College Physics for AP® Courses

Part 1: Chapters 1-17

Senior Contributing Authors

Irina Lyublinskaya, Cuny College Of Staten Island
Gregg Wolfe, Avonworth High School
Douglas Ingram, Texas Christian University
Liza Pujji, Manukau Institute Of Technology
Sudhi Oberoi, Raman Research Institute
Nathan Czuba, Sabio Academy

This book does not contain the chapters 18-34 or the Answers for chapters 18-34 of the original book *College Physics for AP® Courses* by OpenStax.

Chapters 18-34 and the chapter Answers of the original book *The College Physics for AP® Courses* by OpenStax may be found in a book with the ISBN of 9781680920772.

The page numbers for Chapters 1-17 are the same as in the original version of the book.
The page numbers for the Appendices and Index have been recalculated to reflect the removal of the Answers pages for Chapters 18-34.

Table of Contents

PREFACE

Welcome to *College Physics for AP® Courses*, an OpenStax resource. This textbook was written to increase student access to high-quality learning materials, maintaining highest standards of academic rigor at little to no cost.

About OpenStax

OpenStax is a nonprofit based at Rice University, and it's our mission to improve student access to education. Our first openly licensed college textbook was published in 2012, and our library has since scaled to over 25 books for college and AP® courses used by hundreds of thousands of students. Our adaptive learning technology, designed to improve learning outcomes through personalized educational paths, is being piloted in college courses throughout the country. Through our partnerships with philanthropic foundations and our alliance with other educational resource organizations, OpenStax is breaking down the most common barriers to learning and empowering students and instructors to succeed.

About OpenStax Resources

Customization

College Physics for AP® Courses is licensed under a Creative Commons Attribution 4.0 International (CC BY) license, which means that you can distribute, remix, and build upon the content, as long as you provide attribution to OpenStax and its content contributors.

Because our books are openly licensed, you are free to use the entire book or pick and choose the sections that are most relevant to the needs of your course. Feel free to remix the content by assigning your students certain chapters and sections in your syllabus, in the order that you prefer. You can even provide a direct link in your syllabus to the sections in the web view of your book.

Instructors also have the option of creating a customized version of their OpenStax book through the OpenStax Custom platform. The custom version can be made available to students in low-cost print or digital form through their campus bookstore. Visit your book page on openstax.org for a link to your book on OpenStax Custom.

Errata

All OpenStax textbooks undergo a rigorous review process. However, like any professional-grade textbook, errors sometimes occur. Since our books are web based, we can make updates periodically when deemed pedagogically necessary. If you have a correction to suggest, submit it through the link on your book page on openstax.org. Subject matter experts review all errata suggestions. OpenStax is committed to remaining transparent about all updates, so you will also find a list of past errata changes on your book page on openstax.org.

Format

You can access this textbook for free in web view or PDF through openstax.org, and in low-cost print and iBooks editions.

About *College Physics for AP® Courses*

College Physics for AP® Courses is designed to engage students in their exploration of physics and help them apply these concepts to the Advanced Placement® test. Because physics is integral to modern technology and other sciences, the book also includes content that goes beyond the scope of the AP® course to further student understanding. The AP® Connection in each chapter directs students to the material they should focus on for the AP® exam, and what content — although interesting — is not necessarily part of the AP® curriculum. This book is Learning List-approved for AP® Physics courses.

Coverage, Scope, and Alignment to the AP® Curriculum

The current AP® Physics curriculum framework outlines the two full-year physics courses AP® Physics 1: Algebra-Based and AP® Physics 2: Algebra-Based. These two courses replaced the one-year AP® Physics B course, which over the years had become a fast-paced survey of physics facts and formulas that did not provide in-depth conceptual understanding of major physics ideas and the connections between them.

AP® Physics 1 and 2 courses focus on the big ideas typically included in the first and second semesters of an algebra-based, introductory college-level physics course, providing students with the essential knowledge and skills required to support future advanced course work in physics. The AP® Physics 1 curriculum includes mechanics, mechanical waves, sound, and electrostatics. The AP® Physics 2 curriculum focuses on thermodynamics, fluid statics, dynamics, electromagnetism, geometric and physical optics, quantum physics, atomic physics, and nuclear physics. Seven unifying themes of physics called the Big Ideas each include three to seven enduring understandings (EU), which are themselves composed of essential knowledge (EK) that provides details and context for students as they explore physics.

AP® science practices emphasize inquiry-based learning and development of critical thinking and reasoning skills. Inquiry usually uses a series of steps to gain new knowledge, beginning with an observation and following with a hypothesis to explain the observation; then experiments are conducted to test the hypothesis, gather results, and draw conclusions from data. The AP® framework has identified seven major science practices, which can be described by short phrases: using representations and models to communicate information and solve problems; using mathematics appropriately; engaging in questioning; planning and implementing data collection strategies; analyzing and evaluating data; justifying scientific explanations; and connecting concepts. The framework's Learning Objectives merge content (EU and EK) with one or more of the seven science practices that students should develop as they prepare for the AP® Physics exam.

College Physics for AP® Courses is based on the OpenStax *College Physics* text, adapted to focus on the AP curriculum's concepts and practices. Each chapter of OpenStax *College Physics for AP® Courses* begins with a *Connection for AP® Courses* introduction that explains how the content in the chapter sections align to the Big Ideas, enduring understandings, and essential knowledge in the AP® framework. This textbook contains a wealth of information, and the *Connection for AP® Courses* sections will help you distill the required AP® content from material that, although interesting, exceeds the scope of an introductory-level course.

Each section opens with the program's learning objectives as well as the AP® learning objectives and science practices addressed. We have also developed *Real World Connections* features and *Applying the Science Practices* features that highlight concepts, examples, and practices in the framework.

- 1 Introduction: The Nature of Science and Physics
- 2 Kinematics
- 3 Two-Dimensional Kinematics
- 4 Dynamics: Force and Newton's Laws of Motion
- 5 Further Applications of Newton's Laws: Friction, Drag, and Elasticity
- 6 Gravitation and Uniform Circular Motion
- 7 Work, Energy, and Energy Resources
- 8 Linear Momentum and Collisions
- 9 Statics and Torque
- 10 Rotational Motion and Angular Momentum
- 11 Fluid Statics
- 12 Fluid Dynamics and Its Biological and Medical Applications
- 13 Temperature, Kinetic Theory, and the Gas Laws
- 14 Heat and Heat Transfer Methods
- 15 Thermodynamics
- 16 Oscillatory Motion and Waves
- 17 Physics of Hearing
- 18 Electric Charge and Electric Field
- 19 Electric Potential and Electric Field
- 20 Electric Current, Resistance, and Ohm's Law
- 21 Circuits, Bioelectricity, and DC Instruments
- 22 Magnetism
- 23 Electromagnetic Induction, AC Circuits, and Electrical Technologies
- 24 Electromagnetic Waves
- 25 Geometric Optics
- 26 Vision and Optical Instruments
- 27 Wave Optics
- 28 Special Relativity
- 29 Introduction to Quantum Physics
- 30 Atomic Physics
- 31 Radioactivity and Nuclear Physics
- 32 Medical Applications of Nuclear Physics
- 33 Particle Physics
- 34 Frontiers of Physics
- Appendix A: Atomic Masses
- Appendix B: Selected Radioactive Isotopes
- Appendix C: Useful Information
- Appendix D: Glossary of Key Symbols and Notation

Pedagogical Foundation and Features

College Physics for AP® Courses is organized so that topics are introduced conceptually with a steady progression to precise definitions and analytical applications. The analytical, problem-solving aspect is tied back to the conceptual before moving on to another topic. Each introductory chapter, for example, opens with an engaging photograph relevant to the subject of the chapter and interesting applications that are easy for most students to visualize.

- **Connections for AP® Courses** introduce each chapter and explain how its content addresses the AP® curriculum.
- **Worked Examples** Examples start with problems based on real-life situations, then describe a strategy for solving the problem that emphasizes key concepts. The subsequent detailed mathematical solution also includes a follow-up

discussion.

- **Problem-solving Strategies** are presented independently and subsequently appear at crucial points in the text where students can benefit most from them.
- **Misconception Alerts** address common misconceptions that students may bring to class.
- **Take-Home Investigations** provide the opportunity for students to apply or explore what they have learned with a hands-on activity.
- **Real World Connections** highlight important concepts and examples in the AP® framework.
- **Applying the Science Practices** includes activities and challenging questions that engage students while they apply the AP® science practices.
- **Things Great and Small** explains macroscopic phenomena (such as air pressure) with submicroscopic phenomena (such as atoms bouncing off of walls).
- **PhET Explorations** link students to interactive PHeT physics simulations, developed by the University of Colorado, to help them further explore the physics concepts they have learned about in their book module.

Assessment

College Physics for AP® Courses offers a wealth of assessment options, including the following end-of-module problems:

- **Integrated Concept Problems** challenge students to apply both conceptual knowledge and skills to solve a problem.
- **Unreasonable Results** encourage students to solve a problem and then evaluate why the premise or answer to the problem are unrealistic.
- **Construct Your Own Problem** requires students to construct how to solve a particular problem, justify their starting assumptions, show their steps to find the solution to the problem, and finally discuss the meaning of the result.
- **Test Prep for AP® Courses** includes assessment items with the format and rigor found in the AP® exam to help prepare students for the exam.

AP Physics Collection

College Physics for AP® Courses is a part of the AP Physics Collection. The AP Physics Collection is a free, turnkey solution for your AP® Physics course, brought to you through a collaboration between OpenStax and Rice Online Learning. The integrated collection pairs the OpenStax College Physics for AP® Courses text with Concept Trailer videos, instructional videos, problem solution videos, and a correlation guide to help you align all of your content. The instructional videos and problem solution videos were developed by Rice Professor Jason Hafner and AP® Physics teachers Gigi Nevils-Noe and Matt Wilson through Rice Online Learning. You can access all of this free material through the College Physics for AP® Courses page on openstax.org.

Additional Resources

Student and Instructor Resources

We've compiled additional resources for both students and instructors, including Getting Started Guides, an instructor solution manual, and instructional videos. Instructor resources require a verified instructor account, which you can apply for when you log in or create your account on openstax.org. Take advantage of these resources to supplement your OpenStax book.

Partner Resources

OpenStax Partners are our allies in the mission to make high-quality learning materials affordable and accessible to students and instructors everywhere. Their tools integrate seamlessly with our OpenStax titles at a low cost. To access the partner resources for your text, visit your book page on openstax.org.

About the Authors

Senior Contributing Authors

Irina Lyublinskaya, CUNY College of Staten Island
Gregg Wolfe, Avonworth High School
Douglas Ingram, Trinity Christian University
Liza Pujji, Manukau Institute of Technology, New Zealand
Sudhi Oberoi, Visiting Research Student, QuIC Lab, Raman Research Institute, India
Nathan Czuba, Sabio Academy
Julie Kretchman, Science Writer, BS, University of Toronto
John Stoke, Science Writer, MS, University of Chicago
David Anderson, Science Writer, PhD, College of William and Mary
Erika Gasper, Science Writer, MA, University of California, Santa Cruz

Advanced Placement Teacher Reviewers

John Boehringer, Prosper High School
Victor Brazil, Petaluma High School
Michelle Burgess, Avon Lake High School
Bryan Callow, Lindenwold High School
Brian Hastings, Spring Grove Area School District

Alexander Lavy, Xavier High School
Jerome Mass, Glastonbury Public Schools

Faculty Reviewers

John Aiken, Georgia Institute of Technology
Robert Arts, University of Pikeville
Anand Batra, Howard University
Michael Ottinger, Missouri Western State University
James Smith, Caldwell University
Ulrich Zurcher, Cleveland State University

To the AP® Physics Student

The fundamental goal of physics is to discover and understand the "laws" that govern observed phenomena in the world around us. Why study physics? If you plan to become a physicist, the answer is obvious—introductory physics provides the foundation for your career; or if you want to become an engineer, physics provides the basis for the engineering principles used to solve applied and practical problems. For example, after the discovery of the photoelectric effect by physicists, engineers developed photocells that are used in solar panels to convert sunlight to electricity. What if you are an aspiring medical doctor? Although the applications of the laws of physics may not be obvious, their understanding is tremendously valuable. Physics is involved in medical diagnostics, such as x-rays, magnetic resonance imaging (MRI), and ultrasonic blood flow measurements. Medical therapy sometimes directly involves physics; cancer radiotherapy uses ionizing radiation. What if you are planning a nonscience career? Learning physics provides you with a well-rounded education and the ability to make important decisions, such as evaluating the pros and cons of energy production sources or voting on decisions about nuclear waste disposal.

This AP® Physics 1 course begins with **kinematics,** the study of motion without considering its causes. Motion is everywhere: from the vibration of atoms to the planetary revolutions around the Sun. Understanding motion is key to understanding other concepts in physics. You will then study **dynamics,** which considers the forces that affect the motion of moving objects and systems. Newton's laws of motion are the foundation of dynamics. These laws provide an example of the breadth and simplicity of the principles under which nature functions. One of the most remarkable simplifications in physics is that only four distinct forces account for all known phenomena. Your journey will continue as you learn about energy. Energy plays an essential role both in everyday events and in scientific phenomena. You can likely name many forms of energy, from that provided by our foods, to the energy we use to run our cars, to the sunlight that warms us on the beach. The next stop is learning about oscillatory motion and waves. All oscillations involve force and energy: you push a child in a swing to get the motion started and you put energy into a guitar string when you pluck it. Some oscillations create waves. For example, a guitar creates sound waves. You will conclude this first physics course with the study of static electricity and electric currents. Many of the characteristics of static electricity can be explored by rubbing things together. Rubbing creates the spark you get from walking across a wool carpet, for example. Similarly, lightning results from air movements under certain weather conditions.

In the AP® Physics 2 course, you will continue your journey by studying **fluid dynamics,** which explains why rising smoke curls and twists and how the body regulates blood flow. The next stop is **thermodynamics,** the study of heat transfer—energy in transit—that can be used to do work. Basic physical laws govern how heat transfers and its efficiency. Then you will learn more about electric phenomena as you delve into **electromagnetism.** An electric current produces a magnetic field; similarly, a magnetic field produces a current. This phenomenon, known as **magnetic induction,** is essential to our technological society. The generators in cars and nuclear plants use magnetism to generate a current. Other devices that use magnetism to induce currents include pickup coils in electric guitars, transformers of every size, certain microphones, airport security gates, and damping mechanisms on sensitive chemical balances. From electromagnetism you will continue your journey to **optics,** the study of light. You already know that visible light is the type of electromagnetic waves to which our eyes respond. Through vision, light can evoke deep emotions, such as when we view a magnificent sunset or glimpse a rainbow breaking through the clouds. Optics is concerned with the generation and propagation of light. The **quantum mechanics, atomic physics,** and **nuclear physics** are at the end of your journey. These areas of physics have been developed at the end of the 19th and early 20th centuries and deal with submicroscopic objects. Because these objects are smaller than we can observe directly with our senses and generally must be observed with the aid of instruments, parts of these physics areas may seem foreign and bizarre to you at first. However, we have experimentally confirmed most of the ideas in these areas of physics.

AP® Physics is a challenging course. After all, you are taking physics at the introductory college level. You will discover that some concepts are more difficult to understand than others; most students, for example, struggle to understand rotational motion and angular momentum or particle-wave duality. The AP® curriculum promotes depth of understanding over breadth of content, and to make your exploration of topics more manageable, concepts are organized around seven major themes called the **Big Ideas** that apply to all levels of physical systems and interactions between them (see web diagram below). Each Big Idea identifies **enduring understandings** (EU), **essential knowledge** (EK), and **illustrative examples** that support key concepts and content. Simple descriptions define the focus of each Big Idea.

- Big Idea 1: Objects and systems have properties.
- Big Idea 2: Fields explain interactions.
- Big Idea 3: The interactions are described by forces.
- Big Idea 4: Interactions result in changes.
- Big Idea 5: Changes are constrained by conservation laws.
- Big Idea 6: Waves can transfer energy and momentum.
- Big Idea 7: The mathematics of probability can to describe the behavior of complex and quantum mechanical systems.

Doing college work is not easy, but completion of AP® classes is a reliable predictor of college success and prepares you for subsequent courses. The more you engage in the subject, the easier your journey through the curriculum will be. Bring your enthusiasm to class every day along with your notebook, pencil, and calculator. Prepare for class the day before, and review

concepts daily. Form a peer study group and ask your teacher for extra help if necessary. The AP® lab program focuses on more open-ended, student-directed, and inquiry-based lab investigations designed to make you think, ask questions, and analyze data like scientists. You will develop critical thinking and reasoning skills and apply different means of communicating information. By

the time you sit for the AP® exam in May, you will be fluent in the language of physics; because you have been doing real science, you will be ready to show what you have learned. Along the way, you will find the study of the world around us to be one of the most relevant and enjoyable experiences of your high school career.

Irina Lyublinskaya, PhD
Professor of Science Education

To the AP® Physics Teacher

The AP® curriculum was designed to allow instructors flexibility in their approach to teaching the physics courses. *College Physics for AP® Courses* helps you orient students as they delve deeper into the world of physics. Each chapter includes a Connection for AP® Courses introduction that describes the AP® Physics Big Ideas, enduring understandings, and essential knowledge addressed in that chapter.

Each section starts with specific AP® learning objectives and includes essential concepts, illustrative examples, and science practices, along with suggestions for applying the learning objectives through take-home experiments, virtual lab investigations, and activities and questions for preparation and review. At the end of each section, students will find the Test Prep for AP® courses with multiple-choice and open-response questions addressing AP® learning objectives to help them prepare for the AP® exam.

College Physics for AP® Courses has been written to engage students in their exploration of physics and help them relate what they learn in the classroom to their lives outside of it. Physics underlies much of what is happening today in other sciences and in technology. Thus, the book content includes interesting facts and ideas that go beyond the scope of the AP® course. The AP® Connection in each chapter directs students to the material they should focus on for the AP® exam, and what content—although interesting—is not part of the AP® curriculum. Physics is a beautiful and fascinating science. It is in your hands to engage and inspire your students to dive into an amazing world of physics, so they can enjoy it beyond just preparation for the AP® exam.

Irina Lyublinskaya, PhD
Professor of Science Education

The concept map showing major links between Big Ideas and Enduring Understandings is provided below for visual reference.

1   INTRODUCTION: THE NATURE OF SCIENCE AND PHYSICS

Figure 1.1 Galaxies are as immense as atoms are small. Yet the same laws of physics describe both, and all the rest of nature—an indication of the underlying unity in the universe. The laws of physics are surprisingly few in number, implying an underlying simplicity to nature's apparent complexity. (credit: NASA, JPL-Caltech, P. Barmby, Harvard-Smithsonian Center for Astrophysics)

Chapter Outline

1.1. Physics: An Introduction

1.2. Physical Quantities and Units

1.3. Accuracy, Precision, and Significant Figures

1.4. Approximation

Connection for AP® Courses

What is your first reaction when you hear the word "physics"? Did you imagine working through difficult equations or memorizing formulas that seem to have no real use in life outside the physics classroom? Many people come to the subject of physics with a bit of fear. But as you begin your exploration of this broad-ranging subject, you may soon come to realize that physics plays a much larger role in your life than you first thought, no matter your life goals or career choice.

For example, take a look at the image above. This image is of the Andromeda Galaxy, which contains billions of individual stars, huge clouds of gas, and dust. Two smaller galaxies are also visible as bright blue spots in the background. At a staggering 2.5 million light years from Earth, this galaxy is the nearest one to our own galaxy (which is called the Milky Way). The stars and planets that make up Andromeda might seem to be the furthest thing from most people's regular, everyday lives. But Andromeda is a great starting point to think about the forces that hold together the universe. The forces that cause Andromeda to act as it does are the same forces we contend with here on Earth, whether we are planning to send a rocket into space or simply raise the walls for a new home. The same gravity that causes the stars of Andromeda to rotate and revolve also causes water to flow over hydroelectric dams here on Earth. Tonight, take a moment to look up at the stars. The forces out there are the same as the ones here on Earth. Through a study of physics, you may gain a greater understanding of the interconnectedness of everything we can see and know in this universe.

Think now about all of the technological devices that you use on a regular basis. Computers, smart phones, GPS systems, MP3 players, and satellite radio might come to mind. Next, think about the most exciting modern technologies that you have heard about in the news, such as trains that levitate above tracks, "invisibility cloaks" that bend light around them, and microscopic robots that fight cancer cells in our bodies. All of these groundbreaking advancements, commonplace or unbelievable, rely on the principles of physics. Aside from playing a significant role in technology, professionals such as engineers, pilots, physicians, physical therapists, electricians, and computer programmers apply physics concepts in their daily work. For example, a pilot must understand how wind forces affect a flight path and a physical therapist must understand how the muscles in the body experience forces as they move and bend. As you will learn in this text, physics principles are propelling new, exciting technologies, and these principles are applied in a wide range of careers.

In this text, you will begin to explore the history of the formal study of physics, beginning with natural philosophy and the ancient Greeks, and leading up through a review of Sir Isaac Newton and the laws of physics that bear his name. You will also be introduced to the standards scientists use when they study physical quantities and the interrelated system of measurements most of the scientific community uses to communicate in a single mathematical language. Finally, you will study the limits of our ability to be accurate and precise, and the reasons scientists go to painstaking lengths to be as clear as possible regarding their own limitations.

Chapter 1 introduces many fundamental skills and understandings needed for success with the AP® Learning Objectives. While this chapter does not directly address any Big Ideas, its content will allow for a more meaningful understanding when these Big Ideas are addressed in future chapters. For instance, the discussion of models, theories, and laws will assist you in understanding the concept of fields as addressed in Big Idea 2, and the section titled 'The Evolution of Natural Philosophy into Modern Physics' will help prepare you for the statistical topics addressed in Big Idea 7.

This chapter will also prepare you to understand the Science Practices. In explicitly addressing the role of models in representing and communicating scientific phenomena, Section 1.1 supports Science Practice 1. Additionally, anecdotes about historical investigations and the inset on the scientific method will help you to engage in the scientific questioning referenced in Science Practice 3. The appropriate use of mathematics, as called for in Science Practice 2, is a major focus throughout sections 1.2, 1.3, and 1.4.

1.1 Physics: An Introduction

Figure 1.2 The flight formations of migratory birds such as Canada geese are governed by the laws of physics. (credit: David Merrett)

Learning Objectives
By the end of this section, you will be able to: • Explain the difference between a principle and a law. • Explain the difference between a model and a theory.

The physical universe is enormously complex in its detail. Every day, each of us observes a great variety of objects and phenomena. Over the centuries, the curiosity of the human race has led us collectively to explore and catalog a tremendous wealth of information. From the flight of birds to the colors of flowers, from lightning to gravity, from quarks to clusters of galaxies, from the flow of time to the mystery of the creation of the universe, we have asked questions and assembled huge arrays of facts. In the face of all these details, we have discovered that a surprisingly small and unified set of physical laws can explain what we observe. As humans, we make generalizations and seek order. We have found that nature is remarkably cooperative—it exhibits the *underlying order and simplicity* we so value.

It is the underlying order of nature that makes science in general, and physics in particular, so enjoyable to study. For example, what do a bag of chips and a car battery have in common? Both contain energy that can be converted to other forms. The law of conservation of energy (which says that energy can change form but is never lost) ties together such topics as food calories, batteries, heat, light, and watch springs. Understanding this law makes it easier to learn about the various forms energy takes and how they relate to one another. Apparently unrelated topics are connected through broadly applicable physical laws, permitting an understanding beyond just the memorization of lists of facts.

The unifying aspect of physical laws and the basic simplicity of nature form the underlying themes of this text. In learning to apply these laws, you will, of course, study the most important topics in physics. More importantly, you will gain analytical abilities that will enable you to apply these laws far beyond the scope of what can be included in a single book. These analytical skills will help you to excel academically, and they will also help you to think critically in any professional career you choose to pursue. This module discusses the realm of physics (to define what physics is), some applications of physics (to illustrate its relevance to other disciplines), and more precisely what constitutes a physical law (to illuminate the importance of experimentation to theory).

Science and the Realm of Physics

Science consists of the theories and laws that are the general truths of nature as well as the body of knowledge they encompass. Scientists are continually trying to expand this body of knowledge and to perfect the expression of the laws that describe it. **Physics** is concerned with describing the interactions of energy, matter, space, and time, and it is especially interested in what fundamental mechanisms underlie every phenomenon. The concern for describing the basic phenomena in nature essentially defines the *realm of physics*.

Physics aims to describe the function of everything around us, from the movement of tiny charged particles to the motion of people, cars, and spaceships. In fact, almost everything around you can be described quite accurately by the laws of physics. Consider a smart phone (**Figure 1.3**). Physics describes how electricity interacts with the various circuits inside the device. This knowledge helps engineers select the appropriate materials and circuit layout when building the smart phone. Next, consider a GPS system. Physics describes the relationship between the speed of an object, the distance over which it travels, and the time it takes to travel that distance. When you use a GPS device in a vehicle, it utilizes these physics equations to determine the travel time from one location to another.

Figure 1.3 The Apple "iPhone" is a common smart phone with a GPS function. Physics describes the way that electricity flows through the circuits of this device. Engineers use their knowledge of physics to construct an iPhone with features that consumers will enjoy. One specific feature of an iPhone is the GPS function. GPS uses physics equations to determine the driving time between two locations on a map. (credit: @gletham GIS, Social, Mobile Tech Images)

Applications of Physics

You need not be a scientist to use physics. On the contrary, knowledge of physics is useful in everyday situations as well as in nonscientific professions. It can help you understand how microwave ovens work, why metals should not be put into them, and why they might affect pacemakers. (See Figure 1.4 and Figure 1.5.) Physics allows you to understand the hazards of radiation and rationally evaluate these hazards more easily. Physics also explains the reason why a black car radiator helps remove heat in a car engine, and it explains why a white roof helps keep the inside of a house cool. Similarly, the operation of a car's ignition system as well as the transmission of electrical signals through our body's nervous system are much easier to understand when you think about them in terms of basic physics.

Physics is the foundation of many important disciplines and contributes directly to others. Chemistry, for example—since it deals with the interactions of atoms and molecules—is rooted in atomic and molecular physics. Most branches of engineering are applied physics. In architecture, physics is at the heart of structural stability, and is involved in the acoustics, heating, lighting, and cooling of buildings. Parts of geology rely heavily on physics, such as radioactive dating of rocks, earthquake analysis, and heat transfer in the Earth. Some disciplines, such as biophysics and geophysics, are hybrids of physics and other disciplines.

Physics has many applications in the biological sciences. On the microscopic level, it helps describe the properties of cell walls and cell membranes (Figure 1.6 and Figure 1.7). On the macroscopic level, it can explain the heat, work, and power associated with the human body. Physics is involved in medical diagnostics, such as x-rays, magnetic resonance imaging (MRI), and ultrasonic blood flow measurements. Medical therapy sometimes directly involves physics; for example, cancer radiotherapy uses ionizing radiation. Physics can also explain sensory phenomena, such as how musical instruments make sound, how the eye detects color, and how lasers can transmit information.

It is not necessary to formally study all applications of physics. What is most useful is knowledge of the basic laws of physics and a skill in the analytical methods for applying them. The study of physics also can improve your problem-solving skills. Furthermore, physics has retained the most basic aspects of science, so it is used by all of the sciences, and the study of physics makes other sciences easier to understand.

Figure 1.4 The laws of physics help us understand how common appliances work. For example, the laws of physics can help explain how microwave ovens heat up food, and they also help us understand why it is dangerous to place metal objects in a microwave oven. (credit: MoneyBlogNewz)

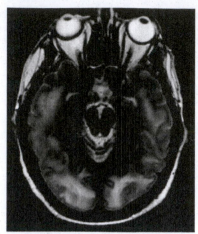

Figure 1.5 These two applications of physics have more in common than meets the eye. Microwave ovens use electromagnetic waves to heat food. Magnetic resonance imaging (MRI) also uses electromagnetic waves to yield an image of the brain, from which the exact location of tumors can be determined. (credit: Rashmi Chawla, Daniel Smith, and Paul E. Marik)

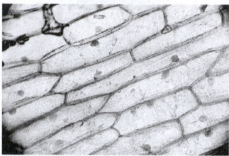

Figure 1.6 Physics, chemistry, and biology help describe the properties of cell walls in plant cells, such as the onion cells seen here. (credit: Umberto Salvagnin)

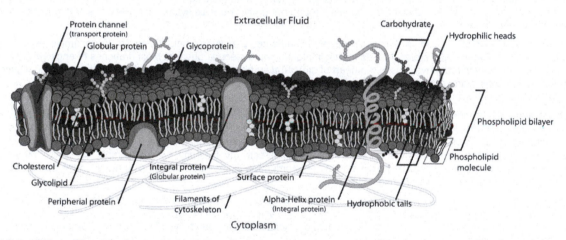

Figure 1.7 An artist's rendition of the the structure of a cell membrane. Membranes form the boundaries of animal cells and are complex in structure and function. Many of the most fundamental properties of life, such as the firing of nerve cells, are related to membranes. The disciplines of biology, chemistry, and physics all help us understand the membranes of animal cells. (credit: Mariana Ruiz)

Models, Theories, and Laws; The Role of Experimentation

The laws of nature are concise descriptions of the universe around us; they are human statements of the underlying laws or rules that all natural processes follow. Such laws are intrinsic to the universe; humans did not create them and so cannot change them. We can only discover and understand them. Their discovery is a very human endeavor, with all the elements of mystery, imagination, struggle, triumph, and disappointment inherent in any creative effort. (See **Figure 1.8** and **Figure 1.9**.) The cornerstone of discovering natural laws is observation; science must describe the universe as it is, not as we may imagine it to be.

Sir Isaac Newton

Figure 1.8 Isaac Newton (1642–1727) was very reluctant to publish his revolutionary work and had to be convinced to do so. In his later years, he stepped down from his academic post and became exchequer of the Royal Mint. He took this post seriously, inventing reeding (or creating ridges) on the edge of coins to prevent unscrupulous people from trimming the silver off of them before using them as currency. (credit: Arthur Shuster and Arthur E. Shipley: *Britain's Heritage of Science*. London, 1917.)

Figure 1.9 Marie Curie (1867–1934) sacrificed monetary assets to help finance her early research and damaged her physical well-being with radiation exposure. She is the only person to win Nobel prizes in both physics and chemistry. One of her daughters also won a Nobel Prize. (credit: Wikimedia Commons)

We all are curious to some extent. We look around, make generalizations, and try to understand what we see—for example, we look up and wonder whether one type of cloud signals an oncoming storm. As we become serious about exploring nature, we become more organized and formal in collecting and analyzing data. We attempt greater precision, perform controlled experiments (if we can), and write down ideas about how the data may be organized and unified. We then formulate models, theories, and laws based on the data we have collected and analyzed to generalize and communicate the results of these experiments.

A **model** is a representation of something that is often too difficult (or impossible) to display directly. While a model is justified with experimental proof, it is only accurate under limited situations. An example is the planetary model of the atom in which electrons are pictured as orbiting the nucleus, analogous to the way planets orbit the Sun. (See **Figure 1.10**.) We cannot observe electron orbits directly, but the mental image helps explain the observations we can make, such as the emission of light from hot gases (atomic spectra). Physicists use models for a variety of purposes. For example, models can help physicists analyze a scenario and perform a calculation, or they can be used to represent a situation in the form of a computer simulation. A **theory** is an explanation for patterns in nature that is supported by scientific evidence and verified multiple times by various groups of researchers. Some theories include models to help visualize phenomena, whereas others do not. Newton's theory of gravity, for example, does not require a model or mental image, because we can observe the objects directly with our own senses. The kinetic theory of gases, on the other hand, is a model in which a gas is viewed as being composed of atoms and molecules. Atoms and molecules are too small to be observed directly with our senses—thus, we picture them mentally to understand what our instruments tell us about the behavior of gases.

A **law** uses concise language to describe a generalized pattern in nature that is supported by scientific evidence and repeated experiments. Often, a law can be expressed in the form of a single mathematical equation. Laws and theories are similar in that they are both scientific statements that result from a tested hypothesis and are supported by scientific evidence. However, the designation *law* is reserved for a concise and very general statement that describes phenomena in nature, such as the law that energy is conserved during any process, or Newton's second law of motion, which relates force, mass, and acceleration by the simple equation $\mathbf{F} = m\mathbf{a}$. A theory, in contrast, is a less concise statement of observed phenomena. For example, the Theory of Evolution and the Theory of Relativity cannot be expressed concisely enough to be considered a law. The biggest difference between a law and a theory is that a theory is much more complex and dynamic. A law describes a single action, whereas a theory explains an entire group of related phenomena. And, whereas a law is a postulate that forms the foundation of the scientific method, a theory is the end result of that process.

Less broadly applicable statements are usually called principles (such as Pascal's principle, which is applicable only in fluids), but the distinction between laws and principles often is not carefully made.

Figure 1.10 What is a model? This planetary model of the atom shows electrons orbiting the nucleus. It is a drawing that we use to form a mental image of the atom that we cannot see directly with our eyes because it is too small.

Models, Theories, and Laws

Models, theories, and laws are used to help scientists analyze the data they have already collected. However, often after a model, theory, or law has been developed, it points scientists toward new discoveries they would not otherwise have made.

The models, theories, and laws we devise sometimes *imply the existence of objects or phenomena as yet unobserved.* These predictions are remarkable triumphs and tributes to the power of science. It is the underlying order in the universe that enables scientists to make such spectacular predictions. However, if *experiment* does not verify our predictions, then the theory or law is wrong, no matter how elegant or convenient it is. Laws can never be known with absolute certainty because it is impossible to perform every imaginable experiment in order to confirm a law in every possible scenario. Physicists operate under the assumption that all scientific laws and theories are valid until a counterexample is observed. If a good-quality, verifiable experiment contradicts a well-established law, then the law must be modified or overthrown completely.

The study of science in general and physics in particular is an adventure much like the exploration of uncharted ocean. Discoveries are made; models, theories, and laws are formulated; and the beauty of the physical universe is made more sublime for the insights gained.

The Scientific Method

As scientists inquire and gather information about the world, they follow a process called the **scientific method**. This process typically begins with an observation and question that the scientist will research. Next, the scientist typically performs some research about the topic and then devises a hypothesis. Then, the scientist will test the hypothesis by performing an experiment. Finally, the scientist analyzes the results of the experiment and draws a conclusion. Note that the scientific method can be applied to many situations that are not limited to science, and this method can be modified to suit the situation.

Consider an example. Let us say that you try to turn on your car, but it will not start. You undoubtedly wonder: Why will the car not start? You can follow a scientific method to answer this question. First off, you may perform some research to determine a variety of reasons why the car will not start. Next, you will state a hypothesis. For example, you may believe that the car is not starting because it has no engine oil. To test this, you open the hood of the car and examine the oil level. You observe that the oil is at an acceptable level, and you thus conclude that the oil level is not contributing to your car issue. To troubleshoot the issue further, you may devise a new hypothesis to test and then repeat the process again.

The Evolution of Natural Philosophy into Modern Physics

Physics was not always a separate and distinct discipline. It remains connected to other sciences to this day. The word *physics* comes from Greek, meaning nature. The study of nature came to be called "natural philosophy." From ancient times through the Renaissance, natural philosophy encompassed many fields, including astronomy, biology, chemistry, physics, mathematics, and medicine. Over the last few centuries, the growth of knowledge has resulted in ever-increasing specialization and branching of natural philosophy into separate fields, with physics retaining the most basic facets. (See **Figure 1.11**, **Figure 1.12**, and **Figure 1.13**.) Physics as it developed from the Renaissance to the end of the 19th century is called **classical physics**. It was transformed into modern physics by revolutionary discoveries made starting at the beginning of the 20th century.

Figure 1.11 Over the centuries, natural philosophy has evolved into more specialized disciplines, as illustrated by the contributions of some of the greatest minds in history. The Greek philosopher **Aristotle** (384–322 B.C.) wrote on a broad range of topics including physics, animals, the soul, politics, and poetry. (credit: Jastrow (2006)/Ludovisi Collection)

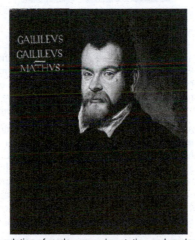

Figure 1.12 **Galileo Galilei** (1564–1642) laid the foundation of modern experimentation and made contributions in mathematics, physics, and astronomy. (credit: Domenico Tintoretto)

Figure 1.13 **Niels Bohr** (1885–1962) made fundamental contributions to the development of quantum mechanics, one part of modern physics. (credit: United States Library of Congress Prints and Photographs Division)

Classical physics is not an exact description of the universe, but it is an excellent approximation under the following conditions: Matter must be moving at speeds less than about 1% of the speed of light, the objects dealt with must be large enough to be seen with a microscope, and only weak gravitational fields, such as the field generated by the Earth, can be involved. Because humans live under such circumstances, classical physics seems intuitively reasonable, while many aspects of modern physics seem bizarre. This is why models are so useful in modern physics—they let us conceptualize phenomena we do not ordinarily experience. We can relate to models in human terms and visualize what happens when objects move at high speeds or imagine what objects too small to observe with our senses might be like. For example, we can understand an atom's properties because we can picture it in our minds, although we have never seen an atom with our eyes. New tools, of course, allow us to better picture phenomena we cannot see. In fact, new instrumentation has allowed us in recent years to actually "picture" the atom.

Limits on the Laws of Classical Physics

For the laws of classical physics to apply, the following criteria must be met: Matter must be moving at speeds less than about 1% of the speed of light, the objects dealt with must be large enough to be seen with a microscope, and only weak gravitational fields (such as the field generated by the Earth) can be involved.

Figure 1.14 Using a scanning tunneling microscope (STM), scientists can see the individual atoms that compose this sheet of gold. (credit: Erwinrossen)

Some of the most spectacular advances in science have been made in modern physics. Many of the laws of classical physics have been modified or rejected, and revolutionary changes in technology, society, and our view of the universe have resulted. Like science fiction, modern physics is filled with fascinating objects beyond our normal experiences, but it has the advantage over science fiction of being very real. Why, then, is the majority of this text devoted to topics of classical physics? There are two main reasons: Classical physics gives an extremely accurate description of the universe under a wide range of everyday circumstances, and knowledge of classical physics is necessary to understand modern physics.

Modern physics itself consists of the two revolutionary theories, relativity and quantum mechanics. These theories deal with the very fast and the very small, respectively. **Relativity** must be used whenever an object is traveling at greater than about 1% of the speed of light or experiences a strong gravitational field such as that near the Sun. **Quantum mechanics** must be used for objects smaller than can be seen with a microscope. The combination of these two theories is *relativistic quantum mechanics,* and it describes the behavior of small objects traveling at high speeds or experiencing a strong gravitational field. Relativistic quantum mechanics is the best universally applicable theory we have. Because of its mathematical complexity, it is used only when necessary, and the other theories are used whenever they will produce sufficiently accurate results. We will find, however, that we can do a great deal of modern physics with the algebra and trigonometry used in this text.

Check Your Understanding

A friend tells you he has learned about a new law of nature. What can you know about the information even before your friend describes the law? How would the information be different if your friend told you he had learned about a scientific theory rather than a law?

Solution
Without knowing the details of the law, you can still infer that the information your friend has learned conforms to the requirements of all laws of nature: it will be a concise description of the universe around us; a statement of the underlying rules that all natural processes follow. If the information had been a theory, you would be able to infer that the information will be a large-scale, broadly applicable generalization.

PhET Explorations: Equation Grapher

Learn about graphing polynomials. The shape of the curve changes as the constants are adjusted. View the curves for the individual terms (e.g. $y = bx$) to see how they add to generate the polynomial curve.

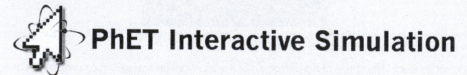

Figure 1.15 Equation Grapher (http://cnx.org/content/m54764/1.2/equation-grapher_en.jar)

1.2 Physical Quantities and Units

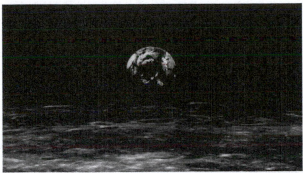

Figure 1.16 The distance from Earth to the Moon may seem immense, but it is just a tiny fraction of the distances from Earth to other celestial bodies. (credit: NASA)

Learning Objectives
By the end of this section, you will be able to: • Perform unit conversions both in the SI and English units. • Explain the most common prefixes in the SI units and be able to write them in scientific notation.

The range of objects and phenomena studied in physics is immense. From the incredibly short lifetime of a nucleus to the age of Earth, from the tiny sizes of sub-nuclear particles to the vast distance to the edges of the known universe, from the force exerted by a jumping flea to the force between Earth and the Sun, there are enough factors of 10 to challenge the imagination of even the most experienced scientist. Giving numerical values for physical quantities and equations for physical principles allows us to understand nature much more deeply than does qualitative description alone. To comprehend these vast ranges, we must also have accepted units in which to express them. And we shall find that (even in the potentially mundane discussion of meters, kilograms, and seconds) a profound simplicity of nature appears—all physical quantities can be expressed as combinations of only four fundamental physical quantities: length, mass, time, and electric current.

We define a **physical quantity** either by *specifying how it is measured* or by *stating how it is calculated* from other measurements. For example, we define distance and time by specifying methods for measuring them, whereas we define *average speed* by stating that it is calculated as distance traveled divided by time of travel.

Measurements of physical quantities are expressed in terms of **units**, which are standardized values. For example, the length of a race, which is a physical quantity, can be expressed in units of meters (for sprinters) or kilometers (for distance runners). Without standardized units, it would be extremely difficult for scientists to express and compare measured values in a meaningful way. (See **Figure 1.17**.)

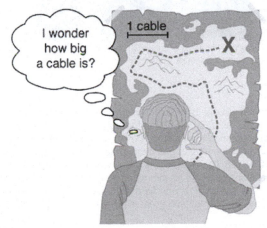

Figure 1.17 Distances given in unknown units are maddeningly useless.

There are two major systems of units used in the world: **SI units** (also known as the metric system) and **English units** (also known as the customary or imperial system). **English units** were historically used in nations once ruled by the British Empire and are still widely used in the United States. Virtually every other country in the world now uses SI units as the standard; the metric system is also the standard system agreed upon by scientists and mathematicians. The acronym "SI" is derived from the French *Système International*.

SI Units: Fundamental and Derived Units

Table 1.1 gives the fundamental SI units that are used throughout this textbook. This text uses non-SI units in a few applications where they are in very common use, such as the measurement of blood pressure in millimeters of mercury (mm Hg). Whenever non-SI units are discussed, they will be tied to SI units through conversions.

Table 1.1 Fundamental SI Units

Length	Mass	Time	Electric Charge
meter (m)	kilogram (kg)	second (s)	coulomb (c)

It is an intriguing fact that some physical quantities are more fundamental than others and that the most fundamental physical quantities can be defined *only* in terms of the procedure used to measure them. The units in which they are measured are thus called **fundamental units**. In this textbook, the fundamental physical quantities are taken to be length, mass, time, and electric charge. (Note that electric current will not be introduced until much later in this text.) All other physical quantities, such as force and electric current, can be expressed as algebraic combinations of length, mass, time, and current (for example, speed is length divided by time); these units are called **derived units**.

Units of Time, Length, and Mass: The Second, Meter, and Kilogram

The Second

The SI unit for time, the **second**(abbreviated s), has a long history. For many years it was defined as 1/86,400 of a mean solar day. More recently, a new standard was adopted to gain greater accuracy and to define the second in terms of a non-varying, or constant, physical phenomenon (because the solar day is getting longer due to very gradual slowing of the Earth's rotation). Cesium atoms can be made to vibrate in a very steady way, and these vibrations can be readily observed and counted. In 1967 the second was redefined as the time required for 9,192,631,770 of these vibrations. (See **Figure 1.18**.) Accuracy in the fundamental units is essential, because all measurements are ultimately expressed in terms of fundamental units and can be no more accurate than are the fundamental units themselves.

Figure 1.18 An atomic clock such as this one uses the vibrations of cesium atoms to keep time to a precision of better than a microsecond per year. The fundamental unit of time, the second, is based on such clocks. This image is looking down from the top of an atomic fountain nearly 30 feet tall! (credit: Steve Jurvetson/Flickr)

The Meter

The SI unit for length is the **meter** (abbreviated m); its definition has also changed over time to become more accurate and precise. The meter was first defined in 1791 as 1/10,000,000 of the distance from the equator to the North Pole. This measurement was improved in 1889 by redefining the meter to be the distance between two engraved lines on a platinum-iridium bar now kept near Paris. By 1960, it had become possible to define the meter even more accurately in terms of the wavelength of light, so it was again redefined as 1,650,763.73 wavelengths of orange light emitted by krypton atoms. In 1983, the meter was given its present definition (partly for greater accuracy) as the distance light travels in a vacuum in 1/299,792,458 of a second. (See **Figure 1.19**.) This change defines the speed of light to be exactly 299,792,458 meters per second. The length of the meter will change if the speed of light is someday measured with greater accuracy.

The Kilogram

The SI unit for mass is the **kilogram** (abbreviated kg); it is defined to be the mass of a platinum-iridium cylinder kept with the old meter standard at the International Bureau of Weights and Measures near Paris. Exact replicas of the standard kilogram are also kept at the United States' National Institute of Standards and Technology, or NIST, located in Gaithersburg, Maryland outside of Washington D.C., and at other locations around the world. The determination of all other masses can be ultimately traced to a comparison with the standard mass.

Light travels a distance of 1 meter
in 1/299,792,458 seconds

Figure 1.19 The meter is defined to be the distance light travels in 1/299,792,458 of a second in a vacuum. Distance traveled is speed multiplied by time.

Electric current and its accompanying unit, the ampere, will be introduced in **Introduction to Electric Current, Resistance, and Ohm's Law** when electricity and magnetism are covered. The initial modules in this textbook are concerned with mechanics, fluids, heat, and waves. In these subjects all pertinent physical quantities can be expressed in terms of the fundamental units of length, mass, and time.

Metric Prefixes

SI units are part of the **metric system**. The metric system is convenient for scientific and engineering calculations because the units are categorized by factors of 10. **Table 1.2** gives metric prefixes and symbols used to denote various factors of 10.

Metric systems have the advantage that conversions of units involve only powers of 10. There are 100 centimeters in a meter, 1000 meters in a kilometer, and so on. In nonmetric systems, such as the system of U.S. customary units, the relationships are not as simple—there are 12 inches in a foot, 5280 feet in a mile, and so on. Another advantage of the metric system is that the same unit can be used over extremely large ranges of values simply by using an appropriate metric prefix. For example, distances in meters are suitable in construction, while distances in kilometers are appropriate for air travel, and the tiny measure of nanometers are convenient in optical design. With the metric system there is no need to invent new units for particular applications.

The term **order of magnitude** refers to the scale of a value expressed in the metric system. Each power of 10 in the metric system represents a different order of magnitude. For example, 10^1, 10^2, 10^3, and so forth are all different orders of magnitude. All quantities that can be expressed as a product of a specific power of 10 are said to be of the *same* order of magnitude. For example, the number 800 can be written as 8×10^2, and the number 450 can be written as 4.5×10^2. Thus, the numbers 800 and 450 are of the same order of magnitude: 10^2. Order of magnitude can be thought of as a ballpark estimate for the scale of a value. The diameter of an atom is on the order of 10^{-9} m, while the diameter of the Sun is on the order of 10^9 m.

The Quest for Microscopic Standards for Basic Units

The fundamental units described in this chapter are those that produce the greatest accuracy and precision in measurement. There is a sense among physicists that, because there is an underlying microscopic substructure to matter, it would be most satisfying to base our standards of measurement on microscopic objects and fundamental physical phenomena such as the speed of light. A microscopic standard has been accomplished for the standard of time, which is based on the oscillations of the cesium atom.

The standard for length was once based on the wavelength of light (a small-scale length) emitted by a certain type of atom, but it has been supplanted by the more precise measurement of the speed of light. If it becomes possible to measure the mass of atoms or a particular arrangement of atoms such as a silicon sphere to greater precision than the kilogram standard, it may become possible to base mass measurements on the small scale. There are also possibilities that electrical phenomena on the small scale may someday allow us to base a unit of charge on the charge of electrons and protons, but at present current and charge are related to large-scale currents and forces between wires.

Table 1.2 Metric Prefixes for Powers of 10 and their Symbols

Prefix	Symbol	Value[1]	Example (some are approximate)			
exa	E	10^{18}	exameter	Em	10^{18} m	distance light travels in a century
peta	P	10^{15}	petasecond	Ps	10^{15} s	30 million years
tera	T	10^{12}	terawatt	TW	10^{12} W	powerful laser output
giga	G	10^{9}	gigahertz	GHz	10^{9} Hz	a microwave frequency
mega	M	10^{6}	megacurie	MCi	10^{6} Ci	high radioactivity
kilo	k	10^{3}	kilometer	km	10^{3} m	about 6/10 mile
hecto	h	10^{2}	hectoliter	hL	10^{2} L	26 gallons
deka	da	10^{1}	dekagram	dag	10^{1} g	teaspoon of butter
—	—	10^{0} (=1)				
deci	d	10^{-1}	deciliter	dL	10^{-1} L	less than half a soda
centi	c	10^{-2}	centimeter	cm	10^{-2} m	fingertip thickness
milli	m	10^{-3}	millimeter	mm	10^{-3} m	flea at its shoulders
micro	µ	10^{-6}	micrometer	µm	10^{-6} m	detail in microscope
nano	n	10^{-9}	nanogram	ng	10^{-9} g	small speck of dust
pico	p	10^{-12}	picofarad	pF	10^{-12} F	small capacitor in radio
femto	f	10^{-15}	femtometer	fm	10^{-15} m	size of a proton
atto	a	10^{-18}	attosecond	as	10^{-18} s	time light crosses an atom

Known Ranges of Length, Mass, and Time

The vastness of the universe and the breadth over which physics applies are illustrated by the wide range of examples of known lengths, masses, and times in **Table 1.3**. Examination of this table will give you some feeling for the range of possible topics and numerical values. (See **Figure 1.20** and **Figure 1.21**.)

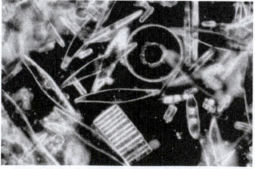

Figure 1.20 Tiny phytoplankton swims among crystals of ice in the Antarctic Sea. They range from a few micrometers to as much as 2 millimeters in length. (credit: Prof. Gordon T. Taylor, Stony Brook University; NOAA Corps Collections)

---

1. See Appendix A for a discussion of powers of 10.

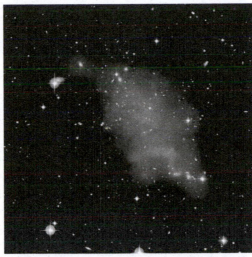

Figure 1.21 Galaxies collide 2.4 billion light years away from Earth. The tremendous range of observable phenomena in nature challenges the imagination. (credit: NASA/CXC/UVic./A. Mahdavi et al. Optical/lensing: CFHT/UVic./H. Hoekstra et al.)

Unit Conversion and Dimensional Analysis

It is often necessary to convert from one type of unit to another. For example, if you are reading a European cookbook, some quantities may be expressed in units of liters and you need to convert them to cups. Or, perhaps you are reading walking directions from one location to another and you are interested in how many miles you will be walking. In this case, you will need to convert units of feet to miles.

Let us consider a simple example of how to convert units. Let us say that we want to convert 80 meters (m) to kilometers (km).

The first thing to do is to list the units that you have and the units that you want to convert to. In this case, we have units in *meters* and we want to convert to *kilometers*.

Next, we need to determine a **conversion factor** relating meters to kilometers. A conversion factor is a ratio expressing how many of one unit are equal to another unit. For example, there are 12 inches in 1 foot, 100 centimeters in 1 meter, 60 seconds in 1 minute, and so on. In this case, we know that there are 1,000 meters in 1 kilometer.

Now we can set up our unit conversion. We will write the units that we have and then multiply them by the conversion factor so that the units cancel out, as shown:

$$80 \, \text{m} \times \frac{1 \, \text{km}}{1000 \, \text{m}} = 0.080 \, \text{km}. \tag{1.1}$$

Note that the unwanted m unit cancels, leaving only the desired km unit. You can use this method to convert between any types of unit.

Click **Appendix C** for a more complete list of conversion factors.

Table 1.3 Approximate Values of Length, Mass, and Time

Lengths in meters		Masses in kilograms (more precise values in parentheses)		Times in seconds (more precise values in parentheses)	
10^{-18}	Present experimental limit to smallest observable detail	10^{-30}	Mass of an electron $(9.11\times10^{-31}\ \text{kg})$	10^{-23}	Time for light to cross a proton
10^{-15}	Diameter of a proton	10^{-27}	Mass of a hydrogen atom $(1.67\times10^{-27}\ \text{kg})$	10^{-22}	Mean life of an extremely unstable nucleus
10^{-14}	Diameter of a uranium nucleus	10^{-15}	Mass of a bacterium	10^{-15}	Time for one oscillation of visible light
10^{-10}	Diameter of a hydrogen atom	10^{-5}	Mass of a mosquito	10^{-13}	Time for one vibration of an atom in a solid
10^{-8}	Thickness of membranes in cells of living organisms	10^{-2}	Mass of a hummingbird	10^{-8}	Time for one oscillation of an FM radio wave
10^{-6}	Wavelength of visible light	1	Mass of a liter of water (about a quart)	10^{-3}	Duration of a nerve impulse
10^{-3}	Size of a grain of sand	10^{2}	Mass of a person	1	Time for one heartbeat
1	Height of a 4-year-old child	10^{3}	Mass of a car	10^{5}	One day $(8.64\times10^{4}\ \text{s})$
10^{2}	Length of a football field	10^{8}	Mass of a large ship	10^{7}	One year (y) $(3.16\times10^{7}\ \text{s})$
10^{4}	Greatest ocean depth	10^{12}	Mass of a large iceberg	10^{9}	About half the life expectancy of a human
10^{7}	Diameter of the Earth	10^{15}	Mass of the nucleus of a comet	10^{11}	Recorded history
10^{11}	Distance from the Earth to the Sun	10^{23}	Mass of the Moon $(7.35\times10^{22}\ \text{kg})$	10^{17}	Age of the Earth
10^{16}	Distance traveled by light in 1 year (a light year)	10^{25}	Mass of the Earth $(5.97\times10^{24}\ \text{kg})$	10^{18}	Age of the universe
10^{21}	Diameter of the Milky Way galaxy	10^{30}	Mass of the Sun $(1.99\times10^{30}\ \text{kg})$		
10^{22}	Distance from the Earth to the nearest large galaxy (Andromeda)	10^{42}	Mass of the Milky Way galaxy (current upper limit)		
10^{26}	Distance from the Earth to the edges of the known universe	10^{53}	Mass of the known universe (current upper limit)		

Example 1.1 Unit Conversions: A Short Drive Home

Suppose that you drive the 10.0 km from your university to home in 20.0 min. Calculate your average speed (a) in kilometers per hour (km/h) and (b) in meters per second (m/s). (Note: Average speed is distance traveled divided by time of travel.)

Strategy

First we calculate the average speed using the given units. Then we can get the average speed into the desired units by picking the correct conversion factor and multiplying by it. The correct conversion factor is the one that cancels the unwanted unit and leaves the desired unit in its place.

Solution for (a)

(1) Calculate average speed. Average speed is distance traveled divided by time of travel. (Take this definition as a given for now—average speed and other motion concepts will be covered in a later module.) In equation form,

$$\text{average speed} = \frac{\text{distance}}{\text{time}}. \tag{1.2}$$

(2) Substitute the given values for distance and time.

$$\text{average speed} = \frac{10.0 \text{ km}}{20.0 \text{ min}} = 0.500 \frac{\text{km}}{\text{min}}. \tag{1.3}$$

(3) Convert km/min to km/h: multiply by the conversion factor that will cancel minutes and leave hours. That conversion factor is 60 min/hr. Thus,

$$\text{average speed} = 0.500 \frac{\text{km}}{\text{min}} \times \frac{60 \text{ min}}{1 \text{ h}} = 30.0 \frac{\text{km}}{\text{h}}. \tag{1.4}$$

Discussion for (a)

To check your answer, consider the following:

(1) Be sure that you have properly cancelled the units in the unit conversion. If you have written the unit conversion factor upside down, the units will not cancel properly in the equation. If you accidentally get the ratio upside down, then the units will not cancel; rather, they will give you the wrong units as follows:

$$\frac{\text{km}}{\text{min}} \times \frac{1 \text{ hr}}{60 \text{ min}} = \frac{1}{60} \frac{\text{km} \cdot \text{hr}}{\text{min}^2}, \tag{1.5}$$

which are obviously not the desired units of km/h.

(2) Check that the units of the final answer are the desired units. The problem asked us to solve for average speed in units of km/h and we have indeed obtained these units.

(3) Check the significant figures. Because each of the values given in the problem has three significant figures, the answer should also have three significant figures. The answer 30.0 km/hr does indeed have three significant figures, so this is appropriate. Note that the significant figures in the conversion factor are not relevant because an hour is *defined* to be 60 minutes, so the precision of the conversion factor is perfect.

(4) Next, check whether the answer is reasonable. Let us consider some information from the problem—if you travel 10 km in a third of an hour (20 min), you would travel three times that far in an hour. The answer does seem reasonable.

Solution for (b)

There are several ways to convert the average speed into meters per second.

(1) Start with the answer to (a) and convert km/h to m/s. Two conversion factors are needed—one to convert hours to seconds, and another to convert kilometers to meters.

(2) Multiplying by these yields

$$\text{Average speed} = 30.0 \frac{\text{km}}{\text{h}} \times \frac{1 \text{ h}}{3{,}600 \text{ s}} \times \frac{1{,}000 \text{ m}}{1 \text{ km}}, \tag{1.6}$$

$$\text{Average speed} = 8.33 \frac{\text{m}}{\text{s}}. \tag{1.7}$$

Discussion for (b)

If we had started with 0.500 km/min, we would have needed different conversion factors, but the answer would have been the same: 8.33 m/s.

You may have noted that the answers in the worked example just covered were given to three digits. Why? When do you need to be concerned about the number of digits in something you calculate? Why not write down all the digits your calculator produces? The module **Accuracy, Precision, and Significant Figures** will help you answer these questions.

Nonstandard Units

While there are numerous types of units that we are all familiar with, there are others that are much more obscure. For example, a **firkin** is a unit of volume that was once used to measure beer. One firkin equals about 34 liters. To learn more about nonstandard units, use a dictionary or encyclopedia to research different "weights and measures." Take note of any unusual units, such as a barleycorn, that are not listed in the text. Think about how the unit is defined and state its relationship to SI units.

Check Your Understanding

Some hummingbirds beat their wings more than 50 times per second. A scientist is measuring the time it takes for a hummingbird to beat its wings once. Which fundamental unit should the scientist use to describe the measurement? Which factor of 10 is the scientist likely to use to describe the motion precisely? Identify the metric prefix that corresponds to this factor of 10.

Solution

The scientist will measure the time between each movement using the fundamental unit of seconds. Because the wings beat so fast, the scientist will probably need to measure in milliseconds, or 10^{-3} seconds. (50 beats per second corresponds to

20 milliseconds per beat.)

One cubic centimeter is equal to one milliliter. What does this tell you about the different units in the SI metric system?

Solution

The fundamental unit of length (meter) is probably used to create the derived unit of volume (liter). The measure of a milliliter is dependent on the measure of a centimeter.

1.3 Accuracy, Precision, and Significant Figures

Figure 1.22 A double-pan mechanical balance is used to compare different masses. Usually an object with unknown mass is placed in one pan and objects of known mass are placed in the other pan. When the bar that connects the two pans is horizontal, then the masses in both pans are equal. The "known masses" are typically metal cylinders of standard mass such as 1 gram, 10 grams, and 100 grams. (credit: Serge Melki)

Figure 1.23 Many mechanical balances, such as double-pan balances, have been replaced by digital scales, which can typically measure the mass of an object more precisely. Whereas a mechanical balance may only read the mass of an object to the nearest tenth of a gram, many digital scales can measure the mass of an object up to the nearest thousandth of a gram. (credit: Karel Jakubec)

Learning Objectives

By the end of this section, you will be able to:

- Determine the appropriate number of significant figures in both addition and subtraction, as well as multiplication and division calculations.
- Calculate the percent uncertainty of a measurement.

Accuracy and Precision of a Measurement

Science is based on observation and experiment—that is, on measurements. **Accuracy** is how close a measurement is to the correct value for that measurement. For example, let us say that you are measuring the length of standard computer paper. The packaging in which you purchased the paper states that it is 11.0 inches long. You measure the length of the paper three times and obtain the following measurements: 11.1 in., 11.2 in., and 10.9 in. These measurements are quite accurate because they are very close to the correct value of 11.0 inches. In contrast, if you had obtained a measurement of 12 inches, your measurement would not be very accurate.

The **precision** of a measurement system is refers to how close the agreement is between repeated measurements (which are repeated under the same conditions). Consider the example of the paper measurements. The precision of the measurements refers to the spread of the measured values. One way to analyze the precision of the measurements would be to determine the range, or difference, between the lowest and the highest measured values. In that case, the lowest value was 10.9 in. and the highest value was 11.2 in. Thus, the measured values deviated from each other by at most 0.3 in. These measurements were relatively precise because they did not vary too much in value. However, if the measured values had been 10.9, 11.1, and 11.9, then the measurements would not be very precise because there would be significant variation from one measurement to another.

The measurements in the paper example are both accurate and precise, but in some cases, measurements are accurate but not precise, or they are precise but not accurate. Let us consider an example of a GPS system that is attempting to locate the position of a restaurant in a city. Think of the restaurant location as existing at the center of a bull's-eye target, and think of each GPS attempt to locate the restaurant as a black dot. In **Figure 1.24**, you can see that the GPS measurements are spread out far apart from each other, but they are all relatively close to the actual location of the restaurant at the center of the target. This indicates a low precision, high accuracy measuring system. However, in **Figure 1.25**, the GPS measurements are concentrated quite closely to one another, but they are far away from the target location. This indicates a high precision, low accuracy measuring system.

Figure 1.24 A GPS system attempts to locate a restaurant at the center of the bull's-eye. The black dots represent each attempt to pinpoint the location of the restaurant. The dots are spread out quite far apart from one another, indicating low precision, but they are each rather close to the actual location of the restaurant, indicating high accuracy. (credit: Dark Evil)

Figure 1.25 In this figure, the dots are concentrated rather closely to one another, indicating high precision, but they are rather far away from the actual location of the restaurant, indicating low accuracy. (credit: Dark Evil)

Accuracy, Precision, and Uncertainty

The degree of accuracy and precision of a measuring system are related to the **uncertainty** in the measurements. Uncertainty is a quantitative measure of how much your measured values deviate from a standard or expected value. If your measurements are not very accurate or precise, then the uncertainty of your values will be very high. In more general terms, uncertainty can be thought of as a disclaimer for your measured values. For example, if someone asked you to provide the mileage on your car, you might say that it is 45,000 miles, plus or minus 500 miles. The plus or minus amount is the uncertainty in your value. That is, you are indicating that the actual mileage of your car might be as low as 44,500 miles or as high as 45,500 miles, or anywhere in between. All measurements contain some amount of uncertainty. In our example of measuring the length of the paper, we might say that the length of the paper is 11 in., plus or minus 0.2 in. The uncertainty in a measurement, A, is often denoted as δA ("delta A"), so the measurement result would be recorded as $A \pm \delta A$. In our paper example, the length of the paper could be expressed as $11 \text{ in.} \pm 0.2$.

The factors contributing to uncertainty in a measurement include:

1.   Limitations of the measuring device,

2. The skill of the person making the measurement,

3. Irregularities in the object being measured,

4. Any other factors that affect the outcome (highly dependent on the situation).

In our example, such factors contributing to the uncertainty could be the following: the smallest division on the ruler is 0.1 in., the person using the ruler has bad eyesight, or one side of the paper is slightly longer than the other. At any rate, the uncertainty in a measurement must be based on a careful consideration of all the factors that might contribute and their possible effects.

Making Connections: Real-World Connections—Fevers or Chills?

Uncertainty is a critical piece of information, both in physics and in many other real-world applications. Imagine you are caring for a sick child. You suspect the child has a fever, so you check his or her temperature with a thermometer. What if the uncertainty of the thermometer were $3.0°C$? If the child's temperature reading was $37.0°C$ (which is normal body temperature), the "true" temperature could be anywhere from a hypothermic $34.0°C$ to a dangerously high $40.0°C$. A thermometer with an uncertainty of $3.0°C$ would be useless.

Percent Uncertainty

One method of expressing uncertainty is as a percent of the measured value. If a measurement A is expressed with uncertainty, δA, the **percent uncertainty** (%unc) is defined to be

$$\% \text{ unc} = \frac{\delta A}{A} \times 100\%.$$

(1.8)

Example 1.2 Calculating Percent Uncertainty: A Bag of Apples

A grocery store sells 5 lb bags of apples. You purchase four bags over the course of a month and weigh the apples each time. You obtain the following measurements:

- Week 1 weight: 4.8 lb
- Week 2 weight: 5.3 lb
- Week 3 weight: 4.9 lb
- Week 4 weight: 5.4 lb

You determine that the weight of the 5 lb bag has an uncertainty of ± 0.4 lb. What is the percent uncertainty of the bag's weight?

Strategy

First, observe that the expected value of the bag's weight, A, is 5 lb. The uncertainty in this value, δA, is 0.4 lb. We can use the following equation to determine the percent uncertainty of the weight:

$$\% \text{ unc} = \frac{\delta A}{A} \times 100\%.$$

(1.9)

Solution

Plug the known values into the equation:

$$\% \text{ unc} = \frac{0.4 \text{ lb}}{5 \text{ lb}} \times 100\% = 8\%.$$

(1.10)

Discussion

We can conclude that the weight of the apple bag is 5 lb $\pm 8\%$. Consider how this percent uncertainty would change if the bag of apples were half as heavy, but the uncertainty in the weight remained the same. Hint for future calculations: when calculating percent uncertainty, always remember that you must multiply the fraction by 100%. If you do not do this, you will have a decimal quantity, not a percent value.

Uncertainties in Calculations

There is an uncertainty in anything calculated from measured quantities. For example, the area of a floor calculated from measurements of its length and width has an uncertainty because the length and width have uncertainties. How big is the uncertainty in something you calculate by multiplication or division? If the measurements going into the calculation have small uncertainties (a few percent or less), then the **method of adding percents** can be used for multiplication or division. This method says that *the percent uncertainty in a quantity calculated by multiplication or division is the sum of the percent uncertainties in the items used to make the calculation*. For example, if a floor has a length of 4.00 m and a width of 3.00 m, with uncertainties of 2% and 1%, respectively, then the area of the floor is 12.0 m^2 and has an uncertainty of 3%.

(Expressed as an area this is 0.36 m^2, which we round to 0.4 m^2 since the area of the floor is given to a tenth of a square meter.)

Check Your Understanding

A high school track coach has just purchased a new stopwatch. The stopwatch manual states that the stopwatch has an uncertainty of $\pm 0.05 \text{ s}$. Runners on the track coach's team regularly clock 100 m sprints of 11.49 s to 15.01 s. At the school's last track meet, the first-place sprinter came in at 12.04 s and the second-place sprinter came in at 12.07 s. Will the coach's new stopwatch be helpful in timing the sprint team? Why or why not?

Solution
No, the uncertainty in the stopwatch is too great to effectively differentiate between the sprint times.

Precision of Measuring Tools and Significant Figures

An important factor in the accuracy and precision of measurements involves the precision of the measuring tool. In general, a precise measuring tool is one that can measure values in very small increments. For example, a standard ruler can measure length to the nearest millimeter, while a caliper can measure length to the nearest 0.01 millimeter. The caliper is a more precise measuring tool because it can measure extremely small differences in length. The more precise the measuring tool, the more precise and accurate the measurements can be.

When we express measured values, we can only list as many digits as we initially measured with our measuring tool. For example, if you use a standard ruler to measure the length of a stick, you may measure it to be 36.7 cm. You could not express this value as 36.71 cm because your measuring tool was not precise enough to measure a hundredth of a centimeter. It should be noted that the last digit in a measured value has been estimated in some way by the person performing the measurement. For example, the person measuring the length of a stick with a ruler notices that the stick length seems to be somewhere in between 36.6 cm and 36.7 cm, and he or she must estimate the value of the last digit. Using the method of **significant figures**, the rule is that *the last digit written down in a measurement is the first digit with some uncertainty*. In order to determine the number of significant digits in a value, start with the first measured value at the left and count the number of digits through the last digit written on the right. For example, the measured value 36.7 cm has three digits, or significant figures. Significant figures indicate the precision of a measuring tool that was used to measure a value.

Zeros

Special consideration is given to zeros when counting significant figures. The zeros in 0.053 are not significant, because they are only placekeepers that locate the decimal point. There are two significant figures in 0.053. The zeros in 10.053 are not placekeepers but are significant—this number has five significant figures. The zeros in 1300 may or may not be significant depending on the style of writing numbers. They could mean the number is known to the last digit, or they could be placekeepers. So 1300 could have two, three, or four significant figures. (To avoid this ambiguity, write 1300 in scientific notation.) *Zeros are significant except when they serve only as placekeepers.*

Check Your Understanding

Determine the number of significant figures in the following measurements:

 a. 0.0009

 b. 15,450.0

 c. 6×10^3

 d. 87.990

 e. 30.42

Solution

(a) 1; the zeros in this number are placekeepers that indicate the decimal point

(b) 6; here, the zeros indicate that a measurement was made to the 0.1 decimal point, so the zeros are significant

(c) 1; the value 10^3 signifies the decimal place, not the number of measured values

(d) 5; the final zero indicates that a measurement was made to the 0.001 decimal point, so it is significant

(e) 4; any zeros located in between significant figures in a number are also significant

Significant Figures in Calculations

When combining measurements with different degrees of accuracy and precision, *the number of significant digits in the final answer can be no greater than the number of significant digits in the least precise measured value*. There are two different rules, one for multiplication and division and the other for addition and subtraction, as discussed below.

1. For multiplication and division: *The result should have the same number of significant figures as the quantity having the least significant figures entering into the calculation.* For example, the area of a circle can be calculated from its radius using $A = \pi r^2$. Let us see how many significant figures the area has if the radius has only two—say, $r = 1.2$ m. Then,

$$A = \pi r^2 = (3.1415927...) \times (1.2 \text{ m})^2 = 4.5238934 \text{ m}^2 \qquad (1.11)$$

is what you would get using a calculator that has an eight-digit output. But because the radius has only two significant figures, it limits the calculated quantity to two significant figures or

$$A = 4.5 \text{ m}^2, \qquad (1.12)$$

even though π is good to at least eight digits.

2. For addition and subtraction: *The answer can contain no more decimal places than the least precise measurement.* Suppose that you buy 7.56 kg of potatoes in a grocery store as measured with a scale with precision 0.01 kg. Then you drop off 6.052 kg of potatoes at your laboratory as measured by a scale with precision 0.001 kg. Finally, you go home and add 13.7 kg of potatoes as measured by a bathroom scale with precision 0.1 kg. How many kilograms of potatoes do you now have, and how many significant figures are appropriate in the answer? The mass is found by simple addition and subtraction:

$$\begin{array}{r} 7.56 \ \ \text{kg} \\ - \ 6.052 \ \text{kg} \\ +13.7 \ \ \ \ \text{kg} \\ \hline 15.208 \ \text{kg} \end{array} = 15.2 \text{ kg}. \qquad (1.13)$$

Next, we identify the least precise measurement: 13.7 kg. This measurement is expressed to the 0.1 decimal place, so our final answer must also be expressed to the 0.1 decimal place. Thus, the answer is rounded to the tenths place, giving us 15.2 kg.

Significant Figures in this Text

In this text, most numbers are assumed to have three significant figures. Furthermore, consistent numbers of significant figures are used in all worked examples. You will note that an answer given to three digits is based on input good to at least three digits, for example. If the input has fewer significant figures, the answer will also have fewer significant figures. Care is also taken that the number of significant figures is reasonable for the situation posed. In some topics, particularly in optics, more accurate numbers are needed and more than three significant figures will be used. Finally, if a number is *exact*, such as the two in the formula for the circumference of a circle, $c = 2\pi r$, it does not affect the number of significant figures in a calculation.

Check Your Understanding

Perform the following calculations and express your answer using the correct number of significant digits.

(a) A woman has two bags weighing 13.5 pounds and one bag with a weight of 10.2 pounds. What is the total weight of the bags?

(b) The force F on an object is equal to its mass m multiplied by its acceleration a. If a wagon with mass 55 kg accelerates at a rate of 0.0255 m/s^2, what is the force on the wagon? (The unit of force is called the newton, and it is expressed with the symbol N.)

Solution

(a) 37.2 pounds; Because the number of bags is an exact value, it is not considered in the significant figures.

(b) 1.4 N; Because the value 55 kg has only two significant figures, the final value must also contain two significant figures.

PhET Explorations: Estimation

Explore size estimation in one, two, and three dimensions! Multiple levels of difficulty allow for progressive skill improvement.

Figure 1.26 Estimation (http://cnx.org/content/m54766/1.7/estimation_en.jar)

1.4 Approximation

Learning Objectives

By the end of this section, you will be able to:

- Make reasonable approximations based on given data.

On many occasions, physicists, other scientists, and engineers need to make **approximations** or "guesstimates" for a particular quantity. What is the distance to a certain destination? What is the approximate density of a given item? About how large a current will there be in a circuit? Many approximate numbers are based on formulae in which the input quantities are known only to a limited accuracy. As you develop problem-solving skills (that can be applied to a variety of fields through a study of physics), you will also develop skills at approximating. You will develop these skills through thinking more quantitatively, and by being willing to take risks. As with any endeavor, experience helps, as well as familiarity with units. These approximations allow us to rule out certain scenarios or unrealistic numbers. Approximations also allow us to challenge others and guide us in our approaches to our scientific world. Let us do two examples to illustrate this concept.

Example 1.3 Approximate the Height of a Building

Can you approximate the height of one of the buildings on your campus, or in your neighborhood? Let us make an approximation based upon the height of a person. In this example, we will calculate the height of a 39-story building.

Strategy

Think about the average height of an adult male. We can approximate the height of the building by scaling up from the height of a person.

Solution

Based on information in the example, we know there are 39 stories in the building. If we use the fact that the height of one story is approximately equal to about the length of two adult humans (each human is about 2 m tall), then we can estimate the total height of the building to be

$$\frac{2 \text{ m}}{1 \text{ person}} \times \frac{2 \text{ person}}{1 \text{ story}} \times 39 \text{ stories} = 156 \text{ m}.$$ (1.14)

Discussion

You can use known quantities to determine an approximate measurement of unknown quantities. If your hand measures 10 cm across, how many hand lengths equal the width of your desk? What other measurements can you approximate besides length?

Example 1.4 Approximating Vast Numbers: a Trillion Dollars

Figure 1.27 A bank stack contains one-hundred $100 bills, and is worth $10,000. How many bank stacks make up a trillion dollars? (credit: Andrew Magill)

The U.S. federal deficit in the 2008 fiscal year was a little greater than $10 trillion. Most of us do not have any concept of how much even one trillion actually is. Suppose that you were given a trillion dollars in $100 bills. If you made 100-bill stacks and used them to evenly cover a football field (between the end zones), make an approximation of how high the money pile would become. (We will use feet/inches rather than meters here because football fields are measured in yards.) One of your friends says 3 in., while another says 10 ft. What do you think?

Strategy

When you imagine the situation, you probably envision thousands of small stacks of 100 wrapped $100 bills, such as you might see in movies or at a bank. Since this is an easy-to-approximate quantity, let us start there. We can find the volume of a stack of 100 bills, find out how many stacks make up one trillion dollars, and then set this volume equal to the area of the football field multiplied by the unknown height.

Solution

(1) Calculate the volume of a stack of 100 bills. The dimensions of a single bill are approximately 3 in. by 6 in. A stack of 100 of these is about 0.5 in. thick. So the total volume of a stack of 100 bills is:

$$\text{volume of stack} = \text{length} \times \text{width} \times \text{height},$$ (1.15)
$$\text{volume of stack} = 6 \text{ in.} \times 3 \text{ in.} \times 0.5 \text{ in.},$$
$$\text{volume of stack} = 9 \text{ in.}^3.$$

(2) Calculate the number of stacks. Note that a trillion dollars is equal to $\$1 \times 10^{12}$, and a stack of one-hundred $100 bills is equal to $\$10,000$, or $\$1 \times 10^4$. The number of stacks you will have is:

$$\$1 \times 10^{12} (\text{a trillion dollars}) / \$1 \times 10^4 \text{ per stack} = 1 \times 10^8 \text{ stacks.}$$ (1.16)

(3) Calculate the area of a football field in square inches. The area of a football field is $100 \text{ yd} \times 50 \text{ yd}$, which gives $5,000 \text{ yd}^2$. Because we are working in inches, we need to convert square yards to square inches:

$$\text{Area} = 5,000 \text{ yd}^2 \times \frac{3 \text{ ft}}{1 \text{ yd}} \times \frac{3 \text{ ft}}{1 \text{ yd}} \times \frac{12 \text{ in.}}{1 \text{ ft}} \times \frac{12 \text{ in.}}{1 \text{ ft}} = 6,480,000 \text{ in.}^2,$$ (1.17)
$$\text{Area} \approx 6 \times 10^6 \text{ in.}^2.$$

This conversion gives us $6 \times 10^6 \text{ in.}^2$ for the area of the field. (Note that we are using only one significant figure in these calculations.)

(4) Calculate the total volume of the bills. The volume of all the $100-bill stacks is

$$9 \text{ in.}^3 / \text{stack} \times 10^8 \text{ stacks} = 9 \times 10^8 \text{ in.}^3.$$

(5) Calculate the height. To determine the height of the bills, use the equation:

$$\text{volume of bills} = \text{area of fiel} \times \text{height of money:}$$ (1.18)
$$\text{Height of money} = \frac{\text{volume of bills}}{\text{area of fiel}},$$
$$\text{Height of money} = \frac{9 \times 10^8 \text{in.}^3}{6 \times 10^6 \text{in.}^2} = 1.33 \times 10^2 \text{ in.,}$$
$$\text{Height of money} \approx 1 \times 10^2 \text{ in.} = 100 \text{ in.}$$

The height of the money will be about 100 in. high. Converting this value to feet gives

$$100 \text{ in.} \times \frac{1 \text{ ft}}{12 \text{ in.}} = 8.33 \text{ ft} \approx 8 \text{ ft.}$$ (1.19)

Discussion

The final approximate value is much higher than the early estimate of 3 in., but the other early estimate of 10 ft (120 in.) was roughly correct. How did the approximation measure up to your first guess? What can this exercise tell you in terms of rough "guesstimates" versus carefully calculated approximations?

Check Your Understanding

Using mental math and your understanding of fundamental units, approximate the area of a regulation basketball court. Describe the process you used to arrive at your final approximation.

Solution

An average male is about two meters tall. It would take approximately 15 men laid out end to end to cover the length, and about 7 to cover the width. That gives an approximate area of 420 m^2.

Glossary

accuracy: the degree to which a measured value agrees with correct value for that measurement

approximation: an estimated value based on prior experience and reasoning

classical physics: physics that was developed from the Renaissance to the end of the 19th century

conversion factor: a ratio expressing how many of one unit are equal to another unit

derived units: units that can be calculated using algebraic combinations of the fundamental units

English units: system of measurement used in the United States; includes units of measurement such as feet, gallons, and pounds

fundamental units: units that can only be expressed relative to the procedure used to measure them

kilogram: the SI unit for mass, abbreviated (kg)

law: a description, using concise language or a mathematical formula, a generalized pattern in nature that is supported by scientific evidence and repeated experiments

meter: the SI unit for length, abbreviated (m)

method of adding percents: the percent uncertainty in a quantity calculated by multiplication or division is the sum of the percent uncertainties in the items used to make the calculation

metric system: a system in which values can be calculated in factors of 10

model: representation of something that is often too difficult (or impossible) to display directly

modern physics: the study of relativity, quantum mechanics, or both

order of magnitude: refers to the size of a quantity as it relates to a power of 10

percent uncertainty: the ratio of the uncertainty of a measurement to the measured value, expressed as a percentage

physical quantity : a characteristic or property of an object that can be measured or calculated from other measurements

physics: the science concerned with describing the interactions of energy, matter, space, and time; it is especially interested in what fundamental mechanisms underlie every phenomenon

precision: the degree to which repeated measurements agree with each other

quantum mechanics: the study of objects smaller than can be seen with a microscope

relativity: the study of objects moving at speeds greater than about 1% of the speed of light, or of objects being affected by a strong gravitational field

scientific method: a method that typically begins with an observation and question that the scientist will research; next, the scientist typically performs some research about the topic and then devises a hypothesis; then, the scientist will test the hypothesis by performing an experiment; finally, the scientist analyzes the results of the experiment and draws a conclusion

second: the SI unit for time, abbreviated (s)

SI units : the international system of units that scientists in most countries have agreed to use; includes units such as meters, liters, and grams

significant figures: express the precision of a measuring tool used to measure a value

theory: an explanation for patterns in nature that is supported by scientific evidence and verified multiple times by various groups of researchers

uncertainty: a quantitative measure of how much your measured values deviate from a standard or expected value

units : a standard used for expressing and comparing measurements

Section Summary

1.1 Physics: An Introduction
- Science seeks to discover and describe the underlying order and simplicity in nature.

- Physics is the most basic of the sciences, concerning itself with energy, matter, space and time, and their interactions.
- Scientific laws and theories express the general truths of nature and the body of knowledge they encompass. These laws of nature are rules that all natural processes appear to follow.

1.2 Physical Quantities and Units

- Physical quantities are a characteristic or property of an object that can be measured or calculated from other measurements.
- Units are standards for expressing and comparing the measurement of physical quantities. All units can be expressed as combinations of four fundamental units.
- The four fundamental units we will use in this text are the meter (for length), the kilogram (for mass), the second (for time), and the ampere (for electric current). These units are part of the metric system, which uses powers of 10 to relate quantities over the vast ranges encountered in nature.
- The four fundamental units are abbreviated as follows: meter, m; kilogram, kg; second, s; and ampere, A. The metric system also uses a standard set of prefixes to denote each order of magnitude greater than or lesser than the fundamental unit itself.
- Unit conversions involve changing a value expressed in one type of unit to another type of unit. This is done by using conversion factors, which are ratios relating equal quantities of different units.

1.3 Accuracy, Precision, and Significant Figures

- Accuracy of a measured value refers to how close a measurement is to the correct value. The uncertainty in a measurement is an estimate of the amount by which the measurement result may differ from this value.
- Precision of measured values refers to how close the agreement is between repeated measurements.
- The precision of a *measuring tool* is related to the size of its measurement increments. The smaller the measurement increment, the more precise the tool.
- Significant figures express the precision of a measuring tool.
- When multiplying or dividing measured values, the final answer can contain only as many significant figures as the least precise value.
- When adding or subtracting measured values, the final answer cannot contain more decimal places than the least precise value.

1.4 Approximation

Scientists often approximate the values of quantities to perform calculations and analyze systems.

Conceptual Questions

1.1 Physics: An Introduction

1. Models are particularly useful in relativity and quantum mechanics, where conditions are outside those normally encountered by humans. What is a model?

2. How does a model differ from a theory?

3. If two different theories describe experimental observations equally well, can one be said to be more valid than the other (assuming both use accepted rules of logic)?

4. What determines the validity of a theory?

5. Certain criteria must be satisfied if a measurement or observation is to be believed. Will the criteria necessarily be as strict for an expected result as for an unexpected result?

6. Can the validity of a model be limited, or must it be universally valid? How does this compare to the required validity of a theory or a law?

7. Classical physics is a good approximation to modern physics under certain circumstances. What are they?

8. When is it *necessary* to use relativistic quantum mechanics?

9. Can classical physics be used to accurately describe a satellite moving at a speed of 7500 m/s? Explain why or why not.

1.2 Physical Quantities and Units

10. Identify some advantages of metric units.

1.3 Accuracy, Precision, and Significant Figures

11. What is the relationship between the accuracy and uncertainty of a measurement?

12. Prescriptions for vision correction are given in units called *diopters* (D). Determine the meaning of that unit. Obtain information (perhaps by calling an optometrist or performing an internet search) on the minimum uncertainty with which corrections in diopters are determined and the accuracy with which corrective lenses can be produced. Discuss the sources of uncertainties in both the prescription and accuracy in the manufacture of lenses.

Problems & Exercises

1.2 Physical Quantities and Units

1. The speed limit on some interstate highways is roughly 100 km/h. (a) What is this in meters per second? (b) How many miles per hour is this?

2. A car is traveling at a speed of 33 m/s. (a) What is its speed in kilometers per hour? (b) Is it exceeding the 90 km/h speed limit?

3. Show that $1.0 \text{ m/s} = 3.6 \text{ km/h}$. Hint: Show the explicit steps involved in converting $1.0 \text{ m/s} = 3.6 \text{ km/h}$.

4. American football is played on a 100-yd-long field, excluding the end zones. How long is the field in meters? (Assume that 1 meter equals 3.281 feet.)

5. Soccer fields vary in size. A large soccer field is 115 m long and 85 m wide. What are its dimensions in feet and inches? (Assume that 1 meter equals 3.281 feet.)

6. What is the height in meters of a person who is 6 ft 1.0 in. tall? (Assume that 1 meter equals 39.37 in.)

7. Mount Everest, at 29,028 feet, is the tallest mountain on the Earth. What is its height in kilometers? (Assume that 1 kilometer equals 3,281 feet.)

8. The speed of sound is measured to be 342 m/s on a certain day. What is this in km/h?

9. Tectonic plates are large segments of the Earth's crust that move slowly. Suppose that one such plate has an average speed of 4.0 cm/year. (a) What distance does it move in 1 s at this speed? (b) What is its speed in kilometers per million years?

10. (a) Refer to **Table 1.3** to determine the average distance between the Earth and the Sun. Then calculate the average speed of the Earth in its orbit in kilometers per second. (b) What is this in meters per second?

1.3 Accuracy, Precision, and Significant Figures

Express your answers to problems in this section to the correct number of significant figures and proper units.

11. Suppose that your bathroom scale reads your mass as 65 kg with a 3% uncertainty. What is the uncertainty in your mass (in kilograms)?

12. A good-quality measuring tape can be off by 0.50 cm over a distance of 20 m. What is its percent uncertainty?

13. (a) A car speedometer has a 5.0% uncertainty. What is the range of possible speeds when it reads 90 km/h? (b) Convert this range to miles per hour. $(1 \text{ km} = 0.6214 \text{ mi})$

14. An infant's pulse rate is measured to be 130 ± 5 beats/min. What is the percent uncertainty in this measurement?

15. (a) Suppose that a person has an average heart rate of 72.0 beats/min. How many beats does he or she have in 2.0 y? (b) In 2.00 y? (c) In 2.000 y?

16. A can contains 375 mL of soda. How much is left after 308 mL is removed?

17. State how many significant figures are proper in the results of the following calculations: (a) $(106.7)(98.2)/(46.210)(1.01)$ (b) $(18.7)^2$ (c) $\left(1.60 \times 10^{-19}\right)(3712)$.

18. (a) How many significant figures are in the numbers 99 and 100? (b) If the uncertainty in each number is 1, what is the percent uncertainty in each? (c) Which is a more meaningful way to express the accuracy of these two numbers, significant figures or percent uncertainties?

19. (a) If your speedometer has an uncertainty of 2.0 km/h at a speed of 90 km/h, what is the percent uncertainty? (b) If it has the same percent uncertainty when it reads 60 km/h, what is the range of speeds you could be going?

20. (a) A person's blood pressure is measured to be $120 \pm 2 \text{ mm Hg}$. What is its percent uncertainty? (b) Assuming the same percent uncertainty, what is the uncertainty in a blood pressure measurement of 80 mm Hg?

21. A person measures his or her heart rate by counting the number of beats in 30 s. If 40 ± 1 beats are counted in $30.0 \pm 0.5 \text{ s}$, what is the heart rate and its uncertainty in beats per minute?

22. What is the area of a circle 3.102 cm in diameter?

23. If a marathon runner averages 9.5 mi/h, how long does it take him or her to run a 26.22 mi marathon?

24. A marathon runner completes a 42.188 km course in 2 h, 30 min, and 12 s. There is an uncertainty of 25 m in the distance traveled and an uncertainty of 1 s in the elapsed time. (a) Calculate the percent uncertainty in the distance. (b) Calculate the uncertainty in the elapsed time. (c) What is the average speed in meters per second? (d) What is the uncertainty in the average speed?

25. The sides of a small rectangular box are measured to be $1.80 \pm 0.01 \text{ cm}$, $2.05 \pm 0.02 \text{ cm}$, and $3.1 \pm 0.1 \text{ cm}$ long. Calculate its volume and uncertainty in cubic centimeters.

26. When non-metric units were used in the United Kingdom, a unit of mass called the *pound-mass* (lbm) was employed, where $1 \text{ lbm} = 0.4539 \text{ kg}$. (a) If there is an uncertainty of 0.0001 kg in the pound-mass unit, what is its percent uncertainty? (b) Based on that percent uncertainty, what mass in pound-mass has an uncertainty of 1 kg when converted to kilograms?

27. The length and width of a rectangular room are measured to be $3.955 \pm 0.005 \text{ m}$ and $3.050 \pm 0.005 \text{ m}$. Calculate the area of the room and its uncertainty in square meters.

28. A car engine moves a piston with a circular cross section of $7.500 \pm 0.002 \text{ cm}$ diameter a distance of $3.250 \pm 0.001 \text{ cm}$ to compress the gas in the cylinder. (a) By what amount is the gas decreased in volume in cubic centimeters? (b) Find the uncertainty in this volume.

1.4 Approximation

29. How many heartbeats are there in a lifetime?

30. A generation is about one-third of a lifetime. Approximately how many generations have passed since the year 0 AD?

31. How many times longer than the mean life of an extremely unstable atomic nucleus is the lifetime of a human? (Hint: The lifetime of an unstable atomic nucleus is on the order of 10^{-22} s.)

32. Calculate the approximate number of atoms in a bacterium. Assume that the average mass of an atom in the bacterium is ten times the mass of a hydrogen atom. (Hint: The mass of a hydrogen atom is on the order of 10^{-27} kg and the mass of a bacterium is on the order of 10^{-15} kg.)

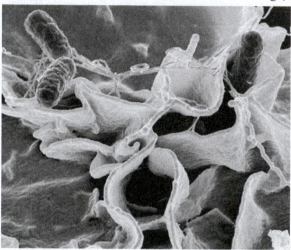

Figure 1.28 This color-enhanced photo shows *Salmonella typhimurium* (red) attacking human cells. These bacteria are commonly known for causing foodborne illness. Can you estimate the number of atoms in each bacterium? (credit: Rocky Mountain Laboratories, NIAID, NIH)

33. Approximately how many atoms thick is a cell membrane, assuming all atoms there average about twice the size of a hydrogen atom?

34. (a) What fraction of Earth's diameter is the greatest ocean depth? (b) The greatest mountain height?

35. (a) Calculate the number of cells in a hummingbird assuming the mass of an average cell is ten times the mass of a bacterium. (b) Making the same assumption, how many cells are there in a human?

36. Assuming one nerve impulse must end before another can begin, what is the maximum firing rate of a nerve in impulses per second?

2 KINEMATICS

Figure 2.1 The motion of an American kestrel through the air can be described by the bird's displacement, speed, velocity, and acceleration. When it flies in a straight line without any change in direction, its motion is said to be one dimensional. (credit: Vince Maidens, Wikimedia Commons)

Chapter Outline

2.1. Displacement

2.2. Vectors, Scalars, and Coordinate Systems

2.3. Time, Velocity, and Speed

2.4. Acceleration

2.5. Motion Equations for Constant Acceleration in One Dimension

2.6. Problem-Solving Basics for One Dimensional Kinematics

2.7. Falling Objects

2.8. Graphical Analysis of One Dimensional Motion

Connection for AP® Courses

Objects are in motion everywhere we look. Everything from a tennis game to a space-probe flyby of the planet Neptune involves motion. When you are resting, your heart moves blood through your veins. Even in inanimate objects, there is a continuous motion in the vibrations of atoms and molecules. Questions about motion are interesting in and of themselves: *How long will it take for a space probe to get to Mars? Where will a football land if it is thrown at a certain angle?*

Understanding motion will not only provide answers to these questions, but will be key to understanding more advanced concepts in physics. For example, the discussion of force in Chapter 4 will not fully make sense until you understand acceleration. This relationship between force and acceleration is also critical to understanding Big Idea 3.

Additionally, this unit will explore the topic of reference frames, a critical component to quantifying how things move. If you have ever waved to a departing friend at a train station, you are likely familiar with this idea. While you see your friend move away from you at a considerable rate, those sitting with her will likely see her as not moving. The effect that the chosen reference frame has on your observations is substantial, and an understanding of this is needed to grasp both Enduring Understanding 3.A and Essential Knowledge 3.A.1.

Our formal study of physics begins with **kinematics,** which is defined as the *study of motion without considering its causes*. In one- and two-dimensional kinematics we will study only the *motion* of a football, for example, without worrying about what forces cause or change its motion. In this chapter, we examine the simplest type of motion—namely, motion along a straight line, or one-dimensional motion. Later, in two-dimensional kinematics, we apply concepts developed here to study motion along curved paths (two- and three-dimensional motion), for example, that of a car rounding a curve.

The content in this chapter supports:

Big Idea 3 The interactions of an object with other objects can be described by forces.

Enduring Understanding 3.A All forces share certain common characteristics when considered by observers in inertial reference frames.

Essential Knowledge 3.A.1 An observer in a particular reference frame can describe the motion of an object using such quantities as position, displacement, distance, velocity, speed, and acceleration.

2.1 Displacement

Figure 2.2 These cyclists in Vietnam can be described by their position relative to buildings and a canal. Their motion can be described by their change in position, or displacement, in the frame of reference. (credit: Suzan Black, Fotopedia)

Learning Objectives

By the end of this section, you will be able to:

- Define position, displacement, distance, and distance traveled in a particular frame of reference.
- Explain the relationship between position and displacement.
- Distinguish between displacement and distance traveled.
- Calculate displacement and distance given initial position, final position, and the path between the two.

The information presented in this section supports the following AP® learning objectives and science practices:

- **3.A.1.1** The student is able to express the motion of an object using narrative, mathematical, and graphical representations. **(S.P. 1.5, 2.1, 2.2)**
- **3.A.1.3** The student is able to analyze experimental data describing the motion of an object and is able to express the results of the analysis using narrative, mathematical, and graphical representations. **(S.P. 5.1)**

Position

In order to describe the motion of an object, you must first be able to describe its **position**—where it is at any particular time. More precisely, you need to specify its position relative to a convenient reference frame. Earth is often used as a reference frame, and we often describe the position of an object as it relates to stationary objects in that reference frame. For example, a rocket launch would be described in terms of the position of the rocket with respect to the Earth as a whole, while a professor's position could be described in terms of where she is in relation to the nearby white board. (See **Figure 2.3**.) In other cases, we use reference frames that are not stationary but are in motion relative to the Earth. To describe the position of a person in an airplane, for example, we use the airplane, not the Earth, as the reference frame. (See **Figure 2.4**.)

Displacement

If an object moves relative to a reference frame (for example, if a professor moves to the right relative to a white board or a passenger moves toward the rear of an airplane), then the object's position changes. This change in position is known as **displacement**. The word "displacement" implies that an object has moved, or has been displaced.

Displacement

Displacement is the *change in position* of an object:

$$\Delta x = x_f - x_0,$$ (2.1)

where Δx is displacement, x_f is the final position, and x_0 is the initial position.

In this text the upper case Greek letter Δ (delta) always means "change in" whatever quantity follows it; thus, Δx means *change in position*. Always solve for displacement by subtracting initial position x_0 from final position x_f.

Note that the SI unit for displacement is the meter (m) (see **Physical Quantities and Units**), but sometimes kilometers, miles, feet, and other units of length are used. Keep in mind that when units other than the meter are used in a problem, you may need to convert them into meters to complete the calculation.

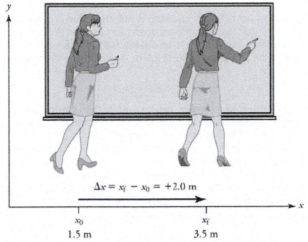

Figure 2.3 A professor paces left and right while lecturing. Her position relative to the blackboard is given by x. The $+2.0\ \text{m}$ displacement of the professor relative to the blackboard is represented by an arrow pointing to the right.

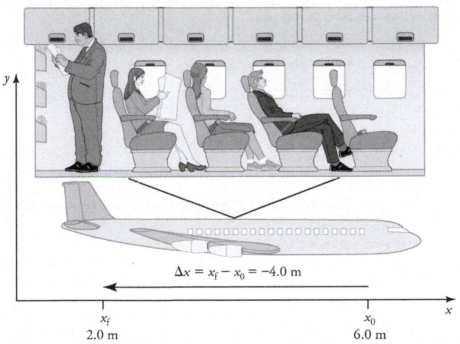

Figure 2.4 A passenger moves from his seat to the back of the plane. His location relative to the airplane is given by x. The −4 m displacement of the passenger relative to the plane is represented by an arrow toward the rear of the plane. Notice that the arrow representing his displacement is twice as long as the arrow representing the displacement of the professor (he moves twice as far) in **Figure 2.3**.

Note that displacement has a direction as well as a magnitude. The professor's displacement is 2.0 m to the right, and the airline passenger's displacement is 4.0 m toward the rear. In one-dimensional motion, direction can be specified with a plus or minus sign. When you begin a problem, you should select which direction is positive (usually that will be to the right or up, but you are free to select positive as being any direction). The professor's initial position is $x_0 = 1.5\ \text{m}$ and her final position is

$x_f = 3.5\ \text{m}$. Thus her displacement is

$$\Delta x = x_f - x_0 = 3.5\ \text{m} - 1.5\ \text{m} = +2.0\ \text{m}. \tag{2.2}$$

In this coordinate system, motion to the right is positive, whereas motion to the left is negative. Similarly, the airplane passenger's initial position is $x_0 = 6.0\ \text{m}$ and his final position is $x_f = 2.0\ \text{m}$, so his displacement is

$$\Delta x = x_f - x_0 = 2.0 \text{ m} - 6.0 \text{ m} = -4.0 \text{ m}. \qquad (2.3)$$

His displacement is negative because his motion is toward the rear of the plane, or in the negative x direction in our coordinate system.

Distance

Although displacement is described in terms of direction, distance is not. **Distance** is defined to be *the magnitude or size of displacement between two positions*. Note that the distance between two positions is not the same as the distance traveled between them. **Distance traveled** is *the total length of the path traveled between two positions*. Distance has no direction and, thus, no sign. For example, the distance the professor walks is 2.0 m. The distance the airplane passenger walks is 4.0 m.

Misconception Alert: Distance Traveled vs. Magnitude of Displacement

It is important to note that the *distance traveled*, however, can be greater than the magnitude of the displacement (by magnitude, we mean just the size of the displacement without regard to its direction; that is, just a number with a unit). For example, the professor could pace back and forth many times, perhaps walking a distance of 150 m during a lecture, yet still end up only 2.0 m to the right of her starting point. In this case her displacement would be +2.0 m, the magnitude of her displacement would be 2.0 m, but the distance she traveled would be 150 m. In kinematics we nearly always deal with displacement and magnitude of displacement, and almost never with distance traveled. One way to think about this is to assume you marked the start of the motion and the end of the motion. The displacement is simply the difference in the position of the two marks and is independent of the path taken in traveling between the two marks. The distance traveled, however, is the total length of the path taken between the two marks.

Check Your Understanding

A cyclist rides 3 km west and then turns around and rides 2 km east. (a) What is her displacement? (b) What distance does she ride? (c) What is the magnitude of her displacement?

Solution

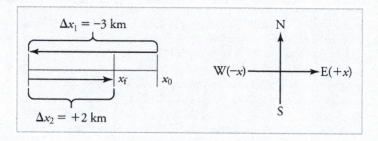

Figure 2.5

(a) The rider's displacement is $\Delta x = x_f - x_0 = -1 \text{ km}$. (The displacement is negative because we take east to be positive and west to be negative.)

(b) The distance traveled is $3 \text{ km} + 2 \text{ km} = 5 \text{ km}$.

(c) The magnitude of the displacement is 1 km.

2.2 Vectors, Scalars, and Coordinate Systems

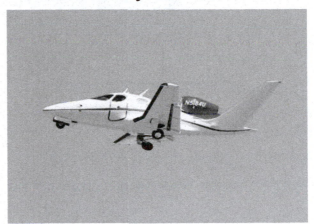

Figure 2.6 The motion of this Eclipse Concept jet can be described in terms of the distance it has traveled (a scalar quantity) or its displacement in a specific direction (a vector quantity). In order to specify the direction of motion, its displacement must be described based on a coordinate system. In this case, it may be convenient to choose motion toward the left as positive motion (it is the forward direction for the plane), although in many cases, the x -coordinate runs from left to right, with motion to the right as positive and motion to the left as negative. (credit: Armchair Aviator, Flickr)

Learning Objectives

By the end of this section, you will be able to:

- Define and distinguish between scalar and vector quantities.
- Assign a coordinate system for a scenario involving one-dimensional motion.

The information presented in this section supports the following AP® learning objectives and science practices:

- **3.A.1.2** The student is able to design an experimental investigation of the motion of an object.

What is the difference between distance and displacement? Whereas displacement is defined by both direction and magnitude, distance is defined only by magnitude. Displacement is an example of a vector quantity. Distance is an example of a scalar quantity. A **vector** is any quantity with both *magnitude and direction*. Other examples of vectors include a velocity of 90 km/h east and a force of 500 newtons straight down.

The direction of a vector in one-dimensional motion is given simply by a plus $(+)$ or minus $(-)$ sign. Vectors are represented graphically by arrows. An arrow used to represent a vector has a length proportional to the vector's magnitude (e.g., the larger the magnitude, the longer the length of the vector) and points in the same direction as the vector.

Some physical quantities, like distance, either have no direction or none is specified. A **scalar** is any quantity that has a magnitude, but no direction. For example, a $20°C$ temperature, the 250 kilocalories (250 Calories) of energy in a candy bar, a 90 km/h speed limit, a person's 1.8 m height, and a distance of 2.0 m are all scalars—quantities with no specified direction. Note, however, that a scalar can be negative, such as a $-20°C$ temperature. In this case, the minus sign indicates a point on a scale rather than a direction. Scalars are never represented by arrows.

Coordinate Systems for One-Dimensional Motion

In order to describe the direction of a vector quantity, you must designate a coordinate system within the reference frame. For one-dimensional motion, this is a simple coordinate system consisting of a one-dimensional coordinate line. In general, when describing horizontal motion, motion to the right is usually considered positive, and motion to the left is considered negative. With vertical motion, motion up is usually positive and motion down is negative. In some cases, however, as with the jet in **Figure 2.6**, it can be more convenient to switch the positive and negative directions. For example, if you are analyzing the motion of falling objects, it can be useful to define downwards as the positive direction. If people in a race are running to the left, it is useful to define left as the positive direction. It does not matter as long as the system is clear and consistent. Once you assign a positive direction and start solving a problem, you cannot change it.

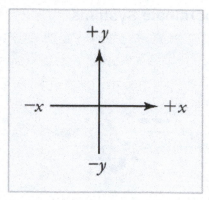

Figure 2.7 It is usually convenient to consider motion upward or to the right as positive $(+)$ and motion downward or to the left as negative $(-)$.

Check Your Understanding

A person's speed can stay the same as he or she rounds a corner and changes direction. Given this information, is speed a scalar or a vector quantity? Explain.

Solution
Speed is a scalar quantity. It does not change at all with direction changes; therefore, it has magnitude only. If it were a vector quantity, it would change as direction changes (even if its magnitude remained constant).

Switching Reference Frames

A fundamental tenet of physics is that information about an event can be gathered from a variety of reference frames. For example, imagine that you are a passenger walking toward the front of a bus. As you walk, your motion is observed by a fellow bus passenger and by an observer standing on the sidewalk.

Both the bus passenger and sidewalk observer will be able to collect information about you. They can determine how far you moved and how much time it took you to do so. However, while you moved at a consistent pace, both observers will get different results. To the passenger sitting on the bus, you moved forward at what one would consider a normal pace, something similar to how quickly you would walk outside on a sunny day. To the sidewalk observer though, you will have moved much quicker. Because the bus is also moving forward, the distance you move forward against the sidewalk each second increases, and the sidewalk observer must conclude that you are moving at a greater pace.

To show that you understand this concept, you will need to create an event and think of a way to view this event from two different frames of reference. In order to ensure that the event is being observed simultaneously from both frames, you will need an assistant to help out. An example of a possible event is to have a friend ride on a skateboard while tossing a ball. How will your friend observe the ball toss, and how will those observations be different from your own?

Your task is to describe your event and the observations of your event from both frames of reference. Answer the following questions below to demonstrate your understanding. For assistance, you can review the information given in the 'Position' paragraph at the start of Section 2.1.

1.   What is your event? What object are both you and your assistant observing?

2.   What do *you* see as the event takes place?

3.   What does *your assistant* see as the event takes place?

4.   How do your reference frames cause you and your assistant to have two different sets of observations?

2.3 Time, Velocity, and Speed

Figure 2.8 The motion of these racing snails can be described by their speeds and their velocities. (credit: tobitasflickr, Flickr)

Learning Objectives

By the end of this section, you will be able to:

- Explain the relationships between instantaneous velocity, average velocity, instantaneous speed, average speed, displacement, and time.
- Calculate velocity and speed given initial position, initial time, final position, and final time.
- Derive a graph of velocity vs. time given a graph of position vs. time.
- Interpret a graph of velocity vs. time.

The information presented in this section supports the following AP® learning objectives and science practices:

- **3.A.1.1** The student is able to express the motion of an object using narrative, mathematical, and graphical representations. **(S.P. 1.5, 2.1, 2.2)**
- **3.A.1.3** The student is able to analyze experimental data describing the motion of an object and is able to express the results of the analysis using narrative, mathematical, and graphical representations. **(S.P. 5.1)**

There is more to motion than distance and displacement. Questions such as, "How long does a foot race take?" and "What was the runner's speed?" cannot be answered without an understanding of other concepts. In this section we add definitions of time, velocity, and speed to expand our description of motion.

Time

As discussed in **Physical Quantities and Units**, the most fundamental physical quantities are defined by how they are measured. This is the case with time. Every measurement of time involves measuring a change in some physical quantity. It may be a number on a digital clock, a heartbeat, or the position of the Sun in the sky. In physics, the definition of time is simple— **time** is *change*, or the interval over which change occurs. It is impossible to know that time has passed unless something changes.

The amount of time or change is calibrated by comparison with a standard. The SI unit for time is the second, abbreviated s. We might, for example, observe that a certain pendulum makes one full swing every 0.75 s. We could then use the pendulum to measure time by counting its swings or, of course, by connecting the pendulum to a clock mechanism that registers time on a dial. This allows us to not only measure the amount of time, but also to determine a sequence of events.

How does time relate to motion? We are usually interested in elapsed time for a particular motion, such as how long it takes an airplane passenger to get from his seat to the back of the plane. To find elapsed time, we note the time at the beginning and end of the motion and subtract the two. For example, a lecture may start at 11:00 A.M. and end at 11:50 A.M., so that the elapsed time would be 50 min. **Elapsed time** Δt is the difference between the ending time and beginning time,

$$\Delta t = t_f - t_0,$$
(2.4)

where Δt is the change in time or elapsed time, t_f is the time at the end of the motion, and t_0 is the time at the beginning of the motion. (As usual, the delta symbol, Δ, means the change in the quantity that follows it.)

Life is simpler if the beginning time t_0 is taken to be zero, as when we use a stopwatch. If we were using a stopwatch, it would simply read zero at the start of the lecture and 50 min at the end. If $t_0 = 0$, then $\Delta t = t_f \equiv t$.

In this text, for simplicity's sake,

- motion starts at time equal to zero $(t_0 = 0)$
- the symbol t is used for elapsed time unless otherwise specified $(\Delta t = t_f \equiv t)$

Velocity

Your notion of velocity is probably the same as its scientific definition. You know that if you have a large displacement in a small amount of time you have a large velocity, and that velocity has units of distance divided by time, such as miles per hour or kilometers per hour.

Average Velocity

Average velocity is *displacement (change in position) divided by the time of travel,*

$$\bar{v} = \frac{\Delta x}{\Delta t} = \frac{x_f - x_0}{t_f - t_0},$$ (2.5)

where $\bar{v}$ is the *average* (indicated by the bar over the v) velocity, Δx is the change in position (or displacement), and x_f and x_0 are the final and beginning positions at times t_f and t_0, respectively. If the starting time t_0 is taken to be zero, then the average velocity is simply

$$\bar{v} = \frac{\Delta x}{t}.$$ (2.6)

Notice that this definition indicates that *velocity is a vector because displacement is a vector*. It has both magnitude and direction. The SI unit for velocity is meters per second or m/s, but many other units, such as km/h, mi/h (also written as mph), and cm/s, are in common use. Suppose, for example, an airplane passenger took 5 seconds to move −4 m (the minus sign indicates that displacement is toward the back of the plane). His average velocity would be

$$\bar{v} = \frac{\Delta x}{t} = \frac{-4 \text{ m}}{5 \text{ s}} = -0.8 \text{ m/s}.$$ (2.7)

The minus sign indicates the average velocity is also toward the rear of the plane.

The average velocity of an object does not tell us anything about what happens to it between the starting point and ending point, however. For example, we cannot tell from average velocity whether the airplane passenger stops momentarily or backs up before he goes to the back of the plane. To get more details, we must consider smaller segments of the trip over smaller time intervals.

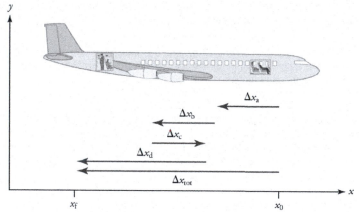

Figure 2.9 A more detailed record of an airplane passenger heading toward the back of the plane, showing smaller segments of his trip.

The smaller the time intervals considered in a motion, the more detailed the information. When we carry this process to its logical conclusion, we are left with an infinitesimally small interval. Over such an interval, the average velocity becomes the *instantaneous velocity* or the *velocity at a specific instant*. A car's speedometer, for example, shows the magnitude (but not the direction) of the instantaneous velocity of the car. (Police give tickets based on instantaneous velocity, but when calculating how long it will take to get from one place to another on a road trip, you need to use average velocity.) **Instantaneous velocity** v is the average velocity at a specific instant in time (or over an infinitesimally small time interval).

Mathematically, finding instantaneous velocity, v, at a precise instant t can involve taking a limit, a calculus operation beyond the scope of this text. However, under many circumstances, we can find precise values for instantaneous velocity without calculus.

Speed

In everyday language, most people use the terms "speed" and "velocity" interchangeably. In physics, however, they do not have the same meaning and they are distinct concepts. One major difference is that speed has no direction. Thus *speed is a scalar*. Just as we need to distinguish between instantaneous velocity and average velocity, we also need to distinguish between instantaneous speed and average speed.

Instantaneous speed is the magnitude of instantaneous velocity. For example, suppose the airplane passenger at one instant had an instantaneous velocity of −3.0 m/s (the minus meaning toward the rear of the plane). At that same time his instantaneous speed was 3.0 m/s. Or suppose that at one time during a shopping trip your instantaneous velocity is 40 km/h due north. Your instantaneous speed at that instant would be 40 km/h—the same magnitude but without a direction. Average speed, however, is very different from average velocity. **Average speed** is the distance traveled divided by elapsed time.

We have noted that distance traveled can be greater than displacement. So average speed can be greater than average velocity, which is displacement divided by time. For example, if you drive to a store and return home in half an hour, and your car's odometer shows the total distance traveled was 6 km, then your average speed was 12 km/h. Your average velocity, however, was zero, because your displacement for the round trip is zero. (Displacement is change in position and, thus, is zero for a round trip.) Thus average speed is *not* simply the magnitude of average velocity.

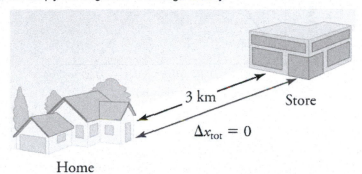

Home

Figure 2.10 During a 30-minute round trip to the store, the total distance traveled is 6 km. The average speed is 12 km/h. The displacement for the round trip is zero, since there was no net change in position. Thus the average velocity is zero.

Another way of visualizing the motion of an object is to use a graph. A plot of position or of velocity as a function of time can be very useful. For example, for this trip to the store, the position, velocity, and speed-vs.-time graphs are displayed in **Figure 2.11**. (Note that these graphs depict a very simplified **model** of the trip. We are assuming that speed is constant during the trip, which is unrealistic given that we'll probably stop at the store. But for simplicity's sake, we will model it with no stops or changes in speed. We are also assuming that the route between the store and the house is a perfectly straight line.)

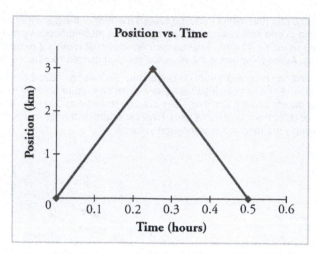

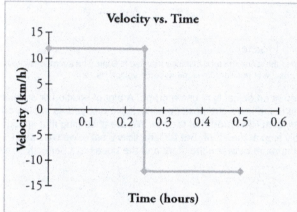

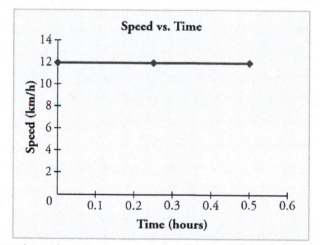

Figure 2.11 Position vs. time, velocity vs. time, and speed vs. time on a trip. Note that the velocity for the return trip is negative.

Making Connections: Take-Home Investigation—Getting a Sense of Speed

If you have spent much time driving, you probably have a good sense of speeds between about 10 and 70 miles per hour. But what are these in meters per second? What do we mean when we say that something is moving at 10 m/s? To get a better sense of what these values really mean, do some observations and calculations on your own:

- calculate typical car speeds in meters per second
- estimate jogging and walking speed by timing yourself; convert the measurements into both m/s and mi/h
- determine the speed of an ant, snail, or falling leaf

A commuter train travels from Baltimore to Washington, DC, and back in 1 hour and 45 minutes. The distance between the two stations is approximately 40 miles. What is (a) the average velocity of the train, and (b) the average speed of the train in m/s?

Solution

(a) The average velocity of the train is zero because $x_f = x_0$; the train ends up at the same place it starts.

(b) The average speed of the train is calculated below. Note that the train travels 40 miles one way and 40 miles back, for a total distance of 80 miles.

$$\frac{\text{distance}}{\text{time}} = \frac{80 \text{ miles}}{105 \text{ minutes}} \tag{2.8}$$

$$\frac{80 \text{ miles}}{105 \text{ minutes}} \times \frac{5280 \text{ feet}}{1 \text{ mile}} \times \frac{1 \text{ meter}}{3.28 \text{ feet}} \times \frac{1 \text{ minute}}{60 \text{ seconds}} = 20 \text{ m/s} \tag{2.9}$$

2.4 Acceleration

Figure 2.12 A plane decelerates, or slows down, as it comes in for landing in St. Maarten. Its acceleration is opposite in direction to its velocity. (credit: Steve Conry, Flickr)

Learning Objectives

By the end of this section, you will be able to:

- Define and distinguish between instantaneous acceleration and average acceleration.
- Calculate acceleration given initial time, initial velocity, final time, and final velocity.

The information presented in this section supports the following AP® learning objectives and science practices:

- **3.A.1.1** The student is able to express the motion of an object using narrative, mathematical, and graphical representations. **(S.P. 1.5, 2.1, 2.2)**
- **3.A.1.3** The student is able to analyze experimental data describing the motion of an object and is able to express the results of the analysis using narrative, mathematical, and graphical representations. **(S.P. 5.1)**

In everyday conversation, to accelerate means to speed up. The accelerator in a car can in fact cause it to speed up. The greater the **acceleration**, the greater the change in velocity over a given time. The formal definition of acceleration is consistent with these notions, but more inclusive.

Average Acceleration

Average Acceleration is *the rate at which velocity changes*,

$$\bar{a} = \frac{\Delta v}{\Delta t} = \frac{v_f - v_0}{t_f - t_0}, \tag{2.10}$$

where $\bar{a}$ is average acceleration, v is velocity, and t is time. (The bar over the a means *average* acceleration.)

Because acceleration is velocity in m/s divided by time in s, the SI units for acceleration are m/s^2, meters per second squared or meters per second per second, which literally means by how many meters per second the velocity changes every second.

Recall that velocity is a vector—it has both magnitude and direction. This means that a change in velocity can be a change in magnitude (or speed), but it can also be a change in *direction*. For example, if a car turns a corner at constant speed, it is accelerating because its direction is changing. The quicker you turn, the greater the acceleration. So there is an acceleration when velocity changes either in magnitude (an increase or decrease in speed) or in direction, or both.

Acceleration as a Vector

Acceleration is a vector in the same direction as the *change* in velocity, Δv. Since velocity is a vector, it can change either in magnitude or in direction. Acceleration is therefore a change in either speed or direction, or both.

Keep in mind that although acceleration is in the direction of the *change* in velocity, it is not always in the direction of *motion*. When an object's acceleration is in the same direction of its motion, the object will speed up. However, when an object's acceleration is opposite to the direction of its motion, the object will slow down. Speeding up and slowing down should not be confused with a positive and negative acceleration. The next two examples should help to make this distinction clear.

Figure 2.13 A subway train in Sao Paulo, Brazil, decelerates as it comes into a station. It is accelerating in a direction opposite to its direction of motion. (credit: Yusuke Kawasaki, Flickr)

Making Connections: Car Motion

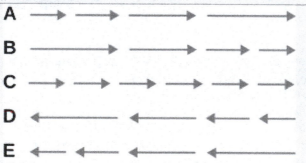

Figure 2.14 Above are arrows representing the motion of five cars (A–E). In all five cases, the positive direction should be considered to the right of the page.

Consider the acceleration and velocity of each car in terms of its direction of travel.

Figure 2.15 Car A is speeding up.

Because the positive direction is considered to the right of the paper, Car A is moving with a positive velocity. Because it is speeding up while moving with a positive velocity, its acceleration is also considered positive.

Figure 2.16 Car B is slowing down.

Because the positive direction is considered to the right of the paper, Car B is also moving with a positive velocity. However, because it is slowing down while moving with a positive velocity, its acceleration is considered negative. (This can be viewed in a mathematical manner as well. If the car was originally moving with a velocity of +25 m/s, it is finishing with a speed less than that, like +5 m/s. Because the change in velocity is negative, the acceleration will be as well.)

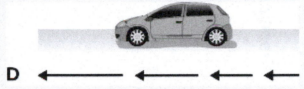

Figure 2.17 Car C has a constant speed.

Because the positive direction is considered to the right of the paper, Car C is moving with a positive velocity. Because all arrows are of the same length, this car is not changing its speed. As a result, its change in velocity is zero, and its acceleration must be zero as well.

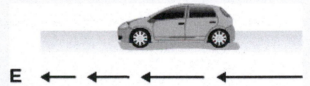

Figure 2.18 Car D is speeding up in the opposite direction of Cars A, B, C.

Because the car is moving opposite to the positive direction, Car D is moving with a negative velocity. Because it is speeding up while moving in a negative direction, its acceleration is negative as well.

Figure 2.19 Car E is slowing down in the same direction as Car D and opposite of Cars A, B, C.

Because it is moving opposite to the positive direction, Car E is moving with a negative velocity as well. However, because it is slowing down while moving in a negative direction, its acceleration is actually positive. As in example B, this may be more easily understood in a mathematical sense. The car is originally moving with a large negative velocity (−25 m/s) but slows to a final velocity that is less negative (−5 m/s). This change in velocity, from −25 m/s to −5 m/s, is actually a positive change ($v_f - v_i = -5 \text{ m/s} - -25 \text{ m/s}$ of 20 m/s. Because the change in velocity is positive, the acceleration must also be positive.

Making Connection - Illustrative Example

The three graphs below are labeled A, B, and C. Each one represents the position of a moving object plotted against time.

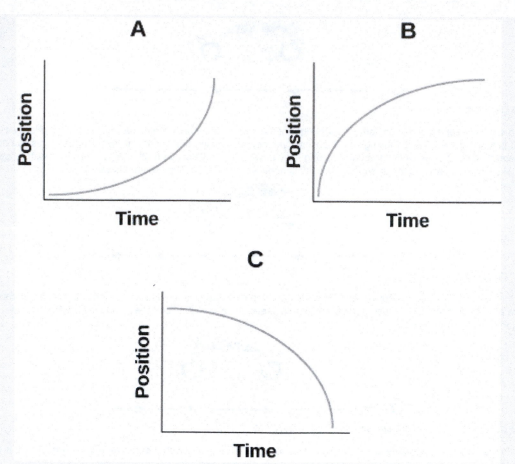

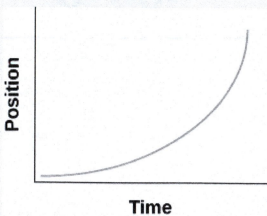

Figure 2.20 Three position and time graphs: A, B, and C.

As we did in the previous example, let's consider the acceleration and velocity of each object in terms of its direction of travel.

Figure 2.21 Graph A of Position (y axis) vs. Time (x axis).

Object A is continually increasing its position in the positive direction. As a result, its velocity is considered positive.

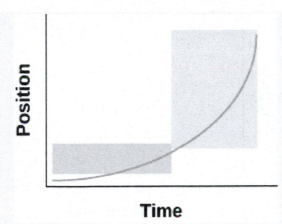

Figure 2.22 Breakdown of Graph A into two separate sections.

During the first portion of time (shaded grey) the position of the object does not change much, resulting in a small positive velocity. During a later portion of time (shaded green) the position of the object changes more, resulting in a larger positive velocity. Because this positive velocity is increasing over time, the acceleration of the object is considered positive.

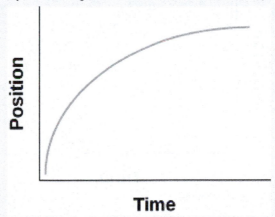

Figure 2.23 Graph B of Position (y axis) vs. Time (x axis).

As in case A, Object B is continually increasing its position in the positive direction. As a result, its velocity is considered positive.

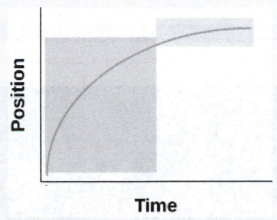

Figure 2.24 Breakdown of Graph B into two separate sections.

During the first portion of time (shaded grey) the position of the object changes a large amount, resulting in a large positive velocity. During a later portion of time (shaded green) the position of the object does not change as much, resulting in a smaller positive velocity. Because this positive velocity is decreasing over time, the acceleration of the object is considered negative.

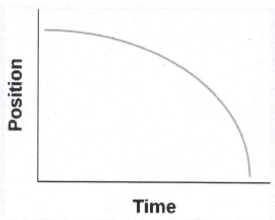

Figure 2.25 Graph C of Position (y axis) vs. Time (x axis).

Object C is continually decreasing its position in the positive direction. As a result, its velocity is considered negative.

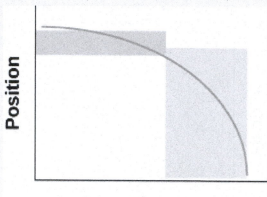

Figure 2.26 Breakdown of Graph C into two separate sections.

During the first portion of time (shaded grey) the position of the object does not change a large amount, resulting in a small negative velocity. During a later portion of time (shaded green) the position of the object changes a much larger amount, resulting in a larger negative velocity. Because the velocity of the object is becoming more negative during the time period, the change in velocity is negative. As a result, the object experiences a negative acceleration.

Example 2.1 Calculating Acceleration: A Racehorse Leaves the Gate

A racehorse coming out of the gate accelerates from rest to a velocity of 15.0 m/s due west in 1.80 s. What is its average acceleration?

Figure 2.27 (credit: Jon Sullivan, PD Photo.org)

Strategy

First we draw a sketch and assign a coordinate system to the problem. This is a simple problem, but it always helps to visualize it. Notice that we assign east as positive and west as negative. Thus, in this case, we have negative velocity.

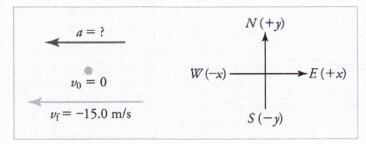

Figure 2.28

We can solve this problem by identifying Δv and Δt from the given information and then calculating the average

acceleration directly from the equation $\bar{a} = \dfrac{\Delta v}{\Delta t} = \dfrac{v_f - v_0}{t_f - t_0}$.

Solution

1. Identify the knowns. $v_0 = 0$, $v_f = -15.0$ m/s (the minus sign indicates direction toward the west), $\Delta t = 1.80$ s.

2. Find the change in velocity. Since the horse is going from zero to -15.0 m/s, its change in velocity equals its final velocity: $\Delta v = v_f = -15.0$ m/s.

3. Plug in the known values (Δv and Δt) and solve for the unknown $\bar{a}$.

$$\bar{a} = \frac{\Delta v}{\Delta t} = \frac{-15.0 \text{ m/s}}{1.80 \text{ s}} = -8.33 \text{ m/s}^2. \tag{2.11}$$

Discussion

The minus sign for acceleration indicates that acceleration is toward the west. An acceleration of 8.33 m/s^2 due west means that the horse increases its velocity by 8.33 m/s due west each second, that is, 8.33 meters per second per second, which we write as 8.33 m/s^2. This is truly an average acceleration, because the ride is not smooth. We shall see later that an acceleration of this magnitude would require the rider to hang on with a force nearly equal to his weight.

Instantaneous Acceleration

Instantaneous acceleration a, or the *acceleration at a specific instant in time*, is obtained by the same process as discussed for instantaneous velocity in **Time, Velocity, and Speed**—that is, by considering an infinitesimally small interval of time. How do we find instantaneous acceleration using only algebra? The answer is that we choose an average acceleration that is representative of the motion. **Figure 2.29** shows graphs of instantaneous acceleration versus time for two very different motions. In **Figure 2.29**(a), the acceleration varies slightly and the average over the entire interval is nearly the same as the instantaneous acceleration at any time. In this case, we should treat this motion as if it had a constant acceleration equal to the average (in this case about 1.8 m/s^2). In **Figure 2.29**(b), the acceleration varies drastically over time. In such situations it is best to consider smaller time intervals and choose an average acceleration for each. For example, we could consider motion over the time intervals from 0 to 1.0 s and from 1.0 to 3.0 s as separate motions with accelerations of $+3.0$ m/s^2 and -2.0 m/s^2, respectively.

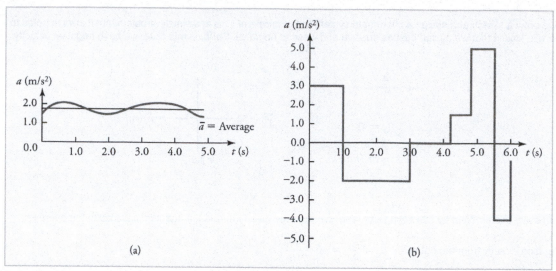

Figure 2.29 Graphs of instantaneous acceleration versus time for two different one-dimensional motions. (a) Here acceleration varies only slightly and is always in the same direction, since it is positive. The average over the interval is nearly the same as the acceleration at any given time. (b) Here the acceleration varies greatly, perhaps representing a package on a post office conveyor belt that is accelerated forward and backward as it bumps along. It is necessary to consider small time intervals (such as from 0 to 1.0 s) with constant or nearly constant acceleration in such a situation.

The next several examples consider the motion of the subway train shown in **Figure 2.30**. In (a) the shuttle moves to the right, and in (b) it moves to the left. The examples are designed to further illustrate aspects of motion and to illustrate some of the reasoning that goes into solving problems.

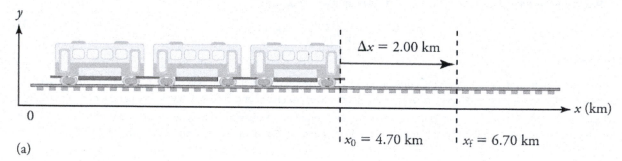

(a)

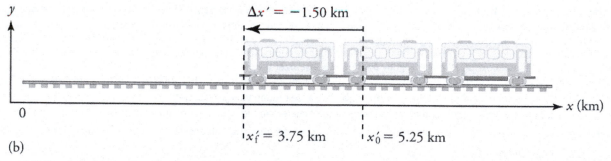

(b)

Figure 2.30 One-dimensional motion of a subway train considered in **Example 2.2, Example 2.3, Example 2.4, Example 2.5, Example 2.6,** and **Example 2.7.** Here we have chosen the x-axis so that + means to the right and − means to the left for displacements, velocities, and accelerations. (a) The subway train moves to the right from x_0 to x_f. Its displacement Δx is +2.0 km. (b) The train moves to the left from x'_0 to x'_f. Its displacement $\Delta x'$ is -1.5 km. (Note that the prime symbol (') is used simply to distinguish between displacement in the two different situations. The distances of travel and the size of the cars are on different scales to fit everything into the diagram.)

Example 2.2 Calculating Displacement: A Subway Train

What are the magnitude and sign of displacements for the motions of the subway train shown in parts (a) and (b) of **Figure 2.30?**

Strategy

A drawing with a coordinate system is already provided, so we don't need to make a sketch, but we should analyze it to make sure we understand what it is showing. Pay particular attention to the coordinate system. To find displacement, we use the equation $\Delta x = x_f - x_0$. This is straightforward since the initial and final positions are given.

Solution

1. Identify the knowns. In the figure we see that $x_f = 6.70$ km and $x_0 = 4.70$ km for part (a), and $x'_f = 3.75$ km and $x'_0 = 5.25$ km for part (b).

2. Solve for displacement in part (a).

$$\Delta x = x_f - x_0 = 6.70 \text{ km} - 4.70 \text{ km} = +2.00 \text{ km} \tag{2.12}$$

3. Solve for displacement in part (b).

$$\Delta x' = x'_f - x'_0 = 3.75 \text{ km} - 5.25 \text{ km} = -1.50 \text{ km} \tag{2.13}$$

Discussion

The direction of the motion in (a) is to the right and therefore its displacement has a positive sign, whereas motion in (b) is to the left and thus has a minus sign.

Example 2.3 Comparing Distance Traveled with Displacement: A Subway Train

What are the distances traveled for the motions shown in parts (a) and (b) of the subway train in **Figure 2.30**?

Strategy

To answer this question, think about the definitions of distance and distance traveled, and how they are related to displacement. Distance between two positions is defined to be the magnitude of displacement, which was found in **Example 2.2**. Distance traveled is the total length of the path traveled between the two positions. (See **Displacement**.) In the case of the subway train shown in **Figure 2.30**, the distance traveled is the same as the distance between the initial and final positions of the train.

Solution

1. The displacement for part (a) was +2.00 km. Therefore, the distance between the initial and final positions was 2.00 km, and the distance traveled was 2.00 km.

2. The displacement for part (b) was -1.5 km. Therefore, the distance between the initial and final positions was 1.50 km, and the distance traveled was 1.50 km.

Discussion

Distance is a scalar. It has magnitude but no sign to indicate direction.

Example 2.4 Calculating Acceleration: A Subway Train Speeding Up

Suppose the train in **Figure 2.30**(a) accelerates from rest to 30.0 km/h in the first 20.0 s of its motion. What is its average acceleration during that time interval?

Strategy

It is worth it at this point to make a simple sketch:

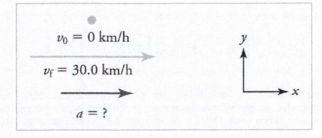

Figure 2.31

This problem involves three steps. First we must determine the change in velocity, then we must determine the change in time, and finally we use these values to calculate the acceleration.

Solution

1. Identify the knowns. $v_0 = 0$ (the trains starts at rest), $v_f = 30.0$ km/h, and $\Delta t = 20.0$ s.

2. Calculate Δv. Since the train starts from rest, its change in velocity is $\Delta v = +30.0$ km/h , where the plus sign means velocity to the right.

3. Plug in known values and solve for the unknown, $\bar{a}$.

$$\bar{a} = \frac{\Delta v}{\Delta t} = \frac{+30.0 \text{ km/h}}{20.0 \text{ s}} \qquad (2.14)$$

4. Since the units are mixed (we have both hours and seconds for time), we need to convert everything into SI units of meters and seconds. (See **Physical Quantities and Units** for more guidance.)

$$\bar{a} = \left(\frac{+30 \text{ km/h}}{20.0 \text{ s}}\right)\left(\frac{10^3 \text{ m}}{1 \text{ km}}\right)\left(\frac{1 \text{ h}}{3600 \text{ s}}\right) = 0.417 \text{ m/s}^2 \qquad (2.15)$$

Discussion

The plus sign means that acceleration is to the right. This is reasonable because the train starts from rest and ends up with a velocity to the right (also positive). So acceleration is in the same direction as the *change* in velocity, as is always the case.

Example 2.5 Calculate Acceleration: A Subway Train Slowing Down

Now suppose that at the end of its trip, the train in **Figure 2.30**(a) slows to a stop from a speed of 30.0 km/h in 8.00 s. What is its average acceleration while stopping?

Strategy

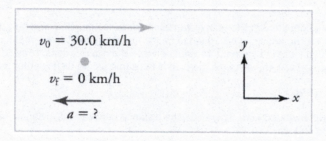

Figure 2.32

In this case, the train is decelerating and its acceleration is negative because it is toward the left. As in the previous example, we must find the change in velocity and the change in time and then solve for acceleration.

Solution

1. Identify the knowns. $v_0 = 30.0$ km/h , $v_f = 0$ km/h (the train is stopped, so its velocity is 0), and $\Delta t = 8.00$ s .

2. Solve for the change in velocity, Δv .

$$\Delta v = v_f - v_0 = 0 - 30.0 \text{ km/h} = -30.0 \text{ km/h} \qquad (2.16)$$

3. Plug in the knowns, Δv and Δt , and solve for $\bar{a}$.

$$\bar{a} = \frac{\Delta v}{\Delta t} = \frac{-30.0 \text{ km/h}}{8.00 \text{ s}} \qquad (2.17)$$

4. Convert the units to meters and seconds.

$$\bar{a} = \frac{\Delta v}{\Delta t} = \left(\frac{-30.0 \text{ km/h}}{8.00 \text{ s}}\right)\left(\frac{10^3 \text{ m}}{1 \text{ km}}\right)\left(\frac{1 \text{ h}}{3600 \text{ s}}\right) = -1.04 \text{ m/s}^2. \qquad (2.18)$$

Discussion

The minus sign indicates that acceleration is to the left. This sign is reasonable because the train initially has a positive velocity in this problem, and a negative acceleration would oppose the motion. Again, acceleration is in the same direction as the *change* in velocity, which is negative here. This acceleration can be called a deceleration because it has a direction opposite to the velocity.

The graphs of position, velocity, and acceleration vs. time for the trains in **Example 2.4** and **Example 2.5** are displayed in **Figure 2.33**. (We have taken the velocity to remain constant from 20 to 40 s, after which the train decelerates.)

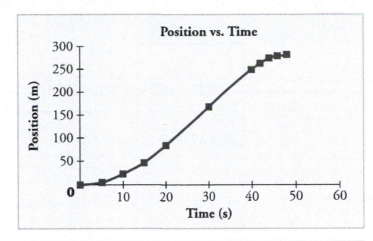

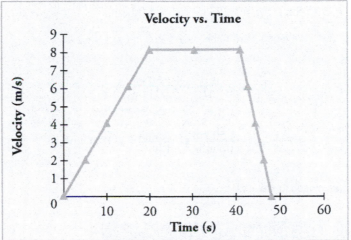

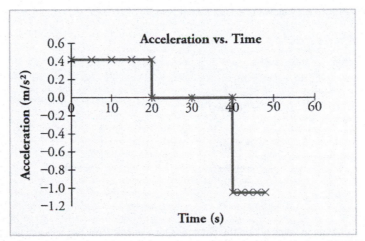

Figure 2.33 (a) Position of the train over time. Notice that the train's position changes slowly at the beginning of the journey, then more and more quickly as it picks up speed. Its position then changes more slowly as it slows down at the end of the journey. In the middle of the journey, while the velocity remains constant, the position changes at a constant rate. (b) Velocity of the train over time. The train's velocity increases as it accelerates at the beginning of the journey. It remains the same in the middle of the journey (where there is no acceleration). It decreases as the train decelerates at the end of the journey. (c) The acceleration of the train over time. The train has positive acceleration as it speeds up at the beginning of the journey. It has no acceleration as it travels at constant velocity in the middle of the journey. Its acceleration is negative as it slows down at the end of the journey.

Example 2.6 Calculating Average Velocity: The Subway Train

What is the average velocity of the train in part b of **Example 2.2**, and shown again below, if it takes 5.00 min to make its trip?

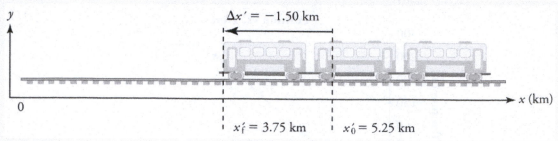

Figure 2.34

Strategy

Average velocity is displacement divided by time. It will be negative here, since the train moves to the left and has a negative displacement.

Solution

1. Identify the knowns. $x'_f = 3.75$ km, $x'_0 = 5.25$ km, $\Delta t = 5.00$ min.

2. Determine displacement, $\Delta x'$. We found $\Delta x'$ to be -1.5 km in **Example 2.2.**

3. Solve for average velocity.

$$\bar{v} = \frac{\Delta x'}{\Delta t} = \frac{-1.50 \text{ km}}{5.00 \text{ min}}$$ (2.19)

4. Convert units.

$$\bar{v} = \frac{\Delta x'}{\Delta t} = \left(\frac{-1.50 \text{ km}}{5.00 \text{ min}}\right)\left(\frac{60 \text{ min}}{1 \text{ h}}\right) = -18.0 \text{ km/h}$$ (2.20)

Discussion

The negative velocity indicates motion to the left.

Example 2.7 Calculating Deceleration: The Subway Train

Finally, suppose the train in **Figure 2.34** slows to a stop from a velocity of 20.0 km/h in 10.0 s. What is its average acceleration?

Strategy

Once again, let's draw a sketch:

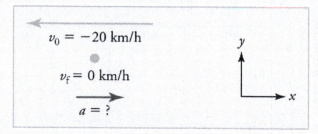

Figure 2.35

As before, we must find the change in velocity and the change in time to calculate average acceleration.

Solution

1. Identify the knowns. $v_0 = -20$ km/h, $v_f = 0$ km/h, $\Delta t = 10.0$ s.

2. Calculate Δv. The change in velocity here is actually positive, since

$$\Delta v = v_f - v_0 = 0 - (-20 \text{ km/h}) = +20 \text{ km/h}.$$ (2.21)

3. Solve for $\bar{a}$.

$$\bar{a} = \frac{\Delta v}{\Delta t} = \frac{+20.0 \text{ km/h}}{10.0 \text{ s}}$$ (2.22)

4. Convert units.

$$\bar{a} = \left(\frac{+20.0 \text{ km/h}}{10.0 \text{ s}}\right)\left(\frac{10^3 \text{ m}}{1 \text{ km}}\right)\left(\frac{1 \text{ h}}{3600 \text{ s}}\right) = +0.556 \text{ m/s}^2 \qquad (2.23)$$

Discussion

The plus sign means that acceleration is to the right. This is reasonable because the train initially has a negative velocity (to the left) in this problem and a positive acceleration opposes the motion (and so it is to the right). Again, acceleration is in the same direction as the *change* in velocity, which is positive here. As in **Example 2.5**, this acceleration can be called a deceleration since it is in the direction opposite to the velocity.

Sign and Direction

Perhaps the most important thing to note about these examples is the signs of the answers. In our chosen coordinate system, plus means the quantity is to the right and minus means it is to the left. This is easy to imagine for displacement and velocity. But it is a little less obvious for acceleration. Most people interpret negative acceleration as the slowing of an object. This was not the case in **Example 2.7**, where a positive acceleration slowed a negative velocity. The crucial distinction was that the acceleration was in the opposite direction from the velocity. In fact, a negative acceleration will *increase* a negative velocity. For example, the train moving to the left in **Figure 2.34** is sped up by an acceleration to the left. In that case, both v and a are negative. The plus and minus signs give the directions of the accelerations. If acceleration has the same sign as the velocity, the object is speeding up. If acceleration has the opposite sign as the velocity, the object is slowing down.

Check Your Understanding

An airplane lands on a runway traveling east. Describe its acceleration.

Solution
If we take east to be positive, then the airplane has negative acceleration, as it is accelerating toward the west. It is also decelerating: its acceleration is opposite in direction to its velocity.

PhET Explorations: Moving Man Simulation

Learn about position, velocity, and acceleration graphs. Move the little man back and forth with the mouse and plot his motion. Set the position, velocity, or acceleration and let the simulation move the man for you.

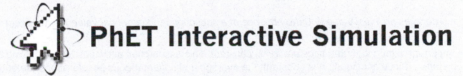

Figure 2.36 Moving Man (http://cnx.org/content/m54772/1.3/moving-man_en.jar)

2.5 Motion Equations for Constant Acceleration in One Dimension

Figure 2.37 Kinematic equations can help us describe and predict the motion of moving objects such as these kayaks racing in Newbury, England. (credit: Barry Skeates, Flickr)

Learning Objectives

By the end of this section, you will be able to:

- Calculate displacement of an object that is not accelerating, given initial position and velocity.
- Calculate final velocity of an accelerating object, given initial velocity, acceleration, and time.
- Calculate displacement and final position of an accelerating object, given initial position, initial velocity, time, and acceleration.

The information presented in this section supports the following AP® learning objectives and science practices:

- **3.A.1.1** The student is able to express the motion of an object using narrative, mathematical, or graphical representations. **(S.P. 1.5, 2.1, 2.2)**
- **3.A.1.3** The student is able to analyze experimental data describing the motion of an object and is able to express the results of the analysis using narrative, mathematical, and graphical representations. **(S.P. 5.1)**

We might know that the greater the acceleration of, say, a car moving away from a stop sign, the greater the displacement in a given time. But we have not developed a specific equation that relates acceleration and displacement. In this section, we develop some convenient equations for kinematic relationships, starting from the definitions of displacement, velocity, and acceleration already covered.

Notation: t, x, v, a

First, let us make some simplifications in notation. Taking the initial time to be zero, as if time is measured with a stopwatch, is a great simplification. Since elapsed time is $\Delta t = t_f - t_0$, taking $t_0 = 0$ means that $\Delta t = t_f$, the final time on the stopwatch. When initial time is taken to be zero, we use the subscript 0 to denote initial values of position and velocity. That is, x_0 is the *initial position* and v_0 *is the initial velocity*. We put no subscripts on the final values. That is, t *is the final time*, x *is the final position*, and v *is the final velocity*. This gives a simpler expression for elapsed time—now, $\Delta t = t$. It also simplifies the expression for displacement, which is now $\Delta x = x - x_0$. Also, it simplifies the expression for change in velocity, which is now $\Delta v = v - v_0$. To summarize, using the simplified notation, with the initial time taken to be zero,

$$\left.\begin{array}{rcl} \Delta t & = & t \\ \Delta x & = & x - x_0 \\ \Delta v & = & v - v_0 \end{array}\right\} \tag{2.24}$$

where *the subscript 0 denotes an initial value and the absence of a subscript denotes a final value* in whatever motion is under consideration.

We now make the important assumption that *acceleration is constant*. This assumption allows us to avoid using calculus to find instantaneous acceleration. Since acceleration is constant, the average and instantaneous accelerations are equal. That is,

$$\bar{a} = a = \text{constant}, \tag{2.25}$$

so we use the symbol a for acceleration at all times. Assuming acceleration to be constant does not seriously limit the situations we can study nor degrade the accuracy of our treatment. For one thing, acceleration *is* constant in a great number of situations. Furthermore, in many other situations we can accurately describe motion by assuming a constant acceleration equal to the average acceleration for that motion. Finally, in motions where acceleration changes drastically, such as a car accelerating to top speed and then braking to a stop, the motion can be considered in separate parts, each of which has its own constant acceleration.

Solving for Displacement (Δx) and Final Position (x) from Average Velocity when Acceleration (a) is Constant

To get our first two new equations, we start with the definition of average velocity:

$$\bar{v} = \frac{\Delta x}{\Delta t}. \tag{2.26}$$

Substituting the simplified notation for Δx and Δt yields

$$\bar{v} = \frac{x - x_0}{t}. \tag{2.27}$$

Solving for x yields

$$x = x_0 + \bar{v}t, \tag{2.28}$$

where the average velocity is

$$\bar{v} = \frac{v_0 + v}{2} \ (\text{constant } a). \tag{2.29}$$

The equation $\bar{v} = \frac{v_0 + v}{2}$ reflects the fact that, when acceleration is constant, v is just the simple average of the initial and final velocities. For example, if you steadily increase your velocity (that is, with constant acceleration) from 30 to 60 km/h, then your average velocity during this steady increase is 45 km/h. Using the equation $\bar{v} = \frac{v_0 + v}{2}$ to check this, we see that

$$\bar{v} = \frac{v_0 + v}{2} = \frac{30 \text{ km/h} + 60 \text{ km/h}}{2} = 45 \text{ km/h,} \qquad (2.30)$$

which seems logical.

Example 2.8 Calculating Displacement: How Far does the Jogger Run?

A jogger runs down a straight stretch of road with an average velocity of 4.00 m/s for 2.00 min. What is his final position, taking his initial position to be zero?

Strategy

Draw a sketch.

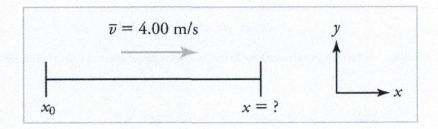

Figure 2.38

The final position x is given by the equation

$$x = x_0 + \bar{v}t. \qquad (2.31)$$

To find x, we identify the values of x_0, $\bar{v}$, and t from the statement of the problem and substitute them into the equation.

Solution

1. Identify the knowns. $\bar{v} = 4.00 \text{ m/s}$, $\Delta t = 2.00 \text{ min}$, and $x_0 = 0 \text{ m}$.

2. Enter the known values into the equation.

$$x = x_0 + \bar{v}t = 0 + (4.00 \text{ m/s})(120 \text{ s}) = 480 \text{ m} \qquad (2.32)$$

Discussion

Velocity and final displacement are both positive, which means they are in the same direction.

The equation $x = x_0 + \bar{v}t$ gives insight into the relationship between displacement, average velocity, and time. It shows, for example, that displacement is a linear function of average velocity. (By linear function, we mean that displacement depends on $\bar{v}$ rather than on $\bar{v}$ raised to some other power, such as $\bar{v}^2$. When graphed, linear functions look like straight lines with a constant slope.) On a car trip, for example, we will get twice as far in a given time if we average 90 km/h than if we average 45 km/h.

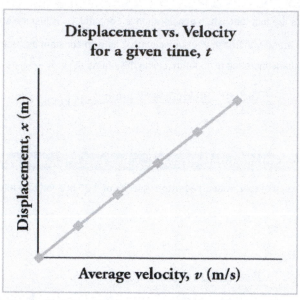

Figure 2.39 There is a linear relationship between displacement and average velocity. For a given time t, an object moving twice as fast as another object will move twice as far as the other object.

Solving for Final Velocity

We can derive another useful equation by manipulating the definition of acceleration.

$$a = \frac{\Delta v}{\Delta t} \qquad (2.33)$$

Substituting the simplified notation for Δv and Δt gives us

$$a = \frac{v - v_0}{t} \text{ (constant } a\text{)}. \qquad (2.34)$$

Solving for v yields

$$v = v_0 + at \text{ (constant } a\text{)}. \qquad (2.35)$$

Example 2.9 Calculating Final Velocity: An Airplane Slowing Down after Landing

An airplane lands with an initial velocity of 70.0 m/s and then decelerates at 1.50 m/s^2 for 40.0 s. What is its final velocity?

Strategy

Draw a sketch. We draw the acceleration vector in the direction opposite the velocity vector because the plane is decelerating.

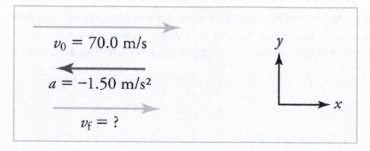

Figure 2.40

Solution

1. Identify the knowns. $v_0 = 70.0$ m/s, $a = -1.50 \text{ m/s}^2$, $t = 40.0$ s.

2. Identify the unknown. In this case, it is final velocity, v_f.

3. Determine which equation to use. We can calculate the final velocity using the equation $v = v_0 + at$.

4. Plug in the known values and solve.

$$v = v_0 + at = 70.0 \text{ m/s} + \left(-1.50 \text{ m/s}^2\right)(40.0 \text{ s}) = 10.0 \text{ m/s} \qquad (2.36)$$

Discussion

The final velocity is much less than the initial velocity, as desired when slowing down, but still positive. With jet engines, reverse thrust could be maintained long enough to stop the plane and start moving it backward. That would be indicated by a negative final velocity, which is not the case here.

$v_0 = 70.0 \text{ m/s}$

$a = -1.5 \text{ m/s}^2$

$t_0 = 0$

$v = 10.0 \text{ m/s}$

$a = -1.50 \text{ m/s}^2$

$t = 40.0 \text{ s}$

Figure 2.41 The airplane lands with an initial velocity of 70.0 m/s and slows to a final velocity of 10.0 m/s before heading for the terminal. Note that the acceleration is negative because its direction is opposite to its velocity, which is positive.

In addition to being useful in problem solving, the equation $v = v_0 + at$ gives us insight into the relationships among velocity, acceleration, and time. From it we can see, for example, that

- final velocity depends on how large the acceleration is and how long it lasts
- if the acceleration is zero, then the final velocity equals the initial velocity $(v = v_0)$, as expected (i.e., velocity is constant)
- if a is negative, then the final velocity is less than the initial velocity

(All of these observations fit our intuition, and it is always useful to examine basic equations in light of our intuition and experiences to check that they do indeed describe nature accurately.)

Making Connections: Real-World Connection

Figure 2.42 The Space Shuttle *Endeavor* blasts off from the Kennedy Space Center in February 2010. (credit: Matthew Simantov, Flickr)

An intercontinental ballistic missile (ICBM) has a larger average acceleration than the Space Shuttle and achieves a greater velocity in the first minute or two of flight (actual ICBM burn times are classified—short-burn-time missiles are more difficult for an enemy to destroy). But the Space Shuttle obtains a greater final velocity, so that it can orbit the earth rather than come directly back down as an ICBM does. The Space Shuttle does this by accelerating for a longer time.

Solving for Final Position When Velocity is Not Constant ($a \neq 0$)

We can combine the equations above to find a third equation that allows us to calculate the final position of an object experiencing constant acceleration. We start with

$$v = v_0 + at. \qquad (2.37)$$

Adding v_0 to each side of this equation and dividing by 2 gives

$$\frac{v_0 + v}{2} = v_0 + \frac{1}{2}at. \qquad (2.38)$$

Since $\frac{v_0 + v}{2} = \bar{v}$ for constant acceleration, then

$$\bar{v} = v_0 + \frac{1}{2}at.$$ (2.39)

Now we substitute this expression for $\bar{v}$ into the equation for displacement, $x = x_0 + \bar{v}t$, yielding

$$x = x_0 + v_0 t + \frac{1}{2}at^2 \text{ (constant } a\text{)}.$$ (2.40)

Example 2.10 Calculating Displacement of an Accelerating Object: Dragsters

Dragsters can achieve average accelerations of 26.0 m/s^2. Suppose such a dragster accelerates from rest at this rate for 5.56 s. How far does it travel in this time?

Figure 2.43 U.S. Army Top Fuel pilot Tony "The Sarge" Schumacher begins a race with a controlled burnout. (credit: Lt. Col. William Thurmond. Photo Courtesy of U.S. Army.)

Strategy

Draw a sketch.

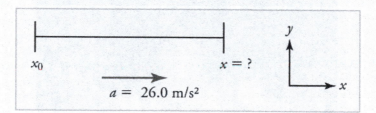

Figure 2.44

We are asked to find displacement, which is x if we take x_0 to be zero. (Think about it like the starting line of a race. It can be anywhere, but we call it 0 and measure all other positions relative to it.) We can use the equation $x = x_0 + v_0 t + \frac{1}{2}at^2$ once we identify v_0, a, and t from the statement of the problem.

Solution

1. Identify the knowns. Starting from rest means that $v_0 = 0$, a is given as 26.0 m/s^2 and t is given as 5.56 s.

2. Plug the known values into the equation to solve for the unknown x:

$$x = x_0 + v_0 t + \frac{1}{2}at^2.$$ (2.41)

Since the initial position and velocity are both zero, this simplifies to

$$x = \frac{1}{2}at^2.$$ (2.42)

Substituting the identified values of a and t gives

$$x = \tfrac{1}{2}(26.0 \text{ m/s}^2)(5.56 \text{ s})^2,$$ (2.43)

yielding

$$x = 402 \text{ m.}$$ (2.44)

Discussion

If we convert 402 m to miles, we find that the distance covered is very close to one quarter of a mile, the standard distance for drag racing. So the answer is reasonable. This is an impressive displacement in only 5.56 s, but top-notch dragsters can do a quarter mile in even less time than this.

What else can we learn by examining the equation $x = x_0 + v_0 t + \tfrac{1}{2}at^2$? We see that:

- displacement depends on the square of the elapsed time when acceleration is not zero. In **Example 2.10**, the dragster covers only one fourth of the total distance in the first half of the elapsed time
- if acceleration is zero, then the initial velocity equals average velocity ($v_0 = \bar{v}$) and $x = x_0 + v_0 t + \tfrac{1}{2}at^2$ becomes

$$x = x_0 + v_0 t$$

Solving for Final Velocity when Velocity Is Not Constant ($a \neq 0$)

A fourth useful equation can be obtained from another algebraic manipulation of previous equations.

If we solve $v = v_0 + at$ for t, we get

$$t = \frac{v - v_0}{a}.$$ (2.45)

Substituting this and $\bar{v} = \frac{v_0 + v}{2}$ into $x = x_0 + \bar{v}t$, we get

$$v^2 = v_0^2 + 2a(x - x_0) \text{ (constant} a).$$ (2.46)

Example 2.11 Calculating Final Velocity: Dragsters

Calculate the final velocity of the dragster in **Example 2.10** without using information about time.

Strategy

Draw a sketch.

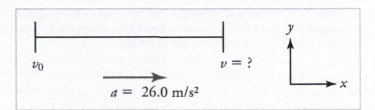

Figure 2.45

The equation $v^2 = v_0^2 + 2a(x - x_0)$ is ideally suited to this task because it relates velocities, acceleration, and displacement, and no time information is required.

Solution

1. Identify the known values. We know that $v_0 = 0$, since the dragster starts from rest. Then we note that $x - x_0 = 402 \text{ m}$ (this was the answer in **Example 2.10**). Finally, the average acceleration was given to be $a = 26.0 \text{ m/s}^2$.

2. Plug the knowns into the equation $v^2 = v_0^2 + 2a(x - x_0)$ and solve for v.

$$v^2 = 0 + 2(26.0 \text{ m/s}^2)(402 \text{ m}).$$ (2.47)

Thus

$$v^2 = 2.09 \times 10^4 \text{ m}^2/\text{s}^2. \tag{2.48}$$

To get v, we take the square root:

$$v = \sqrt{2.09 \times 10^4 \text{ m}^2/\text{s}^2} = 145 \text{ m/s}. \tag{2.49}$$

Discussion

145 m/s is about 522 km/h or about 324 mi/h, but even this breakneck speed is short of the record for the quarter mile. Also, note that a square root has two values; we took the positive value to indicate a velocity in the same direction as the acceleration.

An examination of the equation $v^2 = v_0^2 + 2a(x - x_0)$ can produce further insights into the general relationships among physical quantities:

- The final velocity depends on how large the acceleration is and the distance over which it acts
- For a fixed deceleration, a car that is going twice as fast doesn't simply stop in twice the distance—it takes much further to stop. (This is why we have reduced speed zones near schools.)

Putting Equations Together

In the following examples, we further explore one-dimensional motion, but in situations requiring slightly more algebraic manipulation. The examples also give insight into problem-solving techniques. The box below provides easy reference to the equations needed.

Summary of Kinematic Equations (constant a)

$$x = x_0 + \bar{v}t \tag{2.50}$$

$$\bar{v} = \frac{v_0 + v}{2} \tag{2.51}$$

$$v = v_0 + at \tag{2.52}$$

$$x = x_0 + v_0 t + \frac{1}{2}at^2 \tag{2.53}$$

$$v^2 = v_0^2 + 2a(x - x_0) \tag{2.54}$$

Example 2.12 Calculating Displacement: How Far Does a Car Go When Coming to a Halt?

On dry concrete, a car can decelerate at a rate of 7.00 m/s^2, whereas on wet concrete it can decelerate at only 5.00 m/s^2. Find the distances necessary to stop a car moving at 30.0 m/s (about 110 km/h) (a) on dry concrete and (b) on wet concrete. (c) Repeat both calculations, finding the displacement from the point where the driver sees a traffic light turn red, taking into account his reaction time of 0.500 s to get his foot on the brake.

Strategy

Draw a sketch.

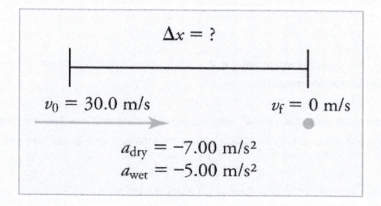

Figure 2.46

In order to determine which equations are best to use, we need to list all of the known values and identify exactly what we need to solve for. We shall do this explicitly in the next several examples, using tables to set them off.

Solution for (a)

1. Identify the knowns and what we want to solve for. We know that $v_0 = 30.0$ m/s ; $v = 0$; $a = -7.00$ m/s^2 (a is negative because it is in a direction opposite to velocity). We take x_0 to be 0. We are looking for displacement Δx , or $x - x_0$.

2. Identify the equation that will help up solve the problem. The best equation to use is

$$v^2 = v_0^2 + 2a(x - x_0). \tag{2.55}$$

This equation is best because it includes only one unknown, x . We know the values of all the other variables in this equation. (There are other equations that would allow us to solve for x , but they require us to know the stopping time, t , which we do not know. We could use them but it would entail additional calculations.)

3. Rearrange the equation to solve for x .

$$x - x_0 = \frac{v^2 - v_0^2}{2a} \tag{2.56}$$

4. Enter known values.

$$x - 0 = \frac{0^2 - (30.0 \text{ m/s})^2}{2(-7.00 \text{ m/s}^2)} \tag{2.57}$$

Thus,

$$x = 64.3 \text{ m on dry concrete.} \tag{2.58}$$

Solution for (b)

This part can be solved in exactly the same manner as Part A. The only difference is that the deceleration is -5.00 m/s^2 . The result is

$$x_{\text{wet}} = 90.0 \text{ m on wet concrete.} \tag{2.59}$$

Solution for (c)

Once the driver reacts, the stopping distance is the same as it is in Parts A and B for dry and wet concrete. So to answer this question, we need to calculate how far the car travels during the reaction time, and then add that to the stopping time. It is reasonable to assume that the velocity remains constant during the driver's reaction time.

1. Identify the knowns and what we want to solve for. We know that $\bar{v} = 30.0$ m/s ; $t_{\text{reaction}} = 0.500$ s ; $a_{\text{reaction}} = 0$. We take $x_{0 - \text{reaction}}$ to be 0. We are looking for x_{reaction} .

2. Identify the best equation to use.

$x = x_0 + \bar{v}t$ works well because the only unknown value is x , which is what we want to solve for.

3. Plug in the knowns to solve the equation.

$$x = 0 + (30.0 \text{ m/s})(0.500 \text{ s}) = 15.0 \text{ m.} \tag{2.60}$$

This means the car travels 15.0 m while the driver reacts, making the total displacements in the two cases of dry and wet concrete 15.0 m greater than if he reacted instantly.

4. Add the displacement during the reaction time to the displacement when braking.

$$x_{\text{braking}} + x_{\text{reaction}} = x_{\text{total}} \tag{2.61}$$

a. 64.3 m + 15.0 m = 79.3 m when dry

b. 90.0 m + 15.0 m = 105 m when wet

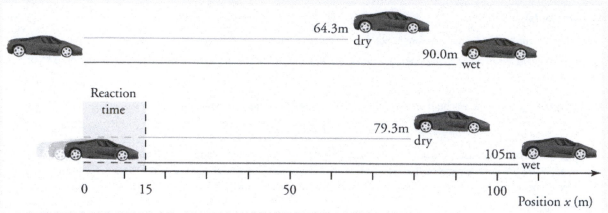

Figure 2.47 The distance necessary to stop a car varies greatly, depending on road conditions and driver reaction time. Shown here are the braking distances for dry and wet pavement, as calculated in this example, for a car initially traveling at 30.0 m/s. Also shown are the total distances traveled from the point where the driver first sees a light turn red, assuming a 0.500 s reaction time.

Discussion

The displacements found in this example seem reasonable for stopping a fast-moving car. It should take longer to stop a car on wet rather than dry pavement. It is interesting that reaction time adds significantly to the displacements. But more important is the general approach to solving problems. We identify the knowns and the quantities to be determined and then find an appropriate equation. There is often more than one way to solve a problem. The various parts of this example can in fact be solved by other methods, but the solutions presented above are the shortest.

Example 2.13 Calculating Time: A Car Merges into Traffic

Suppose a car merges into freeway traffic on a 200-m-long ramp. If its initial velocity is 10.0 m/s and it accelerates at 2.00 m/s^2, how long does it take to travel the 200 m up the ramp? (Such information might be useful to a traffic engineer.)

Strategy

Draw a sketch.

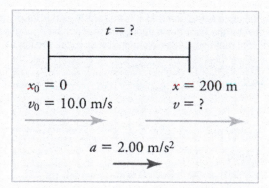

Figure 2.48

We are asked to solve for the time t. As before, we identify the known quantities in order to choose a convenient physical relationship (that is, an equation with one unknown, t).

Solution

1. Identify the knowns and what we want to solve for. We know that $v_0 = 10 \text{ m/s}$; $a = 2.00 \text{ m/s}^2$; and $x = 200 \text{ m}$.

2. We need to solve for t. Choose the best equation. $x = x_0 + v_0 t + \frac{1}{2}at^2$ works best because the only unknown in the equation is the variable t for which we need to solve.

3. We will need to rearrange the equation to solve for t. In this case, it will be easier to plug in the knowns first.

$$200 \text{ m} = 0 \text{ m} + (10.0 \text{ m/s})t + \frac{1}{2}\left(2.00 \text{ m/s}^2\right)t^2 \tag{2.62}$$

4. Simplify the equation. The units of meters (m) cancel because they are in each term. We can get the units of seconds (s)

to cancel by taking $t = t$ s , where t is the magnitude of time and s is the unit. Doing so leaves

$$200 = 10t + t^2.$$ (2.63)

5. Use the quadratic formula to solve for t.

(a) Rearrange the equation to get 0 on one side of the equation.

$$t^2 + 10t - 200 = 0$$ (2.64)

This is a quadratic equation of the form

$$at^2 + bt + c = 0,$$ (2.65)

where the constants are $a = 1.00$, $b = 10.0$, and $c = -200$.

(b) Its solutions are given by the quadratic formula:

$$t = \frac{-b \pm \sqrt{b^2 - 4ac}}{2a}.$$ (2.66)

This yields two solutions for t, which are

$$t = 10.0 \text{ and} -20.0.$$ (2.67)

In this case, then, the time is $t = t$ in seconds, or

$$t = 10.0 \text{ s and} - 20.0 \text{ s}.$$ (2.68)

A negative value for time is unreasonable, since it would mean that the event happened 20 s before the motion began. We can discard that solution. Thus,

$$t = 10.0 \text{ s}.$$ (2.69)

Discussion

Whenever an equation contains an unknown squared, there will be two solutions. In some problems both solutions are meaningful, but in others, such as the above, only one solution is reasonable. The 10.0 s answer seems reasonable for a typical freeway on-ramp.

With the basics of kinematics established, we can go on to many other interesting examples and applications. In the process of developing kinematics, we have also glimpsed a general approach to problem solving that produces both correct answers and insights into physical relationships. **Problem-Solving Basics** discusses problem-solving basics and outlines an approach that will help you succeed in this invaluable task.

Making Connections: Take-Home Experiment—Breaking News

We have been using SI units of meters per second squared to describe some examples of acceleration or deceleration of cars, runners, and trains. To achieve a better feel for these numbers, one can measure the braking deceleration of a car doing a slow (and safe) stop. Recall that, for average acceleration, $\bar{a} = \Delta v / \Delta t$. While traveling in a car, slowly apply the brakes as you come up to a stop sign. Have a passenger note the initial speed in miles per hour and the time taken (in seconds) to stop. From this, calculate the deceleration in miles per hour per second. Convert this to meters per second squared and compare with other decelerations mentioned in this chapter. Calculate the distance traveled in braking.

Check Your Understanding

A manned rocket accelerates at a rate of 20 m/s^2 during launch. How long does it take the rocket reach a velocity of 400 m/s?

Solution

To answer this, choose an equation that allows you to solve for time t, given only a, v_0, and v.

$$v = v_0 + at$$ (2.70)

Rearrange to solve for t.

$$t = \frac{v - v}{a} = \frac{400 \text{ m/s} - 0 \text{ m/s}}{20 \text{ m/s}^2} = 20 \text{ s}$$ (2.71)

2.6 Problem-Solving Basics for One Dimensional Kinematics

Figure 2.49 Problem-solving skills are essential to your success in Physics. (credit: scui3asteveo, Flickr)

Learning Objectives

By the end of this section, you will be able to:

- Apply problem-solving steps and strategies to solve problems of one-dimensional kinematics.
- Apply strategies to determine whether or not the result of a problem is reasonable, and if not, determine the cause.

Problem-solving skills are obviously essential to success in a quantitative course in physics. More importantly, the ability to apply broad physical principles, usually represented by equations, to specific situations is a very powerful form of knowledge. It is much more powerful than memorizing a list of facts. Analytical skills and problem-solving abilities can be applied to new situations, whereas a list of facts cannot be made long enough to contain every possible circumstance. Such analytical skills are useful both for solving problems in this text and for applying physics in everyday and professional life.

Problem-Solving Steps

While there is no simple step-by-step method that works for every problem, the following general procedures facilitate problem solving and make it more meaningful. A certain amount of creativity and insight is required as well.

Step 1

Examine the situation to determine which physical principles are involved. It often helps to *draw a simple sketch* at the outset. You will also need to decide which direction is positive and note that on your sketch. Once you have identified the physical principles, it is much easier to find and apply the equations representing those principles. Although finding the correct equation is essential, keep in mind that equations represent physical principles, laws of nature, and relationships among physical quantities. Without a conceptual understanding of a problem, a numerical solution is meaningless.

Step 2

Make a list of what is given or can be inferred from the problem as stated (identify the knowns). Many problems are stated very succinctly and require some inspection to determine what is known. A sketch can also be very useful at this point. Formally identifying the knowns is of particular importance in applying physics to real-world situations. Remember, "stopped" means velocity is zero, and we often can take initial time and position as zero.

Step 3

Identify exactly what needs to be determined in the problem (identify the unknowns). In complex problems, especially, it is not always obvious what needs to be found or in what sequence. Making a list can help.

Step 4

Find an equation or set of equations that can help you solve the problem. Your list of knowns and unknowns can help here. It is easiest if you can find equations that contain only one unknown—that is, all of the other variables are known, so you can easily solve for the unknown. If the equation contains more than one unknown, then an additional equation is needed to solve the problem. In some problems, several unknowns must be determined to get at the one needed most. In such problems it is especially important to keep physical principles in mind to avoid going astray in a sea of equations. You may have to use two (or more) different equations to get the final answer.

Step 5

Substitute the knowns along with their units into the appropriate equation, and obtain numerical solutions complete with units. This step produces the numerical answer; it also provides a check on units that can help you find errors. If the units of the answer are incorrect, then an error has been made. However, be warned that correct units do not guarantee that the numerical part of the answer is also correct.

Step 6

Check the answer to see if it is reasonable: Does it make sense? This final step is extremely important—the goal of physics is to accurately describe nature. To see if the answer is reasonable, check both its magnitude and its sign, in addition to its units. Your judgment will improve as you solve more and more physics problems, and it will become possible for you to make finer and finer judgments regarding whether nature is adequately described by the answer to a problem. This step brings the problem back to its conceptual meaning. If you can judge whether the answer is reasonable, you have a deeper understanding of physics than just being able to mechanically solve a problem.

When solving problems, we often perform these steps in different order, and we also tend to do several steps simultaneously. There is no rigid procedure that will work every time. Creativity and insight grow with experience, and the basics of problem solving become almost automatic. One way to get practice is to work out the text's examples for yourself as you read. Another is to work as many end-of-section problems as possible, starting with the easiest to build confidence and progressing to the more difficult. Once you become involved in physics, you will see it all around you, and you can begin to apply it to situations you encounter outside the classroom, just as is done in many of the applications in this text.

Unreasonable Results

Physics must describe nature accurately. Some problems have results that are unreasonable because one premise is unreasonable or because certain premises are inconsistent with one another. The physical principle applied correctly then produces an unreasonable result. For example, if a person starting a foot race accelerates at 0.40 m/s^2 for 100 s, his final speed will be 40 m/s (about 150 km/h)—clearly unreasonable because the time of 100 s is an unreasonable premise. The physics is correct in a sense, but there is more to describing nature than just manipulating equations correctly. Checking the result of a problem to see if it is reasonable does more than help uncover errors in problem solving—it also builds intuition in judging whether nature is being accurately described.

Use the following strategies to determine whether an answer is reasonable and, if it is not, to determine what is the cause.

Step 1

Solve the problem using strategies as outlined and in the format followed in the worked examples in the text. In the example given in the preceding paragraph, you would identify the givens as the acceleration and time and use the equation below to find the unknown final velocity. That is,

$$v = v_0 + at = 0 + \left(0.40 \text{ m/s}^2\right)(100 \text{ s}) = 40 \text{ m/s}. \tag{2.72}$$

Step 2

Check to see if the answer is reasonable. Is it too large or too small, or does it have the wrong sign, improper units, …? In this case, you may need to convert meters per second into a more familiar unit, such as miles per hour.

$$\left(\frac{40 \text{ m}}{\text{s}}\right)\left(\frac{3.28 \text{ ft}}{\text{m}}\right)\left(\frac{1 \text{ mi}}{5280 \text{ ft}}\right)\left(\frac{60 \text{ s}}{\text{min}}\right)\left(\frac{60 \text{ min}}{1 \text{ h}}\right) = 89 \text{ mph} \tag{2.73}$$

This velocity is about four times greater than a person can run—so it is too large.

Step 3

If the answer is unreasonable, look for what specifically could cause the identified difficulty. In the example of the runner, there are only two assumptions that are suspect. The acceleration could be too great or the time too long. First look at the acceleration and think about what the number means. If someone accelerates at 0.40 m/s^2, their velocity is increasing by 0.4 m/s each second. Does this seem reasonable? If so, the time must be too long. It is not possible for someone to accelerate at a constant rate of 0.40 m/s^2 for 100 s (almost two minutes).

2.7 Falling Objects

Learning Objectives

By the end of this section, you will be able to:

- Describe the effects of gravity on objects in motion.
- Describe the motion of objects that are in free fall.
- Calculate the position and velocity of objects in free fall.

The information presented in this section supports the following AP® learning objectives and science practices:

- **3.A.1.1** The student is able to express the motion of an object using narrative, mathematical, or graphical representations. **(S.P. 1.5, 2.1, 2.2)**
- **3.A.1.2** The student is able to design an experimental investigation of the motion of an object. **(S.P. 4.2)**
- **3.A.1.3** The student is able to analyze experimental data describing the motion of an object and is able to express the results of the analysis using narrative, mathematical, and graphical representations. **(S.P. 5.1)**

Falling objects form an interesting class of motion problems. For example, we can estimate the depth of a vertical mine shaft by dropping a rock into it and listening for the rock to hit the bottom. By applying the kinematics developed so far to falling objects, we can examine some interesting situations and learn much about gravity in the process.

Gravity

The most remarkable and unexpected fact about falling objects is that, if air resistance and friction are negligible, then in a given location all objects fall toward the center of Earth with the *same constant acceleration, independent of their mass*. This experimentally determined fact is unexpected, because we are so accustomed to the effects of air resistance and friction that we expect light objects to fall slower than heavy ones.

In air   In a vacuum   In a vacuum (the hard way)

Figure 2.50 A hammer and a feather will fall with the same constant acceleration if air resistance is considered negligible. This is a general characteristic of gravity not unique to Earth, as astronaut David R. Scott demonstrated on the Moon in 1971, where the acceleration due to gravity is only 1.67 m/s^2.

In the real world, air resistance can cause a lighter object to fall slower than a heavier object of the same size. A tennis ball will reach the ground after a hard baseball dropped at the same time. (It might be difficult to observe the difference if the height is not large.) Air resistance opposes the motion of an object through the air, while friction between objects—such as between clothes and a laundry chute or between a stone and a pool into which it is dropped—also opposes motion between them. For the ideal situations of these first few chapters, an object *falling without air resistance or friction* is defined to be in **free-fall**.

The force of gravity causes objects to fall toward the center of Earth. The acceleration of free-falling objects is therefore called the **acceleration due to gravity**. The acceleration due to gravity is *constant*, which means we can apply the kinematics equations to any falling object where air resistance and friction are negligible. This opens a broad class of interesting situations to us. The acceleration due to gravity is so important that its magnitude is given its own symbol, g. It is constant at any given location on Earth and has the average value

$$g = 9.80 \text{ m/s}^2.$$

(2.74)

Although g varies from 9.78 m/s^2 to 9.83 m/s^2, depending on latitude, altitude, underlying geological formations, and local topography, the average value of 9.80 m/s^2 will be used in this text unless otherwise specified. The direction of the acceleration due to gravity is *downward (towards the center of Earth)*. In fact, its direction *defines* what we call vertical. Note that whether the acceleration a in the kinematic equations has the value $+g$ or $-g$ depends on how we define our coordinate system. If we define the upward direction as positive, then $a = -g = -9.80 \text{ m/s}^2$, and if we define the downward direction as positive, then $a = g = 9.80 \text{ m/s}^2$.

One-Dimensional Motion Involving Gravity

The best way to see the basic features of motion involving gravity is to start with the simplest situations and then progress toward more complex ones. So we start by considering straight up and down motion with no air resistance or friction. These assumptions mean that the velocity (if there is any) is vertical. If the object is dropped, we know the initial velocity is zero. Once the object has left contact with whatever held or threw it, the object is in free-fall. Under these circumstances, the motion is one-dimensional and has constant acceleration of magnitude g. We will also represent vertical displacement with the symbol y and use x for horizontal displacement.

Kinematic Equations for Objects in Free-Fall where Acceleration = -g

$$v = v_0 - gt$$

(2.75)

$$y = y_0 + v_0t - \frac{1}{2}gt^2$$

(2.76)

$$v^2 = v_0^2 - 2g(y - y_0)$$

(2.77)

Example 2.14 Calculating Position and Velocity of a Falling Object: A Rock Thrown Upward

A person standing on the edge of a high cliff throws a rock straight up with an initial velocity of 13.0 m/s. The rock misses the edge of the cliff as it falls back to Earth. Calculate the position and velocity of the rock 1.00 s, 2.00 s, and 3.00 s after it is thrown, neglecting the effects of air resistance.

Strategy

Draw a sketch.

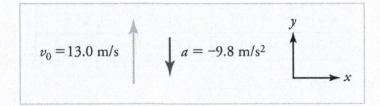

Figure 2.51

We are asked to determine the position y at various times. It is reasonable to take the initial position y_0 to be zero. This problem involves one-dimensional motion in the vertical direction. We use plus and minus signs to indicate direction, with up being positive and down negative. Since up is positive, and the rock is thrown upward, the initial velocity must be positive too. The acceleration due to gravity is downward, so a is negative. It is crucial that the initial velocity and the acceleration due to gravity have opposite signs. Opposite signs indicate that the acceleration due to gravity opposes the initial motion and will slow and eventually reverse it.

Since we are asked for values of position and velocity at three times, we will refer to these as y_1 and v_1; y_2 and v_2; and y_3 and v_3.

Solution for Position y_1

1. Identify the knowns. We know that $y_0 = 0$; $v_0 = 13.0$ m/s; $a = -g = -9.80$ m/s^2; and $t = 1.00$ s.

2. Identify the best equation to use. We will use $y = y_0 + v_0 t + \frac{1}{2}at^2$ because it includes only one unknown, y (or y_1, here), which is the value we want to find.

3. Plug in the known values and solve for y_1.

$$y = 0 + (13.0 \text{ m/s})(1.00 \text{ s}) + \frac{1}{2}\left(-9.80 \text{ m/s}^2\right)(1.00 \text{ s})^2 = 8.10 \text{ m} \tag{2.78}$$

Discussion

The rock is 8.10 m above its starting point at $t = 1.00$ s, since $y_1 > y_0$. It could be *moving* up or down; the only way to tell is to calculate v_1 and find out if it is positive or negative.

Solution for Velocity v_1

1. Identify the knowns. We know that $y_0 = 0$; $v_0 = 13.0$ m/s; $a = -g = -9.80$ m/s^2; and $t = 1.00$ s. We also know from the solution above that $y_1 = 8.10$ m.

2. Identify the best equation to use. The most straightforward is $v = v_0 - gt$ (from $v = v_0 + at$, where $a = \text{gravitational acceleration} = -g$).

3. Plug in the knowns and solve.

$$v_1 = v_0 - gt = 13.0 \text{ m/s} - \left(9.80 \text{ m/s}^2\right)(1.00 \text{ s}) = 3.20 \text{ m/s} \tag{2.79}$$

Discussion

The positive value for v_1 means that the rock is still heading upward at $t = 1.00$ s. However, it has slowed from its original 13.0 m/s, as expected.

Solution for Remaining Times

The procedures for calculating the position and velocity at $t = 2.00$ s and 3.00 s are the same as those above. The results are summarized in **Table 2.1** and illustrated in **Figure 2.52**.

Table 2.1 Results

Time, t	Position, y	Velocity, v	Acceleration, a
1.00 s	8.10 m	3.20 m/s	-9.80 m/s^2
2.00 s	6.40 m	-6.60 m/s	-9.80 m/s^2
3.00 s	-5.10 m	-16.4 m/s	-9.80 m/s^2

Graphing the data helps us understand it more clearly.

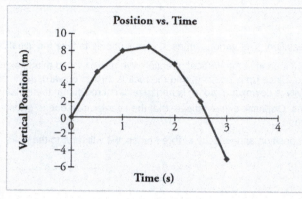

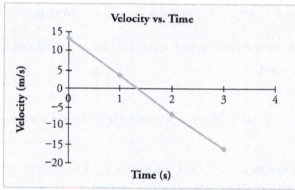

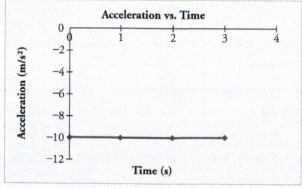

Figure 2.52 Vertical position, vertical velocity, and vertical acceleration vs. time for a rock thrown vertically up at the edge of a cliff. Notice that velocity changes linearly with time and that acceleration is constant. *Misconception Alert!* Notice that the position vs. time graph shows vertical position only. It is easy to get the impression that the graph shows some horizontal motion—the shape of the graph looks like the path of a projectile. But this is not the case; the horizontal axis is *time*, not space. The actual path of the rock in space is straight up, and straight down.

Discussion

The interpretation of these results is important. At 1.00 s the rock is above its starting point and heading upward, since y_1 and v_1 are both positive. At 2.00 s, the rock is still above its starting point, but the negative velocity means it is moving downward. At 3.00 s, both y_3 and v_3 are negative, meaning the rock is below its starting point and continuing to move downward. Notice that when the rock is at its highest point (at 1.5 s), its velocity is zero, but its acceleration is still -9.80 m/s^2. Its acceleration is -9.80 m/s^2 for the whole trip—while it is moving up and while it is moving down. Note that the values for y are the positions (or displacements) of the rock, not the total distances traveled. Finally, note that free-fall applies to upward motion as well as downward. Both have the same acceleration—the acceleration due to gravity, which remains constant the entire time. Astronauts training in the famous Vomit Comet, for example, experience free-fall while arcing up as well as down, as we will discuss in more detail later.

Making Connections: Take-Home Experiment—Reaction Time

A simple experiment can be done to determine your reaction time. Have a friend hold a ruler between your thumb and index finger, separated by about 1 cm. Note the mark on the ruler that is right between your fingers. Have your friend drop the ruler unexpectedly, and try to catch it between your two fingers. Note the new reading on the ruler. Assuming acceleration is that due to gravity, calculate your reaction time. How far would you travel in a car (moving at 30 m/s) if the time it took your foot to go from the gas pedal to the brake was twice this reaction time?

Example 2.15 Calculating Velocity of a Falling Object: A Rock Thrown Down

What happens if the person on the cliff throws the rock straight down, instead of straight up? To explore this question, calculate the velocity of the rock when it is 5.10 m below the starting point, and has been thrown downward with an initial speed of 13.0 m/s.

Strategy

Draw a sketch.

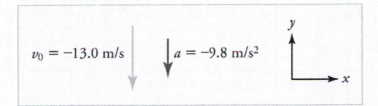

Figure 2.53

Since up is positive, the final position of the rock will be negative because it finishes below the starting point at $y_0 = 0$.

Similarly, the initial velocity is downward and therefore negative, as is the acceleration due to gravity. We expect the final velocity to be negative since the rock will continue to move downward.

Solution

1. Identify the knowns. $y_0 = 0$; $y_1 = -5.10 \text{ m}$; $v_0 = -13.0 \text{ m/s}$; $a = -g = -9.80 \text{ m/s}^2$.

2. Choose the kinematic equation that makes it easiest to solve the problem. The equation $v^2 = v_0^2 + 2a(y - y_0)$ works well because the only unknown in it is v. (We will plug y_1 in for y.)

3. Enter the known values

$$v^2 = (-13.0 \text{ m/s})^2 + 2(-9.80 \text{ m/s}^2)(-5.10 \text{ m} - 0 \text{ m}) = 268.96 \text{ m}^2/\text{s}^2, \tag{2.80}$$

where we have retained extra significant figures because this is an intermediate result.

Taking the square root, and noting that a square root can be positive or negative, gives

$$v = \pm 16.4 \text{ m/s}. \tag{2.81}$$

The negative root is chosen to indicate that the rock is still heading down. Thus,

$$v = -16.4 \text{ m/s}. \tag{2.82}$$

Discussion

Note that *this is exactly the same velocity the rock had at this position when it was thrown straight upward with the same initial speed*. (See **Example 2.14** and **Figure 2.54**(a).) This is not a coincidental result. Because we only consider the acceleration due to gravity in this problem, the *speed* of a falling object depends only on its initial speed and its vertical position relative to the starting point. For example, if the velocity of the rock is calculated at a height of 8.10 m above the starting point (using the method from **Example 2.14**) when the initial velocity is 13.0 m/s straight up, a result of $\pm 3.20 \text{ m/s}$ is obtained. Here both signs are meaningful; the positive value occurs when the rock is at 8.10 m and heading up, and the negative value occurs when the rock is at 8.10 m and heading back down. It has the same *speed* but the opposite direction.

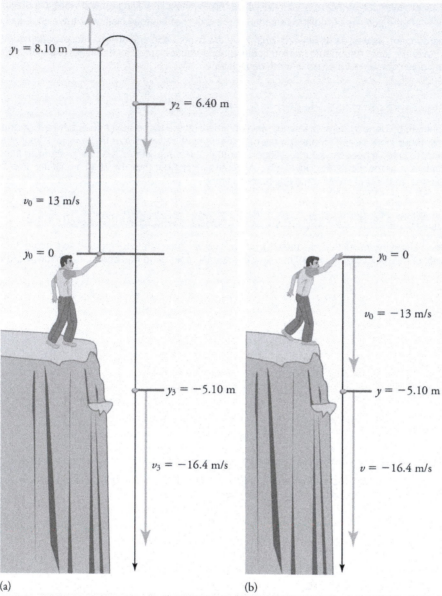

Figure 2.54 (a) A person throws a rock straight up, as explored in Example 2.14. The arrows are velocity vectors at 0, 1.00, 2.00, and 3.00 s. (b) A person throws a rock straight down from a cliff with the same initial speed as before, as in Example 2.15. Note that at the same distance below the point of release, the rock has the same velocity in both cases.

Another way to look at it is this: In **Example 2.14**, the rock is thrown up with an initial velocity of 13.0 m/s. It rises and then falls back down. When its position is $y = 0$ on its way back down, its velocity is -13.0 m/s. That is, it has the same speed on its way down as on its way up. We would then expect its velocity at a position of $y = -5.10 \text{ m}$ to be the same whether we have thrown it upwards at $+13.0 \text{ m/s}$ or thrown it downwards at -13.0 m/s. The velocity of the rock on its way down from $y = 0$ is the same whether we have thrown it up or down to start with, as long as the speed with which it was initially thrown is the same.

Example 2.16 Find *g* from Data on a Falling Object

The acceleration due to gravity on Earth differs slightly from place to place, depending on topography (e.g., whether you are on a hill or in a valley) and subsurface geology (whether there is dense rock like iron ore as opposed to light rock like salt beneath you.) The precise acceleration due to gravity can be calculated from data taken in an introductory physics laboratory course. An object, usually a metal ball for which air resistance is negligible, is dropped and the time it takes to fall a known distance is measured. See, for example, **Figure 2.55**. Very precise results can be produced with this method if sufficient care is taken in measuring the distance fallen and the elapsed time.

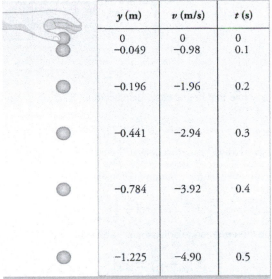

y (m)	v (m/s)	t (s)
0	0	0
−0.049	−0.98	0.1
−0.196	−1.96	0.2
−0.441	−2.94	0.3
−0.784	−3.92	0.4
−1.225	−4.90	0.5

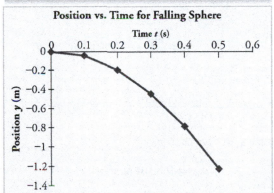

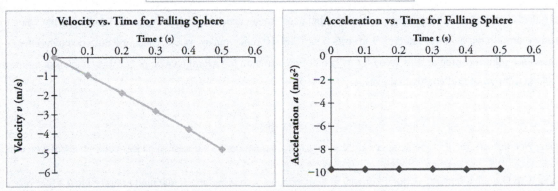

Figure 2.55 Positions and velocities of a metal ball released from rest when air resistance is negligible. Velocity is seen to increase linearly with time while displacement increases with time squared. Acceleration is a constant and is equal to gravitational acceleration.

Suppose the ball falls 1.0000 m in 0.45173 s. Assuming the ball is not affected by air resistance, what is the precise acceleration due to gravity at this location?

Strategy

Draw a sketch.

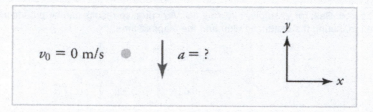

Figure 2.56

We need to solve for acceleration a. Note that in this case, displacement is downward and therefore negative, as is acceleration.

Solution

1. Identify the knowns. $y_0 = 0$; $y = -1.0000$ m; $t = 0.45173$; $v_0 = 0$.

2. Choose the equation that allows you to solve for a using the known values.

$$y = y_0 + v_0 t + \frac{1}{2}at^2 \tag{2.83}$$

3. Substitute 0 for v_0 and rearrange the equation to solve for a. Substituting 0 for v_0 yields

$$y = y_0 + \frac{1}{2}at^2. \tag{2.84}$$

Solving for a gives

$$a = \frac{2(y - y_0)}{t^2}. \tag{2.85}$$

4. Substitute known values yields

$$a = \frac{2(-1.0000 \text{ m} - 0)}{(0.45173 \text{ s})^2} = -9.8010 \text{ m/s}^2, \tag{2.86}$$

so, because $a = -g$ with the directions we have chosen,

$$g = 9.8010 \text{ m/s}^2. \tag{2.87}$$

Discussion

The negative value for a indicates that the gravitational acceleration is downward, as expected. We expect the value to be somewhere around the average value of 9.80 m/s^2, so 9.8010 m/s^2 makes sense. Since the data going into the calculation are relatively precise, this value for g is more precise than the average value of 9.80 m/s^2; it represents the local value for the acceleration due to gravity.

Applying the Science Practices: Finding Acceleration Due to Gravity

While it is well established that the acceleration due to gravity is quite nearly 9.8 m/s^2 at all locations on Earth, you can verify this for yourself with some basic materials.

Your task is to find the acceleration due to gravity at your location. Achieving an acceleration of precisely 9.8 m/s^2 will be difficult. However, with good preparation and attention to detail, you should be able to get close. Before you begin working, consider the following questions.

What measurements will you need to take in order to find the acceleration due to gravity?

What relationships and equations found in this chapter may be useful in calculating the acceleration?

What variables will you need to hold constant?

What materials will you use to record your measurements?

Upon completing these four questions, record your procedure. Once recorded, you may carry out the experiment. If you find that your experiment cannot be carried out, you may revise your procedure.

Once you have found your experimental acceleration, compare it to the assumed value of 9.8 m/s^2. If error exists, what were the likely sources of this error? How could you change your procedure in order to improve the accuracy of your findings?

A chunk of ice breaks off a glacier and falls 30.0 meters before it hits the water. Assuming it falls freely (there is no air resistance), how long does it take to hit the water?

Solution

We know that initial position $y_0 = 0$, final position $y = -30.0 \text{ m}$, and $a = -g = -9.80 \text{ m/s}^2$. We can then use the equation $y = y_0 + v_0 t + \frac{1}{2}at^2$ to solve for t. Inserting $a = -g$, we obtain

$$y = 0 + 0 - \frac{1}{2}gt^2$$

(2.88)

$$t^2 = \frac{2y}{-g}$$

$$t = \pm\sqrt{\frac{2y}{-g}} = \pm\sqrt{\frac{2(-30.0 \text{ m})}{-9.80 \text{ m/s}^2}} = \pm\sqrt{6.12 \text{ s}^2} = 2.47 \text{ s} \approx 2.5 \text{ s}$$

where we take the positive value as the physically relevant answer. Thus, it takes about 2.5 seconds for the piece of ice to hit the water.

PhET Explorations: Equation Grapher

Learn about graphing polynomials. The shape of the curve changes as the constants are adjusted. View the curves for the individual terms (e.g. $y = bx$) to see how they add to generate the polynomial curve.

Figure 2.57 Equation Grapher (http://cnx.org/content/m54775/1.5/equation-grapher_en.jar)

2.8 Graphical Analysis of One Dimensional Motion

Learning Objectives

By the end of this section, you will be able to:

- Describe a straight-line graph in terms of its slope and *y*-intercept.
- Determine average velocity or instantaneous velocity from a graph of position vs. time.
- Determine average or instantaneous acceleration from a graph of velocity vs. time.
- Derive a graph of velocity vs. time from a graph of position vs. time.
- Derive a graph of acceleration vs. time from a graph of velocity vs. time.

A graph, like a picture, is worth a thousand words. Graphs not only contain numerical information; they also reveal relationships between physical quantities. This section uses graphs of displacement, velocity, and acceleration versus time to illustrate one-dimensional kinematics.

Slopes and General Relationships

First note that graphs in this text have perpendicular axes, one horizontal and the other vertical. When two physical quantities are plotted against one another in such a graph, the horizontal axis is usually considered to be an **independent variable** and the vertical axis a **dependent variable**. If we call the horizontal axis the x-axis and the vertical axis the y-axis, as in **Figure 2.58**, a straight-line graph has the general form

$$y = mx + b.$$

(2.89)

Here m is the **slope**, defined to be the rise divided by the run (as seen in the figure) of the straight line. The letter b is used for the **y-intercept**, which is the point at which the line crosses the vertical axis.

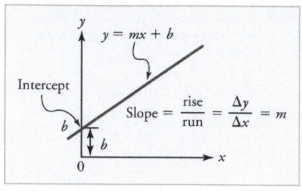

Figure 2.58 A straight-line graph. The equation for a straight line is $y = mx + b$.

Graph of Displacement vs. Time (a = 0, so v is constant)

Time is usually an independent variable that other quantities, such as displacement, depend upon. A graph of displacement versus time would, thus, have x on the vertical axis and t on the horizontal axis. **Figure 2.59** is just such a straight-line graph. It shows a graph of displacement versus time for a jet-powered car on a very flat dry lake bed in Nevada.

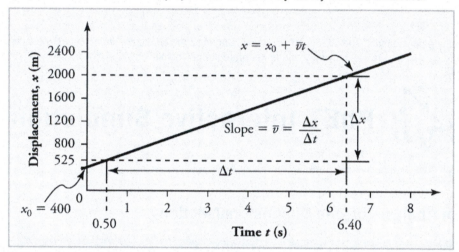

Figure 2.59 Graph of displacement versus time for a jet-powered car on the Bonneville Salt Flats.

Using the relationship between dependent and independent variables, we see that the slope in the graph above is average velocity $\bar{v}$ and the intercept is displacement at time zero—that is, x_0. Substituting these symbols into $y = mx + b$ gives

$$x = \bar{v}t + x_0 \tag{2.90}$$

or

$$x = x_0 + \bar{v}t. \tag{2.91}$$

Thus a graph of displacement versus time gives a general relationship among displacement, velocity, and time, as well as giving detailed numerical information about a specific situation.

The Slope of x vs. t

The slope of the graph of displacement x vs. time t is velocity v .

$$\text{slope} = \frac{\Delta x}{\Delta t} = v \tag{2.92}$$

Notice that this equation is the same as that derived algebraically from other motion equations in **Motion Equations for Constant Acceleration in One Dimension**.

From the figure we can see that the car has a displacement of 400 m at time 0.650 m at t = 1.0 s, and so on. Its displacement at times other than those listed in the table can be read from the graph; furthermore, information about its velocity and acceleration can also be obtained from the graph.

Example 2.17 Determining Average Velocity from a Graph of Displacement versus Time: Jet Car

Find the average velocity of the car whose position is graphed in **Figure 2.59**.

Strategy

The slope of a graph of x vs. t is average velocity, since slope equals rise over run. In this case, rise = change in displacement and run = change in time, so that

$$\text{slope} = \frac{\Delta x}{\Delta t} = \bar{v}. \tag{2.93}$$

Since the slope is constant here, any two points on the graph can be used to find the slope. (Generally speaking, it is most accurate to use two widely separated points on the straight line. This is because any error in reading data from the graph is proportionally smaller if the interval is larger.)

Solution

1. Choose two points on the line. In this case, we choose the points labeled on the graph: (6.4 s, 2000 m) and (0.50 s, 525 m). (Note, however, that you could choose any two points.)

2. Substitute the x and t values of the chosen points into the equation. Remember in calculating change (Δ) we always use final value minus initial value.

$$\bar{v} = \frac{\Delta x}{\Delta t} = \frac{2000 \text{ m} - 525 \text{ m}}{6.4 \text{ s} - 0.50 \text{ s}}, \tag{2.94}$$

yielding

$$\bar{v} = 250 \text{ m/s}. \tag{2.95}$$

Discussion

This is an impressively large land speed (900 km/h, or about 560 mi/h): much greater than the typical highway speed limit of 60 mi/h (27 m/s or 96 km/h), but considerably shy of the record of 343 m/s (1234 km/h or 766 mi/h) set in 1997.

Graphs of Motion when a is constant but $a \neq 0$

The graphs in **Figure 2.60** below represent the motion of the jet-powered car as it accelerates toward its top speed, but only during the time when its acceleration is constant. Time starts at zero for this motion (as if measured with a stopwatch), and the displacement and velocity are initially 200 m and 15 m/s, respectively.

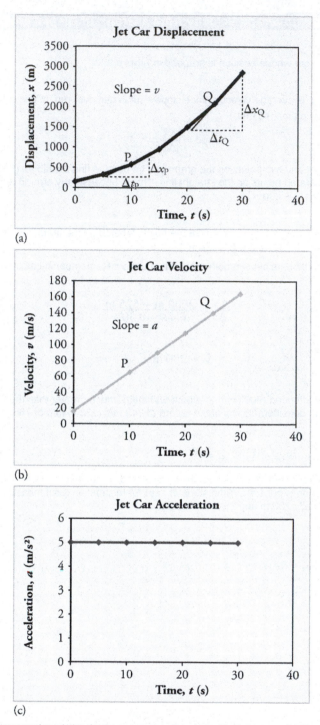

(a)

(b)

(c)

Figure 2.60 Graphs of motion of a jet-powered car during the time span when its acceleration is constant. (a) The slope of an x vs. t graph is velocity. This is shown at two points, and the instantaneous velocities obtained are plotted in the next graph. Instantaneous velocity at any point is the slope of the tangent at that point. (b) The slope of the v vs. t graph is constant for this part of the motion, indicating constant acceleration. (c) Acceleration has the constant value of 5.0 m/s^2 over the time interval plotted.

Figure 2.61 A U.S. Air Force jet car speeds down a track. (credit: Matt Trostle, Flickr)

The graph of displacement versus time in **Figure 2.60**(a) is a curve rather than a straight line. The slope of the curve becomes steeper as time progresses, showing that the velocity is increasing over time. The slope at any point on a displacement-versus-time graph is the instantaneous velocity at that point. It is found by drawing a straight line tangent to the curve at the point of interest and taking the slope of this straight line. Tangent lines are shown for two points in **Figure 2.60**(a). If this is done at every point on the curve and the values are plotted against time, then the graph of velocity versus time shown in **Figure 2.60**(b) is obtained. Furthermore, the slope of the graph of velocity versus time is acceleration, which is shown in **Figure 2.60**(c).

Example 2.18 Determining Instantaneous Velocity from the Slope at a Point: Jet Car

Calculate the velocity of the jet car at a time of 25 s by finding the slope of the x vs. t graph in the graph below.

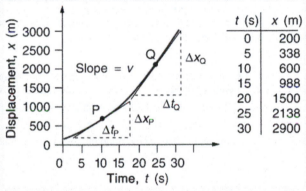

t (s)	x (m)
0	200
5	338
10	600
15	988
20	1500
25	2138
30	2900

Figure 2.62 The slope of an x vs. t graph is velocity. This is shown at two points. Instantaneous velocity at any point is the slope of the tangent at that point.

Strategy

The slope of a curve at a point is equal to the slope of a straight line tangent to the curve at that point. This principle is illustrated in **Figure 2.62**, where Q is the point at $t = 25$ s .

Solution

1. Find the tangent line to the curve at $t = 25$ s .

2. Determine the endpoints of the tangent. These correspond to a position of 1300 m at time 19 s and a position of 3120 m at time 32 s.

3. Plug these endpoints into the equation to solve for the slope, v .

$$\text{slope} = v_Q = \frac{\Delta x_Q}{\Delta t_Q} = \frac{(3120 \text{ m} - 1300 \text{ m})}{(32 \text{ s} - 19 \text{ s})} \qquad (2.96)$$

Thus,

$$v_Q = \frac{1820 \text{ m}}{13 \text{ s}} = 140 \text{ m/s}. \qquad (2.97)$$

Discussion

This is the value given in this figure's table for v at $t = 25$ s. The value of 140 m/s for v_Q is plotted in **Figure 2.62**. The entire graph of v vs. t can be obtained in this fashion.

Carrying this one step further, we note that the slope of a velocity versus time graph is acceleration. Slope is rise divided by run; on a v vs. t graph, rise = change in velocity Δv and run = change in time Δt.

The Slope of v vs. t

The slope of a graph of velocity v vs. time t is acceleration a.

$$\text{slope} = \frac{\Delta v}{\Delta t} = a \tag{2.98}$$

Since the velocity versus time graph in **Figure 2.60**(b) is a straight line, its slope is the same everywhere, implying that acceleration is constant. Acceleration versus time is graphed in **Figure 2.60**(c).

Additional general information can be obtained from **Figure 2.62** and the expression for a straight line, $y = mx + b$.

In this case, the vertical axis y is V, the intercept b is v_0, the slope m is a, and the horizontal axis x is t. Substituting these symbols yields

$$v = v_0 + at. \tag{2.99}$$

A general relationship for velocity, acceleration, and time has again been obtained from a graph. Notice that this equation was also derived algebraically from other motion equations in **Motion Equations for Constant Acceleration in One Dimension**.

It is not accidental that the same equations are obtained by graphical analysis as by algebraic techniques. In fact, an important way to *discover* physical relationships is to measure various physical quantities and then make graphs of one quantity against another to see if they are correlated in any way. Correlations imply physical relationships and might be shown by smooth graphs such as those above. From such graphs, mathematical relationships can sometimes be postulated. Further experiments are then performed to determine the validity of the hypothesized relationships.

Graphs of Motion Where Acceleration is Not Constant

Now consider the motion of the jet car as it goes from 165 m/s to its top velocity of 250 m/s, graphed in **Figure 2.63**. Time again starts at zero, and the initial displacement and velocity are 2900 m and 165 m/s, respectively. (These were the final displacement and velocity of the car in the motion graphed in **Figure 2.60**.) Acceleration gradually decreases from 5.0 m/s^2 to zero when the car hits 250 m/s. The slope of the x vs. t graph increases until $t = 55$ s, after which time the slope is constant. Similarly, velocity increases until 55 s and then becomes constant, since acceleration decreases to zero at 55 s and remains zero afterward.

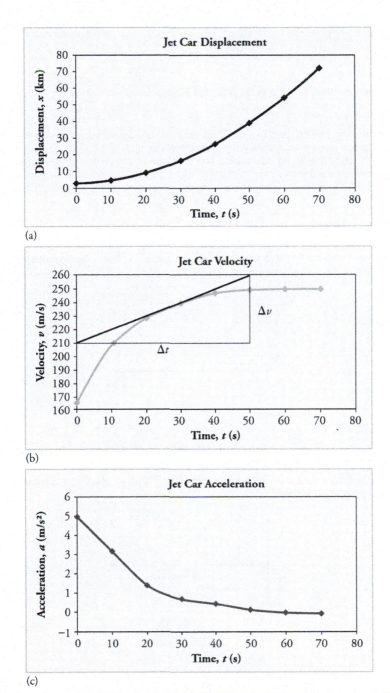

Figure 2.63 Graphs of motion of a jet-powered car as it reaches its top velocity. This motion begins where the motion in **Figure 2.60** ends. (a) The slope of this graph is velocity; it is plotted in the next graph. (b) The velocity gradually approaches its top value. The slope of this graph is acceleration; it is plotted in the final graph. (c) Acceleration gradually declines to zero when velocity becomes constant.

Example 2.19 Calculating Acceleration from a Graph of Velocity versus Time

Calculate the acceleration of the jet car at a time of 25 s by finding the slope of the v vs. t graph in **Figure 2.63**(b).

Strategy

The slope of the curve at $t = 25$ s is equal to the slope of the line tangent at that point, as illustrated in **Figure 2.63**(b).

Solution

Determine endpoints of the tangent line from the figure, and then plug them into the equation to solve for slope, a.

$$\text{slope} = \frac{\Delta v}{\Delta t} = \frac{(260 \text{ m/s} - 210 \text{ m/s})}{(51 \text{ s} - 1.0 \text{ s})} \qquad (2.100)$$

$$a = \frac{50 \text{ m/s}}{50 \text{ s}} = 1.0 \text{ m/s}^2.$$ (2.101)

Discussion

Note that this value for a is consistent with the value plotted in **Figure 2.63**(c) at $t = 25 \text{ s}$.

A graph of displacement versus time can be used to generate a graph of velocity versus time, and a graph of velocity versus time can be used to generate a graph of acceleration versus time. We do this by finding the slope of the graphs at every point. If the graph is linear (i.e., a line with a constant slope), it is easy to find the slope at any point and you have the slope for every point. Graphical analysis of motion can be used to describe both specific and general characteristics of kinematics. Graphs can also be used for other topics in physics. An important aspect of exploring physical relationships is to graph them and look for underlying relationships.

Check Your Understanding

A graph of velocity vs. time of a ship coming into a harbor is shown below. (a) Describe the motion of the ship based on the graph. (b)What would a graph of the ship's acceleration look like?

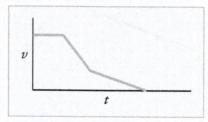

Figure 2.64

Solution

(a) The ship moves at constant velocity and then begins to decelerate at a constant rate. At some point, its deceleration rate decreases. It maintains this lower deceleration rate until it stops moving.

(b) A graph of acceleration vs. time would show zero acceleration in the first leg, large and constant negative acceleration in the second leg, and constant negative acceleration.

Figure 2.65

Glossary

acceleration: the rate of change in velocity; the change in velocity over time

acceleration due to gravity: acceleration of an object as a result of gravity

average acceleration: the change in velocity divided by the time over which it changes

average speed: distance traveled divided by time during which motion occurs

average velocity: displacement divided by time over which displacement occurs

deceleration: acceleration in the direction opposite to velocity; acceleration that results in a decrease in velocity

dependent variable: the variable that is being measured; usually plotted along the y-axis

displacement: the change in position of an object

distance: the magnitude of displacement between two positions

distance traveled: the total length of the path traveled between two positions

elapsed time: the difference between the ending time and beginning time

free-fall: the state of movement that results from gravitational force only

independent variable: the variable that the dependent variable is measured with respect to; usually plotted along the x-axis

instantaneous acceleration: acceleration at a specific point in time

instantaneous speed: magnitude of the instantaneous velocity

instantaneous velocity: velocity at a specific instant, or the average velocity over an infinitesimal time interval

kinematics: the study of motion without considering its causes

model: simplified description that contains only those elements necessary to describe the physics of a physical situation

position: the location of an object at a particular time

scalar: a quantity that is described by magnitude, but not direction

slope: the difference in y-value (the rise) divided by the difference in x-value (the run) of two points on a straight line

time: change, or the interval over which change occurs

vector: a quantity that is described by both magnitude and direction

y-intercept: the y-value when $x = 0$, or when the graph crosses the y-axis

Section Summary

2.1 Displacement
- Kinematics is the study of motion without considering its causes. In this chapter, it is limited to motion along a straight line, called one-dimensional motion.
- Displacement is the change in position of an object.
- In symbols, displacement Δx is defined to be

$$\Delta x = x_f - x_0,$$

where x_0 is the initial position and x_f is the final position. In this text, the Greek letter Δ (delta) always means "change in" whatever quantity follows it. The SI unit for displacement is the meter (m). Displacement has a direction as well as a magnitude.
- When you start a problem, assign which direction will be positive.
- Distance is the magnitude of displacement between two positions.
- Distance traveled is the total length of the path traveled between two positions.

2.2 Vectors, Scalars, and Coordinate Systems
- A vector is any quantity that has magnitude and direction.
- A scalar is any quantity that has magnitude but no direction.
- Displacement and velocity are vectors, whereas distance and speed are scalars.
- In one-dimensional motion, direction is specified by a plus or minus sign to signify left or right, up or down, and the like.

2.3 Time, Velocity, and Speed
- Time is measured in terms of change, and its SI unit is the second (s). Elapsed time for an event is

$$\Delta t = t_f - t_0,$$

where t_f is the final time and t_0 is the initial time. The initial time is often taken to be zero, as if measured with a stopwatch; the elapsed time is then just t.
- Average velocity $\bar{v}$ is defined as displacement divided by the travel time. In symbols, average velocity is

$$\bar{v} = \frac{\Delta x}{\Delta t} = \frac{x_f - x_0}{t_f - t_0}.$$

- The SI unit for velocity is m/s.
- Velocity is a vector and thus has a direction.
- Instantaneous velocity v is the velocity at a specific instant or the average velocity for an infinitesimal interval.
- Instantaneous speed is the magnitude of the instantaneous velocity.
- Instantaneous speed is a scalar quantity, as it has no direction specified.

- Average speed is the total distance traveled divided by the elapsed time. (Average speed is *not* the magnitude of the average velocity.) Speed is a scalar quantity; it has no direction associated with it.

2.4 Acceleration

- Acceleration is the rate at which velocity changes. In symbols, **average acceleration** $\bar{a}$ is

$$\bar{a} = \frac{\Delta v}{\Delta t} = \frac{v_f - v_0}{t_f - t_0}.$$

- The SI unit for acceleration is $\mathrm{m/s}^2$.
- Acceleration is a vector, and thus has a both a magnitude and direction.
- Acceleration can be caused by either a change in the magnitude or the direction of the velocity.
- Instantaneous acceleration a is the acceleration at a specific instant in time.
- Deceleration is an acceleration with a direction opposite to that of the velocity.

2.5 Motion Equations for Constant Acceleration in One Dimension

- To simplify calculations we take acceleration to be constant, so that $\bar{a} = a$ at all times.
- We also take initial time to be zero.
- Initial position and velocity are given a subscript 0; final values have no subscript. Thus,

$$\left.\begin{aligned} \Delta t &= t \\ \Delta x &= x - x_0 \\ \Delta v &= v - v_0 \end{aligned}\right\}$$

- The following kinematic equations for motion with constant a are useful:

$$x = x_0 + \bar{v}\,t$$

$$\bar{v} = \frac{v_0 + v}{2}$$

$$v = v_0 + at$$

$$x = x_0 + v_0 t + \frac{1}{2}at^2$$

$$v^2 = v_0^2 + 2a(x - x_0)$$

- In vertical motion, y is substituted for x.

2.6 Problem-Solving Basics for One Dimensional Kinematics

- *The six basic problem solving steps for physics are:*
  Step 1. Examine the situation to determine which physical principles are involved.

  Step 2. Make a list of what is given or can be inferred from the problem as stated (identify the knowns).

  Step 3. Identify exactly what needs to be determined in the problem (identify the unknowns).

  Step 4. Find an equation or set of equations that can help you solve the problem.

  Step 5. Substitute the knowns along with their units into the appropriate equation, and obtain numerical solutions complete with units.

  Step 6. Check the answer to see if it is reasonable: Does it make sense?

2.7 Falling Objects

- An object in free-fall experiences constant acceleration if air resistance is negligible.
- On Earth, all free-falling objects have an acceleration due to gravity g, which averages

$$g = 9.80 \ \mathrm{m/s}^2.$$

- Whether the acceleration a should be taken as $+g$ or $-g$ is determined by your choice of coordinate system. If you choose the upward direction as positive, $a = -g = -9.80 \ \mathrm{m/s}^2$ is negative. In the opposite case, $a = +g = 9.80 \ \mathrm{m/s}^2$ is positive. Since acceleration is constant, the kinematic equations above can be applied with the appropriate $+g$ or $-g$ substituted for a.
- For objects in free-fall, up is normally taken as positive for displacement, velocity, and acceleration.

2.8 Graphical Analysis of One Dimensional Motion

- Graphs of motion can be used to analyze motion.
- Graphical solutions yield identical solutions to mathematical methods for deriving motion equations.
- The slope of a graph of displacement x vs. time t is velocity v.
- The slope of a graph of velocity v vs. time t graph is acceleration a.
- Average velocity, instantaneous velocity, and acceleration can all be obtained by analyzing graphs.

Conceptual Questions

2.1 Displacement

1. Give an example in which there are clear distinctions among distance traveled, displacement, and magnitude of displacement. Specifically identify each quantity in your example.

2. Under what circumstances does distance traveled equal magnitude of displacement? What is the only case in which magnitude of displacement and displacement are exactly the same?

3. Bacteria move back and forth by using their flagella (structures that look like little tails). Speeds of up to $50 \ \mu\text{m/s} \left(50 \times 10^{-6} \ \text{m/s}\right)$ have been observed. The total distance traveled by a bacterium is large for its size, while its displacement is small. Why is this?

2.2 Vectors, Scalars, and Coordinate Systems

4. A student writes, "*A bird that is diving for prey has a speed of* $-10 \ m/s$." What is wrong with the student's statement? What has the student actually described? Explain.

5. What is the speed of the bird in **Exercise 2.4**?

6. Acceleration is the change in velocity over time. Given this information, is acceleration a vector or a scalar quantity? Explain.

7. A weather forecast states that the temperature is predicted to be $-5°\text{C}$ the following day. Is this temperature a vector or a scalar quantity? Explain.

2.3 Time, Velocity, and Speed

8. Give an example (but not one from the text) of a device used to measure time and identify what change in that device indicates a change in time.

9. There is a distinction between average speed and the magnitude of average velocity. Give an example that illustrates the difference between these two quantities.

10. Does a car's odometer measure position or displacement? Does its speedometer measure speed or velocity?

11. If you divide the total distance traveled on a car trip (as determined by the odometer) by the time for the trip, are you calculating the average speed or the magnitude of the average velocity? Under what circumstances are these two quantities the same?

12. How are instantaneous velocity and instantaneous speed related to one another? How do they differ?

2.4 Acceleration

13. Is it possible for speed to be constant while acceleration is not zero? Give an example of such a situation.

14. Is it possible for velocity to be constant while acceleration is not zero? Explain.

15. Give an example in which velocity is zero yet acceleration is not.

16. If a subway train is moving to the left (has a negative velocity) and then comes to a stop, what is the direction of its acceleration? Is the acceleration positive or negative?

17. Plus and minus signs are used in one-dimensional motion to indicate direction. What is the sign of an acceleration that reduces the magnitude of a negative velocity? Of a positive velocity?

2.6 Problem-Solving Basics for One Dimensional Kinematics

18. What information do you need in order to choose which equation or equations to use to solve a problem? Explain.

19. What is the last thing you should do when solving a problem? Explain.

2.7 Falling Objects

20. What is the acceleration of a rock thrown straight upward on the way up? At the top of its flight? On the way down?

21. An object that is thrown straight up falls back to Earth. This is one-dimensional motion. (a) When is its velocity zero? (b) Does its velocity change direction? (c) Does the acceleration due to gravity have the same sign on the way up as on the way down?

22. Suppose you throw a rock nearly straight up at a coconut in a palm tree, and the rock misses on the way up but hits the coconut on the way down. Neglecting air resistance, how does the speed of the rock when it hits the coconut on the way down compare with what it would have been if it had hit the coconut on the way up? Is it more likely to dislodge the coconut on the way up or down? Explain.

23. If an object is thrown straight up and air resistance is negligible, then its speed when it returns to the starting point is the same as when it was released. If air resistance were not negligible, how would its speed upon return compare with its initial speed? How would the maximum height to which it rises be affected?

24. The severity of a fall depends on your speed when you strike the ground. All factors but the acceleration due to gravity being the same, how many times higher could a safe fall on the Moon be than on Earth (gravitational acceleration on the Moon is about 1/6 that of the Earth)?

25. How many times higher could an astronaut jump on the Moon than on Earth if his takeoff speed is the same in both locations (gravitational acceleration on the Moon is about 1/6 of g on Earth)?

2.8 Graphical Analysis of One Dimensional Motion

26. (a) Explain how you can use the graph of position versus time in **Figure 2.66** to describe the change in velocity over time. Identify (b) the time (t_a, t_b, t_c, t_d, or t_e) at which the instantaneous velocity is greatest, (c) the time at which it is zero, and (d) the time at which it is negative.

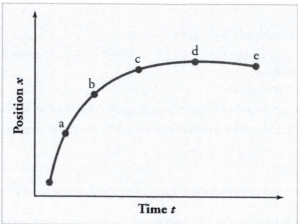

Figure 2.66

27. (a) Sketch a graph of velocity versus time corresponding to the graph of displacement versus time given in **Figure 2.67**. (b) Identify the time or times (t_a, t_b, t_c, etc.) at which the instantaneous velocity is greatest. (c) At which times is it zero? (d) At which times is it negative?

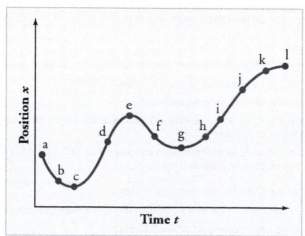

Figure 2.67

28. (a) Explain how you can determine the acceleration over time from a velocity versus time graph such as the one in **Figure 2.68**. (b) Based on the graph, how does acceleration change over time?

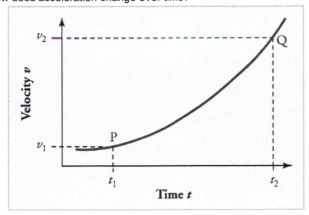

Figure 2.68

29. (a) Sketch a graph of acceleration versus time corresponding to the graph of velocity versus time given in **Figure 2.69**. (b) Identify the time or times (t_a , t_b , t_c , etc.) at which the acceleration is greatest. (c) At which times is it zero? (d) At which times is it negative?

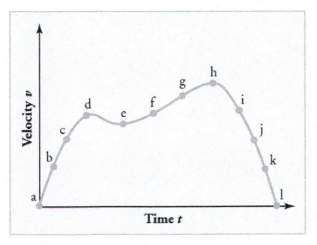

Figure 2.69

30. Consider the velocity vs. time graph of a person in an elevator shown in **Figure 2.70**. Suppose the elevator is initially at rest. It then accelerates for 3 seconds, maintains that velocity for 15 seconds, then decelerates for 5 seconds until it stops. The acceleration for the entire trip is not constant so we cannot use the equations of motion from **Motion Equations for Constant Acceleration in One Dimension** for the complete trip. (We could, however, use them in the three individual sections where acceleration is a constant.) Sketch graphs of (a) position vs. time and (b) acceleration vs. time for this trip.

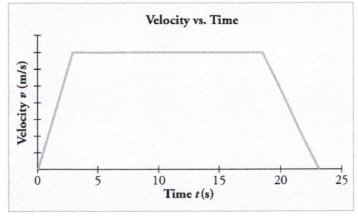

Figure 2.70

31. A cylinder is given a push and then rolls up an inclined plane. If the origin is the starting point, sketch the position, velocity, and acceleration of the cylinder vs. time as it goes up and then down the plane.

Problems & Exercises

2.1 Displacement

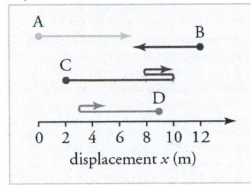

Figure 2.71

1. Find the following for path A in **Figure 2.71**: (a) The distance traveled. (b) The magnitude of the displacement from start to finish. (c) The displacement from start to finish.

2. Find the following for path B in **Figure 2.71**: (a) The distance traveled. (b) The magnitude of the displacement from start to finish. (c) The displacement from start to finish.

3. Find the following for path C in **Figure 2.71**: (a) The distance traveled. (b) The magnitude of the displacement from start to finish. (c) The displacement from start to finish.

4. Find the following for path D in **Figure 2.71**: (a) The distance traveled. (b) The magnitude of the displacement from start to finish. (c) The displacement from start to finish.

2.3 Time, Velocity, and Speed

5. (a) Calculate Earth's average speed relative to the Sun. (b) What is its average velocity over a period of one year?

6. A helicopter blade spins at exactly 100 revolutions per minute. Its tip is 5.00 m from the center of rotation. (a) Calculate the average speed of the blade tip in the helicopter's frame of reference. (b) What is its average velocity over one revolution?

7. The North American and European continents are moving apart at a rate of about 3 cm/y. At this rate how long will it take them to drift 500 km farther apart than they are at present?

8. Land west of the San Andreas fault in southern California is moving at an average velocity of about 6 cm/y northwest relative to land east of the fault. Los Angeles is west of the fault and may thus someday be at the same latitude as San Francisco, which is east of the fault. How far in the future will this occur if the displacement to be made is 590 km northwest, assuming the motion remains constant?

9. On May 26, 1934, a streamlined, stainless steel diesel train called the Zephyr set the world's nonstop long-distance speed record for trains. Its run from Denver to Chicago took 13 hours, 4 minutes, 58 seconds, and was witnessed by more than a million people along the route. The total distance traveled was 1633.8 km. What was its average speed in km/h and m/s?

10. Tidal friction is slowing the rotation of the Earth. As a result, the orbit of the Moon is increasing in radius at a rate of approximately 4 cm/year. Assuming this to be a constant rate, how many years will pass before the radius of the Moon's orbit increases by 3.84×10^6 m (1%)?

11. A student drove to the university from her home and noted that the odometer reading of her car increased by 12.0 km. The trip took 18.0 min. (a) What was her average speed? (b) If the straight-line distance from her home to the university is 10.3 km in a direction $25.0°$ south of east, what was her average velocity? (c) If she returned home by the same path 7 h 30 min after she left, what were her average speed and velocity for the entire trip?

12. The speed of propagation of the action potential (an electrical signal) in a nerve cell depends (inversely) on the diameter of the axon (nerve fiber). If the nerve cell connecting the spinal cord to your feet is 1.1 m long, and the nerve impulse speed is 18 m/s, how long does it take for the nerve signal to travel this distance?

13. Conversations with astronauts on the lunar surface were characterized by a kind of echo in which the earthbound person's voice was so loud in the astronaut's space helmet that it was picked up by the astronaut's microphone and transmitted back to Earth. It is reasonable to assume that the echo time equals the time necessary for the radio wave to travel from the Earth to the Moon and back (that is, neglecting any time delays in the electronic equipment). Calculate the distance from Earth to the Moon given that the echo time was 2.56 s and that radio waves travel at the speed of light $(3.00 \times 10^8 \text{ m/s})$.

14. A football quarterback runs 15.0 m straight down the playing field in 2.50 s. He is then hit and pushed 3.00 m straight backward in 1.75 s. He breaks the tackle and runs straight forward another 21.0 m in 5.20 s. Calculate his average velocity (a) for each of the three intervals and (b) for the entire motion.

15. The planetary model of the atom pictures electrons orbiting the atomic nucleus much as planets orbit the Sun. In this model you can view hydrogen, the simplest atom, as having a single electron in a circular orbit 1.06×10^{-10} m in diameter. (a) If the average speed of the electron in this orbit is known to be 2.20×10^6 m/s, calculate the number of revolutions per second it makes about the nucleus. (b) What is the electron's average velocity?

2.4 Acceleration

16. A cheetah can accelerate from rest to a speed of 30.0 m/s in 7.00 s. What is its acceleration?

17. Professional Application

Dr. John Paul Stapp was U.S. Air Force officer who studied the effects of extreme deceleration on the human body. On December 10, 1954, Stapp rode a rocket sled, accelerating from rest to a top speed of 282 m/s (1015 km/h) in 5.00 s, and was brought jarringly back to rest in only 1.40 s! Calculate his (a) acceleration and (b) deceleration. Express each in multiples of g (9.80 m/s^2) by taking its ratio to the acceleration of gravity.

18. A commuter backs her car out of her garage with an acceleration of 1.40 m/s^2. (a) How long does it take her to reach a speed of 2.00 m/s? (b) If she then brakes to a stop in 0.800 s, what is her deceleration?

19. Assume that an intercontinental ballistic missile goes from rest to a suborbital speed of 6.50 km/s in 60.0 s (the actual speed and time are classified). What is its average acceleration in m/s^2 and in multiples of g (9.80 m/s^2)?

2.5 Motion Equations for Constant Acceleration in One Dimension

20. An Olympic-class sprinter starts a race with an acceleration of 4.50 m/s^2. (a) What is her speed 2.40 s later? (b) Sketch a graph of her position vs. time for this period.

21. A well-thrown ball is caught in a well-padded mitt. If the deceleration of the ball is $2.10 \times 10^4 \text{ m/s}^2$, and 1.85 ms $(1 \text{ ms} = 10^{-3} \text{ s})$ elapses from the time the ball first touches the mitt until it stops, what was the initial velocity of the ball?

22. A bullet in a gun is accelerated from the firing chamber to the end of the barrel at an average rate of $6.20 \times 10^5 \text{ m/s}^2$ for 8.10×10^{-4} s. What is its muzzle velocity (that is, its final velocity)?

23. (a) A light-rail commuter train accelerates at a rate of 1.35 m/s^2. How long does it take to reach its top speed of 80.0 km/h, starting from rest? (b) The same train ordinarily decelerates at a rate of 1.65 m/s^2. How long does it take to come to a stop from its top speed? (c) In emergencies the train can decelerate more rapidly, coming to rest from 80.0 km/h in 8.30 s. What is its emergency deceleration in m/s^2?

24. While entering a freeway, a car accelerates from rest at a rate of 2.40 m/s^2 for 12.0 s. (a) Draw a sketch of the situation. (b) List the knowns in this problem. (c) How far does the car travel in those 12.0 s? To solve this part, first identify the unknown, and then discuss how you chose the appropriate equation to solve for it. After choosing the equation, show your steps in solving for the unknown, check your units, and discuss whether the answer is reasonable. (d) What is the car's final velocity? Solve for this unknown in the same manner as in part (c), showing all steps explicitly.

25. At the end of a race, a runner decelerates from a velocity of 9.00 m/s at a rate of 2.00 m/s^2. (a) How far does she travel in the next 5.00 s? (b) What is her final velocity? (c) Evaluate the result. Does it make sense?

26. Professional Application:

Blood is accelerated from rest to 30.0 cm/s in a distance of 1.80 cm by the left ventricle of the heart. (a) Make a sketch of the situation. (b) List the knowns in this problem. (c) How long does the acceleration take? To solve this part, first identify the unknown, and then discuss how you chose the appropriate equation to solve for it. After choosing the equation, show your steps in solving for the unknown, checking your units. (d) Is the answer reasonable when compared with the time for a heartbeat?

27. In a slap shot, a hockey player accelerates the puck from a velocity of 8.00 m/s to 40.0 m/s in the same direction. If this shot takes 3.33×10^{-2} s, calculate the distance over which the puck accelerates.

28. A powerful motorcycle can accelerate from rest to 26.8 m/s (100 km/h) in only 3.90 s. (a) What is its average acceleration? (b) How far does it travel in that time?

29. Freight trains can produce only relatively small accelerations and decelerations. (a) What is the final velocity of a freight train that accelerates at a rate of 0.0500 m/s^2 for 8.00 min, starting with an initial velocity of 4.00 m/s? (b) If the train can slow down at a rate of 0.550 m/s^2, how long will it take to come to a stop from this velocity? (c) How far will it travel in each case?

30. A fireworks shell is accelerated from rest to a velocity of 65.0 m/s over a distance of 0.250 m. (a) How long did the acceleration last? (b) Calculate the acceleration.

31. A swan on a lake gets airborne by flapping its wings and running on top of the water. (a) If the swan must reach a velocity of 6.00 m/s to take off and it accelerates from rest at an average rate of 0.350 m/s^2, how far will it travel before becoming airborne? (b) How long does this take?

32. Professional Application:

A woodpecker's brain is specially protected from large decelerations by tendon-like attachments inside the skull. While pecking on a tree, the woodpecker's head comes to a stop from an initial velocity of 0.600 m/s in a distance of only 2.00 mm. (a) Find the acceleration in m/s^2 and in multiples of g $(g = 9.80 \text{ m/s}^2)$. (b) Calculate the stopping time. (c) The tendons cradling the brain stretch, making its stopping distance 4.50 mm (greater than the head and, hence, less deceleration of the brain). What is the brain's deceleration, expressed in multiples of g?

33. An unwary football player collides with a padded goalpost while running at a velocity of 7.50 m/s and comes to a full stop after compressing the padding and his body 0.350 m. (a) What is his deceleration? (b) How long does the collision last?

34. In World War II, there were several reported cases of airmen who jumped from their flaming airplanes with no parachute to escape certain death. Some fell about 20,000 feet (6000 m), and some of them survived, with few life-threatening injuries. For these lucky pilots, the tree branches and snow drifts on the ground allowed their deceleration to be relatively small. If we assume that a pilot's speed upon impact was 123 mph (54 m/s), then what was his deceleration? Assume that the trees and snow stopped him over a distance of 3.0 m.

35. Consider a grey squirrel falling out of a tree to the ground. (a) If we ignore air resistance in this case (only for the sake of this problem), determine a squirrel's velocity just before hitting the ground, assuming it fell from a height of 3.0 m. (b) If the squirrel stops in a distance of 2.0 cm through bending its limbs, compare its deceleration with that of the airman in the previous problem.

36. An express train passes through a station. It enters with an initial velocity of 22.0 m/s and decelerates at a rate of 0.150 m/s^2 as it goes through. The station is 210 m long. (a) How long is the nose of the train in the station? (b) How fast is it going when the nose leaves the station? (c) If the train is 130 m long, when does the end of the train leave the station? (d) What is the velocity of the end of the train as it leaves?

37. Dragsters can actually reach a top speed of 145 m/s in only 4.45 s—considerably less time than given in **Example 2.10** and **Example 2.11**. (a) Calculate the average acceleration for such a dragster. (b) Find the final velocity of this dragster starting from rest and accelerating at the rate found in (a) for 402 m (a quarter mile) without using any information on time. (c) Why is the final velocity greater than that used to find the average acceleration? *Hint*: Consider whether the assumption of constant acceleration is valid for a dragster. If not, discuss whether the acceleration would be greater at the beginning or end of the run and what effect that would have on the final velocity.

38. A bicycle racer sprints at the end of a race to clinch a victory. The racer has an initial velocity of 11.5 m/s and accelerates at the rate of 0.500 m/s^2 for 7.00 s. (a) What is his final velocity? (b) The racer continues at this velocity to the finish line. If he was 300 m from the finish line when he started to accelerate, how much time did he save? (c) One other racer was 5.00 m ahead when the winner started to accelerate, but he was unable to accelerate, and traveled at 11.8 m/s until the finish line. How far ahead of him (in meters and in seconds) did the winner finish?

39. In 1967, New Zealander Burt Munro set the world record for an Indian motorcycle, on the Bonneville Salt Flats in Utah, with a maximum speed of 183.58 mi/h. The one-way course was 5.00 mi long. Acceleration rates are often described by the time it takes to reach 60.0 mi/h from rest. If this time was 4.00 s, and Burt accelerated at this rate until he reached his maximum speed, how long did it take Burt to complete the course?

40. (a) A world record was set for the men's 100-m dash in the 2008 Olympic Games in Beijing by Usain Bolt of Jamaica. Bolt "coasted" across the finish line with a time of 9.69 s. If we assume that Bolt accelerated for 3.00 s to reach his maximum speed, and maintained that speed for the rest of the race, calculate his maximum speed and his acceleration. (b) During the same Olympics, Bolt also set the world record in the 200-m dash with a time of 19.30 s. Using the same assumptions as for the 100-m dash, what was his maximum speed for this race?

2.7 Falling Objects

Assume air resistance is negligible unless otherwise stated.

41. Calculate the displacement and velocity at times of (a) 0.500, (b) 1.00, (c) 1.50, and (d) 2.00 s for a ball thrown straight up with an initial velocity of 15.0 m/s. Take the point of release to be $y_0 = 0$.

42. Calculate the displacement and velocity at times of (a) 0.500, (b) 1.00, (c) 1.50, (d) 2.00, and (e) 2.50 s for a rock thrown straight down with an initial velocity of 14.0 m/s from the Verrazano Narrows Bridge in New York City. The roadway of this bridge is 70.0 m above the water.

43. A basketball referee tosses the ball straight up for the starting tip-off. At what velocity must a basketball player leave the ground to rise 1.25 m above the floor in an attempt to get the ball?

44. A rescue helicopter is hovering over a person whose boat has sunk. One of the rescuers throws a life preserver straight down to the victim with an initial velocity of 1.40 m/s and observes that it takes 1.8 s to reach the water. (a) List the knowns in this problem. (b) How high above the water was the preserver released? Note that the downdraft of the helicopter reduces the effects of air resistance on the falling life preserver, so that an acceleration equal to that of gravity is reasonable.

45. A dolphin in an aquatic show jumps straight up out of the water at a velocity of 13.0 m/s. (a) List the knowns in this problem. (b) How high does his body rise above the water? To solve this part, first note that the final velocity is now a known and identify its value. Then identify the unknown, and discuss how you chose the appropriate equation to solve for it. After choosing the equation, show your steps in solving for the unknown, checking units, and discuss whether the answer is reasonable. (c) How long is the dolphin in the air? Neglect any effects due to his size or orientation.

46. A swimmer bounces straight up from a diving board and falls feet first into a pool. She starts with a velocity of 4.00 m/s, and her takeoff point is 1.80 m above the pool. (a) How long are her feet in the air? (b) What is her highest point above the board? (c) What is her velocity when her feet hit the water?

47. (a) Calculate the height of a cliff if it takes 2.35 s for a rock to hit the ground when it is thrown straight up from the cliff with an initial velocity of 8.00 m/s. (b) How long would it take to reach the ground if it is thrown straight down with the same speed?

48. A very strong, but inept, shot putter puts the shot straight up vertically with an initial velocity of 11.0 m/s. How long does he have to get out of the way if the shot was released at a height of 2.20 m, and he is 1.80 m tall?

49. You throw a ball straight up with an initial velocity of 15.0 m/s. It passes a tree branch on the way up at a height of 7.00 m. How much additional time will pass before the ball passes the tree branch on the way back down?

50. A kangaroo can jump over an object 2.50 m high. (a) Calculate its vertical speed when it leaves the ground. (b) How long is it in the air?

51. Standing at the base of one of the cliffs of Mt. Arapiles in Victoria, Australia, a hiker hears a rock break loose from a height of 105 m. He can't see the rock right away but then does, 1.50 s later. (a) How far above the hiker is the rock when he can see it? (b) How much time does he have to move before the rock hits his head?

52. An object is dropped from a height of 75.0 m above ground level. (a) Determine the distance traveled during the first second. (b) Determine the final velocity at which the object hits the ground. (c) Determine the distance traveled during the last second of motion before hitting the ground.

53. There is a 250-m-high cliff at Half Dome in Yosemite National Park in California. Suppose a boulder breaks loose from the top of this cliff. (a) How fast will it be going when it strikes the ground? (b) Assuming a reaction time of 0.300 s, how long will a tourist at the bottom have to get out of the way after hearing the sound of the rock breaking loose (neglecting the height of the tourist, which would become negligible anyway if hit)? The speed of sound is 335 m/s on this day.

54. A ball is thrown straight up. It passes a 2.00-m-high window 7.50 m off the ground on its path up and takes 1.30 s to go past the window. What was the ball's initial velocity?

55. Suppose you drop a rock into a dark well and, using precision equipment, you measure the time for the sound of a splash to return. (a) Neglecting the time required for sound to travel up the well, calculate the distance to the water if the sound returns in 2.0000 s. (b) Now calculate the distance taking into account the time for sound to travel up the well. The speed of sound is 332.00 m/s in this well.

56. A steel ball is dropped onto a hard floor from a height of 1.50 m and rebounds to a height of 1.45 m. (a) Calculate its velocity just before it strikes the floor. (b) Calculate its velocity just after it leaves the floor on its way back up. (c) Calculate its acceleration during contact with the floor if that contact lasts 0.0800 ms $(8.00 \times 10^{-5} \text{ s})$. (d) How much did the ball compress during its collision with the floor, assuming the floor is absolutely rigid?

57. A coin is dropped from a hot-air balloon that is 300 m above the ground and rising at 10.0 m/s upward. For the coin, find (a) the maximum height reached, (b) its position and velocity 4.00 s after being released, and (c) the time before it hits the ground.

58. A soft tennis ball is dropped onto a hard floor from a height of 1.50 m and rebounds to a height of 1.10 m. (a) Calculate its velocity just before it strikes the floor. (b) Calculate its velocity just after it leaves the floor on its way back up. (c) Calculate its acceleration during contact with the floor if that contact lasts 3.50 ms $(3.50 \times 10^{-3} \text{ s})$. (d) How much did the ball compress during its collision with the floor, assuming the floor is absolutely rigid?

2.8 Graphical Analysis of One Dimensional Motion

Note: There is always uncertainty in numbers taken from graphs. If your answers differ from expected values, examine them to see if they are within data extraction uncertainties estimated by you.

59. (a) By taking the slope of the curve in **Figure 2.72**, verify that the velocity of the jet car is 115 m/s at $t = 20$ s. (b) By taking the slope of the curve at any point in **Figure 2.73**, verify that the jet car's acceleration is 5.0 m/s^2.

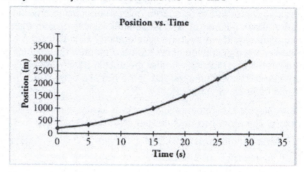

Figure 2.72

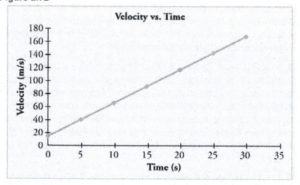

Figure 2.73

60. Using approximate values, calculate the slope of the curve in **Figure 2.74** to verify that the velocity at $t = 10.0$ s is 0.208 m/s. Assume all values are known to 3 significant figures.

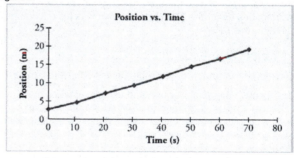

Figure 2.74

61. Using approximate values, calculate the slope of the curve in **Figure 2.74** to verify that the velocity at $t = 30.0$ s is 0.238 m/s. Assume all values are known to 3 significant figures.

62. By taking the slope of the curve in **Figure 2.75**, verify that the acceleration is approximately 3.2 m/s^2 at $t = 10 \text{ s}$.

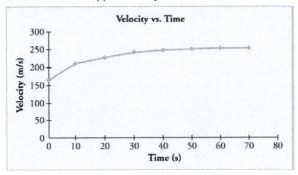

Figure 2.75

63. Construct the displacement graph for the subway shuttle train as shown in **Figure 2.30**(a). Your graph should show the position of the train, in kilometers, from t = 0 to 20 s. You will need to use the information on acceleration and velocity given in the examples for this figure.

64. (a) Take the slope of the curve in **Figure 2.76** to find the jogger's velocity at $t = 2.5 \text{ s}$. (b) Repeat at 7.5 s. These values must be consistent with the graph in **Figure 2.77**.

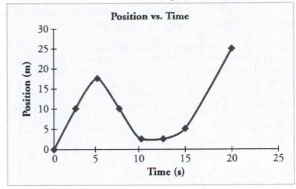

Figure 2.76

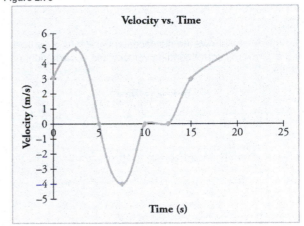

Figure 2.77

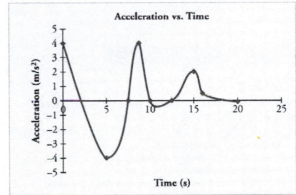

Figure 2.78

65. A graph of $v(t)$ is shown for a world-class track sprinter in a 100-m race. (See **Figure 2.79**). (a) What is his average velocity for the first 4 s? (b) What is his instantaneous velocity at $t = 5$ s ? (c) What is his average acceleration between 0 and 4 s? (d) What is his time for the race?

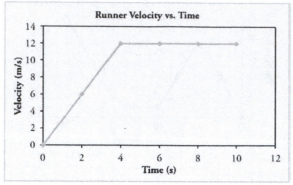

Figure 2.79

66. Figure 2.80 shows the displacement graph for a particle for 5 s. Draw the corresponding velocity and acceleration graphs.

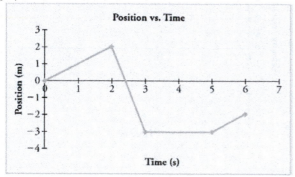

Figure 2.80

Test Prep for AP® Courses

2.1 Displacement

1. Which of the following statements comparing position, distance, and displacement is correct?
  a. An object may record a distance of zero while recording a non-zero displacement.
  b. An object may record a non-zero distance while recording a displacement of zero.
  c. An object may record a non-zero distance while maintaining a position of zero.
  d. An object may record a non-zero displacement while maintaining a position of zero.

2.2 Vectors, Scalars, and Coordinate Systems

2. A student is trying to determine the acceleration of a feather as she drops it to the ground. If the student is looking to achieve a positive velocity and positive acceleration, what is the *most sensible* way to set up her coordinate system?

  a. Her hand should be a coordinate of zero and the upward direction should be considered positive.
  b. Her hand should be a coordinate of zero and the downward direction should be considered positive.
  c. The floor should be a coordinate of zero and the upward direction should be considered positive.
  d. The floor should be a coordinate of zero and the downward direction should be considered positive.

2.3 Time, Velocity, and Speed

3. A group of students has two carts, A and B, with wheels that turn with negligible friction. The two carts travel along a straight horizontal track and eventually collide. Before the collision, cart A travels to the right and cart B is initially at rest. After the collision, the carts stick together.

a. Describe an experimental procedure to determine the velocities of the carts before and after the collision, including all the additional equipment you would need. You may include a labeled diagram of your setup to help in your description. Indicate what measurements you would take and how you would take them. Include enough detail so that another student could carry out your procedure.

b. There will be sources of error in the measurements taken in the experiment both before and after the collision. Which velocity will be more greatly affected by this error: the velocity prior to the collision or the velocity after the collision? Or will both sets of data be affected equally? Justify your answer.

2.4 Acceleration

4.

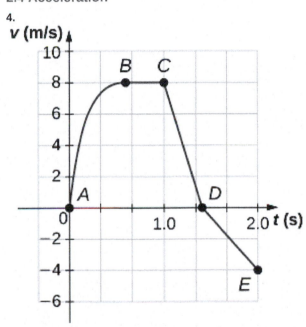

Figure 2.81 Graph showing Velocity vs. Time of a cart. A cart is constrained to move along a straight line. A varying net force along the direction of motion is exerted on the cart. The cart's velocity v as a function of time t is shown in the graph. The five labeled points divide the graph into four sections.

Which of the following correctly ranks the magnitude of the average acceleration of the cart during the four sections of the graph?

a. $a_{CD} > a_{AB} > a_{BC} > a_{DE}$
b. $a_{BC} > a_{AB} > a_{CD} > a_{DE}$
c. $a_{AB} > a_{BC} > a_{DE} > a_{CD}$
d. $a_{CD} > a_{AB} > a_{DE} > a_{BC}$

5. Push a book across a table and observe it slow to a stop.

Draw graphs showing the book's position vs. time and velocity vs. time if the direction of its motion is considered positive.

Draw graphs showing the book's position vs. time and velocity vs. time if the direction of its motion is considered negative.

2.5 Motion Equations for Constant Acceleration in One Dimension

6. A group of students is attempting to determine the average acceleration of a marble released from the top of a long ramp. Below is a set of data representing the marble's position with

respect to time.

Position (cm)	Time (s)
0.0	0.0
0.3	0.5
1.25	1.0
2.8	1.5
5.0	2.0
7.75	2.5
11.3	3.0

Use the data table above to construct a graph determining the acceleration of the marble. Select a set of data points from the table and plot those points on the graph. Fill in the blank column in the table for any quantities you graph other than the given data. Label the axes and indicate the scale for each. Draw a best-fit line or curve through your data points.

Using the best-fit line, determine the value of the marble's acceleration.

2.7 Falling Objects

7. Observing a spacecraft land on a distant asteroid, scientists notice that the craft is falling at a rate of 5 m/s. When it is 100 m closer to the surface of the asteroid, the craft reports a velocity of 8 m/s. According to their data, what is the approximate gravitational acceleration on this asteroid?

a. 0 m/s^2
b. 0.03 m/s^2
c. 0.20 m/s^2
d. 0.65 m/s^2
e. 33 m/s^2

3 TWO-DIMENSIONAL KINEMATICS

Figure 3.1 Everyday motion that we experience is, thankfully, rarely as tortuous as a rollercoaster ride like this—the Dragon Khan in Spain's Universal Port Aventura Amusement Park. However, most motion is in curved, rather than straight-line, paths. Motion along a curved path is two- or three-dimensional motion, and can be described in a similar fashion to one-dimensional motion. (credit: Boris23/Wikimedia Commons)

Chapter Outline

3.1. Kinematics in Two Dimensions: An Introduction

3.2. Vector Addition and Subtraction: Graphical Methods

3.3. Vector Addition and Subtraction: Analytical Methods

3.4. Projectile Motion

3.5. Addition of Velocities

Connection for AP® Courses

Most instances of motion in everyday life involve changes in displacement and velocity that occur in more than one direction. For example, when you take a long road trip, you drive on different roads in different directions for different amounts of time at different speeds. How can these motions all be combined to determine information about the trip such as the total displacement and average velocity? If you kick a ball from ground level at some angle above the horizontal, how can you describe its motion? To what maximum height does the object rise above the ground? How long is the object in the air? How much horizontal distance is covered before the ball lands? To answer questions such as these, we need to describe motion in two dimensions.

Examining two-dimensional motion requires an understanding of both the scalar and the vector quantities associated with the motion. You will learn how to combine vectors to incorporate both the magnitude and direction of vectors into your analysis. You will learn strategies for simplifying the calculations involved by choosing the appropriate reference frame and by treating each dimension of the motion separately as a one-dimensional problem, but you will also see that the motion itself occurs in the same way regardless of your chosen reference frame (Essential Knowledge 3.A.1).

This chapter lays a necessary foundation for examining interactions of objects described by forces (Big Idea 3). Changes in direction result from acceleration, which necessitates force on an object. In this chapter, you will concentrate on describing motion that involves changes in direction. In later chapters, you will apply this understanding as you learn about how forces cause these motions (Enduring Understanding 3.A). The concepts in this chapter support:

Big Idea 3 The interactions of an object with other objects can be described by forces.

Enduring Understanding 3.A All forces share certain common characteristics when considered by observers in inertial reference frames.

Essential Knowledge 3.A.1 An observer in a particular reference frame can describe the motion of an object using such quantities as position, displacement, distance, velocity, speed, and acceleration.

3.1 Kinematics in Two Dimensions: An Introduction

Learning Objectives

By the end of this section, you will be able to:

*   Observe that motion in two dimensions consists of horizontal and vertical components.
*   Understand the independence of horizontal and vertical vectors in two-dimensional motion.

The information presented in this section supports the following AP® learning objectives and science practices:

*   **3.A.1.1** The student is able to express the motion of an object using narrative, mathematical, and graphical representations. **(S.P. 1.5, 2.1, 2.2)**
*   **3.A.1.2** The student is able to design an experimental investigation of the motion of an object. **(S.P. 4.2)**
*   **3.A.1.3** The student is able to analyze experimental data describing the motion of an object and is able to express the results of the analysis using narrative, mathematical, and graphical representations. **(S.P. 5.1)**

Figure 3.2 Walkers and drivers in a city like New York are rarely able to travel in straight lines to reach their destinations. Instead, they must follow roads and sidewalks, making two-dimensional, zigzagged paths. (credit: Margaret W. Carruthers)

Two-Dimensional Motion: Walking in a City

Suppose you want to walk from one point to another in a city with uniform square blocks, as pictured in **Figure 3.3**.

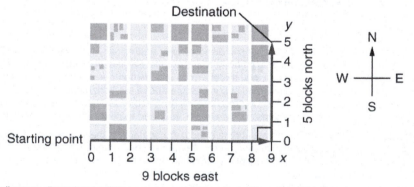

Figure 3.3 A pedestrian walks a two-dimensional path between two points in a city. In this scene, all blocks are square and are the same size.

The straight-line path that a helicopter might fly is blocked to you as a pedestrian, and so you are forced to take a two-dimensional path, such as the one shown. You walk 14 blocks in all, 9 east followed by 5 north. What is the straight-line distance?

An old adage states that the shortest distance between two points is a straight line. The two legs of the trip and the straight-line path form a right triangle, and so the Pythagorean theorem, $a^2 + b^2 = c^2$, can be used to find the straight-line distance.

$$c = \sqrt{a^2 + b^2}$$

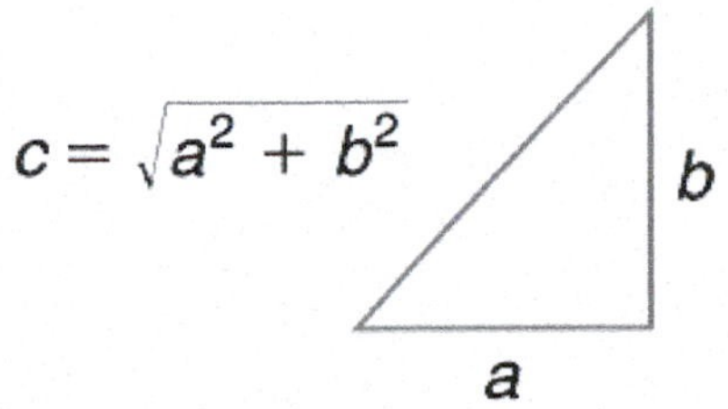

Figure 3.4 The Pythagorean theorem relates the length of the legs of a right triangle, labeled a and b , with the hypotenuse, labeled c . The relationship is given by: $a^2 + b^2 = c^2$. This can be rewritten, solving for c : $c = \sqrt{a^2 + b^2}$.

The hypotenuse of the triangle is the straight-line path, and so in this case its length in units of city blocks is $\sqrt{(9 \text{ blocks})^2 + (5 \text{ blocks})^2} = 10.3 \text{ blocks}$, considerably shorter than the 14 blocks you walked. (Note that we are using three significant figures in the answer. Although it appears that "9" and "5" have only one significant digit, they are discrete numbers. In this case "9 blocks" is the same as "9.0 or 9.00 blocks." We have decided to use three significant figures in the answer in order to show the result more precisely.)

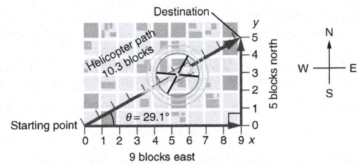

Figure 3.5 The straight-line path followed by a helicopter between the two points is shorter than the 14 blocks walked by the pedestrian. All blocks are square and the same size.

The fact that the straight-line distance (10.3 blocks) in **Figure 3.5** is less than the total distance walked (14 blocks) is one example of a general characteristic of vectors. (Recall that **vectors** are quantities that have both magnitude and direction.)

As for one-dimensional kinematics, we use arrows to represent vectors. The length of the arrow is proportional to the vector's magnitude. The arrow's length is indicated by hash marks in **Figure 3.3** and **Figure 3.5**. The arrow points in the same direction as the vector. For two-dimensional motion, the path of an object can be represented with three vectors: one vector shows the straight-line path between the initial and final points of the motion, one vector shows the horizontal component of the motion, and one vector shows the vertical component of the motion. The horizontal and vertical components of the motion add together to give the straight-line path. For example, observe the three vectors in **Figure 3.5**. The first represents a 9-block displacement east. The second represents a 5-block displacement north. These vectors are added to give the third vector, with a 10.3-block total displacement. The third vector is the straight-line path between the two points. Note that in this example, the vectors that we are adding are perpendicular to each other and thus form a right triangle. This means that we can use the Pythagorean theorem to calculate the magnitude of the total displacement. (Note that we cannot use the Pythagorean theorem to add vectors that are not perpendicular. We will develop techniques for adding vectors having any direction, not just those perpendicular to one another, in **Vector Addition and Subtraction: Graphical Methods** and **Vector Addition and Subtraction: Analytical Methods**.)

The Independence of Perpendicular Motions

The person taking the path shown in **Figure 3.5** walks east and then north (two perpendicular directions). How far he or she walks east is only affected by his or her motion eastward. Similarly, how far he or she walks north is only affected by his or her motion northward.

Independence of Motion

The horizontal and vertical components of two-dimensional motion are independent of each other. Any motion in the horizontal direction does not affect motion in the vertical direction, and vice versa.

This is true in a simple scenario like that of walking in one direction first, followed by another. It is also true of more complicated motion involving movement in two directions at once. For example, let's compare the motions of two baseballs. One baseball is dropped from rest. At the same instant, another is thrown horizontally from the same height and follows a curved path. A stroboscope has captured the positions of the balls at fixed time intervals as they fall.

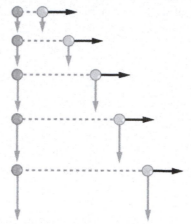

Figure 3.6 This shows the motions of two identical balls—one falls from rest, the other has an initial horizontal velocity. Each subsequent position is an equal time interval. Arrows represent horizontal and vertical velocities at each position. The ball on the right has an initial horizontal velocity, while the ball on the left has no horizontal velocity. Despite the difference in horizontal velocities, the vertical velocities and positions are identical for both balls. This shows that the vertical and horizontal motions are independent.

Applying the Science Practices: Independence of Horizontal and Vertical Motion or Maximum Height and Flight Time

Choose one of the following experiments to design:

Design an experiment to confirm what is shown in **Figure 3.6**, that the vertical motion of the two balls is independent of the horizontal motion. As you think about your experiment, consider the following questions:

- How will you measure the horizontal and vertical positions of each ball over time? What equipment will this require?
- How will you measure the time interval between each of your position measurements? What equipment will this require?
- If you were to create separate graphs of the horizontal velocity for each ball versus time, what do you predict it would look like? Explain.
- If you were to compare graphs of the vertical velocity for each ball versus time, what do you predict it would look like? Explain.
- If there is a significant amount of air resistance, how will that affect each of your graphs?

Design a two-dimensional ballistic motion experiment that demonstrates the relationship between the maximum height reached by an object and the object's time of flight. As you think about your experiment, consider the following questions:

- How will you measure the maximum height reached by your object?
- How can you take advantage of the symmetry of an object in ballistic motion launched from ground level, reaching maximum height, and returning to ground level?
- Will it make a difference if your object has no horizontal component to its velocity? Explain.
- Will you need to measure the time at multiple different positions? Why or why not?
- Predict what a graph of travel time versus maximum height will look like. Will it be linear? Parabolic? Horizontal? Explain the shape of your predicted graph qualitatively or quantitatively.
- If there is a significant amount of air resistance, how will that affect your measurements and your results?

It is remarkable that for each flash of the strobe, the vertical positions of the two balls are the same. This similarity implies that the vertical motion is independent of whether or not the ball is moving horizontally. (Assuming no air resistance, the vertical motion of a falling object is influenced by gravity only, and not by any horizontal forces.) Careful examination of the ball thrown horizontally shows that it travels the same horizontal distance between flashes. This is due to the fact that there are no additional forces on the ball in the horizontal direction after it is thrown. This result means that the horizontal velocity is constant, and affected neither by vertical motion nor by gravity (which is vertical). Note that this case is true only for ideal conditions. In the real world, air resistance will affect the speed of the balls in both directions.

The two-dimensional curved path of the horizontally thrown ball is composed of two independent one-dimensional motions (horizontal and vertical). The key to analyzing such motion, called *projectile motion*, is to *resolve* (break) it into motions along perpendicular directions. Resolving two-dimensional motion into perpendicular components is possible because the components are independent. We shall see how to resolve vectors in **Vector Addition and Subtraction: Graphical Methods** and **Vector Addition and Subtraction: Analytical Methods**. We will find such techniques to be useful in many areas of physics.

PhET Explorations: Ladybug Motion 2D

Learn about position, velocity and acceleration vectors. Move the ladybug by setting the position, velocity or acceleration, and see how the vectors change. Choose linear, circular or elliptical motion, and record and playback the motion to analyze the behavior.

3.2 Vector Addition and Subtraction: Graphical Methods

Learning Objectives

By the end of this section, you will be able to:

- Understand the rules of vector addition, subtraction, and multiplication.
- Apply graphical methods of vector addition and subtraction to determine the displacement of moving objects.

The information presented in this section supports the following AP® learning objectives and science practices:

- **3.A.1.1** The student is able to express the motion of an object using narrative, mathematical, and graphical representations. **(S.P. 1.5, 2.1, 2.2)**
- **3.A.1.3** The student is able to analyze experimental data describing the motion of an object and is able to express the results of the analysis using narrative, mathematical, and graphical representations. **(S.P. 5.1)**

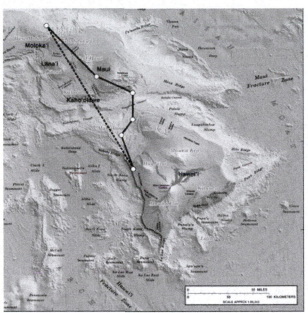

Figure 3.8 Displacement can be determined graphically using a scale map, such as this one of the Hawaiian Islands. A journey from Hawai'i to Moloka'i has a number of legs, or journey segments. These segments can be added graphically with a ruler to determine the total two-dimensional displacement of the journey. (credit: US Geological Survey)

Vectors in Two Dimensions

A **vector** is a quantity that has magnitude and direction. Displacement, velocity, acceleration, and force, for example, are all vectors. In one-dimensional, or straight-line, motion, the direction of a vector can be given simply by a plus or minus sign. In two dimensions (2-d), however, we specify the direction of a vector relative to some reference frame (i.e., coordinate system), using an arrow having length proportional to the vector's magnitude and pointing in the direction of the vector.

Figure 3.9 shows such a *graphical representation of a vector*, using as an example the total displacement for the person walking in a city considered in **Kinematics in Two Dimensions: An Introduction**. We shall use the notation that a boldface symbol, such as $\mathbf{D}$, stands for a vector. Its magnitude is represented by the symbol in italics, D, and its direction by θ.

Vectors in this Text

In this text, we will represent a vector with a boldface variable. For example, we will represent the quantity force with the vector $\mathbf{F}$, which has both magnitude and direction. The magnitude of the vector will be represented by a variable in italics, such as F, and the direction of the variable will be given by an angle θ.

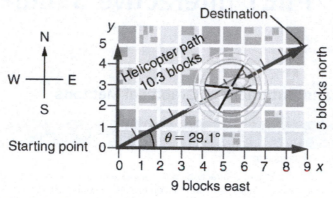

Figure 3.9 A person walks 9 blocks east and 5 blocks north. The displacement is 10.3 blocks at an angle $29.1°$ north of east.

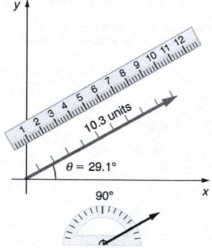

Figure 3.10 To describe the resultant vector for the person walking in a city considered in Figure 3.9 graphically, draw an arrow to represent the total displacement vector D. Using a protractor, draw a line at an angle θ relative to the east-west axis. The length D of the arrow is proportional to the vector's magnitude and is measured along the line with a ruler. In this example, the magnitude D of the vector is 10.3 units, and the direction θ is $29.1°$ north of east.

Vector Addition: Head-to-Tail Method

The **head-to-tail method** is a graphical way to add vectors, described in **Figure 3.11** below and in the steps following. The **tail** of the vector is the starting point of the vector, and the **head** (or tip) of a vector is the final, pointed end of the arrow.

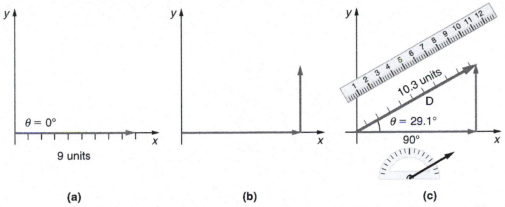

Figure 3.11 Head-to-Tail Method: The head-to-tail method of graphically adding vectors is illustrated for the two displacements of the person walking in a city considered in **Figure 3.9**. (a) Draw a vector representing the displacement to the east. (b) Draw a vector representing the displacement to the north. The tail of this vector should originate from the head of the first, east-pointing vector. (c) Draw a line from the tail of the east-pointing vector to the head of the north-pointing vector to form the sum or **resultant vector D** . The length of the arrow **D** is proportional to the vector's magnitude and is measured to be 10.3 units . Its direction, described as the angle with respect to the east (or horizontal axis) θ is measured with a protractor to be $29.1°$.

Step 1. *Draw an arrow to represent the first vector (9 blocks to the east) using a ruler and protractor.*

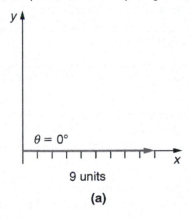

Figure 3.12

Step 2. Now draw an arrow to represent the second vector (5 blocks to the north). *Place the tail of the second vector at the head of the first vector.*

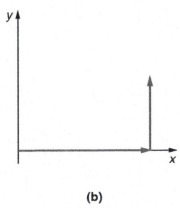

Figure 3.13

Step 3. *If there are more than two vectors, continue this process for each vector to be added. Note that in our example, we have only two vectors, so we have finished placing arrows tip to tail.*

Step 4. *Draw an arrow from the tail of the first vector to the head of the last vector.* This is the **resultant**, or the sum, of the other vectors.

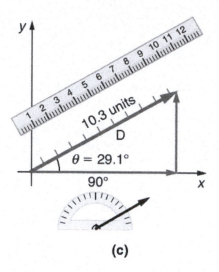

(c)

Figure 3.14

Step 5. To get the **magnitude** of the resultant, *measure its length with a ruler. (Note that in most calculations, we will use the Pythagorean theorem to determine this length.)*

Step 6. To get the **direction** of the resultant, *measure the angle it makes with the reference frame using a protractor. (Note that in most calculations, we will use trigonometric relationships to determine this angle.)*

The graphical addition of vectors is limited in accuracy only by the precision with which the drawings can be made and the precision of the measuring tools. It is valid for any number of vectors.

Example 3.1 Adding Vectors Graphically Using the Head-to-Tail Method: A Woman Takes a Walk

Use the graphical technique for adding vectors to find the total displacement of a person who walks the following three paths (displacements) on a flat field. First, she walks 25.0 m in a direction $49.0°$ north of east. Then, she walks 23.0 m heading $15.0°$ north of east. Finally, she turns and walks 32.0 m in a direction 68.0° south of east.

Strategy

Represent each displacement vector graphically with an arrow, labeling the first $\mathbf{A}$, the second $\mathbf{B}$, and the third $\mathbf{C}$, making the lengths proportional to the distance and the directions as specified relative to an east-west line. The head-to-tail method outlined above will give a way to determine the magnitude and direction of the resultant displacement, denoted $\mathbf{R}$.

Solution

(1) Draw the three displacement vectors.

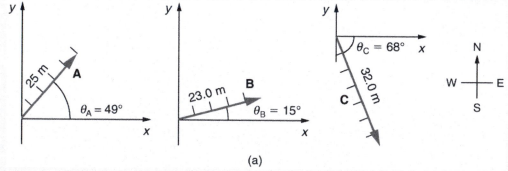

(a)

Figure 3.15

(2) Place the vectors head to tail retaining both their initial magnitude and direction.

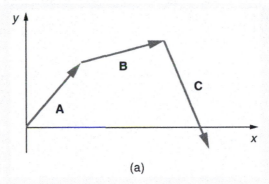

Figure 3.16

(3) Draw the resultant vector, **R** .

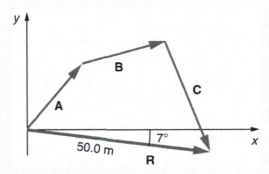

Figure 3.17

(4) Use a ruler to measure the magnitude of **R** , and a protractor to measure the direction of **R** . While the direction of the vector can be specified in many ways, the easiest way is to measure the angle between the vector and the nearest horizontal or vertical axis. Since the resultant vector is south of the eastward pointing axis, we flip the protractor upside down and measure the angle between the eastward axis and the vector.

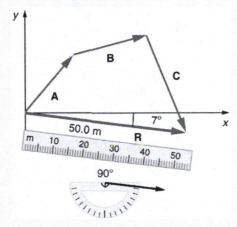

Figure 3.18

In this case, the total displacement **R** is seen to have a magnitude of 50.0 m and to lie in a direction $7.0°$ south of east. By using its magnitude and direction, this vector can be expressed as $R = 50.0$ m and $\theta = 7.0°$ south of east.

Discussion

The head-to-tail graphical method of vector addition works for any number of vectors. It is also important to note that the resultant is independent of the order in which the vectors are added. Therefore, we could add the vectors in any order as illustrated in **Figure 3.19** and we will still get the same solution.

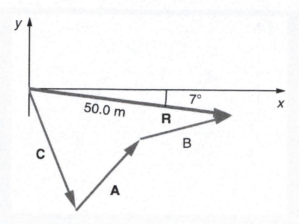

Figure 3.19

Here, we see that when the same vectors are added in a different order, the result is the same. This characteristic is true in every case and is an important characteristic of vectors. Vector addition is **commutative**. Vectors can be added in any order.

$$\mathbf{A} + \mathbf{B} = \mathbf{B} + \mathbf{A}.$$ (3.1)

(This is true for the addition of ordinary numbers as well—you get the same result whether you add $2 + 3$ or $3 + 2$, for example).

Vector Subtraction

Vector subtraction is a straightforward extension of vector addition. To define subtraction (say we want to subtract $\mathbf{B}$ from $\mathbf{A}$, written $\mathbf{A} - \mathbf{B}$, we must first define what we mean by subtraction. The *negative* of a vector $\mathbf{B}$ is defined to be $-\mathbf{B}$; that is, graphically *the negative of any vector has the same magnitude but the opposite direction*, as shown in **Figure 3.20**. In other words, $\mathbf{B}$ has the same length as $-\mathbf{B}$, but points in the opposite direction. Essentially, we just flip the vector so it points in the opposite direction.

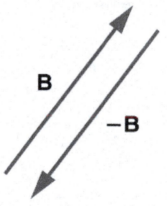

Figure 3.20 The negative of a vector is just another vector of the same magnitude but pointing in the opposite direction. So $\mathbf{B}$ is the negative of $-\mathbf{B}$; it has the same length but opposite direction.

The *subtraction* of vector $\mathbf{B}$ from vector $\mathbf{A}$ is then simply defined to be the addition of $-\mathbf{B}$ to $\mathbf{A}$. Note that vector subtraction is the addition of a negative vector. The order of subtraction does not affect the results.

$$\mathbf{A} - \mathbf{B} = \mathbf{A} + (-\mathbf{B}).$$ (3.2)

This is analogous to the subtraction of scalars (where, for example, $5 - 2 = 5 + (-2)$). Again, the result is independent of the order in which the subtraction is made. When vectors are subtracted graphically, the techniques outlined above are used, as the following example illustrates.

Example 3.2 Subtracting Vectors Graphically: A Woman Sailing a Boat

A woman sailing a boat at night is following directions to a dock. The instructions read to first sail 27.5 m in a direction $66.0°$ north of east from her current location, and then travel 30.0 m in a direction $112°$ north of east (or $22.0°$ west of north). If the woman makes a mistake and travels in the *opposite* direction for the second leg of the trip, where will she end up? Compare this location with the location of the dock.

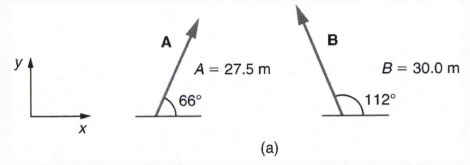

(a)

Figure 3.21

Strategy

We can represent the first leg of the trip with a vector $\mathbf{A}$, and the second leg of the trip with a vector $\mathbf{B}$. The dock is located at a location $\mathbf{A} + \mathbf{B}$. If the woman mistakenly travels in the *opposite* direction for the second leg of the journey, she will travel a distance B (30.0 m) in the direction $180° - 112° = 68°$ south of east. We represent this as $-\mathbf{B}$, as shown below. The vector $-\mathbf{B}$ has the same magnitude as $\mathbf{B}$ but is in the opposite direction. Thus, she will end up at a location $\mathbf{A} + (-\mathbf{B})$, or $\mathbf{A} - \mathbf{B}$.

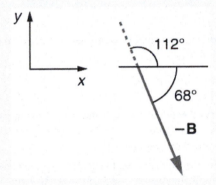

Figure 3.22

We will perform vector addition to compare the location of the dock, $\mathbf{A} + \mathbf{B}$, with the location at which the woman mistakenly arrives, $\mathbf{A} + (-\mathbf{B})$.

Solution

(1) To determine the location at which the woman arrives by accident, draw vectors $\mathbf{A}$ and $-\mathbf{B}$.

(2) Place the vectors head to tail.

(3) Draw the resultant vector $\mathbf{R}$.

(4) Use a ruler and protractor to measure the magnitude and direction of $\mathbf{R}$.

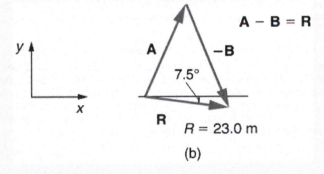

(b)

Figure 3.23

In this case, $R = 23.0$ m and $\theta = 7.5°$ south of east.

(5) To determine the location of the dock, we repeat this method to add vectors $\mathbf{A}$ and $\mathbf{B}$. We obtain the resultant vector $\mathbf{R'}$:

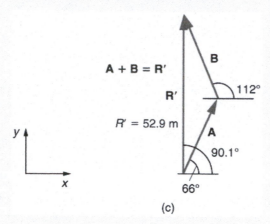

(c)

Figure 3.24

In this case $R = 52.9$ m and $\theta = 90.1°$ north of east.

We can see that the woman will end up a significant distance from the dock if she travels in the opposite direction for the second leg of the trip.

Discussion

Because subtraction of a vector is the same as addition of a vector with the opposite direction, the graphical method of subtracting vectors works the same as for addition.

Multiplication of Vectors and Scalars

If we decided to walk three times as far on the first leg of the trip considered in the preceding example, then we would walk 3×27.5 m, or 82.5 m, in a direction $66.0°$ north of east. This is an example of multiplying a vector by a positive **scalar**. Notice that the magnitude changes, but the direction stays the same.

If the scalar is negative, then multiplying a vector by it changes the vector's magnitude and gives the new vector the *opposite* direction. For example, if you multiply by –2, the magnitude doubles but the direction changes. We can summarize these rules in the following way: When vector $\mathbf{A}$ is multiplied by a scalar c,

- the magnitude of the vector becomes the absolute value of c A,
- if c is positive, the direction of the vector does not change,
- if c is negative, the direction is reversed.

In our case, $c = 3$ and $A = 27.5$ m. Vectors are multiplied by scalars in many situations. Note that division is the inverse of multiplication. For example, dividing by 2 is the same as multiplying by the value (1/2). The rules for multiplication of vectors by scalars are the same for division; simply treat the divisor as a scalar between 0 and 1.

Resolving a Vector into Components

In the examples above, we have been adding vectors to determine the resultant vector. In many cases, however, we will need to do the opposite. We will need to take a single vector and find what other vectors added together produce it. In most cases, this involves determining the perpendicular **components** of a single vector, for example the *x*- *and y*-components, or the north-south and east-west components.

For example, we may know that the total displacement of a person walking in a city is 10.3 blocks in a direction $29.0°$ north of east and want to find out how many blocks east and north had to be walked. This method is called *finding the components (or parts)* of the displacement in the east and north directions, and it is the inverse of the process followed to find the total displacement. It is one example of finding the components of a vector. There are many applications in physics where this is a useful thing to do. We will see this soon in **Projectile Motion**, and much more when we cover **forces** in **Dynamics: Newton's Laws of Motion**. Most of these involve finding components along perpendicular axes (such as north and east), so that right triangles are involved. The analytical techniques presented in **Vector Addition and Subtraction: Analytical Methods** are ideal for finding vector components.

PhET Explorations: Maze Game

Learn about position, velocity, and acceleration in the "Arena of Pain". Use the green arrow to move the ball. Add more walls to the arena to make the game more difficult. Try to make a goal as fast as you can.

3.3 Vector Addition and Subtraction: Analytical Methods

Learning Objectives

By the end of this section, you will be able to:

- Understand the rules of vector addition and subtraction using analytical methods.
- Apply analytical methods to determine vertical and horizontal component vectors.
- Apply analytical methods to determine the magnitude and direction of a resultant vector.

The information presented in this section supports the following AP® learning objectives and science practices:

- **3.A.1.1** The student is able to express the motion of an object using narrative, mathematical, and graphical representations. **(S.P. 1.5, 2.1, 2.2)**

Analytical methods of vector addition and subtraction employ geometry and simple trigonometry rather than the ruler and protractor of graphical methods. Part of the graphical technique is retained, because vectors are still represented by arrows for easy visualization. However, analytical methods are more concise, accurate, and precise than graphical methods, which are limited by the accuracy with which a drawing can be made. Analytical methods are limited only by the accuracy and precision with which physical quantities are known.

Resolving a Vector into Perpendicular Components

Analytical techniques and right triangles go hand-in-hand in physics because (among other things) motions along perpendicular directions are independent. We very often need to separate a vector into perpendicular components. For example, given a vector like $\mathbf{A}$ in **Figure 3.26**, we may wish to find which two perpendicular vectors, $\mathbf{A}_x$ and $\mathbf{A}_y$, add to produce it.

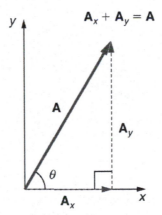

Figure 3.26 The vector $\mathbf{A}$, with its tail at the origin of an x, y-coordinate system, is shown together with its x- and y-components, $\mathbf{A}_x$ and $\mathbf{A}_y$. These vectors form a right triangle. The analytical relationships among these vectors are summarized below.

$\mathbf{A}_x$ and $\mathbf{A}_y$ are defined to be the components of $\mathbf{A}$ along the x- and y-axes. The three vectors $\mathbf{A}$, $\mathbf{A}_x$, and $\mathbf{A}_y$ form a right triangle:

$$\mathbf{A}_x + \mathbf{A_y} = \mathbf{A}. \tag{3.3}$$

Note that this relationship between vector components and the resultant vector holds only for vector quantities (which include both magnitude and direction). The relationship does not apply for the magnitudes alone. For example, if $\mathbf{A}_x = 3$ m east, $\mathbf{A}_y = 4$ m north, and $\mathbf{A} = 5$ m north-east, then it is true that the vectors $\mathbf{A}_x + \mathbf{A_y} = \mathbf{A}$. However, it is *not* true that the sum of the magnitudes of the vectors is also equal. That is,

$$3\,\text{m} + 4\,\text{m} \neq 5\,\text{m} \tag{3.4}$$

Thus,

$$A_x + A_y \neq A \qquad (3.5)$$

If the vector $\mathbf{A}$ is known, then its magnitude A (its length) and its angle θ (its direction) are known. To find A_x and A_y, its x- and y-components, we use the following relationships for a right triangle.

$$A_x = A \cos \theta \qquad (3.6)$$

and

$$A_y = A \sin \theta. \qquad (3.7)$$

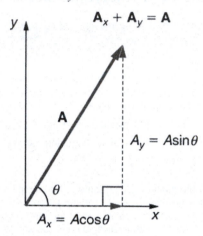

Figure 3.27 The magnitudes of the vector components $\mathbf{A}_x$ and $\mathbf{A}_y$ can be related to the resultant vector $\mathbf{A}$ and the angle θ with trigonometric identities. Here we see that $A_x = A \cos \theta$ and $A_y = A \sin \theta$.

Suppose, for example, that $\mathbf{A}$ is the vector representing the total displacement of the person walking in a city considered in **Kinematics in Two Dimensions: An Introduction** and **Vector Addition and Subtraction: Graphical Methods**.

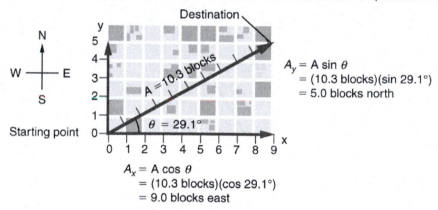

Figure 3.28 We can use the relationships $A_x = A \cos \theta$ and $A_y = A \sin \theta$ to determine the magnitude of the horizontal and vertical component vectors in this example.

Then $A = 10.3$ blocks and $\theta = 29.1°$, so that

$$A_x = A \cos \theta = (10.3 \text{ blocks})(\cos 29.1°) = 9.0 \text{ blocks} \qquad (3.8)$$

$$A_y = A \sin \theta = (10.3 \text{ blocks})(\sin 29.1°) = 5.0 \text{ blocks}. \qquad (3.9)$$

Calculating a Resultant Vector

If the perpendicular components $\mathbf{A}_x$ and $\mathbf{A}_y$ of a vector $\mathbf{A}$ are known, then $\mathbf{A}$ can also be found analytically. To find the magnitude A and direction θ of a vector from its perpendicular components $\mathbf{A}_x$ and $\mathbf{A}_y$, we use the following relationships:

$$A = \sqrt{A_x{}^2 + A_y{}^2} \qquad (3.10)$$

$$\theta = \tan^{-1}(A_y / A_x). \qquad (3.11)$$

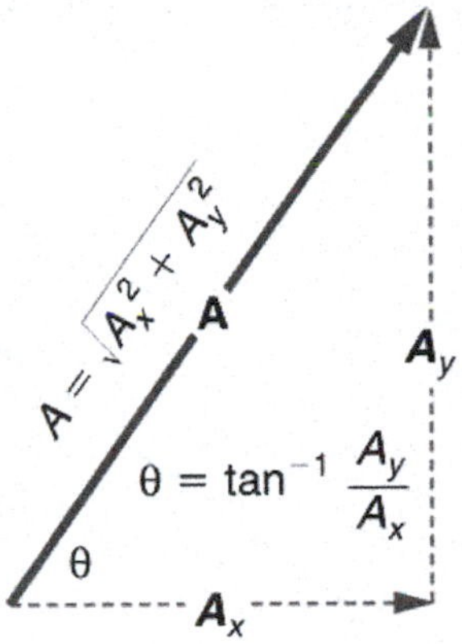

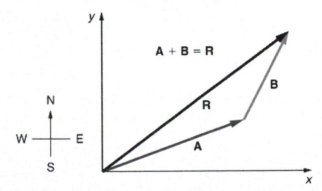

Figure 3.29 The magnitude and direction of the resultant vector can be determined once the horizontal and vertical components A_x and A_y have been determined.

Note that the equation $A = \sqrt{A_x^2 + A_y^2}$ is just the Pythagorean theorem relating the legs of a right triangle to the length of the hypotenuse. For example, if A_x and A_y are 9 and 5 blocks, respectively, then $A = \sqrt{9^2 + 5^2} = 10.3$ blocks, again consistent with the example of the person walking in a city. Finally, the direction is $\theta = \tan^{-1}(5/9) = 29.1°$, as before.

Determining Vectors and Vector Components with Analytical Methods

Equations $A_x = A \cos \theta$ and $A_y = A \sin \theta$ are used to find the perpendicular components of a vector—that is, to go from A and θ to A_x and A_y. Equations $A = \sqrt{A_x^2 + A_y^2}$ and $\theta = \tan^{-1}(A_y / A_x)$ are used to find a vector from its perpendicular components—that is, to go from A_x and A_y to A and θ. Both processes are crucial to analytical methods of vector addition and subtraction.

Adding Vectors Using Analytical Methods

To see how to add vectors using perpendicular components, consider **Figure 3.30**, in which the vectors **A** and **B** are added to produce the resultant **R** .

Figure 3.30 Vectors **A** and **B** are two legs of a walk, and **R** is the resultant or total displacement. You can use analytical methods to determine the magnitude and direction of **R** .

If **A** and **B** represent two legs of a walk (two displacements), then **R** is the total displacement. The person taking the walk ends up at the tip of **R**. There are many ways to arrive at the same point. In particular, the person could have walked first in the x-direction and then in the y-direction. Those paths are the x- and y-components of the resultant, $\mathbf{R}_x$ and $\mathbf{R}_y$. If we know $\mathbf{R}_x$ and $\mathbf{R}_y$, we can find R and θ using the equations $A = \sqrt{A_x^2 + A_y^2}$ and $\theta = \tan^{-1}(A_y / A_x)$. When you use the analytical method of vector addition, you can determine the components or the magnitude and direction of a vector.

Step 1. *Identify the x- and y-axes that will be used in the problem. Then, find the components of each vector to be added along* the chosen perpendicular axes. Use the equations $A_x = A \cos \theta$ and $A_y = A \sin \theta$ to find the components. In **Figure 3.31**, these components are A_x, A_y, B_x, and B_y. The angles that vectors **A** and **B** make with the x-axis are θ_A and θ_B, respectively.

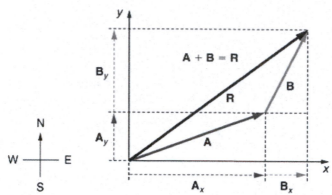

Figure 3.31 To add vectors **A** and **B**, first determine the horizontal and vertical components of each vector. These are the dotted vectors A_x, A_y, B_x and B_y shown in the image.

Step 2. Find the components of the resultant along each axis by adding the components of the individual vectors along that axis. That is, as shown in **Figure 3.32**,

$$R_x = A_x + B_x \tag{3.12}$$

and

$$R_y = A_y + B_y. \tag{3.13}$$

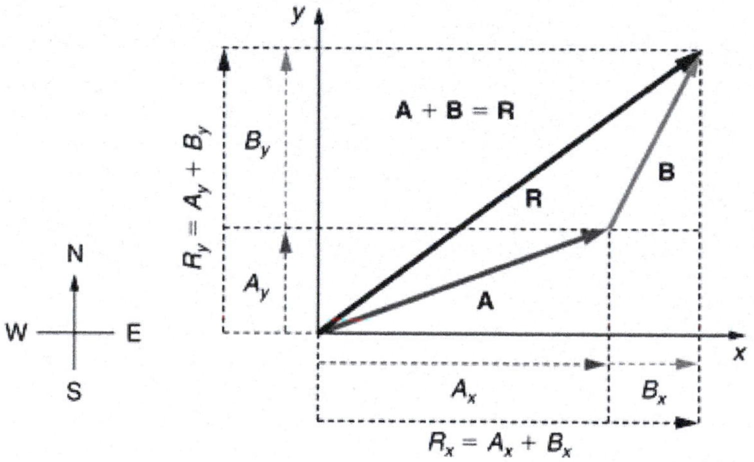

Figure 3.32 The magnitude of the vectors A_x and B_x add to give the magnitude R_x of the resultant vector in the horizontal direction. Similarly, the magnitudes of the vectors A_y and B_y add to give the magnitude R_y of the resultant vector in the vertical direction.

Components along the same axis, say the x-axis, are vectors along the same line and, thus, can be added to one another like ordinary numbers. The same is true for components along the y-axis. (For example, a 9-block eastward walk could be taken in two legs, the first 3 blocks east and the second 6 blocks east, for a total of 9, because they are along the same direction.) So resolving vectors into components along common axes makes it easier to add them. Now that the components of **R** are known, its magnitude and direction can be found.

Step 3. To get the magnitude R of the resultant, use the Pythagorean theorem:

$$R = \sqrt{R_x^2 + R_y^2}. \tag{3.14}$$

Step 4. To get the direction of the resultant:

$$\theta = \tan^{-1}(R_y / R_x). \tag{3.15}$$

The following example illustrates this technique for adding vectors using perpendicular components.

Example 3.3 Adding Vectors Using Analytical Methods

Add the vector **A** to the vector **B** shown in Figure 3.33, using perpendicular components along the x- and y-axes. The x-and y-axes are along the east–west and north–south directions, respectively. Vector **A** represents the first leg of a walk in which a person walks 53.0 m in a direction 20.0° north of east. Vector **B** represents the second leg, a displacement of 34.0 m in a direction 63.0° north of east.

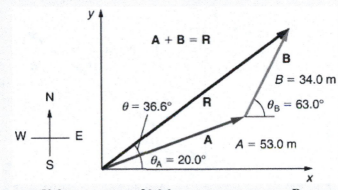

Figure 3.33 Vector **A** has magnitude 53.0 m and direction 20.0 ° north of the x-axis. Vector **B** has magnitude 34.0 m and direction 63.0° north of the x-axis. You can use analytical methods to determine the magnitude and direction of **R** .

Strategy

The components of **A** and **B** along the x- and y-axes represent walking due east and due north to get to the same ending point. Once found, they are combined to produce the resultant.

Solution

Following the method outlined above, we first find the components of **A** and **B** along the x- and y-axes. Note that $A = 53.0 \text{ m}$, $\theta_A = 20.0°$, $B = 34.0 \text{ m}$, and $\theta_B = 63.0°$. We find the x-components by using $A_x = A \cos \theta$, which gives

$$\begin{aligned} A_x &= A \cos \theta_A = (53.0 \text{ m})(\cos 20.0°) \\ &= (53.0 \text{ m})(0.940) = 49.8 \text{ m} \end{aligned} \tag{3.16}$$

and

$$\begin{aligned} B_x &= B \cos \theta_B = (34.0 \text{ m})(\cos 63.0°) \\ &= (34.0 \text{ m})(0.454) = 15.4 \text{ m}. \end{aligned} \tag{3.17}$$

Similarly, the y-components are found using $A_y = A \sin \theta_A$:

$$\begin{aligned} A_y &= A \sin \theta_A = (53.0 \text{ m})(\sin 20.0°) \\ &= (53.0 \text{ m})(0.342) = 18.1 \text{ m} \end{aligned} \tag{3.18}$$

and

$$\begin{aligned} B_y &= B \sin \theta_B = (34.0 \text{ m})(\sin 63.0 °) \\ &= (34.0 \text{ m})(0.891) = 30.3 \text{ m}. \end{aligned} \tag{3.19}$$

The x- and y-components of the resultant are thus

$$R_x = A_x + B_x = 49.8 \text{ m} + 15.4 \text{ m} = 65.2 \text{ m} \tag{3.20}$$

and

$$R_y = A_y + B_y = 18.1 \text{ m} + 30.3 \text{ m} = 48.4 \text{ m}. \tag{3.21}$$

Now we can find the magnitude of the resultant by using the Pythagorean theorem:

$$R = \sqrt{R_x^2 + R_y^2} = \sqrt{(65.2)^2 + (48.4)^2} \text{ m} \tag{3.22}$$

so that

$$R = 81.2 \text{ m}. \tag{3.23}$$

Finally, we find the direction of the resultant:

$$\theta = \tan^{-1}(R_y/R_x) = +\tan^{-1}(48.4/65.2). \tag{3.24}$$

Thus,

$$\theta = \tan^{-1}(0.742) = 36.6\,^\circ. \tag{3.25}$$

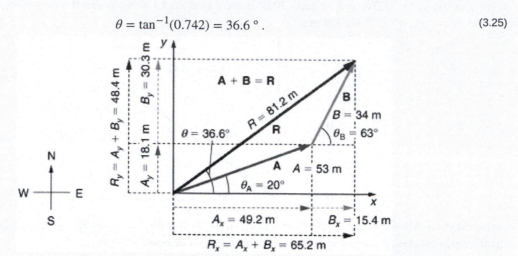

Figure 3.34 Using analytical methods, we see that the magnitude of $\mathbf{R}$ is 81.2 m and its direction is $36.6°$ north of east.

Discussion

This example illustrates the addition of vectors using perpendicular components. Vector subtraction using perpendicular components is very similar—it is just the addition of a negative vector.

Subtraction of vectors is accomplished by the addition of a negative vector. That is, $\mathbf{A} - \mathbf{B} \equiv \mathbf{A} + (-\mathbf{B})$. Thus, *the method for the subtraction of vectors using perpendicular components is identical to that for addition*. The components of $-\mathbf{B}$ are the negatives of the components of $\mathbf{B}$. The *x*- and *y*-components of the resultant $\mathbf{A} - \mathbf{B} = \mathbf{R}$ are thus

$$R_x = A_x + (-B_x) \tag{3.26}$$

and

$$R_y = A_y + (-B_y) \tag{3.27}$$

and the rest of the method outlined above is identical to that for addition. (See **Figure 3.35**.)

Analyzing vectors using perpendicular components is very useful in many areas of physics, because perpendicular quantities are often independent of one another. The next module, **Projectile Motion**, is one of many in which using perpendicular components helps make the picture clear and simplifies the physics.

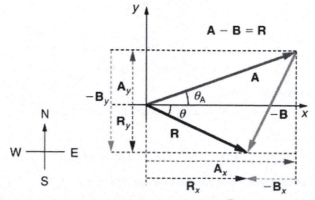

Figure 3.35 The subtraction of the two vectors shown in **Figure 3.30**. The components of $-\mathbf{B}$ are the negatives of the components of $\mathbf{B}$. The method of subtraction is the same as that for addition.

PhET Explorations: Vector Addition

Learn how to add vectors. Drag vectors onto a graph, change their length and angle, and sum them together. The magnitude, angle, and components of each vector can be displayed in several formats.

Figure 3.36 Vector Addition (http://cnx.org/content/m54783/1.2/vector-addition_en.jar)

3.4 Projectile Motion

Learning Objectives
By the end of this section, you will be able to: • Identify and explain the properties of a projectile, such as acceleration due to gravity, range, maximum height, and trajectory. • Determine the location and velocity of a projectile at different points in its trajectory. • Apply the principle of independence of motion to solve projectile motion problems. The information presented in this section supports the following AP® learning objectives: • **3.A.1.1** The student is able to express the motion of an object using narrative, mathematical, and graphical representations. **(S.P. 1.5, 2.1, 2.2)** • **3.A.1.3** The student is able to analyze experimental data describing the motion of an object and is able to express the results of the analysis using narrative, mathematical, and graphical representations. **(S.P. 5.1)**

Projectile motion is the **motion** of an object thrown or projected into the air, subject to only the acceleration of gravity. The object is called a **projectile**, and its path is called its **trajectory**. The motion of falling objects, as covered in **Problem-Solving Basics for One-Dimensional Kinematics**, is a simple one-dimensional type of projectile motion in which there is no horizontal movement. In this section, we consider two-dimensional projectile motion, such as that of a football or other object for which **air resistance** *is negligible*.

The most important fact to remember here is that *motions along perpendicular axes are independent* and thus can be analyzed separately. This fact was discussed in **Kinematics in Two Dimensions: An Introduction**, where vertical and horizontal motions were seen to be independent. The key to analyzing two-dimensional projectile motion is to break it into two motions, one along the horizontal axis and the other along the vertical. (This choice of axes is the most sensible, because acceleration due to gravity is vertical—thus, there will be no acceleration along the horizontal axis when air resistance is negligible.) As is customary, we call the horizontal axis the *x*-axis and the vertical axis the *y*-axis. **Figure 3.37** illustrates the notation for displacement, where **s** is defined to be the total displacement and **x** and **y** are its components along the horizontal and vertical axes, respectively. The magnitudes of these vectors are *s*, *x*, and *y*. (Note that in the last section we used the notation **A** to represent a vector with components $\mathbf{A}_x$ and $\mathbf{A}_y$. If we continued this format, we would call displacement **s** with components $\mathbf{s}_x$ and $\mathbf{s}_y$. However, to simplify the notation, we will simply represent the component vectors as **x** and **y** .)

Of course, to describe motion we must deal with velocity and acceleration, as well as with displacement. We must find their components along the *x*- and *y*-axes, too. We will assume all forces except gravity (such as air resistance and friction, for example) are negligible. The components of acceleration are then very simple: $a_y = -g = -9.80 \text{ m/s}^2$. (Note that this definition assumes that the upwards direction is defined as the positive direction. If you arrange the coordinate system instead such that the downwards direction is positive, then acceleration due to gravity takes a positive value.) Because gravity is vertical, $a_x = 0$. Both accelerations are constant, so the kinematic equations can be used.

Review of Kinematic Equations (constant a)

$$x = x_0 + \bar{v} t \tag{3.28}$$

$$\bar{v} = \frac{v_0 + v}{2} \tag{3.29}$$

$$v = v_0 + at \tag{3.30}$$

$$x = x_0 + v_0 t + \frac{1}{2}at^2 \tag{3.31}$$

$$v^2 = v_0^2 + 2a(x - x_0). \tag{3.32}$$

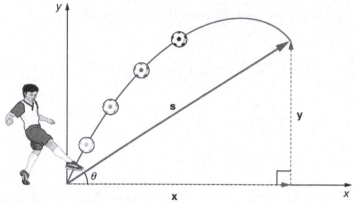

Figure 3.37 The total displacement s of a soccer ball at a point along its path. The vector s has components x and y along the horizontal and vertical axes. Its magnitude is s, and it makes an angle θ with the horizontal.

Given these assumptions, the following steps are then used to analyze projectile motion:

Step 1. *Resolve or break the motion into horizontal and vertical components along the x- and y-axes.* These axes are perpendicular, so $A_x = A \cos \theta$ and $A_y = A \sin \theta$ are used. The magnitude of the components of displacement s along these axes are x and y. The magnitudes of the components of the velocity v are $v_x = v \cos \theta$ and $v_y = v \sin \theta$, where v is the magnitude of the velocity and θ is its direction, as shown in **Figure 3.38**. Initial values are denoted with a subscript 0, as usual.

Step 2. *Treat the motion as two independent one-dimensional motions, one horizontal and the other vertical.* The kinematic equations for horizontal and vertical motion take the following forms:

$$\text{Horizontal Motion}(a_x = 0) \tag{3.33}$$

$$x = x_0 + v_x t \tag{3.34}$$

$$v_x = v_{0x} = v_x = \text{velocity is a constant.} \tag{3.35}$$

$$\text{Vertical Motion(assuming positive is up } a_y = -g = -9.80\text{m/s}^2) \tag{3.36}$$

$$y = y_0 + \frac{1}{2}(v_{0y} + v_y)t \tag{3.37}$$

$$v_y = v_{0y} - gt \tag{3.38}$$

$$y = y_0 + v_{0y}t - \frac{1}{2}gt^2 \tag{3.39}$$

$$v_y^2 = v_{0y}^2 - 2g(y - y_0). \tag{3.40}$$

Step 3. *Solve for the unknowns in the two separate motions—one horizontal and one vertical.* Note that the only common variable between the motions is time t. The problem solving procedures here are the same as for one-dimensional **kinematics** and are illustrated in the solved examples below.

Step 4. *Recombine the two motions to find the total displacement s and velocity v.* Because the x - and y -motions are perpendicular, we determine these vectors by using the techniques outlined in the **Vector Addition and Subtraction: Analytical Methods** and employing $A = \sqrt{A_x^2 + A_y^2}$ and $\theta = \tan^{-1}(A_y / A_x)$ in the following form, where θ is the direction of the displacement s and θ_v is the direction of the velocity v:

Total displacement and velocity

$$s = \sqrt{x^2 + y^2} \tag{3.41}$$

$$\theta = \tan^{-1}(y / x) \tag{3.42}$$

$$v = \sqrt{v_x^2 + v_y^2} \tag{3.43}$$

$$\theta_v = \tan^{-1}(v_y / v_x). \tag{3.44}$$

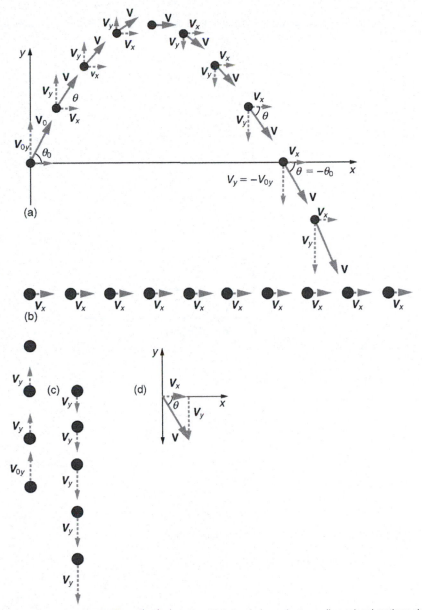

Figure 3.38 (a) We analyze two-dimensional projectile motion by breaking it into two independent one-dimensional motions along the vertical and horizontal axes. (b) The horizontal motion is simple, because $a_x = 0$ and v_x is thus constant. (c) The velocity in the vertical direction begins to decrease as the object rises; at its highest point, the vertical velocity is zero. As the object falls towards the Earth again, the vertical velocity increases again in magnitude but points in the opposite direction to the initial vertical velocity. (d) The x - and y -motions are recombined to give the total velocity at any given point on the trajectory.

Example 3.4 A Fireworks Projectile Explodes High and Away

During a fireworks display, a shell is shot into the air with an initial speed of 70.0 m/s at an angle of $75.0°$ above the horizontal, as illustrated in **Figure 3.39**. The fuse is timed to ignite the shell just as it reaches its highest point above the ground. (a) Calculate the height at which the shell explodes. (b) How much time passed between the launch of the shell and the explosion? (c) What is the horizontal displacement of the shell when it explodes?

Strategy

Because air resistance is negligible for the unexploded shell, the analysis method outlined above can be used. The motion can be broken into horizontal and vertical motions in which $a_x = 0$ and $a_y = -g$. We can then define x_0 and y_0 to be zero and solve for the desired quantities.

Solution for (a)

By "height" we mean the altitude or vertical position y above the starting point. The highest point in any trajectory, called the apex, is reached when $v_y = 0$. Since we know the initial and final velocities as well as the initial position, we use the following equation to find y:

$$v_y^2 = v_{0y}^2 - 2g(y - y_0).$$

(3.45)

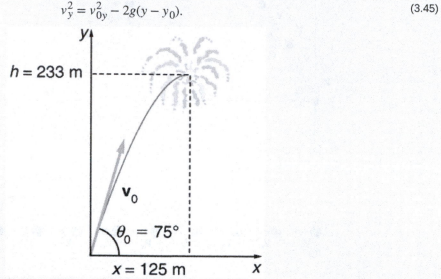

Figure 3.39 The trajectory of a fireworks shell. The fuse is set to explode the shell at the highest point in its trajectory, which is found to be at a height of 233 m and 125 m away horizontally.

Because y_0 and v_y are both zero, the equation simplifies to

$$0 = v_{0y}^2 - 2gy.$$

(3.46)

Solving for y gives

$$y = \frac{v_{0y}^2}{2g}.$$

(3.47)

Now we must find v_{0y}, the component of the initial velocity in the y-direction. It is given by $v_{0y} = v_0 \sin \theta$, where v_{0y} is the initial velocity of 70.0 m/s, and $\theta_0 = 75.0°$ is the initial angle. Thus,

$$v_{0y} = v_0 \sin \theta_0 = (70.0 \text{ m/s})(\sin 75°) = 67.6 \text{ m/s}.$$

(3.48)

and y is

$$y = \frac{(67.6 \text{ m/s})^2}{2(9.80 \text{ m/s}^2)},$$

(3.49)

so that

$$y = 233\text{m}.$$

(3.50)

Discussion for (a)

Note that because up is positive, the initial velocity is positive, as is the maximum height, but the acceleration due to gravity is negative. Note also that the maximum height depends only on the vertical component of the initial velocity, so that any projectile with a 67.6 m/s initial vertical component of velocity will reach a maximum height of 233 m (neglecting air resistance). The numbers in this example are reasonable for large fireworks displays, the shells of which do reach such heights before exploding. In practice, air resistance is not completely negligible, and so the initial velocity would have to be somewhat larger than that given to reach the same height.

Solution for (b)

As in many physics problems, there is more than one way to solve for the time to the highest point. In this case, the easiest method is to use $y = y_0 + \frac{1}{2}(v_{0y} + v_y)t$. Because y_0 is zero, this equation reduces to simply

$$y = \frac{1}{2}(v_{0y} + v_y)t.$$

(3.51)

Note that the final vertical velocity, v_y, at the highest point is zero. Thus,

$$t = \frac{2y}{(v_{0y} + v_y)} = \frac{2(233 \text{ m})}{(67.6 \text{ m/s})} \tag{3.52}$$
$$= 6.90 \text{ s}.$$

Discussion for (b)

This time is also reasonable for large fireworks. When you are able to see the launch of fireworks, you will notice several seconds pass before the shell explodes. (Another way of finding the time is by using $y = y_0 + v_{0y}t - \frac{1}{2}gt^2$, and solving the quadratic equation for t.)

Solution for (c)

Because air resistance is negligible, $a_x = 0$ and the horizontal velocity is constant, as discussed above. The horizontal displacement is horizontal velocity multiplied by time as given by $x = x_0 + v_x t$, where x_0 is equal to zero:

$$x = v_x t, \tag{3.53}$$

where v_x is the x-component of the velocity, which is given by $v_x = v_0 \cos \theta_0$. Now,

$$v_x = v_0 \cos \theta_0 = (70.0 \text{ m/s})(\cos 75.0°) = 18.1 \text{ m/s}. \tag{3.54}$$

The time t for both motions is the same, and so x is

$$x = (18.1 \text{ m/s})(6.90 \text{ s}) = 125 \text{ m}. \tag{3.55}$$

Discussion for (c)

The horizontal motion is a constant velocity in the absence of air resistance. The horizontal displacement found here could be useful in keeping the fireworks fragments from falling on spectators. Once the shell explodes, air resistance has a major effect, and many fragments will land directly below.

In solving part (a) of the preceding example, the expression we found for y is valid for any projectile motion where air resistance is negligible. Call the maximum height $y = h$; then,

$$h = \frac{v_{0y}^2}{2g}. \tag{3.56}$$

This equation defines the *maximum height of a projectile* and depends only on the vertical component of the initial velocity.

Defining a Coordinate System

It is important to set up a coordinate system when analyzing projectile motion. One part of defining the coordinate system is to define an origin for the x and y positions. Often, it is convenient to choose the initial position of the object as the origin such that $x_0 = 0$ and $y_0 = 0$. It is also important to define the positive and negative directions in the x and y directions.

Typically, we define the positive vertical direction as upwards, and the positive horizontal direction is usually the direction of the object's motion. When this is the case, the vertical acceleration, g, takes a negative value (since it is directed downwards towards the Earth). However, it is occasionally useful to define the coordinates differently. For example, if you are analyzing the motion of a ball thrown downwards from the top of a cliff, it may make sense to define the positive direction downwards since the motion of the ball is solely in the downwards direction. If this is the case, g takes a positive value.

Example 3.5 Calculating Projectile Motion: Hot Rock Projectile

Kilauea in Hawaii is the world's most continuously active volcano. Very active volcanoes characteristically eject red-hot rocks and lava rather than smoke and ash. Suppose a large rock is ejected from the volcano with a speed of 25.0 m/s and at an angle $35.0°$ above the horizontal, as shown in **Figure 3.40**. The rock strikes the side of the volcano at an altitude 20.0 m lower than its starting point. (a) Calculate the time it takes the rock to follow this path. (b) What are the magnitude and direction of the rock's velocity at impact?

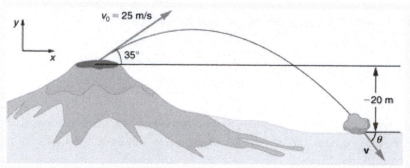

Figure 3.40 The trajectory of a rock ejected from the Kilauea volcano.

Strategy

Again, resolving this two-dimensional motion into two independent one-dimensional motions will allow us to solve for the desired quantities. The time a projectile is in the air is governed by its vertical motion alone. We will solve for t first. While the rock is rising and falling vertically, the horizontal motion continues at a constant velocity. This example asks for the final velocity. Thus, the vertical and horizontal results will be recombined to obtain v and θ_y at the final time t determined in the first part of the example.

Solution for (a)

While the rock is in the air, it rises and then falls to a final position 20.0 m lower than its starting altitude. We can find the time for this by using

$$y = y_0 + v_{0y}t - \frac{1}{2}gt^2. \tag{3.57}$$

If we take the initial position y_0 to be zero, then the final position is $y = -20.0$ m. Now the initial vertical velocity is the vertical component of the initial velocity, found from $v_{0y} = v_0 \sin \theta_0 = (25.0 \text{ m/s})(\sin 35.0°) = 14.3 \text{ m/s}$. Substituting known values yields

$$-20.0 \text{ m} = (14.3 \text{ m/s})t - \left(4.90 \text{ m/s}^2\right)t^2. \tag{3.58}$$

Rearranging terms gives a quadratic equation in t:

$$\left(4.90 \text{ m/s}^2\right)t^2 - (14.3 \text{ m/s})t - (20.0 \text{ m}) = 0. \tag{3.59}$$

This expression is a quadratic equation of the form $at^2 + bt + c = 0$, where the constants are $a = 4.90$, $b = -14.3$, and $c = -20.0$. Its solutions are given by the quadratic formula:

$$t = \frac{-b \pm \sqrt{b^2 - 4ac}}{2a}. \tag{3.60}$$

This equation yields two solutions: $t = 3.96$ and $t = -1.03$. (It is left as an exercise for the reader to verify these solutions.) The time is $t = 3.96$ s or -1.03 s. The negative value of time implies an event before the start of motion, and so we discard it. Thus,

$$t = 3.96 \text{ s}. \tag{3.61}$$

Discussion for (a)

The time for projectile motion is completely determined by the vertical motion. So any projectile that has an initial vertical velocity of 14.3 m/s and lands 20.0 m below its starting altitude will spend 3.96 s in the air.

Solution for (b)

From the information now in hand, we can find the final horizontal and vertical velocities v_x and v_y and combine them to find the total velocity v and the angle θ_0 it makes with the horizontal. Of course, v_x is constant so we can solve for it at any horizontal location. In this case, we chose the starting point since we know both the initial velocity and initial angle. Therefore:

$$v_x = v_0 \cos \theta_0 = (25.0 \text{ m/s})(\cos 35°) = 20.5 \text{ m/s}. \tag{3.62}$$

The final vertical velocity is given by the following equation:

$$v_y = v_{0y} - gt, \tag{3.63}$$

This OpenStax book is available for free at http://cnx.org/content/col11844/1.14   Download for free at https://openstax.org/details/books/college-physics-ap-courses

where v_{0y} was found in part (a) to be 14.3 m/s. Thus,

$$v_y = 14.3 \text{ m/s} - (9.80 \text{ m/s}^2)(3.96 \text{ s}) \tag{3.64}$$

so that

$$v_y = -24.5 \text{ m/s}. \tag{3.65}$$

To find the magnitude of the final velocity v we combine its perpendicular components, using the following equation:

$$v = \sqrt{v_x^2 + v_y^2} = \sqrt{(20.5 \text{ m/s})^2 + (-24.5 \text{ m/s})^2}, \tag{3.66}$$

which gives

$$v = 31.9 \text{ m/s}. \tag{3.67}$$

The direction θ_v is found from the equation:

$$\theta_v = \tan^{-1}(v_y / v_x) \tag{3.68}$$

so that

$$\theta_v = \tan^{-1}(-24.5 / 20.5) = \tan^{-1}(-1.19). \tag{3.69}$$

Thus,

$$\theta_v = -50.1^{\circ}. \tag{3.70}$$

Discussion for (b)

The negative angle means that the velocity is 50.1° below the horizontal. This result is consistent with the fact that the final vertical velocity is negative and hence downward—as you would expect because the final altitude is 20.0 m lower than the initial altitude. (See **Figure 3.40**.)

One of the most important things illustrated by projectile motion is that vertical and horizontal motions are independent of each other. Galileo was the first person to fully comprehend this characteristic. He used it to predict the range of a projectile. On level ground, we define **range** to be the horizontal distance R traveled by a projectile. Galileo and many others were interested in the range of projectiles primarily for military purposes—such as aiming cannons. However, investigating the range of projectiles can shed light on other interesting phenomena, such as the orbits of satellites around the Earth. Let us consider projectile range further.

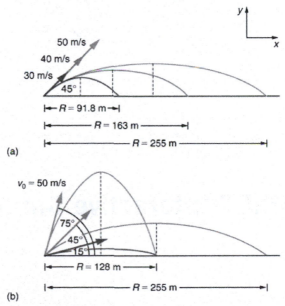

Figure 3.41 Trajectories of projectiles on level ground. (a) The greater the initial speed v_0, the greater the range for a given initial angle. (b) The effect of initial angle θ_0 on the range of a projectile with a given initial speed. Note that the range is the same for 15° and 75°, although the maximum heights of those paths are different.

How does the initial velocity of a projectile affect its range? Obviously, the greater the initial speed v_0, the greater the range, as shown in **Figure 3.41**(a). The initial angle θ_0 also has a dramatic effect on the range, as illustrated in **Figure 3.41**(b). For a fixed initial speed, such as might be produced by a cannon, the maximum range is obtained with $\theta_0 = 45°$. This is true only for conditions neglecting air resistance. If air resistance is considered, the maximum angle is approximately $38°$. Interestingly, for every initial angle except $45°$, there are two angles that give the same range—the sum of those angles is $90°$. The range also depends on the value of the acceleration of gravity g. The lunar astronaut Alan Shepherd was able to drive a golf ball a great distance on the Moon because gravity is weaker there. The range R of a projectile on *level ground* for which air resistance is negligible is given by

$$R = \frac{v_0^2 \sin 2\theta_0}{g},$$ (3.71)

where v_0 is the initial speed and θ_0 is the initial angle relative to the horizontal. The proof of this equation is left as an end-of-chapter problem (hints are given), but it does fit the major features of projectile range as described.

When we speak of the range of a projectile on level ground, we assume that R is very small compared with the circumference of the Earth. If, however, the range is large, the Earth curves away below the projectile and acceleration of gravity changes direction along the path. The range is larger than predicted by the range equation given above because the projectile has farther to fall than it would on level ground. (See **Figure 3.42**.) If the initial speed is great enough, the projectile goes into orbit. This possibility was recognized centuries before it could be accomplished. When an object is in orbit, the Earth curves away from underneath the object at the same rate as it falls. The object thus falls continuously but never hits the surface. These and other aspects of orbital motion, such as the rotation of the Earth, will be covered analytically and in greater depth later in this text.

Once again we see that thinking about one topic, such as the range of a projectile, can lead us to others, such as the Earth orbits. In **Addition of Velocities**, we will examine the addition of velocities, which is another important aspect of two-dimensional kinematics and will also yield insights beyond the immediate topic.

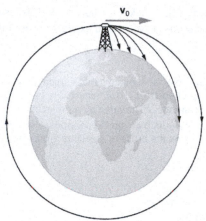

Figure 3.42 Projectile to satellite. In each case shown here, a projectile is launched from a very high tower to avoid air resistance. With increasing initial speed, the range increases and becomes longer than it would be on level ground because the Earth curves away underneath its path. With a large enough initial speed, orbit is achieved.

PhET Explorations: Projectile Motion

Blast a Buick out of a cannon! Learn about projectile motion by firing various objects. Set the angle, initial speed, and mass. Add air resistance. Make a game out of this simulation by trying to hit a target.

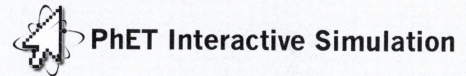

Figure 3.43 Projectile Motion (http://cnx.org/content/m54787/1.4/projectile-motion_en.jar)

3.5 Addition of Velocities

Relative Velocity

If a person rows a boat across a rapidly flowing river and tries to head directly for the other shore, the boat instead moves *diagonally* relative to the shore, as in **Figure 3.44**. The boat does not move in the direction in which it is pointed. The reason, of course, is that the river carries the boat downstream. Similarly, if a small airplane flies overhead in a strong crosswind, you can sometimes see that the plane is not moving in the direction in which it is pointed, as illustrated in **Figure 3.45**. The plane is moving straight ahead relative to the air, but the movement of the air mass relative to the ground carries it sideways.

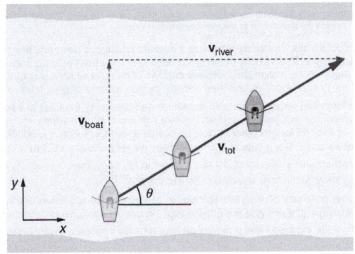

Figure 3.44 A boat trying to head straight across a river will actually move diagonally relative to the shore as shown. Its total velocity (solid arrow) relative to the shore is the sum of its velocity relative to the river plus the velocity of the river relative to the shore.

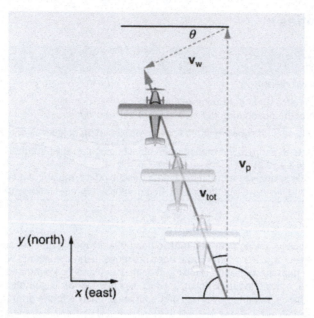

Figure 3.45 An airplane heading straight north is instead carried to the west and slowed down by wind. The plane does not move relative to the ground in the direction it points; rather, it moves in the direction of its total velocity (solid arrow).

In each of these situations, an object has a **velocity** relative to a medium (such as a river) and that medium has a velocity relative to an observer on solid ground. The velocity of the object *relative to the observer* is the sum of these velocity vectors, as indicated in **Figure 3.44** and **Figure 3.45**. These situations are only two of many in which it is useful to add velocities. In this module, we first re-examine how to add velocities and then consider certain aspects of what relative velocity means.

How do we add velocities? Velocity is a vector (it has both magnitude and direction); the rules of **vector addition** discussed in **Vector Addition and Subtraction: Graphical Methods** and **Vector Addition and Subtraction: Analytical Methods** apply to the addition of velocities, just as they do for any other vectors. In one-dimensional motion, the addition of velocities is simple—they add like ordinary numbers. For example, if a field hockey player is moving at 5 m/s straight toward the goal and drives the ball in the same direction with a velocity of 30 m/s relative to her body, then the velocity of the ball is 35 m/s relative to the stationary, profusely sweating goalkeeper standing in front of the goal.

In two-dimensional motion, either graphical or analytical techniques can be used to add velocities. We will concentrate on analytical techniques. The following equations give the relationships between the magnitude and direction of velocity (v and θ) and its components (v_x and v_y) along the x- and y-axes of an appropriately chosen coordinate system:

$$v_x = v \cos \theta \tag{3.72}$$

$$v_y = v \sin \theta \tag{3.73}$$

$$v = \sqrt{v_x^2 + v_y^2} \tag{3.74}$$

$$\theta = \tan^{-1}(v_y / v_x). \tag{3.75}$$

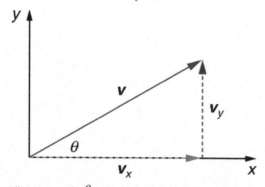

Figure 3.46 The velocity, v, of an object traveling at an angle θ to the horizontal axis is the sum of component vectors $\mathbf{v}_x$ and $\mathbf{v}_y$.

These equations are valid for any vectors and are adapted specifically for velocity. The first two equations are used to find the components of a velocity when its magnitude and direction are known. The last two are used to find the magnitude and direction of velocity when its components are known.

Fill a bathtub half-full of water. Take a toy boat or some other object that floats in water. Unplug the drain so water starts to drain. Try pushing the boat from one side of the tub to the other and perpendicular to the flow of water. Which way do you need to push the boat so that it ends up immediately opposite? Compare the directions of the flow of water, heading of the boat, and actual velocity of the boat.

Example 3.6 Adding Velocities: A Boat on a River

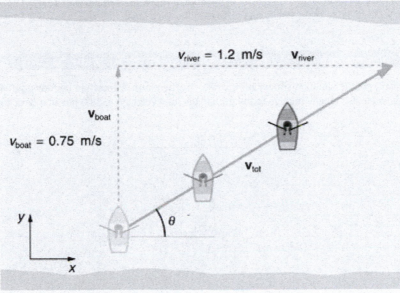

Figure 3.47 A boat attempts to travel straight across a river at a speed 0.75 m/s. The current in the river, however, flows at a speed of 1.20 m/s to the right. What is the total displacement of the boat relative to the shore?

Refer to **Figure 3.47**, which shows a boat trying to go straight across the river. Let us calculate the magnitude and direction of the boat's velocity relative to an observer on the shore, $\mathbf{v}_{tot}$. The velocity of the boat, $\mathbf{v}_{boat}$, is 0.75 m/s in the y-direction relative to the river and the velocity of the river, $\mathbf{v}_{river}$, is 1.20 m/s to the right.

Strategy

We start by choosing a coordinate system with its x-axis parallel to the velocity of the river, as shown in **Figure 3.47**. Because the boat is directed straight toward the other shore, its velocity relative to the water is parallel to the y-axis and perpendicular to the velocity of the river. Thus, we can add the two velocities by using the equations $v_{tot} = \sqrt{v_x^2 + v_y^2}$ and $\theta = \tan^{-1}(v_y/v_x)$ directly.

Solution

The magnitude of the total velocity is

$$v_{tot} = \sqrt{v_x^2 + v_y^2},$$ (3.76)

where

$$v_x = v_{river} = 1.20 \text{ m/s}$$ (3.77)

and

$$v_y = v_{boat} = 0.750 \text{ m/s}.$$ (3.78)

Thus,

$$v_{tot} = \sqrt{(1.20 \text{ m/s})^2 + (0.750 \text{ m/s})^2}$$ (3.79)

yielding

$$v_{tot} = 1.42 \text{ m/s}.$$ (3.80)

The direction of the total velocity θ is given by:

$$\theta = \tan^{-1}(v_y/v_x) = \tan^{-1}(0.750/1.20). \tag{3.81}$$

This equation gives

$$\theta = 32.0°. \tag{3.82}$$

Discussion

Both the magnitude v and the direction θ of the total velocity are consistent with **Figure 3.47**. Note that because the velocity of the river is large compared with the velocity of the boat, it is swept rapidly downstream. This result is evidenced by the small angle (only $32.0°$) the total velocity has relative to the riverbank.

Example 3.7 Calculating Velocity: Wind Velocity Causes an Airplane to Drift

Calculate the wind velocity for the situation shown in **Figure 3.48**. The plane is known to be moving at 45.0 m/s due north relative to the air mass, while its velocity relative to the ground (its total velocity) is 38.0 m/s in a direction $20.0°$ west of north.

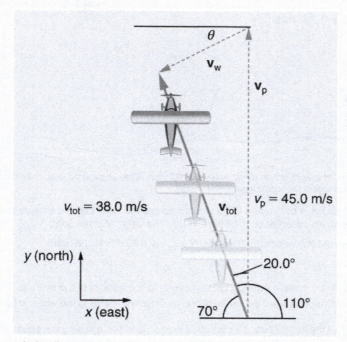

Figure 3.48 An airplane is known to be heading north at 45.0 m/s, though its velocity relative to the ground is 38.0 m/s at an angle west of north. What is the speed and direction of the wind?

Strategy

In this problem, somewhat different from the previous example, we know the total velocity $\mathbf{v}_{tot}$ and that it is the sum of two other velocities, $\mathbf{v}_w$ (the wind) and $\mathbf{v}_p$ (the plane relative to the air mass). The quantity $\mathbf{v}_p$ is known, and we are asked to find $\mathbf{v}_w$. None of the velocities are perpendicular, but it is possible to find their components along a common set of perpendicular axes. If we can find the components of $\mathbf{v}_w$, then we can combine them to solve for its magnitude and direction. As shown in **Figure 3.48**, we choose a coordinate system with its x-axis due east and its y-axis due north (parallel to $\mathbf{v}_p$). (You may wish to look back at the discussion of the addition of vectors using perpendicular components in **Vector Addition and Subtraction: Analytical Methods.**)

Solution

Because $\mathbf{v}_{tot}$ is the vector sum of the $\mathbf{v}_w$ and $\mathbf{v}_p$, its x- and y-components are the sums of the x- and y-components of the wind and plane velocities. Note that the plane only has vertical component of velocity so $v_{px} = 0$ and $v_{py} = v_p$. That is,

$$v_{totx} = v_{wx} \tag{3.83}$$

and

$$v_{\text{tot}y} = v_{\text{w}y} + v_{\text{p}}. \tag{3.84}$$

We can use the first of these two equations to find $v_{\text{w}x}$:

$$v_{\text{w}x} = v_{\text{tot}x} = v_{\text{tot}}\cos 110°. \tag{3.85}$$

Because $v_{\text{tot}} = 38.0 \text{ m / s}$ and $\cos 110° = -0.342$ we have

$$v_{\text{w}x} = (38.0 \text{ m/s})(-0.342) = -13.0 \text{ m/s}. \tag{3.86}$$

The minus sign indicates motion west which is consistent with the diagram.

Now, to find $v_{\text{w}y}$ we note that

$$v_{\text{tot}y} = v_{\text{w}y} + v_{\text{p}} \tag{3.87}$$

Here $v_{\text{tot}y} = v_{\text{tot}}\sin 110°$; thus,

$$v_{\text{w}y} = (38.0 \text{ m/s})(0.940) - 45.0 \text{ m/s} = -9.29 \text{ m/s}. \tag{3.88}$$

This minus sign indicates motion south which is consistent with the diagram.

Now that the perpendicular components of the wind velocity $v_{\text{w}x}$ and $v_{\text{w}y}$ are known, we can find the magnitude and direction of $\mathbf{v_{\text{w}}}$. First, the magnitude is

$$\begin{aligned} v_{\text{w}} &= \sqrt{v_{\text{w}x}^2 + v_{\text{w}y}^2} \\ &= \sqrt{(-13.0 \text{ m/s})^2 + (-9.29 \text{ m/s})^2} \end{aligned} \tag{3.89}$$

so that

$$v_{\text{w}} = 16.0 \text{ m/s}. \tag{3.90}$$

The direction is:

$$\theta = \tan^{-1}(v_{\text{w}y} / v_{\text{w}x}) = \tan^{-1}(-9.29 / -13.0) \tag{3.91}$$

giving

$$\theta = 35.6°. \tag{3.92}$$

Discussion

The wind's speed and direction are consistent with the significant effect the wind has on the total velocity of the plane, as seen in **Figure 3.48**. Because the plane is fighting a strong combination of crosswind and head-wind, it ends up with a total velocity significantly less than its velocity relative to the air mass as well as heading in a different direction.

Note that in both of the last two examples, we were able to make the mathematics easier by choosing a coordinate system with one axis parallel to one of the velocities. We will repeatedly find that choosing an appropriate coordinate system makes problem solving easier. For example, in projectile motion we always use a coordinate system with one axis parallel to gravity.

Relative Velocities and Classical Relativity

When adding velocities, we have been careful to specify that the *velocity is relative to some reference frame*. These velocities are called **relative velocities**. For example, the velocity of an airplane relative to an air mass is different from its velocity relative to the ground. Both are quite different from the velocity of an airplane relative to its passengers (which should be close to zero). Relative velocities are one aspect of **relativity**, which is defined to be the study of how different observers moving relative to each other measure the same phenomenon.

Nearly everyone has heard of relativity and immediately associates it with Albert Einstein (1879–1955), the greatest physicist of the 20th century. Einstein revolutionized our view of nature with his *modern* theory of relativity, which we shall study in later chapters. The relative velocities in this section are actually aspects of classical relativity, first discussed correctly by Galileo and Isaac Newton. **Classical relativity** is limited to situations where speeds are less than about 1% of the speed of light—that is, less than $3{,}000 \text{ km/s}$. Most things we encounter in daily life move slower than this speed.

Let us consider an example of what two different observers see in a situation analyzed long ago by Galileo. Suppose a sailor at the top of a mast on a moving ship drops his binoculars. Where will it hit the deck? Will it hit at the base of the mast, or will it hit behind the mast because the ship is moving forward? The answer is that if air resistance is negligible, the binoculars will hit at the base of the mast at a point directly below its point of release. Now let us consider what two different observers see when the binoculars drop. One observer is on the ship and the other on shore. The binoculars have no horizontal velocity relative to the observer on the ship, and so he sees them fall straight down the mast. (See **Figure 3.49**.) To the observer on shore, the binoculars and the ship have the *same* horizontal velocity, so both move the same distance forward while the binoculars are falling. This observer sees the curved path shown in **Figure 3.49**. Although the paths look different to the different observers, each sees the same result—the binoculars hit at the base of the mast and not behind it. To get the correct description, it is crucial to correctly specify the velocities relative to the observer.

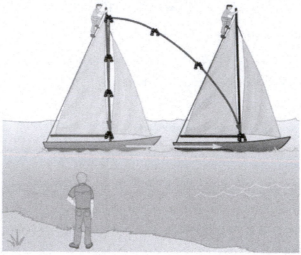

Figure 3.49 Classical relativity. The same motion as viewed by two different observers. An observer on the moving ship sees the binoculars dropped from the top of its mast fall straight down. An observer on shore sees the binoculars take the curved path, moving forward with the ship. Both observers see the binoculars strike the deck at the base of the mast. The initial horizontal velocity is different relative to the two observers. (The ship is shown moving rather fast to emphasize the effect.)

Example 3.8 Calculating Relative Velocity: An Airline Passenger Drops a Coin

An airline passenger drops a coin while the plane is moving at 260 m/s. What is the velocity of the coin when it strikes the floor 1.50 m below its point of release: (a) Measured relative to the plane? (b) Measured relative to the Earth?

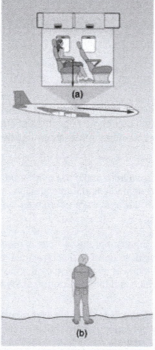

Figure 3.50 The motion of a coin dropped inside an airplane as viewed by two different observers. (a) An observer in the plane sees the coin fall straight down. (b) An observer on the ground sees the coin move almost horizontally.

Strategy

Both problems can be solved with the techniques for falling objects and projectiles. In part (a), the initial velocity of the coin is zero relative to the plane, so the motion is that of a falling object (one-dimensional). In part (b), the initial velocity is 260 m/s horizontal relative to the Earth and gravity is vertical, so this motion is a projectile motion. In both parts, it is best to use a coordinate system with vertical and horizontal axes.

Solution for (a)

Using the given information, we note that the initial velocity and position are zero, and the final position is 1.50 m. The final velocity can be found using the equation:

$$v_y^2 = v_{0y}^2 - 2g(y - y_0).$$

(3.93)

Substituting known values into the equation, we get

$$v_y^2 = 0^2 - 2(9.80 \text{ m/s}^2)(-1.50 \text{ m} - 0 \text{ m}) = 29.4 \text{ m}^2/\text{s}^2$$

(3.94)

yielding

$$v_y = -5.42 \text{ m/s}.$$

(3.95)

We know that the square root of 29.4 has two roots: 5.42 and -5.42. We choose the negative root because we know that the velocity is directed downwards, and we have defined the positive direction to be upwards. There is no initial horizontal velocity relative to the plane and no horizontal acceleration, and so the motion is straight down relative to the plane.

Solution for (b)

Because the initial vertical velocity is zero relative to the ground and vertical motion is independent of horizontal motion, the final vertical velocity for the coin relative to the ground is $v_y = -5.42 \text{ m/s}$, the same as found in part (a). In contrast to part (a), there now is a horizontal component of the velocity. However, since there is no horizontal acceleration, the initial and final horizontal velocities are the same and $v_x = 260 \text{ m/s}$. The x- and y-components of velocity can be combined to find the magnitude of the final velocity:

$$v = \sqrt{v_x^2 + v_y^2}.$$

(3.96)

Thus,

$$v = \sqrt{(260 \text{ m/s})^2 + (-5.42 \text{ m/s})^2}$$

(3.97)

yielding

$$v = 260.06 \text{ m/s}.$$

(3.98)

The direction is given by:

$$\theta = \tan^{-1}(v_y/v_x) = \tan^{-1}(-5.42/260)$$

(3.99)

so that

$$\theta = \tan^{-1}(-0.0208) = -1.19°.$$

(3.100)

Discussion

In part (a), the final velocity relative to the plane is the same as it would be if the coin were dropped from rest on the Earth and fell 1.50 m. This result fits our experience; objects in a plane fall the same way when the plane is flying horizontally as when it is at rest on the ground. This result is also true in moving cars. In part (b), an observer on the ground sees a much different motion for the coin. The plane is moving so fast horizontally to begin with that its final velocity is barely greater than the initial velocity. Once again, we see that in two dimensions, vectors do not add like ordinary numbers—the final velocity v in part (b) is *not* $(260 - 5.42)$ m/s ; rather, it is 260.06 m/s . The velocity's magnitude had to be calculated to five digits to see any difference from that of the airplane. The motions as seen by different observers (one in the plane and one on the ground) in this example are analogous to those discussed for the binoculars dropped from the mast of a moving ship, except that the velocity of the plane is much larger, so that the two observers see *very* different paths. (See **Figure 3.50**.) In addition, both observers see the coin fall 1.50 m vertically, but the one on the ground also sees it move forward 144 m (this calculation is left for the reader). Thus, one observer sees a vertical path, the other a nearly horizontal path.

Making Connections: Relativity and Einstein

Because Einstein was able to clearly define how measurements are made (some involve light) and because the speed of light is the same for all observers, the outcomes are spectacularly unexpected. Time varies with observer, energy is stored as increased mass, and more surprises await.

PhET Explorations: Motion in 2D

Try the new "Ladybug Motion 2D" simulation for the latest updated version. Learn about position, velocity, and acceleration vectors. Move the ball with the mouse or let the simulation move the ball in four types of motion (2 types of linear, simple harmonic, circle).

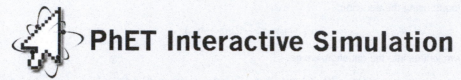

Figure 3.51 Motion in 2D (http://cnx.org/content/m54798/1.2/motion-2d_en.jar)

Glossary

air resistance: a frictional force that slows the motion of objects as they travel through the air; when solving basic physics problems, air resistance is assumed to be zero

analytical method: the method of determining the magnitude and direction of a resultant vector using the Pythagorean theorem and trigonometric identities

classical relativity: the study of relative velocities in situations where speeds are less than about 1% of the speed of light—that is, less than 3000 km/s

commutative: refers to the interchangeability of order in a function; vector addition is commutative because the order in which vectors are added together does not affect the final sum

component (of a 2-d vector): a piece of a vector that points in either the vertical or the horizontal direction; every 2-d vector can be expressed as a sum of two vertical and horizontal vector components

direction (of a vector): the orientation of a vector in space

head (of a vector): the end point of a vector; the location of the tip of the vector's arrowhead; also referred to as the "tip"

head-to-tail method: a method of adding vectors in which the tail of each vector is placed at the head of the previous vector

kinematics: the study of motion without regard to mass or force

magnitude (of a vector): the length or size of a vector; magnitude is a scalar quantity

motion: displacement of an object as a function of time

projectile: an object that travels through the air and experiences only acceleration due to gravity

projectile motion: the motion of an object that is subject only to the acceleration of gravity

range: the maximum horizontal distance that a projectile travels

relative velocity: the velocity of an object as observed from a particular reference frame

relativity: the study of how different observers moving relative to each other measure the same phenomenon

resultant: the sum of two or more vectors

resultant vector: the vector sum of two or more vectors

scalar: a quantity with magnitude but no direction

tail: the start point of a vector; opposite to the head or tip of the arrow

trajectory: the path of a projectile through the air

vector: a quantity that has both magnitude and direction; an arrow used to represent quantities with both magnitude and direction

vector addition: the rules that apply to adding vectors together

velocity: speed in a given direction

Section Summary

3.1 Kinematics in Two Dimensions: An Introduction
- The shortest path between any two points is a straight line. In two dimensions, this path can be represented by a vector with horizontal and vertical components.
- The horizontal and vertical components of a vector are independent of one another. Motion in the horizontal direction does not affect motion in the vertical direction, and vice versa.

3.2 Vector Addition and Subtraction: Graphical Methods
- The **graphical method of adding vectors** $\mathbf{A}$ and $\mathbf{B}$ involves drawing vectors on a graph and adding them using the head-to-tail method. The resultant vector $\mathbf{R}$ is defined such that $\mathbf{A} + \mathbf{B} = \mathbf{R}$. The magnitude and direction of $\mathbf{R}$ are then determined with a ruler and protractor, respectively.
- The **graphical method of subtracting vector** $\mathbf{B}$ from $\mathbf{A}$ involves adding the opposite of vector $\mathbf{B}$, which is defined as $-\mathbf{B}$. In this case, $\mathbf{A} - \mathbf{B} = \mathbf{A} + (-\mathbf{B}) = \mathbf{R}$. Then, the head-to-tail method of addition is followed in the usual way to obtain the resultant vector $\mathbf{R}$.
- Addition of vectors is **commutative** such that $\mathbf{A} + \mathbf{B} = \mathbf{B} + \mathbf{A}$.
- The **head-to-tail method** of adding vectors involves drawing the first vector on a graph and then placing the tail of each subsequent vector at the head of the previous vector. The resultant vector is then drawn from the tail of the first vector to the head of the final vector.
- If a vector $\mathbf{A}$ is multiplied by a scalar quantity c, the magnitude of the product is given by cA. If c is positive, the direction of the product points in the same direction as $\mathbf{A}$; if c is negative, the direction of the product points in the opposite direction as $\mathbf{A}$.

3.3 Vector Addition and Subtraction: Analytical Methods
- The analytical method of vector addition and subtraction involves using the Pythagorean theorem and trigonometric identities to determine the magnitude and direction of a resultant vector.
- The steps to add vectors $\mathbf{A}$ and $\mathbf{B}$ using the analytical method are as follows:

  Step 1: Determine the coordinate system for the vectors. Then, determine the horizontal and vertical components of each vector using the equations

$$\begin{aligned} A_x &= A \cos \theta \\ B_x &= B \cos \theta \end{aligned}$$

  and

$$\begin{aligned} A_y &= A \sin \theta \\ B_y &= B \sin \theta. \end{aligned}$$

  Step 2: Add the horizontal and vertical components of each vector to determine the components R_x and R_y of the resultant vector, $\mathbf{R}$:

$$R_x = A_x + B_x$$

  and

$$R_y = A_y + B_y.$$

  Step 3: Use the Pythagorean theorem to determine the magnitude, R, of the resultant vector $\mathbf{R}$:

$$R = \sqrt{R_x^2 + R_y^2}.$$

  Step 4: Use a trigonometric identity to determine the direction, θ, of $\mathbf{R}$:

$$\theta = \tan^{-1}(R_y / R_x).$$

3.4 Projectile Motion
- Projectile motion is the motion of an object through the air that is subject only to the acceleration of gravity.
- To solve projectile motion problems, perform the following steps:
  1. Determine a coordinate system. Then, resolve the position and/or velocity of the object in the horizontal and vertical components. The components of position $\mathbf{s}$ are given by the quantities x and y, and the components of the velocity $\mathbf{v}$ are given by $v_x = v \cos \theta$ and $v_y = v \sin \theta$, where v is the magnitude of the velocity and θ is its direction.
  2. Analyze the motion of the projectile in the horizontal direction using the following equations:

$$\text{Horizontal motion}(a_x = 0)$$

$$x = x_0 + v_x t$$

$$v_x = v_{0x} = \mathbf{v}_x = \text{velocity is a constant.}$$

3. Analyze the motion of the projectile in the vertical direction using the following equations:

$$\text{Vertical motion}(\text{Assuming positive direction is up; } a_y = -g = -9.80 \text{ m/s}^2)$$

$$y = y_0 + \frac{1}{2}(v_{0y} + v_y)t$$

$$v_y = v_{0y} - gt$$

$$y = y_0 + v_{0y}t - \frac{1}{2}gt^2$$

$$v_y^2 = v_{0y}^2 - 2g(y - y_0).$$

4. Recombine the horizontal and vertical components of location and/or velocity using the following equations:

$$s = \sqrt{x^2 + y^2}$$

$$\theta = \tan^{-1}(y/x)$$

$$v = \sqrt{v_x^2 + v_y^2}$$

$$\theta_v = \tan^{-1}(v_y/v_x).$$

- The maximum height h of a projectile launched with initial vertical velocity v_{0y} is given by

$$h = \frac{v_{0y}^2}{2g}.$$

- The maximum horizontal distance traveled by a projectile is called the **range**. The range R of a projectile on level ground launched at an angle θ_0 above the horizontal with initial speed v_0 is given by

$$R = \frac{v_0^2 \sin 2\theta_0}{g}.$$

3.5 Addition of Velocities

- Velocities in two dimensions are added using the same analytical vector techniques, which are rewritten as

$$v_x = v \cos \theta$$
$$v_y = v \sin \theta$$
$$v = \sqrt{v_x^2 + v_y^2}$$

$$\theta = \tan^{-1}(v_y/v_x).$$

- Relative velocity is the velocity of an object as observed from a particular reference frame, and it varies dramatically with reference frame.
- **Relativity** is the study of how different observers measure the same phenomenon, particularly when the observers move relative to one another. **Classical relativity** is limited to situations where speed is less than about 1% of the speed of light (3000 km/s).

Conceptual Questions

3.2 Vector Addition and Subtraction: Graphical Methods

1. Which of the following is a vector: a person's height, the altitude on Mt. Everest, the age of the Earth, the boiling point of water, the cost of this book, the Earth's population, the acceleration of gravity?

2. Give a specific example of a vector, stating its magnitude, units, and direction.

3. What do vectors and scalars have in common? How do they differ?

4. Two campers in a national park hike from their cabin to the same spot on a lake, each taking a different path, as illustrated below. The total distance traveled along Path 1 is 7.5 km, and that along Path 2 is 8.2 km. What is the final displacement of each camper?

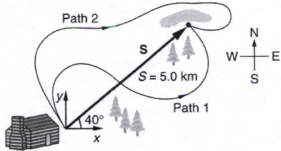

Figure 3.52

5. If an airplane pilot is told to fly 123 km in a straight line to get from San Francisco to Sacramento, explain why he could end up anywhere on the circle shown in **Figure 3.53**. What other information would he need to get to Sacramento?

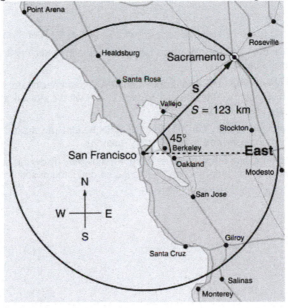

Figure 3.53

6. Suppose you take two steps $\mathbf{A}$ and $\mathbf{B}$ (that is, two nonzero displacements). Under what circumstances can you end up at your starting point? More generally, under what circumstances can two nonzero vectors add to give zero? Is the maximum distance you can end up from the starting point $\mathbf{A} + \mathbf{B}$ the sum of the lengths of the two steps?

7. Explain why it is not possible to add a scalar to a vector.

8. If you take two steps of different sizes, can you end up at your starting point? More generally, can two vectors with different magnitudes ever add to zero? Can three or more?

3.3 Vector Addition and Subtraction: Analytical Methods

9. Suppose you add two vectors $\mathbf{A}$ and $\mathbf{B}$. What relative direction between them produces the resultant with the greatest magnitude? What is the maximum magnitude? What relative direction between them produces the resultant with the smallest magnitude? What is the minimum magnitude?

10. Give an example of a nonzero vector that has a component of zero.

11. Explain why a vector cannot have a component greater than its own magnitude.

12. If the vectors $\mathbf{A}$ and $\mathbf{B}$ are perpendicular, what is the component of $\mathbf{A}$ along the direction of $\mathbf{B}$? What is the component of $\mathbf{B}$ along the direction of $\mathbf{A}$?

3.4 Projectile Motion

13. Answer the following questions for projectile motion on level ground assuming negligible air resistance (the initial angle being neither $0°$ nor $90°$): (a) Is the velocity ever zero? (b) When is the velocity a minimum? A maximum? (c) Can the velocity ever be the same as the initial velocity at a time other than at $t = 0$? (d) Can the speed ever be the same as the initial speed at a time other than at $t = 0$?

14. Answer the following questions for projectile motion on level ground assuming negligible air resistance (the initial angle being neither $0°$ nor $90°$): (a) Is the acceleration ever zero? (b) Is the acceleration ever in the same direction as a component of velocity? (c) Is the acceleration ever opposite in direction to a component of velocity?

15. For a fixed initial speed, the range of a projectile is determined by the angle at which it is fired. For all but the maximum, there are two angles that give the same range. Considering factors that might affect the ability of an archer to hit a target, such as wind, explain why the smaller angle (closer to the horizontal) is preferable. When would it be necessary for the archer to use the larger angle? Why does the punter in a football game use the higher trajectory?

16. During a lecture demonstration, a professor places two coins on the edge of a table. She then flicks one of the coins horizontally off the table, simultaneously nudging the other over the edge. Describe the subsequent motion of the two coins, in particular discussing whether they hit the floor at the same time.

3.5 Addition of Velocities

17. What frame or frames of reference do you instinctively use when driving a car? When flying in a commercial jet airplane?

18. A basketball player dribbling down the court usually keeps his eyes fixed on the players around him. He is moving fast. Why doesn't he need to keep his eyes on the ball?

19. If someone is riding in the back of a pickup truck and throws a softball straight backward, is it possible for the ball to fall straight down as viewed by a person standing at the side of the road? Under what condition would this occur? How would the motion of the ball appear to the person who threw it?

20. The hat of a jogger running at constant velocity falls off the back of his head. Draw a sketch showing the path of the hat in the jogger's frame of reference. Draw its path as viewed by a stationary observer.

21. A clod of dirt falls from the bed of a moving truck. It strikes the ground directly below the end of the truck. What is the direction of its velocity relative to the truck just before it hits? Is this the same as the direction of its velocity relative to ground just before it hits? Explain your answers.

Problems & Exercises

3.2 Vector Addition and Subtraction: Graphical Methods

Use graphical methods to solve these problems. You may assume data taken from graphs is accurate to three digits.

1. Find the following for path A in **Figure 3.54**: (a) the total distance traveled, and (b) the magnitude and direction of the displacement from start to finish.

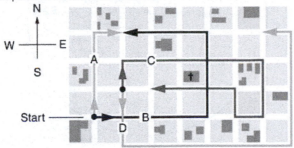

Figure 3.54 The various lines represent paths taken by different people walking in a city. All blocks are 120 m on a side.

2. Find the following for path B in **Figure 3.54**: (a) the total distance traveled, and (b) the magnitude and direction of the displacement from start to finish.

3. Find the north and east components of the displacement for the hikers shown in **Figure 3.52**.

4. Suppose you walk 18.0 m straight west and then 25.0 m straight north. How far are you from your starting point, and what is the compass direction of a line connecting your starting point to your final position? (If you represent the two legs of the walk as vector displacements $\mathbf{A}$ and $\mathbf{B}$, as in **Figure 3.55**, then this problem asks you to find their sum $\mathbf{R} = \mathbf{A} + \mathbf{B}$.)

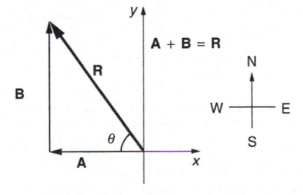

Figure 3.55 The two displacements $\mathbf{A}$ and $\mathbf{B}$ add to give a total displacement $\mathbf{R}$ having magnitude R and direction θ.

5. Suppose you first walk 12.0 m in a direction $20°$ west of north and then 20.0 m in a direction $40.0°$ south of west. How far are you from your starting point, and what is the compass direction of a line connecting your starting point to your final position? (If you represent the two legs of the walk as vector displacements $\mathbf{A}$ and $\mathbf{B}$, as in **Figure 3.56**, then this problem finds their sum $\mathbf{R} = \mathbf{A} + \mathbf{B}$.)

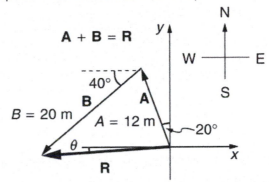

Figure 3.56

6. Repeat the problem above, but reverse the order of the two legs of the walk; show that you get the same final result. That is, you first walk leg $\mathbf{B}$, which is 20.0 m in a direction exactly $40°$ south of west, and then leg $\mathbf{A}$, which is 12.0 m in a direction exactly $20°$ west of north. (This problem shows that $\mathbf{A} + \mathbf{B} = \mathbf{B} + \mathbf{A}$.)

7. (a) Repeat the problem two problems prior, but for the second leg you walk 20.0 m in a direction $40.0°$ north of east (which is equivalent to subtracting $\mathbf{B}$ from $\mathbf{A}$ —that is, to finding $\mathbf{R}' = \mathbf{A} - \mathbf{B}$). (b) Repeat the problem two problems prior, but now you first walk 20.0 m in a direction $40.0°$ south of west and then 12.0 m in a direction $20.0°$ east of south (which is equivalent to subtracting $\mathbf{A}$ from $\mathbf{B}$ —that is, to finding $\mathbf{R}'' = \mathbf{B} - \mathbf{A} = -\mathbf{R}'$). Show that this is the case.

8. Show that the *order* of addition of three vectors does not affect their sum. Show this property by choosing any three vectors $\mathbf{A}$, $\mathbf{B}$, and $\mathbf{C}$, all having different lengths and directions. Find the sum $\mathbf{A} + \mathbf{B} + \mathbf{C}$ then find their sum when added in a different order and show the result is the same. (There are five other orders in which $\mathbf{A}$, $\mathbf{B}$, and $\mathbf{C}$ can be added; choose only one.)

9. Show that the sum of the vectors discussed in **Example 3.2** gives the result shown in **Figure 3.24**.

10. Find the magnitudes of velocities v_A and v_B in **Figure 3.57**

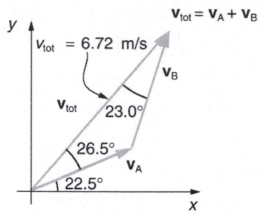

Figure 3.57 The two velocities $\mathbf{v}_A$ and $\mathbf{v}_B$ add to give a total $\mathbf{v}_{tot}$.

11. Find the components of v_{tot} along the x- and y-axes in **Figure 3.57**.

12. Find the components of v_{tot} along a set of perpendicular axes rotated $30°$ counterclockwise relative to those in **Figure 3.57**.

3.3 Vector Addition and Subtraction: Analytical Methods

13. Find the following for path C in **Figure 3.58**: (a) the total distance traveled and (b) the magnitude and direction of the displacement from start to finish. In this part of the problem, explicitly show how you follow the steps of the analytical method of vector addition.

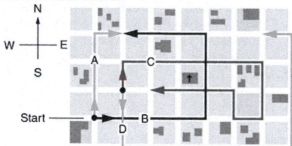

Figure 3.58 The various lines represent paths taken by different people walking in a city. All blocks are 120 m on a side.

14. Find the following for path D in **Figure 3.58**: (a) the total distance traveled and (b) the magnitude and direction of the displacement from start to finish. In this part of the problem, explicitly show how you follow the steps of the analytical method of vector addition.

15. Find the north and east components of the displacement from San Francisco to Sacramento shown in **Figure 3.59**.

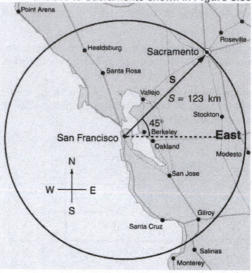

Figure 3.59

16. Solve the following problem using analytical techniques: Suppose you walk 18.0 m straight west and then 25.0 m straight north. How far are you from your starting point, and what is the compass direction of a line connecting your starting point to your final position? (If you represent the two legs of the walk as vector displacements $\mathbf{A}$ and $\mathbf{B}$, as in **Figure 3.60**, then this problem asks you to find their sum $\mathbf{R} = \mathbf{A} + \mathbf{B}$.)

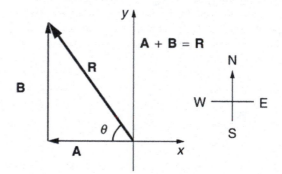

Figure 3.60 The two displacements $\mathbf{A}$ and $\mathbf{B}$ add to give a total displacement $\mathbf{R}$ having magnitude R and direction θ.

Note that you can also solve this graphically. Discuss why the analytical technique for solving this problem is potentially more accurate than the graphical technique.

17. Repeat **Exercise 3.16** using analytical techniques, but reverse the order of the two legs of the walk and show that you get the same final result. (This problem shows that adding them in reverse order gives the same result—that is, $\mathbf{B} + \mathbf{A} = \mathbf{A} + \mathbf{B}$.) Discuss how taking another path to reach the same point might help to overcome an obstacle blocking you other path.

18. You drive 7.50 km in a straight line in a direction $15°$ east of north. (a) Find the distances you would have to drive straight east and then straight north to arrive at the same point. (This determination is equivalent to find the components of the displacement along the east and north directions.) (b) Show that you still arrive at the same point if the east and north legs are reversed in order.

19. Do **Exercise 3.16** again using analytical techniques and change the second leg of the walk to 25.0 m straight south. (This is equivalent to subtracting **B** from **A** —that is, finding $\mathbf{R'} = \mathbf{A} - \mathbf{B}$) (b) Repeat again, but now you first walk 25.0 m north and then 18.0 m east. (This is equivalent to subtract **A** from **B** —that is, to find $\mathbf{A} = \mathbf{B} + \mathbf{C}$. Is that consistent with your result?)

20. A new landowner has a triangular piece of flat land she wishes to fence. Starting at the west corner, she measures the first side to be 80.0 m long and the next to be 105 m. These sides are represented as displacement vectors **A** from **B** in **Figure 3.61**. She then correctly calculates the length and orientation of the third side **C**. What is her result?

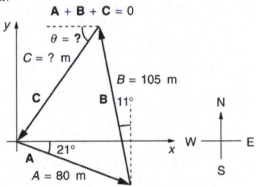

Figure 3.61

21. You fly 32.0 km in a straight line in still air in the direction $35.0°$ south of west. (a) Find the distances you would have to fly straight south and then straight west to arrive at the same point. (This determination is equivalent to finding the components of the displacement along the south and west directions.) (b) Find the distances you would have to fly first in a direction $45.0°$ south of west and then in a direction $45.0°$ west of north. These are the components of the displacement along a different set of axes—one rotated $45°$.

22. A farmer wants to fence off his four-sided plot of flat land. He measures the first three sides, shown as **A**, **B**, and **C** in **Figure 3.62**, and then correctly calculates the length and orientation of the fourth side **D**. What is his result?

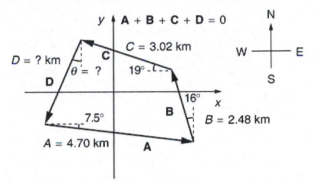

Figure 3.62

23. In an attempt to escape his island, Gilligan builds a raft and sets to sea. The wind shifts a great deal during the day, and he is blown along the following straight lines: 2.50 km $45.0°$ north of west; then 4.70 km $60.0°$ south of east; then 1.30 km $25.0°$ south of west; then 5.10 km straight east; then 1.70 km $5.00°$ east of north; then 7.20 km $55.0°$ south of west; and finally 2.80 km $10.0°$ north of east. What is his final position relative to the island?

24. Suppose a pilot flies 40.0 km in a direction $60°$ north of east and then flies 30.0 km in a direction $15°$ north of east as shown in **Figure 3.63**. Find her total distance R from the starting point and the direction θ of the straight-line path to the final position. Discuss qualitatively how this flight would be altered by a wind from the north and how the effect of the wind would depend on both wind speed and the speed of the plane relative to the air mass.

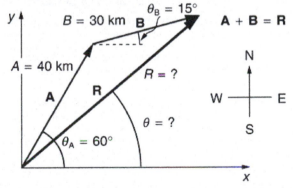

Figure 3.63

3.4 Projectile Motion

25. A projectile is launched at ground level with an initial speed of 50.0 m/s at an angle of $30.0°$ above the horizontal. It strikes a target above the ground 3.00 seconds later. What are the x and y distances from where the projectile was launched to where it lands?

26. A ball is kicked with an initial velocity of 16 m/s in the horizontal direction and 12 m/s in the vertical direction. (a) At what speed does the ball hit the ground? (b) For how long does the ball remain in the air? (c)What maximum height is attained by the ball?

27. A ball is thrown horizontally from the top of a 60.0-m building and lands 100.0 m from the base of the building. Ignore air resistance. (a) How long is the ball in the air? (b) What must have been the initial horizontal component of the velocity? (c) What is the vertical component of the velocity just before the ball hits the ground? (d) What is the velocity (including both the horizontal and vertical components) of the ball just before it hits the ground?

28. (a) A daredevil is attempting to jump his motorcycle over a line of buses parked end to end by driving up a $32°$ ramp at a speed of $40.0 \text{ m/s} \ (144 \text{ km/h})$. How many buses can he clear if the top of the takeoff ramp is at the same height as the bus tops and the buses are 20.0 m long? (b) Discuss what your answer implies about the margin of error in this act—that is, consider how much greater the range is than the horizontal distance he must travel to miss the end of the last bus. (Neglect air resistance.)

29. An archer shoots an arrow at a 75.0 m distant target; the bull's-eye of the target is at same height as the release height of the arrow. (a) At what angle must the arrow be released to hit the bull's-eye if its initial speed is 35.0 m/s? In this part of the problem, explicitly show how you follow the steps involved in solving projectile motion problems. (b) There is a large tree halfway between the archer and the target with an overhanging horizontal branch 3.50 m above the release height of the arrow. Will the arrow go over or under the branch?

30. A rugby player passes the ball 7.00 m across the field, where it is caught at the same height as it left his hand. (a) At what angle was the ball thrown if its initial speed was 12.0 m/s, assuming that the smaller of the two possible angles was used? (b) What other angle gives the same range, and why would it not be used? (c) How long did this pass take?

31. Verify the ranges for the projectiles in Figure 3.41(a) for $\theta = 45°$ and the given initial velocities.

32. Verify the ranges shown for the projectiles in Figure 3.41(b) for an initial velocity of 50 m/s at the given initial angles.

33. The cannon on a battleship can fire a shell a maximum distance of 32.0 km. (a) Calculate the initial velocity of the shell. (b) What maximum height does it reach? (At its highest, the shell is above 60% of the atmosphere—but air resistance is not really negligible as assumed to make this problem easier.) (c) The ocean is not flat, because the Earth is curved. Assume that the radius of the Earth is $6.37 \times 10^3 \text{ km}$. How many meters lower will its surface be 32.0 km from the ship along a horizontal line parallel to the surface at the ship? Does your answer imply that error introduced by the assumption of a flat Earth in projectile motion is significant here?

34. An arrow is shot from a height of 1.5 m toward a cliff of height H. It is shot with a velocity of 30 m/s at an angle of $60°$ above the horizontal. It lands on the top edge of the cliff 4.0 s later. (a) What is the height of the cliff? (b) What is the maximum height reached by the arrow along its trajectory? (c) What is the arrow's impact speed just before hitting the cliff?

35. In the standing broad jump, one squats and then pushes off with the legs to see how far one can jump. Suppose the extension of the legs from the crouch position is 0.600 m and the acceleration achieved from this position is 1.25 times the acceleration due to gravity, g. How far can they jump? State your assumptions. (Increased range can be achieved by swinging the arms in the direction of the jump.)

36. The world long jump record is 8.95 m (Mike Powell, USA, 1991). Treated as a projectile, what is the maximum range obtainable by a person if he has a take-off speed of 9.5 m/s? State your assumptions.

37. Serving at a speed of 170 km/h, a tennis player hits the ball at a height of 2.5 m and an angle θ below the horizontal. The baseline from which the ball is served is 11.9 m from the net, which is 0.91 m high. What is the angle θ such that the ball just crosses the net? Will the ball land in the service box, which has an outermost service line is 6.40 m from the net?

38. A football quarterback is moving straight backward at a speed of 2.00 m/s when he throws a pass to a player 18.0 m straight downfield. (a) If the ball is thrown at an angle of $25°$ relative to the ground and is caught at the same height as it is released, what is its initial speed relative to the ground? (b) How long does it take to get to the receiver? (c) What is its maximum height above its point of release?

39. Gun sights are adjusted to aim high to compensate for the effect of gravity, effectively making the gun accurate only for a specific range. (a) If a gun is sighted to hit targets that are at the same height as the gun and 100.0 m away, how low will the bullet hit if aimed directly at a target 150.0 m away? The muzzle velocity of the bullet is 275 m/s. (b) Discuss qualitatively how a larger muzzle velocity would affect this problem and what would be the effect of air resistance.

40. An eagle is flying horizontally at a speed of 3.00 m/s when the fish in her talons wiggles loose and falls into the lake 5.00 m below. Calculate the velocity of the fish relative to the water when it hits the water.

41. An owl is carrying a mouse to the chicks in its nest. Its position at that time is 4.00 m west and 12.0 m above the center of the 30.0 cm diameter nest. The owl is flying east at 3.50 m/s at an angle $30.0°$ below the horizontal when it accidentally drops the mouse. Is the owl lucky enough to have the mouse hit the nest? To answer this question, calculate the horizontal position of the mouse when it has fallen 12.0 m.

42. Suppose a soccer player kicks the ball from a distance 30 m toward the goal. Find the initial speed of the ball if it just passes over the goal, 2.4 m above the ground, given the initial direction to be $40°$ above the horizontal.

43. Can a goalkeeper at her/ his goal kick a soccer ball into the opponent's goal without the ball touching the ground? The distance will be about 95 m. A goalkeeper can give the ball a speed of 30 m/s.

44. The free throw line in basketball is 4.57 m (15 ft) from the basket, which is 3.05 m (10 ft) above the floor. A player standing on the free throw line throws the ball with an initial speed of 8.15 m/s, releasing it at a height of 2.44 m (8 ft) above the floor. At what angle above the horizontal must the ball be thrown to exactly hit the basket? Note that most players will use a large initial angle rather than a flat shot because it allows for a larger margin of error. Explicitly show how you follow the steps involved in solving projectile motion problems.

45. In 2007, Michael Carter (U.S.) set a world record in the shot put with a throw of 24.77 m. What was the initial speed of the shot if he released it at a height of 2.10 m and threw it at an angle of $38.0°$ above the horizontal? (Although the maximum distance for a projectile on level ground is achieved at $45°$ when air resistance is neglected, the actual angle to achieve maximum range is smaller; thus, $38°$ will give a longer range than $45°$ in the shot put.)

46. A basketball player is running at 5.00 m/s directly toward the basket when he jumps into the air to dunk the ball. He maintains his horizontal velocity. (a) What vertical velocity does he need to rise 0.750 m above the floor? (b) How far from the basket (measured in the horizontal direction) must he start his jump to reach his maximum height at the same time as he reaches the basket?

47. A football player punts the ball at a $45.0°$ angle. Without an effect from the wind, the ball would travel 60.0 m horizontally. (a) What is the initial speed of the ball? (b) When the ball is near its maximum height it experiences a brief gust of wind that reduces its horizontal velocity by 1.50 m/s. What distance does the ball travel horizontally?

48. Prove that the trajectory of a projectile is parabolic, having the form $y = ax + bx^2$. To obtain this expression, solve the equation $x = v_{0x}t$ for t and substitute it into the expression for $y = v_{0y}t - (1/2)gt^2$ (These equations describe the x and y positions of a projectile that starts at the origin.) You should obtain an equation of the form $y = ax + bx^2$ where a and b are constants.

49. Derive $R = \frac{v_0^2 \sin 2\theta_0}{g}$ for the range of a projectile on level ground by finding the time t at which y becomes zero and substituting this value of t into the expression for $x - x_0$, noting that $R = x - x_0$

50. Unreasonable Results (a) Find the maximum range of a super cannon that has a muzzle velocity of 4.0 km/s. (b) What is unreasonable about the range you found? (c) Is the premise unreasonable or is the available equation inapplicable? Explain your answer. (d) If such a muzzle velocity could be obtained, discuss the effects of air resistance, thinning air with altitude, and the curvature of the Earth on the range of the super cannon.

51. Construct Your Own Problem Consider a ball tossed over a fence. Construct a problem in which you calculate the ball's needed initial velocity to just clear the fence. Among the things to determine are; the height of the fence, the distance to the fence from the point of release of the ball, and the height at which the ball is released. You should also consider whether it is possible to choose the initial speed for the ball and just calculate the angle at which it is thrown. Also examine the possibility of multiple solutions given the distances and heights you have chosen.

3.5 Addition of Velocities

52. Bryan Allen pedaled a human-powered aircraft across the English Channel from the cliffs of Dover to Cap Gris-Nez on June 12, 1979. (a) He flew for 169 min at an average velocity of 3.53 m/s in a direction $45°$ south of east. What was his total displacement? (b) Allen encountered a headwind averaging 2.00 m/s almost precisely in the opposite direction of his motion relative to the Earth. What was his average velocity relative to the air? (c) What was his total displacement relative to the air mass?

53. A seagull flies at a velocity of 9.00 m/s straight into the wind. (a) If it takes the bird 20.0 min to travel 6.00 km relative to the Earth, what is the velocity of the wind? (b) If the bird turns around and flies with the wind, how long will he take to return 6.00 km? (c) Discuss how the wind affects the total round-trip time compared to what it would be with no wind.

54. Near the end of a marathon race, the first two runners are separated by a distance of 45.0 m. The front runner has a velocity of 3.50 m/s, and the second a velocity of 4.20 m/s. (a) What is the velocity of the second runner relative to the first? (b) If the front runner is 250 m from the finish line, who will win the race, assuming they run at constant velocity? (c) What distance ahead will the winner be when she crosses the finish line?

55. Verify that the coin dropped by the airline passenger in the **Example 3.8** travels 144 m horizontally while falling 1.50 m in the frame of reference of the Earth.

56. A football quarterback is moving straight backward at a speed of 2.00 m/s when he throws a pass to a player 18.0 m straight downfield. The ball is thrown at an angle of $25.0°$ relative to the ground and is caught at the same height as it is released. What is the initial velocity of the ball *relative to the quarterback* ?

57. A ship sets sail from Rotterdam, The Netherlands, heading due north at 7.00 m/s relative to the water. The local ocean current is 1.50 m/s in a direction $40.0°$ north of east. What is the velocity of the ship relative to the Earth?

58. (a) A jet airplane flying from Darwin, Australia, has an air speed of 260 m/s in a direction $5.0°$ south of west. It is in the jet stream, which is blowing at 35.0 m/s in a direction $15°$ south of east. What is the velocity of the airplane relative to the Earth? (b) Discuss whether your answers are consistent with your expectations for the effect of the wind on the plane's path.

59. (a) In what direction would the ship in **Exercise 3.57** have to travel in order to have a velocity straight north relative to the Earth, assuming its speed relative to the water remains 7.00 m/s ? (b) What would its speed be relative to the Earth?

60. (a) Another airplane is flying in a jet stream that is blowing at 45.0 m/s in a direction $20°$ south of east (as in **Exercise 3.58**). Its direction of motion relative to the Earth is $45.0°$ south of west, while its direction of travel relative to the air is $5.00°$ south of west. What is the airplane's speed relative to the air mass? (b) What is the airplane's speed relative to the Earth?

61. A sandal is dropped from the top of a 15.0-m-high mast on a ship moving at 1.75 m/s due south. Calculate the velocity of the sandal when it hits the deck of the ship: (a) relative to the ship and (b) relative to a stationary observer on shore. (c) Discuss how the answers give a consistent result for the position at which the sandal hits the deck.

62. The velocity of the wind relative to the water is crucial to sailboats. Suppose a sailboat is in an ocean current that has a velocity of 2.20 m/s in a direction $30.0°$ east of north relative to the Earth. It encounters a wind that has a velocity of 4.50 m/s in a direction of $50.0°$ south of west relative to the Earth. What is the velocity of the wind relative to the water?

63. The great astronomer Edwin Hubble discovered that all distant galaxies are receding from our Milky Way Galaxy with velocities proportional to their distances. It appears to an observer on the Earth that we are at the center of an expanding universe. **Figure 3.64** illustrates this for five galaxies lying along a straight line, with the Milky Way Galaxy at the center. Using the data from the figure, calculate the velocities: (a) relative to galaxy 2 and (b) relative to galaxy 5. The results mean that observers on all galaxies will see themselves at the center of the expanding universe, and they would likely be aware of relative velocities, concluding that it is not possible to locate the center of expansion with the given information.

Galaxy 1 300 Mly	Galaxy 2 150 Mly	Galaxy 3 MW	Galaxy 4 190 Mly	Galaxy 5 450 Mly

$v_1 = -4500$ km/s $v_2 = -2200$ km/s $v_4 = 2830$ km/s $v_5 = 6700$ km/s

Figure 3.64 Five galaxies on a straight line, showing their distances and velocities relative to the Milky Way (MW) Galaxy. The distances are in millions of light years (Mly), where a light year is the distance light travels in one year. The velocities are nearly proportional to the distances. The sizes of the galaxies are greatly exaggerated; an average galaxy is about 0.1 Mly across.

64. (a) Use the distance and velocity data in **Figure 3.64** to find the rate of expansion as a function of distance.

(b) If you extrapolate back in time, how long ago would all of the galaxies have been at approximately the same position? The two parts of this problem give you some idea of how the Hubble constant for universal expansion and the time back to the Big Bang are determined, respectively.

65. An athlete crosses a 25-m-wide river by swimming perpendicular to the water current at a speed of 0.5 m/s relative to the water. He reaches the opposite side at a distance 40 m downstream from his starting point. How fast is the water in the river flowing with respect to the ground? What is the speed of the swimmer with respect to a friend at rest on the ground?

66. A ship sailing in the Gulf Stream is heading $25.0°$ west of north at a speed of 4.00 m/s relative to the water. Its velocity relative to the Earth is 4.80 m/s $5.00°$ west of north. What is the velocity of the Gulf Stream? (The velocity obtained is typical for the Gulf Stream a few hundred kilometers off the east coast of the United States.)

67. An ice hockey player is moving at 8.00 m/s when he hits the puck toward the goal. The speed of the puck relative to the player is 29.0 m/s. The line between the center of the goal and the player makes a $90.0°$ angle relative to his path as shown in **Figure 3.65**. What angle must the puck's velocity make relative to the player (in his frame of reference) to hit the center of the goal?

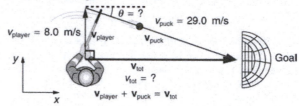

Figure 3.65 An ice hockey player moving across the rink must shoot backward to give the puck a velocity toward the goal.

68. Unreasonable Results Suppose you wish to shoot supplies straight up to astronauts in an orbit 36,000 km above the surface of the Earth. (a) At what velocity must the supplies be launched? (b) What is unreasonable about this velocity? (c) Is there a problem with the relative velocity between the supplies and the astronauts when the supplies reach their maximum height? (d) Is the premise unreasonable or is the available equation inapplicable? Explain your answer.

69. Unreasonable Results A commercial airplane has an air speed of 280 m/s due east and flies with a strong tailwind. It travels 3000 km in a direction $5°$ south of east in 1.50 h. (a) What was the velocity of the plane relative to the ground? (b) Calculate the magnitude and direction of the tailwind's velocity. (c) What is unreasonable about both of these velocities? (d) Which premise is unreasonable?

70. Construct Your Own Problem Consider an airplane headed for a runway in a cross wind. Construct a problem in which you calculate the angle the airplane must fly relative to the air mass in order to have a velocity parallel to the runway. Among the things to consider are the direction of the runway, the wind speed and direction (its velocity) and the speed of the plane relative to the air mass. Also calculate the speed of the airplane relative to the ground. Discuss any last minute maneuvers the pilot might have to perform in order for the plane to land with its wheels pointing straight down the runway.

3.1 Kinematics in Two Dimensions: An

Introduction

1. A ball is thrown at an angle of 45 degrees above the horizontal. Which of the following best describes the

acceleration of the ball from the instant after it leaves the thrower's hand until the time it hits the ground?
a. Always in the same direction as the motion, initially positive and gradually dropping to zero by the time it hits the ground
b. Initially positive in the upward direction, then zero at maximum height, then negative from there until it hits the ground
c. Always in the opposite direction as the motion, initially positive and gradually dropping to zero by the time it hits the ground
d. Always in the downward direction with the same constant value

2. In an experiment, a student launches a ball with an initial horizontal velocity at an elevation 2 meters above ground. The ball follows a parabolic trajectory until it hits the ground. Which of the following accurately describes the graph of the ball's vertical acceleration versus time (taking the downward direction to be negative)?
a. A negative value that does not change with time
b. A gradually increasing negative value (straight line)
c. An increasing rate of negative values over time (parabolic curve)
d. Zero at all times since the initial motion is horizontal

3. A student wishes to design an experiment to show that the acceleration of an object is independent of the object's velocity. To do this, ball A is launched horizontally with some initial speed at an elevation 1.5 meters above the ground, ball B is dropped from rest 1.5 meters above the ground, and ball C is launched vertically with some initial speed at an elevation 1.5 meters above the ground. What information would the student need to collect about each ball in order to test the hypothesis?

3.2 Vector Addition and Subtraction: Graphical Methods

4. A ball is launched vertically upward. The vertical position of the ball is recorded at various points in time in the table shown.

Table 3.1

Height (m)	Time (sec)
0.490	0.1
0.882	0.2
1.176	0.3
1.372	0.4
1.470	0.5
1.470	0.6
1.372	0.7

Which of the following correctly describes the graph of the ball's vertical velocity versus time?

a. Always positive, steadily decreasing
b. Always positive, constant
c. Initially positive, steadily decreasing, becoming negative at the end
d. Initially zero, steadily getting more and more negative

5.

Table 3.2

Height (m)	Time (sec)
0.490	0.1
0.882	0.2
1.176	0.3
1.372	0.4
1.470	0.5
1.470	0.6
1.372	0.7

A ball is launched at an angle of 60 degrees above the horizontal, and the vertical position of the ball is recorded at various points in time in the table shown, assuming the ball was at a height of 0 at time t = 0.
a. Draw a graph of the ball's vertical velocity versus time.
b. Describe the graph of the ball's horizontal velocity.
c. Draw a graph of the ball's vertical acceleration versus time.

3.4 Projectile Motion

6. In an experiment, a student launches a ball with an initial horizontal velocity of 5.00 meters/sec at an elevation 2.00 meters above ground. Draw and clearly label with appropriate values and units a graph of the ball's horizontal velocity vs. time and the ball's vertical velocity vs. time. The graph should cover the motion from the instant after the ball is launched until the instant before it hits the ground. Assume the downward direction is negative for this problem.

4 DYNAMICS: FORCE AND NEWTON'S LAWS OF MOTION

Figure 4.1 Newton's laws of motion describe the motion of the dolphin's path. (credit: Jin Jang)

Chapter Outline

4.1. Development of Force Concept

4.2. Newton's First Law of Motion: Inertia

4.3. Newton's Second Law of Motion: Concept of a System

4.4. Newton's Third Law of Motion: Symmetry in Forces

4.5. Normal, Tension, and Other Examples of Force

4.6. Problem-Solving Strategies

4.7. Further Applications of Newton's Laws of Motion

4.8. Extended Topic: The Four Basic Forces—An Introduction

Connection for AP® Courses

Motion draws our attention. Motion itself can be beautiful, causing us to marvel at the forces needed to achieve spectacular motion, such as that of a jumping dolphin, a leaping pole vaulter, a bird in flight, or an orbiting satellite. The study of motion is kinematics, but kinematics only *describes* the way objects move—their velocity and their acceleration. **Dynamics** considers the forces that affect the motion of moving objects and systems. Newton's laws of motion are the foundation of dynamics. These laws provide an example of the breadth and simplicity of principles under which nature functions. They are also universal laws in that they apply to situations on Earth as well as in space.

Isaac Newton's (1642–1727) laws of motion were just one part of the monumental work that has made him legendary. The development of Newton's laws marks the transition from the Renaissance into the modern era. This transition was characterized by a revolutionary change in the way people thought about the physical universe. For many centuries natural philosophers had debated the nature of the universe based largely on certain rules of logic, with great weight given to the thoughts of earlier classical philosophers such as Aristotle (384–322 BC). Among the many great thinkers who contributed to this change were Newton and Galileo Galilei (1564–1647).

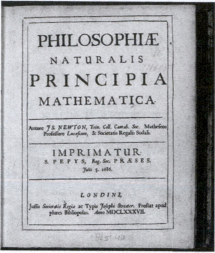

Figure 4.2 Isaac Newton's monumental work, *Philosophiae Naturalis Principia Mathematica*, was published in 1687. It proposed scientific laws that are still used today to describe the motion of objects. (credit: Service commun de la documentation de l'Université de Strasbourg)

Galileo was instrumental in establishing *observation* as the absolute determinant of truth, rather than "logical" argument. Galileo's use of the telescope was his most notable achievement in demonstrating the importance of observation. He discovered moons orbiting Jupiter and made other observations that were inconsistent with certain ancient ideas and religious dogma. For this reason, and because of the manner in which he dealt with those in authority, Galileo was tried by the Inquisition and punished. He spent the final years of his life under a form of house arrest. Because others before Galileo had also made discoveries by *observing* the nature of the universe and because repeated observations verified those of Galileo, his work could not be suppressed or denied. After his death, his work was verified by others, and his ideas were eventually accepted by the church and scientific communities.

Galileo also contributed to the formulation of what is now called Newton's first law of motion. Newton made use of the work of his predecessors, which enabled him to develop laws of motion, discover the law of gravity, invent calculus, and make great contributions to the theories of light and color. It is amazing that many of these developments were made by Newton working alone, without the benefit of the usual interactions that take place among scientists today.

Newton's laws are introduced along with Big Idea 3, that interactions can be described by forces. These laws provide a theoretical basis for studying motion depending on interactions between the objects. In particular, Newton's laws are applicable to all forces in inertial frames of references (Enduring Understanding 3.A). We will find that all forces are vectors; that is, forces always have both a magnitude and a direction (Essential Knowledge 3.A.2). Furthermore, we will learn that all forces are a result of interactions between two or more objects (Essential Knowledge 3.A.3). These interactions between any two objects are described by Newton's third law, stating that the forces exerted on these objects are equal in magnitude and opposite in direction to each other (Essential Knowledge 3.A.4).

We will discover that there is an empirical cause-effect relationship between the net force exerted on an object of mass m and its acceleration, with this relationship described by Newton's second law (Enduring Understanding 3.B). This supports Big Idea 1, that inertial mass is a property of an object or a system. The mass of an object or a system is one of the factors affecting changes in motion when an object or a system interacts with other objects or systems (Essential Knowledge 1.C.1). Another is the net force on an object, which is the vector sum of all the forces exerted on the object (Essential Knowledge 3.B.1). To analyze this, we use free-body diagrams to visualize the forces exerted on a given object in order to find the net force and analyze the object's motion (Essential Knowledge 3.B.2).

Thinking of these objects as systems is a concept introduced in this chapter, where a system is a collection of elements that could be considered as a single object without any internal structure (Essential Knowledge 5.A.1). This will support Big Idea 5, that changes that occur to the system due to interactions are governed by conservation laws. These conservation laws will be the focus of later chapters in this book. They explain whether quantities are conserved in the given system or change due to transfer to or from another system due to interactions between the systems (Enduring Understanding 5.A).

Furthermore, when a situation involves more than one object, it is important to define the system and analyze the motion of a whole system, not its elements, based on analysis of external forces on the system. This supports Big Idea 4, that interactions between systems cause changes in those systems. All kinematics variables in this case describe the motion of the center of mass of the system (Essential Knowledge 4.A.1, Essential Knowledge 4.A.2). The internal forces between the elements of the system do not affect the velocity of the center of mass (Essential Knowledge 4.A.3). The velocity of the center of mass will change only if there is a net external force exerted on the system (Enduring Understanding 4.A).

We will learn that some of these interactions can be explained by the existence of fields extending through space, supporting Big Idea 2. For example, any object that has mass creates a gravitational field in space (Enduring Understanding 2.B). Any material object (one that has mass) placed in the gravitational field will experience gravitational force (Essential Knowledge 2.B.1).

Forces may be categorized as contact or long-distance (Enduring Understanding 3.C). In this chapter we will work with both. An example of a long-distance force is gravitation (Essential Knowledge 3.C.1). Contact forces, such as tension, friction, normal force, and the force of a spring, result from interatomic electric forces at the microscopic level (Essential Knowledge 3.C.4).

It was not until the advent of modern physics early in the twentieth century that it was discovered that Newton's laws of motion produce a good approximation to motion only when the objects are moving at speeds much, much less than the speed of light and when those objects are larger than the size of most molecules (about 10^{-9} m in diameter). These constraints define the realm of classical mechanics, as discussed in **Introduction to the Nature of Science and Physics**. At the beginning of the twentieth century, Albert Einstein (1879–1955) developed the theory of relativity and, along with many other scientists, quantum theory. Quantum theory does not have the constraints present in classical physics. All of the situations we consider in this chapter, and all those preceding the introduction of relativity in **Special Relativity**, are in the realm of classical physics.

The development of special relativity and empirical observations at atomic scales led to the idea that there are four basic forces that account for all known phenomena. These forces are called fundamental (Enduring Understanding 3.G). The properties of gravitational (Essential Knowledge 3.G.1) and electromagnetic (Essential Knowledge 3.G.2) forces are explained in more detail.

Big Idea 1 Objects and systems have properties such as mass and charge. Systems may have internal structure.

Essential Knowledge 1.C.1 Inertial mass is the property of an object or a system that determines how its motion changes when it interacts with other objects or systems.

Big Idea 2 Fields existing in space can be used to explain interactions.

Enduring Understanding 2.A A field associates a value of some physical quantity with every point in space. Field models are useful for describing interactions that occur at a distance (long-range forces) as well as a variety of other physical phenomena.

Essential Knowledge 2.A.1 A vector field gives, as a function of position (and perhaps time), the value of a physical quantity that is described by a vector.

Essential Knowledge 2.A.2 A scalar field gives the value of a physical quantity.

Enduring Understanding 2.B A gravitational field is caused by an object with mass.

Essential Knowledge 2.B.1 A gravitational field g at the location of an object with mass m causes a gravitational force of magnitude mg to be exerted on the object in the direction of the field.

Big Idea 3 The interactions of an object with other objects can be described by forces.

Enduring Understanding 3.A All forces share certain common characteristics when considered by observers in inertial reference frames.

Essential Knowledge 3.A.2 Forces are described by vectors.

Essential Knowledge 3.A.3 A force exerted on an object is always due to the interaction of that object with another object.

Essential Knowledge 3.A.4 If one object exerts a force on a second object, the second object always exerts a force of equal magnitude on the first object in the opposite direction.

Enduring Understanding 3.B Classically, the acceleration of an object interacting with other objects can be predicted by using $a = \sum F / m$.

$$a = \sum F / m .$$

Essential Knowledge 3.B.1 If an object of interest interacts with several other objects, the net force is the vector sum of the individual forces.

Essential Knowledge 3.B.2 Free-body diagrams are useful tools for visualizing the forces being exerted on a single object and writing the equations that represent a physical situation.

Enduring Understanding 3.C At the macroscopic level, forces can be categorized as either long-range (action-at-a-distance) forces or contact forces.

Essential Knowledge 3.C.1 Gravitational force describes the interaction of one object that has mass with another object that has mass.

Essential Knowledge 3.C.4 Contact forces result from the interaction of one object touching another object, and they arise from interatomic electric forces. These forces include tension, friction, normal, spring (Physics 1), and buoyant (Physics 2).

Enduring Understanding 3.G Certain types of forces are considered fundamental.

Essential Knowledge 3.G.1 Gravitational forces are exerted at all scales and dominate at the largest distance and mass scales.

Essential Knowledge 3.G.2 Electromagnetic forces are exerted at all scales and can dominate at the human scale.

Big Idea 4 Interactions between systems can result in changes in those systems.

Enduring Understanding 4.A The acceleration of the center of mass of a system is related to the net force exerted on the system, where $a = \sum F / m$.

$$\text{where } a = \sum F / m .$$

Essential Knowledge 4.A.1 The linear motion of a system can be described by the displacement, velocity, and acceleration of its center of mass.

Essential Knowledge 4.A.2 The acceleration is equal to the rate of change of velocity with time, and velocity is equal to the rate of change of position with time.

Essential Knowledge 4.A.3 Forces that systems exert on each other are due to interactions between objects in the systems. If the interacting objects are parts of the same system, there will be no change in the center-of-mass velocity of that system.

Big Idea 5 Changes that occur as a result of interactions are constrained by conservation laws.

Enduring Understanding 5.A Certain quantities are conserved, in the sense that the changes of those quantities in a given system are always equal to the transfer of that quantity to or from the system by all possible interactions with other systems.

Essential Knowledge 5.A.1 A system is an object or a collection of objects. The objects are treated as having no internal structure.

4.1 Development of Force Concept

Learning Objectives
By the end of this section, you will be able to: • Understand the definition of force. The information presented in this section supports the following AP® learning objectives and science practices: • **3.A.2.1** The student is able to represent forces in diagrams or mathematically using appropriately labeled vectors with magnitude, direction, and units during the analysis of a situation. **(S.P. 1.1)** • **3.A.3.2** The student is able to challenge a claim that an object can exert a force on itself. **(S.P. 6.1)** • **3.A.3.3** The student is able to describe a force as an interaction between two objects and identify both objects for any force. **(S.P. 1.4)** • **3.B.2.1** The student is able to create and use free-body diagrams to analyze physical situations to solve problems with motion qualitatively and quantitatively. **(S.P. 1.1, 1.4, 2.2)**

Dynamics is the study of the forces that cause objects and systems to move. To understand this, we need a working definition of force. Our intuitive definition of **force**—that is, a push or a pull—is a good place to start. We know that a push or pull has both magnitude and direction (therefore, it is a vector quantity) and can vary considerably in each regard. For example, a cannon exerts a strong force on a cannonball that is launched into the air. In contrast, Earth exerts only a tiny downward pull on a flea. Our everyday experiences also give us a good idea of how multiple forces add. If two people push in different directions on a third person, as illustrated in **Figure 4.3**, we might expect the total force to be in the direction shown. Since force is a vector, it adds just like other vectors, as illustrated in **Figure 4.3**(a) for two ice skaters. Forces, like other vectors, are represented by arrows and can be added using the familiar head-to-tail method or by trigonometric methods. These ideas were developed in **Two-Dimensional Kinematics**.

By definition, force is always the result of an interaction of two or more objects. No object possesses force on its own. For example, a cannon does not possess force, but it can exert force on a cannonball. Earth does not possess force on its own, but exerts force on a football or on any other massive object. The skaters in **Figure 4.3** exert force on one another as they interact.

No object can exert force on itself. When you clap your hands, one hand exerts force on the other. When a train accelerates, it exerts force on the track and vice versa. A bowling ball is accelerated by the hand throwing it; once the hand is no longer in contact with the bowling ball, it is no longer accelerating the bowling ball or exerting force on it. The ball continues moving forward due to inertia.

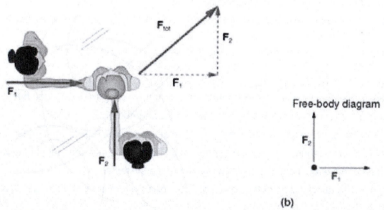

(a)   (b)

Figure 4.3 Part (a) shows an overhead view of two ice skaters pushing on a third. Forces are vectors and add like other vectors, so the total force on the third skater is in the direction shown. In part (b), we see a free-body diagram representing the forces acting on the third skater.

Figure 4.3(b) is our first example of a **free-body diagram**, which is a technique used to illustrate all the **external forces** acting on a body. The body is represented by a single isolated point (or free body), and only those forces acting *on* the body from the outside (external forces) are shown. (These forces are the only ones shown, because only external forces acting on the body affect its motion. We can ignore any internal forces within the body.) Free-body diagrams are very useful in analyzing forces acting on a system and are employed extensively in the study and application of Newton's laws of motion.

A more quantitative definition of force can be based on some standard force, just as distance is measured in units relative to a standard distance. One possibility is to stretch a spring a certain fixed distance, as illustrated in **Figure 4.4**, and use the force it exerts to pull itself back to its relaxed shape—called a *restoring force*—as a standard. The magnitude of all other forces can be stated as multiples of this standard unit of force. Many other possibilities exist for standard forces. (One that we will encounter in **Magnetism** is the magnetic force between two wires carrying electric current.) Some alternative definitions of force will be given later in this chapter.

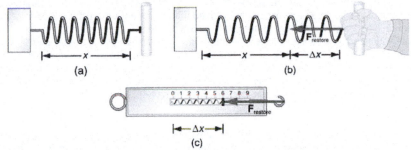

Figure 4.4 The force exerted by a stretched spring can be used as a standard unit of force. (a) This spring has a length x when undistorted. (b) When stretched a distance Δx, the spring exerts a restoring force, $\mathbf{F}_{restore}$, which is reproducible. (c) A spring scale is one device that uses a spring to measure force. The force $\mathbf{F}_{restore}$ is exerted on whatever is attached to the hook. Here $\mathbf{F}_{restore}$ has a magnitude of 6 units in the force standard being employed.

Take-Home Experiment: Force Standards

To investigate force standards and cause and effect, get two identical rubber bands. Hang one rubber band vertically on a hook. Find a small household item that could be attached to the rubber band using a paper clip, and use this item as a weight to investigate the stretch of the rubber band. Measure the amount of stretch produced in the rubber band with one, two, and four of these (identical) items suspended from the rubber band. What is the relationship between the number of items and the amount of stretch? How large a stretch would you expect for the same number of items suspended from two rubber bands? What happens to the amount of stretch of the rubber band (with the weights attached) if the weights are also pushed to the side with a pencil?

4.2 Newton's First Law of Motion: Inertia

Learning Objectives
By the end of this section, you will be able to: • Define mass and inertia. • Understand Newton's first law of motion.

Experience suggests that an object at rest will remain at rest if left alone, and that an object in motion tends to slow down and stop unless some effort is made to keep it moving. What **Newton's first law of motion** states, however, is the following:

Newton's First Law of Motion

There exists an inertial frame of reference such that a body at rest remains at rest, or, if in motion, remains in motion at a constant velocity unless acted on by a net external force.

Note the repeated use of the verb "remains." We can think of this law as preserving the status quo of motion.

The first law of motion postulates the existence of at least one frame of reference which we call an inertial reference frame, relative to which the motion of an object not subject to forces is a straight line at a constant speed. An inertial reference frame is any reference frame that is not itself accelerating. A car traveling at constant velocity is an inertial reference frame. A car slowing down for a stoplight, or speeding up after the light turns green, will be accelerating and is not an inertial reference frame. Finally, when the car goes around a turn, which is due to an acceleration changing the direction of the velocity vector, it is not an inertial reference frame. Note that Newton's laws of motion are only valid for inertial reference frames.

Rather than contradicting our experience, **Newton's first law of motion** states that there must be a *cause* (which is a net external force) *for there to be any change in velocity (either a change in magnitude or direction)* in an inertial reference frame. We will define *net external force* in the next section. An object sliding across a table or floor slows down due to the net force of friction acting on the object. If friction disappeared, would the object still slow down?

The idea of cause and effect is crucial in accurately describing what happens in various situations. For example, consider what happens to an object sliding along a rough horizontal surface. The object quickly grinds to a halt. If we spray the surface with talcum powder to make the surface smoother, the object slides farther. If we make the surface even smoother by rubbing lubricating oil on it, the object slides farther yet. Extrapolating to a frictionless surface, we can imagine the object sliding in a straight line indefinitely. Friction is thus the *cause* of the slowing (consistent with Newton's first law). The object would not slow down at all if friction were completely eliminated. Consider an air hockey table. When the air is turned off, the puck slides only a short distance before friction slows it to a stop. However, when the air is turned on, it creates a nearly frictionless surface, and the puck glides long distances without slowing down. Additionally, if we know enough about the friction, we can accurately predict how quickly the object will slow down. Friction is an external force.

Newton's first law is completely general and can be applied to anything from an object sliding on a table to a satellite in orbit to blood pumped from the heart. Experiments have thoroughly verified that any change in velocity (speed or direction) must be caused by an external force. The idea of *generally applicable or universal laws* is important not only here—it is a basic feature of all laws of physics. Identifying these laws is like recognizing patterns in nature from which further patterns can be discovered. The genius of Galileo, who first developed the idea for the first law, and Newton, who clarified it, was to ask the fundamental question, "What is the cause?" Thinking in terms of cause and effect is a worldview fundamentally different from the typical ancient Greek approach when questions such as "Why does a tiger have stripes?" would have been answered in Aristotelian fashion, "That is the nature of the beast." True perhaps, but not a useful insight.

Mass

The property of a body to remain at rest or to remain in motion with constant velocity is called **inertia**. Newton's first law is often called the **law of inertia**. As we know from experience, some objects have more inertia than others. It is obviously more difficult to change the motion of a large boulder than that of a basketball, for example. The inertia of an object is measured by its **mass**.

An object with a small mass will exhibit less inertia and be more affected by other objects. An object with a large mass will exhibit greater inertia and be less affected by other objects. This inertial mass of an object is a measure of how difficult it is to alter the uniform motion of the object by an external force.

Roughly speaking, mass is a measure of the amount of "stuff" (or matter) in something. The quantity or amount of matter in an object is determined by the numbers of atoms and molecules of various types it contains. Unlike weight, mass does not vary with location. The mass of an object is the same on Earth, in orbit, or on the surface of the Moon. In practice, it is very difficult to count and identify all of the atoms and molecules in an object, so masses are not often determined in this manner. Operationally, the masses of objects are determined by comparison with the standard kilogram.

Check Your Understanding

Which has more mass: a kilogram of cotton balls or a kilogram of gold?

Solution

They are equal. A kilogram of one substance is equal in mass to a kilogram of another substance. The quantities that might differ between them are volume and density.

4.3 Newton's Second Law of Motion: Concept of a System

Learning Objectives

By the end of this section, you will be able to:

- Define net force, external force, and system.
- Understand Newton's second law of motion.
- Apply Newton's second law to determine the weight of an object.

Newton's second law of motion is closely related to Newton's first law of motion. It mathematically states the cause and effect relationship between force and changes in motion. Newton's second law of motion is more quantitative and is used extensively to calculate what happens in situations involving a force. Before we can write down Newton's second law as a simple equation giving the exact relationship of force, mass, and acceleration, we need to sharpen some ideas that have already been mentioned.

First, what do we mean by a change in motion? The answer is that a change in motion is equivalent to a change in velocity. A change in velocity means, by definition, that there is an **acceleration**. Newton's first law says that a net external force causes a change in motion; thus, we see that a *net external force causes acceleration*.

Another question immediately arises. What do we mean by an external force? An intuitive notion of external is correct—an **external force** acts from outside the **system** of interest. For example, in **Figure 4.5**(a) the system of interest is the wagon plus the child in it. The two forces exerted by the other children are external forces. An internal force acts between elements of the system. Again looking at **Figure 4.5**(a), the force the child in the wagon exerts to hang onto the wagon is an internal force between elements of the system of interest. Only external forces affect the motion of a system, according to Newton's first law. (The internal forces actually cancel, as we shall see in the next section.) *You must define the boundaries of the system before you can determine which forces are external.* Sometimes the system is obvious, whereas other times identifying the boundaries of a system is more subtle. The concept of a system is fundamental to many areas of physics, as is the correct application of

Newton's laws. This concept will be revisited many times on our journey through physics.

When we describe the acceleration of a system, we are modeling the system as a single point which contains all of the mass of that system. The point we choose for this is the point about which the system's mass is evenly distributed. For example, in a rigid object, this center of mass is the point where the object will stay balanced even if only supported at this point. For a sphere or disk made of homogenous material, this point is of course at the center. Similarly, for a rod made of homogenous material, the center of mass will be at the midpoint.

For the rider in the wagon in **Figure 4.5**, the center of mass is probably between the rider's hips. Due to internal forces, the rider's hand or hair may accelerate slightly differently, but it is the acceleration of the system's center of mass that interests us. This is true whether the system is a vehicle carrying passengers, a bowl of grapes, or a planet. When we draw a free-body diagram of a system, we represent the system's center of mass with a single point and use vectors to indicate the forces exerted on that center of mass. (See **Figure 4.5**.)

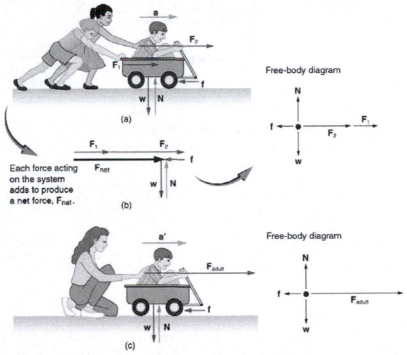

Figure 4.5 Different forces exerted on the same mass produce different accelerations. (a) Two children push a wagon with a child in it. Arrows representing all external forces are shown. The system of interest is the wagon and its rider. The weight $\mathbf{w}$ of the system and the support of the ground $\mathbf{N}$ are also shown for completeness and are assumed to cancel. The vector $\mathbf{f}$ represents the friction acting on the wagon, and it acts to the left, opposing the motion of the wagon. (b) All of the external forces acting on the system add together to produce a net force, $\mathbf{F}_{net}$. The free-body diagram shows all of the forces acting on the system of interest. The dot represents the center of mass of the system. Each force vector extends from this dot. Because there are two forces acting to the right, we draw the vectors collinearly. (c) A larger net external force produces a larger acceleration ($\mathbf{a}' > \mathbf{a}$) when an adult pushes the child.

Now, it seems reasonable that acceleration should be directly proportional to and in the same direction as the net (total) external force acting on a system. This assumption has been verified experimentally and is illustrated in **Figure 4.5**. In part (a), a smaller force causes a smaller acceleration than the larger force illustrated in part (c). For completeness, the vertical forces are also shown; they are assumed to cancel since there is no acceleration in the vertical direction. The vertical forces are the weight $\mathbf{w}$ and the support of the ground $\mathbf{N}$, and the horizontal force $\mathbf{f}$ represents the force of friction. These will be discussed in more detail in later sections. For now, we will define **friction** as a force that opposes the motion past each other of objects that are touching. **Figure 4.5**(b) shows how vectors representing the external forces add together to produce a net force, $\mathbf{F}_{net}$.

To obtain an equation for Newton's second law, we first write the relationship of acceleration and net external force as the proportionality

$$\mathbf{a} \propto \mathbf{F}_{net},$$ (4.1)

where the symbol $\propto$ means "proportional to," and $\mathbf{F}_{net}$ is the **net external force**. (The net external force is the vector sum of all external forces and can be determined graphically, using the head-to-tail method, or analytically, using components. The techniques are the same as for the addition of other vectors, and are covered in **Two-Dimensional Kinematics**.) This proportionality states what we have said in words—*acceleration is directly proportional to the net external force*. Once the system of interest is chosen, it is important to identify the external forces and ignore the internal ones. It is a tremendous simplification not to have to consider the numerous internal forces acting between objects within the system, such as muscular forces within the child's body, let alone the myriad of forces between atoms in the objects, but by doing so, we can easily solve some very complex problems with only minimal error due to our simplification

Now, it also seems reasonable that acceleration should be inversely proportional to the mass of the system. In other words, the larger the mass (the inertia), the smaller the acceleration produced by a given force. And indeed, as illustrated in **Figure 4.6**, the same net external force applied to a car produces a much smaller acceleration than when applied to a basketball. The proportionality is written as

$$\mathbf{a} \propto \frac{1}{m} \tag{4.2}$$

where m is the mass of the system. Experiments have shown that acceleration is exactly inversely proportional to mass, just as it is exactly linearly proportional to the net external force.

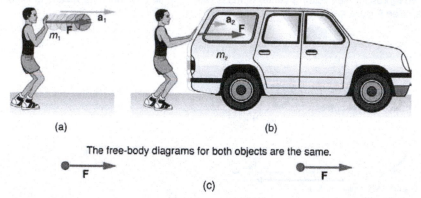

Figure 4.6 The same force exerted on systems of different masses produces different accelerations. (a) A basketball player pushes on a basketball to make a pass. (The effect of gravity on the ball is ignored.) (b) The same player exerts an identical force on a stalled SUV and produces a far smaller acceleration (even if friction is negligible). (c) The free-body diagrams are identical, permitting direct comparison of the two situations. A series of patterns for the free-body diagram will emerge as you do more problems.

Both of these proportionalities have been experimentally verified repeatedly and consistently, for a broad range of systems and scales. Thus, it has been **experimentally** found that the acceleration of an object depends *only* on the net external force and the mass of the object. Combining the two proportionalities just given yields Newton's second law of motion.

Applying the Science Practices: Testing the Relationship Between Mass, Acceleration, and Force

Plan three simple experiments using objects you have at home to test relationships between mass, acceleration, and force.

(a) Design an experiment to test the relationship between mass and acceleration. What will be the independent variable in your experiment? What will be the dependent variable? What controls will you put in place to ensure force is constant?

(b) Design a similar experiment to test the relationship between mass and force. What will be the independent variable in your experiment? What will be the dependent variable? What controls will you put in place to ensure acceleration is constant?

(c) Design a similar experiment to test the relationship between force and acceleration. What will be the independent variable in your experiment? What will be the dependent variable? Will you have any trouble ensuring that the mass is constant?

What did you learn?

Newton's Second Law of Motion

The acceleration of a system is directly proportional to and in the same direction as the net external force acting on the system, and inversely proportional to its mass.

In equation form, Newton's second law of motion is

$$\mathbf{a} = \frac{\mathbf{F}_{net}}{m}. \tag{4.3}$$

This is often written in the more familiar form

$$\mathbf{F}_{net} = m\mathbf{a}. \tag{4.4}$$

When only the magnitude of force and acceleration are considered, this equation is simply

$$F_{net} = ma. \tag{4.5}$$

Although these last two equations are really the same, the first gives more insight into what Newton's second law means. The law is a *cause and effect relationship* among three quantities that is not simply based on their definitions. The validity of the second law is completely based on experimental verification.

First, consider a person on a sled sliding downhill. What is the system in this situation? Try to draw a free-body diagram describing this system, labeling all the forces and their directions. Which of the forces are internal? Which are external?

Next, consider a person on a sled being pushed along level ground by a friend. What is the system in this situation? Try to draw a free-body diagram describing this system, labelling all the forces and their directions. Which of the forces are internal? Which are external?

Units of Force

$\mathbf{F}_{net} = m\mathbf{a}$ is used to define the units of force in terms of the three basic units for mass, length, and time. The SI unit of force is called the **newton** (abbreviated N) and is the force needed to accelerate a 1-kg system at the rate of 1m/s^2. That is, since $\mathbf{F}_{net} = m\mathbf{a}$,

$$1\text{ N} = 1\text{ kg} \cdot \text{m/s}^2. \tag{4.6}$$

While almost the entire world uses the newton for the unit of force, in the United States the most familiar unit of force is the pound (lb), where 1 N = 0.225 lb.

Weight and the Gravitational Force

When an object is dropped, it accelerates toward the center of Earth. Newton's second law states that a net force on an object is responsible for its acceleration. If air resistance is negligible, the net force on a falling object is the gravitational force, commonly called its **weight** $\mathbf{w}$. Weight can be denoted as a vector $\mathbf{w}$ because it has a direction; *down* is, by definition, the direction of gravity, and hence weight is a downward force. The magnitude of weight is denoted as w. Galileo was instrumental in showing that, in the absence of air resistance, all objects fall with the same acceleration g. Using Galileo's result and Newton's second law, we can derive an equation for weight.

Consider an object with mass m falling downward toward Earth. It experiences only the downward force of gravity, which has magnitude w. Newton's second law states that the magnitude of the net external force on an object is $F_{net} = ma$.

Since the object experiences only the downward force of gravity, $F_{net} = w$. We know that the acceleration of an object due to gravity is g, or $a = g$. Substituting these into Newton's second law gives

Weight

This is the equation for *weight*—the gravitational force on a mass m:

$$w = mg. \tag{4.7}$$

Since $g = 9.80\text{ m/s}^2$ on Earth, the weight of a 1.0 kg object on Earth is 9.8 N, as we see:

$$w = mg = (1.0\text{ kg})(9.80\text{ m/s}^2) = 9.8\text{ N}. \tag{4.8}$$

Recall that g can take a positive or negative value, depending on the positive direction in the coordinate system. Be sure to take this into consideration when solving problems with weight.

When the net external force on an object is its weight, we say that it is in **free-fall**. That is, the only force acting on the object is the force of gravity. In the real world, when objects fall downward toward Earth, they are never truly in free-fall because there is always some upward force from the air acting on the object.

The acceleration due to gravity g varies slightly over the surface of Earth, so that the weight of an object depends on location and is not an intrinsic property of the object. Weight varies dramatically if one leaves Earth's surface. On the Moon, for example, the acceleration due to gravity is only 1.67 m/s^2. A 1.0-kg mass thus has a weight of 9.8 N on Earth and only about 1.7 N on the Moon.

The broadest definition of weight in this sense is that *the weight of an object is the gravitational force on it from the nearest large body*, such as Earth, the Moon, the Sun, and so on. This is the most common and useful definition of weight in physics. It differs dramatically, however, from the definition of weight used by NASA and the popular media in relation to space travel and exploration. When they speak of "weightlessness" and "microgravity," they are really referring to the phenomenon we call "free-fall" in physics. We shall use the above definition of weight, and we will make careful distinctions between free-fall and actual weightlessness.

It is important to be aware that weight and mass are very different physical quantities, although they are closely related. Mass is the quantity of matter (how much "stuff") and does not vary in classical physics, whereas weight is the gravitational force and does vary depending on gravity. It is tempting to equate the two, since most of our examples take place on Earth, where the weight of an object only varies a little with the location of the object. Furthermore, the terms *mass* and *weight* are used interchangeably in everyday language; for example, our medical records often show our "weight" in kilograms, but never in the correct units of newtons.

Common Misconceptions: Mass vs. Weight

Mass and weight are often used interchangeably in everyday language. However, in science, these terms are distinctly different from one another. Mass is a measure of how much matter is in an object. The typical measure of mass is the kilogram (or the "slug" in English units). Weight, on the other hand, is a measure of the force of gravity acting on an object. Weight is equal to the mass of an object (m) multiplied by the acceleration due to gravity (g). Like any other force, weight is measured in terms of newtons (or pounds in English units).

Assuming the mass of an object is kept intact, it will remain the same, regardless of its location. However, because weight depends on the acceleration due to gravity, the weight of an object *can change* when the object enters into a region with stronger or weaker gravity. For example, the acceleration due to gravity on the Moon is 1.67 m/s^2 (which is much less than the acceleration due to gravity on Earth, 9.80 m/s^2). If you measured your weight on Earth and then measured your weight on the Moon, you would find that you "weigh" much less, even though you do not look any skinnier. This is because the force of gravity is weaker on the Moon. In fact, when people say that they are "losing weight," they really mean that they are losing "mass" (which in turn causes them to weigh less).

Take-Home Experiment: Mass and Weight

What do bathroom scales measure? When you stand on a bathroom scale, what happens to the scale? It depresses slightly. The scale contains springs that compress in proportion to your weight—similar to rubber bands expanding when pulled. The springs provide a measure of your weight (for an object which is not accelerating). This is a force in newtons (or pounds). In most countries, the measurement is divided by 9.80 to give a reading in mass units of kilograms. The scale measures weight but is calibrated to provide information about mass. While standing on a bathroom scale, push down on a table next to you. What happens to the reading? Why? Would your scale measure the same "mass" on Earth as on the Moon?

Example 4.1 What Acceleration Can a Person Produce when Pushing a Lawn Mower?

Suppose that the net external force (push minus friction) exerted on a lawn mower is 51 N (about 11 lb) parallel to the ground. The mass of the mower is 24 kg. What is its acceleration?

Figure 4.7 The net force on a lawn mower is 51 N to the right. At what rate does the lawn mower accelerate to the right?

Strategy

Since $\mathbf{F}_{\text{net}}$ and m are given, the acceleration can be calculated directly from Newton's second law as stated in $\mathbf{F}_{\text{net}} = m\mathbf{a}$.

Solution

The magnitude of the acceleration a is $a = \dfrac{F_{\text{net}}}{m}$. Entering known values gives

$$a = \frac{51 \text{ N}}{24 \text{ kg}} \tag{4.9}$$

Substituting the units $\text{kg} \cdot \text{m/s}^2$ for N yields

$$a = \frac{51 \text{ kg} \cdot \text{m/s}^2}{24 \text{ kg}} = 2.1 \text{ m/s}^2. \tag{4.10}$$

Discussion

The direction of the acceleration is the same direction as that of the net force, which is parallel to the ground. There is no information given in this example about the individual external forces acting on the system, but we can say something about their relative magnitudes. For example, the force exerted by the person pushing the mower must be greater than the friction opposing the motion (since we know the mower moves forward), and the vertical forces must cancel if there is to be no acceleration in the vertical direction (the mower is moving only horizontally). The acceleration found is small enough to be reasonable for a person pushing a mower. Such an effort would not last too long because the person's top speed would soon be reached.

Example 4.2 What Rocket Thrust Accelerates This Sled?

Prior to manned space flights, rocket sleds were used to test aircraft, missile equipment, and physiological effects on human subjects at high speeds. They consisted of a platform that was mounted on one or two rails and propelled by several rockets. Calculate the magnitude of force exerted by each rocket, called its thrust $\mathbf{T}$, for the four-rocket propulsion system shown in Figure 4.8. The sled's initial acceleration is 49 m/s^2, the mass of the system is 2100 kg, and the force of friction opposing the motion is known to be 650 N.

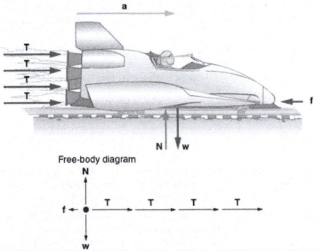

Free-body diagram

Figure 4.8 A sled experiences a rocket thrust that accelerates it to the right. Each rocket creates an identical thrust $\mathbf{T}$. As in other situations where there is only horizontal acceleration, the vertical forces cancel. The ground exerts an upward force $\mathbf{N}$ on the system that is equal in magnitude and opposite in direction to its weight, $\mathbf{w}$. The system here is the sled, its rockets, and rider, so none of the forces *between* these objects are considered. The arrow representing friction ($\mathbf{f}$) is drawn larger than scale.

Strategy

Although there are forces acting vertically and horizontally, we assume the vertical forces cancel since there is no vertical acceleration. This leaves us with only horizontal forces and a simpler one-dimensional problem. Directions are indicated with plus or minus signs, with right taken as the positive direction. See the free-body diagram in the figure.

Solution

Since acceleration, mass, and the force of friction are given, we start with Newton's second law and look for ways to find the thrust of the engines. Since we have defined the direction of the force and acceleration as acting "to the right," we need to consider only the magnitudes of these quantities in the calculations. Hence we begin with

$$F_{\text{net}} = ma, \tag{4.11}$$

where F_{net} is the net force along the horizontal direction. We can see from Figure 4.8 that the engine thrusts add, while friction opposes the thrust. In equation form, the net external force is

$$F_{\text{net}} = 4T - f. \tag{4.12}$$

Substituting this into Newton's second law gives

$$F_{net} = ma = 4T - f.$$ (4.13)

Using a little algebra, we solve for the total thrust $4T$:

$$4T = ma + f.$$ (4.14)

Substituting known values yields

$$4T = ma + f = (2100 \text{ kg})(49 \text{ m/s}^2) + 650 \text{ N}.$$ (4.15)

So the total thrust is

$$4T = 1.0 \times 10^5 \text{ N},$$ (4.16)

and the individual thrusts are

$$T = \frac{1.0 \times 10^5 \text{ N}}{4} = 2.6 \times 10^4 \text{ N}.$$ (4.17)

Discussion

The numbers are quite large, so the result might surprise you. Experiments such as this were performed in the early 1960s to test the limits of human endurance and the setup designed to protect human subjects in jet fighter emergency ejections. Speeds of 1000 km/h were obtained, with accelerations of 45 g 's. (Recall that g , the acceleration due to gravity, is

9.80 m/s^2 . When we say that an acceleration is 45 g 's, it is $45 \times 9.80 \text{ m/s}^2$, which is approximately 440 m/s^2 .) While living subjects are not used any more, land speeds of 10,000 km/h have been obtained with rocket sleds. In this example, as in the preceding one, the system of interest is obvious. We will see in later examples that choosing the system of interest is crucial—and the choice is not always obvious.

Newton's second law of motion is more than a definition; it is a relationship among acceleration, force, and mass. It can help us make predictions. Each of those physical quantities can be defined independently, so the second law tells us something basic and universal about nature. The next section introduces the third and final law of motion.

Applying the Science Practices: Sums of Forces

Recall that forces are vector quantities, and therefore the net force acting on a system should be the vector sum of the forces.

(a) Design an experiment to test this hypothesis. What sort of a system would be appropriate and convenient to have multiple forces applied to it? What features of the system should be held constant? What could be varied? Can forces be arranged in multiple directions so that, while the hypothesis is still tested, the resulting calculations are not too inconvenient?

(b) Another group of students has done such an experiment, using a motion capture system looking down at an air hockey table to measure the motion of the 0.10-kg puck. The table was aligned with the cardinal directions, and a compressed air hose was placed in the center of each side, capable of varying levels of force output and fixed so that it was aimed at the center of the table.

Table 4.1

Forces	Measured acceleration (magnitudes)
3 N north, 4 N west	$48 \pm 4 \text{ m/s}^2$
5 N south, 12 N east	$132 \pm 6 \text{ m/s}^2$
6 N north, 12 N east, 4 N west	$99 \pm 3 \text{ m/s}^2$

Given the data in the table, is the hypothesis confirmed? What were the directions of the accelerations?

4.4 Newton's Third Law of Motion: Symmetry in Forces

Learning Objectives

By the end of this section, you will be able to:

* Understand Newton's third law of motion.
* Apply Newton's third law to define systems and solve problems of motion.

The information presented in this section supports the following AP® learning objectives and science practices:

- **3.A.2.1** The student is able to represent forces in diagrams or mathematically using appropriately labeled vectors with magnitude, direction, and units during the analysis of a situation. **(S.P. 1.1)**
- **3.A.3.1** The student is able to analyze a scenario and make claims (develop arguments, justify assertions) about the forces exerted on an object by other objects for different types of forces or components of forces. **(S.P. 6.4, 7.2)**
- **3.A.3.3** The student is able to describe a force as an interaction between two objects and identify both objects for any force. **(S.P. 1.4)**
- **3.A.4.1** The student is able to construct explanations of physical situations involving the interaction of bodies using Newton's third law and the representation of action-reaction pairs of forces. **(S.P. 1.4, 6.2)**
- **3.A.4.2** The student is able to use Newton's third law to make claims and predictions about the action-reaction pairs of forces when two objects interact. **(S.P. 6.4, 7.2)**
- **3.A.4.3** The student is able to analyze situations involving interactions among several objects by using free-body diagrams that include the application of Newton's third law to identify forces. **(S.P. 1.4)**
- **3.B.2.1** The student is able to create and use free-body diagrams to analyze physical situations to solve problems with motion qualitatively and quantitatively. **(S.P. 1.1, 1.4, 2.2)**
- **4.A.2.1** The student is able to make predictions about the motion of a system based on the fact that acceleration is equal to the change in velocity per unit time, and velocity is equal to the change in position per unit time. **(S.P. 6.4)**
- **4.A.2.2** The student is able to evaluate using given data whether all the forces on a system or whether all the parts of a system have been identified. **(S.P. 5.3)**
- **4.A.3.1** The student is able to apply Newton's second law to systems to calculate the change in the center-of-mass velocity when an external force is exerted on the system. **(S.P. 2.2)**

There is a passage in the musical *Man of la Mancha* that relates to Newton's third law of motion. Sancho, in describing a fight with his wife to Don Quixote, says, "Of course I hit her back, Your Grace, but she's a lot harder than me and you know what they say, 'Whether the stone hits the pitcher or the pitcher hits the stone, it's going to be bad for the pitcher.'" This is exactly what happens whenever one body exerts a force on another—the first also experiences a force (equal in magnitude and opposite in direction). Numerous common experiences, such as stubbing a toe or throwing a ball, confirm this. It is precisely stated in **Newton's third law of motion**.

Newton's Third Law of Motion

Whenever one body exerts a force on a second body, the first body experiences a force that is equal in magnitude and opposite in direction to the force that it exerts.

This law represents a certain *symmetry in nature*: Forces always occur in pairs, and one body cannot exert a force on another without experiencing a force itself. We sometimes refer to this law loosely as "action-reaction," where the force exerted is the action and the force experienced as a consequence is the reaction. Newton's third law has practical uses in analyzing the origin of forces and understanding which forces are external to a system.

We can readily see Newton's third law at work by taking a look at how people move about. Consider a swimmer pushing off from the side of a pool, as illustrated in **Figure 4.9**. She pushes against the pool wall with her feet and accelerates in the direction *opposite* to that of her push. The wall has exerted an equal and opposite force back on the swimmer. You might think that two equal and opposite forces would cancel, but they do not *because they act on different systems*. In this case, there are two systems that we could investigate: the swimmer or the wall. If we select the swimmer to be the system of interest, as in the figure, then $\mathbf{F}_{\text{wall on feet}}$ is an external force on this system and affects its motion. The swimmer moves in the direction of

$\mathbf{F}_{\text{wall on feet}}$. In contrast, the force $\mathbf{F}_{\text{feet on wall}}$ acts on the wall and not on our system of interest. Thus $\mathbf{F}_{\text{feet on wall}}$ does not directly affect the motion of the system and does not cancel $\mathbf{F}_{\text{wall on feet}}$. Note that the swimmer pushes in the direction opposite to that in which she wishes to move. The reaction to her push is thus in the desired direction.

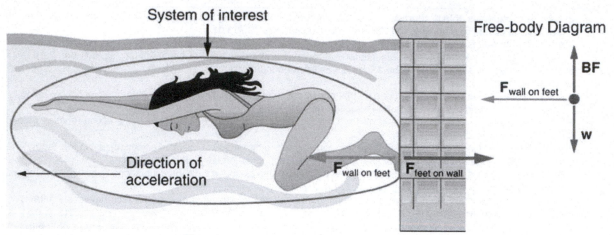

Figure 4.9 When the swimmer exerts a force $\mathbf{F}_{\text{feet on wall}}$ on the wall, she accelerates in the direction opposite to that of her push. This means the net external force on her is in the direction opposite to $\mathbf{F}_{\text{feet on wall}}$. This opposition occurs because, in accordance with Newton's third law of motion, the wall exerts a force $\mathbf{F}_{\text{wall on feet}}$ on her, equal in magnitude but in the direction opposite to the one she exerts on it. The line around the swimmer indicates the system of interest. Note that $\mathbf{F}_{\text{feet on wall}}$ does not act on this system (the swimmer) and, thus, does not cancel $\mathbf{F}_{\text{wall on feet}}$. Thus the free-body diagram shows only $\mathbf{F}_{\text{wall on feet}}$, $\mathbf{w}$, the gravitational force, and $\mathbf{BF}$, the buoyant force of the water supporting the swimmer's weight. The vertical forces $\mathbf{w}$ and $\mathbf{BF}$ cancel since there is no vertical motion.

Similarly, when a person stands on Earth, the Earth exerts a force on the person, pulling the person toward the Earth. As stated by Newton's third law of motion, the person also exerts a force that is equal in magnitude, but opposite in direction, pulling the Earth up toward the person. Since the mass of the Earth is so great, however, and $F = ma$, the acceleration of the Earth toward the person is not noticeable.

Other examples of Newton's third law are easy to find. As a professor paces in front of a whiteboard, she exerts a force backward on the floor. The floor exerts a reaction force forward on the professor that causes her to accelerate forward. Similarly, a car accelerates because the ground pushes forward on the drive wheels in reaction to the drive wheels pushing backward on the ground. You can see evidence of the wheels pushing backward when tires spin on a gravel road and throw rocks backward. In another example, rockets move forward by expelling gas backward at high velocity. This means the rocket exerts a large backward force on the gas in the rocket combustion chamber, and the gas therefore exerts a large reaction force forward on the rocket. This reaction force is called **thrust**. It is a common misconception that rockets propel themselves by pushing on the ground or on the air behind them. They actually work better in a vacuum, where they can more readily expel the exhaust gases. Helicopters similarly create lift by pushing air down, thereby experiencing an upward reaction force. Birds and airplanes also fly by exerting force on air in a direction opposite to that of whatever force they need. For example, the wings of a bird force air downward and backward in order to get lift and move forward. An octopus propels itself in the water by ejecting water through a funnel from its body, similar to a jet ski. In a situation similar to Sancho's, professional cage fighters experience reaction forces when they punch, sometimes breaking their hand by hitting an opponent's body.

Example 4.3 Getting Up To Speed: Choosing the Correct System

A physics professor pushes a cart of demonstration equipment to a lecture hall, as seen in **Figure 4.10**. Her mass is 65.0 kg, the cart's is 12.0 kg, and the equipment's is 7.0 kg. Calculate the acceleration produced when the professor exerts a backward force of 150 N on the floor. All forces opposing the motion, such as friction on the cart's wheels and air resistance, total 24.0 N.

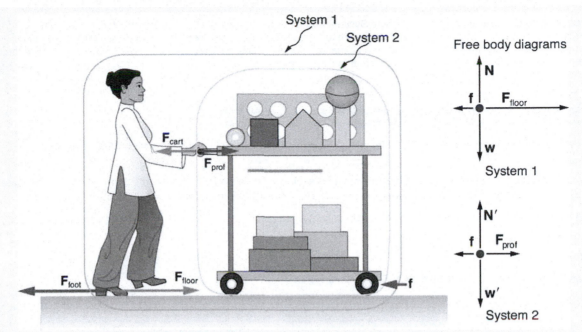

Figure 4.10 A professor pushes a cart of demonstration equipment. The lengths of the arrows are proportional to the magnitudes of the forces (except for $\mathbf{f}$, since it is too small to draw to scale). Different questions are asked in each example; thus, the system of interest must be defined differently for each. System 1 is appropriate for this example, since it asks for the acceleration of the entire group of objects. Only $\mathbf{F}_{\text{floo}}$ and $\mathbf{f}$ are external forces acting on System 1 along the line of motion. All other forces either cancel or act on the outside world. System 2 is chosen for Example 4.4 so that $\mathbf{F}_{\text{prof}}$ will be an external force and enter into Newton's second law. Note that the free-body diagrams, which allow us to apply Newton's second law, vary with the system chosen.

Strategy

Since they accelerate as a unit, we define the system to be the professor, cart, and equipment. This is System 1 in **Figure 4.10**. The professor pushes backward with a force $\mathbf{F}_{\text{foot}}$ of 150 N. According to Newton's third law, the floor exerts a forward reaction force $\mathbf{F}_{\text{floo}}$ of 150 N on System 1. Because all motion is horizontal, we can assume there is no net force in the vertical direction. The problem is therefore one-dimensional along the horizontal direction. As noted, $\mathbf{f}$ opposes the motion and is thus in the opposite direction of $\mathbf{F}_{\text{floo}}$. Note that we do not include the forces $\mathbf{F}_{\text{prof}}$ or $\mathbf{F}_{\text{cart}}$ because these are internal forces, and we do not include $\mathbf{F}_{\text{foot}}$ because it acts on the floor, not on the system. There are no other significant forces acting on System 1. If the net external force can be found from all this information, we can use Newton's second law to find the acceleration as requested. See the free-body diagram in the figure.

Solution

Newton's second law is given by

$$a = \frac{F_{\text{net}}}{m}. \tag{4.18}$$

The net external force on System 1 is deduced from **Figure 4.10** and the discussion above to be

$$F_{\text{net}} = F_{\text{floo}} - f = 150\,\text{N} - 24.0\,\text{N} = 126\,\text{N}. \tag{4.19}$$

The mass of System 1 is

$$m = (65.0 + 12.0 + 7.0)\,\text{kg} = 84\,\text{kg}. \tag{4.20}$$

These values of F_{net} and m produce an acceleration of

$$a = \frac{F_{\text{net}}}{m}, \tag{4.21}$$
$$a = \frac{126\,\text{N}}{84\,\text{kg}} = 1.5\,\text{m/s}^2.$$

Discussion

None of the forces between components of System 1, such as between the professor's hands and the cart, contribute to the net external force because they are internal to System 1. Another way to look at this is to note that forces between components of a system cancel because they are equal in magnitude and opposite in direction. For example, the force exerted by the professor on the cart results in an equal and opposite force back on her. In this case both forces act on the same system and, therefore, cancel. Thus internal forces (between components of a system) cancel. Choosing System 1 was crucial to solving this problem.

Example 4.4 Force on the Cart—Choosing a New System

Calculate the force the professor exerts on the cart in **Figure 4.10** using data from the previous example if needed.

Strategy

If we now define the system of interest to be the cart plus equipment (System 2 in **Figure 4.10**), then the net external force on System 2 is the force the professor exerts on the cart minus friction. The force she exerts on the cart, $\mathbf{F}_{prof}$, is an external force acting on System 2. $\mathbf{F}_{prof}$ was internal to System 1, but it is external to System 2 and will enter Newton's second law for System 2.

Solution

Newton's second law can be used to find $\mathbf{F}_{prof}$. Starting with

$$a = \frac{F_{net}}{m} \tag{4.22}$$

and noting that the magnitude of the net external force on System 2 is

$$F_{net} = F_{prof} - f, \tag{4.23}$$

we solve for F_{prof}, the desired quantity:

$$F_{prof} = F_{net} + f. \tag{4.24}$$

The value of f is given, so we must calculate net F_{net}. That can be done since both the acceleration and mass of System 2 are known. Using Newton's second law we see that

$$F_{net} = ma, \tag{4.25}$$

where the mass of System 2 is 19.0 kg (m = 12.0 kg + 7.0 kg) and its acceleration was found to be $a = 1.5$ m/s^2 in the previous example. Thus,

$$F_{net} = ma, \tag{4.26}$$

$$F_{net} = (19.0 \text{ kg})(1.5 \text{ m/s}^2) = 29 \text{ N}. \tag{4.27}$$

Now we can find the desired force:

$$F_{prof} = F_{net} + f, \tag{4.28}$$

$$F_{prof} = 29 \text{ N} + 24.0 \text{ N} = 53 \text{ N}. \tag{4.29}$$

Discussion

It is interesting that this force is significantly less than the 150-N force the professor exerted backward on the floor. Not all of that 150-N force is transmitted to the cart; some of it accelerates the professor.

The choice of a system is an important analytical step both in solving problems and in thoroughly understanding the physics of the situation (which is not necessarily the same thing).

PhET Explorations: Gravity Force Lab

Visualize the gravitational force that two objects exert on each other. Change properties of the objects in order to see how it changes the gravity force.

Figure 4.11 Gravity Force Lab (http://cnx.org/content/m54849/1.3/gravity-force-lab_en.jar)

4.5 Normal, Tension, and Other Examples of Force

Learning Objectives

By the end of this section, you will be able to:

- Define normal and tension forces.
- Apply Newton's laws of motion to solve problems involving a variety of forces.
- Use trigonometric identities to resolve weight into components.

The information presented in this section supports the following AP® learning objectives and science practices:

- **2.B.1.1** The student is able to apply $F = mg$ to calculate the gravitational force on an object with mass m in a gravitational field of strength g in the context of the effects of a net force on objects and systems. **(S.P. 2.2, 7.2)**
- **3.A.2.1** The student is able to represent forces in diagrams or mathematically using appropriately labeled vectors with magnitude, direction, and units during the analysis of a situation. **(S.P. 1.1)**
- **3.A.3.1** The student is able to analyze a scenario and make claims (develop arguments, justify assertions) about the forces exerted on an object by other objects for different types of forces or components of forces. **(S.P. 6.4, 7.2)**
- **3.A.3.3** The student is able to describe a force as an interaction between two objects and identify both objects for any force. **(S.P. 1.4)**
- **3.A.4.1** The student is able to construct explanations of physical situations involving the interaction of bodies using Newton's third law and the representation of action-reaction pairs of forces. **(S.P. 1.4, 6.2)**
- **3.A.4.2** The student is able to use Newton's third law to make claims and predictions about the action-reaction pairs of forces when two objects interact. **(S.P. 6.4, 7.2)**
- **3.A.4.3** The student is able to analyze situations involving interactions among several objects by using free-body diagrams that include the application of Newton's third law to identify forces. **(S.P. 1.4)**
- **3.B.1.3** The student is able to re-express a free-body diagram representation into a mathematical representation and solve the mathematical representation for the acceleration of the object. **(S.P. 1.5, 2.2)**
- **3.B.2.1** The student is able to create and use free-body diagrams to analyze physical situations to solve problems with motion qualitatively and quantitatively. **(S.P. 1.1, 1.4, 2.2)**

Forces are given many names, such as push, pull, thrust, lift, weight, friction, and tension. Traditionally, forces have been grouped into several categories and given names relating to their source, how they are transmitted, or their effects. The most important of these categories are discussed in this section, together with some interesting applications. Further examples of forces are discussed later in this text.

Normal Force

Weight (also called force of gravity) is a pervasive force that acts at all times and must be counteracted to keep an object from falling. You definitely notice that you must support the weight of a heavy object by pushing up on it when you hold it stationary, as illustrated in **Figure 4.12**(a). But how do inanimate objects like a table support the weight of a mass placed on them, such as shown in **Figure 4.12**(b)? When the bag of dog food is placed on the table, the table actually sags slightly under the load. This would be noticeable if the load were placed on a card table, but even rigid objects deform when a force is applied to them. Unless the object is deformed beyond its limit, it will exert a restoring force much like a deformed spring (or trampoline or diving board). The greater the deformation, the greater the restoring force. So when the load is placed on the table, the table sags until the restoring force becomes as large as the weight of the load. At this point the net external force on the load is zero. That is the situation when the load is stationary on the table. The table sags quickly, and the sag is slight so we do not notice it. But it is similar to the sagging of a trampoline when you climb onto it.

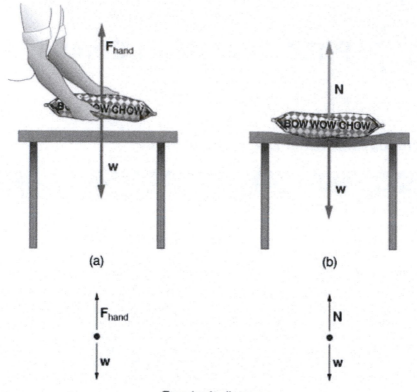

Free-body diagrams

Figure 4.12 (a) The person holding the bag of dog food must supply an upward force $\mathbf{F}_{hand}$ equal in magnitude and opposite in direction to the weight of the food $\mathbf{W}$. (b) The card table sags when the dog food is placed on it, much like a stiff trampoline. Elastic restoring forces in the table grow as it sags until they supply a force $\mathbf{N}$ equal in magnitude and opposite in direction to the weight of the load.

We must conclude that whatever supports a load, be it animate or not, must supply an upward force equal to the weight of the load, as we assumed in a few of the previous examples. If the force supporting a load is perpendicular to the surface of contact between the load and its support, this force is defined to be a **normal force** and here is given the symbol $\mathbf{N}$. (This is not the unit for force N.) The word *normal* means perpendicular to a surface. The normal force can be less than the object's weight if the object is on an incline, as you will see in the next example.

Common Misconception: Normal Force (N) vs. Newton (N)

In this section we have introduced the quantity normal force, which is represented by the variable $\mathbf{N}$. This should not be confused with the symbol for the newton, which is also represented by the letter N. These symbols are particularly important to distinguish because the units of a normal force ($\mathbf{N}$) happen to be newtons (N). For example, the normal force $\mathbf{N}$ that the floor exerts on a chair might be $\mathbf{N} = 100\,\text{N}$. One important difference is that normal force is a vector, while the newton is simply a unit. Be careful not to confuse these letters in your calculations! You will encounter more similarities among variables and units as you proceed in physics. Another example of this is the quantity work (W) and the unit watts (W).

Example 4.5 Weight on an Incline, a Two-Dimensional Problem

Consider the skier on a slope shown in **Figure 4.13**. Her mass including equipment is 60.0 kg. (a) What is her acceleration if friction is negligible? (b) What is her acceleration if friction is known to be 45.0 N?

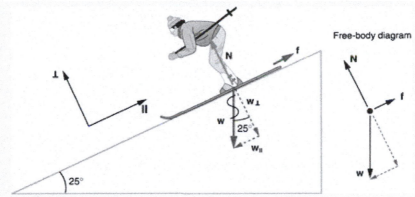

Figure 4.13 Since motion and friction are parallel to the slope, it is most convenient to project all forces onto a coordinate system where one axis is parallel to the slope and the other is perpendicular (axes shown to left of skier). $\mathbf{N}$ is perpendicular to the slope and $\mathbf{f}$ is parallel to the slope, but $\mathbf{W}$ has components along both axes, namely $\mathbf{W}_\perp$ and $\mathbf{W}_\parallel$. $\mathbf{N}$ is equal in magnitude to $\mathbf{W}_\perp$, so that there is no motion perpendicular to the slope, but f is less than $w_\parallel$, so that there is a downslope acceleration (along the parallel axis).

Strategy

This is a two-dimensional problem, since the forces on the skier (the system of interest) are not parallel. The approach we have used in two-dimensional kinematics also works very well here. Choose a convenient coordinate system and project the vectors onto its axes, creating *two* connected *one*-dimensional problems to solve. The most convenient coordinate system for motion on an incline is one that has one coordinate parallel to the slope and one perpendicular to the slope. (Remember that motions along mutually perpendicular axes are independent.) We use the symbols $\perp$ and $\parallel$ to represent perpendicular and parallel, respectively. This choice of axes simplifies this type of problem, because there is no motion perpendicular to the slope and because friction is always parallel to the surface between two objects. The only external forces acting on the system are the skier's weight, friction, and the support of the slope, respectively labeled $\mathbf{w}$, $\mathbf{f}$, and $\mathbf{N}$ in **Figure 4.13**. $\mathbf{N}$ is always perpendicular to the slope, and $\mathbf{f}$ is parallel to it. But $\mathbf{w}$ is not in the direction of either axis, and so the first step we take is to project it into components along the chosen axes, defining $w_\parallel$ to be the component of weight parallel to the slope and $w_\perp$ the component of weight perpendicular to the slope. Once this is done, we can consider the two separate problems of forces parallel to the slope and forces perpendicular to the slope.

Solution

The magnitude of the component of the weight parallel to the slope is $w_\parallel = w \sin (25°) = mg \sin (25°)$, and the magnitude of the component of the weight perpendicular to the slope is $w_\perp = w \cos (25°) = mg \cos (25°)$.

(a) Neglecting friction. Since the acceleration is parallel to the slope, we need only consider forces parallel to the slope. (Forces perpendicular to the slope add to zero, since there is no acceleration in that direction.) The forces parallel to the slope are the amount of the skier's weight parallel to the slope $w_\parallel$ and friction f. Using Newton's second law, with subscripts to denote quantities parallel to the slope,

$$a_\parallel = \frac{F_{\text{net} \parallel}}{m} \qquad (4.30)$$

where $F_{\text{net} \parallel} = w_\parallel = mg \sin (25°)$, assuming no friction for this part, so that

$$a_\parallel = \frac{F_{\text{net} \parallel}}{m} = \frac{mg \sin (25°)}{m} = g \sin (25°) \qquad (4.31)$$

$$(9.80 \text{ m/s}^2)(0.4226) = 4.14 \text{ m/s}^2 \qquad (4.32)$$

is the acceleration.

(b) Including friction. We now have a given value for friction, and we know its direction is parallel to the slope and it opposes motion between surfaces in contact. So the net external force is now

$$F_{\text{net} \parallel} = w_\parallel - f, \qquad (4.33)$$

and substituting this into Newton's second law, $a_\parallel = \dfrac{F_{\text{net} \parallel}}{m}$, gives

$$a_{\|} = \frac{F_{\text{net} \|}}{m} = \frac{w_{\|} - f}{m} = \frac{mg \sin{(25°)} - f}{m}. \tag{4.34}$$

We substitute known values to obtain

$$a_{\|} = \frac{(60.0 \text{ kg})(9.80 \text{ m/s}^2)(0.4226) - 45.0 \text{ N}}{60.0 \text{ kg}}, \tag{4.35}$$

which yields

$$a_{\|} = 3.39 \text{ m/s}^2, \tag{4.36}$$

which is the acceleration parallel to the incline when there is 45.0 N of opposing friction.

Discussion

Since friction always opposes motion between surfaces, the acceleration is smaller when there is friction than when there is none. In fact, it is a general result that if friction on an incline is negligible, then the acceleration down the incline is $a = g \sin\theta$, *regardless of mass*. This is related to the previously discussed fact that all objects fall with the same acceleration in the absence of air resistance. Similarly, all objects, regardless of mass, slide down a frictionless incline with the same acceleration (if the angle is the same).

Resolving Weight into Components

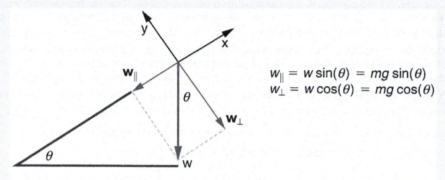

Figure 4.14 An object rests on an incline that makes an angle θ with the horizontal.

When an object rests on an incline that makes an angle θ with the horizontal, the force of gravity acting on the object is divided into two components: a force acting perpendicular to the plane, $\mathbf{w}_{\perp}$, and a force acting parallel to the plane, $\mathbf{w}_{\|}$. The perpendicular force of weight, $\mathbf{w}_{\perp}$, is typically equal in magnitude and opposite in direction to the normal force, $\mathbf{N}$. The force acting parallel to the plane, $\mathbf{w}_{\|}$, causes the object to accelerate down the incline. The force of friction, $\mathbf{f}$, opposes the motion of the object, so it acts upward along the plane.

It is important to be careful when resolving the weight of the object into components. If the angle of the incline is at an angle θ to the horizontal, then the magnitudes of the weight components are

$$w_{\|} = w \sin{(\theta)} = mg \sin{(\theta)} \tag{4.37}$$

and

$$w_{\perp} = w \cos{(\theta)} = mg \cos{(\theta)}. \tag{4.38}$$

Instead of memorizing these equations, it is helpful to be able to determine them from reason. To do this, draw the right triangle formed by the three weight vectors. Notice that the angle θ of the incline is the same as the angle formed between $\mathbf{w}$ and $\mathbf{w}_{\perp}$. Knowing this property, you can use trigonometry to determine the magnitude of the weight components:

$$\cos{(\theta)} = \frac{w_{\perp}}{w} \tag{4.39}$$

$$w_{\perp} = w \cos{(\theta)} = mg \cos{(\theta)}$$

$$\sin{(\theta)} = \frac{w_{\|}}{w} \tag{4.40}$$

$$w_{\|} = w \sin{(\theta)} = mg \sin{(\theta)}$$

Take-Home Experiment: Force Parallel

To investigate how a force parallel to an inclined plane changes, find a rubber band, some objects to hang from the end of the rubber band, and a board you can position at different angles. How much does the rubber band stretch when you hang the object from the end of the board? Now place the board at an angle so that the object slides off when placed on the board. How much does the rubber band extend if it is lined up parallel to the board and used to hold the object stationary on the board? Try two more angles. What does this show?

Tension

A **tension** is a force along the length of a medium, especially a force carried by a flexible medium, such as a rope or cable. The word "tension" comes from a Latin word meaning "to stretch." Not coincidentally, the flexible cords that carry muscle forces to other parts of the body are called *tendons*. Any flexible connector, such as a string, rope, chain, wire, or cable, can exert pulls only parallel to its length; thus, a force carried by a flexible connector is a tension with direction parallel to the connector. It is important to understand that tension is a pull in a connector. In contrast, consider the phrase: "You can't push a rope." The tension force pulls outward along the two ends of a rope.

Consider a person holding a mass on a rope as shown in **Figure 4.15**.

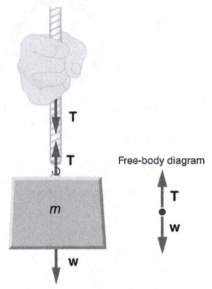

Figure 4.15 When a perfectly flexible connector (one requiring no force to bend it) such as this rope transmits a force $\mathbf{T}$, that force must be parallel to the length of the rope, as shown. The pull such a flexible connector exerts is a tension. Note that the rope pulls with equal force but in opposite directions on the hand and the supported mass (neglecting the weight of the rope). This is an example of Newton's third law. The rope is the medium that carries the equal and opposite forces between the two objects. The tension anywhere in the rope between the hand and the mass is equal. Once you have determined the tension in one location, you have determined the tension at all locations along the rope.

Tension in the rope must equal the weight of the supported mass, as we can prove using Newton's second law. If the 5.00-kg mass in the figure is stationary, then its acceleration is zero, and thus $\mathbf{F}_{\text{net}} = 0$. The only external forces acting on the mass are its weight $\mathbf{w}$ and the tension $\mathbf{T}$ supplied by the rope. Thus,

$$F_{\text{net}} = T - w = 0, \tag{4.41}$$

where T and w are the magnitudes of the tension and weight and their signs indicate direction, with up being positive here. Thus, just as you would expect, the tension equals the weight of the supported mass:

$$T = w = mg. \tag{4.42}$$

For a 5.00-kg mass, then (neglecting the mass of the rope) we see that

$$T = mg = (5.00 \text{ kg})(9.80 \text{ m/s}^2) = 49.0 \text{ N}. \tag{4.43}$$

If we cut the rope and insert a spring, the spring would extend a length corresponding to a force of 49.0 N, providing a direct observation and measure of the tension force in the rope.

Flexible connectors are often used to transmit forces around corners, such as in a hospital traction system, a finger joint, or a bicycle brake cable. If there is no friction, the tension is transmitted undiminished. Only its direction changes, and it is always parallel to the flexible connector. This is illustrated in **Figure 4.16** (a) and (b).

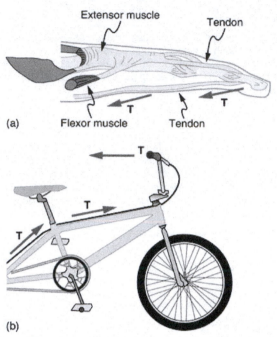

Figure 4.16 (a) Tendons in the finger carry force $\mathbf{T}$ from the muscles to other parts of the finger, usually changing the force's direction, but not its magnitude (the tendons are relatively friction free). (b) The brake cable on a bicycle carries the tension $\mathbf{T}$ from the handlebars to the brake mechanism. Again, the direction but not the magnitude of $\mathbf{T}$ is changed.

Example 4.6 What Is the Tension in a Tightrope?

Calculate the tension in the wire supporting the 70.0-kg tightrope walker shown in **Figure 4.17**.

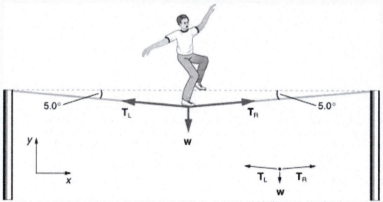

Figure 4.17 The weight of a tightrope walker causes a wire to sag by 5.0 degrees. The system of interest here is the point in the wire at which the tightrope walker is standing.

Strategy

As you can see in the figure, the wire is not perfectly horizontal (it cannot be!), but is bent under the person's weight. Thus, the tension on either side of the person has an upward component that can support his weight. As usual, forces are vectors represented pictorially by arrows having the same directions as the forces and lengths proportional to their magnitudes. The system is the tightrope walker, and the only external forces acting on him are his weight $\mathbf{w}$ and the two tensions $\mathbf{T}_L$ (left tension) and $\mathbf{T}_R$ (right tension), as illustrated. It is reasonable to neglect the weight of the wire itself. The net external force is zero since the system is stationary. A little trigonometry can now be used to find the tensions. One conclusion is possible at the outset—we can see from part (b) of the figure that the magnitudes of the tensions T_L and T_R must be equal. This is because there is no horizontal acceleration in the rope, and the only forces acting to the left and right are T_L and T_R.

Thus, the magnitude of those forces must be equal so that they cancel each other out.

Whenever we have two-dimensional vector problems in which no two vectors are parallel, the easiest method of solution is to pick a convenient coordinate system and project the vectors onto its axes. In this case the best coordinate system has one axis horizontal and the other vertical. We call the horizontal the x -axis and the vertical the y -axis.

Solution

First, we need to resolve the tension vectors into their horizontal and vertical components. It helps to draw a new free-body diagram showing all of the horizontal and vertical components of each force acting on the system.

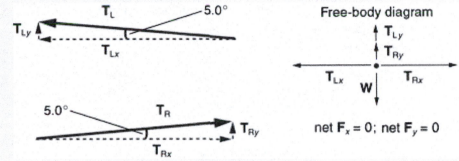

Figure 4.18 When the vectors are projected onto vertical and horizontal axes, their components along those axes must add to zero, since the tightrope walker is stationary. The small angle results in T being much greater than w.

Consider the horizontal components of the forces (denoted with a subscript x):

$$F_{netx} = T_{Lx} - T_{Rx}. \tag{4.44}$$

The net external horizontal force $F_{netx} = 0$, since the person is stationary. Thus,

$$\begin{aligned} F_{netx} = 0 &= T_{Lx} - T_{Rx} \\ T_{Lx} &= T_{Rx}. \end{aligned} \tag{4.45}$$

Now, observe **Figure 4.18**. You can use trigonometry to determine the magnitude of T_L and T_R. Notice that:

$$\begin{aligned} \cos{(5.0°)} &= \frac{T_{Lx}}{T_L} \\ T_{Lx} &= T_L \cos{(5.0°)} \\ \cos{(5.0°)} &= \frac{T_{Rx}}{T_R} \\ T_{Rx} &= T_R \cos{(5.0°)}. \end{aligned} \tag{4.46}$$

Equating T_{Lx} and T_{Rx}:

$$T_L \cos{(5.0°)} = T_R \cos{(5.0°)}. \tag{4.47}$$

Thus,

$$T_L = T_R = T, \tag{4.48}$$

as predicted. Now, considering the vertical components (denoted by a subscript y), we can solve for T. Again, since the person is stationary, Newton's second law implies that net $F_y = 0$. Thus, as illustrated in the free-body diagram in **Figure 4.18**,

$$F_{nety} = T_{Ly} + T_{Ry} - w = 0. \tag{4.49}$$

Observing **Figure 4.18**, we can use trigonometry to determine the relationship between T_{Ly}, T_{Ry}, and T. As we determined from the analysis in the horizontal direction, $T_L = T_R = T$:

$$\begin{aligned} \sin{(5.0°)} &= \frac{T_{Ly}}{T_L} \\ T_{Ly} = T_L \sin{(5.0°)} &= T \sin{(5.0°)} \\ \sin{(5.0°)} &= \frac{T_{Ry}}{T_R} \\ T_{Ry} = T_R \sin{(5.0°)} &= T \sin{(5.0°)}. \end{aligned} \tag{4.50}$$

Now, we can substitute the values for T_{Ly} and T_{Ry}, into the net force equation in the vertical direction:

$$\begin{aligned} F_{nety} &= T_{Ly} + T_{Ry} - w = 0 \\ F_{nety} &= T\sin(5.0°) + T\sin(5.0°) - w = 0 \\ 2T\sin(5.0°) - w &= 0 \\ 2T\sin(5.0°) &= w \end{aligned}$$

(4.51)

and

$$T = \frac{w}{2\sin(5.0°)} = \frac{mg}{2\sin(5.0°)},$$

(4.52)

so that

$$T = \frac{(70.0\text{ kg})(9.80\text{ m/s}^2)}{2(0.0872)},$$

(4.53)

and the tension is

$$T = 3900\text{ N.}$$

(4.54)

Discussion

Note that the vertical tension in the wire acts as a normal force that supports the weight of the tightrope walker. The tension is almost six times the 686-N weight of the tightrope walker. Since the wire is nearly horizontal, the vertical component of its tension is only a small fraction of the tension in the wire. The large horizontal components are in opposite directions and cancel, and so most of the tension in the wire is not used to support the weight of the tightrope walker.

If we wish to *create* a very large tension, all we have to do is exert a force perpendicular to a flexible connector, as illustrated in **Figure 4.19**. As we saw in the last example, the weight of the tightrope walker acted as a force perpendicular to the rope. We saw that the tension in the roped related to the weight of the tightrope walker in the following way:

$$T = \frac{w}{2\sin(\theta)}.$$

(4.55)

We can extend this expression to describe the tension T created when a perpendicular force ($\mathbf{F}_\perp$) is exerted at the middle of a flexible connector:

$$T = \frac{F_\perp}{2\sin(\theta)}.$$

(4.56)

Note that θ is the angle between the horizontal and the bent connector. In this case, T becomes very large as θ approaches zero. Even the relatively small weight of any flexible connector will cause it to sag, since an infinite tension would result if it were horizontal (i.e., $\theta = 0$ and $\sin\theta = 0$). (See **Figure 4.19**.)

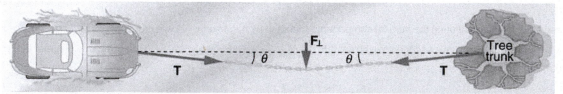

Figure 4.19 We can create a very large tension in the chain by pushing on it perpendicular to its length, as shown. Suppose we wish to pull a car out of the mud when no tow truck is available. Each time the car moves forward, the chain is tightened to keep it as nearly straight as possible. The tension in the chain is given by $T = \frac{F_\perp}{2\sin(\theta)}$; since θ is small, T is very large. This situation is analogous to the tightrope walker shown in **Figure 4.17**, except that the tensions shown here are those transmitted to the car and the tree rather than those acting at the point where $\mathbf{F}_\perp$ is applied.

Figure 4.20 Unless an infinite tension is exerted, any flexible connector—such as the chain at the bottom of the picture—will sag under its own weight, giving a characteristic curve when the weight is evenly distributed along the length. Suspension bridges—such as the Golden Gate Bridge shown in this image—are essentially very heavy flexible connectors. The weight of the bridge is evenly distributed along the length of flexible connectors, usually cables, which take on the characteristic shape. (credit: Leaflet, Wikimedia Commons)

Extended Topic: Real Forces and Inertial Frames

There is another distinction among forces in addition to the types already mentioned. Some forces are real, whereas others are not. *Real forces* are those that have some physical origin, such as the gravitational pull. Contrastingly, *fictitious forces* are those that arise simply because an observer is in an accelerating frame of reference, such as one that rotates (like a merry-go-round) or undergoes linear acceleration (like a car slowing down). For example, if a satellite is heading due north above Earth's northern hemisphere, then to an observer on Earth it will appear to experience a force to the west that has no physical origin. Of course, what is happening here is that Earth is rotating toward the east and moves east under the satellite. In Earth's frame this looks like a westward force on the satellite, or it can be interpreted as a violation of Newton's first law (the law of inertia). An **inertial frame of reference** is one in which all forces are real and, equivalently, one in which Newton's laws have the simple forms given in this chapter.

Earth's rotation is slow enough that Earth is nearly an inertial frame. You ordinarily must perform precise experiments to observe fictitious forces and the slight departures from Newton's laws, such as the effect just described. On the large scale, such as for the rotation of weather systems and ocean currents, the effects can be easily observed.

The crucial factor in determining whether a frame of reference is inertial is whether it accelerates or rotates relative to a known inertial frame. Unless stated otherwise, all phenomena discussed in this text are considered in inertial frames.

All the forces discussed in this section are real forces, but there are a number of other real forces, such as lift and thrust, that are not discussed in this section. They are more specialized, and it is not necessary to discuss every type of force. It is natural, however, to ask where the basic simplicity we seek to find in physics is in the long list of forces. Are some more basic than others? Are some different manifestations of the same underlying force? The answer to both questions is yes, as will be seen in the next (extended) section and in the treatment of modern physics later in the text.

PhET Explorations: Forces in 1 Dimension

Explore the forces at work when you try to push a filing cabinet. Create an applied force and see the resulting friction force and total force acting on the cabinet. Charts show the forces, position, velocity, and acceleration vs. time. View a free-body diagram of all the forces (including gravitational and normal forces).

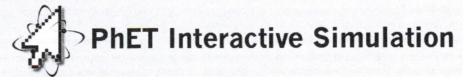

Figure 4.21 Forces in 1 Dimension (http://cnx.org/content/m54857/1.4/forces-1d_en.jar)

4.6 Problem-Solving Strategies

Learning Objectives
By the end of this section, you will be able to:
• Apply a problem-solving procedure to solve problems using Newton's laws of motion

The information presented in this section supports the following AP® learning objectives and science practices:

- **3.A.2.1** The student is able to represent forces in diagrams or mathematically using appropriately labeled vectors with magnitude, direction, and units during the analysis of a situation. **(S.P. 1.1)**
- **3.A.3.3** The student is able to describe a force as an interaction between two objects and identify both objects for any force. **(S.P. 1.4)**
- **3.B.1.1** The student is able to predict the motion of an object subject to forces exerted by several objects using an application of Newton's second law in a variety of physical situations with acceleration in one dimension. **(S.P. 6.4, 7.2)**
- **3.B.1.3** The student is able to re-express a free-body diagram representation into a mathematical representation and solve the mathematical representation for the acceleration of the object. **(S.P. 1.5, 2.2)**
- **3.B.2.1** The student is able to create and use free-body diagrams to analyze physical situations to solve problems with motion qualitatively and quantitatively. **(S.P. 1.1, 1.4, 2.2)**

Success in problem solving is obviously necessary to understand and apply physical principles, not to mention the more immediate need of passing exams. The basics of problem solving, presented earlier in this text, are followed here, but specific strategies useful in applying Newton's laws of motion are emphasized. These techniques also reinforce concepts that are useful in many other areas of physics. Many problem-solving strategies are stated outright in the worked examples, and so the following techniques should reinforce skills you have already begun to develop.

Problem-Solving Strategy for Newton's Laws of Motion

Step 1. As usual, it is first necessary to identify the physical principles involved. *Once it is determined that Newton's laws of motion are involved (if the problem involves forces), it is particularly important to draw a careful sketch of the situation.* Such a sketch is shown in **Figure 4.22**(a). Then, as in **Figure 4.22**(b), use arrows to represent all forces, label them carefully, and make their lengths and directions correspond to the forces they represent (whenever sufficient information exists).

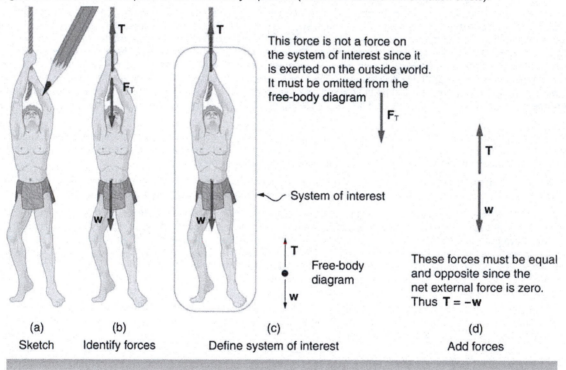

Figure 4.22 (a) A sketch of Tarzan hanging from a vine. (b) Arrows are used to represent all forces. $\mathbf{T}$ is the tension in the vine above Tarzan, $\mathbf{F_T}$ is the force he exerts on the vine, and $\mathbf{W}$ is his weight. All other forces, such as the nudge of a breeze, are assumed negligible. (c) Suppose we are given the ape man's mass and asked to find the tension in the vine. We then define the system of interest as shown and draw a free-body diagram. $\mathbf{F_T}$ is no longer shown, because it is not a force acting on the system of interest; rather, $\mathbf{F_T}$ acts on the outside world. (d) Showing only the arrows, the head-to-tail method of addition is used. It is apparent that $\mathbf{T} = -\mathbf{w}$, if Tarzan is stationary.

Step 2. Identify what needs to be determined and what is known or can be inferred from the problem as stated. That is, make a list of knowns and unknowns. *Then carefully determine the system of interest.* This decision is a crucial step, since Newton's second law involves only external forces. Once the system of interest has been identified, it becomes possible to determine which forces are external and which are internal, a necessary step to employ Newton's second law. (See **Figure 4.22**(c).) Newton's third law may be used to identify whether forces are exerted between components of a system (internal) or between the system and something outside (external). As illustrated earlier in this chapter, the system of interest depends on what question we need to answer. This choice becomes easier with practice, eventually developing into an almost unconscious process. Skill in clearly defining systems will be beneficial in later chapters as well.

A diagram showing the system of interest and all of the external forces is called a **free-body diagram**. Only forces are shown on free-body diagrams, not acceleration or velocity. We have drawn several of these in worked examples. **Figure 4.22**(c) shows a free-body diagram for the system of interest. Note that no internal forces are shown in a free-body diagram.

Step 3. Once a free-body diagram is drawn, *Newton's second law can be applied to solve the problem*. This is done in **Figure 4.22**(d) for a particular situation. In general, once external forces are clearly identified in free-body diagrams, it should be a straightforward task to put them into equation form and solve for the unknown, as done in all previous examples. If the problem is one-dimensional—that is, if all forces are parallel—then they add like scalars. If the problem is two-dimensional, then it must be broken down into a pair of one-dimensional problems. This is done by projecting the force vectors onto a set of axes chosen for convenience. As seen in previous examples, the choice of axes can simplify the problem. For example, when an incline is involved, a set of axes with one axis parallel to the incline and one perpendicular to it is most convenient. It is almost always convenient to make one axis parallel to the direction of motion, if this is known.

Applying Newton's Second Law

Before you write net force equations, it is critical to determine whether the system is accelerating in a particular direction. If the acceleration is zero in a particular direction, then the net force is zero in that direction. Similarly, if the acceleration is nonzero in a particular direction, then the net force is described by the equation: $F_{net} = ma$.

For example, if the system is accelerating in the horizontal direction, but it is not accelerating in the vertical direction, then you will have the following conclusions:

$$F_{net\,x} = ma,$$ (4.57)

$$F_{net\,y} = 0.$$ (4.58)

You will need this information in order to determine unknown forces acting in a system.

Step 4. As always, *check the solution to see whether it is reasonable*. In some cases, this is obvious. For example, it is reasonable to find that friction causes an object to slide down an incline more slowly than when no friction exists. In practice, intuition develops gradually through problem solving, and with experience it becomes progressively easier to judge whether an answer is reasonable. Another way to check your solution is to check the units. If you are solving for force and end up with units of m/s, then you have made a mistake.

4.7 Further Applications of Newton's Laws of Motion

Learning Objectives

By the end of this section, you will be able to:

- Apply problem-solving techniques to solve for quantities in more complex systems of forces.
- Integrate concepts from kinematics to solve problems using Newton's laws of motion.

The information presented in this section supports the following AP® learning objectives and science practices:

- **3.A.2.1** The student is able to represent forces in diagrams or mathematically using appropriately labeled vectors with magnitude, direction, and units during the analysis of a situation. **(S.P. 1.1)**
- **3.A.3.1** The student is able to analyze a scenario and make claims (develop arguments, justify assertions) about the forces exerted on an object by other objects for different types of forces or components of forces. **(S.P. 6.4, 7.2)**
- **3.A.3.3** The student is able to describe a force as an interaction between two objects and identify both objects for any force. **(S.P. 1.4)**
- **3.B.1.1** The student is able to predict the motion of an object subject to forces exerted by several objects using an application of Newton's second law in a variety of physical situations with acceleration in one dimension. **(S.P. 6.4, 7.2)**
- **3.B.1.3** The student is able to re-express a free-body diagram representation into a mathematical representation and solve the mathematical representation for the acceleration of the object. **(S.P. 1.5, 2.2)**
- **3.B.2.1** The student is able to create and use free-body diagrams to analyze physical situations to solve problems with motion qualitatively and quantitatively. **(S.P. 1.1, 1.4, 2.2)**

There are many interesting applications of Newton's laws of motion, a few more of which are presented in this section. These serve also to illustrate some further subtleties of physics and to help build problem-solving skills.

Example 4.7 Drag Force on a Barge

Suppose two tugboats push on a barge at different angles, as shown in **Figure 4.23**. The first tugboat exerts a force of $2.7 \times 10^5 \text{ N}$ in the x-direction, and the second tugboat exerts a force of $3.6 \times 10^5 \text{ N}$ in the y-direction.

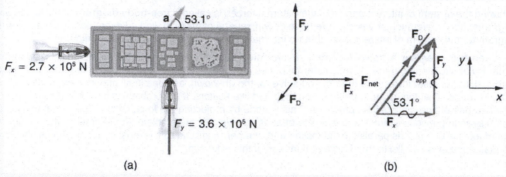

(a)   (b)

Figure 4.23 (a) A view from above of two tugboats pushing on a barge. (b) The free-body diagram for the ship contains only forces acting in the plane of the water. It omits the two vertical forces—the weight of the barge and the buoyant force of the water supporting it cancel and are not shown. Since the applied forces are perpendicular, the x- and y-axes are in the same direction as $\mathbf{F}_x$ and $\mathbf{F}_y$. The problem quickly becomes a one-dimensional problem along the direction of $\mathbf{F}_{app}$, since friction is in the direction opposite to $\mathbf{F}_{app}$.

If the mass of the barge is 5.0×10^6 kg and its acceleration is observed to be 7.5×10^{-2} m/s^2 in the direction shown, what is the drag force of the water on the barge resisting the motion? (Note: drag force is a frictional force exerted by fluids, such as air or water. The drag force opposes the motion of the object.)

Strategy

The directions and magnitudes of acceleration and the applied forces are given in Figure 4.23(a). We will define the total force of the tugboats on the barge as $\mathbf{F}_{app}$ so that:

$$\mathbf{F}_{app} = \mathbf{F}_x + \mathbf{F}_y \tag{4.59}$$

Since the barge is flat bottomed, the drag of the water $\mathbf{F}_D$ will be in the direction opposite to $\mathbf{F}_{app}$, as shown in the free-body diagram in Figure 4.23(b). The system of interest here is the barge, since the forces on *it* are given as well as its acceleration. Our strategy is to find the magnitude and direction of the net applied force $\mathbf{F}_{app}$, and then apply Newton's second law to solve for the drag force $\mathbf{F}_D$.

Solution

Since $\mathbf{F}_x$ and $\mathbf{F}_y$ are perpendicular, the magnitude and direction of $\mathbf{F}_{app}$ are easily found. First, the resultant magnitude is given by the Pythagorean theorem:

$$F_{app} = \sqrt{F_x^2 + F_y^2} \tag{4.60}$$
$$F_{app} = \sqrt{(2.7 \times 10^5 \text{ N})^2 + (3.6 \times 10^5 \text{ N})^2} = 4.5 \times 10^5 \text{ N}.$$

The angle is given by

$$\theta = \tan^{-1}\left(\frac{F_y}{F_x}\right) \tag{4.61}$$
$$\theta = \tan^{-1}\left(\frac{3.6 \times 10^5 \text{ N}}{2.7 \times 10^5 \text{ N}}\right) = 53°,$$

which we know, because of Newton's first law, is the same direction as the acceleration. $\mathbf{F}_D$ is in the opposite direction of $\mathbf{F}_{app}$, since it acts to slow down the acceleration. Therefore, the net external force is in the same direction as $\mathbf{F}_{app}$, but its magnitude is slightly less than $\mathbf{F}_{app}$. The problem is now one-dimensional. From Figure 4.23(b), we can see that

$$F_{net} = F_{app} - F_D. \tag{4.62}$$

But Newton's second law states that

$$F_{net} = ma. \tag{4.63}$$

Thus,

$$F_{app} - F_D = ma. \tag{4.64}$$

This can be solved for the magnitude of the drag force of the water F_D in terms of known quantities:

$$F_D = F_{app} - ma. \tag{4.65}$$

Substituting known values gives

$$F_D = (4.5 \times 10^5 \text{ N}) - (5.0 \times 10^6 \text{ kg})(7.5 \times 10^{-2} \text{ m/s}^2) = 7.5 \times 10^4 \text{ N}. \tag{4.66}$$

The direction of $\mathbf{F}_D$ has already been determined to be in the direction opposite to $\mathbf{F}_{app}$, or at an angle of $53°$ south of west.

Discussion

The numbers used in this example are reasonable for a moderately large barge. It is certainly difficult to obtain larger accelerations with tugboats, and small speeds are desirable to avoid running the barge into the docks. Drag is relatively small for a well-designed hull at low speeds, consistent with the answer to this example, where F_D is less than 1/600th of the weight of the ship.

In the earlier example of a tightrope walker we noted that the tensions in wires supporting a mass were equal only because the angles on either side were equal. Consider the following example, where the angles are not equal; slightly more trigonometry is involved.

Example 4.8 Different Tensions at Different Angles

Consider the traffic light (mass 15.0 kg) suspended from two wires as shown in **Figure 4.24**. Find the tension in each wire, neglecting the masses of the wires.

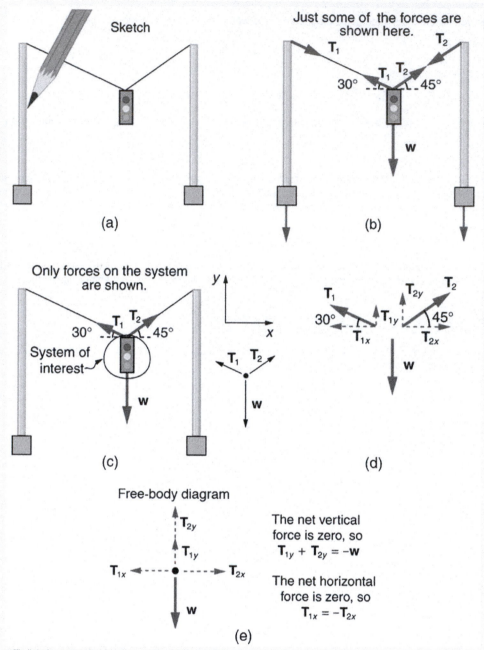

Figure 4.24 A traffic light is suspended from two wires. (b) Some of the forces involved. (c) Only forces acting on the system are shown here. The free-body diagram for the traffic light is also shown. (d) The forces projected onto vertical (*y*) and horizontal (*x*) axes. The horizontal components of the tensions must cancel, and the sum of the vertical components of the tensions must equal the weight of the traffic light. (e) The free-body diagram shows the vertical and horizontal forces acting on the traffic light.

Strategy

The system of interest is the traffic light, and its free-body diagram is shown in **Figure 4.24**(c). The three forces involved are not parallel, and so they must be projected onto a coordinate system. The most convenient coordinate system has one axis vertical and one horizontal, and the vector projections on it are shown in part (d) of the figure. There are two unknowns in this problem (T_1 and T_2), so two equations are needed to find them. These two equations come from applying Newton's second law along the vertical and horizontal axes, noting that the net external force is zero along each axis because acceleration is zero.

Solution

First consider the horizontal or *x*-axis:

$$F_{\text{net}x} = T_{2x} - T_{1x} = 0.$$

(4.67)

Thus, as you might expect,

$$T_{1x} = T_{2x}. \tag{4.68}$$

This gives us the following relationship between T_1 and T_2:

$$T_1 \cos(30°) = T_2 \cos(45°). \tag{4.69}$$

Thus,

$$T_2 = (1.225)T_1. \tag{4.70}$$

Note that T_1 and T_2 are not equal in this case, because the angles on either side are not equal. It is reasonable that T_2 ends up being greater than T_1, because it is exerted more vertically than T_1.

Now consider the force components along the vertical or y-axis:

$$F_{\text{net } y} = T_{1y} + T_{2y} - w = 0. \tag{4.71}$$

This implies

$$T_{1y} + T_{2y} = w. \tag{4.72}$$

Substituting the expressions for the vertical components gives

$$T_1 \sin(30°) + T_2 \sin(45°) = w. \tag{4.73}$$

There are two unknowns in this equation, but substituting the expression for T_2 in terms of T_1 reduces this to one equation with one unknown:

$$T_1(0.500) + (1.225T_1)(0.707) = w = mg, \tag{4.74}$$

which yields

$$(1.366)T_1 = (15.0 \text{ kg})(9.80 \text{ m/s}^2). \tag{4.75}$$

Solving this last equation gives the magnitude of T_1 to be

$$T_1 = 108 \text{ N}. \tag{4.76}$$

Finally, the magnitude of T_2 is determined using the relationship between them, $T_2 = 1.225\ T_1$, found above. Thus we obtain

$$T_2 = 132 \text{ N}. \tag{4.77}$$

Discussion

Both tensions would be larger if both wires were more horizontal, and they will be equal if and only if the angles on either side are the same (as they were in the earlier example of a tightrope walker).

The bathroom scale is an excellent example of a normal force acting on a body. It provides a quantitative reading of how much it must push upward to support the weight of an object. But can you predict what you would see on the dial of a bathroom scale if you stood on it during an elevator ride? Will you see a value greater than your weight when the elevator starts up? What about when the elevator moves upward at a constant speed: will the scale still read more than your weight at rest? Consider the following example.

Example 4.9 What Does the Bathroom Scale Read in an Elevator?

Figure 4.25 shows a 75.0-kg man (weight of about 165 lb) standing on a bathroom scale in an elevator. Calculate the scale reading: (a) if the elevator accelerates upward at a rate of 1.20 m/s^2, and (b) if the elevator moves upward at a constant speed of 1 m/s.

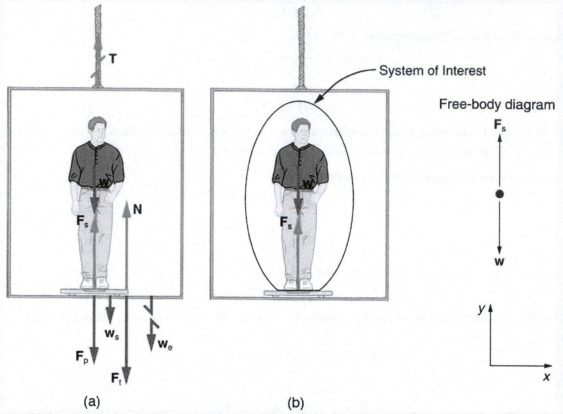

Figure 4.25 (a) The various forces acting when a person stands on a bathroom scale in an elevator. The arrows are approximately correct for when the elevator is accelerating upward—broken arrows represent forces too large to be drawn to scale. $\mathbf{T}$ is the tension in the supporting cable, $\mathbf{w}$ is the weight of the person, $\mathbf{w}_s$ is the weight of the scale, $\mathbf{w}_e$ is the weight of the elevator, $\mathbf{F}_s$ is the force of the scale on the person, $\mathbf{F}_p$ is the force of the person on the scale, $\mathbf{F}_t$ is the force of the scale on the floor of the elevator, and $\mathbf{N}$ is the force of the floor upward on the scale. (b) The free-body diagram shows only the external forces acting on the designated system of interest—the person.

Strategy

If the scale is accurate, its reading will equal F_p, the magnitude of the force the person exerts downward on it. **Figure 4.25**(a) shows the numerous forces acting on the elevator, scale, and person. It makes this one-dimensional problem look much more formidable than if the person is chosen to be the system of interest and a free-body diagram is drawn as in **Figure 4.25**(b). Analysis of the free-body diagram using Newton's laws can produce answers to both parts (a) and (b) of this example, as well as some other questions that might arise. The only forces acting on the person are his weight $\mathbf{w}$ and the upward force of the scale $\mathbf{F}_s$. According to Newton's third law $\mathbf{F}_p$ and $\mathbf{F}_s$ are equal in magnitude and opposite in direction, so that we need to find F_s in order to find what the scale reads. We can do this, as usual, by applying Newton's second law,

$$F_{net} = ma. \tag{4.78}$$

From the free-body diagram we see that $F_{net} = F_s - w$, so that

$$F_s - w = ma. \tag{4.79}$$

Solving for F_s gives an equation with only one unknown:

$$F_s = ma + w, \tag{4.80}$$

or, because $w = mg$, simply

$$F_s = ma + mg. \tag{4.81}$$

No assumptions were made about the acceleration, and so this solution should be valid for a variety of accelerations in addition to the ones in this exercise.

Solution for (a)

In this part of the problem, $a = 1.20 \text{ m/s}^2$, so that

$$F_s = (75.0 \text{ kg})(1.20 \text{ m/s}^2) + (75.0 \text{ kg})(9.80 \text{ m/s}^2),\tag{4.82}$$

yielding

$$F_s = 825 \text{ N}.\tag{4.83}$$

Discussion for (a)

This is about 185 lb. What would the scale have read if he were stationary? Since his acceleration would be zero, the force of the scale would be equal to his weight:

$$
\begin{aligned}
F_{\text{net}} &= ma = 0 = F_s - w \\
F_s &= w = mg \\
F_s &= (75.0 \text{ kg})(9.80 \text{ m/s}^2) \\
F_s &= 735 \text{ N}.
\end{aligned}
\tag{4.84}
$$

So, the scale reading in the elevator is greater than his 735-N (165 lb) weight. This means that the scale is pushing up on the person with a force greater than his weight, as it must in order to accelerate him upward. Clearly, the greater the acceleration of the elevator, the greater the scale reading, consistent with what you feel in rapidly accelerating versus slowly accelerating elevators.

Solution for (b)

Now, what happens when the elevator reaches a constant upward velocity? Will the scale still read more than his weight?

For any constant velocity—up, down, or stationary—acceleration is zero because $a = \dfrac{\Delta v}{\Delta t}$, and $\Delta v = 0$.

Thus,

$$F_s = ma + mg = 0 + mg.\tag{4.85}$$

Now

$$F_s = (75.0 \text{ kg})(9.80 \text{ m/s}^2),\tag{4.86}$$

which gives

$$F_s = 735 \text{ N}.\tag{4.87}$$

Discussion for (b)

The scale reading is 735 N, which equals the person's weight. This will be the case whenever the elevator has a constant velocity—moving up, moving down, or stationary.

The solution to the previous example also applies to an elevator accelerating downward, as mentioned. When an elevator accelerates downward, a is negative, and the scale reading is *less* than the weight of the person, until a constant downward velocity is reached, at which time the scale reading again becomes equal to the person's weight. If the elevator is in free-fall and accelerating downward at g, then the scale reading will be zero and the person will *appear* to be weightless.

Integrating Concepts: Newton's Laws of Motion and Kinematics

Physics is most interesting and most powerful when applied to general situations that involve more than a narrow set of physical principles. Newton's laws of motion can also be integrated with other concepts that have been discussed previously in this text to solve problems of motion. For example, forces produce accelerations, a topic of kinematics, and hence the relevance of earlier chapters. When approaching problems that involve various types of forces, acceleration, velocity, and/or position, use the following steps to approach the problem:

Problem-Solving Strategy

Step 1. *Identify which physical principles are involved.* Listing the givens and the quantities to be calculated will allow you to identify the principles involved.
Step 2. *Solve the problem using strategies outlined in the text.* If these are available for the specific topic, you should refer to them. You should also refer to the sections of the text that deal with a particular topic. The following worked example illustrates how these strategies are applied to an integrated concept problem.

Example 4.10 What Force Must a Soccer Player Exert to Reach Top Speed?

A soccer player starts from rest and accelerates forward, reaching a velocity of 8.00 m/s in 2.50 s. (a) What was his average acceleration? (b) What average force did he exert backward on the ground to achieve this acceleration? The player's mass is 70.0 kg, and air resistance is negligible.

Strategy

1. To solve an *integrated concept problem*, we must first identify the physical principles involved and identify the chapters in which they are found. Part (a) of this example considers *acceleration* along a straight line. This is a topic of *kinematics*. Part (b) deals with *force*, a topic of *dynamics* found in this chapter.

2. The following solutions to each part of the example illustrate how the specific problem-solving strategies are applied. These involve identifying knowns and unknowns, checking to see if the answer is reasonable, and so forth.

Solution for (a)

We are given the initial and final velocities (zero and 8.00 m/s forward); thus, the change in velocity is $\Delta v = 8.00$ m/s . We are given the elapsed time, and so $\Delta t = 2.50$ s . The unknown is acceleration, which can be found from its definition:

$$a = \frac{\Delta v}{\Delta t}. \tag{4.88}$$

Substituting the known values yields

$$\begin{aligned} a &= \frac{8.00 \text{ m/s}}{2.50 \text{ s}} \\ &= 3.20 \text{ m/s}^2. \end{aligned} \tag{4.89}$$

Discussion for (a)

This is an attainable acceleration for an athlete in good condition.

Solution for (b)

Here we are asked to find the average force the player exerts backward to achieve this forward acceleration. Neglecting air resistance, this would be equal in magnitude to the net external force on the player, since this force causes his acceleration. Since we now know the player's acceleration and are given his mass, we can use Newton's second law to find the force exerted. That is,

$$F_{\text{net}} = ma. \tag{4.90}$$

Substituting the known values of m and a gives

$$\begin{aligned} F_{\text{net}} &= (70.0 \text{ kg})(3.20 \text{ m/s}^2) \\ &= 224 \text{ N}. \end{aligned} \tag{4.91}$$

Discussion for (b)

This is about 50 pounds, a reasonable average force.

This worked example illustrates how to apply problem-solving strategies to situations that include topics from different chapters. The first step is to identify the physical principles involved in the problem. The second step is to solve for the unknown using familiar problem-solving strategies. These strategies are found throughout the text, and many worked examples show how to use them for single topics. You will find these techniques for integrated concept problems useful in applications of physics outside of a physics course, such as in your profession, in other science disciplines, and in everyday life. The following problems will build your skills in the broad application of physical principles.

4.8 Extended Topic: The Four Basic Forces—An Introduction

Learning Objectives

By the end of this section, you will be able to:

- Understand the four basic forces that underlie the processes in nature.

The information presented in this section supports the following AP® learning objectives and science practices:

- **3.C.4.1** The student is able to make claims about various contact forces between objects based on the microscopic cause of those forces. **(S.P. 6.1)**
- **3.C.4.2** The student is able to explain contact forces (tension, friction, normal, buoyant, spring) as arising from interatomic electric forces and that they therefore have certain directions. **(S.P. 6.2)**
- **3.G.1.1** The student is able to articulate situations when the gravitational force is the dominant force and when the electromagnetic, weak, and strong forces can be ignored. **(S.P. 7.1)**

One of the most remarkable simplifications in physics is that only four distinct forces account for all known phenomena. In fact, nearly all of the forces we experience directly are due to only one basic force, called the electromagnetic force. (The gravitational force is the only force we experience directly that is not electromagnetic.) This is a tremendous simplification of the myriad of *apparently* different forces we can list, only a few of which were discussed in the previous section. As we will see, the basic forces are all thought to act through the exchange of microscopic carrier particles, and the characteristics of the basic forces are determined by the types of particles exchanged. Action at a distance, such as the gravitational force of Earth on the Moon, is explained by the existence of a **force field** rather than by "physical contact."

The *four basic forces* are the gravitational force, the electromagnetic force, the weak nuclear force, and the strong nuclear force. Their properties are summarized in **Table 4.2**. Since the weak and strong nuclear forces act over an extremely short range, the size of a nucleus or less, we do not experience them directly, although they are crucial to the very structure of matter. These forces determine which nuclei are stable and which decay, and they are the basis of the release of energy in certain nuclear reactions. Nuclear forces determine not only the stability of nuclei, but also the relative abundance of elements in nature. The properties of the nucleus of an atom determine the number of electrons it has and, thus, indirectly determine the chemistry of the atom. More will be said of all of these topics in later chapters.

Concept Connections: The Four Basic Forces

The four basic forces will be encountered in more detail as you progress through the text. The gravitational force is defined in **Uniform Circular Motion and Gravitation**, electric force in **Electric Charge and Electric Field**, magnetic force in **Magnetism**, and nuclear forces in **Radioactivity and Nuclear Physics**. On a macroscopic scale, electromagnetism and gravity are the basis for all forces. The nuclear forces are vital to the substructure of matter, but they are not directly experienced on the macroscopic scale.

Table 4.2 Properties of the Four Basic Forces[1]

Force	Approximate Relative Strengths	Range	Attraction/Repulsion	Carrier Particle
Gravitational	10^{-38}	∞	attractive only	Graviton
Electromagnetic	10^{-2}	∞	attractive and repulsive	Photon
Weak nuclear	10^{-13}	$< 10^{-18}\,\mathrm{m}$	attractive and repulsive	W^+, W^-, Z^0
Strong nuclear	1	$< 10^{-15}\,\mathrm{m}$	attractive and repulsive	gluons

The gravitational force is surprisingly weak—it is only because gravity is always attractive that we notice it at all. Our weight is the gravitational force due to the *entire* Earth acting on us. On the very large scale, as in astronomical systems, the gravitational force is the dominant force determining the motions of moons, planets, stars, and galaxies. The gravitational force also affects the nature of space and time. As we shall see later in the study of general relativity, space is curved in the vicinity of very massive bodies, such as the Sun, and time actually slows down near massive bodies.

Take a good look at the ranges for the four fundamental forces listed in **Table 4.2**. The range of the strong nuclear force, 10^{-15} m, is approximately the size of the nucleus of an atom; the weak nuclear force has an even shorter range. At scales on the order of 10−10 m, approximately the size of an atom, both nuclear forces are completely dominated by the electromagnetic force. Notice that this scale is still utterly tiny compared to our everyday experience. At scales that we do experience daily, electromagnetism tends to be negligible, due to its attractive and repulsive properties canceling each other out. That leaves gravity, which is usually the strongest of the forces at scales above ~10^{-4} m, and hence includes our everyday activities, such as throwing, climbing stairs, and walking.

Electromagnetic forces can be either attractive or repulsive. They are long-range forces, which act over extremely large distances, and they nearly cancel for macroscopic objects. (Remember that it is the *net* external force that is important.) If they did not cancel, electromagnetic forces would completely overwhelm the gravitational force. The electromagnetic force is a combination of electrical forces (such as those that cause static electricity) and magnetic forces (such as those that affect a compass needle). These two forces were thought to be quite distinct until early in the 19th century, when scientists began to discover that they are different manifestations of the same force. This discovery is a classical case of the *unification of forces*. Similarly, friction, tension, and all of the other classes of forces we experience directly (except gravity, of course) are due to electromagnetic interactions of atoms and molecules. It is still convenient to consider these forces separately in specific applications, however, because of the ways they manifest themselves.

Concept Connections: Unifying Forces

Attempts to unify the four basic forces are discussed in relation to elementary particles later in this text. By "unify" we mean finding connections between the forces that show that they are different manifestations of a single force. Even if such unification is achieved, the forces will retain their separate characteristics on the macroscopic scale and may be identical only under extreme conditions such as those existing in the early universe.

---

1. The graviton is a proposed particle, though it has not yet been observed by scientists. See the discussion of gravitational waves later in this section. The particles W^+, W^-, and Z^0 are called vector bosons; these were predicted by theory and first observed in 1983. There are eight types of gluons proposed by scientists, and their existence is indicated by meson exchange in the nuclei of atoms.

Physicists are now exploring whether the four basic forces are in some way related. Attempts to unify all forces into one come under the rubric of Grand Unified Theories (GUTs), with which there has been some success in recent years. It is now known that under conditions of extremely high density and temperature, such as existed in the early universe, the electromagnetic and weak nuclear forces are indistinguishable. They can now be considered to be different manifestations of one force, called the *electroweak* force. So the list of four has been reduced in a sense to only three. Further progress in unifying all forces is proving difficult—especially the inclusion of the gravitational force, which has the special characteristics of affecting the space and time in which the other forces exist.

While the unification of forces will not affect how we discuss forces in this text, it is fascinating that such underlying simplicity exists in the face of the overt complexity of the universe. There is no reason that nature must be simple—it simply is.

Action at a Distance: Concept of a Field

All forces act at a distance. This is obvious for the gravitational force. Earth and the Moon, for example, interact without coming into contact. It is also true for all other forces. Friction, for example, is an electromagnetic force between atoms that may not actually touch. What is it that carries forces between objects? One way to answer this question is to imagine that a **force field** surrounds whatever object creates the force. A second object (often called a *test object*) placed in this field will experience a force that is a function of location and other variables. The field itself is the "thing" that carries the force from one object to another. The field is defined so as to be a characteristic of the object creating it; the field does not depend on the test object placed in it. Earth's gravitational field, for example, is a function of the mass of Earth and the distance from its center, independent of the presence of other masses. The concept of a field is useful because equations can be written for force fields surrounding objects (for gravity, this yields $w = mg$ at Earth's surface), and motions can be calculated from these equations.

(See **Figure 4.26**.)

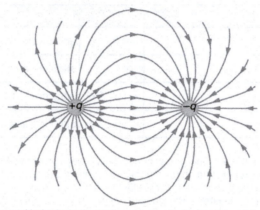

Figure 4.26 The electric force field between a positively charged particle and a negatively charged particle. When a positive test charge is placed in the field, the charge will experience a force in the direction of the force field lines.

Concept Connections: Force Fields

The concept of a *force field* is also used in connection with electric charge and is presented in **Electric Charge and Electric Field**. It is also a useful idea for all the basic forces, as will be seen in **Particle Physics**. Fields help us to visualize forces and how they are transmitted, as well as to describe them with precision and to link forces with subatomic carrier particles.

Making Connections: Vector and Scalar Fields

These fields may be either scalar or vector fields. Gravity and electromagnetism are examples of vector fields. A test object placed in such a field will have both the magnitude and direction of the resulting force on the test object completely defined by the object's location in the field. We will later cover examples of scalar fields, which have a magnitude but no direction.

The field concept has been applied very successfully; we can calculate motions and describe nature to high precision using field equations. As useful as the field concept is, however, it leaves unanswered the question of what carries the force. It has been proposed in recent decades, starting in 1935 with Hideki Yukawa's (1907–1981) work on the strong nuclear force, that all forces are transmitted by the exchange of elementary particles. We can visualize particle exchange as analogous to macroscopic phenomena such as two people passing a basketball back and forth, thereby exerting a repulsive force without touching one another. (See **Figure 4.27**.)

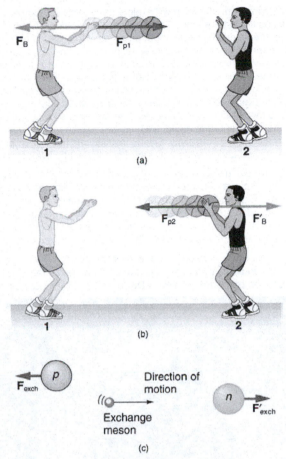

Figure 4.27 The exchange of masses resulting in repulsive forces. (a) The person throwing the basketball exerts a force F_{p1} on it toward the other person and feels a reaction force F_B away from the second person. (b) The person catching the basketball exerts a force F_{p2} on it to stop the ball and feels a reaction force F'_B away from the first person. (c) The analogous exchange of a meson between a proton and a neutron carries the strong nuclear forces F_{exch} and F'_{exch} between them. An attractive force can also be exerted by the exchange of a mass—if person 2 pulled the basketball away from the first person as he tried to retain it, then the force between them would be attractive.

This idea of particle exchange deepens rather than contradicts field concepts. It is more satisfying philosophically to think of something physical actually moving between objects acting at a distance. Table 4.2 lists the exchange or **carrier particles**, both observed and proposed, that carry the four forces. But the real fruit of the particle-exchange proposal is that searches for Yukawa's proposed particle found it *and* a number of others that were completely unexpected, stimulating yet more research. All of this research eventually led to the proposal of quarks as the underlying substructure of matter, which is a basic tenet of GUTs. If successful, these theories will explain not only forces, but also the structure of matter itself. Yet physics is an experimental science, so the test of these theories must lie in the domain of the real world. As of this writing, scientists at the CERN laboratory in Switzerland are starting to test these theories using the world's largest particle accelerator: the Large Hadron Collider. This accelerator (27 km in circumference) allows two high-energy proton beams, traveling in opposite directions, to collide. An energy of 14 trillion electron volts will be available. It is anticipated that some new particles, possibly force carrier particles, will be found. (See Figure 4.28.) One of the force carriers of high interest that researchers hope to detect is the Higgs boson. The observation of its properties might tell us why different particles have different masses.

Figure 4.28 The world's largest particle accelerator spans the border between Switzerland and France. Two beams, traveling in opposite directions close to the speed of light, collide in a tube similar to the central tube shown here. External magnets determine the beam's path. Special detectors will analyze particles created in these collisions. Questions as broad as what is the origin of mass and what was matter like the first few seconds of our universe will be explored. This accelerator began preliminary operation in 2008. (credit: Frank Hommes)

Tiny particles also have wave-like behavior, something we will explore more in a later chapter. To better understand force-carrier particles from another perspective, let us consider gravity. The search for gravitational waves has been going on for a number of years. Almost 100 years ago, Einstein predicted the existence of these waves as part of his general theory of relativity. Gravitational waves are created during the collision of massive stars, in black holes, or in supernova explosions—like shock waves. These gravitational waves will travel through space from such sites much like a pebble dropped into a pond sends out ripples—except these waves move at the speed of light. A detector apparatus has been built in the U.S., consisting of two large installations nearly 3000 km apart—one in Washington state and one in Louisiana! The facility is called the Laser Interferometer Gravitational-Wave Observatory (LIGO). Each installation is designed to use optical lasers to examine any slight shift in the relative positions of two masses due to the effect of gravity waves. The two sites allow simultaneous measurements of these small effects to be separated from other natural phenomena, such as earthquakes. Initial operation of the detectors began in 2002, and work is proceeding on increasing their sensitivity. Similar installations have been built in Italy (VIRGO), Germany (GEO600), and Japan (TAMA300) to provide a worldwide network of gravitational wave detectors.

International collaboration in this area is moving into space with the joint EU/US project LISA (Laser Interferometer Space Antenna). Earthquakes and other Earthly noises will be no problem for these monitoring spacecraft. LISA will complement LIGO by looking at much more massive black holes through the observation of gravitational-wave sources emitting much larger wavelengths. Three satellites will be placed in space above Earth in an equilateral triangle (with 5,000,000-km sides) (**Figure 4.29**). The system will measure the relative positions of each satellite to detect passing gravitational waves. Accuracy to within 10% of the size of an atom will be needed to detect any waves. The launch of this project might be as early as 2018.

"I'm sure LIGO will tell us something about the universe that we didn't know before. The history of science tells us that any time you go where you haven't been before, you usually find something that really shakes the scientific paradigms of the day. Whether gravitational wave astrophysics will do that, only time will tell." —David Reitze, LIGO Input Optics Manager, University of Florida

Figure 4.29 Space-based future experiments for the measurement of gravitational waves. Shown here is a drawing of LISA's orbit. Each satellite of LISA will consist of a laser source and a mass. The lasers will transmit a signal to measure the distance between each satellite's test mass. The relative motion of these masses will provide information about passing gravitational waves. (credit: NASA)

The ideas presented in this section are but a glimpse into topics of modern physics that will be covered in much greater depth in later chapters.

Glossary

acceleration: the rate at which an object's velocity changes over a period of time

carrier particle: a fundamental particle of nature that is surrounded by a characteristic force field; photons are carrier particles of the electromagnetic force

dynamics: the study of how forces affect the motion of objects and systems

external force: a force acting on an object or system that originates outside of the object or system

force: a push or pull on an object with a specific magnitude and direction; can be represented by vectors; can be expressed as a multiple of a standard force

force field: a region in which a test particle will experience a force

free-body diagram: a sketch showing all of the external forces acting on an object or system; the system is represented by a dot, and the forces are represented by vectors extending outward from the dot

free-fall: a situation in which the only force acting on an object is the force due to gravity

friction: a force past each other of objects that are touching; examples include rough surfaces and air resistance

inertia: the tendency of an object to remain at rest or remain in motion

inertial frame of reference: a coordinate system that is not accelerating; all forces acting in an inertial frame of reference are real forces, as opposed to fictitious forces that are observed due to an accelerating frame of reference

law of inertia: see Newton's first law of motion

mass: the quantity of matter in a substance; measured in kilograms

net external force: the vector sum of all external forces acting on an object or system; causes a mass to accelerate

Newton's first law of motion: in an inertial frame of reference, a body at rest remains at rest, or, if in motion, remains in motion at a constant velocity unless acted on by a net external force; also known as the law of inertia

Newton's second law of motion: the net external force $\mathbf{F}_{net}$ on an object with mass m is proportional to and in the same direction as the acceleration of the object, $\mathbf{a}$, and inversely proportional to the mass; defined mathematically as

$$\mathbf{a} = \frac{\mathbf{F}_{net}}{m}$$

Newton's third law of motion: whenever one body exerts a force on a second body, the first body experiences a force that is equal in magnitude and opposite in direction to the force that the first body exerts

normal force: the force that a surface applies to an object to support the weight of the object; acts perpendicular to the surface on which the object rests

system: defined by the boundaries of an object or collection of objects being observed; all forces originating from outside of the system are considered external forces

tension: the pulling force that acts along a medium, especially a stretched flexible connector, such as a rope or cable; when a rope supports the weight of an object, the force on the object due to the rope is called a tension force

thrust: a reaction force that pushes a body forward in response to a backward force; rockets, airplanes, and cars are pushed forward by a thrust reaction force

weight: the force $\mathbf{w}$ due to gravity acting on an object of mass m; defined mathematically as: $w = mg$, where $\mathbf{g}$ is the magnitude and direction of the acceleration due to gravity

Section Summary

4.1 Development of Force Concept
- **Dynamics** is the study of how forces affect the motion of objects.
- **Force** is a push or pull that can be defined in terms of various standards, and it is a vector having both magnitude and direction.
- **External forces** are any outside forces that act on a body. A **free-body diagram** is a drawing of all external forces acting on a body.

4.2 Newton's First Law of Motion: Inertia
- **Newton's first law of motion** states that in an inertial frame of reference a body at rest remains at rest, or, if in motion, remains in motion at a constant velocity unless acted on by a net external force. This is also known as the **law of inertia**.
- **Inertia** is the tendency of an object to remain at rest or remain in motion. Inertia is related to an object's mass.
- **Mass** is the quantity of matter in a substance.

4.3 Newton's Second Law of Motion: Concept of a System
- Acceleration, $\mathbf{a}$, is defined as a change in velocity, meaning a change in its magnitude or direction, or both.
- An external force is one acting on a system from outside the system, as opposed to internal forces, which act between components within the system.
- Newton's second law of motion states that the acceleration of a system is directly proportional to and in the same direction as the net external force acting on the system, and inversely proportional to its mass.
- In equation form, Newton's second law of motion is $\mathbf{a} = \frac{\mathbf{F}_{net}}{m}$.
- This is often written in the more familiar form: $\mathbf{F}_{net} = m\mathbf{a}$.
- The weight $\mathbf{w}$ of an object is defined as the force of gravity acting on an object of mass m. The object experiences an acceleration due to gravity $\mathbf{g}$:

$$\mathbf{w} = m\mathbf{g}.$$

- If the only force acting on an object is due to gravity, the object is in free fall.
- Friction is a force that opposes the motion past each other of objects that are touching.

4.4 Newton's Third Law of Motion: Symmetry in Forces

- **Newton's third law of motion** represents a basic symmetry in nature. It states: Whenever one body exerts a force on a second body, the first body experiences a force that is equal in magnitude and opposite in direction to the force that the first body exerts.
- A **thrust** is a reaction force that pushes a body forward in response to a backward force. Rockets, airplanes, and cars are pushed forward by a thrust reaction force.

4.5 Normal, Tension, and Other Examples of Force

- When objects rest on a surface, the surface applies a force to the object that supports the weight of the object. This supporting force acts perpendicular to and away from the surface. It is called a normal force, $\mathbf{N}$.
- When objects rest on a non-accelerating horizontal surface, the magnitude of the normal force is equal to the weight of the object:

$$N = mg.$$

- When objects rest on an inclined plane that makes an angle θ with the horizontal surface, the weight of the object can be resolved into components that act perpendicular ($\mathbf{w}_\perp$) and parallel ($\mathbf{w}_\parallel$) to the surface of the plane. These components can be calculated using:

$$w_\parallel = w \sin(\theta) = mg \sin(\theta)$$

$$w_\perp = w \cos(\theta) = mg \cos(\theta).$$

- The pulling force that acts along a stretched flexible connector, such as a rope or cable, is called tension, $\mathbf{T}$. When a rope supports the weight of an object that is at rest, the tension in the rope is equal to the weight of the object:

$$T = mg.$$

- In any inertial frame of reference (one that is not accelerated or rotated), Newton's laws have the simple forms given in this chapter and all forces are real forces having a physical origin.

4.6 Problem-Solving Strategies

- To solve problems involving Newton's laws of motion, follow the procedure described:
    1. Draw a sketch of the problem.
    2. Identify known and unknown quantities, and identify the system of interest. Draw a free-body diagram, which is a sketch showing all of the forces acting on an object. The object is represented by a dot, and the forces are represented by vectors extending in different directions from the dot. If vectors act in directions that are not horizontal or vertical, resolve the vectors into horizontal and vertical components and draw them on the free-body diagram.
    3. Write Newton's second law in the horizontal and vertical directions and add the forces acting on the object. If the object does not accelerate in a particular direction (for example, the x-direction) then $F_{\text{net }x} = 0$. If the object does accelerate in that direction, $F_{\text{net }x} = ma$.
    4. Check your answer. Is the answer reasonable? Are the units correct?

4.7 Further Applications of Newton's Laws of Motion

- Newton's laws of motion can be applied in numerous situations to solve problems of motion.
- Some problems will contain multiple force vectors acting in different directions on an object. Be sure to draw diagrams, resolve all force vectors into horizontal and vertical components, and draw a free-body diagram. Always analyze the direction in which an object accelerates so that you can determine whether $F_{\text{net}} = ma$ or $F_{\text{net}} = 0$.

- The normal force on an object is not always equal in magnitude to the weight of the object. If an object is accelerating, the normal force will be less than or greater than the weight of the object. Also, if the object is on an inclined plane, the normal force will always be less than the full weight of the object.
- Some problems will contain various physical quantities, such as forces, acceleration, velocity, or position. You can apply concepts from kinematics and dynamics in order to solve these problems of motion.

4.8 Extended Topic: The Four Basic Forces—An Introduction

- The various types of forces that are categorized for use in many applications are all manifestations of the *four basic forces* in nature.
- The properties of these forces are summarized in Table 4.2.
- Everything we experience directly without sensitive instruments is due to either electromagnetic forces or gravitational forces. The nuclear forces are responsible for the submicroscopic structure of matter, but they are not directly sensed because of their short ranges. Attempts are being made to show all four forces are different manifestations of a single unified force.

• A force field surrounds an object creating a force and is the carrier of that force.

Conceptual Questions

4.1 Development of Force Concept

1. Propose a force standard different from the example of a stretched spring discussed in the text. Your standard must be capable of producing the same force repeatedly.

2. What properties do forces have that allow us to classify them as vectors?

4.2 Newton's First Law of Motion: Inertia

3. How are inertia and mass related?

4. What is the relationship between weight and mass? Which is an intrinsic, unchanging property of a body?

4.3 Newton's Second Law of Motion: Concept of a System

5. Which statement is correct? (a) Net force causes motion. (b) Net force causes change in motion. Explain your answer and give an example.

6. Why can we neglect forces such as those holding a body together when we apply Newton's second law of motion?

7. Explain how the choice of the "system of interest" affects which forces must be considered when applying Newton's second law of motion.

8. Describe a situation in which the net external force on a system is not zero, yet its speed remains constant.

9. A system can have a nonzero velocity while the net external force on it *is* zero. Describe such a situation.

10. A rock is thrown straight up. What is the net external force acting on the rock when it is at the top of its trajectory?

11. (a) Give an example of different net external forces acting on the same system to produce different accelerations. (b) Give an example of the same net external force acting on systems of different masses, producing different accelerations. (c) What law accurately describes both effects? State it in words and as an equation.

12. If the acceleration of a system is zero, are no external forces acting on it? What about internal forces? Explain your answers.

13. If a constant, nonzero force is applied to an object, what can you say about the velocity and acceleration of the object?

14. The gravitational force on the basketball in **Figure 4.6** is ignored. When gravity *is* taken into account, what is the direction of the net external force on the basketball—above horizontal, below horizontal, or still horizontal?

4.4 Newton's Third Law of Motion: Symmetry in Forces

15. When you take off in a jet aircraft, there is a sensation of being pushed back into the seat. Explain why you move backward in the seat—is there really a force backward on you? (The same reasoning explains whiplash injuries, in which the head is apparently thrown backward.)

16. A device used since the 1940s to measure the kick or recoil of the body due to heart beats is the "ballistocardiograph." What physics principle(s) are involved here to measure the force of cardiac contraction? How might we construct such a device?

17. Describe a situation in which one system exerts a force on another and, as a consequence, experiences a force that is equal in magnitude and opposite in direction. Which of Newton's laws of motion apply?

18. Why does an ordinary rifle recoil (kick backward) when fired? The barrel of a recoilless rifle is open at both ends. Describe how Newton's third law applies when one is fired. Can you safely stand close behind one when it is fired?

19. An American football lineman reasons that it is senseless to try to out-push the opposing player, since no matter how hard he pushes he will experience an equal and opposite force from the other player. Use Newton's laws and draw a free-body diagram of an appropriate system to explain how he can still out-push the opposition if he is strong enough.

20. Newton's third law of motion tells us that forces always occur in pairs of equal and opposite magnitude. Explain how the choice of the "system of interest" affects whether one such pair of forces cancels.

4.5 Normal, Tension, and Other Examples of Force

21. If a leg is suspended by a traction setup as shown in **Figure 4.30**, what is the tension in the rope?

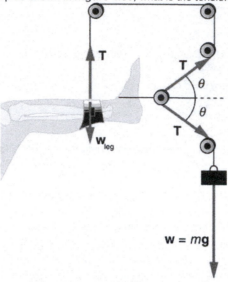

Figure 4.30 A leg is suspended by a traction system in which wires are used to transmit forces. Frictionless pulleys change the direction of the force T without changing its magnitude.

22. In a traction setup for a broken bone, with pulleys and rope available, how might we be able to increase the force along the tibia using the same weight? (See **Figure 4.30**.) (Note that the tibia is the shin bone shown in this image.)

4.7 Further Applications of Newton's Laws of Motion

23. To simulate the apparent weightlessness of space orbit, astronauts are trained in the hold of a cargo aircraft that is accelerating downward at g. Why will they appear to be weightless, as measured by standing on a bathroom scale, in this accelerated frame of reference? Is there any difference between their apparent weightlessness in orbit and in the aircraft?

24. A cartoon shows the toupee coming off the head of an elevator passenger when the elevator rapidly stops during an upward ride. Can this really happen without the person being tied to the floor of the elevator? Explain your answer.

4.8 Extended Topic: The Four Basic Forces—An Introduction

25. Explain, in terms of the properties of the four basic forces, why people notice the gravitational force acting on their bodies if it is such a comparatively weak force.

26. What is the dominant force between astronomical objects? Why are the other three basic forces less significant over these very large distances?

27. Give a detailed example of how the exchange of a particle can result in an *attractive* force. (For example, consider one child pulling a toy out of the hands of another.)

Problems & Exercises

4.3 Newton's Second Law of Motion: Concept of a System

You may assume data taken from illustrations is accurate to three digits.

1. A 63.0-kg sprinter starts a race with an acceleration of $4.20 \ \text{m/s}^2$. What is the net external force on him?

2. If the sprinter from the previous problem accelerates at that rate for 20 m, and then maintains that velocity for the remainder of the 100-m dash, what will be his time for the race?

3. A cleaner pushes a 4.50-kg laundry cart in such a way that the net external force on it is 60.0 N. Calculate the magnitude of its acceleration.

4. Since astronauts in orbit are apparently weightless, a clever method of measuring their masses is needed to monitor their mass gains or losses to adjust diets. One way to do this is to exert a known force on an astronaut and measure the acceleration produced. Suppose a net external force of 50.0 N is exerted and the astronaut's acceleration is measured to be $0.893 \ \text{m/s}^2$. (a) Calculate her mass. (b) By exerting a force on the astronaut, the vehicle in which they orbit experiences an equal and opposite force. Discuss how this would affect the measurement of the astronaut's acceleration. Propose a method in which recoil of the vehicle is avoided.

5. In **Figure 4.7**, the net external force on the 24-kg mower is stated to be 51 N. If the force of friction opposing the motion is 24 N, what force F (in newtons) is the person exerting on the mower? Suppose the mower is moving at 1.5 m/s when the force F is removed. How far will the mower go before stopping?

6. The same rocket sled drawn in **Figure 4.31** is decelerated at a rate of $196 \ \text{m/s}^2$. What force is necessary to produce this deceleration? Assume that the rockets are off. The mass of the system is 2100 kg.

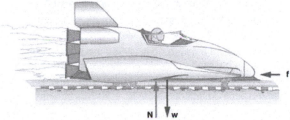

Figure 4.31

7. (a) If the rocket sled shown in **Figure 4.32** starts with only one rocket burning, what is the magnitude of its acceleration? Assume that the mass of the system is 2100 kg, the thrust T is 2.4×10^4 N, and the force of friction opposing the motion is known to be 650 N. (b) Why is the acceleration not one-fourth of what it is with all rockets burning?

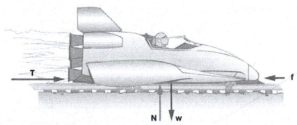

Figure 4.32

8. What is the deceleration of the rocket sled if it comes to rest in 1.1 s from a speed of 1000 km/h? (Such deceleration caused one test subject to black out and have temporary blindness.)

9. Suppose two children push horizontally, but in exactly opposite directions, on a third child in a wagon. The first child exerts a force of 75.0 N, the second a force of 90.0 N, friction is 12.0 N, and the mass of the third child plus wagon is 23.0 kg. (a) What is the system of interest if the acceleration of the child in the wagon is to be calculated? (b) Draw a free-body diagram, including all forces acting on the system. (c) Calculate the acceleration. (d) What would the acceleration be if friction were 15.0 N?

10. A powerful motorcycle can produce an acceleration of $3.50 \ \text{m/s}^2$ while traveling at 90.0 km/h. At that speed the forces resisting motion, including friction and air resistance, total 400 N. (Air resistance is analogous to air friction. It always opposes the motion of an object.) What is the magnitude of the force the motorcycle exerts backward on the ground to produce its acceleration if the mass of the motorcycle with rider is 245 kg?

11. The rocket sled shown in **Figure 4.33** accelerates at a rate of $49.0 \ \text{m/s}^2$. Its passenger has a mass of 75.0 kg. (a) Calculate the horizontal component of the force the seat exerts against his body. Compare this with his weight by using a ratio. (b) Calculate the direction and magnitude of the total force the seat exerts against his body.

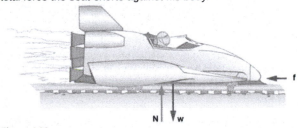

Figure 4.33

12. Repeat the previous problem for the situation in which the rocket sled decelerates at a rate of $201 \ \text{m/s}^2$. In this problem, the forces are exerted by the seat and restraining belts.

13. The weight of an astronaut plus his space suit on the Moon is only 250 N. How much do they weigh on Earth? What is the mass on the Moon? On Earth?

14. Suppose the mass of a fully loaded module in which astronauts take off from the Moon is 10,000 kg. The thrust of its engines is 30,000 N. (a) Calculate its the magnitude of acceleration in a vertical takeoff from the Moon. (b) Could it lift off from Earth? If not, why not? If it could, calculate the magnitude of its acceleration.

4.4 Newton's Third Law of Motion: Symmetry in

Forces

15. What net external force is exerted on a 1100-kg artillery shell fired from a battleship if the shell is accelerated at 2.40×10^4 m/s^2 ? What is the magnitude of the force exerted on the ship by the artillery shell?

16. A brave but inadequate rugby player is being pushed backward by an opposing player who is exerting a force of 800 N on him. The mass of the losing player plus equipment is 90.0 kg, and he is accelerating at 1.20 m/s^2 backward. (a) What is the force of friction between the losing player's feet and the grass? (b) What force does the winning player exert on the ground to move forward if his mass plus equipment is 110 kg? (c) Draw a sketch of the situation showing the system of interest used to solve each part. For this situation, draw a free-body diagram and write the net force equation.

4.5 Normal, Tension, and Other Examples of Force

17. Two teams of nine members each engage in a tug of war. Each of the first team's members has an average mass of 68 kg and exerts an average force of 1350 N horizontally. Each of the second team's members has an average mass of 73 kg and exerts an average force of 1365 N horizontally. (a) What is magnitude of the acceleration of the two teams? (b) What is the tension in the section of rope between the teams?

18. What force does a trampoline have to apply to a 45.0-kg gymnast to accelerate her straight up at 7.50 m/s^2 ? Note that the answer is independent of the velocity of the gymnast—she can be moving either up or down, or be stationary.

19. (a) Calculate the tension in a vertical strand of spider web if a spider of mass 8.00×10^{-5} kg hangs motionless on it. (b) Calculate the tension in a horizontal strand of spider web if the same spider sits motionless in the middle of it much like the tightrope walker in **Figure 4.17**. The strand sags at an angle of $12°$ below the horizontal. Compare this with the tension in the vertical strand (find their ratio).

20. Suppose a 60.0-kg gymnast climbs a rope. (a) What is the tension in the rope if he climbs at a constant speed? (b) What is the tension in the rope if he accelerates upward at a rate of 1.50 m/s^2 ?

21. Show that, as stated in the text, a force $\mathbf{F_\perp}$ exerted on a flexible medium at its center and perpendicular to its length (such as on the tightrope wire in **Figure 4.17**) gives rise to a tension of magnitude $T = \dfrac{F_\perp}{2 \sin (\theta)}$.

22. Consider the baby being weighed in **Figure 4.34**. (a) What is the mass of the child and basket if a scale reading of 55 N is observed? (b) What is the tension T_1 in the cord attaching the baby to the scale? (c) What is the tension T_2 in the cord attaching the scale to the ceiling, if the scale has a mass of 0.500 kg? (d) Draw a sketch of the situation indicating the system of interest used to solve each part. The masses of the cords are negligible.

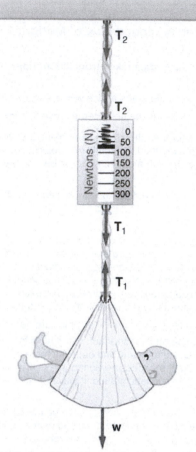

Figure 4.34 A baby is weighed using a spring scale.

4.6 Problem-Solving Strategies

23. A 5.00×10^5-kg rocket is accelerating straight up. Its engines produce 1.250×10^7 N of thrust, and air resistance is 4.50×10^6 N . What is the rocket's acceleration? Explicitly show how you follow the steps in the Problem-Solving Strategy for Newton's laws of motion.

24. The wheels of a midsize car exert a force of 2100 N backward on the road to accelerate the car in the forward direction. If the force of friction including air resistance is 250 N and the acceleration of the car is 1.80 m/s^2 , what is the mass of the car plus its occupants? Explicitly show how you follow the steps in the Problem-Solving Strategy for Newton's laws of motion. For this situation, draw a free-body diagram and write the net force equation.

25. Calculate the force a 70.0-kg high jumper must exert on the ground to produce an upward acceleration 4.00 times the acceleration due to gravity. Explicitly show how you follow the steps in the Problem-Solving Strategy for Newton's laws of motion.

26. When landing after a spectacular somersault, a 40.0-kg gymnast decelerates by pushing straight down on the mat. Calculate the force she must exert if her deceleration is 7.00 times the acceleration due to gravity. Explicitly show how you follow the steps in the Problem-Solving Strategy for Newton's laws of motion.

27. A freight train consists of two 8.00×10^4-kg engines and 45 cars with average masses of 5.50×10^4 kg . (a) What force must each engine exert backward on the track to accelerate the train at a rate of 5.00×10^{-2} m/s^2 if the force of friction is 7.50×10^5 N , assuming the engines exert identical forces? This is not a large frictional force for such a massive system. Rolling friction for trains is small, and consequently trains are very energy-efficient transportation systems. (b) What is the force in the coupling between the 37th and 38th cars (this is the force each exerts on the other), assuming all cars have the same mass and that friction is evenly distributed among all of the cars and engines?

28. Commercial airplanes are sometimes pushed out of the passenger loading area by a tractor. (a) An 1800-kg tractor exerts a force of 1.75×10^4 N backward on the pavement, and the system experiences forces resisting motion that total 2400 N. If the acceleration is 0.150 m/s^2 , what is the mass of the airplane? (b) Calculate the force exerted by the tractor on the airplane, assuming 2200 N of the friction is experienced by the airplane. (c) Draw two sketches showing the systems of interest used to solve each part, including the free-body diagrams for each.

29. A 1100-kg car pulls a boat on a trailer. (a) What total force resists the motion of the car, boat, and trailer, if the car exerts a 1900-N force on the road and produces an acceleration of 0.550 m/s^2 ? The mass of the boat plus trailer is 700 kg. (b) What is the force in the hitch between the car and the trailer if 80% of the resisting forces are experienced by the boat and trailer?

30. (a) Find the magnitudes of the forces $\mathbf{F}_1$ and $\mathbf{F}_2$ that add to give the total force $\mathbf{F}_{\text{tot}}$ shown in **Figure 4.35**. This may be done either graphically or by using trigonometry. (b) Show graphically that the same total force is obtained independent of the order of addition of $\mathbf{F}_1$ and $\mathbf{F}_2$. (c) Find the direction and magnitude of some other pair of vectors that add to give $\mathbf{F}_{\text{tot}}$. Draw these to scale on the same drawing used in part (b) or a similar picture.

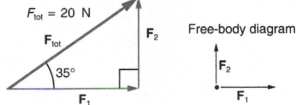

Figure 4.35

31. Two children pull a third child on a snow saucer sled exerting forces $\mathbf{F}_1$ and $\mathbf{F}_2$ as shown from above in **Figure 4.36**. Find the acceleration of the 49.00-kg sled and child system. Note that the direction of the frictional force is unspecified; it will be in the opposite direction of the sum of $\mathbf{F}_1$ and $\mathbf{F}_2$.

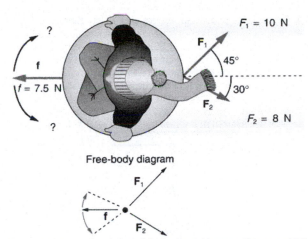

Figure 4.36 An overhead view of the horizontal forces acting on a child's snow saucer sled.

32. Suppose your car was mired deeply in the mud and you wanted to use the method illustrated in **Figure 4.37** to pull it out. (a) What force would you have to exert perpendicular to the center of the rope to produce a force of 12,000 N on the car if the angle is 2.00°? In this part, explicitly show how you follow the steps in the Problem-Solving Strategy for Newton's laws of motion. (b) Real ropes stretch under such forces. What force would be exerted on the car if the angle increases to 7.00° and you still apply the force found in part (a) to its center?

Figure 4.37

33. What force is exerted on the tooth in **Figure 4.38** if the tension in the wire is 25.0 N? Note that the force applied to the tooth is smaller than the tension in the wire, but this is necessitated by practical considerations of how force can be applied in the mouth. Explicitly show how you follow steps in the Problem-Solving Strategy for Newton's laws of motion.

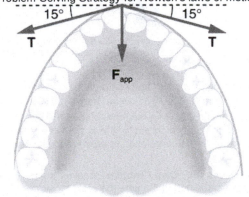

Figure 4.38 Braces are used to apply forces to teeth to realign them. Shown in this figure are the tensions applied by the wire to the protruding tooth. The total force applied to the tooth by the wire, $\mathbf{F}_{\text{app}}$, points straight toward the back of the mouth.

34. Figure 4.39 shows Superhero and Trusty Sidekick hanging motionless from a rope. Superhero's mass is 90.0 kg, while Trusty Sidekick's is 55.0 kg, and the mass of the rope is negligible. (a) Draw a free-body diagram of the situation showing all forces acting on Superhero, Trusty Sidekick, and the rope. (b) Find the tension in the rope above Superhero. (c) Find the tension in the rope between Superhero and Trusty Sidekick. Indicate on your free-body diagram the system of interest used to solve each part.

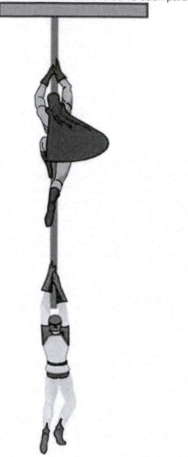

Figure 4.39 Superhero and Trusty Sidekick hang motionless on a rope as they try to figure out what to do next. Will the tension be the same everywhere in the rope?

35. A nurse pushes a cart by exerting a force on the handle at a downward angle $35.0°$ below the horizontal. The loaded cart has a mass of 28.0 kg, and the force of friction is 60.0 N. (a) Draw a free-body diagram for the system of interest. (b) What force must the nurse exert to move at a constant velocity?

36. Construct Your Own Problem Consider the tension in an elevator cable during the time the elevator starts from rest and accelerates its load upward to some cruising velocity. Taking the elevator and its load to be the system of interest, draw a free-body diagram. Then calculate the tension in the cable. Among the things to consider are the mass of the elevator and its load, the final velocity, and the time taken to reach that velocity.

37. Construct Your Own Problem Consider two people pushing a toboggan with four children on it up a snow-covered slope. Construct a problem in which you calculate the acceleration of the toboggan and its load. Include a free-body diagram of the appropriate system of interest as the basis for your analysis. Show vector forces and their components and explain the choice of coordinates. Among the things to be considered are the forces exerted by those pushing, the angle of the slope, and the masses of the toboggan and children.

38. Unreasonable Results (a) Repeat Exercise 4.29, but assume an acceleration of 1.20 m/s^2 is produced. (b) What is unreasonable about the result? (c) Which premise is unreasonable, and why is it unreasonable?

39. Unreasonable Results (a) What is the initial acceleration of a rocket that has a mass of $1.50 \times 10^6 \text{ kg}$ at takeoff, the engines of which produce a thrust of $2.00 \times 10^6 \text{ N}$? Do not neglect gravity. (b) What is unreasonable about the result? (This result has been unintentionally achieved by several real rockets.) (c) Which premise is unreasonable, or which premises are inconsistent? (You may find it useful to compare this problem to the rocket problem earlier in this section.)

4.7 Further Applications of Newton's Laws of Motion

40. A flea jumps by exerting a force of $1.20 \times 10^{-5} \text{ N}$ straight down on the ground. A breeze blowing on the flea parallel to the ground exerts a force of $0.500 \times 10^{-6} \text{ N}$ on the flea. Find the direction and magnitude of the acceleration of the flea if its mass is $6.00 \times 10^{-7} \text{ kg}$. Do not neglect the gravitational force.

41. Two muscles in the back of the leg pull upward on the Achilles tendon, as shown in **Figure 4.40**. (These muscles are called the medial and lateral heads of the gastrocnemius muscle.) Find the magnitude and direction of the total force on the Achilles tendon. What type of movement could be caused by this force?

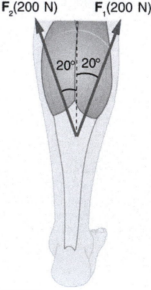

Figure 4.40 Achilles tendon

42. A 76.0-kg person is being pulled away from a burning building as shown in **Figure 4.41**. Calculate the tension in the two ropes if the person is momentarily motionless. Include a free-body diagram in your solution.

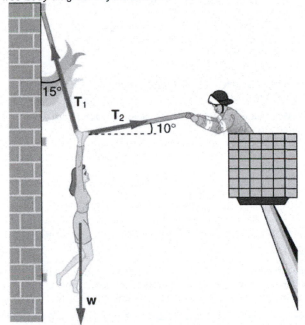

Figure 4.41 The force T_2 needed to hold steady the person being rescued from the fire is less than her weight and less than the force T_1 in the other rope, since the more vertical rope supports a greater part of her weight (a vertical force).

43. Integrated Concepts A 35.0-kg dolphin decelerates from 12.0 to 7.50 m/s in 2.30 s to join another dolphin in play. What average force was exerted to slow him if he was moving horizontally? (The gravitational force is balanced by the buoyant force of the water.)

44. Integrated Concepts When starting a foot race, a 70.0-kg sprinter exerts an average force of 650 N backward on the ground for 0.800 s. (a) What is his final speed? (b) How far does he travel?

45. Integrated Concepts A large rocket has a mass of 2.00×10^6 kg at takeoff, and its engines produce a thrust of 3.50×10^7 N. (a) Find its initial acceleration if it takes off vertically. (b) How long does it take to reach a velocity of 120 km/h straight up, assuming constant mass and thrust? (c) In reality, the mass of a rocket decreases significantly as its fuel is consumed. Describe qualitatively how this affects the acceleration and time for this motion.

46. Integrated Concepts A basketball player jumps straight up for a ball. To do this, he lowers his body 0.300 m and then accelerates through this distance by forcefully straightening his legs. This player leaves the floor with a vertical velocity sufficient to carry him 0.900 m above the floor. (a) Calculate his velocity when he leaves the floor. (b) Calculate his acceleration while he is straightening his legs. He goes from zero to the velocity found in part (a) in a distance of 0.300 m. (c) Calculate the force he exerts on the floor to do this, given that his mass is 110 kg.

47. Integrated Concepts A 2.50-kg fireworks shell is fired straight up from a mortar and reaches a height of 110 m. (a) Neglecting air resistance (a poor assumption, but we will make it for this example), calculate the shell's velocity when it leaves the mortar. (b) The mortar itself is a tube 0.450 m long. Calculate the average acceleration of the shell in the tube as it goes from zero to the velocity found in (a). (c) What is the average force on the shell in the mortar? Express your answer in newtons and as a ratio to the weight of the shell.

48. Integrated Concepts Repeat **Exercise 4.47** for a shell fired at an angle $10.0°$ from the vertical.

49. Integrated Concepts An elevator filled with passengers has a mass of 1700 kg. (a) The elevator accelerates upward from rest at a rate of 1.20 m/s^2 for 1.50 s. Calculate the tension in the cable supporting the elevator. (b) The elevator continues upward at constant velocity for 8.50 s. What is the tension in the cable during this time? (c) The elevator decelerates at a rate of 0.600 m/s^2 for 3.00 s. What is the tension in the cable during deceleration? (d) How high has the elevator moved above its original starting point, and what is its final velocity?

50. Unreasonable Results (a) What is the final velocity of a car originally traveling at 50.0 km/h that decelerates at a rate of 0.400 m/s^2 for 50.0 s? (b) What is unreasonable about the result? (c) Which premise is unreasonable, or which premises are inconsistent?

51. Unreasonable Results A 75.0-kg man stands on a bathroom scale in an elevator that accelerates from rest to 30.0 m/s in 2.00 s. (a) Calculate the scale reading in newtons and compare it with his weight. (The scale exerts an upward force on him equal to its reading.) (b) What is unreasonable about the result? (c) Which premise is unreasonable, or which premises are inconsistent?

4.8 Extended Topic: The Four Basic Forces—An Introduction

52. (a) What is the strength of the weak nuclear force relative to the strong nuclear force? (b) What is the strength of the weak nuclear force relative to the electromagnetic force? Since the weak nuclear force acts at only very short distances, such as inside nuclei, where the strong and electromagnetic forces also act, it might seem surprising that we have any knowledge of it at all. We have such knowledge because the weak nuclear force is responsible for beta decay, a type of nuclear decay not explained by other forces.

53. (a) What is the ratio of the strength of the gravitational force to that of the strong nuclear force? (b) What is the ratio of the strength of the gravitational force to that of the weak nuclear force? (c) What is the ratio of the strength of the gravitational force to that of the electromagnetic force? What do your answers imply about the influence of the gravitational force on atomic nuclei?

54. What is the ratio of the strength of the strong nuclear force to that of the electromagnetic force? Based on this ratio, you might expect that the strong force dominates the nucleus, which is true for small nuclei. Large nuclei, however, have sizes greater than the range of the strong nuclear force. At these sizes, the electromagnetic force begins to affect nuclear stability. These facts will be used to explain nuclear fusion and fission later in this text.

Test Prep for AP® Courses

4.1 Development of Force Concept

1.

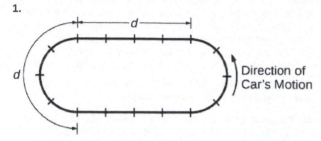

Figure 4.42 The figure above represents a racetrack with semicircular sections connected by straight sections. Each section has length d, and markers along the track are spaced $d/4$ apart. Two people drive cars counterclockwise around the track, as shown. Car X goes around the curves at constant speed vc, increases speed at constant acceleration for half of each straight section to reach a maximum speed of $2vc$, then brakes at constant acceleration for the other half of each straight section to return to speed vc. Car Y also goes around the curves at constant speed vc, increases its speed at constant acceleration for one-fourth of each straight section to reach the same maximum speed $2vc$, stays at that speed for half of each straight section, then brakes at constant acceleration for the remaining fourth of each straight section to return to speed vc.

(a) On the figures below, draw an arrow showing the direction of the net force on each of the cars at the positions noted by the dots. If the net force is zero at any position, label the dot with 0.

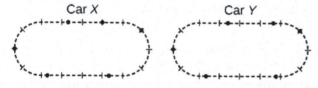

Figure 4.43

The position of the six dots on the Car Y track on the right are as follows:

The first dot on the left center of the track is at the same position as it is on the Car X track.

The second dot is just slight to the right of the Car X dot (less than a dash) past three perpendicular hash marks moving to the right.

The third dot is about one and two-thirds perpendicular hash marks to the right of the center top perpendicular has mark.

The fourth dot is in the same position as the Car X figure (one perpendicular hash mark above the center right perpendicular hash mark).

The fifth dot is about one and two-third perpendicular hash marks to the right of the center bottom perpendicular hash mark.

The sixth dot is in the same position as the Car Y dot (one and two third perpendicular hash marks to the left of the center bottom hash mark).

(b)

i. Indicate which car, if either, completes one trip around the track in less time, and justify your answer qualitatively without using equations.

ii. Justify your answer about which car, if either, completes one trip around the track in less time quantitatively with appropriate equations.

2. Which of the following is an example of a body exerting a force on itself?
   a. a person standing up from a seated position
   b. a car accelerating while driving
   c. both of the above
   d. none of the above

3. A hawk accelerates as it glides in the air. Does the force causing the acceleration come from the hawk itself? Explain.

4. What causes the force that moves a boat forward when someone rows it?
   a. The force is caused by the rower's arms.
   b. The force is caused by an interaction between the oars and gravity.
   c. The force is caused by an interaction between the oars and the water the boat is traveling in.
   d. The force is caused by friction.

4.4 Newton's Third Law of Motion: Symmetry in Forces

5. What object or objects commonly exert forces on the following objects in motion? (a) a soccer ball being kicked, (b) a dolphin jumping, (c) a parachutist drifting to Earth.

6. A ball with a mass of 0.25 kg hits a gym ceiling with a force of 78.0 N. What happens next?
   a. The ball accelerates downward with a force of 80.5 N.
   b. The ball accelerates downward with a force of 78.0 N.
   c. The ball accelerates downward with a force of 2.45 N.
   d. It depends on the height of the ceiling.

7. Which of the following is true?
   a. Earth exerts a force due to gravity on your body, and your body exerts a smaller force on the Earth, because your mass is smaller than the mass of the Earth.
   b. The Moon orbits the Earth because the Earth exerts a force on the Moon and the Moon exerts a force equal in magnitude and direction on the Earth.
   c. A rocket taking off exerts a force on the Earth equal to the force the Earth exerts on the rocket.
   d. An airplane cruising at a constant speed is not affected by gravity.

8. Stationary skater A pushes stationary skater B, who then accelerates at 5.0 m/s^2. Skater A does not move. Since forces act in action-reaction pairs, explain why Skater A did not move?

9. The current in a river exerts a force of 9.0 N on a balloon floating in the river. A wind exerts a force of 13.0 N on the balloon in the opposite direction. Draw a free-body diagram to show the forces acting on the balloon. Use your free-body diagram to predict the effect on the balloon.

10. A force is applied to accelerate an object on a smooth icy surface. When the force stops, which of the following will be true? (Assume zero friction.)
   a. The object's acceleration becomes zero.
   b. The object's speed becomes zero.
   c. The object's acceleration continues to increase at a constant rate.
   d. The object accelerates, but in the opposite direction.

11. A parachutist's fall to Earth is determined by two opposing forces. A gravitational force of 539 N acts on the parachutist. After 2 s, she opens her parachute and experiences an air resistance of 615 N. At what speed is the parachutist falling after 10 s?

12. A flight attendant pushes a cart down the aisle of a plane in flight. In determining the acceleration of the cart relative to the plane, which factor do you not need to consider?
 a.   The friction of the cart's wheels.
 b.   The force with which the flight attendant's feet push on the floor.
 c.   The velocity of the plane.
 d.   The mass of the items in the cart.

13. A landscaper is easing a wheelbarrow full of soil down a hill. Define the system you would analyze and list all the forces that you would need to include to calculate the acceleration of the wheelbarrow.

14. Two water-skiers, with masses of 48 kg and 61 kg, are preparing to be towed behind the same boat. When the boat accelerates, the rope the skiers hold onto accelerates with it and exerts a net force of 290 N on the skiers. At what rate will the skiers accelerate?
 a. 10.8 m/s^2
 b. 2.7 m/s^2
 c. 6.0 m/s^2 and 4.8 m/s^2
 d. 5.3 m/s^2

15. A figure skater has a mass of 40 kg and her partner's mass is 50 kg. She pushes against the ice with a force of 120 N, causing her and her partner to move forward. Calculate the pair's acceleration. Assume that all forces opposing the motion, such as friction and air resistance, total 5.0 N.

4.5 Normal, Tension, and Other Examples of Force

16. An archer shoots an arrow straight up with a force of 24.5 N. The arrow has a mass of 0.4 kg. What is the force of gravity on the arrow?
 a. 9.8 m/s^2
 b.   9.8 N
 c.   61.25 N
 d.   3.9 N

17. A cable raises a mass of 120.0 kg with an acceleration of 1.3 m/s2. What force of tension is in the cable?

18. A child pulls a wagon along a grassy field. Define the system, the pairs of forces at work, and the results.

19. Two teams are engaging in a tug–of–war. The rope suddenly snaps. Which statement is true about the forces involved?
 a.   The forces exerted by the two teams are no longer equal; the teams will accelerate in opposite directions as a result.
 b.   The forces exerted by the players are no longer balanced by the force of tension in the rope; the teams will accelerate in opposite directions as a result.
 c.   The force of gravity balances the forces exerted by the players; the teams will fall as a result
 d.   The force of tension in the rope is transferred to the players; the teams will accelerate in opposite directions as a result.

20. The following free-body diagram represents a toboggan on a hill. What acceleration would you expect, and why?

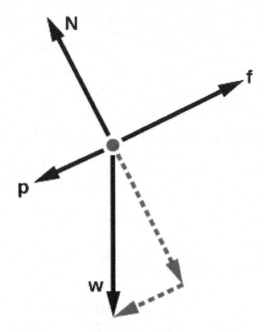

Figure 4.44
 a.   Acceleration down the hill; the force due to being pushed, together with the downhill component of gravity, overcomes the opposing force of friction.
 b.   Acceleration down the hill; friction is less than the opposing component of force due to gravity.
 c.   No movement; friction is greater than the force due to being pushed.
 d.   It depends on how strong the force due to friction is. *p*

21. Draw a free-body diagram to represent the forces acting on a kite on a string that is floating stationary in the air. Label the forces in your diagram.

22. A car is sliding down a hill with a slope of 20°. The mass of the car is 965 kg. When a cable is used to pull the car up the slope, a force of 4215 N is applied. What is the car's acceleration, ignoring friction?

4.6 Problem-Solving Strategies

23. A toboggan with two riders has a total mass of 85.0 kg. A third person is pushing the toboggan with a force of 42.5 N at the top of a hill with an angle of 15°. The force of friction on the toboggan is 31.0 N. Which statement describes an accurate free-body diagram to represent the situation?
 a.   An arrow of magnitude 10.5 N points down the slope of the hill.
 b.   An arrow of magnitude 833 N points straight down.
 c.   An arrow of magnitude 833 N points perpendicular to the slope of the hill.
 d.   An arrow of magnitude 73.5 N points down the slope of the hill.

24. A mass of 2.0 kg is suspended from the ceiling of an elevator by a rope. What is the tension in the rope when the elevator (i) accelerates upward at 1.5 m/s^2? (ii) accelerates downward at 1.5 m/s^2?
 a.   (i) 22.6 N; (ii) 16.6 N
 b.   Because the mass is hanging from the elevator itself, the tension in the rope will not change in either case.
 c.   (i) 22.6 N; (ii) 19.6 N
 d.   (i) 16.6 N; (ii) 19.6 N

25. Which statement is true about drawing free-body diagrams?

a. Drawing a free-body diagram should be the last step in solving a problem about forces.
b. Drawing a free-body diagram helps you compare forces quantitatively.
c. The forces in a free-body diagram should always balance.
d. Drawing a free-body diagram can help you determine the net force.

4.7 Further Applications of Newton's Laws of Motion

26. A basketball player jumps as he shoots the ball. Describe the forces that are acting on the ball and on the basketball player. What are the results?

27. Two people push on a boulder to try to move it. The mass of the boulder is 825 kg. One person pushes north with a force of 64 N. The other pushes west with a force of 38 N. Predict the magnitude of the acceleration of the boulder. Assume that friction is negligible.

28.

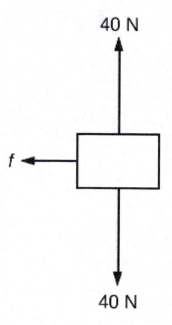

Figure 4.45 The figure shows the forces exerted on a block that is sliding on a horizontal surface: the gravitational force of 40 N, the 40 N normal force exerted by the surface, and a frictional force exerted to the left. The coefficient of friction between the block and the surface is 0.20. The acceleration of the block is most nearly

a. 1.0 m/s^2 to the right
b. 1.0 m/s^2 to the left
c. 2.0 m/s^2 to the right
d. 2.0 m/s^2 to the left

4.8 Extended Topic: The Four Basic Forces—An Introduction

29. Which phenomenon correctly describes the direction and magnitude of normal forces?
a. electromagnetic attraction
b. electromagnetic repulsion
c. gravitational attraction
d. gravitational repulsion

30. Explain which of the four fundamental forces is

responsible for a ball bouncing off the ground after it hits, and why this force has this effect.

31. Which of the basic forces best explains tension in a rope being pulled between two people? Is the acting force causing attraction or repulsion in this instance?
a. gravity; attraction
b. electromagnetic; attraction
c. weak and strong nuclear; attraction
d. weak and strong nuclear; repulsion

32. Explain how interatomic electric forces produce the normal force, and why it has the direction it does.

33. The gravitational force is the weakest of the four basic forces. In which case can the electromagnetic, strong, and weak forces be ignored because the gravitational force is so strongly dominant?
a. a person jumping on a trampoline
b. a rocket blasting off from Earth
c. a log rolling down a hill
d. all of the above

34. Describe a situation in which gravitational force is the dominant force. Why can the other three basic forces be ignored in the situation you described?

5   FURTHER APPLICATIONS OF NEWTON'S LAWS: FRICTION, DRAG, AND ELASTICITY

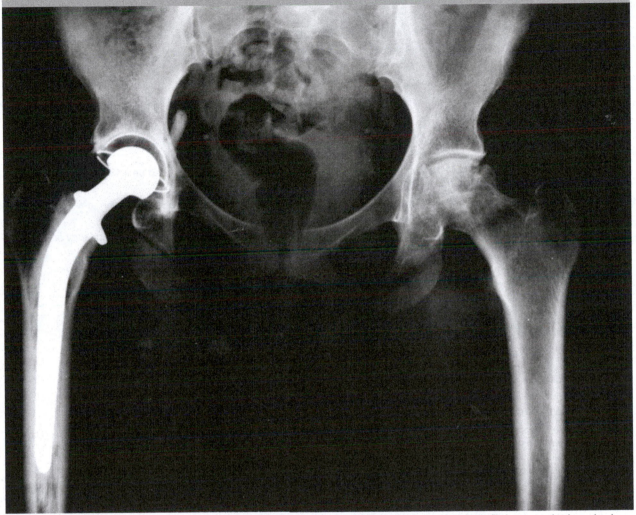

Figure 5.1 Total hip replacement surgery has become a common procedure. The head (or ball) of the patient's femur fits into a cup that has a hard plastic-like inner lining. (credit: National Institutes of Health, via Wikimedia Commons)

Chapter Outline

5.1. Friction

5.2. Drag Forces

5.3. Elasticity: Stress and Strain

Connection for AP® Courses

Have you ever wondered why it is difficult to walk on a smooth surface like ice? The interaction between you and the surface is a result of forces that affect your motion. In the previous chapter, you learned Newton's laws of motion and examined how net force affects the motion, position and shape of an object. Now we will look at some interesting and common forces that will provide further applications of Newton's laws of motion.

The information presented in this chapter supports learning objectives covered under Big Idea 3 of the AP Physics Curriculum Framework, which refer to the nature of forces and their roles in interactions among objects. The chapter discusses examples of specific contact forces, such as friction, air or liquid drag, and elasticity that may affect the motion or shape of an object. It also discusses the nature of forces on both macroscopic and microscopic levels (Enduring Understanding 3.C and Essential Knowledge 3.C.4). In addition, Newton's laws are applied to describe the motion of an object (Enduring Understanding 3.B) and to examine relationships between contact forces and other forces exerted on an object (Enduring Understanding 3.A, 3.A.3 and

Essential Knowledge 3.A.4). The examples in this chapter give you practice in using vector properties of forces (Essential Knowledge 3.A.2) and free-body diagrams (Essential Knowledge 3.B.2) to determine net force (Essential Knowledge 3.B.1).

Big Idea 3 The interactions of an object with other objects can be described by forces.

Enduring Understanding 3.A All forces share certain common characteristics when considered by observers in inertial reference frames.

Essential Knowledge 3.A.2 Forces are described by vectors.

Essential Knowledge 3.A.3 A force exerted on an object is always due to the interaction of that object with another object.

Essential Knowledge 3.A.4 If one object exerts a force on a second object, the second object always exerts a force of equal magnitude on the first object in the opposite direction.

Enduring Understanding 3.B Classically, the acceleration of an object interacting with other objects can be predicted by using

$$\vec{a} = \frac{\sum \vec{F}}{m}.$$

Essential Knowledge 3.B.1 If an object of interest interacts with several other objects, the net force is the vector sum of the individual forces.

Essential Knowledge 3.B.2 Free-body diagrams are useful tools for visualizing forces being exerted on a single object and writing the equations that represent a physical situation.

Enduring Understanding 3.C At the macroscopic level, forces can be categorized as either long-range (action-at-a-distance) forces or contact forces.

Essential Knowledge 3.C.4 Contact forces result from the interaction of one object touching another object, and they arise from interatomic electric forces. These forces include tension, friction, normal, spring (Physics 1), and buoyant (Physics 2).

5.1 Friction

Learning Objectives

By the end of this section, you will be able to:

- Discuss the general characteristics of friction.
- Describe the various types of friction.
- Calculate the magnitudes of static and kinetic frictional forces.

The information presented in this section supports the following AP® learning objectives and science practices:

- **3.C.4.1** The student is able to make claims about various contact forces between objects based on the microscopic cause of those forces. **(S.P. 6.1)**
- **3.C.4.2** The student is able to explain contact forces (tension, friction, normal, buoyant, spring) as arising from interatomic electric forces and that they therefore have certain directions. **(S.P. 6.2)**

Friction is a force that is around us all the time that opposes relative motion between systems in contact but also allows us to move (which you have discovered if you have ever tried to walk on ice). While a common force, the behavior of friction is actually very complicated and is still not completely understood. We have to rely heavily on observations for whatever understandings we can gain. However, we can still deal with its more elementary general characteristics and understand the circumstances in which it behaves.

Friction
___

Friction is a force that opposes relative motion between systems in contact.

One of the simpler characteristics of friction is that it is parallel to the contact surface between systems and always in a direction that opposes motion or attempted motion of the systems relative to each other. If two systems are in contact and moving relative to one another, then the friction between them is called **kinetic friction**. For example, friction slows a hockey puck sliding on ice. But when objects are stationary, **static friction** can act between them; the static friction is usually greater than the kinetic friction between the objects.

Kinetic Friction
___

If two systems are in contact and moving relative to one another, then the friction between them is called kinetic friction.

Imagine, for example, trying to slide a heavy crate across a concrete floor—you may push harder and harder on the crate and not move it at all. This means that the static friction responds to what you do—it increases to be equal to and in the opposite direction of your push. But if you finally push hard enough, the crate seems to slip suddenly and starts to move. Once in motion it is easier to keep it in motion than it was to get it started, indicating that the kinetic friction force is less than the static friction force. If you add mass to the crate, say by placing a box on top of it, you need to push even harder to get it started and also to keep it moving. Furthermore, if you oiled the concrete you would find it to be easier to get the crate started and keep it going (as you might expect).

Figure 5.2 is a crude pictorial representation of how friction occurs at the interface between two objects. Close-up inspection of these surfaces shows them to be rough. So when you push to get an object moving (in this case, a crate), you must raise the object until it can skip along with just the tips of the surface hitting, break off the points, or do both. A considerable force can be resisted by friction with no apparent motion. The harder the surfaces are pushed together (such as if another box is placed on the crate), the more force is needed to move them. Part of the friction is due to adhesive forces between the surface molecules of the two objects, which explain the dependence of friction on the nature of the substances. Adhesion varies with substances in contact and is a complicated aspect of surface physics. Once an object is moving, there are fewer points of contact (fewer molecules adhering), so less force is required to keep the object moving. At small but nonzero speeds, friction is nearly independent of speed.

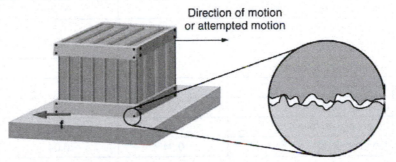

Figure 5.2 Frictional forces, such as f , always oppose motion or attempted motion between objects in contact. Friction arises in part because of the roughness of the surfaces in contact, as seen in the expanded view. In order for the object to move, it must rise to where the peaks can skip along the bottom surface. Thus a force is required just to set the object in motion. Some of the peaks will be broken off, also requiring a force to maintain motion. Much of the friction is actually due to attractive forces between molecules making up the two objects, so that even perfectly smooth surfaces are not friction-free. Such adhesive forces also depend on the substances the surfaces are made of, explaining, for example, why rubber-soled shoes slip less than those with leather soles.

The magnitude of the frictional force has two forms: one for static situations (static friction), the other for when there is motion (kinetic friction).

When there is no motion between the objects, the **magnitude of static friction $\mathbf{f_s}$** is

$$f_s \le \mu_s N, \tag{5.1}$$

where μ_s is the coefficient of static friction and N is the magnitude of the normal force (the force perpendicular to the surface).

Magnitude of Static Friction

Magnitude of static friction f_s is

$$f_s \le \mu_s N, \tag{5.2}$$

where μ_s is the coefficient of static friction and N is the magnitude of the normal force.

The symbol $\le$ means *less than or equal to*, implying that static friction can have a minimum and a maximum value of $\mu_s N$. Static friction is a responsive force that increases to be equal and opposite to whatever force is exerted, up to its maximum limit. Once the applied force exceeds $f_{s(max)}$, the object will move. Thus

$$f_{s(max)} = \mu_s N. \tag{5.3}$$

Once an object is moving, the **magnitude of kinetic friction $\mathbf{f_k}$** is given by

$$f_k = \mu_k N, \tag{5.4}$$

where μ_k is the coefficient of kinetic friction. A system in which $f_k = \mu_k N$ is described as a system in which *friction behaves simply*.

Magnitude of Kinetic Friction

The magnitude of kinetic friction f_k is given by

$$f_k = \mu_k N, \tag{5.5}$$

where μ_k is the coefficient of kinetic friction.

As seen in **Table 5.1**, the coefficients of kinetic friction are less than their static counterparts. That values of μ in **Table 5.1** are stated to only one or, at most, two digits is an indication of the approximate description of friction given by the above two equations.

Table 5.1 Coefficients of Static and Kinetic Friction

System	Static friction μ_s	Kinetic friction μ_k
Rubber on dry concrete	1.0	0.7
Rubber on wet concrete	0.7	0.5
Wood on wood	0.5	0.3
Waxed wood on wet snow	0.14	0.1
Metal on wood	0.5	0.3
Steel on steel (dry)	0.6	0.3
Steel on steel (oiled)	0.05	0.03
Teflon on steel	0.04	0.04
Bone lubricated by synovial fluid	0.016	0.015
Shoes on wood	0.9	0.7
Shoes on ice	0.1	0.05
Ice on ice	0.1	0.03
Steel on ice	0.04	0.02

The equations given earlier include the dependence of friction on materials and the normal force. The direction of friction is always opposite that of motion, parallel to the surface between objects, and perpendicular to the normal force. For example, if the crate you try to push (with a force parallel to the floor) has a mass of 100 kg, then the normal force would be equal to its weight, $W = mg = (100 \text{ kg})(9.80 \text{ m/s}^2) = 980 \text{ N}$, perpendicular to the floor. If the coefficient of static friction is 0.45, you would have to exert a force parallel to the floor greater than $f_{s(\text{max})} = \mu_s N = (0.45)(980 \text{ N}) = 440 \text{ N}$ to move the crate.

Once there is motion, friction is less and the coefficient of kinetic friction might be 0.30, so that a force of only 290 N ($f_k = \mu_k N = (0.30)(980 \text{ N}) = 290 \text{ N}$) would keep it moving at a constant speed. If the floor is lubricated, both coefficients are considerably less than they would be without lubrication. Coefficient of friction is a unit less quantity with a magnitude usually between 0 and 1.0. The coefficient of the friction depends on the two surfaces that are in contact.

Take-Home Experiment

Find a small plastic object (such as a food container) and slide it on a kitchen table by giving it a gentle tap. Now spray water on the table, simulating a light shower of rain. What happens now when you give the object the same-sized tap? Now add a few drops of (vegetable or olive) oil on the surface of the water and give the same tap. What happens now? This latter situation is particularly important for drivers to note, especially after a light rain shower. Why?

Many people have experienced the slipperiness of walking on ice. However, many parts of the body, especially the joints, have much smaller coefficients of friction—often three or four times less than ice. A joint is formed by the ends of two bones, which are connected by thick tissues. The knee joint is formed by the lower leg bone (the tibia) and the thighbone (the femur). The hip is a ball (at the end of the femur) and socket (part of the pelvis) joint. The ends of the bones in the joint are covered by cartilage, which provides a smooth, almost glassy surface. The joints also produce a fluid (synovial fluid) that reduces friction and wear. A damaged or arthritic joint can be replaced by an artificial joint (**Figure 5.3**). These replacements can be made of metals (stainless steel or titanium) or plastic (polyethylene), also with very small coefficients of friction.

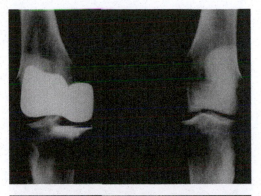

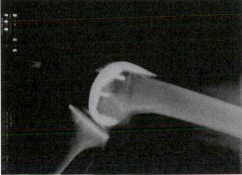

Figure 5.3 Artificial knee replacement is a procedure that has been performed for more than 20 years. In this figure, we see the post-op x rays of the right knee joint replacement. (credit: Mike Baird, Flickr)

Other natural lubricants include saliva produced in our mouths to aid in the swallowing process, and the slippery mucus found between organs in the body, allowing them to move freely past each other during heartbeats, during breathing, and when a person moves. Artificial lubricants are also common in hospitals and doctor's clinics. For example, when ultrasonic imaging is carried out, the gel that couples the transducer to the skin also serves to to lubricate the surface between the transducer and the skin—thereby reducing the coefficient of friction between the two surfaces. This allows the transducer to mover freely over the skin.

Example 5.1 Skiing Exercise

A skier with a mass of 62 kg is sliding down a snowy slope. Find the coefficient of kinetic friction for the skier if friction is known to be 45.0 N.

Strategy

The magnitude of kinetic friction was given in to be 45.0 N. Kinetic friction is related to the normal force N as $f_k = \mu_k N$;

thus, the coefficient of kinetic friction can be found if we can find the normal force of the skier on a slope. The normal force is always perpendicular to the surface, and since there is no motion perpendicular to the surface, the normal force should equal the component of the skier's weight perpendicular to the slope. (See the skier and free-body diagram in **Figure 5.4**.)

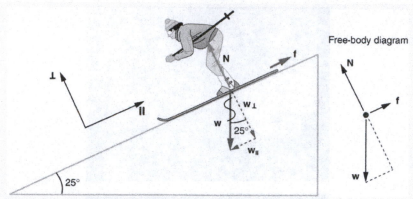

Figure 5.4 The motion of the skier and friction are parallel to the slope and so it is most convenient to project all forces onto a coordinate system where one axis is parallel to the slope and the other is perpendicular (axes shown to left of skier). $\mathbf{N}$ (the normal force) is perpendicular to the slope, and $\mathbf{f}$ (the friction) is parallel to the slope, but $\mathbf{w}$ (the skier's weight) has components along both axes, namely $\mathbf{w}_\perp$ and $\mathbf{W}_{//}$. $\mathbf{N}$ is equal in magnitude to $\mathbf{w}_\perp$, so there is no motion perpendicular to the slope. However, $\mathbf{f}$ is less than $\mathbf{W}_{//}$ in magnitude, so there is acceleration down the slope (along the x-axis).

That is,

$$N = w_\perp = w\cos 25° = mg\cos 25°. \tag{5.6}$$

Substituting this into our expression for kinetic friction, we get

$$f_k = \mu_k mg\cos 25°, \tag{5.7}$$

which can now be solved for the coefficient of kinetic friction μ_k.

Solution

Solving for μ_k gives

$$\mu_k = \frac{f_k}{N} = \frac{f_k}{w\cos 25°} = \frac{f_k}{mg\cos 25°}. \tag{5.8}$$

Substituting known values on the right-hand side of the equation,

$$\mu_k = \frac{45.0\,\text{N}}{(62\,\text{kg})(9.80\,\text{m/s}^2)(0.906)} = 0.082. \tag{5.9}$$

Discussion

This result is a little smaller than the coefficient listed in **Table 5.1** for waxed wood on snow, but it is still reasonable since values of the coefficients of friction can vary greatly. In situations like this, where an object of mass m slides down a slope that makes an angle θ with the horizontal, friction is given by $f_k = \mu_k mg\cos\theta$. All objects will slide down a slope with constant acceleration under these circumstances. Proof of this is left for this chapter's Problems and Exercises.

Take-Home Experiment

An object will slide down an inclined plane at a constant velocity if the net force on the object is zero. We can use this fact to measure the coefficient of kinetic friction between two objects. As shown in **Example 5.1**, the kinetic friction on a slope $f_k = \mu_k mg\cos\theta$. The component of the weight down the slope is equal to $mg\sin\theta$ (see the free-body diagram in **Figure 5.4**). These forces act in opposite directions, so when they have equal magnitude, the acceleration is zero. Writing these out:

$$f_k = Fg_x \tag{5.10}$$
$$\mu_k mg\cos\theta = mg\sin\theta. \tag{5.11}$$

Solving for μ_k, we find that

$$\mu_k = \frac{mg\sin\theta}{mg\cos\theta} = \tan\theta. \tag{5.12}$$

Put a coin on a book and tilt it until the coin slides at a constant velocity down the book. You might need to tap the book lightly to get the coin to move. Measure the angle of tilt relative to the horizontal and find μ_k. Note that the coin will not start to slide at all until an angle greater than θ is attained, since the coefficient of static friction is larger than the coefficient of kinetic friction. Discuss how this may affect the value for μ_k and its uncertainty.

We have discussed that when an object rests on a horizontal surface, there is a normal force supporting it equal in magnitude to its weight. Furthermore, simple friction is always proportional to the normal force.

Making Connections: Submicroscopic Explanations of Friction

The simpler aspects of friction dealt with so far are its macroscopic (large-scale) characteristics. Great strides have been made in the atomic-scale explanation of friction during the past several decades. Researchers are finding that the atomic nature of friction seems to have several fundamental characteristics. These characteristics not only explain some of the simpler aspects of friction—they also hold the potential for the development of nearly friction-free environments that could save hundreds of billions of dollars in energy which is currently being converted (unnecessarily) to heat.

Figure 5.5 illustrates one macroscopic characteristic of friction that is explained by microscopic (small-scale) research. We have noted that friction is proportional to the normal force, but not to the area in contact, a somewhat counterintuitive notion. When two rough surfaces are in contact, the actual contact area is a tiny fraction of the total area since only high spots touch. When a greater normal force is exerted, the actual contact area increases, and it is found that the friction is proportional to this area.

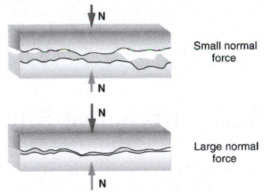

Figure 5.5 Two rough surfaces in contact have a much smaller area of actual contact than their total area. When there is a greater normal force as a result of a greater applied force, the area of actual contact increases as does friction.

But the atomic-scale view promises to explain far more than the simpler features of friction. The mechanism for how heat is generated is now being determined. In other words, why do surfaces get warmer when rubbed? Essentially, atoms are linked with one another to form lattices. When surfaces rub, the surface atoms adhere and cause atomic lattices to vibrate—essentially creating sound waves that penetrate the material. The sound waves diminish with distance and their energy is converted into heat. Chemical reactions that are related to frictional wear can also occur between atoms and molecules on the surfaces. Figure 5.6 shows how the tip of a probe drawn across another material is deformed by atomic-scale friction. The force needed to drag the tip can be measured and is found to be related to shear stress, which will be discussed later in this chapter. The variation in shear stress is remarkable (more than a factor of 10^{12}) and difficult to predict theoretically, but shear stress is yielding a fundamental understanding of a large-scale phenomenon known since ancient times—friction.

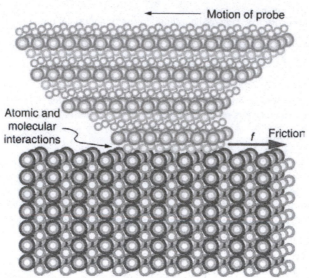

Figure 5.6 The tip of a probe is deformed sideways by frictional force as the probe is dragged across a surface. Measurements of how the force varies for different materials are yielding fundamental insights into the atomic nature of friction.

PhET Explorations: Forces and Motion

Explore the forces at work when you try to push a filing cabinet. Create an applied force and see the resulting friction force and total force acting on the cabinet. Charts show the forces, position, velocity, and acceleration vs. time. Draw a free-body diagram of all the forces (including gravitational and normal forces).

Figure 5.7 Forces and Motion (http://cnx.org/content/m54899/1.3/forces-and-motion_en.jar)

5.2 Drag Forces

Learning Objectives

By the end of this section, you will be able to:

- Define drag force and model it mathematically.
- Discuss the applications of drag force.
- Define terminal velocity.
- Perform calculations to find terminal velocity.

Another interesting force in everyday life is the force of drag on an object when it is moving in a fluid (either a gas or a liquid). You feel the drag force when you move your hand through water. You might also feel it if you move your hand during a strong wind. The faster you move your hand, the harder it is to move. You feel a smaller drag force when you tilt your hand so only the side goes through the air—you have decreased the area of your hand that faces the direction of motion. Like friction, the **drag force** always opposes the motion of an object. Unlike simple friction, the drag force is proportional to some function of the velocity of the object in that fluid. This functionality is complicated and depends upon the shape of the object, its size, its velocity, and the fluid it is in. For most large objects such as bicyclists, cars, and baseballs not moving too slowly, the magnitude of the drag force F_D is found to be proportional to the square of the speed of the object. We can write this relationship mathematically

as $F_D \propto v^2$. When taking into account other factors, this relationship becomes

$$F_\text{D} = \tfrac{1}{2}C\rho A v^2, \qquad\qquad (5.13)$$

where C is the drag coefficient, A is the area of the object facing the fluid, and ρ is the density of the fluid. (Recall that density is mass per unit volume.) This equation can also be written in a more generalized fashion as $F_\text{D} = bv^2$, where b is a constant equivalent to $0.5C\rho A$. We have set the exponent for these equations as 2 because, when an object is moving at high velocity through air, the magnitude of the drag force is proportional to the square of the speed. As we shall see in a few pages on fluid dynamics, for small particles moving at low speeds in a fluid, the exponent is equal to 1.

Drag Force

Drag force F_D is found to be proportional to the square of the speed of the object. Mathematically

$$F_D \propto v^2 \tag{5.14}$$

$$F_D = \frac{1}{2} C \rho A v^2, \tag{5.15}$$

where C is the drag coefficient, A is the area of the object facing the fluid, and ρ is the density of the fluid.

Athletes as well as car designers seek to reduce the drag force to lower their race times. (See **Figure 5.8**). "Aerodynamic" shaping of an automobile can reduce the drag force and so increase a car's gas mileage.

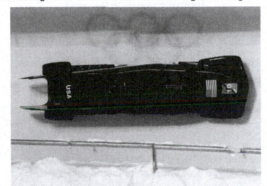

Figure 5.8 From racing cars to bobsled racers, aerodynamic shaping is crucial to achieving top speeds. Bobsleds are designed for speed. They are shaped like a bullet with tapered fins. (credit: U.S. Army, via Wikimedia Commons)

The value of the drag coefficient, C, is determined empirically, usually with the use of a wind tunnel. (See **Figure 5.9**).

Figure 5.9 NASA researchers test a model plane in a wind tunnel. (credit: NASA/Ames)

The drag coefficient can depend upon velocity, but we will assume that it is a constant here. **Table 5.2** lists some typical drag coefficients for a variety of objects. Notice that the drag coefficient is a dimensionless quantity. At highway speeds, over 50% of the power of a car is used to overcome air drag. The most fuel-efficient cruising speed is about 70–80 km/h (about 45–50 mi/h). For this reason, during the 1970s oil crisis in the United States, maximum speeds on highways were set at about 90 km/h (55 mi/h).

Table 5.2 Drag Coefficient Values **Typical values of drag coefficient C.**

Object	C
Airfoil	0.05
Toyota Camry	0.28
Ford Focus	0.32
Honda Civic	0.36
Ferrari Testarossa	0.37
Dodge Ram pickup	0.43
Sphere	0.45
Hummer H2 SUV	0.64
Skydiver (feet first)	0.70
Bicycle	0.90
Skydiver (horizontal)	1.0
Circular flat plate	1.12

Substantial research is under way in the sporting world to minimize drag. The dimples on golf balls are being redesigned as are the clothes that athletes wear. Bicycle racers and some swimmers and runners wear full bodysuits. Australian Cathy Freeman wore a full body suit in the 2000 Sydney Olympics, and won the gold medal for the 400 m race. Many swimmers in the 2008 Beijing Olympics wore (Speedo) body suits; it might have made a difference in breaking many world records (See **Figure 5.10**). Most elite swimmers (and cyclists) shave their body hair. Such innovations can have the effect of slicing away milliseconds in a race, sometimes making the difference between a gold and a silver medal. One consequence is that careful and precise guidelines must be continuously developed to maintain the integrity of the sport.

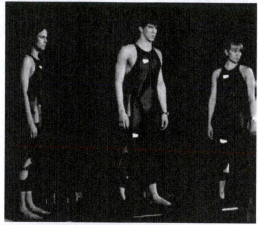

Figure 5.10 Body suits, such as this LZR Racer Suit, have been credited with many world records after their release in 2008. Smoother "skin" and more compression forces on a swimmer's body provide at least 10% less drag. (credit: NASA/Kathy Barnstorff)

Some interesting situations connected to Newton's second law occur when considering the effects of drag forces upon a moving object. For instance, consider a skydiver falling through air under the influence of gravity. The two forces acting on him are the force of gravity and the drag force (ignoring the buoyant force). The downward force of gravity remains constant regardless of the velocity at which the person is moving. However, as the person's velocity increases, the magnitude of the drag force increases until the magnitude of the drag force is equal to the gravitational force, thus producing a net force of zero. A zero net force means that there is no acceleration, as given by Newton's second law. At this point, the person's velocity remains constant and we say that the person has reached his *terminal velocity* (v_t). Since F_D is proportional to the speed, a heavier skydiver must go faster for F_D to equal his weight. Let's see how this works out more quantitatively.

At the terminal velocity,

$$F_{net} = mg - F_D = ma = 0.$$ (5.16)

Thus,

$$mg = F_D.$$ (5.17)

Using the equation for drag force, we have

$$mg = \frac{1}{2}\rho C A v^2.$$
(5.18)

Solving for the velocity, we obtain

$$v = \sqrt{\frac{2mg}{\rho C A}}.$$
(5.19)

Assume the density of air is $\rho = 1.21 \text{ kg/m}^3$. A 75-kg skydiver descending head first will have an area approximately $A = 0.18 \text{ m}^2$ and a drag coefficient of approximately $C = 0.70$. We find that

$$
\begin{aligned}
v &= \sqrt{\frac{2(75 \text{ kg})(9.80 \text{ m/s}^2)}{(1.21 \text{ kg/m}^3)(0.70)(0.18 \text{ m}^2)}} \\
&= 98 \text{ m/s} \\
&= 350 \text{ km/h.}
\end{aligned}
$$
(5.20)

This means a skydiver with a mass of 75 kg achieves a maximum terminal velocity of about 350 km/h while traveling in a pike (head first) position, minimizing the area and his drag. In a spread-eagle position, that terminal velocity may decrease to about 200 km/h as the area increases. This terminal velocity becomes much smaller after the parachute opens.

Take-Home Experiment

This interesting activity examines the effect of weight upon terminal velocity. Gather together some nested coffee filters. Leaving them in their original shape, measure the time it takes for one, two, three, four, and five nested filters to fall to the floor from the same height (roughly 2 m). (Note that, due to the way the filters are nested, drag is constant and only mass varies.) They obtain terminal velocity quite quickly, so find this velocity as a function of mass. Plot the terminal velocity v versus mass. Also plot v^2 versus mass. Which of these relationships is more linear? What can you conclude from these graphs?

Example 5.2 A Terminal Velocity

Find the terminal velocity of an 85-kg skydiver falling in a spread-eagle position.

Strategy

At terminal velocity, $F_{\text{net}} = 0$. Thus the drag force on the skydiver must equal the force of gravity (the person's weight).

Using the equation of drag force, we find $mg = \frac{1}{2}\rho C A v^2$.

Thus the terminal velocity v_t can be written as

$$v_t = \sqrt{\frac{2mg}{\rho C A}}.$$
(5.21)

Solution

All quantities are known except the person's projected area. This is an adult (85 kg) falling spread eagle. We can estimate the frontal area as

$$A = (2 \text{ m})(0.35 \text{ m}) = 0.70 \text{ m}^2.$$
(5.22)

Using our equation for v_t, we find that

$$
\begin{aligned}
v_t &= \sqrt{\frac{2(85 \text{ kg})(9.80 \text{ m/s}^2)}{(1.21 \text{ kg/m}^3)(1.0)(0.70 \text{ m}^2)}} \\
&= 44 \text{ m/s.}
\end{aligned}
$$
(5.23)

Discussion

This result is consistent with the value for v_t mentioned earlier. The 75-kg skydiver going feet first had a $v = 98 \text{ m/s}$. He weighed less but had a smaller frontal area and so a smaller drag due to the air.

The size of the object that is falling through air presents another interesting application of air drag. If you fall from a 5-m high branch of a tree, you will likely get hurt—possibly fracturing a bone. However, a small squirrel does this all the time, without getting hurt. You don't reach a terminal velocity in such a short distance, but the squirrel does.

The following interesting quote on animal size and terminal velocity is from a 1928 essay by a British biologist, J.B.S. Haldane, titled "On Being the Right Size."

To the mouse and any smaller animal, [gravity] presents practically no dangers. You can drop a mouse down a thousand-yard mine shaft; and, on arriving at the bottom, it gets a slight shock and walks away, provided that the ground is fairly soft. A rat is killed, a man is broken, and a horse splashes. For the resistance presented to movement by the air is proportional to the surface of the moving object. Divide an animal's length, breadth, and height each by ten; its weight is reduced to a thousandth, but its surface only to a hundredth. So the resistance to falling in the case of the small animal is relatively ten times greater than the driving force.

The above quadratic dependence of air drag upon velocity does not hold if the object is very small, is going very slow, or is in a denser medium than air. Then we find that the drag force is proportional just to the velocity. This relationship is given by **Stokes' law**, which states that

$$F_s = 6\pi r \eta v, \qquad\qquad (5.24)$$

where r is the radius of the object, η is the viscosity of the fluid, and v is the object's velocity.

Stokes' Law
$$F_s = 6\pi r \eta v, \qquad\qquad (5.25)$$

where r is the radius of the object, η is the viscosity of the fluid, and v is the object's velocity.

Good examples of this law are provided by microorganisms, pollen, and dust particles. Because each of these objects is so small, we find that many of these objects travel unaided only at a constant (terminal) velocity. Terminal velocities for bacteria (size about $1\ \mu m$) can be about $2\ \mu m/s$. To move at a greater speed, many bacteria swim using flagella (organelles shaped like little tails) that are powered by little motors embedded in the cell. Sediment in a lake can move at a greater terminal velocity (about $5\ \mu m/s$), so it can take days to reach the bottom of the lake after being deposited on the surface.

If we compare animals living on land with those in water, you can see how drag has influenced evolution. Fishes, dolphins, and even massive whales are streamlined in shape to reduce drag forces. Birds are streamlined and migratory species that fly large distances often have particular features such as long necks. Flocks of birds fly in the shape of a spear head as the flock forms a streamlined pattern (see **Figure 5.11**). In humans, one important example of streamlining is the shape of sperm, which need to be efficient in their use of energy.

Figure 5.11 Geese fly in a V formation during their long migratory travels. This shape reduces drag and energy consumption for individual birds, and also allows them a better way to communicate. (credit: Julo, Wikimedia Commons)

Galileo's Experiment

Galileo is said to have dropped two objects of different masses from the Tower of Pisa. He measured how long it took each to reach the ground. Since stopwatches weren't readily available, how do you think he measured their fall time? If the objects were the same size, but with different masses, what do you think he should have observed? Would this result be different if done on the Moon?

PhET Explorations: Masses & Springs

A realistic mass and spring laboratory. Hang masses from springs and adjust the spring stiffness and damping. You can even slow time. Transport the lab to different planets. A chart shows the kinetic, potential, and thermal energy for each spring.

Figure 5.12 Masses & Springs (http://cnx.org/content/m54904/1.4/mass-spring-lab_en.jar)

5.3 Elasticity: Stress and Strain

Learning Objectives

By the end of this section, you will be able to:

- State Hooke's law.
- Explain Hooke's law using graphical representation between deformation and applied force.
- Discuss the three types of deformations such as changes in length, sideways shear, and changes in volume.
- Describe with examples the Young's modulus, shear modulus, and bulk modulus.
- Determine the change in length given mass, length, and radius.

We now move from consideration of forces that affect the motion of an object (such as friction and drag) to those that affect an object's shape. If a bulldozer pushes a car into a wall, the car will not move but it will noticeably change shape. A change in shape due to the application of a force is a **deformation**. Even very small forces are known to cause some deformation. For small deformations, two important characteristics are observed. First, the object returns to its original shape when the force is removed—that is, the deformation is elastic for small deformations. Second, the size of the deformation is proportional to the force—that is, for small deformations, Hooke's law is obeyed. In equation form, **Hooke's law** is given by

$$F = k\Delta L,$$ (5.26)

where ΔL is the amount of deformation (the change in length, for example) produced by the force F, and k is a proportionality constant that depends on the shape and composition of the object and the direction of the force. Note that this force is a function of the deformation ΔL —it is not constant as a kinetic friction force is. Rearranging this to

$$\Delta L = \frac{F}{k}$$ (5.27)

makes it clear that the deformation is proportional to the applied force. **Figure 5.13** shows the Hooke's law relationship between the extension ΔL of a spring or of a human bone. For metals or springs, the straight line region in which Hooke's law pertains is much larger. Bones are brittle and the elastic region is small and the fracture abrupt. Eventually a large enough stress to the material will cause it to break or fracture.

Hooke's Law

$$F = k\Delta L,$$ (5.28)

where ΔL is the amount of deformation (the change in length, for example) produced by the force F, and k is a proportionality constant that depends on the shape and composition of the object and the direction of the force.

$$\Delta L = \frac{F}{k}$$ (5.29)

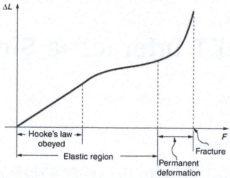

Figure 5.13 A graph of deformation ΔL versus applied force F. The straight segment is the linear region where Hooke's law is obeyed. The slope of the straight region is $\frac{1}{k}$. For larger forces, the graph is curved but the deformation is still elastic— ΔL will return to zero if the force is removed. Still greater forces permanently deform the object until it finally fractures. The shape of the curve near fracture depends on several factors, including how the force F is applied. Note that in this graph the slope increases just before fracture, indicating that a small increase in F is producing a large increase in L near the fracture.

The proportionality constant k depends upon a number of factors for the material. For example, a guitar string made of nylon stretches when it is tightened, and the elongation ΔL is proportional to the force applied (at least for small deformations). Thicker nylon strings and ones made of steel stretch less for the same applied force, implying they have a larger k (see **Figure 5.14**). Finally, all three strings return to their normal lengths when the force is removed, provided the deformation is small. Most materials will behave in this manner if the deformation is less than about 0.1% or about 1 part in 10^3.

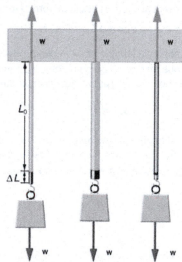

Figure 5.14 The same force, in this case a weight (w), applied to three different guitar strings of identical length produces the three different deformations shown as shaded segments. The string on the left is thin nylon, the one in the middle is thicker nylon, and the one on the right is steel.

Stretch Yourself a Little

How would you go about measuring the proportionality constant k of a rubber band? If a rubber band stretched 3 cm when a 100-g mass was attached to it, then how much would it stretch if two similar rubber bands were attached to the same mass—even if put together in parallel or alternatively if tied together in series?

We now consider three specific types of deformations: changes in length (tension and compression), sideways shear (stress), and changes in volume. All deformations are assumed to be small unless otherwise stated.

Changes in Length—Tension and Compression: Elastic Modulus

A change in length ΔL is produced when a force is applied to a wire or rod parallel to its length L_0, either stretching it (a tension) or compressing it. (See **Figure 5.15**.)

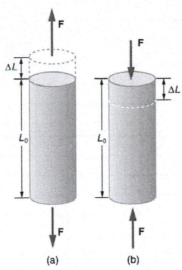

Figure 5.15 (a) Tension. The rod is stretched a length ΔL when a force is applied parallel to its length. (b) Compression. The same rod is compressed by forces with the same magnitude in the opposite direction. For very small deformations and uniform materials, ΔL is approximately the same for the same magnitude of tension or compression. For larger deformations, the cross-sectional area changes as the rod is compressed or stretched.

Experiments have shown that the change in length (ΔL) depends on only a few variables. As already noted, ΔL is proportional to the force F and depends on the substance from which the object is made. Additionally, the change in length is proportional to the original length L_0 and inversely proportional to the cross-sectional area of the wire or rod. For example, a long guitar string will stretch more than a short one, and a thick string will stretch less than a thin one. We can combine all these factors into one equation for ΔL :

$$\Delta L = \frac{1}{Y}\frac{F}{A}L_0, \tag{5.30}$$

where ΔL is the change in length, F the applied force, Y is a factor, called the elastic modulus or Young's modulus, that depends on the substance, A is the cross-sectional area, and L_0 is the original length. Table 5.3 lists values of Y for several materials—those with a large Y are said to have a large **tensile strength** because they deform less for a given tension or compression.

Table 5.3 Elastic Moduli[1]

Material	Young's modulus (tension–compression)Y $(10^9 \ \text{N/m}^2)$	Shear modulus S $(10^9 \ \text{N/m}^2)$	Bulk modulus B $(10^9 \ \text{N/m}^2)$
Aluminum	70	25	75
Bone – tension	16	80	8
Bone – compression	9		
Brass	90	35	75
Brick	15		
Concrete	20		
Glass	70	20	30
Granite	45	20	45
Hair (human)	10		
Hardwood	15	10	
Iron, cast	100	40	90
Lead	16	5	50
Marble	60	20	70
Nylon	5		
Polystyrene	3		
Silk	6		
Spider thread	3		
Steel	210	80	130
Tendon	1		
Acetone			0.7
Ethanol			0.9
Glycerin			4.5
Mercury			25
Water			2.2

Young's moduli are not listed for liquids and gases in Table 5.3 because they cannot be stretched or compressed in only one direction. Note that there is an assumption that the object does not accelerate, so that there are actually two applied forces of magnitude F acting in opposite directions. For example, the strings in Figure 5.15 are being pulled down by a force of magnitude w and held up by the ceiling, which also exerts a force of magnitude w.

Example 5.3 The Stretch of a Long Cable

Suspension cables are used to carry gondolas at ski resorts. (See Figure 5.16) Consider a suspension cable that includes an unsupported span of 3020 m. Calculate the amount of stretch in the steel cable. Assume that the cable has a diameter of 5.6 cm and the maximum tension it can withstand is $3.0 \times 10^6 \ \text{N}$.

---

1. Approximate and average values. Young's moduli Y for tension and compression sometimes differ but are averaged here. Bone has significantly different Young's moduli for tension and compression.

Figure 5.16 Gondolas travel along suspension cables at the Gala Yuzawa ski resort in Japan. (credit: Rudy Herman, Flickr)

Strategy

The force is equal to the maximum tension, or $F = 3.0 \times 10^6 \text{ N}$. The cross-sectional area is $\pi r^2 = 2.46 \times 10^{-3} \text{ m}^2$. The equation $\Delta L = \frac{1}{Y} \frac{F}{A} L_0$ can be used to find the change in length.

Solution

All quantities are known. Thus,

$$
\begin{aligned}
\Delta L &= \left(\frac{1}{210 \times 10^9 \text{ N/m}^2} \right)\left(\frac{3.0 \times 10^6 \text{ N}}{2.46 \times 10^{-3} \text{ m}^2} \right)(3020 \text{ m}) \\
&= 18 \text{ m.}
\end{aligned}
$$

(5.31)

Discussion

This is quite a stretch, but only about 0.6% of the unsupported length. Effects of temperature upon length might be important in these environments.

Bones, on the whole, do not fracture due to tension or compression. Rather they generally fracture due to sideways impact or bending, resulting in the bone shearing or snapping. The behavior of bones under tension and compression is important because it determines the load the bones can carry. Bones are classified as weight-bearing structures such as columns in buildings and trees. Weight-bearing structures have special features; columns in building have steel-reinforcing rods while trees and bones are fibrous. The bones in different parts of the body serve different structural functions and are prone to different stresses. Thus the bone in the top of the femur is arranged in thin sheets separated by marrow while in other places the bones can be cylindrical and filled with marrow or just solid. Overweight people have a tendency toward bone damage due to sustained compressions in bone joints and tendons.

Another biological example of Hooke's law occurs in tendons. Functionally, the tendon (the tissue connecting muscle to bone) must stretch easily at first when a force is applied, but offer a much greater restoring force for a greater strain. **Figure 5.17** shows a stress-strain relationship for a human tendon. Some tendons have a high collagen content so there is relatively little strain, or length change; others, like support tendons (as in the leg) can change length up to 10%. Note that this stress-strain curve is nonlinear, since the slope of the line changes in different regions. In the first part of the stretch called the toe region, the fibers in the tendon begin to align in the direction of the stress—this is called *uncrimping*. In the linear region, the fibrils will be stretched, and in the failure region individual fibers begin to break. A simple model of this relationship can be illustrated by springs in parallel: different springs are activated at different lengths of stretch. Examples of this are given in the problems at end of this chapter. Ligaments (tissue connecting bone to bone) behave in a similar way.

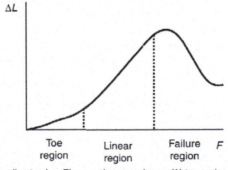

Figure 5.17 Typical stress-strain curve for mammalian tendon. Three regions are shown: (1) toe region (2) linear region, and (3) failure region.

Unlike bones and tendons, which need to be strong as well as elastic, the arteries and lungs need to be very stretchable. The elastic properties of the arteries are essential for blood flow. The pressure in the arteries increases and arterial walls stretch when the blood is pumped out of the heart. When the aortic valve shuts, the pressure in the arteries drops and the arterial walls relax to maintain the blood flow. When you feel your pulse, you are feeling exactly this—the elastic behavior of the arteries as the blood gushes through with each pump of the heart. If the arteries were rigid, you would not feel a pulse. The heart is also an organ with special elastic properties. The lungs expand with muscular effort when we breathe in but relax freely and elastically when we breathe out. Our skins are particularly elastic, especially for the young. A young person can go from 100 kg to 60 kg with no visible sag in their skins. The elasticity of all organs reduces with age. Gradual physiological aging through reduction in elasticity starts in the early 20s.

Example 5.4 Calculating Deformation: How Much Does Your Leg Shorten When You Stand on It?

Calculate the change in length of the upper leg bone (the femur) when a 70.0 kg man supports 62.0 kg of his mass on it, assuming the bone to be equivalent to a uniform rod that is 40.0 cm long and 2.00 cm in radius.

Strategy

The force is equal to the weight supported, or

$$F = mg = (62.0 \text{ kg})(9.80 \text{ m/s}^2) = 607.6 \text{ N}, \tag{5.32}$$

and the cross-sectional area is $\pi r^2 = 1.257 \times 10^{-3} \text{ m}^2$. The equation $\Delta L = \frac{1}{Y}\frac{F}{A}L_0$ can be used to find the change in length.

Solution

All quantities except ΔL are known. Note that the compression value for Young's modulus for bone must be used here. Thus,

$$\begin{aligned} \Delta L &= \left(\frac{1}{9 \times 10^9 \text{ N/m}^2}\right)\left(\frac{607.6 \text{ N}}{1.257 \times 10^{-3} \text{ m}^2}\right)(0.400 \text{ m}) \\ &= 2 \times 10^{-5} \text{ m}. \end{aligned} \tag{5.33}$$

Discussion

This small change in length seems reasonable, consistent with our experience that bones are rigid. In fact, even the rather large forces encountered during strenuous physical activity do not compress or bend bones by large amounts. Although bone is rigid compared with fat or muscle, several of the substances listed in **Table 5.3** have larger values of Young's modulus Y. In other words, they are more rigid and have greater tensile strength.

The equation for change in length is traditionally rearranged and written in the following form:

$$\frac{F}{A} = Y\frac{\Delta L}{L_0}. \tag{5.34}$$

The ratio of force to area, $\frac{F}{A}$, is defined as **stress** (measured in N/m^2), and the ratio of the change in length to length, $\frac{\Delta L}{L_0}$, is defined as **strain** (a unitless quantity). In other words,

$$\text{stress} = Y \times \text{strain}. \tag{5.35}$$

In this form, the equation is analogous to Hooke's law, with stress analogous to force and strain analogous to deformation. If we again rearrange this equation to the form

$$F = YA\frac{\Delta L}{L_0}, \tag{5.36}$$

we see that it is the same as Hooke's law with a proportionality constant

$$k = \frac{YA}{L_0}. \tag{5.37}$$

This general idea—that force and the deformation it causes are proportional for small deformations—applies to changes in length, sideways bending, and changes in volume.

Stress

The ratio of force to area, $\frac{F}{A}$, is defined as stress measured in N/m^2.

The ratio of the change in length to length, $\frac{\Delta L}{L_0}$, is defined as strain (a unitless quantity). In other words,

$$\text{stress} = Y \times \text{strain}. \tag{5.38}$$

Sideways Stress: Shear Modulus

Figure 5.18 illustrates what is meant by a sideways stress or a *shearing force*. Here the deformation is called Δx and it is perpendicular to L_0 , rather than parallel as with tension and compression. Shear deformation behaves similarly to tension and compression and can be described with similar equations. The expression for **shear deformation** is

$$\Delta x = \frac{1}{S}\frac{F}{A}L_0, \tag{5.39}$$

where S is the shear modulus (see **Table 5.3**) and F is the force applied perpendicular to L_0 and parallel to the cross-sectional area A . Again, to keep the object from accelerating, there are actually two equal and opposite forces F applied across opposite faces, as illustrated in **Figure 5.18**. The equation is logical—for example, it is easier to bend a long thin pencil (small A) than a short thick one, and both are more easily bent than similar steel rods (large S).

Shear Deformation

$$\Delta x = \frac{1}{S}\frac{F}{A}L_0, \tag{5.40}$$

where S is the shear modulus and F is the force applied perpendicular to L_0 and parallel to the cross-sectional area A .

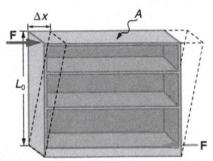

Figure 5.18 Shearing forces are applied perpendicular to the length L_0 and parallel to the area A , producing a deformation Δx . Vertical forces are not shown, but it should be kept in mind that in addition to the two shearing forces, $\mathbf{F}$, there must be supporting forces to keep the object from rotating. The distorting effects of these supporting forces are ignored in this treatment. The weight of the object also is not shown, since it is usually negligible compared with forces large enough to cause significant deformations.

Examination of the shear moduli in **Table 5.3** reveals some telling patterns. For example, shear moduli are less than Young's moduli for most materials. Bone is a remarkable exception. Its shear modulus is not only greater than its Young's modulus, but it is as large as that of steel. This is one reason that bones can be long and relatively thin. Bones can support loads comparable to that of concrete and steel. Most bone fractures are not caused by compression but by excessive twisting and bending.

The spinal column (consisting of 26 vertebral segments separated by discs) provides the main support for the head and upper part of the body. The spinal column has normal curvature for stability, but this curvature can be increased, leading to increased shearing forces on the lower vertebrae. Discs are better at withstanding compressional forces than shear forces. Because the spine is not vertical, the weight of the upper body exerts some of both. Pregnant women and people that are overweight (with large abdomens) need to move their shoulders back to maintain balance, thereby increasing the curvature in their spine and so increasing the shear component of the stress. An increased angle due to more curvature increases the shear forces along the plane. These higher shear forces increase the risk of back injury through ruptured discs. The lumbosacral disc (the wedge shaped disc below the last vertebrae) is particularly at risk because of its location.

The shear moduli for concrete and brick are very small; they are too highly variable to be listed. Concrete used in buildings can withstand compression, as in pillars and arches, but is very poor against shear, as might be encountered in heavily loaded floors or during earthquakes. Modern structures were made possible by the use of steel and steel-reinforced concrete. Almost by definition, liquids and gases have shear moduli near zero, because they flow in response to shearing forces.

Example 5.5 Calculating Force Required to Deform: That Nail Does Not Bend Much Under a Load

Find the mass of the picture hanging from a steel nail as shown in **Figure 5.19**, given that the nail bends only $1.80\ \mu m$. (Assume the shear modulus is known to two significant figures.)

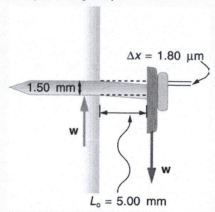

Figure 5.19 Side view of a nail with a picture hung from it. The nail flexes very slightly (shown much larger than actual) because of the shearing effect of the supported weight. Also shown is the upward force of the wall on the nail, illustrating that there are equal and opposite forces applied across opposite cross sections of the nail. See **Example 5.5** for a calculation of the mass of the picture.

Strategy

The force F on the nail (neglecting the nail's own weight) is the weight of the picture w. If we can find w, then the mass of the picture is just $\frac{w}{g}$. The equation $\Delta x = \frac{1}{S}\frac{F}{A}L_0$ can be solved for F.

Solution

Solving the equation $\Delta x = \frac{1}{S}\frac{F}{A}L_0$ for F, we see that all other quantities can be found:

$$F = \frac{SA}{L_0}\Delta x. \tag{5.41}$$

S is found in **Table 5.3** and is $S = 80\times10^9\ \text{N/m}^2$. The radius r is 0.750 mm (as seen in the figure), so the cross-sectional area is

$$A = \pi r^2 = 1.77\times10^{-6}\ \text{m}^2. \tag{5.42}$$

The value for L_0 is also shown in the figure. Thus,

$$F = \frac{(80\times10^9\ \text{N/m}^2)(1.77\times10^{-6}\ \text{m}^2)}{(5.00\times10^{-3}\ \text{m})}(1.80\times10^{-6}\ \text{m}) = 51\ \text{N}. \tag{5.43}$$

This 51 N force is the weight w of the picture, so the picture's mass is

$$m = \frac{w}{g} = \frac{F}{g} = 5.2\ \text{kg}. \tag{5.44}$$

Discussion

This is a fairly massive picture, and it is impressive that the nail flexes only $1.80\ \mu m$ —an amount undetectable to the unaided eye.

Changes in Volume: Bulk Modulus

An object will be compressed in all directions if inward forces are applied evenly on all its surfaces as in **Figure 5.20**. It is relatively easy to compress gases and extremely difficult to compress liquids and solids. For example, air in a wine bottle is compressed when it is corked. But if you try corking a brim-full bottle, you cannot compress the wine—some must be removed if the cork is to be inserted. The reason for these different compressibilities is that atoms and molecules are separated by large empty spaces in gases but packed close together in liquids and solids. To compress a gas, you must force its atoms and molecules closer together. To compress liquids and solids, you must actually compress their atoms and molecules, and very strong electromagnetic forces in them oppose this compression.

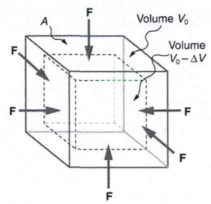

Figure 5.20 An inward force on all surfaces compresses this cube. Its change in volume is proportional to the force per unit area and its original volume, and is related to the compressibility of the substance.

We can describe the compression or volume deformation of an object with an equation. First, we note that a force "applied evenly" is defined to have the same stress, or ratio of force to area $\frac{F}{A}$ on all surfaces. The deformation produced is a change in volume ΔV, which is found to behave very similarly to the shear, tension, and compression previously discussed. (This is not surprising, since a compression of the entire object is equivalent to compressing each of its three dimensions.) The relationship of the change in volume to other physical quantities is given by

$$\Delta V = \frac{1}{B}\frac{F}{A}V_0,\tag{5.45}$$

where B is the bulk modulus (see **Table 5.3**), V_0 is the original volume, and $\frac{F}{A}$ is the force per unit area applied uniformly inward on all surfaces. Note that no bulk moduli are given for gases.

What are some examples of bulk compression of solids and liquids? One practical example is the manufacture of industrial-grade diamonds by compressing carbon with an extremely large force per unit area. The carbon atoms rearrange their crystalline structure into the more tightly packed pattern of diamonds. In nature, a similar process occurs deep underground, where extremely large forces result from the weight of overlying material. Another natural source of large compressive forces is the pressure created by the weight of water, especially in deep parts of the oceans. Water exerts an inward force on all surfaces of a submerged object, and even on the water itself. At great depths, water is measurably compressed, as the following example illustrates.

Example 5.6 Calculating Change in Volume with Deformation: How Much Is Water Compressed at Great Ocean Depths?

Calculate the fractional decrease in volume ($\frac{\Delta V}{V_0}$) for seawater at 5.00 km depth, where the force per unit area is

$5.00 \times 10^7 \text{ N/m}^2$.

Strategy

Equation $\Delta V = \frac{1}{B}\frac{F}{A}V_0$ is the correct physical relationship. All quantities in the equation except $\frac{\Delta V}{V_0}$ are known.

Solution

Solving for the unknown $\frac{\Delta V}{V_0}$ gives

$$\frac{\Delta V}{V_0} = \frac{1}{B}\frac{F}{A}.\tag{5.46}$$

Substituting known values with the value for the bulk modulus B from **Table 5.3**,

$$\begin{aligned}\frac{\Delta V}{V_0} &= \frac{5.00\times10^7 \text{ N/m}^2}{2.2\times10^9 \text{ N/m}^2}\\ &= 0.023 = 2.3\%.\end{aligned}\tag{5.47}$$

Discussion

Although measurable, this is not a significant decrease in volume considering that the force per unit area is about 500 atmospheres (1 million pounds per square foot). Liquids and solids are extraordinarily difficult to compress.

Conversely, very large forces are created by liquids and solids when they try to expand but are constrained from doing so—which is equivalent to compressing them to less than their normal volume. This often occurs when a contained material warms up, since most materials expand when their temperature increases. If the materials are tightly constrained, they deform or break their container. Another very common example occurs when water freezes. Water, unlike most materials, expands when it freezes, and it can easily fracture a boulder, rupture a biological cell, or crack an engine block that gets in its way.

Other types of deformations, such as torsion or twisting, behave analogously to the tension, shear, and bulk deformations considered here.

Glossary

deformation: change in shape due to the application of force

drag force: F_D, found to be proportional to the square of the speed of the object; mathematically

$$F_D \propto v^2$$

$$F_D = \frac{1}{2}C\rho A v^2,$$

where C is the drag coefficient, A is the area of the object facing the fluid, and ρ is the density of the fluid

friction: a force that opposes relative motion or attempts at motion between systems in contact

Hooke's law: proportional relationship between the force F on a material and the deformation ΔL it causes, $F = k\Delta L$

kinetic friction: a force that opposes the motion of two systems that are in contact and moving relative to one another

magnitude of kinetic friction: $f_k = \mu_k N$, where μ_k is the coefficient of kinetic friction

magnitude of static friction: $f_s \leq \mu_s N$, where μ_s is the coefficient of static friction and N is the magnitude of the normal force

shear deformation: deformation perpendicular to the original length of an object

static friction: a force that opposes the motion of two systems that are in contact and are not moving relative to one another

Stokes' law: $F_s = 6\pi r\eta v$, where r is the radius of the object, η is the viscosity of the fluid, and v is the object's velocity

strain: ratio of change in length to original length

stress: ratio of force to area

tensile strength: measure of deformation for a given tension or compression

Section Summary

5.1 Friction
- Friction is a contact force between systems that opposes the motion or attempted motion between them. Simple friction is proportional to the normal force N pushing the systems together. (A normal force is always perpendicular to the contact surface between systems.) Friction depends on both of the materials involved. The magnitude of static friction f_s between systems stationary relative to one another is given by

$$f_s \leq \mu_s N,$$

where μ_s is the coefficient of static friction, which depends on both of the materials.
- The kinetic friction force f_k between systems moving relative to one another is given by

$$f_k = \mu_k N,$$

where μ_k is the coefficient of kinetic friction, which also depends on both materials.

5.2 Drag Forces

- Drag forces acting on an object moving in a fluid oppose the motion. For larger objects (such as a baseball) moving at a velocity v in air, the drag force is given by

$$F_\text{D} = \tfrac{1}{2} C\rho A v^2,$$

where C is the drag coefficient (typical values are given in **Table 5.2**), A is the area of the object facing the fluid, and ρ is the fluid density.

- For small objects (such as a bacterium) moving in a denser medium (such as water), the drag force is given by Stokes' law,

$$F_\text{s} = 6\pi\eta r v,$$

where r is the radius of the object, η is the fluid viscosity, and v is the object's velocity.

5.3 Elasticity: Stress and Strain

- Hooke's law is given by

$$F = k\Delta L,$$

where ΔL is the amount of deformation (the change in length), F is the applied force, and k is a proportionality constant that depends on the shape and composition of the object and the direction of the force. The relationship between the deformation and the applied force can also be written as

$$\Delta L = \frac{1}{Y}\frac{F}{A}L_0,$$

where Y is *Young's modulus*, which depends on the substance, A is the cross-sectional area, and L_0 is the original length.

- The ratio of force to area, $\frac{F}{A}$, is defined as *stress*, measured in N/m^2.

- The ratio of the change in length to length, $\frac{\Delta L}{L_0}$, is defined as *strain* (a unitless quantity). In other words,

$$\text{stress} = Y \times \text{strain}.$$

- The expression for shear deformation is

$$\Delta x = \frac{1}{S}\frac{F}{A}L_0,$$

where S is the shear modulus and F is the force applied perpendicular to L_0 and parallel to the cross-sectional area A.

- The relationship of the change in volume to other physical quantities is given by

$$\Delta V = \frac{1}{B}\frac{F}{A}V_0,$$

where B is the bulk modulus, V_0 is the original volume, and $\frac{F}{A}$ is the force per unit area applied uniformly inward on all surfaces.

Conceptual Questions

5.1 Friction

1. Define normal force. What is its relationship to friction when friction behaves simply?

2. The glue on a piece of tape can exert forces. Can these forces be a type of simple friction? Explain, considering especially that tape can stick to vertical walls and even to ceilings.

3. When you learn to drive, you discover that you need to let up slightly on the brake pedal as you come to a stop or the car will stop with a jerk. Explain this in terms of the relationship between static and kinetic friction.

4. When you push a piece of chalk across a chalkboard, it sometimes screeches because it rapidly alternates between slipping and sticking to the board. Describe this process in more detail, in particular explaining how it is related to the fact that kinetic friction is less than static friction. (The same slip-grab process occurs when tires screech on pavement.)

5.2 Drag Forces

5. Athletes such as swimmers and bicyclists wear body suits in competition. Formulate a list of pros and cons of such suits.

6. Two expressions were used for the drag force experienced by a moving object in a liquid. One depended upon the speed, while the other was proportional to the square of the speed. In which types of motion would each of these expressions be more applicable than the other one?

7. As cars travel, oil and gasoline leaks onto the road surface. If a light rain falls, what does this do to the control of the car? Does a heavy rain make any difference?

8. Why can a squirrel jump from a tree branch to the ground and run away undamaged, while a human could break a bone in such a fall?

5.3 Elasticity: Stress and Strain

9. The elastic properties of the arteries are essential for blood flow. Explain the importance of this in terms of the characteristics of the flow of blood (pulsating or continuous).

10. What are you feeling when you feel your pulse? Measure your pulse rate for 10 s and for 1 min. Is there a factor of 6 difference?

11. Examine different types of shoes, including sports shoes and thongs. In terms of physics, why are the bottom surfaces designed as they are? What differences will dry and wet conditions make for these surfaces?

12. Would you expect your height to be different depending upon the time of day? Why or why not?

13. Why can a squirrel jump from a tree branch to the ground and run away undamaged, while a human could break a bone in such a fall?

14. Explain why pregnant women often suffer from back strain late in their pregnancy.

15. An old carpenter's trick to keep nails from bending when they are pounded into hard materials is to grip the center of the nail firmly with pliers. Why does this help?

16. When a glass bottle full of vinegar warms up, both the vinegar and the glass expand, but vinegar expands significantly more with temperature than glass. The bottle will break if it was filled to its tightly capped lid. Explain why, and also explain how a pocket of air above the vinegar would prevent the break. (This is the function of the air above liquids in glass containers.)

Problems & Exercises

5.1 Friction

1. A physics major is cooking breakfast when he notices that the frictional force between his steel spatula and his Teflon frying pan is only 0.200 N. Knowing the coefficient of kinetic friction between the two materials, he quickly calculates the normal force. What is it?

2. (a) When rebuilding her car's engine, a physics major must exert 300 N of force to insert a dry steel piston into a steel cylinder. What is the magnitude of the normal force between the piston and cylinder? (b) What is the magnitude of the force would she have to exert if the steel parts were oiled?

3. (a) What is the maximum frictional force in the knee joint of a person who supports 66.0 kg of her mass on that knee? (b) During strenuous exercise it is possible to exert forces to the joints that are easily ten times greater than the weight being supported. What is the maximum force of friction under such conditions? The frictional forces in joints are relatively small in all circumstances except when the joints deteriorate, such as from injury or arthritis. Increased frictional forces can cause further damage and pain.

4. Suppose you have a 120-kg wooden crate resting on a wood floor. (a) What maximum force can you exert horizontally on the crate without moving it? (b) If you continue to exert this force once the crate starts to slip, what will the magnitude of its acceleration then be?

5. (a) If half of the weight of a small 1.00×10^3 kg utility truck is supported by its two drive wheels, what is the magnitude of the maximum acceleration it can achieve on dry concrete? (b) Will a metal cabinet lying on the wooden bed of the truck slip if it accelerates at this rate? (c) Solve both problems assuming the truck has four-wheel drive.

6. A team of eight dogs pulls a sled with waxed wood runners on wet snow (mush!). The dogs have average masses of 19.0 kg, and the loaded sled with its rider has a mass of 210 kg. (a) Calculate the magnitude of the acceleration starting from rest if each dog exerts an average force of 185 N backward on the snow. (b) What is the magnitude of the acceleration once the sled starts to move? (c) For both situations, calculate the magnitude of the force in the coupling between the dogs and the sled.

7. Consider the 65.0-kg ice skater being pushed by two others shown in **Figure 5.21**. (a) Find the direction and magnitude of $\mathbf{F}_{tot}$, the total force exerted on her by the others, given that the magnitudes F_1 and F_2 are 26.4 N and 18.6 N, respectively. (b) What is her initial acceleration if she is initially stationary and wearing steel-bladed skates that point in the direction of $\mathbf{F}_{tot}$? (c) What is her acceleration assuming she is already moving in the direction of $\mathbf{F}_{tot}$? (Remember that friction always acts in the direction opposite that of motion or attempted motion between surfaces in contact.)

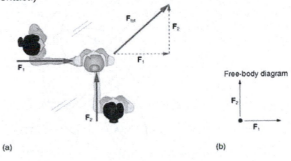

(a)   (b)

Figure 5.21

8. Show that the acceleration of any object down a frictionless incline that makes an angle θ with the horizontal is $a = g \sin \theta$. (Note that this acceleration is independent of mass.)

9. Show that the acceleration of any object down an incline where friction behaves simply (that is, where $f_k = \mu_k N$) is $a = g(\sin \theta - \mu_k \cos \theta)$. Note that the acceleration is independent of mass and reduces to the expression found in the previous problem when friction becomes negligibly small $(\mu_k = 0)$.

10. Calculate the deceleration of a snow boarder going up a $5.0°$, slope assuming the coefficient of friction for waxed wood on wet snow. The result of **Exercise 5.9** may be useful, but be careful to consider the fact that the snow boarder is going uphill. Explicitly show how you follow the steps in **Problem-Solving Strategies**.

11. (a) Calculate the acceleration of a skier heading down a $10.0°$ slope, assuming the coefficient of friction for waxed wood on wet snow. (b) Find the angle of the slope down which this skier could coast at a constant velocity. You can neglect air resistance in both parts, and you will find the result of **Exercise 5.9** to be useful. Explicitly show how you follow the steps in the **Problem-Solving Strategies**.

12. If an object is to rest on an incline without slipping, then friction must equal the component of the weight of the object parallel to the incline. This requires greater and greater friction for steeper slopes. Show that the maximum angle of an incline above the horizontal for which an object will not slide down is $\theta = \tan^{-1} \mu_s$. You may use the result of the previous problem. Assume that $a = 0$ and that static friction has reached its maximum value.

13. Calculate the maximum deceleration of a car that is heading down a $6°$ slope (one that makes an angle of $6°$ with the horizontal) under the following road conditions. You may assume that the weight of the car is evenly distributed on all four tires and that the coefficient of static friction is involved—that is, the tires are not allowed to slip during the deceleration. (Ignore rolling.) Calculate for a car: (a) On dry concrete. (b) On wet concrete. (c) On ice, assuming that $\mu_s = 0.100$, the same as for shoes on ice.

14. Calculate the maximum acceleration of a car that is heading up a $4°$ slope (one that makes an angle of $4°$ with the horizontal) under the following road conditions. Assume that only half the weight of the car is supported by the two drive wheels and that the coefficient of static friction is involved—that is, the tires are not allowed to slip during the acceleration. (Ignore rolling.) (a) On dry concrete. (b) On wet concrete. (c) On ice, assuming that $\mu_s = 0.100$, the same as for shoes on ice.

15. Repeat **Exercise 5.14** for a car with four-wheel drive.

16. A freight train consists of two 8.00×10^5-kg engines and 45 cars with average masses of $5.50 \times 10^5 \, \text{kg}$. (a) What force must each engine exert backward on the track to accelerate the train at a rate of $5.00 \times 10^{-2} \, \text{m/s}^2$ if the force of friction is $7.50 \times 10^5 \, \text{N}$, assuming the engines exert identical forces? This is not a large frictional force for such a massive system. Rolling friction for trains is small, and consequently trains are very energy-efficient transportation systems. (b) What is the magnitude of the force in the coupling between the 37th and 38th cars (this is the force each exerts on the other), assuming all cars have the same mass and that friction is evenly distributed among all of the cars and engines?

17. Consider the 52.0-kg mountain climber in **Figure 5.22**. (a) Find the tension in the rope and the force that the mountain climber must exert with her feet on the vertical rock face to remain stationary. Assume that the force is exerted parallel to her legs. Also, assume negligible force exerted by her arms. (b) What is the minimum coefficient of friction between her shoes and the cliff?

Figure 5.22 Part of the climber's weight is supported by her rope and part by friction between her feet and the rock face.

18. A contestant in a winter sporting event pushes a 45.0-kg block of ice across a frozen lake as shown in **Figure 5.23**(a). (a) Calculate the minimum force F he must exert to get the block moving. (b) What is the magnitude of its acceleration once it starts to move, if that force is maintained?

19. Repeat **Exercise 5.18** with the contestant pulling the block of ice with a rope over his shoulder at the same angle above the horizontal as shown in **Figure 5.23**(b).

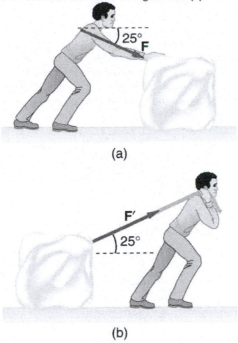

(a)

(b)

Figure 5.23 Which method of sliding a block of ice requires less force—(a) pushing or (b) pulling at the same angle above the horizontal?

5.2 Drag Forces

20. The terminal velocity of a person falling in air depends upon the weight and the area of the person facing the fluid. Find the terminal velocity (in meters per second and kilometers per hour) of an 80.0-kg skydiver falling in a pike (headfirst) position with a surface area of 0.140 m^2 .

21. A 60-kg and a 90-kg skydiver jump from an airplane at an altitude of 6000 m, both falling in the pike position. Make some assumption on their frontal areas and calculate their terminal velocities. How long will it take for each skydiver to reach the ground (assuming the time to reach terminal velocity is small)? Assume all values are accurate to three significant digits.

22. A 560-g squirrel with a surface area of 930 cm^2 falls from a 5.0-m tree to the ground. Estimate its terminal velocity. (Use a drag coefficient for a horizontal skydiver.) What will be the velocity of a 56-kg person hitting the ground, assuming no drag contribution in such a short distance?

23. To maintain a constant speed, the force provided by a car's engine must equal the drag force plus the force of friction of the road (the rolling resistance). (a) What are the magnitudes of drag forces at 70 km/h and 100 km/h for a Toyota Camry? (Drag area is 0.70 m^2) (b) What is the magnitude of drag force at 70 km/h and 100 km/h for a Hummer H2? (Drag area is 2.44 m^2) Assume all values are accurate to three significant digits.

24. By what factor does the drag force on a car increase as it goes from 65 to 110 km/h?

25. Calculate the speed a spherical rain drop would achieve falling from 5.00 km (a) in the absence of air drag (b) with air drag. Take the size across of the drop to be 4 mm, the density to be $1.00 \times 10^3 \text{ kg/m}^3$, and the surface area to be πr^2 .

26. Using Stokes' law, verify that the units for viscosity are kilograms per meter per second.

27. Find the terminal velocity of a spherical bacterium (diameter 2.00 μm) falling in water. You will first need to note that the drag force is equal to the weight at terminal velocity. Take the density of the bacterium to be $1.10 \times 10^3 \text{ kg/m}^3$.

28. Stokes' law describes sedimentation of particles in liquids and can be used to measure viscosity. Particles in liquids achieve terminal velocity quickly. One can measure the time it takes for a particle to fall a certain distance and then use Stokes' law to calculate the viscosity of the liquid. Suppose a steel ball bearing (density $7.8 \times 10^3 \text{ kg/m}^3$, diameter 3.0 mm) is dropped in a container of motor oil. It takes 12 s to fall a distance of 0.60 m. Calculate the viscosity of the oil.

5.3 Elasticity: Stress and Strain

29. During a circus act, one performer swings upside down hanging from a trapeze holding another, also upside-down, performer by the legs. If the upward force on the lower performer is three times her weight, how much do the bones (the femurs) in her upper legs stretch? You may assume each is equivalent to a uniform rod 35.0 cm long and 1.80 cm in radius. Her mass is 60.0 kg.

30. During a wrestling match, a 150 kg wrestler briefly stands on one hand during a maneuver designed to perplex his already moribund adversary. By how much does the upper arm bone shorten in length? The bone can be represented by a uniform rod 38.0 cm in length and 2.10 cm in radius.

31. (a) The "lead" in pencils is a graphite composition with a Young's modulus of about $1 \times 10^9 \text{ N / m}^2$. Calculate the change in length of the lead in an automatic pencil if you tap it straight into the pencil with a force of 4.0 N. The lead is 0.50 mm in diameter and 60 mm long. (b) Is the answer reasonable? That is, does it seem to be consistent with what you have observed when using pencils?

32. TV broadcast antennas are the tallest artificial structures on Earth. In 1987, a 72.0-kg physicist placed himself and 400 kg of equipment at the top of one 610-m high antenna to perform gravity experiments. By how much was the antenna compressed, if we consider it to be equivalent to a steel cylinder 0.150 m in radius?

33. (a) By how much does a 65.0-kg mountain climber stretch her 0.800-cm diameter nylon rope when she hangs 35.0 m below a rock outcropping? (b) Does the answer seem to be consistent with what you have observed for nylon ropes? Would it make sense if the rope were actually a bungee cord?

34. A 20.0-m tall hollow aluminum flagpole is equivalent in strength to a solid cylinder 4.00 cm in diameter. A strong wind bends the pole much as a horizontal force of 900 N exerted at the top would. How far to the side does the top of the pole flex?

35. As an oil well is drilled, each new section of drill pipe supports its own weight and that of the pipe and drill bit beneath it. Calculate the stretch in a new 6.00 m length of steel pipe that supports 3.00 km of pipe having a mass of 20.0 kg/m and a 100-kg drill bit. The pipe is equivalent in strength to a solid cylinder 5.00 cm in diameter.

36. Calculate the force a piano tuner applies to stretch a steel piano wire 8.00 mm, if the wire is originally 0.850 mm in diameter and 1.35 m long.

37. A vertebra is subjected to a shearing force of 500 N. Find the shear deformation, taking the vertebra to be a cylinder 3.00 cm high and 4.00 cm in diameter.

38. A disk between vertebrae in the spine is subjected to a shearing force of 600 N. Find its shear deformation, taking it to have the shear modulus of $1 \times 10^9 \, \text{N}/\text{m}^2$. The disk is equivalent to a solid cylinder 0.700 cm high and 4.00 cm in diameter.

39. When using a pencil eraser, you exert a vertical force of 6.00 N at a distance of 2.00 cm from the hardwood-eraser joint. The pencil is 6.00 mm in diameter and is held at an angle of $20.0°$ to the horizontal. (a) By how much does the wood flex perpendicular to its length? (b) How much is it compressed lengthwise?

40. To consider the effect of wires hung on poles, we take data from **Example 4.8**, in which tensions in wires supporting a traffic light were calculated. The left wire made an angle $30.0°$ below the horizontal with the top of its pole and carried a tension of 108 N. The 12.0 m tall hollow aluminum pole is equivalent in strength to a 4.50 cm diameter solid cylinder. (a) How far is it bent to the side? (b) By how much is it compressed?

41. A farmer making grape juice fills a glass bottle to the brim and caps it tightly. The juice expands more than the glass when it warms up, in such a way that the volume increases by 0.2% (that is, $\Delta V / V_0 = 2 \times 10^{-3}$) relative to the space available. Calculate the magnitude of the normal force exerted by the juice per square centimeter if its bulk modulus is $1.8 \times 10^9 \, \text{N}/\text{m}^2$, assuming the bottle does not break. In view of your answer, do you think the bottle will survive?

42. (a) When water freezes, its volume increases by 9.05% (that is, $\Delta V / V_0 = 9.05 \times 10^{-2}$). What force per unit area is water capable of exerting on a container when it freezes? (It is acceptable to use the bulk modulus of water in this problem.) (b) Is it surprising that such forces can fracture engine blocks, boulders, and the like?

43. This problem returns to the tightrope walker studied in **Example 4.6**, who created a tension of $3.94 \times 10^3 \, \text{N}$ in a wire making an angle $5.0°$ below the horizontal with each supporting pole. Calculate how much this tension stretches the steel wire if it was originally 15 m long and 0.50 cm in diameter.

44. The pole in **Figure 5.24** is at a $90.0°$ bend in a power line and is therefore subjected to more shear force than poles in straight parts of the line. The tension in each line is $4.00 \times 10^4 \, \text{N}$, at the angles shown. The pole is 15.0 m tall, has an 18.0 cm diameter, and can be considered to have half the strength of hardwood. (a) Calculate the compression of the pole. (b) Find how much it bends and in what direction. (c) Find the tension in a guy wire used to keep the pole straight if it is attached to the top of the pole at an angle of $30.0°$ with the vertical. (Clearly, the guy wire must be in the opposite direction of the bend.)

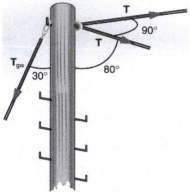

Figure 5.24 This telephone pole is at a $90°$ bend in a power line. A guy wire is attached to the top of the pole at an angle of $30°$ with the vertical.

Test Prep for AP® Courses

5.1 Friction

1. When a force of 20 N is applied to a stationary box weighing 40 N, the box does not move. This means the coefficient of static friction

a. is equal to 0.5.
b. is greater than 0.5.
c. is less than 0.5.
d. cannot be determined.

2. A 2-kg block slides down a ramp which is at an incline of 25°. If the frictional force is 4.86 N, what is the coefficient of friction? At what incline will the box slide at a constant velocity? Assume g = 10 m/s^2.

3. A block is given a short push and then slides with constant friction across a horizontal floor. Which statement best explains the direction of the force that friction applies on the moving block?

  a. Friction will be in the same direction as the block's motion because molecular interactions between the block and the floor will deform the block in the direction of motion.

  b. Friction will be in the same direction as the block's motion because thermal energy generated at the interface between the block and the floor adds kinetic energy to the block.

  c. Friction will be in the opposite direction of the block's motion because molecular interactions between the block and the floor will deform the block in the opposite direction of motion.

  d. Friction will be in the opposite direction of the block's motion because thermal energy generated at the interface between the block and the floor converts some of the block's kinetic energy to potential energy.

4. A student pushes a cardboard box across a carpeted floor and afterwards notices that the bottom of the box feels warm. Explain how interactions between molecules in the cardboard and molecules in the carpet produced this heat.

6   GRAVITATION AND UNIFORM CIRCULAR MOTION

Figure 6.1 This Australian Grand Prix Formula 1 race car moves in a circular path as it makes the turn. Its wheels also spin rapidly—the latter completing many revolutions, the former only part of one (a circular arc). The same physical principles are involved in each. (credit: Richard Munckton)

Chapter Outline

6.1. Rotation Angle and Angular Velocity

6.2. Centripetal Acceleration

6.3. Centripetal Force

6.4. Fictitious Forces and Non-inertial Frames: The Coriolis Force

6.5. Newton's Universal Law of Gravitation

6.6. Satellites and Kepler's Laws: An Argument for Simplicity

Connection for AP® Courses

Many motions, such as the arc of a bird's flight or Earth's path around the Sun, are curved. Recall that Newton's first law tells us that motion is along a straight line at constant speed unless there is a net external force. We will therefore study not only motion along curves, but also the forces that cause it, including gravitational forces. This chapter supports Big Idea 3 that interactions between objects are described by forces, and thus change in motion is a result of a net force exerted on an object. In this chapter, this idea is applied to uniform circular motion. In some ways, this chapter is a continuation of **Dynamics: Newton's Laws of Motion** as we study more applications of Newton's laws of motion.

This chapter deals with the simplest form of curved motion, **uniform circular motion**, which is motion in a circular path at constant speed. As an object moves on a circular path, the magnitude of its velocity remains constant, but the direction of the velocity is changing. This means there is an acceleration that we will refer to as a "centripetal" acceleration caused by a net external force, also called the "centripetal" force (Enduring Understanding 3.B). The centripetal force is the net force totaling all external forces acting on the object (Essential Knowledge 3.B.1). In order to determine the net force, a free-body diagram may be useful (Essential Knowledge 3.B.2).

Studying this topic illustrates most of the concepts associated with rotational motion and leads to many new topics we group under the name *rotation*. This motion can be described using kinematics variables (Essential Knowledge 3.A.1), but in addition to linear variables, we will introduce angular variables. We use various ways to describe motion, namely, verbally, algebraically and graphically (Learning Objective 3.A.1.1). Pure *rotational motion* occurs when points in an object move in circular paths centered on one point. Pure *translational motion* is motion with no rotation. Some motion combines both types, such as a rotating hockey puck moving over ice. Some combinations of both types of motion are conveniently described with fictitious forces which appear as a result of using a non-inertial frame of reference (Enduring Understanding 3.A).

Furthermore, the properties of uniform circular motion can be applied to the motion of massive objects in a gravitational field. Thus, this chapter supports Big Idea 1 that gravitational mass is an important property of an object or a system.

We have experimental evidence that gravitational and inertial masses are equal (Enduring Understanding 1.C), and that gravitational mass is a measure of the strength of the gravitational interaction (Essential Knowledge 1.C.2). Therefore, this chapter will support Big Idea 2 that fields existing in space can be used to explain interactions, because any massive object creates a gravitational field in space (Enduring Understanding 2.B). Mathematically, we use Newton's universal law of gravitation to provide a model for the gravitational interaction between two massive objects (Essential Knowledge 2.B.2). We will discover that this model describes the interaction of one object with mass with another object with mass (Essential Knowledge 3.C.1), and also that gravitational force is a long-range force (Enduring Understanding 3.C).

The concepts in this chapter support:

Big Idea 1 Objects and systems have properties such as mass and charge. Systems may have internal structure.

Enduring Understanding 1.C Objects and systems have properties of inertial mass and gravitational mass that are experimentally verified to be the same and that satisfy conservation principles.

Essential Knowledge 1.C.2 Gravitational mass is the property of an object or a system that determines the strength of the gravitational interaction with other objects, systems, or gravitational fields.

Essential Knowledge 1.C.3 Objects and systems have properties of inertial mass and gravitational mass that are experimentally verified to be the same and that satisfy conservation principles.

Big Idea 2 Fields existing in space can be used to explain interactions.

Enduring Understanding 2.B A gravitational field is caused by an object with mass.

Essential Knowledge 2.B.2. The gravitational field caused by a spherically symmetric object with mass is radial and, outside the object, varies as the inverse square of the radial distance from the center of that object.

Big Idea 3 The interactions of an object with other objects can be described by forces.

Enduring Understanding 3.A All forces share certain common characteristics when considered by observers in inertial reference frames.

Essential Knowledge 3.A.1. An observer in a particular reference frame can describe the motion of an object using such quantities as position, displacement, distance, velocity, speed, and acceleration.

Essential Knowledge 3.A.3. A force exerted on an object is always due to the interaction of that object with another object.

Enduring Understanding 3.B Classically, the acceleration of an object interacting with other objects can be predicted by using $a = \sum F / m$.

Essential Knowledge 3.B.1 If an object of interest interacts with several other objects, the net force is the vector sum of the individual forces.

Essential Knowledge 3.B.2 Free-body diagrams are useful tools for visualizing forces being exerted on a single object and writing the equations that represent a physical situation.

Enduring Understanding 3.C At the macroscopic level, forces can be categorized as either long-range (action-at-a-distance) forces or contact forces.

Essential Knowledge 3.C.1. Gravitational force describes the interaction of one object that has mass with another object that has mass.

6.1 Rotation Angle and Angular Velocity

Learning Objectives

By the end of this section, you will be able to:

- Define arc length, rotation angle, radius of curvature, and angular velocity.
- Calculate the angular velocity of a car wheel spin.

In **Kinematics**, we studied motion along a straight line and introduced such concepts as displacement, velocity, and acceleration.

Two-Dimensional Kinematics dealt with motion in two dimensions. Projectile motion is a special case of two-dimensional kinematics in which the object is projected into the air, while being subject to the gravitational force, and lands a distance away. In this chapter, we consider situations where the object does not land but moves in a curve. We begin the study of uniform circular motion by defining two angular quantities needed to describe rotational motion.

Rotation Angle

When objects rotate about some axis—for example, when the CD (compact disc) in **Figure 6.2** rotates about its center—each point in the object follows a circular arc. Consider a line from the center of the CD to its edge. Each **pit** used to record sound along this line moves through the same angle in the same amount of time. The rotation angle is the amount of rotation and is analogous to linear distance. We define the **rotation angle** $\Delta\theta$ to be the ratio of the arc length to the radius of curvature:

$$\Delta\theta = \frac{\Delta s}{r}. \tag{6.1}$$

Figure 6.2 All points on a CD travel in circular arcs. The pits along a line from the center to the edge all move through the same angle $\Delta\theta$ in a time Δt.

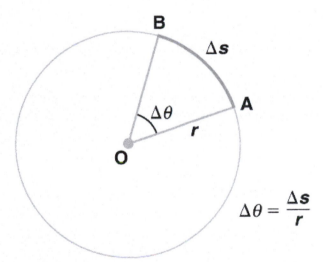

Figure 6.3 The radius of a circle is rotated through an angle $\Delta\theta$. The arc length Δs is described on the circumference.

The **arc length** Δs is the distance traveled along a circular path as shown in **Figure 6.3** Note that r is the **radius of curvature** of the circular path.

We know that for one complete revolution, the arc length is the circumference of a circle of radius r. The circumference of a circle is $2\pi r$. Thus for one complete revolution the rotation angle is

$$\Delta\theta = \frac{2\pi r}{r} = 2\pi. \tag{6.2}$$

This result is the basis for defining the units used to measure rotation angles, $\Delta\theta$ to be **radians** (rad), defined so that

$$2\pi \, \text{rad} = 1 \, \text{revolution}. \tag{6.3}$$

A comparison of some useful angles expressed in both degrees and radians is shown in **Table 6.1**.

Table 6.1 Comparison of Angular Units

Degree Measures	Radian Measure
30°	$\frac{\pi}{6}$
60°	$\frac{\pi}{3}$
90°	$\frac{\pi}{2}$
120°	$\frac{2\pi}{3}$
135°	$\frac{3\pi}{4}$
180°	π

$$\Delta\theta = \frac{\Delta s_1}{r_1}$$

$$\Delta\theta = \frac{\Delta s_2}{r_2}$$

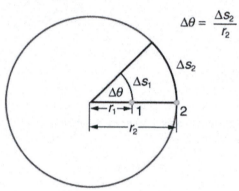

Figure 6.4 Points 1 and 2 rotate through the same angle ($\Delta\theta$), but point 2 moves through a greater arc length (Δs) because it is at a greater distance from the center of rotation (r).

If $\Delta\theta = 2\pi$ rad, then the CD has made one complete revolution, and every point on the CD is back at its original position. Because there are $360°$ in a circle or one revolution, the relationship between radians and degrees is thus

$$2\pi \text{ rad} = 360° \tag{6.4}$$

so that

$$1 \text{ rad} = \frac{360°}{2\pi} \approx 57.3°. \tag{6.5}$$

Angular Velocity

How fast is an object rotating? We define **angular velocity** ω as the rate of change of an angle. In symbols, this is

$$\omega = \frac{\Delta\theta}{\Delta t}, \tag{6.6}$$

where an angular rotation $\Delta\theta$ takes place in a time Δt. The greater the rotation angle in a given amount of time, the greater the angular velocity. The units for angular velocity are radians per second (rad/s).

Angular velocity ω is analogous to linear velocity v. To get the precise relationship between angular and linear velocity, we again consider a pit on the rotating CD. This pit moves an arc length Δs in a time Δt, and so it has a linear velocity

$$v = \frac{\Delta s}{\Delta t}. \tag{6.7}$$

From $\Delta\theta = \frac{\Delta s}{r}$ we see that $\Delta s = r\Delta\theta$. Substituting this into the expression for v gives

$$v = \frac{r\Delta\theta}{\Delta t} = r\omega. \tag{6.8}$$

We write this relationship in two different ways and gain two different insights:

$$v = r\omega \text{ or } \omega = \frac{v}{r}.$$

(6.9)

The first relationship in $v = r\omega$ or $\omega = \frac{v}{r}$ states that the linear velocity v is proportional to the distance from the center of rotation, thus, it is largest for a point on the rim (largest r), as you might expect. We can also call this linear speed v of a point on the rim the *tangential speed*. The second relationship in $v = r\omega$ or $\omega = \frac{v}{r}$ can be illustrated by considering the tire of a moving car. Note that the speed of a point on the rim of the tire is the same as the speed v of the car. See **Figure 6.5**. So the faster the car moves, the faster the tire spins—large v means a large ω, because $v = r\omega$. Similarly, a larger-radius tire rotating at the same angular velocity (ω) will produce a greater linear speed (v) for the car.

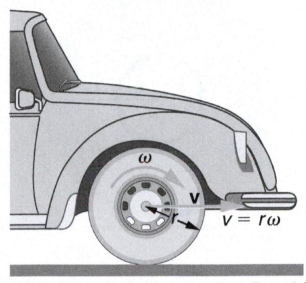

Figure 6.5 A car moving at a velocity v to the right has a tire rotating with an angular velocity ω. The speed of the tread of the tire relative to the axle is v, the same as if the car were jacked up. Thus the car moves forward at linear velocity $v = r\omega$, where r is the tire radius. A larger angular velocity for the tire means a greater velocity for the car.

Example 6.1 How Fast Does a Car Tire Spin?

Calculate the angular velocity of a 0.300 m radius car tire when the car travels at 15.0 m/s (about 54 km/h). See **Figure 6.5**.

Strategy

Because the linear speed of the tire rim is the same as the speed of the car, we have $v = 15.0$ m/s. The radius of the tire is given to be $r = 0.300$ m. Knowing v and r, we can use the second relationship in $v = r\omega$, $\omega = \frac{v}{r}$ to calculate the angular velocity.

Solution

To calculate the angular velocity, we will use the following relationship:

$$\omega = \frac{v}{r}.$$

(6.10)

Substituting the knowns,

$$\omega = \frac{15.0 \text{ m/s}}{0.300 \text{ m}} = 50.0 \text{ rad/s}.$$

(6.11)

Discussion

When we cancel units in the above calculation, we get 50.0/s. But the angular velocity must have units of rad/s. Because radians are actually unitless (radians are defined as a ratio of distance), we can simply insert them into the answer for the angular velocity. Also note that if an earth mover with much larger tires, say 1.20 m in radius, were moving at the same speed of 15.0 m/s, its tires would rotate more slowly. They would have an angular velocity

$$\omega = (15.0 \text{ m/s}) / (1.20 \text{ m}) = 12.5 \text{ rad/s}.$$

(6.12)

Both ω and v have directions (hence they are angular and linear *velocities*, respectively). Angular velocity has only two directions with respect to the axis of rotation—it is either clockwise or counterclockwise. Linear velocity is tangent to the path, as illustrated in **Figure 6.6**.

Tie an object to the end of a string and swing it around in a horizontal circle above your head (swing at your wrist). Maintain uniform speed as the object swings and measure the angular velocity of the motion. What is the approximate speed of the object? Identify a point close to your hand and take appropriate measurements to calculate the linear speed at this point. Identify other circular motions and measure their angular velocities.

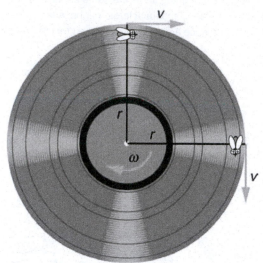

Figure 6.6 As an object moves in a circle, here a fly on the edge of an old-fashioned vinyl record, its instantaneous velocity is always tangent to the circle. The direction of the angular velocity is clockwise in this case.

PhET Explorations: Ladybug Revolution

Figure 6.7 Ladybug Revolution (http://cnx.org/content/m54992/1.2/rotation_en.jar)

Join the ladybug in an exploration of rotational motion. Rotate the merry-go-round to change its angle, or choose a constant angular velocity or angular acceleration. Explore how circular motion relates to the bug's x,y position, velocity, and acceleration using vectors or graphs.

6.2 Centripetal Acceleration

Learning Objectives

By the end of this section, you will be able to:

- Establish the expression for centripetal acceleration.
- Explain the centrifuge.

We know from kinematics that acceleration is a change in velocity, either in its magnitude or in its direction, or both. In uniform circular motion, the direction of the velocity changes constantly, so there is always an associated acceleration, even though the magnitude of the velocity might be constant. You experience this acceleration yourself when you turn a corner in your car. (If you hold the wheel steady during a turn and move at constant speed, you are in uniform circular motion.) What you notice is a sideways acceleration because you and the car are changing direction. The sharper the curve and the greater your speed, the more noticeable this acceleration will become. In this section we examine the direction and magnitude of that acceleration.

Figure 6.8 shows an object moving in a circular path at constant speed. The direction of the instantaneous velocity is shown at two points along the path. Acceleration is in the direction of the change in velocity, which points directly toward the center of rotation (the center of the circular path). This pointing is shown with the vector diagram in the figure. We call the acceleration of an object moving in uniform circular motion (resulting from a net external force) the **centripetal acceleration**(a_c); centripetal means "toward the center" or "center seeking."

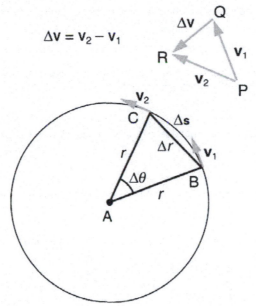

Figure 6.8 The directions of the velocity of an object at two different points are shown, and the change in velocity $\Delta \mathbf{v}$ is seen to point directly toward the center of curvature. (See small inset.) Because $\mathbf{a}_c = \Delta \mathbf{v} / \Delta t$, the acceleration is also toward the center; $\mathbf{a}_c$ is called centripetal acceleration. (Because $\Delta \theta$ is very small, the arc length Δs is equal to the chord length Δr for small time differences.)

The direction of centripetal acceleration is toward the center of curvature, but what is its magnitude? Note that the triangle formed by the velocity vectors and the one formed by the radii r and Δs are similar. Both the triangles ABC and PQR are isosceles triangles (two equal sides). The two equal sides of the velocity vector triangle are the speeds $v_1 = v_2 = v$. Using the properties of two similar triangles, we obtain

$$\frac{\Delta v}{v} = \frac{\Delta s}{r}. \tag{6.13}$$

Acceleration is $\Delta v / \Delta t$, and so we first solve this expression for Δv:

$$\Delta v = \frac{v}{r} \Delta s. \tag{6.14}$$

Then we divide this by Δt, yielding

$$\frac{\Delta v}{\Delta t} = \frac{v}{r} \times \frac{\Delta s}{\Delta t}. \tag{6.15}$$

Finally, noting that $\Delta v / \Delta t = a_c$ and that $\Delta s / \Delta t = v$, the linear or tangential speed, we see that the magnitude of the centripetal acceleration is

$$a_c = \frac{v^2}{r}, \tag{6.16}$$

which is the acceleration of an object in a circle of radius r at a speed v. So, centripetal acceleration is greater at high speeds and in sharp curves (smaller radius), as you have noticed when driving a car. But it is a bit surprising that a_c is proportional to speed squared, implying, for example, that it is four times as hard to take a curve at 100 km/h than at 50 km/h. A sharp corner has a small radius, so that a_c is greater for tighter turns, as you have probably noticed.

It is also useful to express a_c in terms of angular velocity. Substituting $v = r\omega$ into the above expression, we find $a_c = (r\omega)^2 / r = r\omega^2$. We can express the magnitude of centripetal acceleration using either of two equations:

$$a_c = \frac{v^2}{r}; \ a_c = r\omega^2. \tag{6.17}$$

Recall that the direction of a_c is toward the center. You may use whichever expression is more convenient, as illustrated in examples below.

A **centrifuge** (see **Figure 6.9**b) is a rotating device used to separate specimens of different densities. High centripetal acceleration significantly decreases the time it takes for separation to occur, and makes separation possible with small samples. Centrifuges are used in a variety of applications in science and medicine, including the separation of single cell suspensions such as bacteria, viruses, and blood cells from a liquid medium and the separation of macromolecules, such as DNA and protein, from a solution. Centrifuges are often rated in terms of their centripetal acceleration relative to acceleration due to gravity (g); maximum centripetal acceleration of several hundred thousand g is possible in a vacuum. Human centrifuges, extremely large centrifuges, have been used to test the tolerance of astronauts to the effects of accelerations larger than that of Earth's gravity.

Example 6.2 How Does the Centripetal Acceleration of a Car Around a Curve Compare with That Due to Gravity?

What is the magnitude of the centripetal acceleration of a car following a curve of radius 500 m at a speed of 25.0 m/s (about 90 km/h)? Compare the acceleration with that due to gravity for this fairly gentle curve taken at highway speed. See Figure 6.9(a).

Strategy

Because v and r are given, the first expression in $a_c = \frac{v^2}{r}; a_c = r\omega^2$ is the most convenient to use.

Solution

Entering the given values of $v = 25.0$ m/s and $r = 500$ m into the first expression for a_c gives

$$a_c = \frac{v^2}{r} = \frac{(25.0 \text{ m/s})^2}{500 \text{ m}} = 1.25 \text{ m/s}^2. \tag{6.18}$$

Discussion

To compare this with the acceleration due to gravity $(g = 9.80 \text{ m/s}^2)$, we take the ratio of $a_c/g = \left(1.25 \text{ m/s}^2\right)/\left(9.80 \text{ m/s}^2\right) = 0.128$. Thus, $a_c = 0.128$ g and is noticeable especially if you were not wearing a seat belt.

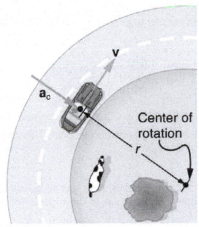

(a) Car around corner

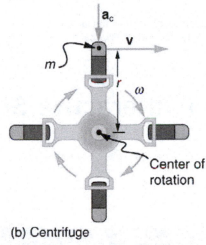

(b) Centrifuge

Figure 6.9 (a) The car following a circular path at constant speed is accelerated perpendicular to its velocity, as shown. The magnitude of this centripetal acceleration is found in Example 6.2. (b) A particle of mass in a centrifuge is rotating at constant angular velocity . It must be accelerated perpendicular to its velocity or it would continue in a straight line. The magnitude of the necessary acceleration is found in Example 6.3.

Example 6.3 How Big Is the Centripetal Acceleration in an Ultracentrifuge?

Calculate the centripetal acceleration of a point 7.50 cm from the axis of an **ultracentrifuge** spinning at 7.5×10^4 rev/min. Determine the ratio of this acceleration to that due to gravity. See **Figure 6.9(b)**.

Strategy

The term rev/min stands for revolutions per minute. By converting this to radians per second, we obtain the angular velocity ω. Because r is given, we can use the second expression in the equation $a_c = \frac{v^2}{r}$; $a_c = r\omega^2$ to calculate the centripetal acceleration.

Solution

To convert 7.50×10^4 rev / min to radians per second, we use the facts that one revolution is $2\pi \text{rad}$ and one minute is 60.0 s. Thus,

$$\omega = 7.50 \times 10^4 \ \frac{\text{rev}}{\text{min}} \times \frac{2\pi \text{ rad}}{1 \text{ rev}} \times \frac{1 \text{ min}}{60.0 \text{ s}} = 7854 \text{ rad/s.} \tag{6.19}$$

Now the centripetal acceleration is given by the second expression in $a_c = \frac{v^2}{r}$; $a_c = r\omega^2$ as

$$a_c = r\omega^2. \tag{6.20}$$

Converting 7.50 cm to meters and substituting known values gives

$$a_c = (0.0750 \text{ m})(7854 \text{ rad/s})^2 = 4.63 \times 10^6 \text{ m/s}^2. \tag{6.21}$$

Note that the unitless radians are discarded in order to get the correct units for centripetal acceleration. Taking the ratio of a_c to g yields

$$\frac{a_c}{g} = \frac{4.63 \times 10^6}{9.80} = 4.72 \times 10^5.$$
(6.22)

Discussion

This last result means that the centripetal acceleration is 472,000 times as strong as g. It is no wonder that such high ω centrifuges are called ultracentrifuges. The extremely large accelerations involved greatly decrease the time needed to cause the sedimentation of blood cells or other materials.

Of course, a net external force is needed to cause any acceleration, just as Newton proposed in his second law of motion. So a net external force is needed to cause a centripetal acceleration. In **Centripetal Force**, we will consider the forces involved in circular motion.

PhET Explorations: Ladybug Motion 2D

Learn about position, velocity and acceleration vectors. Move the ladybug by setting the position, velocity or acceleration, and see how the vectors change. Choose linear, circular or elliptical motion, and record and playback the motion to analyze the behavior.

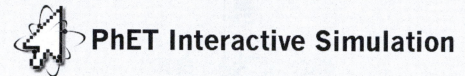

Figure 6.10 Ladybug Motion 2D (http://cnx.org/content/m54995/1.3/ladybug-motion-2d_en.jar)

6.3 Centripetal Force

Learning Objectives
By the end of this section, you will be able to:
• Calculate coefficient of friction on a car tire.   • Calculate ideal speed and angle of a car on a turn.

Any force or combination of forces can cause a centripetal or radial acceleration. Just a few examples are the tension in the rope on a tether ball, the force of Earth's gravity on the Moon, friction between roller skates and a rink floor, a banked roadway's force on a car, and forces on the tube of a spinning centrifuge.

Any net force causing uniform circular motion is called a **centripetal force**. The direction of a centripetal force is toward the center of curvature, the same as the direction of centripetal acceleration. According to Newton's second law of motion, net force is mass times acceleration: net $\mathbf{F} = ma$. For uniform circular motion, the acceleration is the centripetal acceleration— $a = a_c$. Thus, the magnitude of centripetal force $\mathbf{F}_c$ is

$$\mathbf{F}_c = ma_c.$$
(6.23)

By using the expressions for centripetal acceleration a_c from $a_c = \frac{v^2}{r}$; $a_c = r\omega^2$, we get two expressions for the centripetal force $\mathbf{F}_c$ in terms of mass, velocity, angular velocity, and radius of curvature:

$$F_c = m\frac{v^2}{r}; \; F_c = mr\omega^2.$$
(6.24)

You may use whichever expression for centripetal force is more convenient. Centripetal force $\mathbf{F}_c$ is always perpendicular to the path and pointing to the center of curvature, because $\mathbf{a}_c$ is perpendicular to the velocity and pointing to the center of curvature.

Note that if you solve the first expression for r, you get

$$r = \frac{mv^2}{F_c}.$$
(6.25)

This implies that for a given mass and velocity, a large centripetal force causes a small radius of curvature—that is, a tight curve.

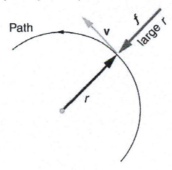

$f = \mathbf{F}_c$ is parallel to $\mathbf{a}_c$ since $\mathbf{F}_c = m\mathbf{a}_c$

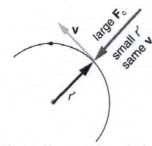

Figure 6.11 The frictional force supplies the centripetal force and is numerically equal to it. Centripetal force is perpendicular to velocity and causes uniform circular motion. The larger the F_c, the smaller the radius of curvature r and the sharper the curve. The second curve has the same v, but a larger F_c produces a smaller r'.

Example 6.4 What Coefficient of Friction Do Car Tires Need on a Flat Curve?

(a) Calculate the centripetal force exerted on a 900 kg car that negotiates a 500 m radius curve at 25.0 m/s.

(b) Assuming an unbanked curve, find the minimum static coefficient of friction, between the tires and the road, static friction being the reason that keeps the car from slipping (see **Figure 6.12**).

Strategy and Solution for (a)

We know that $F_c = \frac{mv^2}{r}$. Thus,

$$F_c = \frac{mv^2}{r} = \frac{(900 \text{ kg})(25.0 \text{ m/s})^2}{(500 \text{ m})} = 1125 \text{ N}.$$ (6.26)

Strategy for (b)

Figure 6.12 shows the forces acting on the car on an unbanked (level ground) curve. Friction is to the left, keeping the car from slipping, and because it is the only horizontal force acting on the car, the friction is the centripetal force in this case. We know that the maximum static friction (at which the tires roll but do not slip) is $\mu_s N$, where μ_s is the static coefficient of friction and N is the normal force. The normal force equals the car's weight on level ground, so that $N = mg$. Thus the centripetal force in this situation is

$$F_c = f = \mu_s N = \mu_s mg.$$ (6.27)

Now we have a relationship between centripetal force and the coefficient of friction. Using the first expression for F_c from the equation

(6.28)

$$\left.\begin{array}{l} F_c = m\dfrac{v^2}{r} \\ F_c = mr\omega^2 \end{array}\right\},$$

$$m\frac{v^2}{r} = \mu_s mg.$$ (6.29)

We solve this for μ_s, noting that mass cancels, and obtain

$$\mu_s = \frac{v^2}{rg}.$$

(6.30)

Solution for (b)

Substituting the knowns,

$$\mu_s = \frac{(25.0 \text{ m/s})^2}{(500 \text{ m})(9.80 \text{ m/s}^2)} = 0.13.$$

(6.31)

(Because coefficients of friction are approximate, the answer is given to only two digits.)

Discussion

We could also solve part (a) using the first expression in $\left. \begin{array}{c} F_c = m\frac{v^2}{r} \\ F_c = mr\omega^2 \end{array} \right\}$, because *m*, *v*, and *r* are given. The coefficient

of friction found in part (b) is much smaller than is typically found between tires and roads. The car will still negotiate the curve if the coefficient is greater than 0.13, because static friction is a responsive force, being able to assume a value less than but no more than $\mu_s N$. A higher coefficient would also allow the car to negotiate the curve at a higher speed, but if the coefficient of friction is less, the safe speed would be less than 25 m/s. Note that mass cancels, implying that in this example, it does not matter how heavily loaded the car is to negotiate the turn. Mass cancels because friction is assumed proportional to the normal force, which in turn is proportional to mass. If the surface of the road were banked, the normal force would be less as will be discussed below.

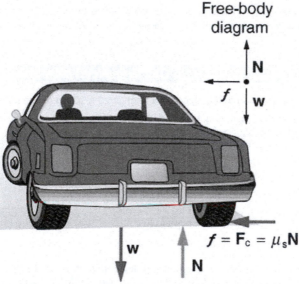

Figure 6.12 This car on level ground is moving away and turning to the left. The centripetal force causing the car to turn in a circular path is due to friction between the tires and the road. A minimum coefficient of friction is needed, or the car will move in a larger-radius curve and leave the roadway.

Let us now consider **banked curves**, where the slope of the road helps you negotiate the curve. See **Figure 6.13**. The greater the angle θ, the faster you can take the curve. Race tracks for bikes as well as cars, for example, often have steeply banked curves. In an "ideally banked curve," the angle θ is such that you can negotiate the curve at a certain speed without the aid of friction between the tires and the road. We will derive an expression for θ for an ideally banked curve and consider an example related to it.

For **ideal banking**, the net external force equals the horizontal centripetal force in the absence of friction. The components of the normal force N in the horizontal and vertical directions must equal the centripetal force and the weight of the car, respectively. In cases in which forces are not parallel, it is most convenient to consider components along perpendicular axes—in this case, the vertical and horizontal directions.

Figure 6.13 shows a free body diagram for a car on a frictionless banked curve. If the angle θ is ideal for the speed and radius, then the net external force will equal the necessary centripetal force. The only two external forces acting on the car are its weight $\mathbf{w}$ and the normal force of the road $\mathbf{N}$. (A frictionless surface can only exert a force perpendicular to the surface—that is, a normal force.) These two forces must add to give a net external force that is horizontal toward the center of curvature and has magnitude mv^2/r. Because this is the crucial force and it is horizontal, we use a coordinate system with vertical and horizontal axes. Only the normal force has a horizontal component, and so this must equal the centripetal force—that is,

$$N \sin \theta = \frac{mv^2}{r}.$$

(6.32)

Because the car does not leave the surface of the road, the net vertical force must be zero, meaning that the vertical components of the two external forces must be equal in magnitude and opposite in direction. From the figure, we see that the vertical component of the normal force is $N \cos \theta$, and the only other vertical force is the car's weight. These must be equal in magnitude; thus,

$$N \cos \theta = mg.$$

(6.33)

Now we can combine the last two equations to eliminate N and get an expression for θ, as desired. Solving the second equation for $N = mg / (\cos \theta)$, and substituting this into the first yields

$$mg\frac{\sin \theta}{\cos \theta} = \frac{mv^2}{r}$$

(6.34)

$$mg \tan(\theta) = \frac{mv^2}{r}$$

(6.35)

$$\tan \theta = \frac{v^2}{rg}.$$

Taking the inverse tangent gives

$$\theta = \tan^{-1}\left(\frac{v^2}{rg}\right) \text{ (ideally banked curve, no friction).}$$

(6.36)

This expression can be understood by considering how θ depends on v and r. A large θ will be obtained for a large v and a small r. That is, roads must be steeply banked for high speeds and sharp curves. Friction helps, because it allows you to take the curve at greater or lower speed than if the curve is frictionless. Note that θ does not depend on the mass of the vehicle.

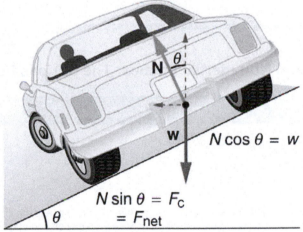

$$N \cos \theta = w$$
$$N \sin \theta = F_c$$
$$= F_{net}$$

Figure 6.13 The car on this banked curve is moving away and turning to the left.

Example 6.5 What Is the Ideal Speed to Take a Steeply Banked Tight Curve?

Curves on some test tracks and race courses, such as the Daytona International Speedway in Florida, are very steeply banked. This banking, with the aid of tire friction and very stable car configurations, allows the curves to be taken at very high speed. To illustrate, calculate the speed at which a 100 m radius curve banked at 65.0° should be driven if the road is frictionless.

Strategy

We first note that all terms in the expression for the ideal angle of a banked curve except for speed are known; thus, we need only rearrange it so that speed appears on the left-hand side and then substitute known quantities.

Solution

Starting with

$$\tan \theta = \frac{v^2}{rg}$$

(6.37)

we get

$$v = (rg \tan \theta)^{1/2}. \tag{6.38}$$

Noting that tan 65.0° = 2.14, we obtain

$$v = \left[(100 \text{ m})(9.80 \text{ m/s}^2)(2.14)\right]^{1/2} \tag{6.39}$$
$$= 45.8 \text{ m/s}.$$

Discussion

This is just about 165 km/h, consistent with a very steeply banked and rather sharp curve. Tire friction enables a vehicle to take the curve at significantly higher speeds.

Calculations similar to those in the preceding examples can be performed for a host of interesting situations in which centripetal force is involved—a number of these are presented in this chapter's Problems and Exercises.

Take-Home Experiment

Ask a friend or relative to swing a golf club or a tennis racquet. Take appropriate measurements to estimate the centripetal acceleration of the end of the club or racquet. You may choose to do this in slow motion.

PhET Explorations: Gravity and Orbits

Move the sun, earth, moon and space station to see how it affects their gravitational forces and orbital paths. Visualize the sizes and distances between different heavenly bodies, and turn off gravity to see what would happen without it!

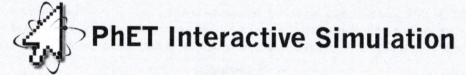

Figure 6.14 Gravity and Orbits (http://cnx.org/content/m55002/1.2/gravity-and-orbits_en.jar)

6.4 Fictitious Forces and Non-inertial Frames: The Coriolis Force

Learning Objectives
By the end of this section, you will be able to: • Discuss the inertial frame of reference. • Discuss the non-inertial frame of reference. • Describe the effects of the Coriolis force.

What do taking off in a jet airplane, turning a corner in a car, riding a merry-go-round, and the circular motion of a tropical cyclone have in common? Each exhibits fictitious forces—unreal forces that arise from motion and may *seem* real, because the observer's frame of reference is accelerating or rotating.

When taking off in a jet, most people would agree it feels as if you are being pushed back into the seat as the airplane accelerates down the runway. Yet a physicist would say that *you* tend to remain stationary while the *seat* pushes forward on you, and there is no real force backward on you. An even more common experience occurs when you make a tight curve in your car—say, to the right. You feel as if you are thrown (that is, *forced*) toward the left relative to the car. Again, a physicist would say that *you* are going in a straight line but the *car* moves to the right, and there is no real force on you to the left. Recall Newton's first law.

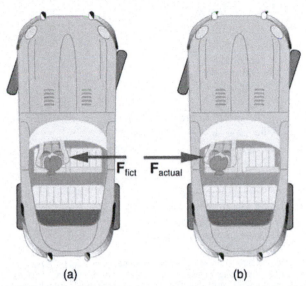

(a)   (b)

Figure 6.15 (a) The car driver feels herself forced to the left relative to the car when she makes a right turn. This is a fictitious force arising from the use of the car as a frame of reference. (b) In the Earth's frame of reference, the driver moves in a straight line, obeying Newton's first law, and the car moves to the right. There is no real force to the left on the driver relative to Earth. There is a real force to the right on the car to make it turn.

We can reconcile these points of view by examining the frames of reference used. Let us concentrate on people in a car. Passengers instinctively use the car as a frame of reference, while a physicist uses Earth. The physicist chooses Earth because it is very nearly an inertial frame of reference—one in which all forces are real (that is, in which all forces have an identifiable physical origin). In such a frame of reference, Newton's laws of motion take the form given in **Dynamics: Newton's Laws of Motion** The car is a **non-inertial frame of reference** because it is accelerated to the side. The force to the left sensed by car passengers is a **fictitious force** having no physical origin. There is nothing real pushing them left—the car, as well as the driver, is actually accelerating to the right.

Let us now take a mental ride on a merry-go-round—specifically, a rapidly rotating playground merry-go-round. You take the merry-go-round to be your frame of reference because you rotate together. In that non-inertial frame, you feel a fictitious force, named **centrifugal force (**not to be confused with centripetal force), trying to throw you off. You must hang on tightly to counteract the centrifugal force. In Earth's frame of reference, there is no force trying to throw you off. Rather you must hang on to make yourself go in a circle because otherwise you would go in a straight line, right off the merry-go-round.

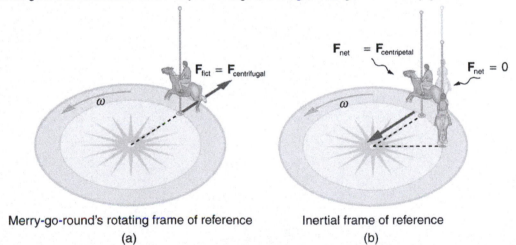

Merry-go-round's rotating frame of reference   Inertial frame of reference
(a)   (b)

Figure 6.16 (a) A rider on a merry-go-round feels as if he is being thrown off. This fictitious force is called the centrifugal force—it explains the rider's motion in the rotating frame of reference. (b) In an inertial frame of reference and according to Newton's laws, it is his inertia that carries him off and not a real force (the unshaded rider has $F_{net} = 0$ and heads in a straight line). A real force, $F_{centripetal}$, is needed to cause a circular path.

This inertial effect, carrying you away from the center of rotation if there is no centripetal force to cause circular motion, is put to good use in centrifuges (see **Figure 6.17**). A centrifuge spins a sample very rapidly, as mentioned earlier in this chapter. Viewed from the rotating frame of reference, the fictitious centrifugal force throws particles outward, hastening their sedimentation. The greater the angular velocity, the greater the centrifugal force. But what really happens is that the inertia of the particles carries them along a line tangent to the circle while the test tube is forced in a circular path by a centripetal force.

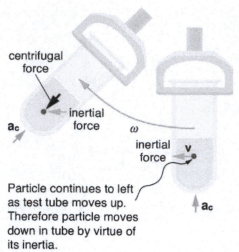

Figure 6.17 Centrifuges use inertia to perform their task. Particles in the fluid sediment come out because their inertia carries them away from the center of rotation. The large angular velocity of the centrifuge quickens the sedimentation. Ultimately, the particles will come into contact with the test tube walls, which will then supply the centripetal force needed to make them move in a circle of constant radius.

Let us now consider what happens if something moves in a frame of reference that rotates. For example, what if you slide a ball directly away from the center of the merry-go-round, as shown in **Figure 6.18**? The ball follows a straight path relative to Earth (assuming negligible friction) and a path curved to the right on the merry-go-round's surface. A person standing next to the merry-go-round sees the ball moving straight and the merry-go-round rotating underneath it. In the merry-go-round's frame of reference, we explain the apparent curve to the right by using a fictitious force, called the **Coriolis force**, that causes the ball to curve to the right. The fictitious Coriolis force can be used by anyone in that frame of reference to explain why objects follow curved paths and allows us to apply Newton's Laws in non-inertial frames of reference.

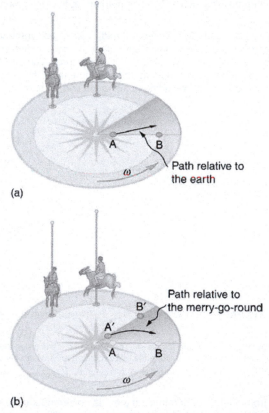

Figure 6.18 Looking down on the counterclockwise rotation of a merry-go-round, we see that a ball slid straight toward the edge follows a path curved to the right. The person slides the ball toward point B, starting at point A. Both points rotate to the shaded positions (A' and B') shown in the time that the ball follows the curved path in the rotating frame and a straight path in Earth's frame.

Up until now, we have considered Earth to be an inertial frame of reference with little or no worry about effects due to its rotation. Yet such effects *do* exist—in the rotation of weather systems, for example. Most consequences of Earth's rotation can be qualitatively understood by analogy with the merry-go-round. Viewed from above the North Pole, Earth rotates counterclockwise, as does the merry-go-round in **Figure 6.18**. As on the merry-go-round, any motion in Earth's northern hemisphere experiences a

Coriolis force to the right. Just the opposite occurs in the southern hemisphere; there, the force is to the left. Because Earth's angular velocity is small, the Coriolis force is usually negligible, but for large-scale motions, such as wind patterns, it has substantial effects.

The Coriolis force causes hurricanes in the northern hemisphere to rotate in the counterclockwise direction, while the tropical cyclones (what hurricanes are called below the equator) in the southern hemisphere rotate in the clockwise direction. The terms hurricane, typhoon, and tropical storm are regionally-specific names for tropical cyclones, storm systems characterized by low pressure centers, strong winds, and heavy rains. Figure 6.19 helps show how these rotations take place. Air flows toward any region of low pressure, and tropical cyclones contain particularly low pressures. Thus winds flow toward the center of a tropical cyclone or a low-pressure weather system at the surface. In the northern hemisphere, these inward winds are deflected to the right, as shown in the figure, producing a counterclockwise circulation at the surface for low-pressure zones of any type. Low pressure at the surface is associated with rising air, which also produces cooling and cloud formation, making low-pressure patterns quite visible from space. Conversely, wind circulation around high-pressure zones is clockwise in the northern hemisphere but is less visible because high pressure is associated with sinking air, producing clear skies.

The rotation of tropical cyclones and the path of a ball on a merry-go-round can just as well be explained by inertia and the rotation of the system underneath. When non-inertial frames are used, fictitious forces, such as the Coriolis force, must be invented to explain the curved path. There is no identifiable physical source for these fictitious forces. In an inertial frame, inertia explains the path, and no force is found to be without an identifiable source. Either view allows us to describe nature, but a view in an inertial frame is the simplest and truest, in the sense that all forces have real origins and explanations.

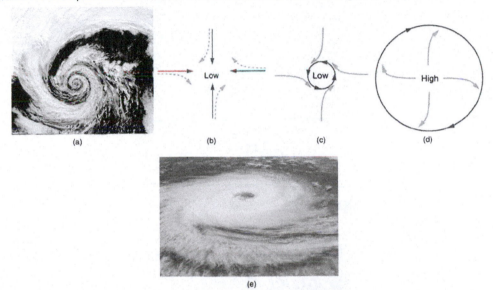

Figure 6.19 (a) The counterclockwise rotation of this northern hemisphere hurricane is a major consequence of the Coriolis force. (credit: NASA) (b) Without the Coriolis force, air would flow straight into a low-pressure zone, such as that found in tropical cyclones. (c) The Coriolis force deflects the winds to the right, producing a counterclockwise rotation. (d) Wind flowing away from a high-pressure zone is also deflected to the right, producing a clockwise rotation. (e) The opposite direction of rotation is produced by the Coriolis force in the southern hemisphere, leading to tropical cyclones. (credit: NASA)

6.5 Newton's Universal Law of Gravitation

Learning Objectives

By the end of this section, you will be able to:

- Explain Earth's gravitational force.
- Describe the gravitational effect of the Moon on Earth.
- Discuss weightlessness in space.
- Understand the Cavendish experiment.

The information presented in this section supports the following AP® learning objectives and science practices:

- **2.B.2.1** The student is able to apply $g = \dfrac{GM}{r^2}$ to calculate the gravitational field due to an object with mass M, where the field is a vector directed toward the center of the object of mass M. **(S.P. 2.2)**
- **2.B.2.2** The student is able to approximate a numerical value of the gravitational field (g) near the surface of an object from its radius and mass relative to those of the Earth or other reference objects. **(S.P. 2.2)**
- **3.A.3.4.** The student is able to make claims about the force on an object due to the presence of other objects with the same property: mass, electric charge. **(S.P. 6.1, 6.4)**

What do aching feet, a falling apple, and the orbit of the Moon have in common? Each is caused by the gravitational force. Our feet are strained by supporting our weight—the force of Earth's gravity on us. An apple falls from a tree because of the same force acting a few meters above Earth's surface. And the Moon orbits Earth because gravity is able to supply the necessary centripetal force at a distance of hundreds of millions of meters. In fact, the same force causes planets to orbit the Sun, stars to orbit the center of the galaxy, and galaxies to cluster together. Gravity is another example of underlying simplicity in nature. It is the weakest of the four basic forces found in nature, and in some ways the least understood. It is a force that acts at a distance, without physical contact, and is expressed by a formula that is valid everywhere in the universe, for masses and distances that vary from the tiny to the immense.

Sir Isaac Newton was the first scientist to precisely define the gravitational force, and to show that it could explain both falling bodies and astronomical motions. See Figure 6.20. But Newton was not the first to suspect that the same force caused both our weight and the motion of planets. His forerunner Galileo Galilei had contended that falling bodies and planetary motions had the same cause. Some of Newton's contemporaries, such as Robert Hooke, Christopher Wren, and Edmund Halley, had also made some progress toward understanding gravitation. But Newton was the first to propose an exact mathematical form and to use that form to show that the motion of heavenly bodies should be conic sections—circles, ellipses, parabolas, and hyperbolas. This theoretical prediction was a major triumph—it had been known for some time that moons, planets, and comets follow such paths, but no one had been able to propose a mechanism that caused them to follow these paths and not others. This was one of the earliest examples of a theory derived from empirical evidence doing more than merely describing those empirical results; it made claims about the fundamental workings of the universe.

Figure 6.20 According to early accounts, Newton was inspired to make the connection between falling bodies and astronomical motions when he saw an apple fall from a tree and realized that if the gravitational force could extend above the ground to a tree, it might also reach the Sun. The inspiration of Newton's apple is a part of worldwide folklore and may even be based in fact. Great importance is attached to it because Newton's universal law of gravitation and his laws of motion answered very old questions about nature and gave tremendous support to the notion of underlying simplicity and unity in nature. Scientists still expect underlying simplicity to emerge from their ongoing inquiries into nature.

The gravitational force is relatively simple. It is always attractive, and it depends only on the masses involved and the distance between them. Stated in modern language, **Newton's universal law of gravitation** states that every particle in the universe attracts every other particle with a force along a line joining them. The force is directly proportional to the product of their masses and inversely proportional to the square of the distance between them.

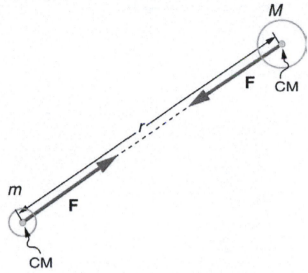

Figure 6.21 Gravitational attraction is along a line joining the centers of mass of these two bodies. The magnitude of the force is the same on each, consistent with Newton's third law.

Misconception Alert

The magnitude of the force on each object (one has larger mass than the other) is the same, consistent with Newton's third law.

The bodies we are dealing with tend to be large. To simplify the situation we assume that the body acts as if its entire mass is concentrated at one specific point called the **center of mass** (CM), which will be further explored in **Linear Momentum and Collisions**. For two bodies having masses m and M with a distance r between their centers of mass, the equation for Newton's universal law of gravitation is

$$F = G\frac{mM}{r^2},$$

(6.40)

where F is the magnitude of the gravitational force and G is a proportionality factor called the **gravitational constant**. G is a universal gravitational constant—that is, it is thought to be the same everywhere in the universe. It has been measured experimentally to be

$$G = 6.673 \times 10^{-11} \frac{\text{N} \cdot \text{m}^2}{\text{kg}^2}$$

(6.41)

in SI units. Note that the units of G are such that a force in newtons is obtained from $F = G\frac{mM}{r^2}$, when considering masses in kilograms and distance in meters. For example, two 1.000 kg masses separated by 1.000 m will experience a gravitational attraction of 6.673×10^{-11} N. This is an extraordinarily small force. The small magnitude of the gravitational force is consistent with everyday experience. We are unaware that even large objects like mountains exert gravitational forces on us. In fact, our body weight is the force of attraction of the *entire Earth* on us with a mass of 6×10^{24} kg.

The experiment to measure G was first performed by Cavendish, and is explained in more detail later. The fundamental concept it is based on is having a known mass on a spring with a known force (or spring) constant. Then, a second known mass is placed at multiple known distances from the first, and the amount of stretch in the spring resulting from the gravitational attraction of the two masses is measured.

Recall that the acceleration due to gravity g is about 9.80 m/s^2 on Earth. We can now determine why this is so. The weight of an object mg is the gravitational force between it and Earth. Substituting mg for F in Newton's universal law of gravitation gives

$$mg = G\frac{mM}{r^2},$$

(6.42)

where m is the mass of the object, M is the mass of Earth, and r is the distance to the center of Earth (the distance between the centers of mass of the object and Earth). See **Figure 6.22**. The mass m of the object cancels, leaving an equation for g:

$$g = G\frac{M}{r^2}.$$

(6.43)

Substituting known values for Earth's mass and radius (to three significant figures),

$$g = \left(6.67 \times 10^{-11} \frac{\text{N} \cdot \text{m}^2}{\text{kg}^2}\right) \times \frac{5.98 \times 10^{24} \text{ kg}}{(6.38 \times 10^6 \text{ m})^2},$$

(6.44)

and we obtain a value for the acceleration of a falling body:

$$g = 9.80 \text{ m/s}^2.$$

(6.45)

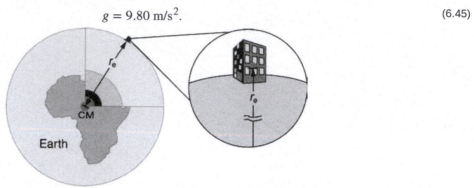

Figure 6.22 The distance between the centers of mass of Earth and an object on its surface is very nearly the same as the radius of Earth, because Earth is so much larger than the object.

This is the expected value *and is independent of the body's mass*. Newton's law of gravitation takes Galileo's observation that all masses fall with the same acceleration a step further, explaining the observation in terms of a force that causes objects to fall—in fact, in terms of a universally existing force of attraction between masses.

Gravitational Mass and Inertial Mass

Notice that, in Equation 6.40, the mass of the objects under consideration is directly proportional to the gravitational force. More mass means greater forces, and vice versa. However, we have already seen the concept of mass before in a different context.

In Chapter 4, you read that mass is a measure of inertia. However, we normally measure the mass of an object by measuring the force of gravity (F) on it.

How do we know that inertial mass is identical to gravitational mass? Assume that we compare the mass of two objects. The objects have inertial masses m_1 and m_2. If the objects balance each other on a pan balance, we can conclude that they have the same gravitational mass, that is, that they experience the same force due to gravity, F. Using Newton's second law of motion, $F = ma$, we can write $m_1 a_1 = m_2 a_2$.

If we can show that the two objects experience the same acceleration due to gravity, we can conclude that $m_1 = m_2$, that is, that the objects' inertial masses are equal.

In fact, Galileo and others conducted experiments to show that, when factors such as wind resistance are kept constant, all objects, regardless of their mass, experience the same acceleration due to gravity. Galileo is famously said to have dropped two balls of different masses off the leaning tower of Pisa to demonstrate this. The balls accelerated at the same rate. Since acceleration due to gravity is constant for all objects on Earth, regardless of their mass or composition, i.e., $a_1 = a_2$, then $m_1 = m_2$. Thus, we can conclude that inertial mass is identical to gravitational mass. This allows us to calculate the acceleration of free fall due to gravity, such as in the orbits of planets.

Take-Home Experiment

Take a marble, a ball, and a spoon and drop them from the same height. Do they hit the floor at the same time? If you drop a piece of paper as well, does it behave like the other objects? Explain your observations.

Making Connections: Gravitation, Other Forces, and General Relativity

Attempts are still being made to understand the gravitational force. As we shall see in **Particle Physics**, modern physics is exploring the connections of gravity to other forces, space, and time. General relativity alters our view of gravitation, leading us to think of gravitation as bending space and time.

Applying the Science Practices: All Objects Have Gravitational Fields

We can use the formula developed above, $g = \dfrac{GM}{r^2}$, to calculate the gravitational fields of other objects.

For example, the Moon has a radius of 1.7×10^6 m and a mass of 7.3×10^{22} kg. The gravitational field on the surface of the

Moon can be expressed as

$$g = G\frac{M}{r^2}$$

$$= \left(6.67\times10^{-11}\frac{N{\cdot}m^2}{kg^2}\right)\times\frac{7.3\times10^{22}\ kg}{\left(1.7\times10^6\ m\right)^2}$$

$$= 1.685\ m/s^2$$

This is about 1/6 of the gravity on Earth, which seems reasonable, since the Moon has a much smaller mass than Earth does.

A person has a mass of 50 kg. The gravitational field 1.0 m from the person's center of mass can be expressed as

$$g = G\frac{M}{r^2}$$

$$= \left(6.67\times10^{-11}\frac{N{\cdot}m^2}{kg^2}\right)\times\frac{50\ kg}{(1\ m)^2}$$

$$= 3.34\times10^{-9}\ m/s^2$$

This is less than one millionth of the gravitational field at the surface of Earth.

In the following example, we make a comparison similar to one made by Newton himself. He noted that if the gravitational force caused the Moon to orbit Earth, then the acceleration due to gravity should equal the centripetal acceleration of the Moon in its orbit. Newton found that the two accelerations agreed "pretty nearly."

Example 6.6 Earth's Gravitational Force Is the Centripetal Force Making the Moon Move in a Curved Path

(a) Find the acceleration due to Earth's gravity at the distance of the Moon.

(b) Calculate the centripetal acceleration needed to keep the Moon in its orbit (assuming a circular orbit about a fixed Earth), and compare it with the value of the acceleration due to Earth's gravity that you have just found.

Strategy for (a)

This calculation is the same as the one finding the acceleration due to gravity at Earth's surface, except that r is the distance from the center of Earth to the center of the Moon. The radius of the Moon's nearly circular orbit is $3.84\times10^8\ m$.

Solution for (a)

Substituting known values into the expression for g found above, remembering that M is the mass of Earth not the Moon, yields

$$g = G\frac{M}{r^2} = \left(6.67\times10^{-11}\frac{N\cdot m^2}{kg^2}\right)\times\frac{5.98\times10^{24}\ kg}{(3.84\times10^8\ m)^2} \tag{6.46}$$
$$= 2.70\times10^{-3}\ m/s.^2$$

Strategy for (b)

Centripetal acceleration can be calculated using either form of

$$\left.\begin{array}{l}a_c = \frac{v^2}{r}\\ a_c = r\omega^2\end{array}\right\}. \tag{6.47}$$

We choose to use the second form:

$$a_c = r\omega^2, \tag{6.48}$$

where ω is the angular velocity of the Moon about Earth.

Solution for (b)

Given that the period (the time it takes to make one complete rotation) of the Moon's orbit is 27.3 days, (d) and using

$$1 \text{ d} \times 24\frac{\text{hr}}{\text{d}} \times 60\frac{\text{min}}{\text{hr}} \times 60\frac{\text{s}}{\text{min}} = 86,400 \text{ s} \tag{6.49}$$

we see that

$$\omega = \frac{\Delta\theta}{\Delta t} = \frac{2\pi \text{ rad}}{(27.3 \text{ d})(86,400 \text{ s/d})} = 2.66 \times 10^{-6}\frac{\text{rad}}{\text{s}}. \tag{6.50}$$

The centripetal acceleration is

$$\begin{aligned} a_c &= r\omega^2 = (3.84 \times 10^8 \text{ m})(2.66 \times 10^{-6} \text{ rad/s})^2 \\ &= 2.72 \times 10^{-3} \text{ m/s.}^2 \end{aligned} \tag{6.51}$$

The direction of the acceleration is toward the center of the Earth.

Discussion

The centripetal acceleration of the Moon found in (b) differs by less than 1% from the acceleration due to Earth's gravity found in (a). This agreement is approximate because the Moon's orbit is slightly elliptical, and Earth is not stationary (rather the Earth-Moon system rotates about its center of mass, which is located some 1700 km below Earth's surface). The clear implication is that Earth's gravitational force causes the Moon to orbit Earth.

Why does Earth not remain stationary as the Moon orbits it? This is because, as expected from Newton's third law, if Earth exerts a force on the Moon, then the Moon should exert an equal and opposite force on Earth (see **Figure 6.23**). We do not sense the Moon's effect on Earth's motion, because the Moon's gravity moves our bodies right along with Earth but there are other signs on Earth that clearly show the effect of the Moon's gravitational force as discussed in **Satellites and Kepler's Laws: An Argument for Simplicity**.

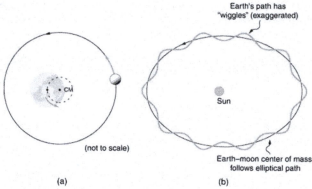

Figure 6.23 (a) Earth and the Moon rotate approximately once a month around their common center of mass. (b) Their center of mass orbits the Sun in an elliptical orbit, but Earth's path around the Sun has "wiggles" in it. Similar wiggles in the paths of stars have been observed and are considered direct evidence of planets orbiting those stars. This is important because the planets' reflected light is often too dim to be observed.

Tides

Ocean tides are one very observable result of the Moon's gravity acting on Earth. **Figure 6.24** is a simplified drawing of the Moon's position relative to the tides. Because water easily flows on Earth's surface, a high tide is created on the side of Earth nearest to the Moon, where the Moon's gravitational pull is strongest. Why is there also a high tide on the opposite side of Earth? The answer is that Earth is pulled toward the Moon more than the water on the far side, because Earth is closer to the Moon. So the water on the side of Earth closest to the Moon is pulled away from Earth, and Earth is pulled away from water on the far side. As Earth rotates, the tidal bulge (an effect of the tidal forces between an orbiting natural satellite and the primary planet that it orbits) keeps its orientation with the Moon. Thus there are two tides per day (the actual tidal period is about 12 hours and 25.2 minutes), because the Moon moves in its orbit each day as well).

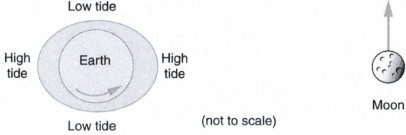

Figure 6.24 The Moon causes ocean tides by attracting the water on the near side more than Earth, and by attracting Earth more than the water on the far side. The distances and sizes are not to scale. For this simplified representation of the Earth-Moon system, there are two high and two low tides per day at any location, because Earth rotates under the tidal bulge.

The Sun also affects tides, although it has about half the effect of the Moon. However, the largest tides, called spring tides, occur when Earth, the Moon, and the Sun are aligned. The smallest tides, called neap tides, occur when the Sun is at a $90°$ angle to the Earth-Moon alignment.

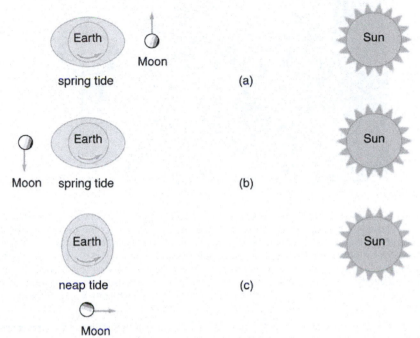

Figure 6.25 (a, b) Spring tides: The highest tides occur when Earth, the Moon, and the Sun are aligned. (c) Neap tide: The lowest tides occur when the Sun lies at $90°$ to the Earth-Moon alignment. Note that this figure is not drawn to scale.

Tides are not unique to Earth but occur in many astronomical systems. The most extreme tides occur where the gravitational force is the strongest and varies most rapidly, such as near black holes (see **Figure 6.26**). A few likely candidates for black holes have been observed in our galaxy. These have masses greater than the Sun but have diameters only a few kilometers across. The tidal forces near them are so great that they can actually tear matter from a companion star.

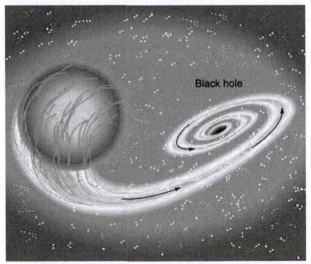

Figure 6.26 A black hole is an object with such strong gravity that not even light can escape it. This black hole was created by the supernova of one star in a two-star system. The tidal forces created by the black hole are so great that it tears matter from the companion star. This matter is compressed and heated as it is sucked into the black hole, creating light and X-rays observable from Earth.

"Weightlessness" and Microgravity

In contrast to the tremendous gravitational force near black holes is the apparent gravitational field experienced by astronauts orbiting Earth. What is the effect of "weightlessness" upon an astronaut who is in orbit for months? Or what about the effect of weightlessness upon plant growth? Weightlessness doesn't mean that an astronaut is not being acted upon by the gravitational force. There is no "zero gravity" in an astronaut's orbit. The term just means that the astronaut is in free-fall, accelerating with the acceleration due to gravity. If an elevator cable breaks, the passengers inside will be in free fall and will experience weightlessness. You can experience short periods of weightlessness in some rides in amusement parks.

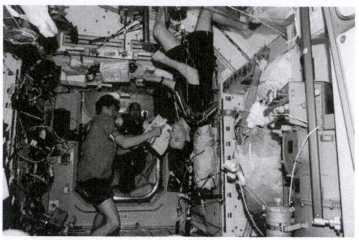

Figure 6.27 Astronauts experiencing weightlessness on board the International Space Station. (credit: NASA)

Microgravity refers to an environment in which the apparent net acceleration of a body is small compared with that produced by Earth at its surface. Many interesting biology and physics topics have been studied over the past three decades in the presence of microgravity. Of immediate concern is the effect on astronauts of extended times in outer space, such as at the International Space Station. Researchers have observed that muscles will atrophy (waste away) in this environment. There is also a corresponding loss of bone mass. Study continues on cardiovascular adaptation to space flight. On Earth, blood pressure is usually higher in the feet than in the head, because the higher column of blood exerts a downward force on it, due to gravity. When standing, 70% of your blood is below the level of the heart, while in a horizontal position, just the opposite occurs. What difference does the absence of this pressure differential have upon the heart?

Some findings in human physiology in space can be clinically important to the management of diseases back on Earth. On a somewhat negative note, spaceflight is known to affect the human immune system, possibly making the crew members more vulnerable to infectious diseases. Experiments flown in space also have shown that some bacteria grow faster in microgravity than they do on Earth. However, on a positive note, studies indicate that microbial antibiotic production can increase by a factor of two in space-grown cultures. One hopes to be able to understand these mechanisms so that similar successes can be achieved on the ground. In another area of physics space research, inorganic crystals and protein crystals have been grown in outer space that have much higher quality than any grown on Earth, so crystallography studies on their structure can yield much better results.

Plants have evolved with the stimulus of gravity and with gravity sensors. Roots grow downward and shoots grow upward. Plants might be able to provide a life support system for long duration space missions by regenerating the atmosphere, purifying water, and producing food. Some studies have indicated that plant growth and development are not affected by gravity, but there is still uncertainty about structural changes in plants grown in a microgravity environment.

The Cavendish Experiment: Then and Now

As previously noted, the universal gravitational constant G is determined experimentally. This definition was first done accurately by Henry Cavendish (1731–1810), an English scientist, in 1798, more than 100 years after Newton published his universal law of gravitation. The measurement of G is very basic and important because it determines the strength of one of the four forces in nature. Cavendish's experiment was very difficult because he measured the tiny gravitational attraction between two ordinary-sized masses (tens of kilograms at most), using apparatus like that in **Figure 6.28**. Remarkably, his value for G differs by less than 1% from the best modern value.

One important consequence of knowing G was that an accurate value for Earth's mass could finally be obtained. This was done by measuring the acceleration due to gravity as accurately as possible and then calculating the mass of Earth M from the relationship Newton's universal law of gravitation gives

$$mg = G\frac{mM}{r^2}, \tag{6.52}$$

where m is the mass of the object, M is the mass of Earth, and r is the distance to the center of Earth (the distance between the centers of mass of the object and Earth). See **Figure 6.21**. The mass m of the object cancels, leaving an equation for g:

$$g = G\frac{M}{r^2}. \tag{6.53}$$

Rearranging to solve for M yields

$$M = \frac{gr^2}{G}. \tag{6.54}$$

So M can be calculated because all quantities on the right, including the radius of Earth r, are known from direct measurements. We shall see in **Satellites and Kepler's Laws: An Argument for Simplicity** that knowing G also allows for the determination of astronomical masses. Interestingly, of all the fundamental constants in physics, G is by far the least well determined.

The Cavendish experiment is also used to explore other aspects of gravity. One of the most interesting questions is whether the gravitational force depends on substance as well as mass—for example, whether one kilogram of lead exerts the same gravitational pull as one kilogram of water. A Hungarian scientist named Roland von Eötvös pioneered this inquiry early in the 20th century. He found, with an accuracy of five parts per billion, that the gravitational force does not depend on the substance. Such experiments continue today, and have improved upon Eötvös' measurements. Cavendish-type experiments such as those of Eric Adelberger and others at the University of Washington, have also put severe limits on the possibility of a fifth force and have verified a major prediction of general relativity—that gravitational energy contributes to rest mass. Ongoing measurements there use a torsion balance and a parallel plate (not spheres, as Cavendish used) to examine how Newton's law of gravitation works over sub-millimeter distances. On this small-scale, do gravitational effects depart from the inverse square law? So far, no deviation has been observed.

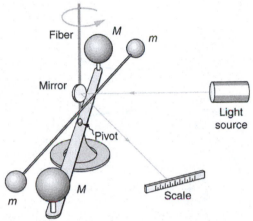

Figure 6.28 Cavendish used an apparatus like this to measure the gravitational attraction between the two suspended spheres (m) and the two on the stand (M) by observing the amount of torsion (twisting) created in the fiber. Distance between the masses can be varied to check the dependence of the force on distance. Modern experiments of this type continue to explore gravity.

6.6 Satellites and Kepler's Laws: An Argument for Simplicity

Learning Objectives

By the end of this section, you will be able to:

- State Kepler's laws of planetary motion.
- Derive Kepler's third law for circular orbits.
- Discuss the Ptolemaic model of the universe.

Examples of gravitational orbits abound. Hundreds of artificial satellites orbit Earth together with thousands of pieces of debris. The Moon's orbit about Earth has intrigued humans from time immemorial. The orbits of planets, asteroids, meteors, and comets about the Sun are no less interesting. If we look further, we see almost unimaginable numbers of stars, galaxies, and other celestial objects orbiting one another and interacting through gravity.

All these motions are governed by gravitational force, and it is possible to describe them to various degrees of precision. Precise descriptions of complex systems must be made with large computers. However, we can describe an important class of orbits without the use of computers, and we shall find it instructive to study them. These orbits have the following characteristics:

1. *A small mass m orbits a much larger mass M*. This allows us to view the motion as if M were stationary—in fact, as if from an inertial frame of reference placed on M —without significant error. Mass m is the satellite of M, if the orbit is gravitationally bound.

2. *The system is isolated from other masses*. This allows us to neglect any small effects due to outside masses.

The conditions are satisfied, to good approximation, by Earth's satellites (including the Moon), by objects orbiting the Sun, and by the satellites of other planets. Historically, planets were studied first, and there is a classical set of three laws, called Kepler's laws of planetary motion, that describe the orbits of all bodies satisfying the two previous conditions (not just planets in our solar system). These descriptive laws are named for the German astronomer Johannes Kepler (1571–1630), who devised them after careful study (over some 20 years) of a large amount of meticulously recorded observations of planetary motion done by Tycho Brahe (1546–1601). Such careful collection and detailed recording of methods and data are hallmarks of good science. Data constitute the evidence from which new interpretations and meanings can be constructed.

Kepler's Laws of Planetary Motion

Kepler's First Law

The orbit of each planet about the Sun is an ellipse with the Sun at one focus.

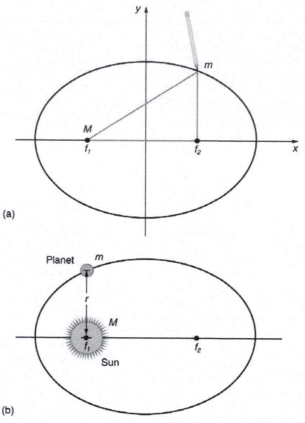

(a)

(b)

Figure 6.29 (a) An ellipse is a closed curve such that the sum of the distances from a point on the curve to the two foci (f_1 and f_2) is a constant. You can draw an ellipse as shown by putting a pin at each focus, and then placing a string around a pencil and the pins and tracing a line on paper. A circle is a special case of an ellipse in which the two foci coincide (thus any point on the circle is the same distance from the center). (b) For any closed gravitational orbit, m follows an elliptical path with M at one focus. Kepler's first law states this fact for planets orbiting the Sun.

Kepler's Second Law

Each planet moves so that an imaginary line drawn from the Sun to the planet sweeps out equal areas in equal times (see Figure 6.30).

Kepler's Third Law

The ratio of the squares of the periods of any two planets about the Sun is equal to the ratio of the cubes of their average distances from the Sun. In equation form, this is

$$\frac{T_1^{\,2}}{T_2^{\,2}} = \frac{r_1^{\,3}}{r_2^{\,3}},$$

(6.55)

where T is the period (time for one orbit) and r is the average radius. This equation is valid only for comparing two small masses orbiting the same large one. Most importantly, this is a descriptive equation only, giving no information as to the cause of the equality.

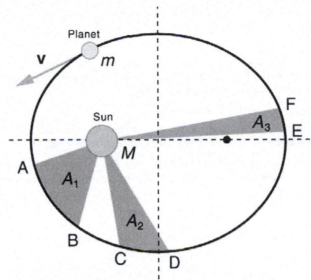

Figure 6.30 The shaded regions have equal areas. It takes equal times for m to go from A to B, from C to D, and from E to F. The mass m moves fastest when it is closest to M. Kepler's second law was originally devised for planets orbiting the Sun, but it has broader validity.

Note again that while, for historical reasons, Kepler's laws are stated for planets orbiting the Sun, they are actually valid for all bodies satisfying the two previously stated conditions.

Example 6.7 Find the Time for One Orbit of an Earth Satellite

Given that the Moon orbits Earth each 27.3 d and that it is an average distance of 3.84×10^8 m from the center of Earth, calculate the period of an artificial satellite orbiting at an average altitude of 1500 km above Earth's surface.

Strategy

The period, or time for one orbit, is related to the radius of the orbit by Kepler's third law, given in mathematical form in $\frac{T_1^2}{T_2^2} = \frac{r_1^3}{r_2^3}$. Let us use the subscript 1 for the Moon and the subscript 2 for the satellite. We are asked to find T_2. The

given information tells us that the orbital radius of the Moon is $r_1 = 3.84 \times 10^8$ m , and that the period of the Moon is $T_1 = 27.3$ d . The height of the artificial satellite above Earth's surface is given, and so we must add the radius of Earth (6380 km) to get $r_2 = (1500 + 6380)$ km $= 7880$ km . Now all quantities are known, and so T_2 can be found.

Solution

Kepler's third law is

$$\frac{T_1^2}{T_2^2} = \frac{r_1^3}{r_2^3}. \tag{6.56}$$

To solve for T_2 , we cross-multiply and take the square root, yielding

$$T_2^2 = T_1^2 \left(\frac{r_2}{r_1}\right)^3 \tag{6.57}$$

$$T_2 = T_1 \left(\frac{r_2}{r_1}\right)^{3/2}. \tag{6.58}$$

Substituting known values yields

$$T_2 = 27.3 \, \text{d} \times \frac{24.0 \, \text{h}}{\text{d}} \times \left(\frac{7880 \, \text{km}}{3.84 \times 10^5 \, \text{km}}\right)^{3/2} \tag{6.59}$$

$$= 1.93 \, \text{h}.$$

Discussion This is a reasonable period for a satellite in a fairly low orbit. It is interesting that any satellite at this altitude will orbit in the same amount of time. This fact is related to the condition that the satellite's mass is small compared with that of

Earth.

People immediately search for deeper meaning when broadly applicable laws, like Kepler's, are discovered. It was Newton who took the next giant step when he proposed the law of universal gravitation. While Kepler was able to discover *what* was happening, Newton discovered that gravitational force was the cause.

Derivation of Kepler's Third Law for Circular Orbits

We shall derive Kepler's third law, starting with Newton's laws of motion and his universal law of gravitation. The point is to demonstrate that the force of gravity is the cause for Kepler's laws (although we will only derive the third one).

Let us consider a circular orbit of a small mass m around a large mass M, satisfying the two conditions stated at the beginning of this section. Gravity supplies the centripetal force to mass m. Starting with Newton's second law applied to circular motion,

$$F_{\text{net}} = ma_c = m\frac{v^2}{r}. \tag{6.60}$$

The net external force on mass m is gravity, and so we substitute the force of gravity for F_{net}:

$$G\frac{mM}{r^2} = m\frac{v^2}{r}. \tag{6.61}$$

The mass m cancels, yielding

$$G\frac{M}{r} = v^2. \tag{6.62}$$

The fact that m cancels out is another aspect of the oft-noted fact that at a given location all masses fall with the same acceleration. Here we see that at a given orbital radius r, all masses orbit at the same speed. (This was implied by the result of the preceding worked example.) Now, to get at Kepler's third law, we must get the period T into the equation. By definition, period T is the time for one complete orbit. Now the average speed v is the circumference divided by the period—that is,

$$v = \frac{2\pi r}{T}. \tag{6.63}$$

Substituting this into the previous equation gives

$$G\frac{M}{r} = \frac{4\pi^2 r^2}{T^2}. \tag{6.64}$$

Solving for T^2 yields

$$T^2 = \frac{4\pi^2}{GM}r^3. \tag{6.65}$$

Using subscripts 1 and 2 to denote two different satellites, and taking the ratio of the last equation for satellite 1 to satellite 2 yields

$$\frac{T_1^2}{T_2^2} = \frac{r_1^3}{r_2^3}. \tag{6.66}$$

This is Kepler's third law. Note that Kepler's third law is valid only for comparing satellites of the same parent body, because only then does the mass of the parent body M cancel.

Now consider what we get if we solve $T^2 = \frac{4\pi^2}{GM}r^3$ for the ratio r^3/T^2. We obtain a relationship that can be used to determine the mass M of a parent body from the orbits of its satellites:

$$\frac{r^3}{T^2} = \frac{G}{4\pi^2}M. \tag{6.67}$$

If r and T are known for a satellite, then the mass M of the parent can be calculated. This principle has been used extensively to find the masses of heavenly bodies that have satellites. Furthermore, the ratio r^3/T^2 should be a constant for all satellites of the same parent body (because $r^3/T^2 = GM/4\pi^2$). (See Table 6.2).

It is clear from Table 6.2 that the ratio of r^3/T^2 is constant, at least to the third digit, for all listed satellites of the Sun, and for those of Jupiter. Small variations in that ratio have two causes—uncertainties in the r and T data, and perturbations of the orbits due to other bodies. Interestingly, those perturbations can be—and have been—used to predict the location of new planets and moons. This is another verification of Newton's universal law of gravitation.

Making Connections: General Relativity and Mercury

Newton's universal law of gravitation is modified by Einstein's general theory of relativity, as we shall see in **Particle Physics**. Newton's gravity is not seriously in error—it was and still is an extremely good approximation for most situations. Einstein's modification is most noticeable in extremely large gravitational fields, such as near black holes. However, general relativity also explains such phenomena as small but long-known deviations of the orbit of the planet Mercury from classical predictions.

The Case for Simplicity

The development of the universal law of gravitation by Newton played a pivotal role in the history of ideas. While it is beyond the scope of this text to cover that history in any detail, we note some important points. The definition of planet set in 2006 by the International Astronomical Union (IAU) states that in the solar system, a planet is a celestial body that:

1. is in orbit around the Sun,

2. has sufficient mass to assume hydrostatic equilibrium and

3. has cleared the neighborhood around its orbit.

A non-satellite body fulfilling only the first two of the above criteria is classified as "dwarf planet."

In 2006, Pluto was demoted to a 'dwarf planet' after scientists revised their definition of what constitutes a "true" planet.

Table 6.2 Orbital Data and Kepler's Third Law

Parent	Satellite	Average orbital radius r(km)	Period $T(y)$	r^3 / T^2 (km^3 / y^2)
Earth	Moon	3.84×10^5	0.07481	1.01×10^{19}
Sun	Mercury	5.79×10^7	0.2409	3.34×10^{24}
	Venus	1.082×10^8	0.6150	3.35×10^{24}
	Earth	1.496×10^8	1.000	3.35×10^{24}
	Mars	2.279×10^8	1.881	3.35×10^{24}
	Jupiter	7.783×10^8	11.86	3.35×10^{24}
	Saturn	1.427×10^9	29.46	3.35×10^{24}
	Neptune	4.497×10^9	164.8	3.35×10^{24}
	Pluto	5.90×10^9	248.3	3.33×10^{24}
Jupiter	Io	4.22×10^5	0.00485 (1.77 d)	3.19×10^{21}
	Europa	6.71×10^5	0.00972 (3.55 d)	3.20×10^{21}
	Ganymede	1.07×10^6	0.0196 (7.16 d)	3.19×10^{21}
	Callisto	1.88×10^6	0.0457 (16.19 d)	3.20×10^{21}

The universal law of gravitation is a good example of a physical principle that is very broadly applicable. That single equation for the gravitational force describes all situations in which gravity acts. It gives a cause for a vast number of effects, such as the orbits of the planets and moons in the solar system. It epitomizes the underlying unity and simplicity of physics.

Before the discoveries of Kepler, Copernicus, Galileo, Newton, and others, the solar system was thought to revolve around Earth as shown in **Figure 6.31**(a). This is called the Ptolemaic view, for the Greek philosopher who lived in the second century AD. This model is characterized by a list of facts for the motions of planets with no cause and effect explanation. There tended to be a different rule for each heavenly body and a general lack of simplicity.

Figure 6.31(b) represents the modern or Copernican model. In this model, a small set of rules and a single underlying force explain not only all motions in the solar system, but all other situations involving gravity. The breadth and simplicity of the laws of physics are compelling. As our knowledge of nature has grown, the basic simplicity of its laws has become ever more evident.

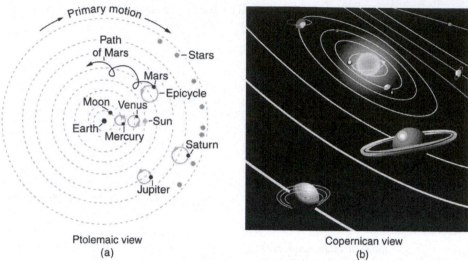

Ptolemaic view
(a)

Copernican view
(b)

Figure 6.31 (a) The Ptolemaic model of the universe has Earth at the center with the Moon, the planets, the Sun, and the stars revolving about it in complex superpositions of circular paths. This geocentric model, which can be made progressively more accurate by adding more circles, is purely descriptive, containing no hints as to what are the causes of these motions. (b) The Copernican model has the Sun at the center of the solar system. It is fully explained by a small number of laws of physics, including Newton's universal law of gravitation.

Glossary

angular velocity: ω , the rate of change of the angle with which an object moves on a circular path

arc length: Δs , the distance traveled by an object along a circular path

banked curve: the curve in a road that is sloping in a manner that helps a vehicle negotiate the curve

center of mass: the point where the entire mass of an object can be thought to be concentrated

centrifugal force: a fictitious force that tends to throw an object off when the object is rotating in a non-inertial frame of reference

centripetal acceleration: the acceleration of an object moving in a circle, directed toward the center

centripetal force: any net force causing uniform circular motion

Coriolis force: the fictitious force causing the apparent deflection of moving objects when viewed in a rotating frame of reference

fictitious force: a force having no physical origin

gravitational constant, G: a proportionality factor used in the equation for Newton's universal law of gravitation; it is a universal constant—that is, it is thought to be the same everywhere in the universe

ideal angle: the angle at which a car can turn safely on a steep curve, which is in proportion to the ideal speed

ideal banking: the sloping of a curve in a road, where the angle of the slope allows the vehicle to negotiate the curve at a certain speed without the aid of friction between the tires and the road; the net external force on the vehicle equals the horizontal centripetal force in the absence of friction

ideal speed: the maximum safe speed at which a vehicle can turn on a curve without the aid of friction between the tire and the road

microgravity: an environment in which the apparent net acceleration of a body is small compared with that produced by Earth at its surface

Newton's universal law of gravitation: every particle in the universe attracts every other particle with a force along a line joining them; the force is directly proportional to the product of their masses and inversely proportional to the square of the distance between them

non-inertial frame of reference: an accelerated frame of reference

pit: a tiny indentation on the spiral track moulded into the top of the polycarbonate layer of CD

radians: a unit of angle measurement

radius of curvature: radius of a circular path

rotation angle: the ratio of the arc length to the radius of curvature on a circular path:

$$\Delta\theta = \frac{\Delta s}{r}$$

ultracentrifuge: a centrifuge optimized for spinning a rotor at very high speeds

uniform circular motion: the motion of an object in a circular path at constant speed

Section Summary

6.1 Rotation Angle and Angular Velocity
- Uniform circular motion is motion in a circle at constant speed. The rotation angle $\Delta\theta$ is defined as the ratio of the arc length to the radius of curvature:

$$\Delta\theta = \frac{\Delta s}{r},$$

  where arc length Δs is distance traveled along a circular path and r is the radius of curvature of the circular path. The quantity $\Delta\theta$ is measured in units of radians (rad), for which

$$2\pi \text{ rad} = 360° = 1 \text{ revolution.}$$
- The conversion between radians and degrees is $1 \text{ rad} = 57.3°$.
- Angular velocity ω is the rate of change of an angle,

$$\omega = \frac{\Delta\theta}{\Delta t},$$

  where a rotation $\Delta\theta$ takes place in a time Δt. The units of angular velocity are radians per second (rad/s). Linear velocity v and angular velocity ω are related by

$$v = r\omega \text{ or } \omega = \frac{v}{r}.$$

6.2 Centripetal Acceleration
- Centripetal acceleration a_c is the acceleration experienced while in uniform circular motion. It always points toward the center of rotation. It is perpendicular to the linear velocity v and has the magnitude

$$a_c = \frac{v^2}{r}; \ a_c = r\omega^2.$$
- The unit of centripetal acceleration is m/s^2.

6.3 Centripetal Force
- Centripetal force F_c is any force causing uniform circular motion. It is a "center-seeking" force that always points toward the center of rotation. It is perpendicular to linear velocity v and has magnitude

$$F_c = ma_c,$$

  which can also be expressed as

$$\left.\begin{array}{c} F_c = m\frac{v^2}{r} \\ \text{or} \\ F_c = mr\omega^2 \end{array}\right\},$$

6.4 Fictitious Forces and Non-inertial Frames: The Coriolis Force
- Rotating and accelerated frames of reference are non-inertial.
- Fictitious forces, such as the Coriolis force, are needed to explain motion in such frames.

6.5 Newton's Universal Law of Gravitation
- Newton's universal law of gravitation: Every particle in the universe attracts every other particle with a force along a line joining them. The force is directly proportional to the product of their masses and inversely proportional to the square of the distance between them. In equation form, this is

$$F = G\frac{mM}{r^2},$$

  where F is the magnitude of the gravitational force. G is the gravitational constant, given by

$$G = 6.673 \times 10^{-11} \text{ N} \cdot \text{m}^2/\text{kg}^2.$$

- Newton's law of gravitation applies universally.

6.6 Satellites and Kepler's Laws: An Argument for Simplicity

- Kepler's laws are stated for a small mass m orbiting a larger mass M in near-isolation. Kepler's laws of planetary motion are then as follows:

Kepler's first law

The orbit of each planet about the Sun is an ellipse with the Sun at one focus.

Kepler's second law

Each planet moves so that an imaginary line drawn from the Sun to the planet sweeps out equal areas in equal times.

Kepler's third law

The ratio of the squares of the periods of any two planets about the Sun is equal to the ratio of the cubes of their average distances from the Sun:

$$\frac{T_1^2}{T_2^2} = \frac{r_1^3}{r_2^3},$$

where T is the period (time for one orbit) and r is the average radius of the orbit.

- The period and radius of a satellite's orbit about a larger body M are related by

$$T^2 = \frac{4\pi^2}{GM}r^3$$

or

$$\frac{r^3}{T^2} = \frac{G}{4\pi^2}M.$$

Conceptual Questions

6.1 Rotation Angle and Angular Velocity

1. There is an analogy between rotational and linear physical quantities. What rotational quantities are analogous to distance and velocity?

6.2 Centripetal Acceleration

2. Can centripetal acceleration change the speed of circular motion? Explain.

6.3 Centripetal Force

3. If you wish to reduce the stress (which is related to centripetal force) on high-speed tires, would you use large- or small-diameter tires? Explain.

4. Define centripetal force. Can any type of force (for example, tension, gravitational force, friction, and so on) be a centripetal force? Can any combination of forces be a centripetal force?

5. If centripetal force is directed toward the center, why do you feel that you are 'thrown' away from the center as a car goes around a curve? Explain.

6. Race car drivers routinely cut corners as shown in **Figure 6.32**. Explain how this allows the curve to be taken at the greatest speed.

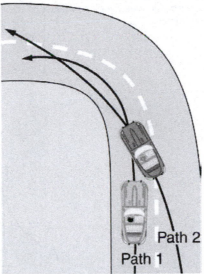

Figure 6.32 Two paths around a race track curve are shown. Race car drivers will take the inside path (called cutting the corner) whenever possible because it allows them to take the curve at the highest speed.

7. A number of amusement parks have rides that make vertical loops like the one shown in **Figure 6.33**. For safety, the cars are attached to the rails in such a way that they cannot fall off. If the car goes over the top at just the right speed, gravity alone will supply the centripetal force. What other force acts and what is its direction if:

(a) The car goes over the top at faster than this speed?

(b) The car goes over the top at slower than this speed?

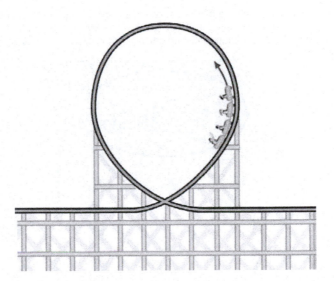

Figure 6.33 Amusement rides with a vertical loop are an example of a form of curved motion.

8. What is the direction of the force exerted by the car on the passenger as the car goes over the top of the amusement ride pictured in **Figure 6.33** under the following circumstances:

(a) The car goes over the top at such a speed that the gravitational force is the only force acting?

(b) The car goes over the top faster than this speed?

(c) The car goes over the top slower than this speed?

9. As a skater forms a circle, what force is responsible for making her turn? Use a free body diagram in your answer.

10. Suppose a child is riding on a merry-go-round at a distance about halfway between its center and edge. She has a lunch box resting on wax paper, so that there is very little friction between it and the merry-go-round. Which path shown in **Figure 6.34** will the lunch box take when she lets go? The lunch box leaves a trail in the dust on the merry-go-round. Is that trail straight, curved to the left, or curved to the right? Explain your answer.

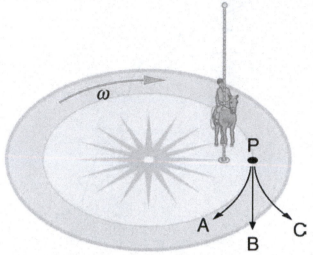

Merry-go-round's rotating frame of reference

Figure 6.34 A child riding on a merry-go-round releases her lunch box at point P. This is a view from above the clockwise rotation. Assuming it slides with negligible friction, will it follow path A, B, or C, as viewed from Earth's frame of reference? What will be the shape of the path it leaves in the dust on the merry-go-round?

11. Do you feel yourself thrown to either side when you negotiate a curve that is ideally banked for your car's speed? What is the direction of the force exerted on you by the car seat?

12. Suppose a mass is moving in a circular path on a frictionless table as shown in figure. In the Earth's frame of reference, there is no centrifugal force pulling the mass away from the centre of rotation, yet there is a very real force stretching the string attaching the mass to the nail. Using concepts related to centripetal force and Newton's third law, explain what force stretches the string, identifying its physical origin.

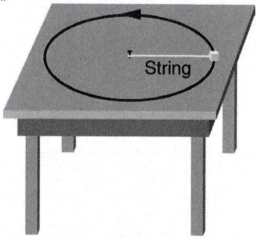

Figure 6.35 A mass attached to a nail on a frictionless table moves in a circular path. The force stretching the string is real and not fictional. What is the physical origin of the force on the string?

6.4 Fictitious Forces and Non-inertial Frames: The Coriolis Force

13. When a toilet is flushed or a sink is drained, the water (and other material) begins to rotate about the drain on the way down. Assuming no initial rotation and a flow initially directly straight toward the drain, explain what causes the rotation and which direction it has in the northern hemisphere. (Note that this is a small effect and in most toilets the rotation is caused by directional water jets.) Would the direction of rotation reverse if water were forced up the drain?

14. Is there a real force that throws water from clothes during the spin cycle of a washing machine? Explain how the water is removed.

15. In one amusement park ride, riders enter a large vertical barrel and stand against the wall on its horizontal floor. The barrel is spun up and the floor drops away. Riders feel as if they are pinned to the wall by a force something like the gravitational force. This is a fictitious force sensed and used by the riders to explain events in the rotating frame of reference of the barrel. Explain in an inertial frame of reference (Earth is nearly one) what pins the riders to the wall, and identify all of the real forces acting on them.

16. Action at a distance, such as is the case for gravity, was once thought to be illogical and therefore untrue. What is the ultimate determinant of the truth in physics, and why was this action ultimately accepted?

17. Two friends are having a conversation. Anna says a satellite in orbit is in freefall because the satellite keeps falling toward Earth. Tom says a satellite in orbit is not in freefall because the acceleration due to gravity is not 9.80 m/s^2. Who do you agree with and why?

18. A non-rotating frame of reference placed at the center of the Sun is very nearly an inertial one. Why is it not exactly an inertial frame?

6.5 Newton's Universal Law of Gravitation

19. Action at a distance, such as is the case for gravity, was once thought to be illogical and therefore untrue. What is the ultimate determinant of the truth in physics, and why was this action ultimately accepted?

20. Two friends are having a conversation. Anna says a satellite in orbit is in freefall because the satellite keeps falling toward Earth. Tom says a satellite in orbit is not in freefall because the acceleration due to gravity is not 9.80 m/s^2. Who do you agree with and why?

21. Draw a free body diagram for a satellite in an elliptical orbit showing why its speed increases as it approaches its parent body and decreases as it moves away.

22. Newton's laws of motion and gravity were among the first to convincingly demonstrate the underlying simplicity and unity in nature. Many other examples have since been discovered, and we now expect to find such underlying order in complex situations. Is there proof that such order will always be found in new explorations?

6.6 Satellites and Kepler's Laws: An Argument for Simplicity

23. In what frame(s) of reference are Kepler's laws valid? Are Kepler's laws purely descriptive, or do they contain causal information?

Problems & Exercises

6.1 Rotation Angle and Angular Velocity

1. Semi-trailer trucks have an odometer on one hub of a trailer wheel. The hub is weighted so that it does not rotate, but it contains gears to count the number of wheel revolutions—it then calculates the distance traveled. If the wheel has a 1.15 m diameter and goes through 200,000 rotations, how many kilometers should the odometer read?

2. Microwave ovens rotate at a rate of about 6 rev/min. What is this in revolutions per second? What is the angular velocity in radians per second?

3. An automobile with 0.260 m radius tires travels 80,000 km before wearing them out. How many revolutions do the tires make, neglecting any backing up and any change in radius due to wear?

4. (a) What is the period of rotation of Earth in seconds? (b) What is the angular velocity of Earth? (c) Given that Earth has a radius of 6.4×10^6 m at its equator, what is the linear velocity at Earth's surface?

5. A baseball pitcher brings his arm forward during a pitch, rotating the forearm about the elbow. If the velocity of the ball in the pitcher's hand is 35.0 m/s and the ball is 0.300 m from the elbow joint, what is the angular velocity of the forearm?

6. In lacrosse, a ball is thrown from a net on the end of a stick by rotating the stick and forearm about the elbow. If the angular velocity of the ball about the elbow joint is 30.0 rad/s and the ball is 1.30 m from the elbow joint, what is the velocity of the ball?

7. A truck with 0.420-m-radius tires travels at 32.0 m/s. What is the angular velocity of the rotating tires in radians per second? What is this in rev/min?

8. Integrated Concepts When kicking a football, the kicker rotates his leg about the hip joint.

(a) If the velocity of the tip of the kicker's shoe is 35.0 m/s and the hip joint is 1.05 m from the tip of the shoe, what is the shoe tip's angular velocity?

(b) The shoe is in contact with the initially stationary 0.500 kg football for 20.0 ms. What average force is exerted on the football to give it a velocity of 20.0 m/s?

(c) Find the maximum range of the football, neglecting air resistance.

9. Construct Your Own Problem

Consider an amusement park ride in which participants are rotated about a vertical axis in a cylinder with vertical walls. Once the angular velocity reaches its full value, the floor drops away and friction between the walls and the riders prevents them from sliding down. Construct a problem in which you calculate the necessary angular velocity that assures the riders will not slide down the wall. Include a free body diagram of a single rider. Among the variables to consider are the radius of the cylinder and the coefficients of friction between the riders' clothing and the wall.

6.2 Centripetal Acceleration

10. A fairground ride spins its occupants inside a flying saucer-shaped container. If the horizontal circular path the riders follow has an 8.00 m radius, at how many revolutions per minute will the riders be subjected to a centripetal acceleration whose magnitude is 1.50 times that due to gravity?

11. A runner taking part in the 200 m dash must run around the end of a track that has a circular arc with a radius of curvature of 30 m. If he completes the 200 m dash in 23.2 s and runs at constant speed throughout the race, what is the magnitude of his centripetal acceleration as he runs the curved portion of the track?

12. Taking the age of Earth to be about 4×10^9 years and assuming its orbital radius of 1.5×10^{11} m has not changed and is circular, calculate the approximate total distance Earth has traveled since its birth (in a frame of reference stationary with respect to the Sun).

13. The propeller of a World War II fighter plane is 2.30 m in diameter.

(a) What is its angular velocity in radians per second if it spins at 1200 rev/min?

(b) What is the linear speed of its tip at this angular velocity if the plane is stationary on the tarmac?

(c) What is the centripetal acceleration of the propeller tip under these conditions? Calculate it in meters per second squared and convert to multiples of g.

14. An ordinary workshop grindstone has a radius of 7.50 cm and rotates at 6500 rev/min.

(a) Calculate the magnitude of the centripetal acceleration at its edge in meters per second squared and convert it to multiples of g.

(b) What is the linear speed of a point on its edge?

15. Helicopter blades withstand tremendous stresses. In addition to supporting the weight of a helicopter, they are spun at rapid rates and experience large centripetal accelerations, especially at the tip.

(a) Calculate the magnitude of the centripetal acceleration at the tip of a 4.00 m long helicopter blade that rotates at 300 rev/min.

(b) Compare the linear speed of the tip with the speed of sound (taken to be 340 m/s).

16. Olympic ice skaters are able to spin at about 5 rev/s.

(a) What is their angular velocity in radians per second?

(b) What is the centripetal acceleration of the skater's nose if it is 0.120 m from the axis of rotation?

(c) An exceptional skater named Dick Button was able to spin much faster in the 1950s than anyone since—at about 9 rev/s. What was the centripetal acceleration of the tip of his nose, assuming it is at 0.120 m radius?

(d) Comment on the magnitudes of the accelerations found. It is reputed that Button ruptured small blood vessels during his spins.

17. What percentage of the acceleration at Earth's surface is the acceleration due to gravity at the position of a satellite located 300 km above Earth?

18. Verify that the linear speed of an ultracentrifuge is about 0.50 km/s, and Earth in its orbit is about 30 km/s by calculating:

(a) The linear speed of a point on an ultracentrifuge 0.100 m from its center, rotating at 50,000 rev/min.

(b) The linear speed of Earth in its orbit about the Sun (use data from the text on the radius of Earth's orbit and approximate it as being circular).

19. A rotating space station is said to create "artificial gravity"—a loosely-defined term used for an acceleration that would be crudely similar to gravity. The outer wall of the rotating space station would become a floor for the astronauts, and centripetal acceleration supplied by the floor would allow astronauts to exercise and maintain muscle and bone strength more naturally than in non-rotating space environments. If the space station is 200 m in diameter, what angular velocity would produce an "artificial gravity" of 9.80 m/s^2 at the rim?

20. At takeoff, a commercial jet has a 60.0 m/s speed. Its tires have a diameter of 0.850 m.

(a) At how many rev/min are the tires rotating?

(b) What is the centripetal acceleration at the edge of the tire?

(c) With what force must a determined $1.00 \times 10^{-15} \text{ kg}$ bacterium cling to the rim?

(d) Take the ratio of this force to the bacterium's weight.

21. Integrated Concepts

Riders in an amusement park ride shaped like a Viking ship hung from a large pivot are rotated back and forth like a rigid pendulum. Sometime near the middle of the ride, the ship is momentarily motionless at the top of its circular arc. The ship then swings down under the influence of gravity.

(a) Assuming negligible friction, find the speed of the riders at the bottom of its arc, given the system's center of mass travels in an arc having a radius of 14.0 m and the riders are near the center of mass.

(b) What is the centripetal acceleration at the bottom of the arc?

(c) Draw a free body diagram of the forces acting on a rider at the bottom of the arc.

(d) Find the force exerted by the ride on a 60.0 kg rider and compare it to her weight.

(e) Discuss whether the answer seems reasonable.

22. Unreasonable Results

A mother pushes her child on a swing so that his speed is 9.00 m/s at the lowest point of his path. The swing is suspended 2.00 m above the child's center of mass.

(a) What is the magnitude of the centripetal acceleration of the child at the low point?

(b) What is the magnitude of the force the child exerts on the seat if his mass is 18.0 kg?

(c) What is unreasonable about these results?

(d) Which premises are unreasonable or inconsistent?

6.3 Centripetal Force

23. (a) A 22.0 kg child is riding a playground merry-go-round that is rotating at 40.0 rev/min. What centripetal force must she exert to stay on if she is 1.25 m from its center?

(b) What centripetal force does she need to stay on an amusement park merry-go-round that rotates at 3.00 rev/min if she is 8.00 m from its center?

(c) Compare each force with her weight.

24. Calculate the centripetal force on the end of a 100 m (radius) wind turbine blade that is rotating at 0.5 rev/s. Assume the mass is 4 kg.

25. What is the ideal banking angle for a gentle turn of 1.20 km radius on a highway with a 105 km/h speed limit (about 65 mi/h), assuming everyone travels at the limit?

26. What is the ideal speed to take a 100 m radius curve banked at a 20.0° angle?

27. (a) What is the radius of a bobsled turn banked at 75.0° and taken at 30.0 m/s, assuming it is ideally banked?

(b) Calculate the centripetal acceleration.

(c) Does this acceleration seem large to you?

28. Part of riding a bicycle involves leaning at the correct angle when making a turn, as seen in **Figure 6.36**. To be stable, the force exerted by the ground must be on a line going through the center of gravity. The force on the bicycle wheel can be resolved into two perpendicular components—friction parallel to the road (this must supply the centripetal force), and the vertical normal force (which must equal the system's weight).

(a) Show that θ (as defined in the figure) is related to the speed v and radius of curvature r of the turn in the same way as for an ideally banked roadway—that is,

$$\theta = \tan^{-1} v^2 \big/ rg$$

(b) Calculate θ for a 12.0 m/s turn of radius 30.0 m (as in a race).

Free-body diagram

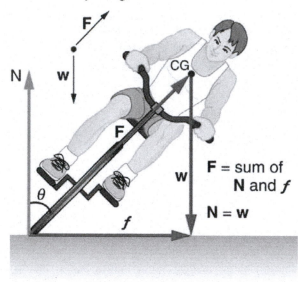

$$F = \text{sum of } N \text{ and } f$$
$$N = w$$

Figure 6.36 A bicyclist negotiating a turn on level ground must lean at the correct angle—the ability to do this becomes instinctive. The force of the ground on the wheel needs to be on a line through the center of gravity. The net external force on the system is the centripetal force. The vertical component of the force on the wheel cancels the weight of the system while its horizontal component must supply the centripetal force. This process produces a relationship among the angle θ, the speed v, and the radius of curvature r of the turn similar to that for the ideal banking of roadways.

29. A large centrifuge, like the one shown in **Figure 6.37**(a), is used to expose aspiring astronauts to accelerations similar to those experienced in rocket launches and atmospheric reentries.

(a) At what angular velocity is the centripetal acceleration $10\ g$ if the rider is 15.0 m from the center of rotation?

(b) The rider's cage hangs on a pivot at the end of the arm, allowing it to swing outward during rotation as shown in **Figure 6.37**(b). At what angle θ below the horizontal will the cage hang when the centripetal acceleration is $10\ g$? (Hint: The arm supplies centripetal force and supports the weight of the cage. Draw a free body diagram of the forces to see what the angle θ should be.)

(a) NASA centrifuge and ride

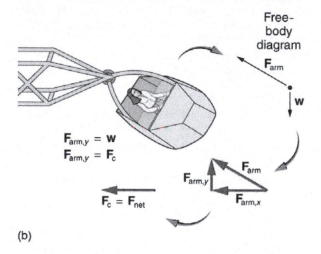

(b)

Figure 6.37 (a) NASA centrifuge used to subject trainees to accelerations similar to those experienced in rocket launches and reentries. (credit: NASA) (b) Rider in cage showing how the cage pivots outward during rotation. This allows the total force exerted on the rider by the cage to be along its axis at all times.

30. Integrated Concepts

If a car takes a banked curve at less than the ideal speed, friction is needed to keep it from sliding toward the inside of the curve (a real problem on icy mountain roads). (a) Calculate the ideal speed to take a 100 m radius curve banked at 15.0º. (b) What is the minimum coefficient of friction needed for a frightened driver to take the same curve at 20.0 km/h?

31. Modern roller coasters have vertical loops like the one shown in Figure 6.38. The radius of curvature is smaller at the top than on the sides so that the downward centripetal acceleration at the top will be greater than the acceleration due to gravity, keeping the passengers pressed firmly into their seats. What is the speed of the roller coaster at the top of the loop if the radius of curvature there is 15.0 m and the downward acceleration of the car is 1.50 g?

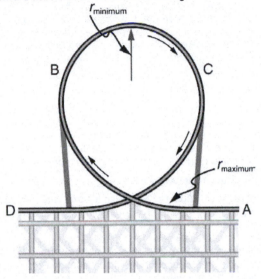

Figure 6.38 Teardrop-shaped loops are used in the latest roller coasters so that the radius of curvature gradually decreases to a minimum at the top. This means that the centripetal acceleration builds from zero to a maximum at the top and gradually decreases again. A circular loop would cause a jolting change in acceleration at entry, a disadvantage discovered long ago in railroad curve design. With a small radius of curvature at the top, the centripetal acceleration can more easily be kept greater than g so that the passengers do not lose contact with their seats nor do they need seat belts to keep them in place.

32. Unreasonable Results

(a) Calculate the minimum coefficient of friction needed for a car to negotiate an unbanked 50.0 m radius curve at 30.0 m/s.

(b) What is unreasonable about the result?

(c) Which premises are unreasonable or inconsistent?

6.5 Newton's Universal Law of Gravitation

33. (a) Calculate Earth's mass given the acceleration due to gravity at the North Pole is 9.830 m/s^2 and the radius of the Earth is 6371 km from center to pole.

(b) Compare this with the accepted value of $5.979 \times 10^{24} \text{ kg}$.

34. (a) Calculate the magnitude of the acceleration due to gravity on the surface of Earth due to the Moon.

(b) Calculate the magnitude of the acceleration due to gravity at Earth due to the Sun.

(c) Take the ratio of the Moon's acceleration to the Sun's and comment on why the tides are predominantly due to the Moon in spite of this number.

35. (a) What is the acceleration due to gravity on the surface of the Moon?

(b) On the surface of Mars? The mass of Mars is $6.418 \times 10^{23} \text{ kg}$ and its radius is $3.38 \times 10^6 \text{ m}$.

36. (a) Calculate the acceleration due to gravity on the surface of the Sun.

(b) By what factor would your weight increase if you could stand on the Sun? (Never mind that you cannot.)

37. The Moon and Earth rotate about their common center of mass, which is located about 4700 km from the center of Earth. (This is 1690 km below the surface.)

(a) Calculate the magnitude of the acceleration due to the Moon's gravity at that point.

(b) Calculate the magnitude of the centripetal acceleration of the center of Earth as it rotates about that point once each lunar month (about 27.3 d) and compare it with the acceleration found in part (a). Comment on whether or not they are equal and why they should or should not be.

38. Solve part (b) of Example 6.6 using $a_c = v^2 / r$.

39. Astrology, that unlikely and vague pseudoscience, makes much of the position of the planets at the moment of one's birth. The only known force a planet exerts on Earth is gravitational.

(a) Calculate the magnitude of the gravitational force exerted on a 4.20 kg baby by a 100 kg father 0.200 m away at birth (he is assisting, so he is close to the child).

(b) Calculate the magnitude of the force on the baby due to Jupiter if it is at its closest distance to Earth, some $6.29 \times 10^{11} \text{ m}$ away. How does the force of Jupiter on the baby compare to the force of the father on the baby? Other objects in the room and the hospital building also exert similar gravitational forces. (Of course, there could be an unknown force acting, but scientists first need to be convinced that there is even an effect, much less that an unknown force causes it.)

40. The existence of the dwarf planet Pluto was proposed based on irregularities in Neptune's orbit. Pluto was subsequently discovered near its predicted position. But it now appears that the discovery was fortuitous, because Pluto is small and the irregularities in Neptune's orbit were not well known. To illustrate that Pluto has a minor effect on the orbit of Neptune compared with the closest planet to Neptune:

(a) Calculate the acceleration due to gravity at Neptune due to Pluto when they are $4.50 \times 10^{12} \text{ m}$ apart, as they are at present. The mass of Pluto is $1.4 \times 10^{22} \text{ kg}$.

(b) Calculate the acceleration due to gravity at Neptune due to Uranus, presently about $2.50 \times 10^{12} \text{ m}$ apart, and compare it with that due to Pluto. The mass of Uranus is $8.62 \times 10^{25} \text{ kg}$.

41. (a) The Sun orbits the Milky Way galaxy once each 2.60×10^8 y , with a roughly circular orbit averaging 3.00×10^4 light years in radius. (A light year is the distance traveled by light in 1 y.) Calculate the centripetal acceleration of the Sun in its galactic orbit. Does your result support the contention that a nearly inertial frame of reference can be located at the Sun?

(b) Calculate the average speed of the Sun in its galactic orbit. Does the answer surprise you?

42. Unreasonable Result

A mountain 10.0 km from a person exerts a gravitational force on him equal to 2.00% of his weight.

(a) Calculate the mass of the mountain.

(b) Compare the mountain's mass with that of Earth.

(c) What is unreasonable about these results?

(d) Which premises are unreasonable or inconsistent? (Note that accurate gravitational measurements can easily detect the effect of nearby mountains and variations in local geology.)

6.6 Satellites and Kepler's Laws: An Argument for Simplicity

43. A geosynchronous Earth satellite is one that has an orbital period of precisely 1 day. Such orbits are useful for communication and weather observation because the satellite remains above the same point on Earth (provided it orbits in the equatorial plane in the same direction as Earth's rotation). Calculate the radius of such an orbit based on the data for the moon in **Table 6.2**.

44. Calculate the mass of the Sun based on data for Earth's orbit and compare the value obtained with the Sun's actual mass.

45. Find the mass of Jupiter based on data for the orbit of one of its moons, and compare your result with its actual mass.

46. Find the ratio of the mass of Jupiter to that of Earth based on data in **Table 6.2**.

47. Astronomical observations of our Milky Way galaxy indicate that it has a mass of about 8.0×10^{11} solar masses.

A star orbiting on the galaxy's periphery is about 6.0×10^4 light years from its center. (a) What should the orbital period of that star be? (b) If its period is 6.0×10^7 instead, what is the mass of the galaxy? Such calculations are used to imply the existence of "dark matter" in the universe and have indicated, for example, the existence of very massive black holes at the centers of some galaxies.

48. Integrated Concepts

Space debris left from old satellites and their launchers is becoming a hazard to other satellites. (a) Calculate the speed of a satellite in an orbit 900 km above Earth's surface. (b) Suppose a loose rivet is in an orbit of the same radius that intersects the satellite's orbit at an angle of $90°$ relative to Earth. What is the velocity of the rivet relative to the satellite just before striking it? (c) Given the rivet is 3.00 mm in size, how long will its collision with the satellite last? (d) If its mass is 0.500 g, what is the average force it exerts on the satellite? (e) How much energy in joules is generated by the collision? (The satellite's velocity does not change appreciably, because its mass is much greater than the rivet's.)

49. Unreasonable Results

(a) Based on Kepler's laws and information on the orbital characteristics of the Moon, calculate the orbital radius for an Earth satellite having a period of 1.00 h. (b) What is unreasonable about this result? (c) What is unreasonable or inconsistent about the premise of a 1.00 h orbit?

50. Construct Your Own Problem

On February 14, 2000, the NEAR spacecraft was successfully inserted into orbit around Eros, becoming the first artificial satellite of an asteroid. Construct a problem in which you determine the orbital speed for a satellite near Eros. You will need to find the mass of the asteroid and consider such things as a safe distance for the orbit. Although Eros is not spherical, calculate the acceleration due to gravity on its surface at a point an average distance from its center of mass. Your instructor may also wish to have you calculate the escape velocity from this point on Eros.

Test Prep for AP® Courses

6.5 Newton's Universal Law of Gravitation

1. Jupiter has a mass approximately 300 times greater than Earth's and a radius about 11 times greater. How will the gravitational acceleration at the surface of Jupiter compare to that at the surface of the Earth?
   a. Greater
   b. Less
   c. About the same
   d. Not enough information

2. Given Newton's universal law of gravitation (Equation 6.40), under what circumstances is the force due to gravity maximized?

3. In the formula $g = \dfrac{GM}{r^2}$, what does G represent?

a. The acceleration due to gravity
b. A gravitational constant that is the same everywhere in the universe
c. A gravitational constant that is inversely proportional to the radius
d. The factor by which you multiply the inertial mass to obtain the gravitational mass

4. Saturn's moon Titan has a radius of 2.58×10^6 m and a measured gravitational field of 1.35 m/s^2. What is its mass?

5. A recently discovered planet has a mass twice as great as Earth's and a radius twice as large as Earth's. What will be the approximate size of its gravitational field?

a. 19 m/s^2
b. 4.9 m/s^2
c. 2.5 m/s^2
d. 9.8 m/s^2

6. 4. Earth is 1.5×10^{11} m from the Sun. Mercury is 5.7×10^{10} m from the Sun. How does the gravitational field of the Sun on Mercury (g_{SM}) compare to the gravitational field of the Sun on Earth (g_{SE})?

7   WORK, ENERGY, AND ENERGY RESOURCES

Figure 7.1 How many forms of energy can you identify in this photograph of a wind farm in Iowa? (credit: Jürgen from Sandesneben, Germany, Wikimedia Commons)

Chapter Outline

7.1. Work: The Scientific Definition

7.2. Kinetic Energy and the Work-Energy Theorem

7.3. Gravitational Potential Energy

7.4. Conservative Forces and Potential Energy

7.5. Nonconservative Forces

7.6. Conservation of Energy

7.7. Power

7.8. Work, Energy, and Power in Humans

7.9. World Energy Use

Connection for AP® Courses

Energy plays an essential role both in everyday events and in scientific phenomena. You can no doubt name many forms of energy, from that provided by our foods to the energy we use to run our cars and the sunlight that warms us on the beach. You can also cite examples of what people call "energy" that may not be scientific, such as someone having an energetic personality. Not only does energy have many interesting forms, it is involved in almost all phenomena, and is one of the most important concepts of physics.

There is no simple and accurate scientific definition for energy. Energy is characterized by its many forms and the fact that it is conserved. We can loosely define energy as the ability to do **work**, admitting that in some circumstances not all energy is available to do work. Because of the association of energy with work, we begin the chapter with a discussion of work. Work is intimately related to energy and how energy moves from one system to another or changes form. The work-energy theorem supports Big Idea 3, that interactions between objects are described by forces. In particular, exerting a force on an object may do work on it, changing it's energy (Enduring Understanding 3.E). The work-energy theorem, introduced in this chapter, establishes the relationship between work done on an object by an external force and changes in the object's kinetic energy (Essential Knowledge 3.E.1).

Similarly, systems can do work on each other, supporting Big Idea 4, that interactions between systems can result in changes in those systems—in this case, changes in the total energy of the system (Enduring Understanding 4.C). The total energy of the system is the sum of its kinetic energy, potential energy, and microscopic internal energy (Essential Knowledge 4.C.1). In this chapter students learn how to calculate kinetic, gravitational, and elastic potential energy in order to determine the total mechanical energy of a system. The transfer of mechanical energy into or out of a system is equal to the work done on the system by an external force with a nonzero component parallel to the displacement (Essential Knowledge 4.C.2).

An important aspect of energy is that the total amount of energy in the universe is constant. Energy can change forms, but it cannot appear from nothing or disappear without a trace. Energy is thus one of a handful of physical quantities that we say is "conserved." **Conservation of energy** (as physicists call the principle that energy can neither be created nor destroyed) is based on experiment. Even as scientists discovered new forms of energy, conservation of energy has always been found to apply. Perhaps the most dramatic example of this was supplied by Einstein when he suggested that mass is equivalent to energy (his famous equation $E = mc^2$). This is one of the most important applications of Big Idea 5, that changes that occur as a result of interactions are constrained by conservation laws. Specifically, there are many situations where conservation of energy (Enduring Understanding 5.B) is both a useful concept and starting point for calculations related to the system. Note, however, that conservation doesn't necessarily mean that energy in a system doesn't change. Energy may be transferred into or out of the system, and the change must be equal to the amount transferred (Enduring Understanding 5.A). This may occur if there is an external force or a transfer between external objects and the system (Essential Knowledge 5.A.3). Energy is one of the fundamental quantities that are conserved for all systems (Essential Knowledge 5.A.2). The chapter introduces concepts of kinetic energy and potential energy. Kinetic energy is introduced as an energy of motion that can be changed by the amount of work done by an external force. Potential energy can only exist when objects interact with each other via conservative forces according to classical physics (Essential Knowledge 5.B.3). Because of this, a single object can only have kinetic energy and no potential energy (Essential Knowledge 5.B.1). The chapter also introduces the idea that the energy transfer is equal to the work done on the system by external forces and the rate of energy transfer is defined as power (Essential Knowledge 5.B.5).

From a societal viewpoint, energy is one of the major building blocks of modern civilization. Energy resources are key limiting factors to economic growth. The world use of energy resources, especially oil, continues to grow, with ominous consequences economically, socially, politically, and environmentally. We will briefly examine the world's energy use patterns at the end of this chapter.

The concepts in this chapter support:

Big Idea 3 The interactions of an object with other objects can be described by forces.

Enduring Understanding 3.E A force exerted on an object can change the kinetic energy of the object.

Essential Knowledge 3.E.1 The change in the kinetic energy of an object depends on the force exerted on the object and on the displacement of the object during the interval that the force is exerted.

Big Idea 4 Interactions between systems can result in changes in those systems.

Enduring Understanding 4.C Interactions with other objects or systems can change the total energy of a system.

Essential Knowledge 4.C.1 The energy of a system includes its kinetic energy, potential energy, and microscopic internal energy. Examples should include gravitational potential energy, elastic potential energy, and kinetic energy.

Essential Knowledge 4.C.2 Mechanical energy (the sum of kinetic and potential energy) is transferred into or out of a system when an external force is exerted on a system such that a component of the force is parallel to its displacement. The process through which the energy is transferred is called work.

Big Idea 5 Changes that occur as a result of interactions are constrained by conservation laws.

Enduring Understanding 5.A Certain quantities are conserved, in the sense that the changes of those quantities in a given system are always equal to the transfer of that quantity to or from the system by all possible interactions with other systems.

Essential Knowledge 5.A.2 For all systems under all circumstances, energy, charge, linear momentum, and angular momentum are conserved.

Essential Knowledge 5.A.3 An interaction can be either a force exerted by objects outside the system or the transfer of some quantity with objects outside the system.

Enduring Understanding 5.B The energy of a system is conserved.

Essential Knowledge 5.B.1 Classically, an object can only have kinetic energy since potential energy requires an interaction between two or more objects.

Essential Knowledge 5.B.3 A system with internal structure can have potential energy. Potential energy exists within a system if the objects within that system interact with conservative forces.

Essential Knowledge 5.B.5 Energy can be transferred by an external force exerted on an object or system that moves the object or system through a distance; this energy transfer is called work. Energy transfer in mechanical or electrical systems may occur at different rates. Power is defined as the rate of energy transfer into, out of, or within a system.

7.1 Work: The Scientific Definition

Learning Objectives

By the end of this section, you will be able to:

- Explain how an object must be displaced for a force on it to do work.
- Explain how relative directions of force and displacement of an object determine whether the work done on the object is positive, negative, or zero.

The information presented in this section supports the following AP® learning objectives and science practices:

- **5.B.5.1** The student is able to design an experiment and analyze data to examine how a force exerted on an object or system does work on the object or system as it moves through a distance. **(S.P. 4.2, 5.1)**

- **5.B.5.2** The student is able to design an experiment and analyze graphical data in which interpretations of the area under a force-distance curve are needed to determine the work done on or by the object or system. **(S.P. 4.5, 5.1)**
- **5.B.5.3** The student is able to predict and calculate from graphical data the energy transfer to or work done on an object or system from information about a force exerted on the object or system through a distance. **(S.P. 1.5, 2.2, 6.4)**

What It Means to Do Work

The scientific definition of work differs in some ways from its everyday meaning. Certain things we think of as hard work, such as writing an exam or carrying a heavy load on level ground, are not work as defined by a scientist. The scientific definition of work reveals its relationship to energy—whenever work is done, energy is transferred.

For work, in the scientific sense, to be done on an object, a force must be exerted on that object and there must be displacement of that object in the direction of the force.

Formally, the **work** done on a system by a constant force is defined to be *the product of the component of the force in the direction of motion and the distance through which the force acts*. For a constant force, this is expressed in equation form as

$$W = |\mathbf{F}| (\cos \theta) |\mathbf{d}|,\qquad(7.1)$$

where W is work, $\mathbf{d}$ is the displacement of the system, and θ is the angle between the force vector $\mathbf{F}$ and the displacement vector $\mathbf{d}$, as in **Figure 7.2**. We can also write this as

$$W = Fd \cos \theta.\qquad(7.2)$$

To find the work done on a system that undergoes motion that is not one-way or that is in two or three dimensions, we divide the motion into one-way one-dimensional segments and add up the work done over each segment.

What is Work?

The work done on a system by a constant force is *the product of the component of the force in the direction of motion times the distance through which the force acts*. For one-way motion in one dimension, this is expressed in equation form as

$$W = Fd \cos \theta,\qquad(7.3)$$

where W is work, F is the magnitude of the force on the system, d is the magnitude of the displacement of the system, and θ is the angle between the force vector $\mathbf{F}$ and the displacement vector $\mathbf{d}$.

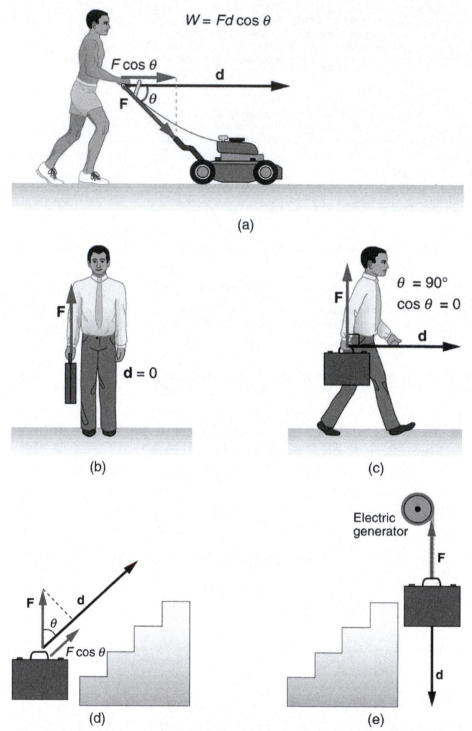

Figure 7.2 Examples of work. (a) The work done by the force $\mathbf{F}$ on this lawn mower is $Fd \cos \theta$. Note that $F \cos \theta$ is the component of the force in the direction of motion. (b) A person holding a briefcase does no work on it, because there is no displacement. No energy is transferred to or from the briefcase. (c) The person moving the briefcase horizontally at a constant speed does no work on it, and transfers no energy to it. (d) Work *is* done on the briefcase by carrying it up stairs at constant speed, because there is necessarily a component of force $\mathbf{F}$ in the direction of the motion. Energy is transferred to the briefcase and could in turn be used to do work. (e) When the briefcase is lowered, energy is transferred out of the briefcase and into an electric generator. Here the work done on the briefcase by the generator is negative, removing energy from the briefcase, because $\mathbf{F}$ and $\mathbf{d}$ are in opposite directions.

To examine what the definition of work means, let us consider the other situations shown in **Figure 7.2**. The person holding the briefcase in **Figure 7.2**(b) does no work, for example. Here $d = 0$, so $W = 0$. Why is it you get tired just holding a load? The answer is that your muscles are doing work against one another, *but they are doing no work on the system of interest* (the "briefcase-Earth system"—see **Gravitational Potential Energy** for more details). There must be displacement for work to be done, and there must be a component of the force in the direction of the motion. For example, the person carrying the briefcase on level ground in **Figure 7.2**(c) does no work on it, because the force is perpendicular to the motion. That is, $\cos 90° = 0$, and so $W = 0$.

In contrast, when a force exerted on the system has a component in the direction of motion, such as in **Figure 7.2**(d), work *is* done—energy is transferred to the briefcase. Finally, in **Figure 7.2**(e), energy is transferred from the briefcase to a generator. There are two good ways to interpret this energy transfer. One interpretation is that the briefcase's weight does work on the generator, giving it energy. The other interpretation is that the generator does negative work on the briefcase, thus removing energy from it. The drawing shows the latter, with the force from the generator upward on the briefcase, and the displacement downward. This makes $\theta = 180°$, and $\cos 180° = -1$; therefore, W is negative.

Real World Connections: When Work Happens

Note that work as we define it is not the same as effort. You can push against a concrete wall all you want, but you won't move it. While the pushing represents effort on your part, the fact that you have not changed the wall's state in any way indicates that you haven't done work. If you did somehow push the wall over, this would indicate a change in the wall's state, and therefore you would have done work.

This can also be shown with **Figure 7.2**(a): as you push a lawnmower against friction, both you and friction are changing the lawnmower's state. However, only the component of the force parallel to the movement is changing the lawnmower's state. The component perpendicular to the motion is trying to push the lawnmower straight into Earth; the lawnmower does not move into Earth, and therefore the lawnmower's state is not changing in the direction of Earth.

Similarly, in **Figure 7.2**(c), both your hand and gravity are exerting force on the briefcase. However, they are both acting perpendicular to the direction of motion, hence they are not changing the condition of the briefcase and do no work. However, if the briefcase were dropped, then its displacement would be parallel to the force of gravity, which would do work on it, changing its state (it would fall to the ground).

Calculating Work

Work and energy have the same units. From the definition of work, we see that those units are force times distance. Thus, in SI units, work and energy are measured in **newton-meters**. A newton-meter is given the special name **joule** (J), and $1 \text{ J} = 1 \text{ N} \cdot \text{m} = 1 \text{ kg} \cdot \text{m}^2/\text{s}^2$. One joule is not a large amount of energy; it would lift a small 100-gram apple a distance of about 1 meter.

Example 7.1 Calculating the Work You Do to Push a Lawn Mower Across a Large Lawn

How much work is done on the lawn mower by the person in **Figure 7.2**(a) if he exerts a constant force of 75.0 N at an angle $35°$ below the horizontal and pushes the mower 25.0 m on level ground? Convert the amount of work from joules to kilocalories and compare it with this person's average daily intake of $10,000 \text{ kJ}$ (about 2400 kcal) of food energy. One *calorie* (1 cal) of heat is the amount required to warm 1 g of water by $1°\text{C}$, and is equivalent to 4.184 J, while one *food calorie* (1 kcal) is equivalent to 4184 J.

Strategy

We can solve this problem by substituting the given values into the definition of work done on a system, stated in the equation $W = Fd \cos \theta$. The force, angle, and displacement are given, so that only the work W is unknown.

Solution

The equation for the work is

$$W = Fd \cos \theta. \tag{7.4}$$

Substituting the known values gives

$$\begin{aligned} W &= (75.0 \text{ N})(25.0 \text{ m}) \cos (35.0°) \\ &= 1536 \text{ J} = 1.54 \times 10^3 \text{ J}. \end{aligned} \tag{7.5}$$

Converting the work in joules to kilocalories yields $W = (1536 \text{ J})(1 \text{ kcal} / 4184 \text{ J}) = 0.367 \text{ kcal}$. The ratio of the work done to the daily consumption is

$$\frac{W}{2400 \text{ kcal}} = 1.53 \times 10^{-4}. \tag{7.6}$$

Discussion

This ratio is a tiny fraction of what the person consumes, but it is typical. Very little of the energy released in the consumption of food is used to do work. Even when we "work" all day long, less than 10% of our food energy intake is used to do work and more than 90% is converted to thermal energy or stored as chemical energy in fat.

Applying the Science Practices: Boxes on Floors

Plan and design an experiment to determine how much work you do on a box when you are pushing it over different floor surfaces. Make sure your experiment can help you answer the following questions: What happens on different surfaces? What happens if you take different routes across the same surface? Do you get different results with two people pushing on perpendicular surfaces of the box? What if you vary the mass in the box? Remember to think about both your effort in any given instant (a proxy for force exerted) and the total work you do. Also, when planning your experiments, remember that in any given set of trials you should only change one variable.

You should find that you have to exert more effort on surfaces that will create more friction with the box, though you might be surprised by which surfaces the box slides across easily. Longer routes result in your doing more work, even though the box ends up in the same place. Two people pushing on perpendicular sides do less work for their total effort, due to the forces and displacement not being parallel. A more massive box will take more effort to move.

Applying the Science Practices: Force-Displacement Diagrams

Suppose you are given two carts and a track to run them on, a motion detector, a force sensor, and a computer that can record the data from the two sensors. Plan and design an experiment to measure the work done on one of the carts, and compare your results to the work-energy theorem. Note that the motion detector can measure both displacement and velocity versus time, while the force sensor measures force over time, and the carts have known masses. Recall that the work-energy theorem states that the work done on a system (force over displacement) should equal the change in kinetic energy. In your experimental design, describe and compare two possible ways to calculate the work done.

Sample Response: One possible technique is to set up the motion detector at one end of the track, and have the computer record both displacement and velocity over time. Then attach the force sensor to one of the carts, and use this cart, through the force sensor, to push the second cart toward the motion detector. Calculate the difference between the final and initial kinetic energies (the kinetic energies after and before the push), and compare this to the area of a graph of force versus displacement for the duration of the push. They should be the same.

7.2 Kinetic Energy and the Work-Energy Theorem

Learning Objectives

By the end of this section, you will be able to:

- Explain work as a transfer of energy and net work as the work done by the net force.
- Explain and apply the work-energy theorem.

The information presented in this section supports the following AP® learning objectives and science practices:

- **3.E.1.1** The student is able to make predictions about the changes in kinetic energy of an object based on considerations of the direction of the net force on the object as the object moves. **(S.P. 6.4, 7.2)**
- **3.E.1.2** The student is able to use net force and velocity vectors to determine qualitatively whether kinetic energy of an object would increase, decrease, or remain unchanged. **(S.P. 1.4)**
- **3.E.1.3** The student is able to use force and velocity vectors to determine qualitatively or quantitatively the net force exerted on an object and qualitatively whether kinetic energy of that object would increase, decrease, or remain unchanged. **(S.P. 1.4, 2.2)**
- **3.E.1.4** The student is able to apply mathematical routines to determine the change in kinetic energy of an object given the forces on the object and the displacement of the object. **(S.P. 2.2)**
- **4.C.1.1** The student is able to calculate the total energy of a system and justify the mathematical routines used in the calculation of component types of energy within the system whose sum is the total energy. **(S.P. 1.4, 2.1, 2.2)**
- **4.C.2.1** The student is able to make predictions about the changes in the mechanical energy of a system when a component of an external force acts parallel or antiparallel to the direction of the displacement of the center of mass. **(S.P. 6.4)**
- **4.C.2.2** The student is able to apply the concepts of conservation of energy and the work-energy theorem to determine qualitatively and/or quantitatively that work done on a two-object system in linear motion will change the kinetic energy of the center of mass of the system, the potential energy of the systems, and/or the internal energy of the system. **(S.P. 1.4, 2.2, 7.2)**
- **5.B.5.3** The student is able to predict and calculate from graphical data the energy transfer to or work done on an object or system from information about a force exerted on the object or system through a distance. **(S.P. 1.5, 2.2, 6.4)**

Work Transfers Energy

What happens to the work done on a system? Energy is transferred into the system, but in what form? Does it remain in the system or move on? The answers depend on the situation. For example, if the lawn mower in **Figure 7.2**(a) is pushed just hard enough to keep it going at a constant speed, then energy put into the mower by the person is removed continuously by friction, and eventually leaves the system in the form of heat transfer. In contrast, work done on the briefcase by the person carrying it up stairs in **Figure 7.2**(d) is stored in the briefcase-Earth system and can be recovered at any time, as shown in **Figure 7.2**(e). In fact, the building of the pyramids in ancient Egypt is an example of storing energy in a system by doing work on the system.

Some of the energy imparted to the stone blocks in lifting them during construction of the pyramids remains in the stone-Earth system and has the potential to do work.

In this section we begin the study of various types of work and forms of energy. We will find that some types of work leave the energy of a system constant, for example, whereas others change the system in some way, such as making it move. We will also develop definitions of important forms of energy, such as the energy of motion.

Net Work and the Work-Energy Theorem

We know from the study of Newton's laws in **Dynamics: Force and Newton's Laws of Motion** that net force causes acceleration. We will see in this section that work done by the net force gives a system energy of motion, and in the process we will also find an expression for the energy of motion.

Let us start by considering the total, or net, work done on a system. Net work is defined to be the sum of work done by all external forces—that is, **net work** is the work done by the net external force $\mathbf{F}_{net}$. In equation form, this is

$W_{net} = F_{net}d \cos \theta$ where θ is the angle between the force vector and the displacement vector.

Figure 7.3(a) shows a graph of force versus displacement for the component of the force in the direction of the displacement—that is, an $F \cos \theta$ vs. d graph. In this case, $F \cos \theta$ is constant. You can see that the area under the graph is $Fd \cos \theta$, or the work done. **Figure 7.3**(b) shows a more general process where the force varies. The area under the curve is divided into strips, each having an average force $(F \cos \theta)_{i(ave)}$. The work done is $(F \cos \theta)_{i(ave)}d_i$ for each strip, and the total work done is the sum of the W_i. Thus the total work done is the total area under the curve, a useful property to which we shall refer later.

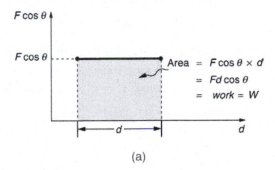

(a)

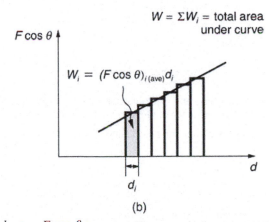

(b)

Figure 7.3 (a) A graph of $F \cos \theta$ vs. d, when $F \cos \theta$ is constant. The area under the curve represents the work done by the force. (b) A graph of $F \cos \theta$ vs. d in which the force varies. The work done for each interval is the area of each strip; thus, the total area under the curve equals the total work done.

Real World Connections: Work and Direction

Consider driving in a car. While moving, you have forward velocity and therefore kinetic energy. When you hit the brakes, they exert a force opposite to your direction of motion (acting through the wheels). The brakes do work on your car and reduce the kinetic energy. Similarly, when you accelerate, the engine (acting through the wheels) exerts a force in the direction of motion. The engine does work on your car, and increases the kinetic energy. Finally, if you go around a corner at a constant speed, you have the same kinetic energy both before and after the corner. The force exerted by the engine was perpendicular to the direction of motion, and therefore did no work and did not change the kinetic energy.

Net work will be simpler to examine if we consider a one-dimensional situation where a force is used to accelerate an object in a direction parallel to its initial velocity. Such a situation occurs for the package on the roller belt conveyor system shown in **Figure**

7.4.

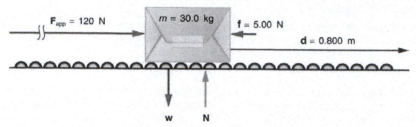

Figure 7.4 A package on a roller belt is pushed horizontally through a distance $\mathbf{d}$.

The force of gravity and the normal force acting on the package are perpendicular to the displacement and do no work. Moreover, they are also equal in magnitude and opposite in direction so they cancel in calculating the net force. The net force arises solely from the horizontal applied force $\mathbf{F}_{app}$ and the horizontal friction force $\mathbf{f}$. Thus, as expected, the net force is parallel to the displacement, so that $\theta = 0°$ and $\cos \theta = 1$, and the net work is given by

$$W_{net} = F_{net}d. \qquad (7.7)$$

The effect of the net force $\mathbf{F}_{net}$ is to accelerate the package from v_0 to v. The kinetic energy of the package increases, indicating that the net work done on the system is positive. (See Example 7.2.) By using Newton's second law, and doing some algebra, we can reach an interesting conclusion. Substituting $F_{net} = ma$ from Newton's second law gives

$$W_{net} = mad. \qquad (7.8)$$

To get a relationship between net work and the speed given to a system by the net force acting on it, we take $d = x - x_0$ and use the equation studied in **Motion Equations for Constant Acceleration in One Dimension** for the change in speed over a distance d if the acceleration has the constant value a; namely, $v^2 = v_0^2 + 2ad$ (note that a appears in the expression for the net work). Solving for acceleration gives $a = \dfrac{v^2 - v_0^2}{2d}$. When a is substituted into the preceding expression for W_{net}, we obtain

$$W_{net} = m\left(\frac{v^2 - v_0^2}{2d}\right)d. \qquad (7.9)$$

The d cancels, and we rearrange this to obtain

$$W = \frac{1}{2}mv^2 - \frac{1}{2}mv_0^2. \qquad (7.10)$$

This expression is called the **work-energy theorem**, and it actually applies *in general* (even for forces that vary in direction and magnitude), although we have derived it for the special case of a constant force parallel to the displacement. The theorem implies that the net work on a system equals the change in the quantity $\frac{1}{2}mv^2$. This quantity is our first example of a form of energy.

The Work-Energy Theorem

The net work on a system equals the change in the quantity $\frac{1}{2}mv^2$.

$$W_{net} = \frac{1}{2}mv^2 - \frac{1}{2}mv_0^2 \qquad (7.11)$$

The quantity $\frac{1}{2}mv^2$ in the work-energy theorem is defined to be the translational **kinetic energy** (KE) of a mass m moving at a speed v. (*Translational* kinetic energy is distinct from *rotational* kinetic energy, which is considered later.) In equation form, the translational kinetic energy,

$$KE = \frac{1}{2}mv^2, \qquad (7.12)$$

is the energy associated with translational motion. Kinetic energy is a form of energy associated with the motion of a particle, single body, or system of objects moving together.

We are aware that it takes energy to get an object, like a car or the package in **Figure 7.4**, up to speed, but it may be a bit surprising that kinetic energy is proportional to speed squared. This proportionality means, for example, that a car traveling at 100 km/h has four times the kinetic energy it has at 50 km/h, helping to explain why high-speed collisions are so devastating. We will now consider a series of examples to illustrate various aspects of work and energy.

Applying the Science Practices: Cars on a Hill

Assemble a ramp suitable for rolling some toy cars up or down. Then plan a series of experiments to determine how the direction of a force relative to the velocity of an object alters the kinetic energy of the object. Note that gravity will be pointing down in all cases. What happens if you start the car at the top? How about at the bottom, with an initial velocity that is increasing? If your ramp is wide enough, what happens if you send the toy car straight across? Does varying the surface of the ramp change your results?

Sample Response: When the toy car is going down the ramp, with a component of gravity in the same direction, the kinetic energy increases. Sending the car up the ramp decreases the kinetic energy, as gravity is opposing the motion. Sending the car sideways should result in little to no change. If you have a surface that generates more friction than a smooth surface (carpet), note that the friction always opposed the motion, and hence decreases the kinetic energy.

Example 7.2 Calculating the Kinetic Energy of a Package

Suppose a 30.0-kg package on the roller belt conveyor system in **Figure 7.4** is moving at 0.500 m/s. What is its kinetic energy?

Strategy

Because the mass m and speed v are given, the kinetic energy can be calculated from its definition as given in the equation $KE = \frac{1}{2}mv^2$.

Solution

The kinetic energy is given by

$$KE = \frac{1}{2}mv^2. \tag{7.13}$$

Entering known values gives

$$KE = 0.5(30.0 \text{ kg})(0.500 \text{ m/s})^2, \tag{7.14}$$

which yields

$$KE = 3.75 \text{ kg} \cdot \text{m}^2/\text{s}^2 = 3.75 \text{ J}. \tag{7.15}$$

Discussion

Note that the unit of kinetic energy is the joule, the same as the unit of work, as mentioned when work was first defined. It is also interesting that, although this is a fairly massive package, its kinetic energy is not large at this relatively low speed. This fact is consistent with the observation that people can move packages like this without exhausting themselves.

Real World Connections: Center of Mass

Suppose we have two experimental carts, of equal mass, latched together on a track with a compressed spring between them. When the latch is released, the spring does 10 J of work on the carts (we'll see how in a couple of sections). The carts move relative to the spring, which is the center of mass of the system. However, the center of mass stays fixed. How can we consider the kinetic energy of this system?

By the work-energy theorem, the work done by the spring on the carts must turn into kinetic energy. So this system has 10 J of kinetic energy. The total kinetic energy of the system is the kinetic energy of the center of mass of the system relative to the fixed origin plus the kinetic energy of each cart relative to the center of mass. We know that the center of mass relative to the fixed origin does not move, and therefore all of the kinetic energy must be distributed among the carts relative to the center of mass. Since the carts have equal mass, they each receive an equal amount of kinetic energy, so each cart has 5.0 J of kinetic energy.

In our example, the forces between the spring and each cart are internal to the system. According to Newton's third law, these internal forces will cancel since they are equal and opposite in direction. However, this does not imply that these internal forces will not do work. Thus, the change in kinetic energy of the system is caused by work done by the force of the spring, and results in the motion of the two carts relative to the center of mass.

Example 7.3 Determining the Work to Accelerate a Package

Suppose that you push on the 30.0-kg package in **Figure 7.4** with a constant force of 120 N through a distance of 0.800 m, and that the opposing friction force averages 5.00 N.

(a) Calculate the net work done on the package. (b) Solve the same problem as in part (a), this time by finding the work done by each force that contributes to the net force.

Strategy and Concept for (a)

This is a motion in one dimension problem, because the downward force (from the weight of the package) and the normal force have equal magnitude and opposite direction, so that they cancel in calculating the net force, while the applied force, friction, and the displacement are all horizontal. (See **Figure 7.4**.) As expected, the net work is the net force times distance.

Solution for (a)

The net force is the push force minus friction, or $F_{net} = 120 \text{ N} - 5.00 \text{ N} = 115 \text{ N}$. Thus the net work is

$$\begin{aligned} W_{net} &= F_{net}d = (115 \text{ N})(0.800 \text{ m}) \\ &= 92.0 \text{ N} \cdot \text{m} = 92.0 \text{ J}. \end{aligned} \tag{7.16}$$

Discussion for (a)

This value is the net work done on the package. The person actually does more work than this, because friction opposes the motion. Friction does negative work and removes some of the energy the person expends and converts it to thermal energy. The net work equals the sum of the work done by each individual force.

Strategy and Concept for (b)

The forces acting on the package are gravity, the normal force, the force of friction, and the applied force. The normal force and force of gravity are each perpendicular to the displacement, and therefore do no work.

Solution for (b)

The applied force does work.

$$\begin{aligned} W_{app} &= F_{app}d \cos(0°) = F_{app}d \\ &= (120 \text{ N})(0.800 \text{ m}) \\ &= 96.0 \text{ J} \end{aligned} \tag{7.17}$$

The friction force and displacement are in opposite directions, so that $\theta = 180°$, and the work done by friction is

$$\begin{aligned} W_{fr} &= F_{fr}d \cos(180°) = -F_{fr}d \\ &= -(5.00 \text{ N})(0.800 \text{ m}) \\ &= -4.00 \text{ J}. \end{aligned} \tag{7.18}$$

So the amounts of work done by gravity, by the normal force, by the applied force, and by friction are, respectively,

$$\begin{aligned} W_{gr} &= 0, \\ W_{N} &= 0, \\ W_{app} &= 96.0 \text{ J}, \\ W_{fr} &= -4.00 \text{ J}. \end{aligned} \tag{7.19}$$

The total work done as the sum of the work done by each force is then seen to be

$$W_{total} = W_{gr} + W_{N} + W_{app} + W_{fr} = 92.0 \text{ J}. \tag{7.20}$$

Discussion for (b)

The calculated total work W_{total} as the sum of the work by each force agrees, as expected, with the work W_{net} done by the net force. The work done by a collection of forces acting on an object can be calculated by either approach.

Example 7.4 Determining Speed from Work and Energy

Find the speed of the package in **Figure 7.4** at the end of the push, using work and energy concepts.

Strategy

Here the work-energy theorem can be used, because we have just calculated the net work, W_{net}, and the initial kinetic energy, $\frac{1}{2}mv_0^2$. These calculations allow us to find the final kinetic energy, $\frac{1}{2}mv^2$, and thus the final speed v.

Solution

The work-energy theorem in equation form is

$$W_{\text{net}} = \tfrac{1}{2}mv^2 - \tfrac{1}{2}mv_0{}^2. \tag{7.21}$$

Solving for $\tfrac{1}{2}mv^2$ gives

$$\tfrac{1}{2}mv^2 = W_{\text{net}} + \tfrac{1}{2}mv_0{}^2. \tag{7.22}$$

Thus,

$$\tfrac{1}{2}mv^2 = 92.0 \text{ J} + 3.75 \text{ J} = 95.75 \text{ J}. \tag{7.23}$$

Solving for the final speed as requested and entering known values gives

$$v = \sqrt{\frac{2(95.75 \text{ J})}{m}} = \sqrt{\frac{191.5 \text{ kg} \cdot \text{m}^2/\text{s}^2}{30.0 \text{ kg}}} \tag{7.24}$$
$$= 2.53 \text{ m/s}.$$

Discussion

Using work and energy, we not only arrive at an answer, we see that the final kinetic energy is the sum of the initial kinetic energy and the net work done on the package. This means that the work indeed adds to the energy of the package.

Example 7.5 Work and Energy Can Reveal Distance, Too

How far does the package in **Figure 7.4** coast after the push, assuming friction remains constant? Use work and energy considerations.

Strategy

We know that once the person stops pushing, friction will bring the package to rest. In terms of energy, friction does negative work until it has removed all of the package's kinetic energy. The work done by friction is the force of friction times the distance traveled times the cosine of the angle between the friction force and displacement; hence, this gives us a way of finding the distance traveled after the person stops pushing.

Solution

The normal force and force of gravity cancel in calculating the net force. The horizontal friction force is then the net force, and it acts opposite to the displacement, so $\theta = 180°$. To reduce the kinetic energy of the package to zero, the work W_{fr} by friction must be minus the kinetic energy that the package started with plus what the package accumulated due to the pushing. Thus $W_{\text{fr}} = -95.75 \text{ J}$. Furthermore, $W_{\text{fr}} = fd' \cos\theta = -fd'$, where d' is the distance it takes to stop. Thus,

$$d' = -\frac{W_{\text{fr}}}{f} = -\frac{-95.75 \text{ J}}{5.00 \text{ N}}, \tag{7.25}$$

and so

$$d' = 19.2 \text{ m}. \tag{7.26}$$

Discussion

This is a reasonable distance for a package to coast on a relatively friction-free conveyor system. Note that the work done by friction is negative (the force is in the opposite direction of motion), so it removes the kinetic energy.

Some of the examples in this section can be solved without considering energy, but at the expense of missing out on gaining insights about what work and energy are doing in this situation. On the whole, solutions involving energy are generally shorter and easier than those using kinematics and dynamics alone.

7.3 Gravitational Potential Energy

Learning Objectives

By the end of this section, you will be able to:

- Explain gravitational potential energy in terms of work done against gravity.
- Show that the gravitational potential energy of an object of mass m at height h on Earth is given by $PE_g = mgh$.
- Show how knowledge of potential energy as a function of position can be used to simplify calculations and explain physical phenomena.

> The information presented in this section supports the following AP® learning objectives and science practices:
>
> - **4.C.1.1** The student is able to calculate the total energy of a system and justify the mathematical routines used in the calculation of component types of energy within the system whose sum is the total energy. **(S.P. 1.4, 2.1, 2.2)**
> - **5.B.1.1** The student is able to set up a representation or model showing that a single object can only have kinetic energy and use information about that object to calculate its kinetic energy. **(S.P. 1.4, 2.2)**
> - **5.B.1.2** The student is able to translate between a representation of a single object, which can only have kinetic energy, and a system that includes the object, which may have both kinetic and potential energies. **(S.P. 1.5)**

Work Done Against Gravity

Climbing stairs and lifting objects is work in both the scientific and everyday sense—it is work done against the gravitational force. When there is work, there is a transformation of energy. The work done against the gravitational force goes into an important form of stored energy that we will explore in this section.

Let us calculate the work done in lifting an object of mass m through a height h, such as in **Figure 7.5**. If the object is lifted straight up at constant speed, then the force needed to lift it is equal to its weight mg. The work done on the mass is then

$W = Fd = mgh$. We define this to be the **gravitational potential energy** (PE_g) put into (or gained by) the object-Earth

system. This energy is associated with the state of separation between two objects that attract each other by the gravitational force. For convenience, we refer to this as the PE_g gained by the object, recognizing that this is energy stored in the

gravitational field of Earth. Why do we use the word "system"? Potential energy is a property of a system rather than of a single object—due to its physical position. An object's gravitational potential is due to its position relative to the surroundings within the Earth-object system. The force applied to the object is an external force, from outside the system. When it does positive work it increases the gravitational potential energy of the system. Because gravitational potential energy depends on relative position, we need a reference level at which to set the potential energy equal to 0. We usually choose this point to be Earth's surface, but this point is arbitrary; what is important is the *difference* in gravitational potential energy, because this difference is what relates to the work done. The difference in gravitational potential energy of an object (in the Earth-object system) between two rungs of a ladder will be the same for the first two rungs as for the last two rungs.

Converting Between Potential Energy and Kinetic Energy

Gravitational potential energy may be converted to other forms of energy, such as kinetic energy. If we release the mass, gravitational force will do an amount of work equal to mgh on it, thereby increasing its kinetic energy by that same amount (by

the work-energy theorem). We will find it more useful to consider just the conversion of PE_g to KE without explicitly

considering the intermediate step of work. (See **Example 7.7**.) This shortcut makes it is easier to solve problems using energy (if possible) rather than explicitly using forces.

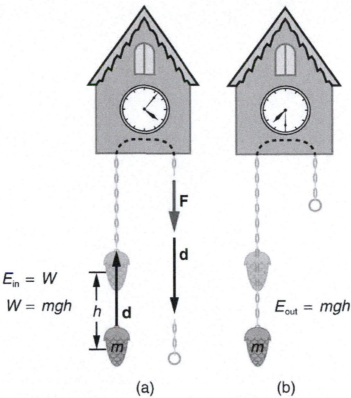

Figure 7.5 (a) The work done to lift the weight is stored in the mass-Earth system as gravitational potential energy. (b) As the weight moves downward, this gravitational potential energy is transferred to the cuckoo clock.

More precisely, we define the *change* in gravitational potential energy ΔPE_g to be

$$\Delta\text{PE}_g = mgh, \tag{7.27}$$

where, for simplicity, we denote the change in height by h rather than the usual Δh. Note that h is positive when the final height is greater than the initial height, and vice versa. For example, if a 0.500-kg mass hung from a cuckoo clock is raised 1.00 m, then its change in gravitational potential energy is

$$
\begin{aligned}
mgh &= (0.500 \text{ kg})(9.80 \text{ m/s}^2)(1.00 \text{ m}) \\
&= 4.90 \text{ kg} \cdot \text{m}^2/\text{s}^2 = 4.90 \text{ J}.
\end{aligned}
\tag{7.28}
$$

Note that the units of gravitational potential energy turn out to be joules, the same as for work and other forms of energy. As the clock runs, the mass is lowered. We can think of the mass as gradually giving up its 4.90 J of gravitational potential energy, *without directly considering the force of gravity that does the work*.

Using Potential Energy to Simplify Calculations

The equation $\Delta\text{PE}_g = mgh$ applies for any path that has a change in height of h, not just when the mass is lifted straight up.

(See **Figure 7.6**.) It is much easier to calculate mgh (a simple multiplication) than it is to calculate the work done along a complicated path. The idea of gravitational potential energy has the double advantage that it is very broadly applicable and it makes calculations easier. From now on, we will consider that any change in vertical position h of a mass m is accompanied by a change in gravitational potential energy mgh, and we will avoid the equivalent but more difficult task of calculating work done by or against the gravitational force.

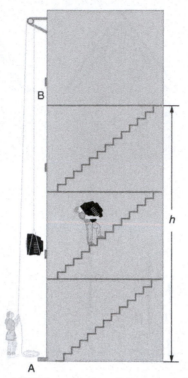

Figure 7.6 The change in gravitational potential energy $\left(\Delta PE_g\right)$ between points A and B is independent of the path. $\Delta PE_g = mgh$ for any path between the two points. Gravity is one of a small class of forces where the work done by or against the force depends only on the starting and ending points, not on the path between them.

Example 7.6 The Force to Stop Falling

A 60.0-kg person jumps onto the floor from a height of 3.00 m. If he lands stiffly (with his knee joints compressing by 0.500 cm), calculate the force on the knee joints.

Strategy

This person's energy is brought to zero in this situation by the work done on him by the floor as he stops. The initial PE_g is transformed into KE as he falls. The work done by the floor reduces this kinetic energy to zero.

Solution

The work done on the person by the floor as he stops is given by

$$W = Fd \cos \theta = -Fd, \tag{7.29}$$

with a minus sign because the displacement while stopping and the force from floor are in opposite directions $(\cos \theta = \cos 180° = -1)$. The floor removes energy from the system, so it does negative work.

The kinetic energy the person has upon reaching the floor is the amount of potential energy lost by falling through height h :

$$KE = -\Delta PE_g = -mgh, \tag{7.30}$$

The distance d that the person's knees bend is much smaller than the height h of the fall, so the additional change in gravitational potential energy during the knee bend is ignored.

The work W done by the floor on the person stops the person and brings the person's kinetic energy to zero:

$$W = -KE = mgh. \tag{7.31}$$

Combining this equation with the expression for W gives

$$-Fd = mgh. \tag{7.32}$$

Recalling that h is negative because the person fell *down*, the force on the knee joints is given by

$$F = -\frac{mgh}{d} = -\frac{(60.0 \text{ kg})(9.80 \text{ m/s}^2)(-3.00 \text{ m})}{5.00 \times 10^{-3} \text{ m}} = 3.53 \times 10^5 \text{ N}. \tag{7.33}$$

Discussion

Such a large force (500 times more than the person's weight) over the short impact time is enough to break bones. A much better way to cushion the shock is by bending the legs or rolling on the ground, increasing the time over which the force acts. A bending motion of 0.5 m this way yields a force 100 times smaller than in the example. A kangaroo's hopping shows this method in action. The kangaroo is the only large animal to use hopping for locomotion, but the shock in hopping is cushioned by the bending of its hind legs in each jump.(See **Figure 7.7**.)

Figure 7.7 The work done by the ground upon the kangaroo reduces its kinetic energy to zero as it lands. However, by applying the force of the ground on the hind legs over a longer distance, the impact on the bones is reduced. (credit: Chris Samuel, Flickr)

Example 7.7 Finding the Speed of a Roller Coaster from its Height

(a) What is the final speed of the roller coaster shown in **Figure 7.8** if it starts from rest at the top of the 20.0 m hill and work done by frictional forces is negligible? (b) What is its final speed (again assuming negligible friction) if its initial speed is 5.00 m/s?

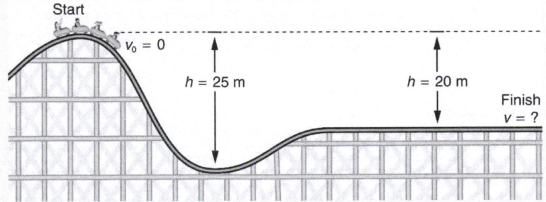

Figure 7.8 The speed of a roller coaster increases as gravity pulls it downhill and is greatest at its lowest point. Viewed in terms of energy, the roller-coaster-Earth system's gravitational potential energy is converted to kinetic energy. If work done by friction is negligible, all ΔPE_g is converted to KE.

Strategy

The roller coaster loses potential energy as it goes downhill. We neglect friction, so that the remaining force exerted by the track is the normal force, which is perpendicular to the direction of motion and does no work. The net work on the roller coaster is then done by gravity alone. The *loss* of gravitational potential energy from moving *downward* through a distance h equals the *gain* in kinetic energy. This can be written in equation form as $-\Delta PE_g = \Delta KE$. Using the equations for

PE_g and KE, we can solve for the final speed v, which is the desired quantity.

Solution for (a)

Here the initial kinetic energy is zero, so that $\Delta \text{KE} = \frac{1}{2}mv^2$. The equation for change in potential energy states that $\Delta \text{PE}_g = mgh$. Since h is negative in this case, we will rewrite this as $\Delta \text{PE}_g = -mg \mid h \mid$ to show the minus sign clearly. Thus,

$$-\Delta \text{PE}_g = \Delta \text{KE} \tag{7.34}$$

becomes

$$mg \mid h \mid = \frac{1}{2}mv^2. \tag{7.35}$$

Solving for v, we find that mass cancels and that

$$v = \sqrt{2g \mid h \mid}. \tag{7.36}$$

Substituting known values,

$$\begin{aligned} v &= \sqrt{2(9.80 \text{ m/s}^2)(20.0 \text{ m})} \\ &= 19.8 \text{ m/s}. \end{aligned} \tag{7.37}$$

Solution for (b)

Again $-\Delta \text{PE}_g = \Delta \text{KE}$. In this case there is initial kinetic energy, so $\Delta \text{KE} = \frac{1}{2}mv^2 - \frac{1}{2}mv_0^2$. Thus,

$$mg \mid h \mid = \frac{1}{2}mv^2 - \frac{1}{2}mv_0^2. \tag{7.38}$$

Rearranging gives

$$\frac{1}{2}mv^2 = mg \mid h \mid + \frac{1}{2}mv_0^2. \tag{7.39}$$

This means that the final kinetic energy is the sum of the initial kinetic energy and the gravitational potential energy. Mass again cancels, and

$$v = \sqrt{2g \mid h \mid + v_0^2}. \tag{7.40}$$

This equation is very similar to the kinematics equation $v = \sqrt{v_0^2 + 2ad}$, but it is more general—the kinematics equation is valid only for constant acceleration, whereas our equation above is valid for any path regardless of whether the object moves with a constant acceleration. Now, substituting known values gives

$$\begin{aligned} v &= \sqrt{2(9.80 \text{ m/s}^2)(20.0 \text{ m}) + (5.00 \text{ m/s})^2} \\ &= 20.4 \text{ m/s}. \end{aligned} \tag{7.41}$$

Discussion and Implications

First, note that mass cancels. This is quite consistent with observations made in **Falling Objects** that all objects fall at the same rate if friction is negligible. Second, only the speed of the roller coaster is considered; there is no information about its direction at any point. This reveals another general truth. When friction is negligible, the speed of a falling body depends only on its initial speed and height, and not on its mass or the path taken. For example, the roller coaster will have the same final speed whether it falls 20.0 m straight down or takes a more complicated path like the one in the figure. Third, and perhaps unexpectedly, the final speed in part (b) is greater than in part (a), but by far less than 5.00 m/s. Finally, note that speed can be found at *any* height along the way by simply using the appropriate value of h at the point of interest.

We have seen that work done by or against the gravitational force depends only on the starting and ending points, and not on the path between, allowing us to define the simplifying concept of gravitational potential energy. We can do the same thing for a few other forces, and we will see that this leads to a formal definition of the law of conservation of energy.

Making Connections: Take-Home Investigation—Converting Potential to Kinetic Energy

One can study the conversion of gravitational potential energy into kinetic energy in this experiment. On a smooth, level surface, use a ruler of the kind that has a groove running along its length and a book to make an incline (see **Figure 7.9**).

Place a marble at the 10-cm position on the ruler and let it roll down the ruler. When it hits the level surface, measure the time it takes to roll one meter. Now place the marble at the 20-cm and the 30-cm positions and again measure the times it takes to roll 1 m on the level surface. Find the velocity of the marble on the level surface for all three positions. Plot velocity squared versus the distance traveled by the marble. What is the shape of each plot? If the shape is a straight line, the plot shows that the marble's kinetic energy at the bottom is proportional to its potential energy at the release point.

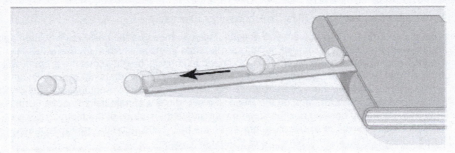

Figure 7.9 A marble rolls down a ruler, and its speed on the level surface is measured.

7.4 Conservative Forces and Potential Energy

Learning Objectives

By the end of this section, you will be able to:

- Define conservative force, potential energy, and mechanical energy.
- Explain the potential energy of a spring in terms of its compression when Hooke's law applies.
- Use the work-energy theorem to show how having only conservative forces leads to conservation of mechanical energy.

The information presented in this section supports the following AP® learning objectives and science practices:

- **4.C.1.1** The student is able to calculate the total energy of a system and justify the mathematical routines used in the calculation of component types of energy within the system whose sum is the total energy. **(S.P. 1.4, 2.1, 2.2)**
- **4.C.2.1** The student is able to make predictions about the changes in the mechanical energy of a system when a component of an external force acts parallel or antiparallel to the direction of the displacement of the center of mass. **(S.P. 6.4)**
- **5.B.1.1** The student is able to set up a representation or model showing that a single object can only have kinetic energy and use information about that object to calculate its kinetic energy. **(S.P. 1.4, 2.2)**
- **5.B.1.2** The student is able to translate between a representation of a single object, which can only have kinetic energy, and a system that includes the object, which may have both kinetic and potential energies. **(S.P. 1.5)**
- **5.B.3.1** The student is able to describe and make qualitative and/or quantitative predictions about everyday examples of systems with internal potential energy. **(S.P. 2.2, 6.4, 7.2)**
- **5.B.3.2** The student is able to make quantitative calculations of the internal potential energy of a system from a description or diagram of that system. **(S.P. 1.4, 2.2)**
- **5.B.3.3** The student is able to apply mathematical reasoning to create a description of the internal potential energy of a system from a description or diagram of the objects and interactions in that system. **(S.P. 1.4, 2.2)**

Potential Energy and Conservative Forces

Work is done by a force, and some forces, such as weight, have special characteristics. A **conservative force** is one, like the gravitational force, for which work done by or against it depends only on the starting and ending points of a motion and not on the path taken. We can define a **potential energy** (PE) for any conservative force, just as we did for the gravitational force. For example, when you wind up a toy, an egg timer, or an old-fashioned watch, you do work against its spring and store energy in it. (We treat these springs as ideal, in that we assume there is no friction and no production of thermal energy.) This stored energy is recoverable as work, and it is useful to think of it as potential energy contained in the spring. Indeed, the reason that the spring has this characteristic is that its force is *conservative*. That is, a conservative force results in stored or potential energy. Gravitational potential energy is one example, as is the energy stored in a spring. We will also see how conservative forces are related to the conservation of energy.

Potential Energy and Conservative Forces

Potential energy is the energy a system has due to position, shape, or configuration. It is stored energy that is completely recoverable.

A conservative force is one for which work done by or against it depends only on the starting and ending points of a motion and not on the path taken.

We can define a potential energy (PE) for any conservative force. The work done against a conservative force to reach a final configuration depends on the configuration, not the path followed, and is the potential energy added.

Real World Connections: Energy of a Bowling Ball

How much energy does a bowling ball have? (Just think about it for a minute.)

If you are thinking that you need more information, you're right. If we can measure the ball's velocity, then determining its kinetic energy is simple. Note that this does require defining a reference frame in which to measure the velocity. Determining the ball's potential energy also requires more information. You need to know its height above the ground, which requires a reference frame of the ground. Without the ground—in other words, Earth—the ball does not classically have potential energy. Potential energy comes from the interaction between the ball and the ground. Another way of thinking about this is to compare the ball's potential energy on Earth and on the Moon. A bowling ball a certain height above Earth is going to have more potential energy than the same bowling ball the same height above the surface of the Moon, because Earth has greater mass than the Moon and therefore exerts more gravity on the ball. Thus, potential energy requires a system of at least two objects, or an object with an internal structure of at least two parts.

Potential Energy of a Spring

First, let us obtain an expression for the potential energy stored in a spring (PE_s). We calculate the work done to stretch or compress a spring that obeys Hooke's law. (Hooke's law was examined in **Elasticity: Stress and Strain**, and states that the magnitude of force F on the spring and the resulting deformation ΔL are proportional, $F = k\Delta L$.) (See **Figure 7.10**.) For our spring, we will replace ΔL (the amount of deformation produced by a force F) by the distance x that the spring is stretched or compressed along its length. So the force needed to stretch the spring has magnitude $F = kx$, where k is the spring's force constant. The force increases linearly from 0 at the start to kx in the fully stretched position. The average force is $kx/2$. Thus the work done in stretching or compressing the spring is $W_s = Fd = \left(\frac{kx}{2}\right)x = \frac{1}{2}kx^2$. Alternatively, we noted in **Kinetic Energy and the Work-Energy Theorem** that the area under a graph of F vs. x is the work done by the force. In **Figure 7.10**(c) we see that this area is also $\frac{1}{2}kx^2$. We therefore define the **potential energy of a spring**, PE_s, to be

$$PE_s = \frac{1}{2}kx^2, \tag{7.42}$$

where k is the spring's force constant and x is the displacement from its undeformed position. The potential energy represents the work done *on* the spring and the energy stored in it as a result of stretching or compressing it a distance x. The potential energy of the spring PE_s does not depend on the path taken; it depends only on the stretch or squeeze x in the final configuration.

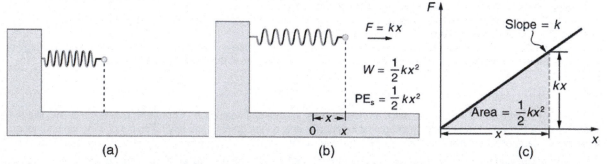

Figure 7.10 (a) An undeformed spring has no PE_s stored in it. (b) The force needed to stretch (or compress) the spring a distance x has a magnitude $F = kx$, and the work done to stretch (or compress) it is $\frac{1}{2}kx^2$. Because the force is conservative, this work is stored as potential energy (PE_s) in the spring, and it can be fully recovered. (c) A graph of F vs. x has a slope of k, and the area under the graph is $\frac{1}{2}kx^2$. Thus the work done or potential energy stored is $\frac{1}{2}kx^2$.

The equation $PE_s = \frac{1}{2}kx^2$ has general validity beyond the special case for which it was derived. Potential energy can be stored in any elastic medium by deforming it. Indeed, the general definition of **potential energy** is energy due to position, shape, or configuration. For shape or position deformations, stored energy is $\text{PE}_s = \frac{1}{2}kx^2$, where k is the force constant of the particular system and x is its deformation. Another example is seen in **Figure 7.11** for a guitar string.

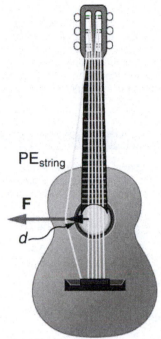

PE$_{\text{string}}$

F

d

Figure 7.11 Work is done to deform the guitar string, giving it potential energy. When released, the potential energy is converted to kinetic energy and back to potential as the string oscillates back and forth. A very small fraction is dissipated as sound energy, slowly removing energy from the string.

Conservation of Mechanical Energy

Let us now consider what form the work-energy theorem takes when only conservative forces are involved. This will lead us to the conservation of energy principle. The work-energy theorem states that the net work done by all forces acting on a system equals its change in kinetic energy. In equation form, this is

$$W_{\text{net}} = \frac{1}{2}mv^2 - \frac{1}{2}mv_0{}^2 = \Delta\text{KE}. \tag{7.43}$$

If only conservative forces act, then

$$W_{\text{net}} = W_{\text{c}}, \tag{7.44}$$

where W_{c} is the total work done by all conservative forces. Thus,

$$W_{\text{c}} = \Delta\text{KE}. \tag{7.45}$$

Now, if the conservative force, such as the gravitational force or a spring force, does work, the system loses potential energy. That is, $W_{\text{c}} = -\Delta\text{PE}$. Therefore,

$$-\Delta\text{PE} = \Delta\text{KE} \tag{7.46}$$

or

$$\Delta\text{KE} + \Delta\text{PE} = 0. \tag{7.47}$$

This equation means that the total kinetic and potential energy is constant for any process involving only conservative forces. That is,

$$\left.\begin{array}{l} \text{KE} + \text{PE} = \text{constant} \\ \text{or} \\ \text{KE}_i + \text{PE}_i = \text{KE}_f + \text{PE}_f \end{array}\right\}(\text{conservative forces only}), \tag{7.48}$$

where i and f denote initial and final values. This equation is a form of the work-energy theorem for conservative forces; it is known as the **conservation of mechanical energy** principle. Remember that this applies to the extent that all the forces are conservative, so that friction is negligible. The total kinetic plus potential energy of a system is defined to be its **mechanical energy**, $(\text{KE} + \text{PE})$. In a system that experiences only conservative forces, there is a potential energy associated with each force, and the energy only changes form between KE and the various types of PE, with the total energy remaining constant.

The internal energy of a system is the sum of the kinetic energies of all of its elements, plus the potential energy due to all of the interactions due to conservative forces between all of the elements.

Real World Connections

Consider a wind-up toy, such as a car. It uses a spring system to store energy. The amount of energy stored depends only on how many times it is wound, not how quickly or slowly the winding happens. Similarly, a dart gun using compressed air stores energy in its internal structure. In this case, the energy stored inside depends only on how many times it is pumped, not how quickly or slowly the pumping is done. The total energy put into the system, whether through winding or pumping, is equal to the total energy conserved in the system (minus any energy loss in the system due to interactions between its parts, such as air leaks in the dart gun). Since the internal energy of the system is conserved, you can calculate the amount of stored energy by measuring the kinetic energy of the system (the moving car or dart) when the potential energy is released.

Example 7.8 Using Conservation of Mechanical Energy to Calculate the Speed of a Toy Car

A 0.100-kg toy car is propelled by a compressed spring, as shown in **Figure 7.12**. The car follows a track that rises 0.180 m above the starting point. The spring is compressed 4.00 cm and has a force constant of 250.0 N/m. Assuming work done by friction to be negligible, find (a) how fast the car is going before it starts up the slope and (b) how fast it is going at the top of the slope.

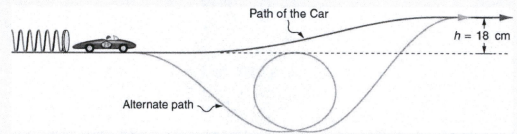

Figure 7.12 A toy car is pushed by a compressed spring and coasts up a slope. Assuming negligible friction, the potential energy in the spring is first completely converted to kinetic energy, and then to a combination of kinetic and gravitational potential energy as the car rises. The details of the path are unimportant because all forces are conservative—the car would have the same final speed if it took the alternate path shown.

Strategy

The spring force and the gravitational force are conservative forces, so conservation of mechanical energy can be used. Thus,

$$\mathrm{KE_i + PE_i = KE_f + PE_f} \tag{7.49}$$

or

$$\tfrac{1}{2}mv_i^2 + mgh_i + \tfrac{1}{2}kx_i^2 = \tfrac{1}{2}mv_f^2 + mgh_f + \tfrac{1}{2}kx_f^2, \tag{7.50}$$

where h is the height (vertical position) and x is the compression of the spring. This general statement looks complex but becomes much simpler when we start considering specific situations. First, we must identify the initial and final conditions in a problem; then, we enter them into the last equation to solve for an unknown.

Solution for (a)

This part of the problem is limited to conditions just before the car is released and just after it leaves the spring. Take the initial height to be zero, so that both h_i and h_f are zero. Furthermore, the initial speed v_i is zero and the final compression of the spring x_f is zero, and so several terms in the conservation of mechanical energy equation are zero and it simplifies to

$$\tfrac{1}{2}kx_i^2 = \tfrac{1}{2}mv_f^2. \tag{7.51}$$

In other words, the initial potential energy in the spring is converted completely to kinetic energy in the absence of friction. Solving for the final speed and entering known values yields

$$
\begin{aligned}
v_f &= \sqrt{\tfrac{k}{m}}x_i \\
&= \sqrt{\tfrac{250.0\ \mathrm{N/m}}{0.100\ \mathrm{kg}}}(0.0400\ \mathrm{m}) \\
&= 2.00\ \mathrm{m/s}.
\end{aligned}
\tag{7.52}
$$

Solution for (b)

One method of finding the speed at the top of the slope is to consider conditions just before the car is released and just after it reaches the top of the slope, completely ignoring everything in between. Doing the same type of analysis to find which terms are zero, the conservation of mechanical energy becomes

$$\frac{1}{2}kx_i^2 = \frac{1}{2}mv_f^2 + mgh_f.$$ (7.53)

This form of the equation means that the spring's initial potential energy is converted partly to gravitational potential energy and partly to kinetic energy. The final speed at the top of the slope will be less than at the bottom. Solving for v_f and

substituting known values gives

(7.54)

$$
\begin{aligned}
v_f &= \sqrt{\frac{kx_i^2}{m} - 2gh_f} \\
&= \sqrt{\left(\frac{250.0 \text{ N/m}}{0.100 \text{ kg}}\right)(0.0400 \text{ m})^2 - 2(9.80 \text{ m/s}^2)(0.180 \text{ m})} \\
&= 0.687 \text{ m/s.}
\end{aligned}
$$

Discussion

Another way to solve this problem is to realize that the car's kinetic energy before it goes up the slope is converted partly to potential energy—that is, to take the final conditions in part (a) to be the initial conditions in part (b).

Applying the Science Practices: Potential Energy in a Spring

Suppose you are running an experiment in which two 250 g carts connected by a spring (with spring constant 120 N/m) are run into a solid block, and the compression of the spring is measured. In one run of this experiment, the spring was measured to compress from its rest length of 5.0 cm to a minimum length of 2.0 cm. What was the potential energy stored in this system?

Answer

Note that the change in length of the spring is 3.0 cm. Hence we can apply Equation 7.42 to find that the potential energy is $PE = (1/2)(120 \text{ N/m})(0.030 \text{ m})^2 = 0.0541$ J.

Note that, for conservative forces, we do not directly calculate the work they do; rather, we consider their effects through their corresponding potential energies, just as we did in **Example 7.8**. Note also that we do not consider details of the path taken—only the starting and ending points are important (as long as the path is not impossible). This assumption is usually a tremendous simplification, because the path may be complicated and forces may vary along the way.

PhET Explorations: Energy Skate Park

Learn about conservation of energy with a skater dude! Build tracks, ramps and jumps for the skater and view the kinetic energy, potential energy and friction as he moves. You can also take the skater to different planets or even space!

Figure 7.13 Energy Skate Park (http://cnx.org/content/m55076/1.5/energy-skate-park_en.jar)

7.5 Nonconservative Forces

Learning Objectives
By the end of this section, you will be able to: • Define nonconservative forces and explain how they affect mechanical energy. • Show how the principle of conservation of energy can be applied by treating the conservative forces in terms of their potential energies and any nonconservative forces in terms of the work they do. The information presented in this section supports the following AP® learning objectives and science practices: • **4.C.1.2** The student is able to predict changes in the total energy of a system due to changes in position and speed of

objects or frictional interactions within the system. **(S.P. 6.4)**
- **4.C.2.1** The student is able to make predictions about the changes in the mechanical energy of a system when a component of an external force acts parallel or antiparallel to the direction of the displacement of the center of mass. **(S.P. 6.4)**

Nonconservative Forces and Friction

Forces are either conservative or nonconservative. Conservative forces were discussed in **Conservative Forces and Potential Energy**. A **nonconservative force** is one for which work depends on the path taken. Friction is a good example of a nonconservative force. As illustrated in **Figure 7.14**, work done against friction depends on the length of the path between the starting and ending points. Because of this dependence on path, there is no potential energy associated with nonconservative forces. An important characteristic is that the work done by a nonconservative force *adds or removes mechanical energy from a system*. **Friction**, for example, creates **thermal energy** that dissipates, removing energy from the system. Furthermore, even if the thermal energy is retained or captured, it cannot be fully converted back to work, so it is lost or not recoverable in that sense as well.

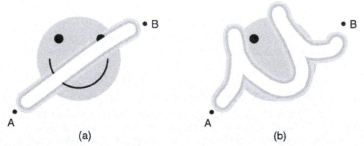

Figure 7.14 The amount of the happy face erased depends on the path taken by the eraser between points A and B, as does the work done against friction. Less work is done and less of the face is erased for the path in (a) than for the path in (b). The force here is friction, and most of the work goes into thermal energy that subsequently leaves the system (the happy face plus the eraser). The energy expended cannot be fully recovered.

How Nonconservative Forces Affect Mechanical Energy

Mechanical energy *may* not be conserved when nonconservative forces act. For example, when a car is brought to a stop by friction on level ground, it loses kinetic energy, which is dissipated as thermal energy, reducing its mechanical energy. **Figure 7.15** compares the effects of conservative and nonconservative forces. We often choose to understand simpler systems such as that described in **Figure 7.15**(a) first before studying more complicated systems as in **Figure 7.15**(b).

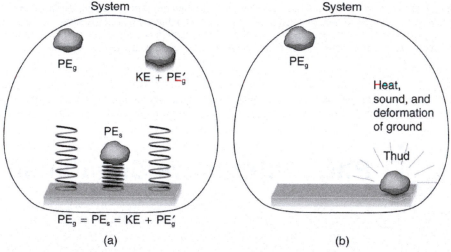

Figure 7.15 Comparison of the effects of conservative and nonconservative forces on the mechanical energy of a system. (a) A system with only conservative forces. When a rock is dropped onto a spring, its mechanical energy remains constant (neglecting air resistance) because the force in the spring is conservative. The spring can propel the rock back to its original height, where it once again has only potential energy due to gravity. (b) A system with nonconservative forces. When the same rock is dropped onto the ground, it is stopped by nonconservative forces that dissipate its mechanical energy as thermal energy, sound, and surface distortion. The rock has lost mechanical energy.

How the Work-Energy Theorem Applies

Now let us consider what form the work-energy theorem takes when both conservative and nonconservative forces act. We will see that the work done by nonconservative forces equals the change in the mechanical energy of a system. As noted in **Kinetic Energy and the Work-Energy Theorem**, the work-energy theorem states that the net work on a system equals the change in its kinetic energy, or $W_{\text{net}} = \Delta \text{KE}$. The net work is the sum of the work by nonconservative forces plus the work by conservative forces. That is,

$$W_{net} = W_{nc} + W_c, \tag{7.55}$$

so that

$$W_{nc} + W_c = \Delta KE, \tag{7.56}$$

where W_{nc} is the total work done by all nonconservative forces and W_c is the total work done by all conservative forces.

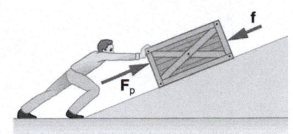

Figure 7.16 A person pushes a crate up a ramp, doing work on the crate. Friction and gravitational force (not shown) also do work on the crate; both forces oppose the person's push. As the crate is pushed up the ramp, it gains mechanical energy, implying that the work done by the person is greater than the work done by friction.

Consider **Figure 7.16**, in which a person pushes a crate up a ramp and is opposed by friction. As in the previous section, we note that work done by a conservative force comes from a loss of gravitational potential energy, so that $W_c = -\Delta PE$.

Substituting this equation into the previous one and solving for W_{nc} gives

$$W_{nc} = \Delta KE + \Delta PE. \tag{7.57}$$

This equation means that the total mechanical energy $(KE + PE)$ changes by exactly the amount of work done by nonconservative forces. In **Figure 7.16**, this is the work done by the person minus the work done by friction. So even if energy is not conserved for the system of interest (such as the crate), we know that an equal amount of work was done to cause the change in total mechanical energy.

We rearrange $W_{nc} = \Delta KE + \Delta PE$ to obtain

$$KE_i + PE_i + W_{nc} = KE_f + PE_f. \tag{7.58}$$

This means that the amount of work done by nonconservative forces adds to the mechanical energy of a system. If W_{nc} is positive, then mechanical energy is increased, such as when the person pushes the crate up the ramp in **Figure 7.16**. If W_{nc} is negative, then mechanical energy is decreased, such as when the rock hits the ground in **Figure 7.15**(b). If W_{nc} is zero, then mechanical energy is conserved, and nonconservative forces are balanced. For example, when you push a lawn mower at constant speed on level ground, your work done is removed by the work of friction, and the mower has a constant energy.

Applying Energy Conservation with Nonconservative Forces

When no change in potential energy occurs, applying $KE_i + PE_i + W_{nc} = KE_f + PE_f$ amounts to applying the work-energy theorem by setting the change in kinetic energy to be equal to the net work done on the system, which in the most general case includes both conservative and nonconservative forces. But when seeking instead to find a change in total mechanical energy in situations that involve changes in both potential and kinetic energy, the previous equation $KE_i + PE_i + W_{nc} = KE_f + PE_f$ says that you can start by finding the change in mechanical energy that would have resulted from just the conservative forces, including the potential energy changes, and add to it the work done, with the proper sign, by any nonconservative forces involved.

Example 7.9 Calculating Distance Traveled: How Far a Baseball Player Slides

Consider the situation shown in **Figure 7.17**, where a baseball player slides to a stop on level ground. Using energy considerations, calculate the distance the 65.0-kg baseball player slides, given that his initial speed is 6.00 m/s and the force of friction against him is a constant 450 N.

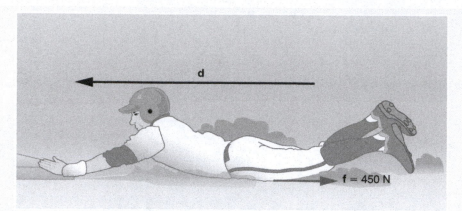

Figure 7.17 The baseball player slides to a stop in a distance d . In the process, friction removes the player's kinetic energy by doing an amount of work fd equal to the initial kinetic energy.

Strategy

Friction stops the player by converting his kinetic energy into other forms, including thermal energy. In terms of the work-energy theorem, the work done by friction, which is negative, is added to the initial kinetic energy to reduce it to zero. The work done by friction is negative, because $\mathbf{f}$ is in the opposite direction of the motion (that is, $\theta = 180°$, and so $\cos \theta = -1$). Thus $W_{\text{nc}} = -fd$. The equation simplifies to

$$\frac{1}{2}mv_{\text{i}}^2 - fd = 0 \tag{7.59}$$

or

$$fd = \frac{1}{2}mv_{\text{i}}^2. \tag{7.60}$$

This equation can now be solved for the distance d .

Solution

Solving the previous equation for d and substituting known values yields

$$\begin{aligned} d &= \frac{mv_{\text{i}}^2}{2f} \\ &= \frac{(65.0 \text{ kg})(6.00 \text{ m/s})^2}{(2)(450 \text{ N})} \\ &= 2.60 \text{ m}. \end{aligned} \tag{7.61}$$

Discussion

The most important point of this example is that the amount of nonconservative work equals the change in mechanical energy. For example, you must work harder to stop a truck, with its large mechanical energy, than to stop a mosquito.

Example 7.10 Calculating Distance Traveled: Sliding Up an Incline

Suppose that the player from **Example 7.9** is running up a hill having a $5.00°$ incline upward with a surface similar to that in the baseball stadium. The player slides with the same initial speed, and the frictional force is still 450 N. Determine how far he slides.

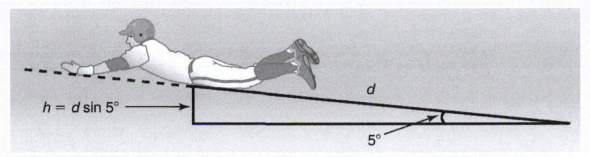

Figure 7.18 The same baseball player slides to a stop on a $5.00°$ slope.

Strategy

In this case, the work done by the nonconservative friction force on the player reduces the mechanical energy he has from his kinetic energy at zero height, to the final mechanical energy he has by moving through distance d to reach height h along the hill, with $h = d \sin 5.00°$. This is expressed by the equation

$$KE + PE_i + W_{nc} = KE_f + PE_f. \tag{7.62}$$

Solution

The work done by friction is again $W_{nc} = -fd$; initially the potential energy is $PE_i = mg \cdot 0 = 0$ and the kinetic energy

is $KE_i = \frac{1}{2}mv_i^2$; the final energy contributions are $KE_f = 0$ for the kinetic energy and $PE_f = mgh = mgd \sin \theta$ for the potential energy.

Substituting these values gives

$$\frac{1}{2}mv_i^2 + 0 + \left(-fd\right) = 0 + mgd \sin \theta. \tag{7.63}$$

Solve this for d to obtain

$$\begin{aligned} d &= \frac{\left(\frac{1}{2}\right)mv_i^2}{f + mg \sin \theta} \\ &= \frac{(0.5)(65.0 \text{ kg})(6.00 \text{ m/s})^2}{450 \text{ N} + (65.0 \text{ kg})(9.80 \text{ m/s}^2) \sin (5.00°)} \\ &= 2.31 \text{ m.} \end{aligned} \tag{7.64}$$

Discussion

As might have been expected, the player slides a shorter distance by sliding uphill. Note that the problem could also have been solved in terms of the forces directly and the work energy theorem, instead of using the potential energy. This method would have required combining the normal force and force of gravity vectors, which no longer cancel each other because they point in different directions, and friction, to find the net force. You could then use the net force and the net work to find the distance d that reduces the kinetic energy to zero. By applying conservation of energy and using the potential energy instead, we need only consider the gravitational potential energy mgh, without combining and resolving force vectors. This simplifies the solution considerably.

Making Connections: Take-Home Investigation—Determining Friction from the Stopping Distance

This experiment involves the conversion of gravitational potential energy into thermal energy. Use the ruler, book, and marble from **Making Connections: Take-Home Investigation—Converting Potential to Kinetic Energy**. In addition, you will need a foam cup with a small hole in the side, as shown in **Figure 7.19**. From the 10-cm position on the ruler, let the marble roll into the cup positioned at the bottom of the ruler. Measure the distance d the cup moves before stopping. What forces caused it to stop? What happened to the kinetic energy of the marble at the bottom of the ruler? Next, place the marble at the 20-cm and the 30-cm positions and again measure the distance the cup moves after the marble enters it. Plot the distance the cup moves versus the initial marble position on the ruler. Is this relationship linear?

With some simple assumptions, you can use these data to find the coefficient of kinetic friction μ_k of the cup on the table.

The force of friction f on the cup is $\mu_k N$, where the normal force N is just the weight of the cup plus the marble. The normal force and force of gravity do no work because they are perpendicular to the displacement of the cup, which moves horizontally. The work done by friction is fd. You will need the mass of the marble as well to calculate its initial kinetic energy.

It is interesting to do the above experiment also with a steel marble (or ball bearing). Releasing it from the same positions on the ruler as you did with the glass marble, is the velocity of this steel marble the same as the velocity of the marble at the bottom of the ruler? Is the distance the cup moves proportional to the mass of the steel and glass marbles?

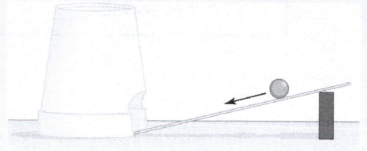

Figure 7.19 Rolling a marble down a ruler into a foam cup.

PhET Explorations: The Ramp

Explore forces, energy and work as you push household objects up and down a ramp. Lower and raise the ramp to see how the angle of inclination affects the parallel forces acting on the file cabinet. Graphs show forces, energy and work.

Figure 7.20 The Ramp (http://cnx.org/content/m55047/1.5/the-ramp_en.jar)

7.6 Conservation of Energy

Learning Objectives
By the end of this section, you will be able to: • Explain the law of the conservation of energy. • Describe some of the many forms of energy. • Define efficiency of an energy conversion process as the fraction left as useful energy or work, rather than being transformed, for example, into thermal energy. The information presented in this section supports the following AP® learning objectives and science practices: • **4.C.1.2** The student is able to predict changes in the total energy of a system due to changes in position and speed of objects or frictional interactions within the system. **(S.P. 6.4)** • **4.C.2.1** The student is able to make predictions about the changes in the mechanical energy of a system when a component of an external force acts parallel or antiparallel to the direction of the displacement of the center of mass. **(S.P. 6.4)** • **4.C.2.2** The student is able to apply the concepts of conservation of energy and the work-energy theorem to determine qualitatively and/or quantitatively that work done on a two-object system in linear motion will change the kinetic energy of the center of mass of the system, the potential energy of the systems, and/or the internal energy of the system. **(S.P. 1.4, 2.2, 7.2)** • **5.A.2.1** The student is able to define open and closed systems for everyday situations and apply conservation concepts for energy, charge, and linear momentum to those situations. **(S.P. 6.4, 7.2)** • **5.B.5.4** The student is able to make claims about the interaction between a system and its environment in which the environment exerts a force on the system, thus doing work on the system and changing the energy of the system (kinetic energy plus potential energy). **(S.P. 6.4, 7.2)** • **5.B.5.5** The student is able to predict and calculate the energy transfer to (i.e., the work done on) an object or system from information about a force exerted on the object or system through a distance. **(S.P. 2.2, 6.4)**

Law of Conservation of Energy

Energy, as we have noted, is conserved, making it one of the most important physical quantities in nature. The **law of**

conservation of energy can be stated as follows:

Total energy is constant in any process. It may change in form or be transferred from one system to another, but the total remains the same.

We have explored some forms of energy and some ways it can be transferred from one system to another. This exploration led to the definition of two major types of energy—mechanical energy $(KE + PE)$ and energy transferred via work done by nonconservative forces (W_{nc}). But energy takes *many* other forms, manifesting itself in *many* different ways, and we need to be able to deal with all of these before we can write an equation for the above general statement of the conservation of energy.

Other Forms of Energy than Mechanical Energy

At this point, we deal with all other forms of energy by lumping them into a single group called other energy (OE). Then we can state the conservation of energy in equation form as

$$KE_i + PE_i + W_{nc} + OE_i = KE_f + PE_f + OE_f. \tag{7.65}$$

All types of energy and work can be included in this very general statement of conservation of energy. Kinetic energy is KE, work done by a conservative force is represented by PE, work done by nonconservative forces is W_{nc}, and all other energies are included as OE. This equation applies to all previous examples; in those situations OE was constant, and so it subtracted out and was not directly considered.

Making Connections: Usefulness of the Energy Conservation Principle

The fact that energy is conserved and has many forms makes it very important. You will find that energy is discussed in many contexts, because it is involved in all processes. It will also become apparent that many situations are best understood in terms of energy and that problems are often most easily conceptualized and solved by considering energy.

When does OE play a role? One example occurs when a person eats. Food is oxidized with the release of carbon dioxide, water, and energy. Some of this chemical energy is converted to kinetic energy when the person moves, to potential energy when the person changes altitude, and to thermal energy (another form of OE).

Some of the Many Forms of Energy

What are some other forms of energy? You can probably name a number of forms of energy not yet discussed. Many of these will be covered in later chapters, but let us detail a few here. **Electrical energy** is a common form that is converted to many other forms and does work in a wide range of practical situations. Fuels, such as gasoline and food, carry **chemical energy** that can be transferred to a system through oxidation. Chemical fuel can also produce electrical energy, such as in batteries. Batteries can in turn produce light, which is a very pure form of energy. Most energy sources on Earth are in fact stored energy from the energy we receive from the Sun. We sometimes refer to this as **radiant energy**, or electromagnetic radiation, which includes visible light, infrared, and ultraviolet radiation. **Nuclear energy** comes from processes that convert measurable amounts of mass into energy. Nuclear energy is transformed into the energy of sunlight, into electrical energy in power plants, and into the energy of the heat transfer and blast in weapons. Atoms and molecules inside all objects are in random motion. This internal mechanical energy from the random motions is called **thermal energy**, because it is related to the temperature of the object. These and all other forms of energy can be converted into one another and can do work.

Real World Connections: Open or Closed System?

Consider whether the following systems are open or closed: a car, a spring-operated dart gun, and the system shown in **Figure 7.15(a)**.

A car is not a closed system. You add energy in the form of more gas in the tank (or charging the batteries), and energy is lost due to air resistance and friction.

A spring-operated dart gun is not a closed system. You have to initially compress the spring. Once that has been done, however, the dart gun and dart can be treated as a closed system. All of the energy remains in the system consisting of these two objects.

Figure 7.15(a) is an example of a closed system, once it has been started. All of the energy in the system remains there; none is brought in from outside or leaves.

Table 7.1 gives the amount of energy stored, used, or released from various objects and in various phenomena. The range of energies and the variety of types and situations is impressive.

Problem-Solving Strategies for Energy

You will find the following problem-solving strategies useful whenever you deal with energy. The strategies help in organizing and reinforcing energy concepts. In fact, they are used in the examples presented in this chapter. The familiar general problem-solving strategies presented earlier—involving identifying physical principles, knowns, and unknowns, checking units, and so on—continue to be relevant here.

Step 1. Determine the system of interest and identify what information is given and what quantity is to be calculated. A sketch will help.

Step 2. Examine all the forces involved and determine whether you know or are given the potential energy from the work done by the forces. Then use step 3 or step 4.

Step 3. If you know the potential energies for the forces that enter into the problem, then forces are all conservative, and you can apply conservation of mechanical energy simply in terms of potential and kinetic energy. The equation expressing conservation of energy is

$$KE_i + PE_i = KE_f + PE_f. \tag{7.66}$$

Step 4. If you know the potential energy for only some of the forces, possibly because some of them are nonconservative and do not have a potential energy, or if there are other energies that are not easily treated in terms of force and work, then the conservation of energy law in its most general form must be used.

$$KE_i + PE_i + W_{nc} + OE_i = KE_f + PE_f + OE_f. \tag{7.67}$$

In most problems, one or more of the terms is zero, simplifying its solution. Do not calculate W_c, the work done by conservative forces; it is already incorporated in the PE terms.

Step 5. You have already identified the types of work and energy involved (in step 2). Before solving for the unknown, *eliminate terms wherever possible* to simplify the algebra. For example, choose $h = 0$ at either the initial or final point, so that PE_g is zero there. Then solve for the unknown in the customary manner.

Step 6. *Check the answer to see if it is reasonable.* Once you have solved a problem, reexamine the forms of work and energy to see if you have set up the conservation of energy equation correctly. For example, work done against friction should be negative, potential energy at the bottom of a hill should be less than that at the top, and so on. Also check to see that the numerical value obtained is reasonable. For example, the final speed of a skateboarder who coasts down a 3-m-high ramp could reasonably be 20 km/h, but *not* 80 km/h.

Transformation of Energy

The transformation of energy from one form into others is happening all the time. The chemical energy in food is converted into thermal energy through metabolism; light energy is converted into chemical energy through photosynthesis. In a larger example, the chemical energy contained in coal is converted into thermal energy as it burns to turn water into steam in a boiler. This thermal energy in the steam in turn is converted to mechanical energy as it spins a turbine, which is connected to a generator to produce electrical energy. (In all of these examples, not all of the initial energy is converted into the forms mentioned. This important point is discussed later in this section.)

Another example of energy conversion occurs in a solar cell. Sunlight impinging on a solar cell (see **Figure 7.21**) produces electricity, which in turn can be used to run an electric motor. Energy is converted from the primary source of solar energy into electrical energy and then into mechanical energy.

Figure 7.21 Solar energy is converted into electrical energy by solar cells, which is used to run a motor in this solar-power aircraft. (credit: NASA)

Table 7.1 Energy of Various Objects and Phenomena

Object/phenomenon	Energy in joules
Big Bang	10^{68}
Energy released in a supernova	10^{44}
Fusion of all the hydrogen in Earth's oceans	10^{34}
Annual world energy use	4×10^{20}
Large fusion bomb (9 megaton)	3.8×10^{16}
1 kg hydrogen (fusion to helium)	6.4×10^{14}
1 kg uranium (nuclear fission)	8.0×10^{13}
Hiroshima-size fission bomb (10 kiloton)	4.2×10^{13}
90,000-ton aircraft carrier at 30 knots	1.1×10^{10}
1 barrel crude oil	5.9×10^{9}
1 ton TNT	4.2×10^{9}
1 gallon of gasoline	1.2×10^{8}
Daily home electricity use (developed countries)	7×10^{7}
Daily adult food intake (recommended)	1.2×10^{7}
1000-kg car at 90 km/h	3.1×10^{5}
1 g fat (9.3 kcal)	3.9×10^{4}
ATP hydrolysis reaction	3.2×10^{4}
1 g carbohydrate (4.1 kcal)	1.7×10^{4}
1 g protein (4.1 kcal)	1.7×10^{4}
Tennis ball at 100 km/h	22
Mosquito $\left(10^{-2} \text{ g at } 0.5 \text{ m/s}\right)$	1.3×10^{-6}
Single electron in a TV tube beam	4.0×10^{-15}
Energy to break one DNA strand	10^{-19}

Efficiency

Even though energy is conserved in an energy conversion process, the output of *useful energy* or work will be less than the energy input. The **efficiency** Eff of an energy conversion process is defined as

$$\text{Efficien} \quad (Eff) = \frac{\text{useful energy or work output}}{\text{total energy input}} = \frac{W_{\text{out}}}{E_{\text{in}}}. \tag{7.68}$$

Table 7.2 lists some efficiencies of mechanical devices and human activities. In a coal-fired power plant, for example, about 40% of the chemical energy in the coal becomes useful electrical energy. The other 60% transforms into other (perhaps less useful) energy forms, such as thermal energy, which is then released to the environment through combustion gases and cooling towers.

Table 7.2 Efficiency of the Human Body and Mechanical Devices

Activity/device	Efficiency (%)[1]
Cycling and climbing	20
Swimming, surface	2
Swimming, submerged	4
Shoveling	3
Weightlifting	9
Steam engine	17
Gasoline engine	30
Diesel engine	35
Nuclear power plant	35
Coal power plant	42
Electric motor	98
Compact fluorescent light	20
Gas heater (residential)	90
Solar cell	10

PhET Explorations: Masses and Springs

A realistic mass and spring laboratory. Hang masses from springs and adjust the spring stiffness and damping. You can even slow time. Transport the lab to different planets. A chart shows the kinetic, potential, and thermal energies for each spring.

Figure 7.22 Masses and Springs (http://cnx.org/content/m55049/1.4/mass-spring-lab_en.jar)

7.7 Power

Learning Objectives

By the end of this section, you will be able to:

- Calculate power by calculating changes in energy over time.
- Examine power consumption and calculations of the cost of energy consumed.

What is Power?

Power—the word conjures up many images: a professional football player muscling aside his opponent, a dragster roaring away from the starting line, a volcano blowing its lava into the atmosphere, or a rocket blasting off, as in **Figure 7.23**.

---

1. Representative values

Figure 7.23 This powerful rocket on the Space Shuttle *Endeavor* did work and consumed energy at a very high rate. (credit: NASA)

These images of power have in common the rapid performance of work, consistent with the scientific definition of **power** (P) as the rate at which work is done.

Power

Power is the rate at which work is done.

$$P = \frac{W}{t}$$

(7.69)

The SI unit for power is the **watt** (W), where 1 watt equals 1 joule/second $(1 \text{ W} = 1 \text{ J/s})$.

Because work is energy transfer, power is also the rate at which energy is expended. A 60-W light bulb, for example, expends 60 J of energy per second. Great power means a large amount of work or energy developed in a short time. For example, when a powerful car accelerates rapidly, it does a large amount of work and consumes a large amount of fuel in a short time.

Calculating Power from Energy

Example 7.11 Calculating the Power to Climb Stairs

What is the power output for a 60.0-kg woman who runs up a 3.00 m high flight of stairs in 3.50 s, starting from rest but having a final speed of 2.00 m/s? (See **Figure 7.24**.)

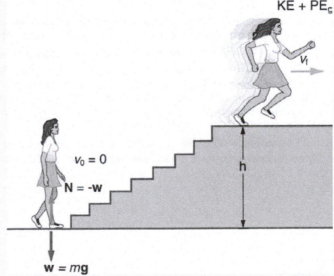

Figure 7.24 When this woman runs upstairs starting from rest, she converts the chemical energy originally from food into kinetic energy and gravitational potential energy. Her power output depends on how fast she does this.

Strategy and Concept

The work going into mechanical energy is $W = \text{KE} + \text{PE}$. At the bottom of the stairs, we take both KE and PE_g as initially zero; thus, $W = \text{KE}_f + \text{PE}_g = \frac{1}{2}mv_f^2 + mgh$, where h is the vertical height of the stairs. Because all terms are given, we can calculate W and then divide it by time to get power.

Solution

Substituting the expression for W into the definition of power given in the previous equation, $P = W/t$ yields

$$P = \frac{W}{t} = \frac{\frac{1}{2}mv_f^2 + mgh}{t}. \tag{7.70}$$

Entering known values yields

$$\begin{aligned} P &= \frac{0.5(60.0 \text{ kg})(2.00 \text{ m/s})^2 + (60.0 \text{ kg})(9.80 \text{ m/s}^2)(3.00 \text{ m})}{3.50 \text{ s}} \\ &= \frac{120 \text{ J} + 1764 \text{ J}}{3.50 \text{ s}} \\ &= 538 \text{ W}. \end{aligned} \tag{7.71}$$

Discussion

The woman does 1764 J of work to move up the stairs compared with only 120 J to increase her kinetic energy; thus, most of her power output is required for climbing rather than accelerating.

It is impressive that this woman's useful power output is slightly less than 1 **horsepower** $(1 \text{ hp} = 746 \text{ W})$! People can generate more than a horsepower with their leg muscles for short periods of time by rapidly converting available blood sugar and oxygen into work output. (A horse can put out 1 hp for hours on end.) Once oxygen is depleted, power output decreases and the person begins to breathe rapidly to obtain oxygen to metabolize more food—this is known as the *aerobic* stage of exercise. If the woman climbed the stairs slowly, then her power output would be much less, although the amount of work done would be the same.

Making Connections: Take-Home Investigation—Measure Your Power Rating

Determine your own power rating by measuring the time it takes you to climb a flight of stairs. We will ignore the gain in kinetic energy, as the above example showed that it was a small portion of the energy gain. Don't expect that your output will be more than about 0.5 hp.

Examples of Power

Examples of power are limited only by the imagination, because there are as many types as there are forms of work and energy. (See **Table 7.3** for some examples.) Sunlight reaching Earth's surface carries a maximum power of about 1.3 kilowatts per square meter (kW/m^2). A tiny fraction of this is retained by Earth over the long term. Our consumption rate of fossil fuels is far greater than the rate at which they are stored, so it is inevitable that they will be depleted. Power implies that energy is transferred, perhaps changing form. It is never possible to change one form completely into another without losing some of it as thermal energy. For example, a 60-W incandescent bulb converts only 5 W of electrical power to light, with 55 W dissipating into thermal energy. Furthermore, the typical electric power plant converts only 35 to 40% of its fuel into electricity. The remainder becomes a huge amount of thermal energy that must be dispersed as heat transfer, as rapidly as it is created. A coal-fired power plant may produce 1000 megawatts; 1 megawatt (MW) is 10^6 W of electric power. But the power plant consumes chemical energy at a rate of about 2500 MW, creating heat transfer to the surroundings at a rate of 1500 MW. (See **Figure 7.25**.)

Figure 7.25 Tremendous amounts of electric power are generated by coal-fired power plants such as this one in China, but an even larger amount of power goes into heat transfer to the surroundings. The large cooling towers here are needed to transfer heat as rapidly as it is produced. The transfer of heat is not unique to coal plants but is an unavoidable consequence of generating electric power from any fuel—nuclear, coal, oil, natural gas, or the like. (credit: Kleinolive, Wikimedia Commons)

Table 7.3 Power Output or Consumption

Object or Phenomenon	Power in Watts
Supernova (at peak)	5×10^{37}
Milky Way galaxy	10^{37}
Crab Nebula pulsar	10^{28}
The Sun	4×10^{26}
Volcanic eruption (maximum)	4×10^{15}
Lightning bolt	2×10^{12}
Nuclear power plant (total electric and heat transfer)	3×10^{9}
Aircraft carrier (total useful and heat transfer)	10^{8}
Dragster (total useful and heat transfer)	2×10^{6}
Car (total useful and heat transfer)	8×10^{4}
Football player (total useful and heat transfer)	5×10^{3}
Clothes dryer	4×10^{3}
Person at rest (all heat transfer)	100
Typical incandescent light bulb (total useful and heat transfer)	60
Heart, person at rest (total useful and heat transfer)	8
Electric clock	3
Pocket calculator	10^{-3}

Power and Energy Consumption

We usually have to pay for the energy we use. It is interesting and easy to estimate the cost of energy for an electrical appliance if its power consumption rate and time used are known. The higher the power consumption rate and the longer the appliance is used, the greater the cost of that appliance. The power consumption rate is $P = W/t = E/t$, where E is the energy supplied by the electricity company. So the energy consumed over a time t is

$$E = Pt. \tag{7.72}$$

Electricity bills state the energy used in units of **kilowatt-hours** $(\text{kW} \cdot \text{h}),$ which is the product of power in kilowatts and time in hours. This unit is convenient because electrical power consumption at the kilowatt level for hours at a time is typical.

Example 7.12 Calculating Energy Costs

What is the cost of running a 0.200-kW computer 6.00 h per day for 30.0 d if the cost of electricity is $0.120 per $\text{kW} \cdot \text{h}$?

Strategy

Cost is based on energy consumed; thus, we must find E from $E = Pt$ and then calculate the cost. Because electrical energy is expressed in $\text{kW} \cdot \text{h}$, at the start of a problem such as this it is convenient to convert the units into kW and hours.

Solution

The energy consumed in $\text{kW} \cdot \text{h}$ is

$$\begin{aligned} E &= Pt = (0.200 \, \text{kW})(6.00 \, \text{h/d})(30.0 \, \text{d}) \\ &= 36.0 \, \text{kW} \cdot \text{h}, \end{aligned} \tag{7.73}$$

and the cost is simply given by

$$\text{cost} = (36.0 \, \text{kW} \cdot \text{h})(\$0.120 \, \text{per kW} \cdot \text{h}) = \$4.32 \, \text{per month}. \tag{7.74}$$

Discussion

The cost of using the computer in this example is neither exorbitant nor negligible. It is clear that the cost is a combination of power and time. When both are high, such as for an air conditioner in the summer, the cost is high.

The motivation to save energy has become more compelling with its ever-increasing price. Armed with the knowledge that energy consumed is the product of power and time, you can estimate costs for yourself and make the necessary value judgments about where to save energy. Either power or time must be reduced. It is most cost-effective to limit the use of high-power devices that normally operate for long periods of time, such as water heaters and air conditioners. This would not include relatively high power devices like toasters, because they are on only a few minutes per day. It would also not include electric clocks, in spite of their 24-hour-per-day usage, because they are very low power devices. It is sometimes possible to use devices that have greater efficiencies—that is, devices that consume less power to accomplish the same task. One example is the compact fluorescent light bulb, which produces over four times more light per watt of power consumed than its incandescent cousin.

Modern civilization depends on energy, but current levels of energy consumption and production are not sustainable. The likelihood of a link between global warming and fossil fuel use (with its concomitant production of carbon dioxide), has made reduction in energy use as well as a shift to non-fossil fuels of the utmost importance. Even though energy in an isolated system is a conserved quantity, the final result of most energy transformations is waste heat transfer to the environment, which is no longer useful for doing work. As we will discuss in more detail in **Thermodynamics**, the potential for energy to produce useful work has been "degraded" in the energy transformation.

7.8 Work, Energy, and Power in Humans

Learning Objectives

By the end of this section, you will be able to:

- Explain the human body's consumption of energy when at rest versus when engaged in activities that do useful work.
- Calculate the conversion of chemical energy in food into useful work.

Energy Conversion in Humans

Our own bodies, like all living organisms, are energy conversion machines. Conservation of energy implies that the chemical energy stored in food is converted into work, thermal energy, and/or stored as chemical energy in fatty tissue. (See **Figure 7.26**.) The fraction going into each form depends both on how much we eat and on our level of physical activity. If we eat more than is needed to do work and stay warm, the remainder goes into body fat.

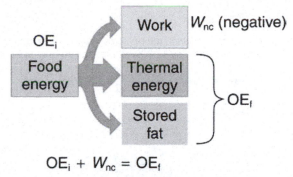

Figure 7.26 Energy consumed by humans is converted to work, thermal energy, and stored fat. By far the largest fraction goes to thermal energy, although the fraction varies depending on the type of physical activity.

Power Consumed at Rest

The *rate* at which the body uses food energy to sustain life and to do different activities is called the **metabolic rate**. The total energy conversion rate of a person *at rest* is called the **basal metabolic rate** (BMR) and is divided among various systems in the body, as shown in **Table 7.4**. The largest fraction goes to the liver and spleen, with the brain coming next. Of course, during vigorous exercise, the energy consumption of the skeletal muscles and heart increase markedly. About 75% of the calories burned in a day go into these basic functions. The BMR is a function of age, gender, total body weight, and amount of muscle mass (which burns more calories than body fat). Athletes have a greater BMR due to this last factor.

Table 7.4 Basal Metabolic Rates (BMR)

Organ	Power consumed at rest (W)	Oxygen consumption (mL/min)	Percent of BMR
Liver & spleen	23	67	27
Brain	16	47	19
Skeletal muscle	15	45	18
Kidney	9	26	10
Heart	6	17	7
Other	16	48	19
Totals	85 W	250 mL/min	100%

Energy consumption is directly proportional to oxygen consumption because the digestive process is basically one of oxidizing food. We can measure the energy people use during various activities by measuring their oxygen use. (See **Figure 7.27**.) Approximately 20 kJ of energy are produced for each liter of oxygen consumed, independent of the type of food. **Table 7.5** shows energy and oxygen consumption rates (power expended) for a variety of activities.

Power of Doing Useful Work

Work done by a person is sometimes called **useful work**, which is *work done on the outside world*, such as lifting weights. Useful work requires a force exerted through a distance on the outside world, and so it excludes internal work, such as that done by the heart when pumping blood. Useful work does include that done in climbing stairs or accelerating to a full run, because these are accomplished by exerting forces on the outside world. Forces exerted by the body are nonconservative, so that they can change the mechanical energy ($KE + PE$) of the system worked upon, and this is often the goal. A baseball player throwing a ball, for example, increases both the ball's kinetic and potential energy.

If a person needs more energy than they consume, such as when doing vigorous work, the body must draw upon the chemical energy stored in fat. So exercise can be helpful in losing fat. However, the amount of exercise needed to produce a loss in fat, or to burn off extra calories consumed that day, can be large, as **Example 7.13** illustrates.

Example 7.13 Calculating Weight Loss from Exercising

If a person who normally requires an average of 12,000 kJ (3000 kcal) of food energy per day consumes 13,000 kJ per day, he will steadily gain weight. How much bicycling per day is required to work off this extra 1000 kJ?

Solution

Table 7.5 states that 400 W are used when cycling at a moderate speed. The time required to work off 1000 kJ at this rate is then

$$\text{Time} = \frac{\text{energy}}{\left(\frac{\text{energy}}{\text{time}}\right)} = \frac{1000 \text{ kJ}}{400 \text{ W}} = 2500 \text{ s} = 42 \text{ min.} \tag{7.75}$$

Discussion

If this person uses more energy than he or she consumes, the person's body will obtain the needed energy by metabolizing body fat. If the person uses 13,000 kJ but consumes only 12,000 kJ, then the amount of fat loss will be

$$\text{Fat loss} = (1000 \text{ kJ})\left(\frac{1.0 \text{ g fat}}{39 \text{ kJ}}\right) = 26 \text{ g,} \qquad (7.76)$$

assuming the energy content of fat to be 39 kJ/g.

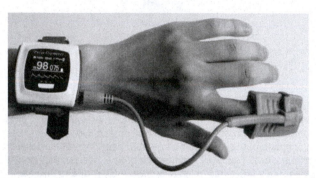

Figure 7.27 A pulse oxymeter is an apparatus that measures the amount of oxygen in blood. Oxymeters can be used to determine a person's metabolic rate, which is the rate at which food energy is converted to another form. Such measurements can indicate the level of athletic conditioning as well as certain medical problems. (credit: UusiAjaja, Wikimedia Commons)

Table 7.5 Energy and Oxygen Consumption Rates[2] (Power)

Activity	Energy consumption in watts	Oxygen consumption in liters O_2/min
Sleeping	83	0.24
Sitting at rest	120	0.34
Standing relaxed	125	0.36
Sitting in class	210	0.60
Walking (5 km/h)	280	0.80
Cycling (13–18 km/h)	400	1.14
Shivering	425	1.21
Playing tennis	440	1.26
Swimming breaststroke	475	1.36
Ice skating (14.5 km/h)	545	1.56
Climbing stairs (116/min)	685	1.96
Cycling (21 km/h)	700	2.00
Running cross-country	740	2.12
Playing basketball	800	2.28
Cycling, professional racer	1855	5.30
Sprinting	2415	6.90

All bodily functions, from thinking to lifting weights, require energy. (See **Figure 7.28**.) The many small muscle actions accompanying all quiet activity, from sleeping to head scratching, ultimately become thermal energy, as do less visible muscle actions by the heart, lungs, and digestive tract. Shivering, in fact, is an involuntary response to low body temperature that pits muscles against one another to produce thermal energy in the body (and do no work). The kidneys and liver consume a surprising amount of energy, but the biggest surprise of all it that a full 25% of all energy consumed by the body is used to maintain electrical potentials in all living cells. (Nerve cells use this electrical potential in nerve impulses.) This bioelectrical energy ultimately becomes mostly thermal energy, but some is utilized to power chemical processes such as in the kidneys and liver, and in fat production.

---

2. for an average 76-kg male

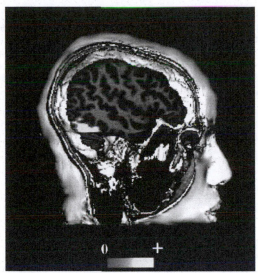

Figure 7.28 This fMRI scan shows an increased level of energy consumption in the vision center of the brain. Here, the patient was being asked to recognize faces. (credit: NIH via Wikimedia Commons)

7.9 World Energy Use

Learning Objectives
By the end of this section, you will be able to: • Describe the distinction between renewable and nonrenewable energy sources. • Explain why the inevitable conversion of energy to less useful forms makes it necessary to conserve energy resources.

Energy is an important ingredient in all phases of society. We live in a very interdependent world, and access to adequate and reliable energy resources is crucial for economic growth and for maintaining the quality of our lives. But current levels of energy consumption and production are not sustainable. About 40% of the world's energy comes from oil, and much of that goes to transportation uses. Oil prices are dependent as much upon new (or foreseen) discoveries as they are upon political events and situations around the world. The U.S., with 4.5% of the world's population, consumes 24% of the world's oil production per year; 66% of that oil is imported!

Renewable and Nonrenewable Energy Sources

The principal energy resources used in the world are shown in **Figure 7.29**. The fuel mix has changed over the years but now is dominated by oil, although natural gas and solar contributions are increasing. **Renewable forms of energy** are those sources that cannot be used up, such as water, wind, solar, and biomass. About 85% of our energy comes from nonrenewable **fossil fuels**—oil, natural gas, coal. The likelihood of a link between global warming and fossil fuel use, with its production of carbon dioxide through combustion, has made, in the eyes of many scientists, a shift to non-fossil fuels of utmost importance—but it will not be easy.

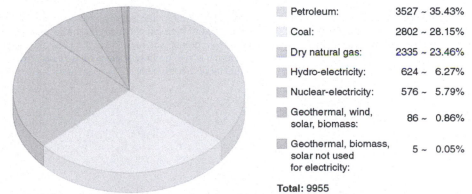

Petroleum:	3527 ~ 35.43%
Coal:	2802 ~ 28.15%
Dry natural gas:	2335 ~ 23.46%
Hydro-electricity:	624 ~ 6.27%
Nuclear-electricity:	576 ~ 5.79%
Geothermal, wind, solar, biomass:	86 ~ 0.86%
Geothermal, biomass, solar not used for electricity:	5 ~ 0.05%

Total: 9955

Figure 7.29 World energy consumption by source, in billions of kilowatt-hours: 2006. (credit: KVDP)

The World's Growing Energy Needs

World energy consumption continues to rise, especially in the developing countries. (See **Figure 7.30**.) Global demand for energy has tripled in the past 50 years and might triple again in the next 30 years. While much of this growth will come from the rapidly booming economies of China and India, many of the developed countries, especially those in Europe, are hoping to meet their energy needs by expanding the use of renewable sources. Although presently only a small percentage, renewable energy is growing very fast, especially wind energy. For example, Germany plans to meet 20% of its electricity and 10% of its overall energy needs with renewable resources by the year 2020. (See **Figure 7.31**.) Energy is a key constraint in the rapid economic growth of China and India. In 2003, China surpassed Japan as the world's second largest consumer of oil. However, over 1/3 of this is imported. Unlike most Western countries, coal dominates the commercial energy resources of China, accounting for 2/3 of its energy consumption. In 2009 China surpassed the United States as the largest generator of CO_2. In India, the main energy resources are biomass (wood and dung) and coal. Half of India's oil is imported. About 70% of India's electricity is generated by highly polluting coal. Yet there are sizeable strides being made in renewable energy. India has a rapidly growing wind energy base, and it has the largest solar cooking program in the world.

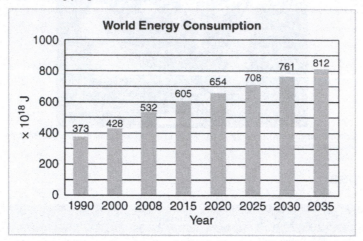

Figure 7.30 Past and projected world energy use (source: Based on data from U.S. Energy Information Administration, 2011)

Figure 7.31 Solar cell arrays at a power plant in Steindorf, Germany (credit: Michael Betke, Flickr)

Table 7.6 displays the 2006 commercial energy mix by country for some of the prime energy users in the world. While non-renewable sources dominate, some countries get a sizeable percentage of their electricity from renewable resources. For example, about 67% of New Zealand's electricity demand is met by hydroelectric. Only 10% of the U.S. electricity is generated by renewable resources, primarily hydroelectric. It is difficult to determine total contributions of renewable energy in some countries with a large rural population, so these percentages in this table are left blank.

Table 7.6 Energy Consumption—Selected Countries (2006)

Country	Consumption, in EJ (10^{18} J)	Oil	Natural Gas	Coal	Nuclear	Hydro	Other Renewables	Electricity Use per capita (kWh/yr)	Energy Use per capita (GJ/yr)
Australia	5.4	34%	17%	44%	0%	3%	1%	10000	260
Brazil	9.6	48%	7%	5%	1%	35%	2%	2000	50
China	63	22%	3%	69%	1%	6%		1500	35
Egypt	2.4	50%	41%	1%	0%	6%		990	32
Germany	16	37%	24%	24%	11%	1%	3%	6400	173
India	15	34%	7%	52%	1%	5%		470	13
Indonesia	4.9	51%	26%	16%	0%	2%	3%	420	22
Japan	24	48%	14%	21%	12%	4%	1%	7100	176
New Zealand	0.44	32%	26%	6%	0%	11%	19%	8500	102
Russia	31	19%	53%	16%	5%	6%		5700	202
U.S.	105	40%	23%	22%	8%	3%	1%	12500	340
World	**432**	**39%**	**23%**	**24%**	**6%**	**6%**	**2%**	**2600**	**71**

Energy and Economic Well-being

The last two columns in this table examine the energy and electricity use per capita. Economic well-being is dependent upon energy use, and in most countries higher standards of living, as measured by GDP (gross domestic product) per capita, are matched by higher levels of energy consumption per capita. This is borne out in **Figure 7.32**. Increased efficiency of energy use will change this dependency. A global problem is balancing energy resource development against the harmful effects upon the environment in its extraction and use.

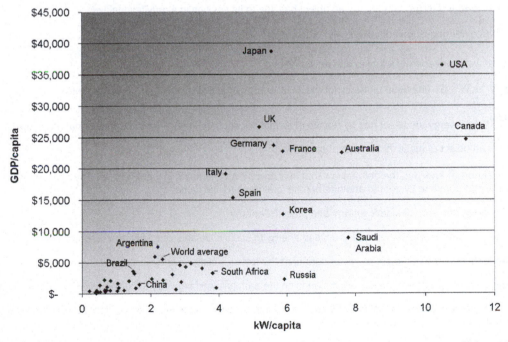

Figure 7.32 Power consumption per capita versus GDP per capita for various countries. Note the increase in energy usage with increasing GDP. (2007, credit: Frank van Mierlo, Wikimedia Commons)

Conserving Energy

As we finish this chapter on energy and work, it is relevant to draw some distinctions between two sometimes misunderstood terms in the area of energy use. As has been mentioned elsewhere, the "law of the conservation of energy" is a very useful principle in analyzing physical processes. It is a statement that cannot be proven from basic principles, but is a very good bookkeeping device, and no exceptions have ever been found. It states that the total amount of energy in an isolated system will always remain constant. Related to this principle, but remarkably different from it, is the important philosophy of energy conservation. This concept has to do with seeking to decrease the amount of energy used by an individual or group through (1) reduced activities (e.g., turning down thermostats, driving fewer kilometers) and/or (2) increasing conversion efficiencies in the performance of a particular task—such as developing and using more efficient room heaters, cars that have greater miles-per-gallon ratings, energy-efficient compact fluorescent lights, etc.

Since energy in an isolated system is not destroyed or created or generated, one might wonder why we need to be concerned about our energy resources, since energy is a conserved quantity. The problem is that the final result of most energy transformations is waste heat transfer to the environment and conversion to energy forms no longer useful for doing work. To state it in another way, the potential for energy to produce useful work has been "degraded" in the energy transformation. (This will be discussed in more detail in **Thermodynamics**.)

Glossary

basal metabolic rate: the total energy conversion rate of a person at rest

chemical energy: the energy in a substance stored in the bonds between atoms and molecules that can be released in a chemical reaction

conservation of mechanical energy: the rule that the sum of the kinetic energies and potential energies remains constant if only conservative forces act on and within a system

conservative force: a force that does the same work for any given initial and final configuration, regardless of the path followed

efficiency: a measure of the effectiveness of the input of energy to do work; useful energy or work divided by the total input of energy

electrical energy: the energy carried by a flow of charge

energy: the ability to do work

fossil fuels: oil, natural gas, and coal

friction: the force between surfaces that opposes one sliding on the other; friction changes mechanical energy into thermal energy

gravitational potential energy: the energy an object has due to its position in a gravitational field

horsepower: an older non-SI unit of power, with $1 \text{ hp} = 746 \text{ W}$

joule: SI unit of work and energy, equal to one newton-meter

kilowatt-hour: $(\text{kW} \cdot \text{h})$ unit used primarily for electrical energy provided by electric utility companies

kinetic energy: the energy an object has by reason of its motion, equal to $\frac{1}{2}mv^2$ for the translational (i.e., non-rotational) motion of an object of mass m moving at speed v

law of conservation of energy: the general law that total energy is constant in any process; energy may change in form or be transferred from one system to another, but the total remains the same

mechanical energy: the sum of kinetic energy and potential energy

metabolic rate: the rate at which the body uses food energy to sustain life and to do different activities

net work: work done by the net force, or vector sum of all the forces, acting on an object

nonconservative force: a force whose work depends on the path followed between the given initial and final configurations

nuclear energy: energy released by changes within atomic nuclei, such as the fusion of two light nuclei or the fission of a heavy nucleus

potential energy: energy due to position, shape, or configuration

potential energy of a spring: the stored energy of a spring as a function of its displacement; when Hooke's law applies, it is given by the expression $\frac{1}{2}kx^2$ where x is the distance the spring is compressed or extended and k is the spring constant

power: the rate at which work is done

radiant energy: the energy carried by electromagnetic waves

renewable forms of energy: those sources that cannot be used up, such as water, wind, solar, and biomass

thermal energy: the energy within an object due to the random motion of its atoms and molecules that accounts for the object's temperature

useful work: work done on an external system

watt: (W) SI unit of power, with $1 \text{ W} = 1 \text{ J/s}$

work: the transfer of energy by a force that causes an object to be displaced; the product of the component of the force in the direction of the displacement and the magnitude of the displacement

work-energy theorem: the result, based on Newton's laws, that the net work done on an object is equal to its change in kinetic energy

Section Summary

7.1 Work: The Scientific Definition
- Work is the transfer of energy by a force acting on an object as it is displaced.
- The work W that a force $\mathbf{F}$ does on an object is the product of the magnitude F of the force, times the magnitude d of the displacement, times the cosine of the angle θ between them. In symbols,

$$W = Fd \cos \theta.$$

- The SI unit for work and energy is the joule (J), where $1 \text{ J} = 1 \text{ N} \cdot \text{m} = 1 \text{ kg} \cdot \text{m}^2/\text{s}^2$.
- The work done by a force is zero if the displacement is either zero or perpendicular to the force.
- The work done is positive if the force and displacement have the same direction, and negative if they have opposite direction.

7.2 Kinetic Energy and the Work-Energy Theorem
- The net work W_{net} is the work done by the net force acting on an object.
- Work done on an object transfers energy to the object.
- The translational kinetic energy of an object of mass m moving at speed v is $\text{KE} = \frac{1}{2}mv^2$.
- The work-energy theorem states that the net work W_{net} on a system changes its kinetic energy,

$$W_{\text{net}} = \frac{1}{2}mv^2 - \frac{1}{2}mv_0^2.$$

7.3 Gravitational Potential Energy
- Work done against gravity in lifting an object becomes potential energy of the object-Earth system.
- The change in gravitational potential energy, ΔPE_g, is $\Delta\text{PE}_g = mgh$, with h being the increase in height and g the acceleration due to gravity.
- The gravitational potential energy of an object near Earth's surface is due to its position in the mass-Earth system. Only differences in gravitational potential energy, ΔPE_g, have physical significance.
- As an object descends without friction, its gravitational potential energy changes into kinetic energy corresponding to increasing speed, so that $\Delta\text{KE} = -\Delta\text{PE}_g$.

7.4 Conservative Forces and Potential Energy
- A conservative force is one for which work depends only on the starting and ending points of a motion, not on the path taken.
- We can define potential energy (PE) for any conservative force, just as we defined PE_g for the gravitational force.
- The potential energy of a spring is $\text{PE}_s = \frac{1}{2}kx^2$, where k is the spring's force constant and x is the displacement from its undeformed position.
- Mechanical energy is defined to be $\text{KE} + \text{PE}$ for a conservative force.
- When only conservative forces act on and within a system, the total mechanical energy is constant. In equation form,

$$\left. \begin{array}{c} KE + PE = \text{constant} \\ \text{or} \\ KE_i + PE_i = KE_f + PE_f \end{array} \right\}$$

where i and f denote initial and final values. This is known as the conservation of mechanical energy.

7.5 Nonconservative Forces
- A nonconservative force is one for which work depends on the path.
- Friction is an example of a nonconservative force that changes mechanical energy into thermal energy.
- Work W_{nc} done by a nonconservative force changes the mechanical energy of a system. In equation form,

$W_{nc} = \Delta KE + \Delta PE$ or, equivalently, $KE_i + PE_i + W_{nc} = KE_f + PE_f$.

- When both conservative and nonconservative forces act, energy conservation can be applied and used to calculate motion in terms of the known potential energies of the conservative forces and the work done by nonconservative forces, instead of finding the net work from the net force, or having to directly apply Newton's laws.

7.6 Conservation of Energy
- The law of conservation of energy states that the total energy is constant in any process. Energy may change in form or be transferred from one system to another, but the total remains the same.
- When all forms of energy are considered, conservation of energy is written in equation form as
$KE_i + PE_i + W_{nc} + OE_i = KE_f + PE_f + OE_f$, where OE is all **other forms of energy** besides mechanical energy.
- Commonly encountered forms of energy include electric energy, chemical energy, radiant energy, nuclear energy, and thermal energy.
- Energy is often utilized to do work, but it is not possible to convert all the energy of a system to work.
- The efficiency Eff of a machine or human is defined to be $Eff = \dfrac{W_{out}}{E_{in}}$, where W_{out} is useful work output and E_{in} is the energy consumed.

7.7 Power
- Power is the rate at which work is done, or in equation form, for the average power P for work W done over a time t,
$P = W/t$.
- The SI unit for power is the watt (W), where $1\ W = 1\ J/s$.
- The power of many devices such as electric motors is also often expressed in horsepower (hp), where $1\ hp = 746\ W$.

7.8 Work, Energy, and Power in Humans
- The human body converts energy stored in food into work, thermal energy, and/or chemical energy that is stored in fatty tissue.
- The *rate* at which the body uses food energy to sustain life and to do different activities is called the metabolic rate, and the corresponding rate when at rest is called the basal metabolic rate (BMR)
- The energy included in the basal metabolic rate is divided among various systems in the body, with the largest fraction going to the liver and spleen, and the brain coming next.
- About 75% of food calories are used to sustain basic body functions included in the basal metabolic rate.
- The energy consumption of people during various activities can be determined by measuring their oxygen use, because the digestive process is basically one of oxidizing food.

7.9 World Energy Use
- The relative use of different fuels to provide energy has changed over the years, but fuel use is currently dominated by oil, although natural gas and solar contributions are increasing.
- Although non-renewable sources dominate, some countries meet a sizeable percentage of their electricity needs from renewable resources.
- The United States obtains only about 10% of its energy from renewable sources, mostly hydroelectric power.
- Economic well-being is dependent upon energy use, and in most countries higher standards of living, as measured by GDP (Gross Domestic Product) per capita, are matched by higher levels of energy consumption per capita.
- Even though, in accordance with the law of conservation of energy, energy can never be created or destroyed, energy that can be used to do work is always partly converted to less useful forms, such as waste heat to the environment, in all of our uses of energy for practical purposes.

Conceptual Questions

7.1 Work: The Scientific Definition

1. Give an example of something we think of as work in everyday circumstances that is not work in the scientific sense. Is energy transferred or changed in form in your example? If so, explain how this is accomplished without doing work.

2. Give an example of a situation in which there is a force and a displacement, but the force does no work. Explain why it does no work.

3. Describe a situation in which a force is exerted for a long time but does no work. Explain.

7.2 Kinetic Energy and the Work-Energy Theorem

4. The person in **Figure 7.33** does work on the lawn mower. Under what conditions would the mower gain energy? Under what conditions would it lose energy?

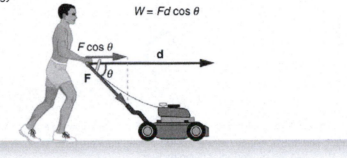

Figure 7.33

5. Work done on a system puts energy into it. Work done by a system removes energy from it. Give an example for each statement.

6. When solving for speed in **Example 7.4**, we kept only the positive root. Why?

7.3 Gravitational Potential Energy

7. In **Example 7.7**, we calculated the final speed of a roller coaster that descended 20 m in height and had an initial speed of 5 m/s downhill. Suppose the roller coaster had had an initial speed of 5 m/s uphill instead, and it coasted uphill, stopped, and then rolled back down to a final point 20 m below the start. We would find in that case that its final speed is the same as its initial. Explain in terms of conservation of energy.

8. Does the work you do on a book when you lift it onto a shelf depend on the path taken? On the time taken? On the height of the shelf? On the mass of the book?

7.4 Conservative Forces and Potential Energy

9. What is a conservative force?

10. The force exerted by a diving board is conservative, provided the internal friction is negligible. Assuming friction is negligible, describe changes in the potential energy of a diving board as a swimmer dives from it, starting just before the swimmer steps on the board until just after his feet leave it.

11. Define mechanical energy. What is the relationship of mechanical energy to nonconservative forces? What happens to mechanical energy if only conservative forces act?

12. What is the relationship of potential energy to conservative force?

7.6 Conservation of Energy

13. Consider the following scenario. A car for which friction is *not* negligible accelerates from rest down a hill, running out of gasoline after a short distance. The driver lets the car coast farther down the hill, then up and over a small crest. He then coasts down that hill into a gas station, where he brakes to a stop and fills the tank with gasoline. Identify the forms of energy the car has, and how they are changed and transferred in this series of events. (See **Figure 7.34**.)

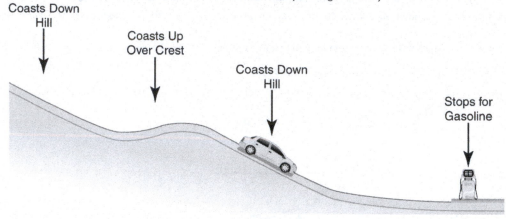

Figure 7.34 A car experiencing non-negligible friction coasts down a hill, over a small crest, then downhill again, and comes to a stop at a gas station.

14. Describe the energy transfers and transformations for a javelin, starting from the point at which an athlete picks up the javelin and ending when the javelin is stuck into the ground after being thrown.

15. Do devices with efficiencies of less than one violate the law of conservation of energy? Explain.

16. List four different forms or types of energy. Give one example of a conversion from each of these forms to another form.

17. List the energy conversions that occur when riding a bicycle.

7.7 Power

18. Most electrical appliances are rated in watts. Does this rating depend on how long the appliance is on? (When off, it is a zero-watt device.) Explain in terms of the definition of power.

19. Explain, in terms of the definition of power, why energy consumption is sometimes listed in kilowatt-hours rather than joules. What is the relationship between these two energy units?

20. A spark of static electricity, such as that you might receive from a doorknob on a cold dry day, may carry a few hundred watts of power. Explain why you are not injured by such a spark.

7.8 Work, Energy, and Power in Humans

21. Explain why it is easier to climb a mountain on a zigzag path rather than one straight up the side. Is your increase in gravitational potential energy the same in both cases? Is your energy consumption the same in both?

22. Do you do work on the outside world when you rub your hands together to warm them? What is the efficiency of this activity?

23. Shivering is an involuntary response to lowered body temperature. What is the efficiency of the body when shivering, and is this a desirable value?

24. Discuss the relative effectiveness of dieting and exercise in losing weight, noting that most athletic activities consume food energy at a rate of 400 to 500 W, while a single cup of yogurt can contain 1360 kJ (325 kcal). Specifically, is it likely that exercise alone will be sufficient to lose weight? You may wish to consider that regular exercise may increase the metabolic rate, whereas protracted dieting may reduce it.

7.9 World Energy Use

25. What is the difference between energy conservation and the law of conservation of energy? Give some examples of each.

26. If the efficiency of a coal-fired electrical generating plant is 35%, then what do we mean when we say that energy is a conserved quantity?

Problems & Exercises

7.1 Work: The Scientific Definition

1. How much work does a supermarket checkout attendant do on a can of soup he pushes 0.600 m horizontally with a force of 5.00 N? Express your answer in joules and kilocalories.

2. A 75.0-kg person climbs stairs, gaining 2.50 meters in height. Find the work done to accomplish this task.

3. (a) Calculate the work done on a 1500-kg elevator car by its cable to lift it 40.0 m at constant speed, assuming friction averages 100 N. (b) What is the work done on the lift by the gravitational force in this process? (c) What is the total work done on the lift?

4. Suppose a car travels 108 km at a speed of 30.0 m/s, and uses 2.0 gal of gasoline. Only 30% of the gasoline goes into useful work by the force that keeps the car moving at constant speed despite friction. (See **Table 7.1** for the energy content of gasoline.) (a) What is the magnitude of the force exerted to keep the car moving at constant speed? (b) If the required force is directly proportional to speed, how many gallons will be used to drive 108 km at a speed of 28.0 m/s?

5. Calculate the work done by an 85.0-kg man who pushes a crate 4.00 m up along a ramp that makes an angle of $20.0°$ with the horizontal. (See **Figure 7.35**.) He exerts a force of 500 N on the crate parallel to the ramp and moves at a constant speed. Be certain to include the work he does on the crate *and* on his body to get up the ramp.

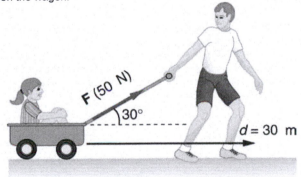

Figure 7.35 A man pushes a crate up a ramp.

6. How much work is done by the boy pulling his sister 30.0 m in a wagon as shown in **Figure 7.36**? Assume no friction acts on the wagon.

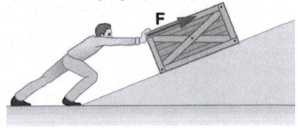

Figure 7.36 The boy does work on the system of the wagon and the child when he pulls them as shown.

7. A shopper pushes a grocery cart 20.0 m at constant speed on level ground, against a 35.0 N frictional force. He pushes in a direction $25.0°$ below the horizontal. (a) What is the work done on the cart by friction? (b) What is the work done on the cart by the gravitational force? (c) What is the work done on the cart by the shopper? (d) Find the force the shopper exerts, using energy considerations. (e) What is the total work done on the cart?

8. Suppose the ski patrol lowers a rescue sled and victim, having a total mass of 90.0 kg, down a $60.0°$ slope at constant speed, as shown in **Figure 7.37**. The coefficient of friction between the sled and the snow is 0.100. (a) How much work is done by friction as the sled moves 30.0 m along the hill? (b) How much work is done by the rope on the sled in this distance? (c) What is the work done by the gravitational force on the sled? (d) What is the total work done?

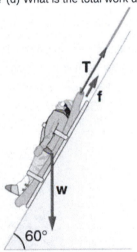

Figure 7.37 A rescue sled and victim are lowered down a steep slope.

7.2 Kinetic Energy and the Work-Energy Theorem

9. Compare the kinetic energy of a 20,000-kg truck moving at 110 km/h with that of an 80.0-kg astronaut in orbit moving at 27,500 km/h.

10. (a) How fast must a 3000-kg elephant move to have the same kinetic energy as a 65.0-kg sprinter running at 10.0 m/s? (b) Discuss how the larger energies needed for the movement of larger animals would relate to metabolic rates.

11. Confirm the value given for the kinetic energy of an aircraft carrier in **Table 7.1**. You will need to look up the definition of a nautical mile (1 knot = 1 nautical mile/h).

12. (a) Calculate the force needed to bring a 950-kg car to rest from a speed of 90.0 km/h in a distance of 120 m (a fairly typical distance for a non-panic stop). (b) Suppose instead the car hits a concrete abutment at full speed and is brought to a stop in 2.00 m. Calculate the force exerted on the car and compare it with the force found in part (a).

13. A car's bumper is designed to withstand a 4.0-km/h (1.1-m/s) collision with an immovable object without damage to the body of the car. The bumper cushions the shock by absorbing the force over a distance. Calculate the magnitude of the average force on a bumper that collapses 0.200 m while bringing a 900-kg car to rest from an initial speed of 1.1 m/s.

14. Boxing gloves are padded to lessen the force of a blow. (a) Calculate the force exerted by a boxing glove on an opponent's face, if the glove and face compress 7.50 cm during a blow in which the 7.00-kg arm and glove are brought to rest from an initial speed of 10.0 m/s. (b) Calculate the force exerted by an identical blow in the gory old days when no gloves were used and the knuckles and face would compress only 2.00 cm. (c) Discuss the magnitude of the force with glove on. Does it seem high enough to cause damage even though it is lower than the force with no glove?

15. Using energy considerations, calculate the average force a 60.0-kg sprinter exerts backward on the track to accelerate from 2.00 to 8.00 m/s in a distance of 25.0 m, if he encounters a headwind that exerts an average force of 30.0 N against him.

7.3 Gravitational Potential Energy

16. A hydroelectric power facility (see **Figure 7.38**) converts the gravitational potential energy of water behind a dam to electric energy. (a) What is the gravitational potential energy relative to the generators of a lake of volume 50.0 km^3 ($\text{mass} = 5.00 \times 10^{13} \text{ kg}$), given that the lake has an average height of 40.0 m above the generators? (b) Compare this with the energy stored in a 9-megaton fusion bomb.

Figure 7.38 Hydroelectric facility (credit: Denis Belevich, Wikimedia Commons)

17. (a) How much gravitational potential energy (relative to the ground on which it is built) is stored in the Great Pyramid of Cheops, given that its mass is about $7 \times 10^9 \text{ kg}$ and its center of mass is 36.5 m above the surrounding ground? (b) How does this energy compare with the daily food intake of a person?

18. Suppose a 350-g kookaburra (a large kingfisher bird) picks up a 75-g snake and raises it 2.5 m from the ground to a branch. (a) How much work did the bird do on the snake? (b) How much work did it do to raise its own center of mass to the branch?

19. In **Example 7.7**, we found that the speed of a roller coaster that had descended 20.0 m was only slightly greater when it had an initial speed of 5.00 m/s than when it started from rest. This implies that $\Delta\text{PE} \gg \text{KE}_i$. Confirm this statement by taking the ratio of ΔPE to KE_i. (Note that mass cancels.)

20. A 100-g toy car is propelled by a compressed spring that starts it moving. The car follows the curved track in **Figure 7.39**. Show that the final speed of the toy car is 0.687 m/s if its initial speed is 2.00 m/s and it coasts up the frictionless slope, gaining 0.180 m in altitude.

Figure 7.39 A toy car moves up a sloped track. (credit: Leszek Leszczynski, Flickr)

21. In a downhill ski race, surprisingly, little advantage is gained by getting a running start. (This is because the initial kinetic energy is small compared with the gain in gravitational potential energy on even small hills.) To demonstrate this, find the final speed and the time taken for a skier who skies 70.0 m along a $30°$ slope neglecting friction: (a) Starting from rest. (b) Starting with an initial speed of 2.50 m/s. (c) Does the answer surprise you? Discuss why it is still advantageous to get a running start in very competitive events.

7.4 Conservative Forces and Potential Energy

22. A 5.00×10^5-kg subway train is brought to a stop from a speed of 0.500 m/s in 0.400 m by a large spring bumper at the end of its track. What is the force constant k of the spring?

23. A pogo stick has a spring with a force constant of $2.50 \times 10^4 \text{ N/m}$, which can be compressed 12.0 cm. To what maximum height can a child jump on the stick using only the energy in the spring, if the child and stick have a total mass of 40.0 kg? Explicitly show how you follow the steps in the **Problem-Solving Strategies for Energy**.

7.5 Nonconservative Forces

24. A 60.0-kg skier with an initial speed of 12.0 m/s coasts up a 2.50-m-high rise as shown in **Figure 7.40**. Find her final speed at the top, given that the coefficient of friction between her skis and the snow is 0.0800. (Hint: Find the distance traveled up the incline assuming a straight-line path as shown in the figure.)

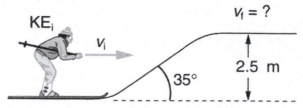

Figure 7.40 The skier's initial kinetic energy is partially used in coasting to the top of a rise.

25. (a) How high a hill can a car coast up (engine disengaged) if work done by friction is negligible and its initial speed is 110 km/h? (b) If, in actuality, a 750-kg car with an initial speed of 110 km/h is observed to coast up a hill to a height 22.0 m above its starting point, how much thermal energy was generated by friction? (c) What is the average force of friction if the hill has a slope $2.5°$ above the horizontal?

7.6 Conservation of Energy

26. Using values from Table 7.1, how many DNA molecules could be broken by the energy carried by a single electron in the beam of an old-fashioned TV tube? (These electrons were not dangerous in themselves, but they did create dangerous x rays. Later model tube TVs had shielding that absorbed x rays before they escaped and exposed viewers.)

27. Using energy considerations and assuming negligible air resistance, show that a rock thrown from a bridge 20.0 m above water with an initial speed of 15.0 m/s strikes the water with a speed of 24.8 m/s independent of the direction thrown.

28. If the energy in fusion bombs were used to supply the energy needs of the world, how many of the 9-megaton variety would be needed for a year's supply of energy (using data from Table 7.1)? This is not as far-fetched as it may sound—there are thousands of nuclear bombs, and their energy can be trapped in underground explosions and converted to electricity, as natural geothermal energy is.

29. (a) Use of hydrogen fusion to supply energy is a dream that may be realized in the next century. Fusion would be a relatively clean and almost limitless supply of energy, as can be seen from Table 7.1. To illustrate this, calculate how many years the present energy needs of the world could be supplied by one millionth of the oceans' hydrogen fusion energy. (b) How does this time compare with historically significant events, such as the duration of stable economic systems?

7.7 Power

30. The Crab Nebula (see Figure 7.41) pulsar is the remnant of a supernova that occurred in A.D. 1054. Using data from Table 7.3, calculate the approximate factor by which the power output of this astronomical object has declined since its explosion.

Figure 7.41 Crab Nebula (credit: ESO, via Wikimedia Commons)

31. Suppose a star 1000 times brighter than our Sun (that is, emitting 1000 times the power) suddenly goes supernova. Using data from Table 7.3: (a) By what factor does its power output increase? (b) How many times brighter than our entire Milky Way galaxy is the supernova? (c) Based on your answers, discuss whether it should be possible to observe supernovas in distant galaxies. Note that there are on the order of 10^{11} observable galaxies, the average brightness of which is somewhat less than our own galaxy.

32. A person in good physical condition can put out 100 W of useful power for several hours at a stretch, perhaps by pedaling a mechanism that drives an electric generator. Neglecting any problems of generator efficiency and practical considerations such as resting time: (a) How many people would it take to run a 4.00-kW electric clothes dryer? (b) How many people would it take to replace a large electric power plant that generates 800 MW?

33. What is the cost of operating a 3.00-W electric clock for a year if the cost of electricity is $0.0900 per $kW \cdot h$?

34. A large household air conditioner may consume 15.0 kW of power. What is the cost of operating this air conditioner 3.00 h per day for 30.0 d if the cost of electricity is $0.110 per $kW \cdot h$?

35. (a) What is the average power consumption in watts of an appliance that uses $5.00 \, kW \cdot h$ of energy per day? (b) How many joules of energy does this appliance consume in a year?

36. (a) What is the average useful power output of a person who does 6.00×10^6 J of useful work in 8.00 h? (b) Working at this rate, how long will it take this person to lift 2000 kg of bricks 1.50 m to a platform? (Work done to lift his body can be omitted because it is not considered useful output here.)

37. A 500-kg dragster accelerates from rest to a final speed of 110 m/s in 400 m (about a quarter of a mile) and encounters an average frictional force of 1200 N. What is its average power output in watts and horsepower if this takes 7.30 s?

38. (a) How long will it take an 850-kg car with a useful power output of 40.0 hp (1 hp = 746 W) to reach a speed of 15.0 m/s, neglecting friction? (b) How long will this acceleration take if the car also climbs a 3.00-m-high hill in the process?

39. (a) Find the useful power output of an elevator motor that lifts a 2500-kg load a height of 35.0 m in 12.0 s, if it also increases the speed from rest to 4.00 m/s. Note that the total mass of the counterbalanced system is 10,000 kg—so that only 2500 kg is raised in height, but the full 10,000 kg is accelerated. (b) What does it cost, if electricity is $0.0900 per $kW \cdot h$?

40. (a) What is the available energy content, in joules, of a battery that operates a 2.00-W electric clock for 18 months?

(b) How long can a battery that can supply 8.00×10^4 J run a pocket calculator that consumes energy at the rate of 1.00×10^{-3} W?

41. (a) How long would it take a 1.50×10^5 -kg airplane with engines that produce 100 MW of power to reach a speed of 250 m/s and an altitude of 12.0 km if air resistance were negligible? (b) If it actually takes 900 s, what is the power? (c) Given this power, what is the average force of air resistance if the airplane takes 1200 s? (Hint: You must find the distance the plane travels in 1200 s assuming constant acceleration.)

42. Calculate the power output needed for a 950-kg car to climb a $2.00°$ slope at a constant 30.0 m/s while encountering wind resistance and friction totaling 600 N. Explicitly show how you follow the steps in the **Problem-Solving Strategies for Energy**.

43. (a) Calculate the power per square meter reaching Earth's upper atmosphere from the Sun. (Take the power output of the Sun to be 4.00×10^{26} W.) (b) Part of this is absorbed and reflected by the atmosphere, so that a maximum of 1.30 kW/m^2 reaches Earth's surface. Calculate the area in km^2 of solar energy collectors needed to replace an electric power plant that generates 750 MW if the collectors convert an average of 2.00% of the maximum power into electricity. (This small conversion efficiency is due to the devices themselves, and the fact that the sun is directly overhead only briefly.) With the same assumptions, what area would be needed to meet the United States' energy needs $(1.05 \times 10^{20}$ J)? Australia's energy needs $(5.4 \times 10^{18}$ J)? China's energy needs $(6.3 \times 10^{19}$ J)? (These energy consumption values are from 2006.)

7.8 Work, Energy, and Power in Humans

44. (a) How long can you rapidly climb stairs (116/min) on the 93.0 kcal of energy in a 10.0-g pat of butter? (b) How many flights is this if each flight has 16 stairs?

45. (a) What is the power output in watts and horsepower of a 70.0-kg sprinter who accelerates from rest to 10.0 m/s in 3.00 s? (b) Considering the amount of power generated, do you think a well-trained athlete could do this repetitively for long periods of time?

46. Calculate the power output in watts and horsepower of a shot-putter who takes 1.20 s to accelerate the 7.27-kg shot from rest to 14.0 m/s, while raising it 0.800 m. (Do not include the power produced to accelerate his body.)

Figure 7.42 Shot putter at the Dornoch Highland Gathering in 2007. (credit: John Haslam, Flickr)

47. (a) What is the efficiency of an out-of-condition professor who does 2.10×10^5 J of useful work while metabolizing 500 kcal of food energy? (b) How many food calories would a well-conditioned athlete metabolize in doing the same work with an efficiency of 20%?

48. Energy that is not utilized for work or heat transfer is converted to the chemical energy of body fat containing about 39 kJ/g. How many grams of fat will you gain if you eat 10,000 kJ (about 2500 kcal) one day and do nothing but sit relaxed for 16.0 h and sleep for the other 8.00 h? Use data from **Table 7.5** for the energy consumption rates of these activities.

49. Using data from **Table 7.5**, calculate the daily energy needs of a person who sleeps for 7.00 h, walks for 2.00 h, attends classes for 4.00 h, cycles for 2.00 h, sits relaxed for 3.00 h, and studies for 6.00 h. (Studying consumes energy at the same rate as sitting in class.)

50. What is the efficiency of a subject on a treadmill who puts out work at the rate of 100 W while consuming oxygen at the rate of 2.00 L/min? (Hint: See **Table 7.5**.)

51. Shoveling snow can be extremely taxing because the arms have such a low efficiency in this activity. Suppose a person shoveling a footpath metabolizes food at the rate of 800 W. (a) What is her useful power output? (b) How long will it take her to lift 3000 kg of snow 1.20 m? (This could be the amount of heavy snow on 20 m of footpath.) (c) How much waste heat transfer in kilojoules will she generate in the process?

52. Very large forces are produced in joints when a person jumps from some height to the ground. (a) Calculate the magnitude of the force produced if an 80.0-kg person jumps from a 0.600–m-high ledge and lands stiffly, compressing joint material 1.50 cm as a result. (Be certain to include the weight of the person.) (b) In practice the knees bend almost involuntarily to help extend the distance over which you stop. Calculate the magnitude of the force produced if the stopping distance is 0.300 m. (c) Compare both forces with the weight of the person.

53. Jogging on hard surfaces with insufficiently padded shoes produces large forces in the feet and legs. (a) Calculate the magnitude of the force needed to stop the downward motion of a jogger's leg, if his leg has a mass of 13.0 kg, a speed of 6.00 m/s, and stops in a distance of 1.50 cm. (Be certain to include the weight of the 75.0-kg jogger's body.) (b) Compare this force with the weight of the jogger.

54. (a) Calculate the energy in kJ used by a 55.0-kg woman who does 50 deep knee bends in which her center of mass is lowered and raised 0.400 m. (She does work in both directions.) You may assume her efficiency is 20%. (b) What is the average power consumption rate in watts if she does this in 3.00 min?

55. Kanellos Kanellopoulos flew 119 km from Crete to Santorini, Greece, on April 23, 1988, in the *Daedalus 88*, an aircraft powered by a bicycle-type drive mechanism (see **Figure 7.43**). His useful power output for the 234-min trip was about 350 W. Using the efficiency for cycling from **Table 7.2**, calculate the food energy in kilojoules he metabolized during the flight.

Figure 7.43 The Daedalus 88 in flight. (credit: NASA photo by Beasley)

56. The swimmer shown in **Figure 7.44** exerts an average horizontal backward force of 80.0 N with his arm during each 1.80 m long stroke. (a) What is his work output in each stroke? (b) Calculate the power output of his arms if he does 120 strokes per minute.

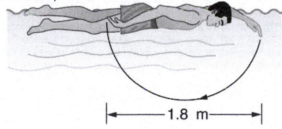

Figure 7.44

57. Mountain climbers carry bottled oxygen when at very high altitudes. (a) Assuming that a mountain climber uses oxygen at twice the rate for climbing 116 stairs per minute (because of low air temperature and winds), calculate how many liters of oxygen a climber would need for 10.0 h of climbing. (These are liters at sea level.) Note that only 40% of the inhaled oxygen is utilized; the rest is exhaled. (b) How much useful work does the climber do if he and his equipment have a mass of 90.0 kg and he gains 1000 m of altitude? (c) What is his efficiency for the 10.0-h climb?

58. The awe-inspiring Great Pyramid of Cheops was built more than 4500 years ago. Its square base, originally 230 m on a side, covered 13.1 acres, and it was 146 m high, with a mass of about 7×10^9 kg. (The pyramid's dimensions are slightly different today due to quarrying and some sagging.) Historians estimate that 20,000 workers spent 20 years to construct it, working 12-hour days, 330 days per year. (a) Calculate the gravitational potential energy stored in the pyramid, given its center of mass is at one-fourth its height. (b) Only a fraction of the workers lifted blocks; most were involved in support services such as building ramps (see **Figure 7.45**), bringing food and water, and hauling blocks to the site. Calculate the efficiency of the workers who did the lifting, assuming there were 1000 of them and they consumed food energy at the rate of 300 kcal/h. What does your answer imply about how much of their work went into block-lifting, versus how much work went into friction and lifting and lowering their own bodies? (c) Calculate the mass of food that had to be supplied each day, assuming that the average worker required 3600 kcal per day and that their diet was 5% protein, 60% carbohydrate, and 35% fat. (These proportions neglect the mass of bulk and nondigestible materials consumed.)

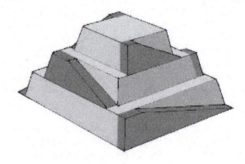

Figure 7.45 Ancient pyramids were probably constructed using ramps as simple machines. (credit: Franck Monnier, Wikimedia Commons)

59. (a) How long can you play tennis on the 800 kJ (about 200 kcal) of energy in a candy bar? (b) Does this seem like a long time? Discuss why exercise is necessary but may not be sufficient to cause a person to lose weight.

7.9 World Energy Use

60. Integrated Concepts

(a) Calculate the force the woman in **Figure 7.46** exerts to do a push-up at constant speed, taking all data to be known to three digits. (b) How much work does she do if her center of mass rises 0.240 m? (c) What is her useful power output if she does 25 push-ups in 1 min? (Should work done lowering her body be included? See the discussion of useful work in **Work, Energy, and Power in Humans**.

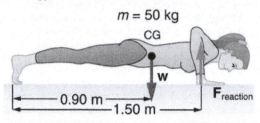

Figure 7.46 Forces involved in doing push-ups. The woman's weight acts as a force exerted downward on her center of gravity (CG).

61. Integrated Concepts

A 75.0-kg cross-country skier is climbing a $3.0°$ slope at a constant speed of 2.00 m/s and encounters air resistance of 25.0 N. Find his power output for work done against the gravitational force and air resistance. (b) What average force does he exert backward on the snow to accomplish this? (c) If he continues to exert this force and to experience the same air resistance when he reaches a level area, how long will it take him to reach a velocity of 10.0 m/s?

62. Integrated Concepts

The 70.0-kg swimmer in **Figure 7.44** starts a race with an initial velocity of 1.25 m/s and exerts an average force of 80.0 N backward with his arms during each 1.80 m long stroke. (a) What is his initial acceleration if water resistance is 45.0 N? (b) What is the subsequent average resistance force from the water during the 5.00 s it takes him to reach his top velocity of 2.50 m/s? (c) Discuss whether water resistance seems to increase linearly with velocity.

63. Integrated Concepts

A toy gun uses a spring with a force constant of 300 N/m to propel a 10.0-g steel ball. If the spring is compressed 7.00 cm and friction is negligible: (a) How much force is needed to compress the spring? (b) To what maximum height can the ball be shot? (c) At what angles above the horizontal may a child aim to hit a target 3.00 m away at the same height as the gun? (d) What is the gun's maximum range on level ground?

64. Integrated Concepts

(a) What force must be supplied by an elevator cable to produce an acceleration of 0.800 m/s^2 against a 200-N frictional force, if the mass of the loaded elevator is 1500 kg? (b) How much work is done by the cable in lifting the elevator 20.0 m? (c) What is the final speed of the elevator if it starts from rest? (d) How much work went into thermal energy?

65. Unreasonable Results

A car advertisement claims that its 900-kg car accelerated from rest to 30.0 m/s and drove 100 km, gaining 3.00 km in altitude, on 1.0 gal of gasoline. The average force of friction including air resistance was 700 N. Assume all values are known to three significant figures. (a) Calculate the car's efficiency. (b) What is unreasonable about the result? (c) Which premise is unreasonable, or which premises are inconsistent?

66. Unreasonable Results

Body fat is metabolized, supplying 9.30 kcal/g, when dietary intake is less than needed to fuel metabolism. The manufacturers of an exercise bicycle claim that you can lose 0.500 kg of fat per day by vigorously exercising for 2.00 h per day on their machine. (a) How many kcal are supplied by the metabolization of 0.500 kg of fat? (b) Calculate the kcal/min that you would have to utilize to metabolize fat at the rate of 0.500 kg in 2.00 h. (c) What is unreasonable about the results? (d) Which premise is unreasonable, or which premises are inconsistent?

67. Construct Your Own Problem

Consider a person climbing and descending stairs. Construct a problem in which you calculate the long-term rate at which stairs can be climbed considering the mass of the person, his ability to generate power with his legs, and the height of a single stair step. Also consider why the same person can descend stairs at a faster rate for a nearly unlimited time in spite of the fact that very similar forces are exerted going down as going up. (This points to a fundamentally different process for descending versus climbing stairs.)

68. Construct Your Own Problem

Consider humans generating electricity by pedaling a device similar to a stationary bicycle. Construct a problem in which you determine the number of people it would take to replace a large electrical generation facility. Among the things to consider are the power output that is reasonable using the legs, rest time, and the need for electricity 24 hours per day. Discuss the practical implications of your results.

69. Integrated Concepts

A 105-kg basketball player crouches down 0.400 m while waiting to jump. After exerting a force on the floor through this 0.400 m, his feet leave the floor and his center of gravity rises 0.950 m above its normal standing erect position. (a) Using energy considerations, calculate his velocity when he leaves the floor. (b) What average force did he exert on the floor? (Do not neglect the force to support his weight as well as that to accelerate him.) (c) What was his power output during the acceleration phase?

7.1 Work: The Scientific Definition

1. Given **Table 7.7** about how much force does the rocket engine exert on the 3.0-kg payload?

Table 7.7

Distance traveled with rocket engine firing (m)	Payload final velocity (m/s)
500	310
490	300
1020	450
505	312

a. 150 N
b. 300 N
c. 450 N
d. 600 N

2. You have a cart track, a cart, several masses, and a position-sensing pulley. Design an experiment to examine how the force exerted on the cart does work as it moves through a distance.

3. Look at **Figure 7.10**(c). You compress a spring by x, and then release it. Next you compress the spring by $2x$. How much more work did you do the second time than the first?
a. Half as much
b. The same
c. Twice as much
d. Four times as much

4. You have a cart track, two carts, several masses, a position-sensing pulley, and a piece of carpet (a rough surface) that will fit over the track. Design an experiment to examine how the force exerted on the cart does work as the cart moves through a distance.

5. A crane is lifting construction materials from the ground to an elevation of 60 m. Over the first 10 m, the motor linearly increases the force it exerts from 0 to 10 kN. It exerts that constant force for the next 40 m, and then winds down to 0 N again over the last 10 m, as shown in the figure. What is the total work done on the construction materials?

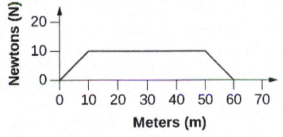

Figure 7.47
a. 500 kJ
b. 600 kJ
c. 300 kJ
d. 18 MJ

7.2 Kinetic Energy and the Work-Energy Theorem

6. A toy car is going around a loop-the-loop. Gravity _____ the kinetic energy on the upward side of the loop, _____ the kinetic energy at the top, and _____ the kinetic energy on the downward side of the loop.
a. increases, decreases, has no effect on
b. decreases, has no effect on, increases
c. increases, has no effect on, decreases
d. decreases, increases, has no effect on

7. A roller coaster is set up with a track in the form of a perfect cosine. Describe and graph what happens to the kinetic energy of a cart as it goes through the first full period of the track.

8. If wind is blowing horizontally toward a car with an angle of 30 degrees from the direction of travel, the kinetic energy will _____. If the wind is blowing at a car at 135 degrees from the direction of travel, the kinetic energy will _____.
a. increase, increase
b. increase, decrease
c. decrease, increase
d. decrease, decrease

9. In what direction relative to the direction of travel can a force act on a car (traveling on level ground), and not change the kinetic energy? Can you give examples of such forces?

10. A 2000-kg airplane is coming in for a landing, with a velocity 5 degrees below the horizontal and a drag force of 40 kN acting directly rearward. Kinetic energy will _____ due to the net force of _____.
a. increase, 20 kN
b. decrease, 40 kN
c. increase, 45 kN
d. decrease, 45 kN

11. You are participating in the Iditarod, and your sled dogs are pulling you across a frozen lake with a force of 1200 N while a 300 N wind is blowing at you at 135 degrees from your direction of travel. What is the net force, and will your kinetic energy increase or decrease?

12. A model drag car is being accelerated along its track from rest by a motor with a force of 75 N, but there is a drag force of 30 N due to the track. What is the kinetic energy after 2 m of travel?
a. 90 J
b. 150 J
c. 210 J
d. 60 J

13. You are launching a 2-kg potato out of a potato cannon. The cannon is 1.5 m long and is aimed 30 degrees above the horizontal. It exerts a 50 N force on the potato. What is the kinetic energy of the potato as it leaves the muzzle of the potato cannon?

14. When the force acting on an object is parallel to the direction of the motion of the center of mass, the mechanical energy _____. When the force acting on an object is antiparallel to the direction of the center of mass, the mechanical energy _____.
a. increases, increases
b. increases, decreases
c. decreases, increases
d. decreases, decreases

15. Describe a system in which the main forces acting are parallel or antiparallel to the center of mass, and justify your answer.

16. A child is pulling two red wagons, with the second one tied to the first by a (non-stretching) rope. Each wagon has a mass of 10 kg. If the child exerts a force of 30 N for 5.0 m, how much has the kinetic energy of the two-wagon system

changed?
  a. 300 J
  b. 150 J
  c. 75 J
  d. 60 J

17. A child has two red wagons, with the rear one tied to the front by a (non-stretching) rope. If the child pushes on the rear wagon, what happens to the kinetic energy of each of the wagons, and the two-wagon system?

18. Draw a graph of the force parallel to displacement exerted on a stunt motorcycle going through a loop-the-loop versus the distance traveled around the loop. Explain the net change in energy.

7.3 Gravitational Potential Energy

19. A 1.0 kg baseball is flying at 10 m/s. How much kinetic energy does it have? Potential energy?
  a. 10 J, 20 J
  b. 50 J, 20 J
  c. unknown, 50 J
  d. 50 J, unknown

20. A 2.0-kg potato has been launched out of a potato cannon at 9.0 m/s. What is the kinetic energy? If you then learn that it is 4.0 m above the ground, what is the total mechanical energy relative to the ground?
  a. 78 J, 3 J
  b. 160 J, 81 J
  c. 81 J, 160 J
  d. 81 J, 3 J

21. You have a 120-g yo-yo that you are swinging at 0.9 m/s. How much energy does it have? How high can it get above the lowest point of the swing without your doing any additional work, on Earth? How high could it get on the Moon, where gravity is 1/6 Earth's?

7.4 Conservative Forces and Potential Energy

22. Two 4.0 kg masses are connected to each other by a spring with a force constant of 25 N/m and a rest length of 1.0 m. If the spring has been compressed to 0.80 m in length and the masses are traveling toward each other at 0.50 m/s (each), what is the total energy in the system?
  a. 1.0 J
  b. 1.5 J
  c. 9.0 J
  d. 8.0 J

23. A spring with a force constant of 5000 N/m and a rest length of 3.0 m is used in a catapult. When compressed to 1.0 m, it is used to launch a 50 kg rock. However, there is an error in the release mechanism, so the rock gets launched almost straight up. How high does it go, and how fast is it going when it hits the ground?

24. What information do you need to calculate the kinetic energy and potential energy of a spring? Potential energy due to gravity? How many objects do you need information about for each of these cases?

25. You are loading a toy dart gun, which has two settings, the more powerful with the spring compressed twice as far as the lower setting. If it takes 5.0 J of work to compress the dart gun to the lower setting, how much work does it take for the higher setting?
  a. 20 J
  b. 10 J
  c. 2.5 J
  d. 40 J

26. Describe a system you use daily with internal potential energy.

27. Old-fashioned pendulum clocks are powered by masses that need to be wound back to the top of the clock about once a week to counteract energy lost due to friction and to the chimes. One particular clock has three masses: 4.0 kg, 4.0 kg, and 6.0 kg. They can drop 1.3 meters. How much energy does the clock use in a week?
  a. 51 J
  b. 76 J
  c. 127 J
  d. 178 J

28. A water tower stores not only water, but (at least part of) the energy to move the water. How much? Make reasonable estimates for how much water is in the tower, and other quantities you need.

29. Old-fashioned pocket watches needed to be wound daily so they wouldn't run down and lose time, due to the friction in the internal components. This required a large number of turns of the winding key, but not much force per turn, and it was possible to overwind and break the watch. How was the energy stored?
  a. A small mass raised a long distance
  b. A large mass raised a short distance
  c. A weak spring deformed a long way
  d. A strong spring deformed a short way

30. Some of the very first clocks invented in China were powered by water. Describe how you think this was done.

7.5 Nonconservative Forces

31. You are in a room in a basement with a smooth concrete floor (friction force equals 40 N) and a nice rug (friction force equals 55 N) that is 3 m by 4 m. However, you have to push a very heavy box from one corner of the rug to the opposite corner of the rug. Will you do more work against friction going around the floor or across the rug, and how much extra?
  a. Across the rug is 275 J extra
  b. Around the floor is 5 J extra
  c. Across the rug is 5 J extra
  d. Around the floor is 280 J extra

32. In the Appalachians, along the interstate, there are ramps of loose gravel for semis that have had their brakes fail to drive into to stop. Design an experiment to measure how effective this would be.

7.6 Conservation of Energy

33. You do 30 J of work to load a toy dart gun. However, the dart is 10 cm long and feels a frictional force of 10 N while going through the dart gun's barrel. What is the kinetic energy of the fired dart?
  a. 30 J
  b. 29 J
  c. 28 J
  d. 27 J

34. When an object is lifted by a crane, it begins and ends its motion at rest. The same is true of an object pushed across a rough surface. Explain why this happens. What are the differences between these systems?

35. A child has two red wagons, with the rear one tied to the front by a stretchy rope (a spring). If the child pulls on the front wagon, the _____ increases.

   a. kinetic energy of the wagons
   b. potential energy stored in the spring
   c. both A and B
   d. not enough information

36. A child has two red wagons, with the rear one tied to the front by a stretchy rope (a spring). If the child pulls on the front wagon, the energy stored in the system increases. How do the relative amounts of potential and kinetic energy in this system change over time?

37. Which of the following are closed systems?
   a. Earth
   b. a car
   c. a frictionless pendulum
   d. a mass on a spring in a vacuum

38. Describe a real-world example of a closed system.

39. A 5.0-kg rock falls off of a 10 m cliff. If air resistance exerts a force of 10 N, what is the kinetic energy when the rock hits the ground?
   a. 400 J
   b. 12.6 m/s
   c. 100 J
   d. 500 J

40. Hydroelectricity is generated by storing water behind a dam, and then letting some of it run through generators in the dam to turn them. If the system is the water, what is the environment that is doing work on it? If a dam has water 100 m deep behind it, how much energy was generated if 10,000 kg of water exited the dam at 2.0 m/s?

41. Before railroads were invented, goods often traveled along canals, with mules pulling barges from the bank. If a mule is exerting a 1200 N force for 10 km, and the rope connecting the mule to the barge is at a 20 degree angle from the direction of travel, how much work did the mule do on the barge?
   a. 12 MJ
   b. 11 MJ
   c. 4.1 MJ
   d. 6 MJ

42. Describe an instance today in which you did work, by the scientific definition. Then calculate how much work you did in that instance, showing your work.

8 LINEAR MOMENTUM AND COLLISIONS

Figure 8.1 Each rugby player has great momentum, which will affect the outcome of their collisions with each other and the ground. (credit: ozzzie, Flickr)

Chapter Outline

8.1. Linear Momentum and Force

8.2. Impulse

8.3. Conservation of Momentum

8.4. Elastic Collisions in One Dimension

8.5. Inelastic Collisions in One Dimension

8.6. Collisions of Point Masses in Two Dimensions

8.7. Introduction to Rocket Propulsion

Connection for AP® courses

In this chapter, you will learn about the concept of momentum and the relationship between momentum and force (both vector quantities) applied over a time interval. Have you ever considered why a glass dropped on a tile floor will often break, but a glass dropped on carpet will often remain intact? Both involve changes in momentum, but the actual collision with the floor is different in each case, just as an automobile collision without the benefit of an airbag can have a significantly different outcome than one with an airbag.

You will learn that the interaction of objects (like a glass and the floor or two automobiles) results in forces, which in turn result in changes in the momentum of each object. At the same time, you will see how the law of momentum conservation can be applied to a system to help determine the outcome of a collision.

The content in this chapter supports:

Big Idea 3 The interactions of an object with other objects can be described by forces.

Enduring Understanding 3.D A force exerted on an object can change the momentum of the object.

Essential Knowledge 3.D.2 The change in momentum of an object occurs over a time interval.

Big Idea 4: Interactions between systems can result in changes in those systems.

Enduring Understanding 4.B Interactions with other objects or systems can change the total linear momentum of a system.

Essential Knowledge 4.B.1 The change in linear momentum for a constant-mass system is the product of the mass of the system and the change in velocity of the center of mass.

Big Idea 5 Changes that occur as a result of interactions are constrained by conservation laws.

Enduring Understanding 5.A Certain quantities are conserved, in the sense that the changes of those quantities in a given system are always equal to the transfer of that quantity to or from the system by all possible interactions with other systems.

Essential Knowledge 5.A.2 For all systems under all circumstances, energy, charge, linear momentum, and angular momentum are conserved.

Essential Knowledge 5.D.1 In a collision between objects, linear momentum is conserved. In an elastic collision, kinetic energy is the same before and after.

Essential Knowledge 5.D.2 In a collision between objects, linear momentum is conserved. In an inelastic collision, kinetic energy is not the same before and after the collision.

8.1 Linear Momentum and Force

Learning Objectives

By the end of this section, you will be able to:

- Define linear momentum.
- Explain the relationship between linear momentum and force.
- State Newton's second law of motion in terms of linear momentum.
- Calculate linear momentum given mass and velocity.

The information presented in this section supports the following AP® learning objectives and science practices:

- **3.D.1.1** The student is able to justify the selection of data needed to determine the relationship between the direction of the force acting on an object and the change in momentum caused by that force. **(S.P. 4.1)**

Linear Momentum

The scientific definition of linear momentum is consistent with most people's intuitive understanding of momentum: a large, fast-moving object has greater momentum than a smaller, slower object. **Linear momentum** is defined as the product of a system's mass multiplied by its velocity. In symbols, linear momentum is expressed as

$$\mathbf{p} = m\mathbf{v}. \tag{8.1}$$

Momentum is directly proportional to the object's mass and also its velocity. Thus the greater an object's mass or the greater its velocity, the greater its momentum. Momentum $\mathbf{p}$ is a vector having the same direction as the velocity $\mathbf{v}$. The SI unit for momentum is $\text{kg} \cdot \text{m/s}$.

Linear Momentum

Linear momentum is defined as the product of a system's mass multiplied by its velocity:

$$\mathbf{p} = m\mathbf{v}. \tag{8.2}$$

Example 8.1 Calculating Momentum: A Football Player and a Football

(a) Calculate the momentum of a 110-kg football player running at 8.00 m/s. (b) Compare the player's momentum with the momentum of a hard-thrown 0.410-kg football that has a speed of 25.0 m/s.

Strategy

No information is given regarding direction, and so we can calculate only the magnitude of the momentum, p. (As usual, a symbol that is in italics is a magnitude, whereas one that is italicized, boldfaced, and has an arrow is a vector.) In both parts of this example, the magnitude of momentum can be calculated directly from the definition of momentum given in the equation, which becomes

$$p = mv \tag{8.3}$$

when only magnitudes are considered.

Solution for (a)

To determine the momentum of the player, substitute the known values for the player's mass and speed into the equation.

$$p_{\text{player}} = (110 \text{ kg})(8.00 \text{ m/s}) = 880 \text{ kg} \cdot \text{m/s} \tag{8.4}$$

Solution for (b)

To determine the momentum of the ball, substitute the known values for the ball's mass and speed into the equation.

$$p_{\text{ball}} = (0.410 \text{ kg})(25.0 \text{ m/s}) = 10.3 \text{ kg} \cdot \text{m/s} \tag{8.5}$$

The ratio of the player's momentum to that of the ball is

$$\frac{p_{\text{player}}}{p_{\text{ball}}} = \frac{880}{10.3} = 85.9.$$

(8.6)

Discussion

Although the ball has greater velocity, the player has a much greater mass. Thus the momentum of the player is much greater than the momentum of the football, as you might guess. As a result, the player's motion is only slightly affected if he catches the ball. We shall quantify what happens in such collisions in terms of momentum in later sections.

Momentum and Newton's Second Law

The importance of momentum, unlike the importance of energy, was recognized early in the development of classical physics. Momentum was deemed so important that it was called the "quantity of motion." Newton actually stated his **second law of motion** in terms of momentum: The net external force equals the change in momentum of a system divided by the time over which it changes. Using symbols, this law is

$$\mathbf{F}_{\text{net}} = \frac{\Delta \mathbf{p}}{\Delta t},$$

(8.7)

where $\mathbf{F}_{\text{net}}$ is the net external force, $\Delta \mathbf{p}$ is the change in momentum, and Δt is the change in time.

Newton's Second Law of Motion in Terms of Momentum

The net external force equals the change in momentum of a system divided by the time over which it changes.

$$\mathbf{F}_{\text{net}} = \frac{\Delta \mathbf{p}}{\Delta t}$$

(8.8)

Making Connections: Force and Momentum

Force and momentum are intimately related. Force acting over time can change momentum, and Newton's second law of motion, can be stated in its most broadly applicable form in terms of momentum. Momentum continues to be a key concept in the study of atomic and subatomic particles in quantum mechanics.

This statement of Newton's second law of motion includes the more familiar $\mathbf{F}_{\text{net}} = m\mathbf{a}$ as a special case. We can derive this form as follows. First, note that the change in momentum $\Delta \mathbf{p}$ is given by

$$\Delta \mathbf{p} = \Delta(m\mathbf{v}).$$

(8.9)

If the mass of the system is constant, then

$$\Delta(m\mathbf{v}) = m\Delta\mathbf{v}.$$

(8.10)

So that for constant mass, Newton's second law of motion becomes

$$\mathbf{F}_{\text{net}} = \frac{\Delta \mathbf{p}}{\Delta t} = \frac{m\Delta\mathbf{v}}{\Delta t}.$$

(8.11)

Because $\frac{\Delta\mathbf{v}}{\Delta t} = \mathbf{a}$, we get the familiar equation

$$\mathbf{F}_{\text{net}} = m\mathbf{a}$$

(8.12)

when the mass of the system is constant.

Newton's second law of motion stated in terms of momentum is more generally applicable because it can be applied to systems where the mass is changing, such as rockets, as well as to systems of constant mass. We will consider systems with varying mass in some detail; however, the relationship between momentum and force remains useful when mass is constant, such as in the following example.

Example 8.2 Calculating Force: Venus Williams' Racquet

During the 2007 French Open, Venus Williams hit the fastest recorded serve in a premier women's match, reaching a speed of 58 m/s (209 km/h). What is the average force exerted on the 0.057-kg tennis ball by Venus Williams' racquet, assuming that the ball's speed just after impact is 58 m/s, that the initial horizontal component of the velocity before impact is negligible, and that the ball remained in contact with the racquet for 5.0 ms (milliseconds)?

Strategy

This problem involves only one dimension because the ball starts from having no horizontal velocity component before impact. Newton's second law stated in terms of momentum is then written as

$$\mathbf{F}_{net} = \frac{\Delta \mathbf{p}}{\Delta t}. \tag{8.13}$$

As noted above, when mass is constant, the change in momentum is given by

$$\Delta p = m\Delta v = m(v_f - v_i). \tag{8.14}$$

In this example, the velocity just after impact and the change in time are given; thus, once Δp is calculated, $F_{net} = \frac{\Delta p}{\Delta t}$ can be used to find the force.

Solution

To determine the change in momentum, substitute the values for the initial and final velocities into the equation above.

$$
\begin{aligned}
\Delta p &= m(v_f - v_i) \\
&= (0.057 \text{ kg})(58 \text{ m/s} - 0 \text{ m/s}) \\
&= 3.306 \text{ kg} \cdot \text{m/s} \approx 3.3 \text{ kg} \cdot \text{m/s}
\end{aligned}
\tag{8.15}
$$

Now the magnitude of the net external force can determined by using $F_{net} = \frac{\Delta p}{\Delta t}$:

$$
\begin{aligned}
F_{net} &= \frac{\Delta p}{\Delta t} = \frac{3.306 \text{ kg} \cdot \text{m/s}}{5.0 \times 10^{-3} \text{ s}} \\
&= 661 \text{ N} \approx 660 \text{ N},
\end{aligned}
\tag{8.16}
$$

where we have retained only two significant figures in the final step.

Discussion

This quantity was the average force exerted by Venus Williams' racquet on the tennis ball during its brief impact (note that the ball also experienced the 0.56-N force of gravity, but that force was not due to the racquet). This problem could also be solved by first finding the acceleration and then using $F_{net} = ma$, but one additional step would be required compared with the strategy used in this example.

Making Connections: Illustrative Example

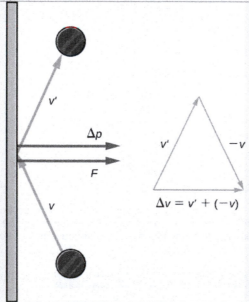

Figure 8.2 A puck has an elastic, glancing collision with the edge of an air hockey table.

In **Figure 8.2**, a puck is shown colliding with the edge of an air hockey table at a glancing angle. During the collision, the edge of the table exerts a force **F** on the puck, and the velocity of the puck changes as a result of the collision. The change in momentum is found by the equation:

$$\Delta p = m\Delta v = mv' - mv = m(v' + (-v)) \tag{8.17}$$

As shown, the direction of the change in velocity is the same as the direction of the change in momentum, which in turn is in the same direction as the force exerted by the edge of the table. Note that there is only a horizontal change in velocity. There is no difference in the vertical components of the initial and final velocity vectors; therefore, there is no vertical component to the change in velocity vector or the change in momentum vector. This is consistent with the fact that the force exerted by the edge of the table is purely in the horizontal direction.

8.2 Impulse

Learning Objectives

By the end of this section, you will be able to:

- Define impulse.
- Describe effects of impulses in everyday life.
- Determine the average effective force using graphical representation.
- Calculate average force and impulse given mass, velocity, and time.

The information presented in this section supports the following AP® learning objectives and science practices:

- **3.D.2.1** The student is able to justify the selection of routines for the calculation of the relationships between changes in momentum of an object, average force, impulse, and time of interaction. **(S.P. 2.1)**
- **3.D.2.2** The student is able to predict the change in momentum of an object from the average force exerted on the object and the interval of time during which the force is exerted. **(S.P. 6.4)**
- **3.D.2.3** The student is able to analyze data to characterize the change in momentum of an object from the average force exerted on the object and the interval of time during which the force is exerted. **(S.P. 5.1)**
- **3.D.2.4** The student is able to design a plan for collecting data to investigate the relationship between changes in momentum and the average force exerted on an object over time. **(S.P. 4.1)**
- **4.B.2.1** The student is able to apply mathematical routines to calculate the change in momentum of a system by analyzing the average force exerted over a certain time on the system. **(S.P. 2.2)**
- **4.B.2.2** The student is able to perform analysis on data presented as a force-time graph and predict the change in momentum of a system. **(S.P. 5.1)**

The effect of a force on an object depends on how long it acts, as well as how great the force is. In **Example 8.1**, a very large force acting for a short time had a great effect on the momentum of the tennis ball. A small force could cause the same **change in momentum**, but it would have to act for a much longer time. For example, if the ball were thrown upward, the gravitational force (which is much smaller than the tennis racquet's force) would eventually reverse the momentum of the ball. Quantitatively, the effect we are talking about is the change in momentum $\Delta \mathbf{p}$.

By rearranging the equation $\mathbf{F}_{net} = \dfrac{\Delta \mathbf{p}}{\Delta t}$ to be

$$\Delta \mathbf{p} = \mathbf{F}_{net}\Delta t, \tag{8.18}$$

we can see how the change in momentum equals the average net external force multiplied by the time this force acts. The quantity $\mathbf{F}_{net}\,\Delta t$ is given the name **impulse**. Impulse is the same as the change in momentum.

Impulse: Change in Momentum

Change in momentum equals the average net external force multiplied by the time this force acts.

$$\Delta \mathbf{p} = \mathbf{F}_{net}\Delta t \tag{8.19}$$

The quantity $\mathbf{F}_{net}\,\Delta t$ is given the name impulse.

There are many ways in which an understanding of impulse can save lives, or at least limbs. The dashboard padding in a car, and certainly the airbags, allow the net force on the occupants in the car to act over a much longer time when there is a sudden stop. The momentum change is the same for an occupant, whether an air bag is deployed or not, but the force (to bring the occupant to a stop) will be much less if it acts over a larger time. Cars today have many plastic components. One advantage of plastics is their lighter weight, which results in better gas mileage. Another advantage is that a car will crumple in a collision, especially in the event of a head-on collision. A longer collision time means the force on the car will be less. Deaths during car races decreased dramatically when the rigid frames of racing cars were replaced with parts that could crumple or collapse in the event of an accident.

Bones in a body will fracture if the force on them is too large. If you jump onto the floor from a table, the force on your legs can be immense if you land stiff-legged on a hard surface. Rolling on the ground after jumping from the table, or landing with a parachute, extends the time over which the force (on you from the ground) acts.

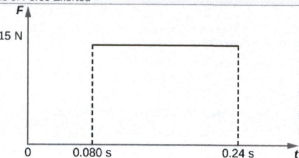

Figure 8.3 This is a graph showing the force exerted by a fixed barrier on a block versus time.

A 1.2-kg block slides across a horizontal, frictionless surface with a constant speed of 3.0 m/s before striking a fixed barrier and coming to a stop. In **Figure 8.3**, the force exerted by the barrier is assumed to be a constant 15 N during the 0.24-s collision. The impulse can be calculated using the area under the curve.

$$\Delta p = F\Delta t = (15 \ \text{N})(0.24 \ \text{s}) = \ 3.6 \ \text{kg} \bullet \text{m/s} \tag{8.20}$$

Note that the initial momentum of the block is:

$$p_{initial} = mv_{initial} = (1.2 \ \text{kg})(-3.0 \ \text{m/s}) = \ -3.6 \ \text{kg} \bullet \text{m/s} \tag{8.21}$$

We are assuming that the initial velocity is –3.0 m/s. We have established that the force exerted by the barrier is in the positive direction, so the initial velocity of the block must be in the negative direction. Since the final momentum of the block is zero, the impulse is equal to the change in momentum of the block.

Suppose that, instead of striking a fixed barrier, the block is instead stopped by a spring. Consider the force exerted by the spring over the time interval from the beginning of the collision until the block comes to rest.

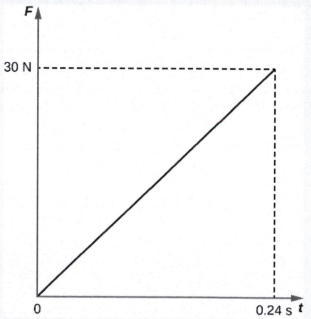

Figure 8.4 This is a graph showing the force exerted by a spring on a block versus time.

In this case, the impulse can be calculated again using the area under the curve (the area of a triangle):

$$\Delta p = \frac{1}{2}(\text{base})(\text{height}) = \frac{1}{2}(0.24 \ \text{s})(30 \ \text{N}) = \ 3.6 \ \text{kg} \bullet \text{m/s} \tag{8.22}$$

Again, this is equal to the difference between the initial and final momentum of the block, so the impulse is equal to the change in momentum.

Example 8.3 Calculating Magnitudes of Impulses: Two Billiard Balls Striking a Rigid Wall

Two identical billiard balls strike a rigid wall with the same speed, and are reflected without any change of speed. The first ball strikes perpendicular to the wall. The second ball strikes the wall at an angle of $30°$ from the perpendicular, and bounces off at an angle of $30°$ from perpendicular to the wall.

(a) Determine the direction of the force on the wall due to each ball.

(b) Calculate the ratio of the magnitudes of impulses on the two balls by the wall.

Strategy for (a)

In order to determine the force on the wall, consider the force on the ball due to the wall using Newton's second law and then apply Newton's third law to determine the direction. Assume the x-axis to be normal to the wall and to be positive in the initial direction of motion. Choose the y-axis to be along the wall in the plane of the second ball's motion. The momentum direction and the velocity direction are the same.

Solution for (a)

The first ball bounces directly into the wall and exerts a force on it in the $+x$ direction. Therefore the wall exerts a force on the ball in the $-x$ direction. The second ball continues with the same momentum component in the y direction, but reverses its x-component of momentum, as seen by sketching a diagram of the angles involved and keeping in mind the proportionality between velocity and momentum.

These changes mean the change in momentum for both balls is in the $-x$ direction, so the force of the wall on each ball is along the $-x$ direction.

Strategy for (b)

Calculate the change in momentum for each ball, which is equal to the impulse imparted to the ball.

Solution for (b)

Let u be the speed of each ball before and after collision with the wall, and m the mass of each ball. Choose the x-axis and y-axis as previously described, and consider the change in momentum of the first ball which strikes perpendicular to the wall.

$$p_{xi} = mu; \; p_{yi} = 0 \tag{8.23}$$

$$p_{xf} = -mu; \; p_{yf} = 0 \tag{8.24}$$

Impulse is the change in momentum vector. Therefore the x-component of impulse is equal to $-2mu$ and the y-component of impulse is equal to zero.

Now consider the change in momentum of the second ball.

$$p_{xi} = mu \cos 30°; \; p_{yi} = -mu \sin 30° \tag{8.25}$$

$$p_{xf} = -mu \cos 30°; \; p_{yf} = -mu \sin 30° \tag{8.26}$$

It should be noted here that while p_x changes sign after the collision, p_y does not. Therefore the x-component of impulse is equal to $-2mu \cos 30°$ and the y-component of impulse is equal to zero.

The ratio of the magnitudes of the impulse imparted to the balls is

$$\frac{2mu}{2mu \cos 30°} = \frac{2}{\sqrt{3}} = 1.155. \tag{8.27}$$

Discussion

The direction of impulse and force is the same as in the case of (a); it is normal to the wall and along the negative x-direction. Making use of Newton's third law, the force on the wall due to each ball is normal to the wall along the positive x-direction.

Our definition of impulse includes an assumption that the force is constant over the time interval Δt. *Forces are usually not constant.* Forces vary considerably even during the brief time intervals considered. It is, however, possible to find an average effective force F_{eff} that produces the same result as the corresponding time-varying force. **Figure 8.5** shows a graph of what an actual force looks like as a function of time for a ball bouncing off the floor. The area under the curve has units of momentum and is equal to the impulse or change in momentum between times t_1 and t_2. That area is equal to the area inside the rectangle bounded by F_{eff}, t_1, and t_2. Thus the impulses and their effects are the same for both the actual and effective forces.

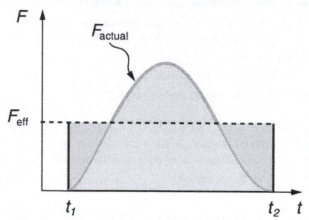

Figure 8.5 A graph of force versus time with time along the x-axis and force along the y-axis for an actual force and an equivalent effective force. The areas under the two curves are equal.

Making Connections: Baseball

In most real-life collisions, the forces acting on an object are not constant. For example, when a bat strikes a baseball, the force is very small at the beginning of the collision since only a small portion of the ball is initially in contact with the bat. As the collision continues, the ball deforms so that a greater fraction of the ball is in contact with the bat, resulting in a greater force. As the ball begins to leave the bat, the force drops to zero, much like the force curve in **Figure 8.5**. Although the changing force is difficult to precisely calculate at each instant, the average force can be estimated very well in most cases.

Suppose that a 150-g baseball experiences an average force of 480 N in a direction opposite the initial 32 m/s speed of the baseball over a time interval of 0.017 s. What is the final velocity of the baseball after the collision?

$$\Delta p = F \Delta t = (480)(0.017) = \quad 8.16 \quad \text{kg} \bullet \text{m/s} \tag{8.28}$$

$$mv_f - mv_i = 8.16 \quad \text{kg} \bullet \text{m/s} \tag{8.29}$$

$$(0.150 \text{ kg})v_f - (0.150 \text{ kg})(-32 \text{ m/s}) = 8.16 \quad \text{kg} \bullet \text{m/s} \tag{8.30}$$

$$v_f = 22 \text{ m/s} \tag{8.31}$$

Note in the above example that the initial velocity of the baseball prior to the collision is negative, consistent with the assumption we initially made that the force exerted by the bat is positive and in the direction opposite the initial velocity of the baseball. In this case, even though the force acting on the baseball varies with time, the average force is a good approximation of the effective force acting on the ball for the purposes of calculating the impulse and the change in momentum.

Making Connections: Take-Home Investigation—Hand Movement and Impulse

Try catching a ball while "giving" with the ball, pulling your hands toward your body. Then, try catching a ball while keeping your hands still. Hit water in a tub with your full palm. After the water has settled, hit the water again by diving your hand with your fingers first into the water. (Your full palm represents a swimmer doing a belly flop and your diving hand represents a swimmer doing a dive.) Explain what happens in each case and why. Which orientations would you advise people to avoid and why?

Making Connections: Constant Force and Constant Acceleration

The assumption of a constant force in the definition of impulse is analogous to the assumption of a constant acceleration in kinematics. In both cases, nature is adequately described without the use of calculus.

Applying the Science Practices: Verifying the Relationship between Force and Change in Linear Momentum

Design an experiment in order to experimentally verify the relationship between the impulse of a force and change in linear momentum. For simplicity, it would be best to ensure that frictional forces are very small or zero in your experiment so that the effect of friction can be neglected. As you design your experiment, consider the following:

- Would it be easier to analyze a one-dimensional collision or a two-dimensional collision?
- How will you measure the force?
- Should you have two objects in motion or one object bouncing off a rigid surface?
- How will you measure the duration of the collision?
- How will you measure the initial and final velocities of the object(s)?

- Would it be easier to analyze an elastic or inelastic collision?
- Should you verify the relationship mathematically or graphically?

8.3 Conservation of Momentum

<table>
<tr><td colspan="2">Learning Objectives</td></tr>
</table>

By the end of this section, you will be able to:

- Describe the law of conservation of linear momentum.
- Derive an expression for the conservation of momentum.
- Explain conservation of momentum with examples.
- Explain the law of conservation of momentum as it relates to atomic and subatomic particles.

The information presented in this section supports the following AP® learning objectives and science practices:

- **5.A.2.1** The student is able to define open and closed systems for everyday situations and apply conservation concepts for energy, charge, and linear momentum to those situations. **(S.P. 6.4, 7.2)**
- **5.D.1.4** The student is able to design an experimental test of an application of the principle of the conservation of linear momentum, predict an outcome of the experiment using the principle, analyze data generated by that experiment whose uncertainties are expressed numerically, and evaluate the match between the prediction and the outcome. **(S.P. 4.2, 5.1, 5.3, 6.4)**
- **5.D.2.1** The student is able to qualitatively predict, in terms of linear momentum and kinetic energy, how the outcome of a collision between two objects changes depending on whether the collision is elastic or inelastic. **(S.P. 6.4, 7.2)**
- **5.D.2.2** The student is able to plan data collection strategies to test the law of conservation of momentum in a two-object collision that is elastic or inelastic and analyze the resulting data graphically. **(S.P.4.1, 4.2, 5.1)**
- **5.D.3.1** The student is able to predict the velocity of the center of mass of a system when there is no interaction outside of the system but there is an interaction within the system (i.e., the student simply recognizes that interactions within a system do not affect the center of mass motion of the system and is able to determine that there is no external force). **(S.P. 6.4)**

Momentum is an important quantity because it is conserved. Yet it was not conserved in the examples in Impulse and Linear Momentum and Force, where large changes in momentum were produced by forces acting on the system of interest. Under what circumstances is momentum conserved?

The answer to this question entails considering a sufficiently large system. It is always possible to find a larger system in which total momentum is constant, even if momentum changes for components of the system. If a football player runs into the goalpost in the end zone, there will be a force on him that causes him to bounce backward. However, the Earth also recoils —conserving momentum—because of the force applied to it through the goalpost. Because Earth is many orders of magnitude more massive than the player, its recoil is immeasurably small and can be neglected in any practical sense, but it is real nevertheless.

Consider what happens if the masses of two colliding objects are more similar than the masses of a football player and Earth—for example, one car bumping into another, as shown in Figure 8.6. Both cars are coasting in the same direction when the lead car (labeled m_2) is bumped by the trailing car (labeled m_1). The only unbalanced force on each car is the force of the collision. (Assume that the effects due to friction are negligible.) Car 1 slows down as a result of the collision, losing some momentum, while car 2 speeds up and gains some momentum. We shall now show that the total momentum of the two-car system remains constant.

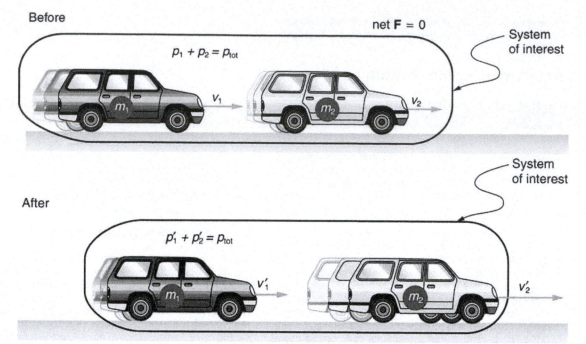

Figure 8.6 A car of mass m_1 moving with a velocity of v_1 bumps into another car of mass m_2 and velocity v_2 that it is following. As a result, the first car slows down to a velocity of v'_1 and the second speeds up to a velocity of v'_2. The momentum of each car is changed, but the total momentum p_{tot} of the two cars is the same before and after the collision (if you assume friction is negligible).

Using the definition of impulse, the change in momentum of car 1 is given by

$$\Delta p_1 = F_1 \Delta t, \tag{8.32}$$

where F_1 is the force on car 1 due to car 2, and Δt is the time the force acts (the duration of the collision). Intuitively, it seems obvious that the collision time is the same for both cars, but it is only true for objects traveling at ordinary speeds. This assumption must be modified for objects travelling near the speed of light, without affecting the result that momentum is conserved.

Similarly, the change in momentum of car 2 is

$$\Delta p_2 = F_2 \Delta t, \tag{8.33}$$

where F_2 is the force on car 2 due to car 1, and we assume the duration of the collision Δt is the same for both cars. We know from Newton's third law that $F_2 = -F_1$, and so

$$\Delta p_2 = -F_1 \Delta t = -\Delta p_1. \tag{8.34}$$

Thus, the changes in momentum are equal and opposite, and

$$\Delta p_1 + \Delta p_2 = 0. \tag{8.35}$$

Because the changes in momentum add to zero, the total momentum of the two-car system is constant. That is,

$$p_1 + p_2 = \text{constant}, \tag{8.36}$$
$$p_1 + p_2 = p'_1 + p'_2, \tag{8.37}$$

where p'_1 and p'_2 are the momenta of cars 1 and 2 after the collision. (We often use primes to denote the final state.)

This result—that momentum is conserved—has validity far beyond the preceding one-dimensional case. It can be similarly shown that total momentum is conserved for any isolated system, with any number of objects in it. In equation form, the **conservation of momentum principle** for an isolated system is written

$$\mathbf{P}_{tot} = \text{constant}, \tag{8.38}$$

or

$$\mathbf{P}_{tot} = \mathbf{P}'_{tot}, \tag{8.39}$$

where $\mathbf{P}_{tot}$ is the total momentum (the sum of the momenta of the individual objects in the system) and $\mathbf{P}'_{tot}$ is the total momentum some time later. (The total momentum can be shown to be the momentum of the center of mass of the system.) An **isolated system** is defined to be one for which the net external force is zero $(\mathbf{F}_{net} = 0)$.

Conservation of Momentum Principle

$$\mathbf{p}_{tot} = \text{constant}$$
$$\mathbf{p}_{tot} = \mathbf{p}'_{tot} \quad \text{(isolated system)}$$

(8.40)

Isolated System

An isolated system is defined to be one for which the net external force is zero $(\mathbf{F}_{net} = 0)$.

Making Connections: Cart Collisions

Consider two air carts with equal mass *(m)* on a linear track. The first cart moves with a speed *v* towards the second cart, which is initially at rest. We will take the initial direction of motion of the first cart as the positive direction.

The momentum of the system will be conserved in the collision. If the collision is elastic, then the first cart will stop after the collision. Conservation of momentum therefore tells us that the second cart will have a final velocity *v* after the collision in the same direction as the initial velocity of the first cart.

The kinetic energy of the system will be conserved since the masses are equal and the final velocity of cart 2 is equal to the initial velocity of cart 1. What would a graph of total momentum vs. time look like in this case? What would a graph of total kinetic energy vs. time look like in this case?

Consider the center of mass of this system as the frame of reference. As cart 1 approaches cart 2, the center of mass remains exactly halfway between the two carts. The center of mass moves toward the stationary cart 2 at a speed $\frac{v}{2}$. After the collision, the center of mass continues moving in the same direction, away from (now stationary) cart 1 at a speed $\frac{v}{2}$.

How would a graph of center-of-mass velocity vs. time compare to a graph of momentum vs. time?

Suppose instead that the two carts move with equal speeds *v* in opposite directions towards the center of mass. Again, they have an elastic collision, so after the collision, they exchange velocities (each cart moving in the opposite direction of its initial motion with the same speed). As the two carts approach, the center of mass is exactly between the two carts, at the point where they will collide. In this case, how would a graph of center-of-mass velocity vs. time compare to a graph of the momentum of the system vs. time?

Let us return to the example where the first cart is moving with a speed *v* toward the second cart, initially at rest. Suppose the second cart has some putty on one end so that, when the collision occurs, the two carts stick together in an inelastic collision. In this case, conservation of momentum tells us that the final velocity of the two-cart system will be half the initial velocity of the first cart, in the same direction as the first cart's initial motion. Kinetic energy will not be conserved in this case, however. Compared to the moving cart before the collision, the overall moving mass after the collision is doubled, but the velocity is halved.

The initial kinetic energy of the system is:

$$k_i = \tfrac{1}{2}mv^2(\text{1st cart}) + 0(\text{2nd cart}) = \tfrac{1}{2}mv^2$$

(8.41)

The final kinetic energy of the two carts (2*m*) moving together (at speed *v*/2) is:

$$k_f = \tfrac{1}{2}(2m)\left(\tfrac{v}{2}\right)^2 = \tfrac{1}{4}mv^2$$

(8.42)

What would a graph of total momentum vs. time look like in this case? What would a graph of total kinetic energy vs. time look like in this case?

Consider the center of mass of this system. As cart 1 approaches cart 2, the center of mass remains exactly halfway between the two carts. The center of mass moves toward the stationary cart 2 at a speed $\frac{v}{2}$. After the collision, the two carts move together at a speed $\frac{v}{2}$. How would a graph of center-of-mass velocity vs. time compare to a graph of momentum vs. time?

Suppose instead that the two carts move with equal speeds *v* in opposite directions towards the center of mass. They have putty on the end of each cart so that they stick together after the collision. As the two carts approach, the center of mass is exactly between the two carts, at the point where they will collide. In this case, how would a graph of center-of-mass velocity vs. time compare to a graph of the momentum of the system vs. time?

Perhaps an easier way to see that momentum is conserved for an isolated system is to consider Newton's second law in terms of momentum, $\mathbf{F}_{\text{net}} = \dfrac{\Delta \mathbf{p}_{\text{tot}}}{\Delta t}$. For an isolated system, $(\mathbf{F}_{\text{net}} = 0)$; thus, $\Delta \mathbf{p}_{\text{tot}} = 0$, and $\mathbf{p}_{\text{tot}}$ is constant.

We have noted that the three length dimensions in nature—x, y, and z—are independent, and it is interesting to note that momentum can be conserved in different ways along each dimension. For example, during projectile motion and where air resistance is negligible, momentum is conserved in the horizontal direction because horizontal forces are zero and momentum is unchanged. But along the vertical direction, the net vertical force is not zero and the momentum of the projectile is not conserved. (See **Figure 8.7**.) However, if the momentum of the projectile-Earth system is considered in the vertical direction, we find that the total momentum is conserved.

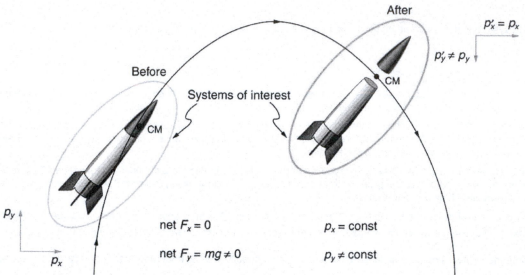

Figure 8.7 The horizontal component of a projectile's momentum is conserved if air resistance is negligible, even in this case where a space probe separates. The forces causing the separation are internal to the system, so that the net external horizontal force $F_{x-\text{net}}$ is still zero. The vertical component of the momentum is not conserved, because the net vertical force $F_{y-\text{net}}$ is not zero. In the vertical direction, the space probe-Earth system needs to be considered and we find that the total momentum is conserved. The center of mass of the space probe takes the same path it would if the separation did not occur.

The conservation of momentum principle can be applied to systems as different as a comet striking Earth and a gas containing huge numbers of atoms and molecules. Conservation of momentum is violated only when the net external force is not zero. But another larger system can always be considered in which momentum is conserved by simply including the source of the external force. For example, in the collision of two cars considered above, the two-car system conserves momentum while each one-car system does not.

Making Connections: Take-Home Investigation—Drop of Tennis Ball and a Basketball

Hold a tennis ball side by side and in contact with a basketball. Drop the balls together. (Be careful!) What happens? Explain your observations. Now hold the tennis ball above and in contact with the basketball. What happened? Explain your observations. What do you think will happen if the basketball ball is held above and in contact with the tennis ball?

Making Connections: Take-Home Investigation—Two Tennis Balls in a Ballistic Trajectory

Tie two tennis balls together with a string about a foot long. Hold one ball and let the other hang down and throw it in a ballistic trajectory. Explain your observations. Now mark the center of the string with bright ink or attach a brightly colored sticker to it and throw again. What happened? Explain your observations.

Some aquatic animals such as jellyfish move around based on the principles of conservation of momentum. A jellyfish fills its umbrella section with water and then pushes the water out resulting in motion in the opposite direction to that of the jet of water. Squids propel themselves in a similar manner but, in contrast with jellyfish, are able to control the direction in which they move by aiming their nozzle forward or backward. Typical squids can move at speeds of 8 to 12 km/h.

The ballistocardiograph (BCG) was a diagnostic tool used in the second half of the 20th century to study the strength of the heart. About once a second, your heart beats, forcing blood into the aorta. A force in the opposite direction is exerted on the rest of your body (recall Newton's third law). A ballistocardiograph is a device that can measure this reaction force. This measurement is done by using a sensor (resting on the person) or by using a moving table suspended from the ceiling. This technique can gather information on the strength of the heart beat and the volume of blood passing from the heart. However, the electrocardiogram (ECG or EKG) and the echocardiogram (cardiac ECHO or ECHO; a technique that uses ultrasound to see an image of the heart) are more widely used in the practice of cardiology.

Design an experiment to verify the conservation of linear momentum in a one-dimensional collision, both elastic and inelastic. For simplicity, try to ensure that friction is minimized so that it has a negligible effect on your experiment. As you consider your experiment, consider the following questions:

- Predict how the final momentum of the system will compare to the initial momentum of the system that you will measure. Justify your prediction.
- How will you measure the momentum of each object?
- Should you have two objects in motion or one object bouncing off a rigid surface?
- Should you verify the relationship mathematically or graphically?
- How will you estimate the uncertainty of your measurements? How will you express this uncertainty in your data?

When you have completed each experiment, compare the outcome to your prediction about the initial and final momentum of the system and evaluate your results.

Conservation of momentum is quite useful in describing collisions. Momentum is crucial to our understanding of atomic and subatomic particles because much of what we know about these particles comes from collision experiments.

Subatomic Collisions and Momentum

The conservation of momentum principle not only applies to the macroscopic objects, it is also essential to our explorations of atomic and subatomic particles. Giant machines hurl subatomic particles at one another, and researchers evaluate the results by assuming conservation of momentum (among other things).

On the small scale, we find that particles and their properties are invisible to the naked eye but can be measured with our instruments, and models of these subatomic particles can be constructed to describe the results. Momentum is found to be a property of all subatomic particles including massless particles such as photons that compose light. Momentum being a property of particles hints that momentum may have an identity beyond the description of an object's mass multiplied by the object's velocity. Indeed, momentum relates to wave properties and plays a fundamental role in what measurements are taken and how we take these measurements. Furthermore, we find that the conservation of momentum principle is valid when considering systems of particles. We use this principle to analyze the masses and other properties of previously undetected particles, such as the nucleus of an atom and the existence of quarks that make up particles of nuclei. **Figure 8.8** below illustrates how a particle scattering backward from another implies that its target is massive and dense. Experiments seeking evidence that **quarks** make up protons (one type of particle that makes up nuclei) scattered high-energy electrons off of protons (nuclei of hydrogen atoms). Electrons occasionally scattered straight backward in a manner that implied a very small and very dense particle makes up the proton—this observation is considered nearly direct evidence of quarks. The analysis was based partly on the same conservation of momentum principle that works so well on the large scale.

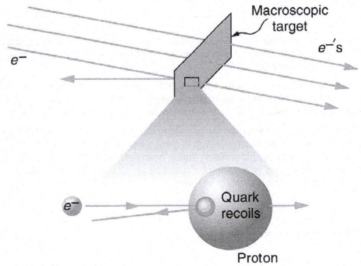

Figure 8.8 A subatomic particle scatters straight backward from a target particle. In experiments seeking evidence for quarks, electrons were observed to occasionally scatter straight backward from a proton.

8.4 Elastic Collisions in One Dimension

Learning Objectives

By the end of this section, you will be able to:

- Describe an elastic collision of two objects in one dimension.
- Define internal kinetic energy.
- Derive an expression for conservation of internal kinetic energy in a one-dimensional collision.
- Determine the final velocities in an elastic collision given masses and initial velocities.

The information presented in this section supports the following AP® learning objectives and science practices:

- **4.B.1.1** The student is able to calculate the change in linear momentum of a two-object system with constant mass in linear motion from a representation of the system (data, graphs, etc.). **(S.P. 1.4, 2.2)**
- **4.B.1.2** The student is able to analyze data to find the change in linear momentum for a constant-mass system using the product of the mass and the change in velocity of the center of mass. **(S.P. 5.1)**
- **5.A.2.1** The student is able to define open and closed systems for everyday situations and apply conservation concepts for energy, charge, and linear momentum to those situations. **(S.P. 6.4, 7.2)**
- **5.D.1.1** The student is able to make qualitative predictions about natural phenomena based on conservation of linear momentum and restoration of kinetic energy in elastic collisions. **(S.P. 6.4, 7.2)**
- **5.D.1.2** The student is able to apply the principles of conservation of momentum and restoration of kinetic energy to reconcile a situation that appears to be isolated and elastic, but in which data indicate that linear momentum and kinetic energy are not the same after the interaction, by refining a scientific question to identify interactions that have not been considered. Students will be expected to solve qualitatively and/or quantitatively for one-dimensional situations and only qualitatively in two-dimensional situations. **(S.P. 2.2, 3.2, 5.1, 5.3)**
- **5.D.1.5** The student is able to classify a given collision situation as elastic or inelastic, justify the selection of conservation of linear momentum and restoration of kinetic energy as the appropriate principles for analyzing an elastic collision, solve for missing variables, and calculate their values. **(S.P. 2.1, 2.2)**
- **5.D.1.6** The student is able to make predictions of the dynamical properties of a system undergoing a collision by application of the principle of linear momentum conservation and the principle of the conservation of energy in situations in which an elastic collision may also be assumed. **(S.P. 6.4)**
- **5.D.1.7** The student is able to classify a given collision situation as elastic or inelastic, justify the selection of conservation of linear momentum and restoration of kinetic energy as the appropriate principles for analyzing an elastic collision, solve for missing variables, and calculate their values. **(S.P. 2.1, 2.2)**
- **5.D.2.1** The student is able to qualitatively predict, in terms of linear momentum and kinetic energy, how the outcome of a collision between two objects changes depending on whether the collision is elastic or inelastic. **(S.P. 6.4, 7.2)**
- **5.D.2.2** The student is able to plan data collection strategies to test the law of conservation of momentum in a two-object collision that is elastic or inelastic and analyze the resulting data graphically. **(S.P.4.1, 4.2, 5.1)**
- **5.D.3.2** The student is able to make predictions about the velocity of the center of mass for interactions within a defined one-dimensional system. **(S.P. 6.4)**

Let us consider various types of two-object collisions. These collisions are the easiest to analyze, and they illustrate many of the physical principles involved in collisions. The conservation of momentum principle is very useful here, and it can be used whenever the net external force on a system is zero.

We start with the elastic collision of two objects moving along the same line—a one-dimensional problem. An **elastic collision** is one that also conserves internal kinetic energy. **Internal kinetic energy** is the sum of the kinetic energies of the objects in the system. **Figure 8.9** illustrates an elastic collision in which internal kinetic energy and momentum are conserved.

Truly elastic collisions can only be achieved with subatomic particles, such as electrons striking nuclei. Macroscopic collisions can be very nearly, but not quite, elastic—some kinetic energy is always converted into other forms of energy such as heat transfer due to friction and sound. One macroscopic collision that is nearly elastic is that of two steel blocks on ice. Another nearly elastic collision is that between two carts with spring bumpers on an air track. Icy surfaces and air tracks are nearly frictionless, more readily allowing nearly elastic collisions on them.

Elastic Collision

An **elastic collision** is one that conserves internal kinetic energy.

Internal Kinetic Energy

Internal kinetic energy is the sum of the kinetic energies of the objects in the system.

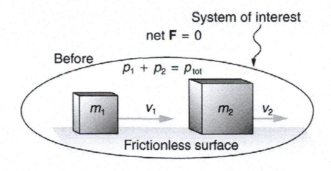

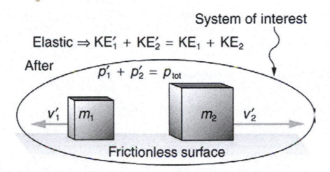

Figure 8.9 An elastic one-dimensional two-object collision. Momentum and internal kinetic energy are conserved.

Now, to solve problems involving one-dimensional elastic collisions between two objects we can use the equations for conservation of momentum and conservation of internal kinetic energy. First, the equation for conservation of momentum for two objects in a one-dimensional collision is

$$p_1 + p_2 = p'_1 + p'_2 \quad (F_{\text{net}} = 0) \tag{8.43}$$

or

$$m_1 v_1 + m_2 v_2 = m_1 v'_1 + m_2 v'_2 \quad (F_{\text{net}} = 0), \tag{8.44}$$

where the primes (') indicate values after the collision. By definition, an elastic collision conserves internal kinetic energy, and so the sum of kinetic energies before the collision equals the sum after the collision. Thus,

$$\tfrac{1}{2} m_1 v_1{}^2 + \tfrac{1}{2} m_2 v_2{}^2 = \tfrac{1}{2} m_1 v'_1{}^2 + \tfrac{1}{2} m_2 v'_2{}^2 \quad \text{(two-object elastic collision)} \tag{8.45}$$

expresses the equation for conservation of internal kinetic energy in a one-dimensional collision.

Making Connections: Collisions

Suppose data are collected on a collision between two masses sliding across a frictionless surface. Mass A (1.0 kg) moves with a velocity of +12 m/s, and mass B (2.0 kg) moves with a velocity of −12 m/s. The two masses collide and stick together after the collision. The table below shows the measured velocities of each mass at times before and after the collision:

Table 8.1

Time (s)	Velocity A (m/s)	Velocity B (m/s)
0	+12	−12
1.0 s	+12	−12
2.0 s	−4.0	−4.0
3.0 s	−4.0	−4.0

The total mass of the system is 3.0 kg. The velocity of the center of mass of this system can be determined from the conservation of momentum. Consider the system before the collision:

$$(m_A + m_B)v_{cm} = m_A v_A + m_B v_B \tag{8.46}$$

$$(3.0)v_{cm} = (1)(12) + (2)(-12) \tag{8.47}$$

$$v_{cm} = -4.0 \text{ m/s} \tag{8.48}$$

After the collision, the center-of-mass velocity is the same:

$$(m_A + m_B)v_{cm} = (m_A + m_B)v_{final} \tag{8.49}$$

$$(3.0)v_{cm} = (3)(-4.0) \tag{8.50}$$

$$v_{cm} = -4.0 \text{ m/s} \tag{8.51}$$

The total momentum of the system before the collision is:

$$m_A v_A + m_B v_B = (1)(12) + (2)(-12) = -12 \text{ kg} \bullet \text{m/s} \tag{8.52}$$

The total momentum of the system after the collision is:

$$(m_A + m_B)v_{final} = (3)(-4) = -12 \text{ kg} \bullet \text{m/s} \tag{8.53}$$

Thus, the change in momentum of the system is zero when measured this way. We get a similar result when we calculate the momentum using the center-of-mass velocity. Since the center-of-mass velocity is the same both before and after the collision, we calculate the same momentum for the system using this method both before and after the collision.

Example 8.4 Calculating Velocities Following an Elastic Collision

Calculate the velocities of two objects following an elastic collision, given that

$$m_1 = 0.500 \text{ kg}, \ m_2 = 3.50 \text{ kg}, \ v_1 = 4.00 \text{ m/s, and } \ v_2 = 0. \tag{8.54}$$

Strategy and Concept

First, visualize what the initial conditions mean—a small object strikes a larger object that is initially at rest. This situation is slightly simpler than the situation shown in **Figure 8.9** where both objects are initially moving. We are asked to find two unknowns (the final velocities v'_1 and v'_2). To find two unknowns, we must use two independent equations. Because this collision is elastic, we can use the above two equations. Both can be simplified by the fact that object 2 is initially at rest, and thus $v_2 = 0$. Once we simplify these equations, we combine them algebraically to solve for the unknowns.

Solution

For this problem, note that $v_2 = 0$ and use conservation of momentum. Thus,

$$p_1 = p'_1 + p'_2 \tag{8.55}$$

or

$$m_1 v_1 = m_1 v'_1 + m_2 v'_2. \tag{8.56}$$

Using conservation of internal kinetic energy and that $v_2 = 0$,

$$\frac{1}{2}m_1 v_1^2 = \frac{1}{2}m_1 v'_1{}^2 + \frac{1}{2}m_2 v'_2{}^2. \tag{8.57}$$

Solving the first equation (momentum equation) for v'_2, we obtain

$$v'_2 = \frac{m_1}{m_2}(v_1 - v'_1). \tag{8.58}$$

Substituting this expression into the second equation (internal kinetic energy equation) eliminates the variable v'_2, leaving only v'_1 as an unknown (the algebra is left as an exercise for the reader). There are two solutions to any quadratic equation; in this example, they are

$$v'_1 = 4.00 \text{ m/s} \tag{8.59}$$

and

$$v'_1 = -3.00 \text{ m/s}. \tag{8.60}$$

As noted when quadratic equations were encountered in earlier chapters, both solutions may or may not be meaningful. In this case, the first solution is the same as the initial condition. The first solution thus represents the situation before the collision and is discarded. The second solution ($v'_1 = -3.00$ m/s) is negative, meaning that the first object bounces backward. When this negative value of v'_1 is used to find the velocity of the second object after the collision, we get

$$v'_2 = \frac{m_1}{m_2}(v_1 - v'_1) = \frac{0.500 \text{ kg}}{3.50 \text{ kg}}[4.00 - (-3.00)] \text{ m/s} \tag{8.61}$$

or

$$v'_2 = 1.00 \text{ m/s}. \tag{8.62}$$

Discussion

The result of this example is intuitively reasonable. A small object strikes a larger one at rest and bounces backward. The larger one is knocked forward, but with a low speed. (This is like a compact car bouncing backward off a full-size SUV that is initially at rest.) As a check, try calculating the internal kinetic energy before and after the collision. You will see that the internal kinetic energy is unchanged at 4.00 J. Also check the total momentum before and after the collision; you will find it, too, is unchanged.

The equations for conservation of momentum and internal kinetic energy as written above can be used to describe any one-dimensional elastic collision of two objects. These equations can be extended to more objects if needed.

Making Connections: Take-Home Investigation—Ice Cubes and Elastic Collision

Find a few ice cubes which are about the same size and a smooth kitchen tabletop or a table with a glass top. Place the ice cubes on the surface several centimeters away from each other. Flick one ice cube toward a stationary ice cube and observe the path and velocities of the ice cubes after the collision. Try to avoid edge-on collisions and collisions with rotating ice cubes. Have you created approximately elastic collisions? Explain the speeds and directions of the ice cubes using momentum.

PhET Explorations: Collision Lab

Investigate collisions on an air hockey table. Set up your own experiments: vary the number of discs, masses and initial conditions. Is momentum conserved? Is kinetic energy conserved? Vary the elasticity and see what happens.

Figure 8.10 Collision Lab (http://cnx.org/content/m55171/1.7/collision-lab_en.jar)

8.5 Inelastic Collisions in One Dimension

Learning Objectives

By the end of this section, you will be able to:

- Define inelastic collision.
- Explain perfectly inelastic collisions.
- Apply an understanding of collisions to sports.
- Determine recoil velocity and loss in kinetic energy given mass and initial velocity.

The information presented in this section supports the following AP® learning objectives and science practices:

- **4.B.1.1** The student is able to calculate the change in linear momentum of a two-object system with constant mass in linear motion from a representation of the system (data, graphs, etc.). **(S.P. 1.4, 2.2)**
- **5.A.2.1** The student is able to define open and closed systems for everyday situations and apply conservation concepts for energy, charge, and linear momentum to those situations. **(S.P. 6.4, 7.2)**
- **5.D.1.3** The student is able to apply mathematical routines appropriately to problems involving elastic collisions in one dimension and justify the selection of those mathematical routines based on conservation of momentum and restoration of kinetic energy. **(S.P. 2.1, 2.2)**
- **5.D.1.5** The student is able to classify a given collision situation as elastic or inelastic, justify the selection of conservation of linear momentum and restoration of kinetic energy as the appropriate principles for analyzing an elastic collision, solve for missing variables, and calculate their values. **(S.P. 2.1, 2.2)**
- **5.D.2.1** The student is able to qualitatively predict, in terms of linear momentum and kinetic energy, how the outcome of a collision between two objects changes depending on whether the collision is elastic or inelastic. **(S.P. 6.4, 7.2)**
- **5.D.2.2** The student is able to plan data collection strategies to test the law of conservation of momentum in a two-object collision that is elastic or inelastic and analyze the resulting data graphically. **(S.P.4.1, 4.2, 5.1)**
- **5.D.2.3** The student is able to apply the conservation of linear momentum to a closed system of objects involved in an

inelastic collision to predict the change in kinetic energy. **(S.P. 6.4, 7.2)**
- **5.D.2.4** The student is able to analyze data that verify conservation of momentum in collisions with and without an external friction force. **(S.P. 4.1, 4.2, 4.4, 5.1, 5.3)**
- **5.D.2.5** The student is able to classify a given collision situation as elastic or inelastic, justify the selection of conservation of linear momentum as the appropriate solution method for an inelastic collision, recognize that there is a common final velocity for the colliding objects in the totally inelastic case, solve for missing variables, and calculate their values. **(S.P. 2.1 2.2)**
- **5.D.2.6** The student is able to apply the conservation of linear momentum to an isolated system of objects involved in an inelastic collision to predict the change in kinetic energy. **(S.P. 6.4, 7.2)**

We have seen that in an elastic collision, internal kinetic energy is conserved. An **inelastic collision** is one in which the internal kinetic energy changes (it is not conserved). This lack of conservation means that the forces between colliding objects may remove or add internal kinetic energy. Work done by internal forces may change the forms of energy within a system. For inelastic collisions, such as when colliding objects stick together, this internal work may transform some internal kinetic energy into heat transfer. Or it may convert stored energy into internal kinetic energy, such as when exploding bolts separate a satellite from its launch vehicle.

Inelastic Collision

An inelastic collision is one in which the internal kinetic energy changes (it is not conserved).

Figure 8.11 shows an example of an inelastic collision. Two objects that have equal masses head toward one another at equal speeds and then stick together. Their total internal kinetic energy is initially $\frac{1}{2}mv^2 + \frac{1}{2}mv^2 = mv^2$. The two objects come to rest after sticking together, conserving momentum. But the internal kinetic energy is zero after the collision. A collision in which the objects stick together is sometimes called a **perfectly inelastic collision** because it reduces internal kinetic energy more than does any other type of inelastic collision. In fact, such a collision reduces internal kinetic energy to the minimum it can have while still conserving momentum.

Perfectly Inelastic Collision

A collision in which the objects stick together is sometimes called "perfectly inelastic."

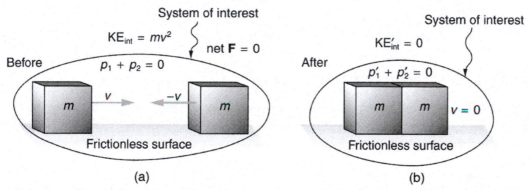

Figure 8.11 An inelastic one-dimensional two-object collision. Momentum is conserved, but internal kinetic energy is not conserved. (a) Two objects of equal mass initially head directly toward one another at the same speed. (b) The objects stick together (a perfectly inelastic collision), and so their final velocity is zero. The internal kinetic energy of the system changes in any inelastic collision and is reduced to zero in this example.

Example 8.5 Calculating Velocity and Change in Kinetic Energy: Inelastic Collision of a Puck and a Goalie

(a) Find the recoil velocity of a 70.0-kg ice hockey goalie, originally at rest, who catches a 0.150-kg hockey puck slapped at him at a velocity of 35.0 m/s. (b) How much kinetic energy is lost during the collision? Assume friction between the ice and the puck-goalie system is negligible. (See **Figure 8.12**)

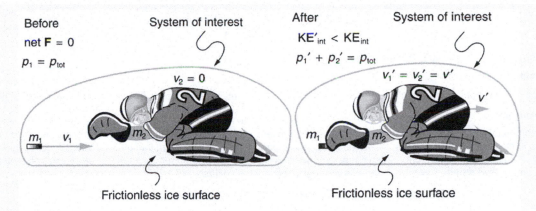

Figure 8.12 An ice hockey goalie catches a hockey puck and recoils backward. The initial kinetic energy of the puck is almost entirely converted to thermal energy and sound in this inelastic collision.

Strategy

Momentum is conserved because the net external force on the puck-goalie system is zero. We can thus use conservation of momentum to find the final velocity of the puck and goalie system. Note that the initial velocity of the goalie is zero and that the final velocity of the puck and goalie are the same. Once the final velocity is found, the kinetic energies can be calculated before and after the collision and compared as requested.

Solution for (a)

Momentum is conserved because the net external force on the puck-goalie system is zero.

Conservation of momentum is

$$p_1 + p_2 = p'_1 + p'_2 \tag{8.63}$$

or

$$m_1 v_1 + m_2 v_2 = m_1 v'_1 + m_2 v'_2. \tag{8.64}$$

Because the goalie is initially at rest, we know $v_2 = 0$. Because the goalie catches the puck, the final velocities are equal, or $v'_1 = v'_2 = v'$. Thus, the conservation of momentum equation simplifies to

$$m_1 v_1 = (m_1 + m_2)v'. \tag{8.65}$$

Solving for v' yields

$$v' = \frac{m_1}{m_1 + m_2} v_1. \tag{8.66}$$

Entering known values in this equation, we get

$$v' = \left(\frac{0.150 \text{ kg}}{70.0 \text{ kg} + 0.150 \text{ kg}}\right)(35.0 \text{ m/s}) = 7.48 \times 10^{-2} \text{ m/s}. \tag{8.67}$$

Discussion for (a)

This recoil velocity is small and in the same direction as the puck's original velocity, as we might expect.

Solution for (b)

Before the collision, the internal kinetic energy KE_{int} of the system is that of the hockey puck, because the goalie is initially at rest. Therefore, KE_{int} is initially

$$\begin{aligned} \text{KE}_{\text{int}} &= \tfrac{1}{2}mv^2 = \tfrac{1}{2}(0.150 \text{ kg})(35.0 \text{ m/s})^2 \\ &= 91.9 \text{ J}. \end{aligned} \tag{8.68}$$

After the collision, the internal kinetic energy is

$$\begin{aligned} \text{KE}'_{\text{int}} &= \tfrac{1}{2}(m + M)v^2 = \tfrac{1}{2}(70.15 \text{ kg})\left(7.48 \times 10^{-2} \text{ m/s}\right)^2 \\ &= 0.196 \text{ J}. \end{aligned} \tag{8.69}$$

The change in internal kinetic energy is thus

$$KE'_{int} - KE_{int} = 0.196 \text{ J} - 91.9 \text{ J}$$
$$= -91.7 \text{ J} \tag{8.70}$$

where the minus sign indicates that the energy was lost.

Discussion for (b)

Nearly all of the initial internal kinetic energy is lost in this perfectly inelastic collision. KE_{int} is mostly converted to thermal energy and sound.

During some collisions, the objects do not stick together and less of the internal kinetic energy is removed—such as happens in most automobile accidents. Alternatively, stored energy may be converted into internal kinetic energy during a collision. **Figure 8.13** shows a one-dimensional example in which two carts on an air track collide, releasing potential energy from a compressed spring. **Example 8.6** deals with data from such a collision.

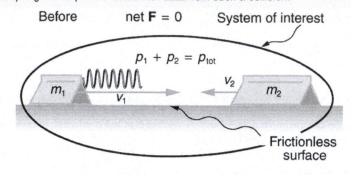

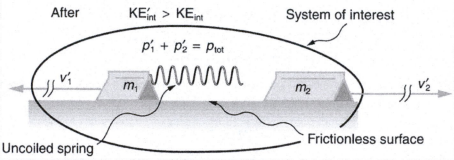

Figure 8.13 An air track is nearly frictionless, so that momentum is conserved. Motion is one-dimensional. In this collision, examined in **Example 8.6**, the potential energy of a compressed spring is released during the collision and is converted to internal kinetic energy.

Collisions are particularly important in sports and the sporting and leisure industry utilizes elastic and inelastic collisions. Let us look briefly at tennis. Recall that in a collision, it is momentum and not force that is important. So, a heavier tennis racquet will have the advantage over a lighter one. This conclusion also holds true for other sports—a lightweight bat (such as a softball bat) cannot hit a hardball very far.

The location of the impact of the tennis ball on the racquet is also important, as is the part of the stroke during which the impact occurs. A smooth motion results in the maximizing of the velocity of the ball after impact and reduces sports injuries such as tennis elbow. A tennis player tries to hit the ball on the "sweet spot" on the racquet, where the vibration and impact are minimized and the ball is able to be given more velocity. Sports science and technologies also use physics concepts such as momentum and rotational motion and vibrations.

Take-Home Experiment—Bouncing of Tennis Ball

1. Find a racquet (a tennis, badminton, or other racquet will do). Place the racquet on the floor and stand on the handle. Drop a tennis ball on the strings from a measured height. Measure how high the ball bounces. Now ask a friend to hold the racquet firmly by the handle and drop a tennis ball from the same measured height above the racquet. Measure how high the ball bounces and observe what happens to your friend's hand during the collision. Explain your observations and measurements.

2. The coefficient of restitution (c) is a measure of the elasticity of a collision between a ball and an object, and is defined as the ratio of the speeds after and before the collision. A perfectly elastic collision has a c of 1. For a ball bouncing off the floor (or a racquet on the floor), c can be shown to be $c = (h/H)^{1/2}$ where h is the height to which the ball bounces and H is the height from which the ball is dropped. Determine c for the cases in Part 1 and

for the case of a tennis ball bouncing off a concrete or wooden floor ($c = 0.85$ for new tennis balls used on a tennis court).

Example 8.6 Calculating Final Velocity and Energy Release: Two Carts Collide

In the collision pictured in **Figure 8.13**, two carts collide inelastically. Cart 1 (denoted m_1 carries a spring which is initially compressed. During the collision, the spring releases its potential energy and converts it to internal kinetic energy. The mass of cart 1 and the spring is 0.350 kg, and the cart and the spring together have an initial velocity of 2.00 m/s . Cart 2 (denoted m_2 in **Figure 8.13**) has a mass of 0.500 kg and an initial velocity of -0.500 m/s . After the collision, cart 1 is observed to recoil with a velocity of -4.00 m/s . (a) What is the final velocity of cart 2? (b) How much energy was released by the spring (assuming all of it was converted into internal kinetic energy)?

Strategy

We can use conservation of momentum to find the final velocity of cart 2, because $F_{net} = 0$ (the track is frictionless and the force of the spring is internal). Once this velocity is determined, we can compare the internal kinetic energy before and after the collision to see how much energy was released by the spring.

Solution for (a)

As before, the equation for conservation of momentum in a two-object system is

$$m_1 v_1 + m_2 v_2 = m_1 v'_1 + m_2 v'_2. \tag{8.71}$$

The only unknown in this equation is v'_2. Solving for v'_2 and substituting known values into the previous equation yields

$$\begin{aligned} v'_2 &= \frac{m_1 v_1 + m_2 v_2 - m_1 v'_1}{m_2} \\ &= \frac{(0.350 \text{ kg})(2.00 \text{ m/s}) + (0.500 \text{ kg})(-0.500 \text{ m/s})}{0.500 \text{ kg}} - \frac{(0.350 \text{ kg})(-4.00 \text{ m/s})}{0.500 \text{ kg}} \\ &= 3.70 \text{ m/s}. \end{aligned} \tag{8.72}$$

Solution for (b)

The internal kinetic energy before the collision is

$$\begin{aligned} KE_{int} &= \tfrac{1}{2}m_1 v_1^2 + \tfrac{1}{2}m_2 v_2^2 \\ &= \tfrac{1}{2}(0.350 \text{ kg})(2.00 \text{ m/s})^2 + \tfrac{1}{2}(0.500 \text{ kg})(- 0.500 \text{ m/s})^2 \\ &= 0.763 \text{ J}. \end{aligned} \tag{8.73}$$

After the collision, the internal kinetic energy is

$$\begin{aligned} KE'_{int} &= \tfrac{1}{2}m_1 v'^2_1 + \tfrac{1}{2}m_2 v'^2_2 \\ &= \tfrac{1}{2}(0.350 \text{ kg})(-4.00 \text{ m/s})^2 + \tfrac{1}{2}(0.500 \text{ kg})(3.70 \text{ m/s})^2 \\ &= 6.22 \text{ J}. \end{aligned} \tag{8.74}$$

The change in internal kinetic energy is thus

$$\begin{aligned} KE'_{int} - KE_{int} &= 6.22 \text{ J} - 0.763 \text{ J} \\ &= 5.46 \text{ J}. \end{aligned} \tag{8.75}$$

Discussion

The final velocity of cart 2 is large and positive, meaning that it is moving to the right after the collision. The internal kinetic energy in this collision increases by 5.46 J. That energy was released by the spring.

8.6 Collisions of Point Masses in Two Dimensions

Learning Objectives

By the end of this section, you will be able to:

- Discuss two-dimensional collisions as an extension of one-dimensional analysis.

- Define point masses.
- Derive an expression for conservation of momentum along the x-axis and y-axis.
- Describe elastic collisions of two objects with equal mass.
- Determine the magnitude and direction of the final velocity given initial velocity and scattering angle.

The information presented in this section supports the following AP® learning objectives and science practices:

- **5.D.1.2** The student is able to apply the principles of conservation of momentum and restoration of kinetic energy to reconcile a situation that appears to be isolated and elastic, but in which data indicate that linear momentum and kinetic energy are not the same after the interaction, by refining a scientific question to identify interactions that have not been considered. Students will be expected to solve qualitatively and/or quantitatively for one-dimensional situations and only qualitatively in two-dimensional situations.
- **5.D.3.3** The student is able to make predictions about the velocity of the center of mass for interactions within a defined two-dimensional system.

In the previous two sections, we considered only one-dimensional collisions; during such collisions, the incoming and outgoing velocities are all along the same line. But what about collisions, such as those between billiard balls, in which objects scatter to the side? These are two-dimensional collisions, and we shall see that their study is an extension of the one-dimensional analysis already presented. The approach taken (similar to the approach in discussing two-dimensional kinematics and dynamics) is to choose a convenient coordinate system and resolve the motion into components along perpendicular axes. Resolving the motion yields a pair of one-dimensional problems to be solved simultaneously.

One complication arising in two-dimensional collisions is that the objects might rotate before or after their collision. For example, if two ice skaters hook arms as they pass by one another, they will spin in circles. We will not consider such rotation until later, and so for now we arrange things so that no rotation is possible. To avoid rotation, we consider only the scattering of **point masses**—that is, structureless particles that cannot rotate or spin.

We start by assuming that $\mathbf{F}_{\text{net}} = 0$, so that momentum $\mathbf{p}$ is conserved. The simplest collision is one in which one of the particles is initially at rest. (See **Figure 8.14**.) The best choice for a coordinate system is one with an axis parallel to the velocity of the incoming particle, as shown in **Figure 8.14**. Because momentum is conserved, the components of momentum along the x- and y-axes (p_x and p_y) will also be conserved, but with the chosen coordinate system, p_y is initially zero and p_x is the momentum of the incoming particle. Both facts simplify the analysis. (Even with the simplifying assumptions of point masses, one particle initially at rest, and a convenient coordinate system, we still gain new insights into nature from the analysis of two-dimensional collisions.)

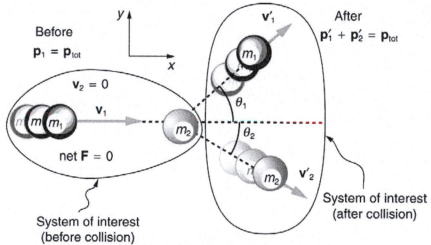

Figure 8.14 A two-dimensional collision with the coordinate system chosen so that m_2 is initially at rest and v_1 is parallel to the x-axis. This coordinate system is sometimes called the laboratory coordinate system, because many scattering experiments have a target that is stationary in the laboratory, while particles are scattered from it to determine the particles that make-up the target and how they are bound together. The particles may not be observed directly, but their initial and final velocities are.

Along the x-axis, the equation for conservation of momentum is

$$p_{1x} + p_{2x} = p'_{1x} + p'_{2x}. \tag{8.76}$$

Where the subscripts denote the particles and axes and the primes denote the situation after the collision. In terms of masses and velocities, this equation is

$$m_1 v_{1x} + m_2 v_{2x} = m_1 v'_{1x} + m_2 v'_{2x}. \tag{8.77}$$

But because particle 2 is initially at rest, this equation becomes

$$m_1 v_{1x} = m_1 v'_{1x} + m_2 v'_{2x}. \tag{8.78}$$

The components of the velocities along the x-axis have the form $v \cos \theta$. Because particle 1 initially moves along the x-axis, we find $v_{1x} = v_1$.

Conservation of momentum along the x-axis gives the following equation:

$$m_1 v_1 = m_1 v'_1 \cos \theta_1 + m_2 v'_2 \cos \theta_2, \tag{8.79}$$

where θ_1 and θ_2 are as shown in **Figure 8.14**.

Conservation of Momentum along the x-axis

$$m_1 v_1 = m_1 v'_1 \cos \theta_1 + m_2 v'_2 \cos \theta_2 \tag{8.80}$$

Along the y-axis, the equation for conservation of momentum is

$$p_{1y} + p_{2y} = p'_{1y} + p'_{2y} \tag{8.81}$$

or

$$m_1 v_{1y} + m_2 v_{2y} = m_1 v'_{1y} + m_2 v'_{2y}. \tag{8.82}$$

But v_{1y} is zero, because particle 1 initially moves along the x-axis. Because particle 2 is initially at rest, v_{2y} is also zero. The equation for conservation of momentum along the y-axis becomes

$$0 = m_1 v'_{1y} + m_2 v'_{2y}. \tag{8.83}$$

The components of the velocities along the y-axis have the form $v \sin \theta$.

Thus, conservation of momentum along the y-axis gives the following equation:

$$0 = m_1 v'_1 \sin \theta_1 + m_2 v'_2 \sin \theta_2. \tag{8.84}$$

Conservation of Momentum along the y-axis

$$0 = m_1 v'_1 \sin \theta_1 + m_2 v'_2 \sin \theta_2 \tag{8.85}$$

The equations of conservation of momentum along the x-axis and y-axis are very useful in analyzing two-dimensional collisions of particles, where one is originally stationary (a common laboratory situation). But two equations can only be used to find two unknowns, and so other data may be necessary when collision experiments are used to explore nature at the subatomic level.

Making Connections: Real World Connections

We have seen, in one-dimensional collisions when momentum is conserved, that the center-of-mass velocity of the system remains unchanged as a result of the collision. If you calculate the momentum and center-of-mass velocity before the collision, you will get the same answer as if you calculate both quantities after the collision. This logic also works for two-dimensional collisions.

For example, consider two cars of equal mass. Car A is driving east (+x-direction) with a speed of 40 m/s. Car B is driving north (+y-direction) with a speed of 80 m/s. What is the velocity of the center-of-mass of this system before and after an inelastic collision, in which the cars move together as one mass after the collision?

Since both cars have equal mass, the center-of-mass velocity components are just the average of the components of the individual velocities before the collision. The x-component of the center of mass velocity is 20 m/s, and the y-component is 40 m/s.

Using momentum conservation for the collision in both the x-component and y-component yields similar answers:

$$m(40) + m(0) = (2m)v_{\text{fina}}(x) \tag{8.86}$$

$$v_{\text{fina}}(x) = 20 \text{ m/s} \tag{8.87}$$

$$m(0) + m(80) = (2m)v_{\text{fina}}(y) \tag{8.88}$$

$$v_{\text{fina}}(y) = 40 \text{ m/s} \tag{8.89}$$

Since the two masses move together after the collision, the velocity of this combined object is equal to the center-of-mass velocity. Thus, the center-of-mass velocity before and after the collision is identical, even in two-dimensional collisions, when momentum is conserved.

Example 8.7 Determining the Final Velocity of an Unseen Object from the Scattering of Another Object

Suppose the following experiment is performed. A 0.250-kg object (m_1) is slid on a frictionless surface into a dark room, where it strikes an initially stationary object with mass of 0.400 kg (m_2). The 0.250-kg object emerges from the room at an angle of $45.0°$ with its incoming direction.

The speed of the 0.250-kg object is originally 2.00 m/s and is 1.50 m/s after the collision. Calculate the magnitude and direction of the velocity $(v'_2$ and $\theta_2)$ of the 0.400-kg object after the collision.

Strategy

Momentum is conserved because the surface is frictionless. The coordinate system shown in **Figure 8.15** is one in which m_2 is originally at rest and the initial velocity is parallel to the x-axis, so that conservation of momentum along the x- and y-axes is applicable.

Everything is known in these equations except v'_2 and θ_2, which are precisely the quantities we wish to find. We can find two unknowns because we have two independent equations: the equations describing the conservation of momentum in the x- and y-directions.

Solution

Solving $m_1 v_1 = m_1 v'_1 \cos\theta_1 + m_2 v'_2 \cos\theta_2$ for $v'_2 \cos\theta_2$ and $0 = m_1 v'_1 \sin\theta_1 + m_2 v'_2 \sin\theta_2$ for $v'_2 \sin\theta_2$ and taking the ratio yields an equation (in which θ_2 is the only unknown quantity. Applying the identity $\left(\tan\theta = \frac{\sin\theta}{\cos\theta}\right)$, we obtain:

$$\tan\theta_2 = \frac{v'_1 \sin\theta_1}{v'_1 \cos\theta_1 - v_1}. \tag{8.90}$$

Entering known values into the previous equation gives

$$\tan\theta_2 = \frac{(1.50 \text{ m/s})(0.7071)}{(1.50 \text{ m/s})(0.7071) - 2.00 \text{ m/s}} = -1.129. \tag{8.91}$$

Thus,

$$\theta_2 = \tan^{-1}(-1.129) = 311.5° \approx 312°. \tag{8.92}$$

Angles are defined as positive in the counter clockwise direction, so this angle indicates that m_2 is scattered to the right in **Figure 8.15**, as expected (this angle is in the fourth quadrant). Either equation for the x- or y-axis can now be used to solve for v'_2, but the latter equation is easiest because it has fewer terms.

$$v'_2 = -\frac{m_1}{m_2} v'_1 \frac{\sin\theta_1}{\sin\theta_2} \tag{8.93}$$

Entering known values into this equation gives

$$v'_2 = -\left(\frac{0.250 \text{ kg}}{0.400 \text{ kg}}\right)(1.50 \text{ m/s})\left(\frac{0.7071}{-0.7485}\right). \tag{8.94}$$

Thus,

$$v'_2 = 0.886 \text{ m/s}. \tag{8.95}$$

Discussion

It is instructive to calculate the internal kinetic energy of this two-object system before and after the collision. (This calculation is left as an end-of-chapter problem.) If you do this calculation, you will find that the internal kinetic energy is less after the collision, and so the collision is inelastic. This type of result makes a physicist want to explore the system further.

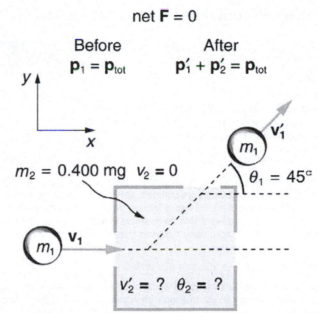

Figure 8.15 A collision taking place in a dark room is explored in Example 8.7. The incoming object m_1 is scattered by an initially stationary object. Only the stationary object's mass m_2 is known. By measuring the angle and speed at which m_1 emerges from the room, it is possible to calculate the magnitude and direction of the initially stationary object's velocity after the collision.

Elastic Collisions of Two Objects with Equal Mass

Some interesting situations arise when the two colliding objects have equal mass and the collision is elastic. This situation is nearly the case with colliding billiard balls, and precisely the case with some subatomic particle collisions. We can thus get a mental image of a collision of subatomic particles by thinking about billiards (or pool). (Refer to **Figure 8.14** for masses and angles.) First, an elastic collision conserves internal kinetic energy. Again, let us assume object 2 (m_2) is initially at rest. Then, the internal kinetic energy before and after the collision of two objects that have equal masses is

$$\frac{1}{2}m{v_1}^2 = \frac{1}{2}m{v'_1}^2 + \frac{1}{2}m{v'_2}^2.$$ (8.96)

Because the masses are equal, $m_1 = m_2 = m$. Algebraic manipulation (left to the reader) of conservation of momentum in the x- and y-directions can show that

$$\frac{1}{2}m{v_1}^2 = \frac{1}{2}m{v'_1}^2 + \frac{1}{2}m{v'_2}^2 + m{v'_1}v'_2 \, \cos(\theta_1 - \theta_2).$$ (8.97)

(Remember that θ_2 is negative here.) The two preceding equations can both be true only if

$$m{v'_1}v'_2 \, \cos(\theta_1 - \theta_2) = 0.$$ (8.98)

There are three ways that this term can be zero. They are

- $v'_1 = 0$: head-on collision; incoming ball stops
- $v'_2 = 0$: no collision; incoming ball continues unaffected
- $\cos(\theta_1 - \theta_2) = 0$: angle of separation $(\theta_1 - \theta_2)$ is $90°$ after the collision

All three of these ways are familiar occurrences in billiards and pool, although most of us try to avoid the second. If you play enough pool, you will notice that the angle between the balls is very close to $90°$ after the collision, although it will vary from this value if a great deal of spin is placed on the ball. (Large spin carries in extra energy and a quantity called *angular momentum*, which must also be conserved.) The assumption that the scattering of billiard balls is elastic is reasonable based on the correctness of the three results it produces. This assumption also implies that, to a good approximation, momentum is conserved for the two-ball system in billiards and pool. The problems below explore these and other characteristics of two-dimensional collisions.

Connections to Nuclear and Particle Physics

Two-dimensional collision experiments have revealed much of what we know about subatomic particles, as we shall see in **Medical Applications of Nuclear Physics** and **Particle Physics**. Ernest Rutherford, for example, discovered the nature of the atomic nucleus from such experiments.

8.7 Introduction to Rocket Propulsion

Rockets range in size from fireworks so small that ordinary people use them to immense Saturn Vs that once propelled massive payloads toward the Moon. The propulsion of all rockets, jet engines, deflating balloons, and even squids and octopuses is explained by the same physical principle—Newton's third law of motion. Matter is forcefully ejected from a system, producing an equal and opposite reaction on what remains. Another common example is the recoil of a gun. The gun exerts a force on a bullet to accelerate it and consequently experiences an equal and opposite force, causing the gun's recoil or kick.

Making Connections: Take-Home Experiment—Propulsion of a Balloon

Hold a balloon and fill it with air. Then, let the balloon go. In which direction does the air come out of the balloon and in which direction does the balloon get propelled? If you fill the balloon with water and then let the balloon go, does the balloon's direction change? Explain your answer.

Figure 8.16 shows a rocket accelerating straight up. In part (a), the rocket has a mass m and a velocity v relative to Earth, and hence a momentum mv. In part (b), a time Δt has elapsed in which the rocket has ejected a mass Δm of hot gas at a velocity v_e relative to the rocket. The remainder of the mass $(m - \Delta m)$ now has a greater velocity $(v + \Delta v)$. The momentum of the entire system (rocket plus expelled gas) has actually decreased because the force of gravity has acted for a time Δt, producing a negative impulse $\Delta p = -mg\Delta t$. (Remember that impulse is the net external force on a system multiplied by the time it acts, and it equals the change in momentum of the system.) So, the center of mass of the system is in free fall but, by rapidly expelling mass, part of the system can accelerate upward. It is a commonly held misconception that the rocket exhaust pushes on the ground. If we consider thrust; that is, the force exerted on the rocket by the exhaust gases, then a rocket's thrust is greater in outer space than in the atmosphere or on the launch pad. In fact, gases are easier to expel into a vacuum.

By calculating the change in momentum for the entire system over Δt, and equating this change to the impulse, the following expression can be shown to be a good approximation for the acceleration of the rocket.

$$a = \frac{v_e}{m} \frac{\Delta m}{\Delta t} - g \tag{8.99}$$

"The rocket" is that part of the system remaining after the gas is ejected, and g is the acceleration due to gravity.

Acceleration of a Rocket

Acceleration of a rocket is

$$a = \frac{v_e}{m} \frac{\Delta m}{\Delta t} - g, \tag{8.100}$$

where a is the acceleration of the rocket, v_e is the escape velocity, m is the mass of the rocket, Δm is the mass of the ejected gas, and Δt is the time in which the gas is ejected.

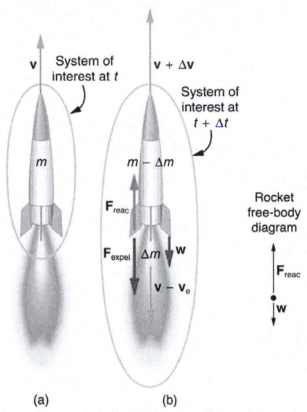

Figure 8.16 (a) This rocket has a mass m and an upward velocity v . The net external force on the system is $-mg$, if air resistance is neglected.

(b) A time Δt later the system has two main parts, the ejected gas and the remainder of the rocket. The reaction force on the rocket is what overcomes the gravitational force and accelerates it upward.

A rocket's acceleration depends on three major factors, consistent with the equation for acceleration of a rocket . First, the greater the exhaust velocity of the gases relative to the rocket, v_e , the greater the acceleration is. The practical limit for v_e is about 2.5×10^3 m/s for conventional (non-nuclear) hot-gas propulsion systems. The second factor is the rate at which mass is ejected from the rocket. This is the factor $\Delta m / \Delta t$ in the equation. The quantity $(\Delta m / \Delta t) v_e$, with units of newtons, is called "thrust." The faster the rocket burns its fuel, the greater its thrust, and the greater its acceleration. The third factor is the mass m of the rocket. The smaller the mass is (all other factors being the same), the greater the acceleration. The rocket mass m decreases dramatically during flight because most of the rocket is fuel to begin with, so that acceleration increases continuously, reaching a maximum just before the fuel is exhausted.

Factors Affecting a Rocket's Acceleration
- The greater the exhaust velocity v_e of the gases relative to the rocket, the greater the acceleration.

- The faster the rocket burns its fuel, the greater its acceleration.
- The smaller the rocket's mass (all other factors being the same), the greater the acceleration.

Example 8.8 Calculating Acceleration: Initial Acceleration of a Moon Launch

A Saturn V's mass at liftoff was 2.80×10^6 kg , its fuel-burn rate was 1.40×10^4 kg/s , and the exhaust velocity was 2.40×10^3 m/s . Calculate its initial acceleration.

Strategy

This problem is a straightforward application of the expression for acceleration because a is the unknown and all of the terms on the right side of the equation are given.

Solution

Substituting the given values into the equation for acceleration yields

$$a = \frac{v_e}{m}\frac{\Delta m}{\Delta t} - g \qquad\qquad\qquad\qquad (8.101)$$

$$= \frac{2.40\times10^3 \text{ m/s}}{2.80\times10^6 \text{ kg}}\left(1.40\times10^4 \text{ kg/s}\right) - 9.80 \text{ m/s}^2$$

$$= 2.20 \text{ m/s}^2.$$

Discussion

This value is fairly small, even for an initial acceleration. The acceleration does increase steadily as the rocket burns fuel, because m decreases while v_e and $\frac{\Delta m}{\Delta t}$ remain constant. Knowing this acceleration and the mass of the rocket, you can show that the thrust of the engines was $3.36\times10^7 \text{ N}$.

To achieve the high speeds needed to hop continents, obtain orbit, or escape Earth's gravity altogether, the mass of the rocket other than fuel must be as small as possible. It can be shown that, in the absence of air resistance and neglecting gravity, the final velocity of a one-stage rocket initially at rest is

$$v = v_e \ln\frac{m_0}{m_r}, \qquad\qquad\qquad\qquad (8.102)$$

where $\ln(m_0/m_r)$ is the natural logarithm of the ratio of the initial mass of the rocket (m_0) to what is left (m_r) after all of the fuel is exhausted. (Note that v is actually the change in velocity, so the equation can be used for any segment of the flight. If we start from rest, the change in velocity equals the final velocity.) For example, let us calculate the mass ratio needed to escape Earth's gravity starting from rest, given that the escape velocity from Earth is about 11.2×10^3 m/s , and assuming an exhaust velocity $v_e = 2.5\times10^3$ m/s .

$$\ln\frac{m_0}{m_r} = \frac{v}{v_e} = \frac{11.2\times10^3 \text{ m/s}}{2.5\times10^3 \text{ m/s}} = 4.48 \qquad\qquad\qquad\qquad (8.103)$$

Solving for m_0/m_r gives

$$\frac{m_0}{m_r} = e^{4.48} = 88. \qquad\qquad\qquad\qquad (8.104)$$

Thus, the mass of the rocket is

$$m_r = \frac{m_0}{88}. \qquad\qquad\qquad\qquad (8.105)$$

This result means that only $1/88$ of the mass is left when the fuel is burnt, and $87/88$ of the initial mass was fuel. Expressed as percentages, 98.9% of the rocket is fuel, while payload, engines, fuel tanks, and other components make up only 1.10%. Taking air resistance and gravitational force into account, the mass m_r remaining can only be about $m_0/180$. It is difficult to build a rocket in which the fuel has a mass 180 times everything else. The solution is multistage rockets. Each stage only needs to achieve part of the final velocity and is discarded after it burns its fuel. The result is that each successive stage can have smaller engines and more payload relative to its fuel. Once out of the atmosphere, the ratio of payload to fuel becomes more favorable, too.

The space shuttle was an attempt at an economical vehicle with some reusable parts, such as the solid fuel boosters and the craft itself. (See Figure 8.17) The shuttle's need to be operated by humans, however, made it at least as costly for launching satellites as expendable, unmanned rockets. Ideally, the shuttle would only have been used when human activities were required for the success of a mission, such as the repair of the Hubble space telescope. Rockets with satellites can also be launched from airplanes. Using airplanes has the double advantage that the initial velocity is significantly above zero and a rocket can avoid most of the atmosphere's resistance.

Figure 8.17 The space shuttle had a number of reusable parts. Solid fuel boosters on either side were recovered and refueled after each flight, and the entire orbiter returned to Earth for use in subsequent flights. The large liquid fuel tank was expended. The space shuttle was a complex assemblage of technologies, employing both solid and liquid fuel and pioneering ceramic tiles as reentry heat shields. As a result, it permitted multiple launches as opposed to single-use rockets. (credit: NASA)

PhET Explorations: Lunar Lander

Can you avoid the boulder field and land safely, just before your fuel runs out, as Neil Armstrong did in 1969? Our version of this classic video game accurately simulates the real motion of the lunar lander with the correct mass, thrust, fuel consumption rate, and lunar gravity. The real lunar lander is very hard to control.

Figure 8.18 Lunar Lander (http://cnx.org/content/m55174/1.2/lunar-lander_en.jar)

Glossary

change in momentum: the difference between the final and initial momentum; the mass times the change in velocity

conservation of momentum principle: when the net external force is zero, the total momentum of the system is conserved or constant

elastic collision: a collision that also conserves internal kinetic energy

impulse: the average net external force times the time it acts; equal to the change in momentum

inelastic collision: a collision in which internal kinetic energy is not conserved

internal kinetic energy: the sum of the kinetic energies of the objects in a system

isolated system: a system in which the net external force is zero

linear momentum: the product of mass and velocity

perfectly inelastic collision: a collision in which the colliding objects stick together

point masses: structureless particles with no rotation or spin

quark: fundamental constituent of matter and an elementary particle

second law of motion: physical law that states that the net external force equals the change in momentum of a system divided by the time over which it changes

Section Summary

8.1 Linear Momentum and Force
- Linear momentum (*momentum* for brevity) is defined as the product of a system's mass multiplied by its velocity.
- In symbols, linear momentum $\mathbf{p}$ is defined to be

$$\mathbf{p} = m\mathbf{v},$$

where m is the mass of the system and $\mathbf{v}$ is its velocity.
- The SI unit for momentum is $\text{kg} \cdot \text{m/s}$.
- Newton's second law of motion in terms of momentum states that the net external force equals the change in momentum of a system divided by the time over which it changes.
- In symbols, Newton's second law of motion is defined to be

$$\mathbf{F}_{\text{net}} = \frac{\Delta \mathbf{p}}{\Delta t},$$

$\mathbf{F}_{\text{net}}$ is the net external force, $\Delta \mathbf{p}$ is the change in momentum, and Δt is the change time.

8.2 Impulse
- Impulse, or change in momentum, equals the average net external force multiplied by the time this force acts:

$$\Delta \mathbf{p} = \mathbf{F}_{\text{net}} \Delta t.$$

- Forces are usually not constant over a period of time.

8.3 Conservation of Momentum
- The conservation of momentum principle is written

$$\mathbf{p}_{\text{tot}} = \text{constant}$$

or

$$\mathbf{p}_{\text{tot}} = \mathbf{p}'_{\text{tot}} \quad (\text{isolated system}),$$

$\mathbf{p}_{\text{tot}}$ is the initial total momentum and $\mathbf{p}'_{\text{tot}}$ is the total momentum some time later.
- An isolated system is defined to be one for which the net external force is zero $\left(\mathbf{F}_{\text{net}} = 0 \right)$.
- During projectile motion and where air resistance is negligible, momentum is conserved in the horizontal direction because horizontal forces are zero.
- Conservation of momentum applies only when the net external force is zero.
- The conservation of momentum principle is valid when considering systems of particles.

8.4 Elastic Collisions in One Dimension
- An elastic collision is one that conserves internal kinetic energy.
- Conservation of kinetic energy and momentum together allow the final velocities to be calculated in terms of initial velocities and masses in one dimensional two-body collisions.

8.5 Inelastic Collisions in One Dimension
- An inelastic collision is one in which the internal kinetic energy changes (it is not conserved).
- A collision in which the objects stick together is sometimes called perfectly inelastic because it reduces internal kinetic energy more than does any other type of inelastic collision.
- Sports science and technologies also use physics concepts such as momentum and rotational motion and vibrations.

8.6 Collisions of Point Masses in Two Dimensions
- The approach to two-dimensional collisions is to choose a convenient coordinate system and break the motion into components along perpendicular axes. Choose a coordinate system with the x-axis parallel to the velocity of the incoming particle.
- Two-dimensional collisions of point masses where mass 2 is initially at rest conserve momentum along the initial direction of mass 1 (the x-axis), stated by $m_1 v_1 = m_1 v'_1 \cos \theta_1 + m_2 v'_2 \cos \theta_2$ and along the direction perpendicular to the initial direction (the y-axis) stated by $0 = m_1 v'_{1y} + m_2 v'_{2y}$.
- The internal kinetic before and after the collision of two objects that have equal masses is

$$\frac{1}{2} m v_1^2 = \frac{1}{2} m v'_1^2 + \frac{1}{2} m v'_2^2 + m v'_1 v'_2 \cos(\theta_1 - \theta_2).$$

- Point masses are structureless particles that cannot spin.

8.7 Introduction to Rocket Propulsion
- Newton's third law of motion states that to every action, there is an equal and opposite reaction.
- Acceleration of a rocket is $a = \frac{v_e}{m} \frac{\Delta m}{\Delta t} - g$.
- A rocket's acceleration depends on three main factors. They are
    1. The greater the exhaust velocity of the gases, the greater the acceleration.
    2. The faster the rocket burns its fuel, the greater its acceleration.
    3. The smaller the rocket's mass, the greater the acceleration.

Conceptual Questions

8.1 Linear Momentum and Force

1. An object that has a small mass and an object that has a large mass have the same momentum. Which object has the largest kinetic energy?

2. An object that has a small mass and an object that has a large mass have the same kinetic energy. Which mass has the largest momentum?

3. Professional Application

Football coaches advise players to block, hit, and tackle with their feet on the ground rather than by leaping through the air. Using the concepts of momentum, work, and energy, explain how a football player can be more effective with his feet on the ground.

4. How can a small force impart the same momentum to an object as a large force?

8.2 Impulse

5. Professional Application

Explain in terms of impulse how padding reduces forces in a collision. State this in terms of a real example, such as the advantages of a carpeted vs. tile floor for a day care center.

6. While jumping on a trampoline, sometimes you land on your back and other times on your feet. In which case can you reach a greater height and why?

7. Professional Application

Tennis racquets have "sweet spots." If the ball hits a sweet spot then the player's arm is not jarred as much as it would be otherwise. Explain why this is the case.

8.3 Conservation of Momentum

8. Professional Application

If you dive into water, you reach greater depths than if you do a belly flop. Explain this difference in depth using the concept of conservation of energy. Explain this difference in depth using what you have learned in this chapter.

9. Under what circumstances is momentum conserved?

10. Can momentum be conserved for a system if there are external forces acting on the system? If so, under what conditions? If not, why not?

11. Momentum for a system can be conserved in one direction while not being conserved in another. What is the angle between the directions? Give an example.

12. Professional Application

Explain in terms of momentum and Newton's laws how a car's air resistance is due in part to the fact that it pushes air in its direction of motion.

13. Can objects in a system have momentum while the momentum of the system is zero? Explain your answer.

14. Must the total energy of a system be conserved whenever its momentum is conserved? Explain why or why not.

8.4 Elastic Collisions in One Dimension

15. What is an elastic collision?

8.5 Inelastic Collisions in One Dimension

16. What is an inelastic collision? What is a perfectly inelastic collision?

17. Mixed-pair ice skaters performing in a show are standing motionless at arms length just before starting a routine. They reach out, clasp hands, and pull themselves together by only using their arms. Assuming there is no friction between the blades of their skates and the ice, what is their velocity after their bodies meet?

18. A small pickup truck that has a camper shell slowly coasts toward a red light with negligible friction. Two dogs in the back of the truck are moving and making various inelastic collisions with each other and the walls. What is the effect of the dogs on the motion of the center of mass of the system (truck plus entire load)? What is their effect on the motion of the truck?

8.6 Collisions of Point Masses in Two Dimensions

19. Figure 8.19 shows a cube at rest and a small object heading toward it. (a) Describe the directions (angle θ_1) at which the small object can emerge after colliding elastically with the cube. How does θ_1 depend on b, the so-called impact parameter? Ignore any effects that might be due to rotation after the collision, and assume that the cube is much more massive than the small object. (b) Answer the same questions if the small object instead collides with a massive sphere.

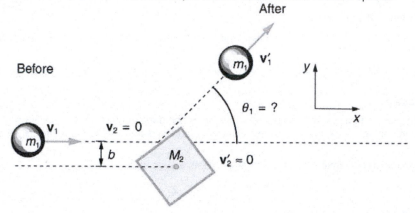

Figure 8.19 A small object approaches a collision with a much more massive cube, after which its velocity has the direction θ_1. The angles at which the small object can be scattered are determined by the shape of the object it strikes and the impact parameter b.

8.7 Introduction to Rocket Propulsion

20. Professional Application

Suppose a fireworks shell explodes, breaking into three large pieces for which air resistance is negligible. How is the motion of the center of mass affected by the explosion? How would it be affected if the pieces experienced significantly more air resistance than the intact shell?

21. Professional Application

During a visit to the International Space Station, an astronaut was positioned motionless in the center of the station, out of reach of any solid object on which he could exert a force. Suggest a method by which he could move himself away from this position, and explain the physics involved.

22. Professional Application

It is possible for the velocity of a rocket to be greater than the exhaust velocity of the gases it ejects. When that is the case, the gas velocity and gas momentum are in the same direction as that of the rocket. How is the rocket still able to obtain thrust by ejecting the gases?

Problems & Exercises

8.1 Linear Momentum and Force

1. (a) Calculate the momentum of a 2000-kg elephant charging a hunter at a speed of 7.50 m/s. (b) Compare the elephant's momentum with the momentum of a 0.0400-kg tranquilizer dart fired at a speed of 600 m/s. (c) What is the momentum of the 90.0-kg hunter running at 7.40 m/s after missing the elephant?

2. (a) What is the mass of a large ship that has a momentum of $1.60 \times 10^9 \text{ kg} \cdot \text{m/s}$, when the ship is moving at a speed of 48.0 km/h? (b) Compare the ship's momentum to the momentum of a 1100-kg artillery shell fired at a speed of 1200 m/s.

3. (a) At what speed would a 2.00×10^4-kg airplane have to fly to have a momentum of $1.60 \times 10^9 \text{ kg} \cdot \text{m/s}$ (the same as the ship's momentum in the problem above)? (b) What is the plane's momentum when it is taking off at a speed of 60.0 m/s? (c) If the ship is an aircraft carrier that launches these airplanes with a catapult, discuss the implications of your answer to (b) as it relates to recoil effects of the catapult on the ship.

4. (a) What is the momentum of a garbage truck that is $1.20 \times 10^4 \text{ kg}$ and is moving at 10.0 m/s? (b) At what speed would an 8.00-kg trash can have the same momentum as the truck?

5. A runaway train car that has a mass of 15,000 kg travels at a speed of 5.4 m/s down a track. Compute the time required for a force of 1500 N to bring the car to rest.

6. The mass of Earth is $5.972 \times 10^{24} \text{ kg}$ and its orbital radius is an average of $1.496 \times 10^{11} \text{ m}$. Calculate its linear momentum.

8.2 Impulse

7. A bullet is accelerated down the barrel of a gun by hot gases produced in the combustion of gun powder. What is the average force exerted on a 0.0300-kg bullet to accelerate it to a speed of 600 m/s in a time of 2.00 ms (milliseconds)?

8. Professional Application

A car moving at 10 m/s crashes into a tree and stops in 0.26 s. Calculate the force the seat belt exerts on a passenger in the car to bring him to a halt. The mass of the passenger is 70 kg.

9. A person slaps her leg with her hand, bringing her hand to rest in 2.50 milliseconds from an initial speed of 4.00 m/s. (a) What is the average force exerted on the leg, taking the effective mass of the hand and forearm to be 1.50 kg? (b) Would the force be any different if the woman clapped her hands together at the same speed and brought them to rest in the same time? Explain why or why not.

10. Professional Application

A professional boxer hits his opponent with a 1000-N horizontal blow that lasts for 0.150 s. (a) Calculate the impulse imparted by this blow. (b) What is the opponent's final velocity, if his mass is 105 kg and he is motionless in midair when struck near his center of mass? (c) Calculate the recoil velocity of the opponent's 10.0-kg head if hit in this manner, assuming the head does not initially transfer significant momentum to the boxer's body. (d) Discuss the implications of your answers for parts (b) and (c).

11. Professional Application

Suppose a child drives a bumper car head on into the side rail, which exerts a force of 4000 N on the car for 0.200 s. (a) What impulse is imparted by this force? (b) Find the final velocity of the bumper car if its initial velocity was 2.80 m/s and the car plus driver have a mass of 200 kg. You may neglect friction between the car and floor.

12. Professional Application

One hazard of space travel is debris left by previous missions. There are several thousand objects orbiting Earth that are large enough to be detected by radar, but there are far greater numbers of very small objects, such as flakes of paint. Calculate the force exerted by a 0.100-mg chip of paint that strikes a spacecraft window at a relative speed of $4.00 \times 10^3 \text{ m/s}$, given the collision lasts $6.00 \times 10^{-8} \text{ s}$.

13. Professional Application

A 75.0-kg person is riding in a car moving at 20.0 m/s when the car runs into a bridge abutment. (a) Calculate the average force on the person if he is stopped by a padded dashboard that compresses an average of 1.00 cm. (b) Calculate the average force on the person if he is stopped by an air bag that compresses an average of 15.0 cm.

14. Professional Application

Military rifles have a mechanism for reducing the recoil forces of the gun on the person firing it. An internal part recoils over a relatively large distance and is stopped by damping mechanisms in the gun. The larger distance reduces the average force needed to stop the internal part. (a) Calculate the recoil velocity of a 1.00-kg plunger that directly interacts with a 0.0200-kg bullet fired at 600 m/s from the gun. (b) If this part is stopped over a distance of 20.0 cm, what average force is exerted upon it by the gun? (c) Compare this to the force exerted on the gun if the bullet is accelerated to its velocity in 10.0 ms (milliseconds).

15. A cruise ship with a mass of $1.00 \times 10^7 \text{ kg}$ strikes a pier at a speed of 0.750 m/s. It comes to rest 6.00 m later, damaging the ship, the pier, and the tugboat captain's finances. Calculate the average force exerted on the pier using the concept of impulse. (Hint: First calculate the time it took to bring the ship to rest.)

16. Calculate the final speed of a 110-kg rugby player who is initially running at 8.00 m/s but collides head-on with a padded goalpost and experiences a backward force of $1.76 \times 10^4 \text{ N}$ for $5.50 \times 10^{-2} \text{ s}$.

17. Water from a fire hose is directed horizontally against a wall at a rate of 50.0 kg/s and a speed of 42.0 m/s. Calculate the magnitude of the force exerted on the wall, assuming the water's horizontal momentum is reduced to zero.

18. A 0.450-kg hammer is moving horizontally at 7.00 m/s when it strikes a nail and comes to rest after driving the nail 1.00 cm into a board. (a) Calculate the duration of the impact. (b) What was the average force exerted on the nail?

19. Starting with the definitions of momentum and kinetic energy, derive an equation for the kinetic energy of a particle expressed as a function of its momentum.

20. A ball with an initial velocity of 10 m/s moves at an angle $60°$ above the $+x$-direction. The ball hits a vertical wall and bounces off so that it is moving $60°$ above the $-x$-direction with the same speed. What is the impulse delivered by the wall?

21. When serving a tennis ball, a player hits the ball when its velocity is zero (at the highest point of a vertical toss). The racquet exerts a force of 540 N on the ball for 5.00 ms, giving it a final velocity of 45.0 m/s. Using these data, find the mass of the ball.

22. A punter drops a ball from rest vertically 1 meter down onto his foot. The ball leaves the foot with a speed of 18 m/s at an angle $55°$ above the horizontal. What is the impulse delivered by the foot (magnitude and direction)?

8.3 Conservation of Momentum

23. Professional Application

Train cars are coupled together by being bumped into one another. Suppose two loaded train cars are moving toward one another, the first having a mass of 150,000 kg and a velocity of 0.300 m/s, and the second having a mass of 110,000 kg and a velocity of -0.120 m/s. (The minus indicates direction of motion.) What is their final velocity?

24. Suppose a clay model of a koala bear has a mass of 0.200 kg and slides on ice at a speed of 0.750 m/s. It runs into another clay model, which is initially motionless and has a mass of 0.350 kg. Both being soft clay, they naturally stick together. What is their final velocity?

25. Professional Application

Consider the following question: *A car moving at 10 m/s crashes into a tree and stops in 0.26 s. Calculate the force the seatbelt exerts on a passenger in the car to bring him to a halt. The mass of the passenger is 70 kg.* Would the answer to this question be different if the car with the 70-kg passenger had collided with a car that has a mass equal to and is traveling in the opposite direction and at the same speed? Explain your answer.

26. What is the velocity of a 900-kg car initially moving at 30.0 m/s, just after it hits a 150-kg deer initially running at 12.0 m/s in the same direction? Assume the deer remains on the car.

27. A 1.80-kg falcon catches a 0.650-kg dove from behind in midair. What is their velocity after impact if the falcon's velocity is initially 28.0 m/s and the dove's velocity is 7.00 m/s in the same direction?

8.4 Elastic Collisions in One Dimension

28. Two identical objects (such as billiard balls) have a one-dimensional collision in which one is initially motionless. After the collision, the moving object is stationary and the other moves with the same speed as the other originally had. Show that both momentum and kinetic energy are conserved.

29. Professional Application

Two manned satellites approach one another at a relative speed of 0.250 m/s, intending to dock. The first has a mass of 4.00×10^3 kg, and the second a mass of 7.50×10^3 kg. If the two satellites collide elastically rather than dock, what is their final relative velocity?

30. A 70.0-kg ice hockey goalie, originally at rest, catches a 0.150-kg hockey puck slapped at him at a velocity of 35.0 m/s. Suppose the goalie and the ice puck have an elastic collision and the puck is reflected back in the direction from which it came. What would their final velocities be in this case?

8.5 Inelastic Collisions in One Dimension

31. A 0.240-kg billiard ball that is moving at 3.00 m/s strikes the bumper of a pool table and bounces straight back at 2.40 m/s (80% of its original speed). The collision lasts 0.0150 s. (a) Calculate the average force exerted on the ball by the bumper. (b) How much kinetic energy in joules is lost during the collision? (c) What percent of the original energy is left?

32. During an ice show, a 60.0-kg skater leaps into the air and is caught by an initially stationary 75.0-kg skater. (a) What is their final velocity assuming negligible friction and that the 60.0-kg skater's original horizontal velocity is 4.00 m/s? (b) How much kinetic energy is lost?

33. Professional Application

Using mass and speed data from Example 8.1 and assuming that the football player catches the ball with his feet off the ground with both of them moving horizontally, calculate: (a) the final velocity if the ball and player are going in the same direction and (b) the loss of kinetic energy in this case. (c) Repeat parts (a) and (b) for the situation in which the ball and the player are going in opposite directions. Might the loss of kinetic energy be related to how much it hurts to catch the pass?

34. A battleship that is 6.00×10^7 kg and is originally at rest fires a 1100-kg artillery shell horizontally with a velocity of 575 m/s. (a) If the shell is fired straight aft (toward the rear of the ship), there will be negligible friction opposing the ship's recoil. Calculate its recoil velocity. (b) Calculate the increase in internal kinetic energy (that is, for the ship and the shell). This energy is less than the energy released by the gun powder—significant heat transfer occurs.

35. Professional Application

Two manned satellites approaching one another, at a relative speed of 0.250 m/s, intending to dock. The first has a mass of 4.00×10^3 kg, and the second a mass of 7.50×10^3 kg.

(a) Calculate the final velocity (after docking) by using the frame of reference in which the first satellite was originally at rest. (b) What is the loss of kinetic energy in this inelastic collision? (c) Repeat both parts by using the frame of reference in which the second satellite was originally at rest. Explain why the change in velocity is different in the two frames, whereas the change in kinetic energy is the same in both.

36. Professional Application

A 30,000-kg freight car is coasting at 0.850 m/s with negligible friction under a hopper that dumps 110,000 kg of scrap metal into it. (a) What is the final velocity of the loaded freight car? (b) How much kinetic energy is lost?

37. Professional Application

Space probes may be separated from their launchers by exploding bolts. (They bolt away from one another.) Suppose a 4800-kg satellite uses this method to separate from the 1500-kg remains of its launcher, and that 5000 J of kinetic energy is supplied to the two parts. What are their subsequent velocities using the frame of reference in which they were at rest before separation?

38. A 0.0250-kg bullet is accelerated from rest to a speed of 550 m/s in a 3.00-kg rifle. The pain of the rifle's kick is much worse if you hold the gun loosely a few centimeters from your shoulder rather than holding it tightly against your shoulder. (a) Calculate the recoil velocity of the rifle if it is held loosely away from the shoulder. (b) How much kinetic energy does the rifle gain? (c) What is the recoil velocity if the rifle is held tightly against the shoulder, making the effective mass 28.0 kg? (d) How much kinetic energy is transferred to the rifle-shoulder combination? The pain is related to the amount of kinetic energy, which is significantly less in this latter situation. (e) Calculate the momentum of a 110-kg football player running at 8.00 m/s. Compare the player's momentum with the momentum of a hard-thrown 0.410-kg football that has a speed of 25.0 m/s. Discuss its relationship to this problem.

39. Professional Application

One of the waste products of a nuclear reactor is plutonium-239 $\left(^{239}\text{Pu}\right)$. This nucleus is radioactive and decays by splitting into a helium-4 nucleus and a uranium-235 nucleus $\left(^{4}\text{He} + {}^{235}\text{U}\right)$, the latter of which is also radioactive and will itself decay some time later. The energy emitted in the plutonium decay is 8.40×10^{-13} J and is entirely converted to kinetic energy of the helium and uranium nuclei. The mass of the helium nucleus is 6.68×10^{-27} kg, while that of the uranium is 3.92×10^{-25} kg (note that the ratio of the masses is 4 to 235). (a) Calculate the velocities of the two nuclei, assuming the plutonium nucleus is originally at rest. (b) How much kinetic energy does each nucleus carry away? Note that the data given here are accurate to three digits only.

40. Professional Application

The Moon's craters are remnants of meteorite collisions. Suppose a fairly large asteroid that has a mass of 5.00×10^{12} kg (about a kilometer across) strikes the Moon at a speed of 15.0 km/s. (a) At what speed does the Moon recoil after the perfectly inelastic collision (the mass of the Moon is 7.36×10^{22} kg) ? (b) How much kinetic energy is lost in the collision? Such an event may have been observed by medieval English monks who reported observing a red glow and subsequent haze about the Moon. (c) In October 2009, NASA crashed a rocket into the Moon, and analyzed the plume produced by the impact. (Significant amounts of water were detected.) Answer part (a) and (b) for this real-life experiment. The mass of the rocket was 2000 kg and its speed upon impact was 9000 km/h. How does the plume produced alter these results?

41. Professional Application

Two football players collide head-on in midair while trying to catch a thrown football. The first player is 95.0 kg and has an initial velocity of 6.00 m/s, while the second player is 115 kg and has an initial velocity of –3.50 m/s. What is their velocity just after impact if they cling together?

42. What is the speed of a garbage truck that is 1.20×10^{4} kg and is initially moving at 25.0 m/s just after it hits and adheres to a trash can that is 80.0 kg and is initially at rest?

43. During a circus act, an elderly performer thrills the crowd by catching a cannon ball shot at him. The cannon ball has a mass of 10.0 kg and the horizontal component of its velocity is 8.00 m/s when the 65.0-kg performer catches it. If the performer is on nearly frictionless roller skates, what is his recoil velocity?

44. (a) During an ice skating performance, an initially motionless 80.0-kg clown throws a fake barbell away. The clown's ice skates allow her to recoil frictionlessly. If the clown recoils with a velocity of 0.500 m/s and the barbell is thrown with a velocity of 10.0 m/s, what is the mass of the barbell? (b) How much kinetic energy is gained by this maneuver? (c) Where does the kinetic energy come from?

8.6 Collisions of Point Masses in Two Dimensions

45. Two identical pucks collide on an air hockey table. One puck was originally at rest. (a) If the incoming puck has a speed of 6.00 m/s and scatters to an angle of $30.0°$,what is the velocity (magnitude and direction) of the second puck? (You may use the result that $\theta_1 - \theta_2 = 90°$ for elastic collisions of objects that have identical masses.) (b) Confirm that the collision is elastic.

46. Confirm that the results of the example **Example 8.7** do conserve momentum in both the x - and y -directions.

47. A 3000-kg cannon is mounted so that it can recoil only in the horizontal direction. (a) Calculate its recoil velocity when it fires a 15.0-kg shell at 480 m/s at an angle of $20.0°$ above the horizontal. (b) What is the kinetic energy of the cannon? This energy is dissipated as heat transfer in shock absorbers that stop its recoil. (c) What happens to the vertical component of momentum that is imparted to the cannon when it is fired?

48. Professional Application

A 5.50-kg bowling ball moving at 9.00 m/s collides with a 0.850-kg bowling pin, which is scattered at an angle of $85.0°$ to the initial direction of the bowling ball and with a speed of 15.0 m/s. (a) Calculate the final velocity (magnitude and direction) of the bowling ball. (b) Is the collision elastic? (c) Linear kinetic energy is greater after the collision. Discuss how spin on the ball might be converted to linear kinetic energy in the collision.

49. Professional Application

Ernest Rutherford (the first New Zealander to be awarded the Nobel Prize in Chemistry) demonstrated that nuclei were very small and dense by scattering helium-4 nuclei $\left(^4\text{He}\right)$ from gold-197 nuclei $\left(^{197}\text{Au}\right)$. The energy of the incoming helium nucleus was 8.00×10^{-13} J , and the masses of the helium and gold nuclei were 6.68×10^{-27} kg and 3.29×10^{-25} kg , respectively (note that their mass ratio is 4 to 197). (a) If a helium nucleus scatters to an angle of $120°$ during an elastic collision with a gold nucleus, calculate the helium nucleus's final speed and the final velocity (magnitude and direction) of the gold nucleus. (b) What is the final kinetic energy of the helium nucleus?

50. Professional Application

Two cars collide at an icy intersection and stick together afterward. The first car has a mass of 1200 kg and is approaching at 8.00 m/s due south. The second car has a mass of 850 kg and is approaching at 17.0 m/s due west. (a) Calculate the final velocity (magnitude and direction) of the cars. (b) How much kinetic energy is lost in the collision? (This energy goes into deformation of the cars.) Note that because both cars have an initial velocity, you cannot use the equations for conservation of momentum along the x -axis and y -axis; instead, you must look for other simplifying aspects.

51. Starting with equations

$m_1 v_1 = m_1 v'_1 \cos\theta_1 + m_2 v'_2 \cos\theta_2$ and

$0 = m_1 v'_1 \sin\theta_1 + m_2 v'_2 \sin\theta_2$ for conservation of momentum in the x - and y -directions and assuming that one object is originally stationary, prove that for an elastic collision of two objects of equal masses,

$\frac{1}{2}mv_1^2 = \frac{1}{2}mv'_1{}^2 + \frac{1}{2}mv'_2{}^2 + mv'_1 v'_2 \cos\left(\theta_1 - \theta_2\right)$

as discussed in the text.

52. Integrated Concepts

A 90.0-kg ice hockey player hits a 0.150-kg puck, giving the puck a velocity of 45.0 m/s. If both are initially at rest and if the ice is frictionless, how far does the player recoil in the time it takes the puck to reach the goal 15.0 m away?

8.7 Introduction to Rocket Propulsion

53. Professional Application

Antiballistic missiles (ABMs) are designed to have very large accelerations so that they may intercept fast-moving incoming missiles in the short time available. What is the takeoff acceleration of a 10,000-kg ABM that expels 196 kg of gas per second at an exhaust velocity of 2.50×10^3 m/s?

54. Professional Application

What is the acceleration of a 5000-kg rocket taking off from the Moon, where the acceleration due to gravity is only 1.6 m/s^2 , if the rocket expels 8.00 kg of gas per second at an exhaust velocity of 2.20×10^3 m/s?

55. Professional Application

Calculate the increase in velocity of a 4000-kg space probe that expels 3500 kg of its mass at an exhaust velocity of 2.00×10^3 m/s . You may assume the gravitational force is negligible at the probe's location.

56. Professional Application

Ion-propulsion rockets have been proposed for use in space. They employ atomic ionization techniques and nuclear energy sources to produce extremely high exhaust velocities, perhaps as great as 8.00×10^6 m/s . These techniques allow a much more favorable payload-to-fuel ratio. To illustrate this fact: (a) Calculate the increase in velocity of a 20,000-kg space probe that expels only 40.0-kg of its mass at the given exhaust velocity. (b) These engines are usually designed to produce a very small thrust for a very long time—the type of engine that might be useful on a trip to the outer planets, for example. Calculate the acceleration of such an engine if it expels 4.50×10^{-6} kg/s at the given velocity, assuming the acceleration due to gravity is negligible.

57. Derive the equation for the vertical acceleration of a rocket.

58. Professional Application

(a) Calculate the maximum rate at which a rocket can expel gases if its acceleration cannot exceed seven times that of gravity. The mass of the rocket just as it runs out of fuel is 75,000-kg, and its exhaust velocity is 2.40×10^3 m/s . Assume that the acceleration of gravity is the same as on Earth's surface $\left(9.80 \text{ m/s}^2\right)$. (b) Why might it be necessary to limit the acceleration of a rocket?

59. Given the following data for a fire extinguisher-toy wagon rocket experiment, calculate the average exhaust velocity of the gases expelled from the extinguisher. Starting from rest, the final velocity is 10.0 m/s. The total mass is initially 75.0 kg and is 70.0 kg after the extinguisher is fired.

60. How much of a single-stage rocket that is 100,000 kg can be anything but fuel if the rocket is to have a final speed of 8.00 km/s , given that it expels gases at an exhaust velocity of 2.20×10^3 m/s?

61. Professional Application

(a) A 5.00-kg squid initially at rest ejects 0.250-kg of fluid with a velocity of 10.0 m/s. What is the recoil velocity of the squid if the ejection is done in 0.100 s and there is a 5.00-N frictional force opposing the squid's movement. (b) How much energy is lost to work done against friction?

62. Unreasonable Results

Squids have been reported to jump from the ocean and travel 30.0 m (measured horizontally) before re-entering the water. (a) Calculate the initial speed of the squid if it leaves the water at an angle of $20.0°$, assuming negligible lift from the air and negligible air resistance. (b) The squid propels itself by squirting water. What fraction of its mass would it have to eject in order to achieve the speed found in the previous part? The water is ejected at 12.0 m/s ; gravitational force and friction are neglected. (c) What is unreasonable about the results? (d) Which premise is unreasonable, or which premises are inconsistent?

63. Construct Your Own Problem

Consider an astronaut in deep space cut free from her space ship and needing to get back to it. The astronaut has a few packages that she can throw away to move herself toward the ship. Construct a problem in which you calculate the time it takes her to get back by throwing all the packages at one time compared to throwing them one at a time. Among the things to be considered are the masses involved, the force she can exert on the packages through some distance, and the distance to the ship.

64. Construct Your Own Problem

Consider an artillery projectile striking armor plating. Construct a problem in which you find the force exerted by the projectile on the plate. Among the things to be considered are the mass and speed of the projectile and the distance over which its speed is reduced. Your instructor may also wish for you to consider the relative merits of depleted uranium versus lead projectiles based on the greater density of uranium.

Test Prep for AP® Courses

8.1 Linear Momentum and Force

1. A boy standing on a frictionless ice rink is initially at rest. He throws a snowball in the +x-direction, and it travels on a ballistic trajectory, hitting the ground some distance away. Which of the following is true about the boy while he is in the act of throwing the snowball?
  a. He feels an upward force to compensate for the downward trajectory of the snowball.
  b. He feels a backward force exerted by the snowball he is throwing.
  c. He feels no net force.
  d. He feels a forward force, the same force that propels the snowball.

2. A 150-g baseball is initially moving 80 mi/h in the –x-direction. After colliding with a baseball bat for 20 ms, the baseball moves 80 mi/h in the +x-direction. What is the magnitude and direction of the average force exerted by the bat on the baseball?

8.2 Impulse

3. A 1.0-kg ball of putty is released from rest and falls vertically 1.5 m until it strikes a hard floor, where it comes to rest in a 0.045-s time interval. What is the magnitude and direction of the average force exerted on the ball by the floor during the collision?
  a. 33 N, up
  b. 120 N, up
  c. 120 N, down
  d. 240 N, down

4. A 75-g ball is dropped from rest from a height of 2.2 m. It bounces off the floor and rebounds to a maximum height of 1.7 m. If the ball is in contact with the floor for 0.024 s, what is the magnitude and direction of the average force exerted on the ball by the floor during the collision?

5. A 2.4-kg ceramic bowl falls to the floor. During the 0.018-s impact, the bowl experiences an average force of 750 N from the floor. The bowl is at rest after the impact. From what initial height did the bowl fall?

  a. 1.6 m
  b. 2.8 m
  c. 3.2 m
  d. 5.6 m

6. Whether or not an object (such as a plate, glass, or bone) breaks upon impact depends on the average force exerted on that object by the surface. When a 1.2-kg glass figure hits the floor, it will break if it experiences an average force of 330 N. When it hits a tile floor, the glass comes to a stop in 0.015 s. From what minimum height must the glass fall to experience sufficient force to break? How would your answer change if the figure were falling to a padded or carpeted surface? Explain.

7. A 2.5-kg block slides across a frictionless table toward a horizontal spring. As the block bounces off the spring, a probe measures the velocity of the block (initially negative, moving away from the probe) over time as follows:

Table 8.2

Velocity (m/s)	Time (s)
−12.0	0
−10.0	0.10
−6.0	0.20
0	0.30
6.0	0.40
10.0	0.50
12.0	0.60

What is the average force exerted on the block by the spring over the entire 0.60-s time interval of the collision?

  a. 50 N
  b. 60 N
  c. 100 N
  d. 120 N

8. During an automobile crash test, the average force exerted by a solid wall on a 2500-kg car that hits the wall is measured to be 740,000 N over a 0.22-s time interval. What was the initial speed of the car prior to the collision, assuming the car is at rest at the end of the time interval?

9. A test car is driving toward a solid crash-test barrier with a speed of 45 mi/h. Two seconds prior to impact, the car begins to brake, but it is still moving when it hits the wall. After the collision with the wall, the car crumples somewhat and comes to a complete stop. In order to estimate the average force exerted by the wall on the car, what information would you need to collect?
   a. The (negative) acceleration of the car before it hits the wall and the distance the car travels while braking.
   b. The (negative) acceleration of the car before it hits the wall and the velocity of the car just before impact.
   c. The velocity of the car just before impact and the duration of the collision with the wall.
   d. The duration of the collision with the wall and the distance the car travels while braking.

10. Design an experiment to verify the relationship between the average force exerted on an object and the change in momentum of that object. As part of your explanation, list the equipment you would use and describe your experimental setup. What would you measure and how? How exactly would you verify the relationship? Explain.

11. A 22-g puck hits the wall of an air hockey table perpendicular to the wall with an initial speed of 14 m/s. The puck is in contact with the wall for 0.0055 s, and it rebounds from the wall with a speed of 14 m/s in the opposite direction. What is the magnitude of the average force exerted by the wall on the puck?
   a. 0.308 N
   b. 0.616 N
   c. 56 N
   d. 112 N

12. A 22-g puck hits the wall of an air hockey table perpendicular to the wall with an initial speed of 7 m/s. The puck is in contact with the wall for 0.011 s, and the wall exerts an average force of 28 N on the puck during that time. Calculate the magnitude and direction of the change in momentum of the puck.

13.

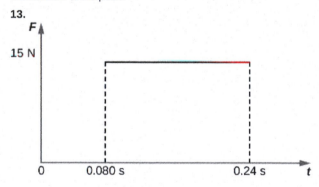

Figure 8.20 This is a graph showing the force exerted by a rigid wall versus time. The graph in **Figure 8.20** represents the force exerted on a particle during a collision. What is the magnitude of the change in momentum of the particle as a result of the collision?
   a. 1.2 kg • m/s
   b. 2.4 kg • m/s
   c. 3.6 kg • m/s
   d. 4.8 kg • m/s

14.

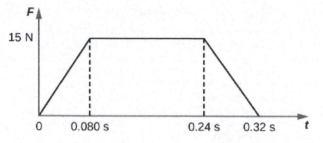

Figure 8.21 This is a graph showing the force exerted by a rigid wall versus time. The graph in **Figure 8.21** represents the force exerted on a particle during a collision. What is the magnitude of the change in momentum of the particle as a result of the collision?

8.3 Conservation of Momentum

15. Which of the following is an example of an open system?
   a. Two air cars colliding on a track elastically.
   b. Two air cars colliding on a track and sticking together.
   c. A bullet being fired into a hanging wooden block and becoming embedded in the block, with the system then acting as a ballistic pendulum.
   d. A bullet being fired into a hillside and becoming buried in the earth.

16. A 40-kg girl runs across a mat with a speed of 5.0 m/s and jumps onto a 120-kg hanging platform initially at rest, causing the girl and platform to swing back and forth like a pendulum together after her jump. What is the combined velocity of the girl and platform after the jump? What is the combined momentum of the girl and platform both before and after the collision?

A 50-kg boy runs across a mat with a speed of 6.0 m/s and collides with a soft barrier on the wall, rebounding off the wall and falling to the ground. The boy is at rest after the collision. What is the momentum of the boy before and after the collision? Is momentum conserved in this collision? Explain. Which of these is an example of an open system and which is an example of a closed system? Explain your answer.

17. A student sets up an experiment to measure the momentum of a system of two air cars, A and B, of equal mass, moving on a linear, frictionless track. Before the collision, car A has a certain speed, and car B is at rest. Which of the following will be true about the total momentum of the two cars?
   a. It will be greater before the collision.
   b. It will be equal before and after the collision.
   c. It will be greater after the collision.
   d. The answer depends on whether the collision is elastic or inelastic.

18. A group of students has two carts, A and B, with wheels that turn with negligible friction. The carts can travel along a straight horizontal track. Cart A has known mass m_A. The students are asked to use a one-dimensional collision between the carts to determine the mass of cart B. Before the collision, cart A travels to the right and cart B is initially at rest. After the collision, the carts stick together.

a. Describe an experimental procedure to determine the velocities of the carts before and after a collision, including all the additional equipment you would need. You may include a labeled diagram of your setup to help in your description. Indicate what measurements you would take and how you would take them. Include enough detail so that another student could carry out your procedure.

b. There will be sources of error in the measurements taken in the experiment, both before and after the collision. For your experimental procedure, will the uncertainty in the calculated value of the mass of cart *B* be affected more by the error in the measurements taken before the collision or by those taken after the collision, or will it be equally affected by both sets of measurements? Justify your answer.

A group of students took measurements for one collision. A graph of the students' data is shown below.

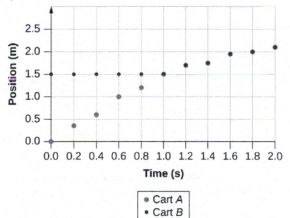

● Cart *A*
● Cart *B*

Figure 8.22 The image shows a graph with position in meters on the vertical axis and time in seconds on the horizontal axis.

c. Given m_A = 0.50 kg, use the graph to calculate the mass of cart *B*. Explicitly indicate the principles used in your calculations.

d. The students are now asked to Consider the kinetic energy changes in an inelastic collision, specifically whether the initial values of one of the physical quantities affect the fraction of mechanical energy dissipated in the collision. How could you modify the experiment to investigate this question? Be sure to explicitly describe the calculations you would make, specifying all equations you would use (but do not actually do any algebra or arithmetic).

19. Cart A is moving with an initial velocity +v (in the positive direction) toward cart B, initially at rest. Both carts have equal mass and are on a frictionless surface. Which of the following statements correctly characterizes the velocity of the center of mass of the system before and after the collision?

a. $\frac{+v}{2}$ before, $\frac{-v}{2}$ after

b. $\frac{+v}{2}$ before, 0 after

c. $\frac{+v}{2}$ before, $\frac{+v}{2}$ after

d. 0 before, 0 after

20. Cart A is moving with a velocity of +10 m/s toward cart B, which is moving with a velocity of +4 m/s. Both carts have equal mass and are moving on a frictionless surface. The two carts have an inelastic collision and stick together after the

collision. Calculate the velocity of the center of mass of the system before and after the collision. If there were friction present in this problem, how would this external force affect the center-of-mass velocity both before and after the collision?

8.4 Elastic Collisions in One Dimension

21. Two cars (A and B) of mass 1.5 kg collide. Car A is initially moving at 12 m/s, and car B is initially moving in the same direction with a speed of 6 m/s. The two cars are moving along a straight line before and after the collision. What will be the change in momentum of this system after the collision?
   a. −27 kg • m/s
   b. zero
   c. +27 kg • m/s
   d. It depends on whether the collision is elastic or inelastic.

22. Two cars (A and B) of mass 1.5 kg collide. Car A is initially moving at 24 m/s, and car B is initially moving in the opposite direction with a speed of 12 m/s. The two cars are moving along a straight line before and after the collision. (a) If the two cars have an elastic collision, calculate the change in momentum of the two-car system. (b) If the two cars have a completely inelastic collision, calculate the change in momentum of the two-car system.

23. Puck A (200 g) slides across a frictionless surface to collide with puck B (800 g), initially at rest. The velocity of each puck is measured during the experiment as follows:

Table 8.3

Time	Velocity A	Velocity B
0	+8.0 m/s	0
1.0 s	+8.0 m/s	0
2.0 s	−2.0 m/s	+2.5 m/s
3.0 s	−2.0 m/s	+2.5 m/s

What is the change in momentum of the center of mass of the system as a result of the collision?

   a. +1.6 kg•m/s
   b. +0.8 kg•m/s
   c. 0
   d. −1.6 kg•m/s

24. For the table above, calculate the center-of-mass velocity of the system both before and after the collision, then calculate the center-of-mass momentum of the system both before and after the collision. From this, determine the change in the momentum of the system as a result of the collision.

25. Two cars (A and B) of equal mass have an elastic collision. Prior to the collision, car A is moving at 15 m/s in the +x-direction, and car B is moving at 10 m/s in the −x-direction. Assuming that both cars continue moving along the x-axis after the collision, what will be the velocity of car A after the collision?
   a. same as the original 15 m/s speed, opposite direction
   b. equal to car B's velocity prior to the collision
   c. equal to the average of the two velocities, in its original direction
   d. equal to the average of the two velocities, in the opposite direction

26. Two cars (A and B) of equal mass have an elastic

collision. Prior to the collision, car A is moving at 20 m/s in the +x-direction, and car B is moving at 10 m/s in the –x-direction. Assuming that both cars continue moving along the x-axis after the collision, what will be the velocities of each car after the collision?

27. A rubber ball is dropped from rest at a fixed height. It bounces off a hard floor and rebounds upward, but it only reaches 90% of its original fixed height. What is the best way to explain the loss of kinetic energy of the ball during the collision?
  a. Energy was required to deform the ball's shape during the collision with the floor.
  b. Energy was lost due to work done by the ball pushing on the floor during the collision.
  c. Energy was lost due to friction between the ball and the floor.
  d. Energy was lost due to the work done by gravity during the motion.

28. A tennis ball strikes a wall with an initial speed of 15 m/s. The ball bounces off the wall but rebounds with slightly less speed (14 m/s) after the collision. Explain (a) what else changed its momentum in response to the ball's change in momentum so that overall momentum is conserved, and (b) how some of the ball's kinetic energy was lost.

29. Two objects, A and B, have equal mass. Prior to the collision, mass A is moving 10 m/s in the +x-direction, and mass B is moving 4 m/s in the +x-direction. Which of the following results represents an inelastic collision between A and B?
  a. After the collision, mass A is at rest, and mass B moves 14 m/s in the +x-direction.
  b. After the collision, mass A moves 4 m/s in the –x-direction, and mass B moves 18 m/s in the +x-direction.
  c. After the collision, the two masses stick together and move 7 m/s in the +x-direction.
  d. After the collision, mass A moves 4 m/s in the +x-direction, and mass B moves 10 m/s in the +x-direction.

30. Mass A is three times more massive than mass B. Mass A is initially moving 12 m/s in the +x-direction. Mass B is initially moving 12 m/s in the –x-direction. Assuming that the collision is elastic, calculate the final velocity of both masses after the collision. Show that your results are consistent with conservation of momentum and conservation of kinetic energy.

31. Two objects (A and B) of equal mass collide elastically. Mass A is initially moving 5.0 m/s in the +x-direction prior to the collision. Mass B is initially moving 3.0 m/s in the –x-direction prior to the collision. After the collision, mass A will be moving with a velocity of 3.0 m/s in the –x-direction. What will be the velocity of mass B after the collision?
  a. 3.0 m/s in the +x-direction
  b. 5.0 m/s in the +x-direction
  c. 3.0 m/s in the –x-direction
  d. 5.0 m/s in the –x-direction

32. Two objects (A and B) of equal mass collide elastically. Mass A is initially moving 4.0 m/s in the +x-direction prior to the collision. Mass B is initially moving 8.0 m/s in the –x-direction prior to the collision. After the collision, mass A will be moving with a velocity of 8.0 m/s in the –x-direction. (a) Use the principle of conservation of momentum to predict the velocity of mass B after the collision. (b) Use the fact that kinetic energy is conserved in elastic collisions to predict the velocity of mass B after the collision.

33. Two objects of equal mass collide. Object A is initially moving in the +x-direction with a speed of 12 m/s, and object B is initially at rest. After the collision, object A is at rest, and object B is moving away with some unknown velocity. There are no external forces acting on the system of two masses. What statement can we make about this collision?
  a. Both momentum and kinetic energy are conserved.
  b. Momentum is conserved, but kinetic energy is not conserved.
  c. Neither momentum nor kinetic energy is conserved.
  d. More information is needed in order to determine which is conserved.

34. Two objects of equal mass collide. Object A is initially moving with a velocity of 15 m/s in the +x-direction, and object B is initially at rest. After the collision, object A is at rest. There are no external forces acting on the system of two masses. (a) Use momentum conservation to deduce the velocity of object B after the collision. (b) Is this collision elastic? Justify your answer.

35. Which of the following statements is true about an inelastic collision?
  a. Momentum is conserved, and kinetic energy is conserved.
  b. Momentum is conserved, and kinetic energy is not conserved.
  c. Momentum is not conserved, and kinetic energy is conserved.
  d. Momentum is not conserved, and kinetic energy is not conserved.

36. Explain how the momentum and kinetic energy of a system of two colliding objects changes as a result of (a) an elastic collision and (b) an inelastic collision.

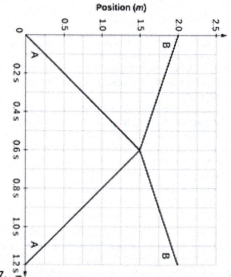

37.
This figure shows the positions of two colliding objects measured before, during, and after a collision. Mass A is 1.0 kg. Mass B is 3.0 kg. Which of the following statements is true?
  a. This is an elastic collision, with a total momentum of 0 kg • m/s.
  b. This is an elastic collision, with a total momentum of 1.67 kg • m/s.
  c. This is an inelastic collision, with a total momentum of 0 kg • m/s.
  d. This is an inelastic collision, with a total momentum of 1.67 kg • m/s.

38. For the above graph, determine the initial and final momentum for both objects, assuming mass A is 1.0 kg and mass B is 3.0 kg. Also, determine the initial and final kinetic energies for both objects. Based on your results, explain whether momentum is conserved in this collision, and state whether the collision is elastic or inelastic.

39. Mass A (1.0 kg) slides across a frictionless surface with a velocity of 8 m/s in the positive direction. Mass B (3.0 kg) is initially at rest. The two objects collide and stick together. What will be the change in the center-of-mass velocity of the system as a result of the collision?
a. There will be no change in the center-of-mass velocity.
b. The center-of-mass velocity will decrease by 2 m/s.
c. The center-of-mass velocity will decrease by 6 m/s.
d. The center-of-mass velocity will decrease by 8 m/s.

40. Mass A (1.0 kg) slides across a frictionless surface with a velocity of 4 m/s in the positive direction. Mass B (1.0 kg) slides across the same surface in the opposite direction with a velocity of −8 m/s. The two objects collide and stick together after the collision. Predict how the center-of-mass velocity will change as a result of the collision, and explain your prediction. Calculate the center-of-mass velocity of the system both before and after the collision and explain why it remains the same or why it has changed.

8.5 Inelastic Collisions in One Dimension

41. Mass A (2.0 kg) has an initial velocity of 4 m/s in the +x-direction. Mass B (2.0 kg) has an initial velocity of 5 m/s in the −x-direction. If the two masses have an elastic collision, what will be the final velocities of the masses after the collision?
a. Both will move 0.5 m/s in the −x-direction.
b. Mass A will stop; mass B will move 9 m/s in the +x-direction.
c. Mass B will stop; mass A will move 9 m/s in the −x-direction.
d. Mass A will move 5 m/s in the −x-direction; mass B will move 4 m/s in the +x-direction.

42. Mass A has an initial velocity of 22 m/s in the +x-direction. Mass B is three times more massive than mass A and has an initial velocity of 22 m/s in the −x-direction. If the two masses have an elastic collision, what will be the final velocities of the masses after the collision?

43. Mass A (2.0 kg) is moving with an initial velocity of 15 m/s in the +x-direction, and it collides with mass B (5.0 kg), initially at rest. After the collision, the two objects stick together and move as one. What is the change in kinetic energy of the system as a result of the collision?
a. no change
b. decrease by 225 J
c. decrease by 161 J
d. decrease by 64 J

44. Mass A (2.0 kg) is moving with an initial velocity of 15 m/s in the +x-direction, and it collides with mass B (4.0 kg), initially moving 7.0 m/s in the +x-direction. After the collision, the two objects stick together and move as one. What is the change in kinetic energy of the system as a result of the collision?

45. Mass A slides across a rough table with an initial velocity of 12 m/s in the +x-direction. By the time mass A collides with mass B (a stationary object with equal mass), mass A has slowed to 10 m/s. After the collision, the two objects stick together and move as one. Immediately after the collision, the velocity of the system is measured to be 5 m/s in the +x-direction, and the system eventually slides to a stop.

Which of the following statements is true about this motion?
a. Momentum is conserved during the collision, but it is not conserved during the motion before and after the collision.
b. Momentum is not conserved at any time during this analysis.
c. Momentum is conserved at all times during this analysis.
d. Momentum is not conserved during the collision, but it is conserved during the motion before and after the collision.

46. Mass A is initially moving with a velocity of 12 m/s in the +x-direction. Mass B is twice as massive as mass A and is initially at rest. After the two objects collide, the two masses move together as one with a velocity of 4 m/s in the +x-direction. Is momentum conserved in this collision?

47. Mass A is initially moving with a velocity of 24 m/s in the +x-direction. Mass B is twice as massive as mass A and is initially at rest. The two objects experience a totally inelastic collision. What is the final speed of both objects after the collision?
a. A is not moving; B is moving 24 m/s in the +x-direction.
b. Neither A nor B is moving.
c. A is moving 24 m/s in the −x-direction. B is not moving.
d. Both A and B are moving together 8 m/s in the +x-direction.

48. Mass A is initially moving with some unknown velocity in the +x-direction. Mass B is twice as massive as mass A and initially at rest. The two objects collide, and after the collision, they move together with a speed of 6 m/s in the +x-direction. (a) Is this collision elastic or inelastic? Explain. (b) Determine the initial velocity of mass A.

49. Mass A is initially moving with a velocity of 2 m/s in the +x-direction. Mass B is initially moving with a velocity of 6 m/s in the −x-direction. The two objects have equal masses. After they collide, mass A moves with a speed of 4 m/s in the −x-direction. What is the final velocity of mass B after the collision?
a. 6 m/s in the +x-direction
b. 4 m/s in the +x-direction
c. zero
d. 4 m/s in the −x-direction

50. Mass A is initially moving with a velocity of 15 m/s in the +x-direction. Mass B is twice as massive and is initially moving with a velocity of 10 m/s in the −x-direction. The two objects collide, and after the collision, mass A moves with a speed of 15 m/s in the −x-direction. (a) What is the final velocity of mass B after the collision? (b) Calculate the change in kinetic energy as a result of the collision, assuming mass A is 5.0 kg.

8.6 Collisions of Point Masses in Two Dimensions

51. Two cars of equal mass approach an intersection. Car A is moving east at a speed of 45 m/s. Car B is moving south at a speed of 35 m/s. They collide inelastically and stick together after the collision, moving as one object. Which of the following statements is true about the center-of-mass velocity of this system?

a. The center-of-mass velocity will decrease after the
    collision as a result of lost energy (but not drop to zero).
b. The center-of-mass velocity will remain the same after
    the collision since momentum is conserved.
c. The center-of-mass velocity will drop to zero since the
    two objects stick together.
d. The magnitude of the center-of-mass velocity will
    remain the same, but the direction of the velocity will
    change.

52. Car A has a mass of 2000 kg and approaches an
intersection with a velocity of 38 m/s directed to the east. Car
B has a mass of 3500 kg and approaches the intersection
with a velocity of 53 m/s directed 63° north of east. The two
cars collide and stick together after the collision. Will the
center-of-mass velocity change as a result of the collision?
Explain why or why not. Calculate the center-of-mass velocity
before and after the collision.

9 STATICS AND TORQUE

Figure 9.1 On a short time scale, rocks like these in Australia's Kings Canyon are static, or motionless relative to the Earth. (credit: freeaussiestock.com)

Chapter Outline

9.1. The First Condition for Equilibrium

9.2. The Second Condition for Equilibrium

9.3. Stability

9.4. Applications of Statics, Including Problem-Solving Strategies

9.5. Simple Machines

9.6. Forces and Torques in Muscles and Joints

Connection for AP® Courses

What might desks, bridges, buildings, trees, and mountains have in common? What do these objects have in common with a car moving at a constant velocity? While it may be apparent that the objects in the first group are all motionless relative to Earth, they also share something with the moving car and all objects moving at a constant velocity. All of these objects, stationary and moving, share an acceleration of zero. How can this be? Consider Newton's second law, $F = ma$. When acceleration is zero, as is the case for both stationary objects and objects moving at a constant velocity, the net external force must also be zero (Big Idea 3). Forces are acting on both stationary objects and on objects moving at a constant velocity, but the forces are balanced. That is, they are in *equilibrium*. In equilibrium, the net force is zero.

The first two sections of this chapter will focus on the two conditions necessary for equilibrium. They will not only help you to distinguish between stationary bridges and cars moving at constant velocity, but will introduce a second equilibrium condition, this time involving rotation. As you explore the second equilibrium condition, you will learn about torque, in support of both Enduring Understanding 3.F and Essential Knowledge 3.F.1. Much like a force, torque provides the capability for acceleration; however, with careful attention, torques may also be balanced and equilibrium can be reached.

The remainder of this chapter will discuss a variety of interesting equilibrium applications. From the art of balancing, to simple machines, to the muscles in your body, the world around you relies upon the principles of equilibrium to remain stable. This chapter will help you to see just how closely related these events truly are.

The content in this chapter supports:

Big Idea 3 The interactions of an object with other objects can be described by forces.

Enduring Understanding 3.F A force exerted on an object can cause a torque on that object.

Essential Knowledge 3.F.1 Only the force component perpendicular to the line connecting the axis of rotation and the point of application of the force results in a torque about that axis.

9.1 The First Condition for Equilibrium

The first condition necessary to achieve equilibrium is the one already mentioned: the net external force on the system must be zero. Expressed as an equation, this is simply

$$\text{net } \mathbf{F} = 0 \tag{9.1}$$

Note that if net F is zero, then the net external force in *any* direction is zero. For example, the net external forces along the typical *x*- and *y*-axes are zero. This is written as

$$\text{net } F_x = 0 \text{ and } F_y = 0 \tag{9.2}$$

Figure 9.2 and **Figure 9.3** illustrate situations where net $F = 0$ for both **static equilibrium** (motionless), and **dynamic equilibrium** (constant velocity).

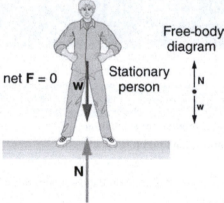

Figure 9.2 This motionless person is in static equilibrium. The forces acting on him add up to zero. Both forces are vertical in this case.

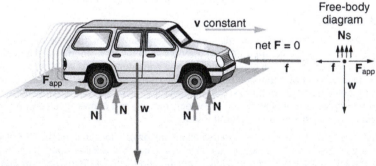

Figure 9.3 This car is in dynamic equilibrium because it is moving at constant velocity. There are horizontal and vertical forces, but the net external force in any direction is zero. The applied force F_{app} between the tires and the road is balanced by air friction, and the weight of the car is supported by the normal forces, here shown to be equal for all four tires.

However, it is not sufficient for the net external force of a system to be zero for a system to be in equilibrium. Consider the two situations illustrated in **Figure 9.4** and **Figure 9.5** where forces are applied to an ice hockey stick lying flat on ice. The net external force is zero in both situations shown in the figure; but in one case, equilibrium is achieved, whereas in the other, it is not. In **Figure 9.4**, the ice hockey stick remains motionless. But in **Figure 9.5**, with the same forces applied in different places, the stick experiences accelerated rotation. Therefore, we know that the point at which a force is applied is another factor in determining whether or not equilibrium is achieved. This will be explored further in the next section.

Equilibrium: remains stationary

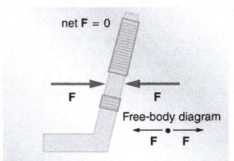

Figure 9.4 An ice hockey stick lying flat on ice with two equal and opposite horizontal forces applied to it. Friction is negligible, and the gravitational force is balanced by the support of the ice (a normal force). Thus, $\text{net } F = 0$. Equilibrium is achieved, which is static equilibrium in this case.

Nonequilibrium: rotation accelerates

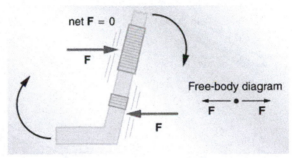

Figure 9.5 The same forces are applied at other points and the stick rotates—in fact, it experiences an accelerated rotation. Here $\text{net } F = 0$ but the system is *not* at equilibrium. Hence, the $\text{net } F = 0$ is a necessary—but not sufficient—condition for achieving equilibrium.

PhET Explorations: Torque

Investigate how torque causes an object to rotate. Discover the relationships between angular acceleration, moment of inertia, angular momentum and torque.

Figure 9.6 Torque (http://cnx.org/content/m55176/1.2/torque_en.jar)

9.2 The Second Condition for Equilibrium

Learning Objectives
By the end of this section, you will be able to: • State the second condition that is necessary to achieve equilibrium. • Explain torque and the factors on which it depends. • Describe the role of torque in rotational mechanics. The information presented in this section supports the following AP® learning objectives and science practices: • **3.F.1.1** The student is able to use representations of the relationship between force and torque. **(S.P. 1.4)** • **3.F.1.2** The student is able to compare the torques on an object caused by various forces. **(S.P. 1.4)** • **3.F.1.3** The student is able to estimate the torque on an object caused by various forces in comparison to other situations. **(S.P. 2.3)**

Torque

The second condition necessary to achieve equilibrium involves avoiding accelerated rotation (maintaining a constant angular velocity. A rotating body or system can be in equilibrium if its rate of rotation is constant and remains unchanged by the forces acting on it. To understand what factors affect rotation, let us think about what happens when you open an ordinary door by rotating it on its hinges.

Several familiar factors determine how effective you are in opening the door. See **Figure 9.7**. First of all, the larger the force, the more effective it is in opening the door—obviously, the harder you push, the more rapidly the door opens. Also, the point at which you push is crucial. If you apply your force too close to the hinges, the door will open slowly, if at all. Most people have been embarrassed by making this mistake and bumping up against a door when it did not open as quickly as expected. Finally, the direction in which you push is also important. The most effective direction is perpendicular to the door—we push in this direction almost instinctively.

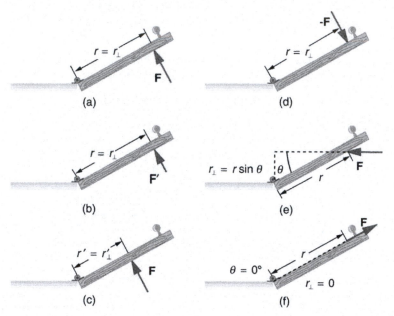

Figure 9.7 Torque is the turning or twisting effectiveness of a force, illustrated here for door rotation on its hinges (as viewed from overhead). Torque has both magnitude and direction. (a) Counterclockwise torque is produced by this force, which means that the door will rotate in a counterclockwise due to $\mathbf{F}$. Note that $r_\perp$ is the perpendicular distance of the pivot from the line of action of the force. (b) A smaller counterclockwise torque is produced by a smaller force $\mathbf{F'}$ acting at the same distance from the hinges (the pivot point). (c) The same force as in (a) produces a smaller counterclockwise torque when applied at a smaller distance from the hinges. (d) The same force as in (a), but acting in the opposite direction, produces a clockwise torque. (e) A smaller counterclockwise torque is produced by the same magnitude force acting at the same point but in a different direction. Here, θ is less than $90°$. (f) Torque is zero here since the force just pulls on the hinges, producing no rotation. In this case, $\theta = 0°$.

The magnitude, direction, and point of application of the force are incorporated into the definition of the physical quantity called torque. **Torque** is the rotational equivalent of a force. It is a measure of the effectiveness of a force in changing or accelerating a rotation (changing the angular velocity over a period of time). In equation form, the magnitude of torque is defined to be

$$\tau = rF \sin \theta \tag{9.3}$$

where τ (the Greek letter tau) is the symbol for torque, r is the distance from the pivot point to the point where the force is applied, F is the magnitude of the force, and θ is the angle between the force and the vector directed from the point of application to the pivot point, as seen in **Figure 9.7** and **Figure 9.8**. An alternative expression for torque is given in terms of the **perpendicular lever arm** $r_\perp$ as shown in **Figure 9.7** and **Figure 9.8**, which is defined as

$$r_\perp = r \sin \theta \tag{9.4}$$

so that

$$\tau = r_\perp F. \tag{9.5}$$

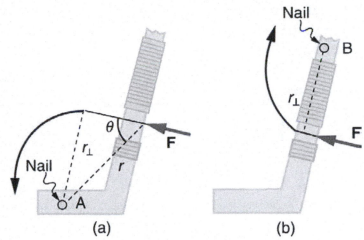

Figure 9.8 A force applied to an object can produce a torque, which depends on the location of the pivot point. (a) The three factors r, F, and θ for pivot point A on a body are shown here— r is the distance from the chosen pivot point to the point where the force F is applied, and θ is the angle between F and the vector directed from the point of application to the pivot point. If the object can rotate around point A, it will rotate counterclockwise. This means that torque is counterclockwise relative to pivot A. (b) In this case, point B is the pivot point. The torque from the applied force will cause a clockwise rotation around point B, and so it is a clockwise torque relative to B.

The perpendicular lever arm $r_\perp$ is the shortest distance from the pivot point to the line along which F acts; it is shown as a dashed line in **Figure 9.7** and **Figure 9.8**. Note that the line segment that defines the distance $r_\perp$ is perpendicular to F, as its name implies. It is sometimes easier to find or visualize $r_\perp$ than to find both r and θ. In such cases, it may be more convenient to use $\tau = r_\perp F$ rather than $\tau = rF \sin \theta$ for torque, but both are equally valid.

The **SI unit of torque** is newtons times meters, usually written as $N \cdot m$. For example, if you push perpendicular to the door with a force of 40 N at a distance of 0.800 m from the hinges, you exert a torque of $32\ N\cdot m (0.800\ m \times 40\ N \times \sin 90°)$ relative to the hinges. If you reduce the force to 20 N, the torque is reduced to $16\ N\cdot m$, and so on.

The torque is always calculated with reference to some chosen pivot point. For the same applied force, a different choice for the location of the pivot will give you a different value for the torque, since both r and θ depend on the location of the pivot. Any point in any object can be chosen to calculate the torque about that point. The object may not actually pivot about the chosen "pivot point."

Note that for rotation in a plane, torque has two possible directions. Torque is either clockwise or counterclockwise relative to the chosen pivot point, as illustrated for points B and A, respectively, in **Figure 9.8**. If the object can rotate about point A, it will rotate counterclockwise, which means that the torque for the force is shown as counterclockwise relative to A. But if the object can rotate about point B, it will rotate clockwise, which means the torque for the force shown is clockwise relative to B. Also, the magnitude of the torque is greater when the lever arm is longer.

Making Connections: Pivoting Block

A solid block of length *d* is pinned to a wall on its right end. Three forces act on the block as shown below: **F**A, **F**B, and **F**C. While all three forces are of equal magnitude, and all three are equal distances away from the pivot point, all three forces will create a different torque upon the object.

FA is vectored perpendicular to its distance from the pivot point; as a result, the magnitude of its torque can be found by the equation $\tau = F_A * d$. Vector **F**B is parallel to the line connecting the point of application of force and the pivot point. As a result, it does not provide an ability to rotate the object and, subsequently, its torque is zero. **F**C, however, is directed at an angle θ to the line connecting the point of application of force and the pivot point. In this instance, only the component perpendicular to this line is exerting a torque. This component, labeled $F_\perp$, can be found using the equation $F_\perp = F_C \sin\theta$. The component of the force parallel to this line, labeled $F_\parallel$, does not provide an ability to rotate the object and, as a result, does not provide a torque. Therefore, the resulting torque created by **F**C is $\tau = F_\perp * d$.

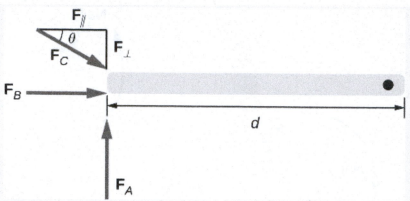

Figure 9.9 Forces on a block pinned to a wall. A solid block of length *d* is pinned to a wall on its right end. Three forces act on the block: F$_A$, F$_B$, and F$_C$.

Now, *the second condition necessary to achieve equilibrium* is that *the net external torque on a system must be zero*. An external torque is one that is created by an external force. You can choose the point around which the torque is calculated. The point can be the physical pivot point of a system or any other point in space—but it must be the same point for all torques. If the second condition (net external torque on a system is zero) is satisfied for one choice of pivot point, it will also hold true for any other choice of pivot point in or out of the system of interest. (This is true only in an inertial frame of reference.) The second condition necessary to achieve equilibrium is stated in equation form as

$$\text{net } \tau = 0 \tag{9.6}$$

where net means total. Torques, which are in opposite directions are assigned opposite signs. A common convention is to call counterclockwise (ccw) torques positive and clockwise (cw) torques negative.

When two children balance a seesaw as shown in **Figure 9.10**, they satisfy the two conditions for equilibrium. Most people have perfect intuition about seesaws, knowing that the lighter child must sit farther from the pivot and that a heavier child can keep a lighter one off the ground indefinitely.

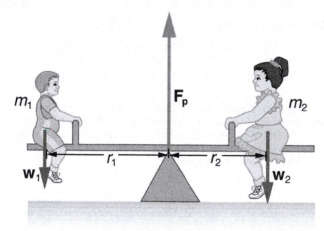

Figure 9.10 Two children balancing a seesaw satisfy both conditions for equilibrium. The lighter child sits farther from the pivot to create a torque equal in magnitude to that of the heavier child.

Example 9.1 She Saw Torques On A Seesaw

The two children shown in **Figure 9.10** are balanced on a seesaw of negligible mass. (This assumption is made to keep the example simple—more involved examples will follow.) The first child has a mass of 26.0 kg and sits 1.60 m from the pivot.(a) If the second child has a mass of 32.0 kg, how far is she from the pivot? (b) What is F_p, the supporting force exerted by the pivot?

Strategy

Both conditions for equilibrium must be satisfied. In part (a), we are asked for a distance; thus, the second condition (regarding torques) must be used, since the first (regarding only forces) has no distances in it. To apply the second condition for equilibrium, we first identify the system of interest to be the seesaw plus the two children. We take the supporting pivot to be the point about which the torques are calculated. We then identify all external forces acting on the system.

Solution (a)

The three external forces acting on the system are the weights of the two children and the supporting force of the pivot. Let us examine the torque produced by each. Torque is defined to be

$$\tau = rF \sin \theta. \tag{9.7}$$

Here $\theta = 90°$, so that $\sin \theta = 1$ for all three forces. That means $r_\perp = r$ for all three. The torques exerted by the three forces are first,

$$\tau_1 = r_1 w_1 \tag{9.8}$$

second,

$$\tau_2 = -r_2 w_2 \tag{9.9}$$

and third,

$$\begin{aligned} \tau_p &= r_p F_p \\ &= 0 \cdot F_p \\ &= 0. \end{aligned} \tag{9.10}$$

Note that a minus sign has been inserted into the second equation because this torque is clockwise and is therefore negative by convention. Since F_p acts directly on the pivot point, the distance r_p is zero. A force acting on the pivot cannot cause a rotation, just as pushing directly on the hinges of a door will not cause it to rotate. Now, the second condition for equilibrium is that the sum of the torques on both children is zero. Therefore

$$\tau_2 = -\tau_1, \tag{9.11}$$

or

$$r_2 w_2 = r_1 w_1. \tag{9.12}$$

Weight is mass times the acceleration due to gravity. Entering mg for w, we get

$$r_2 m_2 g = r_1 m_1 g. \tag{9.13}$$

Solve this for the unknown r_2:

$$r_2 = r_1 \frac{m_1}{m_2}. \tag{9.14}$$

The quantities on the right side of the equation are known; thus, r_2 is

$$r_2 = (1.60 \text{ m}) \frac{26.0 \text{ kg}}{32.0 \text{ kg}} = 1.30 \text{ m}. \tag{9.15}$$

As expected, the heavier child must sit closer to the pivot (1.30 m versus 1.60 m) to balance the seesaw.

Solution (b)

This part asks for a force F_p. The easiest way to find it is to use the first condition for equilibrium, which is

$$\text{net } \mathbf{F} = 0. \tag{9.16}$$

The forces are all vertical, so that we are dealing with a one-dimensional problem along the vertical axis; hence, the condition can be written as

$$\text{net } F_y = 0 \tag{9.17}$$

where we again call the vertical axis the y-axis. Choosing upward to be the positive direction, and using plus and minus signs to indicate the directions of the forces, we see that

$$F_p - w_1 - w_2 = 0. \tag{9.18}$$

This equation yields what might have been guessed at the beginning:

$$F_p = w_1 + w_2. \tag{9.19}$$

So, the pivot supplies a supporting force equal to the total weight of the system:

$$F_p = m_1 g + m_2 g. \tag{9.20}$$

Entering known values gives

$$F_{\text{p}} = (26.0 \text{ kg})(9.80 \text{ m/s}^2) + (32.0 \text{ kg})(9.80 \text{ m/s}^2) \qquad\qquad (9.21)$$
$$= 568 \text{ N}.$$

Discussion

The two results make intuitive sense. The heavier child sits closer to the pivot. The pivot supports the weight of the two children. Part (b) can also be solved using the second condition for equilibrium, since both distances are known, but only if the pivot point is chosen to be somewhere other than the location of the seesaw's actual pivot!

Several aspects of the preceding example have broad implications. First, the choice of the pivot as the point around which torques are calculated simplified the problem. Since F_{p} is exerted on the pivot point, its lever arm is zero. Hence, the torque exerted by the supporting force F_{p} is zero relative to that pivot point. The second condition for equilibrium holds for any choice of pivot point, and so we choose the pivot point to simplify the solution of the problem.

Second, the acceleration due to gravity canceled in this problem, and we were left with a ratio of masses. *This will not always be the case*. Always enter the correct forces—do not jump ahead to enter some ratio of masses.

Third, the weight of each child is distributed over an area of the seesaw, yet we treated the weights as if each force were exerted at a single point. This is not an approximation—the distances r_1 and r_2 are the distances to points directly below the **center of gravity** of each child. As we shall see in the next section, the mass and weight of a system can act as if they are located at a single point.

Finally, note that the concept of torque has an importance beyond static equilibrium. *Torque plays the same role in rotational motion that force plays in linear motion.* We will examine this in the next chapter.

Take-Home Experiment

Take a piece of modeling clay and put it on a table, then mash a cylinder down into it so that a ruler can balance on the round side of the cylinder while everything remains still. Put a penny 8 cm away from the pivot. Where would you need to put two pennies to balance? Three pennies?

9.3 Stability

Learning Objectives

By the end of this section, you will be able to:

- State the types of equilibrium.
- Describe stable and unstable equilibriums.
- Describe neutral equilibrium.

The information presented in this section supports the following AP® learning objectives and science practices:

- **3.F.1.1** The student is able to use representations of the relationship between force and torque. **(S.P. 1.4)**
- **3.F.1.2** The student is able to compare the torques on an object caused by various forces. **(S.P. 1.4)**
- **3.F.1.3** The student is able to estimate the torque on an object caused by various forces in comparison to other situations. **(S.P. 2.3)**
- **3.F.1.4** The student is able to design an experiment and analyze data testing a question about torques in a balanced rigid system. **(S.P. 4.1, 4.2, 5.1)**
- **3.F.1.5** The student is able to calculate torques on a two-dimensional system in static equilibrium, by examining a representation or model (such as a diagram or physical construction). **(S.P. 1.4, 2.2)**

It is one thing to have a system in equilibrium; it is quite another for it to be stable. The toy doll perched on the man's hand in **Figure 9.11**, for example, is not in stable equilibrium. There are *three types of equilibrium*: *stable*, *unstable*, and *neutral*. Figures throughout this module illustrate various examples.

Figure 9.11 presents a balanced system, such as the toy doll on the man's hand, which has its center of gravity (cg) directly over the pivot, so that the torque of the total weight is zero. This is equivalent to having the torques of the individual parts balanced about the pivot point, in this case the hand. The cgs of the arms, legs, head, and torso are labeled with smaller type.

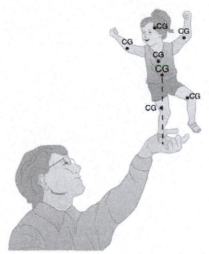

Figure 9.11 A man balances a toy doll on one hand.

A system is said to be in **stable equilibrium** if, when displaced from equilibrium, it experiences a net force or torque in a direction opposite to the direction of the displacement. For example, a marble at the bottom of a bowl will experience a *restoring* force when displaced from its equilibrium position. This force moves it back toward the equilibrium position. Most systems are in stable equilibrium, especially for small displacements. For another example of stable equilibrium, see the pencil in **Figure 9.12**.

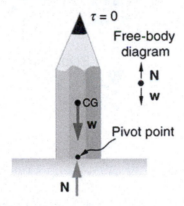

Figure 9.12 This pencil is in the condition of equilibrium. The net force on the pencil is zero and the total torque about any pivot is zero.

A system is in **unstable equilibrium** if, when displaced, it experiences a net force or torque in the *same* direction as the displacement from equilibrium. A system in unstable equilibrium accelerates away from its equilibrium position if displaced even slightly. An obvious example is a ball resting on top of a hill. Once displaced, it accelerates away from the crest. See the next several figures for examples of unstable equilibrium.

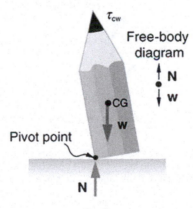

Figure 9.13 If the pencil is displaced slightly to the side (counterclockwise), it is no longer in equilibrium. Its weight produces a clockwise torque that returns the pencil to its equilibrium position.

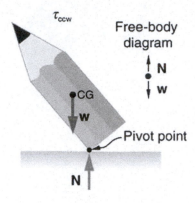

Figure 9.14 If the pencil is displaced too far, the torque caused by its weight changes direction to counterclockwise and causes the displacement to increase.

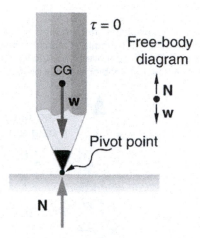

Figure 9.15 This figure shows unstable equilibrium, although both conditions for equilibrium are satisfied.

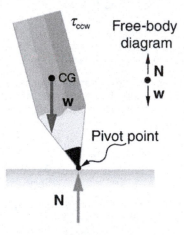

Figure 9.16 If the pencil is displaced even slightly, a torque is created by its weight that is in the same direction as the displacement, causing the displacement to increase.

A system is in **neutral equilibrium** if its equilibrium is independent of displacements from its original position. A marble on a flat horizontal surface is an example. Combinations of these situations are possible. For example, a marble on a saddle is stable for displacements toward the front or back of the saddle and unstable for displacements to the side. **Figure 9.17** shows another example of neutral equilibrium.

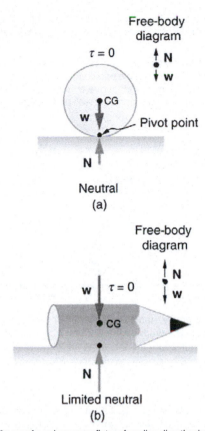

Figure 9.17 (a) Here we see neutral equilibrium. The cg of a sphere on a flat surface lies directly above the point of support, independent of the position on the surface. The sphere is therefore in equilibrium in any location, and if displaced, it will remain put. (b) Because it has a circular cross section, the pencil is in neutral equilibrium for displacements perpendicular to its length.

When we consider how far a system in stable equilibrium can be displaced before it becomes unstable, we find that some systems in stable equilibrium are more stable than others. The pencil in **Figure 9.12** and the person in **Figure 9.18**(a) are in stable equilibrium, but become unstable for relatively small displacements to the side. The critical point is reached when the cg is no longer *above* the base of support. Additionally, since the cg of a person's body is above the pivots in the hips, displacements must be quickly controlled. This control is a central nervous system function that is developed when we learn to hold our bodies erect as infants. For increased stability while standing, the feet should be spread apart, giving a larger base of support. Stability is also increased by lowering one's center of gravity by bending the knees, as when a football player prepares to receive a ball or braces themselves for a tackle. A cane, a crutch, or a walker increases the stability of the user, even more as the base of support widens. Usually, the cg of a female is lower (closer to the ground) than a male. Young children have their center of gravity between their shoulders, which increases the challenge of learning to walk.

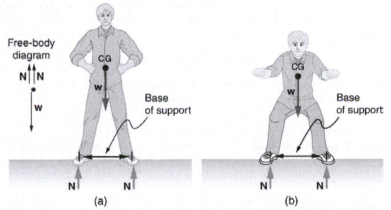

Figure 9.18 (a) The center of gravity of an adult is above the hip joints (one of the main pivots in the body) and lies between two narrowly-separated feet. Like a pencil standing on its eraser, this person is in stable equilibrium in relation to sideways displacements, but relatively small displacements take his cg outside the base of support and make him unstable. Humans are less stable relative to forward and backward displacements because the feet are not very long. Muscles are used extensively to balance the body in the front-to-back direction. (b) While bending in the manner shown, stability is increased by lowering the center of gravity. Stability is also increased if the base is expanded by placing the feet farther apart.

Animals such as chickens have easier systems to control. **Figure 9.19** shows that the cg of a chicken lies below its hip joints and between its widely separated and broad feet. Even relatively large displacements of the chicken's cg are stable and result in restoring forces and torques that return the cg to its equilibrium position with little effort on the chicken's part. Not all birds are like chickens, of course. Some birds, such as the flamingo, have balance systems that are almost as sophisticated as that of humans.

Figure 9.19 shows that the cg of a chicken is below the hip joints and lies above a broad base of support formed by widely-separated and large feet. Hence, the chicken is in very stable equilibrium, since a relatively large displacement is needed to render it unstable. The body of the chicken is supported from above by the hips and acts as a pendulum between the hips. Therefore, the chicken is stable for front-to-back displacements as well as for side-to-side displacements.

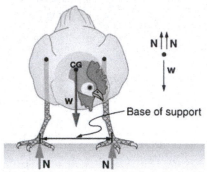

Figure 9.19 The center of gravity of a chicken is below the hip joints. The chicken is in stable equilibrium. The body of the chicken is supported from above by the hips and acts as a pendulum between them.

Engineers and architects strive to achieve extremely stable equilibriums for buildings and other systems that must withstand wind, earthquakes, and other forces that displace them from equilibrium. Although the examples in this section emphasize gravitational forces, the basic conditions for equilibrium are the same for all types of forces. The net external force must be zero, and the net torque must also be zero.

Take-Home Experiment

Stand straight with your heels, back, and head against a wall. Bend forward from your waist, keeping your heels and bottom against the wall, to touch your toes. Can you do this without toppling over? Explain why and what you need to do to be able to touch your toes without losing your balance. Is it easier for a woman to do this?

9.4 Applications of Statics, Including Problem-Solving Strategies

Learning Objectives

By the end of this section, you will be able to:

- Discuss the applications of statics in real life.
- State and discuss various problem-solving strategies in statics.

The information presented in this section supports the following AP® learning objectives and science practices:

- **3.F.1.1** The student is able to use representations of the relationship between force and torque. **(S.P. 1.4)**
- **3.F.1.2** The student is able to compare the torques on an object caused by various forces. **(S.P. 1.4)**
- **3.F.1.3** The student is able to estimate the torque on an object caused by various forces in comparison to other situations. **(S.P. 2.3)**
- **3.F.1.4** The student is able to design an experiment and analyze data testing a question about torques in a balanced rigid system. **(S.P. 4.1, 4.2, 5.1)**
- **3.F.1.5** The student is able to calculate torques on a two-dimensional system in static equilibrium, by examining a representation or model (such as a diagram or physical construction). **(S.P. 1.4, 2.2)**

Statics can be applied to a variety of situations, ranging from raising a drawbridge to bad posture and back strain. We begin with a discussion of problem-solving strategies specifically used for statics. Since statics is a special case of Newton's laws, both the general problem-solving strategies and the special strategies for Newton's laws, discussed in **Problem-Solving Strategies**, still apply.

Problem-Solving Strategy: Static Equilibrium Situations

1. The first step is to determine whether or not the system is in **static equilibrium**. This condition is always the case when the *acceleration of the system is zero and accelerated rotation does not occur*.

2. It is particularly important to *draw a free body diagram for the system of interest*. Carefully label all forces, and note their relative magnitudes, directions, and points of application whenever these are known.

3. Solve the problem by applying either or both of the conditions for equilibrium (represented by the equations

---

net $F = 0$ and net $\tau = 0$, depending on the list of known and unknown factors. If the second condition is involved, *choose the pivot point to simplify the solution.* Any pivot point can be chosen, but the most useful ones cause torques by unknown forces to be zero. (Torque is zero if the force is applied at the pivot (then $r = 0$), or along a line through the pivot point (then $\theta = 0$)). Always choose a convenient coordinate system for projecting forces.

4. *Check the solution to see if it is reasonable* by examining the magnitude, direction, and units of the answer. The importance of this last step never diminishes, although in unfamiliar applications, it is usually more difficult to judge reasonableness. These judgments become progressively easier with experience.

Now let us apply this problem-solving strategy for the pole vaulter shown in the three figures below. The pole is uniform and has a mass of 5.00 kg. In **Figure 9.20**, the pole's cg lies halfway between the vaulter's hands. It seems reasonable that the force exerted by each hand is equal to half the weight of the pole, or 24.5 N. This obviously satisfies the first condition for equilibrium (net $F = 0$). The second condition (net $\tau = 0$) is also satisfied, as we can see by choosing the cg to be the pivot point. The weight exerts no torque about a pivot point located at the cg, since it is applied at that point and its lever arm is zero. The equal forces exerted by the hands are equidistant from the chosen pivot, and so they exert equal and opposite torques. Similar arguments hold for other systems where supporting forces are exerted symmetrically about the cg. For example, the four legs of a uniform table each support one-fourth of its weight.

In **Figure 9.20**, a pole vaulter holding a pole with its cg halfway between his hands is shown. Each hand exerts a force equal to half the weight of the pole, $F_R = F_L = w/2$. (b) The pole vaulter moves the pole to his left, and the forces that the hands exert are no longer equal. See **Figure 9.20**. If the pole is held with its cg to the left of the person, then he must push down with his right hand and up with his left. The forces he exerts are larger here because they are in opposite directions and the cg is at a long distance from either hand.

Similar observations can be made using a meter stick held at different locations along its length.

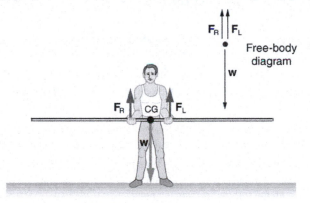

Figure 9.20 A pole vaulter holds a pole horizontally with both hands.

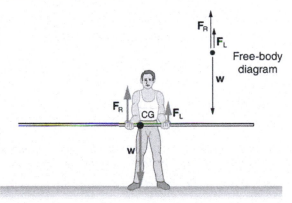

Figure 9.21 A pole vaulter is holding a pole horizontally with both hands. The center of gravity is near his right hand.

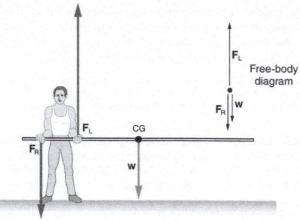

Figure 9.22 A pole vaulter is holding a pole horizontally with both hands. The center of gravity is to the left side of the vaulter.

If the pole vaulter holds the pole as shown in **Figure 9.210**, the situation is not as simple. The total force he exerts is still equal to the weight of the pole, but it is not evenly divided between his hands. (If $F_L = F_R$, then the torques about the cg would not be equal since the lever arms are different.) Logically, the right hand should support more weight, since it is closer to the cg. In fact, if the right hand is moved directly under the cg, it will support all the weight. This situation is exactly analogous to two people carrying a load; the one closer to the cg carries more of its weight. Finding the forces F_L and F_R is straightforward, as the next example shows.

If the pole vaulter holds the pole from near the end of the pole (**Figure 9.22**), the direction of the force applied by the right hand of the vaulter reverses its direction.

Example 9.2 What Force Is Needed to Support a Weight Held Near Its CG?

For the situation shown in **Figure 9.20**, calculate: (a) F_R , the force exerted by the right hand, and (b) F_L , the force exerted by the left hand. The hands are 0.900 m apart, and the cg of the pole is 0.600 m from the left hand.

Strategy

Figure 9.20 includes a free body diagram for the pole, the system of interest. There is not enough information to use the first condition for equilibrium (net $F = 0$), since two of the three forces are unknown and the hand forces cannot be assumed to be equal in this case. There is enough information to use the second condition for equilibrium (net $\tau = 0$) if the pivot point is chosen to be at either hand, thereby making the torque from that hand zero. We choose to locate the pivot at the left hand in this part of the problem, to eliminate the torque from the left hand.

Solution for (a)

There are now only two nonzero torques, those from the gravitational force (τ_W) and from the push or pull of the right hand (τ_R). Stating the second condition in terms of clockwise and counterclockwise torques,

$$\text{net } \tau_{cw} = -\text{net } \tau_{ccw}. \tag{9.22}$$

or the algebraic sum of the torques is zero.

Here this is

$$\tau_R = -\tau_w \tag{9.23}$$

since the weight of the pole creates a counterclockwise torque and the right hand counters with a clockwise toque. Using the definition of torque, $\tau = rF \sin \theta$, noting that $\theta = 90°$, and substituting known values, we obtain

$$(0.900 \text{ m})(F_R) = (0.600 \text{ m})(mg). \tag{9.24}$$

Thus,

$$F_R = (0.667)(5.00 \text{ kg})(9.80 \text{ m/s}^2) \tag{9.25}$$
$$= 32.7 \text{ N}.$$

Solution for (b)

The first condition for equilibrium is based on the free body diagram in the figure. This implies that by Newton's second law:

$$F_L + F_R - mg = 0 \tag{9.26}$$

From this we can conclude:

$$F_L + F_R = w = mg \tag{9.27}$$

Solving for F_L, we obtain

$$
\begin{aligned}
F_L &= mg - F_R \\
&= mg - 32.7 \text{ N} \\
&= (5.00 \text{ kg})(9.80 \text{ m/s}^2) - 32.7 \text{ N} \\
&= 16.3 \text{ N}
\end{aligned}
\tag{9.28}
$$

Discussion

F_L is seen to be exactly half of F_R, as we might have guessed, since F_L is applied twice as far from the cg as F_R.

If the pole vaulter holds the pole as he might at the start of a run, shown in **Figure 9.22**, the forces change again. Both are considerably greater, and one force reverses direction.

Take-Home Experiment

This is an experiment to perform while standing in a bus or a train. Stand facing sideways. How do you move your body to readjust the distribution of your mass as the bus accelerates and decelerates? Now stand facing forward. How do you move your body to readjust the distribution of your mass as the bus accelerates and decelerates? Why is it easier and safer to stand facing sideways rather than forward? Note: For your safety (and those around you), make sure you are holding onto something while you carry out this activity!

PhET Explorations: Balancing Act

Play with objects on a teeter totter to learn about balance. Test what you've learned by trying the Balance Challenge game.

Figure 9.23 Balancing Act (http://phet.colorado.edu/en/simulation/balancing-act)

9.5 Simple Machines

Learning Objectives
By the end of this section, you will be able to: • Describe different simple machines.   • Calculate the mechanical advantage.   The information presented in this section supports the following AP® learning objectives and science practices: • **3.F.1.1** The student is able to use representations of the relationship between force and torque. **(S.P. 1.4)**   • **3.F.1.2** The student is able to compare the torques on an object caused by various forces. **(S.P. 1.4)**   • **3.F.1.3** The student is able to estimate the torque on an object caused by various forces in comparison to other situations. **(S.P. 2.3)**   • **3.F.1.5** The student is able to calculate torques on a two-dimensional system in static equilibrium, by examining a representation or model (such as a diagram or physical construction). **(S.P. 1.4, 2.2)**

Simple machines are devices that can be used to multiply or augment a force that we apply – often at the expense of a distance through which we apply the force. The word for "machine" comes from the Greek word meaning "to help make things easier." Levers, gears, pulleys, wedges, and screws are some examples of machines. Energy is still conserved for these devices because a machine cannot do more work than the energy put into it. However, machines can reduce the input force that is needed to perform the job. The ratio of output to input force magnitudes for any simple machine is called its **mechanical advantage** (MA).

$$\mathrm{MA} = \frac{F_\mathrm{o}}{F_\mathrm{i}} \qquad\qquad (9.29)$$

One of the simplest machines is the lever, which is a rigid bar pivoted at a fixed place called the fulcrum. Torques are involved in levers, since there is rotation about a pivot point. Distances from the physical pivot of the lever are crucial, and we can obtain a useful expression for the MA in terms of these distances.

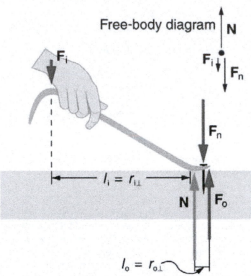

Figure 9.24 A nail puller is a lever with a large mechanical advantage. The external forces on the nail puller are represented by solid arrows. The force that the nail puller applies to the nail ($\mathbf{F}_\mathrm{o}$) is not a force on the nail puller. The reaction force the nail exerts back on the puller ($\mathbf{F}_\mathrm{n}$) is an external force and is equal and opposite to $\mathbf{F}_\mathrm{o}$. The perpendicular lever arms of the input and output forces are l_i and l_0 .

Figure 9.24 shows a lever type that is used as a nail puller. Crowbars, seesaws, and other such levers are all analogous to this one. $\mathbf{F}_\mathrm{i}$ is the input force and $\mathbf{F}_\mathrm{o}$ is the output force. There are three vertical forces acting on the nail puller (the system of interest) – these are $\mathbf{F}_\mathrm{i}$, $\mathbf{F}_\mathrm{o}$, and $\mathbf{N}$. $\mathbf{F}_\mathrm{n}$ is the reaction force back on the system, equal and opposite to $\mathbf{F}_\mathrm{o}$. (Note that $\mathbf{F}_\mathrm{o}$ is not a force on the system.) $\mathbf{N}$ is the normal force upon the lever, and its torque is zero since it is exerted at the pivot. The torques due to $\mathbf{F}_\mathrm{i}$ and $\mathbf{F}_\mathrm{n}$ must be equal to each other if the nail is not moving, to satisfy the second condition for equilibrium (net $\tau = 0$). (In order for the nail to actually move, the torque due to $\mathbf{F}_\mathrm{i}$ must be ever-so-slightly greater than torque due to $\mathbf{F}_\mathrm{n}$.) Hence,

$$l_\mathrm{i} F_\mathrm{i} = l_\mathrm{o} F_\mathrm{o} \qquad\qquad (9.30)$$

where l_i and l_o are the distances from where the input and output forces are applied to the pivot, as shown in the figure. Rearranging the last equation gives

$$\frac{F_\mathrm{o}}{F_\mathrm{i}} = \frac{l_\mathrm{i}}{l_\mathrm{o}}. \qquad\qquad (9.31)$$

What interests us most here is that the magnitude of the force exerted by the nail puller, F_o, is much greater than the magnitude of the input force applied to the puller at the other end, F_i. For the nail puller,

$$\mathrm{MA} = \frac{F_\mathrm{o}}{F_\mathrm{i}} = \frac{l_\mathrm{i}}{l_\mathrm{o}}. \qquad\qquad (9.32)$$

This equation is true for levers in general. For the nail puller, the MA is certainly greater than one. The longer the handle on the nail puller, the greater the force you can exert with it.

Two other types of levers that differ slightly from the nail puller are a wheelbarrow and a shovel, shown in **Figure 9.25**. All these lever types are similar in that only three forces are involved – the input force, the output force, and the force on the pivot – and thus their MAs are given by $\mathrm{MA} = \dfrac{F_\mathrm{o}}{F_\mathrm{i}}$ and $\mathrm{MA} = \dfrac{d_1}{d_2}$, with distances being measured relative to the physical pivot. The wheelbarrow and shovel differ from the nail puller because both the input and output forces are on the same side of the pivot.

In the case of the wheelbarrow, the output force or load is between the pivot (the wheel's axle) and the input or applied force. In the case of the shovel, the input force is between the pivot (at the end of the handle) and the load, but the input lever arm is shorter than the output lever arm. In this case, the MA is less than one.

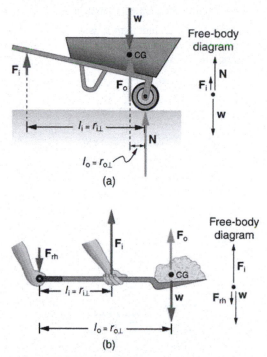

Figure 9.25 (a) In the case of the wheelbarrow, the output force or load is between the pivot and the input force. The pivot is the wheel's axle. Here, the output force is greater than the input force. Thus, a wheelbarrow enables you to lift much heavier loads than you could with your body alone. (b) In the case of the shovel, the input force is between the pivot and the load, but the input lever arm is shorter than the output lever arm. The pivot is at the handle held by the right hand. Here, the output force (supporting the shovel's load) is less than the input force (from the hand nearest the load), because the input is exerted closer to the pivot than is the output.

Example 9.3 What is the Advantage for the Wheelbarrow?

In the wheelbarrow of **Figure 9.25**, the load has a perpendicular lever arm of 7.50 cm, while the hands have a perpendicular lever arm of 1.02 m. (a) What upward force must you exert to support the wheelbarrow and its load if their combined mass is 45.0 kg? (b) What force does the wheelbarrow exert on the ground?

Strategy

Here, we use the concept of mechanical advantage.

Solution

(a) In this case, $\dfrac{F_o}{F_i} = \dfrac{l_i}{l_o}$ becomes

$$F_i = F_o \frac{l_o}{l_i}. \tag{9.33}$$

Adding values into this equation yields

$$F_i = (45.0 \text{ kg})(9.80 \text{ m/s}^2)\frac{0.075 \text{ m}}{1.02 \text{ m}} = 32.4 \text{ N.} \tag{9.34}$$

The free-body diagram (see **Figure 9.25**) gives the following normal force: $F_i + N = W$. Therefore,

$N = (45.0 \text{ kg})(9.80 \text{ m/s}^2) - 32.4 \text{ N} = 409 \text{ N}$. N is the normal force acting on the wheel; by Newton's third law, the

force the wheel exerts on the ground is 409 N.

Discussion

An even longer handle would reduce the force needed to lift the load. The MA here is $\text{MA} = 1.02/0.0750 = 13.6$.

Another very simple machine is the inclined plane. Pushing a cart up a plane is easier than lifting the same cart straight up to the top using a ladder, because the applied force is less. However, the work done in both cases (assuming the work done by friction is negligible) is the same. Inclined lanes or ramps were probably used during the construction of the Egyptian pyramids to move large blocks of stone to the top.

A crank is a lever that can be rotated $360°$ about its pivot, as shown in **Figure 9.26**. Such a machine may not look like a lever, but the physics of its actions remain the same. The MA for a crank is simply the ratio of the radii r_i / r_0. Wheels and gears have this simple expression for their MAs too. The MA can be greater than 1, as it is for the crank, or less than 1, as it is for the simplified car axle driving the wheels, as shown. If the axle's radius is 2.0 cm and the wheel's radius is 24.0 cm, then $MA = 2.0 / 24.0 = 0.083$ and the axle would have to exert a force of $12,000 \text{ N}$ on the wheel to enable it to exert a force of 1000 N on the ground.

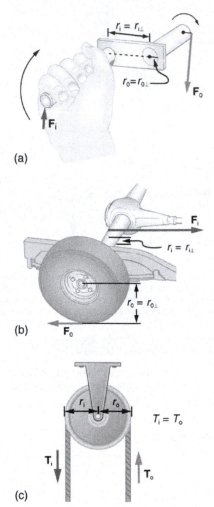

(a)

(b)

(c)

Figure 9.26 (a) A crank is a type of lever that can be rotated $360°$ about its pivot. Cranks are usually designed to have a large MA. (b) A simplified automobile axle drives a wheel, which has a much larger diameter than the axle. The MA is less than 1. (c) An ordinary pulley is used to lift a heavy load. The pulley changes the direction of the force T exerted by the cord without changing its magnitude. Hence, this machine has an MA of 1.

An ordinary pulley has an MA of 1; it only changes the direction of the force and not its magnitude. Combinations of pulleys, such as those illustrated in **Figure 9.27**, are used to multiply force. If the pulleys are friction-free, then the force output is approximately an integral multiple of the tension in the cable. The number of cables pulling directly upward on the system of interest, as illustrated in the figures given below, is approximately the MA of the pulley system. Since each attachment applies an external force in approximately the same direction as the others, they add, producing a total force that is nearly an integral multiple of the input force T.

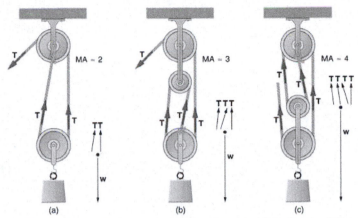

Figure 9.27 (a) The combination of pulleys is used to multiply force. The force is an integral multiple of tension if the pulleys are frictionless. This pulley system has two cables attached to its load, thus applying a force of approximately $2T$. This machine has $MA \approx 2$. (b) Three pulleys are used to lift a load in such a way that the mechanical advantage is about 3. Effectively, there are three cables attached to the load. (c) This pulley system applies a force of $4T$, so that it has $MA \approx 4$. Effectively, four cables are pulling on the system of interest.

9.6 Forces and Torques in Muscles and Joints

Learning Objectives

By the end of this section, you will be able to:

- Explain the forces exerted by muscles.
- State how a bad posture causes back strain.
- Discuss the benefits of skeletal muscles attached close to joints.
- Discuss various complexities in the real system of muscles, bones, and joints.

The information presented in this section supports the following AP® learning objectives and science practices:

- **3.F.1.1** The student is able to use representations of the relationship between force and torque. **(S.P. 1.4)**
- **3.F.1.2** The student is able to compare the torques on an object caused by various forces. **(S.P. 1.4)**
- **3.F.1.3** The student is able to estimate the torque on an object caused by various forces in comparison to other situations. **(S.P. 4.1, 4.2, 5.1)**
- **3.F.1.5** The student is able to calculate torques on a two-dimensional system in static equilibrium, by examining a representation or model (such as a diagram or physical construction). **(S.P. 1.4, 2.2)**

Muscles, bones, and joints are some of the most interesting applications of statics. There are some surprises. Muscles, for example, exert far greater forces than we might think. **Figure 9.28** shows a forearm holding a book and a schematic diagram of an analogous lever system. The schematic is a good approximation for the forearm, which looks more complicated than it is, and we can get some insight into the way typical muscle systems function by analyzing it.

Muscles can only contract, so they occur in pairs. In the arm, the biceps muscle is a flexor—that is, it closes the limb. The triceps muscle is an extensor that opens the limb. This configuration is typical of skeletal muscles, bones, and joints in humans and other vertebrates. Most skeletal muscles exert much larger forces within the body than the limbs apply to the outside world. The reason is clear once we realize that most muscles are attached to bones via tendons close to joints, causing these systems to have mechanical advantages much less than one. Viewing them as simple machines, the input force is much greater than the output force, as seen in **Figure 9.28**.

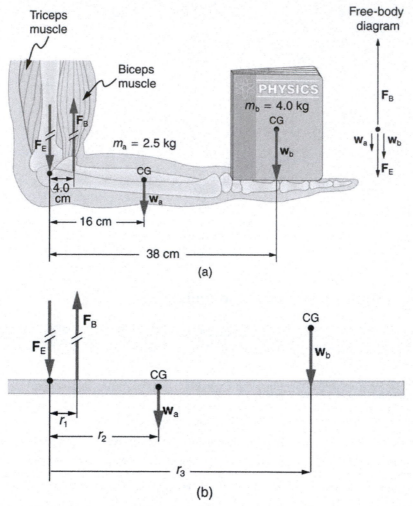

Figure 9.28 (a) The figure shows the forearm of a person holding a book. The biceps exert a force F_B to support the weight of the forearm and the book. The triceps are assumed to be relaxed. (b) Here, you can view an approximately equivalent mechanical system with the pivot at the elbow joint as seen in **Example 9.4.**

Example 9.4 Muscles Exert Bigger Forces Than You Might Think

Calculate the force the biceps muscle must exert to hold the forearm and its load as shown in **Figure 9.28**, and compare this force with the weight of the forearm plus its load. You may take the data in the figure to be accurate to three significant figures.

Strategy

There are four forces acting on the forearm and its load (the system of interest). The magnitude of the force of the biceps is F_B; that of the elbow joint is F_E; that of the weights of the forearm is w_a, and its load is w_b. Two of these are unknown (F_B and F_E), so that the first condition for equilibrium cannot by itself yield F_B. But if we use the second condition and choose the pivot to be at the elbow, then the torque due to F_E is zero, and the only unknown becomes F_B.

Solution

The torques created by the weights are clockwise relative to the pivot, while the torque created by the biceps is counterclockwise; thus, the second condition for equilibrium (net $\tau = 0$) becomes

$$r_2 w_a + r_3 w_b = r_1 F_B. \tag{9.35}$$

Note that $\sin \theta = 1$ for all forces, since $\theta = 90°$ for all forces. This equation can easily be solved for F_B in terms of known quantities, yielding

$$F_B = \frac{r_2 w_a + r_3 w_b}{r_1}. \tag{9.36}$$

Entering the known values gives

$$F_B = \frac{(0.160 \text{ m})(2.50 \text{ kg})(9.80 \text{ m/s}^2) + (0.380 \text{ m})(4.00 \text{ kg})(9.80 \text{ m/s}^2)}{0.0400 \text{ m}} \tag{9.37}$$

which yields

$$F_B = 470 \text{ N}. \tag{9.38}$$

Now, the combined weight of the arm and its load is $(6.50 \text{ kg})(9.80 \text{ m/s}^2) = 63.7 \text{ N}$, so that the ratio of the force exerted by the biceps to the total weight is

$$\frac{F_B}{w_a + w_b} = \frac{470}{63.7} = 7.38. \tag{9.39}$$

Discussion

This means that the biceps muscle is exerting a force 7.38 times the weight supported.

In the above example of the biceps muscle, the angle between the forearm and upper arm is 90°. If this angle changes, the force exerted by the biceps muscle also changes. In addition, the length of the biceps muscle changes. The force the biceps muscle can exert depends upon its length; it is smaller when it is shorter than when it is stretched.

Very large forces are also created in the joints. In the previous example, the downward force F_E exerted by the humerus at the elbow joint equals 407 N, or 6.38 times the total weight supported. (The calculation of F_E is straightforward and is left as an end-of-chapter problem.) Because muscles can contract, but not expand beyond their resting length, joints and muscles often exert forces that act in opposite directions and thus subtract. (In the above example, the upward force of the muscle minus the downward force of the joint equals the weight supported—that is, $470 \text{ N} - 407 \text{ N} = 63 \text{ N}$, approximately equal to the weight supported.) Forces in muscles and joints are largest when their load is a long distance from the joint, as the book is in the previous example.

In racquet sports such as tennis the constant extension of the arm during game play creates large forces in this way. The mass times the lever arm of a tennis racquet is an important factor, and many players use the heaviest racquet they can handle. It is no wonder that joint deterioration and damage to the tendons in the elbow, such as "tennis elbow," can result from repetitive motion, undue torques, and possibly poor racquet selection in such sports. Various tried techniques for holding and using a racquet or bat or stick not only increases sporting prowess but can minimize fatigue and long-term damage to the body. For example, tennis balls correctly hit at the "sweet spot" on the racquet will result in little vibration or impact force being felt in the racquet and the body—less torque as explained in **Collisions of Extended Bodies in Two Dimensions**. Twisting the hand to provide top spin on the ball or using an extended rigid elbow in a backhand stroke can also aggravate the tendons in the elbow.

Training coaches and physical therapists use the knowledge of relationships between forces and torques in the treatment of muscles and joints. In physical therapy, an exercise routine can apply a particular force and torque which can, over a period of time, revive muscles and joints. Some exercises are designed to be carried out under water, because this requires greater forces to be exerted, further strengthening muscles. However, connecting tissues in the limbs, such as tendons and cartilage as well as joints are sometimes damaged by the large forces they carry. Often, this is due to accidents, but heavily muscled athletes, such as weightlifters, can tear muscles and connecting tissue through effort alone.

The back is considerably more complicated than the arm or leg, with various muscles and joints between vertebrae, all having mechanical advantages less than 1. Back muscles must, therefore, exert very large forces, which are borne by the spinal column. Discs crushed by mere exertion are very common. The jaw is somewhat exceptional—the masseter muscles that close the jaw have a mechanical advantage greater than 1 for the back teeth, allowing us to exert very large forces with them. A cause of stress headaches is persistent clenching of teeth where the sustained large force translates into fatigue in muscles around the skull.

Figure 9.29 shows how bad posture causes back strain. In part (a), we see a person with good posture. Note that her upper body's cg is directly above the pivot point in the hips, which in turn is directly above the base of support at her feet. Because of this, her upper body's weight exerts no torque about the hips. The only force needed is a vertical force at the hips equal to the weight supported. No muscle action is required, since the bones are rigid and transmit this force from the floor. This is a position of unstable equilibrium, but only small forces are needed to bring the upper body back to vertical if it is slightly displaced. Bad posture is shown in part (b); we see that the upper body's cg is in front of the pivot in the hips. This creates a clockwise torque around the hips that is counteracted by muscles in the lower back. These muscles must exert large forces, since they have typically small mechanical advantages. (In other words, the perpendicular lever arm for the muscles is much smaller than for the cg.) Poor posture can also cause muscle strain for people sitting at their desks using computers. Special chairs are available that allow the body's CG to be more easily situated above the seat, to reduce back pain. Prolonged muscle action produces muscle strain. Note that the cg of the entire body is still directly above the base of support in part (b) of **Figure 9.29**. This is compulsory; otherwise the person would not be in equilibrium. We lean forward for the same reason when carrying a load on our backs, to the side when carrying a load in one arm, and backward when carrying a load in front of us, as seen in **Figure 9.30**.

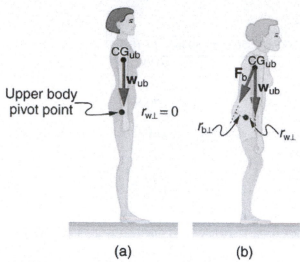

Figure 9.29 (a) Good posture places the upper body's cg over the pivots in the hips, eliminating the need for muscle action to balance the body. (b) Poor posture requires exertion by the back muscles to counteract the clockwise torque produced around the pivot by the upper body's weight. The back muscles have a small effective perpendicular lever arm, $r_{b\perp}$, and must therefore exert a large force $\mathbf{F}_b$. Note that the legs lean backward to keep the cg of the entire body above the base of support in the feet.

You have probably been warned against lifting objects with your back. This action, even more than bad posture, can cause muscle strain and damage discs and vertebrae, since abnormally large forces are created in the back muscles and spine.

Figure 9.30 People adjust their stance to maintain balance. (a) A father carrying his son piggyback leans forward to position their overall cg above the base of support at his feet. (b) A student carrying a shoulder bag leans to the side to keep the overall cg over his feet. (c) Another student carrying a load of books in her arms leans backward for the same reason.

Example 9.5 Do Not Lift with Your Back

Consider the person lifting a heavy box with his back, shown in **Figure 9.31**. (a) Calculate the magnitude of the force F_B — in the back muscles that is needed to support the upper body plus the box and compare this with his weight. The mass of the upper body is 55.0 kg and the mass of the box is 30.0 kg. (b) Calculate the magnitude and direction of the force $\mathbf{F}_V$ — exerted by the vertebrae on the spine at the indicated pivot point. Again, data in the figure may be taken to be accurate to three significant figures.

Strategy

By now, we sense that the second condition for equilibrium is a good place to start, and inspection of the known values confirms that it can be used to solve for F_B — if the pivot is chosen to be at the hips. The torques created by $\mathbf{w}_{ub}$ and $\mathbf{w}_{box}$ — are clockwise, while that created by $\mathbf{F}_B$ — is counterclockwise.

Solution for (a)

Using the perpendicular lever arms given in the figure, the second condition for equilibrium (net $\tau = 0$) becomes

$$(0.350 \text{ m})(55.0 \text{ kg})(9.80 \text{ m/s}^2) + (0.500 \text{ m})(30.0 \text{ kg})(9.80 \text{ m/s}^2) = (0.0800 \text{ m})F_B. \qquad (9.40)$$

Solving for F_B yields

$$F_B = 4.20 \times 10^3 \text{ N}. \tag{9.41}$$

The ratio of the force the back muscles exert to the weight of the upper body plus its load is

$$\frac{F_B}{w_{ub} + w_{box}} = \frac{4200 \text{ N}}{833 \text{ N}} = 5.04. \tag{9.42}$$

This force is considerably larger than it would be if the load were not present.

Solution for (b)

More important in terms of its damage potential is the force on the vertebrae $\mathbf{F_V}$. The first condition for equilibrium (net $\mathbf{F} = 0$) can be used to find its magnitude and direction. Using y for vertical and x for horizontal, the condition for the net external forces along those axes to be zero

$$\text{net } F_y = 0 \text{ and net } F_x = 0. \tag{9.43}$$

Starting with the vertical (y) components, this yields

$$F_{Vy} - w_{ub} - w_{box} - F_B \sin 29.0° = 0. \tag{9.44}$$

Thus,

$$\begin{aligned} F_{Vy} &= w_{ub} + w_{box} + F_B \sin 29.0° \\ &= 833 \text{ N} + (4200 \text{ N}) \sin 29.0° \end{aligned} \tag{9.45}$$

yielding

$$F_{Vy} = 2.87 \times 10^3 \text{ N}. \tag{9.46}$$

Similarly, for the horizontal (x) components,

$$F_{Vx} - F_B \cos 29.0° = 0 \tag{9.47}$$

yielding

$$F_{Vx} = 3.67 \times 10^3 \text{ N}. \tag{9.48}$$

The magnitude of $\mathbf{F_V}$ is given by the Pythagorean theorem:

$$F_V = \sqrt{F_{Vx}^2 + F_{Vy}^2} = 4.66 \times 10^3 \text{ N}. \tag{9.49}$$

The direction of $\mathbf{F_V}$ is

$$\theta = \tan^{-1}\left(\frac{F_{Vy}}{F_{Vx}}\right) = 38.0°. \tag{9.50}$$

Note that the ratio of F_V to the weight supported is

$$\frac{F_V}{w_{ub} + w_{box}} = \frac{4660 \text{ N}}{833 \text{ N}} = 5.59. \tag{9.51}$$

Discussion

This force is about 5.6 times greater than it would be if the person were standing erect. The trouble with the back is not so much that the forces are large—because similar forces are created in our hips, knees, and ankles—but that our spines are relatively weak. Proper lifting, performed with the back erect and using the legs to raise the body and load, creates much smaller forces in the back—in this case, about 5.6 times smaller.

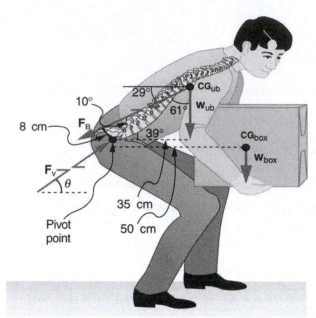

Figure 9.31 This figure shows that large forces are exerted by the back muscles and experienced in the vertebrae when a person lifts with their back, since these muscles have small effective perpendicular lever arms. The data shown here are analyzed in the preceding example, **Example 9.5.**

What are the benefits of having most skeletal muscles attached so close to joints? One advantage is speed because small muscle contractions can produce large movements of limbs in a short period of time. Other advantages are flexibility and agility, made possible by the large numbers of joints and the ranges over which they function. For example, it is difficult to imagine a system with biceps muscles attached at the wrist that would be capable of the broad range of movement we vertebrates possess.

There are some interesting complexities in real systems of muscles, bones, and joints. For instance, the pivot point in many joints changes location as the joint is flexed, so that the perpendicular lever arms and the mechanical advantage of the system change, too. Thus the force the biceps muscle must exert to hold up a book varies as the forearm is flexed. Similar mechanisms operate in the legs, which explain, for example, why there is less leg strain when a bicycle seat is set at the proper height. The methods employed in this section give a reasonable description of real systems provided enough is known about the dimensions of the system. There are many other interesting examples of force and torque in the body—a few of these are the subject of end-of-chapter problems.

Glossary

center of gravity: the point where the total weight of the body is assumed to be concentrated

dynamic equilibrium: a state of equilibrium in which the net external force and torque on a system moving with constant velocity are zero

mechanical advantage: the ratio of output to input forces for any simple machine

neutral equilibrium: a state of equilibrium that is independent of a system's displacements from its original position

perpendicular lever arm: the shortest distance from the pivot point to the line along which $\mathbf{F}$ lies

SI units of torque: newton times meters, usually written as N·m

stable equilibrium: a system, when displaced, experiences a net force or torque in a direction opposite to the direction of the displacement

static equilibrium: a state of equilibrium in which the net external force and torque acting on a system is zero

static equilibrium: equilibrium in which the acceleration of the system is zero and accelerated rotation does not occur

torque: turning or twisting effectiveness of a force

unstable equilibrium: a system, when displaced, experiences a net force or torque in the same direction as the displacement from equilibrium

Section Summary

9.1 The First Condition for Equilibrium

- Statics is the study of forces in equilibrium.
- Two conditions must be met to achieve equilibrium, which is defined to be motion without linear or rotational acceleration.
- The first condition necessary to achieve equilibrium is that the net external force on the system must be zero, so that net $\mathbf{F} = 0$.

9.2 The Second Condition for Equilibrium

- The second condition assures those torques are also balanced. Torque is the rotational equivalent of a force in producing a rotation and is defined to be

$$\tau = rF \sin \theta$$

where τ is torque, r is the distance from the pivot point to the point where the force is applied, F is the magnitude of the force, and θ is the angle between $\mathbf{F}$ and the vector directed from the point where the force acts to the pivot point. The perpendicular lever arm $r_\perp$ is defined to be

$$r_\perp = r \sin \theta$$

so that

$$\tau = r_\perp F.$$

- The perpendicular lever arm $r_\perp$ is the shortest distance from the pivot point to the line along which F acts. The SI unit for torque is newton-meter (N·m). The second condition necessary to achieve equilibrium is that the net external torque on a system must be zero:

$$\text{net } \tau = 0$$

By convention, counterclockwise torques are positive, and clockwise torques are negative.

9.3 Stability

- A system is said to be in stable equilibrium if, when displaced from equilibrium, it experiences a net force or torque in a direction opposite the direction of the displacement.
- A system is in unstable equilibrium if, when displaced from equilibrium, it experiences a net force or torque in the same direction as the displacement from equilibrium.
- A system is in neutral equilibrium if its equilibrium is independent of displacements from its original position.

9.4 Applications of Statics, Including Problem-Solving Strategies

- Statics can be applied to a variety of situations, ranging from raising a drawbridge to bad posture and back strain. We have discussed the problem-solving strategies specifically useful for statics. Statics is a special case of Newton's laws, both the general problem-solving strategies and the special strategies for Newton's laws, discussed in **Problem-Solving Strategies**, still apply.

9.5 Simple Machines

- Simple machines are devices that can be used to multiply or augment a force that we apply – often at the expense of a distance through which we have to apply the force.
- The ratio of output to input forces for any simple machine is called its mechanical advantage
- A few simple machines are the lever, nail puller, wheelbarrow, crank, etc.

9.6 Forces and Torques in Muscles and Joints

- Statics plays an important part in understanding everyday strains in our muscles and bones.
- Many lever systems in the body have a mechanical advantage of significantly less than one, as many of our muscles are attached close to joints.
- Someone with good posture stands or sits in such a way that the person's center of gravity lies directly above the pivot point in the hips, thereby avoiding back strain and damage to disks.

Conceptual Questions

9.1 The First Condition for Equilibrium

1. What can you say about the velocity of a moving body that is in dynamic equilibrium? Draw a sketch of such a body using clearly labeled arrows to represent all external forces on the body.

2. Under what conditions can a rotating body be in equilibrium? Give an example.

9.2 The Second Condition for Equilibrium

3. What three factors affect the torque created by a force relative to a specific pivot point?

4. A wrecking ball is being used to knock down a building. One tall unsupported concrete wall remains standing. If the wrecking ball hits the wall near the top, is the wall more likely to fall over by rotating at its base or by falling straight down? Explain your answer. How is it most likely to fall if it is struck with the same force at its base? Note that this depends on how firmly the wall is attached at its base.

5. Mechanics sometimes put a length of pipe over the handle of a wrench when trying to remove a very tight bolt. How does this help? (It is also hazardous since it can break the bolt.)

9.3 Stability

6. A round pencil lying on its side as in **Figure 9.14** is in neutral equilibrium relative to displacements perpendicular to its length. What is its stability relative to displacements parallel to its length?

7. Explain the need for tall towers on a suspension bridge to ensure stable equilibrium.

9.4 Applications of Statics, Including Problem-Solving Strategies

8. When visiting some countries, you may see a person balancing a load on the head. Explain why the center of mass of the load needs to be directly above the person's neck vertebrae.

9.5 Simple Machines

9. Scissors are like a double-lever system. Which of the simple machines in **Figure 9.24** and **Figure 9.25** is most analogous to scissors?

10. Suppose you pull a nail at a constant rate using a nail puller as shown in **Figure 9.24**. Is the nail puller in equilibrium? What if you pull the nail with some acceleration – is the nail puller in equilibrium then? In which case is the force applied to the nail puller larger and why?

11. Why are the forces exerted on the outside world by the limbs of our bodies usually much smaller than the forces exerted by muscles inside the body?

12. Explain why the forces in our joints are several times larger than the forces we exert on the outside world with our limbs. Can these forces be even greater than muscle forces (see previous Question)?

9.6 Forces and Torques in Muscles and Joints

13. Why are the forces exerted on the outside world by the limbs of our bodies usually much smaller than the forces exerted by muscles inside the body?

14. Explain why the forces in our joints are several times larger than the forces we exert on the outside world with our limbs. Can these forces be even greater than muscle forces?

15. Certain types of dinosaurs were bipedal (walked on two legs). What is a good reason that these creatures invariably had long tails if they had long necks?

16. Swimmers and athletes during competition need to go through certain postures at the beginning of the race. Consider the balance of the person and why start-offs are so important for races.

17. If the maximum force the biceps muscle can exert is 1000 N, can we pick up an object that weighs 1000 N? Explain your answer.

18. Suppose the biceps muscle was attached through tendons to the upper arm close to the elbow and the forearm near the wrist. What would be the advantages and disadvantages of this type of construction for the motion of the arm?

19. Explain one of the reasons why pregnant women often suffer from back strain late in their pregnancy.

Problems & Exercises

9.2 The Second Condition for Equilibrium

1. (a) When opening a door, you push on it perpendicularly with a force of 55.0 N at a distance of 0.850m from the hinges. What torque are you exerting relative to the hinges? (b) Does it matter if you push at the same height as the hinges?

2. When tightening a bolt, you push perpendicularly on a wrench with a force of 165 N at a distance of 0.140 m from the center of the bolt. (a) How much torque are you exerting in newton × meters (relative to the center of the bolt)? (b) Convert this torque to footpounds.

3. Two children push on opposite sides of a door during play. Both push horizontally and perpendicular to the door. One child pushes with a force of 17.5 N at a distance of 0.600 m from the hinges, and the second child pushes at a distance of 0.450 m. What force must the second child exert to keep the door from moving? Assume friction is negligible.

4. Use the second condition for equilibrium $(\text{net } \tau = 0)$ to calculate F_{p} in **Example 9.1**, employing any data given or solved for in part (a) of the example.

5. Repeat the seesaw problem in **Example 9.1** with the center of mass of the seesaw 0.160 m to the left of the pivot (on the side of the lighter child) and assuming a mass of 12.0 kg for the seesaw. The other data given in the example remain unchanged. Explicitly show how you follow the steps in the Problem-Solving Strategy for static equilibrium.

9.3 Stability

6. Suppose a horse leans against a wall as in **Figure 9.32**. Calculate the force exerted on the wall assuming that force is horizontal while using the data in the schematic representation of the situation. Note that the force exerted on the wall is equal in magnitude and opposite in direction to the force exerted on the horse, keeping it in equilibrium. The total mass of the horse and rider is 500 kg. Take the data to be accurate to three digits.

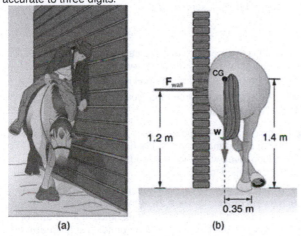

(a)   (b)

Figure 9.32

7. Two children of mass 20 kg and 30 kg sit balanced on a seesaw with the pivot point located at the center of the seesaw. If the children are separated by a distance of 3 m, at what distance from the pivot point is the small child sitting in order to maintain the balance?

8. (a) Calculate the magnitude and direction of the force on each foot of the horse in **Figure 9.32** (two are on the ground), assuming the center of mass of the horse is midway between the feet. The total mass of the horse and rider is 500kg. (b) What is the minimum coefficient of friction between the hooves and ground? Note that the force exerted by the wall is horizontal.

9. A person carries a plank of wood 2 m long with one hand pushing down on it at one end with a force F_1 and the other hand holding it up at 50 cm from the end of the plank with force F_2. If the plank has a mass of 20 kg and its center of gravity is at the middle of the plank, what are the magnitudes of the forces F_1 and F_2?

10. A 17.0-m-high and 11.0-m-long wall under construction and its bracing are shown in **Figure 9.33**. The wall is in stable equilibrium without the bracing but can pivot at its base. Calculate the force exerted by each of the 10 braces if a strong wind exerts a horizontal force of 650 N on each square meter of the wall. Assume that the net force from the wind acts at a height halfway up the wall and that all braces exert equal forces parallel to their lengths. Neglect the thickness of the wall.

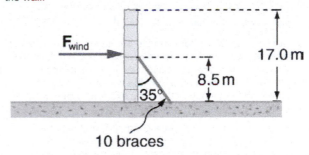

Figure 9.33

11. (a) What force must be exerted by the wind to support a 2.50-kg chicken in the position shown in **Figure 9.34**? (b) What is the ratio of this force to the chicken's weight? (c) Does this support the contention that the chicken has a relatively stable construction?

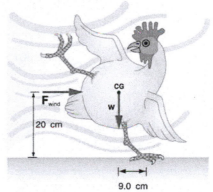

Figure 9.34

12. Suppose the weight of the drawbridge in **Figure 9.35** is supported entirely by its hinges and the opposite shore, so that its cables are slack. (a) What fraction of the weight is supported by the opposite shore if the point of support is directly beneath the cable attachments? (b) What is the direction and magnitude of the force the hinges exert on the bridge under these circumstances? The mass of the bridge is 2500 kg.

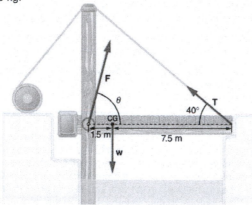

Figure 9.35 A small drawbridge, showing the forces on the hinges (F), its weight (W), and the tension in its wires (T).

13. Suppose a 900-kg car is on the bridge in **Figure 9.35** with its center of mass halfway between the hinges and the cable attachments. (The bridge is supported by the cables and hinges only.) (a) Find the force in the cables. (b) Find the direction and magnitude of the force exerted by the hinges on the bridge.

14. A sandwich board advertising sign is constructed as shown in **Figure 9.36**. The sign's mass is 8.00 kg. (a) Calculate the tension in the chain assuming no friction between the legs and the sidewalk. (b) What force is exerted by each side on the hinge?

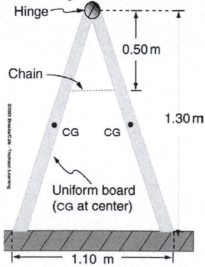

Figure 9.36 A sandwich board advertising sign demonstrates tension.

15. (a) What minimum coefficient of friction is needed between the legs and the ground to keep the sign in **Figure 9.36** in the position shown if the chain breaks? (b) What force is exerted by each side on the hinge?

16. A gymnast is attempting to perform splits. From the information given in **Figure 9.37**, calculate the magnitude and direction of the force exerted on each foot by the floor.

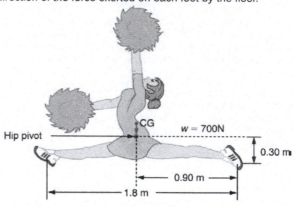

Figure 9.37 A gymnast performs full split. The center of gravity and the various distances from it are shown.

9.4 Applications of Statics, Including Problem-Solving Strategies

17. To get up on the roof, a person (mass 70.0 kg) places a 6.00-m aluminum ladder (mass 10.0 kg) against the house on a concrete pad with the base of the ladder 2.00 m from the house. The ladder rests against a plastic rain gutter, which we can assume to be frictionless. The center of mass of the ladder is 2 m from the bottom. The person is standing 3 m from the bottom. What are the magnitudes of the forces on the ladder at the top and bottom?

18. In **Figure 9.22**, the cg of the pole held by the pole vaulter is 2.00 m from the left hand, and the hands are 0.700 m apart. Calculate the force exerted by (a) his right hand and (b) his left hand. (c) If each hand supports half the weight of the pole in **Figure 9.20**, show that the second condition for equilibrium (net $\tau = 0$) is satisfied for a pivot other than the one located at the center of gravity of the pole. Explicitly show how you follow the steps in the Problem-Solving Strategy for static equilibrium described above.

9.5 Simple Machines

19. What is the mechanical advantage of a nail puller—similar to the one shown in **Figure 9.24** —where you exert a force 45 cm from the pivot and the nail is 1.8 cm on the other side? What minimum force must you exert to apply a force of 1250 N to the nail?

20. Suppose you needed to raise a 250-kg mower a distance of 6.0 cm above the ground to change a tire. If you had a 2.0-m long lever, where would you place the fulcrum if your force was limited to 300 N?

21. a) What is the mechanical advantage of a wheelbarrow, such as the one in **Figure 9.25**, if the center of gravity of the wheelbarrow and its load has a perpendicular lever arm of 5.50 cm, while the hands have a perpendicular lever arm of 1.02 m? (b) What upward force should you exert to support the wheelbarrow and its load if their combined mass is 55.0 kg? (c) What force does the wheel exert on the ground?

22. A typical car has an axle with 1.10 cm radius driving a tire with a radius of 27.5 cm. What is its mechanical advantage assuming the very simplified model in **Figure 9.26(b)**?

23. What force does the nail puller in **Exercise 9.19** exert on the supporting surface? The nail puller has a mass of 2.10 kg.

24. If you used an ideal pulley of the type shown in **Figure 9.27**(a) to support a car engine of mass 115 kg, (a) What would be the tension in the rope? (b) What force must the ceiling supply, assuming you pull straight down on the rope? Neglect the pulley system's mass.

25. Repeat **Exercise 9.24** for the pulley shown in **Figure 9.27**(c), assuming you pull straight up on the rope. The pulley system's mass is 7.00 kg.

9.6 Forces and Torques in Muscles and Joints

26. Verify that the force in the elbow joint in **Example 9.4** is 407 N, as stated in the text.

27. Two muscles in the back of the leg pull on the Achilles tendon as shown in **Figure 9.38**. What total force do they exert?

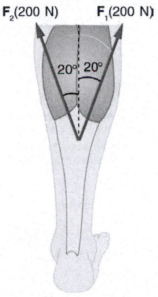

F₂(200 N)   **F₁(200 N)**

Figure 9.38 The Achilles tendon of the posterior leg serves to attach plantaris, gastrocnemius, and soleus muscles to calcaneus bone.

28. The upper leg muscle (quadriceps) exerts a force of 1250 N, which is carried by a tendon over the kneecap (the patella) at the angles shown in **Figure 9.39**. Find the direction and magnitude of the force exerted by the kneecap on the upper leg bone (the femur).

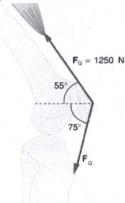

F_Q = 1250 N

Figure 9.39 The knee joint works like a hinge to bend and straighten the lower leg. It permits a person to sit, stand, and pivot.

29. A device for exercising the upper leg muscle is shown in **Figure 9.40**, together with a schematic representation of an equivalent lever system. Calculate the force exerted by the upper leg muscle to lift the mass at a constant speed. Explicitly show how you follow the steps in the Problem-Solving Strategy for static equilibrium in **Applications of Statistics, Including Problem-Solving Strategies**.

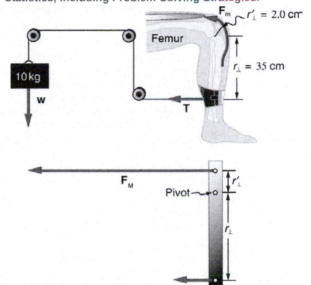

Figure 9.40 A mass is connected by pulleys and wires to the ankle in this exercise device.

30. A person working at a drafting board may hold her head as shown in **Figure 9.41**, requiring muscle action to support the head. The three major acting forces are shown. Calculate the direction and magnitude of the force supplied by the upper vertebrae $\mathbf{F_V}$ to hold the head stationary, assuming that this force acts along a line through the center of mass as do the weight and muscle force.

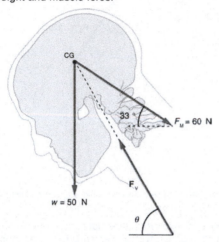

Figure 9.41

31. We analyzed the biceps muscle example with the angle between forearm and upper arm set at $90°$. Using the same numbers as in **Example 9.4**, find the force exerted by the biceps muscle when the angle is $120°$ and the forearm is in a downward position.

32. Even when the head is held erect, as in **Figure 9.42**, its center of mass is not directly over the principal point of support (the atlanto-occipital joint). The muscles at the back of the neck should therefore exert a force to keep the head erect. That is why your head falls forward when you fall asleep in the class. (a) Calculate the force exerted by these muscles using the information in the figure. (b) What is the force exerted by the pivot on the head?

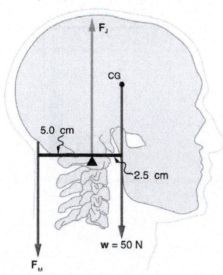

Figure 9.42 The center of mass of the head lies in front of its major point of support, requiring muscle action to hold the head erect. A simplified lever system is shown.

33. A 75-kg man stands on his toes by exerting an upward force through the Achilles tendon, as in **Figure 9.43**. (a) What is the force in the Achilles tendon if he stands on one foot? (b) Calculate the force at the pivot of the simplified lever system shown—that force is representative of forces in the ankle joint.

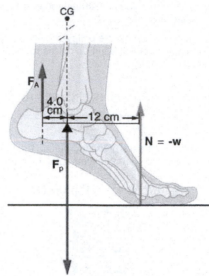

Figure 9.43 The muscles in the back of the leg pull the Achilles tendon when one stands on one's toes. A simplified lever system is shown.

34. A father lifts his child as shown in **Figure 9.44**. What force should the upper leg muscle exert to lift the child at a constant speed?

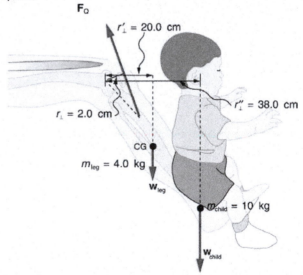

Figure 9.44 A child being lifted by a father's lower leg.

35. Unlike most of the other muscles in our bodies, the masseter muscle in the jaw, as illustrated in **Figure 9.45**, is attached relatively far from the joint, enabling large forces to be exerted by the back teeth. (a) Using the information in the figure, calculate the force exerted by the lower teeth on the bullet. (b) Calculate the force on the joint.

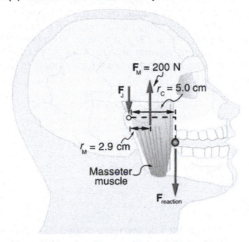

Figure 9.45 A person clenching a bullet between his teeth.

36. Integrated Concepts

Suppose we replace the 4.0-kg book in **Exercise 9.31** of the biceps muscle with an elastic exercise rope that obeys Hooke's Law. Assume its force constant $k = 600$ N/m . (a) How much is the rope stretched (past equilibrium) to provide the same force F_B as in this example? Assume the rope is held in the hand at the same location as the book. (b) What force is on the biceps muscle if the exercise rope is pulled straight up so that the forearm makes an angle of $25°$ with the horizontal? Assume the biceps muscle is still perpendicular to the forearm.

37. (a) What force should the woman in **Figure 9.46** exert on the floor with each hand to do a push-up? Assume that she moves up at a constant speed. (b) The triceps muscle at the back of her upper arm has an effective lever arm of 1.75 cm, and she exerts force on the floor at a horizontal distance of 20.0 cm from the elbow joint. Calculate the magnitude of the force in each triceps muscle, and compare it to her weight. (c) How much work does she do if her center of mass rises 0.240 m? (d) What is her useful power output if she does 25 pushups in one minute?

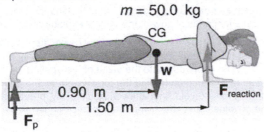

Figure 9.46 A woman doing pushups.

Test Prep for AP® Courses

9.2 The Second Condition for Equilibrium

1. Which of the following is not an example of an object

38. You have just planted a sturdy 2-m-tall palm tree in your front lawn for your mother's birthday. Your brother kicks a 500 g ball, which hits the top of the tree at a speed of 5 m/s and stays in contact with it for 10 ms. The ball falls to the ground near the base of the tree and the recoil of the tree is minimal. (a) What is the force on the tree? (b) The length of the sturdy section of the root is only 20 cm. Furthermore, the soil around the roots is loose and we can assume that an effective force is applied at the tip of the 20 cm length. What is the effective force exerted by the end of the tip of the root to keep the tree from toppling? Assume the tree will be uprooted rather than bend. (c) What could you have done to ensure that the tree does not uproot easily?

39. Unreasonable Results

Suppose two children are using a uniform seesaw that is 3.00 m long and has its center of mass over the pivot. The first child has a mass of 30.0 kg and sits 1.40 m from the pivot. (a) Calculate where the second 18.0 kg child must sit to balance the seesaw. (b) What is unreasonable about the result? (c) Which premise is unreasonable, or which premises are inconsistent?

40. Construct Your Own Problem

Consider a method for measuring the mass of a person's arm in anatomical studies. The subject lies on her back, extends her relaxed arm to the side and two scales are placed below the arm. One is placed under the elbow and the other under the back of her hand. Construct a problem in which you calculate the mass of the arm and find its center of mass based on the scale readings and the distances of the scales from the shoulder joint. You must include a free body diagram of the arm to direct the analysis. Consider changing the position of the scale under the hand to provide more information, if needed. You may wish to consult references to obtain reasonable mass values.

undergoing a torque?

a.   A car is rounding a bend at a constant speed.
b.   A merry-go-round increases from rest to a constant
     rotational speed.
c.   A pendulum swings back and forth.
d.   A bowling ball rolls down a bowling alley.

2. Five forces of equal magnitude, labeled **A–E**, are applied to
the object shown below. If the object is anchored at point **P**,
which force will provide the greatest torque?

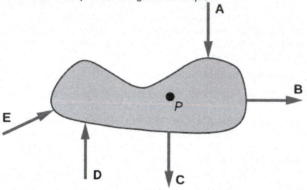

Figure 9.47 Five forces acting on an object.
a.   Force **A**
b.   Force **B**
c.   Force **C**
d.   Force **D**
e.   Force **E**

9.3 Stability

3. Using the concept of torque, explain why a traffic cone
placed on its base is in stable equilibrium, while a traffic cone
placed on its tip is in unstable equilibrium.

9.4 Applications of Statics, Including Problem-Solving Strategies

4. A child sits on the end of a playground see-saw. Which of
the following values is the most appropriate estimate of the
torque created by the child?
a.   6 N•m
b.   60 N•m
c.   600 N•m
d.   6000 N•m

5. A group of students is stacking a set of identical books,
each one overhanging the one below it by 1 inch. They would
like to estimate how many books they could place on top of
each other before the stack tipped. What information below
would they need to know to make this calculation?

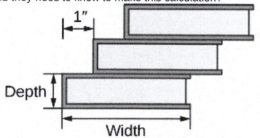

Figure 9.48 3 overlapping stacked books.
I.   The mass of each book
II.   The width of each book
III. The depth of each book

a.   I only
b.   I and II only
c.   I and III only
d.   II only
e.   I, II, and III

6. A 10 N board of uniform density is 5 meters long. It is
supported on the left by a string bearing a 3 N upward force.
In order to prevent the string from breaking, a person must
place an upward force of 7 N at a position along the bottom
surface of the board. At what distance from its left edge would
they need to place this force in order for the board to be in
static equilibrium?

a. $\frac{3}{7}$ m

b. $\frac{5}{2}$ m

c. $\frac{25}{7}$ m

d. $\frac{30}{7}$ m

e.   5 m

7. A bridge is supported by two piers located 20 meters apart.
Both the left and right piers provide an upward force on the
bridge, labeled F_L and F_R respectively.
a.   If a 1000 kg car comes to rest at a point 5 meters from
     the left pier, how much force will the bridge provide to
     the left and right piers?
b.   How will F_L and F_R change as the car drives to the right
     side of the bridge?

8. An object of unknown mass is provided to a student.
Without using a scale, design an experimental procedure
detailing how the magnitude of this mass could be
experimentally found. Your explanation must include the
concept of torque and all steps should be provided in an
orderly sequence. You may include a labeled diagram of your
setup to help in your description. Include enough detail so
that another student could carry out your procedure.

9.5 Simple Machines

9. As a young student, you likely learned that simple
machines are capable of increasing the ability to lift and move
objects. Now, as an educated AP Physics student, you are
aware that this capability is governed by the relationship
between force and torque.

In the space below, explain why torque is integral to the
increase in force created by a simple machine. You may use
an example or diagram to assist in your explanation. Be sure
to cite the mechanical advantage in your explanation as well.

10. Figure 9.24(a) shows a wheelbarrow being lifted by an
applied force F_i. If the wheelbarrow is filled with twenty bricks
massing 3 kg each, estimate the value of the applied force F_i.
Provide an explanation behind the total weight **w** and any
reasoning toward your final answer. Additionally, provide a
range of values over which you feel the force could exist.

9.6 Forces and Torques in Muscles and Joints

11. When you use your hand to raise a 20 lb dumbbell in a
curling motion, the force on your bicep muscle is not equal to
20 lb.

a. Compare the size of the force placed on your bicep muscle to the force of the 20 lb dumbbell lifted by your hand. Using the concept of torque, which force is greater and explain why the two forces are not identical.

b. Does the force placed on your bicep muscle change as you curl the weight closer toward your body? (In other words, is the force on your muscle different when your forearm is 90° to your upper arm than when it is 45° to your upper arm?) Explain your answer using torque.

10 ROTATIONAL MOTION AND ANGULAR MOMENTUM

Figure 10.1 The mention of a tornado conjures up images of raw destructive power. Tornadoes blow houses away as if they were made of paper and have been known to pierce tree trunks with pieces of straw. They descend from clouds in funnel-like shapes that spin violently, particularly at the bottom where they are most narrow, producing winds as high as 500 km/h. (credit: Daphne Zaras, U.S. National Oceanic and Atmospheric Administration)

Chapter Outline

10.1. Angular Acceleration

10.2. Kinematics of Rotational Motion

10.3. Dynamics of Rotational Motion: Rotational Inertia

10.4. Rotational Kinetic Energy: Work and Energy Revisited

10.5. Angular Momentum and Its Conservation

10.6. Collisions of Extended Bodies in Two Dimensions

10.7. Gyroscopic Effects: Vector Aspects of Angular Momentum

Connection for AP® Courses

Why do tornados spin? And why do tornados spin so rapidly? The answer is that the air masses that produce tornados are themselves rotating, and when the radii of the air masses decrease, their rate of rotation increases. An ice skater increases her spin in an exactly analogous manner, as seen in **Figure 10.2.** The skater starts her rotation with outstretched limbs and increases her rate of spin by pulling them in toward her body. The same physics describes the exhilarating spin of a skater and the wrenching force of a tornado. We will find that this is another example of the importance of conservation laws and their role in determining how changes happen in a system, supporting Big Idea 5. The idea that a change of a conserved quantity is always equal to the transfer of that quantity between interacting systems (Enduring Understanding 5.A) is presented for both energy and angular momentum (Enduring Understanding 5.E). The conservation of **angular momentum** in relation to the external net **torque** (Essential Knowledge 5.E.1) parallels that of linear momentum conservation in relation to the external net force. The concept of **rotational inertia** is introduced, a concept that takes into account not only the mass of an object or a system, but also the distribution of mass within the object or system. Therefore, changes in the rotational inertia of a system could lead to changes in the motion (Essential Knowledge 5.E.2) of the system. We shall see that all important aspects of rotational motion either have already been defined for linear motion or have exact analogues in linear motion.

Clearly, therefore, force, energy, and power are associated with rotational motion. This supports Big Idea 3, that interactions are described by forces. The ability of forces to cause torques (Enduring Understanding 3.F) is extended to the interactions between objects that result in nonzero net torque. This nonzero net torque in turn causes changes in the rotational motion of an object (Essential Knowledge 3.F.2) and results in changes of the angular momentum of an object (Essential Knowledge 3.F.3).

Similarly, Big Idea 4, that interactions between systems cause changes in those systems, is supported by the empirical observation that when torques are exerted on rigid bodies these torques cause changes in the angular momentum of the system (Enduring Understanding 4.D).

Again, there is a clear analogy between linear and rotational motion in this interaction. Both the angular kinematics variables (angular displacement, angular velocity, and **angular acceleration**) and the dynamics variables (torque and angular momentum) are vectors with direction depending on whether the rotation is clockwise or counterclockwise with respect to an axis of rotation (Essential Knowledge 4.D.1). The angular momentum of the system can change due to interactions (Essential Knowledge 4.D.2). This change is defined as the product of the average torque and the time interval during which torque is exerted (Essential Knowledge 4.D.3), analogous to the impulse-momentum theorem for linear motion.

The concepts in this chapter support:

Big Idea 3. The interactions of an object with other objects can be described by forces.

Enduring Understanding 3.F. A force exerted on an object can cause a torque on that object.

Extended Knowledge 3.F.2. The presence of a net torque along any axis will cause a rigid system to change its rotational motion or an object to change its rotational motion about that axis.

Extended Knowledge 3.F.3. A torque exerted on an object can change the angular momentum of an object.

Big Idea 4. Interactions between systems can result in changes in those systems.

Enduring Understanding 4.D. A net torque exerted on a system by other objects or systems will change the angular momentum of the system.

Extended Knowledge 4.D.1. Torque, angular velocity, angular acceleration, and angular momentum are vectors and can be characterized as positive or negative depending upon whether they give rise to or correspond to counterclockwise or clockwise rotation with respect to an axis.

Extended Knowledge 4.D.2. The angular momentum of a system may change due to interactions with other objects or systems.

Extended Knowledge 4.D.3. The change in angular momentum is given by the product of the average torque and the time interval during which the torque is exerted.

Big Idea 5. Changes that occur as a result of interactions are constrained by conservation laws.

Enduring Understanding 5.A. Certain quantities are conserved, in the sense that the changes of those quantities in a given system are always equal to the transfer of that quantity to or from the system by all possible interactions with other systems.

Extended Knowledge 5.A.2. For all systems under all circumstances, energy, charge, linear momentum, and angular momentum are conserved.

Enduring Understanding 5.E. The angular momentum of a system is conserved.

Extended Knowledge 5.E.1. If the net external torque exerted on the system is zero, the angular momentum of the system does not change.

Extended Knowledge 5.E.2. The angular momentum of a system is determined by the locations and velocities of the objects that make up the system. The rotational inertia of an object or system depends upon the distribution of mass within the object or system. Changes in the radius of a system or in the distribution of mass within the system result in changes in the system's rotational inertia, and hence in its angular velocity and linear speed for a given angular momentum. Examples should include elliptical orbits in an Earth-satellite system. Mathematical expressions for the moments of inertia will be provided where needed. Students will not be expected to know the parallel axis theorem.

Figure 10.2 This figure skater increases her rate of spin by pulling her arms and her extended leg closer to her axis of rotation. (credit: Luu, Wikimedia Commons)

10.1 Angular Acceleration

Learning Objectives
By the end of this section, you will be able to: • Describe uniform circular motion. • Explain nonuniform circular motion. • Calculate angular acceleration of an object. • Observe the link between linear and angular acceleration.

Uniform Circular Motion and Gravitation discussed only uniform circular motion, which is motion in a circle at constant speed and, hence, constant angular velocity. Recall that angular velocity ω was defined as the time rate of change of angle θ:

$$\omega = \frac{\Delta\theta}{\Delta t}, \tag{10.1}$$

where θ is the angle of rotation as seen in **Figure 10.3**. The relationship between angular velocity ω and linear velocity v was also defined in **Rotation Angle and Angular Velocity** as

$$v = r\omega \tag{10.2}$$

or

$$\omega = \frac{v}{r}, \tag{10.3}$$

where r is the radius of curvature, also seen in **Figure 10.3**. According to the sign convention, the counter clockwise direction is considered as positive direction and clockwise direction as negative

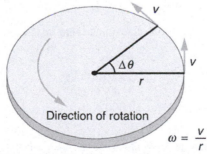

Figure 10.3 This figure shows uniform circular motion and some of its defined quantities.

Angular velocity is not constant when a skater pulls in her arms, when a child starts up a merry-go-round from rest, or when a computer's hard disk slows to a halt when switched off. In all these cases, there is an **angular acceleration**, in which ω

changes. The faster the change occurs, the greater the angular acceleration. Angular acceleration α is defined as the rate of change of angular velocity. In equation form, angular acceleration is expressed as follows:

$$\alpha = \frac{\Delta\omega}{\Delta t},$$ (10.4)

where $\Delta\omega$ is the **change in angular velocity** and Δt is the change in time. The units of angular acceleration are $(\text{rad/s})/\text{s}$, or rad/s^2. If ω increases, then α is positive. If ω decreases, then α is negative.

Example 10.1 Calculating the Angular Acceleration and Deceleration of a Bike Wheel

Suppose a teenager puts her bicycle on its back and starts the rear wheel spinning from rest to a final angular velocity of 250 rpm in 5.00 s. (a) Calculate the angular acceleration in rad/s^2. (b) If she now slams on the brakes, causing an angular acceleration of $-87.3\ \text{rad/s}^2$, how long does it take the wheel to stop?

Strategy for (a)

The angular acceleration can be found directly from its definition in $\alpha = \frac{\Delta\omega}{\Delta t}$ because the final angular velocity and time are given. We see that $\Delta\omega$ is 250 rpm and Δt is 5.00 s.

Solution for (a)

Entering known information into the definition of angular acceleration, we get

$$\begin{aligned} \alpha &= \frac{\Delta\omega}{\Delta t} \\ &= \frac{250\ \text{rpm}}{5.00\ \text{s}}. \end{aligned}$$ (10.5)

Because $\Delta\omega$ is in revolutions per minute (rpm) and we want the standard units of rad/s^2 for angular acceleration, we need to convert $\Delta\omega$ from rpm to rad/s:

$$\begin{aligned} \Delta\omega &= 250\frac{\text{rev}}{\text{min}} \cdot \frac{2\pi\ \text{rad}}{\text{rev}} \cdot \frac{1\ \text{min}}{60\ \text{sec}} \\ &= 26.2\frac{\text{rad}}{\text{s}}. \end{aligned}$$ (10.6)

Entering this quantity into the expression for α, we get

$$\begin{aligned} \alpha &= \frac{\Delta\omega}{\Delta t} \\ &= \frac{26.2\ \text{rad/s}}{5.00\ \text{s}} \\ &= 5.24\ \text{rad/s}^2. \end{aligned}$$ (10.7)

Strategy for (b)

In this part, we know the angular acceleration and the initial angular velocity. We can find the stoppage time by using the definition of angular acceleration and solving for Δt, yielding

$$\Delta t = \frac{\Delta\omega}{\alpha}.$$ (10.8)

Solution for (b)

Here the angular velocity decreases from 26.2 rad/s (250 rpm) to zero, so that $\Delta\omega$ is $-26.2\ \text{rad/s}$, and α is given to be $-87.3\ \text{rad/s}^2$. Thus,

$$\begin{aligned} \Delta t &= \frac{-26.2\ \text{rad/s}}{-87.3\ \text{rad/s}^2} \\ &= 0.300\ \text{s}. \end{aligned}$$ (10.9)

Discussion

Note that the angular acceleration as the girl spins the wheel is small and positive; it takes 5 s to produce an appreciable angular velocity. When she hits the brake, the angular acceleration is large and negative. The angular velocity quickly goes to zero. In both cases, the relationships are analogous to what happens with linear motion. For example, there is a large deceleration when you crash into a brick wall—the velocity change is large in a short time interval.

If the bicycle in the preceding example had been on its wheels instead of upside-down, it would first have accelerated along the ground and then come to a stop. This connection between circular motion and linear motion needs to be explored. For example, it would be useful to know how linear and angular acceleration are related. In circular motion, linear acceleration is *tangent* to the circle at the point of interest, as seen in **Figure 10.4**. Thus, linear acceleration is called **tangential acceleration** a_t.

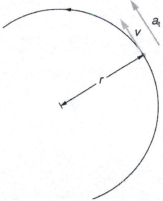

Figure 10.4 In circular motion, linear acceleration a, occurs as the magnitude of the velocity changes: a is tangent to the motion. In the context of circular motion, linear acceleration is also called tangential acceleration a_t.

Linear or tangential acceleration refers to changes in the magnitude of velocity but not its direction. We know from **Uniform Circular Motion and Gravitation** that in circular motion centripetal acceleration, a_c, refers to changes in the direction of the velocity but not its magnitude. An object undergoing circular motion experiences centripetal acceleration, as seen in **Figure 10.5**. Thus, a_t and a_c are perpendicular and independent of one another. Tangential acceleration a_t is directly related to the angular acceleration α and is linked to an increase or decrease in the velocity, but not its direction.

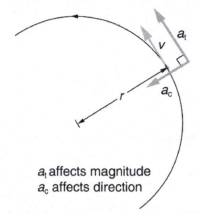

a_t affects magnitude
a_c affects direction

Figure 10.5 Centripetal acceleration a_c occurs as the direction of velocity changes; it is perpendicular to the circular motion. Centripetal and tangential acceleration are thus perpendicular to each other.

Now we can find the exact relationship between linear acceleration a_t and angular acceleration α. Because linear acceleration is proportional to a change in the magnitude of the velocity, it is defined (as it was in **One-Dimensional Kinematics**) to be

$$a_t = \frac{\Delta v}{\Delta t}.$$ (10.10)

For circular motion, note that $v = r\omega$, so that

$$a_t = \frac{\Delta(r\omega)}{\Delta t}.$$ (10.11)

The radius r is constant for circular motion, and so $\Delta(r\omega) = r(\Delta\omega)$. Thus,

$$a_t = r\frac{\Delta\omega}{\Delta t}.$$ (10.12)

By definition, $\alpha = \frac{\Delta\omega}{\Delta t}$. Thus,

$$a_t = r\alpha,$$ (10.13)

or

$$\alpha = \frac{a_t}{r}. \tag{10.14}$$

These equations mean that linear acceleration and angular acceleration are directly proportional. The greater the angular acceleration is, the larger the linear (tangential) acceleration is, and vice versa. For example, the greater the angular acceleration of a car's drive wheels, the greater the acceleration of the car. The radius also matters. For example, the smaller a wheel, the smaller its linear acceleration for a given angular acceleration α.

Example 10.2 Calculating the Angular Acceleration of a Motorcycle Wheel

A powerful motorcycle can accelerate from 0 to 30.0 m/s (about 108 km/h) in 4.20 s. What is the angular acceleration of its 0.320-m-radius wheels? (See Figure 10.6.)

Figure 10.6 The linear acceleration of a motorcycle is accompanied by an angular acceleration of its wheels.

Strategy

We are given information about the linear velocities of the motorcycle. Thus, we can find its linear acceleration a_t. Then, the expression $\alpha = \frac{a_t}{r}$ can be used to find the angular acceleration.

Solution

The linear acceleration is

$$\begin{aligned} a_t &= \frac{\Delta v}{\Delta t} \\ &= \frac{30.0 \text{ m/s}}{4.20 \text{ s}} \\ &= 7.14 \text{ m/s}^2. \end{aligned} \tag{10.15}$$

We also know the radius of the wheels. Entering the values for a_t and r into $\alpha = \frac{a_t}{r}$, we get

$$\begin{aligned} \alpha &= \frac{a_t}{r} \\ &= \frac{7.14 \text{ m/s}^2}{0.320 \text{ m}} \\ &= 22.3 \text{ rad/s}^2. \end{aligned} \tag{10.16}$$

Discussion

Units of radians are dimensionless and appear in any relationship between angular and linear quantities.

So far, we have defined three rotational quantities— θ, ω, and α. These quantities are analogous to the translational quantities x, v, and a. Table 10.1 displays rotational quantities, the analogous translational quantities, and the relationships between them.

Table 10.1 Rotational and Translational Quantities

Rotational	Translational	Relationship
θ	x	$\theta = \frac{x}{r}$
ω	v	$\omega = \frac{v}{r}$
α	a	$\alpha = \frac{a_t}{r}$

Making Connections: Take-Home Experiment

Sit down with your feet on the ground on a chair that rotates. Lift one of your legs such that it is unbent (straightened out). Using the other leg, begin to rotate yourself by pushing on the ground. Stop using your leg to push the ground but allow the chair to rotate. From the origin where you began, sketch the angle, angular velocity, and angular acceleration of your leg as a function of time in the form of three separate graphs. Estimate the magnitudes of these quantities.

Check Your Understanding

Angular acceleration is a vector, having both magnitude and direction. How do we denote its magnitude and direction? Illustrate with an example.

Solution

The magnitude of angular acceleration is α and its most common units are rad/s^2. The direction of angular acceleration along a fixed axis is denoted by a + or a – sign, just as the direction of linear acceleration in one dimension is denoted by a + or a – sign. For example, consider a gymnast doing a forward flip. Her angular momentum would be parallel to the mat and to her left. The magnitude of her angular acceleration would be proportional to her angular velocity (spin rate) and her moment of inertia about her spin axis.

PhET Explorations: Ladybug Revolution

Join the ladybug in an exploration of rotational motion. Rotate the merry-go-round to change its angle, or choose a constant angular velocity or angular acceleration. Explore how circular motion relates to the bug's x,y position, velocity, and acceleration using vectors or graphs.

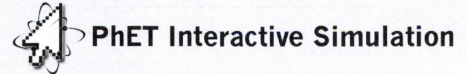

Figure 10.7 Ladybug Revolution (http://cnx.org/content/m55183/1.2/rotation_en.jar)

10.2 Kinematics of Rotational Motion

Learning Objectives

By the end of this section, you will be able to:

- Observe the kinematics of rotational motion.
- Derive rotational kinematic equations.
- Evaluate problem solving strategies for rotational kinematics.

Just by using our intuition, we can begin to see how rotational quantities like θ, ω, and α are related to one another. For example, if a motorcycle wheel has a large angular acceleration for a fairly long time, it ends up spinning rapidly and rotates through many revolutions. In more technical terms, if the wheel's angular acceleration α is large for a long period of time t, then the final angular velocity ω and angle of rotation θ are large. The wheel's rotational motion is exactly analogous to the fact that the motorcycle's large translational acceleration produces a large final velocity, and the distance traveled will also be large.

Kinematics is the description of motion. The **kinematics of rotational motion** describes the relationships among rotation angle, angular velocity, angular acceleration, and time. Let us start by finding an equation relating ω, α, and t. To determine this equation, we recall a familiar kinematic equation for translational, or straight-line, motion:

$$v = v_0 + at \quad \text{(constant } a\text{)} \tag{10.17}$$

Note that in rotational motion $a = a_{\text{t}}$, and we shall use the symbol a for tangential or linear acceleration from now on. As in linear kinematics, we assume a is constant, which means that angular acceleration α is also a constant, because $a = r\alpha$. Now, let us substitute $v = r\omega$ and $a = r\alpha$ into the linear equation above:

$$r\omega = r\omega_0 + r\alpha t. \tag{10.18}$$

The radius r cancels in the equation, yielding

$$\omega = \omega_0 + at \quad \text{(constant } a\text{)}, \tag{10.19}$$

where ω_0 is the initial angular velocity. This last equation is a *kinematic relationship* among ω, α, and t —that is, it describes their relationship without reference to forces or masses that may affect rotation. It is also precisely analogous in form to its translational counterpart.

Making Connections

Kinematics for rotational motion is completely analogous to translational kinematics, first presented in **One-Dimensional Kinematics**. Kinematics is concerned with the description of motion without regard to force or mass. We will find that translational kinematic quantities, such as displacement, velocity, and acceleration have direct analogs in rotational motion.

Starting with the four kinematic equations we developed in **One-Dimensional Kinematics**, we can derive the following four rotational kinematic equations (presented together with their translational counterparts):

Table 10.2 Rotational Kinematic Equations

Rotational	Translational	
$\theta = \bar{\omega}t$	$x = \bar{v}t$	
$\omega = \omega_0 + \alpha t$	$v = v_0 + at$	(constant α, a)
$\theta = \omega_0 t + \frac{1}{2}\alpha t^2$	$x = v_0 t + \frac{1}{2}at^2$	(constant α, a)
$\omega^2 = \omega_0^2 + 2\alpha\theta$	$v^2 = v_0^2 + 2ax$	(constant α, a)

In these equations, the subscript 0 denotes initial values (θ_0, x_0, and t_0 are initial values), and the average angular velocity $\bar{\omega}$ and average velocity $\bar{v}$ are defined as follows:

$$\bar{\omega} = \frac{\omega_0 + \omega}{2} \text{ and } \bar{v} = \frac{v_0 + v}{2}. \tag{10.20}$$

The equations given above in **Table 10.2** can be used to solve any rotational or translational kinematics problem in which a and α are constant.

Problem-Solving Strategy for Rotational Kinematics

1. *Examine the situation to determine that rotational kinematics (rotational motion) is involved*. Rotation must be involved, but without the need to consider forces or masses that affect the motion.

2. *Identify exactly what needs to be determined in the problem (identify the unknowns)*. A sketch of the situation is useful.

3. *Make a list of what is given or can be inferred from the problem as stated (identify the knowns)*.

4. *Solve the appropriate equation or equations for the quantity to be determined (the unknown)*. It can be useful to think in terms of a translational analog because by now you are familiar with such motion.

5. *Substitute the known values along with their units into the appropriate equation, and obtain numerical solutions complete with units*. Be sure to use units of radians for angles.

6. *Check your answer to see if it is reasonable: Does your answer make sense?*

Example 10.3 Calculating the Acceleration of a Fishing Reel

A deep-sea fisherman hooks a big fish that swims away from the boat pulling the fishing line from his fishing reel. The whole system is initially at rest and the fishing line unwinds from the reel at a radius of 4.50 cm from its axis of rotation. The reel is given an angular acceleration of 110 rad/s^2 for 2.00 s as seen in **Figure 10.8**.

(a) What is the final angular velocity of the reel?

(b) At what speed is fishing line leaving the reel after 2.00 s elapses?

(c) How many revolutions does the reel make?

(d) How many meters of fishing line come off the reel in this time?

Strategy

In each part of this example, the strategy is the same as it was for solving problems in linear kinematics. In particular, known values are identified and a relationship is then sought that can be used to solve for the unknown.

Solution for (a)

Here α and t are given and ω needs to be determined. The most straightforward equation to use is $\omega = \omega_0 + \alpha t$ because the unknown is already on one side and all other terms are known. That equation states that

$$\omega = \omega_0 + \alpha t. \tag{10.21}$$

We are also given that $\omega_0 = 0$ (it starts from rest), so that

$$\omega = 0 + \left(110 \text{ rad/s}^2\right)(2.00\text{s}) = 220 \text{ rad/s}. \tag{10.22}$$

Solution for (b)

Now that ω is known, the speed v can most easily be found using the relationship

$$v = r\omega, \tag{10.23}$$

where the radius r of the reel is given to be 4.50 cm; thus,

$$v = (0.0450 \text{ m})(220 \text{ rad/s}) = 9.90 \text{ m/s}. \tag{10.24}$$

Note again that radians must always be used in any calculation relating linear and angular quantities. Also, because radians are dimensionless, we have $\text{m} \times \text{rad} = \text{m}$.

Solution for (c)

Here, we are asked to find the number of revolutions. Because $1 \text{ rev} = 2\pi \text{ rad}$, we can find the number of revolutions by finding θ in radians. We are given α and t, and we know ω_0 is zero, so that θ can be obtained using

$\theta = \omega_0 t + \frac{1}{2}\alpha t^2$.

$$\begin{aligned}\theta &= \omega_0 t + \frac{1}{2}\alpha t^2 \\ &= 0 + (0.500)\left(110 \text{ rad/s}^2\right)(2.00 \text{ s})^2 = 220 \text{ rad}.\end{aligned} \tag{10.25}$$

Converting radians to revolutions gives

$$\theta = (220 \text{ rad})\frac{1 \text{ rev}}{2\pi \text{ rad}} = 35.0 \text{ rev}. \tag{10.26}$$

Solution for (d)

The number of meters of fishing line is x, which can be obtained through its relationship with θ:

$$x = r\theta = (0.0450 \text{ m})(220 \text{ rad}) = 9.90 \text{ m}. \tag{10.27}$$

Discussion

This example illustrates that relationships among rotational quantities are highly analogous to those among linear quantities. We also see in this example how linear and rotational quantities are connected. The answers to the questions are realistic. After unwinding for two seconds, the reel is found to spin at 220 rad/s, which is 2100 rpm. (No wonder reels sometimes make high-pitched sounds.) The amount of fishing line played out is 9.90 m, about right for when the big fish bites.

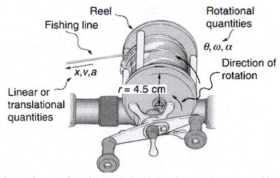

Figure 10.8 Fishing line coming off a rotating reel moves linearly. **Example 10.3** and **Example 10.4** consider relationships between rotational and linear quantities associated with a fishing reel.

Example 10.4 Calculating the Duration When the Fishing Reel Slows Down and Stops

Now let us consider what happens if the fisherman applies a brake to the spinning reel, achieving an angular acceleration of -300 rad/s^2. How long does it take the reel to come to a stop?

Strategy

We are asked to find the time t for the reel to come to a stop. The initial and final conditions are different from those in the previous problem, which involved the same fishing reel. Now we see that the initial angular velocity is $\omega_0 = 220 \text{ rad/s}$ and the final angular velocity ω is zero. The angular acceleration is given to be $\alpha = -300 \text{ rad/s}^2$. Examining the available equations, we see all quantities but t are known in $\omega = \omega_0 + \alpha t$, making it easiest to use this equation.

Solution

The equation states

$$\omega = \omega_0 + \alpha t. \tag{10.28}$$

We solve the equation algebraically for t, and then substitute the known values as usual, yielding

$$t = \frac{\omega - \omega_0}{\alpha} = \frac{0 - 220 \text{ rad/s}}{-300 \text{ rad/s}^2} = 0.733 \text{ s}. \tag{10.29}$$

Discussion

Note that care must be taken with the signs that indicate the directions of various quantities. Also, note that the time to stop the reel is fairly small because the acceleration is rather large. Fishing lines sometimes snap because of the accelerations involved, and fishermen often let the fish swim for a while before applying brakes on the reel. A tired fish will be slower, requiring a smaller acceleration.

Example 10.5 Calculating the Slow Acceleration of Trains and Their Wheels

Large freight trains accelerate very slowly. Suppose one such train accelerates from rest, giving its 0.350-m-radius wheels an angular acceleration of 0.250 rad/s^2. After the wheels have made 200 revolutions (assume no slippage): (a) How far has the train moved down the track? (b) What are the final angular velocity of the wheels and the linear velocity of the train?

Strategy

In part (a), we are asked to find x, and in (b) we are asked to find ω and v. We are given the number of revolutions θ, the radius of the wheels r, and the angular acceleration α.

Solution for (a)

The distance x is very easily found from the relationship between distance and rotation angle:

$$\theta = \frac{x}{r}. \tag{10.30}$$

Solving this equation for x yields

$$x = r\theta. \tag{10.31}$$

Before using this equation, we must convert the number of revolutions into radians, because we are dealing with a relationship between linear and rotational quantities:

$$\theta = (200 \text{ rev})\frac{2\pi \text{ rad}}{1 \text{ rev}} = 1257 \text{ rad}. \tag{10.32}$$

Now we can substitute the known values into $x = r\theta$ to find the distance the train moved down the track:

$$x = r\theta = (0.350 \text{ m})(1257 \text{ rad}) = 440 \text{ m}. \tag{10.33}$$

Solution for (b)

We cannot use any equation that incorporates t to find ω, because the equation would have at least two unknown values. The equation $\omega^2 = \omega_0^2 + 2\alpha\theta$ will work, because we know the values for all variables except ω:

$$\omega^2 = \omega_0^2 + 2\alpha\theta \tag{10.34}$$

Taking the square root of this equation and entering the known values gives

$$\omega = \left[0 + 2(0.250 \text{ rad/s}^2)(1257 \text{ rad})\right]^{1/2}$$ (10.35)
$$= 25.1 \text{ rad/s}.$$

We can find the linear velocity of the train, v, through its relationship to ω:

$$v = r\omega = (0.350 \text{ m})(25.1 \text{ rad/s}) = 8.77 \text{ m/s}.$$ (10.36)

Discussion

The distance traveled is fairly large and the final velocity is fairly slow (just under 32 km/h).

There is translational motion even for something spinning in place, as the following example illustrates. **Figure 10.9** shows a fly on the edge of a rotating microwave oven plate. The example below calculates the total distance it travels.

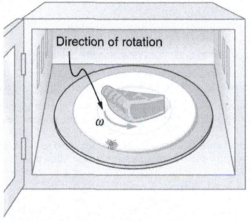

Figure 10.9 The image shows a microwave plate. The fly makes revolutions while the food is heated (along with the fly).

Example 10.6 Calculating the Distance Traveled by a Fly on the Edge of a Microwave Oven Plate

A person decides to use a microwave oven to reheat some lunch. In the process, a fly accidentally flies into the microwave and lands on the outer edge of the rotating plate and remains there. If the plate has a radius of 0.15 m and rotates at 6.0 rpm, calculate the total distance traveled by the fly during a 2.0-min cooking period. (Ignore the start-up and slow-down times.)

Strategy

First, find the total number of revolutions θ, and then the linear distance x traveled. $\theta = \bar{\omega}t$ can be used to find θ because $\bar{\omega}$ is given to be 6.0 rpm.

Solution

Entering known values into $\theta = \bar{\omega}t$ gives

$$\theta = \bar{\omega}t = (6.0 \text{ rpm})(2.0 \text{ min}) = 12 \text{ rev}.$$ (10.37)

As always, it is necessary to convert revolutions to radians before calculating a linear quantity like x from an angular quantity like θ:

$$\theta = (12 \text{ rev})\left(\frac{2\pi \text{ rad}}{1 \text{ rev}}\right) = 75.4 \text{ rad}.$$ (10.38)

Now, using the relationship between x and θ, we can determine the distance traveled:

$$x = r\theta = (0.15 \text{ m})(75.4 \text{ rad}) = 11 \text{ m}.$$ (10.39)

Discussion

Quite a trip (if it survives)! Note that this distance is the total distance traveled by the fly. Displacement is actually zero for complete revolutions because they bring the fly back to its original position. The distinction between total distance traveled and displacement was first noted in **One-Dimensional Kinematics**.

Rotational kinematics has many useful relationships, often expressed in equation form. Are these relationships laws of physics or are they simply descriptive? (Hint: the same question applies to linear kinematics.)

Solution

Rotational kinematics (just like linear kinematics) is descriptive and does not represent laws of nature. With kinematics, we can describe many things to great precision but kinematics does not consider causes. For example, a large angular acceleration describes a very rapid change in angular velocity without any consideration of its cause.

10.3 Dynamics of Rotational Motion: Rotational Inertia

Learning Objectives

By the end of this section, you will be able to:

- Understand the relationship between force, mass, and acceleration.
- Study the turning effect of force.
- Study the analogy between force and torque, mass and moment of inertia, and linear acceleration and angular acceleration.

The information presented in this section supports the following AP® learning objectives and science practices:

- **4.D.1.1** The student is able to describe a representation and use it to analyze a situation in which several forces exerted on a rotating system of rigidly connected objects change the angular velocity and angular momentum of the system. **(S.P. 1.2, 1.4)**
- **4.D.1.2** The student is able to plan data collection strategies designed to establish that torque, angular velocity, angular acceleration, and angular momentum can be predicted accurately when the variables are treated as being clockwise or counterclockwise with respect to a well-defined axis of rotation, and refine the research question based on the examination of data. **(S.P. 3.2, 4.1, 5.1, 5.3)**
- **5.E.2.1** The student is able to describe or calculate the angular momentum and rotational inertia of a system in terms of the locations and velocities of objects that make up the system. Students are expected to do qualitative reasoning with compound objects. Students are expected to do calculations with a fixed set of extended objects and point masses. **(S.P. 2.2)**

If you have ever spun a bike wheel or pushed a merry-go-round, you know that force is needed to change angular velocity as seen in **Figure 10.10**. In fact, your intuition is reliable in predicting many of the factors that are involved. For example, we know that a door opens slowly if we push too close to its hinges. Furthermore, we know that the more massive the door, the more slowly it opens. The first example implies that the farther the force is applied from the pivot, the greater the angular acceleration; another implication is that angular acceleration is inversely proportional to mass. These relationships should seem very similar to the familiar relationships among force, mass, and acceleration embodied in Newton's second law of motion. There are, in fact, precise rotational analogs to both force and mass.

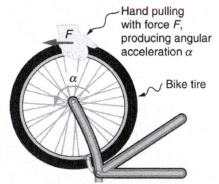

Hand pulling with force F, producing angular acceleration α

Bike tire

Figure 10.10 Force is required to spin the bike wheel. The greater the force, the greater the angular acceleration produced. The more massive the wheel, the smaller the angular acceleration. If you push on a spoke closer to the axle, the angular acceleration will be smaller.

To develop the precise relationship among force, mass, radius, and angular acceleration, consider what happens if we exert a force F on a point mass m that is at a distance r from a pivot point, as shown in **Figure 10.11**. Because the force is perpendicular to r, an acceleration $a = \frac{F}{m}$ is obtained in the direction of F. We can rearrange this equation such that $F = ma$ and then look for ways to relate this expression to expressions for rotational quantities. We note that $a = r\alpha$, and we substitute this expression into $F = ma$, yielding

$$F = mr\alpha. \tag{10.40}$$

Recall that **torque** is the turning effectiveness of a force. In this case, because $\mathbf{F}$ is perpendicular to r, torque is simply $\tau = Fr$. So, if we multiply both sides of the equation above by r, we get torque on the left-hand side. That is,

$$rF = mr^2\alpha \tag{10.41}$$

or

$$\tau = mr^2\alpha. \tag{10.42}$$

This last equation is the rotational analog of Newton's second law ($F = ma$), where torque is analogous to force, angular acceleration is analogous to translational acceleration, and mr^2 is analogous to mass (or inertia). The quantity mr^2 is called the **rotational inertia** or **moment of inertia** of a point mass m a distance r from the center of rotation.

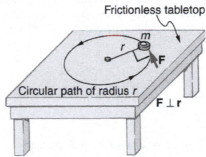

Figure 10.11 An object is supported by a horizontal frictionless table and is attached to a pivot point by a cord that supplies centripetal force. A force F is applied to the object perpendicular to the radius r, causing it to accelerate about the pivot point. The force is kept perpendicular to r.

Making Connections: Rotational Motion Dynamics

Dynamics for rotational motion is completely analogous to linear or translational dynamics. Dynamics is concerned with force and mass and their effects on motion. For rotational motion, we will find direct analogs to force and mass that behave just as we would expect from our earlier experiences.

Rotational Inertia and Moment of Inertia

Before we can consider the rotation of anything other than a point mass like the one in **Figure 10.11**, we must extend the idea of rotational inertia to all types of objects. To expand our concept of rotational inertia, we define the **moment of inertia** I of an object to be the sum of mr^2 for all the point masses of which it is composed. That is, $I = \sum mr^2$. Here I is analogous to m in translational motion. Because of the distance r, the moment of inertia for any object depends on the chosen axis. Actually, calculating I is beyond the scope of this text except for one simple case—that of a hoop, which has all its mass at the same distance from its axis. A hoop's moment of inertia around its axis is therefore MR^2, where M is its total mass and R its radius. (We use M and R for an entire object to distinguish them from m and r for point masses.) In all other cases, we must consult **Figure 10.12** (note that the table is piece of artwork that has shapes as well as formulae) for formulas for I that have been derived from integration over the continuous body. Note that I has units of mass multiplied by distance squared ($\text{kg} \cdot \text{m}^2$), as we might expect from its definition.

The general relationship among torque, moment of inertia, and angular acceleration is

$$\text{net } \tau = I\alpha \tag{10.43}$$

or

$$\alpha = \frac{\text{net } \tau}{I}, \tag{10.44}$$

where net τ is the total torque from all forces relative to a chosen axis. For simplicity, we will only consider torques exerted by forces in the plane of the rotation. Such torques are either positive or negative and add like ordinary numbers. The relationship in $\tau = I\alpha$, $\alpha = \frac{\text{net } \tau}{I}$ is the rotational analog to Newton's second law and is very generally applicable. This equation is actually valid for *any* torque, applied to *any* object, relative to *any* axis.

As we might expect, the larger the torque is, the larger the angular acceleration is. For example, the harder a child pushes on a merry-go-round, the faster it accelerates. Furthermore, the more massive a merry-go-round, the slower it accelerates for the same torque. The basic relationship between moment of inertia and angular acceleration is that the larger the moment of inertia, the smaller is the angular acceleration. But there is an additional twist. The moment of inertia depends not only on the mass of an object, but also on its *distribution* of mass relative to the axis around which it rotates. For example, it will be much easier to accelerate a merry-go-round full of children if they stand close to its axis than if they all stand at the outer edge. The mass is the same in both cases; but the moment of inertia is much larger when the children are at the edge.

Take-Home Experiment

Cut out a circle that has about a 10 cm radius from stiff cardboard. Near the edge of the circle, write numbers 1 to 12 like hours on a clock face. Position the circle so that it can rotate freely about a horizontal axis through its center, like a wheel. (You could loosely nail the circle to a wall.) Hold the circle stationary and with the number 12 positioned at the top, attach a lump of blue putty (sticky material used for fixing posters to walls) at the number 3. How large does the lump need to be to just rotate the circle? Describe how you can change the moment of inertia of the circle. How does this change affect the amount of blue putty needed at the number 3 to just rotate the circle? Change the circle's moment of inertia and then try rotating the circle by using different amounts of blue putty. Repeat this process several times.

In what direction did the circle rotate when you added putty at the number 3 (clockwise or counterclockwise)? In which of these directions was the resulting angular velocity? Was the angular velocity constant? What can we say about the direction (clockwise or counterclockwise) of the angular acceleration? How could you change the placement of the putty to create angular velocity in the opposite direction?

Problem-Solving Strategy for Rotational Dynamics

1. *Examine the situation to determine that torque and mass are involved in the rotation.* Draw a careful sketch of the situation.

2. *Determine the system of interest.*

3. *Draw a free body diagram.* That is, draw and label all external forces acting on the system of interest.

4. *Apply* $net \ \tau = I\alpha, \ \alpha = \dfrac{net \ \tau}{I}$, *the rotational equivalent of Newton's second law, to solve the problem.* Care must be taken to use the correct moment of inertia and to consider the torque about the point of rotation.

5. *As always, check the solution to see if it is reasonable.*

Making Connections

In statics, the net torque is zero, and there is no angular acceleration. In rotational motion, net torque is the cause of angular acceleration, exactly as in Newton's second law of motion for rotation.

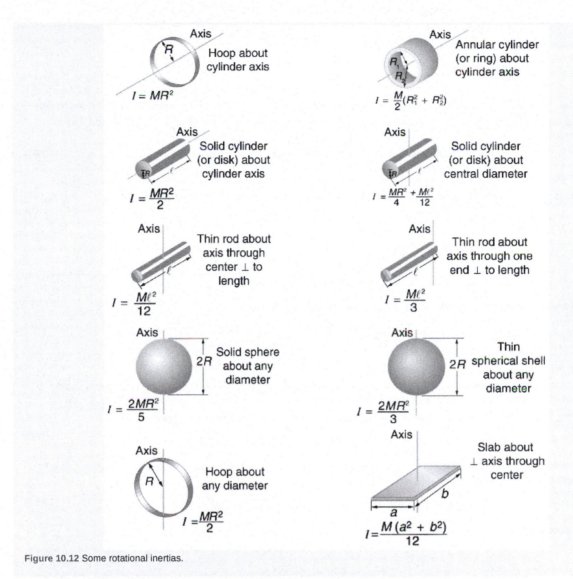

$I = MR^2$ — Hoop about cylinder axis

$I = \frac{M}{2}(R_1^2 + R_2^2)$ — Annular cylinder (or ring) about cylinder axis

$I = \frac{MR^2}{2}$ — Solid cylinder (or disk) about cylinder axis

$I = \frac{MR^2}{4} + \frac{M\ell^2}{12}$ — Solid cylinder (or disk) about central diameter

$I = \frac{M\ell^2}{12}$ — Thin rod about axis through center ⊥ to length

$I = \frac{M\ell^2}{3}$ — Thin rod about axis through one end ⊥ to length

$I = \frac{2MR^2}{5}$ — Solid sphere about any diameter

$I = \frac{2MR^2}{3}$ — Thin spherical shell about any diameter

$I = \frac{MR^2}{2}$ — Hoop about any diameter

$I = \frac{M(a^2 + b^2)}{12}$ — Slab about ⊥ axis through center

Figure 10.12 Some rotational inertias.

Example 10.7 Calculating the Effect of Mass Distribution on a Merry-Go-Round

Consider the father pushing a playground merry-go-round in Figure 10.13. He exerts a force of 250 N at the edge of the 50.0-kg merry-go-round, which has a 1.50 m radius. Calculate the angular acceleration produced (a) when no one is on the merry-go-round and (b) when an 18.0-kg child sits 1.25 m away from the center. Consider the merry-go-round itself to be a uniform disk with negligible retarding friction.

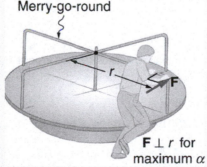

Merry-go-round

$\mathbf{F} \perp r$ for maximum α

Figure 10.13 A father pushes a playground merry-go-round at its edge and perpendicular to its radius to achieve maximum torque.

Strategy

Angular acceleration is given directly by the expression $\alpha = \dfrac{\text{net } \tau}{I}$:

$$\alpha = \frac{\tau}{I}. \tag{10.45}$$

To solve for α, we must first calculate the torque τ (which is the same in both cases) and moment of inertia I (which is greater in the second case). To find the torque, we note that the applied force is perpendicular to the radius and friction is negligible, so that

$$\tau = rF \sin \theta = (1.50 \text{ m})(250 \text{ N}) = 375 \text{ N} \cdot \text{m}. \tag{10.46}$$

Solution for (a)

The moment of inertia of a solid disk about this axis is given in **Figure 10.12** to be

$$\frac{1}{2}MR^2, \tag{10.47}$$

where $M = 50.0 \text{ kg}$ and $R = 1.50 \text{ m}$, so that

$$I = (0.500)(50.0 \text{ kg})(1.50 \text{ m})^2 = 56.25 \text{ kg} \cdot \text{m}^2. \tag{10.48}$$

Now, after we substitute the known values, we find the angular acceleration to be

$$\alpha = \frac{\tau}{I} = \frac{375 \text{ N} \cdot \text{m}}{56.25 \text{ kg} \cdot \text{m}^2} = 6.67\frac{\text{rad}}{\text{s}^2}. \tag{10.49}$$

Solution for (b)

We expect the angular acceleration for the system to be less in this part, because the moment of inertia is greater when the child is on the merry-go-round. To find the total moment of inertia I, we first find the child's moment of inertia I_c by considering the child to be equivalent to a point mass at a distance of 1.25 m from the axis. Then,

$$I_c = MR^2 = (18.0 \text{ kg})(1.25 \text{ m})^2 = 28.13 \text{ kg} \cdot \text{m}^2. \tag{10.50}$$

The total moment of inertia is the sum of moments of inertia of the merry-go-round and the child (about the same axis). To justify this sum to yourself, examine the definition of I:

$$I = 28.13 \text{ kg} \cdot \text{m}^2 + 56.25 \text{ kg} \cdot \text{m}^2 = 84.38 \text{ kg} \cdot \text{m}^2. \tag{10.51}$$

Substituting known values into the equation for α gives

$$\alpha = \frac{\tau}{I} = \frac{375 \text{ N} \cdot \text{m}}{84.38 \text{ kg} \cdot \text{m}^2} = 4.44\frac{\text{rad}}{\text{s}^2}. \tag{10.52}$$

Discussion

The angular acceleration is less when the child is on the merry-go-round than when the merry-go-round is empty, as expected. The angular accelerations found are quite large, partly due to the fact that friction was considered to be negligible. If, for example, the father kept pushing perpendicularly for 2.00 s, he would give the merry-go-round an angular velocity of 13.3 rad/s when it is empty but only 8.89 rad/s when the child is on it. In terms of revolutions per second, these angular velocities are 2.12 rev/s and 1.41 rev/s, respectively. The father would end up running at about 50 km/h in the first case. Summer Olympics, here he comes! Confirmation of these numbers is left as an exercise for the reader.

Making Connections: Multiple Forces on One System

A large potter's wheel has a diameter of 60.0 cm and a mass of 8.0 kg. It is powered by a 20.0 N motor acting on the outer edge. There is also a brake capable of exerting a 15.0 N force at a radius of 12.0 cm from the axis of rotation, on the underside.

What is the angular acceleration when the motor is in use?

The torque is found by $\tau = rF \sin \theta = (0.300 \text{ m})(20.0 \text{ N}) = 6.00 \text{ N} \cdot \text{m}$.

The moment of inertia is calculated as $I = \frac{1}{2}MR^2 = \frac{1}{2}(8.0 \text{ kg})(0.300 \text{ m})^2 = 0.36 \text{ kg} \cdot \text{m}^2$.

Thus, the angular acceleration would be $\alpha = \frac{\tau}{I} = \frac{6.00 \text{ N} \cdot \text{m}}{0.36 \text{ kg} \cdot \text{m}^2} = 17 \text{ rad/s}^2$.

Note that the friction is always acting in a direction opposite to the rotation that is currently happening in this system. If the potter makes a mistake and has both the brake and motor on simultaneously, the friction force of the brake will exert a torque opposite that of the motor.

The torque from the brake is $\tau = rF\sin\theta = (0.120\,\text{m})(15.0\,\text{N}) = 1.80\,\text{N·m}$.

Thus, the net torque is $6.00\,\text{N·m} - 1.80\,\text{N·m} = 4.20\,\text{N·m}$.

And the angular acceleration is $\alpha = \dfrac{\tau}{I} = \dfrac{4.20\,\text{N·m}}{0.36\,\text{kg·m}^2} = 12\,\text{rad/s}^2$.

Check Your Understanding

Torque is the analog of force and moment of inertia is the analog of mass. Force and mass are physical quantities that depend on only one factor. For example, mass is related solely to the numbers of atoms of various types in an object. Are torque and moment of inertia similarly simple?

Solution
No. Torque depends on three factors: force magnitude, force direction, and point of application. Moment of inertia depends on both mass and its distribution relative to the axis of rotation. So, while the analogies are precise, these rotational quantities depend on more factors.

10.4 Rotational Kinetic Energy: Work and Energy Revisited

Learning Objectives

By the end of this section, you will be able to:

- Derive the equation for rotational work.
- Calculate rotational kinetic energy.
- Demonstrate the law of conservation of energy.

The information presented in this section supports the following AP® learning objectives and science practices:

- **3.F.2.1** The student is able to make predictions about the change in the angular velocity about an axis for an object when forces exerted on the object cause a torque about that axis. **(S.P. 6.4)**
- **3.F.2.2** The student is able to plan data collection and analysis strategies designed to test the relationship between a torque exerted on an object and the change in angular velocity of that object about an axis. **(S.P. 4.1, 4.2, 5.1)**

In this module, we will learn about work and energy associated with rotational motion. **Figure 10.14** shows a worker using an electric grindstone propelled by a motor. Sparks are flying, and noise and vibration are created as layers of steel are pared from the pole. The stone continues to turn even after the motor is turned off, but it is eventually brought to a stop by friction. Clearly, the motor had to work to get the stone spinning. This work went into heat, light, sound, vibration, and considerable **rotational kinetic energy**.

Figure 10.14 The motor works in spinning the grindstone, giving it rotational kinetic energy. That energy is then converted to heat, light, sound, and vibration. (credit: U.S. Navy photo by Mass Communication Specialist Seaman Zachary David Bell)

Work must be done to rotate objects such as grindstones or merry-go-rounds. Work was defined in **Uniform Circular Motion and Gravitation** for translational motion, and we can build on that knowledge when considering work done in rotational motion. The simplest rotational situation is one in which the net force is exerted perpendicular to the radius of a disk (as shown in **Figure 10.15**) and remains perpendicular as the disk starts to rotate. The force is parallel to the displacement, and so the net work done is the product of the force times the arc length traveled:

$$\text{net } W = (\text{net } F)\Delta s. \tag{10.53}$$

To get torque and other rotational quantities into the equation, we multiply and divide the right-hand side of the equation by r, and gather terms:

$$\text{net } W = (r \text{ net } F)\frac{\Delta s}{r}.$$

(10.54)

We recognize that $r \text{ net } F = \text{net } \tau$ and $\Delta s / r = \theta$, so that

$$\text{net } W = (\text{net } \tau)\theta.$$

(10.55)

This equation is the expression for rotational work. It is very similar to the familiar definition of translational work as force multiplied by distance. Here, torque is analogous to force, and angle is analogous to distance. The equation $\text{net } W = (\text{net } \tau)\theta$ is valid in general, even though it was derived for a special case.

To get an expression for rotational kinetic energy, we must again perform some algebraic manipulations. The first step is to note that $\text{net } \tau = I\alpha$, so that

$$\text{net } W = I\alpha\theta.$$

(10.56)

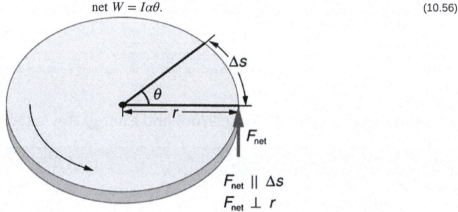

Figure 10.15 The net force on this disk is kept perpendicular to its radius as the force causes the disk to rotate. The net work done is thus $(\text{net } F)\Delta s$. The net work goes into rotational kinetic energy.

Making Connections

qWork and energy in rotational motion are completely analogous to work and energy in translational motion, first presented in **Uniform Circular Motion and Gravitation**.

Now, we solve one of the rotational kinematics equations for $\alpha\theta$. We start with the equation

$$\omega^2 = \omega_0^2 + 2\alpha\theta.$$

(10.57)

Next, we solve for $\alpha\theta$:

$$\alpha\theta = \frac{\omega^2 - \omega_0^2}{2}.$$

(10.58)

Substituting this into the equation for net W and gathering terms yields

$$\text{net } W = \frac{1}{2}I\omega^2 - \frac{1}{2}I\omega_0^2.$$

(10.59)

This equation is the **work-energy theorem** for rotational motion only. As you may recall, net work changes the kinetic energy of a system. Through an analogy with translational motion, we define the term $\left(\frac{1}{2}\right)I\omega^2$ to be **rotational kinetic energy** KE_{rot} for an object with a moment of inertia I and an angular velocity ω:

$$KE_{rot} = \frac{1}{2}I\omega^2.$$

(10.60)

The expression for rotational kinetic energy is exactly analogous to translational kinetic energy, with I being analogous to m and ω to v. Rotational kinetic energy has important effects. Flywheels, for example, can be used to store large amounts of rotational kinetic energy in a vehicle, as seen in **Figure 10.16**.

Figure 10.16 Experimental vehicles, such as this bus, have been constructed in which rotational kinetic energy is stored in a large flywheel. When the bus goes down a hill, its transmission converts its gravitational potential energy into KE_{rot}. It can also convert translational kinetic energy, when the bus stops, into KE_{rot}. The flywheel's energy can then be used to accelerate, to go up another hill, or to keep the bus from going against friction.

Example 10.8 Calculating the Work and Energy for Spinning a Grindstone

Consider a person who spins a large grindstone by placing her hand on its edge and exerting a force through part of a revolution as shown in **Figure 10.17**. In this example, we verify that the work done by the torque she exerts equals the change in rotational energy. (a) How much work is done if she exerts a force of 200 N through a rotation of $1.00 \text{ rad}(57.3°)$? The force is kept perpendicular to the grindstone's 0.320-m radius at the point of application, and the effects of friction are negligible. (b) What is the final angular velocity if the grindstone has a mass of 85.0 kg? (c) What is the final rotational kinetic energy? (It should equal the work.)

Strategy

To find the work, we can use the equation net $W = (\text{net } \tau)\theta$. We have enough information to calculate the torque and are given the rotation angle. In the second part, we can find the final angular velocity using one of the kinematic relationships. In the last part, we can calculate the rotational kinetic energy from its expression in $KE_{rot} = \frac{1}{2}I\omega^2$.

Solution for (a)

The net work is expressed in the equation

$$\text{net } W = (\text{net } \tau)\theta, \tag{10.61}$$

where net τ is the applied force multiplied by the radius (rF) because there is no retarding friction, and the force is perpendicular to r. The angle θ is given. Substituting the given values in the equation above yields

$$\begin{aligned} \text{net } W &= rF\theta = (0.320 \text{ m})(200 \text{ N})(1.00 \text{ rad}) \\ &= 64.0 \text{ N} \cdot \text{m}. \end{aligned} \tag{10.62}$$

Noting that $1 \text{ N} \cdot \text{m} = 1 \text{ J}$,

$$\text{net } W = 64.0 \text{ J}. \tag{10.63}$$

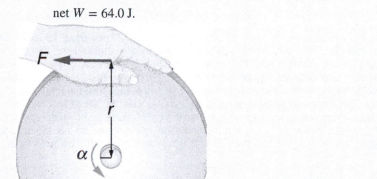

Figure 10.17 A large grindstone is given a spin by a person grasping its outer edge.

Solution for (b)

To find ω from the given information requires more than one step. We start with the kinematic relationship in the equation

$$\omega^2 = \omega_0^2 + 2\alpha\theta. \tag{10.64}$$

Note that $\omega_0 = 0$ because we start from rest. Taking the square root of the resulting equation gives

$$\omega = (2\alpha\theta)^{1/2}. \tag{10.65}$$

Now we need to find α. One possibility is

$$\alpha = \frac{\text{net } \tau}{I}, \tag{10.66}$$

where the torque is

$$\text{net } \tau = rF = (0.320 \text{ m})(200 \text{ N}) = 64.0 \text{ N} \cdot \text{m}. \tag{10.67}$$

The formula for the moment of inertia for a disk is found in **Figure 10.12**:

$$I = \tfrac{1}{2}MR^2 = 0.5(85.0 \text{ kg})(0.320 \text{ m})^2 = 4.352 \text{ kg} \cdot \text{m}^2. \tag{10.68}$$

Substituting the values of torque and moment of inertia into the expression for α, we obtain

$$\alpha = \frac{64.0 \text{ N} \cdot \text{m}}{4.352 \text{ kg} \cdot \text{m}^2} = 14.7\frac{\text{rad}}{\text{s}^2}. \tag{10.69}$$

Now, substitute this value and the given value for θ into the above expression for ω:

$$\omega = (2\alpha\theta)^{1/2} = \left[2\left(14.7\frac{\text{rad}}{\text{s}^2}\right)(1.00 \text{ rad})\right]^{1/2} = 5.42\frac{\text{rad}}{\text{s}}. \tag{10.70}$$

Solution for (c)

The final rotational kinetic energy is

$$KE_{\text{rot}} = \tfrac{1}{2}I\omega^2. \tag{10.71}$$

Both I and ω were found above. Thus,

$$KE_{\text{rot}} = (0.5)\left(4.352 \text{ kg} \cdot \text{m}^2\right)(5.42 \text{ rad/s})^2 = 64.0 \text{ J}. \tag{10.72}$$

Discussion

The final rotational kinetic energy equals the work done by the torque, which confirms that the work done went into rotational kinetic energy. We could, in fact, have used an expression for energy instead of a kinematic relation to solve part (b). We will do this in later examples.

Helicopter pilots are quite familiar with rotational kinetic energy. They know, for example, that a point of no return will be reached if they allow their blades to slow below a critical angular velocity during flight. The blades lose lift, and it is impossible to immediately get the blades spinning fast enough to regain it. Rotational kinetic energy must be supplied to the blades to get them to rotate faster, and enough energy cannot be supplied in time to avoid a crash. Because of weight limitations, helicopter engines are too small to supply both the energy needed for lift and to replenish the rotational kinetic energy of the blades once they have slowed down. The rotational kinetic energy is put into them before takeoff and must not be allowed to drop below this crucial level. One possible way to avoid a crash is to use the gravitational potential energy of the helicopter to replenish the rotational kinetic energy of the blades by losing altitude and aligning the blades so that the helicopter is spun up in the descent. Of course, if the helicopter's altitude is too low, then there is insufficient time for the blade to regain lift before reaching the ground.

Take-Home Experiment

Rotational motion can be observed in wrenches, clocks, wheels or spools on axels, and seesaws. Choose an object or system that exhibits rotational motion and plan an experiment to test how torque affects angular velocity. How will you create and measure different amounts of torque? How will you measure angular velocity? Remember that

$\text{net } \tau = I\alpha$, $I \propto mr^2$, and $\omega = \frac{v}{r}$.

Problem-Solving Strategy for Rotational Energy

1. *Determine that energy or work is involved in the rotation.*
2. *Determine the system of interest.* A sketch usually helps.
3. *Analyze the situation to determine the types of work and energy involved.*

4. *For closed systems, mechanical energy is conserved.* That is, $\mathrm{KE_i + PE_i = KE_f + PE_f}$. Note that $\mathrm{KE_i}$ and $\mathrm{KE_f}$ may each include translational and rotational contributions.

5. *For open systems,* mechanical energy may not be conserved, and other forms of energy (referred to previously as OE), such as heat transfer, may enter or leave the system. Determine what they are, and calculate them as necessary.

6. *Eliminate terms wherever possible to simplify the algebra.*

7. *Check the answer to see if it is reasonable.*

Example 10.9 Calculating Helicopter Energies

A typical small rescue helicopter, similar to the one in **Figure 10.18**, has four blades, each is 4.00 m long and has a mass of 50.0 kg. The blades can be approximated as thin rods that rotate about one end of an axis perpendicular to their length. The helicopter has a total loaded mass of 1000 kg. (a) Calculate the rotational kinetic energy in the blades when they rotate at 300 rpm. (b) Calculate the translational kinetic energy of the helicopter when it flies at 20.0 m/s, and compare it with the rotational energy in the blades. (c) To what height could the helicopter be raised if all of the rotational kinetic energy could be used to lift it?

Strategy

Rotational and translational kinetic energies can be calculated from their definitions. The last part of the problem relates to the idea that energy can change form, in this case from rotational kinetic energy to gravitational potential energy.

Solution for (a)

The rotational kinetic energy is

$$\mathrm{KE_{rot}} = \tfrac{1}{2}I\omega^2. \tag{10.73}$$

We must convert the angular velocity to radians per second and calculate the moment of inertia before we can find $\mathrm{KE_{rot}}$. The angular velocity ω is

$$\omega = \frac{300 \text{ rev}}{1.00 \text{ min}} \cdot \frac{2\pi \text{ rad}}{1 \text{ rev}} \cdot \frac{1.00 \text{ min}}{60.0 \text{ s}} = 31.4\frac{\text{rad}}{\text{s}}. \tag{10.74}$$

The moment of inertia of one blade will be that of a thin rod rotated about its end, found in **Figure 10.12**. The total I is four times this moment of inertia, because there are four blades. Thus,

$$I = 4\frac{M\ell^2}{3} = 4\times\frac{(50.0 \text{ kg})(4.00 \text{ m})^2}{3} = 1067 \text{ kg} \cdot \text{m}^2. \tag{10.75}$$

Entering ω and I into the expression for rotational kinetic energy gives

$$\begin{aligned} \mathrm{KE_{rot}} &= 0.5(1067 \text{ kg} \cdot \text{m}^2)(31.4 \text{ rad/s})^2 \\ &= 5.26\times10^5 \text{ J} \end{aligned} \tag{10.76}$$

Solution for (b)

Translational kinetic energy was defined in **Uniform Circular Motion and Gravitation**. Entering the given values of mass and velocity, we obtain

$$\mathrm{KE_{trans}} = \tfrac{1}{2}mv^2 = (0.5)(1000 \text{ kg})(20.0 \text{ m/s})^2 = 2.00\times10^5 \text{ J}. \tag{10.77}$$

To compare kinetic energies, we take the ratio of translational kinetic energy to rotational kinetic energy. This ratio is

$$\frac{2.00\times10^5 \text{ J}}{5.26\times10^5 \text{ J}} = 0.380. \tag{10.78}$$

Solution for (c)

At the maximum height, all rotational kinetic energy will have been converted to gravitational energy. To find this height, we equate those two energies:

$$\mathrm{KE_{rot} = PE_{grav}} \tag{10.79}$$

or

$$\tfrac{1}{2}I\omega^2 = mgh. \tag{10.80}$$

We now solve for h and substitute known values into the resulting equation

$$h = \frac{\frac{1}{2}I\omega^2}{mg} = \frac{5.26\times10^5 \text{ J}}{(1000 \text{ kg})(9.80 \text{ m/s}^2)} = 53.7 \text{ m}. \tag{10.81}$$

Discussion

The ratio of translational energy to rotational kinetic energy is only 0.380. This ratio tells us that most of the kinetic energy of the helicopter is in its spinning blades—something you probably would not suspect. The 53.7 m height to which the helicopter could be raised with the rotational kinetic energy is also impressive, again emphasizing the amount of rotational kinetic energy in the blades.

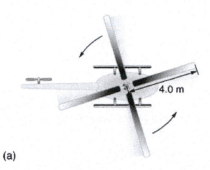

(a)

(b)

Figure 10.18 The first image shows how helicopters store large amounts of rotational kinetic energy in their blades. This energy must be put into the blades before takeoff and maintained until the end of the flight. The engines do not have enough power to simultaneously provide lift and put significant rotational energy into the blades. The second image shows a helicopter from the Auckland Westpac Rescue Helicopter Service. Over 50,000 lives have been saved since its operations beginning in 1973. Here, a water rescue operation is shown. (credit: 111 Emergency, Flickr)

Making Connections

Conservation of energy includes rotational motion, because rotational kinetic energy is another form of KE. **Uniform Circular Motion and Gravitation** has a detailed treatment of conservation of energy.

How Thick Is the Soup? Or Why Don't All Objects Roll Downhill at the Same Rate?

One of the quality controls in a tomato soup factory consists of rolling filled cans down a ramp. If they roll too fast, the soup is too thin. Why should cans of identical size and mass roll down an incline at different rates? And why should the thickest soup roll the slowest?

The easiest way to answer these questions is to consider energy. Suppose each can starts down the ramp from rest. Each can starting from rest means each starts with the same gravitational potential energy PE_{grav}, which is converted entirely to KE, provided each rolls without slipping. KE, however, can take the form of KE_{trans} or KE_{rot}, and total KE is the sum of the two. If a can rolls down a ramp, it puts part of its energy into rotation, leaving less for translation. Thus, the can goes slower than it would if it slid down. Furthermore, the thin soup does not rotate, whereas the thick soup does, because it sticks to the can. The thick soup thus puts more of the can's original gravitational potential energy into rotation than the thin soup, and the can rolls more slowly, as seen in **Figure 10.19**.

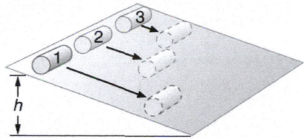

Figure 10.19 Three cans of soup with identical masses race down an incline. The first can has a low friction coating and does not roll but just slides down the incline. It wins because it converts its entire PE into translational KE. The second and third cans both roll down the incline without slipping. The second can contains thin soup and comes in second because part of its initial PE goes into rotating the can (but not the thin soup). The third can contains thick soup. It comes in third because the soup rotates along with the can, taking even more of the initial PE for rotational KE, leaving less for translational KE.

Assuming no losses due to friction, there is only one force doing work—gravity. Therefore the total work done is the change in kinetic energy. As the cans start moving, the potential energy is changing into kinetic energy. Conservation of energy gives

$$\text{PE}_\text{i} = \text{KE}_\text{f}. \tag{10.82}$$

More specifically,

$$\text{PE}_\text{grav} = \text{KE}_\text{trans} + \text{KE}_\text{rot} \tag{10.83}$$

or

$$mgh = \tfrac{1}{2}mv^2 + \tfrac{1}{2}I\omega^2. \tag{10.84}$$

So, the initial mgh is divided between translational kinetic energy and rotational kinetic energy; and the greater I is, the less energy goes into translation. If the can slides down without friction, then $\omega = 0$ and all the energy goes into translation; thus, the can goes faster.

Take-Home Experiment

Locate several cans each containing different types of food. First, predict which can will win the race down an inclined plane and explain why. See if your prediction is correct. You could also do this experiment by collecting several empty cylindrical containers of the same size and filling them with different materials such as wet or dry sand.

Example 10.10 Calculating the Speed of a Cylinder Rolling Down an Incline

Calculate the final speed of a solid cylinder that rolls down a 2.00-m-high incline. The cylinder starts from rest, has a mass of 0.750 kg, and has a radius of 4.00 cm.

Strategy

We can solve for the final velocity using conservation of energy, but we must first express rotational quantities in terms of translational quantities to end up with v as the only unknown.

Solution

Conservation of energy for this situation is written as described above:

$$mgh = \tfrac{1}{2}mv^2 + \tfrac{1}{2}I\omega^2. \tag{10.85}$$

Before we can solve for v , we must get an expression for I from **Figure 10.12**. Because v and ω are related (note here that the cylinder is rolling without slipping), we must also substitute the relationship $\omega = v / R$ into the expression. These substitutions yield

$$mgh = \tfrac{1}{2}mv^2 + \tfrac{1}{2}\left(\tfrac{1}{2}mR^2\right)\left(\frac{v^2}{R^2}\right). \tag{10.86}$$

Interestingly, the cylinder's radius R and mass m cancel, yielding

$$gh = \tfrac{1}{2}v^2 + \tfrac{1}{4}v^2 = \tfrac{3}{4}v^2. \tag{10.87}$$

Solving algebraically, the equation for the final velocity v gives

$$v = \left(\frac{4gh}{3}\right)^{1/2}.$$ (10.88)

Substituting known values into the resulting expression yields

$$v = \left[\frac{4\left(9.80 \text{ m/s}^2\right)(2.00 \text{ m})}{3}\right]^{1/2} = 5.11 \text{ m/s}.$$ (10.89)

Discussion

Because m and R cancel, the result $v = \left(\frac{4}{3}gh\right)^{1/2}$ is valid for any solid cylinder, implying that all solid cylinders will roll down an incline at the same rate independent of their masses and sizes. (Rolling cylinders down inclines is what Galileo actually did to show that objects fall at the same rate independent of mass.) Note that if the cylinder slid without friction down the incline without rolling, then the entire gravitational potential energy would go into translational kinetic energy. Thus, $\frac{1}{2}mv^2 = mgh$ and $v = (2gh)^{1/2}$, which is 22% greater than $(4gh/3)^{1/2}$. That is, the cylinder would go faster at the bottom.

Check Your Understanding

Analogy of Rotational and Translational Kinetic Energy

Is rotational kinetic energy completely analogous to translational kinetic energy? What, if any, are their differences? Give an example of each type of kinetic energy.

Solution

Yes, rotational and translational kinetic energy are exact analogs. They both are the energy of motion involved with the coordinated (non-random) movement of mass relative to some reference frame. The only difference between rotational and translational kinetic energy is that translational is straight line motion while rotational is not. An example of both kinetic and translational kinetic energy is found in a bike tire while being ridden down a bike path. The rotational motion of the tire means it has rotational kinetic energy while the movement of the bike along the path means the tire also has translational kinetic energy. If you were to lift the front wheel of the bike and spin it while the bike is stationary, then the wheel would have only rotational kinetic energy relative to the Earth.

PhET Explorations: My Solar System

Build your own system of heavenly bodies and watch the gravitational ballet. With this orbit simulator, you can set initial positions, velocities, and masses of 2, 3, or 4 bodies, and then see them orbit each other.

Figure 10.20 My Solar System (http://cnx.org/content/m55188/1.5/my-solar-system_en.jar)

10.5 Angular Momentum and Its Conservation

Learning Objectives

By the end of this section, you will be able to:

- Understand the analogy between angular momentum and linear momentum.
- Observe the relationship between torque and angular momentum.
- Apply the law of conservation of angular momentum.

The information presented in this section supports the following AP® learning objectives and science practices:

- **4.D.2.1** The student is able to describe a model of a rotational system and use that model to analyze a situation in which angular momentum changes due to interaction with other objects or systems. **(S.P. 1.2, 1.4)**
- **4.D.2.2** The student is able to plan a data collection and analysis strategy to determine the change in angular momentum of a system and relate it to interactions with other objects and systems. **(S.P. 2.2)**
- **4.D.3.1** The student is able to use appropriate mathematical routines to calculate values for initial or final angular

momentum, or change in angular momentum of a system, or average torque or time during which the torque is exerted in analyzing a situation involving torque and angular momentum. **(S.P. 2.2)**
- **4.D.3.2** The student is able to plan a data collection strategy designed to test the relationship between the change in angular momentum of a system and the product of the average torque applied to the system and the time interval during which the torque is exerted. **(S.P. 4.1, 4.2)**
- **5.E.1.1** The student is able to make qualitative predictions about the angular momentum of a system for a situation in which there is no net external torque. **(S.P. 6.4, 7.2)**
- **5.E.1.2** The student is able to make calculations of quantities related to the angular momentum of a system when the net external torque on the system is zero. **(S.P. 2.1, 2.2)**
- **5.E.2.1** The student is able to describe or calculate the angular momentum and rotational inertia of a system in terms of the locations and velocities of objects that make up the system. Students are expected to do qualitative reasoning with compound objects. Students are expected to do calculations with a fixed set of extended objects and point masses. **(S.P. 2.2)**

Why does Earth keep on spinning? What started it spinning to begin with? And how does an ice skater manage to spin faster and faster simply by pulling her arms in? Why does she not have to exert a torque to spin faster? Questions like these have answers based in angular momentum, the rotational analog to linear momentum.

By now the pattern is clear—every rotational phenomenon has a direct translational analog. It seems quite reasonable, then, to define **angular momentum** L as

$$L = I\omega. \tag{10.90}$$

This equation is an analog to the definition of linear momentum as $p = mv$. Units for linear momentum are $\text{kg} \cdot \text{m/s}$ while units for angular momentum are $\text{kg} \cdot \text{m}^2/\text{s}$. As we would expect, an object that has a large moment of inertia I, such as Earth, has a very large angular momentum. An object that has a large angular velocity ω, such as a centrifuge, also has a rather large angular momentum.

Making Connections

Angular momentum is completely analogous to linear momentum, first presented in **Uniform Circular Motion and Gravitation**. It has the same implications in terms of carrying rotation forward, and it is conserved when the net external torque is zero. Angular momentum, like linear momentum, is also a property of the atoms and subatomic particles.

Example 10.11 Calculating Angular Momentum of the Earth

Strategy

No information is given in the statement of the problem; so we must look up pertinent data before we can calculate $L = I\omega$. First, according to **Figure 10.12**, the formula for the moment of inertia of a sphere is

$$I = \frac{2MR^2}{5} \tag{10.91}$$

so that

$$L = I\omega = \frac{2MR^2\omega}{5}. \tag{10.92}$$

Earth's mass M is 5.979×10^{24} kg and its radius R is 6.376×10^6 m. The Earth's angular velocity ω is, of course, exactly one revolution per day, but we must covert ω to radians per second to do the calculation in SI units.

Solution

Substituting known information into the expression for L and converting ω to radians per second gives

$$\begin{aligned} L &= 0.4\left(5.979 \times 10^{24}\ \text{kg}\right)\left(6.376 \times 10^6\ \text{m}\right)^2\left(\frac{1\ \text{rev}}{\text{d}}\right) \\ &= 9.72 \times 10^{37}\ \text{kg} \cdot \text{m}^2 \cdot \text{rev/d}. \end{aligned} \tag{10.93}$$

Substituting 2π rad for 1 rev and 8.64×10^4 s for 1 day gives

$$\begin{aligned} L &= \left(9.72 \times 10^{37}\ \text{kg} \cdot \text{m}^2\right)\left(\frac{2\pi\ \text{rad/rev}}{8.64 \times 10^4\ \text{s/d}}\right)(1\ \text{rev/d}) \\ &= 7.07 \times 10^{33}\ \text{kg} \cdot \text{m}^2/\text{s}. \end{aligned} \tag{10.94}$$

Discussion

This number is large, demonstrating that Earth, as expected, has a tremendous angular momentum. The answer is approximate, because we have assumed a constant density for Earth in order to estimate its moment of inertia.

When you push a merry-go-round, spin a bike wheel, or open a door, you exert a torque. If the torque you exert is greater than opposing torques, then the rotation accelerates, and angular momentum increases. The greater the net torque, the more rapid the increase in L. The relationship between torque and angular momentum is

$$\text{net } \tau = \frac{\Delta L}{\Delta t}. \tag{10.95}$$

This expression is exactly analogous to the relationship between force and linear momentum, $F = \Delta p / \Delta t$. The equation $\text{net } \tau = \frac{\Delta L}{\Delta t}$ is very fundamental and broadly applicable. It is, in fact, the rotational form of Newton's second law.

Example 10.12 Calculating the Torque Putting Angular Momentum Into a Lazy Susan

Figure 10.21 shows a Lazy Susan food tray being rotated by a person in quest of sustenance. Suppose the person exerts a 2.50 N force perpendicular to the lazy Susan's 0.260-m radius for 0.150 s. (a) What is the final angular momentum of the lazy Susan if it starts from rest, assuming friction is negligible? (b) What is the final angular velocity of the lazy Susan, given that its mass is 4.00 kg and assuming its moment of inertia is that of a disk?

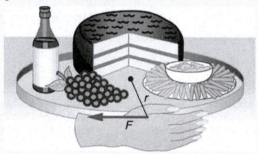

Figure 10.21 A partygoer exerts a torque on a lazy Susan to make it rotate. The equation $\text{net } \tau = \frac{\Delta L}{\Delta t}$ gives the relationship between torque and the angular momentum produced.

Strategy

We can find the angular momentum by solving $\text{net } \tau = \frac{\Delta L}{\Delta t}$ for ΔL, and using the given information to calculate the torque. The final angular momentum equals the change in angular momentum, because the lazy Susan starts from rest. That is, $\Delta L = L$. To find the final velocity, we must calculate ω from the definition of L in $L = I\omega$.

Solution for (a)

Solving $\text{net } \tau = \frac{\Delta L}{\Delta t}$ for ΔL gives

$$\Delta L = (\text{net } \tau)\Delta t. \tag{10.96}$$

Because the force is perpendicular to r, we see that $\text{net } \tau = rF$, so that

$$\begin{aligned} L &= rF\Delta t = (0.260 \text{ m})(2.50 \text{ N})(0.150 \text{ s}) \\ &= 9.75 \times 10^{-2} \text{ kg} \cdot \text{m}^2/\text{s}. \end{aligned} \tag{10.97}$$

Solution for (b)

The final angular velocity can be calculated from the definition of angular momentum,

$$L = I\omega. \tag{10.98}$$

Solving for ω and substituting the formula for the moment of inertia of a disk into the resulting equation gives

$$\omega = \frac{L}{I} = \frac{L}{\frac{1}{2}MR^2}. \tag{10.99}$$

And substituting known values into the preceding equation yields

$$\omega = \frac{9.75 \times 10^{-2} \text{ kg} \cdot \text{m}^2/\text{s}}{(0.500)(4.00 \text{ kg})(0.260 \text{ m})} = 0.721 \text{ rad/s}. \qquad (10.100)$$

Discussion

Note that the imparted angular momentum does not depend on any property of the object but only on torque and time. The final angular velocity is equivalent to one revolution in 8.71 s (determination of the time period is left as an exercise for the reader), which is about right for a lazy Susan.

Take-Home Experiment

Plan an experiment to analyze changes to a system's angular momentum. Choose a system capable of rotational motion such as a lazy Susan or a merry-go-round. Predict how the angular momentum of this system will change when you add an object to the lazy Susan or jump onto the merry-go-round. What variables can you control? What are you measuring? In other words, what are your independent and dependent variables? Are there any independent variables that it would be useful to keep constant (angular velocity, perhaps)? Collect data in order to calculate or estimate the angular momentum of your system when in motion. What do you observe? Collect data in order to calculate the change in angular momentum as a result of the interaction you performed.

Using your data, how does the angular momentum vary with the size and location of an object added to the rotating system?

Example 10.13 Calculating the Torque in a Kick

The person whose leg is shown in **Figure 10.22** kicks his leg by exerting a 2000-N force with his upper leg muscle. The effective perpendicular lever arm is 2.20 cm. Given the moment of inertia of the lower leg is $1.25 \text{ kg} \cdot \text{m}^2$, (a) find the angular acceleration of the leg. (b) Neglecting the gravitational force, what is the rotational kinetic energy of the leg after it has rotated through $57.3°$ (1.00 rad)?

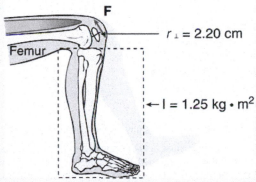

Figure 10.22 The muscle in the upper leg gives the lower leg an angular acceleration and imparts rotational kinetic energy to it by exerting a torque about the knee. $\mathbf{F}$ is a vector that is perpendicular to r. This example examines the situation.

Strategy

The angular acceleration can be found using the rotational analog to Newton's second law, or $\alpha = \text{net } \tau / I$. The moment of inertia I is given and the torque can be found easily from the given force and perpendicular lever arm. Once the angular acceleration α is known, the final angular velocity and rotational kinetic energy can be calculated.

Solution to (a)

From the rotational analog to Newton's second law, the angular acceleration α is

$$\alpha = \frac{\text{net } \tau}{I}. \qquad (10.101)$$

Because the force and the perpendicular lever arm are given and the leg is vertical so that its weight does not create a torque, the net torque is thus

$$
\begin{aligned}
\text{net } \tau &= r_\perp F \\
&= (0.0220 \text{ m})(2000 \text{ N}) \\
&= 44.0 \text{ N} \cdot \text{m}.
\end{aligned}
\qquad (10.102)
$$

Substituting this value for the torque and the given value for the moment of inertia into the expression for α gives

$$\alpha = \frac{44.0 \text{ N} \cdot \text{m}}{1.25 \text{ kg} \cdot \text{m}^2} = 35.2 \text{ rad/s}^2.$$ (10.103)

Solution to (b)

The final angular velocity can be calculated from the kinematic expression

$$\omega^2 = \omega_0{}^2 + 2\alpha\theta$$ (10.104)

or

$$\omega^2 = 2\alpha\theta$$ (10.105)

because the initial angular velocity is zero. The kinetic energy of rotation is

$$\text{KE}_{\text{rot}} = \frac{1}{2}I\omega^2$$ (10.106)

so it is most convenient to use the value of ω^2 just found and the given value for the moment of inertia. The kinetic energy is then

$$\begin{aligned} \text{KE}_{\text{rot}} &= 0.5\left(1.25 \text{ kg} \cdot \text{m}^2\right)\left(70.4 \text{ rad}^2/\text{s}^2\right) \\ &= 44.0 \text{ J} \end{aligned}$$ (10.107)

Discussion

These values are reasonable for a person kicking his leg starting from the position shown. The weight of the leg can be neglected in part (a) because it exerts no torque when the center of gravity of the lower leg is directly beneath the pivot in the knee. In part (b), the force exerted by the upper leg is so large that its torque is much greater than that created by the weight of the lower leg as it rotates. The rotational kinetic energy given to the lower leg is enough that it could give a ball a significant velocity by transferring some of this energy in a kick.

Making Connections: Conservation Laws

Angular momentum, like energy and linear momentum, is conserved. This universally applicable law is another sign of underlying unity in physical laws. Angular momentum is conserved when net external torque is zero, just as linear momentum is conserved when the net external force is zero.

Conservation of Angular Momentum

We can now understand why Earth keeps on spinning. As we saw in the previous example, $\Delta L = (\text{net } \tau)\Delta t$. This equation means that, to change angular momentum, a torque must act over some period of time. Because Earth has a large angular momentum, a large torque acting over a long time is needed to change its rate of spin. So what external torques are there? Tidal friction exerts torque that is slowing Earth's rotation, but tens of millions of years must pass before the change is very significant. Recent research indicates the length of the day was 18 h some 900 million years ago. Only the tides exert significant retarding torques on Earth, and so it will continue to spin, although ever more slowly, for many billions of years.

What we have here is, in fact, another conservation law. If the net torque is *zero*, then angular momentum is constant or *conserved*. We can see this rigorously by considering net $\tau = \dfrac{\Delta L}{\Delta t}$ for the situation in which the net torque is zero. In that case,

$$\text{net}\tau = 0$$ (10.108)

implying that

$$\frac{\Delta L}{\Delta t} = 0.$$ (10.109)

If the change in angular momentum ΔL is zero, then the angular momentum is constant; thus,

$$L = \text{constant} \ (\text{net } \tau = 0)$$ (10.110)

or

$$L = L'(\text{net}\tau = 0).$$ (10.111)

These expressions are the **law of conservation of angular momentum**. Conservation laws are as scarce as they are important.

An example of conservation of angular momentum is seen in **Figure 10.23**, in which an ice skater is executing a spin. The net torque on her is very close to zero, because there is relatively little friction between her skates and the ice and because the friction is exerted very close to the pivot point. (Both F and r are small, and so τ is negligibly small.) Consequently, she can spin for quite some time. She can do something else, too. She can increase her rate of spin by pulling her arms and legs in. Why does pulling her arms and legs in increase her rate of spin? The answer is that her angular momentum is constant, so that

$$L = L'.$$ (10.112)

Expressing this equation in terms of the moment of inertia,

$$I\omega = I'\omega',$$ (10.113)

where the primed quantities refer to conditions after she has pulled in her arms and reduced her moment of inertia. Because I' is smaller, the angular velocity ω' must increase to keep the angular momentum constant. The change can be dramatic, as the following example shows.

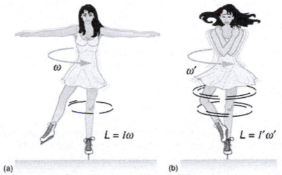

Figure 10.23 (a) An ice skater is spinning on the tip of her skate with her arms extended. Her angular momentum is conserved because the net torque on her is negligibly small. In the next image, her rate of spin increases greatly when she pulls in her arms, decreasing her moment of inertia. The work she does to pull in her arms results in an increase in rotational kinetic energy.

Example 10.14 Calculating the Angular Momentum of a Spinning Skater

Suppose an ice skater, such as the one in Figure 10.23, is spinning at 0.800 rev/ s with her arms extended. She has a moment of inertia of $2.34 \text{ kg} \cdot \text{m}^2$ with her arms extended and of $0.363 \text{ kg} \cdot \text{m}^2$ with her arms close to her body. (These moments of inertia are based on reasonable assumptions about a 60.0-kg skater.) (a) What is her angular velocity in revolutions per second after she pulls in her arms? (b) What is her rotational kinetic energy before and after she does this?

Strategy

In the first part of the problem, we are looking for the skater's angular velocity ω' after she has pulled in her arms. To find this quantity, we use the conservation of angular momentum and note that the moments of inertia and initial angular velocity are given. To find the initial and final kinetic energies, we use the definition of rotational kinetic energy given by

$$KE_{\text{rot}} = \tfrac{1}{2}I\omega^2.$$ (10.114)

Solution for (a)

Because torque is negligible (as discussed above), the conservation of angular momentum given in $I\omega = I'\omega'$ is applicable. Thus,

$$L = L'$$ (10.115)

or

$$I\omega = I'\omega'$$ (10.116)

Solving for ω' and substituting known values into the resulting equation gives

$$\omega' = \frac{I}{I'}\omega = \left(\frac{2.34 \text{ kg} \cdot \text{m}^2}{0.363 \text{ kg} \cdot \text{m}^2}\right)(0.800 \text{ rev/s})$$ (10.117)
$$= 5.16 \text{ rev/s}.$$

Solution for (b)

Rotational kinetic energy is given by

$$KE_{\text{rot}} = \tfrac{1}{2}I\omega^2.$$ (10.118)

The initial value is found by substituting known values into the equation and converting the angular velocity to rad/s:

$$KE_{rot} = (0.5)(2.34 \text{ kg} \cdot \text{m}^2)((0.800 \text{ rev/s})(2\pi \text{ rad/rev}))^2$$
$$= 29.6 \text{ J}.$$

(10.119)

The final rotational kinetic energy is

$$KE_{rot}' = \frac{1}{2}I'\omega'^2.$$

(10.120)

Substituting known values into this equation gives

$$KE_{rot}' = (0.5)(0.363 \text{ kg} \cdot \text{m}^2)[(5.16 \text{ rev/s})(2\pi \text{ rad/rev})]^2$$
$$= 191 \text{ J}.$$

(10.121)

Discussion

In both parts, there is an impressive increase. First, the final angular velocity is large, although most world-class skaters can achieve spin rates about this great. Second, the final kinetic energy is much greater than the initial kinetic energy. The increase in rotational kinetic energy comes from work done by the skater in pulling in her arms. This work is internal work that depletes some of the skater's food energy.

There are several other examples of objects that increase their rate of spin because something reduced their moment of inertia. Tornadoes are one example. Storm systems that create tornadoes are slowly rotating. When the radius of rotation narrows, even in a local region, angular velocity increases, sometimes to the furious level of a tornado. Earth is another example. Our planet was born from a huge cloud of gas and dust, the rotation of which came from turbulence in an even larger cloud. Gravitational forces caused the cloud to contract, and the rotation rate increased as a result. (See **Figure 10.24**.)

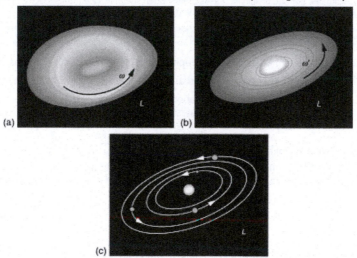

Figure 10.24 The Solar System coalesced from a cloud of gas and dust that was originally rotating. The orbital motions and spins of the planets are in the same direction as the original spin and conserve the angular momentum of the parent cloud.

In case of human motion, one would not expect angular momentum to be conserved when a body interacts with the environment as its foot pushes off the ground. Astronauts floating in space aboard the International Space Station have no angular momentum relative to the inside of the ship if they are motionless. Their bodies will continue to have this zero value no matter how they twist about as long as they do not give themselves a push off the side of the vessel.

Check Your Undestanding

Is angular momentum completely analogous to linear momentum? What, if any, are their differences?

Solution
Yes, angular and linear momentums are completely analogous. While they are exact analogs they have different units and are not directly inter-convertible like forms of energy are.

10.6 Collisions of Extended Bodies in Two Dimensions

Learning Objectives

By the end of this section, you will be able to:

- Observe collisions of extended bodies in two dimensions.
- Examine collisions at the point of percussion.

The information presented in this section supports the following AP® learning objectives and science practices:

- **3.F.3.1** The student is able to predict the behavior of rotational collision situations by the same processes that are used to analyze linear collision situations using an analogy between impulse and change of linear momentum and angular impulse and change of angular momentum. **(S.P. 6.4, 7.2)**
- **3.F.3.2** In an unfamiliar context or using representations beyond equations, the student is able to justify the selection of a mathematical routine to solve for the change in angular momentum of an object caused by torques exerted on the object. **(S.P. 2.1)**
- **3.F.3.3** The student is able to plan data collection and analysis strategies designed to test the relationship between torques exerted on an object and the change in angular momentum of that object. **(S.P. 4.1, 4.2,, 5.1, 5.3)**
- **4.D.2.1** The student is able to describe a model of a rotational system and use that model to analyze a situation in which angular momentum changes due to interaction with other objects or systems. **(S.P. 1.2, 1.4)**
- **4.D.2.2** The student is able to plan a data collection and analysis strategy to determine the change in angular momentum of a system and relate it to interactions with other objects and systems. **(S.P. 2.2)**

Bowling pins are sent flying and spinning when hit by a bowling ball—angular momentum as well as linear momentum and energy have been imparted to the pins. (See **Figure 10.25**). Many collisions involve angular momentum. Cars, for example, may spin and collide on ice or a wet surface. Baseball pitchers throw curves by putting spin on the baseball. A tennis player can put a lot of top spin on the tennis ball which causes it to dive down onto the court once it crosses the net. We now take a brief look at what happens when objects that can rotate collide.

Consider the relatively simple collision shown in **Figure 10.26**, in which a disk strikes and adheres to an initially motionless stick nailed at one end to a frictionless surface. After the collision, the two rotate about the nail. There is an unbalanced external force on the system at the nail. This force exerts no torque because its lever arm r is zero. Angular momentum is therefore conserved in the collision. Kinetic energy is not conserved, because the collision is inelastic. It is possible that momentum is not conserved either because the force at the nail may have a component in the direction of the disk's initial velocity. Let us examine a case of rotation in a collision in **Example 10.15**.

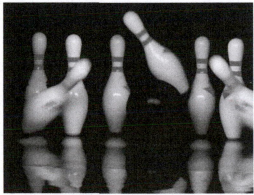

Figure 10.25 The bowling ball causes the pins to fly, some of them spinning violently. (credit: Tinou Bao, Flickr)

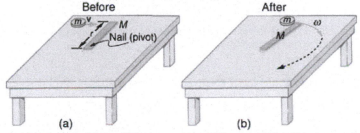

Figure 10.26 (a) A disk slides toward a motionless stick on a frictionless surface. (b) The disk hits the stick at one end and adheres to it, and they rotate together, pivoting around the nail. Angular momentum is conserved for this inelastic collision because the surface is frictionless and the unbalanced external force at the nail exerts no torque.

Example 10.15 Rotation in a Collision

Suppose the disk in **Figure 10.26** has a mass of 50.0 g and an initial velocity of 30.0 m/s when it strikes the stick that is 1.20 m long and 2.00 kg.

(a) What is the angular velocity of the two after the collision?

(b) What is the kinetic energy before and after the collision?

(c) What is the total linear momentum before and after the collision?

Strategy for (a)

We can answer the first question using conservation of angular momentum as noted. Because angular momentum is $I\omega$, we can solve for angular velocity.

Solution for (a)

Conservation of angular momentum states

$$L = L', \tag{10.122}$$

where primed quantities stand for conditions after the collision and both momenta are calculated relative to the pivot point. The initial angular momentum of the system of stick-disk is that of the disk just before it strikes the stick. That is,

$$L = I\omega, \tag{10.123}$$

where I is the moment of inertia of the disk and ω is its angular velocity around the pivot point. Now, $I = mr^2$ (taking the disk to be approximately a point mass) and $\omega = v/r$, so that

$$L = mr^2 \frac{v}{r} = mvr. \tag{10.124}$$

After the collision,

$$L' = I\prime\omega\prime. \tag{10.125}$$

It is ω' that we wish to find. Conservation of angular momentum gives

$$I'\omega' = mvr. \tag{10.126}$$

Rearranging the equation yields

$$\omega' = \frac{mvr}{I'} \tag{10.127}$$

where I' is the moment of inertia of the stick and disk stuck together, which is the sum of their individual moments of inertia about the nail. **Figure 10.12** gives the formula for a rod rotating around one end to be $I = Mr^2/3$. Thus,

$$I\prime = mr^2 + \frac{Mr^2}{3} = \left(m + \frac{M}{3}\right)r^2. \tag{10.128}$$

Entering known values in this equation yields,

$$I' = (0.0500 \text{ kg} + 0.667 \text{ kg})(1.20 \text{ m})^2 = 1.032 \text{ kg} \cdot \text{m}^2. \tag{10.129}$$

The value of I' is now entered into the expression for ω', which yields

$$\omega' = \frac{mvr}{I'} = \frac{(0.0500 \text{ kg})(30.0 \text{ m/s})(1.20 \text{ m})}{1.032 \text{ kg} \cdot \text{m}^2} \tag{10.130}$$

$$= 1.744 \text{ rad/s} \approx 1.74 \text{ rad/s}.$$

Strategy for (b)

The kinetic energy before the collision is the incoming disk's translational kinetic energy, and after the collision, it is the rotational kinetic energy of the two stuck together.

Solution for (b)

First, we calculate the translational kinetic energy by entering given values for the mass and speed of the incoming disk.

$$\text{KE} = \frac{1}{2}mv^2 = (0.500)(0.0500 \text{ kg})(30.0 \text{ m/s})^2 = 22.5 \text{ J} \tag{10.131}$$

After the collision, the rotational kinetic energy can be found because we now know the final angular velocity and the final moment of inertia. Thus, entering the values into the rotational kinetic energy equation gives

$$\text{KE}' = \frac{1}{2}I'\omega'^2 = (0.5)\left(1.032 \text{ kg} \cdot \text{m}^2\right)\left(1.744\frac{\text{rad}}{\text{s}}\right)^2 \tag{10.132}$$

$$= 1.57 \text{ J}.$$

Strategy for (c)

The linear momentum before the collision is that of the disk. After the collision, it is the sum of the disk's momentum and that of the center of mass of the stick.

Solution of (c)

Before the collision, then, linear momentum is

$$p = mv = (0.0500 \text{ kg})(30.0 \text{ m/s}) = 1.50 \text{ kg} \cdot \text{m/s}. \tag{10.133}$$

After the collision, the disk and the stick's center of mass move in the same direction. The total linear momentum is that of the disk moving at a new velocity $v' = r\omega'$ plus that of the stick's center of mass,

which moves at half this speed because $v_{\text{CM}} = \left(\frac{r}{2}\right)\omega' = \frac{v'}{2}$. Thus,

$$p' = mv' + Mv_{\text{CM}} = mv' + \frac{Mv'}{2}. \tag{10.134}$$

Gathering similar terms in the equation yields,

$$p' = \left(m + \frac{M}{2}\right)v' \tag{10.135}$$

so that

$$p' = \left(m + \frac{M}{2}\right)r\omega'. \tag{10.136}$$

Substituting known values into the equation,

$$p' = (1.050 \text{ kg})(1.20 \text{ m})(1.744 \text{ rad/s}) = 2.20 \text{ kg} \cdot \text{m/s}. \tag{10.137}$$

Discussion

First note that the kinetic energy is less after the collision, as predicted, because the collision is inelastic. More surprising is that the momentum after the collision is actually greater than before the collision. This result can be understood if you consider how the nail affects the stick and vice versa. Apparently, the stick pushes backward on the nail when first struck by the disk. The nail's reaction (consistent with Newton's third law) is to push forward on the stick, imparting momentum to it in the same direction in which the disk was initially moving, thereby increasing the momentum of the system.

Applying the Science Practices: Rotational Collisions

When the disk in **Example 10.15** strikes the stick, it exerts a torque on the stick. It is this torque that changes the angular momentum of the stick. A greater torque would produce a greater increase in angular momentum. As you saw in **Example 10.12**, the relationship between net torque and angular momentum can be expressed as $\tau = \frac{\Delta L}{\Delta t}$, where ΔL represents

the change in angular momentum and Δt represents the time interval that it took for the angular momentum to change.

How can you test this? Design an experiment in which you apply different measurable torques to a simple system. The torque applied to the stick can be varied by changing the mass of the disk, the initial velocity of the disk, or the radius of the impact of the disk on the stick. How will you measure changes to angular momentum? Recall that angular momentum is defined as $L = I\omega$.

The above example has other implications. For example, what would happen if the disk hit very close to the nail? Obviously, a force would be exerted on the nail in the forward direction. So, when the stick is struck at the end farthest from the nail, a backward force is exerted on the nail, and when it is hit at the end nearest the nail, a forward force is exerted on the nail. Thus, striking it at a certain point in between produces no force on the nail. This intermediate point is known as the *percussion point*.

An analogous situation occurs in tennis as seen in **Figure 10.27**. If you hit a ball with the end of your racquet, the handle is pulled away from your hand. If you hit a ball much farther down, for example, on the shaft of the racquet, the handle is pushed into your palm. And if you hit the ball at the racquet's percussion point (what some people call the "sweet spot"), then little or *no* force is exerted on your hand, and there is less vibration, reducing chances of a tennis elbow. The same effect occurs for a baseball bat.

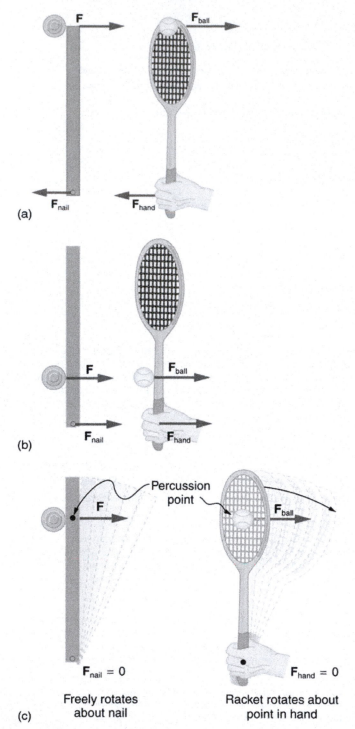

Figure 10.27 A disk hitting a stick is compared to a tennis ball being hit by a racquet. (a) When the ball strikes the racquet near the end, a backward force is exerted on the hand. (b) When the racquet is struck much farther down, a forward force is exerted on the hand. (c) When the racquet is struck at the percussion point, no force is delivered to the hand.

Check Your Understanding

Is rotational kinetic energy a vector? Justify your answer.

Solution
No, energy is always scalar whether motion is involved or not. No form of energy has a direction in space and you can see that rotational kinetic energy does not depend on the direction of motion just as linear kinetic energy is independent of the direction of motion.

10.7 Gyroscopic Effects: Vector Aspects of Angular Momentum

Learning Objectives

By the end of this section, you will be able to:

- Describe the **right-hand rule** to find the direction of angular velocity, momentum, and torque.
- Explain the gyroscopic effect.
- Study how Earth acts like a gigantic gyroscope.

The information presented in this section supports the following AP® learning objectives and science practices:

- **4.D.3.1** The student is able to use appropriate mathematical routines to calculate values for initial or final angular momentum, or change in angular momentum of a system, or average torque or time during which the torque is exerted in analyzing a situation involving torque and angular momentum. **(S.P. 2.2)**
- **4.D.3.2** The student is able to plan a data collection strategy designed to test the relationship between the change in angular momentum of a system and the product of the average torque applied to the system and the time interval during which the torque is exerted. **(S.P. 4.1, 4.2)**

Angular momentum is a vector and, therefore, *has direction as well as magnitude*. Torque affects both the direction and the magnitude of angular momentum. What is the direction of the angular momentum of a rotating object like the disk in **Figure 10.28**? The figure shows the **right-hand rule** used to find the direction of both angular momentum and angular velocity. Both $\mathbf{L}$ and ω are vectors—each has direction and magnitude. Both can be represented by arrows. The right-hand rule defines both to be perpendicular to the plane of rotation in the direction shown. Because angular momentum is related to angular velocity by $\mathbf{L} = I\omega$, the direction of $\mathbf{L}$ is the same as the direction of ω. Notice in the figure that both point along the axis of rotation.

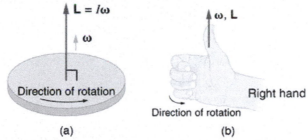

(a)   (b)

Figure 10.28 Figure (a) shows a disk is rotating counterclockwise when viewed from above. Figure (b) shows the right-hand rule. The direction of angular velocity ω size and angular momentum $\mathbf{L}$ are defined to be the direction in which the thumb of your right hand points when you curl your fingers in the direction of the disk's rotation as shown.

Now, recall that torque changes angular momentum as expressed by

$$\text{net } \boldsymbol{\tau} = \frac{\Delta \mathbf{L}}{\Delta t}. \tag{10.138}$$

This equation means that the direction of $\Delta \mathbf{L}$ is the same as the direction of the torque $\boldsymbol{\tau}$ that creates it. This result is illustrated in **Figure 10.29**, which shows the direction of torque and the angular momentum it creates.

Let us now consider a bicycle wheel with a couple of handles attached to it, as shown in **Figure 10.30**. (This device is popular in demonstrations among physicists, because it does unexpected things.) With the wheel rotating as shown, its angular momentum is to the woman's left. Suppose the person holding the wheel tries to rotate it as in the figure. Her natural expectation is that the wheel will rotate in the direction she pushes it—but what happens is quite different. The forces exerted create a torque that is horizontal toward the person, as shown in **Figure 10.30**(a). This torque creates a change in angular momentum $\mathbf{L}$ in the same direction, perpendicular to the original angular momentum $\mathbf{L}$, thus changing the direction of $\mathbf{L}$ but not the magnitude of $\mathbf{L}$. **Figure 10.30** shows how $\Delta \mathbf{L}$ and $\mathbf{L}$ add, giving a new angular momentum with direction that is inclined more toward the person than before. The axis of the wheel has thus moved *perpendicular to the forces exerted on it*, instead of in the expected direction.

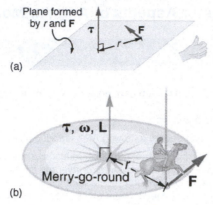

Figure 10.29 In figure (a), the torque is perpendicular to the plane formed by r and $\mathbf{F}$ and is the direction your right thumb would point to if you curled your fingers in the direction of $\mathbf{F}$. Figure (b) shows that the direction of the torque is the same as that of the angular momentum it produces.

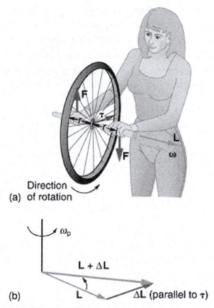

Figure 10.30 In figure (a), a person holding the spinning bike wheel lifts it with her right hand and pushes down with her left hand in an attempt to rotate the wheel. This action creates a torque directly toward her. This torque causes a change in angular momentum $\Delta \mathbf{L}$ in exactly the same direction. Figure (b) shows a vector diagram depicting how $\Delta \mathbf{L}$ and $\mathbf{L}$ add, producing a new angular momentum pointing more toward the person. The wheel moves toward the person, perpendicular to the forces she exerts on it.

Applying Science Practices: Angular Momentum and Torque

You have seen that change in angular momentum depends on the average torque applied and the time interval during which the torque is applied. Plan an experiment similar to the one shown in **Figure 10.30** to test the relationship between the change in angular momentum of a system and the product of the average torque applied to the system and the time interval during which the torque is exerted. What would you use as your test system? How could you measure applied torque? What observations could you make to help you analyze changes in angular momentum? Remember that, since angular momentum is a vector, changes can relate to its magnitude or its direction.

This same logic explains the behavior of gyroscopes. **Figure 10.31** shows the two forces acting on a spinning gyroscope. The torque produced is perpendicular to the angular momentum, thus the direction of the torque is changed, but not its magnitude. The gyroscope *precesses* around a vertical axis, since the torque is always horizontal and perpendicular to $\mathbf{L}$. If the gyroscope is *not* spinning, it acquires angular momentum in the direction of the torque ($\mathbf{L} = \Delta \mathbf{L}$), and it rotates around a horizontal axis, falling over just as we would expect.

Earth itself acts like a gigantic gyroscope. Its angular momentum is along its axis and points at Polaris, the North Star. But Earth is slowly precessing (once in about 26,000 years) due to the torque of the Sun and the Moon on its nonspherical shape.

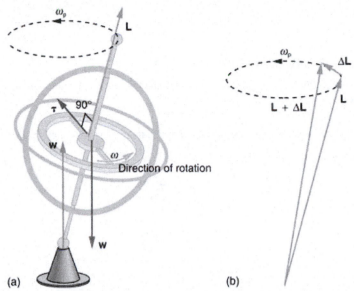

Figure 10.31 As seen in figure (a), the forces on a spinning gyroscope are its weight and the supporting force from the stand. These forces create a horizontal torque on the gyroscope, which create a change in angular momentum $\Delta\mathbf{L}$ that is also horizontal. In figure (b), $\Delta\mathbf{L}$ and $\mathbf{L}$ add to produce a new angular momentum with the same magnitude, but different direction, so that the gyroscope precesses in the direction shown instead of falling over.

Check Your Understanding

Rotational kinetic energy is associated with angular momentum? Does that mean that rotational kinetic energy is a vector?

Solution
No, energy is always a scalar whether motion is involved or not. No form of energy has a direction in space and you can see that rotational kinetic energy does not depend on the direction of motion just as linear kinetic energy is independent of the direction of motion.

Glossary

angular acceleration: the rate of change of angular velocity with time

angular momentum: the product of moment of inertia and angular velocity

change in angular velocity: the difference between final and initial values of angular velocity

kinematics of rotational motion: describes the relationships among rotation angle, angular velocity, angular acceleration, and time

law of conservation of angular momentum: angular momentum is conserved, i.e., the initial angular momentum is equal to the final angular momentum when no external torque is applied to the system

moment of inertia: mass times the square of perpendicular distance from the rotation axis; for a point mass, it is $I = mr^2$ and, because any object can be built up from a collection of point masses, this relationship is the basis for all other moments of inertia

right-hand rule: direction of angular velocity ω and angular momentum L in which the thumb of your right hand points when you curl your fingers in the direction of the disk's rotation

rotational inertia: resistance to change of rotation. The more rotational inertia an object has, the harder it is to rotate

rotational kinetic energy: the kinetic energy due to the rotation of an object. This is part of its total kinetic energy

tangential acceleration: the acceleration in a direction tangent to the circle at the point of interest in circular motion

torque: the turning effectiveness of a force

work-energy theorem: if one or more external forces act upon a rigid object, causing its kinetic energy to change from KE_1 to KE_2, then the work W done by the net force is equal to the change in kinetic energy

Section Summary

10.1 Angular Acceleration

- Uniform circular motion is the motion with a constant angular velocity $\omega = \frac{\Delta\theta}{\Delta t}$.
- In non-uniform circular motion, the velocity changes with time and the rate of change of angular velocity (i.e. angular acceleration) is $\alpha = \frac{\Delta\omega}{\Delta t}$.
- Linear or tangential acceleration refers to changes in the magnitude of velocity but not its direction, given as $a_t = \frac{\Delta v}{\Delta t}$.
- For circular motion, note that $v = r\omega$, so that

$$a_t = \frac{\Delta(r\omega)}{\Delta t}.$$

- The radius r is constant for circular motion, and so $\Delta(r\omega) = r\Delta\omega$. Thus,

$$a_t = r\frac{\Delta\omega}{\Delta t}.$$

- By definition, $\Delta\omega/\Delta t = \alpha$. Thus,

$$a_t = r\alpha$$

or

$$\alpha = \frac{a_t}{r}.$$

10.2 Kinematics of Rotational Motion

- Kinematics is the description of motion.
- The kinematics of rotational motion describes the relationships among rotation angle, angular velocity, angular acceleration, and time.
- Starting with the four kinematic equations we developed in the **One-Dimensional Kinematics**, we can derive the four rotational kinematic equations (presented together with their translational counterparts) seen in **Table 10.2**.
- In these equations, the subscript 0 denotes initial values (x_0 and t_0 are initial values), and the average angular velocity $\bar{\omega}$ and average velocity $\bar{v}$ are defined as follows:

$$\bar{\omega} = \frac{\omega_0 + \omega}{2} \text{ and } \bar{v} = \frac{v_0 + v}{2}.$$

10.3 Dynamics of Rotational Motion: Rotational Inertia

- The farther the force is applied from the pivot, the greater is the angular acceleration; angular acceleration is inversely proportional to mass.
- If we exert a force F on a point mass m that is at a distance r from a pivot point and because the force is perpendicular to r, an acceleration $a = F/m$ is obtained in the direction of F. We can rearrange this equation such that

$$F = ma,$$

and then look for ways to relate this expression to expressions for rotational quantities. We note that $a = r\alpha$, and we substitute this expression into $F=ma$, yielding

$$F=mr\alpha$$

- Torque is the turning effectiveness of a force. In this case, because F is perpendicular to r, torque is simply $\tau = rF$. If we multiply both sides of the equation above by r, we get torque on the left-hand side. That is,

$$rF = mr^2\alpha$$

or

$$\tau = mr^2\alpha.$$

- The moment of inertia I of an object is the sum of MR^2 for all the point masses of which it is composed. That is,

$$I = \sum mr^2.$$

- The general relationship among torque, moment of inertia, and angular acceleration is

$$\tau = I\alpha$$

or

$$\alpha = \frac{\text{net } \tau}{I}.$$

10.4 Rotational Kinetic Energy: Work and Energy Revisited

- The rotational kinetic energy KE_{rot} for an object with a moment of inertia I and an angular velocity ω is given by

$$KE_{rot} = \frac{1}{2} I \omega^2.$$

- Helicopters store large amounts of rotational kinetic energy in their blades. This energy must be put into the blades before takeoff and maintained until the end of the flight. The engines do not have enough power to simultaneously provide lift and put significant rotational energy into the blades.
- Work and energy in rotational motion are completely analogous to work and energy in translational motion.
- The equation for the **work-energy theorem** for rotational motion is,

$$net\ W = \frac{1}{2} I \omega^2 - \frac{1}{2} I \omega_0^2.$$

10.5 Angular Momentum and Its Conservation

- Every rotational phenomenon has a direct translational analog , likewise angular momentum L can be defined as $L = I\omega$.
- This equation is an analog to the definition of linear momentum as $p = mv$. The relationship between torque and angular momentum is $net\ \tau = \frac{\Delta L}{\Delta t}$.
- Angular momentum, like energy and linear momentum, is conserved. This universally applicable law is another sign of underlying unity in physical laws. Angular momentum is conserved when net external torque is zero, just as linear momentum is conserved when the net external force is zero.

10.6 Collisions of Extended Bodies in Two Dimensions

- Angular momentum L is analogous to linear momentum and is given by $L = I\omega$.
- Angular momentum is changed by torque, following the relationship $net\ \tau = \frac{\Delta L}{\Delta t}$.
- Angular momentum is conserved if the net torque is zero $L = constant$ (net $\tau = 0$) or $L = L'$ (net $\tau = 0$) . This equation is known as the law of conservation of angular momentum, which may be conserved in collisions.

10.7 Gyroscopic Effects: Vector Aspects of Angular Momentum

- Torque is perpendicular to the plane formed by r and $\mathbf{F}$ and is the direction your right thumb would point if you curled the fingers of your right hand in the direction of $\mathbf{F}$. The direction of the torque is thus the same as that of the angular momentum it produces.
- The gyroscope precesses around a vertical axis, since the torque is always horizontal and perpendicular to $\mathbf{L}$. If the gyroscope is not spinning, it acquires angular momentum in the direction of the torque ($\mathbf{L} = \Delta\mathbf{L}$), and it rotates about a horizontal axis, falling over just as we would expect.
- Earth itself acts like a gigantic gyroscope. Its angular momentum is along its axis and points at Polaris, the North Star.

Conceptual Questions

10.1 Angular Acceleration

1. Analogies exist between rotational and translational physical quantities. Identify the rotational term analogous to each of the following: acceleration, force, mass, work, translational kinetic energy, linear momentum, impulse.

2. Explain why centripetal acceleration changes the direction of velocity in circular motion but not its magnitude.

3. In circular motion, a tangential acceleration can change the magnitude of the velocity but not its direction. Explain your answer.

4. Suppose a piece of food is on the edge of a rotating microwave oven plate. Does it experience nonzero tangential acceleration, centripetal acceleration, or both when: (a) The plate starts to spin? (b) The plate rotates at constant angular velocity? (c) The plate slows to a halt?

10.3 Dynamics of Rotational Motion: Rotational Inertia

5. The moment of inertia of a long rod spun around an axis through one end perpendicular to its length is $ML^2/3$. Why is this moment of inertia greater than it would be if you spun a point mass M at the location of the center of mass of the rod (at $L/2$)? (That would be $ML^2/4$.)

6. Why is the moment of inertia of a hoop that has a mass M and a radius R greater than the moment of inertia of a disk that has the same mass and radius? Why is the moment of inertia of a spherical shell that has a mass M and a radius R greater than that of a solid sphere that has the same mass and radius?

7. Give an example in which a small force exerts a large torque. Give another example in which a large force exerts a small torque.

8. While reducing the mass of a racing bike, the greatest benefit is realized from reducing the mass of the tires and wheel rims. Why does this allow a racer to achieve greater accelerations than would an identical reduction in the mass of the bicycle's frame?

Figure 10.32 The image shows a side view of a racing bicycle. Can you see evidence in the design of the wheels on this racing bicycle that their moment of inertia has been purposely reduced? (credit: Jesús Rodriguez)

9. A ball slides up a frictionless ramp. It is then rolled without slipping and with the same initial velocity up another frictionless ramp (with the same slope angle). In which case does it reach a greater height, and why?

10.4 Rotational Kinetic Energy: Work and Energy Revisited

10. Describe the energy transformations involved when a yo-yo is thrown downward and then climbs back up its string to be caught in the user's hand.

11. What energy transformations are involved when a dragster engine is revved, its clutch let out rapidly, its tires spun, and it starts to accelerate forward? Describe the source and transformation of energy at each step.

12. The Earth has more rotational kinetic energy now than did the cloud of gas and dust from which it formed. Where did this energy come from?

Figure 10.33 An immense cloud of rotating gas and dust contracted under the influence of gravity to form the Earth and in the process rotational kinetic energy increased. (credit: NASA)

10.5 Angular Momentum and Its Conservation

13. When you start the engine of your car with the transmission in neutral, you notice that the car rocks in the opposite sense of the engine's rotation. Explain in terms of conservation of angular momentum. Is the angular momentum of the car conserved for long (for more than a few seconds)?

14. Suppose a child walks from the outer edge of a rotating merry-go round to the inside. Does the angular velocity of the merry-go-round increase, decrease, or remain the same? Explain your answer.

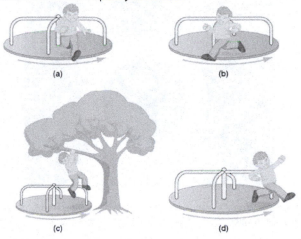

(a)

(b)

(c)

(d)

Figure 10.34 A child may jump off a merry-go-round in a variety of directions.

15. Suppose a child gets off a rotating merry-go-round. Does the angular velocity of the merry-go-round increase, decrease, or remain the same if: (a) He jumps off radially? (b) He jumps backward to land motionless? (c) He jumps straight up and hangs onto an overhead tree branch? (d) He jumps off forward, tangential to the edge? Explain your answers. (Refer to **Figure 10.34**).

16. Helicopters have a small propeller on their tail to keep them from rotating in the opposite direction of their main lifting blades. Explain in terms of Newton's third law why the helicopter body rotates in the opposite direction to the blades.

17. Whenever a helicopter has two sets of lifting blades, they rotate in opposite directions (and there will be no tail propeller). Explain why it is best to have the blades rotate in opposite directions.

18. Describe how work is done by a skater pulling in her arms during a spin. In particular, identify the force she exerts on each arm to pull it in and the distance each moves, noting that a component of the force is in the direction moved. Why is angular momentum not increased by this action?

19. When there is a global heating trend on Earth, the atmosphere expands and the length of the day increases very slightly. Explain why the length of a day increases.

20. Nearly all conventional piston engines have flywheels on them to smooth out engine vibrations caused by the thrust of individual piston firings. Why does the flywheel have this effect?

21. Jet turbines spin rapidly. They are designed to fly apart if something makes them seize suddenly, rather than transfer angular momentum to the plane's wing, possibly tearing it off. Explain how flying apart conserves angular momentum without transferring it to the wing.

22. An astronaut tightens a bolt on a satellite in orbit. He rotates in a direction opposite to that of the bolt, and the satellite rotates in the same direction as the bolt. Explain why. If a handhold is available on the satellite, can this counter-rotation be prevented? Explain your answer.

23. Competitive divers pull their limbs in and curl up their bodies when they do flips. Just before entering the water, they fully extend their limbs to enter straight down. Explain the effect of both actions on their angular velocities. Also explain the effect on their angular momenta.

ω large

ω' small

Figure 10.35 The diver spins rapidly when curled up and slows when she extends her limbs before entering the water.

24. Draw a free body diagram to show how a diver gains angular momentum when leaving the diving board.

25. In terms of angular momentum, what is the advantage of giving a football or a rifle bullet a spin when throwing or releasing it?

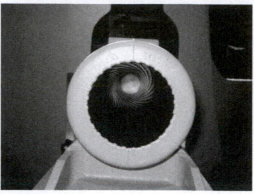

Figure 10.36 The image shows a view down the barrel of a cannon, emphasizing its rifling. Rifling in the barrel of a canon causes the projectile to spin just as is the case for rifles (hence the name for the grooves in the barrel). (credit: Elsie esq., Flickr)

10.6 Collisions of Extended Bodies in Two Dimensions

26. Describe two different collisions—one in which angular momentum is conserved, and the other in which it is not. Which condition determines whether or not angular momentum is conserved in a collision?

27. Suppose an ice hockey puck strikes a hockey stick that lies flat on the ice and is free to move in any direction. Which quantities are likely to be conserved: angular momentum, linear momentum, or kinetic energy (assuming the puck and stick are very resilient)?

28. While driving his motorcycle at highway speed, a physics student notices that pulling back lightly on the right handlebar tips the cycle to the left and produces a left turn. Explain why this happens.

10.7 Gyroscopic Effects: Vector Aspects of Angular Momentum

29. While driving his motorcycle at highway speed, a physics student notices that pulling back lightly on the right handlebar tips the cycle to the left and produces a left turn. Explain why this happens.

30. Gyroscopes used in guidance systems to indicate directions in space must have an angular momentum that does not change in direction. Yet they are often subjected to large forces and accelerations. How can the direction of their angular momentum be constant when they are accelerated?

Problems & Exercises

10.1 Angular Acceleration

1. At its peak, a tornado is 60.0 m in diameter and carries 500 km/h winds. What is its angular velocity in revolutions per second?

2. Integrated Concepts

An ultracentrifuge accelerates from rest to 100,000 rpm in 2.00 min. (a) What is its angular acceleration in rad/s^2? (b) What is the tangential acceleration of a point 9.50 cm from the axis of rotation? (c) What is the radial acceleration in m/s^2 and multiples of g of this point at full rpm?

3. Integrated Concepts

You have a grindstone (a disk) that is 90.0 kg, has a 0.340-m radius, and is turning at 90.0 rpm, and you press a steel axe against it with a radial force of 20.0 N. (a) Assuming the kinetic coefficient of friction between steel and stone is 0.20, calculate the angular acceleration of the grindstone. (b) How many turns will the stone make before coming to rest?

4. Unreasonable Results

You are told that a basketball player spins the ball with an angular acceleration of 100 rad/s^2. (a) What is the ball's final angular velocity if the ball starts from rest and the acceleration lasts 2.00 s? (b) What is unreasonable about the result? (c) Which premises are unreasonable or inconsistent?

10.2 Kinematics of Rotational Motion

5. With the aid of a string, a gyroscope is accelerated from rest to 32 rad/s in 0.40 s.

(a) What is its angular acceleration in rad/s²?

(b) How many revolutions does it go through in the process?

6. Suppose a piece of dust finds itself on a CD. If the spin rate of the CD is 500 rpm, and the piece of dust is 4.3 cm from the center, what is the total distance traveled by the dust in 3 minutes? (Ignore accelerations due to getting the CD rotating.)

7. A gyroscope slows from an initial rate of 32.0 rad/s at a rate of 0.700 rad/s^2.

(a) How long does it take to come to rest?

(b) How many revolutions does it make before stopping?

8. During a very quick stop, a car decelerates at 7.00 m/s^2.

(a) What is the angular acceleration of its 0.280-m-radius tires, assuming they do not slip on the pavement?

(b) How many revolutions do the tires make before coming to rest, given their initial angular velocity is 95.0 rad/s?

(c) How long does the car take to stop completely?

(d) What distance does the car travel in this time?

(e) What was the car's initial velocity?

(f) Do the values obtained seem reasonable, considering that this stop happens very quickly?

Figure 10.37 Yo-yos are amusing toys that display significant physics and are engineered to enhance performance based on physical laws. (credit: Beyond Neon, Flickr)

9. Everyday application: Suppose a yo-yo has a center shaft that has a 0.250 cm radius and that its string is being pulled.

(a) If the string is stationary and the yo-yo accelerates away from it at a rate of 1.50 m/s^2, what is the angular acceleration of the yo-yo?

(b) What is the angular velocity after 0.750 s if it starts from rest?

(c) The outside radius of the yo-yo is 3.50 cm. What is the tangential acceleration of a point on its edge?

10.3 Dynamics of Rotational Motion: Rotational Inertia

10. This problem considers additional aspects of example **Calculating the Effect of Mass Distribution on a Merry-Go-Round.** (a) How long does it take the father to give the merry-go-round an angular velocity of 1.50 rad/s? (b) How many revolutions must he go through to generate this velocity? (c) If he exerts a slowing force of 300 N at a radius of 1.35 m, how long would it take him to stop them?

11. Calculate the moment of inertia of a skater given the following information. (a) The 60.0-kg skater is approximated as a cylinder that has a 0.110-m radius. (b) The skater with arms extended is approximately a cylinder that is 52.5 kg, has a 0.110-m radius, and has two 0.900-m-long arms which are 3.75 kg each and extend straight out from the cylinder like rods rotated about their ends.

12. The triceps muscle in the back of the upper arm extends the forearm. This muscle in a professional boxer exerts a force of $2.00 \times 10^3 \text{ N}$ with an effective perpendicular lever arm of 3.00 cm, producing an angular acceleration of the forearm of 120 rad/s^2. What is the moment of inertia of the boxer's forearm?

13. A soccer player extends her lower leg in a kicking motion by exerting a force with the muscle above the knee in the front of her leg. She produces an angular acceleration of 30.00 rad/s^2 and her lower leg has a moment of inertia of $0.750 \text{ kg} \cdot \text{m}^2$. What is the force exerted by the muscle if its effective perpendicular lever arm is 1.90 cm?

14. Suppose you exert a force of 180 N tangential to a 0.280-m-radius 75.0-kg grindstone (a solid disk).

(a)What torque is exerted? (b) What is the angular acceleration assuming negligible opposing friction? (c) What is the angular acceleration if there is an opposing frictional force of 20.0 N exerted 1.50 cm from the axis?

15. Consider the 12.0 kg motorcycle wheel shown in **Figure 10.38**. Assume it to be approximately an annular ring with an inner radius of 0.280 m and an outer radius of 0.330 m. The motorcycle is on its center stand, so that the wheel can spin freely. (a) If the drive chain exerts a force of 2200 N at a radius of 5.00 cm, what is the angular acceleration of the wheel? (b) What is the tangential acceleration of a point on the outer edge of the tire? (c) How long, starting from rest, does it take to reach an angular velocity of 80.0 rad/s?

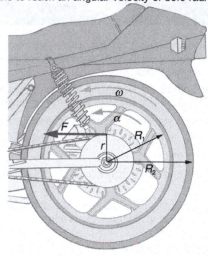

Figure 10.38 A motorcycle wheel has a moment of inertia approximately that of an annular ring.

16. Zorch, an archenemy of Superman, decides to slow Earth's rotation to once per 28.0 h by exerting an opposing force at and parallel to the equator. Superman is not immediately concerned, because he knows Zorch can only exert a force of $4.00 \times 10^7 \text{ N}$ (a little greater than a Saturn V rocket's thrust). How long must Zorch push with this force to accomplish his goal? (This period gives Superman time to devote to other villains.) Explicitly show how you follow the steps found in **Problem-Solving Strategy for Rotational Dynamics**.

17. An automobile engine can produce 200 N · m of torque. Calculate the angular acceleration produced if 95.0% of this torque is applied to the drive shaft, axle, and rear wheels of a car, given the following information. The car is suspended so that the wheels can turn freely. Each wheel acts like a 15.0 kg disk that has a 0.180 m radius. The walls of each tire act like a 2.00-kg annular ring that has inside radius of 0.180 m and outside radius of 0.320 m. The tread of each tire acts like a 10.0-kg hoop of radius 0.330 m. The 14.0-kg axle acts like a rod that has a 2.00-cm radius. The 30.0-kg drive shaft acts like a rod that has a 3.20-cm radius.

18. Starting with the formula for the moment of inertia of a rod rotated around an axis through one end perpendicular to its length $\left(I = M\ell^2/3 \right)$, prove that the moment of inertia of a rod rotated about an axis through its center perpendicular to its length is $I = M\ell^2/12$. You will find the graphics in **Figure 10.12** useful in visualizing these rotations.

19. Unreasonable Results

A gymnast doing a forward flip lands on the mat and exerts a 500-N · m torque to slow and then reverse her angular velocity. Her initial angular velocity is 10.0 rad/s, and her moment of inertia is $0.050 \text{ kg} \cdot \text{m}^2$. (a) What time is required for her to exactly reverse her spin? (b) What is unreasonable about the result? (c) Which premises are unreasonable or inconsistent?

20. Unreasonable Results

An advertisement claims that an 800-kg car is aided by its 20.0-kg flywheel, which can accelerate the car from rest to a speed of 30.0 m/s. The flywheel is a disk with a 0.150-m radius. (a) Calculate the angular velocity the flywheel must have if 95.0% of its rotational energy is used to get the car up to speed. (b) What is unreasonable about the result? (c) Which premise is unreasonable or which premises are inconsistent?

10.4 Rotational Kinetic Energy: Work and Energy Revisited

21. This problem considers energy and work aspects of **Example 10.7**—use data from that example as needed. (a) Calculate the rotational kinetic energy in the merry-go-round plus child when they have an angular velocity of 20.0 rpm. (b) Using energy considerations, find the number of revolutions the father will have to push to achieve this angular velocity starting from rest. (c) Again, using energy considerations, calculate the force the father must exert to stop the merry-go-round in two revolutions

22. What is the final velocity of a hoop that rolls without slipping down a 5.00-m-high hill, starting from rest?

23. (a) Calculate the rotational kinetic energy of Earth on its axis. (b) What is the rotational kinetic energy of Earth in its orbit around the Sun?

24. Calculate the rotational kinetic energy in the motorcycle wheel (**Figure 10.38**) if its angular velocity is 120 rad/s. Assume M = 12.0 kg, R_1 = 0.280 m, and R_2 = 0.330 m.

25. A baseball pitcher throws the ball in a motion where there is rotation of the forearm about the elbow joint as well as other movements. If the linear velocity of the ball relative to the elbow joint is 20.0 m/s at a distance of 0.480 m from the joint and the moment of inertia of the forearm is $0.500 \text{ kg} \cdot \text{m}^2$, what is the rotational kinetic energy of the forearm?

26. While punting a football, a kicker rotates his leg about the hip joint. The moment of inertia of the leg is $3.75 \text{ kg} \cdot \text{m}^2$ and its rotational kinetic energy is 175 J. (a) What is the angular velocity of the leg? (b) What is the velocity of tip of the punter's shoe if it is 1.05 m from the hip joint? (c) Explain how the football can be given a velocity greater than the tip of the shoe (necessary for a decent kick distance).

27. A bus contains a 1500 kg flywheel (a disk that has a 0.600 m radius) and has a total mass of 10,000 kg. (a) Calculate the angular velocity the flywheel must have to contain enough energy to take the bus from rest to a speed of 20.0 m/s, assuming 90.0% of the rotational kinetic energy can be transformed into translational energy. (b) How high a hill can the bus climb with this stored energy and still have a speed of 3.00 m/s at the top of the hill? Explicitly show how you follow the steps in the **Problem-Solving Strategy for Rotational Energy**.

28. A ball with an initial velocity of 8.00 m/s rolls up a hill without slipping. Treating the ball as a spherical shell, calculate the vertical height it reaches. (b) Repeat the calculation for the same ball if it slides up the hill without rolling.

29. While exercising in a fitness center, a man lies face down on a bench and lifts a weight with one lower leg by contacting the muscles in the back of the upper leg. (a) Find the angular acceleration produced given the mass lifted is 10.0 kg at a distance of 28.0 cm from the knee joint, the moment of inertia of the lower leg is $0.900 \text{ kg} \cdot \text{m}^2$, the muscle force is 1500 N, and its effective perpendicular lever arm is 3.00 cm. (b) How much work is done if the leg rotates through an angle of $20.0°$ with a constant force exerted by the muscle?

30. To develop muscle tone, a woman lifts a 2.00-kg weight held in her hand. She uses her biceps muscle to flex the lower arm through an angle of $60.0°$. (a) What is the angular acceleration if the weight is 24.0 cm from the elbow joint, her forearm has a moment of inertia of $0.250 \text{ kg} \cdot \text{m}^2$, and the net force she exerts is 750 N at an effective perpendicular lever arm of 2.00 cm? (b) How much work does she do?

31. Consider two cylinders that start down identical inclines from rest except that one is frictionless. Thus one cylinder rolls without slipping, while the other slides frictionlessly without rolling. They both travel a short distance at the bottom and then start up another incline. (a) Show that they both reach the same height on the other incline, and that this height is equal to their original height. (b) Find the ratio of the time the rolling cylinder takes to reach the height on the second incline to the time the sliding cylinder takes to reach the height on the second incline. (c) Explain why the time for the rolling motion is greater than that for the sliding motion.

32. What is the moment of inertia of an object that rolls without slipping down a 2.00-m-high incline starting from rest, and has a final velocity of 6.00 m/s? Express the moment of inertia as a multiple of MR^2, where M is the mass of the object and R is its radius.

33. Suppose a 200-kg motorcycle has two wheels like, the one described in **Example 10.15** and is heading toward a hill at a speed of 30.0 m/s. (a) How high can it coast up the hill, if you neglect friction? (b) How much energy is lost to friction if the motorcycle only gains an altitude of 35.0 m before coming to rest?

34. In softball, the pitcher throws with the arm fully extended (straight at the elbow). In a fast pitch the ball leaves the hand with a speed of 139 km/h. (a) Find the rotational kinetic energy of the pitcher's arm given its moment of inertia is $0.720 \text{ kg} \cdot \text{m}^2$ and the ball leaves the hand at a distance of 0.600 m from the pivot at the shoulder. (b) What force did the muscles exert to cause the arm to rotate if their effective perpendicular lever arm is 4.00 cm and the ball is 0.156 kg?

35. Construct Your Own Problem

Consider the work done by a spinning skater pulling her arms in to increase her rate of spin. Construct a problem in which you calculate the work done with a "force multiplied by distance" calculation and compare it to the skater's increase in kinetic energy.

10.5 Angular Momentum and Its Conservation

36. (a) Calculate the angular momentum of the Earth in its orbit around the Sun.

(b) Compare this angular momentum with the angular momentum of Earth on its axis.

37. (a) What is the angular momentum of the Moon in its orbit around Earth?

(b) How does this angular momentum compare with the angular momentum of the Moon on its axis? Remember that the Moon keeps one side toward Earth at all times.

(c) Discuss whether the values found in parts (a) and (b) seem consistent with the fact that tidal effects with Earth have caused the Moon to rotate with one side always facing Earth.

38. Suppose you start an antique car by exerting a force of 300 N on its crank for 0.250 s. What angular momentum is given to the engine if the handle of the crank is 0.300 m from the pivot and the force is exerted to create maximum torque the entire time?

39. A playground merry-go-round has a mass of 120 kg and a radius of 1.80 m and it is rotating with an angular velocity of 0.500 rev/s. What is its angular velocity after a 22.0-kg child gets onto it by grabbing its outer edge? The child is initially at rest.

40. Three children are riding on the edge of a merry-go-round that is 100 kg, has a 1.60-m radius, and is spinning at 20.0 rpm. The children have masses of 22.0, 28.0, and 33.0 kg. If the child who has a mass of 28.0 kg moves to the center of the merry-go-round, what is the new angular velocity in rpm?

41. (a) Calculate the angular momentum of an ice skater spinning at 6.00 rev/s given his moment of inertia is $0.400 \text{ kg} \cdot \text{m}^2$. (b) He reduces his rate of spin (his angular velocity) by extending his arms and increasing his moment of inertia. Find the value of his moment of inertia if his angular velocity decreases to 1.25 rev/s. (c) Suppose instead he keeps his arms in and allows friction of the ice to slow him to 3.00 rev/s. What average torque was exerted if this takes 15.0 s?

42. Consider the Earth-Moon system. Construct a problem in which you calculate the total angular momentum of the system including the spins of the Earth and the Moon on their axes and the orbital angular momentum of the Earth-Moon system in its nearly monthly rotation. Calculate what happens to the Moon's orbital radius if the Earth's rotation decreases due to tidal drag. Among the things to be considered are the amount by which the Earth's rotation slows and the fact that the Moon will continue to have one side always facing the Earth.

10.6 Collisions of Extended Bodies in Two Dimensions

43. Repeat **Example 10.15** in which the disk strikes and adheres to the stick 0.100 m from the nail.

44. Repeat **Example 10.15** in which the disk originally spins clockwise at 1000 rpm and has a radius of 1.50 cm.

45. Twin skaters approach one another as shown in **Figure 10.39** and lock hands. (a) Calculate their final angular velocity, given each had an initial speed of 2.50 m/s relative to the ice. Each has a mass of 70.0 kg, and each has a center of mass located 0.800 m from their locked hands. You may approximate their moments of inertia to be that of point masses at this radius. (b) Compare the initial kinetic energy and final kinetic energy.

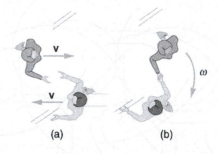

(a)   (b)

Figure 10.39 Twin skaters approach each other with identical speeds. Then, the skaters lock hands and spin.

46. Suppose a 0.250-kg ball is thrown at 15.0 m/s to a motionless person standing on ice who catches it with an outstretched arm as shown in **Figure 10.40**.

(a) Calculate the final linear velocity of the person, given his mass is 70.0 kg.

(b) What is his angular velocity if each arm is 5.00 kg? You may treat the ball as a point mass and treat the person's arms as uniform rods (each has a length of 0.900 m) and the rest of his body as a uniform cylinder of radius 0.180 m. Neglect the effect of the ball on his center of mass so that his center of mass remains in his geometrical center.

(c) Compare the initial and final total kinetic energies.

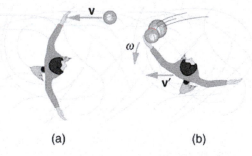

(a)   (b)

Figure 10.40 The figure shows the overhead view of a person standing motionless on ice about to catch a ball. Both arms are outstretched. After catching the ball, the skater recoils and rotates.

47. Repeat **Example 10.15** in which the stick is free to have translational motion as well as rotational motion.

Test Prep for AP® Courses

10.3 Dynamics of Rotational Motion: Rotational Inertia

1. A piece of wood can be carved by spinning it on a motorized lathe and holding a sharp chisel to the edge of the wood as it spins. How does the angular velocity of a piece of wood with a radius of 0.2 m spinning on a lathe change when

10.7 Gyroscopic Effects: Vector Aspects of Angular Momentum

48. Integrated Concepts

The axis of Earth makes a 23.5° angle with a direction perpendicular to the plane of Earth's orbit. As shown in **Figure 10.41**, this axis precesses, making one complete rotation in 25,780 y.

(a) Calculate the change in angular momentum in half this time.

(b) What is the average torque producing this change in angular momentum?

(c) If this torque were created by a single force (it is not) acting at the most effective point on the equator, what would its magnitude be?

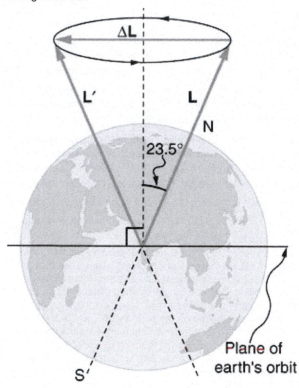

Figure 10.41 The Earth's axis slowly precesses, always making an angle of 23.5° with the direction perpendicular to the plane of Earth's orbit. The change in angular momentum for the two shown positions is quite large, although the magnitude L is unchanged.

a chisel is held to the wood's edge with a force of 50 N?
  a. It increases by 0.1 N•m multiplied by the moment of inertia of the wood.
  b. It decreases by 0.1 N•m divided by the moment of inertia of the wood-and-lathe system.
  c. It decreases by 0.1 N•m multiplied by the moment of inertia of the wood.
  d. It decreases by 0.1 m/s².

2. A Ferris wheel is loaded with people in the chairs at the following positions: 4 o'clock, 1 o'clock, 9 o'clock, and 6 o'clock. As the wheel begins to turn, what forces are acting on the system? How will each force affect the angular velocity and angular momentum?

3. A lever is placed on a fulcrum. A rock is placed on the left end of the lever and a downward (clockwise) force is applied to the right end of the lever. What measurements would be most effective to help you determine the angular momentum of the system? (Assume the lever itself has negligible mass.)
   a. the angular velocity and mass of the rock
   b. the angular velocity and mass of the rock, and the radius of the lever
   c. the velocity of the force, the radius of the lever, and the mass of the rock
   d. the mass of the rock, the length of the lever on both sides of the fulcrum, and the force applied on the right side of the lever

4. You can use the following setup to determine angular acceleration and angular momentum: A lever is placed on a fulcrum. A rock is placed on the left end of the lever and a known downward (clockwise) force is applied to the right end of the lever. What calculations would you perform? How would you account for gravity in your calculations?

5. Consider two sizes of disk, both of mass M. One size of disk has radius R; the other has radius $2R$. System A consists of two of the larger disks rigidly connected to each other with a common axis of rotation. System B consists of one of the larger disks and a number of the smaller disks rigidly connected with a common axis of rotation. If the moment of inertia for system A equals the moment of inertia for system B, how many of the smaller disks are in system B?
   a. 1
   b. 2
   c. 3
   d. 4

6. How do you arrange these objects so that the resulting system has the maximum possible moment of inertia? What is that moment of inertia?

10.4 Rotational Kinetic Energy: Work and Energy Revisited

7. Gear A, which turns clockwise, meshes with gear B, which turns counterclockwise. When more force is applied through gear A, torque is created. How does the angular velocity of gear B change as a result?
   a. It increases in magnitude.
   b. It decreases in magnitude.
   c. It changes direction.
   d. It stays the same.

8. Which will cause a greater increase in the angular velocity of a disk: doubling the torque applied or halving the radius at which the torque is applied? Explain.

9. Which measure would not be useful to help you determine the change in angular velocity when the torque on a fishing reel is increased?
   a. the radius of the reel
   b. the amount of line that unspools
   c. the angular momentum of the fishing line
   d. the time it takes the line to unspool

10. What data could you collect to study the change in angular velocity when two people push a merry-go-round instead of one, providing twice as much torque? How would you use the data you collect?

10.5 Angular Momentum and Its Conservation

11. Which rotational system would be best to use as a model to measure how angular momentum changes when forces on the system are changed?
   a. a fishing reel
   b. a planet and its moon
   c. a figure skater spinning
   d. a person's lower leg

12. You are collecting data to study changes in the angular momentum of a bicycle wheel when a force is applied to it. Which of the following measurements would be least helpful to you?
   a. the time for which the force is applied
   b. the radius at which the force is applied
   c. the angular velocity of the wheel when the force is applied
   d. the direction of the force

13. Which torque applied to a disk with radius 7.0 cm for 3.5 s will produce an angular momentum of 25 N•m•s?
   a. 7.1 N•m
   b. 357.1 N•m
   c. 3.6 N•m
   d. 612.5 N•m

14. Which of the following would be the best way to produce measurable amounts of torque on a system to test the relationship between the angular momentum of the system, the average torque applied to the system, and the time for which the torque is applied?
   a. having different numbers of people push on a merry-go-round
   b. placing known masses on one end of a seesaw
   c. touching the outer edge of a bicycle wheel to a treadmill that is moving at different speeds
   d. hanging known masses from a string that is wound around a spool suspended horizontally on an axle

15.

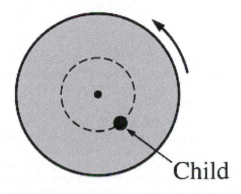

Figure 10.42 A curved arrow lies at the side of a gray disk. There is a point at the center of the disk, and around the point there is a dashed circle. There is a point labeled "Child" on the dashed circle. Below the disc is a label saying "Top View". The diagram above shows a top view of a child of mass M on a circular platform of mass $2M$ that is rotating counterclockwise. Assume the platform rotates without friction. Which of the following describes an action by the child that will increase the angular speed of the platform-child system and why?

a. The child moves toward the center of the platform, increasing the total angular momentum of the system.
b. The child moves toward the center of the platform, decreasing the rotational inertia of the system.
c. The child moves away from the center of the platform, increasing the total angular momentum of the system.
d. The child moves away from the center of the platform, decreasing the rotational inertia of the system.

16.

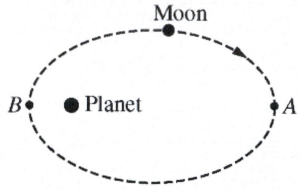

Figure 10.43 A point labeled "Moon" lies on a dashed ellipse. Two other points, labeled "A" and "B", lie at opposite ends of the ellipse. A point labeled "Planet" lies inside the ellipse. A moon is in an elliptical orbit about a planet as shown above. At point A the moon has speed u_A and is at distance R_A from the planet. At point B the moon has speed u_B. Has the moon's angular momentum changed? Explain your answer.

17. A hamster sits 0.10 m from the center of a lazy Susan of negligible mass. The wheel has an angular velocity of 1.0 rev/s. How will the angular velocity of the lazy Susan change if the hamster walks to 0.30 m from the center of rotation? Assume zero friction and no external torque.
a. It will speed up to 2.0 rev/s.
b. It will speed up to 9.0 rev/s.
c. It will slow to 0.01 rev/s.
d. It will slow to 0.02 rev/s.

18. Earth has a mass of 6.0×10^{24} kg, a radius of 6.4×10^6 m, and an angular velocity of 1.2×10^{-5} rev/s. How would the planet's angular velocity change if a layer of Earth with mass 1.0×10^{23} kg broke off of the Earth, decreasing Earth's radius by 0.2×10^6 m? Assume no friction.

19. Consider system A, consisting of two disks of radius R, with both rotating clockwise. Now consider system B, consisting of one disk of radius R rotating counterclockwise and another disk of radius $2R$ rotating clockwise. All of the disks have the same mass, and all have the same magnitude of angular velocity.

Which system has the greatest angular momentum?

a. A
b. B
c. They're equal.
d. Not enough information

20. Assume that a baseball bat being swung at 3π rad/s by a batting machine is equivalent to a 1.1 m thin rod with a mass of 1.0 kg. How fast would a 0.15 kg baseball that squarely hits the very tip of the bat have to be going for the net angular momentum of the bat-ball system to be zero?

10.6 Collisions of Extended Bodies in Two

Dimensions

21. A box with a mass of 2.0 kg rests on one end of a seesaw. The seesaw is 6.0 m long, and we can assume it has negligible mass. Approximately what angular momentum will the box have if someone with a mass of 65 kg sits on the other end of the seesaw quickly, with a velocity of 1.2 m/s?
a. 702 kg•m^2/s
b. 39 kg•m^2/s
c. 18 kg•m^2/s
d. 1.2 kg•m^2/s

22. A spinner in a board game can be thought of as a thin rod that spins about an axis at its center. The spinner in a certain game is 12 cm long and has a mass of 10 g. How will its angular velocity change when it is flicked at one end with a force equivalent to 15 g travelling at 5.0 m/s if all the energy of the collision is transferred to the spinner? (You can use the table in **Figure 10.12** to estimate the rotational inertia of the spinner.)

23. A cyclist pedals to exert a torque on the rear wheel of the bicycle. When the cyclist changes to a higher gear, the torque increases. Which of the following would be the most effective strategy to help you determine the change in angular momentum of the bicycle wheel?
a. multiplying the ratio between the two torques by the mass of the bicycle and rider
b. adding the two torques together, and multiplying by the time for which both torques are applied
c. multiplying the difference in the two torques by the time for which the new torque is applied
d. multiplying both torques by the mass of the bicycle and rider

24. An electric screwdriver has two speeds, each of which exerts a different torque on a screw. Describe what calculations you could use to help you compare the angular momentum of a screw at each speed. What measurements would you need to make in order to calculate this?

25. Why is it important to consider the shape of an object when determining the object's angular momentum?
a. The shape determines the location of the center of mass. The location of the center of mass in turn determines the angular velocity of the object.
b. The shape helps you determine the location of the object's outer edge, where rotational velocity will be greatest.
c. The shape helps you determine the location of the center of rotation.
d. The shape determines the location of the center of mass. The location of the center of mass contributes to the object's rotational inertia, which contributes to its angular momentum.

26. How could you collect and analyze data to test the difference between the torques provided by two speeds on a tabletop fan?

27. Describe a rotational system you could use to demonstrate the effect on the system's angular momentum of applying different amounts of external torque.

28. How could you use simple equipment such as balls and string to study the changes in angular momentum of a system when it interacts with another system?

10.7 Gyroscopic Effects: Vector Aspects of Angular Momentum

29. A globe (model of the Earth) is a hollow sphere with a radius of 16 cm. By wrapping a cord around the equator of a globe and pulling on it, a person exerts a torque on the globe of 120 N • m for 1.2 s. What angular momentum does the globe have after 1.2 s?

30. How could you use a fishing reel to test the relationship between the torque applied to a system, the time for which the torque was applied, and the resulting angular momentum of the system? How would you measure angular momentum?

11 FLUID STATICS

Figure 11.1 The fluid essential to all life has a beauty of its own. It also helps support the weight of this swimmer. (credit: Terren, Wikimedia Commons)

Chapter Outline

11.1. What Is a Fluid?

11.2. Density

11.3. Pressure

11.4. Variation of Pressure with Depth in a Fluid

11.5. Pascal's Principle

11.6. Gauge Pressure, Absolute Pressure, and Pressure Measurement

11.7. Archimedes' Principle

11.8. Cohesion and Adhesion in Liquids: Surface Tension and Capillary Action

11.9. Pressures in the Body

Connection for AP® Courses

Much of what we value in life is fluid: a breath of fresh winter air; the water we drink, swim in, and bathe in; the blood in our veins. But what exactly is a fluid? Can we understand fluids with the laws already presented, or will new laws emerge from their study?

As you read this chapter, you will learn how the arrangement and interaction of the particles—atoms and molecules—that make up a fluid define many of its macroscopic characteristics, like density and pressure (Big Idea 1, Enduring Understanding 1.E, Essential Knowledge 1.E.1). While the number of particles in a fluid is often immense, you will be able to use a probabilistic approach in order to explain how a fluid affects its environment in a variety of ways (Big Idea 7, Enduring Understanding 7.A, Essential Knowledge 7.A.1). From the pressure a fluid places on the walls of a hydraulic system to a variety of biological and medical applications, understanding a fluid's properties begins with understanding its internal structure.

Big Idea 1 Objects and systems have properties such as mass and charge. Systems may have internal structure.

Enduring Understanding 1.E Materials have many macroscopic properties that result from the arrangement and interactions of the atoms and molecules that make up the material.

Essential Knowledge 1.E.1 Matter has a property called density.

Big Idea 7 The mathematics of probability can be used to describe the behavior of complex systems and to interpret the behavior of quantum mechanical systems.

Enduring Understanding 7.A The properties of an ideal gas can be explained in terms of a small number of macroscopic variables including temperature and pressure.

Essential Knowledge 7.A.1 The pressure of a system determines the force that the system exerts on the walls of its container and is a measure of the average change in the momentum or impulse of the molecules colliding with the walls of the container. The pressure also exists inside the system itself, not just at the walls of the container.

11.1 What Is a Fluid?

Matter most commonly exists as a solid, liquid, or gas; these states are known as the three common *phases of matter*. Solids have a definite shape and a specific volume, liquids have a definite volume but their shape changes depending on the container in which they are held, and gases have neither a definite shape nor a specific volume as their molecules move to fill the container in which they are held. (See **Figure 11.2.**) Liquids and gases are considered to be fluids because they yield to shearing forces, whereas solids resist them. Note that the extent to which fluids yield to shearing forces (and hence flow easily and quickly) depends on a quantity called the viscosity which is discussed in detail in **Viscosity and Laminar Flow; Poiseuille's Law**. We can understand the phases of matter and what constitutes a fluid by considering the forces between atoms that make up matter in the three phases.

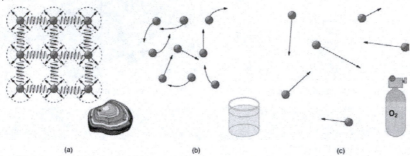

(a)   (b)   (c)

Figure 11.2 (a) Atoms in a solid always have the same neighbors, held near home by forces represented here by springs. These atoms are essentially in contact with one another. A rock is an example of a solid. This rock retains its shape because of the forces holding its atoms together. (b) Atoms in a liquid are also in close contact but can slide over one another. Forces between them strongly resist attempts to push them closer together and also hold them in close contact. Water is an example of a liquid. Water can flow, but it also remains in an open container because of the forces between its atoms. (c) Atoms in a gas are separated by distances that are considerably larger than the size of the atoms themselves, and they move about freely. A gas must be held in a closed container to prevent it from moving out freely.

Atoms in *solids* are in close contact, with forces between them that allow the atoms to vibrate but not to change positions with neighboring atoms. (These forces can be thought of as springs that can be stretched or compressed, but not easily broken.) Thus a solid *resists* all types of stress. A solid cannot be easily deformed because the atoms that make up the solid are not able to move about freely. Solids also resist compression, because their atoms form part of a lattice structure in which the atoms are a relatively fixed distance apart. Under compression, the atoms would be forced into one another. Most of the examples we have studied so far have involved solid objects which deform very little when stressed.

Connections: Submicroscopic Explanation of Solids and Liquids

Atomic and molecular characteristics explain and underlie the macroscopic characteristics of solids and fluids. This submicroscopic explanation is one theme of this text and is highlighted in the Things Great and Small features in **Conservation of Momentum**. See, for example, microscopic description of collisions and momentum or microscopic description of pressure in a gas. This present section is devoted entirely to the submicroscopic explanation of solids and liquids.

In contrast, *liquids* deform easily when stressed and do not spring back to their original shape once the force is removed because the atoms are free to slide about and change neighbors—that is, they *flow* (so they are a type of fluid), with the molecules held together by their mutual attraction. When a liquid is placed in a container with no lid on, it remains in the container (providing the container has no holes below the surface of the liquid!). Because the atoms are closely packed, liquids, like solids, resist compression.

Atoms in *gases* are separated by distances that are large compared with the size of the atoms. The forces between gas atoms are therefore very weak, except when the atoms collide with one another. Gases thus not only flow (and are therefore considered to be fluids) but they are relatively easy to compress because there is much space and little force between atoms. When placed in an open container gases, unlike liquids, will escape. The major distinction is that gases are easily compressed, whereas liquids are not. We shall generally refer to both gases and liquids simply as **fluids**, and make a distinction between them only when they behave differently.

PhET Explorations: States of Matter—Basics

Heat, cool, and compress atoms and molecules and watch as they change between solid, liquid, and gas phases.

Figure 11.3 States of Matter: Basics (http://cnx.org/content/m55205/1.2/states-of-matter-basics_en.jar)

11.2 Density

Learning Objectives

By the end of this section, you will be able to:

- Define density.
- Calculate the mass of a reservoir from its density.
- Compare and contrast the densities of various substances.

The information presented in this section supports the following AP® learning objectives and science practices:

- **1.E.1.1** The student is able to predict the densities, differences in densities, or changes in densities under different conditions for natural phenomena and design an investigation to verify the prediction. **(S.P. 6.2, 6.4)**
- **1.E.1.2** The student is able to select from experimental data the information necessary to determine the density of an object and/or compare densities of several objects. **(S.P. 4.1, 6.4)**

Which weighs more, a ton of feathers or a ton of bricks? This old riddle plays with the distinction between mass and density. A ton is a ton, of course; but bricks have much greater density than feathers, and so we are tempted to think of them as heavier. (See **Figure 11.4**.)

Density, as you will see, is an important characteristic of substances. It is crucial, for example, in determining whether an object sinks or floats in a fluid. Density is the mass per unit volume of a substance or object. In equation form, density is defined as

$$\rho = \frac{m}{V},$$

(11.1)

where the Greek letter ρ (rho) is the symbol for density, m is the mass, and V is the volume occupied by the substance.

Density

Density is mass per unit volume.

$$\rho = \frac{m}{V},$$

(11.2)

where ρ is the symbol for density, m is the mass, and V is the volume occupied by the substance.

In the riddle regarding the feathers and bricks, the masses are the same, but the volume occupied by the feathers is much greater, since their density is much lower. The SI unit of density is kg/m^3, representative values are given in **Table 11.1**. The metric system was originally devised so that water would have a density of 1 g/cm^3, equivalent to 10^3 kg/m^3. Thus the basic mass unit, the kilogram, was first devised to be the mass of 1000 mL of water, which has a volume of 1000 cm^3.

Table 11.1 Densities of Various Substances

Substance	$\rho(10^3 \text{ kg/m}^3 \text{ or g/mL})$	Substance	$\rho(10^3 \text{ kg/m}^3 \text{ or g/mL})$	Substance	$\rho(10^3 \text{ kg/m}^3 \text{ or g/mL})$
Solids		**Liquids**		**Gases**	
Aluminum	2.7	Water (4ºC)	1.000	Air	1.29×10^{-3}
Brass	8.44	Blood	1.05	Carbon dioxide	1.98×10^{-3}
Copper (average)	8.8	Sea water	1.025	Carbon monoxide	1.25×10^{-3}
Gold	19.32	Mercury	13.6	Hydrogen	0.090×10^{-3}
Iron or steel	7.8	Ethyl alcohol	0.79	Helium	0.18×10^{-3}
Lead	11.3	Petrol	0.68	Methane	0.72×10^{-3}
Polystyrene	0.10	Glycerin	1.26	Nitrogen	1.25×10^{-3}
Tungsten	19.30	Olive oil	0.92	Nitrous oxide	1.98×10^{-3}
Uranium	18.70			Oxygen	1.43×10^{-3}
Concrete	2.30–3.0			Steam (100° C)	0.60×10^{-3}
Cork	0.24				
Glass, common (average)	2.6				
Granite	2.7				
Earth's crust	3.3				
Wood	0.3–0.9				
Ice (0°C)	0.917				
Bone	1.7–2.0				

Figure 11.4 A ton of feathers and a ton of bricks have the same mass, but the feathers make a much bigger pile because they have a much lower density.

As you can see by examining **Table 11.1**, the density of an object may help identify its composition. The density of gold, for example, is about 2.5 times the density of iron, which is about 2.5 times the density of aluminum. Density also reveals something about the phase of the matter and its substructure. Notice that the densities of liquids and solids are roughly comparable, consistent with the fact that their atoms are in close contact. The densities of gases are much less than those of liquids and solids, because the atoms in gases are separated by large amounts of empty space.

Take-Home Experiment Sugar and Salt

A pile of sugar and a pile of salt look pretty similar, but which weighs more? If the volumes of both piles are the same, any difference in mass is due to their different densities (including the air space between crystals). Which do you think has the greater density? What values did you find? What method did you use to determine these values?

Example 11.1 Calculating the Mass of a Reservoir From Its Volume

A reservoir has a surface area of 50.0 km^2 and an average depth of 40.0 m. What mass of water is held behind the dam? (See **Figure 11.5** for a view of a large reservoir—the Three Gorges Dam site on the Yangtze River in central China.)

Strategy

We can calculate the volume V of the reservoir from its dimensions, and find the density of water ρ in **Table 11.1**. Then the mass m can be found from the definition of density

$$\rho = \frac{m}{V}. \tag{11.3}$$

Solution

Solving equation $\rho = m / V$ for m gives $m = \rho V$.

The volume V of the reservoir is its surface area A times its average depth h:

$$
\begin{aligned}
V &= Ah = \left(50.0 \text{ km}^2\right)(40.0 \text{ m}) \\
&= \left[\left(50.0 \text{ km}^2\right)\left(\frac{10^3 \text{ m}}{1 \text{ km}}\right)^2\right](40.0 \text{ m}) = 2.00 \times 10^9 \text{ m}^3
\end{aligned}
\tag{11.4}
$$

The density of water ρ from **Table 11.1** is $1.000 \times 10^3 \text{ kg/m}^3$. Substituting V and ρ into the expression for mass gives

$$
\begin{aligned}
m &= \left(1.00 \times 10^3 \text{ kg/m}^3\right)\left(2.00 \times 10^9 \text{ m}^3\right) \\
&= 2.00 \times 10^{12} \text{ kg}.
\end{aligned}
\tag{11.5}
$$

Discussion

A large reservoir contains a very large mass of water. In this example, the weight of the water in the reservoir is $mg = 1.96 \times 10^{13} \text{ N}$, where g is the acceleration due to the Earth's gravity (about 9.80 m/s^2). It is reasonable to ask whether the dam must supply a force equal to this tremendous weight. The answer is no. As we shall see in the following sections, the force the dam must supply can be much smaller than the weight of the water it holds back.

Figure 11.5 Three Gorges Dam in central China. When completed in 2008, this became the world's largest hydroelectric plant, generating power equivalent to that generated by 22 average-sized nuclear power plants. The concrete dam is 181 m high and 2.3 km across. The reservoir made by this dam is 660 km long. Over 1 million people were displaced by the creation of the reservoir. (credit: Le Grand Portage)

11.3 Pressure

Learning Objectives

By the end of this section, you will be able to:

- Define pressure.
- Explain the relationship between pressure and force.
- Calculate force given pressure and area.

The information presented in this section supports the following AP® learning objectives and science practices:

- **7.A.1.1** The student is able to make claims about how the pressure of an ideal gas is connected to the force exerted by molecules on the walls of the container, and how changes in pressure affect the thermal equilibrium of the system. **(S.P. 6.4, 7.2)**

You have no doubt heard the word **pressure** being used in relation to blood (high or low blood pressure) and in relation to the weather (high- and low-pressure weather systems). These are only two of many examples of pressures in fluids. Pressure P is defined as

$$P = \frac{F}{A} \tag{11.6}$$

where F is a force applied to an area A that is perpendicular to the force.

Pressure

Pressure is defined as the force divided by the area perpendicular to the force over which the force is applied, or

$$P = \frac{F}{A}. \tag{11.7}$$

A given force can have a significantly different effect depending on the area over which the force is exerted, as shown in **Figure 11.6**. The SI unit for pressure is the *pascal*, where

$$1 \text{ Pa} = 1 \text{ N/m}^2. \tag{11.8}$$

In addition to the pascal, there are many other units for pressure that are in common use. In meteorology, atmospheric pressure is often described in units of millibar (mb), where

$$100 \text{ mb} = 1 \times 10^4 \text{ Pa}. \tag{11.9}$$

Pounds per square inch $\left(\text{lb/in}^2 \text{ or psi} \right)$ is still sometimes used as a measure of tire pressure, and millimeters of mercury (mm Hg) is still often used in the measurement of blood pressure. Pressure is defined for all states of matter but is particularly important when discussing fluids.

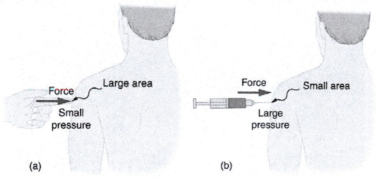

Figure 11.6 (a) While the person being poked with the finger might be irritated, the force has little lasting effect. (b) In contrast, the same force applied to an area the size of the sharp end of a needle is great enough to break the skin.

Making Connections: Gas Particles in a Container

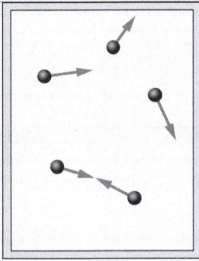

Figure 11.7 Gas particles vibrate within a closed container.

Imagine a closed container full of quickly vibrating gas particles. As the particles rapidly move around the container, they will repeatedly strike each other and the walls of the container.

When the particles strike the walls, a few interesting changes will occur.

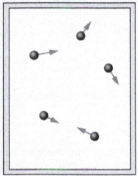 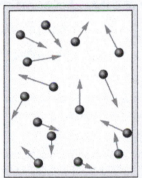

Figure 11.8 The figure on the left shows gas particles traveling within a closed container. As the particles travel, they collide with each other and with the container walls. The figure on the right shows an increased number of gas particles within the container, which will result in an increased number of collisions.

Each time the particles strike the walls of this container, they will apply a force to the container walls. An increase in gas particles will result in more collisions, and a greater force will be applied. The increased force will result in an increased pressure on the container walls, as the areas of the container walls remain constant.

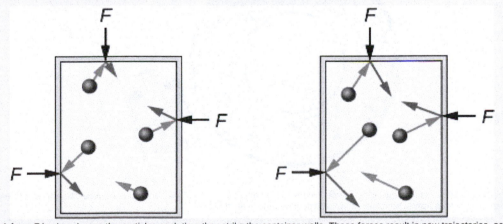

Figure 11.9 A force **F** is placed upon the particles each time they strike the container walls. These forces result in new trajectories, as shown by the red arrows. Because the particles in the figure on the right are moving more quickly, they experience a larger force than those shown in the figure on the left.

If the speed of the particles is increased, then each particle will experience a greater change in momentum when it strikes a container wall. Just like a fast-moving tennis ball recoiling off a hard surface, the greater the particle's momentum, the more force it will experience when it collides. (For verification, see the impulse-momentum theorem described in Chapter 8.)

However, the more interesting change will be at the wall itself. Due to Newton's third law, it is not only the force on the particle that will increase, but the force on the container will increase as well! While not all particles will move with the same velocity, or strike the wall in the same way, they will experience an average change in momentum upon each collision. The force that these particles impart to the container walls is a good measure of this average change in momentum. Both of these relationships will be useful in Chapter 12, as you consider the ideal gas law. For now, it is good to recognize that laws commonly used to understand macroscopic phenomena can be applied to phenomena at the particle level as well.

Example 11.2 Calculating Force Exerted by the Air: What Force Does a Pressure Exert?

An astronaut is working outside the International Space Station where the atmospheric pressure is essentially zero. The pressure gauge on her air tank reads 6.90×10^6 Pa. What force does the air inside the tank exert on the flat end of the cylindrical tank, a disk 0.150 m in diameter?

Strategy

We can find the force exerted from the definition of pressure given in $P = \dfrac{F}{A}$, provided we can find the area A acted upon.

Solution

By rearranging the definition of pressure to solve for force, we see that

$$F = PA. \tag{11.10}$$

Here, the pressure P is given, as is the area of the end of the cylinder A, given by $A = \pi r^2$. Thus,

$$
\begin{aligned}
F &= \left(6.90 \times 10^6 \text{ N/m}^2\right)(3.14)(0.0750 \text{ m})^2 \\
&= 1.22 \times 10^5 \text{ N.}
\end{aligned}
\tag{11.11}
$$

Discussion

Wow! No wonder the tank must be strong. Since we found $F = PA$, we see that the force exerted by a pressure is directly proportional to the area acted upon as well as the pressure itself.

The force exerted on the end of the tank is perpendicular to its inside surface. This direction is because the force is exerted by a static or stationary fluid. We have already seen that fluids cannot *withstand* shearing (sideways) forces; they cannot *exert* shearing forces, either. Fluid pressure has no direction, being a scalar quantity. The forces due to pressure have well-defined directions: they are always exerted perpendicular to any surface. (See the tire in **Figure 11.10**, for example.) Finally, note that pressure is exerted on all surfaces. Swimmers, as well as the tire, feel pressure on all sides. (See **Figure 11.11**.)

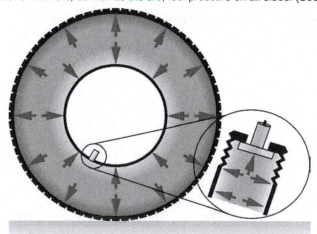

Figure 11.10 Pressure inside this tire exerts forces perpendicular to all surfaces it contacts. The arrows give representative directions and magnitudes of the forces exerted at various points. Note that static fluids do not exert shearing forces.

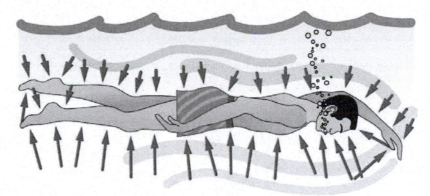

Figure 11.11 Pressure is exerted on all sides of this swimmer, since the water would flow into the space he occupies if he were not there. The arrows represent the directions and magnitudes of the forces exerted at various points on the swimmer. Note that the forces are larger underneath, due to greater depth, giving a net upward or buoyant force that is balanced by the weight of the swimmer.

Making Connections: Pressure

Figure 11.10 and **Figure 11.11** both show pressure at the barrier between an object and a fluid. Note that this pressure also exists within the fluid itself. Just as particles will create a force when colliding with the swimmer in **Figure 11.11**, they will do the same each time they strike each other. These forces can be represented by arrows, whose vectors show the resulting direction of particle movement. The same factors that determine the magnitude of pressure upon the fluid barrier will determine the magnitude of pressure within the fluid itself. These factors will be discussed in Chapter 13.

PhET Explorations: Gas Properties

Pump gas molecules to a box and see what happens as you change the volume, add or remove heat, change gravity, and more. Measure the temperature and pressure, and discover how the properties of the gas vary in relation to each other.

Figure 11.12 Gas Properties (http://cnx.org/content/m55209/1.4/gas-properties_en.jar)

11.4 Variation of Pressure with Depth in a Fluid

Learning Objectives
By the end of this section, you will be able to: • Define pressure in terms of weight. • Explain the variation of pressure with depth in a fluid. • Calculate density given pressure and altitude.

If your ears have ever popped on a plane flight or ached during a deep dive in a swimming pool, you have experienced the effect of depth on pressure in a fluid. At the Earth's surface, the air pressure exerted on you is a result of the weight of air above you. This pressure is reduced as you climb up in altitude and the weight of air above you decreases. Under water, the pressure exerted on you increases with increasing depth. In this case, the pressure being exerted upon you is a result of both the weight of water above you *and* that of the atmosphere above you. You may notice an air pressure change on an elevator ride that transports you many stories, but you need only dive a meter or so below the surface of a pool to feel a pressure increase. The difference is that water is much denser than air, about 775 times as dense.

Consider the container in **Figure 11.13**. Its bottom supports the weight of the fluid in it. Let us calculate the pressure exerted on the bottom by the weight of the fluid. That **pressure** is the weight of the fluid mg divided by the area A supporting it (the area of the bottom of the container):

$$P = \frac{mg}{A}.$$

(11.12)

We can find the mass of the fluid from its volume and density:

$$m = \rho V.$$

(11.13)

The volume of the fluid V is related to the dimensions of the container. It is

$$V = Ah, \tag{11.14}$$

where A is the cross-sectional area and h is the depth. Combining the last two equations gives

$$m = \rho Ah. \tag{11.15}$$

If we enter this into the expression for pressure, we obtain

$$P = \frac{(\rho Ah)g}{A}. \tag{11.16}$$

The area cancels, and rearranging the variables yields

$$P = h\rho g. \tag{11.17}$$

This value is the *pressure due to the weight of a fluid*. The equation has general validity beyond the special conditions under which it is derived here. Even if the container were not there, the surrounding fluid would still exert this pressure, keeping the fluid static. Thus the equation $P = h\rho g$ represents the pressure due to the weight of any fluid of *average density* ρ at any depth h below its surface. For liquids, which are nearly incompressible, this equation holds to great depths. For gases, which are quite compressible, one can apply this equation as long as the density changes are small over the depth considered. **Example 11.4** illustrates this situation.

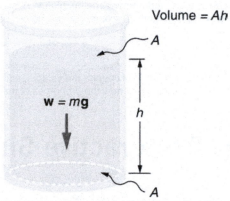

Figure 11.13 The bottom of this container supports the entire weight of the fluid in it. The vertical sides cannot exert an upward force on the fluid (since it cannot withstand a shearing force), and so the bottom must support it all.

Example 11.3 Calculating the Average Pressure and Force Exerted: What Force Must a Dam Withstand?

In **Example 11.1**, we calculated the mass of water in a large reservoir. We will now consider the pressure and force acting on the dam retaining water. (See **Figure 11.14**.) The dam is 500 m wide, and the water is 80.0 m deep at the dam. (a) What is the average pressure on the dam due to the water? (b) Calculate the force exerted against the dam and compare it with the weight of water in the dam (previously found to be 1.96×10^{13} N).

Strategy for (a)

The average pressure $\bar{P}$ due to the weight of the water is the pressure at the average depth $\bar{h}$ of 40.0 m, since pressure increases linearly with depth.

Solution for (a)

The average pressure due to the weight of a fluid is

$$\bar{P} = \bar{h}\rho g. \tag{11.18}$$

Entering the density of water from **Table 11.1** and taking $\bar{h}$ to be the average depth of 40.0 m, we obtain

$$
\begin{aligned}
\bar{P} &= (40.0 \text{ m})\left(10^3 \frac{\text{kg}}{\text{m}^3}\right)\left(9.80 \frac{\text{m}}{\text{s}^2}\right) \\
&= 3.92 \times 10^5 \frac{\text{N}}{\text{m}^2} = 392 \text{ kPa.}
\end{aligned} \tag{11.19}
$$

Strategy for (b)

The force exerted on the dam by the water is the average pressure times the area of contact:

$$F = \bar{P} A. \tag{11.20}$$

Solution for (b)

We have already found the value for $\bar{P}$. The area of the dam is $A = 80.0 \text{ m} \times 500 \text{ m} = 4.00 \times 10^4 \text{ m}^2$, so that

$$
\begin{aligned}
F &= (3.92 \times 10^5 \text{ N/m}^2)(4.00 \times 10^4 \text{ m}^2) \\
&= 1.57 \times 10^{10} \text{ N}.
\end{aligned} \tag{11.21}
$$

Discussion

Although this force seems large, it is small compared with the $1.96 \times 10^{13} \text{ N}$ weight of the water in the reservoir—in fact, it is only 0.0800% of the weight. Note that the pressure found in part (a) is completely independent of the width and length of the lake—it depends only on its average depth at the dam. Thus the force depends only on the water's average depth and the dimensions of the dam, *not* on the horizontal extent of the reservoir. In the diagram, the thickness of the dam increases with depth to balance the increasing force due to the increasing pressure.epth to balance the increasing force due to the increasing pressure.

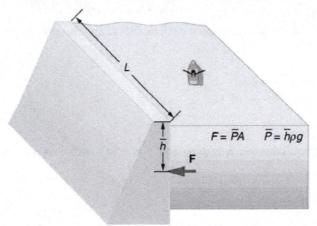

Figure 11.14 The dam must withstand the force exerted against it by the water it retains. This force is small compared with the weight of the water behind the dam.

Atmospheric pressure is another example of pressure due to the weight of a fluid, in this case due to the weight of *air* above a given height. The atmospheric pressure at the Earth's surface varies a little due to the large-scale flow of the atmosphere induced by the Earth's rotation (this creates weather "highs" and "lows"). However, the average pressure at sea level is given by the *standard atmospheric pressure* P_{atm}, measured to be

$$1 \text{ atmosphere (atm)} = P_{\text{atm}} = 1.01 \times 10^5 \text{ N/m}^2 = 101 \text{ kPa}. \tag{11.22}$$

This relationship means that, on average, at sea level, a column of air above 1.00 m^2 of the Earth's surface has a weight of $1.01 \times 10^5 \text{ N}$, equivalent to 1 atm. (See **Figure 11.15**.)

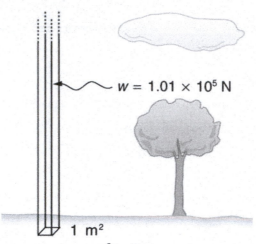

Figure 11.15 Atmospheric pressure at sea level averages 1.01×10^5 Pa (equivalent to 1 atm), since the column of air over this 1 m^2, extending to the top of the atmosphere, weighs 1.01×10^5 N.

Example 11.4 Calculating Average Density: How Dense Is the Air?

Calculate the average density of the atmosphere, given that it extends to an altitude of 120 km. Compare this density with that of air listed in **Table 11.1**.

Strategy

If we solve $P = h\rho g$ for density, we see that

$$\bar{\rho} = \frac{P}{hg}. \tag{11.23}$$

We then take P to be atmospheric pressure, h is given, and g is known, and so we can use this to calculate $\bar{\rho}$.

Solution

Entering known values into the expression for $\bar{\rho}$ yields

$$\bar{\rho} = \frac{1.01 \times 10^5 \ \text{N/m}^2}{(120 \times 10^3 \ \text{m})(9.80 \ \text{m/s}^2)} = 8.59 \times 10^{-2} \ \text{kg/m}^3. \tag{11.24}$$

Discussion

This result is the average density of air between the Earth's surface and the top of the Earth's atmosphere, which essentially ends at 120 km. The density of air at sea level is given in **Table 11.1** as $1.29 \ \text{kg/m}^3$ —about 15 times its average value. Because air is so compressible, its density has its highest value near the Earth's surface and declines rapidly with altitude.

Example 11.5 Calculating Depth Below the Surface of Water: What Depth of Water Creates the Same Pressure as the Entire Atmosphere?

Calculate the depth below the surface of water at which the pressure due to the weight of the water equals 1.00 atm.

Strategy

We begin by solving the equation $P = h\rho g$ for depth h:

$$h = \frac{P}{\rho g}. \tag{11.25}$$

Then we take P to be 1.00 atm and ρ to be the density of the water that creates the pressure.

Solution

Entering the known values into the expression for h gives

$$h = \frac{1.01 \times 10^5 \ \text{N/m}^2}{(1.00 \times 10^3 \ \text{kg/m}^3)(9.80 \ \text{m/s}^2)} = 10.3 \ \text{m}. \tag{11.26}$$

Discussion

Just 10.3 m of water creates the same pressure as 120 km of air. Since water is nearly incompressible, we can neglect any change in its density over this depth.

What do you suppose is the *total* pressure at a depth of 10.3 m in a swimming pool? Does the atmospheric pressure on the water's surface affect the pressure below? The answer is yes. This seems only logical, since both the water's weight and the atmosphere's weight must be supported. So the *total* pressure at a depth of 10.3 m is 2 atm—half from the water above and half from the air above. We shall see in **Pascal's Principle** that fluid pressures always add in this way.

11.5 Pascal's Principle

Learning Objectives
By the end of this section, you will be able to: • Define pressure. • State Pascal's principle. • Understand applications of Pascal's principle. • Derive relationships between forces in a hydraulic system.

Pressure is defined as force per unit area. Can pressure be increased in a fluid by pushing directly on the fluid? Yes, but it is much easier if the fluid is enclosed. The heart, for example, increases blood pressure by pushing directly on the blood in an enclosed system (valves closed in a chamber). If you try to push on a fluid in an open system, such as a river, the fluid flows away. An enclosed fluid cannot flow away, and so pressure is more easily increased by an applied force.

What happens to a pressure in an enclosed fluid? Since atoms in a fluid are free to move about, they transmit the pressure to all parts of the fluid and to the walls of the container. Remarkably, the pressure is transmitted *undiminished*. This phenomenon is called **Pascal's principle**, because it was first clearly stated by the French philosopher and scientist Blaise Pascal (1623–1662): A change in pressure applied to an enclosed fluid is transmitted undiminished to all portions of the fluid and to the walls of its container.

Pascal's Principle

A change in pressure applied to an enclosed fluid is transmitted undiminished to all portions of the fluid and to the walls of its container.

Pascal's principle, an experimentally verified fact, is what makes pressure so important in fluids. Since a change in pressure is transmitted undiminished in an enclosed fluid, we often know more about pressure than other physical quantities in fluids. Moreover, Pascal's principle implies that *the total pressure in a fluid is the sum of the pressures from different sources*. We shall find this fact—that pressures add—very useful.

Blaise Pascal had an interesting life in that he was home-schooled by his father who removed all of the mathematics textbooks from his house and forbade him to study mathematics until the age of 15. This, of course, raised the boy's curiosity, and by the age of 12, he started to teach himself geometry. Despite this early deprivation, Pascal went on to make major contributions in the mathematical fields of probability theory, number theory, and geometry. He is also well known for being the inventor of the first mechanical digital calculator, in addition to his contributions in the field of fluid statics.

Application of Pascal's Principle

One of the most important technological applications of Pascal's principle is found in a *hydraulic system*, which is an enclosed fluid system used to exert forces. The most common hydraulic systems are those that operate car brakes. Let us first consider the simple hydraulic system shown in **Figure 11.16**.

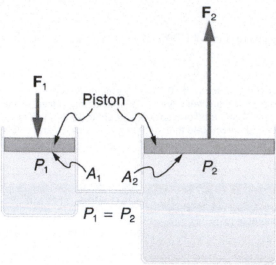

Figure 11.16 A typical hydraulic system with two fluid-filled cylinders, capped with pistons and connected by a tube called a hydraulic line. A downward force $\mathbf{F}_1$ on the left piston creates a pressure that is transmitted undiminished to all parts of the enclosed fluid. This results in an upward force $\mathbf{F}_2$ on the right piston that is larger than $\mathbf{F}_1$ because the right piston has a larger area.

Relationship Between Forces in a Hydraulic System

We can derive a relationship between the forces in the simple hydraulic system shown in **Figure 11.16** by applying Pascal's principle. Note first that the two pistons in the system are at the same height, and so there will be no difference in pressure due to a difference in depth. Now the pressure due to F_1 acting on area A_1 is simply $P_1 = \dfrac{F_1}{A_1}$, as defined by $P = \dfrac{F}{A}$. According to Pascal's principle, this pressure is transmitted undiminished throughout the fluid and to all walls of the container. Thus, a pressure P_2 is felt at the other piston that is equal to P_1. That is $P_1 = P_2$.

But since $P_2 = \dfrac{F_2}{A_2}$, we see that $\dfrac{F_1}{A_1} = \dfrac{F_2}{A_2}$.

This equation relates the ratios of force to area in any hydraulic system, providing the pistons are at the same vertical height and that friction in the system is negligible. Hydraulic systems can increase or decrease the force applied to them. To make the force larger, the pressure is applied to a larger area. For example, if a 100-N force is applied to the left cylinder in **Figure 11.16** and the right one has an area five times greater, then the force out is 500 N. Hydraulic systems are analogous to simple levers, but they have the advantage that pressure can be sent through tortuously curved lines to several places at once.

Example 11.6 Calculating Force of Slave Cylinders: Pascal Puts on the Brakes

Consider the automobile hydraulic system shown in **Figure 11.17**.

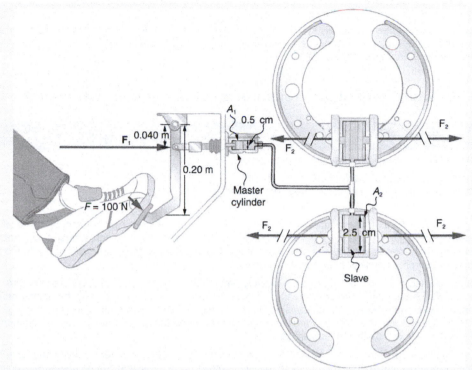

Figure 11.17 Hydraulic brakes use Pascal's principle. The driver exerts a force of 100 N on the brake pedal. This force is increased by the simple lever and again by the hydraulic system. Each of the identical slave cylinders receives the same pressure and, therefore, creates the same force output F_2. The circular cross-sectional areas of the master and slave cylinders are represented by A_1 and A_2, respectively

A force of 100 N is applied to the brake pedal, which acts on the cylinder—called the master—through a lever. A force of 500 N is exerted on the master cylinder. (The reader can verify that the force is 500 N using techniques of statics from **Applications of Statics, Including Problem-Solving Strategies.**) Pressure created in the master cylinder is transmitted to four so-called slave cylinders. The master cylinder has a diameter of 0.500 cm, and each slave cylinder has a diameter of 2.50 cm. Calculate the force F_2 created at each of the slave cylinders.

Strategy

We are given the force F_1 that is applied to the master cylinder. The cross-sectional areas A_1 and A_2 can be calculated from their given diameters. Then $\frac{F_1}{A_1} = \frac{F_2}{A_2}$ can be used to find the force F_2. Manipulate this algebraically to get F_2 on one side and substitute known values:

Solution

Pascal's principle applied to hydraulic systems is given by $\frac{F_1}{A_1} = \frac{F_2}{A_2}$:

$$F_2 = \frac{A_2}{A_1}F_1 = \frac{\pi r_2^2}{\pi r_1^2}F_1 = \frac{(1.25 \text{ cm})^2}{(0.250 \text{ cm})^2}\times 500 \text{ N} = 1.25\times 10^4 \text{ N}.$$

(11.27)

Discussion

This value is the force exerted by each of the four slave cylinders. Note that we can add as many slave cylinders as we wish. If each has a 2.50-cm diameter, each will exert 1.25×10^4 N.

A simple hydraulic system, such as a simple machine, can increase force but cannot do more work than done on it. Work is force times distance moved, and the slave cylinder moves through a smaller distance than the master cylinder. Furthermore, the more slaves added, the smaller the distance each moves. Many hydraulic systems—such as power brakes and those in bulldozers—have a motorized pump that actually does most of the work in the system. The movement of the legs of a spider is achieved partly by hydraulics. Using hydraulics, a jumping spider can create a force that makes it capable of jumping 25 times its length!

Conservation of energy applied to a hydraulic system tells us that the system cannot do more work than is done on it. Work transfers energy, and so the work output cannot exceed the work input. Power brakes and other similar hydraulic systems use pumps to supply extra energy when needed.

11.6 Gauge Pressure, Absolute Pressure, and Pressure Measurement

Learning Objectives

By the end of this section, you will be able to:

- Define gauge pressure and absolute pressure.
- Understand the working of aneroid and open-tube barometers.

If you limp into a gas station with a nearly flat tire, you will notice the tire gauge on the airline reads nearly zero when you begin to fill it. In fact, if there were a gaping hole in your tire, the gauge would read zero, even though atmospheric pressure exists in the tire. Why does the gauge read zero? There is no mystery here. Tire gauges are simply designed to read zero at atmospheric pressure and positive when pressure is greater than atmospheric.

Similarly, atmospheric pressure adds to blood pressure in every part of the circulatory system. (As noted in **Pascal's Principle**, the total pressure in a fluid is the sum of the pressures from different sources—here, the heart and the atmosphere.) But atmospheric pressure has no net effect on blood flow since it adds to the pressure coming out of the heart and going back into it, too. What is important is how much *greater* blood pressure is than atmospheric pressure. Blood pressure measurements, like tire pressures, are thus made relative to atmospheric pressure.

In brief, it is very common for pressure gauges to ignore atmospheric pressure—that is, to read zero at atmospheric pressure. We therefore define **gauge pressure** to be the pressure relative to atmospheric pressure. Gauge pressure is positive for pressures above atmospheric pressure, and negative for pressures below it.

Gauge Pressure

Gauge pressure is the pressure relative to atmospheric pressure. Gauge pressure is positive for pressures above atmospheric pressure, and negative for pressures below it.

In fact, atmospheric pressure does add to the pressure in any fluid not enclosed in a rigid container. This happens because of Pascal's principle. The total pressure, or **absolute pressure**, is thus the sum of gauge pressure and atmospheric pressure: $P_{abs} = P_g + P_{atm}$ where P_{abs} is absolute pressure, P_g is gauge pressure, and P_{atm} is atmospheric pressure. For example, if your tire gauge reads 34 psi (pounds per square inch), then the absolute pressure is 34 psi plus 14.7 psi (P_{atm} in psi), or 48.7 psi (equivalent to 336 kPa).

Absolute Pressure

Absolute pressure is the sum of gauge pressure and atmospheric pressure.

For reasons we will explore later, in most cases the absolute pressure in fluids cannot be negative. Fluids push rather than pull, so the smallest absolute pressure is zero. (A negative absolute pressure is a pull.) Thus the smallest possible gauge pressure is $P_g = -P_{atm}$ (this makes P_{abs} zero). There is no theoretical limit to how large a gauge pressure can be.

There are a host of devices for measuring pressure, ranging from tire gauges to blood pressure cuffs. Pascal's principle is of major importance in these devices. The undiminished transmission of pressure through a fluid allows precise remote sensing of pressures. Remote sensing is often more convenient than putting a measuring device into a system, such as a person's artery.

Figure 11.18 shows one of the many types of mechanical pressure gauges in use today. In all mechanical pressure gauges, pressure results in a force that is converted (or transduced) into some type of readout.

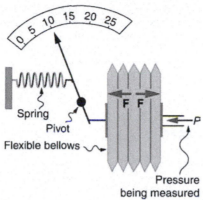

Figure 11.18 This aneroid gauge utilizes flexible bellows connected to a mechanical indicator to measure pressure.

An entire class of gauges uses the property that pressure due to the weight of a fluid is given by $P = h\rho g$. Consider the U-shaped tube shown in **Figure 11.19**, for example. This simple tube is called a *manometer*. In **Figure 11.19**(a), both sides of the tube are open to the atmosphere. Atmospheric pressure therefore pushes down on each side equally so its effect cancels. If the fluid is deeper on one side, there is a greater pressure on the deeper side, and the fluid flows away from that side until the depths are equal.

Let us examine how a manometer is used to measure pressure. Suppose one side of the U-tube is connected to some source of pressure P_{abs} such as the toy balloon in **Figure 11.19**(b) or the vacuum-packed peanut jar shown in **Figure 11.19**(c). Pressure is transmitted undiminished to the manometer, and the fluid levels are no longer equal. In **Figure 11.19**(b), P_{abs} is greater than atmospheric pressure, whereas in **Figure 11.19**(c), P_{abs} is less than atmospheric pressure. In both cases, P_{abs} differs from atmospheric pressure by an amount $h\rho g$, where ρ is the density of the fluid in the manometer. In **Figure 11.19**(b), P_{abs} can support a column of fluid of height h, and so it must exert a pressure $h\rho g$ greater than atmospheric pressure (the gauge pressure P_g is positive). In **Figure 11.19**(c), atmospheric pressure can support a column of fluid of height h, and so P_{abs} is less than atmospheric pressure by an amount $h\rho g$ (the gauge pressure P_g is negative). A manometer with one side open to the atmosphere is an ideal device for measuring gauge pressures. The gauge pressure is $P_g = h\rho g$ and is found by measuring h.

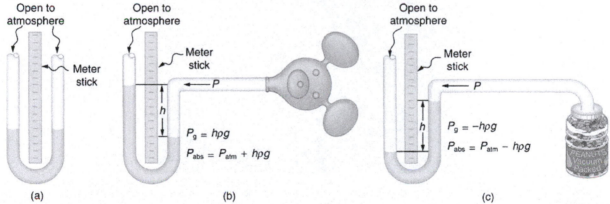

Figure 11.19 An open-tube manometer has one side open to the atmosphere. (a) Fluid depth must be the same on both sides, or the pressure each side exerts at the bottom will be unequal and there will be flow from the deeper side. (b) A positive gauge pressure $P_g = h\rho g$ transmitted to one side of the manometer can support a column of fluid of height h. (c) Similarly, atmospheric pressure is greater than a negative gauge pressure P_g by an amount $h\rho g$. The jar's rigidity prevents atmospheric pressure from being transmitted to the peanuts.

Mercury manometers are often used to measure arterial blood pressure. An inflatable cuff is placed on the upper arm as shown in **Figure 11.20**. By squeezing the bulb, the person making the measurement exerts pressure, which is transmitted undiminished to both the main artery in the arm and the manometer. When this applied pressure exceeds blood pressure, blood flow below the cuff is cut off. The person making the measurement then slowly lowers the applied pressure and listens for blood flow to resume. Blood pressure pulsates because of the pumping action of the heart, reaching a maximum, called **systolic pressure**, and a minimum, called **diastolic pressure**, with each heartbeat. Systolic pressure is measured by noting the value of h when blood flow first begins as cuff pressure is lowered. Diastolic pressure is measured by noting h when blood flows without interruption.

The typical blood pressure of a young adult raises the mercury to a height of 120 mm at systolic and 80 mm at diastolic. This is commonly quoted as 120 over 80, or 120/80. The first pressure is representative of the maximum output of the heart; the second is due to the elasticity of the arteries in maintaining the pressure between beats. The density of the mercury fluid in the manometer is 13.6 times greater than water, so the height of the fluid will be 1/13.6 of that in a water manometer. This reduced height can make measurements difficult, so mercury manometers are used to measure larger pressures, such as blood pressure. The density of mercury is such that $1.0 \text{ mm Hg} = 133 \text{ Pa}$.

Systolic Pressure

Systolic pressure is the maximum blood pressure.

Diastolic Pressure

Diastolic pressure is the minimum blood pressure.

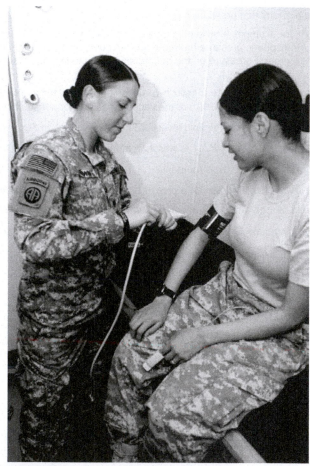

Figure 11.20 In routine blood pressure measurements, an inflatable cuff is placed on the upper arm at the same level as the heart. Blood flow is detected just below the cuff, and corresponding pressures are transmitted to a mercury-filled manometer. (credit: U.S. Army photo by Spc. Micah E. Clare\4TH BCT)

Example 11.7 Calculating Height of IV Bag: Blood Pressure and Intravenous Infusions

Intravenous infusions are usually made with the help of the gravitational force. Assuming that the density of the fluid being administered is 1.00 g/ml, at what height should the IV bag be placed above the entry point so that the fluid just enters the vein if the blood pressure in the vein is 18 mm Hg above atmospheric pressure? Assume that the IV bag is collapsible.

Strategy for (a)

For the fluid to just enter the vein, its pressure at entry must exceed the blood pressure in the vein (18 mm Hg above atmospheric pressure). We therefore need to find the height of fluid that corresponds to this gauge pressure.

Solution

We first need to convert the pressure into SI units. Since $1.0 \text{ mm Hg} = 133 \text{ Pa}$,

$$P = 18 \text{ mm Hg} \times \frac{133 \text{ Pa}}{1.0 \text{ mm Hg}} = 2400 \text{ Pa}. \tag{11.28}$$

Rearranging $P_g = h\rho g$ for h gives $h = \dfrac{P_g}{\rho g}$. Substituting known values into this equation gives

$$\begin{aligned} h &= \frac{2400 \text{ N/m}^2}{\left(1.0 \times 10^3 \text{ kg/m}^3\right)\left(9.80 \text{ m/s}^2\right)} \\ &= 0.24 \text{ m}. \end{aligned} \tag{11.29}$$

Discussion

The IV bag must be placed at 0.24 m above the entry point into the arm for the fluid to just enter the arm. Generally, IV bags are placed higher than this. You may have noticed that the bags used for blood collection are placed below the donor to allow blood to flow easily from the arm to the bag, which is the opposite direction of flow than required in the example presented here.

A *barometer* is a device that measures atmospheric pressure. A mercury barometer is shown in **Figure 11.21**. This device measures atmospheric pressure, rather than gauge pressure, because there is a nearly pure vacuum above the mercury in the tube. The height of the mercury is such that $h\rho g = P_{\text{atm}}$. When atmospheric pressure varies, the mercury rises or falls, giving important clues to weather forecasters. The barometer can also be used as an altimeter, since average atmospheric pressure varies with altitude. Mercury barometers and manometers are so common that units of mm Hg are often quoted for atmospheric pressure and blood pressures. **Table 11.2** gives conversion factors for some of the more commonly used units of pressure.

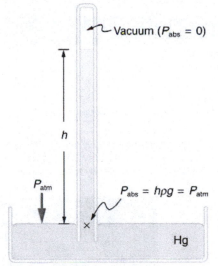

Figure 11.21 A mercury barometer measures atmospheric pressure. The pressure due to the mercury's weight, $h\rho g$, equals atmospheric pressure. The atmosphere is able to force mercury in the tube to a height h because the pressure above the mercury is zero.

Table 11.2 Conversion Factors for Various Pressure Units

Conversion to N/m² (Pa)	Conversion from atm
$1.0 \text{ atm} = 1.013 \times 10^5 \text{ N/m}^2$	$1.0 \text{ atm} = 1.013 \times 10^5 \text{ N/m}^2$
$1.0 \text{ dyne/cm}^2 = 0.10 \text{ N/m}^2$	$1.0 \text{ atm} = 1.013 \times 10^6 \text{ dyne/cm}^2$
$1.0 \text{ kg/cm}^2 = 9.8 \times 10^4 \text{ N/m}^2$	$1.0 \text{ atm} = 1.013 \text{ kg/cm}^2$
$1.0 \text{ lb/in.}^2 = 6.90 \times 10^3 \text{ N/m}^2$	$1.0 \text{ atm} = 14.7 \text{ lb/in.}^2$
$1.0 \text{ mm Hg} = 133 \text{ N/m}^2$	$1.0 \text{ atm} = 760 \text{ mm Hg}$
$1.0 \text{ cm Hg} = 1.33 \times 10^3 \text{ N/m}^2$	$1.0 \text{ atm} = 76.0 \text{ cm Hg}$
$1.0 \text{ cm water} = 98.1 \text{ N/m}^2$	$1.0 \text{ atm} = 1.03 \times 10^3 \text{ cm water}$
$1.0 \text{ bar} = 1.000 \times 10^5 \text{ N/m}^2$	$1.0 \text{ atm} = 1.013 \text{ bar}$
$1.0 \text{ millibar} = 1.000 \times 10^2 \text{ N/m}^2$	$1.0 \text{ atm} = 1013 \text{ millibar}$

11.7 Archimedes' Principle

Learning Objectives
By the end of this section, you will be able to: • Define buoyant force. • State Archimedes' principle. • Understand why objects float or sink. • Understand the relationship between density and Archimedes' principle.

When you rise from lounging in a warm bath, your arms feel strangely heavy. This is because you no longer have the buoyant support of the water. Where does this buoyant force come from? Why is it that some things float and others do not? Do objects that sink get any support at all from the fluid? Is your body buoyed by the atmosphere, or are only helium balloons affected? (See **Figure 11.22**.)

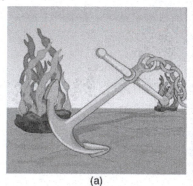  

| (a) | (b) | (c) |

Figure 11.22 (a) Even objects that sink, like this anchor, are partly supported by water when submerged. (b) Submarines have adjustable density (ballast tanks) so that they may float or sink as desired. (credit: Allied Navy) (c) Helium-filled balloons tug upward on their strings, demonstrating air's buoyant effect. (credit: Crystl)

Answers to all these questions, and many others, are based on the fact that pressure increases with depth in a fluid. This means that the upward force on the bottom of an object in a fluid is greater than the downward force on the top of the object. There is a net upward, or **buoyant force** on any object in any fluid. (See **Figure 11.23**.) If the buoyant force is greater than the object's weight, the object will rise to the surface and float. If the buoyant force is less than the object's weight, the object will sink. If the buoyant force equals the object's weight, the object will remain suspended at that depth. The buoyant force is always present whether the object floats, sinks, or is suspended in a fluid.

Buoyant Force

The buoyant force is the net upward force on any object in any fluid.

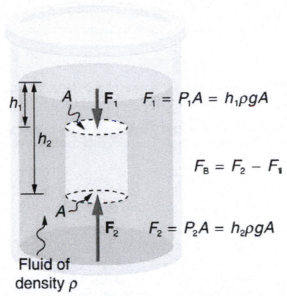

Figure 11.23 Pressure due to the weight of a fluid increases with depth since $P = h\rho g$. This pressure and associated upward force on the bottom of the cylinder are greater than the downward force on the top of the cylinder. Their difference is the buoyant force $\mathbf{F_B}$. (Horizontal forces cancel.)

Just how great is this buoyant force? To answer this question, think about what happens when a submerged object is removed from a fluid, as in **Figure 11.24**.

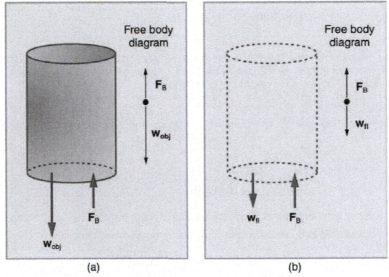

Figure 11.24 (a) An object submerged in a fluid experiences a buoyant force F_B. If F_B is greater than the weight of the object, the object will rise. If F_B is less than the weight of the object, the object will sink. (b) If the object is removed, it is replaced by fluid having weight w_{fl}. Since this weight is supported by surrounding fluid, the buoyant force must equal the weight of the fluid displaced. That is, $F_B = w_{fl}$, a statement of Archimedes' principle.

The space it occupied is filled by fluid having a weight w_{fl}. This weight is supported by the surrounding fluid, and so the buoyant force must equal w_{fl}, the weight of the fluid displaced by the object. It is a tribute to the genius of the Greek mathematician and inventor Archimedes (ca. 287–212 B.C.) that he stated this principle long before concepts of force were well established. Stated in words, **Archimedes' principle** is as follows: The buoyant force on an object equals the weight of the fluid it displaces. In equation form, Archimedes' principle is

$$F_B = w_{fl},$$ (11.30)

where F_B is the buoyant force and w_{fl} is the weight of the fluid displaced by the object. Archimedes' principle is valid in general, for any object in any fluid, whether partially or totally submerged.

Archimedes' Principle

According to this principle the buoyant force on an object equals the weight of the fluid it displaces. In equation form, Archimedes' principle is

$$F_B = w_{fl},$$ (11.31)

where F_B is the buoyant force and w_{fl} is the weight of the fluid displaced by the object.

Humm ... High-tech body swimsuits were introduced in 2008 in preparation for the Beijing Olympics. One concern (and international rule) was that these suits should not provide any buoyancy advantage. How do you think that this rule could be verified?

Making Connections: Take-Home Investigation

The density of aluminum foil is 2.7 times the density of water. Take a piece of foil, roll it up into a ball and drop it into water. Does it sink? Why or why not? Can you make it sink?

Floating and Sinking

Drop a lump of clay in water. It will sink. Then mold the lump of clay into the shape of a boat, and it will float. Because of its shape, the boat displaces more water than the lump and experiences a greater buoyant force. The same is true of steel ships.

Example 11.8 Calculating buoyant force: dependency on shape

(a) Calculate the buoyant force on 10,000 metric tons $(1.00 \times 10^7 \text{ kg})$ of solid steel completely submerged in water, and compare this with the steel's weight. (b) What is the maximum buoyant force that water could exert on this same steel if it were shaped into a boat that could displace $1.00 \times 10^5 \text{ m}^3$ of water?

Strategy for (a)

To find the buoyant force, we must find the weight of water displaced. We can do this by using the densities of water and steel given in **Table 11.1**. We note that, since the steel is completely submerged, its volume and the water's volume are the same. Once we know the volume of water, we can find its mass and weight.

Solution for (a)

First, we use the definition of density $\rho = \frac{m}{V}$ to find the steel's volume, and then we substitute values for mass and density. This gives

$$V_{st} = \frac{m_{st}}{\rho_{st}} = \frac{1.00 \times 10^7 \text{ kg}}{7.8 \times 10^3 \text{ kg/m}^3} = 1.28 \times 10^3 \text{ m}^3.$$ (11.32)

Because the steel is completely submerged, this is also the volume of water displaced, V_w. We can now find the mass of water displaced from the relationship between its volume and density, both of which are known. This gives

$$\begin{aligned} m_w &= \rho_w V_w = (1.000 \times 10^3 \text{ kg/m}^3)(1.28 \times 10^3 \text{ m}^3) \\ &= 1.28 \times 10^6 \text{ kg}. \end{aligned}$$ (11.33)

By Archimedes' principle, the weight of water displaced is $m_w g$, so the buoyant force is

$$\begin{aligned} F_B &= w_w = m_w g = \left(1.28 \times 10^6 \text{ kg}\right)\left(9.80 \text{ m/s}^2\right) \\ &= 1.3 \times 10^7 \text{ N}. \end{aligned}$$ (11.34)

The steel's weight is $m_w g = 9.80 \times 10^7 \text{ N}$, which is much greater than the buoyant force, so the steel will remain submerged. Note that the buoyant force is rounded to two digits because the density of steel is given to only two digits.

Strategy for (b)

Here we are given the maximum volume of water the steel boat can displace. The buoyant force is the weight of this volume of water.

Solution for (b)

The mass of water displaced is found from its relationship to density and volume, both of which are known. That is,

$$m_w = \rho_w V_w = \left(1.000 \times 10^3 \text{ kg/m}^3\right)\left(1.00 \times 10^5 \text{ m}^3\right) \tag{11.35}$$
$$= 1.00 \times 10^8 \text{ kg}.$$

The maximum buoyant force is the weight of this much water, or

$$F_B = w_w = m_w g = \left(1.00 \times 10^8 \text{ kg}\right)\left(9.80 \text{ m/s}^2\right) \tag{11.36}$$
$$= 9.80 \times 10^8 \text{ N}.$$

Discussion

The maximum buoyant force is ten times the weight of the steel, meaning the ship can carry a load nine times its own weight without sinking.

Making Connections: Take-Home Investigation

A piece of household aluminum foil is 0.016 mm thick. Use a piece of foil that measures 10 cm by 15 cm. (a) What is the mass of this amount of foil? (b) If the foil is folded to give it four sides, and paper clips or washers are added to this "boat," what shape of the boat would allow it to hold the most "cargo" when placed in water? Test your prediction.

Density and Archimedes' Principle

Density plays a crucial role in Archimedes' principle. The average density of an object is what ultimately determines whether it floats. If its average density is less than that of the surrounding fluid, it will float. This is because the fluid, having a higher density, contains more mass and hence more weight in the same volume. The buoyant force, which equals the weight of the fluid displaced, is thus greater than the weight of the object. Likewise, an object denser than the fluid will sink.

The extent to which a floating object is submerged depends on how the object's density is related to that of the fluid. In **Figure 11.25**, for example, the unloaded ship has a lower density and less of it is submerged compared with the same ship loaded. We can derive a quantitative expression for the fraction submerged by considering density. The fraction submerged is the ratio of the volume submerged to the volume of the object, or

$$\text{fraction submerged} = \frac{V_{sub}}{V_{obj}} = \frac{V_{fl}}{V_{obj}}. \tag{11.37}$$

The volume submerged equals the volume of fluid displaced, which we call V_{fl}. Now we can obtain the relationship between the densities by substituting $\rho = \frac{m}{V}$ into the expression. This gives

$$\frac{V_{fl}}{V_{obj}} = \frac{m_{fl}/\rho_{fl}}{m_{obj}/\bar{\rho}_{obj}}, \tag{11.38}$$

where $\bar{\rho}_{obj}$ is the average density of the object and ρ_{fl} is the density of the fluid. Since the object floats, its mass and that of the displaced fluid are equal, and so they cancel from the equation, leaving

$$\text{fraction submerged} = \frac{\bar{\rho}_{obj}}{\rho_{fl}}. \tag{11.39}$$

(a)   (b)

Figure 11.25 An unloaded ship (a) floats higher in the water than a loaded ship (b).

We use this last relationship to measure densities. This is done by measuring the fraction of a floating object that is submerged—for example, with a hydrometer. It is useful to define the ratio of the density of an object to a fluid (usually water) as **specific gravity**:

$$\text{specific gravity} = \frac{\bar{\rho}}{\rho_w},$$

(11.40)

where $\bar{\rho}$ is the average density of the object or substance and ρ_w is the density of water at 4.00°C. Specific gravity is dimensionless, independent of whatever units are used for ρ. If an object floats, its specific gravity is less than one. If it sinks, its specific gravity is greater than one. Moreover, the fraction of a floating object that is submerged equals its specific gravity. If an object's specific gravity is exactly 1, then it will remain suspended in the fluid, neither sinking nor floating. Scuba divers try to obtain this state so that they can hover in the water. We measure the specific gravity of fluids, such as battery acid, radiator fluid, and urine, as an indicator of their condition. One device for measuring specific gravity is shown in **Figure 11.26**.

Specific Gravity
_____

Specific gravity is the ratio of the density of an object to a fluid (usually water).

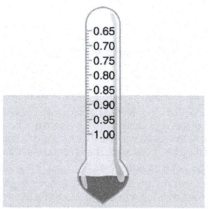

Figure 11.26 This hydrometer is floating in a fluid of specific gravity 0.87. The glass hydrometer is filled with air and weighted with lead at the bottom. It floats highest in the densest fluids and has been calibrated and labeled so that specific gravity can be read from it directly.

Example 11.9 Calculating Average Density: Floating Woman

Suppose a 60.0-kg woman floats in freshwater with 97.0% of her volume submerged when her lungs are full of air. What is her average density?

Strategy

We can find the woman's density by solving the equation

$$\text{fraction submerged} = \frac{\bar{\rho}_{obj}}{\rho_{fl}}$$

(11.41)

for the density of the object. This yields

$$\bar{\rho}_{obj} = \bar{\rho}_{person} = (\text{fraction submerged}) \cdot \rho_{fl}.$$

(11.42)

We know both the fraction submerged and the density of water, and so we can calculate the woman's density.

Solution

Entering the known values into the expression for her density, we obtain

$$\bar{\rho}_{person} = 0.970 \cdot \left(10^3 \frac{\text{kg}}{\text{m}^3}\right) = 970 \frac{\text{kg}}{\text{m}^3}.$$

(11.43)

Discussion

Her density is less than the fluid density. We expect this because she floats. Body density is one indicator of a person's percent body fat, of interest in medical diagnostics and athletic training. (See **Figure 11.27**.)

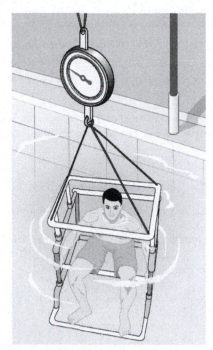

Figure 11.27 Subject in a "fat tank," where he is weighed while completely submerged as part of a body density determination. The subject must completely empty his lungs and hold a metal weight in order to sink. Corrections are made for the residual air in his lungs (measured separately) and the metal weight. His corrected submerged weight, his weight in air, and pinch tests of strategic fatty areas are used to calculate his percent body fat.

There are many obvious examples of lower-density objects or substances floating in higher-density fluids—oil on water, a hot-air balloon, a bit of cork in wine, an iceberg, and hot wax in a "lava lamp," to name a few. Less obvious examples include lava rising in a volcano and mountain ranges floating on the higher-density crust and mantle beneath them. Even seemingly solid Earth has fluid characteristics.

More Density Measurements

One of the most common techniques for determining density is shown in **Figure 11.28**.

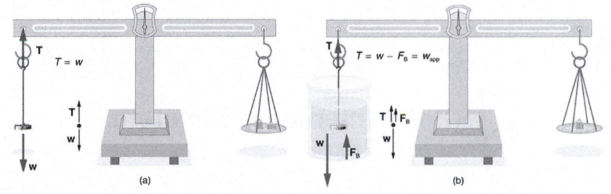

Figure 11.28 (a) A coin is weighed in air. (b) The apparent weight of the coin is determined while it is completely submerged in a fluid of known density. These two measurements are used to calculate the density of the coin.

An object, here a coin, is weighed in air and then weighed again while submerged in a liquid. The density of the coin, an indication of its authenticity, can be calculated if the fluid density is known. This same technique can also be used to determine the density of the fluid if the density of the coin is known. All of these calculations are based on Archimedes' principle.

Archimedes' principle states that the buoyant force on the object equals the weight of the fluid displaced. This, in turn, means that the object *appears* to weigh less when submerged; we call this measurement the object's *apparent weight*. The object suffers an *apparent weight loss* equal to the weight of the fluid displaced. Alternatively, on balances that measure mass, the object suffers an *apparent mass loss* equal to the mass of fluid displaced. That is

$$\text{apparent weight loss} = \text{weight of fluid displace} \tag{11.44}$$

or

$$\text{apparent mass loss} = \text{mass of fluid displaced} \tag{11.45}$$

The next example illustrates the use of this technique.

Example 11.10 Calculating Density: Is the Coin Authentic?

The mass of an ancient Greek coin is determined in air to be 8.630 g. When the coin is submerged in water as shown in Figure 11.28, its apparent mass is 7.800 g. Calculate its density, given that water has a density of 1.000 g/cm^3 and that effects caused by the wire suspending the coin are negligible.

Strategy

To calculate the coin's density, we need its mass (which is given) and its volume. The volume of the coin equals the volume of water displaced. The volume of water displaced V_w can be found by solving the equation for density $\rho = \frac{m}{V}$ for V .

Solution

The volume of water is $V_w = \frac{m_w}{\rho_w}$ where m_w is the mass of water displaced. As noted, the mass of the water displaced equals the apparent mass loss, which is $m_w = 8.630 \text{ g} - 7.800 \text{ g} = 0.830 \text{ g}$. Thus the volume of water is

$V_w = \dfrac{0.830 \text{ g}}{1.000 \text{ g/cm}^3} = 0.830 \text{ cm}^3$. This is also the volume of the coin, since it is completely submerged. We can now find the density of the coin using the definition of density:

$$\rho_c = \frac{m_c}{V_c} = \frac{8.630 \text{ g}}{0.830 \text{ cm}^3} = 10.4 \text{ g/cm}^3. \qquad (11.46)$$

Discussion

You can see from Table 11.1 that this density is very close to that of pure silver, appropriate for this type of ancient coin. Most modern counterfeits are not pure silver.

This brings us back to Archimedes' principle and how it came into being. As the story goes, the king of Syracuse gave Archimedes the task of determining whether the royal crown maker was supplying a crown of pure gold. The purity of gold is difficult to determine by color (it can be diluted with other metals and still look as yellow as pure gold), and other analytical techniques had not yet been conceived. Even ancient peoples, however, realized that the density of gold was greater than that of any other then-known substance. Archimedes purportedly agonized over his task and had his inspiration one day while at the public baths, pondering the support the water gave his body. He came up with his now-famous principle, saw how to apply it to determine density, and ran naked down the streets of Syracuse crying "Eureka!" (Greek for "I have found it"). Similar behavior can be observed in contemporary physicists from time to time!

PhET Explorations: Buoyancy

When will objects float and when will they sink? Learn how buoyancy works with blocks. Arrows show the applied forces, and you can modify the properties of the blocks and the fluid.

Figure 11.29 Buoyancy (http://cnx.org/content/m55215/1.2/buoyancy_en.jar)

11.8 Cohesion and Adhesion in Liquids: Surface Tension and Capillary Action

Learning Objectives

By the end of this section, you will be able to:

- Understand cohesive and adhesive forces.
- Define surface tension.
- Understand capillary action.

Cohesion and Adhesion in Liquids

Children blow soap bubbles and play in the spray of a sprinkler on a hot summer day. (See Figure 11.30.) An underwater spider keeps his air supply in a shiny bubble he carries wrapped around him. A technician draws blood into a small-diameter tube just by touching it to a drop on a pricked finger. A premature infant struggles to inflate her lungs. What is the common thread? All these activities are dominated by the attractive forces between atoms and molecules in liquids—both within a liquid and between the liquid and its surroundings.

Attractive forces between molecules of the same type are called **cohesive forces**. Liquids can, for example, be held in open containers because cohesive forces hold the molecules together. Attractive forces between molecules of different types are called **adhesive forces**. Such forces cause liquid drops to cling to window panes, for example. In this section we examine effects directly attributable to cohesive and adhesive forces in liquids.

Cohesive Forces

Attractive forces between molecules of the same type are called cohesive forces.

Adhesive Forces

Attractive forces between molecules of different types are called adhesive forces.

Figure 11.30 The soap bubbles in this photograph are caused by cohesive forces among molecules in liquids. (credit: Steve Ford Elliott)

Surface Tension

Cohesive forces between molecules cause the surface of a liquid to contract to the smallest possible surface area. This general effect is called **surface tension**. Molecules on the surface are pulled inward by cohesive forces, reducing the surface area. Molecules inside the liquid experience zero net force, since they have neighbors on all sides.

Surface Tension

Cohesive forces between molecules cause the surface of a liquid to contract to the smallest possible surface area. This general effect is called surface tension.

Making Connections: Surface Tension

Forces between atoms and molecules underlie the macroscopic effect called surface tension. These attractive forces pull the molecules closer together and tend to minimize the surface area. This is another example of a submicroscopic explanation for a macroscopic phenomenon.

The model of a liquid surface acting like a stretched elastic sheet can effectively explain surface tension effects. For example, some insects can walk on water (as opposed to floating in it) as we would walk on a trampoline—they dent the surface as shown in **Figure 11.31**(a). **Figure 11.31**(b) shows another example, where a needle rests on a water surface. The iron needle cannot, and does not, float, because its density is greater than that of water. Rather, its weight is supported by forces in the stretched surface that try to make the surface smaller or flatter. If the needle were placed point down on the surface, its weight acting on a smaller area would break the surface, and it would sink.

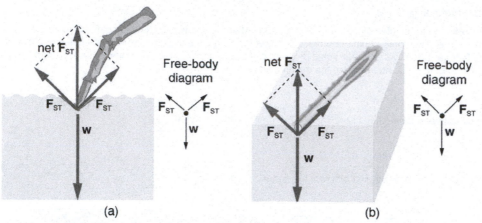

Figure 11.31 Surface tension supporting the weight of an insect and an iron needle, both of which rest on the surface without penetrating it. They are not floating; rather, they are supported by the surface of the liquid. (a) An insect leg dents the water surface. F_{ST} is a restoring force (surface tension) parallel to the surface. (b) An iron needle similarly dents a water surface until the restoring force (surface tension) grows to equal its weight.

Surface tension is proportional to the strength of the cohesive force, which varies with the type of liquid. Surface tension γ is defined to be the force F per unit length L exerted by a stretched liquid membrane:

$$\gamma = \frac{F}{L}. \tag{11.47}$$

Table 11.3 lists values of γ for some liquids. For the insect of **Figure 11.31**(a), its weight w is supported by the upward components of the surface tension force: $w = \gamma L \sin \theta$, where L is the circumference of the insect's foot in contact with the water. **Figure 11.32** shows one way to measure surface tension. The liquid film exerts a force on the movable wire in an attempt to reduce its surface area. The magnitude of this force depends on the surface tension of the liquid and can be measured accurately.

Surface tension is the reason why liquids form bubbles and droplets. The inward surface tension force causes bubbles to be approximately spherical and raises the pressure of the gas trapped inside relative to atmospheric pressure outside. It can be shown that the gauge pressure P inside a spherical bubble is given by

$$P = \frac{4\gamma}{r}, \tag{11.48}$$

where r is the radius of the bubble. Thus the pressure inside a bubble is greatest when the bubble is the smallest. Another bit of evidence for this is illustrated in **Figure 11.33**. When air is allowed to flow between two balloons of unequal size, the smaller balloon tends to collapse, filling the larger balloon.

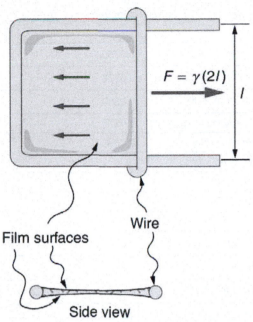

Figure 11.32 Sliding wire device used for measuring surface tension; the device exerts a force to reduce the film's surface area. The force needed to hold the wire in place is $F = \gamma L = \gamma(2l)$, since there are *two* liquid surfaces attached to the wire. This force remains nearly constant as the film is stretched, until the film approaches its breaking point.

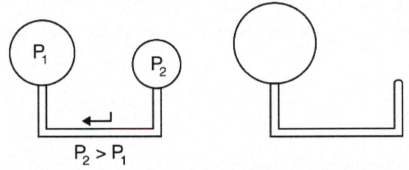

Figure 11.33 With the valve closed, two balloons of different sizes are attached to each end of a tube. Upon opening the valve, the smaller balloon decreases in size with the air moving to fill the larger balloon. The pressure in a spherical balloon is inversely proportional to its radius, so that the smaller balloon has a greater internal pressure than the larger balloon, resulting in this flow.

Table 11.3 Surface Tension of Some Liquids[1]

Liquid	Surface tension γ(N/m)
Water at $0°C$	0.0756
Water at $20°C$	0.0728
Water at $100°C$	0.0589
Soapy water (typical)	0.0370
Ethyl alcohol	0.0223
Glycerin	0.0631
Mercury	0.465
Olive oil	0.032
Tissue fluids (typical)	0.050
Blood, whole at $37°C$	0.058
Blood plasma at $37°C$	0.073
Gold at $1070°C$	1.000
Oxygen at $-193°C$	0.0157
Helium at $-269°C$	0.00012

Example 11.11 Surface Tension: Pressure Inside a Bubble

Calculate the gauge pressure inside a soap bubble 2.00×10^{-4} m in radius using the surface tension for soapy water in **Table 11.3**. Convert this pressure to mm Hg.

Strategy

The radius is given and the surface tension can be found in **Table 11.3**, and so P can be found directly from the equation $P = \frac{4\gamma}{r}$.

Solution

Substituting r and γ into the equation $P = \frac{4\gamma}{r}$, we obtain

$$P = \frac{4\gamma}{r} = \frac{4(0.037 \text{ N/m})}{2.00 \times 10^{-4} \text{ m}} = 740 \text{ N/m}^2 = 740 \text{ Pa}. \tag{11.49}$$

We use a conversion factor to get this into units of mm Hg:

$$P = \left(740 \text{ N/m}^2\right)\frac{1.00 \text{ mm Hg}}{133 \text{ N/m}^2} = 5.56 \text{ mm Hg}. \tag{11.50}$$

Discussion

Note that if a hole were to be made in the bubble, the air would be forced out, the bubble would decrease in radius, and the pressure inside would *increase* to atmospheric pressure (760 mm Hg).

Our lungs contain hundreds of millions of mucus-lined sacs called *alveoli*, which are very similar in size, and about 0.1 mm in diameter. (See **Figure 11.34**.) You can exhale without muscle action by allowing surface tension to contract these sacs. Medical patients whose breathing is aided by a positive pressure respirator have air blown into the lungs, but are generally allowed to exhale on their own. Even if there is paralysis, surface tension in the alveoli will expel air from the lungs. Since pressure increases as the radii of the alveoli decrease, an occasional deep cleansing breath is needed to fully reinflate the alveoli. Respirators are programmed to do this and we find it natural, as do our companion dogs and cats, to take a cleansing breath before settling into a nap.

---

1. At 20ºC unless otherwise stated.

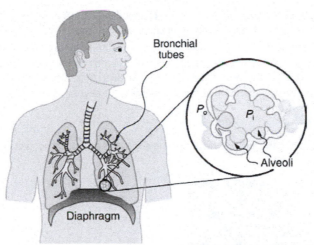

Figure 11.34 Bronchial tubes in the lungs branch into ever-smaller structures, finally ending in alveoli. The alveoli act like tiny bubbles. The surface tension of their mucous lining aids in exhalation and can prevent inhalation if too great.

The tension in the walls of the alveoli results from the membrane tissue and a liquid on the walls of the alveoli containing a long lipoprotein that acts as a surfactant (a surface-tension reducing substance). The need for the surfactant results from the tendency of small alveoli to collapse and the air to fill into the larger alveoli making them even larger (as demonstrated in **Figure 11.33**). During inhalation, the lipoprotein molecules are pulled apart and the wall tension increases as the radius increases (increased surface tension). During exhalation, the molecules slide back together and the surface tension decreases, helping to prevent a collapse of the alveoli. The surfactant therefore serves to change the wall tension so that small alveoli don't collapse and large alveoli are prevented from expanding too much. This tension change is a unique property of these surfactants, and is not shared by detergents (which simply lower surface tension). (See **Figure 11.35**.)

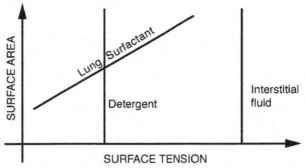

Figure 11.35 Surface tension as a function of surface area. The surface tension for lung surfactant decreases with decreasing area. This ensures that small alveoli don't collapse and large alveoli are not able to over expand.

If water gets into the lungs, the surface tension is too great and you cannot inhale. This is a severe problem in resuscitating drowning victims. A similar problem occurs in newborn infants who are born without this surfactant—their lungs are very difficult to inflate. This condition is known as *hyaline membrane disease* and is a leading cause of death for infants, particularly in premature births. Some success has been achieved in treating hyaline membrane disease by spraying a surfactant into the infant's breathing passages. Emphysema produces the opposite problem with alveoli. Alveolar walls of emphysema victims deteriorate, and the sacs combine to form larger sacs. Because pressure produced by surface tension decreases with increasing radius, these larger sacs produce smaller pressure, reducing the ability of emphysema victims to exhale. A common test for emphysema is to measure the pressure and volume of air that can be exhaled.

Making Connections: Take-Home Investigation

(1) Try floating a sewing needle on water. In order for this activity to work, the needle needs to be very clean as even the oil from your fingers can be sufficient to affect the surface properties of the needle. (2) Place the bristles of a paint brush into water. Pull the brush out and notice that for a short while, the bristles will stick together. The surface tension of the water surrounding the bristles is sufficient to hold the bristles together. As the bristles dry out, the surface tension effect dissipates. (3) Place a loop of thread on the surface of still water in such a way that all of the thread is in contact with the water. Note the shape of the loop. Now place a drop of detergent into the middle of the loop. What happens to the shape of the loop? Why? (4) Sprinkle pepper onto the surface of water. Add a drop of detergent. What happens? Why? (5) Float two matches parallel to each other and add a drop of detergent between them. What happens? Note: For each new experiment, the water needs to be replaced and the bowl washed to free it of any residual detergent.

Adhesion and Capillary Action

Why is it that water beads up on a waxed car but does not on bare paint? The answer is that the adhesive forces between water and wax are much smaller than those between water and paint. Competition between the forces of adhesion and cohesion are important in the macroscopic behavior of liquids. An important factor in studying the roles of these two forces is the angle θ between the tangent to the liquid surface and the surface. (See **Figure 11.36**.) The **contact angle** θ is directly related to the relative strength of the cohesive and adhesive forces. The larger the strength of the cohesive force relative to the adhesive force, the larger θ is, and the more the liquid tends to form a droplet. The smaller θ is, the smaller the relative strength, so that the adhesive force is able to flatten the drop. **Table 11.4** lists contact angles for several combinations of liquids and solids.

Contact Angle
_____

The angle θ between the tangent to the liquid surface and the surface is called the contact angle.

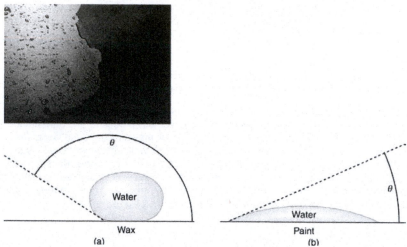

Figure 11.36 In the photograph, water beads on the waxed car paint and flattens on the unwaxed paint. (a) Water forms beads on the waxed surface because the cohesive forces responsible for surface tension are larger than the adhesive forces, which tend to flatten the drop. (b) Water beads on bare paint are flattened considerably because the adhesive forces between water and paint are strong, overcoming surface tension. The contact angle θ is directly related to the relative strengths of the cohesive and adhesive forces. The larger θ is, the larger the ratio of cohesive to adhesive forces. (credit: P. P. Urone)

One important phenomenon related to the relative strength of cohesive and adhesive forces is **capillary action**—the tendency of a fluid to be raised or suppressed in a narrow tube, or *capillary tube*. This action causes blood to be drawn into a small-diameter tube when the tube touches a drop.

Capillary Action
_____

The tendency of a fluid to be raised or suppressed in a narrow tube, or capillary tube, is called capillary action.

If a capillary tube is placed vertically into a liquid, as shown in **Figure 11.37**, capillary action will raise or suppress the liquid inside the tube depending on the combination of substances. The actual effect depends on the relative strength of the cohesive and adhesive forces and, thus, the contact angle θ given in the table. If θ is less than $90°$, then the fluid will be raised; if θ is greater than $90°$, it will be suppressed. Mercury, for example, has a very large surface tension and a large contact angle with glass. When placed in a tube, the surface of a column of mercury curves downward, somewhat like a drop. The curved surface of a fluid in a tube is called a **meniscus**. The tendency of surface tension is always to reduce the surface area. Surface tension thus flattens the curved liquid surface in a capillary tube. This results in a downward force in mercury and an upward force in water, as seen in **Figure 11.37**.

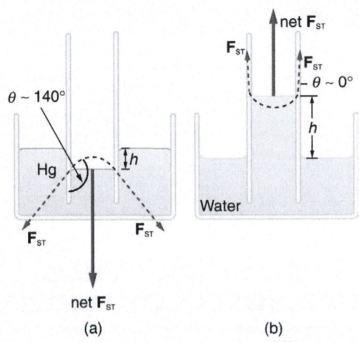

Figure 11.37 (a) Mercury is suppressed in a glass tube because its contact angle is greater than $90°$. Surface tension exerts a downward force as it flattens the mercury, suppressing it in the tube. The dashed line shows the shape the mercury surface would have without the flattening effect of surface tension. (b) Water is raised in a glass tube because its contact angle is nearly $0°$. Surface tension therefore exerts an upward force when it flattens the surface to reduce its area.

Table 11.4 Contact Angles of Some Substances

Interface	Contact angle Θ
Mercury–glass	$140°$
Water–glass	$0°$
Water–paraffin	$107°$
Water–silver	$90°$
Organic liquids (most)–glass	$0°$
Ethyl alcohol–glass	$0°$
Kerosene–glass	$26°$

Capillary action can move liquids horizontally over very large distances, but the height to which it can raise or suppress a liquid in a tube is limited by its weight. It can be shown that this height h is given by

$$h = \frac{2\gamma \cos \theta}{\rho g r}.$$

(11.51)

If we look at the different factors in this expression, we might see how it makes good sense. The height is directly proportional to the surface tension γ, which is its direct cause. Furthermore, the height is inversely proportional to tube radius—the smaller the radius r, the higher the fluid can be raised, since a smaller tube holds less mass. The height is also inversely proportional to fluid density ρ, since a larger density means a greater mass in the same volume. (See **Figure 11.38**.)

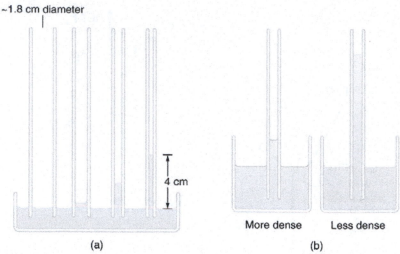

Figure 11.38 (a) Capillary action depends on the radius of a tube. The smaller the tube, the greater the height reached. The height is negligible for large-radius tubes. (b) A denser fluid in the same tube rises to a smaller height, all other factors being the same.

Example 11.12 Calculating Radius of a Capillary Tube: Capillary Action: Tree Sap

Can capillary action be solely responsible for sap rising in trees? To answer this question, calculate the radius of a capillary tube that would raise sap 100 m to the top of a giant redwood, assuming that sap's density is 1050 kg/m^3, its contact angle is zero, and its surface tension is the same as that of water at $20.0° \text{ C}$.

Strategy

The height to which a liquid will rise as a result of capillary action is given by $h = \dfrac{2\gamma \cos \theta}{\rho g r}$, and every quantity is known except for r.

Solution

Solving for r and substituting known values produces

$$r = \frac{2\gamma \cos \theta}{\rho g h} = \frac{2(0.0728 \text{ N/m})\cos(0°)}{(1050 \text{ kg/m}^3)(9.80 \text{ m/s}^2)(100 \text{ m})} \qquad (11.52)$$

$$= 1.41 \times 10^{-7} \text{ m}.$$

Discussion

This result is unreasonable. Sap in trees moves through the *xylem*, which forms tubes with radii as small as $2.5 \times 10^{-5} \text{ m}$. This value is about 180 times as large as the radius found necessary here to raise sap 100 m. This means that capillary action alone cannot be solely responsible for sap getting to the tops of trees.

How *does* sap get to the tops of tall trees? (Recall that a column of water can only rise to a height of 10 m when there is a vacuum at the top—see Example 11.5.) The question has not been completely resolved, but it appears that it is pulled up like a chain held together by cohesive forces. As each molecule of sap enters a leaf and evaporates (a process called transpiration), the entire chain is pulled up a notch. So a negative pressure created by water evaporation must be present to pull the sap up through the xylem vessels. In most situations, *fluids can push but can exert only negligible pull*, because the cohesive forces seem to be too small to hold the molecules tightly together. But in this case, the cohesive force of water molecules provides a very strong pull. Figure 11.39 shows one device for studying negative pressure. Some experiments have demonstrated that negative pressures sufficient to pull sap to the tops of the tallest trees *can* be achieved.

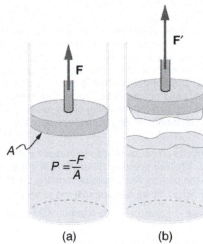

Figure 11.39 (a) When the piston is raised, it stretches the liquid slightly, putting it under tension and creating a negative absolute pressure $P = -F/A$. (b) The liquid eventually separates, giving an experimental limit to negative pressure in this liquid.

11.9 Pressures in the Body

Learning Objectives

By the end of this section, you will be able to:

- Explain the concept of pressure in the human body.
- Explain systolic and diastolic blood pressures.
- Describe pressures in the eye, lungs, spinal column, bladder, and skeletal system.

Pressure in the Body

Next to taking a person's temperature and weight, measuring blood pressure is the most common of all medical examinations. Control of high blood pressure is largely responsible for the significant decreases in heart attack and stroke fatalities achieved in the last three decades. The pressures in various parts of the body can be measured and often provide valuable medical indicators. In this section, we consider a few examples together with some of the physics that accompanies them.

Table 11.5 lists some of the measured pressures in mm Hg, the units most commonly quoted.

Table 11.5 Typical Pressures in Humans

Body system	Gauge pressure in mm Hg
Blood pressures in large arteries (resting)	
Maximum (systolic)	100–140
Minimum (diastolic)	60–90
Blood pressure in large veins	4–15
Eye	12–24
Brain and spinal fluid (lying down)	5–12
Bladder	
While filling	0–25
When full	100–150
Chest cavity between lungs and ribs	−8 to −4
Inside lungs	−2 to +3
Digestive tract	
Esophagus	−2
Stomach	0–20
Intestines	10–20
Middle ear	<1

Blood Pressure

Common arterial blood pressure measurements typically produce values of 120 mm Hg and 80 mm Hg, respectively, for systolic and diastolic pressures. Both pressures have health implications. When systolic pressure is chronically high, the risk of stroke and heart attack is increased. If, however, it is too low, fainting is a problem. **Systolic pressure** increases dramatically during exercise to increase blood flow and returns to normal afterward. This change produces no ill effects and, in fact, may be beneficial to the tone of the circulatory system. **Diastolic pressure** can be an indicator of fluid balance. When low, it may indicate that a person is hemorrhaging internally and needs a transfusion. Conversely, high diastolic pressure indicates a ballooning of the blood vessels, which may be due to the transfusion of too much fluid into the circulatory system. High diastolic pressure is also an indication that blood vessels are not dilating properly to pass blood through. This can seriously strain the heart in its attempt to pump blood.

Blood leaves the heart at about 120 mm Hg but its pressure continues to decrease (to almost 0) as it goes from the aorta to smaller arteries to small veins (see **Figure 11.40**). The pressure differences in the circulation system are caused by blood flow through the system as well as the position of the person. For a person standing up, the pressure in the feet will be larger than at the heart due to the weight of the blood $(P = h\rho g)$. If we assume that the distance between the heart and the feet of a person in an upright position is 1.4 m, then the increase in pressure in the feet relative to that in the heart (for a static column of blood) is given by

$$\Delta P = \Delta h\rho g = (1.4 \text{ m})\left(1050 \text{ kg/m}^3\right)\left(9.80 \text{ m/s}^2\right) = 1.4 \times 10^4 \text{ Pa} = 108 \text{ mm Hg.} \tag{11.53}$$

Increase in Pressure in the Feet of a Person
$$\Delta P = \Delta h\rho g = (1.4 \text{ m})\left(1050 \text{ kg/m}^3\right)\left(9.80 \text{ m/s}^2\right) = 1.4 \times 10^4 \text{ Pa} = 108 \text{ mm Hg.} \tag{11.54}$$

Standing a long time can lead to an accumulation of blood in the legs and swelling. This is the reason why soldiers who are required to stand still for long periods of time have been known to faint. Elastic bandages around the calf can help prevent this accumulation and can also help provide increased pressure to enable the veins to send blood back up to the heart. For similar reasons, doctors recommend tight stockings for long-haul flights.

Blood pressure may also be measured in the major veins, the heart chambers, arteries to the brain, and the lungs. But these pressures are usually only monitored during surgery or for patients in intensive care since the measurements are invasive. To obtain these pressure measurements, qualified health care workers thread thin tubes, called catheters, into appropriate locations to transmit pressures to external measuring devices.

The heart consists of two pumps—the right side forcing blood through the lungs and the left causing blood to flow through the rest of the body (**Figure 11.40**). Right-heart failure, for example, results in a rise in the pressure in the vena cavae and a drop in pressure in the arteries to the lungs. Left-heart failure results in a rise in the pressure entering the left side of the heart and a drop in aortal pressure. Implications of these and other pressures on flow in the circulatory system will be discussed in more detail in **Fluid Dynamics and Its Biological and Medical Applications**.

Two Pumps of the Heart

The heart consists of two pumps—the right side forcing blood through the lungs and the left causing blood to flow through the rest of the body.

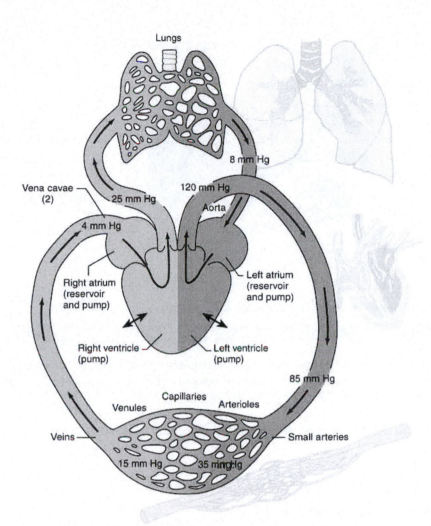

Figure 11.40 Schematic of the circulatory system showing typical pressures. The two pumps in the heart increase pressure and that pressure is reduced as the blood flows through the body. Long-term deviations from these pressures have medical implications discussed in some detail in the **Fluid Dynamics and Its Biological and Medical Applications**. Only aortal or arterial blood pressure can be measured noninvasively.

Pressure in the Eye

The shape of the eye is maintained by fluid pressure, called **intraocular pressure**, which is normally in the range of 12.0 to 24.0 mm Hg. When the circulation of fluid in the eye is blocked, it can lead to a buildup in pressure, a condition called **glaucoma**. The net pressure can become as great as 85.0 mm Hg, an abnormally large pressure that can permanently damage the optic nerve.

To get an idea of the force involved, suppose the back of the eye has an area of 6.0 cm^2, and the net pressure is 85.0 mm Hg.

Force is given by $F = PA$. To get F in newtons, we convert the area to m^2 ($1 \text{ m}^2 = 10^4 \text{ cm}^2$). Then we calculate as follows:

$$F = h\rho g A = (85.0 \times 10^{-3} \text{ m})(13.6 \times 10^3 \text{ kg/m}^3)(9.80 \text{ m/s}^2)(6.0 \times 10^{-4} \text{ m}^2) = 6.8 \text{ N}. \tag{11.55}$$

Eye Pressure

The shape of the eye is maintained by fluid pressure, called intraocular pressure. When the circulation of fluid in the eye is blocked, it can lead to a buildup in pressure, a condition called glaucoma. The force is calculated as

$$F = h\rho g A = (85.0 \times 10^{-3} \text{ m})(13.6 \times 10^3 \text{ kg/m}^3)(9.80 \text{ m/s}^2)(6.0 \times 10^{-4} \text{ m}^2) = 6.8 \text{ N}. \tag{11.56}$$

This force is the weight of about a 680-g mass. A mass of 680 g resting on the eye (imagine 1.5 lb resting on your eye) would be sufficient to cause it damage. (A normal force here would be the weight of about 120 g, less than one-quarter of our initial value.)

People over 40 years of age are at greatest risk of developing glaucoma and should have their intraocular pressure tested routinely. Most measurements involve exerting a force on the (anesthetized) eye over some area (a pressure) and observing the eye's response. A noncontact approach uses a puff of air and a measurement is made of the force needed to indent the eye (**Figure 11.41**). If the intraocular pressure is high, the eye will deform less and rebound more vigorously than normal. Excessive intraocular pressures can be detected reliably and sometimes controlled effectively.

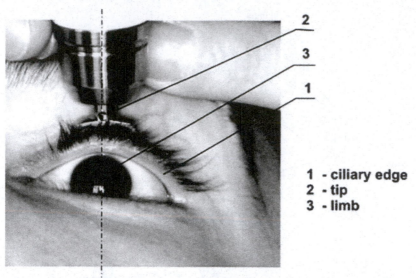

1 - ciliary edge
2 - tip
3 - limb

Figure 11.41 The intraocular eye pressure can be read with a tonometer. (credit: DevelopAll at the Wikipedia Project.)

Example 11.13 Calculating Gauge Pressure and Depth: Damage to the Eardrum

Suppose a 3.00-N force can rupture an eardrum. (a) If the eardrum has an area of 1.00 cm^2, calculate the maximum tolerable gauge pressure on the eardrum in newtons per meter squared and convert it to millimeters of mercury. (b) At what depth in freshwater would this person's eardrum rupture, assuming the gauge pressure in the middle ear is zero?

Strategy for (a)

The pressure can be found directly from its definition since we know the force and area. We are looking for the gauge pressure.

Solution for (a)

$$P_g = F / A = 3.00 \text{ N} / (1.00 \times 10^{-4} \text{ m}^2) = 3.00 \times 10^4 \text{ N/m}^2. \tag{11.57}$$

We now need to convert this to units of mm Hg:

$$P_g = 3.0 \times 10^4 \text{ N/m}^2 \left(\frac{1.0 \text{ mm Hg}}{133 \text{ N/m}^2} \right) = 226 \text{ mm Hg}. \tag{11.58}$$

Strategy for (b)

Here we will use the fact that the water pressure varies linearly with depth h below the surface.

Solution for (b)

$P = h\rho g$ and therefore $h = P / \rho g$. Using the value above for P, we have

$$h = \frac{3.0 \times 10^4 \text{ N/m}^2}{(1.00 \times 10^3 \text{ kg/m}^3)(9.80 \text{ m/s}^2)} = 3.06 \text{ m}. \tag{11.59}$$

Discussion

Similarly, increased pressure exerted upon the eardrum from the middle ear can arise when an infection causes a fluid buildup.

Pressure Associated with the Lungs

The pressure inside the lungs increases and decreases with each breath. The pressure drops to below atmospheric pressure (negative gauge pressure) when you inhale, causing air to flow into the lungs. It increases above atmospheric pressure (positive gauge pressure) when you exhale, forcing air out.

Lung pressure is controlled by several mechanisms. Muscle action in the diaphragm and rib cage is necessary for inhalation; this muscle action increases the volume of the lungs thereby reducing the pressure within them **Figure 11.42**. Surface tension in the alveoli creates a positive pressure opposing inhalation. (See **Cohesion and Adhesion in Liquids: Surface Tension and Capillary Action.**) You can exhale without muscle action by letting surface tension in the alveoli create its own positive pressure. Muscle action can add to this positive pressure to produce forced exhalation, such as when you blow up a balloon, blow out a candle, or cough.

The lungs, in fact, would collapse due to the surface tension in the alveoli, if they were not attached to the inside of the chest wall by liquid adhesion. The gauge pressure in the liquid attaching the lungs to the inside of the chest wall is thus negative, ranging from -4 to -8 mm Hg during exhalation and inhalation, respectively. If air is allowed to enter the chest cavity, it breaks the attachment, and one or both lungs may collapse. Suction is applied to the chest cavity of surgery patients and trauma victims to reestablish negative pressure and inflate the lungs.

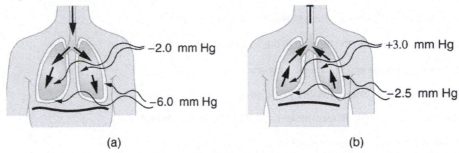

Figure 11.42 (a) During inhalation, muscles expand the chest, and the diaphragm moves downward, reducing pressure inside the lungs to less than atmospheric (negative gauge pressure). Pressure between the lungs and chest wall is even lower to overcome the positive pressure created by surface tension in the lungs. (b) During gentle exhalation, the muscles simply relax and surface tension in the alveoli creates a positive pressure inside the lungs, forcing air out. Pressure between the chest wall and lungs remains negative to keep them attached to the chest wall, but it is less negative than during inhalation.

Other Pressures in the Body

Spinal Column and Skull

Normally, there is a 5- to12-mm Hg pressure in the fluid surrounding the brain and filling the spinal column. This cerebrospinal fluid serves many purposes, one of which is to supply flotation to the brain. The buoyant force supplied by the fluid nearly equals the weight of the brain, since their densities are nearly equal. If there is a loss of fluid, the brain rests on the inside of the skull, causing severe headaches, constricted blood flow, and serious damage. Spinal fluid pressure is measured by means of a needle inserted between vertebrae that transmits the pressure to a suitable measuring device.

Bladder Pressure

This bodily pressure is one of which we are often aware. In fact, there is a relationship between our awareness of this pressure and a subsequent increase in it. Bladder pressure climbs steadily from zero to about 25 mm Hg as the bladder fills to its normal capacity of 500 cm^3. This pressure triggers the **micturition reflex**, which stimulates the feeling of needing to urinate. What is more, it also causes muscles around the bladder to contract, raising the pressure to over 100 mm Hg, accentuating the sensation. Coughing, straining, tensing in cold weather, wearing tight clothes, and experiencing simple nervous tension all can increase bladder pressure and trigger this reflex. So can the weight of a pregnant woman's fetus, especially if it is kicking vigorously or pushing down with its head! Bladder pressure can be measured by a catheter or by inserting a needle through the bladder wall and transmitting the pressure to an appropriate measuring device. One hazard of high bladder pressure (sometimes created by an obstruction), is that such pressure can force urine back into the kidneys, causing potentially severe damage.

Pressures in the Skeletal System

These pressures are the largest in the body, due both to the high values of initial force, and the small areas to which this force is applied, such as in the joints.. For example, when a person lifts an object improperly, a force of 5000 N may be created between vertebrae in the spine, and this may be applied to an area as small as 10 cm^2. The pressure created is

$$P = F / A = (5000 \text{ N}) / (10^{-3} \text{ m}^2) = 5.0 \times 10^6 \text{ N/m}^2$$

or about 50 atm! This pressure can damage both the spinal discs (the cartilage between vertebrae), as well as the bony vertebrae themselves. Even under normal circumstances, forces between vertebrae in the spine are large enough to create pressures of several atmospheres. Most causes of excessive pressure in the skeletal system can be avoided by lifting properly and avoiding extreme physical activity. (See **Forces and Torques in Muscles and Joints**.)

There are many other interesting and medically significant pressures in the body. For example, pressure caused by various muscle actions drives food and waste through the digestive system. Stomach pressure behaves much like bladder pressure and is tied to the sensation of hunger. Pressure in the relaxed esophagus is normally negative because pressure in the chest cavity is normally negative. Positive pressure in the stomach may thus force acid into the esophagus, causing "heartburn." Pressure in the middle ear can result in significant force on the eardrum if it differs greatly from atmospheric pressure, such as while scuba diving. The decrease in external pressure is also noticeable during plane flights (due to a decrease in the weight of air above relative to that at the Earth's surface). The Eustachian tubes connect the middle ear to the throat and allow us to equalize pressure in the middle ear to avoid an imbalance of force on the eardrum.

Many pressures in the human body are associated with the flow of fluids. Fluid flow will be discussed in detail in the **Fluid Dynamics and Its Biological and Medical Applications**.

Glossary

absolute pressure: the sum of gauge pressure and atmospheric pressure

adhesive forces: the attractive forces between molecules of different types

Archimedes' principle: the buoyant force on an object equals the weight of the fluid it displaces

buoyant force: the net upward force on any object in any fluid

capillary action: the tendency of a fluid to be raised or lowered in a narrow tube

cohesive forces: the attractive forces between molecules of the same type

contact angle: the angle θ between the tangent to the liquid surface and the surface

density: the mass per unit volume of a substance or object

diastolic pressure: the minimum blood pressure in the artery

diastolic pressure: minimum arterial blood pressure; indicator for the fluid balance

fluids: liquids and gases; a fluid is a state of matter that yields to shearing forces

gauge pressure: the pressure relative to atmospheric pressure

glaucoma: condition caused by the buildup of fluid pressure in the eye

intraocular pressure: fluid pressure in the eye

micturition reflex: stimulates the feeling of needing to urinate, triggered by bladder pressure

Pascal's Principle: a change in pressure applied to an enclosed fluid is transmitted undiminished to all portions of the fluid and to the walls of its container

pressure: the force per unit area perpendicular to the force, over which the force acts

pressure: the weight of the fluid divided by the area supporting it

specific gravity: the ratio of the density of an object to a fluid (usually water)

surface tension: the cohesive forces between molecules which cause the surface of a liquid to contract to the smallest possible surface area

systolic pressure: the maximum blood pressure in the artery

systolic pressure: maximum arterial blood pressure; indicator for the blood flow

Section Summary

11.1 What Is a Fluid?
*   A fluid is a state of matter that yields to sideways or shearing forces. Liquids and gases are both fluids. Fluid statics is the physics of stationary fluids.

11.2 Density
*   Density is the mass per unit volume of a substance or object. In equation form, density is defined as

$$\rho = \frac{m}{V}.$$

*   The SI unit of density is kg/m^3.

11.3 Pressure
*   Pressure is the force per unit perpendicular area over which the force is applied. In equation form, pressure is defined as

$$P = \frac{F}{A}.$$

*   The SI unit of pressure is pascal and $1\ \text{Pa} = 1\ \text{N/m}^2$.

11.4 Variation of Pressure with Depth in a Fluid
*   Pressure is the weight of the fluid mg divided by the area A supporting it (the area of the bottom of the container):

$$P = \frac{mg}{A}.$$

*   Pressure due to the weight of a liquid is given by

$$P = h\rho g,$$

where P is the pressure, h is the height of the liquid, ρ is the density of the liquid, and g is the acceleration due to gravity.

11.5 Pascal's Principle
- Pressure is force per unit area.
- A change in pressure applied to an enclosed fluid is transmitted undiminished to all portions of the fluid and to the walls of its container.
- A hydraulic system is an enclosed fluid system used to exert forces.

11.6 Gauge Pressure, Absolute Pressure, and Pressure Measurement
- Gauge pressure is the pressure relative to atmospheric pressure.
- Absolute pressure is the sum of gauge pressure and atmospheric pressure.
- Aneroid gauge measures pressure using a bellows-and-spring arrangement connected to the pointer of a calibrated scale.
- Open-tube manometers have U-shaped tubes and one end is always open. It is used to measure pressure.
- A mercury barometer is a device that measures atmospheric pressure.

11.7 Archimedes' Principle
- Buoyant force is the net upward force on any object in any fluid. If the buoyant force is greater than the object's weight, the object will rise to the surface and float. If the buoyant force is less than the object's weight, the object will sink. If the buoyant force equals the object's weight, the object will remain suspended at that depth. The buoyant force is always present whether the object floats, sinks, or is suspended in a fluid.
- Archimedes' principle states that the buoyant force on an object equals the weight of the fluid it displaces.
- Specific gravity is the ratio of the density of an object to a fluid (usually water).

11.8 Cohesion and Adhesion in Liquids: Surface Tension and Capillary Action
- Attractive forces between molecules of the same type are called cohesive forces.
- Attractive forces between molecules of different types are called adhesive forces.
- Cohesive forces between molecules cause the surface of a liquid to contract to the smallest possible surface area. This general effect is called surface tension.
- Capillary action is the tendency of a fluid to be raised or suppressed in a narrow tube, or capillary tube which is due to the relative strength of cohesive and adhesive forces.

11.9 Pressures in the Body
- Measuring blood pressure is among the most common of all medical examinations.
- The pressures in various parts of the body can be measured and often provide valuable medical indicators.
- The shape of the eye is maintained by fluid pressure, called intraocular pressure.
- When the circulation of fluid in the eye is blocked, it can lead to a buildup in pressure, a condition called glaucoma.
- Some of the other pressures in the body are spinal and skull pressures, bladder pressure, pressures in the skeletal system.

Conceptual Questions

11.1 What Is a Fluid?

1. What physical characteristic distinguishes a fluid from a solid?

2. Which of the following substances are fluids at room temperature: air, mercury, water, glass?

3. Why are gases easier to compress than liquids and solids?

4. How do gases differ from liquids?

11.2 Density

5. Approximately how does the density of air vary with altitude?

6. Give an example in which density is used to identify the substance composing an object. Would information in addition to average density be needed to identify the substances in an object composed of more than one material?

7. Figure 11.43 shows a glass of ice water filled to the brim. Will the water overflow when the ice melts? Explain your answer.

Figure 11.43

11.3 Pressure

8. How is pressure related to the sharpness of a knife and its ability to cut?

9. Why does a dull hypodermic needle hurt more than a sharp one?

10. The outward force on one end of an air tank was calculated in **Example 11.2**. How is this force balanced? (The tank does not accelerate, so the force must be balanced.)

11. Why is force exerted by static fluids always perpendicular to a surface?

12. In a remote location near the North Pole, an iceberg floats in a lake. Next to the lake (assume it is not frozen) sits a comparably sized glacier sitting on land. If both chunks of ice should melt due to rising global temperatures (and the melted ice all goes into the lake), which ice chunk would give the greatest increase in the level of the lake water, if any?

13. How do jogging on soft ground and wearing padded shoes reduce the pressures to which the feet and legs are subjected?

14. Toe dancing (as in ballet) is much harder on toes than normal dancing or walking. Explain in terms of pressure.

15. How do you convert pressure units like millimeters of mercury, centimeters of water, and inches of mercury into units like newtons per meter squared without resorting to a table of pressure conversion factors?

11.4 Variation of Pressure with Depth in a Fluid

16. Atmospheric pressure exerts a large force (equal to the weight of the atmosphere above your body—about 10 tons) on the top of your body when you are lying on the beach sunbathing. Why are you able to get up?

17. Why does atmospheric pressure decrease more rapidly than linearly with altitude?

18. What are two reasons why mercury rather than water is used in barometers?

19. Figure 11.44 shows how sandbags placed around a leak outside a river levee can effectively stop the flow of water under the levee. Explain how the small amount of water inside the column formed by the sandbags is able to balance the much larger body of water behind the levee.

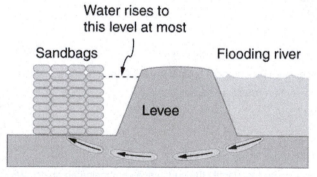

Figure 11.44 Because the river level is very high, it has started to leak under the levee. Sandbags are placed around the leak, and the water held by them rises until it is the same level as the river, at which point the water there stops rising.

20. Why is it difficult to swim under water in the Great Salt Lake?

21. Is there a net force on a dam due to atmospheric pressure? Explain your answer.

22. Does atmospheric pressure add to the gas pressure in a rigid tank? In a toy balloon? When, in general, does atmospheric pressure *not* affect the total pressure in a fluid?

23. You can break a strong wine bottle by pounding a cork into it with your fist, but the cork must press directly against the liquid filling the bottle—there can be no air between the cork and liquid. Explain why the bottle breaks, and why it will not if there is air between the cork and liquid.

11.5 Pascal's Principle

24. Suppose the master cylinder in a hydraulic system is at a greater height than the slave cylinder. Explain how this will affect the force produced at the slave cylinder.

11.6 Gauge Pressure, Absolute Pressure, and Pressure Measurement

25. Explain why the fluid reaches equal levels on either side of a manometer if both sides are open to the atmosphere, even if the tubes are of different diameters.

26. Figure 11.20 shows how a common measurement of arterial blood pressure is made. Is there any effect on the measured pressure if the manometer is lowered? What is the effect of raising the arm above the shoulder? What is the effect of placing the cuff on the upper leg with the person standing? Explain your answers in terms of pressure created by the weight of a fluid.

27. Considering the magnitude of typical arterial blood pressures, why are mercury rather than water manometers used for these measurements?

11.7 Archimedes' Principle

28. More force is required to pull the plug in a full bathtub than when it is empty. Does this contradict Archimedes' principle? Explain your answer.

29. Do fluids exert buoyant forces in a "weightless" environment, such as in the space shuttle? Explain your answer.

30. Will the same ship float higher in salt water than in freshwater? Explain your answer.

31. Marbles dropped into a partially filled bathtub sink to the bottom. Part of their weight is supported by buoyant force, yet the downward force on the bottom of the tub increases by exactly the weight of the marbles. Explain why.

11.8 Cohesion and Adhesion in Liquids: Surface Tension and Capillary Action

32. The density of oil is less than that of water, yet a loaded oil tanker sits lower in the water than an empty one. Why?

33. Is surface tension due to cohesive or adhesive forces, or both?

34. Is capillary action due to cohesive or adhesive forces, or both?

35. Birds such as ducks, geese, and swans have greater densities than water, yet they are able to sit on its surface. Explain this ability, noting that water does not wet their feathers and that they cannot sit on soapy water.

36. Water beads up on an oily sunbather, but not on her neighbor, whose skin is not oiled. Explain in terms of cohesive and adhesive forces.

37. Could capillary action be used to move fluids in a "weightless" environment, such as in an orbiting space probe?

38. What effect does capillary action have on the reading of a manometer with uniform diameter? Explain your answer.

39. Pressure between the inside chest wall and the outside of the lungs normally remains negative. Explain how pressure inside the lungs can become positive (to cause exhalation) without muscle action.

Problems & Exercises

11.2 Density

1. Gold is sold by the troy ounce (31.103 g). What is the volume of 1 troy ounce of pure gold?

2. Mercury is commonly supplied in flasks containing 34.5 kg (about 76 lb). What is the volume in liters of this much mercury?

3. (a) What is the mass of a deep breath of air having a volume of 2.00 L? (b) Discuss the effect taking such a breath has on your body's volume and density.

4. A straightforward method of finding the density of an object is to measure its mass and then measure its volume by submerging it in a graduated cylinder. What is the density of a 240-g rock that displaces $89.0 \, \text{cm}^3$ of water? (Note that the accuracy and practical applications of this technique are more limited than a variety of others that are based on Archimedes' principle.)

5. Suppose you have a coffee mug with a circular cross section and vertical sides (uniform radius). What is its inside radius if it holds 375 g of coffee when filled to a depth of 7.50 cm? Assume coffee has the same density as water.

6. (a) A rectangular gasoline tank can hold 50.0 kg of gasoline when full. What is the depth of the tank if it is 0.500-m wide by 0.900-m long? (b) Discuss whether this gas tank has a reasonable volume for a passenger car.

7. A trash compactor can reduce the volume of its contents to 0.350 their original value. Neglecting the mass of air expelled, by what factor is the density of the rubbish increased?

8. A 2.50-kg steel gasoline can holds 20.0 L of gasoline when full. What is the average density of the full gas can, taking into account the volume occupied by steel as well as by gasoline?

9. What is the density of 18.0-karat gold that is a mixture of 18 parts gold, 5 parts silver, and 1 part copper? (These values are parts by mass, not volume.) Assume that this is a simple mixture having an average density equal to the weighted densities of its constituents.

10. There is relatively little empty space between atoms in solids and liquids, so that the average density of an atom is about the same as matter on a macroscopic scale—approximately $10^3 \, \text{kg/m}^3$. The nucleus of an atom has a radius about 10^{-5} that of the atom and contains nearly all the mass of the entire atom. (a) What is the approximate density of a nucleus? (b) One remnant of a supernova, called a neutron star, can have the density of a nucleus. What would be the radius of a neutron star with a mass 10 times that of our Sun (the radius of the Sun is $7 \times 10^8 \, \text{m}$)?

11.3 Pressure

11. As a woman walks, her entire weight is momentarily placed on one heel of her high-heeled shoes. Calculate the pressure exerted on the floor by the heel if it has an area of $1.50 \, \text{cm}^2$ and the woman's mass is 55.0 kg. Express the pressure in Pa. (In the early days of commercial flight, women were not allowed to wear high-heeled shoes because aircraft floors were too thin to withstand such large pressures.)

12. The pressure exerted by a phonograph needle on a record is surprisingly large. If the equivalent of 1.00 g is supported by a needle, the tip of which is a circle 0.200 mm in radius, what pressure is exerted on the record in N/m^2 ?

13. Nail tips exert tremendous pressures when they are hit by hammers because they exert a large force over a small area. What force must be exerted on a nail with a circular tip of 1.00 mm diameter to create a pressure of $3.00 \times 10^9 \, \text{N/m}^2$? (This high pressure is possible because the hammer striking the nail is brought to rest in such a short distance.)

11.4 Variation of Pressure with Depth in a Fluid

14. What depth of mercury creates a pressure of 1.00 atm?

15. The greatest ocean depths on the Earth are found in the Marianas Trench near the Philippines. Calculate the pressure due to the ocean at the bottom of this trench, given its depth is 11.0 km and assuming the density of seawater is constant all the way down.

16. Verify that the SI unit of $h\rho g$ is N/m^2 .

17. Water towers store water above the level of consumers for times of heavy use, eliminating the need for high-speed pumps. How high above a user must the water level be to create a gauge pressure of $3.00 \times 10^5 \, \text{N/m}^2$?

18. The aqueous humor in a person's eye is exerting a force of 0.300 N on the 1.10-cm^2 area of the cornea. (a) What pressure is this in mm Hg? (b) Is this value within the normal range for pressures in the eye?

19. How much force is exerted on one side of an 8.50 cm by 11.0 cm sheet of paper by the atmosphere? How can the paper withstand such a force?

20. What pressure is exerted on the bottom of a 0.500-m-wide by 0.900-m-long gas tank that can hold 50.0 kg of gasoline by the weight of the gasoline in it when it is full?

21. Calculate the average pressure exerted on the palm of a shot-putter's hand by the shot if the area of contact is $50.0 \, \text{cm}^2$ and he exerts a force of 800 N on it. Express the pressure in N/m^2 and compare it with the $1.00 \times 10^6 \, \text{Pa}$ pressures sometimes encountered in the skeletal system.

22. The left side of the heart creates a pressure of 120 mm Hg by exerting a force directly on the blood over an effective area of $15.0 \, \text{cm}^2$. What force does it exert to accomplish this?

23. Show that the total force on a rectangular dam due to the water behind it increases with the *square* of the water depth. In particular, show that this force is given by $F = \rho g h^2 L / 2$, where ρ is the density of water, h is its depth at the dam, and L is the length of the dam. You may assume the face of the dam is vertical. (Hint: Calculate the average pressure exerted and multiply this by the area in contact with the water. (See **Figure 11.45**.)

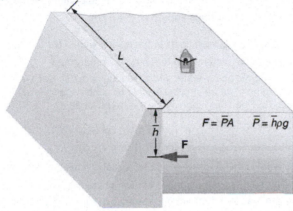

Figure 11.45

11.5 Pascal's Principle

24. How much pressure is transmitted in the hydraulic system considered in **Example 11.6**? Express your answer in pascals and in atmospheres.

25. What force must be exerted on the master cylinder of a hydraulic lift to support the weight of a 2000-kg car (a large car) resting on the slave cylinder? The master cylinder has a 2.00-cm diameter and the slave has a 24.0-cm diameter.

26. A crass host pours the remnants of several bottles of wine into a jug after a party. He then inserts a cork with a 2.00-cm diameter into the bottle, placing it in direct contact with the wine. He is amazed when he pounds the cork into place and the bottom of the jug (with a 14.0-cm diameter) breaks away. Calculate the extra force exerted against the bottom if he pounded the cork with a 120-N force.

27. A certain hydraulic system is designed to exert a force 100 times as large as the one put into it. (a) What must be the ratio of the area of the slave cylinder to the area of the master cylinder? (b) What must be the ratio of their diameters? (c) By what factor is the distance through which the output force moves reduced relative to the distance through which the input force moves? Assume no losses to friction.

28. (a) Verify that work input equals work output for a hydraulic system assuming no losses to friction. Do this by showing that the distance the output force moves is reduced by the same factor that the output force is increased. Assume the volume of the fluid is constant. (b) What effect would friction within the fluid and between components in the system have on the output force? How would this depend on whether or not the fluid is moving?

11.6 Gauge Pressure, Absolute Pressure, and Pressure Measurement

29. Find the gauge and absolute pressures in the balloon and peanut jar shown in **Figure 11.19**, assuming the manometer connected to the balloon uses water whereas the manometer connected to the jar contains mercury. Express in units of centimeters of water for the balloon and millimeters of mercury for the jar, taking $h = 0.0500$ m for each.

30. (a) Convert normal blood pressure readings of 120 over 80 mm Hg to newtons per meter squared using the relationship for pressure due to the weight of a fluid $(P = h\rho g)$ rather than a conversion factor. (b) Discuss why blood pressures for an infant could be smaller than those for an adult. Specifically, consider the smaller height to which blood must be pumped.

31. How tall must a water-filled manometer be to measure blood pressures as high as 300 mm Hg?

32. Pressure cookers have been around for more than 300 years, although their use has strongly declined in recent years (early models had a nasty habit of exploding). How much force must the latches holding the lid onto a pressure cooker be able to withstand if the circular lid is 25.0 cm in diameter and the gauge pressure inside is 300 atm? Neglect the weight of the lid.

33. Suppose you measure a standing person's blood pressure by placing the cuff on his leg 0.500 m below the heart. Calculate the pressure you would observe (in units of mm Hg) if the pressure at the heart were 120 over 80 mm Hg. Assume that there is no loss of pressure due to resistance in the circulatory system (a reasonable assumption, since major arteries are large).

34. A submarine is stranded on the bottom of the ocean with its hatch 25.0 m below the surface. Calculate the force needed to open the hatch from the inside, given it is circular and 0.450 m in diameter. Air pressure inside the submarine is 1.00 atm.

35. Assuming bicycle tires are perfectly flexible and support the weight of bicycle and rider by pressure alone, calculate the total area of the tires in contact with the ground. The bicycle plus rider has a mass of 80.0 kg, and the gauge pressure in the tires is 3.50×10^5 Pa.

11.7 Archimedes' Principle

36. What fraction of ice is submerged when it floats in freshwater, given the density of water at 0°C is very close to 1000 kg/m^3?

37. Logs sometimes float vertically in a lake because one end has become water-logged and denser than the other. What is the average density of a uniform-diameter log that floats with 20.0% of its length above water?

38. Find the density of a fluid in which a hydrometer having a density of 0.750 g/mL floats with 92.0% of its volume submerged.

39. If your body has a density of 995 kg/m^3, what fraction of you will be submerged when floating gently in: (a) Freshwater? (b) Salt water, which has a density of 1027 kg/m^3?

40. Bird bones have air pockets in them to reduce their weight—this also gives them an average density significantly less than that of the bones of other animals. Suppose an ornithologist weighs a bird bone in air and in water and finds its mass is 45.0 g and its apparent mass when submerged is 3.60 g (the bone is watertight). (a) What mass of water is displaced? (b) What is the volume of the bone? (c) What is its average density?

41. A rock with a mass of 540 g in air is found to have an apparent mass of 342 g when submerged in water. (a) What mass of water is displaced? (b) What is the volume of the rock? (c) What is its average density? Is this consistent with the value for granite?

42. Archimedes' principle can be used to calculate the density of a fluid as well as that of a solid. Suppose a chunk of iron with a mass of 390.0 g in air is found to have an apparent mass of 350.5 g when completely submerged in an unknown liquid. (a) What mass of fluid does the iron displace? (b) What is the volume of iron, using its density as given in **Table 11.1** (c) Calculate the fluid's density and identify it.

43. In an immersion measurement of a woman's density, she is found to have a mass of 62.0 kg in air and an apparent mass of 0.0850 kg when completely submerged with lungs empty. (a) What mass of water does she displace? (b) What is her volume? (c) Calculate her density. (d) If her lung capacity is 1.75 L, is she able to float without treading water with her lungs filled with air?

44. Some fish have a density slightly less than that of water and must exert a force (swim) to stay submerged. What force must an 85.0-kg grouper exert to stay submerged in salt water if its body density is 1015 kg/m^3 ?

45. (a) Calculate the buoyant force on a 2.00-L helium balloon. (b) Given the mass of the rubber in the balloon is 1.50 g, what is the net vertical force on the balloon if it is let go? You can neglect the volume of the rubber.

46. (a) What is the density of a woman who floats in freshwater with 4.00% of her volume above the surface? This could be measured by placing her in a tank with marks on the side to measure how much water she displaces when floating and when held under water (briefly). (b) What percent of her volume is above the surface when she floats in seawater?

47. A certain man has a mass of 80 kg and a density of 955 kg/m^3 (excluding the air in his lungs). (a) Calculate his volume. (b) Find the buoyant force air exerts on him. (c) What is the ratio of the buoyant force to his weight?

48. A simple compass can be made by placing a small bar magnet on a cork floating in water. (a) What fraction of a plain cork will be submerged when floating in water? (b) If the cork has a mass of 10.0 g and a 20.0-g magnet is placed on it, what fraction of the cork will be submerged? (c) Will the bar magnet and cork float in ethyl alcohol?

49. What fraction of an iron anchor's weight will be supported by buoyant force when submerged in saltwater?

50. Scurrilous con artists have been known to represent gold-plated tungsten ingots as pure gold and sell them to the greedy at prices much below gold value but deservedly far above the cost of tungsten. With what accuracy must you be able to measure the mass of such an ingot in and out of water to tell that it is almost pure tungsten rather than pure gold?

51. A twin-sized air mattress used for camping has dimensions of 100 cm by 200 cm by 15 cm when blown up. The weight of the mattress is 2 kg. How heavy a person could the air mattress hold if it is placed in freshwater?

52. Referring to **Figure 11.24**, prove that the buoyant force on the cylinder is equal to the weight of the fluid displaced (Archimedes' principle). You may assume that the buoyant force is $F_2 - F_1$ and that the ends of the cylinder have equal areas A . Note that the volume of the cylinder (and that of the fluid it displaces) equals $(h_2 - h_1)A$.

53. (a) A 75.0-kg man floats in freshwater with 3.00% of his volume above water when his lungs are empty, and 5.00% of his volume above water when his lungs are full. Calculate the volume of air he inhales—called his lung capacity—in liters. (b) Does this lung volume seem reasonable?

11.8 Cohesion and Adhesion in Liquids: Surface Tension and Capillary Action

54. What is the pressure inside an alveolus having a radius of 2.50×10^{-4} m if the surface tension of the fluid-lined wall is the same as for soapy water? You may assume the pressure is the same as that created by a spherical bubble.

55. (a) The pressure inside an alveolus with a 2.00×10^{-4} -m radius is $1.40\times10^3 \text{ Pa}$, due to its fluid-lined walls. Assuming the alveolus acts like a spherical bubble, what is the surface tension of the fluid? (b) Identify the likely fluid. (You may need to extrapolate between values in **Table 11.3**.)

56. What is the gauge pressure in millimeters of mercury inside a soap bubble 0.100 m in diameter?

57. Calculate the force on the slide wire in **Figure 11.32** if it is 3.50 cm long and the fluid is ethyl alcohol.

58. **Figure 11.38**(a) shows the effect of tube radius on the height to which capillary action can raise a fluid. (a) Calculate the height h for water in a glass tube with a radius of 0.900 cm—a rather large tube like the one on the left. (b) What is the radius of the glass tube on the right if it raises water to 4.00 cm?

59. We stated in **Example 11.12** that a xylem tube is of radius 2.50×10^{-5} m . Verify that such a tube raises sap less than a meter by finding h for it, making the same assumptions that sap's density is 1050 kg/m^3 , its contact angle is zero, and its surface tension is the same as that of water at $20.0° \text{ C}$.

60. What fluid is in the device shown in **Figure 11.32** if the force is $3.16\times10^{-3} \text{ N}$ and the length of the wire is 2.50 cm? Calculate the surface tension γ and find a likely match from **Table 11.3**.

61. If the gauge pressure inside a rubber balloon with a 10.0-cm radius is 1.50 cm of water, what is the effective surface tension of the balloon?

62. Calculate the gauge pressures inside 2.00-cm-radius bubbles of water, alcohol, and soapy water. Which liquid forms the most stable bubbles, neglecting any effects of evaporation?

63. Suppose water is raised by capillary action to a height of 5.00 cm in a glass tube. (a) To what height will it be raised in a paraffin tube of the same radius? (b) In a silver tube of the same radius?

64. Calculate the contact angle θ for olive oil if capillary action raises it to a height of 7.07 cm in a glass tube with a radius of 0.100 mm. Is this value consistent with that for most organic liquids?

65. When two soap bubbles touch, the larger is inflated by the smaller until they form a single bubble. (a) What is the gauge pressure inside a soap bubble with a 1.50-cm radius? (b) Inside a 4.00-cm-radius soap bubble? (c) Inside the single bubble they form if no air is lost when they touch?

66. Calculate the ratio of the heights to which water and mercury are raised by capillary action in the same glass tube.

67. What is the ratio of heights to which ethyl alcohol and water are raised by capillary action in the same glass tube?

11.9 Pressures in the Body

68. During forced exhalation, such as when blowing up a balloon, the diaphragm and chest muscles create a pressure of 60.0 mm Hg between the lungs and chest wall. What force in newtons does this pressure create on the 600 cm^2 surface area of the diaphragm?

69. You can chew through very tough objects with your incisors because they exert a large force on the small area of a pointed tooth. What pressure in pascals can you create by exerting a force of 500 N with your tooth on an area of 1.00 mm^2 ?

70. One way to force air into an unconscious person's lungs is to squeeze on a balloon appropriately connected to the subject. What force must you exert on the balloon with your hands to create a gauge pressure of 4.00 cm water, assuming you squeeze on an effective area of 50.0 cm^2 ?

71. Heroes in movies hide beneath water and breathe through a hollow reed (villains never catch on to this trick). In practice, you cannot inhale in this manner if your lungs are more than 60.0 cm below the surface. What is the maximum negative gauge pressure you can create in your lungs on dry land, assuming you can achieve -3.00 cm water pressure with your lungs 60.0 cm below the surface?

72. Gauge pressure in the fluid surrounding an infant's brain may rise as high as 85.0 mm Hg (5 to 12 mm Hg is normal), creating an outward force large enough to make the skull grow abnormally large. (a) Calculate this outward force in newtons on each side of an infant's skull if the effective area of each side is 70.0 cm^2 . (b) What is the net force acting on the skull?

73. A full-term fetus typically has a mass of 3.50 kg. (a) What pressure does the weight of such a fetus create if it rests on the mother's bladder, supported on an area of 90.0 cm^2 ?

(b) Convert this pressure to millimeters of mercury and determine if it alone is great enough to trigger the micturition reflex (it will add to any pressure already existing in the bladder).

74. If the pressure in the esophagus is -2.00 mm Hg while that in the stomach is $+20.0 \text{ mm Hg}$, to what height could stomach fluid rise in the esophagus, assuming a density of 1.10 g/mL? (This movement will not occur if the muscle closing the lower end of the esophagus is working properly.)

75. Pressure in the spinal fluid is measured as shown in **Figure 11.46**. If the pressure in the spinal fluid is 10.0 mm Hg: (a) What is the reading of the water manometer in cm water? (b) What is the reading if the person sits up, placing the top of the fluid 60 cm above the tap? The fluid density is 1.05 g/mL.

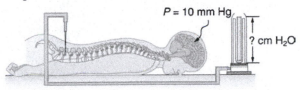

Figure 11.46 A water manometer used to measure pressure in the spinal fluid. The height of the fluid in the manometer is measured relative to the spinal column, and the manometer is open to the atmosphere. The measured pressure will be considerably greater if the person sits up.

76. Calculate the maximum force in newtons exerted by the blood on an aneurysm, or ballooning, in a major artery, given the maximum blood pressure for this person is 150 mm Hg and the effective area of the aneurysm is 20.0 cm^2 . Note that this force is great enough to cause further enlargement and subsequently greater force on the ever-thinner vessel wall.

77. During heavy lifting, a disk between spinal vertebrae is subjected to a 5000-N compressional force. (a) What pressure is created, assuming that the disk has a uniform circular cross section 2.00 cm in radius? (b) What deformation is produced if the disk is 0.800 cm thick and has a Young's modulus of $1.5 \times 10^9 \text{ N/m}^2$?

78. When a person sits erect, increasing the vertical position of their brain by 36.0 cm, the heart must continue to pump blood to the brain at the same rate. (a) What is the gain in gravitational potential energy for 100 mL of blood raised 36.0 cm? (b) What is the drop in pressure, neglecting any losses due to friction? (c) Discuss how the gain in gravitational potential energy and the decrease in pressure are related.

79. (a) How high will water rise in a glass capillary tube with a 0.500-mm radius? (b) How much gravitational potential energy does the water gain? (c) Discuss possible sources of this energy.

80. A negative pressure of 25.0 atm can sometimes be achieved with the device in **Figure 11.47** before the water separates. (a) To what height could such a negative gauge pressure raise water? (b) How much would a steel wire of the same diameter and length as this capillary stretch if suspended from above?

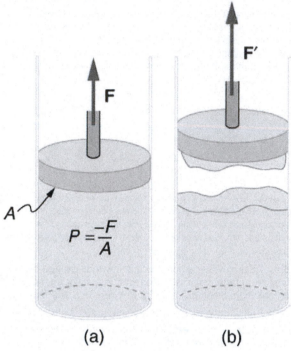

(a)   (b)

Figure 11.47 (a) When the piston is raised, it stretches the liquid slightly, putting it under tension and creating a negative absolute pressure $P = -F / A$ (b) The liquid eventually separates, giving an experimental limit to negative pressure in this liquid.

81. Suppose you hit a steel nail with a 0.500-kg hammer, initially moving at 15.0 m/s and brought to rest in 2.80 mm. (a) What average force is exerted on the nail? (b) How much is the nail compressed if it is 2.50 mm in diameter and 6.00-cm long? (c) What pressure is created on the 1.00-mm-diameter tip of the nail?

82. Calculate the pressure due to the ocean at the bottom of the Marianas Trench near the Philippines, given its depth is 11.0 km and assuming the density of sea water is constant all the way down. (b) Calculate the percent decrease in volume of sea water due to such a pressure, assuming its bulk modulus is the same as water and is constant. (c) What would be the percent increase in its density? Is the assumption of constant density valid? Will the actual pressure be greater or smaller than that calculated under this assumption?

83. The hydraulic system of a backhoe is used to lift a load as shown in **Figure 11.48**. (a) Calculate the force F the slave cylinder must exert to support the 400-kg load and the 150-kg brace and shovel. (b) What is the pressure in the hydraulic fluid if the slave cylinder is 2.50 cm in diameter? (c) What force would you have to exert on a lever with a mechanical advantage of 5.00 acting on a master cylinder 0.800 cm in diameter to create this pressure?

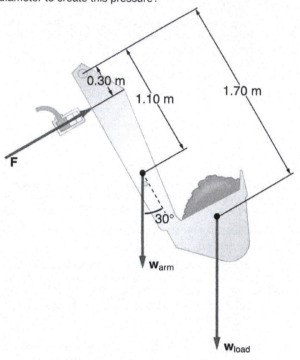

Figure 11.48 Hydraulic and mechanical lever systems are used in heavy machinery such as this back hoe.

84. Some miners wish to remove water from a mine shaft. A pipe is lowered to the water 90 m below, and a negative pressure is applied to raise the water. (a) Calculate the pressure needed to raise the water. (b) What is unreasonable about this pressure? (c) What is unreasonable about the premise?

85. You are pumping up a bicycle tire with a hand pump, the piston of which has a 2.00-cm radius.

(a) What force in newtons must you exert to create a pressure of $6.90 \times 10^5 \text{ Pa}$ (b) What is unreasonable about this (a) result? (c) Which premises are unreasonable or inconsistent?

86. Consider a group of people trying to stay afloat after their boat strikes a log in a lake. Construct a problem in which you calculate the number of people that can cling to the log and keep their heads out of the water. Among the variables to be considered are the size and density of the log, and what is needed to keep a person's head and arms above water without swimming or treading water.

87. The alveoli in emphysema victims are damaged and effectively form larger sacs. Construct a problem in which you calculate the loss of pressure due to surface tension in the alveoli because of their larger average diameters. (Part of the lung's ability to expel air results from pressure created by surface tension in the alveoli.) Among the things to consider are the normal surface tension of the fluid lining the alveoli, the average alveolar radius in normal individuals and its average in emphysema sufferers.

Test Prep for AP® Courses

11.2 Density

1. An under-inflated volleyball is pumped full of air so that its radius increases by 10%. Ignoring the mass of the air inserted into the ball, what will happen to the volleyball's density?

   a. The density of the volleyball will increase by approximately 25%.

   b. The density of the volleyball will increase by approximately 10%.

   c. The density of the volleyball will decrease by approximately 10%.

   d. The density of the volleyball will decrease by approximately 17%.

   e. The density of the volleyball will decrease by approximately 25%.

2. A piece of aluminum foil has a known surface density of 15 g/cm^2. If a 100-gram hollow cube were constructed using this foil, determine the approximate side length of this cube.

   a. 1.05 cm

   b. 1.10 cm

   c. 2.6 cm

   d. 6.67 cm

   e. 15 cm

3. A cube of polystyrene measuring 10 cm per side lies partially submerged in a large container of water.

   a. If 90% of the polystyrene floats above the surface of the water, what is the density of the polystyrene? (Note: The density of water is 1000 kg/m^3.)

   b. A 0.5 kg mass is placed on the block of polystyrene. What percentage of the block now remains above water?

   c. The water is poured out of the container and replaced with ethyl alcohol (density = 790 kg/m^3).

      i. Will the block be able to remain partially submerged in this new fluid? Explain.

      ii. Will the block be able to remain partially submerged in this new fluid with the 0.5 kg mass placed on top? Explain.

   d. Without using a container of water, explain how you could determine the density of the polystyrene mentioned above if the material instead were spherical.

4. Four spheres are hung from a variety of different springs. The table below describes the characteristics of both the spheres and the springs from which they are hung. Use this information to rank the density of each sphere from least to greatest. Show work supporting your ranking.

Table 11.6

Material Type	Radius of Sphere	Stretch of Spring (from equilibrium)	Spring Constant
A	10 cm	5 cm	2 N/m
B	5 cm	8 cm	8 N/m
C	8 cm	10 cm	6 N/m
D	8 cm	12 cm	10 N/m

Rank the densities of the objects listed above, from greatest to least. Show work supporting your ranking.

11.3 Pressure

5. A cylindrical drum of radius 0.5 m is used to hold 400 liters of petroleum ether (density = .68 g/mL or 680 kg/m^3).

(Note: 1 liter = 0.001 m^3)

   a. Determine the amount of pressure applied to the walls of the drum if the petroleum ether fills the drum to its top.

   b. Determine the amount of pressure applied to the floor of the drum if the petroleum ether fills the drum to its top.

   c. If the drum were redesigned to hold 800 liters of petroleum ether:

      i. How would the pressure on the walls change? Would it increase, decrease, or stay the same?

      ii. How would the pressure on the floor change? Would it increase, decrease, or stay the same?

12 FLUID DYNAMICS AND ITS BIOLOGICAL AND MEDICAL APPLICATIONS

Figure 12.1 Many fluids are flowing in this scene. Water from the hose and smoke from the fire are visible flows. Less visible are the flow of air and the flow of fluids on the ground and within the people fighting the fire. Explore all types of flow, such as visible, implied, turbulent, laminar, and so on, present in this scene. Make a list and discuss the relative energies involved in the various flows, including the level of confidence in your estimates. (credit: Andrew Magill, Flickr)

Chapter Outline

Connection for AP® Courses

How do planes fly? How do we model blood flow? How do sprayers work for paints or aerosols? What is the purpose of a water tower? To answer these questions, we will examine fluid dynamics. The equations governing fluid dynamics are derived from the same equations that represent energy conservation. One of the most powerful equations in fluid dynamics is Bernoulli's equation, which governs the relationship between fluid pressure, kinetic energy, and potential energy (Essential Knowledge 5.B.10). We will see how Bernoulli's equation explains the pressure difference that provides lift for airplanes and provides the means for fluids (like water or paint or perfume) to move in useful ways.

The content in this chapter supports:

Big Idea 5 Changes that occur as a result of interactions are constrained by conservation laws.

Enduring Understanding 5.B The energy of a system is conserved.

Essential Knowledge 5.B.10 Bernoulli's equation describes the conservation of energy in a fluid flow.

Enduring Understanding 5.F Classically, the mass of a system is conserved.

Essential Knowledge 5.F.1 The continuity equation describes conservation of mass flow rate in fluids.

12.1 Flow Rate and Its Relation to Velocity

Learning Objectives

By the end of this section, you will be able to:

- Calculate flow rate.
- Define units of volume.
- Describe incompressible fluids.
- Explain the consequences of the equation of continuity.

The information presented in this section supports the following AP® learning objectives and science practices:

- **5.F.1.1** The student is able to make calculations of quantities related to flow of a fluid, using mass conservation principles (the continuity equation). **(S.P. 6.4, 7.2)**

Flow rate Q is defined to be the volume of fluid passing by some location through an area during a period of time, as seen in Figure 12.2. In symbols, this can be written as

$$Q = \frac{V}{t}, \tag{12.1}$$

where V is the volume and t is the elapsed time.

The SI unit for flow rate is m^3/s, but a number of other units for Q are in common use. For example, the heart of a resting adult pumps blood at a rate of 5.00 liters per minute (L/min). Note that a **liter** (L) is 1/1000 of a cubic meter or 1000 cubic centimeters ($10^{-3} \ m^3$ or $10^3 \ cm^3$). In this text we shall use whatever metric units are most convenient for a given situation.

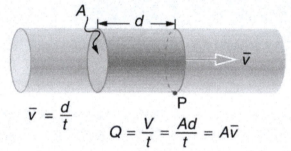

Figure 12.2 Flow rate is the volume of fluid per unit time flowing past a point through the area A. Here the shaded cylinder of fluid flows past point P in a uniform pipe in time t. The volume of the cylinder is Ad and the average velocity is $\bar{v} = d/t$ so that the flow rate is $Q = Ad/t = A\bar{v}$.

Example 12.1 Calculating Volume from Flow Rate: The Heart Pumps a Lot of Blood in a Lifetime

How many cubic meters of blood does the heart pump in a 75-year lifetime, assuming the average flow rate is 5.00 L/min?

Strategy

Time and flow rate Q are given, and so the volume V can be calculated from the definition of flow rate.

Solution

Solving $Q = V/t$ for volume gives

$$V = Qt. \tag{12.2}$$

Substituting known values yields

$$
\begin{aligned}
V &= \left(\frac{5.00 \ L}{1 \ min}\right)(75 \ y)\left(\frac{1 \ m^3}{10^3 \ L}\right)\left(5.26 \times 10^5 \frac{min}{y}\right) \\
&= 2.0 \times 10^5 \ m^3.
\end{aligned}
\tag{12.3}
$$

Discussion

This amount is about 200,000 tons of blood. For comparison, this value is equivalent to about 200 times the volume of water contained in a 6-lane 50-m lap pool.

Flow rate and velocity are related, but quite different, physical quantities. To make the distinction clear, think about the flow rate of a river. The greater the velocity of the water, the greater the flow rate of the river. But flow rate also depends on the size of the river. A rapid mountain stream carries far less water than the Amazon River in Brazil, for example. The precise relationship between flow rate Q and velocity $\bar{v}$ is

$$Q = A\bar{v}, \tag{12.4}$$

where A is the cross-sectional area and $\bar{v}$ is the average velocity. This equation seems logical enough. The relationship tells us that flow rate is directly proportional to both the magnitude of the average velocity (hereafter referred to as the speed) and the size of a river, pipe, or other conduit. The larger the conduit, the greater its cross-sectional area. **Figure 12.2** illustrates how this relationship is obtained. The shaded cylinder has a volume

$$V = Ad, \tag{12.5}$$

which flows past the point P in a time t. Dividing both sides of this relationship by t gives

$$\frac{V}{t} = \frac{Ad}{t}. \tag{12.6}$$

We note that $Q = V/t$ and the average speed is $\bar{v} = d/t$. Thus the equation becomes $Q = A\bar{v}$.

Figure 12.3 shows an incompressible fluid flowing along a pipe of decreasing radius. Because the fluid is incompressible, the same amount of fluid must flow past any point in the tube in a given time to ensure continuity of flow. In this case, because the cross-sectional area of the pipe decreases, the velocity must necessarily increase. This logic can be extended to say that the flow rate must be the same at all points along the pipe. In particular, for points 1 and 2,

$$\left.\begin{array}{r} Q_1 = Q_2 \\ A_1 \bar{v}_1 = A_2 \bar{v}_2 \end{array}\right\}. \tag{12.7}$$

This is called the equation of continuity and is valid for any incompressible fluid. The consequences of the equation of continuity can be observed when water flows from a hose into a narrow spray nozzle: it emerges with a large speed—that is the purpose of the nozzle. Conversely, when a river empties into one end of a reservoir, the water slows considerably, perhaps picking up speed again when it leaves the other end of the reservoir. In other words, speed increases when cross-sectional area decreases, and speed decreases when cross-sectional area increases.

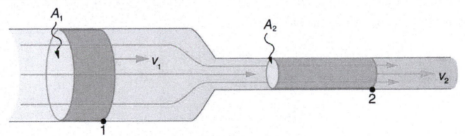

Figure 12.3 When a tube narrows, the same volume occupies a greater length. For the same volume to pass points 1 and 2 in a given time, the speed must be greater at point 2. The process is exactly reversible. If the fluid flows in the opposite direction, its speed will decrease when the tube widens. (Note that the relative volumes of the two cylinders and the corresponding velocity vector arrows are not drawn to scale.)

Since liquids are essentially incompressible, the equation of continuity is valid for all liquids. However, gases are compressible, and so the equation must be applied with caution to gases if they are subjected to compression or expansion.

Making Connections: Incompressible Fluid

The continuity equation tells us that the flow rate must be the same throughout an incompressible fluid. The flow rate, Q, has units of volume per unit time (m^3/s). Another way to think about it would be as a conservation principle, that the volume of fluid flowing past any point in a given amount of time must be conserved throughout the fluid.

For incompressible fluids, we can also say that the mass flowing past any point in a given amount of time must also be conserved. That is because the mass of a given volume of fluid is just the density of the fluid multiplied by the volume:

$$m = \rho V \tag{12.8}$$

When we say a fluid is incompressible, we mean that the density of the fluid does not change. Every cubic meter of fluid has the same number of particles. There is no room to add more particles, nor is the fluid allowed to expand so that the particles will spread out. Since the density is constant, we can express the conservation principle as follows for any two regions of fluid flow, starting with the continuity equation:

$$\frac{V_1}{t_1} = \frac{V_2}{t_2} \tag{12.9}$$

$$\frac{\rho V_1}{t_1} = \frac{\rho V_2}{t_2} \tag{12.10}$$

$$\frac{m_1}{t_1} = \frac{m_2}{t_2} \tag{12.11}$$

More generally, we say that the mass flow rate $\left(\frac{\Delta m}{\Delta t}\right)$ is conserved.

Example 12.2 Calculating Fluid Speed: Speed Increases When a Tube Narrows

A nozzle with a radius of 0.250 cm is attached to a garden hose with a radius of 0.900 cm. The flow rate through hose and nozzle is 0.500 L/s. Calculate the speed of the water (a) in the hose and (b) in the nozzle.

Strategy

We can use the relationship between flow rate and speed to find both velocities. We will use the subscript 1 for the hose and 2 for the nozzle.

Solution for (a)

First, we solve $Q = A\bar{v}$ for v_1 and note that the cross-sectional area is $A = \pi r^2$, yielding

$$\bar{v}_1 = \frac{Q}{A_1} = \frac{Q}{\pi r_1^2}. \tag{12.12}$$

Substituting known values and making appropriate unit conversions yields

$$\bar{v}_1 = \frac{(0.500 \text{ L/s})(10^{-3} \text{ m}^3/\text{L})}{\pi(9.00\times10^{-3} \text{ m})^2} = 1.96 \text{ m/s}. \tag{12.13}$$

Solution for (b)

We could repeat this calculation to find the speed in the nozzle $\bar{v}_2$, but we will use the equation of continuity to give a somewhat different insight. Using the equation which states

$$A_1\bar{v}_1 = A_2\bar{v}_2, \tag{12.14}$$

solving for $\bar{v}_2$ and substituting πr^2 for the cross-sectional area yields

$$\bar{v}_2 = \frac{A_1}{A_2}\bar{v}_1 = \frac{\pi r_1^2}{\pi r_2^2}\bar{v}_1 = \frac{r_1^2}{r_2^2}\bar{v}_1. \tag{12.15}$$

Substituting known values,

$$\bar{v}_2 = \frac{(0.900 \text{ cm})^2}{(0.250 \text{ cm})^2}1.96 \text{ m/s} = 25.5 \text{ m/s}. \tag{12.16}$$

Discussion

A speed of 1.96 m/s is about right for water emerging from a nozzleless hose. The nozzle produces a considerably faster stream merely by constricting the flow to a narrower tube.

Making Connections: Different-Sized Pipes

For incompressible fluids, the density of the fluid remains constant throughout, no matter the flow rate or the size of the opening through which the fluid flows. We say that, to ensure continuity of flow, the amount of fluid that flows past any point is constant. That amount can be measured by either volume or mass.

Flow rate has units of volume/time (m^3/s or L/s). Mass flow rate $\left(\frac{\Delta m}{\Delta t}\right)$ has units of mass/time (kg/s) and can be calculated from the flow rate by using the density:

$$m = \rho V \tag{12.17}$$

The average mass flow rate can be found from the flow rate:

$$\frac{\Delta m}{\Delta t} = \frac{m}{t} = \frac{\rho V}{t} = \rho Q = \rho A v \tag{12.18}$$

Suppose that crude oil with a density of 880 kg/m^3 is flowing through a pipe with a diameter of 55 cm and a speed of 1.8 m/s. Calculate the new speed of the crude oil when the pipe narrows to a new diameter of 31 cm, and calculate the mass flow rate in both sections of the pipe, assuming the density of the oil is constant throughout the pipe.

Solution: To calculate the new speed, we simply use the continuity equation.

Since the cross section of a pipe is a circle, the area of each cross section can be found as follows:

For the larger pipe:

$$A_1 = \pi\left(\frac{d_1}{2}\right)^2 = \pi(0.275)^2 = 0.238\,\text{m}^2 \tag{12.19}$$

For the smaller pipe:

$$A_2 = \pi(0.155)^2 = 0.0755\,\text{m}^2 \tag{12.20}$$

So the larger part of the pipe (A_1) has a cross-sectional area of 0.238 m^2, and the smaller part of the pipe (A_2) has a cross-sectional area of 0.0755 m^2. The continuity equation tells us that the oil will flow faster through the portion of the pipe with the smaller cross-sectional area. Using the continuity equation, we get

$$A_1 v_1 = A_2 v_2 \tag{12.21}$$

$$v_2 = \left(\frac{A_1}{A_2}\right)v_1 = \left(\frac{0.238}{0.0755}\right)(1.8) = 5.7\ \text{m}\Big/\text{s} \tag{12.22}$$

So we find that the oil is flowing at a speed of 1.8 m/s through the larger section of the pipe (A_1), and it is flowing much faster (5.7 m/s) through the smaller section (A_2).

The mass flow rate in both sections should be the same.

For the larger portion of the pipe:

$$\left(\frac{\Delta m}{\Delta t}\right)_1 = \rho A_1 v_1 = (880)(0.238)(1.8) = 380\ \text{kg}\Big/\text{s} \tag{12.23}$$

For the smaller portion of the pipe:

$$\left(\frac{\Delta m}{\Delta t}\right)_2 = \rho A_2 v_2 = (880)(0.75538)(5.7) = 380\ \text{kg}\Big/\text{s} \tag{12.24}$$

And so mass is conserved throughout the pipe. Every second, 380 kg of oil flows out of the larger portion of the pipe, and 380 kg of oil flows into the smaller portion of the pipe.

The solution to the last part of the example shows that speed is inversely proportional to the *square* of the radius of the tube, making for large effects when radius varies. We can blow out a candle at quite a distance, for example, by pursing our lips, whereas blowing on a candle with our mouth wide open is quite ineffective.

In many situations, including in the cardiovascular system, branching of the flow occurs. The blood is pumped from the heart into arteries that subdivide into smaller arteries (arterioles) which branch into very fine vessels called capillaries. In this situation, continuity of flow is maintained but it is the *sum* of the flow rates in each of the branches in any portion along the tube that is maintained. The equation of continuity in a more general form becomes

$$n_1 A_1 \bar{v}_1 = n_2 A_2 \bar{v}_2, \tag{12.25}$$

where n_1 and n_2 are the number of branches in each of the sections along the tube.

Example 12.3 Calculating Flow Speed and Vessel Diameter: Branching in the Cardiovascular System

The aorta is the principal blood vessel through which blood leaves the heart in order to circulate around the body. (a) Calculate the average speed of the blood in the aorta if the flow rate is 5.0 L/min. The aorta has a radius of 10 mm. (b) Blood also flows through smaller blood vessels known as capillaries. When the rate of blood flow in the aorta is 5.0 L/min, the speed of blood in the capillaries is about 0.33 mm/s. Given that the average diameter of a capillary is $8.0\ \mu\text{m}$, calculate the number of capillaries in the blood circulatory system.

Strategy

We can use $Q = A\bar{v}$ to calculate the speed of flow in the aorta and then use the general form of the equation of continuity to calculate the number of capillaries as all of the other variables are known.

Solution for (a)

The flow rate is given by $Q = A\bar{v}$ or $\bar{v} = \dfrac{Q}{\pi r^2}$ for a cylindrical vessel.

Substituting the known values (converted to units of meters and seconds) gives

$$\bar{v} = \frac{(5.0 \text{ L/min})(10^{-3} \text{ m}^3/\text{L})(1 \text{ min/60 s})}{\pi(0.010 \text{ m})^2} = 0.27 \text{ m/s}. \tag{12.26}$$

Solution for (b)

Using $n_1 A_1 \bar{v}_1 = n_2 A_2 \bar{v}_1$, assigning the subscript 1 to the aorta and 2 to the capillaries, and solving for n_2 (the

number of capillaries) gives $n_2 = \dfrac{n_1 A_1 \bar{v}_1}{A_2 \bar{v}_2}$. Converting all quantities to units of meters and seconds and substituting into

the equation above gives

$$n_2 = \frac{(1)(\pi)\left(10 \times 10^{-3} \text{ m}\right)^2 (0.27 \text{ m/s})}{(\pi)\left(4.0 \times 10^{-6} \text{ m}\right)^2 \left(0.33 \times 10^{-3} \text{ m/s}\right)} = 5.0 \times 10^9 \text{ capillaries.} \tag{12.27}$$

Discussion

Note that the speed of flow in the capillaries is considerably reduced relative to the speed in the aorta due to the significant increase in the total cross-sectional area at the capillaries. This low speed is to allow sufficient time for effective exchange to occur although it is equally important for the flow not to become stationary in order to avoid the possibility of clotting. Does this large number of capillaries in the body seem reasonable? In active muscle, one finds about 200 capillaries per mm^3, or about 200×10^6 per 1 kg of muscle. For 20 kg of muscle, this amounts to about 4×10^9 capillaries.

Making Connections: Syringes

A horizontally oriented hypodermic syringe has a barrel diameter of 1.2 cm and a needle diameter of 2.4 mm. A plunger pushes liquid in the barrel at a rate of 4.0 mm/s. Calculate the flow rate of liquid in both parts of the syringe (in mL/s) and the velocity of the liquid emerging from the needle.

Solution:

First, calculate the area of both parts of the syringe:

$$A_1 = \pi\left(\frac{d_1}{2}\right)^2 = \pi(0.006)^2 = 1.13 \times 10^{-4} \text{ m}^2 \tag{12.28}$$

$$A_2 = \pi\left(\frac{d_2}{2}\right)^2 = \pi(0.0012)^2 = 4.52 \times 10^{-6} \text{ m}^2 \tag{12.29}$$

Next, we can use the continuity equation to find the velocity of the liquid in the smaller part of the barrel (v_2):

$$A_1 v_1 = A_2 v_2 \tag{12.30}$$

$$v_2 = \left(\frac{A_1}{A_2}\right) v_1 \tag{12.31}$$

$$v_2 = \left(\frac{1.13 \times 10^{-4}}{4.52 \times 10^{-6}}\right)(0.004) = 0.10 \text{ m}\Big/\text{s} \tag{12.32}$$

Double-check the numbers to be sure that the flow rate in both parts of the syringe is the same:

$$Q_1 = A_1 v_1 = \left(1.13 \times 10^{-4}\right)(0.004) = 4.52 \times 10^{-7} \text{ m}^3\Big/\text{s} \tag{12.33}$$

$$Q_2 = A_2 v_2 = \left(4.52 \times 10^{-6}\right)(0.10) = 4.52 \times 10^{-7} \text{ m}^3\Big/\text{s} \tag{12.34}$$

Finally, by converting to mL/s:

$$\left(\frac{4.52\times10^{-7}\ \ m^3}{1\ \ s}\right)\left(\frac{10^6\ \ mL}{1\ \ m^3}\right) = 0.452\ \ mL\bigg/\ s \tag{12.35}$$

12.2 Bernoulli's Equation

When a fluid flows into a narrower channel, its speed increases. That means its kinetic energy also increases. Where does that change in kinetic energy come from? The increased kinetic energy comes from the net work done on the fluid to push it into the channel and the work done on the fluid by the gravitational force, if the fluid changes vertical position. Recall the work-energy theorem,

$$W_{net} = \frac{1}{2}mv^2 - \frac{1}{2}mv_0^2. \tag{12.36}$$

There is a pressure difference when the channel narrows. This pressure difference results in a net force on the fluid: recall that pressure times area equals force. The net work done increases the fluid's kinetic energy. As a result, the *pressure will drop in a rapidly-moving fluid*, whether or not the fluid is confined to a tube.

There are a number of common examples of pressure dropping in rapidly-moving fluids. Shower curtains have a disagreeable habit of bulging into the shower stall when the shower is on. The high-velocity stream of water and air creates a region of lower pressure inside the shower, and standard atmospheric pressure on the other side. The pressure difference results in a net force inward pushing the curtain in. You may also have noticed that when passing a truck on the highway, your car tends to veer toward it. The reason is the same—the high velocity of the air between the car and the truck creates a region of lower pressure, and the vehicles are pushed together by greater pressure on the outside. (See **Figure 12.4**.) This effect was observed as far back as the mid-1800s, when it was found that trains passing in opposite directions tipped precariously toward one another.

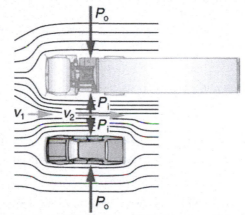

Figure 12.4 An overhead view of a car passing a truck on a highway. Air passing between the vehicles flows in a narrower channel and must increase its speed (v_2 is greater than v_1), causing the pressure between them to drop (P_i is less than P_o). Greater pressure on the outside pushes the car and truck together.

Making Connections: Take-Home Investigation with a Sheet of Paper

Hold the short edge of a sheet of paper parallel to your mouth with one hand on each side of your mouth. The page should slant downward over your hands. Blow over the top of the page. Describe what happens and explain the reason for this behavior.

Bernoulli's Equation

The relationship between pressure and velocity in fluids is described quantitatively by **Bernoulli's equation**, named after its discoverer, the Swiss scientist Daniel Bernoulli (1700–1782). Bernoulli's equation states that for an incompressible, frictionless fluid, the following sum is constant:

$$P + \frac{1}{2}\rho v^2 + \rho gh = \text{constant,} \tag{12.37}$$

where P is the absolute pressure, ρ is the fluid density, v is the velocity of the fluid, h is the height above some reference point, and g is the acceleration due to gravity. If we follow a small volume of fluid along its path, various quantities in the sum may change, but the total remains constant. Let the subscripts 1 and 2 refer to any two points along the path that the bit of fluid follows; Bernoulli's equation becomes

$$P_1 + \frac{1}{2}\rho v_1^2 + \rho gh_1 = P_2 + \frac{1}{2}\rho v_2^2 + \rho gh_2. \tag{12.38}$$

Bernoulli's equation is a form of the conservation of energy principle. Note that the second and third terms are the kinetic and potential energy with m replaced by ρ. In fact, each term in the equation has units of energy per unit volume. We can prove this for the second term by substituting $\rho = m/V$ into it and gathering terms:

$$\frac{1}{2}\rho v^2 = \frac{\frac{1}{2}mv^2}{V} = \frac{\text{KE}}{V}. \tag{12.39}$$

So $\frac{1}{2}\rho v^2$ is the kinetic energy per unit volume. Making the same substitution into the third term in the equation, we find

$$\rho gh = \frac{mgh}{V} = \frac{\text{PE}_g}{V}, \tag{12.40}$$

so ρgh is the gravitational potential energy per unit volume. Note that pressure P has units of energy per unit volume, too. Since $P = F/A$, its units are N/m^2. If we multiply these by m/m, we obtain $\text{N} \cdot \text{m/m}^3 = \text{J/m}^3$, or energy per unit volume. Bernoulli's equation is, in fact, just a convenient statement of conservation of energy for an incompressible fluid in the absence of friction.

Making Connections: Conservation of Energy

Conservation of energy applied to fluid flow produces Bernoulli's equation. The net work done by the fluid's pressure results in changes in the fluid's KE and PE_g per unit volume. If other forms of energy are involved in fluid flow, Bernoulli's equation can be modified to take these forms into account. Such forms of energy include thermal energy dissipated because of fluid viscosity.

The general form of Bernoulli's equation has three terms in it, and it is broadly applicable. To understand it better, we will look at a number of specific situations that simplify and illustrate its use and meaning.

Bernoulli's Equation for Static Fluids

Let us first consider the very simple situation where the fluid is static—that is, $v_1 = v_2 = 0$. Bernoulli's equation in that case is

$$P_1 + \rho gh_1 = P_2 + \rho gh_2. \tag{12.41}$$

We can further simplify the equation by taking $h_2 = 0$ (we can always choose some height to be zero, just as we often have done for other situations involving the gravitational force, and take all other heights to be relative to this). In that case, we get

$$P_2 = P_1 + \rho gh_1. \tag{12.42}$$

This equation tells us that, in static fluids, pressure increases with depth. As we go from point 1 to point 2 in the fluid, the depth increases by h_1, and consequently, P_2 is greater than P_1 by an amount ρgh_1. In the very simplest case, P_1 is zero at the top of the fluid, and we get the familiar relationship $P = \rho gh$. (Recall that $P = \rho gh$ and $\Delta\text{PE}_g = mgh$.) Bernoulli's equation includes the fact that the pressure due to the weight of a fluid is ρgh. Although we introduce Bernoulli's equation for fluid flow, it includes much of what we studied for static fluids in the preceding chapter.

Bernoulli's Principle—Bernoulli's Equation at Constant Depth

Another important situation is one in which the fluid moves but its depth is constant—that is, $h_1 = h_2$. Under that condition, Bernoulli's equation becomes

$$P_1 + \frac{1}{2}\rho v_1^2 = P_2 + \frac{1}{2}\rho v_2^2. \tag{12.43}$$

Situations in which fluid flows at a constant depth are so important that this equation is often called **Bernoulli's principle**. It is Bernoulli's equation for fluids at constant depth. (Note again that this applies to a small volume of fluid as we follow it along its path.) As we have just discussed, pressure drops as speed increases in a moving fluid. We can see this from Bernoulli's principle. For example, if v_2 is greater than v_1 in the equation, then P_2 must be less than P_1 for the equality to hold.

Example 12.4 Calculating Pressure: Pressure Drops as a Fluid Speeds Up

In **Example 12.2**, we found that the speed of water in a hose increased from 1.96 m/s to 25.5 m/s going from the hose to the nozzle. Calculate the pressure in the hose, given that the absolute pressure in the nozzle is $1.01 \times 10^5 \text{ N/m}^2$ (atmospheric, as it must be) and assuming level, frictionless flow.

Strategy

Level flow means constant depth, so Bernoulli's principle applies. We use the subscript 1 for values in the hose and 2 for those in the nozzle. We are thus asked to find P_1.

Solution

Solving Bernoulli's principle for P_1 yields

$$P_1 = P_2 + \frac{1}{2}\rho v_2^2 - \frac{1}{2}\rho v_1^2 = P_2 + \frac{1}{2}\rho(v_2^2 - v_1^2). \tag{12.44}$$

Substituting known values,

$$\begin{aligned} P_1 &= 1.01 \times 10^5 \text{ N/m}^2 \\ &\quad + \frac{1}{2}(10^3 \text{ kg/m}^3)\left[(25.5 \text{ m/s})^2 - (1.96 \text{ m/s})^2\right] \\ &= 4.24 \times 10^5 \text{ N/m}^2. \end{aligned} \tag{12.45}$$

Discussion

This absolute pressure in the hose is greater than in the nozzle, as expected since v is greater in the nozzle. The pressure P_2 in the nozzle must be atmospheric since it emerges into the atmosphere without other changes in conditions.

Applications of Bernoulli's Principle

There are a number of devices and situations in which fluid flows at a constant height and, thus, can be analyzed with Bernoulli's principle.

Entrainment

People have long put the Bernoulli principle to work by using reduced pressure in high-velocity fluids to move things about. With a higher pressure on the outside, the high-velocity fluid forces other fluids into the stream. This process is called *entrainment*. Entrainment devices have been in use since ancient times, particularly as pumps to raise water small heights, as in draining swamps, fields, or other low-lying areas. Some other devices that use the concept of entrainment are shown in **Figure 12.5**.

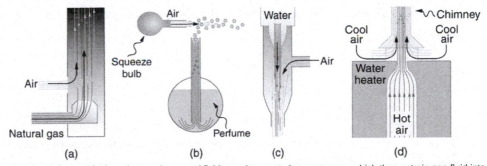

Figure 12.5 Examples of entrainment devices that use increased fluid speed to create low pressures, which then entrain one fluid into another. (a) A Bunsen burner uses an adjustable gas nozzle, entraining air for proper combustion. (b) An atomizer uses a squeeze bulb to create a jet of air that entrains drops of perfume. Paint sprayers and carburetors use very similar techniques to move their respective liquids. (c) A common aspirator uses a high-speed stream of water to create a region of lower pressure. Aspirators may be used as suction pumps in dental and surgical situations or for draining a flooded basement or producing a reduced pressure in a vessel. (d) The chimney of a water heater is designed to entrain air into the pipe leading through the ceiling.

Wings and Sails

The airplane wing is a beautiful example of Bernoulli's principle in action. **Figure 12.6**(a) shows the characteristic shape of a wing. The wing is tilted upward at a small angle and the upper surface is longer, causing air to flow faster over it. The pressure on top of the wing is therefore reduced, creating a net upward force or lift. (Wings can also gain lift by pushing air downward, utilizing the conservation of momentum principle. The deflected air molecules result in an upward force on the wing — Newton's third law.) Sails also have the characteristic shape of a wing. (See **Figure 12.6**(b).) The pressure on the front side of the sail, P_{front}, is lower than the pressure on the back of the sail, P_{back}. This results in a forward force and even allows you to sail into the wind.

Making Connections: Take-Home Investigation with Two Strips of Paper

For a good illustration of Bernoulli's principle, make two strips of paper, each about 15 cm long and 4 cm wide. Hold the small end of one strip up to your lips and let it drape over your finger. Blow across the paper. What happens? Now hold two strips of paper up to your lips, separated by your fingers. Blow between the strips. What happens?

Velocity measurement

Figure 12.7 shows two devices that measure fluid velocity based on Bernoulli's principle. The manometer in **Figure 12.7**(a) is connected to two tubes that are small enough not to appreciably disturb the flow. The tube facing the oncoming fluid creates a dead spot having zero velocity ($v_1 = 0$) in front of it, while fluid passing the other tube has velocity v_2. This means that Bernoulli's principle as stated in $P_1 + \frac{1}{2}\rho v_1^2 = P_2 + \frac{1}{2}\rho v_2^2$ becomes

$$P_1 = P_2 + \frac{1}{2}\rho v_2^2. \tag{12.46}$$

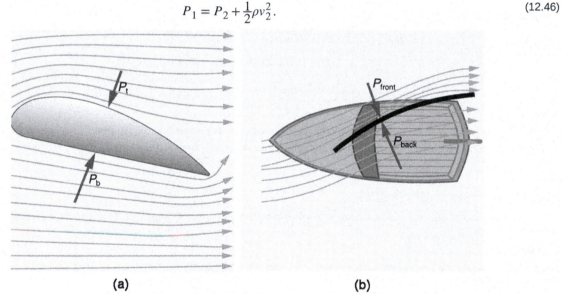

(a)   **(b)**

Figure 12.6 (a) The Bernoulli principle helps explain lift generated by a wing. (b) Sails use the same technique to generate part of their thrust.

Thus pressure P_2 over the second opening is reduced by $\frac{1}{2}\rho v_2^2$, and so the fluid in the manometer rises by h on the side connected to the second opening, where

$$h \propto \frac{1}{2}\rho v_2^2. \tag{12.47}$$

(Recall that the symbol $\propto$ means "proportional to.") Solving for v_2, we see that

$$v_2 \propto \sqrt{h}. \tag{12.48}$$

Figure 12.7(b) shows a version of this device that is in common use for measuring various fluid velocities; such devices are frequently used as air speed indicators in aircraft.

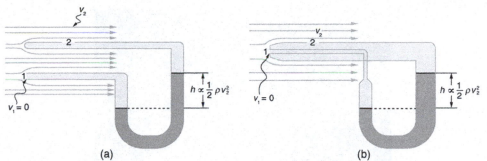

Figure 12.7 Measurement of fluid speed based on Bernoulli's principle. (a) A manometer is connected to two tubes that are close together and small enough not to disturb the flow. Tube 1 is open at the end facing the flow. A dead spot having zero speed is created there. Tube 2 has an opening on the side, and so the fluid has a speed v across the opening; thus, pressure there drops. The difference in pressure at the manometer is $\frac{1}{2}\rho v_2^2$, and so h is proportional to $\frac{1}{2}\rho v_2^2$. (b) This type of velocity measuring device is a Prandtl tube, also known as a pitot tube.

12.3 The Most General Applications of Bernoulli's Equation

Torricelli's Theorem

Figure 12.8 shows water gushing from a large tube through a dam. What is its speed as it emerges? Interestingly, if resistance is negligible, the speed is just what it would be if the water fell a distance h from the surface of the reservoir; the water's speed is independent of the size of the opening. Let us check this out. Bernoulli's equation must be used since the depth is not constant. We consider water flowing from the surface (point 1) to the tube's outlet (point 2). Bernoulli's equation as stated in previously is

$$P_1 + \frac{1}{2}\rho v_1^2 + \rho g h_1 = P_2 + \frac{1}{2}\rho v_2^2 + \rho g h_2. \tag{12.49}$$

Both P_1 and P_2 equal atmospheric pressure (P_1 is atmospheric pressure because it is the pressure at the top of the reservoir. P_2 must be atmospheric pressure, since the emerging water is surrounded by the atmosphere and cannot have a pressure different from atmospheric pressure.) and subtract out of the equation, leaving

$$\frac{1}{2}\rho v_1^2 + \rho g h_1 = \frac{1}{2}\rho v_2^2 + \rho g h_2. \tag{12.50}$$

Solving this equation for v_2^2, noting that the density ρ cancels (because the fluid is incompressible), yields

$$v_2^2 = v_1^2 + 2g(h_1 - h_2). \tag{12.51}$$

We let $h = h_1 - h_2$; the equation then becomes

$$v_2^2 = v_1^2 + 2gh \tag{12.52}$$

where h is the height dropped by the water. This is simply a kinematic equation for any object falling a distance h with negligible resistance. In fluids, this last equation is called *Torricelli's theorem*. Note that the result is independent of the velocity's direction, just as we found when applying conservation of energy to falling objects.

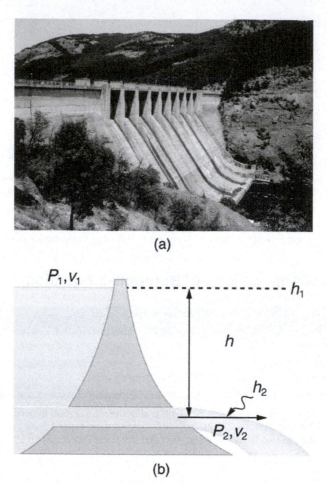

Figure 12.8 (a) Water gushes from the base of the Studen Kladenetz dam in Bulgaria. (credit: Kiril Kapustin; http://www.ImagesFromBulgaria.com) (b) In the absence of significant resistance, water flows from the reservoir with the same speed it would have if it fell the distance h without friction. This is an example of Torricelli's theorem.

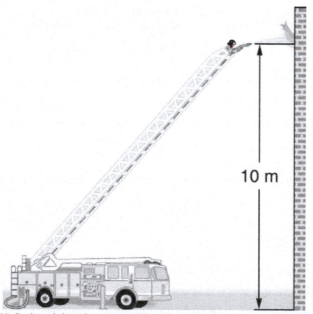

Figure 12.9 Pressure in the nozzle of this fire hose is less than at ground level for two reasons: the water has to go uphill to get to the nozzle, and speed increases in the nozzle. In spite of its lowered pressure, the water can exert a large force on anything it strikes, by virtue of its kinetic energy. Pressure in the water stream becomes equal to atmospheric pressure once it emerges into the air.

All preceding applications of Bernoulli's equation involved simplifying conditions, such as constant height or constant pressure. The next example is a more general application of Bernoulli's equation in which pressure, velocity, and height all change. (See

Figure 12.9.)

Example 12.5 Calculating Pressure: A Fire Hose Nozzle

Fire hoses used in major structure fires have inside diameters of 6.40 cm. Suppose such a hose carries a flow of 40.0 L/s starting at a gauge pressure of 1.62×10^6 N/m^2. The hose goes 10.0 m up a ladder to a nozzle having an inside diameter of 3.00 cm. Assuming negligible resistance, what is the pressure in the nozzle?

Strategy

Here we must use Bernoulli's equation to solve for the pressure, since depth is not constant.

Solution

Bernoulli's equation states

$$P_1 + \frac{1}{2}\rho v_1^2 + \rho g h_1 = P_2 + \frac{1}{2}\rho v_2^2 + \rho g h_2, \tag{12.53}$$

where the subscripts 1 and 2 refer to the initial conditions at ground level and the final conditions inside the nozzle, respectively. We must first find the speeds v_1 and v_2. Since $Q = A_1 v_1$, we get

$$v_1 = \frac{Q}{A_1} = \frac{40.0 \times 10^{-3} \text{ m}^3/\text{s}}{\pi(3.20 \times 10^{-2} \text{ m})^2} = 12.4 \text{ m/s}. \tag{12.54}$$

Similarly, we find

$$v_2 = 56.6 \text{ m/s}. \tag{12.55}$$

(This rather large speed is helpful in reaching the fire.) Now, taking h_1 to be zero, we solve Bernoulli's equation for P_2:

$$P_2 = P_1 + \frac{1}{2}\rho\left(v_1^2 - v_2^2\right) - \rho g h_2. \tag{12.56}$$

Substituting known values yields

$$P_2 = 1.62 \times 10^6 \text{ N/m}^2 + \frac{1}{2}(1000 \text{ kg/m}^3)\left[(12.4 \text{ m/s})^2 - (56.6 \text{ m/s})^2\right] - (1000 \text{ kg/m}^3)(9.80 \text{ m/s}^2)(10.0 \text{ m}) = 0.$$

$$\tag{12.57}$$

Discussion

This value is a gauge pressure, since the initial pressure was given as a gauge pressure. Thus the nozzle pressure equals atmospheric pressure, as it must because the water exits into the atmosphere without changes in its conditions.

Making Connections: Squirt Toy

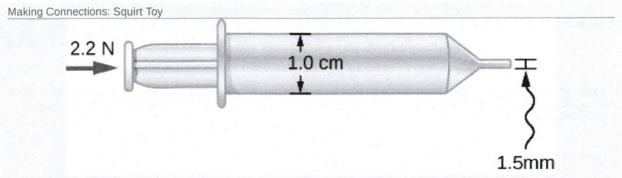

Figure 12.10

A horizontally oriented squirt toy contains a 1.0-cm-diameter barrel for the water. A 2.2-N force on the plunger forces water down the barrel and into a 1.5-mm-diameter opening at the end of the squirt gun. In addition to the force pushing on the plunger, pressure from the atmosphere is also present at both ends of the gun, pushing the plunger in and also pushing the water back in to the narrow opening at the other end. Assuming that the water is moving very slowly in the barrel, with what speed does it emerge from the toy?

Solution

First, find the cross-sectional areas for each part of the toy. The wider part is

$$A_1 = \pi r_1^2 = \pi\left(\frac{d_1}{2}\right)^2 = \pi(0.005)^2 = 7.85 \times 10^{-5} \text{ m}^2 \tag{12.58}$$

Next, we will find the area of the narrower part of the toy:

$$A_2 = \pi r_2^2 = \pi\left(\frac{d_2}{2}\right)^2 = \pi(0.0015)^2 = 7.07 \times 10^{-6} \text{ m}^2 \tag{12.59}$$

The pressure pushing on the barrel is equal to the sum of the pressure from the atmosphere ($1.0 \text{ atm} = 101,300 \text{ N/m}^2$) and the pressure created by the 2.2-N force.

$$P_1 = 101,300 + \left(\frac{\text{Force}}{A_1}\right) \tag{12.60}$$

$$P_1 = 101,300 + \left(\frac{2.2 \text{ N}}{7.85 \times 10^{-5} \text{ m}^2}\right) = 129,300 \text{ N}\Big/ \text{m}^2 \tag{12.61}$$

The pressure pushing on the smaller end of the toy is simply the pressure from the atmosphere:

$$P_2 = 101,300 \text{ N}\Big/ \text{m}^2 \tag{12.62}$$

Since the gun is oriented horizontally ($h_1 = h_2$), we can ignore the potential energy term in Bernoulli's equation, so the equation becomes:

$$P_1 + \tfrac{1}{2}\rho v_1{}^2 = P_2 + \tfrac{1}{2}\rho v_2{}^2 \tag{12.63}$$

The problem states that the water is moving very slowly in the barrel. That means we can make the approximation that $v_1 \approx 0$, which we will justify mathematically.

$$129,300 + (0.500)(1000)(0)^2 = 101,300 + (0.500)(1000)v_2^2 \tag{12.64}$$

$$v_2^2 = \frac{(129,300 - 101,300)}{500} \tag{12.65}$$

$$v_2^2 = \frac{(129,300 - 101,300)}{500} \tag{12.66}$$

$$v_2 = 7.5 \text{ m/s} \tag{12.67}$$

How accurate is our assumption that the water velocity in the barrel is approximately zero? Check using the continuity equation:

$$v_1 = \left(\frac{A_2}{A_1}\right)v_2 = \left(\frac{7.07 \times 10^{-6}}{7.85 \times 10^{-5}}\right)(7.5) = 0.17 \text{ m}\Big/ \text{s} \tag{12.68}$$

How does the kinetic energy per unit volume term for water in the barrel fit into Bernoulli's equation?

$$129,300 + (0.500)(1000)(0.17)^2 = 101,300 + (0.500)(1000)(7.5)^2 \tag{12.69}$$

$$129,300 + 14 = 101,300 + 28,000 \tag{12.70}$$

As you can see, the kinetic energy per unit volume term for water in the barrel is very small (14) compared to the other terms (which are all at least 1000 times larger). Another way to look at this is to consider the ratio of the two terms that represent kinetic energy per unit volume:

$$\frac{K_2}{K_1} = \frac{\tfrac{1}{2}\rho v_2^2}{\tfrac{1}{2}\rho v_1^2} = \frac{v_2^2}{v_1^2} \tag{12.71}$$

Remember that from the continuity equation

$$\frac{v_2}{v_1} = \frac{A_2}{A_1} = \frac{\pi\left(\frac{d_2}{2}\right)^2}{\pi\left(\frac{d_1}{2}\right)^2} = \frac{d_2^2}{d_1^2} \tag{12.72}$$

Thus, the ratio of the kinetic energy per unit volume terms depends on the fourth power of the ratio of the diameters:

$$\frac{K_2}{K_1} = \left(\frac{v_2}{v_1}\right)^2 = \left(\frac{d_2^2}{d_1^2}\right)^2 = \left(\frac{d_2}{d_1}\right)^4 \tag{12.73}$$

In this case, the diameter of the barrel (d_2) is 6.7 times larger than the diameter of the opening at the end of the toy (d_1), which makes the kinetic energy per unit volume term for water in the barrel $(6.7)^4 \approx 2000$ times smaller. We can usually neglect such small terms in addition or subtraction without a significant loss of accuracy.

Power in Fluid Flow

Power is the *rate* at which work is done or energy in any form is used or supplied. To see the relationship of power to fluid flow, consider Bernoulli's equation:

$$P + \frac{1}{2}\rho v^2 + \rho g h = \text{constant}. \tag{12.74}$$

All three terms have units of energy per unit volume, as discussed in the previous section. Now, considering units, if we multiply energy per unit volume by flow rate (volume per unit time), we get units of power. That is, $(E/V)(V/t) = E/t$. This means that if we multiply Bernoulli's equation by flow rate Q, we get power. In equation form, this is

$$\left(P + \frac{1}{2}\rho v^2 + \rho g h\right)Q = \text{power}. \tag{12.75}$$

Each term has a clear physical meaning. For example, PQ is the power supplied to a fluid, perhaps by a pump, to give it its pressure P. Similarly, $\frac{1}{2}\rho v^2 Q$ is the power supplied to a fluid to give it its kinetic energy. And $\rho g h Q$ is the power going to gravitational potential energy.

Making Connections: Power

Power is defined as the rate of energy transferred, or E/t. Fluid flow involves several types of power. Each type of power is identified with a specific type of energy being expended or changed in form.

Example 12.6 Calculating Power in a Moving Fluid

Suppose the fire hose in the previous example is fed by a pump that receives water through a hose with a 6.40-cm diameter coming from a hydrant with a pressure of $0.700 \times 10^6 \text{ N/m}^2$. What power does the pump supply to the water?

Strategy

Here we must consider energy forms as well as how they relate to fluid flow. Since the input and output hoses have the same diameters and are at the same height, the pump does not change the speed of the water nor its height, and so the water's kinetic energy and gravitational potential energy are unchanged. That means the pump only supplies power to increase water pressure by $0.92 \times 10^6 \text{ N/m}^2$ (from $0.700 \times 10^6 \text{ N/m}^2$ to $1.62 \times 10^6 \text{ N/m}^2$).

Solution

As discussed above, the power associated with pressure is

$$
\begin{aligned}
\text{power} &= PQ \\
&= \left(0.920 \times 10^6 \text{ N/m}^2\right)\left(40.0 \times 10^{-3} \text{ m}^3/\text{s}\right). \\
&= 3.68 \times 10^4 \text{ W} = 36.8 \text{ kW}
\end{aligned}
\tag{12.76}
$$

Discussion

Such a substantial amount of power requires a large pump, such as is found on some fire trucks. (This kilowatt value converts to about 50 hp.) The pump in this example increases only the water's pressure. If a pump—such as the heart—directly increases velocity and height as well as pressure, we would have to calculate all three terms to find the power it supplies.

12.4 Viscosity and Laminar Flow; Poiseuille's Law

Learning Objectives

By the end of this section, you will be able to:

- Define laminar flow and turbulent flow.
- Explain what viscosity is.

- Calculate flow and resistance with Poiseuille's law.
- Explain how pressure drops due to resistance.

Laminar Flow and Viscosity

When you pour yourself a glass of juice, the liquid flows freely and quickly. But when you pour syrup on your pancakes, that liquid flows slowly and sticks to the pitcher. The difference is fluid friction, both within the fluid itself and between the fluid and its surroundings. We call this property of fluids *viscosity*. Juice has low viscosity, whereas syrup has high viscosity. In the previous sections we have considered ideal fluids with little or no viscosity. In this section, we will investigate what factors, including viscosity, affect the rate of fluid flow.

The precise definition of viscosity is based on *laminar*, or nonturbulent, flow. Before we can define viscosity, then, we need to define laminar flow and turbulent flow. **Figure 12.11** shows both types of flow. **Laminar** flow is characterized by the smooth flow of the fluid in layers that do not mix. Turbulent flow, or **turbulence**, is characterized by eddies and swirls that mix layers of fluid together.

Figure 12.11 Smoke rises smoothly for a while and then begins to form swirls and eddies. The smooth flow is called laminar flow, whereas the swirls and eddies typify turbulent flow. If you watch the smoke (being careful not to breathe on it), you will notice that it rises more rapidly when flowing smoothly than after it becomes turbulent, implying that turbulence poses more resistance to flow. (credit: Creativity103)

Figure 12.12 shows schematically how laminar and turbulent flow differ. Layers flow without mixing when flow is laminar. When there is turbulence, the layers mix, and there are significant velocities in directions other than the overall direction of flow. The lines that are shown in many illustrations are the paths followed by small volumes of fluids. These are called *streamlines*. Streamlines are smooth and continuous when flow is laminar, but break up and mix when flow is turbulent. Turbulence has two main causes. First, any obstruction or sharp corner, such as in a faucet, creates turbulence by imparting velocities perpendicular to the flow. Second, high speeds cause turbulence. The drag both between adjacent layers of fluid and between the fluid and its surroundings forms swirls and eddies, if the speed is great enough. We shall concentrate on laminar flow for the remainder of this section, leaving certain aspects of turbulence for later sections.

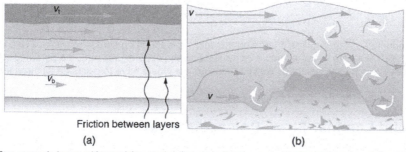

Friction between layers

(a)   (b)

Figure 12.12 (a) Laminar flow occurs in layers without mixing. Notice that viscosity causes drag between layers as well as with the fixed surface. (b) An obstruction in the vessel produces turbulence. Turbulent flow mixes the fluid. There is more interaction, greater heating, and more resistance than in laminar flow.

Making Connections: Take-Home Experiment: Go Down to the River

Try dropping simultaneously two sticks into a flowing river, one near the edge of the river and one near the middle. Which one travels faster? Why?

Figure 12.13 shows how viscosity is measured for a fluid. Two parallel plates have the specific fluid between them. The bottom plate is held fixed, while the top plate is moved to the right, dragging fluid with it. The layer (or lamina) of fluid in contact with either plate does not move relative to the plate, and so the top layer moves at v while the bottom layer remains at rest. Each successive layer from the top down exerts a force on the one below it, trying to drag it along, producing a continuous variation in speed from v to 0 as shown. Care is taken to insure that the flow is laminar; that is, the layers do not mix. The motion in **Figure 12.13** is like a continuous shearing motion. Fluids have zero shear strength, but the *rate* at which they are sheared is related to the same geometrical factors A and L as is shear deformation for solids.

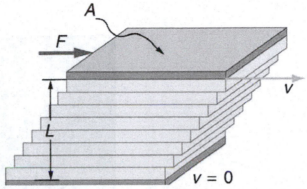

Figure 12.13 The graphic shows laminar flow of fluid between two plates of area A. The bottom plate is fixed. When the top plate is pushed to the right, it drags the fluid along with it.

A force F is required to keep the top plate in **Figure 12.13** moving at a constant velocity v, and experiments have shown that this force depends on four factors. First, F is directly proportional to v (until the speed is so high that turbulence occurs—then a much larger force is needed, and it has a more complicated dependence on v). Second, F is proportional to the area A of the plate. This relationship seems reasonable, since A is directly proportional to the amount of fluid being moved. Third, F is inversely proportional to the distance between the plates L. This relationship is also reasonable; L is like a lever arm, and the greater the lever arm, the less force that is needed. Fourth, F is directly proportional to *the coefficient of viscosity*, η. The greater the viscosity, the greater the force required. These dependencies are combined into the equation

$$F = \eta \frac{vA}{L},$$

(12.77)

which gives us a working definition of fluid **viscosity** η. Solving for η gives

$$\eta = \frac{FL}{vA},$$

(12.78)

which defines viscosity in terms of how it is measured. The SI unit of viscosity is $N \cdot m/[(m/s)m^2] = (N/m^2)s$ or $Pa \cdot s$.

Table 12.1 lists the coefficients of viscosity for various fluids.

Viscosity varies from one fluid to another by several orders of magnitude. As you might expect, the viscosities of gases are much less than those of liquids, and these viscosities are often temperature dependent. The viscosity of blood can be reduced by aspirin consumption, allowing it to flow more easily around the body. (When used over the long term in low doses, aspirin can help prevent heart attacks, and reduce the risk of blood clotting.)

Laminar Flow Confined to Tubes—Poiseuille's Law

What causes flow? The answer, not surprisingly, is pressure difference. In fact, there is a very simple relationship between horizontal flow and pressure. Flow rate Q is in the direction from high to low pressure. The greater the pressure differential between two points, the greater the flow rate. This relationship can be stated as

$$Q = \frac{P_2 - P_1}{R},$$

(12.79)

where P_1 and P_2 are the pressures at two points, such as at either end of a tube, and R is the resistance to flow. The resistance R includes everything, except pressure, that affects flow rate. For example, R is greater for a long tube than for a short one. The greater the viscosity of a fluid, the greater the value of R. Turbulence greatly increases R, whereas increasing the diameter of a tube decreases R.

If viscosity is zero, the fluid is frictionless and the resistance to flow is also zero. Comparing frictionless flow in a tube to viscous flow, as in **Figure 12.14**, we see that for a viscous fluid, speed is greatest at midstream because of drag at the boundaries. We can see the effect of viscosity in a Bunsen burner flame, even though the viscosity of natural gas is small.

The resistance R to laminar flow of an incompressible fluid having viscosity η through a horizontal tube of uniform radius r and length l, such as the one in **Figure 12.15**, is given by

$$R = \frac{8\eta l}{\pi r^4}. \tag{12.80}$$

This equation is called **Poiseuille's law for resistance** after the French scientist J. L. Poiseuille (1799–1869), who derived it in an attempt to understand the flow of blood, an often turbulent fluid.

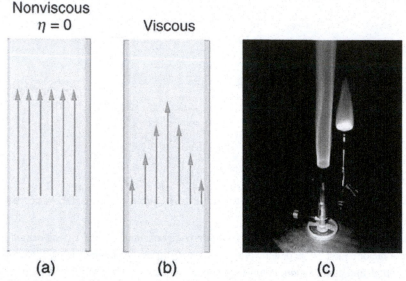

(a)   **(b)**   **(c)**

Figure 12.14 (a) If fluid flow in a tube has negligible resistance, the speed is the same all across the tube. (b) When a viscous fluid flows through a tube, its speed at the walls is zero, increasing steadily to its maximum at the center of the tube. (c) The shape of the Bunsen burner flame is due to the velocity profile across the tube. (credit: Jason Woodhead)

Let us examine Poiseuille's expression for R to see if it makes good intuitive sense. We see that resistance is directly proportional to both fluid viscosity η and the length l of a tube. After all, both of these directly affect the amount of friction encountered—the greater either is, the greater the resistance and the smaller the flow. The radius r of a tube affects the resistance, which again makes sense, because the greater the radius, the greater the flow (all other factors remaining the same). But it is surprising that r is raised to the *fourth* power in Poiseuille's law. This exponent means that any change in the radius of a tube has a very large effect on resistance. For example, doubling the radius of a tube decreases resistance by a factor of $2^4 = 16$.

Taken together, $Q = \frac{P_2 - P_1}{R}$ and $R = \frac{8\eta l}{\pi r^4}$ give the following expression for flow rate:

$$Q = \frac{(P_2 - P_1)\pi r^4}{8\eta l}. \tag{12.81}$$

This equation describes laminar flow through a tube. It is sometimes called Poiseuille's law for laminar flow, or simply **Poiseuille's law**.

Example 12.7 Using Flow Rate: Plaque Deposits Reduce Blood Flow

Suppose the flow rate of blood in a coronary artery has been reduced to half its normal value by plaque deposits. By what factor has the radius of the artery been reduced, assuming no turbulence occurs?

Strategy

Assuming laminar flow, Poiseuille's law states that

$$Q = \frac{(P_2 - P_1)\pi r^4}{8\eta l}. \tag{12.82}$$

We need to compare the artery radius before and after the flow rate reduction.

Solution

With a constant pressure difference assumed and the same length and viscosity, along the artery we have

$$\frac{Q_1}{r_1^4} = \frac{Q_2}{r_2^4}.$$

(12.83)

So, given that $Q_2 = 0.5Q_1$, we find that $r_2^4 = 0.5r_1^4$.

Therefore, $r_2 = (0.5)^{0.25}r_1 = 0.841r_1$, a decrease in the artery radius of 16%.

Discussion

This decrease in radius is surprisingly small for this situation. To restore the blood flow in spite of this buildup would require an increase in the pressure difference $(P_2 - P_1)$ of a factor of two, with subsequent strain on the heart.

Table 12.1 Coefficients of Viscosity of Various Fluids

Fluid	Temperature (°C)	Viscosity η (mPa·s)
Gases		
Air	0	0.0171
	20	0.0181
	40	0.0190
	100	0.0218
Ammonia	20	0.00974
Carbon dioxide	20	0.0147
Helium	20	0.0196
Hydrogen	0	0.0090
Mercury	20	0.0450
Oxygen	20	0.0203
Steam	100	0.0130
Liquids		
Water	0	1.792
	20	1.002
	37	0.6947
	40	0.653
	100	0.282
Whole blood[1]	20	3.015
	37	2.084
Blood plasma[2]	20	1.810
	37	1.257
Ethyl alcohol	20	1.20
Methanol	20	0.584
Oil (heavy machine)	20	660
Oil (motor, SAE 10)	30	200
Oil (olive)	20	138
Glycerin	20	1500
Honey	20	2000–10000
Maple Syrup	20	2000–3000
Milk	20	3.0
Oil (Corn)	20	65

The circulatory system provides many examples of Poiseuille's law in action—with blood flow regulated by changes in vessel size and blood pressure. Blood vessels are not rigid but elastic. Adjustments to blood flow are primarily made by varying the size of the vessels, since the resistance is so sensitive to the radius. During vigorous exercise, blood vessels are selectively dilated to important muscles and organs and blood pressure increases. This creates both greater overall blood flow and increased flow to specific areas. Conversely, decreases in vessel radii, perhaps from plaques in the arteries, can greatly reduce blood flow. If a vessel's radius is reduced by only 5% (to 0.95 of its original value), the flow rate is reduced to about $(0.95)^4 = 0.81$ of its original value. A 19% decrease in flow is caused by a 5% decrease in radius. The body may compensate by increasing blood pressure by 19%, but this presents hazards to the heart and any vessel that has weakened walls. Another example comes from automobile engine oil. If you have a car with an oil pressure gauge, you may notice that oil pressure is high when the engine is cold. Motor oil has greater viscosity when cold than when warm, and so pressure must be greater to pump the same amount of cold oil.

---

1. The ratios of the viscosities of blood to water are nearly constant between 0°C and 37°C.
2. See note on Whole Blood.

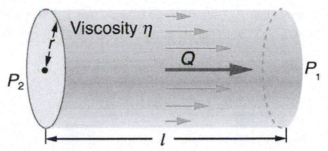

Figure 12.15 Poiseuille's law applies to laminar flow of an incompressible fluid of viscosity η through a tube of length l and radius r. The direction of flow is from greater to lower pressure. Flow rate Q is directly proportional to the pressure difference $P_2 - P_1$, and inversely proportional to the length l of the tube and viscosity η of the fluid. Flow rate increases with r^4, the fourth power of the radius.

Example 12.8 What Pressure Produces This Flow Rate?

An intravenous (IV) system is supplying saline solution to a patient at the rate of $0.120 \text{ cm}^3/\text{s}$ through a needle of radius 0.150 mm and length 2.50 cm. What pressure is needed at the entrance of the needle to cause this flow, assuming the viscosity of the saline solution to be the same as that of water? The gauge pressure of the blood in the patient's vein is 8.00 mm Hg. (Assume that the temperature is $20°C$.)

Strategy

Assuming laminar flow, Poiseuille's law applies. This is given by

$$Q = \frac{(P_2 - P_1)\pi r^4}{8\eta l},$$ (12.84)

where P_2 is the pressure at the entrance of the needle and P_1 is the pressure in the vein. The only unknown is P_2.

Solution

Solving for P_2 yields

$$P_2 = \frac{8\eta l}{\pi r^4}Q + P_1.$$ (12.85)

P_1 is given as 8.00 mm Hg, which converts to $1.066 \times 10^3 \text{ N/m}^2$. Substituting this and the other known values yields

$$
\begin{aligned}
P_2 &= \left[\frac{8(1.00 \times 10^{-3} \text{ N} \cdot \text{s/m}^2)(2.50 \times 10^{-2} \text{ m})}{\pi(0.150 \times 10^{-3} \text{ m})^4}\right](1.20 \times 10^{-7} \text{ m}^3/\text{s}) + 1.066 \times 10^3 \text{ N/m}^2 \\
&= 1.62 \times 10^4 \text{ N/m}^2.
\end{aligned}
$$ (12.86)

Discussion

This pressure could be supplied by an IV bottle with the surface of the saline solution 1.61 m above the entrance to the needle (this is left for you to solve in this chapter's Problems and Exercises), assuming that there is negligible pressure drop in the tubing leading to the needle.

Flow and Resistance as Causes of Pressure Drops

You may have noticed that water pressure in your home might be lower than normal on hot summer days when there is more use. This pressure drop occurs in the water main before it reaches your home. Let us consider flow through the water main as illustrated in **Figure 12.16**. We can understand why the pressure P_1 to the home drops during times of heavy use by rearranging

$$Q = \frac{P_2 - P_1}{R}$$ (12.87)

to

$$P_2 - P_1 = RQ,$$ (12.88)

where, in this case, P_2 is the pressure at the water works and R is the resistance of the water main. During times of heavy use, the flow rate Q is large. This means that $P_2 - P_1$ must also be large. Thus P_1 must decrease. It is correct to think of flow and resistance as causing the pressure to drop from P_2 to P_1. $P_2 - P_1 = RQ$ is valid for both laminar and turbulent flows.

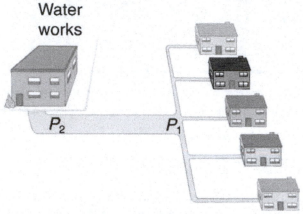

Figure 12.16 During times of heavy use, there is a significant pressure drop in a water main, and P_1 supplied to users is significantly less than P_2 created at the water works. If the flow is very small, then the pressure drop is negligible, and $P_2 \approx P_1$.

We can use $P_2 - P_1 = RQ$ to analyze pressure drops occurring in more complex systems in which the tube radius is not the same everywhere. Resistance will be much greater in narrow places, such as an obstructed coronary artery. For a given flow rate Q, the pressure drop will be greatest where the tube is most narrow. This is how water faucets control flow. Additionally, R is greatly increased by turbulence, and a constriction that creates turbulence greatly reduces the pressure downstream. Plaque in an artery reduces pressure and hence flow, both by its resistance and by the turbulence it creates.

Figure 12.17 is a schematic of the human circulatory system, showing average blood pressures in its major parts for an adult at rest. Pressure created by the heart's two pumps, the right and left ventricles, is reduced by the resistance of the blood vessels as the blood flows through them. The left ventricle increases arterial blood pressure that drives the flow of blood through all parts of the body except the lungs. The right ventricle receives the lower pressure blood from two major veins and pumps it through the lungs for gas exchange with atmospheric gases – the disposal of carbon dioxide from the blood and the replenishment of oxygen. Only one major organ is shown schematically, with typical branching of arteries to ever smaller vessels, the smallest of which are the capillaries, and rejoining of small veins into larger ones. Similar branching takes place in a variety of organs in the body, and the circulatory system has considerable flexibility in flow regulation to these organs by the dilation and constriction of the arteries leading to them and the capillaries within them. The sensitivity of flow to tube radius makes this flexibility possible over a large range of flow rates.

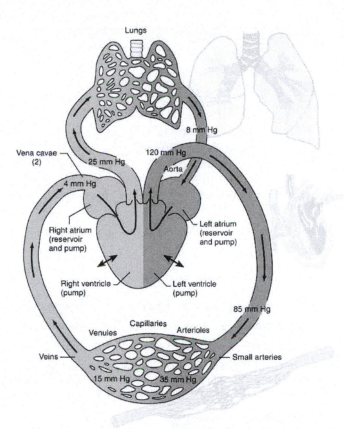

Figure 12.17 Schematic of the circulatory system. Pressure difference is created by the two pumps in the heart and is reduced by resistance in the vessels. Branching of vessels into capillaries allows blood to reach individual cells and exchange substances, such as oxygen and waste products, with them. The system has an impressive ability to regulate flow to individual organs, accomplished largely by varying vessel diameters.

Each branching of larger vessels into smaller vessels increases the total cross-sectional area of the tubes through which the blood flows. For example, an artery with a cross section of 1 cm^2 may branch into 20 smaller arteries, each with cross sections of 0.5 cm^2, with a total of 10 cm^2. In that manner, the resistance of the branchings is reduced so that pressure is not entirely lost. Moreover, because $Q = A\bar{v}$ and A increases through branching, the average velocity of the blood in the smaller vessels is reduced. The blood velocity in the aorta ($\text{diameter} = 1 \text{ cm}$) is about 25 cm/s, while in the capillaries ($20\mu\text{m}$ in diameter) the velocity is about 1 mm/s. This reduced velocity allows the blood to exchange substances with the cells in the capillaries and alveoli in particular.

12.5 The Onset of Turbulence

Learning Objectives

By the end of this section, you will be able to:

- Calculate Reynolds number.
- Use the Reynolds number for a system to determine whether it is laminar or turbulent.

Sometimes we can predict if flow will be laminar or turbulent. We know that flow in a very smooth tube or around a smooth, streamlined object will be laminar at low velocity. We also know that at high velocity, even flow in a smooth tube or around a smooth object will experience turbulence. In between, it is more difficult to predict. In fact, at intermediate velocities, flow may oscillate back and forth indefinitely between laminar and turbulent.

An occlusion, or narrowing, of an artery, such as shown in **Figure 12.18**, is likely to cause turbulence because of the irregularity of the blockage, as well as the complexity of blood as a fluid. Turbulence in the circulatory system is noisy and can sometimes be detected with a stethoscope, such as when measuring diastolic pressure in the upper arm's partially collapsed brachial artery. These turbulent sounds, at the onset of blood flow when the cuff pressure becomes sufficiently small, are called *Korotkoff sounds*. Aneurysms, or ballooning of arteries, create significant turbulence and can sometimes be detected with a stethoscope. Heart murmurs, consistent with their name, are sounds produced by turbulent flow around damaged and insufficiently closed heart valves. Ultrasound can also be used to detect turbulence as a medical indicator in a process analogous to Doppler-shift radar used to detect storms.

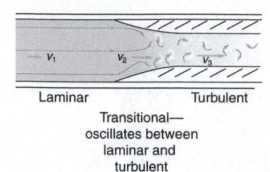

Laminar   Turbulent

Transitional—
oscillates between
laminar and
turbulent

Figure 12.18 Flow is laminar in the large part of this blood vessel and turbulent in the part narrowed by plaque, where velocity is high. In the transition region, the flow can oscillate chaotically between laminar and turbulent flow.

An indicator called the **Reynolds number** N_R can reveal whether flow is laminar or turbulent. For flow in a tube of uniform diameter, the Reynolds number is defined as

$$N_R = \frac{2\rho v r}{\eta}\text{(fl w in tube),} \tag{12.89}$$

where ρ is the fluid density, v its speed, η its viscosity, and r the tube radius. The Reynolds number is a unitless quantity. Experiments have revealed that N_R is related to the onset of turbulence. For N_R below about 2000, flow is laminar. For N_R above about 3000, flow is turbulent. For values of N_R between about 2000 and 3000, flow is unstable—that is, it can be laminar, but small obstructions and surface roughness can make it turbulent, and it may oscillate randomly between being laminar and turbulent. The blood flow through most of the body is a quiet, laminar flow. The exception is in the aorta, where the speed of the blood flow rises above a critical value of 35 m/s and becomes turbulent.

Example 12.9 Is This Flow Laminar or Turbulent?

Calculate the Reynolds number for flow in the needle considered in **Example 12.8** to verify the assumption that the flow is laminar. Assume that the density of the saline solution is 1025 kg/m^3.

Strategy

We have all of the information needed, except the fluid speed v, which can be calculated from $\bar{v} = Q/A = 1.70 \text{ m/s}$ (verification of this is in this chapter's Problems and Exercises).

Solution

Entering the known values into $N_R = \frac{2\rho v r}{\eta}$ gives

$$\begin{aligned} N_R &= \frac{2\rho v r}{\eta} \\ &= \frac{2(1025 \text{ kg/m}^3)(1.70 \text{ m/s})(0.150\times10^{-3} \text{ m})}{1.00\times10^{-3} \text{ N} \cdot \text{s/m}^2} \\ &= 523. \end{aligned} \tag{12.90}$$

Discussion

Since N_R is well below 2000, the flow should indeed be laminar.

Take-Home Experiment: Inhalation

Under the conditions of normal activity, an adult inhales about 1 L of air during each inhalation. With the aid of a watch, determine the time for one of your own inhalations by timing several breaths and dividing the total length by the number of breaths. Calculate the average flow rate Q of air traveling through the trachea during each inhalation.

The topic of chaos has become quite popular over the last few decades. A system is defined to be *chaotic* when its behavior is so sensitive to some factor that it is extremely difficult to predict. The field of *chaos* is the study of chaotic behavior. A good example of chaotic behavior is the flow of a fluid with a Reynolds number between 2000 and 3000. Whether or not the flow is turbulent is difficult, but not impossible, to predict—the difficulty lies in the extremely sensitive dependence on factors like roughness and obstructions on the nature of the flow. A tiny variation in one factor has an exaggerated (or nonlinear) effect on the flow. Phenomena as disparate as turbulence, the orbit of Pluto, and the onset of irregular heartbeats are chaotic and can be analyzed with similar techniques.

12.6 Motion of an Object in a Viscous Fluid

Learning Objectives

By the end of this section, you will be able to:

- Calculate the Reynolds number for an object moving through a fluid.
- Explain whether the Reynolds number indicates laminar or turbulent flow.
- Describe the conditions under which an object has a terminal speed.

A moving object in a viscous fluid is equivalent to a stationary object in a flowing fluid stream. (For example, when you ride a bicycle at 10 m/s in still air, you feel the air in your face exactly as if you were stationary in a 10-m/s wind.) Flow of the stationary fluid around a moving object may be laminar, turbulent, or a combination of the two. Just as with flow in tubes, it is possible to predict when a moving object creates turbulence. We use another form of the Reynolds number N'_R, defined for an object moving in a fluid to be

$$N'_R = \frac{\rho v L}{\eta} \text{(object in fluid)} \tag{12.91}$$

where L is a characteristic length of the object (a sphere's diameter, for example), ρ the fluid density, η its viscosity, and v the object's speed in the fluid. If N'_R is less than about 1, flow around the object can be laminar, particularly if the object has a smooth shape. The transition to turbulent flow occurs for N'_R between 1 and about 10, depending on surface roughness and so on. Depending on the surface, there can be a *turbulent wake* behind the object with some laminar flow over its surface. For an N'_R between 10 and 10^6, the flow may be either laminar or turbulent and may oscillate between the two. For N'_R greater than about 10^6, the flow is entirely turbulent, even at the surface of the object. (See Figure 12.19.) Laminar flow occurs mostly when the objects in the fluid are small, such as raindrops, pollen, and blood cells in plasma.

Example 12.10 Does a Ball Have a Turbulent Wake?

Calculate the Reynolds number N'_R for a ball with a 7.40-cm diameter thrown at 40.0 m/s.

Strategy

We can use $N'_R = \frac{\rho v L}{\eta}$ to calculate N'_R, since all values in it are either given or can be found in tables of density and viscosity.

Solution

Substituting values into the equation for N'_R yields

$$N'_R = \frac{\rho v L}{\eta} = \frac{(1.29 \text{ kg/m}^3)(40.0 \text{ m/s})(0.0740 \text{ m})}{1.81 \times 10^{-5} \, 1.00 \text{ Pa} \cdot \text{s}} \tag{12.92}$$
$$= 2.11 \times 10^5.$$

Discussion

This value is sufficiently high to imply a turbulent wake. Most large objects, such as airplanes and sailboats, create significant turbulence as they move. As noted before, the Bernoulli principle gives only qualitatively-correct results in such situations.

One of the consequences of viscosity is a resistance force called **viscous drag** F_V that is exerted on a moving object. This force typically depends on the object's speed (in contrast with simple friction). Experiments have shown that for laminar flow (N'_R less than about one) viscous drag is proportional to speed, whereas for N'_R between about 10 and 10^6, viscous drag is proportional to speed squared. (This relationship is a strong dependence and is pertinent to bicycle racing, where even a small headwind causes significantly increased drag on the racer. Cyclists take turns being the leader in the pack for this reason.) For N'_R greater than 10^6, drag increases dramatically and behaves with greater complexity. For laminar flow around a sphere, F_V is proportional to fluid viscosity η, the object's characteristic size L, and its speed v. All of which makes sense—the more viscous the fluid and the larger the object, the more drag we expect. Recall Stoke's law $F_S = 6\pi r\eta v$. For the special case of a small sphere of radius R moving slowly in a fluid of viscosity η , the drag force F_S is given by

$$F_S = 6\pi R\eta v. \tag{12.93}$$

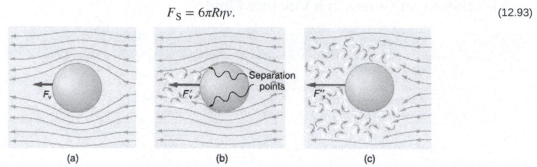

(a)   (b)   (c)

Figure 12.19 (a) Motion of this sphere to the right is equivalent to fluid flow to the left. Here the flow is laminar with N'_R less than 1. There is a force, called viscous drag F_V , to the left on the ball due to the fluid's viscosity. (b) At a higher speed, the flow becomes partially turbulent, creating a wake starting where the flow lines separate from the surface. Pressure in the wake is less than in front of the sphere, because fluid speed is less, creating a net force to the left F'_V that is significantly greater than for laminar flow. Here N'_R is greater than 10. (c) At much higher speeds, where N'_R is greater than 10^6 , flow becomes turbulent everywhere on the surface and behind the sphere. Drag increases dramatically.

An interesting consequence of the increase in F_V with speed is that an object falling through a fluid will not continue to accelerate indefinitely (as it would if we neglect air resistance, for example). Instead, viscous drag increases, slowing acceleration, until a critical speed, called the **terminal speed**, is reached and the acceleration of the object becomes zero. Once this happens, the object continues to fall at constant speed (the terminal speed). This is the case for particles of sand falling in the ocean, cells falling in a centrifuge, and sky divers falling through the air. **Figure 12.20** shows some of the factors that affect terminal speed. There is a viscous drag on the object that depends on the viscosity of the fluid and the size of the object. But there is also a buoyant force that depends on the density of the object relative to the fluid. Terminal speed will be greatest for low-viscosity fluids and objects with high densities and small sizes. Thus a skydiver falls more slowly with outspread limbs than when they are in a pike position—head first with hands at their side and legs together.

Take-Home Experiment: Don't Lose Your Marbles

By measuring the terminal speed of a slowly moving sphere in a viscous fluid, one can find the viscosity of that fluid (at that temperature). It can be difficult to find small ball bearings around the house, but a small marble will do. Gather two or three fluids (syrup, motor oil, honey, olive oil, etc.) and a thick, tall clear glass or vase. Drop the marble into the center of the fluid and time its fall (after letting it drop a little to reach its terminal speed). Compare your values for the terminal speed and see if they are inversely proportional to the viscosities as listed in **Table 12.1**. Does it make a difference if the marble is dropped near the side of the glass?

Knowledge of terminal speed is useful for estimating sedimentation rates of small particles. We know from watching mud settle out of dirty water that sedimentation is usually a slow process. Centrifuges are used to speed sedimentation by creating accelerated frames in which gravitational acceleration is replaced by centripetal acceleration, which can be much greater, increasing the terminal speed.

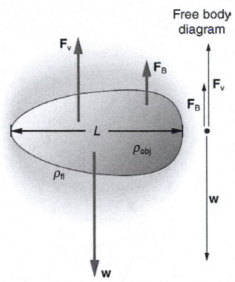

Figure 12.20 There are three forces acting on an object falling through a viscous fluid: its weight w , the viscous drag F_V , and the buoyant force $\mathbf{F_B}$.

12.7 Molecular Transport Phenomena: Diffusion, Osmosis, and Related Processes

Learning Objectives
By the end of this section, you will be able to: • Define diffusion, osmosis, dialysis, and active transport. • Calculate diffusion rates.

Diffusion

There is something fishy about the ice cube from your freezer—how did it pick up those food odors? How does soaking a sprained ankle in Epsom salt reduce swelling? The answer to these questions are related to atomic and molecular transport phenomena—another mode of fluid motion. Atoms and molecules are in constant motion at any temperature. In fluids they move about randomly even in the absence of macroscopic flow. This motion is called a random walk and is illustrated in **Figure 12.21**. **Diffusion** is the movement of substances due to random thermal molecular motion. Fluids, like fish fumes or odors entering ice cubes, can even diffuse through solids.

Diffusion is a slow process over macroscopic distances. The densities of common materials are great enough that molecules cannot travel very far before having a collision that can scatter them in any direction, including straight backward. It can be shown that the average distance x_{rms} that a molecule travels is proportional to the square root of time:

$$x_{\mathrm{rms}} = \sqrt{2Dt},$$

(12.94)

where x_{rms} stands for the **root-mean-square distance** and is the statistical average for the process. The quantity D is the diffusion constant for the particular molecule in a specific medium. **Table 12.2** lists representative values of D for various substances, in units of m^2/s .

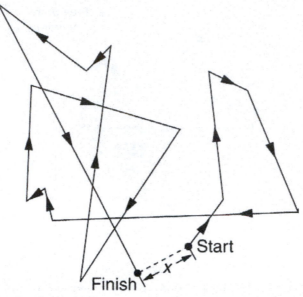

Figure 12.21 The random thermal motion of a molecule in a fluid in time t . This type of motion is called a random walk.

Table 12.2 Diffusion Constants for Various Molecules[3]

Diffusing molecule	Medium	D (m^2/s)
Hydrogen (H_2)	Air	6.4×10^{-5}
Oxygen (O_2)	Air	1.8×10^{-5}
Oxygen (O_2)	Water	1.0×10^{-9}
Glucose $(C_6 H_{12} O_6)$	Water	6.7×10^{-10}
Hemoglobin	Water	6.9×10^{-11}
DNA	Water	1.3×10^{-12}

Note that D gets progressively smaller for more massive molecules. This decrease is because the average molecular speed at a given temperature is inversely proportional to molecular mass. Thus the more massive molecules diffuse more slowly. Another interesting point is that D for oxygen in air is much greater than D for oxygen in water. In water, an oxygen molecule makes many more collisions in its random walk and is slowed considerably. In water, an oxygen molecule moves only about $40 \ \mu m$ in

1 s. (Each molecule actually collides about 10^{10} times per second!). Finally, note that diffusion constants increase with temperature, because average molecular speed increases with temperature. This is because the average kinetic energy of molecules, $\frac{1}{2}mv^2$, is proportional to absolute temperature.

Example 12.11 Calculating Diffusion: How Long Does Glucose Diffusion Take?

Calculate the average time it takes a glucose molecule to move 1.0 cm in water.

Strategy

We can use $x_{rms} = \sqrt{2Dt}$, the expression for the average distance moved in time t , and solve it for t . All other quantities are known.

Solution

Solving for t and substituting known values yields

---

3. At 20°C and 1 atm

$$t = \frac{x^2_{\text{rms}}}{2D} = \frac{(0.010 \text{ m})^2}{2(6.7 \times 10^{-10} \text{ m}^2/\text{s})}$$

$$= 7.5 \times 10^4 \text{ s} = 21 \text{ h}.$$

(12.95)

Discussion

This is a remarkably long time for glucose to move a mere centimeter! For this reason, we stir sugar into water rather than waiting for it to diffuse.

Because diffusion is typically very slow, its most important effects occur over small distances. For example, the cornea of the eye gets most of its oxygen by diffusion through the thin tear layer covering it.

The Rate and Direction of Diffusion

If you very carefully place a drop of food coloring in a still glass of water, it will slowly diffuse into the colorless surroundings until its concentration is the same everywhere. This type of diffusion is called free diffusion, because there are no barriers inhibiting it. Let us examine its direction and rate. Molecular motion is random in direction, and so simple chance dictates that more molecules will move out of a region of high concentration than into it. The net rate of diffusion is higher initially than after the process is partially completed. (See Figure 12.22.)

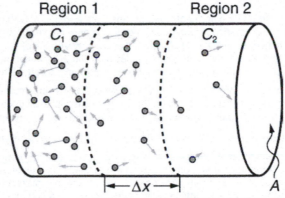

Figure 12.22 Diffusion proceeds from a region of higher concentration to a lower one. The net rate of movement is proportional to the difference in concentration.

The net rate of diffusion is proportional to the concentration difference. Many more molecules will leave a region of high concentration than will enter it from a region of low concentration. In fact, if the concentrations were the same, there would be *no* net movement. The net rate of diffusion is also proportional to the diffusion constant D, which is determined experimentally. The farther a molecule can diffuse in a given time, the more likely it is to leave the region of high concentration. Many of the factors that affect the rate are hidden in the diffusion constant D. For example, temperature and cohesive and adhesive forces all affect values of D.

Diffusion is the dominant mechanism by which the exchange of nutrients and waste products occur between the blood and tissue, and between air and blood in the lungs. In the evolutionary process, as organisms became larger, they needed quicker methods of transportation than net diffusion, because of the larger distances involved in the transport, leading to the development of circulatory systems. Less sophisticated, single-celled organisms still rely totally on diffusion for the removal of waste products and the uptake of nutrients.

Osmosis and Dialysis—Diffusion across Membranes

Some of the most interesting examples of diffusion occur through barriers that affect the rates of diffusion. For example, when you soak a swollen ankle in Epsom salt, water diffuses through your skin. Many substances regularly move through cell membranes; oxygen moves in, carbon dioxide moves out, nutrients go in, and wastes go out, for example. Because membranes are thin structures (typically 6.5×10^{-9} to 10×10^{-9} m across) diffusion rates through them can be high. Diffusion through membranes is an important method of transport.

Membranes are generally selectively permeable, or **semipermeable**. (See Figure 12.23.) One type of semipermeable membrane has small pores that allow only small molecules to pass through. In other types of membranes, the molecules may actually dissolve in the membrane or react with molecules in the membrane while moving across. Membrane function, in fact, is the subject of much current research, involving not only physiology but also chemistry and physics.

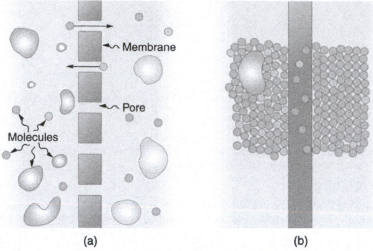

Figure 12.23 (a) A semipermeable membrane with small pores that allow only small molecules to pass through. (b) Certain molecules dissolve in this membrane and diffuse across it.

Osmosis is the transport of water through a semipermeable membrane from a region of high concentration to a region of low concentration. Osmosis is driven by the imbalance in water concentration. For example, water is more concentrated in your body than in Epsom salt. When you soak a swollen ankle in Epsom salt, the water moves out of your body into the lower-concentration region in the salt. Similarly, **dialysis** is the transport of any other molecule through a semipermeable membrane due to its concentration difference. Both osmosis and dialysis are used by the kidneys to cleanse the blood.

Osmosis can create a substantial pressure. Consider what happens if osmosis continues for some time, as illustrated in **Figure 12.24**. Water moves by osmosis from the left into the region on the right, where it is less concentrated, causing the solution on the right to rise. This movement will continue until the pressure ρgh created by the extra height of fluid on the right is large enough to stop further osmosis. This pressure is called a *back pressure*. The back pressure ρgh that stops osmosis is also called the **relative osmotic pressure** if neither solution is pure water, and it is called the **osmotic pressure** if one solution is pure water. Osmotic pressure can be large, depending on the size of the concentration difference. For example, if pure water and sea water are separated by a semipermeable membrane that passes no salt, osmotic pressure will be 25.9 atm. This value means that water will diffuse through the membrane until the salt water surface rises 268 m above the pure-water surface! One example of pressure created by osmosis is turgor in plants (many wilt when too dry). Turgor describes the condition of a plant in which the fluid in a cell exerts a pressure against the cell wall. This pressure gives the plant support. Dialysis can similarly cause substantial pressures.

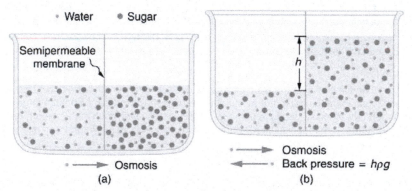

Figure 12.24 (a) Two sugar-water solutions of different concentrations, separated by a semipermeable membrane that passes water but not sugar. Osmosis will be to the right, since water is less concentrated there. (b) The fluid level rises until the back pressure ρgh equals the relative osmotic pressure; then, the net transfer of water is zero.

Reverse osmosis and **reverse dialysis** (also called filtration) are processes that occur when back pressure is sufficient to reverse the normal direction of substances through membranes. Back pressure can be created naturally as on the right side of **Figure 12.24**. (A piston can also create this pressure.) Reverse osmosis can be used to desalinate water by simply forcing it through a membrane that will not pass salt. Similarly, reverse dialysis can be used to filter out any substance that a given membrane will not pass.

One further example of the movement of substances through membranes deserves mention. We sometimes find that substances pass in the direction opposite to what we expect. Cypress tree roots, for example, extract pure water from salt water, although osmosis would move it in the opposite direction. This is not reverse osmosis, because there is no back pressure to cause it. What is happening is called **active transport**, a process in which a living membrane expends energy to move substances across it. Many living membranes move water and other substances by active transport. The kidneys, for example, not only use osmosis and dialysis—they also employ significant active transport to move substances into and out of blood. In fact, it is estimated that at least 25% of the body's energy is expended on active transport of substances at the cellular level. The study of active transport carries us into the realms of microbiology, biophysics, and biochemistry and it is a fascinating application of the laws of nature to living structures.

Glossary

active transport: the process in which a living membrane expends energy to move substances across

Bernoulli's equation: the equation resulting from applying conservation of energy to an incompressible frictionless fluid: $P + 1/2pv^2 + pgh$ = constant , through the fluid

Bernoulli's principle: Bernoulli's equation applied at constant depth: $P_1 + 1/2pv_1^2 = P_2 + 1/2pv_2^2$

dialysis: the transport of any molecule other than water through a semipermeable membrane from a region of high concentration to one of low concentration

diffusion: the movement of substances due to random thermal molecular motion

flow rate: abbreviated Q, it is the volume V that flows past a particular point during a time t, or $Q = V/t$

fluid dynamics: the physics of fluids in motion

laminar: a type of fluid flow in which layers do not mix

liter: a unit of volume, equal to 10^{-3} m^3

osmosis: the transport of water through a semipermeable membrane from a region of high concentration to one of low concentration

osmotic pressure: the back pressure which stops the osmotic process if one solution is pure water

Poiseuille's law: the rate of laminar flow of an incompressible fluid in a tube: $Q = (P_2 - P_1)\pi r^4/8\eta l$

Poiseuille's law for resistance: the resistance to laminar flow of an incompressible fluid in a tube: $R = 8\eta l/\pi r^4$

relative osmotic pressure: the back pressure which stops the osmotic process if neither solution is pure water

reverse dialysis: the process that occurs when back pressure is sufficient to reverse the normal direction of dialysis through membranes

reverse osmosis: the process that occurs when back pressure is sufficient to reverse the normal direction of osmosis through membranes

Reynolds number: a dimensionless parameter that can reveal whether a particular flow is laminar or turbulent

semipermeable: a type of membrane that allows only certain small molecules to pass through

terminal speed: the speed at which the viscous drag of an object falling in a viscous fluid is equal to the other forces acting on the object (such as gravity), so that the acceleration of the object is zero

turbulence: fluid flow in which layers mix together via eddies and swirls

viscosity: the friction in a fluid, defined in terms of the friction between layers

viscous drag: a resistance force exerted on a moving object, with a nontrivial dependence on velocity

Section Summary

12.1 Flow Rate and Its Relation to Velocity

- Flow rate Q is defined to be the volume V flowing past a point in time t, or $Q = \dfrac{V}{t}$ where V is volume and t is time.

- The SI unit of volume is m^3 .

- Another common unit is the liter (L), which is 10^{-3} m^3 .

- Flow rate and velocity are related by $Q = A\bar{v}$ where A is the cross-sectional area of the flow and $\bar{v}$ is its average velocity.
- For incompressible fluids, flow rate at various points is constant. That is,

$$\left.\begin{array}{c} Q_1 = Q_2 \\ A_1 \bar{v}_1 = A_2 \bar{v}_2 \\ n_1 A_1 \bar{v}_1 = n_2 A_2 \bar{v}_2 \end{array}\right\}.$$

12.2 Bernoulli's Equation

- Bernoulli's equation states that the sum on each side of the following equation is constant, or the same at any two points in an incompressible frictionless fluid:

$$P_1 + \frac{1}{2}\rho v_1^2 + \rho g h_1 = P_2 + \frac{1}{2}\rho v_2^2 + \rho g h_2.$$

- Bernoulli's principle is Bernoulli's equation applied to situations in which depth is constant. The terms involving depth (or height h) subtract out, yielding

$$P_1 + \frac{1}{2}\rho v_1^2 = P_2 + \frac{1}{2}\rho v_2^2.$$

- Bernoulli's principle has many applications, including entrainment, wings and sails, and velocity measurement.

12.3 The Most General Applications of Bernoulli's Equation

- Power in fluid flow is given by the equation $\left(P_1 + \frac{1}{2}\rho v^2 + \rho g h\right)Q = \text{power},$ where the first term is power associated with pressure, the second is power associated with velocity, and the third is power associated with height.

12.4 Viscosity and Laminar Flow; Poiseuille's Law

- Laminar flow is characterized by smooth flow of the fluid in layers that do not mix.
- Turbulence is characterized by eddies and swirls that mix layers of fluid together.
- Fluid viscosity η is due to friction within a fluid. Representative values are given in Table 12.1. Viscosity has units of $(\text{N/m}^2)\text{s}$ or $\text{Pa}\cdot\text{s}$.
- Flow is proportional to pressure difference and inversely proportional to resistance:

$$Q = \frac{P_2 - P_1}{R}.$$

- For laminar flow in a tube, Poiseuille's law for resistance states that

$$R = \frac{8\eta l}{\pi r^4}.$$

- Poiseuille's law for flow in a tube is

$$Q = \frac{(P_2 - P_1)\pi r^4}{8\eta l}.$$

- The pressure drop caused by flow and resistance is given by

$$P_2 - P_1 = RQ.$$

12.5 The Onset of Turbulence

- The Reynolds number N_R can reveal whether flow is laminar or turbulent. It is

$$N_R = \frac{2\rho v r}{\eta}.$$

- For N_R below about 2000, flow is laminar. For N_R above about 3000, flow is turbulent. For values of N_R between 2000 and 3000, it may be either or both.

12.6 Motion of an Object in a Viscous Fluid

- When an object moves in a fluid, there is a different form of the Reynolds number $N'_R = \frac{\rho v L}{\eta} (\text{object in fluid})$ which indicates whether flow is laminar or turbulent.
- For N'_R less than about one, flow is laminar.
- For N'_R greater than 10^6, flow is entirely turbulent.

12.7 Molecular Transport Phenomena: Diffusion, Osmosis, and Related Processes

- Diffusion is the movement of substances due to random thermal molecular motion.
- The average distance x_{rms} a molecule travels by diffusion in a given amount of time is given by

$$x_{\text{rms}} = \sqrt{2Dt},$$

where D is the diffusion constant, representative values of which are found in **Table 12.2**.

- Osmosis is the transport of water through a semipermeable membrane from a region of high concentration to a region of low concentration.
- Dialysis is the transport of any other molecule through a semipermeable membrane due to its concentration difference.
- Both processes can be reversed by back pressure.
- Active transport is a process in which a living membrane expends energy to move substances across it.

Conceptual Questions

12.1 Flow Rate and Its Relation to Velocity

1. What is the difference between flow rate and fluid velocity? How are they related?

2. Many figures in the text show streamlines. Explain why fluid velocity is greatest where streamlines are closest together. (Hint: Consider the relationship between fluid velocity and the cross-sectional area through which it flows.)

3. Identify some substances that are incompressible and some that are not.

12.2 Bernoulli's Equation

4. You can squirt water a considerably greater distance by placing your thumb over the end of a garden hose and then releasing, than by leaving it completely uncovered. Explain how this works.

5. Water is shot nearly vertically upward in a decorative fountain and the stream is observed to broaden as it rises. Conversely, a stream of water falling straight down from a faucet narrows. Explain why, and discuss whether surface tension enhances or reduces the effect in each case.

6. Look back to **Figure 12.4**. Answer the following two questions. Why is P_o less than atmospheric? Why is P_o greater than P_i?

7. Give an example of entrainment not mentioned in the text.

8. Many entrainment devices have a constriction, called a Venturi, such as shown in **Figure 12.25**. How does this bolster entrainment?

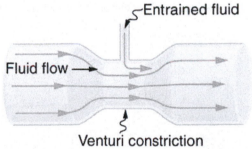

Figure 12.25 A tube with a narrow segment designed to enhance entrainment is called a Venturi. These are very commonly used in carburetors and aspirators.

9. Some chimney pipes have a T-shape, with a crosspiece on top that helps draw up gases whenever there is even a slight breeze. Explain how this works in terms of Bernoulli's principle.

10. Is there a limit to the height to which an entrainment device can raise a fluid? Explain your answer.

11. Why is it preferable for airplanes to take off into the wind rather than with the wind?

12. Roofs are sometimes pushed off vertically during a tropical cyclone, and buildings sometimes explode outward when hit by a tornado. Use Bernoulli's principle to explain these phenomena.

13. Why does a sailboat need a keel?

14. It is dangerous to stand close to railroad tracks when a rapidly moving commuter train passes. Explain why atmospheric pressure would push you toward the moving train.

15. Water pressure inside a hose nozzle can be less than atmospheric pressure due to the Bernoulli effect. Explain in terms of energy how the water can emerge from the nozzle against the opposing atmospheric pressure.

16. A perfume bottle or atomizer sprays a fluid that is in the bottle. (**Figure 12.26.**) How does the fluid rise up in the vertical tube in the bottle?

Figure 12.26 Atomizer: perfume bottle with tube to carry perfume up through the bottle. (credit: Antonia Foy, Flickr)

17. If you lower the window on a car while moving, an empty plastic bag can sometimes fly out the window. Why does this happen?

12.3 The Most General Applications of Bernoulli's Equation

18. Based on Bernoulli's equation, what are three forms of energy in a fluid? (Note that these forms are conservative, unlike heat transfer and other dissipative forms not included in Bernoulli's equation.)

19. Water that has emerged from a hose into the atmosphere has a gauge pressure of zero. Why? When you put your hand in front of the emerging stream you feel a force, yet the water's gauge pressure is zero. Explain where the force comes from in terms of energy.

20. The old rubber boot shown in **Figure 12.27** has two leaks. To what maximum height can the water squirt from Leak 1? How does the velocity of water emerging from Leak 2 differ from that of leak 1? Explain your responses in terms of energy.

Figure 12.27 Water emerges from two leaks in an old boot.

21. Water pressure inside a hose nozzle can be less than atmospheric pressure due to the Bernoulli effect. Explain in terms of energy how the water can emerge from the nozzle against the opposing atmospheric pressure.

12.4 Viscosity and Laminar Flow; Poiseuille's Law

22. Explain why the viscosity of a liquid decreases with temperature—that is, how might increased temperature reduce the effects of cohesive forces in a liquid? Also explain why the viscosity of a gas increases with temperature—that is, how does increased gas temperature create more collisions between atoms and molecules?

23. When paddling a canoe upstream, it is wisest to travel as near to the shore as possible. When canoeing downstream, it may be best to stay near the middle. Explain why.

24. Why does flow decrease in your shower when someone flushes the toilet?

25. Plumbing usually includes air-filled tubes near water faucets, as shown in **Figure 12.28**. Explain why they are needed and how they work.

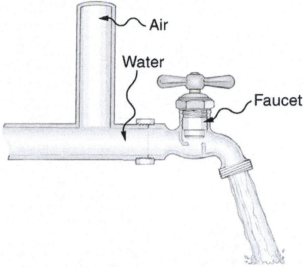

Figure 12.28 The vertical tube near the water tap remains full of air and serves a useful purpose.

12.5 The Onset of Turbulence

26. Doppler ultrasound can be used to measure the speed of blood in the body. If there is a partial constriction of an artery, where would you expect blood speed to be greatest, at or nearby the constriction? What are the two distinct causes of higher resistance in the constriction?

27. Sink drains often have a device such as that shown in **Figure 12.29** to help speed the flow of water. How does this work?

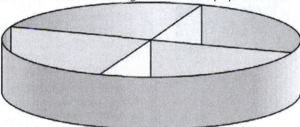

Figure 12.29 You will find devices such as this in many drains. They significantly increase flow rate.

28. Some ceiling fans have decorative wicker reeds on their blades. Discuss whether these fans are as quiet and efficient as those with smooth blades.

12.6 Motion of an Object in a Viscous Fluid

29. What direction will a helium balloon move inside a car that is slowing down—toward the front or back? Explain your answer.

30. Will identical raindrops fall more rapidly in $5°\,C$ air or $25°\,C$ air, neglecting any differences in air density? Explain your answer.

31. If you took two marbles of different sizes, what would you expect to observe about the relative magnitudes of their terminal velocities?

12.7 Molecular Transport Phenomena: Diffusion, Osmosis, and Related Processes

32. Why would you expect the rate of diffusion to increase with temperature? Can you give an example, such as the fact that you can dissolve sugar more rapidly in hot water?

33. How are osmosis and dialysis similar? How do they differ?

Problems & Exercises

12.1 Flow Rate and Its Relation to Velocity

1. What is the average flow rate in cm^3/s of gasoline to the engine of a car traveling at 100 km/h if it averages 10.0 km/L?

2. The heart of a resting adult pumps blood at a rate of 5.00 L/min. (a) Convert this to cm^3/s. (b) What is this rate in m^3/s?

3. Blood is pumped from the heart at a rate of 5.0 L/min into the aorta (of radius 1.0 cm). Determine the speed of blood through the aorta.

4. Blood is flowing through an artery of radius 2 mm at a rate of 40 cm/s. Determine the flow rate and the volume that passes through the artery in a period of 30 s.

5. The Huka Falls on the Waikato River is one of New Zealand's most visited natural tourist attractions (see **Figure 12.30**). On average the river has a flow rate of about 300,000 L/s. At the gorge, the river narrows to 20 m wide and averages 20 m deep. (a) What is the average speed of the river in the gorge? (b) What is the average speed of the water in the river downstream of the falls when it widens to 60 m and its depth increases to an average of 40 m?

Figure 12.30 The Huka Falls in Taupo, New Zealand, demonstrate flow rate. (credit: RaviGogna, Flickr)

6. A major artery with a cross-sectional area of $1.00\ cm^2$ branches into 18 smaller arteries, each with an average cross-sectional area of $0.400\ cm^2$. By what factor is the average velocity of the blood reduced when it passes into these branches?

7. (a) As blood passes through the capillary bed in an organ, the capillaries join to form venules (small veins). If the blood speed increases by a factor of 4.00 and the total cross-sectional area of the venules is $10.0\ cm^2$, what is the total cross-sectional area of the capillaries feeding these venules? (b) How many capillaries are involved if their average diameter is $10.0\ \mu m$?

8. The human circulation system has approximately 1×10^9 capillary vessels. Each vessel has a diameter of about $8\ \mu m$. Assuming cardiac output is 5 L/min, determine the average velocity of blood flow through each capillary vessel.

9. (a) Estimate the time it would take to fill a private swimming pool with a capacity of 80,000 L using a garden hose delivering 60 L/min. (b) How long would it take to fill if you could divert a moderate size river, flowing at $5000\ m^3/s$, into it?

10. The flow rate of blood through a 2.00×10^{-6}-m -radius capillary is $3.80 \times 10^9\ cm^3/s$. (a) What is the speed of the blood flow? (This small speed allows time for diffusion of materials to and from the blood.) (b) Assuming all the blood in the body passes through capillaries, how many of them must there be to carry a total flow of $90.0\ cm^3/s$? (The large number obtained is an overestimate, but it is still reasonable.)

11. (a) What is the fluid speed in a fire hose with a 9.00-cm diameter carrying 80.0 L of water per second? (b) What is the flow rate in cubic meters per second? (c) Would your answers be different if salt water replaced the fresh water in the fire hose?

12. The main uptake air duct of a forced air gas heater is 0.300 m in diameter. What is the average speed of air in the duct if it carries a volume equal to that of the house's interior every 15 min? The inside volume of the house is equivalent to a rectangular solid 13.0 m wide by 20.0 m long by 2.75 m high.

13. Water is moving at a velocity of 2.00 m/s through a hose with an internal diameter of 1.60 cm. (a) What is the flow rate in liters per second? (b) The fluid velocity in this hose's nozzle is 15.0 m/s. What is the nozzle's inside diameter?

14. Prove that the speed of an incompressible fluid through a constriction, such as in a Venturi tube, increases by a factor equal to the square of the factor by which the diameter decreases. (The converse applies for flow out of a constriction into a larger-diameter region.)

15. Water emerges straight down from a faucet with a 1.80-cm diameter at a speed of 0.500 m/s. (Because of the construction of the faucet, there is no variation in speed across the stream.) (a) What is the flow rate in cm^3/s? (b) What is the diameter of the stream 0.200 m below the faucet? Neglect any effects due to surface tension.

16. Unreasonable Results

A mountain stream is 10.0 m wide and averages 2.00 m in depth. During the spring runoff, the flow in the stream reaches $100,000\ m^3/s$. (a) What is the average velocity of the stream under these conditions? (b) What is unreasonable about this velocity? (c) What is unreasonable or inconsistent about the premises?

12.2 Bernoulli's Equation

17. Verify that pressure has units of energy per unit volume.

18. Suppose you have a wind speed gauge like the pitot tube shown in **Example 12.2**(b). By what factor must wind speed increase to double the value of h in the manometer? Is this independent of the moving fluid and the fluid in the manometer?

19. If the pressure reading of your pitot tube is 15.0 mm Hg at a speed of 200 km/h, what will it be at 700 km/h at the same altitude?

20. Calculate the maximum height to which water could be squirted with the hose in **Example 12.2** example if it: (a) Emerges from the nozzle. (b) Emerges with the nozzle removed, assuming the same flow rate.

21. Every few years, winds in Boulder, Colorado, attain sustained speeds of 45.0 m/s (about 100 mi/h) when the jet stream descends during early spring. Approximately what is the force due to the Bernoulli effect on a roof having an area of 220 m^2 ? Typical air density in Boulder is 1.14 kg/m^3, and the corresponding atmospheric pressure is $8.89 \times 10^4 \text{ N/m}^2$. (Bernoulli's principle as stated in the text assumes laminar flow. Using the principle here produces only an approximate result, because there is significant turbulence.)

22. (a) Calculate the approximate force on a square meter of sail, given the horizontal velocity of the wind is 6.00 m/s parallel to its front surface and 3.50 m/s along its back surface. Take the density of air to be 1.29 kg/m^3. (The calculation, based on Bernoulli's principle, is approximate due to the effects of turbulence.) (b) Discuss whether this force is great enough to be effective for propelling a sailboat.

23. (a) What is the pressure drop due to the Bernoulli effect as water goes into a 3.00-cm-diameter nozzle from a 9.00-cm-diameter fire hose while carrying a flow of 40.0 L/s? (b) To what maximum height above the nozzle can this water rise? (The actual height will be significantly smaller due to air resistance.)

24. (a) Using Bernoulli's equation, show that the measured fluid speed v for a pitot tube, like the one in **Figure 12.7**(b), is given by $v = \left(\dfrac{2\rho'gh}{\rho}\right)^{1/2}$,

where h is the height of the manometer fluid, ρ' is the density of the manometer fluid, ρ is the density of the moving fluid, and g is the acceleration due to gravity. (Note that v is indeed proportional to the square root of h, as stated in the text.) (b) Calculate v for moving air if a mercury manometer's h is 0.200 m.

12.3 The Most General Applications of Bernoulli's Equation

25. Hoover Dam on the Colorado River is the highest dam in the United States at 221 m, with an output of 1300 MW. The dam generates electricity with water taken from a depth of 150 m and an average flow rate of $650 \text{ m}^3/\text{s}$. (a) Calculate the power in this flow. (b) What is the ratio of this power to the facility's average of 680 MW?

26. A frequently quoted rule of thumb in aircraft design is that wings should produce about 1000 N of lift per square meter of wing. (The fact that a wing has a top and bottom surface does not double its area.) (a) At takeoff, an aircraft travels at 60.0 m/s, so that the air speed relative to the bottom of the wing is 60.0 m/s. Given the sea level density of air to be 1.29 kg/m^3, how fast must it move over the upper surface to create the ideal lift? (b) How fast must air move over the upper surface at a cruising speed of 245 m/s and at an altitude where air density is one-fourth that at sea level? (Note that this is not all of the aircraft's lift—some comes from the body of the plane, some from engine thrust, and so on. Furthermore, Bernoulli's principle gives an approximate answer because flow over the wing creates turbulence.)

27. The left ventricle of a resting adult's heart pumps blood at a flow rate of $83.0 \text{ cm}^3/\text{s}$, increasing its pressure by 110 mm Hg, its speed from zero to 30.0 cm/s, and its height by 5.00 cm. (All numbers are averaged over the entire heartbeat.) Calculate the total power output of the left ventricle. Note that most of the power is used to increase blood pressure.

28. A sump pump (used to drain water from the basement of houses built below the water table) is draining a flooded basement at the rate of 0.750 L/s, with an output pressure of $3.00 \times 10^5 \text{ N/m}^2$. (a) The water enters a hose with a 3.00-cm inside diameter and rises 2.50 m above the pump. What is its pressure at this point? (b) The hose goes over the foundation wall, losing 0.500 m in height, and widens to 4.00 cm in diameter. What is the pressure now? You may neglect frictional losses in both parts of the problem.

12.4 Viscosity and Laminar Flow; Poiseuille's Law

29. (a) Calculate the retarding force due to the viscosity of the air layer between a cart and a level air track given the following information—air temperature is $20° \text{ C}$, the cart is moving at 0.400 m/s, its surface area is $2.50 \times 10^{-2} \text{ m}^2$, and the thickness of the air layer is $6.00 \times 10^{-5} \text{ m}$. (b) What is the ratio of this force to the weight of the 0.300-kg cart?

30. What force is needed to pull one microscope slide over another at a speed of 1.00 cm/s, if there is a 0.500-mm-thick layer of $20° \text{ C}$ water between them and the contact area is 8.00 cm^2 ?

31. A glucose solution being administered with an IV has a flow rate of $4.00 \text{ cm}^3/\text{min}$. What will the new flow rate be if the glucose is replaced by whole blood having the same density but a viscosity 2.50 times that of the glucose? All other factors remain constant.

32. The pressure drop along a length of artery is 100 Pa, the radius is 10 mm, and the flow is laminar. The average speed of the blood is 15 mm/s. (a) What is the net force on the blood in this section of artery? (b) What is the power expended maintaining the flow?

33. A small artery has a length of $1.1 \times 10^{-3} \text{ m}$ and a radius of $2.5 \times 10^{-5} \text{ m}$. If the pressure drop across the artery is 1.3 kPa, what is the flow rate through the artery? (Assume that the temperature is $37° \text{ C}$.)

34. Fluid originally flows through a tube at a rate of $100 \text{ cm}^3/\text{s}$. To illustrate the sensitivity of flow rate to various factors, calculate the new flow rate for the following changes with all other factors remaining the same as in the original conditions. (a) Pressure difference increases by a factor of 1.50. (b) A new fluid with 3.00 times greater viscosity is substituted. (c) The tube is replaced by one having 4.00 times the length. (d) Another tube is used with a radius 0.100 times the original. (e) Yet another tube is substituted with a radius 0.100 times the original and half the length, *and* the pressure difference is increased by a factor of 1.50.

35. The arterioles (small arteries) leading to an organ, constrict in order to decrease flow to the organ. To shut down an organ, blood flow is reduced naturally to 1.00% of its original value. By what factor did the radii of the arterioles constrict? Penguins do this when they stand on ice to reduce the blood flow to their feet.

36. Angioplasty is a technique in which arteries partially blocked with plaque are dilated to increase blood flow. By what factor must the radius of an artery be increased in order to increase blood flow by a factor of 10?

37. (a) Suppose a blood vessel's radius is decreased to 90.0% of its original value by plaque deposits and the body compensates by increasing the pressure difference along the vessel to keep the flow rate constant. By what factor must the pressure difference increase? (b) If turbulence is created by the obstruction, what additional effect would it have on the flow rate?

38. A spherical particle falling at a terminal speed in a liquid must have the gravitational force balanced by the drag force and the buoyant force. The buoyant force is equal to the weight of the displaced fluid, while the drag force is assumed to be given by Stokes Law, $F_s = 6\pi r \eta v$. Show that the terminal speed is given by $v = \dfrac{2R^2 g}{9\eta}(\rho_s - \rho_1)$,

where R is the radius of the sphere, ρ_s is its density, and ρ_1 is the density of the fluid and η the coefficient of viscosity.

39. Using the equation of the previous problem, find the viscosity of motor oil in which a steel ball of radius 0.8 mm falls with a terminal speed of 4.32 cm/s. The densities of the ball and the oil are 7.86 and 0.88 g/mL, respectively.

40. A skydiver will reach a terminal velocity when the air drag equals their weight. For a skydiver with high speed and a large body, turbulence is a factor. The drag force then is approximately proportional to the square of the velocity. Taking the drag force to be $F_D = \dfrac{1}{2}\rho A v^2$ and setting this equal to the person's weight, find the terminal speed for a person falling "spread eagle." Find both a formula and a number for v_t, with assumptions as to size.

41. A layer of oil 1.50 mm thick is placed between two microscope slides. Researchers find that a force of $5.50\times10^{-4} \text{ N}$ is required to glide one over the other at a speed of 1.00 cm/s when their contact area is 6.00 cm^2. What is the oil's viscosity? What type of oil might it be?

42. (a) Verify that a 19.0% decrease in laminar flow through a tube is caused by a 5.00% decrease in radius, assuming that all other factors remain constant, as stated in the text. (b) What increase in flow is obtained from a 5.00% increase in radius, again assuming all other factors remain constant?

43. Example 12.8 dealt with the flow of saline solution in an IV system. (a) Verify that a pressure of $1.62\times10^4 \text{ N/m}^2$ is created at a depth of 1.61 m in a saline solution, assuming its density to be that of sea water. (b) Calculate the new flow rate if the height of the saline solution is decreased to 1.50 m. (c) At what height would the direction of flow be reversed? (This reversal can be a problem when patients stand up.)

44. When physicians diagnose arterial blockages, they quote the reduction in flow rate. If the flow rate in an artery has been reduced to 10.0% of its normal value by a blood clot and the average pressure difference has increased by 20.0%, by what factor has the clot reduced the radius of the artery?

45. During a marathon race, a runner's blood flow increases to 10.0 times her resting rate. Her blood's viscosity has dropped to 95.0% of its normal value, and the blood pressure difference across the circulatory system has increased by 50.0%. By what factor has the average radii of her blood vessels increased?

46. Water supplied to a house by a water main has a pressure of $3.00\times10^5 \text{ N/m}^2$ early on a summer day when neighborhood use is low. This pressure produces a flow of 20.0 L/min through a garden hose. Later in the day, pressure at the exit of the water main and entrance to the house drops, and a flow of only 8.00 L/min is obtained through the same hose. (a) What pressure is now being supplied to the house, assuming resistance is constant? (b) By what factor did the flow rate in the water main increase in order to cause this decrease in delivered pressure? The pressure at the entrance of the water main is $5.00\times10^5 \text{ N/m}^2$, and the original flow rate was 200 L/min. (c) How many more users are there, assuming each would consume 20.0 L/min in the morning?

47. An oil gusher shoots crude oil 25.0 m into the air through a pipe with a 0.100-m diameter. Neglecting air resistance but not the resistance of the pipe, and assuming laminar flow, calculate the gauge pressure at the entrance of the 50.0-m-long vertical pipe. Take the density of the oil to be 900 kg/m^3 and its viscosity to be $1.00 \text{ (N/m}^2) \cdot \text{s}$ (or $1.00 \text{ Pa} \cdot \text{s}$). Note that you must take into account the pressure due to the 50.0-m column of oil in the pipe.

48. Concrete is pumped from a cement mixer to the place it is being laid, instead of being carried in wheelbarrows. The flow rate is 200.0 L/min through a 50.0-m-long, 8.00-cm-diameter hose, and the pressure at the pump is $8.00\times10^6 \text{ N/m}^2$. (a) Calculate the resistance of the hose. (b) What is the viscosity of the concrete, assuming the flow is laminar? (c) How much power is being supplied, assuming the point of use is at the same level as the pump? You may neglect the power supplied to increase the concrete's velocity.

49. Construct Your Own Problem

Consider a coronary artery constricted by arteriosclerosis. Construct a problem in which you calculate the amount by which the diameter of the artery is decreased, based on an assessment of the decrease in flow rate.

50. Consider a river that spreads out in a delta region on its way to the sea. Construct a problem in which you calculate the average speed at which water moves in the delta region, based on the speed at which it was moving up river. Among the things to consider are the size and flow rate of the river before it spreads out and its size once it has spread out. You can construct the problem for the river spreading out into one large river or into multiple smaller rivers.

12.5 The Onset of Turbulence

51. Verify that the flow of oil is laminar (barely) for an oil gusher that shoots crude oil 25.0 m into the air through a pipe with a 0.100-m diameter. The vertical pipe is 50 m long. Take the density of the oil to be 900 kg/m^3 and its viscosity to be $1.00 \, (\text{N/m}^2) \cdot \text{s}$ (or $1.00 \, \text{Pa} \cdot \text{s}$).

52. Show that the Reynolds number N_R is unitless by substituting units for all the quantities in its definition and cancelling.

53. Calculate the Reynolds numbers for the flow of water through (a) a nozzle with a radius of 0.250 cm and (b) a garden hose with a radius of 0.900 cm, when the nozzle is attached to the hose. The flow rate through hose and nozzle is 0.500 L/s. Can the flow in either possibly be laminar?

54. A fire hose has an inside diameter of 6.40 cm. Suppose such a hose carries a flow of 40.0 L/s starting at a gauge pressure of $1.62 \times 10^6 \text{ N/m}^2$. The hose goes 10.0 m up a ladder to a nozzle having an inside diameter of 3.00 cm. Calculate the Reynolds numbers for flow in the fire hose and nozzle to show that the flow in each must be turbulent.

55. Concrete is pumped from a cement mixer to the place it is being laid, instead of being carried in wheelbarrows. The flow rate is 200.0 L/min through a 50.0-m-long, 8.00-cm-diameter hose, and the pressure at the pump is $8.00 \times 10^6 \text{ N/m}^2$. Verify that the flow of concrete is laminar taking concrete's viscosity to be $48.0 \, (\text{N/m}^2) \cdot \text{s}$, and given its density is 2300 kg/m^3.

56. At what flow rate might turbulence begin to develop in a water main with a 0.200-m diameter? Assume a $20° \text{ C}$ temperature.

57. What is the greatest average speed of blood flow at $37° \text{ C}$ in an artery of radius 2.00 mm if the flow is to remain laminar? What is the corresponding flow rate? Take the density of blood to be $1025 \text{ kg}/\text{m}^3$.

58. In **Take-Home Experiment: Inhalation**, we measured the average flow rate Q of air traveling through the trachea during each inhalation. Now calculate the average air speed in meters per second through your trachea during each inhalation. The radius of the trachea in adult humans is approximately 10^{-2} m. From the data above, calculate the Reynolds number for the air flow in the trachea during inhalation. Do you expect the air flow to be laminar or turbulent?

59. Gasoline is piped underground from refineries to major users. The flow rate is $3.00 \times 10^{-2} \text{ m}^3/\text{s}$ (about 500 gal/min), the viscosity of gasoline is $1.00 \times 10^{-3} \, (\text{N/m}^2) \cdot \text{s}$, and its density is 680 kg/m^3. (a) What minimum diameter must the pipe have if the Reynolds number is to be less than 2000? (b) What pressure difference must be maintained along each kilometer of the pipe to maintain this flow rate?

60. Assuming that blood is an ideal fluid, calculate the critical flow rate at which turbulence is a certainty in the aorta. Take the diameter of the aorta to be 2.50 cm. (Turbulence will actually occur at lower average flow rates, because blood is not an ideal fluid. Furthermore, since blood flow pulses, turbulence may occur during only the high-velocity part of each heartbeat.)

61. Unreasonable Results

A fairly large garden hose has an internal radius of 0.600 cm and a length of 23.0 m. The nozzleless horizontal hose is attached to a faucet, and it delivers 50.0 L/s. (a) What water pressure is supplied by the faucet? (b) What is unreasonable about this pressure? (c) What is unreasonable about the premise? (d) What is the Reynolds number for the given flow? (Take the viscosity of water as $1.005 \times 10^{-3} \, \left(\text{N}/\text{m}^2\right) \cdot \text{s}$.)

12.7 Molecular Transport Phenomena: Diffusion, Osmosis, and Related Processes

62. You can smell perfume very shortly after opening the bottle. To show that it is not reaching your nose by diffusion, calculate the average distance a perfume molecule moves in one second in air, given its diffusion constant D to be $1.00 \times 10^{-6} \text{ m}^2/\text{s}$.

63. What is the ratio of the average distances that oxygen will diffuse in a given time in air and water? Why is this distance less in water (equivalently, why is D less in water)?

64. Oxygen reaches the veinless cornea of the eye by diffusing through its tear layer, which is 0.500-mm thick. How long does it take the average oxygen molecule to do this?

65. (a) Find the average time required for an oxygen molecule to diffuse through a 0.200-mm-thick tear layer on the cornea. (b) How much time is required to diffuse 0.500 cm^3 of oxygen to the cornea if its surface area is 1.00 cm^2?

66. Suppose hydrogen and oxygen are diffusing through air. A small amount of each is released simultaneously. How much time passes before the hydrogen is 1.00 s ahead of the oxygen? Such differences in arrival times are used as an analytical tool in gas chromatography.

12.1 Flow Rate and Its Relation to Velocity

1. Water flows through a small horizontal pipe with a speed of 12 m/s into a larger part of the pipe for which the diameter of the pipe is doubled. What is the speed of the water in the larger part of the pipe?
   a.   0.75 m/s
   b.   3.0 m/s
   c.   6.0 m/s
   d.   12 m/s

2. A popular pool toy allows the user to push a plunger to compress water in a barrel with a diameter of 3.0 cm. The water is compressed with a speed of 1.2 m/s and emerges from a small opening with a speed of 15 m/s. What is the diameter of the opening? Assume the toy is oriented horizontally.

12.2 Bernoulli's Equation

3. At what depth beneath the surface of the lake is the pressure in the water equal to twice atmospheric pressure?
   a.   10 m
   b.   100 m
   c.   1000 m
   d.   10,000 m

4. A pump provides pressure to the lower end of a long pipeline that supplies water from a reservoir to a house located on a hill 150 m vertically upward from the lower end of the pipe (where the water is initially at rest before being pumped). The pipeline has a constant diameter of 3.5 cm, and the upper end of the pipeline is open to the atmosphere. What pressure must the pump provide for water to flow from the upper end of the pipeline at a rate of 5.0 m/s?

5. According to Bernoulli's equation, if the pressure in a given fluid is constant and the kinetic energy per unit volume of a fluid increases, which of the following is true?
   a.   The potential energy per unit volume of the fluid decreases.
   b.   The potential energy per unit volume of the fluid increases.
   c.   The fluid must no longer be considered incompressible.
   d.   The flow rate of the fluid increases.

6. Consider the following circumstances within a fluid, and determine the answer using Bernoulli's equation. (a) The pressure and kinetic energy per unit volume along a fluid path increases. What must be true about the potential energy per unit volume of the fluid along the fluid path? Explain. (b) The pressure along a fluid path increases, and the kinetic energy per unit volume remains constant. What must be true about the potential energy per unit volume of the fluid along the fluid path? Explain.

12.3 The Most General Applications of Bernoulli's Equation

7. A horizontally oriented pipe has a diameter of 5.6 cm and is filled with water. The pipe draws water from a reservoir that is initially at rest. A manually operated plunger provides a force of 440 N in the pipe. Assuming that the other end of the pipe is open to the air, with what speed does the water emerge from the pipe?
   a.   12 m/s
   b.   19 m/s
   c.   150 m/s
   d.   190 m/s

8. A 3.5-cm-diameter pipe contains a pumping mechanism that provides a force of 320 N to push water up into a tall building. Upon entering the piston mechanism, the water is flowing at a rate of 2.5 m/s. The water is then pumped to a level 21 m higher where the other end of the pipe is open to the air. With what speed does water leave the pipe?

9. A large container of water is open to the air, and it develops a hole of area 10 cm^2 at a point 5 m below the surface of the water. What is the flow rate (m^3/s) of the water emerging from this hole?
   a.   99 m^3/s
   b.   9.9 m^3/s
   c.   0.099 m^3/s
   d.   0.0099 m^3/s

10. A pipe is tapered so that the large end has a diameter twice as large as the small end. What must be the gauge pressure (the difference between pressure at the large end and pressure at the small end) in order for water to emerge from the small end with a speed of 12 m/s if the small end is elevated 8 m above the large end of the pipe?

13 TEMPERATURE, KINETIC THEORY, AND THE GAS LAWS

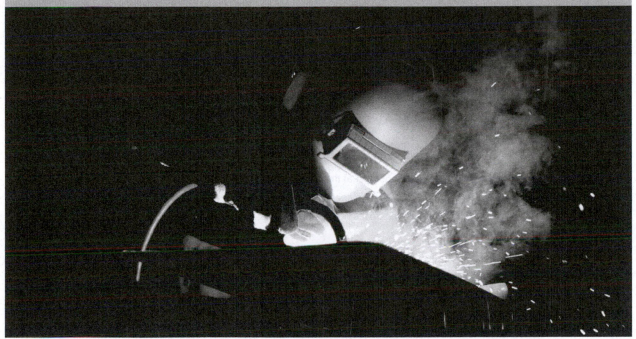

Figure 13.1 The welder's gloves and helmet protect him from the electric arc that transfers enough thermal energy to melt the rod, spray sparks, and burn the retina of an unprotected eye. The thermal energy can be felt on exposed skin a few meters away, and its light can be seen for kilometers. (credit: Kevin S. O'Brien/U.S. Navy)

Chapter Outline

13.1. Temperature

13.2. Thermal Expansion of Solids and Liquids

13.3. The Ideal Gas Law

13.4. Kinetic Theory: Atomic and Molecular Explanation of Pressure and Temperature

13.5. Phase Changes

13.6. Humidity, Evaporation, and Boiling

Connection for AP® Courses

Heat is something familiar to each of us. We feel the warmth of the summer sun, the chill of a clear summer night, the heat of coffee after a winter stroll, and the cooling effect of our sweat. Heat transfer is maintained by temperature differences. Manifestations of heat transfer—the movement of heat energy from one place or material to another—are apparent throughout the universe. Heat from beneath Earth's surface is brought to the surface in flows of incandescent lava. The Sun warms Earth's surface and is the source of much of the energy we find on it. Rising levels of atmospheric carbon dioxide threaten to trap more of the Sun's energy, perhaps fundamentally altering the ecosphere. In space, supernovas explode, briefly radiating more heat than an entire galaxy does.

In this chapter several concepts related to heat are discussed – what is heat, how is it related to temperature, what are heat's effects, and how is it related to other forms of energy and to work. We will find that, in spite of the richness of the phenomena, there is a small set of underlying physical principles that unite the subjects and tie them to other fields.

Figure 13.2 Alcohol thermometer with red dye. (credit: Chemical Engineer, Wikimedia Commons).

In a typical thermometer like this one, the alcohol, with a red dye, expands more rapidly than the glass containing it. When the thermometer's temperature increases, the liquid from the bulb is forced into the narrow tube, producing a large change in the length of the column for a small change in temperature.

The learning objectives covered under Big Idea 7 of the AP Physics Curriculum Framework are supported in this chapter through descriptive, algebraic, and graphical representations. Complex systems with internal structure can be described by both microscopic and macroscopic quantities. Big Idea 7 corresponds to the use of probability to describe the behavior of such systems by reducing large number of microscopic quantities to a small number of macroscopic quantities. The macroscopic quantities for an ideal gas, including temperature and pressure, are explained in this chapter (Enduring Understanding 7.A). The temperature represents average kinetic energy of gas molecules (Essential Knowledge 7.A.2). The pressure of a system determines the force that the system exerts on the walls of its container and is a measure of the average change in the momentum or impulse of the molecules colliding with the walls of the container (Essential Knowledge 7.A.1). The pressure also exists inside the system itself, not just at the walls of the container. This chapter discussed the "ideal gas law" which relates the temperature, pressure, and volume of an ideal gas using a simple equation (Essential Knowledge 7.A.3).

Big Idea 7 The mathematics of probability can be used to describe the behavior of complex systems and to interpret the behavior of quantum mechanical systems.

Enduring Understanding 7.A The properties of an ideal gas can be explained in terms of a small number of macroscopic variables including temperature and pressure.

Essential Knowledge 7.A.1 The pressure of a system determines the force that the system exerts on the walls of its container and is a measure of the average change in the momentum or impulse of the molecules colliding with the walls of the container. The pressure also exists inside the system itself, not just at the walls of the container.

Essential Knowledge 7.A.2 The temperature of a system characterizes the average kinetic energy of its molecules.

Essential Knowledge 7.A.3 In an ideal gas, the macroscopic (average) pressure (P), temperature (T), and volume (V), are related by the equation $PV = nRT$.

13.1 Temperature

Learning Objectives
By the end of this section, you will be able to: • Define temperature. • Convert temperatures between the Celsius, Fahrenheit, and Kelvin scales. • Define thermal equilibrium. • State the zeroth law of thermodynamics.

The concept of temperature has evolved from the common concepts of hot and cold. Human perception of what feels hot or cold is a relative one. For example, if you place one hand in hot water and the other in cold water, and then place both hands in tepid water, the tepid water will feel cool to the hand that was in hot water, and warm to the one that was in cold water. The scientific definition of temperature is less ambiguous than your senses of hot and cold. **Temperature** is operationally defined to be what we measure with a thermometer. (Many physical quantities are defined solely in terms of how they are measured. We shall see later how temperature is related to the kinetic energies of atoms and molecules, a more physical explanation.) Two accurate thermometers, one placed in hot water and the other in cold water, will show the hot water to have a higher temperature. If they are then placed in the tepid water, both will give identical readings (within measurement uncertainties). In this section, we discuss temperature, its measurement by thermometers, and its relationship to thermal equilibrium. Again, temperature is the quantity measured by a thermometer.

Misconception Alert: Human Perception vs. Reality
_____

On a cold winter morning, the wood on a porch feels warmer than the metal of your bike. The wood and bicycle are in thermal equilibrium with the outside air, and are thus the same temperature. They _feel_ different because of the difference in the way that they conduct heat away from your skin. The metal conducts heat away from your body faster than the wood does (see more about conductivity in **Conduction**). This is just one example demonstrating that the human sense of hot and cold is not determined by temperature alone.

Another factor that affects our perception of temperature is humidity. Most people feel much hotter on hot, humid days than on hot, dry days. This is because on humid days, sweat does not evaporate from the skin as efficiently as it does on dry days. It is the evaporation of sweat (or water from a sprinkler or pool) that cools us off.

Any physical property that depends on temperature, and whose response to temperature is reproducible, can be used as the basis of a thermometer. Because many physical properties depend on temperature, the variety of thermometers is remarkable. For example, volume increases with temperature for most substances. This property is the basis for the common alcohol thermometer, the old mercury thermometer, and the bimetallic strip (**Figure 13.3**). Other properties used to measure temperature include electrical resistance and color, as shown in **Figure 13.4**, and the emission of infrared radiation, as shown in **Figure 13.5**.

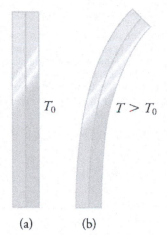

Figure 13.3 The curvature of a bimetallic strip depends on temperature. (a) The strip is straight at the starting temperature, where its two components have the same length. (b) At a higher temperature, this strip bends to the right, because the metal on the left has expanded more than the metal on the right.

Figure 13.4 Each of the six squares on this plastic (liquid crystal) thermometer contains a film of a different heat-sensitive liquid crystal material. Below $95°F$, all six squares are black. When the plastic thermometer is exposed to temperature that increases to $95°F$, the first liquid crystal square changes color. When the temperature increases above $96.8°F$ the second liquid crystal square also changes color, and so forth. (credit: Arkrishna, Wikimedia Commons)

Figure 13.5 Fireman Jason Ormand uses a pyrometer to check the temperature of an aircraft carrier's ventilation system. Infrared radiation (whose emission varies with temperature) from the vent is measured and a temperature readout is quickly produced. Infrared measurements are also frequently used as a measure of body temperature. These modern thermometers, placed in the ear canal, are more accurate than alcohol thermometers placed under the tongue or in the armpit. (credit: Lamel J. Hinton/U.S. Navy)

Temperature Scales

Thermometers are used to measure temperature according to well-defined scales of measurement, which use pre-defined reference points to help compare quantities. The three most common temperature scales are the Fahrenheit, Celsius, and Kelvin scales. A temperature scale can be created by identifying two easily reproducible temperatures. The freezing and boiling temperatures of water at standard atmospheric pressure are commonly used.

The **Celsius** scale (which replaced the slightly different *centigrade* scale) has the freezing point of water at $0°C$ and the boiling point at $100°C$. Its unit is the **degree Celsius** ($°C$). On the **Fahrenheit** scale (still the most frequently used in the United States), the freezing point of water is at $32°F$ and the boiling point is at $212°F$. The unit of temperature on this scale is the **degree Fahrenheit** ($°F$). Note that a temperature difference of one degree Celsius is greater than a temperature difference of one degree Fahrenheit. Only 100 Celsius degrees span the same range as 180 Fahrenheit degrees, thus one degree on the

Celsius scale is 1.8 times larger than one degree on the Fahrenheit scale $180/100 = 9/5$.

The **Kelvin** scale is the temperature scale that is commonly used in science. It is an *absolute temperature* scale defined to have 0 K at the lowest possible temperature, called **absolute zero**. The official temperature unit on this scale is the *kelvin*, which is abbreviated K, and is not accompanied by a degree sign. The freezing and boiling points of water are 273.15 K and 373.15 K, respectively. Thus, the magnitude of temperature differences is the same in units of kelvins and degrees Celsius. Unlike other temperature scales, the Kelvin scale is an absolute scale. It is used extensively in scientific work because a number of physical quantities, such as the volume of an ideal gas, are directly related to absolute temperature. The kelvin is the SI unit used in scientific work.

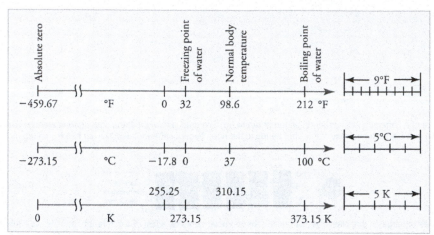

Figure 13.6 Relationships between the Fahrenheit, Celsius, and Kelvin temperature scales, rounded to the nearest degree. The relative sizes of the scales are also shown.

The relationships between the three common temperature scales is shown in **Figure 13.6**. Temperatures on these scales can be converted using the equations in **Table 13.1**.

Table 13.1 Temperature Conversions

To convert from ...	Use this equation ...	Also written as ...
Celsius to Fahrenheit	$T(°F) = \frac{9}{5}T(°C) + 32$	$T_{°F} = \frac{9}{5}T_{°C} + 32$
Fahrenheit to Celsius	$T(°C) = \frac{5}{9}(T(°F) - 32)$	$T_{°C} = \frac{5}{9}(T_{°F} - 32)$
Celsius to Kelvin	$T(K) = T(°C) + 273.15$	$T_K = T_{°C} + 273.15$
Kelvin to Celsius	$T(°C) = T(K) - 273.15$	$T_{°C} = T_K - 273.15$
Fahrenheit to Kelvin	$T(K) = \frac{5}{9}(T(°F) - 32) + 273.15$	$T_K = \frac{5}{9}(T_{°F} - 32) + 273.15$
Kelvin to Fahrenheit	$T(°F) = \frac{9}{5}(T(K) - 273.15) + 32$	$T_{°F} = \frac{9}{5}(T_K - 273.15) + 32$

Notice that the conversions between Fahrenheit and Kelvin look quite complicated. In fact, they are simple combinations of the conversions between Fahrenheit and Celsius, and the conversions between Celsius and Kelvin.

Example 13.1 Converting between Temperature Scales: Room Temperature

"Room temperature" is generally defined to be $25°C$. (a) What is room temperature in $°F$? (b) What is it in K?

Strategy

To answer these questions, all we need to do is choose the correct conversion equations and plug in the known values.

Solution for (a)

1. Choose the right equation. To convert from $°C$ to $°F$, use the equation

$$T_{°F} = \frac{9}{5}T_{°C} + 32.$$ (13.1)

2. Plug the known value into the equation and solve:

$$T_{°F} = \frac{9}{5}25°C + 32 = 77°F.$$

(13.2)

Solution for (b)

1. Choose the right equation. To convert from $°C$ to K, use the equation

$$T_K = T_{°C} + 273.15.$$

(13.3)

2. Plug the known value into the equation and solve:

$$T_K = 25°C + 273.15 = 298 \text{ K.}$$

(13.4)

Example 13.2 Converting between Temperature Scales: the Reaumur Scale

The Reaumur scale is a temperature scale that was used widely in Europe in the 18th and 19th centuries. On the Reaumur temperature scale, the freezing point of water is $0°R$ and the boiling temperature is $80°R$. If "room temperature" is $25°C$ on the Celsius scale, what is it on the Reaumur scale?

Strategy

To answer this question, we must compare the Reaumur scale to the Celsius scale. The difference between the freezing point and boiling point of water on the Reaumur scale is $80°R$. On the Celsius scale it is $100°C$. Therefore $100°C = 80°R$. Both scales start at $0°$ for freezing, so we can derive a simple formula to convert between temperatures on the two scales.

Solution

1. Derive a formula to convert from one scale to the other:

$$T_{°R} = \frac{0.8°R}{°C} \times T_{°C}.$$

(13.5)

2. Plug the known value into the equation and solve:

$$T_{°R} = \frac{0.8°R}{°C} \times 25°C = 20°R.$$

(13.6)

Temperature Ranges in the Universe

Figure 13.8 shows the wide range of temperatures found in the universe. Human beings have been known to survive with body temperatures within a small range, from $24°C$ to $44°C$ ($75°F$ to $111°F$). The average normal body temperature is usually given as $37.0°C$ ($98.6°F$), and variations in this temperature can indicate a medical condition: a fever, an infection, a tumor, or circulatory problems (see **Figure 13.7**).

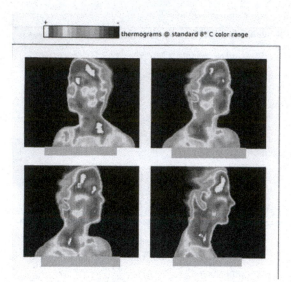

Figure 13.7 This image of radiation from a person's body (an infrared thermograph) shows the location of temperature abnormalities in the upper body. Dark blue corresponds to cold areas and red to white corresponds to hot areas. An elevated temperature might be an indication of malignant tissue (a cancerous tumor in the breast, for example), while a depressed temperature might be due to a decline in blood flow from a clot. In this case, the abnormalities are caused by a condition called hyperhidrosis. (credit: Porcelina81, Wikimedia Commons)

The lowest temperatures ever recorded have been measured during laboratory experiments: 4.5×10^{-10} K at the Massachusetts Institute of Technology (USA), and 1.0×10^{-10} K at Helsinki University of Technology (Finland). In comparison, the coldest recorded place on Earth's surface is Vostok, Antarctica at 183 K $(-89°C)$, and the coldest place (outside the lab) known in the universe is the Boomerang Nebula, with a temperature of 1 K.

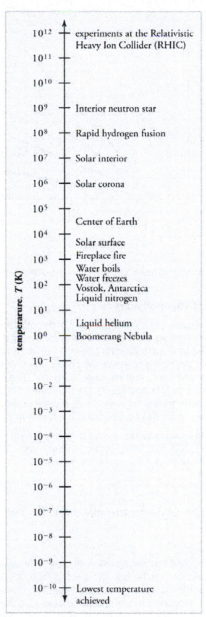

Figure 13.8 Each increment on this logarithmic scale indicates an increase by a factor of ten, and thus illustrates the tremendous range of temperatures in nature. Note that zero on a logarithmic scale would occur off the bottom of the page at infinity.

Making Connections: Absolute Zero

What is absolute zero? Absolute zero is the temperature at which all molecular motion has ceased. The concept of absolute zero arises from the behavior of gases. **Figure 13.9** shows how the pressure of gases at a constant volume decreases as temperature decreases. Various scientists have noted that the pressures of gases extrapolate to zero at the same temperature, $-273.15°C$. This extrapolation implies that there is a lowest temperature. This temperature is called *absolute zero*. Today we know that most gases first liquefy and then freeze, and it is not actually possible to reach absolute zero. The numerical value of absolute zero temperature is $-273.15°C$ or 0 K.

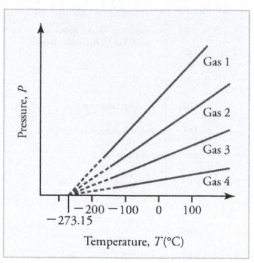

Figure 13.9 Graph of pressure versus temperature for various gases kept at a constant volume. Note that all of the graphs extrapolate to zero pressure at the same temperature.

Thermal Equilibrium and the Zeroth Law of Thermodynamics

Thermometers actually take their *own* temperature, not the temperature of the object they are measuring. This raises the question of how we can be certain that a thermometer measures the temperature of the object with which it is in contact. It is based on the fact that any two systems placed in *thermal contact* (meaning heat transfer can occur between them) will reach the same temperature. That is, heat will flow from the hotter object to the cooler one until they have exactly the same temperature. The objects are then in **thermal equilibrium**, and no further changes will occur. The systems interact and change because their temperatures differ, and the changes stop once their temperatures are the same. Thus, if enough time is allowed for this transfer of heat to run its course, the temperature a thermometer registers *does* represent the system with which it is in thermal equilibrium. Thermal equilibrium is established when two bodies are in contact with each other and can freely exchange energy.

Furthermore, experimentation has shown that if two systems, A and B, are in thermal equilibrium with each another, and B is in thermal equilibrium with a third system C, then A is also in thermal equilibrium with C. This conclusion may seem obvious, because all three have the same temperature, but it is basic to thermodynamics. It is called the **zeroth law of thermodynamics**.

The Zeroth Law of Thermodynamics

If two systems, A and B, are in thermal equilibrium with each other, and B is in thermal equilibrium with a third system, C, then A is also in thermal equilibrium with C.

This law was postulated in the 1930s, after the first and second laws of thermodynamics had been developed and named. It is called the *zeroth law* because it comes logically before the first and second laws (discussed in **Thermodynamics**). An example of this law in action is seen in babies in incubators: babies in incubators normally have very few clothes on, so to an observer they look as if they may not be warm enough. However, the temperature of the air, the cot, and the baby is the same, because they are in thermal equilibrium, which is accomplished by maintaining air temperature to keep the baby comfortable.

Check Your Understanding

Does the temperature of a body depend on its size?

Solution
No, the system can be divided into smaller parts each of which is at the same temperature. We say that the temperature is an *intensive* quantity. Intensive quantities are independent of size.

13.2 Thermal Expansion of Solids and Liquids

Learning Objectives

By the end of this section, you will be able to:

- Define and describe thermal expansion.
- Calculate the linear expansion of an object given its initial length, change in temperature, and coefficient of linear expansion.
- Calculate the volume expansion of an object given its initial volume, change in temperature, and coefficient of volume expansion.

- Calculate thermal stress on an object given its original volume, temperature change, volume change, and bulk modulus.

Figure 13.10 Thermal expansion joints like these in the Auckland Harbour Bridge in New Zealand allow bridges to change length without buckling. (credit: Ingolfson, Wikimedia Commons)

The expansion of alcohol in a thermometer is one of many commonly encountered examples of **thermal expansion**, the change in size or volume of a given mass with temperature. Hot air rises because its volume increases, which causes the hot air's density to be smaller than the density of surrounding air, causing a buoyant (upward) force on the hot air. The same happens in all liquids and gases, driving natural heat transfer upwards in homes, oceans, and weather systems. Solids also undergo thermal expansion. Railroad tracks and bridges, for example, have expansion joints to allow them to freely expand and contract with temperature changes.

What are the basic properties of thermal expansion? First, thermal expansion is clearly related to temperature change. The greater the temperature change, the more a bimetallic strip will bend. Second, it depends on the material. In a thermometer, for example, the expansion of alcohol is much greater than the expansion of the glass containing it.

What is the underlying cause of thermal expansion? As is discussed in **Kinetic Theory: Atomic and Molecular Explanation of Pressure and Temperature**, an increase in temperature implies an increase in the kinetic energy of the individual atoms. In a solid, unlike in a gas, the atoms or molecules are closely packed together, but their kinetic energy (in the form of small, rapid vibrations) pushes neighboring atoms or molecules apart from each other. This neighbor-to-neighbor pushing results in a slightly greater distance, on average, between neighbors, and adds up to a larger size for the whole body. For most substances under ordinary conditions, there is no preferred direction, and an increase in temperature will increase the solid's size by a certain fraction in each dimension.

Linear Thermal Expansion—Thermal Expansion in One Dimension

The change in length ΔL is proportional to length L. The dependence of thermal expansion on temperature, substance, and length is summarized in the equation

$$\Delta L = \alpha L \Delta T,$$

$$(13.7)$$

where ΔL is the change in length L, ΔT is the change in temperature, and α is the **coefficient of linear expansion**, which varies slightly with temperature.

Table 13.2 lists representative values of the coefficient of linear expansion, which may have units of $1/^\circ\text{C}$ or 1/K. Because the size of a kelvin and a degree Celsius are the same, both α and ΔT can be expressed in units of kelvins or degrees Celsius. The equation $\Delta L = \alpha L \Delta T$ is accurate for small changes in temperature and can be used for large changes in temperature if an average value of α is used.

Table 13.2 Thermal Expansion Coefficients at $20°C$ [1]

Material	Coefficient of linear expansion $\alpha(1/°C)$	Coefficient of volume expansion $\beta(1/°C)$
Solids		
Aluminum	25×10^{-6}	75×10^{-6}
Brass	19×10^{-6}	56×10^{-6}
Copper	17×10^{-6}	51×10^{-6}
Gold	14×10^{-6}	42×10^{-6}
Iron or Steel	12×10^{-6}	35×10^{-6}
Invar (Nickel-iron alloy)	0.9×10^{-6}	2.7×10^{-6}
Lead	29×10^{-6}	87×10^{-6}
Silver	18×10^{-6}	54×10^{-6}
Glass (ordinary)	9×10^{-6}	27×10^{-6}
Glass (Pyrex®)	3×10^{-6}	9×10^{-6}
Quartz	0.4×10^{-6}	1×10^{-6}
Concrete, Brick	$\sim12\times10^{-6}$	$\sim36\times10^{-6}$
Marble (average)	2.5×10^{-6}	7.5×10^{-6}
Liquids		
Ether		1650×10^{-6}
Ethyl alcohol		1100×10^{-6}
Petrol		950×10^{-6}
Glycerin		500×10^{-6}
Mercury		180×10^{-6}
Water		210×10^{-6}
Gases		
Air and most other gases at atmospheric pressure		3400×10^{-6}

Example 13.3 Calculating Linear Thermal Expansion: The Golden Gate Bridge

The main span of San Francisco's Golden Gate Bridge is 1275 m long at its coldest. The bridge is exposed to temperatures ranging from $-15°C$ to $40°C$. What is its change in length between these temperatures? Assume that the bridge is made entirely of steel.

Strategy

Use the equation for linear thermal expansion $\Delta L = \alpha L\Delta T$ to calculate the change in length, ΔL. Use the coefficient of linear expansion, α, for steel from Table 13.2, and note that the change in temperature, ΔT, is $55°C$.

Solution

Plug all of the known values into the equation to solve for ΔL.

---

1. Values for liquids and gases are approximate.

$$\Delta L = \alpha L \Delta T = \left(\frac{12 \times 10^{-6}}{°C}\right)(1275 \text{ m})(55°C) = 0.84 \text{ m}. \tag{13.8}$$

Discussion

Although not large compared with the length of the bridge, this change in length is observable. It is generally spread over many expansion joints so that the expansion at each joint is small.

Thermal Expansion in Two and Three Dimensions

Objects expand in all dimensions, as illustrated in **Figure 13.11**. That is, their areas and volumes, as well as their lengths, increase with temperature. Holes also get larger with temperature. If you cut a hole in a metal plate, the remaining material will expand exactly as it would if the plug was still in place. The plug would get bigger, and so the hole must get bigger too. (Think of the ring of neighboring atoms or molecules on the wall of the hole as pushing each other farther apart as temperature increases. Obviously, the ring of neighbors must get slightly larger, so the hole gets slightly larger).

Thermal Expansion in Two Dimensions

For small temperature changes, the change in area ΔA is given by

$$\Delta A = 2\alpha A \Delta T, \tag{13.9}$$

where ΔA is the change in area A, ΔT is the change in temperature, and α is the coefficient of linear expansion, which varies slightly with temperature.

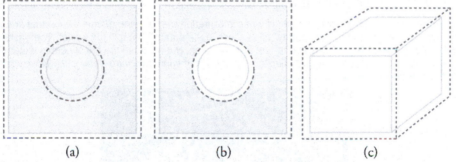

(a)　　　　　　(b)　　　　　　(c)

Figure 13.11 In general, objects expand in all directions as temperature increases. In these drawings, the original boundaries of the objects are shown with solid lines, and the expanded boundaries with dashed lines. (a) Area increases because both length and width increase. The area of a circular plug also increases. (b) If the plug is removed, the hole it leaves becomes larger with increasing temperature, just as if the expanding plug were still in place. (c) Volume also increases, because all three dimensions increase.

Thermal Expansion in Three Dimensions

The change in volume ΔV is very nearly $\Delta V = 3\alpha V \Delta T$. This equation is usually written as

$$\Delta V = \beta V \Delta T, \tag{13.10}$$

where β is the **coefficient of volume expansion** and $\beta \approx 3\alpha$. Note that the values of β in **Table 13.2** are almost exactly equal to 3α.

In general, objects will expand with increasing temperature. Water is the most important exception to this rule. Water expands with increasing temperature (its density *decreases*) when it is at temperatures greater than $4°C (40°F)$. However, it expands with *decreasing* temperature when it is between $+4°C$ and $0°C$ ($40°F$ to $32°F$). Water is densest at $+4°C$. (See **Figure 13.12**.) Perhaps the most striking effect of this phenomenon is the freezing of water in a pond. When water near the surface cools down to $4°C$ it is denser than the remaining water and thus will sink to the bottom. This "turnover" results in a layer of warmer water near the surface, which is then cooled. Eventually the pond has a uniform temperature of $4°C$. If the temperature in the surface layer drops below $4°C$, the water is less dense than the water below, and thus stays near the top. As a result, the pond surface can completely freeze over. The ice on top of liquid water provides an insulating layer from winter's harsh exterior air temperatures. Fish and other aquatic life can survive in $4°C$ water beneath ice, due to this unusual characteristic of water. It also produces circulation of water in the pond that is necessary for a healthy ecosystem of the body of water.

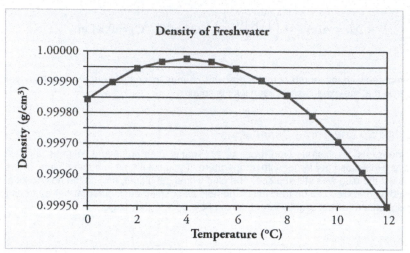

Figure 13.12 The density of water as a function of temperature. Note that the thermal expansion is actually very small. The maximum density at $+4°C$ is only *0.0075%* greater than the density at $2°C$, and *0.012%* greater than that at $0°C$.

Making Connections: Real-World Connections—Filling the Tank

Differences in the thermal expansion of materials can lead to interesting effects at the gas station. One example is the dripping of gasoline from a freshly filled tank on a hot day. Gasoline starts out at the temperature of the ground under the gas station, which is cooler than the air temperature above. The gasoline cools the steel tank when it is filled. Both gasoline and steel tank expand as they warm to air temperature, but gasoline expands much more than steel, and so it may overflow.

This difference in expansion can also cause problems when interpreting the gasoline gauge. The actual amount (mass) of gasoline left in the tank when the gauge hits "empty" is a lot less in the summer than in the winter. The gasoline has the same volume as it does in the winter when the "add fuel" light goes on, but because the gasoline has expanded, there is less mass. If you are used to getting another 40 miles on "empty" in the winter, beware—you will probably run out much more quickly in the summer.

Figure 13.13 Because the gas expands more than the gas tank with increasing temperature, you can't drive as many miles on "empty" in the summer as you can in the winter. (credit: Hector Alejandro, Flickr)

Example 13.4 Calculating Thermal Expansion: Gas vs. Gas Tank

Suppose your 60.0-L (15.9-gal) steel gasoline tank is full of gas, so both the tank and the gasoline have a temperature of $15.0°C$. How much gasoline has spilled by the time they warm to $35.0°C$?

Strategy

The tank and gasoline increase in volume, but the gasoline increases more, so the amount spilled is the difference in their volume changes. (The gasoline tank can be treated as solid steel.) We can use the equation for volume expansion to calculate the change in volume of the gasoline and of the tank.

Solution

1. Use the equation for volume expansion to calculate the increase in volume of the steel tank:

$$\Delta V_s = \beta_s V_s \Delta T. \qquad (13.11)$$

2. The increase in volume of the gasoline is given by this equation:

$$\Delta V_{gas} = \beta_{gas} V_{gas} \Delta T. \tag{13.12}$$

3. Find the difference in volume to determine the amount spilled as

$$V_{spill} = \Delta V_{gas} - \Delta V_s. \tag{13.13}$$

Alternatively, we can combine these three equations into a single equation. (Note that the original volumes are equal.)

$$
\begin{aligned}
V_{spill} &= (\beta_{gas} - \beta_s)V\Delta T \\
&= \left[(950 - 35)\times 10^{-6}/°C\right](60.0\ L)(20.0°C) \\
&= 1.10\ L.
\end{aligned}
\tag{13.14}
$$

Discussion

This amount is significant, particularly for a 60.0-L tank. The effect is so striking because the gasoline and steel expand quickly. The rate of change in thermal properties is discussed in **Heat and Heat Transfer Methods**.

If you try to cap the tank tightly to prevent overflow, you will find that it leaks anyway, either around the cap or by bursting the tank. Tightly constricting the expanding gas is equivalent to compressing it, and both liquids and solids resist being compressed with extremely large forces. To avoid rupturing rigid containers, these containers have air gaps, which allow them to expand and contract without stressing them.

Thermal Stress

Thermal stress is created by thermal expansion or contraction (see **Elasticity: Stress and Strain** for a discussion of stress and strain). Thermal stress can be destructive, such as when expanding gasoline ruptures a tank. It can also be useful, for example, when two parts are joined together by heating one in manufacturing, then slipping it over the other and allowing the combination to cool. Thermal stress can explain many phenomena, such as the weathering of rocks and pavement by the expansion of ice when it freezes.

Example 13.5 Calculating Thermal Stress: Gas Pressure

What pressure would be created in the gasoline tank considered in **Example 13.4**, if the gasoline increases in temperature from $15.0°C$ to $35.0°C$ without being allowed to expand? Assume that the bulk modulus B for gasoline is $1.00\times 10^9\ N/m^2$. (For more on bulk modulus, see **Elasticity: Stress and Strain**.)

Strategy

To solve this problem, we must use the following equation, which relates a change in volume ΔV to pressure:

$$\Delta V = \frac{1}{B}\frac{F}{A}V_0, \tag{13.15}$$

where F/A is pressure, V_0 is the original volume, and B is the bulk modulus of the material involved. We will use the amount spilled in **Example 13.4** as the change in volume, ΔV.

Solution

1. Rearrange the equation for calculating pressure:

$$P = \frac{F}{A} = \frac{\Delta V}{V_0}B. \tag{13.16}$$

2. Insert the known values. The bulk modulus for gasoline is $B = 1.00\times 10^9\ N/m^2$. In the previous example, the change in volume $\Delta V = 1.10\ L$ is the amount that would spill. Here, $V_0 = 60.0\ L$ is the original volume of the gasoline. Substituting these values into the equation, we obtain

$$P = \frac{1.10\ L}{60.0\ L}(1.00\times 10^9\ Pa) = 1.83\times 10^7\ Pa. \tag{13.17}$$

Discussion

This pressure is about $2500\ lb/in^2$, *much* more than a gasoline tank can handle.

Forces and pressures created by thermal stress are typically as great as that in the example above. Railroad tracks and roadways can buckle on hot days if they lack sufficient expansion joints. (See **Figure 13.14**.) Power lines sag more in the summer than in the winter, and will snap in cold weather if there is insufficient slack. Cracks open and close in plaster walls as a house warms and cools. Glass cooking pans will crack if cooled rapidly or unevenly, because of differential contraction and the stresses it creates. (Pyrex® is less susceptible because of its small coefficient of thermal expansion.) Nuclear reactor pressure vessels are threatened by overly rapid cooling, and although none have failed, several have been cooled faster than considered desirable. Biological cells are ruptured when foods are frozen, detracting from their taste. Repeated thawing and freezing accentuate the damage. Even the oceans can be affected. A significant portion of the rise in sea level that is resulting from global warming is due to the thermal expansion of sea water.

Figure 13.14 Thermal stress contributes to the formation of potholes. (credit: Editor5807, Wikimedia Commons)

Metal is regularly used in the human body for hip and knee implants. Most implants need to be replaced over time because, among other things, metal does not bond with bone. Researchers are trying to find better metal coatings that would allow metal-to-bone bonding. One challenge is to find a coating that has an expansion coefficient similar to that of metal. If the expansion coefficients are too different, the thermal stresses during the manufacturing process lead to cracks at the coating-metal interface.

Another example of thermal stress is found in the mouth. Dental fillings can expand differently from tooth enamel. It can give pain when eating ice cream or having a hot drink. Cracks might occur in the filling. Metal fillings (gold, silver, etc.) are being replaced by composite fillings (porcelain), which have smaller coefficients of expansion, and are closer to those of teeth.

Check Your Understanding

Two blocks, A and B, are made of the same material. Block A has dimensions $l \times w \times h = L \times 2L \times L$ and Block B has dimensions $2L \times 2L \times 2L$. If the temperature changes, what is (a) the change in the volume of the two blocks, (b) the change in the cross-sectional area $l \times w$, and (c) the change in the height h of the two blocks?

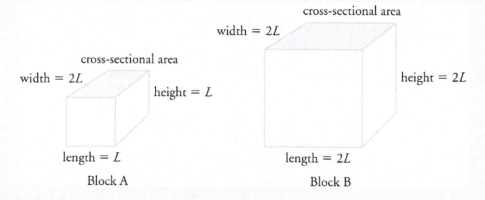

Figure 13.15

Solution

(a) The change in volume is proportional to the original volume. Block A has a volume of $L \times 2L \times L = 2L^3$. Block B has a volume of $2L \times 2L \times 2L = 8L^3$, which is 4 times that of Block A. Thus the change in volume of Block B should be 4 times the change in volume of Block A.

(b) The change in area is proportional to the area. The cross-sectional area of Block A is $L \times 2L = 2L^2$, while that of Block B is $2L \times 2L = 4L^2$. Because cross-sectional area of Block B is twice that of Block A, the change in the cross-sectional area of Block B is twice that of Block A.

(c) The change in height is proportional to the original height. Because the original height of Block B is twice that of A, the change in the height of Block B is twice that of Block A.

13.3 The Ideal Gas Law

Figure 13.16 The air inside this hot air balloon flying over Putrajaya, Malaysia, is hotter than the ambient air. As a result, the balloon experiences a buoyant force pushing it upward. (credit: Kevin Poh, Flickr)

In this section, we continue to explore the thermal behavior of gases. In particular, we examine the characteristics of atoms and molecules that compose gases. (Most gases, for example nitrogen, N_2, and oxygen, O_2, are composed of two or more atoms.

We will primarily use the term "molecule" in discussing a gas because the term can also be applied to monatomic gases, such as helium.)

Gases are easily compressed. We can see evidence of this in Table 13.2, where you will note that gases have the *largest* coefficients of volume expansion. The large coefficients mean that gases expand and contract very rapidly with temperature changes. In addition, you will note that most gases expand at the *same* rate, or have the same β. This raises the question as to why gases should all act in nearly the same way, when liquids and solids have widely varying expansion rates.

The answer lies in the large separation of atoms and molecules in gases, compared to their sizes, as illustrated in Figure 13.17. Because atoms and molecules have large separations, forces between them can be ignored, except when they collide with each other during collisions. The motion of atoms and molecules (at temperatures well above the boiling temperature) is fast, such that the gas occupies all of the accessible volume and the expansion of gases is rapid. In contrast, in liquids and solids, atoms and molecules are closer together and are quite sensitive to the forces between them.

Figure 13.17 Atoms and molecules in a gas are typically widely separated, as shown. Because the forces between them are quite weak at these distances, the properties of a gas depend more on the number of atoms per unit volume and on temperature than on the type of atom.

To get some idea of how pressure, temperature, and volume of a gas are related to one another, consider what happens when you pump air into an initially deflated tire. The tire's volume first increases in direct proportion to the amount of air injected, without much increase in the tire pressure. Once the tire has expanded to nearly its full size, the walls limit volume expansion. If we continue to pump air into it, the pressure increases. The pressure will further increase when the car is driven and the tires move. Most manufacturers specify optimal tire pressure for cold tires. (See **Figure 13.18**.)

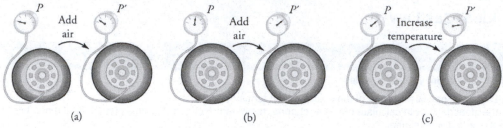

Figure 13.18 (a) When air is pumped into a deflated tire, its volume first increases without much increase in pressure. (b) When the tire is filled to a certain point, the tire walls resist further expansion and the pressure increases with more air. (c) Once the tire is inflated, its pressure increases with temperature.

At room temperatures, collisions between atoms and molecules can be ignored. In this case, the gas is called an ideal gas, in which case the relationship between the pressure, volume, and temperature is given by the equation of state called the ideal gas law.

Ideal Gas Law

The **ideal gas law** states that

$$PV = NkT,$$ (13.18)

where P is the absolute pressure of a gas, V is the volume it occupies, N is the number of atoms and molecules in the gas, and T is its absolute temperature. The constant k is called the **Boltzmann constant** in honor of Austrian physicist Ludwig Boltzmann (1844–1906) and has the value

$$k = 1.38 \times 10^{-23} \, \text{J/K}.$$ (13.19)

The ideal gas law can be derived from basic principles, but was originally deduced from experimental measurements of Charles' law (that volume occupied by a gas is proportional to temperature at a fixed pressure) and from Boyle's law (that for a fixed temperature, the product PV is a constant). In the ideal gas model, the volume occupied by its atoms and molecules is a negligible fraction of V. The ideal gas law describes the behavior of real gases under most conditions. (Note, for example, that N is the total number of atoms and molecules, independent of the type of gas.)

Let us see how the ideal gas law is consistent with the behavior of filling the tire when it is pumped slowly and the temperature is constant. At first, the pressure P is essentially equal to atmospheric pressure, and the volume V increases in direct proportion to the number of atoms and molecules N put into the tire. Once the volume of the tire is constant, the equation $PV = NkT$ predicts that the pressure should increase in proportion to *the number N of atoms and molecules*.

Example 13.6 Calculating Pressure Changes Due to Temperature Changes: Tire Pressure

Suppose your bicycle tire is fully inflated, with an absolute pressure of 7.00×10^5 Pa (a gauge pressure of just under $90.0 \, \text{lb/in}^2$) at a temperature of $18.0°C$. What is the pressure after its temperature has risen to $35.0°C$? Assume that there are no appreciable leaks or changes in volume.

Strategy

The pressure in the tire is changing only because of changes in temperature. First we need to identify what we know and what we want to know, and then identify an equation to solve for the unknown.

We know the initial pressure $P_0 = 7.00 \times 10^5$ Pa, the initial temperature $T_0 = 18.0°C$, and the final temperature $T_f = 35.0°C$. We must find the final pressure P_f. How can we use the equation $PV = NkT$? At first, it may seem that not enough information is given, because the volume V and number of atoms N are not specified. What we can do is use the equation twice: $P_0 V_0 = NkT_0$ and $P_f V_f = NkT_f$. If we divide $P_f V_f$ by $P_0 V_0$ we can come up with an equation that allows us to solve for P_f.

$$\frac{P_f V_f}{P_0 V_0} = \frac{N_f k T_f}{N_0 k T_0}$$ (13.20)

Since the volume is constant, V_f and V_0 are the same and they cancel out. The same is true for N_f and N_0, and k, which is a constant. Therefore,

$$\frac{P_f}{P_0} = \frac{T_f}{T_0}. \tag{13.21}$$

We can then rearrange this to solve for P_f:

$$P_f = P_0 \frac{T_f}{T_0}, \tag{13.22}$$

where the temperature must be in units of kelvins, because T_0 and T_f are absolute temperatures.

Solution

1. Convert temperatures from Celsius to Kelvin.

$$T_0 = (18.0 + 273)\text{K} = 291\ \text{K} \tag{13.23}$$
$$T_f = (35.0 + 273)\text{K} = 308\ \text{K}$$

2. Substitute the known values into the equation.

$$P_f = P_0 \frac{T_f}{T_0} = 7.00 \times 10^5\ \text{Pa} \left(\frac{308\ \text{K}}{291\ \text{K}}\right) = 7.41 \times 10^5\ \text{Pa} \tag{13.24}$$

Discussion

The final temperature is about 6% greater than the original temperature, so the final pressure is about 6% greater as well. Note that *absolute* pressure and *absolute* temperature must be used in the ideal gas law.

Making Connections: Take-Home Experiment—Refrigerating a Balloon

---

Inflate a balloon at room temperature. Leave the inflated balloon in the refrigerator overnight. What happens to the balloon, and why?

Example 13.7 Calculating the Number of Molecules in a Cubic Meter of Gas

How many molecules are in a typical object, such as gas in a tire or water in a drink? We can use the ideal gas law to give us an idea of how large N typically is.

Calculate the number of molecules in a cubic meter of gas at standard temperature and pressure (STP), which is defined to be $0°\text{C}$ and atmospheric pressure.

Strategy

Because pressure, volume, and temperature are all specified, we can use the ideal gas law $PV = NkT$, to find N.

Solution

1. Identify the knowns.

$$
\begin{aligned}
T &= 0°\text{C} = 273\ \text{K} \\
P &= 1.01 \times 10^5\ \text{Pa} \\
V &= 1.00\ \text{m}^3 \\
k &= 1.38 \times 10^{-23}\ \text{J/K}
\end{aligned}
\tag{13.25}
$$

2. Identify the unknown: number of molecules, N.

3. Rearrange the ideal gas law to solve for N.

$$
\begin{aligned}
PV &= NkT \\
N &= \frac{PV}{kT}
\end{aligned}
\tag{13.26}
$$

4. Substitute the known values into the equation and solve for N.

$$N = \frac{PV}{kT} = \frac{\left(1.01 \times 10^5 \text{ Pa}\right)\left(1.00 \text{ m}^3\right)}{\left(1.38 \times 10^{-23} \text{ J/K}\right)(273 \text{ K})} = 2.68 \times 10^{25} \text{ molecules} \tag{13.27}$$

Discussion

This number is undeniably large, considering that a gas is mostly empty space. N is huge, even in small volumes. For example, 1 cm^3 of a gas at STP has 2.68×10^{19} molecules in it. Once again, note that N is the same for all types or mixtures of gases.

Moles and Avogadro's Number

It is sometimes convenient to work with a unit other than molecules when measuring the amount of substance. A **mole** (abbreviated mol) is defined to be the amount of a substance that contains as many atoms or molecules as there are atoms in exactly 12 grams (0.012 kg) of carbon-12. The actual number of atoms or molecules in one mole is called **Avogadro's number** (N_A), in recognition of Italian scientist Amedeo Avogadro (1776–1856). He developed the concept of the mole, based on the hypothesis that equal volumes of gas, at the same pressure and temperature, contain equal numbers of molecules. That is, the number is independent of the type of gas. This hypothesis has been confirmed, and the value of Avogadro's number is

$$N_A = 6.02 \times 10^{23} \text{ mol}^{-1}. \tag{13.28}$$

Avogadro's Number

One mole always contains 6.02×10^{23} particles (atoms or molecules), independent of the element or substance. A mole of any substance has a mass in grams equal to its molecular mass, which can be calculated from the atomic masses given in the periodic table of elements.

$$N_A = 6.02 \times 10^{23} \text{ mol}^{-1} \tag{13.29}$$

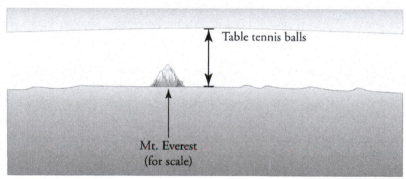

Figure 13.19 How big is a mole? On a macroscopic level, one mole of table tennis balls would cover the Earth to a depth of about 40 km.

Check Your Understanding

The active ingredient in a Tylenol pill is 325 mg of acetaminophen $(C_8 H_9 NO_2)$. Find the number of active molecules of acetaminophen in a single pill.

Solution

We first need to calculate the molar mass (the mass of one mole) of acetaminophen. To do this, we need to multiply the number of atoms of each element by the element's atomic mass.

$$\text{(8 moles of carbon)(12 grams/mole)} + \text{(9 moles hydrogen)(1 gram/mole)} \tag{13.30}$$
$$+\text{(1 mole nitrogen)(14 grams/mole)} + \text{(2 moles oxygen)(16 grams/mole)} = 151 \text{ g}$$

Then we need to calculate the number of moles in 325 mg.

$$\left(\frac{325 \text{ mg}}{151 \text{ grams/mole}}\right)\left(\frac{1 \text{ gram}}{1000 \text{ mg}}\right) = 2.15 \times 10^{-3} \text{ moles} \tag{13.31}$$

Then use Avogadro's number to calculate the number of molecules.

$$N = \left(2.15 \times 10^{-3} \text{ moles}\right)\left(6.02 \times 10^{23} \text{ molecules/mole}\right) = 1.30 \times 10^{21} \text{ molecules} \tag{13.32}$$

Example 13.8 Calculating Moles per Cubic Meter and Liters per Mole

Calculate: (a) the number of moles in 1.00 m^3 of gas at STP, and (b) the number of liters of gas per mole.

Strategy and Solution

(a) We are asked to find the number of moles per cubic meter, and we know from **Example 13.7** that the number of molecules per cubic meter at STP is 2.68×10^{25}. The number of moles can be found by dividing the number of molecules by Avogadro's number. We let n stand for the number of moles,

$$n \text{ mol/m}^3 = \frac{N \text{ molecules/m}^3}{6.02 \times 10^{23} \text{ molecules/mol}} = \frac{2.68 \times 10^{25} \text{ molecules/m}^3}{6.02 \times 10^{23} \text{ molecules/mol}} = 44.5 \text{ mol/m}^3. \tag{13.33}$$

(b) Using the value obtained for the number of moles in a cubic meter, and converting cubic meters to liters, we obtain

$$\frac{\left(10^3 \text{ L/m}^3\right)}{44.5 \text{ mol/m}^3} = 22.5 \text{ L/mol}. \tag{13.34}$$

Discussion

This value is very close to the accepted value of 22.4 L/mol. The slight difference is due to rounding errors caused by using three-digit input. Again this number is the same for all gases. In other words, it is independent of the gas.

The (average) molar weight of air (approximately 80% N_2 and 20% O_2 is $M = 28.8$ g. Thus the mass of one cubic meter of air is 1.28 kg. If a living room has dimensions $5 \text{ m} \times 5 \text{ m} \times 3 \text{ m}$, the mass of air inside the room is 96 kg, which is the typical mass of a human.

Check Your Understanding

The density of air at standard conditions $(P = 1 \text{ atm and } T = 20°C)$ is 1.28 kg/m^3. At what pressure is the density 0.64 kg/m^3 if the temperature and number of molecules are kept constant?

Solution

The best way to approach this question is to think about what is happening. If the density drops to half its original value and no molecules are lost, then the volume must double. If we look at the equation $PV = NkT$, we see that when the temperature is constant, the pressure is inversely proportional to volume. Therefore, if the volume doubles, the pressure must drop to half its original value, and $P_f = 0.50$ atm.

The Ideal Gas Law Restated Using Moles

A very common expression of the ideal gas law uses the number of moles, n, rather than the number of atoms and molecules, N. We start from the ideal gas law,

$$PV = NkT, \tag{13.35}$$

and multiply and divide the equation by Avogadro's number N_A. This gives

$$PV = \frac{N}{N_A} N_A kT. \tag{13.36}$$

Note that $n = N/N_A$ is the number of moles. We define the universal gas constant $R = N_A k$, and obtain the ideal gas law in terms of moles.

Ideal Gas Law (in terms of moles)

The ideal gas law (in terms of moles) is

$$PV = nRT. \tag{13.37}$$

The numerical value of R in SI units is

$$R = N_A k = \left(6.02 \times 10^{23} \text{ mol}^{-1}\right)\left(1.38 \times 10^{-23} \text{ J/K}\right) = 8.31 \text{ J/mol} \cdot \text{K}. \tag{13.38}$$

In other units,

$$R = 1.99 \text{ cal/mol} \cdot K \qquad (13.39)$$
$$R = 0.0821 \text{ L} \cdot \text{atm/mol} \cdot K.$$

You can use whichever value of R is most convenient for a particular problem.

Example 13.9 Calculating Number of Moles: Gas in a Bike Tire

How many moles of gas are in a bike tire with a volume of 2.00×10^{-3} m^3(2.00 L), a pressure of 7.00×10^5 Pa (a gauge pressure of just under 90.0 lb/in^2), and at a temperature of $18.0°C$?

Strategy

Identify the knowns and unknowns, and choose an equation to solve for the unknown. In this case, we solve the ideal gas law, $PV = nRT$, for the number of moles n .

Solution

1. Identify the knowns.

$$\begin{aligned} P &= 7.00 \times 10^5 \text{ Pa} \\ V &= 2.00 \times 10^{-3} \text{ m}^3 \\ T &= 18.0°C = 291 \text{ K} \\ R &= 8.31 \text{ J/mol} \cdot K \end{aligned} \qquad (13.40)$$

2. Rearrange the equation to solve for n and substitute known values.

$$n = \frac{PV}{RT} = \frac{\left(7.00 \times 10^5 \text{ Pa}\right)\left(2.00 \times 10^{-3} \text{ m}^3\right)}{(8.31 \text{ J/mol} \cdot K)(291 \text{ K})} \qquad (13.41)$$
$$= 0.579 \text{ mol}$$

Discussion

The most convenient choice for R in this case is 8.31 J/mol $\cdot$ K, because our known quantities are in SI units. The pressure and temperature are obtained from the initial conditions in **Example 13.6**, but we would get the same answer if we used the final values.

The ideal gas law can be considered to be another manifestation of the law of conservation of energy (see **Conservation of Energy**). Work done on a gas results in an increase in its energy, increasing pressure and/or temperature, or decreasing volume. This increased energy can also be viewed as increased internal kinetic energy, given the gas's atoms and molecules.

The Ideal Gas Law and Energy

Let us now examine the role of energy in the behavior of gases. When you inflate a bike tire by hand, you do work by repeatedly exerting a force through a distance. This energy goes into increasing the pressure of air inside the tire and increasing the temperature of the pump and the air.

The ideal gas law is closely related to energy: the units on both sides are joules. The right-hand side of the ideal gas law in $PV = NkT$ is NkT . This term is roughly the amount of translational kinetic energy of N atoms or molecules at an absolute temperature T , as we shall see formally in **Kinetic Theory: Atomic and Molecular Explanation of Pressure and Temperature**. The left-hand side of the ideal gas law is PV , which also has the units of joules. We know from our study of fluids that pressure is one type of potential energy per unit volume, so pressure multiplied by volume is energy. The important point is that there is energy in a gas related to both its pressure and its volume. The energy can be changed when the gas is doing work as it expands—something we explore in **Heat and Heat Transfer Methods**—similar to what occurs in gasoline or steam engines and turbines.

Problem-Solving Strategy: The Ideal Gas Law

Step 1 Examine the situation to determine that an ideal gas is involved. Most gases are nearly ideal.

Step 2 Make a list of what quantities are given, or can be inferred from the problem as stated (identify the known quantities). Convert known values into proper SI units (K for temperature, Pa for pressure, m^3 for volume, molecules for N , and moles for n).

Step 3 Identify exactly what needs to be determined in the problem (identify the unknown quantities). A written list is useful.

Step 4 Determine whether the number of molecules or the number of moles is known, in order to decide which form of the ideal gas law to use. The first form is $PV = NkT$ and involves N, the number of atoms or molecules. The second form is $PV = nRT$ and involves n, the number of moles.

Step 5 Solve the ideal gas law for the quantity to be determined (the unknown quantity). You may need to take a ratio of final states to initial states to eliminate the unknown quantities that are kept fixed.

Step 6 Substitute the known quantities, along with their units, into the appropriate equation, and obtain numerical solutions complete with units. Be certain to use absolute temperature and absolute pressure.

Step 7 Check the answer to see if it is reasonable: Does it make sense?

Check Your Understanding

Liquids and solids have densities about 1000 times greater than gases. Explain how this implies that the distances between atoms and molecules in gases are about 10 times greater than the size of their atoms and molecules.

Solution
Atoms and molecules are close together in solids and liquids. In gases they are separated by empty space. Thus gases have lower densities than liquids and solids. Density is mass per unit volume, and volume is related to the size of a body (such as a sphere) cubed. So if the distance between atoms and molecules increases by a factor of 10, then the volume occupied increases by a factor of 1000, and the density decreases by a factor of 1000.

13.4 Kinetic Theory: Atomic and Molecular Explanation of Pressure and Temperature

Learning Objectives

By the end of this section, you will be able to:

- Express the ideal gas law in terms of molecular mass and velocity.
- Define thermal energy.
- Calculate the kinetic energy of a gas molecule, given its temperature.
- Describe the relationship between the temperature of a gas and the kinetic energy of atoms and molecules.
- Describe the distribution of speeds of molecules in a gas.

The information presented in this section supports the following AP® learning objectives and science practices:

- **7.A.1.2** Treating a gas molecule as an object (i.e., ignoring its internal structure), the student is able to analyze qualitatively the collisions with a container wall and determine the cause of pressure and at thermal equilibrium to quantitatively calculate the pressure, force, or area for a thermodynamic problem given two of the variables. **(S.P. 1.4, 2.2)**
- **7.A.2.1** The student is able to qualitatively connect the average of all kinetic energies of molecules in a system to the temperature of the system. **(S.P. 7.1)**
- **7.A.2.2** The student is able to connect the statistical distribution of microscopic kinetic energies of molecules to the macroscopic temperature of the system and to relate this to thermodynamic processes. **(S.P. 7.1)**

We have developed macroscopic definitions of pressure and temperature. Pressure is the force divided by the area on which the force is exerted, and temperature is measured with a thermometer. We gain a better understanding of pressure and temperature from the kinetic theory of gases, which assumes that atoms and molecules are in continuous random motion.

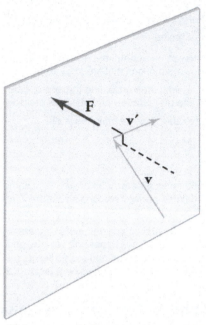

Figure 13.20 When a molecule collides with a rigid wall, the component of its momentum perpendicular to the wall is reversed. A force is thus exerted on the wall, creating pressure.

Figure 13.20 shows an elastic collision of a gas molecule with the wall of a container, so that it exerts a force on the wall (by Newton's third law). Because a huge number of molecules will collide with the wall in a short time, we observe an average force per unit area. These collisions are the source of pressure in a gas. As the number of molecules increases, the number of collisions and thus the pressure increase. Similarly, the gas pressure is higher if the average velocity of molecules is higher. The actual relationship is derived in the **Things Great and Small** feature below. The following relationship is found:

$$PV = \frac{1}{3}Nm\overline{v^2},$$

(13.42)

where P is the pressure (average force per unit area), V is the volume of gas in the container, N is the number of molecules

in the container, m is the mass of a molecule, and $\overline{v^2}$ is the average of the molecular speed squared.

What can we learn from this atomic and molecular version of the ideal gas law? We can derive a relationship between temperature and the average translational kinetic energy of molecules in a gas. Recall the previous expression of the ideal gas law:

$$PV = NkT.$$

(13.43)

Equating the right-hand side of this equation with the right-hand side of $PV = \frac{1}{3}Nm\overline{v^2}$ gives

$$\frac{1}{3}Nm\overline{v^2} = NkT.$$

(13.44)

Making Connections: Things Great and Small—Atomic and Molecular Origin of Pressure in a Gas

Figure 13.21 shows a box filled with a gas. We know from our previous discussions that putting more gas into the box produces greater pressure, and that increasing the temperature of the gas also produces a greater pressure. But why should increasing the temperature of the gas increase the pressure in the box? A look at the atomic and molecular scale gives us some answers, and an alternative expression for the ideal gas law.

The figure shows an expanded view of an elastic collision of a gas molecule with the wall of a container. Calculating the average force exerted by such molecules will lead us to the ideal gas law, and to the connection between temperature and molecular kinetic energy. We assume that a molecule is small compared with the separation of molecules in the gas, and that its interaction with other molecules can be ignored. We also assume the wall is rigid and that the molecule's direction changes, but that its speed remains constant (and hence its kinetic energy and the magnitude of its momentum remain constant as well). This assumption is not always valid, but the same result is obtained with a more detailed description of the molecule's exchange of energy and momentum with the wall.

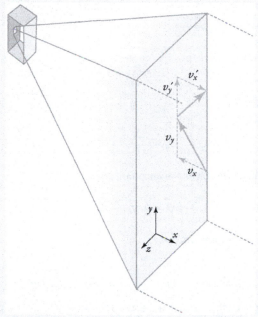

Figure 13.21 Gas in a box exerts an outward pressure on its walls. A molecule colliding with a rigid wall has the direction of its velocity and momentum in the x-direction reversed. This direction is perpendicular to the wall. The components of its velocity momentum in the y- and z-directions are not changed, which means there is no force parallel to the wall.

If the molecule's velocity changes in the x-direction, its momentum changes from $-mv_x$ to $+mv_x$. Thus, its change in momentum is $\Delta mv = +mv_x - (-mv_x) = 2mv_x$. The force exerted on the molecule is given by

$$F = \frac{\Delta p}{\Delta t} = \frac{2mv_x}{\Delta t}.$$ (13.45)

There is no force between the wall and the molecule until the molecule hits the wall. During the short time of the collision, the force between the molecule and wall is relatively large. We are looking for an average force; we take Δt to be the average time between collisions of the molecule with this wall. It is the time it would take the molecule to go across the box and back (a distance $2l$) at a speed of v_x. Thus $\Delta t = 2l / v_x$, and the expression for the force becomes

$$F = \frac{2mv_x}{2l / v_x} = \frac{mv_x^2}{l}.$$ (13.46)

This force is due to *one* molecule. We multiply by the number of molecules N and use their average squared velocity to find the force

$$F = N\frac{m\overline{v_x^2}}{l},$$ (13.47)

where the bar over a quantity means its average value. We would like to have the force in terms of the speed v, rather than the x-component of the velocity. We note that the total velocity squared is the sum of the squares of its components, so that

$$\overline{v^2} = \overline{v_x^2} + \overline{v_y^2} + \overline{v_z^2}.$$ (13.48)

Because the velocities are random, their average components in all directions are the same:

$$\overline{v_x^2} = \overline{v_y^2} = \overline{v_z^2}.$$ (13.49)

Thus,

$$\overline{v^2} = 3\overline{v_x^2},$$ (13.50)

or

$$\overline{v_x^2} = \frac{1}{3}\overline{v^2}.$$ (13.51)

Substituting $\frac{1}{3}\overline{v^2}$ into the expression for F gives

$$F = N\frac{m\overline{v^2}}{3l}.$$
(13.52)

The pressure is F/A, so that we obtain

$$P = \frac{F}{A} = N\frac{m\overline{v^2}}{3Al} = \frac{1}{3}\frac{Nm\overline{v^2}}{V},$$
(13.53)

where we used $V = Al$ for the volume. This gives the important result.

$$PV = \frac{1}{3}Nm\overline{v^2}$$
(13.54)

This equation is another expression of the ideal gas law.

We can get the average kinetic energy of a molecule, $\frac{1}{2}mv^2$, from the right-hand side of the equation by canceling N and multiplying by 3/2. This calculation produces the result that the average kinetic energy of a molecule is directly related to absolute temperature.

$$\overline{KE} = \frac{1}{2}m\overline{v^2} = \frac{3}{2}kT$$
(13.55)

The average translational kinetic energy of a molecule, $\overline{KE}$, is called **thermal energy.** The equation $\overline{KE} = \frac{1}{2}m\overline{v^2} = \frac{3}{2}kT$ is a molecular interpretation of temperature, and it has been found to be valid for gases and reasonably accurate in liquids and solids. It is another definition of temperature based on an expression of the molecular energy.

It is sometimes useful to rearrange $\overline{KE} = \frac{1}{2}m\overline{v^2} = \frac{3}{2}kT$, and solve for the average speed of molecules in a gas in terms of temperature,

$$\sqrt{\overline{v^2}} = v_{rms} = \sqrt{\frac{3kT}{m}},$$
(13.56)

where v_{rms} stands for root-mean-square (rms) speed.

Example 13.10 Calculating Kinetic Energy and Speed of a Gas Molecule

(a) What is the average kinetic energy of a gas molecule at $20.0°C$ (room temperature)? (b) Find the rms speed of a nitrogen molecule (N_2) at this temperature.

Strategy for (a)

The known in the equation for the average kinetic energy is the temperature.

$$\overline{KE} = \frac{1}{2}m\overline{v^2} = \frac{3}{2}kT$$
(13.57)

Before substituting values into this equation, we must convert the given temperature to kelvins. This conversion gives $T = (20.0 + 273) \text{ K} = 293 \text{ K}.$

Solution for (a)

The temperature alone is sufficient to find the average translational kinetic energy. Substituting the temperature into the translational kinetic energy equation gives

$$\overline{KE} = \frac{3}{2}kT = \frac{3}{2}\left(1.38\times10^{-23} \text{ J/K}\right)(293 \text{ K}) = 6.07\times10^{-21} \text{ J}.$$
(13.58)

Strategy for (b)

Finding the rms speed of a nitrogen molecule involves a straightforward calculation using the equation

$$\sqrt{\bar{v^2}} = v_{rms} = \sqrt{\frac{3kT}{m}}, \tag{13.59}$$

but we must first find the mass of a nitrogen molecule. Using the molecular mass of nitrogen N_2 from the periodic table,

$$m = \frac{2(14.0067)\times 10^{-3}\ \text{kg/mol}}{6.02\times 10^{23}\ \text{mol}^{-1}} = 4.65\times 10^{-26}\ \text{kg}. \tag{13.60}$$

Solution for (b)

Substituting this mass and the value for k into the equation for v_{rms} yields

$$v_{rms} = \sqrt{\frac{3kT}{m}} = \sqrt{\frac{3\left(1.38\times 10^{-23}\ \text{J/K}\right)(293\ \text{K})}{4.65\times 10^{-26}\ \text{kg}}} = 511\ \text{m/s}. \tag{13.61}$$

Discussion

Note that the average kinetic energy of the molecule is independent of the type of molecule. The average translational kinetic energy depends only on absolute temperature. The kinetic energy is very small compared to macroscopic energies, so that we do not feel when an air molecule is hitting our skin. The rms velocity of the nitrogen molecule is surprisingly large. These large molecular velocities do not yield macroscopic movement of air, since the molecules move in all directions with equal likelihood. The *mean free path* (the distance a molecule can move on average between collisions) of molecules in air is very small, and so the molecules move rapidly but do not get very far in a second. The high value for rms speed is reflected in the speed of sound, however, which is about 340 m/s at room temperature. The faster the rms speed of air molecules, the faster that sound vibrations can be transferred through the air. The speed of sound increases with temperature and is greater in gases with small molecular masses, such as helium. (See **Figure 13.22**.)

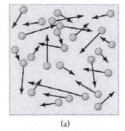 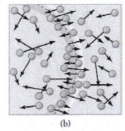

(a)   (b)

Figure 13.22 (a) There are many molecules moving so fast in an ordinary gas that they collide a billion times every second. (b) Individual molecules do not move very far in a small amount of time, but disturbances like sound waves are transmitted at speeds related to the molecular speeds.

Making Connections: Historical Note—Kinetic Theory of Gases

The kinetic theory of gases was developed by Daniel Bernoulli (1700–1782), who is best known in physics for his work on fluid flow (hydrodynamics). Bernoulli's work predates the atomistic view of matter established by Dalton.

Distribution of Molecular Speeds

The motion of molecules in a gas is random in magnitude and direction for individual molecules, but a gas of many molecules has a predictable distribution of molecular speeds. This distribution is called the *Maxwell-Boltzmann distribution*, after its originators, who calculated it based on kinetic theory, and has since been confirmed experimentally. (See **Figure 13.23**.) The distribution has a long tail, because a few molecules may go several times the rms speed. The most probable speed v_p is less than the rms speed v_{rms}. **Figure 13.24** shows that the curve is shifted to higher speeds at higher temperatures, with a broader range of speeds.

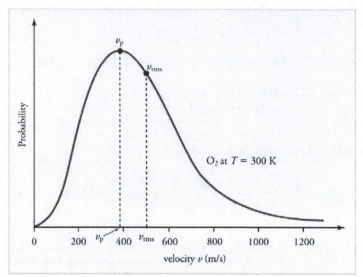

Figure 13.23 The Maxwell-Boltzmann distribution of molecular speeds in an ideal gas. The most likely speed v_p is less than the rms speed v_{rms}. Although very high speeds are possible, only a tiny fraction of the molecules have speeds that are an order of magnitude greater than v_{rms}.

The distribution of thermal speeds depends strongly on temperature. As temperature increases, the speeds are shifted to higher values and the distribution is broadened.

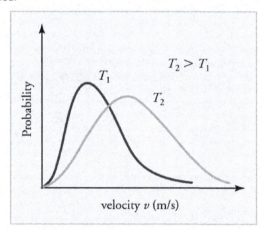

Figure 13.24 The Maxwell-Boltzmann distribution is shifted to higher speeds and is broadened at higher temperatures.

What is the implication of the change in distribution with temperature shown in **Figure 13.24** for humans? All other things being equal, if a person has a fever, he or she is likely to lose more water molecules, particularly from linings along moist cavities such as the lungs and mouth, creating a dry sensation in the mouth.

Example 13.11 Calculating Temperature: Escape Velocity of Helium Atoms

In order to escape Earth's gravity, an object near the top of the atmosphere (at an altitude of 100 km) must travel away from Earth at 11.1 km/s. This speed is called the *escape velocity*. At what temperature would helium atoms have an rms speed equal to the escape velocity?

Strategy

Identify the knowns and unknowns and determine which equations to use to solve the problem.

Solution

1. Identify the knowns: v is the escape velocity, 11.1 km/s.

2. Identify the unknowns: We need to solve for temperature, T. We also need to solve for the mass m of the helium atom.

3. Determine which equations are needed.

   • To solve for mass m of the helium atom, we can use information from the periodic table:

$$m = \frac{\text{molar mass}}{\text{number of atoms per mole}}. \tag{13.62}$$

- To solve for temperature T, we can rearrange either

$$\overline{KE} = \frac{1}{2}m\overline{v^2} = \frac{3}{2}kT \qquad (13.63)$$

or

$$\sqrt{\overline{v^2}} = v_{rms} = \sqrt{\frac{3kT}{m}} \qquad (13.64)$$

to yield

$$T = \frac{m\overline{v^2}}{3k}, \qquad (13.65)$$

where k is the Boltzmann constant and m is the mass of a helium atom.

4. Plug the known values into the equations and solve for the unknowns.

$$m = \frac{\text{molar mass}}{\text{number of atoms per mole}} = \frac{4.0026 \times 10^{-3}\ \text{kg/mol}}{6.02 \times 10^{23}\ \text{mol}} = 6.65 \times 10^{-27}\ \text{kg} \qquad (13.66)$$

$$T = \frac{\left(6.65 \times 10^{-27}\ \text{kg}\right)\left(11.1 \times 10^{3}\ \text{m/s}\right)^2}{3\left(1.38 \times 10^{-23}\ \text{J/K}\right)} = 1.98 \times 10^4\ \text{K} \qquad (13.67)$$

Discussion

This temperature is much higher than atmospheric temperature, which is approximately 250 K ($-25°C$ or $-10°F$) at high altitude. Very few helium atoms are left in the atmosphere, but there were many when the atmosphere was formed. The reason for the loss of helium atoms is that there are a small number of helium atoms with speeds higher than Earth's escape velocity even at normal temperatures. The speed of a helium atom changes from one instant to the next, so that at any instant, there is a small, but nonzero chance that the speed is greater than the escape speed and the molecule escapes from Earth's gravitational pull. Heavier molecules, such as oxygen, nitrogen, and water (very little of which reach a very high altitude), have smaller rms speeds, and so it is much less likely that any of them will have speeds greater than the escape velocity. In fact, so few have speeds above the escape velocity that billions of years are required to lose significant amounts of the atmosphere. **Figure 13.25** shows the impact of a lack of an atmosphere on the Moon. Because the gravitational pull of the Moon is much weaker, it has lost almost its entire atmosphere. The comparison between Earth and the Moon is discussed in this chapter's Problems and Exercises.

Figure 13.25 This photograph of Apollo 17 Commander Eugene Cernan driving the lunar rover on the Moon in 1972 looks as though it was taken at night with a large spotlight. In fact, the light is coming from the Sun. Because the acceleration due to gravity on the Moon is so low (about 1/6 that of Earth), the Moon's escape velocity is much smaller. As a result, gas molecules escape very easily from the Moon, leaving it with virtually no atmosphere. Even during the daytime, the sky is black because there is no gas to scatter sunlight. (credit: Harrison H. Schmitt/NASA)

Check Your Understanding

If you consider a very small object such as a grain of pollen, in a gas, then the number of atoms and molecules striking its surface would also be relatively small. Would the grain of pollen experience any fluctuations in pressure due to statistical fluctuations in the number of gas atoms and molecules striking it in a given amount of time?

Solution

Yes. Such fluctuations actually occur for a body of any size in a gas, but since the numbers of atoms and molecules are immense for macroscopic bodies, the fluctuations are a tiny percentage of the number of collisions, and the averages spoken of in this section vary imperceptibly. Roughly speaking the fluctuations are proportional to the inverse square root of the number of collisions, so for small bodies they can become significant. This was actually observed in the 19th century for pollen grains in water, and is known as the Brownian effect.

PhET Explorations: Gas Properties

Pump gas molecules into a box and see what happens as you change the volume, add or remove heat, change gravity, and more. Measure the temperature and pressure, and discover how the properties of the gas vary in relation to each other.

Figure 13.26 Gas Properties (http://phet.colorado.edu/en/simulation/gas-properties)

13.5 Phase Changes

Learning Objectives
By the end of this section, you will be able to: • Interpret a phase diagram.   • State Dalton's law.   • Identify and describe the triple point of a gas from its phase diagram.   • Describe the state of equilibrium between a liquid and a gas, a liquid and a solid, and a gas and a solid.

Up to now, we have considered the behavior of ideal gases. Real gases are like ideal gases at high temperatures. At lower temperatures, however, the interactions between the molecules and their volumes cannot be ignored. The molecules are very close (condensation occurs) and there is a dramatic decrease in volume, as seen in **Figure 13.27**. The substance changes from a gas to a liquid. When a liquid is cooled to even lower temperatures, it becomes a solid. The volume never reaches zero because of the finite volume of the molecules.

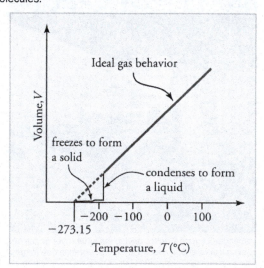

Figure 13.27 A sketch of volume versus temperature for a real gas at constant pressure. The linear (straight line) part of the graph represents ideal gas behavior—volume and temperature are directly and positively related and the line extrapolates to zero volume at $-273.15°C$, or absolute zero. When the gas becomes a liquid, however, the volume actually decreases precipitously at the liquefaction point. The volume decreases slightly once the substance is solid, but it never becomes zero.

High pressure may also cause a gas to change phase to a liquid. Carbon dioxide, for example, is a gas at room temperature and atmospheric pressure, but becomes a liquid under sufficiently high pressure. If the pressure is reduced, the temperature drops and the liquid carbon dioxide solidifies into a snow-like substance at the temperature $-78°C$. Solid CO_2 is called "dry ice." Another example of a gas that can be in a liquid phase is liquid nitrogen (LN_2). LN_2 is made by liquefaction of atmospheric air (through compression and cooling). It boils at 77 K $(-196°C)$ at atmospheric pressure. LN_2 is useful as a refrigerant and allows for the preservation of blood, sperm, and other biological materials. It is also used to reduce noise in electronic sensors and equipment, and to help cool down their current-carrying wires. In dermatology, LN_2 is used to freeze and painlessly remove warts and other growths from the skin.

PV Diagrams

We can examine aspects of the behavior of a substance by plotting a graph of pressure versus volume, called a **PV diagram**. When the substance behaves like an ideal gas, the ideal gas law describes the relationship between its pressure and volume. That is,

$$PV = NkT \text{ (ideal gas).}$$ (13.68)

Now, assuming the number of molecules and the temperature are fixed,

$$PV = \text{constant (ideal gas, constant temperature).}$$ (13.69)

For example, the volume of the gas will decrease as the pressure increases. If you plot the relationship $PV = \text{constant}$ on a PV diagram, you find a hyperbola. **Figure 13.28** shows a graph of pressure versus volume. The hyperbolas represent ideal-gas behavior at various fixed temperatures, and are called *isotherms*. At lower temperatures, the curves begin to look less like hyperbolas—the gas is not behaving ideally and may even contain liquid. There is a **critical point**—that is, a **critical temperature**—above which liquid cannot exist. At sufficiently high pressure above the critical point, the gas will have the density of a liquid but will not condense. Carbon dioxide, for example, cannot be liquefied at a temperature above $31.0°C$. **Critical pressure** is the minimum pressure needed for liquid to exist at the critical temperature. **Table 13.3** lists representative critical temperatures and pressures.

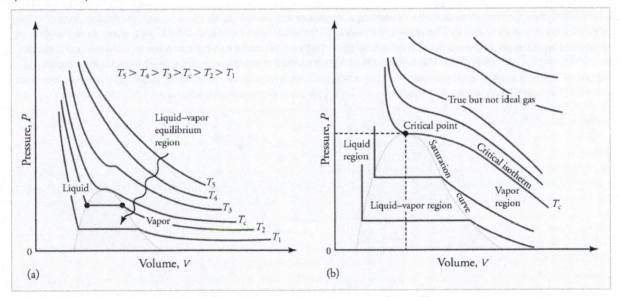

Figure 13.28 PV diagrams. (a) Each curve (isotherm) represents the relationship between P and V at a fixed temperature; the upper curves are at higher temperatures. The lower curves are not hyperbolas, because the gas is no longer an ideal gas. (b) An expanded portion of the PV diagram for low temperatures, where the phase can change from a gas to a liquid. The term "vapor" refers to the gas phase when it exists at a temperature below the boiling temperature.

Table 13.3 Critical Temperatures and Pressures

Substance	Critical temperature		Critical pressure	
	K	**°C**	**Pa**	**atm**
Water	647.4	374.3	22.12×10^6	219.0
Sulfur dioxide	430.7	157.6	7.88×10^6	78.0
Ammonia	405.5	132.4	11.28×10^6	111.7
Carbon dioxide	304.2	31.1	7.39×10^6	73.2
Oxygen	154.8	−118.4	5.08×10^6	50.3
Nitrogen	126.2	−146.9	3.39×10^6	33.6
Hydrogen	33.3	−239.9	1.30×10^6	12.9
Helium	5.3	−267.9	0.229×10^6	2.27

Phase Diagrams

The plots of pressure versus temperatures provide considerable insight into thermal properties of substances. There are well-defined regions on these graphs that correspond to various phases of matter, so PT graphs are called **phase diagrams**. **Figure 13.29** shows the phase diagram for water. Using the graph, if you know the pressure and temperature you can determine the phase of water. The solid lines—boundaries between phases—indicate temperatures and pressures at which the phases coexist (that is, they exist together in ratios, depending on pressure and temperature). For example, the boiling point of water is $100°C$ at 1.00 atm. As the pressure increases, the boiling temperature rises steadily to $374°C$ at a pressure of 218 atm. A pressure cooker (or even a covered pot) will cook food faster because the water can exist as a liquid at temperatures greater than $100°C$ without all boiling away. The curve ends at a point called the *critical point*, because at higher temperatures the liquid phase does not exist at any pressure. The critical point occurs at the critical temperature, as you can see for water from **Table 13.3**. The critical temperature for oxygen is $-118°C$, so oxygen cannot be liquefied above this temperature.

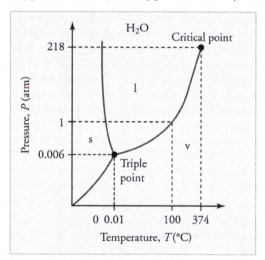

Figure 13.29 The phase diagram (PT graph) for water. Note that the axes are nonlinear and the graph is not to scale. This graph is simplified—there are several other exotic phases of ice at higher pressures.

Similarly, the curve between the solid and liquid regions in **Figure 13.29** gives the melting temperature at various pressures. For example, the melting point is $0°C$ at 1.00 atm, as expected. Note that, at a fixed temperature, you can change the phase from solid (ice) to liquid (water) by increasing the pressure. Ice melts from pressure in the hands of a snowball maker. From the phase diagram, we can also say that the melting temperature of ice rises with increased pressure. When a car is driven over snow, the increased pressure from the tires melts the snowflakes; afterwards the water refreezes and forms an ice layer.

At sufficiently low pressures there is no liquid phase, but the substance can exist as either gas or solid. For water, there is no liquid phase at pressures below 0.00600 atm. The phase change from solid to gas is called **sublimation**. It accounts for large losses of snow pack that never make it into a river, the routine automatic defrosting of a freezer, and the freeze-drying process applied to many foods. Carbon dioxide, on the other hand, sublimates at standard atmospheric pressure of 1 atm. (The solid form of CO_2 is known as dry ice because it does not melt. Instead, it moves directly from the solid to the gas state.)

All three curves on the phase diagram meet at a single point, the **triple point**, where all three phases exist in equilibrium. For water, the triple point occurs at 273.16 K $(0.01°C)$, and is a more accurate calibration temperature than the melting point of water at 1.00 atm, or 273.15 K $(0.0°C)$. See **Table 13.4** for the triple point values of other substances.

Equilibrium

Liquid and gas phases are in equilibrium at the boiling temperature. (See **Figure 13.30**.) If a substance is in a closed container at the boiling point, then the liquid is boiling and the gas is condensing at the same rate without net change in their relative amount. Molecules in the liquid escape as a gas at the same rate at which gas molecules stick to the liquid, or form droplets and become part of the liquid phase. The combination of temperature and pressure has to be "just right"; if the temperature and pressure are increased, equilibrium is maintained by the same increase of boiling and condensation rates.

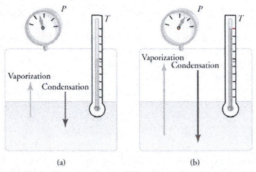

Figure 13.30 Equilibrium between liquid and gas at two different boiling points inside a closed container. (a) The rates of boiling and condensation are equal at this combination of temperature and pressure, so the liquid and gas phases are in equilibrium. (b) At a higher temperature, the boiling rate is faster and the rates at which molecules leave the liquid and enter the gas are also faster. Because there are more molecules in the gas, the gas pressure is higher and the rate at which gas molecules condense and enter the liquid is faster. As a result the gas and liquid are in equilibrium at this higher temperature.

Table 13.4 Triple Point Temperatures and Pressures

Substance	Temperature		Pressure	
	K	°C	Pa	atm
Water	273.16	0.01	$6.10×10^2$	0.00600
Carbon dioxide	216.55	−56.60	$5.16×10^5$	5.11
Sulfur dioxide	197.68	−75.47	$1.67×10^3$	0.0167
Ammonia	195.40	−77.75	$6.06×10^3$	0.0600
Nitrogen	63.18	−210.0	$1.25×10^4$	0.124
Oxygen	54.36	−218.8	$1.52×10^2$	0.00151
Hydrogen	13.84	−259.3	$7.04×10^3$	0.0697

One example of equilibrium between liquid and gas is that of water and steam at $100°C$ and 1.00 atm. This temperature is the boiling point at that pressure, so they should exist in equilibrium. Why does an open pot of water at $100°C$ boil completely away? The gas surrounding an open pot is not pure water: it is mixed with air. If pure water and steam are in a closed container at $100°C$ and 1.00 atm, they would coexist—but with air over the pot, there are fewer water molecules to condense, and water boils. What about water at $20.0°C$ and 1.00 atm? This temperature and pressure correspond to the liquid region, yet an open glass of water at this temperature will completely evaporate. Again, the gas around it is air and not pure water vapor, so that the reduced evaporation rate is greater than the condensation rate of water from dry air. If the glass is sealed, then the liquid phase remains. We call the gas phase a **vapor** when it exists, as it does for water at $20.0°C$, at a temperature below the boiling temperature.

Explain why a cup of water (or soda) with ice cubes stays at $0°C$, even on a hot summer day.

Solution
The ice and liquid water are in thermal equilibrium, so that the temperature stays at the freezing temperature as long as ice remains in the liquid. (Once all of the ice melts, the water temperature will start to rise.)

Vapor Pressure, Partial Pressure, and Dalton's Law

Vapor pressure is defined as the pressure at which a gas coexists with its solid or liquid phase. Vapor pressure is created by faster molecules that break away from the liquid or solid and enter the gas phase. The vapor pressure of a substance depends on both the substance and its temperature—an increase in temperature increases the vapor pressure.

Partial pressure is defined as the pressure a gas would create if it occupied the total volume available. In a mixture of gases, *the total pressure is the sum of partial pressures of the component gases*, assuming ideal gas behavior and no chemical reactions between the components. This law is known as **Dalton's law of partial pressures**, after the English scientist John Dalton (1766–1844), who proposed it. Dalton's law is based on kinetic theory, where each gas creates its pressure by molecular collisions, independent of other gases present. It is consistent with the fact that pressures add according to **Pascal's Principle**. Thus water evaporates and ice sublimates when their vapor pressures exceed the partial pressure of water vapor in the surrounding mixture of gases. If their vapor pressures are less than the partial pressure of water vapor in the surrounding gas, liquid droplets or ice crystals (frost) form.

Is energy transfer involved in a phase change? If so, will energy have to be supplied to change phase from solid to liquid and liquid to gas? What about gas to liquid and liquid to solid? Why do they spray the orange trees with water in Florida when the temperatures are near or just below freezing?

Solution
Yes, energy transfer is involved in a phase change. We know that atoms and molecules in solids and liquids are bound to each other because we know that force is required to separate them. So in a phase change from solid to liquid and liquid to gas, a force must be exerted, perhaps by collision, to separate atoms and molecules. Force exerted through a distance is work, and energy is needed to do work to go from solid to liquid and liquid to gas. This is intuitively consistent with the need for energy to melt ice or boil water. The converse is also true. Going from gas to liquid or liquid to solid involves atoms and molecules pushing together, doing work and releasing energy.

PhET Explorations: States of Matter—Basics

Heat, cool, and compress atoms and molecules and watch as they change between solid, liquid, and gas phases.

Figure 13.31 States of Matter: Basics (http://phet.colorado.edu/en/simulation/states-of-matter-basics)

13.6 Humidity, Evaporation, and Boiling

Learning Objectives

By the end of this section, you will be able to:

- Explain the relationship between vapor pressure of water and the capacity of air to hold water vapor.
- Explain the relationship between relative humidity and partial pressure of water vapor in the air.
- Calculate vapor density using vapor pressure.
- Calculate humidity and dew point.

Figure 13.32 Dew drops like these, on a banana leaf photographed just after sunrise, form when the air temperature drops to or below the dew point. At the dew point, the air can no longer hold all of the water vapor it held at higher temperatures, and some of the water condenses to form droplets. (credit: Aaron Escobar, Flickr)

The expression "it's not the heat, it's the humidity" makes a valid point. We keep cool in hot weather by evaporating sweat from our skin and water from our breathing passages. Because evaporation is inhibited by high humidity, we feel hotter at a given temperature when the humidity is high. Low humidity, on the other hand, can cause discomfort from excessive drying of mucous membranes and can lead to an increased risk of respiratory infections.

When we say humidity, we really mean **relative humidity**. Relative humidity tells us how much water vapor is in the air compared with the maximum possible. At its maximum, denoted as **saturation**, the relative humidity is 100%, and evaporation is inhibited. The amount of water vapor the air can hold depends on its temperature. For example, relative humidity rises in the evening, as air temperature declines, sometimes reaching the **dew point**. At the dew point temperature, relative humidity is 100%, and fog may result from the condensation of water droplets if they are small enough to stay in suspension. Conversely, if you wish to dry something (perhaps your hair), it is more effective to blow hot air over it rather than cold air, because, among other things, hot air can hold more water vapor.

The capacity of air to hold water vapor is based on vapor pressure of water. The liquid and solid phases are continuously giving off vapor because some of the molecules have high enough speeds to enter the gas phase; see **Figure 13.33**(a). If a lid is placed over the container, as in **Figure 13.33**(b), evaporation continues, increasing the pressure, until sufficient vapor has built up for condensation to balance evaporation. Then equilibrium has been achieved, and the vapor pressure is equal to the partial pressure of water in the container. Vapor pressure increases with temperature because molecular speeds are higher as temperature increases. **Table 13.5** gives representative values of water vapor pressure over a range of temperatures.

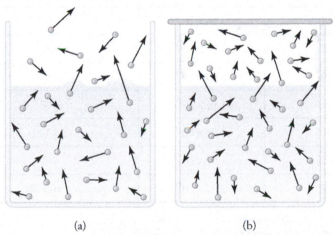

(a)   (b)

Figure 13.33 (a) Because of the distribution of speeds and kinetic energies, some water molecules can break away to the vapor phase even at temperatures below the ordinary boiling point. (b) If the container is sealed, evaporation will continue until there is enough vapor density for the condensation rate to equal the evaporation rate. This vapor density and the partial pressure it creates are the saturation values. They increase with temperature and are independent of the presence of other gases, such as air. They depend only on the vapor pressure of water.

Relative humidity is related to the partial pressure of water vapor in the air. At 100% humidity, the partial pressure is equal to the vapor pressure, and no more water can enter the vapor phase. If the partial pressure is less than the vapor pressure, then evaporation will take place, as humidity is less than 100%. If the partial pressure is greater than the vapor pressure, condensation takes place. The capacity of air to "hold" water vapor is determined by the vapor pressure of water and has nothing to do with the properties of air.

Table 13.5 Saturation Vapor Density of Water

Temperature ($^{\circ}$C)	Vapor pressure (Pa)	Saturation vapor density (g/m^3)
−50	4.0	0.039
−20	1.04×10^2	0.89
−10	2.60×10^2	2.36
0	6.10×10^2	4.84
5	8.68×10^2	6.80
10	1.19×10^3	9.40
15	1.69×10^3	12.8
20	2.33×10^3	17.2
25	3.17×10^3	23.0
30	4.24×10^3	30.4
37	6.31×10^3	44.0
40	7.34×10^3	51.1
50	1.23×10^4	82.4
60	1.99×10^4	130
70	3.12×10^4	197
80	4.73×10^4	294
90	7.01×10^4	418
95	8.59×10^4	505
100	$\mathbf{1.01 \times 10^5}$	**598**
120	1.99×10^5	1095
150	4.76×10^5	2430
200	1.55×10^6	7090
220	2.32×10^6	10,200

Example 13.12 Calculating Density Using Vapor Pressure

Table 13.5 gives the vapor pressure of water at 20.0°C as 2.33×10^3 Pa. Use the ideal gas law to calculate the density of water vapor in g / m^3 that would create a partial pressure equal to this vapor pressure. Compare the result with the saturation vapor density given in the table.

Strategy

To solve this problem, we need to break it down into a two steps. The partial pressure follows the ideal gas law,

$$PV = nRT, \tag{13.70}$$

where n is the number of moles. If we solve this equation for n / V to calculate the number of moles per cubic meter, we can then convert this quantity to grams per cubic meter as requested. To do this, we need to use the molecular mass of water, which is given in the periodic table.

Solution

1. Identify the knowns and convert them to the proper units:

   a.   temperature $T = 20°C=293$ K

   b.   vapor pressure P of water at $20°C$ is $2.33×10^3$ Pa

   c.   molecular mass of water is 18.0 g/mol

2. Solve the ideal gas law for n/V.

$$\frac{n}{V} = \frac{P}{RT} \tag{13.71}$$

3. Substitute known values into the equation and solve for n/V.

$$\frac{n}{V} = \frac{P}{RT} = \frac{2.33×10^3 \text{ Pa}}{(8.31 \text{ J/mol} \cdot \text{K})(293 \text{ K})} = 0.957 \text{ mol/m}^3 \tag{13.72}$$

4. Convert the density in moles per cubic meter to grams per cubic meter.

$$\rho = \left(0.957\frac{\text{mol}}{\text{m}^3}\right)\left(\frac{18.0 \text{ g}}{\text{mol}}\right) = 17.2 \text{ g/m}^3 \tag{13.73}$$

Discussion

The density is obtained by assuming a pressure equal to the vapor pressure of water at $20.0°C$. The density found is identical to the value in **Table 13.5**, which means that a vapor density of 17.2 g/m^3 at $20.0°C$ creates a partial pressure of $2.33×10^3$ Pa, equal to the vapor pressure of water at that temperature. If the partial pressure is equal to the vapor pressure, then the liquid and vapor phases are in equilibrium, and the relative humidity is 100%. Thus, there can be no more than 17.2 g of water vapor per m^3 at $20.0°C$, so that this value is the saturation vapor density at that temperature. This example illustrates how water vapor behaves like an ideal gas: the pressure and density are consistent with the ideal gas law (assuming the density in the table is correct). The saturation vapor densities listed in **Table 13.5** are the maximum amounts of water vapor that air can hold at various temperatures.

Percent Relative Humidity

We define **percent relative humidity** as the ratio of vapor density to saturation vapor density, or

$$\text{percent relative humidity} = \frac{\text{vapor density}}{\text{saturation vapor density}}×100 \tag{13.74}$$

We can use this and the data in **Table 13.5** to do a variety of interesting calculations, keeping in mind that relative humidity is based on the comparison of the partial pressure of water vapor in air and ice.

Example 13.13 Calculating Humidity and Dew Point

(a) Calculate the percent relative humidity on a day when the temperature is $25.0°C$ and the air contains 9.40 g of water vapor per m^3. (b) At what temperature will this air reach 100% relative humidity (the saturation density)? This temperature is the dew point. (c) What is the humidity when the air temperature is $25.0°C$ and the dew point is $-10.0°C$?

Strategy and Solution

(a) Percent relative humidity is defined as the ratio of vapor density to saturation vapor density.

$$\text{percent relative humidity} = \frac{\text{vapor density}}{\text{saturation vapor density}}×100 \tag{13.75}$$

The first is given to be 9.40 g/m^3, and the second is found in **Table 13.5** to be 23.0 g/m^3. Thus,

$$\text{percent relative humidity} = \frac{9.40 \text{ g/m}^3}{23.0 \text{ g/m}^3}×100 = 40.9.\% \tag{13.76}$$

(b) The air contains 9.40 g/m^3 of water vapor. The relative humidity will be 100% at a temperature where 9.40 g/m^3 is the saturation density. Inspection of **Table 13.5** reveals this to be the case at $10.0°C$, where the relative humidity will be 100%. That temperature is called the dew point for air with this concentration of water vapor.

(c) Here, the dew point temperature is given to be $-10.0°C$. Using **Table 13.5**, we see that the vapor density is 2.36 g/m^3, because this value is the saturation vapor density at $-10.0°C$. The saturation vapor density at $25.0°C$ is seen to be 23.0 g/m^3. Thus, the relative humidity at $25.0°C$ is

$$\text{percent relative humidity} = \frac{2.36 \text{ g/m}^3}{23.0 \text{ g/m}^3} \times 100 = 10.3\%. \tag{13.77}$$

Discussion

The importance of dew point is that air temperature cannot drop below $10.0°C$ in part (b), or $-10.0°C$ in part (c), without water vapor condensing out of the air. If condensation occurs, considerable transfer of heat occurs (discussed in **Heat and Heat Transfer Methods**), which prevents the temperature from further dropping. When dew points are below $0°C$, freezing temperatures are a greater possibility, which explains why farmers keep track of the dew point. Low humidity in deserts means low dew-point temperatures. Thus condensation is unlikely. If the temperature drops, vapor does not condense in liquid drops. Because no heat is released into the air, the air temperature drops more rapidly compared to air with higher humidity. Likewise, at high temperatures, liquid droplets do not evaporate, so that no heat is removed from the gas to the liquid phase. This explains the large range of temperature in arid regions.

Why does water boil at $100°C$? You will note from **Table 13.5** that the vapor pressure of water at $100°C$ is 1.01×10^5 Pa, or 1.00 atm. Thus, it can evaporate without limit at this temperature and pressure. But why does it form bubbles when it boils? This is because water ordinarily contains significant amounts of dissolved air and other impurities, which are observed as small bubbles of air in a glass of water. If a bubble starts out at the bottom of the container at $20°C$, it contains water vapor (about 2.30%). The pressure inside the bubble is fixed at 1.00 atm (we ignore the slight pressure exerted by the water around it). As the temperature rises, the amount of air in the bubble stays the same, but the water vapor increases; the bubble expands to keep the pressure at 1.00 atm. At $100°C$, water vapor enters the bubble continuously since the partial pressure of water is equal to 1.00 atm in equilibrium. It cannot reach this pressure, however, since the bubble also contains air and total pressure is 1.00 atm. The bubble grows in size and thereby increases the buoyant force. The bubble breaks away and rises rapidly to the surface—we call this boiling! (See **Figure 13.34**.)

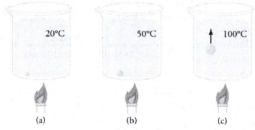

Figure 13.34 (a) An air bubble in water starts out saturated with water vapor at $20°C$. (b) As the temperature rises, water vapor enters the bubble because its vapor pressure increases. The bubble expands to keep its pressure at 1.00 atm. (c) At $100°C$, water vapor enters the bubble continuously because water's vapor pressure exceeds its partial pressure in the bubble, which must be less than 1.00 atm. The bubble grows and rises to the surface.

Check Your Understanding

Freeze drying is a process in which substances, such as foods, are dried by placing them in a vacuum chamber and lowering the atmospheric pressure around them. How does the lowered atmospheric pressure speed the drying process, and why does it cause the temperature of the food to drop?

Solution

Decreased the atmospheric pressure results in decreased partial pressure of water, hence a lower humidity. So evaporation of water from food, for example, will be enhanced. The molecules of water most likely to break away from the food will be those with the greatest velocities. Those remaining thus have a lower average velocity and a lower temperature. This can (and does) result in the freezing and drying of the food; hence the process is aptly named freeze drying.

Watch different types of molecules form a solid, liquid, or gas. Add or remove heat and watch the phase change. Change the temperature or volume of a container and see a pressure-temperature diagram respond in real time. Relate the interaction potential to the forces between molecules.

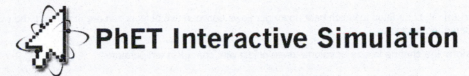

Figure 13.35 States of Matter: Basics (http://phet.colorado.edu/en/simulation/states-of-matter)

Glossary

absolute zero: the lowest possible temperature; the temperature at which all molecular motion ceases

Avogadro's number: N_A , the number of molecules or atoms in one mole of a substance; $N_A = 6.02 \times 10^{23}$ particles/mole

Boltzmann constant: k , a physical constant that relates energy to temperature; $k = 1.38 \times 10^{-23}$ J/K

Celsius scale: temperature scale in which the freezing point of water is $0°C$ and the boiling point of water is $100°C$

coefficient of linear expansion: α , the change in length, per unit length, per $1°C$ change in temperature; a constant used in the calculation of linear expansion; the coefficient of linear expansion depends on the material and to some degree on the temperature of the material

coefficient of volume expansion: β , the change in volume, per unit volume, per $1°C$ change in temperature

critical point: the temperature above which a liquid cannot exist

critical pressure: the minimum pressure needed for a liquid to exist at the critical temperature

critical temperature: the temperature above which a liquid cannot exist

Dalton's law of partial pressures: the physical law that states that the total pressure of a gas is the sum of partial pressures of the component gases

degree Celsius: unit on the Celsius temperature scale

degree Fahrenheit: unit on the Fahrenheit temperature scale

dew point: the temperature at which relative humidity is 100%; the temperature at which water starts to condense out of the air

Fahrenheit scale: temperature scale in which the freezing point of water is $32°F$ and the boiling point of water is $212°F$

ideal gas law: the physical law that relates the pressure and volume of a gas to the number of gas molecules or number of moles of gas and the temperature of the gas

Kelvin scale: temperature scale in which 0 K is the lowest possible temperature, representing absolute zero

mole: the quantity of a substance whose mass (in grams) is equal to its molecular mass

partial pressure: the pressure a gas would create if it occupied the total volume of space available

percent relative humidity: the ratio of vapor density to saturation vapor density

phase diagram: a graph of pressure vs. temperature of a particular substance, showing at which pressures and temperatures the three phases of the substance occur

PV diagram: a graph of pressure vs. volume

relative humidity: the amount of water in the air relative to the maximum amount the air can hold

saturation: the condition of 100% relative humidity

sublimation: the phase change from solid to gas

temperature: the quantity measured by a thermometer

thermal energy: $\overline{\text{KE}}$, the average translational kinetic energy of a molecule

thermal equilibrium: the condition in which heat no longer flows between two objects that are in contact; the two objects have the same temperature

thermal expansion: the change in size or volume of an object with change in temperature

thermal stress: stress caused by thermal expansion or contraction

triple point: the pressure and temperature at which a substance exists in equilibrium as a solid, liquid, and gas

vapor: a gas at a temperature below the boiling temperature

vapor pressure: the pressure at which a gas coexists with its solid or liquid phase

zeroth law of thermodynamics: law that states that if two objects are in thermal equilibrium, and a third object is in thermal equilibrium with one of those objects, it is also in thermal equilibrium with the other object

Section Summary

13.1 Temperature
- Temperature is the quantity measured by a thermometer.
- Temperature is related to the average kinetic energy of atoms and molecules in a system.
- Absolute zero is the temperature at which there is no molecular motion.
- There are three main temperature scales: Celsius, Fahrenheit, and Kelvin.
- Temperatures on one scale can be converted to temperatures on another scale using the following equations:

$$T_{°F} = \frac{9}{5}T_{°C} + 32$$

$$T_{°C} = \frac{5}{9}(T_{°F} - 32)$$

$$T_K = T_{°C} + 273.15$$

$$T_{°C} = T_K - 273.15$$

- Systems are in thermal equilibrium when they have the same temperature.
- Thermal equilibrium occurs when two bodies are in contact with each other and can freely exchange energy.
- The zeroth law of thermodynamics states that when two systems, A and B, are in thermal equilibrium with each other, and B is in thermal equilibrium with a third system, C, then A is also in thermal equilibrium with C.

13.2 Thermal Expansion of Solids and Liquids
- Thermal expansion is the increase, or decrease, of the size (length, area, or volume) of a body due to a change in temperature.
- Thermal expansion is large for gases, and relatively small, but not negligible, for liquids and solids.
- Linear thermal expansion is

$$\Delta L = \alpha L \Delta T,$$

where ΔL is the change in length L , ΔT is the change in temperature, and α is the coefficient of linear expansion, which varies slightly with temperature.
- The change in area due to thermal expansion is

$$\Delta A = 2\alpha A \Delta T,$$

where ΔA is the change in area.
- The change in volume due to thermal expansion is

$$\Delta V = \beta V \Delta T,$$

where β is the coefficient of volume expansion and $\beta \approx 3\alpha$. Thermal stress is created when thermal expansion is constrained.

13.3 The Ideal Gas Law
- The ideal gas law relates the pressure and volume of a gas to the number of gas molecules and the temperature of the gas.

- The ideal gas law can be written in terms of the number of molecules of gas:

$$PV = NkT,$$

where P is pressure, V is volume, T is temperature, N is number of molecules, and k is the Boltzmann constant

$$k = 1.38 \times 10^{-23} \text{ J/K}.$$

- A mole is the number of atoms in a 12-g sample of carbon-12.
- The number of molecules in a mole is called Avogadro's number N_A,

$$N_A = 6.02 \times 10^{23} \text{ mol}^{-1}.$$

- A mole of any substance has a mass in grams equal to its molecular weight, which can be determined from the periodic table of elements.
- The ideal gas law can also be written and solved in terms of the number of moles of gas:

$$PV = nRT,$$

where n is number of moles and R is the universal gas constant,

$$R = 8.31 \text{ J/mol} \cdot \text{K}.$$

- The ideal gas law is generally valid at temperatures well above the boiling temperature.

13.4 Kinetic Theory: Atomic and Molecular Explanation of Pressure and Temperature

- Kinetic theory is the atomistic description of gases as well as liquids and solids.
- Kinetic theory models the properties of matter in terms of continuous random motion of atoms and molecules.
- The ideal gas law can also be expressed as

$$PV = \frac{1}{3}Nm\overline{v^2},$$

where P is the pressure (average force per unit area), V is the volume of gas in the container, N is the number of molecules in the container, m is the mass of a molecule, and $\overline{v^2}$ is the average of the molecular speed squared.

- Thermal energy is defined to be the average translational kinetic energy $\overline{KE}$ of an atom or molecule.
- The temperature of gases is proportional to the average translational kinetic energy of atoms and molecules.

$$\overline{KE} = \frac{1}{2}m\overline{v^2} = \frac{3}{2}kT$$

or

$$\sqrt{\overline{v^2}} = v_{rms} = \sqrt{\frac{3kT}{m}}.$$

- The motion of individual molecules in a gas is random in magnitude and direction. However, a gas of many molecules has a predictable distribution of molecular speeds, known as the *Maxwell-Boltzmann distribution*.

13.5 Phase Changes

- Most substances have three distinct phases: gas, liquid, and solid.
- Phase changes among the various phases of matter depend on temperature and pressure.
- The existence of the three phases with respect to pressure and temperature can be described in a phase diagram.
- Two phases coexist (i.e., they are in thermal equilibrium) at a set of pressures and temperatures. These are described as a line on a phase diagram.
- The three phases coexist at a single pressure and temperature. This is known as the triple point and is described by a single point on a phase diagram.
- A gas at a temperature below its boiling point is called a vapor.
- Vapor pressure is the pressure at which a gas coexists with its solid or liquid phase.
- Partial pressure is the pressure a gas would create if it existed alone.
- Dalton's law states that the total pressure is the sum of the partial pressures of all of the gases present.

13.6 Humidity, Evaporation, and Boiling

- Relative humidity is the fraction of water vapor in a gas compared to the saturation value.
- The saturation vapor density can be determined from the vapor pressure for a given temperature.
- Percent relative humidity is defined to be

$$\text{percent relative humidity} = \frac{\text{vapor density}}{\text{saturation vapor density}} \times 100.$$

- The dew point is the temperature at which air reaches 100% relative humidity.

Conceptual Questions

13.1 Temperature

1. What does it mean to say that two systems are in thermal equilibrium?

2. Give an example of a physical property that varies with temperature and describe how it is used to measure temperature.

3. When a cold alcohol thermometer is placed in a hot liquid, the column of alcohol goes *down* slightly before going up. Explain why.

4. If you add boiling water to a cup at room temperature, what would you expect the final equilibrium temperature of the unit to be? You will need to include the surroundings as part of the system. Consider the zeroth law of thermodynamics.

13.2 Thermal Expansion of Solids and Liquids

5. Thermal stresses caused by uneven cooling can easily break glass cookware. Explain why Pyrex®, a glass with a small coefficient of linear expansion, is less susceptible.

6. Water expands significantly when it freezes: a volume increase of about 9% occurs. As a result of this expansion and because of the formation and growth of crystals as water freezes, anywhere from 10% to 30% of biological cells are burst when animal or plant material is frozen. Discuss the implications of this cell damage for the prospect of preserving human bodies by freezing so that they can be thawed at some future date when it is hoped that all diseases are curable.

7. One method of getting a tight fit, say of a metal peg in a hole in a metal block, is to manufacture the peg slightly larger than the hole. The peg is then inserted when at a different temperature than the block. Should the block be hotter or colder than the peg during insertion? Explain your answer.

8. Does it really help to run hot water over a tight metal lid on a glass jar before trying to open it? Explain your answer.

9. Liquids and solids expand with increasing temperature, because the kinetic energy of a body's atoms and molecules increases. Explain why some materials *shrink* with increasing temperature.

13.3 The Ideal Gas Law

10. Find out the human population of Earth. Is there a mole of people inhabiting Earth? If the average mass of a person is 60 kg, calculate the mass of a mole of people. How does the mass of a mole of people compare with the mass of Earth?

11. Under what circumstances would you expect a gas to behave significantly differently than predicted by the ideal gas law?

12. A constant-volume gas thermometer contains a fixed amount of gas. What property of the gas is measured to indicate its temperature?

13.4 Kinetic Theory: Atomic and Molecular Explanation of Pressure and Temperature

13. How is momentum related to the pressure exerted by a gas? Explain on the atomic and molecular level, considering the behavior of atoms and molecules.

13.5 Phase Changes

14. A pressure cooker contains water and steam in equilibrium at a pressure greater than atmospheric pressure. How does this greater pressure increase cooking speed?

15. Why does condensation form most rapidly on the coldest object in a room—for example, on a glass of ice water?

16. What is the vapor pressure of solid carbon dioxide (dry ice) at $-78.5°C$?

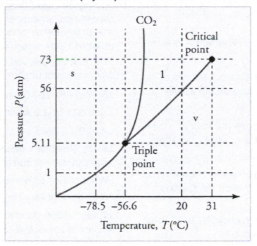

Figure 13.36 The phase diagram for carbon dioxide. The axes are nonlinear, and the graph is not to scale. Dry ice is solid carbon dioxide and has a sublimation temperature of $-78.5°C$.

17. Can carbon dioxide be liquefied at room temperature ($20°C$)? If so, how? If not, why not? (See **Figure 13.36**.)

18. Oxygen cannot be liquefied at room temperature by placing it under a large enough pressure to force its molecules together. Explain why this is.

19. What is the distinction between gas and vapor?

13.6 Humidity, Evaporation, and Boiling

20. Because humidity depends only on water's vapor pressure and temperature, are the saturation vapor densities listed in **Table 13.5** valid in an atmosphere of helium at a pressure of $1.01×10^5$ N/m^2 , rather than air? Are those values affected by altitude on Earth?

21. Why does a beaker of $40.0°C$ water placed in a vacuum chamber start to boil as the chamber is evacuated (air is pumped out of the chamber)? At what pressure does the boiling begin? Would food cook any faster in such a beaker?

22. Why does rubbing alcohol evaporate much more rapidly than water at STP (standard temperature and pressure)?

Problems & Exercises

13.1 Temperature

1. What is the Fahrenheit temperature of a person with a $39.0°C$ fever?

2. Frost damage to most plants occurs at temperatures of $28.0°F$ or lower. What is this temperature on the Kelvin scale?

3. To conserve energy, room temperatures are kept at $68.0°F$ in the winter and $78.0°F$ in the summer. What are these temperatures on the Celsius scale?

4. A tungsten light bulb filament may operate at 2900 K. What is its Fahrenheit temperature? What is this on the Celsius scale?

5. The surface temperature of the Sun is about 5750 K. What is this temperature on the Fahrenheit scale?

6. One of the hottest temperatures ever recorded on the surface of Earth was $134°F$ in Death Valley, CA. What is this temperature in Celsius degrees? What is this temperature in Kelvin?

7. (a) Suppose a cold front blows into your locale and drops the temperature by 40.0 Fahrenheit degrees. How many degrees Celsius does the temperature decrease when there is a $40.0°F$ decrease in temperature? (b) Show that any change in temperature in Fahrenheit degrees is nine-fifths the change in Celsius degrees.

8. (a) At what temperature do the Fahrenheit and Celsius scales have the same numerical value? (b) At what temperature do the Fahrenheit and Kelvin scales have the same numerical value?

13.2 Thermal Expansion of Solids and Liquids

9. The height of the Washington Monument is measured to be 170 m on a day when the temperature is $35.0°C$. What will its height be on a day when the temperature falls to $-10.0°C$? Although the monument is made of limestone, assume that its thermal coefficient of expansion is the same as marble's.

10. How much taller does the Eiffel Tower become at the end of a day when the temperature has increased by $15°C$? Its original height is 321 m and you can assume it is made of steel.

11. What is the change in length of a 3.00-cm-long column of mercury if its temperature changes from $37.0°C$ to $40.0°C$, assuming the mercury is unconstrained?

12. How large an expansion gap should be left between steel railroad rails if they may reach a maximum temperature $35.0°C$ greater than when they were laid? Their original length is 10.0 m.

13. You are looking to purchase a small piece of land in Hong Kong. The price is "only" $60,000 per square meter! The land title says the dimensions are $20 \text{ m} \times 30 \text{ m}$. By how much would the total price change if you measured the parcel with a steel tape measure on a day when the temperature was $20°C$ above normal?

14. Global warming will produce rising sea levels partly due to melting ice caps but also due to the expansion of water as average ocean temperatures rise. To get some idea of the size of this effect, calculate the change in length of a column of water 1.00 km high for a temperature increase of $1.00°C$. Note that this calculation is only approximate because ocean warming is not uniform with depth.

15. Show that 60.0 L of gasoline originally at $15.0°C$ will expand to 61.1 L when it warms to $35.0°C$, as claimed in **Example 13.4**.

16. (a) Suppose a meter stick made of steel and one made of invar (an alloy of iron and nickel) are the same length at $0°C$. What is their difference in length at $22.0°C$? (b) Repeat the calculation for two 30.0-m-long surveyor's tapes.

17. (a) If a 500-mL glass beaker is filled to the brim with ethyl alcohol at a temperature of $5.00°C$, how much will overflow when its temperature reaches $22.0°C$? (b) How much less water would overflow under the same conditions?

18. Most automobiles have a coolant reservoir to catch radiator fluid that may overflow when the engine is hot. A radiator is made of copper and is filled to its 16.0-L capacity when at $10.0°C$. What volume of radiator fluid will overflow when the radiator and fluid reach their $95.0°C$ operating temperature, given that the fluid's volume coefficient of expansion is $\beta = 400 \times 10^{-6}/°C$? Note that this coefficient is approximate, because most car radiators have operating temperatures of greater than $95.0°C$.

19. A physicist makes a cup of instant coffee and notices that, as the coffee cools, its level drops 3.00 mm in the glass cup. Show that this decrease cannot be due to thermal contraction by calculating the decrease in level if the 350 cm^3 of coffee is in a 7.00-cm-diameter cup and decreases in temperature from $95.0°C$ to $45.0°C$. (Most of the drop in level is actually due to escaping bubbles of air.)

20. (a) The density of water at $0°C$ is very nearly 1000 kg/m^3 (it is actually 999.84 kg/m^3), whereas the density of ice at $0°C$ is 917 kg/m^3. Calculate the pressure necessary to keep ice from expanding when it freezes, neglecting the effect such a large pressure would have on the freezing temperature. (This problem gives you only an indication of how large the forces associated with freezing water might be.) (b) What are the implications of this result for biological cells that are frozen?

21. Show that $\beta \approx 3\alpha$, by calculating the change in volume ΔV of a cube with sides of length L.

13.3 The Ideal Gas Law

22. The gauge pressure in your car tires is $2.50 \times 10^5 \text{ N/m}^2$ at a temperature of $35.0°C$ when you drive it onto a ferry boat to Alaska. What is their gauge pressure later, when their temperature has dropped to $-40.0°C$?

23. Convert an absolute pressure of 7.00×10^5 N/m^2 to gauge pressure in lb/in^2. (This value was stated to be just less than 90.0 lb/in^2 in **Example 13.9**. Is it?)

24. Suppose a gas-filled incandescent light bulb is manufactured so that the gas inside the bulb is at atmospheric pressure when the bulb has a temperature of $20.0°C$. (a) Find the gauge pressure inside such a bulb when it is hot, assuming its average temperature is $60.0°C$ (an approximation) and neglecting any change in volume due to thermal expansion or gas leaks. (b) The actual final pressure for the light bulb will be less than calculated in part (a) because the glass bulb will expand. What will the actual final pressure be, taking this into account? Is this a negligible difference?

25. Large helium-filled balloons are used to lift scientific equipment to high altitudes. (a) What is the pressure inside such a balloon if it starts out at sea level with a temperature of $10.0°C$ and rises to an altitude where its volume is twenty times the original volume and its temperature is $-50.0°C$? (b) What is the gauge pressure? (Assume atmospheric pressure is constant.)

26. Confirm that the units of nRT are those of energy for each value of R: (a) 8.31 J/mol $\cdot$ K, (b) 1.99 cal/mol $\cdot$ K, and (c) 0.0821 L $\cdot$ atm/mol $\cdot$ K.

27. In the text, it was shown that $N/V = 2.68 \times 10^{25}$ m^{-3} for gas at STP. (a) Show that this quantity is equivalent to $N/V = 2.68 \times 10^{19}$ cm^{-3}, as stated. (b) About how many atoms are there in one μm^3 (a cubic micrometer) at STP? (c) What does your answer to part (b) imply about the separation of atoms and molecules?

28. Calculate the number of moles in the 2.00-L volume of air in the lungs of the average person. Note that the air is at $37.0°C$ (body temperature).

29. An airplane passenger has 100 cm^3 of air in his stomach just before the plane takes off from a sea-level airport. What volume will the air have at cruising altitude if cabin pressure drops to 7.50×10^4 N/m^2?

30. (a) What is the volume (in km^3) of Avogadro's number of sand grains if each grain is a cube and has sides that are 1.0 mm long? (b) How many kilometers of beaches in length would this cover if the beach averages 100 m in width and 10.0 m in depth? Neglect air spaces between grains.

31. An expensive vacuum system can achieve a pressure as low as 1.00×10^{-7} N/m^2 at $20°C$. How many atoms are there in a cubic centimeter at this pressure and temperature?

32. The number density of gas atoms at a certain location in the space above our planet is about 1.00×10^{11} m^{-3}, and the pressure is 2.75×10^{-10} N/m^2 in this space. What is the temperature there?

33. A bicycle tire has a pressure of 7.00×10^5 N/m^2 at a temperature of $18.0°C$ and contains 2.00 L of gas. What will its pressure be if you let out an amount of air that has a volume of 100 cm^3 at atmospheric pressure? Assume tire temperature and volume remain constant.

34. A high-pressure gas cylinder contains 50.0 L of toxic gas at a pressure of 1.40×10^7 N/m^2 and a temperature of $25.0°C$. Its valve leaks after the cylinder is dropped. The cylinder is cooled to dry ice temperature ($-78.5°C$) to reduce the leak rate and pressure so that it can be safely repaired. (a) What is the final pressure in the tank, assuming a negligible amount of gas leaks while being cooled and that there is no phase change? (b) What is the final pressure if one-tenth of the gas escapes? (c) To what temperature must the tank be cooled to reduce the pressure to 1.00 atm (assuming the gas does not change phase and that there is no leakage during cooling)? (d) Does cooling the tank appear to be a practical solution?

35. Find the number of moles in 2.00 L of gas at $35.0°C$ and under 7.41×10^7 N/m^2 of pressure.

36. Calculate the depth to which Avogadro's number of table tennis balls would cover Earth. Each ball has a diameter of 3.75 cm. Assume the space between balls adds an extra 25.0% to their volume and assume they are not crushed by their own weight.

37. (a) What is the gauge pressure in a $25.0°C$ car tire containing 3.60 mol of gas in a 30.0 L volume? (b) What will its gauge pressure be if you add 1.00 L of gas originally at atmospheric pressure and $25.0°C$? Assume the temperature returns to $25.0°C$ and the volume remains constant.

38. (a) In the deep space between galaxies, the density of atoms is as low as 10^6 atoms/m^3, and the temperature is a frigid 2.7 K. What is the pressure? (b) What volume (in m^3) is occupied by 1 mol of gas? (c) If this volume is a cube, what is the length of its sides in kilometers?

13.4 Kinetic Theory: Atomic and Molecular Explanation of Pressure and Temperature

39. Some incandescent light bulbs are filled with argon gas. What is v_{rms} for argon atoms near the filament, assuming their temperature is 2500 K?

40. Average atomic and molecular speeds (v_{rms}) are large, even at low temperatures. What is v_{rms} for helium atoms at 5.00 K, just one degree above helium's liquefaction temperature?

41. (a) What is the average kinetic energy in joules of hydrogen atoms on the $5500°C$ surface of the Sun? (b) What is the average kinetic energy of helium atoms in a region of the solar corona where the temperature is 6.00×10^5 K?

42. The escape velocity of any object from Earth is 11.2 km/s. (a) Express this speed in m/s and km/h. (b) At what temperature would oxygen molecules (molecular mass is equal to 32.0 g/mol) have an average velocity v_{rms} equal to Earth's escape velocity of 11.1 km/s?

43. The escape velocity from the Moon is much smaller than from Earth and is only 2.38 km/s. At what temperature would hydrogen molecules (molecular mass is equal to 2.016 g/mol) have an average velocity v_{rms} equal to the Moon's escape velocity?

44. Nuclear fusion, the energy source of the Sun, hydrogen bombs, and fusion reactors, occurs much more readily when the average kinetic energy of the atoms is high—that is, at high temperatures. Suppose you want the atoms in your fusion experiment to have average kinetic energies of 6.40×10^{-14} J . What temperature is needed?

45. Suppose that the average velocity (v_{rms}) of carbon dioxide molecules (molecular mass is equal to 44.0 g/mol) in a flame is found to be 1.05×10^5 m/s . What temperature does this represent?

46. Hydrogen molecules (molecular mass is equal to 2.016 g/mol) have an average velocity v_{rms} equal to 193 m/s. What is the temperature?

47. Much of the gas near the Sun is atomic hydrogen. Its temperature would have to be 1.5×10^7 K for the average velocity v_{rms} to equal the escape velocity from the Sun. What is that velocity?

48. There are two important isotopes of uranium— ^{235}U and ^{238}U ; these isotopes are nearly identical chemically but have different atomic masses. Only ^{235}U is very useful in nuclear reactors. One of the techniques for separating them (gas diffusion) is based on the different average velocities v_{rms} of uranium hexafluoride gas, UF_6 . (a) The molecular masses for ^{235}U UF_6 and ^{238}U UF_6 are 349.0 g/mol and 352.0 g/mol, respectively. What is the ratio of their average velocities? (b) At what temperature would their average velocities differ by 1.00 m/s? (c) Do your answers in this problem imply that this technique may be difficult?

13.6 Humidity, Evaporation, and Boiling

49. Dry air is 78.1% nitrogen. What is the partial pressure of nitrogen when the atmospheric pressure is 1.01×10^5 N/m^2?

50. (a) What is the vapor pressure of water at $20.0°C$? (b) What percentage of atmospheric pressure does this correspond to? (c) What percent of $20.0°C$ air is water vapor if it has 100% relative humidity? (The density of dry air at $20.0°C$ is 1.20 kg/m^3 .)

51. Pressure cookers increase cooking speed by raising the boiling temperature of water above its value at atmospheric pressure. (a) What pressure is necessary to raise the boiling point to $120.0°C$? (b) What gauge pressure does this correspond to?

52. (a) At what temperature does water boil at an altitude of 1500 m (about 5000 ft) on a day when atmospheric pressure is 8.59×10^4 N/m^2? (b) What about at an altitude of 3000 m (about 10,000 ft) when atmospheric pressure is 7.00×10^4 N/m^2?

53. What is the atmospheric pressure on top of Mt. Everest on a day when water boils there at a temperature of $70.0°C$?

54. At a spot in the high Andes, water boils at $80.0°C$, greatly reducing the cooking speed of potatoes, for example. What is atmospheric pressure at this location?

55. What is the relative humidity on a $25.0°C$ day when the air contains 18.0 g/m^3 of water vapor?

56. What is the density of water vapor in g/m^3 on a hot dry day in the desert when the temperature is $40.0°C$ and the relative humidity is 6.00%?

57. A deep-sea diver should breathe a gas mixture that has the same oxygen partial pressure as at sea level, where dry air contains 20.9% oxygen and has a total pressure of 1.01×10^5 N/m^2 . (a) What is the partial pressure of oxygen at sea level? (b) If the diver breathes a gas mixture at a pressure of 2.00×10^6 N/m^2 , what percent oxygen should it be to have the same oxygen partial pressure as at sea level?

58. The vapor pressure of water at $40.0°C$ is 7.34×10^3 N/m^2 . Using the ideal gas law, calculate the density of water vapor in g/m^3 that creates a partial pressure equal to this vapor pressure. The result should be the same as the saturation vapor density at that temperature $(51.1$ g/m$^3)$.

59. Air in human lungs has a temperature of $37.0°C$ and a saturation vapor density of 44.0 g/m^3 . (a) If 2.00 L of air is exhaled and very dry air inhaled, what is the maximum loss of water vapor by the person? (b) Calculate the partial pressure of water vapor having this density, and compare it with the vapor pressure of 6.31×10^3 N/m^2 .

60. If the relative humidity is 90.0% on a muggy summer morning when the temperature is $20.0°C$, what will it be later in the day when the temperature is $30.0°C$, assuming the water vapor density remains constant?

61. Late on an autumn day, the relative humidity is 45.0% and the temperature is $20.0°C$. What will the relative humidity be that evening when the temperature has dropped to $10.0°C$, assuming constant water vapor density?

62. Atmospheric pressure atop Mt. Everest is 3.30×10^4 N/m^2 . (a) What is the partial pressure of oxygen there if it is 20.9% of the air? (b) What percent oxygen should a mountain climber breathe so that its partial pressure is the same as at sea level, where atmospheric pressure is 1.01×10^5 N/m^2? (c) One of the most severe problems for those climbing very high mountains is the extreme drying of breathing passages. Why does this drying occur?

63. What is the dew point (the temperature at which 100% relative humidity would occur) on a day when relative humidity is 39.0% at a temperature of $20.0°C$?

64. On a certain day, the temperature is $25.0°C$ and the relative humidity is 90.0%. How many grams of water must condense out of each cubic meter of air if the temperature falls to $15.0°C$? Such a drop in temperature can, thus, produce heavy dew or fog.

65. Integrated Concepts

The boiling point of water increases with depth because pressure increases with depth. At what depth will fresh water have a boiling point of $150°C$, if the surface of the water is at sea level?

66. Integrated Concepts

(a) At what depth in fresh water is the critical pressure of water reached, given that the surface is at sea level? (b) At what temperature will this water boil? (c) Is a significantly higher temperature needed to boil water at a greater depth?

67. Integrated Concepts

To get an idea of the small effect that temperature has on Archimedes' principle, calculate the fraction of a copper block's weight that is supported by the buoyant force in $0°C$ water and compare this fraction with the fraction supported in $95.0°C$ water.

68. Integrated Concepts

If you want to cook in water at $150°C$, you need a pressure cooker that can withstand the necessary pressure. (a) What pressure is required for the boiling point of water to be this high? (b) If the lid of the pressure cooker is a disk 25.0 cm in diameter, what force must it be able to withstand at this pressure?

69. Unreasonable Results

(a) How many moles per cubic meter of an ideal gas are there at a pressure of $1.00×10^{14}$ N/m^2 and at $0°C$? (b) What is unreasonable about this result? (c) Which premise or assumption is responsible?

70. Unreasonable Results

(a) An automobile mechanic claims that an aluminum rod fits loosely into its hole on an aluminum engine block because the engine is hot and the rod is cold. If the hole is 10.0% bigger in diameter than the $22.0°C$ rod, at what temperature will the rod be the same size as the hole? (b) What is unreasonable about this temperature? (c) Which premise is responsible?

71. Unreasonable Results

The temperature inside a supernova explosion is said to be $2.00×10^{13}$ K . (a) What would the average velocity v_{rms} of hydrogen atoms be? (b) What is unreasonable about this velocity? (c) Which premise or assumption is responsible?

72. Unreasonable Results

Suppose the relative humidity is 80% on a day when the temperature is $30.0°C$. (a) What will the relative humidity be if the air cools to $25.0°C$ and the vapor density remains constant? (b) What is unreasonable about this result? (c) Which premise is responsible?

Test Prep for AP® Courses

13.3 The Ideal Gas Law

1. A fixed amount of ideal gas is kept in a container of fixed volume. The absolute pressure P, in pascals, of the gas is plotted as a function of its temperature T, in degrees Celsius. Which of the following are properties of a best fit curve to the data? Select *two* answers.

a. Having a positive slope
b. Passing through the origin
c. Having zero pressure at a certain negative temperature
d. Approaching zero pressure as temperature approaches infinity

2.

Figure 13.37 This figure shows a clear plastic container with a movable piston that contains a fixed amount of gas. A group of students is asked to determine whether the gas is ideal. The students design and conduct an experiment. They measure the three quantities recorded in the data table below.

Table 13.6

Trial	Absolute Gas Pressure (x10m^5 Pa)	Volume (m^3)	Temp. (K)		
1	1.1	0.020	270		
2	1.4	0.016	270		
3	1.9	0.012	270		
4	2.2	0.010	270		
5	2.8	0.008	270		
6	1.2	0.020	290		
7	1.5	0.016	290		
8	2.0	0.012	290		
9	2.4	0.010	290		
10	3.0	0.008	290		
11	1.3	0.020	310		
12	1.6	0.016	310		
13	2.1	0.012	310		
14	2.6	0.010	310		
15	3.2	0.008	310		

a. Select a set of data points from the table and plot those points on a graph to determine whether the gas exhibits properties of an ideal gas. Fill in blank columns in the table for any quantities you graph other than the given data. Label the axes and indicate the scale for each. Draw a best-fit line or curve through your data points.

b. Indicate whether the gas exhibits properties of an ideal gas, and explain what characteristic of your graph provides the evidence.

c. The students repeat their experiment with an identical container that contains half as much gas. They take data for the same values of volume and temperature as in the table. Would the new data result in a different conclusion about whether the gas is ideal? Justify your answer in terms of interactions between the molecules of the gas and the container walls.

13.4 Kinetic Theory: Atomic and Molecular Explanation of Pressure and Temperature

3. Two samples of ideal gas in separate containers have the same number of molecules and the same temperature, but the molecular mass of gas X is greater than that of gas Y. Which of the following correctly compares the average speed of the molecules of the gases and the average force the gases exert on their respective containers?

Table 13.7

	Average Speed of Molecules	Average Force on Container
(a)	Greater for gas X	Greater for gas X
(b)	Greater for gas X	The forces cannot be compared without knowing the volumes of the gases.
(c)	Greater for gas Y	Greater for gas Y
(d)	Greater for gas Y	The forces cannot be compared without knowing the volumes of the gases.

4. How will the average kinetic energy of a gas molecule change if its temperature is increased from 20ºC to 313ºC?

a. It will become sixteen times its original value.
b. It will become four times its original value
c. It will become double its original value
d. It will remain unchanged.

5.

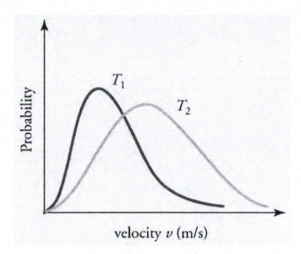

Figure 13.38 This graph shows the Maxwell-Boltzmann distribution of molecular speeds in an ideal gas for two temperatures, T_1 and T_2. Which of the following statements is false?

   a. T_1 is lower than T_2
   b.   The *rms* speed at T_1 is higher than that at T_2.
   c.   The peak of each graph shows the most probable speed at the corresponding temperature.
   d.   None of the above.

6. Suppose you have gas in a cylinder with a movable piston which has an area of 0.40 m^2. The pressure of the gas is 150 Pa when the height of the piston is 0.02 m. Find the force exerted by the gas on the piston. How does this force change if the piston is moved to a height of 0.03 m? Assume temperature remains constant.

7. What is the average kinetic energy of a nitrogen molecule (N_2) if its *rms* speed is 560 m/s? At what temperature is this *rms* speed achieved?

8. What will be the ratio of kinetic energies and *rms* speeds of a nitrogen molecule and a helium atom at the same temperature?

14   HEAT AND HEAT TRANSFER METHODS

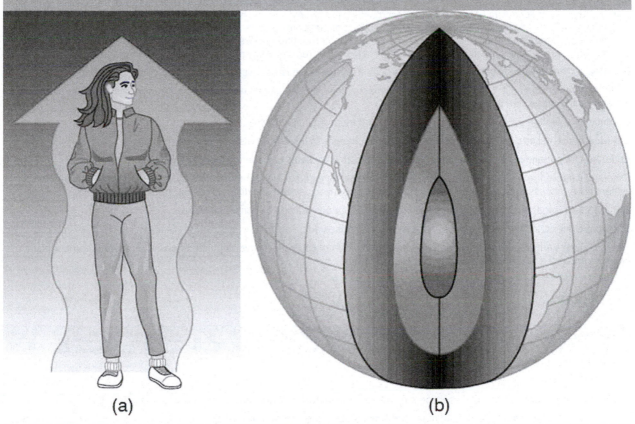

(a)   (b)

Figure 14.1 (a) The chilling effect of a clear breezy night is produced by the wind and by radiative heat transfer to cold outer space. (b) There was once great controversy about the Earth's age, but it is now generally accepted to be about 4.5 billion years old. Much of the debate is centered on the Earth's molten interior. According to our understanding of heat transfer, if the Earth is really that old, its center should have cooled off long ago. The discovery of radioactivity in rocks revealed the source of energy that keeps the Earth's interior molten, despite heat transfer to the surface, and from there to cold outer space.

Chapter Outline

14.1. Heat

14.2. Temperature Change and Heat Capacity

14.3. Phase Change and Latent Heat

14.4. Heat Transfer Methods

14.5. Conduction

14.6. Convection

14.7. Radiation

Connection for AP® Courses

Heat is one of the most intriguing of the many ways in which energy goes from one place to another. Heat is often hidden, as it only exists when energy is in transit, and the methods of transfer are distinctly different. Energy transfer by heat touches every aspect of our lives, and helps us to understand how the universe functions. It explains the chill you feel on a clear breezy night, and why Earth's core has yet to cool.

In this chapter, the ideas of temperature and thermal energy are used to examine and define heat, how heat is affected by the thermal properties of materials, and how the various mechanisms of heat transfer function. These topics are fundamental and practical, and will be returned to in future chapters. Big Idea 4 of the AP® Physics Curriculum Framework is supported by a discussion of how systems interact through energy transfer by heat and how this leads to changes in the energy of each system. Big Idea 5 is supported by exploration of the law of energy conservation that governs any changes in the energy of a system. Heat involves the transfer of thermal energy, or internal energy, from one system to another or to its surroundings, and so leads to a change in the internal energy of the system. This is analogous to the way work transfers mechanical energy to a mass to change its kinetic or potential energy. However, heat occurs as a spontaneous process, in which thermal energy is transferred from a higher temperature system to a lower temperature system.

Big Idea 1 is supported by examination of the internal structure of systems, which determines the nature of those energy changes and the mechanism of heat transfer. Macroscopic properties, such as heat capacity, latent heat, and thermal conductivity, depend on the arrangements and interactions of the atoms or molecules in a substance. The arrangement of these particles also determines whether thermal energy will be transferred through direct physical contact between systems (conduction), through the motion of fluids with different temperatures (convection), or through emission or absorption of radiation.

Big Idea 1 Objects and systems have properties such as mass and charge. Systems may have internal structure.

Enduring Understanding 1.E Materials have many macroscopic properties that result from the arrangement and interactions of the atoms and molecules that make up the material.

Essential Knowledge 1.E.3 Matter has a property called thermal conductivity.

Big Idea 4 Interactions between systems can result in changes in those systems.

Enduring Understanding 4.C Interactions with other objects or systems can change the total energy of a system.

Essential Knowledge 4.C.3 Energy is transferred spontaneously from a higher temperature system to a lower temperature system. The process through which energy is transferred between systems at different temperatures is called heat.

Big Idea 5 Changes that occur as a result of interactions are constrained by conservation laws.

Enduring Understanding 5.B The energy of a system is conserved.

Essential Knowledge 5.B.6 Energy can be transferred by thermal processes involving differences in temperature; the amount of energy transferred in this process of transfer is called heat.

14.1 Heat

Learning Objectives
By the end of this section, you will be able to: • Define heat as transfer of energy.   The information presented in this section supports the following AP® learning objectives and science practices: • **4.C.3.1** The student is able to make predictions about the direction of energy transfer due to temperature differences based on interactions at the microscopic level. **(S.P. 6.1)**

In **Work, Energy, and Energy Resources**, we defined work as force times distance and learned that work done on an object changes its kinetic energy. We also saw in **Temperature, Kinetic Theory, and the Gas Laws** that temperature is proportional to the (average) kinetic energy of atoms and molecules. We say that a thermal system has a certain internal energy: its internal energy is higher if the temperature is higher. If two objects at different temperatures are brought in contact with each other, energy is transferred from the hotter to the colder object until equilibrium is reached and the bodies reach thermal equilibrium (i.e., they are at the same temperature). No work is done by either object, because no force acts through a distance. The transfer of energy is caused by the temperature difference, and ceases once the temperatures are equal. These observations lead to the following definition of **heat**: Heat is the spontaneous transfer of energy due to a temperature difference.

As noted in **Temperature, Kinetic Theory, and the Gas Laws**, heat is often confused with temperature. For example, we may say the heat was unbearable, when we actually mean that the temperature was high. Heat is a form of energy, whereas temperature is not. The misconception arises because we are sensitive to the flow of heat, rather than the temperature.

Owing to the fact that heat is a form of energy, it has the SI unit of *joule* (J). The *calorie* (cal) is a common unit of energy, defined as the energy needed to change the temperature of 1.00 g of water by $1.00°C$ —specifically, between $14.5°C$ and $15.5°C$, since there is a slight temperature dependence. Perhaps the most common unit of heat is the **kilocalorie** (kcal), which is the energy needed to change the temperature of 1.00 kg of water by $1.00°C$. Since mass is most often specified in kilograms, kilocalorie is commonly used. Food calories (given the notation Cal, and sometimes called "big calorie") are actually kilocalories ($1 \text{ kilocalorie} = 1000 \text{ calories}$), a fact not easily determined from package labeling.

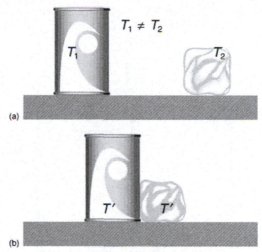

Figure 14.2 In figure (a) the soft drink and the ice have different temperatures, T_1 and T_2, and are not in thermal equilibrium. In figure (b), when the soft drink and ice are allowed to interact, energy is transferred until they reach the same temperature T', achieving equilibrium. Heat transfer occurs due to the difference in temperatures. In fact, since the soft drink and ice are both in contact with the surrounding air and bench, the equilibrium temperature will be the same for both.

Making Connections: Heat Interpreted at the Molecular Level

What is observed as a change in temperature of two macroscopic objects in contact, such as a warm can of liquid and an ice cube, consists of the transfer of kinetic energy from particles (atoms or molecules) with greater kinetic energy to those with lower kinetic energy. In this respect, the process can be viewed in terms of collisions, as described through classical mechanics. Consider the particles in two substances at different temperatures. The particles of each substance move with a range of speeds that are distributed around a mean value, $\bar{v}$. The temperature of each substance is defined in terms of the average kinetic energy of its particles, $\frac{1}{2}m\bar{v}^2$. The simplest mathematical description of this is for an ideal gas, and is given by the following equation:

$$T = \frac{2(\frac{1}{2}m\bar{v}^2)}{3k},$$

(14.1)

where k is Boltzmann's constant ($k = 1.38 \times 10^{-23}$ J/K). The equations for non-ideal gases, liquids, and solids are more complicated, but the general relation between the kinetic energies of the particles and the overall temperature of the substance still holds: the particles in the substance with the higher temperature have greater average kinetic energies than do the particles of a substance with a lower temperature.

When the two substances are in thermal contact, the particles of both substances can collide with each other. In the vast majority of collisions, a particle with greater kinetic energy will transfer some of its energy to a particle with lower kinetic energy. By giving up this energy, the average kinetic energy of this particle is reduced, and therefore, the temperature of the substance associated with that particle decreases slightly. Similarly, the average kinetic energy of the particle in the second substance increases through the collision, causing that substance's temperature to increase by a minuscule amount. In this way, through a vast number of particle collisions, thermal energy is transferred macroscopically from the substance with greater temperature (that is, greater internal energy) to the substance with lower temperature (lower internal energy).

Macroscopically, heat appears to transfer thermal energy spontaneously in only one direction. When interpreted at the microscopic level, the transfer of kinetic energy between particles occurs in both directions. This is because some of the particles in the low-temperature substance have higher kinetic energies than the particles in the high-temperature substance, so that some of the energy transfer is in the direction from the lower temperature substance to the higher temperature substance. However, much more of the energy is transferred in the other direction. When thermal equilibrium is reached, the energy transfer in either direction is, on average, the same, so that there is no further change in the internal energy, or temperature, of either substance.

Mechanical Equivalent of Heat

It is also possible to change the temperature of a substance by doing work. Work can transfer energy into or out of a system. This realization helped establish the fact that heat is a form of energy. James Prescott Joule (1818–1889) performed many experiments to establish the **mechanical equivalent of heat**—*the work needed to produce the same effects as heat transfer*. In terms of the units used for these two terms, the best modern value for this equivalence is

$$1.000 \text{ kcal} = 4186 \text{ J}.$$

(14.2)

We consider this equation as the conversion between two different units of energy.

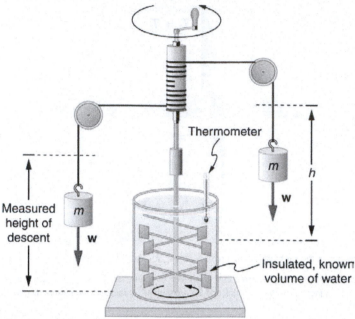

Figure 14.3 Schematic depiction of Joule's experiment that established the equivalence of heat and work.

The figure above shows one of Joule's most famous experimental setups for demonstrating the mechanical equivalent of heat. It demonstrated that work and heat can produce the same effects, and helped establish the principle of conservation of energy. Gravitational potential energy (PE) (work done by the gravitational force) is converted into kinetic energy (KE), and then randomized by viscosity and turbulence into increased average kinetic energy of atoms and molecules in the system, producing a temperature increase. His contributions to the field of thermodynamics were so significant that the SI unit of energy was named after him.

Heat added or removed from a system changes its internal energy and thus its temperature. Such a temperature increase is observed while cooking. However, adding heat does not necessarily increase the temperature. An example is melting of ice; that is, when a substance changes from one phase to another. Work done on the system or by the system can also change the internal energy of the system. Joule demonstrated that the temperature of a system can be increased by stirring. If an ice cube is rubbed against a rough surface, work is done by the frictional force. A system has a well-defined internal energy, but we cannot say that it has a certain "heat content" or "work content". We use the phrase "heat transfer" to emphasize its nature.

Check Your Understanding

Two samples (A and B) of the same substance are kept in a lab. Someone adds 10 kilojoules (kJ) of heat to one sample, while 10 kJ of work is done on the other sample. How can you tell to which sample the heat was added?

Solution
Heat and work both change the internal energy of the substance. However, the properties of the sample only depend on the internal energy so that it is impossible to tell whether heat was added to sample A or B.

14.2 Temperature Change and Heat Capacity

Learning Objectives

By the end of this section, you will be able to:

- Observe heat transfer and change in temperature and mass.
- Calculate final temperature after heat transfer between two objects.

One of the major effects of heat transfer is temperature change: heating increases the temperature while cooling decreases it. We assume that there is no phase change and that no work is done on or by the system. Experiments show that the transferred heat depends on three factors—the change in temperature, the mass of the system, and the substance and phase of the substance.

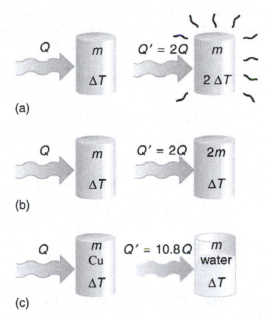

Figure 14.4 The heat Q transferred to cause a temperature change depends on the magnitude of the temperature change, the mass of the system, and the substance and phase involved. (a) The amount of heat transferred is directly proportional to the temperature change. To double the temperature change of a mass m , you need to add twice the heat. (b) The amount of heat transferred is also directly proportional to the mass. To cause an equivalent temperature change in a doubled mass, you need to add twice the heat. (c) The amount of heat transferred depends on the substance and its phase. If it takes an amount Q of heat to cause a temperature change ΔT in a given mass of copper, it will take 10.8 times that amount of heat to cause the equivalent temperature change in the same mass of water assuming no phase change in either substance.

The dependence on temperature change and mass are easily understood. Owing to the fact that the (average) kinetic energy of an atom or molecule is proportional to the absolute temperature, the internal energy of a system is proportional to the absolute temperature and the number of atoms or molecules. Owing to the fact that the transferred heat is equal to the change in the internal energy, the heat is proportional to the mass of the substance and the temperature change. The transferred heat also depends on the substance so that, for example, the heat necessary to raise the temperature is less for alcohol than for water. For the same substance, the transferred heat also depends on the phase (gas, liquid, or solid).

Heat Transfer and Temperature Change

The quantitative relationship between heat transfer and temperature change contains all three factors:

$$Q = mc\Delta T, \tag{14.3}$$

where Q is the symbol for heat transfer, m is the mass of the substance, and ΔT is the change in temperature. The symbol c stands for **specific heat** and depends on the material and phase. The specific heat is the amount of heat necessary to change the temperature of 1.00 kg of mass by $1.00°C$. The specific heat c is a property of the substance; its SI unit is $J/(kg \cdot K)$ or $J/(kg·°C)$. Recall that the temperature change (ΔT) is the same in units of kelvin and degrees Celsius. If heat transfer is measured in kilocalories, then *the unit of specific heat* is $kcal/(kg·°C)$.

Values of specific heat must generally be looked up in tables, because there is no simple way to calculate them. In general, the specific heat also depends on the temperature. Table 14.1 lists representative values of specific heat for various substances. Except for gases, the temperature and volume dependence of the specific heat of most substances is weak. We see from this table that the specific heat of water is five times that of glass and ten times that of iron, which means that it takes five times as much heat to raise the temperature of water the same amount as for glass and ten times as much heat to raise the temperature of water as for iron. In fact, water has one of the largest specific heats of any material, which is important for sustaining life on Earth.

Example 14.1 Calculating the Required Heat: Heating Water in an Aluminum Pan

A 0.500 kg aluminum pan on a stove is used to heat 0.250 liters of water from $20.0°C$ to $80.0°C$. (a) How much heat is required? What percentage of the heat is used to raise the temperature of (b) the pan and (c) the water?

Strategy

The pan and the water are always at the same temperature. When you put the pan on the stove, the temperature of the water and the pan is increased by the same amount. We use the equation for the heat transfer for the given temperature change and mass of water and aluminum. The specific heat values for water and aluminum are given in **Table 14.1**.

Solution

Because water is in thermal contact with the aluminum, the pan and the water are at the same temperature.

1. Calculate the temperature difference:

$$\Delta T = T_f - T_i = 60.0°C. \tag{14.4}$$

2. Calculate the mass of water. Because the density of water is 1000 kg/m^3, one liter of water has a mass of 1 kg, and the mass of 0.250 liters of water is $m_w = 0.250 \text{ kg}$.

3. Calculate the heat transferred to the water. Use the specific heat of water in **Table 14.1**:

$$Q_w = m_w c_w \Delta T = (0.250 \text{ kg})(4186 \text{ J/kg°C})(60.0°C) = 62.8 \text{ kJ}. \tag{14.5}$$

4. Calculate the heat transferred to the aluminum. Use the specific heat for aluminum in **Table 14.1**:

$$Q_{Al} = m_{Al} c_{Al} \Delta T = (0.500 \text{ kg})(900 \text{ J/kg°C})(60.0°C) = 27.0 \times 10^4 \text{ J} = 27.0 \text{ kJ}. \tag{14.6}$$

5. Compare the percentage of heat going into the pan versus that going into the water. First, find the total transferred heat:

$$Q_{Total} = Q_W + Q_{Al} = 62.8 \text{ kJ} + 27.0 \text{ kJ} = 89.8 \text{ kJ}. \tag{14.7}$$

Thus, the amount of heat going into heating the pan is

$$\frac{27.0 \text{ kJ}}{89.8 \text{ kJ}} \times 100\% = 30.1\%, \tag{14.8}$$

and the amount going into heating the water is

$$\frac{62.8 \text{ kJ}}{89.8 \text{ kJ}} \times 100\% = 69.9\%. \tag{14.9}$$

Discussion

In this example, the heat transferred to the container is a significant fraction of the total transferred heat. Although the mass of the pan is twice that of the water, the specific heat of water is over four times greater than that of aluminum. Therefore, it takes a bit more than twice the heat to achieve the given temperature change for the water as compared to the aluminum pan.

Figure 14.5 The smoking brakes on this truck are a visible evidence of the mechanical equivalent of heat.

Example 14.2 Calculating the Temperature Increase from the Work Done on a Substance: Truck Brakes Overheat on Downhill Runs

Truck brakes used to control speed on a downhill run do work, converting gravitational potential energy into increased internal energy (higher temperature) of the brake material. This conversion prevents the gravitational potential energy from being converted into kinetic energy of the truck. The problem is that the mass of the truck is large compared with that of the brake material absorbing the energy, and the temperature increase may occur too fast for sufficient heat to transfer from the brakes to the environment.

Calculate the temperature increase of 100 kg of brake material with an average specific heat of $800 \text{ J/kg} \cdot {}^\circ\text{C}$ if the material retains 10% of the energy from a 10,000-kg truck descending 75.0 m (in vertical displacement) at a constant speed.

Strategy

If the brakes are not applied, gravitational potential energy is converted into kinetic energy. When brakes are applied, gravitational potential energy is converted into internal energy of the brake material. We first calculate the gravitational potential energy (Mgh) that the entire truck loses in its descent and then find the temperature increase produced in the brake material alone.

Solution

1. Calculate the change in gravitational potential energy as the truck goes downhill

$$Mgh = (10{,}000 \text{ kg})(9.80 \text{ m/s}^2)(75.0 \text{ m}) = 7.35 \times 10^6 \text{ J}. \tag{14.10}$$

2. Calculate the temperature from the heat transferred using $Q = Mgh$ and

$$\Delta T = \frac{Q}{mc}, \tag{14.11}$$

where m is the mass of the brake material. Insert the values $m = 100 \text{ kg}$ and $c = 800 \text{ J/kg} \cdot {}^\circ\text{C}$ to find

$$\Delta T = \frac{\left(7.35 \times 10^6 \text{ J}\right)}{(100 \text{ kg})(800 \text{ J/kg}{}^\circ\text{C})} = 92{}^\circ\text{C}. \tag{14.12}$$

Discussion

This temperature is close to the boiling point of water. If the truck had been traveling for some time, then just before the descent, the brake temperature would likely be higher than the ambient temperature. The temperature increase in the descent would likely raise the temperature of the brake material above the boiling point of water, so this technique is not practical. However, the same idea underlies the recent hybrid technology of cars, where mechanical energy (gravitational potential energy) is converted by the brakes into electrical energy (battery).

Table 14.1 Specific Heats[1] of Various Substances

Substances	Specific heat (c)	
Solids	J/kg·°C	kcal/kg·°C[2]
Aluminum	900	0.215
Asbestos	800	0.19
Concrete, granite (average)	840	0.20
Copper	387	0.0924
Glass	840	0.20
Gold	129	0.0308
Human body (average at 37 °C)	3500	0.83
Ice (average, -50°C to 0°C)	2090	0.50
Iron, steel	452	0.108
Lead	128	0.0305
Silver	235	0.0562
Wood	1700	0.4
Liquids		
Benzene	1740	0.415
Ethanol	2450	0.586
Glycerin	2410	0.576
Mercury	139	0.0333
Water (15.0 °C)	4186	1.000
Gases [3]		
Air (dry)	721 (1015)	0.172 (0.242)
Ammonia	1670 (2190)	0.399 (0.523)
Carbon dioxide	638 (833)	0.152 (0.199)
Nitrogen	739 (1040)	0.177 (0.248)
Oxygen	651 (913)	0.156 (0.218)
Steam (100°C)	1520 (2020)	0.363 (0.482)

Note that **Example 14.2** is an illustration of the mechanical equivalent of heat. Alternatively, the temperature increase could be produced by a blow torch instead of mechanically.

Example 14.3 Calculating the Final Temperature When Heat Is Transferred Between Two Bodies: Pouring Cold Water in a Hot Pan

Suppose you pour 0.250 kg of $20.0°C$ water (about a cup) into a 0.500-kg aluminum pan off the stove with a temperature of $150°C$. Assume that the pan is placed on an insulated pad and that a negligible amount of water boils off. What is the temperature when the water and pan reach thermal equilibrium a short time later?

Strategy

The pan is placed on an insulated pad so that little heat transfer occurs with the surroundings. Originally the pan and water are not in thermal equilibrium: the pan is at a higher temperature than the water. Heat transfer then restores thermal equilibrium once the water and pan are in contact. Because heat transfer between the pan and water takes place rapidly, the mass of evaporated water is negligible and the magnitude of the heat lost by the pan is equal to the heat gained by the water. The exchange of heat stops once a thermal equilibrium between the pan and the water is achieved. The heat exchange can be written as $|Q_{hot}| = Q_{cold}$.

---

1. The values for solids and liquids are at constant volume and at $25°C$, except as noted.

2. These values are identical in units of $cal/g·°C$.

3. c_v at constant volume and at $20.0°C$, except as noted, and at 1.00 atm average pressure. Values in parentheses are c_p

at a constant pressure of 1.00 atm.

Solution

1. Use the equation for heat transfer $Q = mc\Delta T$ to express the heat lost by the aluminum pan in terms of the mass of the pan, the specific heat of aluminum, the initial temperature of the pan, and the final temperature:

$$Q_{hot} = m_{Al}c_{Al}(T_f - 150°C).$$ (14.13)

2. Express the heat gained by the water in terms of the mass of the water, the specific heat of water, the initial temperature of the water and the final temperature:

$$Q_{cold} = m_W c_W (T_f - 20.0°C).$$ (14.14)

3. Note that $Q_{hot} < 0$ and $Q_{cold} > 0$ and that they must sum to zero because the heat lost by the hot pan must be the same as the heat gained by the cold water:

$$Q_{cold} + Q_{hot} = 0,$$ (14.15)
$$Q_{cold} = -Q_{hot},$$
$$m_W c_W (T_f - 20.0°C) = -m_{Al} c_{Al}(T_f - 150°C.)$$

4. This an equation for the unknown final temperature, T_f

5. Bring all terms involving T_f on the left hand side and all other terms on the right hand side. Solve for T_f,

$$T_f = \frac{m_{Al}c_{Al}(150°C) + m_W c_W(20.0°C)}{m_{Al}c_{Al} + m_W c_W},$$ (14.16)

and insert the numerical values:

$$T_f = \frac{(0.500 \text{ kg})(900 \text{ J/kg°C})(150°C)+(0.250 \text{ kg})(4186 \text{ J/kg°C})(20.0°C)}{(0.500 \text{ kg})(900 \text{ J/kg°C}) + (0.250 \text{ kg})(4186 \text{ J/kg°C})}$$ (14.17)

$$= \frac{88430 \text{ J}}{1496.5 \text{ J/°C}}$$

$$= 59.1°C.$$

Discussion

This is a typical *calorimetry* problem—two bodies at different temperatures are brought in contact with each other and exchange heat until a common temperature is reached. Why is the final temperature so much closer to $20.0°C$ than $150°C$? The reason is that water has a greater specific heat than most common substances and thus undergoes a small temperature change for a given heat transfer. A large body of water, such as a lake, requires a large amount of heat to increase its temperature appreciably. This explains why the temperature of a lake stays relatively constant during a day even when the temperature change of the air is large. However, the water temperature does change over longer times (e.g., summer to winter).

Take-Home Experiment: Temperature Change of Land and Water

What heats faster, land or water?

To study differences in heat capacity:

- Place equal masses of dry sand (or soil) and water at the same temperature into two small jars. (The average density of soil or sand is about 1.6 times that of water, so you can achieve approximately equal masses by using 50% more water by volume.)
- Heat both (using an oven or a heat lamp) for the same amount of time.
- Record the final temperature of the two masses.
- Now bring both jars to the same temperature by heating for a longer period of time.
- Remove the jars from the heat source and measure their temperature every 5 minutes for about 30 minutes.

Which sample cools off the fastest? This activity replicates the phenomena responsible for land breezes and sea breezes.

Check Your Understanding

If 25 kJ is necessary to raise the temperature of a block from $25°C$ to $30°C$, how much heat is necessary to heat the block from $45°C$ to $50°C$?

Solution
The heat transfer depends only on the temperature difference. Since the temperature differences are the same in both cases, the same 25 kJ is necessary in the second case.

14.3 Phase Change and Latent Heat

So far we have discussed temperature change due to heat transfer. No temperature change occurs from heat transfer if ice melts and becomes liquid water (i.e., during a phase change). For example, consider water dripping from icicles melting on a roof warmed by the Sun. Conversely, water freezes in an ice tray cooled by lower-temperature surroundings.

Figure 14.6 Heat from the air transfers to the ice causing it to melt. (credit: Mike Brand)

Energy is required to melt a solid because the cohesive bonds between the molecules in the solid must be broken apart such that, in the liquid, the molecules can move around at comparable kinetic energies; thus, there is no rise in temperature. Similarly, energy is needed to vaporize a liquid, because molecules in a liquid interact with each other via attractive forces. There is no temperature change until a phase change is complete. The temperature of a cup of soda initially at $0°C$ stays at $0°C$ until all the ice has melted. Conversely, energy is released during freezing and condensation, usually in the form of thermal energy. Work is done by cohesive forces when molecules are brought together. The corresponding energy must be given off (dissipated) to allow them to stay together Figure 14.7.

The energy involved in a phase change depends on two major factors: the number and strength of bonds or force pairs. The number of bonds is proportional to the number of molecules and thus to the mass of the sample. The strength of forces depends on the type of molecules. The heat Q required to change the phase of a sample of mass m is given by

$$Q = mL_f \text{ (melting/freezing)},$$ (14.18)

$$Q = mL_v \text{ (vaporization/condensation)},$$ (14.19)

where the latent heat of fusion, L_f, and latent heat of vaporization, L_v, are material constants that are determined experimentally. See (Table 14.2).

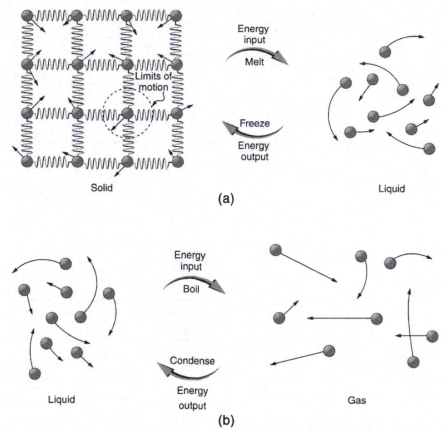

Figure 14.7 (a) Energy is required to partially overcome the attractive forces between molecules in a solid to form a liquid. That same energy must be removed for freezing to take place. (b) Molecules are separated by large distances when going from liquid to vapor, requiring significant energy to overcome molecular attraction. The same energy must be removed for condensation to take place. There is no temperature change until a phase change is complete.

Latent heat is measured in units of J/kg. Both L_f and L_v depend on the substance, particularly on the strength of its molecular forces as noted earlier. L_f and L_v are collectively called **latent heat coefficients**. They are *latent*, or hidden, because in phase changes, energy enters or leaves a system without causing a temperature change in the system; so, in effect, the energy is hidden. **Table 14.2** lists representative values of L_f and L_v, together with melting and boiling points.

The table shows that significant amounts of energy are involved in phase changes. Let us look, for example, at how much energy is needed to melt a kilogram of ice at $0°C$ to produce a kilogram of water at $0°C$. Using the equation for a change in temperature and the value for water from **Table 14.2**, we find that $Q = mL_f = (1.0 \text{ kg})(334 \text{ kJ/kg}) = 334 \text{ kJ}$ is the energy to melt a kilogram of ice. This is a lot of energy as it represents the same amount of energy needed to raise the temperature of 1 kg of liquid water from $0°C$ to $79.8°C$. Even more energy is required to vaporize water; it would take 2256 kJ to change 1 kg of liquid water at the normal boiling point ($100°C$ at atmospheric pressure) to steam (water vapor). This example shows that the energy for a phase change is enormous compared to energy associated with temperature changes without a phase change.

Table 14.2 Heats of Fusion and Vaporization [4]

Substance	Melting point (°C)	L_f		Boiling point (°C)	L_v	
		kJ/kg	kcal/kg		kJ/kg	kcal/kg
Helium	−269.7	5.23	1.25	−268.9	20.9	4.99
Hydrogen	−259.3	58.6	14.0	−252.9	452	108
Nitrogen	−210.0	25.5	6.09	−195.8	201	48.0
Oxygen	−218.8	13.8	3.30	−183.0	213	50.9
Ethanol	−114	104	24.9	78.3	854	204
Ammonia	−75		108	−33.4	1370	327
Mercury	−38.9	11.8	2.82	357	272	65.0
Water	0.00	334	79.8	100.0	2256[5]	539[6]
Sulfur	119	38.1	9.10	444.6	326	77.9
Lead	327	24.5	5.85	1750	871	208
Antimony	631	165	39.4	1440	561	134
Aluminum	660	380	90	2450	11400	2720
Silver	961	88.3	21.1	2193	2336	558
Gold	1063	64.5	15.4	2660	1578	377
Copper	1083	134	32.0	2595	5069	1211
Uranium	1133	84	20	3900	1900	454
Tungsten	3410	184	44	5900	4810	1150

Phase changes can have a tremendous stabilizing effect even on temperatures that are not near the melting and boiling points, because evaporation and condensation (conversion of a gas into a liquid state) occur even at temperatures below the boiling point. Take, for example, the fact that air temperatures in humid climates rarely go above $35.0°C$, which is because most heat transfer goes into evaporating water into the air. Similarly, temperatures in humid weather rarely fall below the dew point because enormous heat is released when water vapor condenses.

We examine the effects of phase change more precisely by considering adding heat into a sample of ice at $−20°C$ (Figure 14.8). The temperature of the ice rises linearly, absorbing heat at a constant rate of $0.50\ \mathrm{cal/g \cdot °C}$ until it reaches $0°C$. Once at this temperature, the ice begins to melt until all the ice has melted, absorbing 79.8 cal/g of heat. The temperature remains constant at $0°C$ during this phase change. Once all the ice has melted, the temperature of the liquid water rises, absorbing heat at a new constant rate of $1.00\ \mathrm{cal/g \cdot °C}$. At $100°C$, the water begins to boil and the temperature again remains constant while the water absorbs 539 cal/g of heat during this phase change. When all the liquid has become steam vapor, the temperature rises again, absorbing heat at a rate of $0.482\ \mathrm{cal/g \cdot °C}$.

---

4. Values quoted at the normal melting and boiling temperatures at standard atmospheric pressure (1 atm).
5. At $37.0°C$ (body temperature), the heat of vaporization L_v for water is 2430 kJ/kg or 580 kcal/kg
6. At $37.0°C$ (body temperature), the heat of vaporization L_v for water is 2430 kJ/kg or 580 kcal/kg

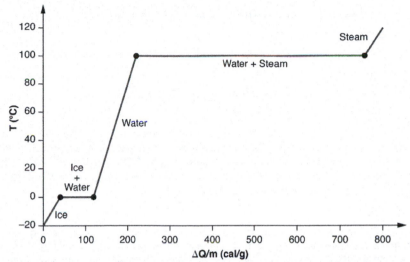

Figure 14.8 A graph of temperature versus energy added. The system is constructed so that no vapor evaporates while ice warms to become liquid water, and so that, when vaporization occurs, the vapor remains in of the system. The long stretches of constant temperature values at $0°C$ and $100°C$ reflect the large latent heat of melting and vaporization, respectively.

Water can evaporate at temperatures below the boiling point. More energy is required than at the boiling point, because the kinetic energy of water molecules at temperatures below $100°C$ is less than that at $100°C$, hence less energy is available from random thermal motions. Take, for example, the fact that, at body temperature, perspiration from the skin requires a heat input of 2428 kJ/kg, which is about 10 percent higher than the latent heat of vaporization at $100°C$. This heat comes from the skin, and thus provides an effective cooling mechanism in hot weather. High humidity inhibits evaporation, so that body temperature might rise, leaving unevaporated sweat on your brow.

Example 14.4 Calculate Final Temperature from Phase Change: Cooling Soda with Ice Cubes

Three ice cubes are used to chill a soda at $20°C$ with mass $m_{soda} = 0.25$ kg. The ice is at $0°C$ and each ice cube has a mass of 6.0 g. Assume that the soda is kept in a foam container so that heat loss can be ignored. Assume the soda has the same heat capacity as water. Find the final temperature when all ice has melted.

Strategy

The ice cubes are at the melting temperature of $0°C$. Heat is transferred from the soda to the ice for melting. Melting of ice occurs in two steps: first the phase change occurs and solid (ice) transforms into liquid water at the melting temperature, then the temperature of this water rises. Melting yields water at $0°C$, so more heat is transferred from the soda to this water until the water plus soda system reaches thermal equilibrium,

$$Q_{ice} = -Q_{soda}. \tag{14.20}$$

The heat transferred to the ice is $Q_{ice} = m_{ice}L_f + m_{ice}c_W(T_f - 0°C)$. The heat given off by the soda is $Q_{soda} = m_{soda}c_W(T_f - 20°C)$. Since no heat is lost, $Q_{ice} = -Q_{soda}$, so that

$$m_{ice}L_f + m_{ice}c_W(T_f - 0°C) = -m_{soda}c_W(T_f - 20°C). \tag{14.21}$$

Bring all terms involving T_f on the left-hand-side and all other terms on the right-hand-side. Solve for the unknown quantity T_f:

$$T_f = \frac{m_{soda}c_W(20°C) - m_{ice}L_f}{(m_{soda} + m_{ice})c_W}. \tag{14.22}$$

Solution

1. Identify the known quantities. The mass of ice is $m_{ice} = 3 \times 6.0$ g $= 0.018$ kg and the mass of soda is $m_{soda} = 0.25$ kg.

2. Calculate the terms in the numerator:
$$m_{soda}c_W(20°C) = (0.25 \text{ kg})(4186 \text{ J/kg·°C})(20°C) = 20{,}930 \text{ J} \tag{14.23}$$
and

$$m_{\text{ice}}L_f = (0.018 \text{ kg})(334{,}000 \text{ J/kg}) = 6012 \text{ J}. \tag{14.24}$$

3. Calculate the denominator:

$$(m_{\text{soda}} + m_{\text{ice}})c_W = (0.25 \text{ kg} + 0.018 \text{ kg})(4186 \text{ K/(kg·°C)} = 1122 \text{ J/°C}. \tag{14.25}$$

4. Calculate the final temperature:

$$T_f = \frac{20{,}930 \text{ J} - 6012 \text{ J}}{1122 \text{ J/°C}} = 13°\text{C}. \tag{14.26}$$

Discussion

This example illustrates the enormous energies involved during a phase change. The mass of ice is about 7 percent the mass of water but leads to a noticeable change in the temperature of soda. Although we assumed that the ice was at the freezing temperature, this is incorrect: the typical temperature is $-6°\text{C}$. However, this correction gives a final temperature that is essentially identical to the result we found. Can you explain why?

We have seen that vaporization requires heat transfer to a liquid from the surroundings, so that energy is released by the surroundings. Condensation is the reverse process, increasing the temperature of the surroundings. This increase may seem surprising, since we associate condensation with cold objects—the glass in the figure, for example. However, energy must be removed from the condensing molecules to make a vapor condense. The energy is exactly the same as that required to make the phase change in the other direction, from liquid to vapor, and so it can be calculated from $Q = mL_v$.

Figure 14.9 Condensation forms on this glass of iced tea because the temperature of the nearby air is reduced to below the dew point. The air cannot hold as much water as it did at room temperature, and so water condenses. Energy is released when the water condenses, speeding the melting of the ice in the glass. (credit: Jenny Downing)

Real-World Application

Energy is also released when a liquid freezes. This phenomenon is used by fruit growers in Florida to protect oranges when the temperature is close to the freezing point $(0°\text{C})$. Growers spray water on the plants in orchards so that the water freezes and heat is released to the growing oranges on the trees. This prevents the temperature inside the orange from dropping below freezing, which would damage the fruit.

Figure 14.10 The ice on these trees released large amounts of energy when it froze, helping to prevent the temperature of the trees from dropping below $0°C$. Water is intentionally sprayed on orchards to help prevent hard frosts. (credit: Hermann Hammer)

Sublimation is the transition from solid to vapor phase. You may have noticed that snow can disappear into thin air without a trace of liquid water, or the disappearance of ice cubes in a freezer. The reverse is also true: Frost can form on very cold windows without going through the liquid stage. A popular effect is the making of "smoke" from dry ice, which is solid carbon dioxide. Sublimation occurs because the equilibrium vapor pressure of solids is not zero. Certain air fresheners use the sublimation of a solid to inject a perfume into the room. Moth balls are a slightly toxic example of a phenol (an organic compound) that sublimates, while some solids, such as osmium tetroxide, are so toxic that they must be kept in sealed containers to prevent human exposure to their sublimation-produced vapors.

Figure 14.11 Direct transitions between solid and vapor are common, sometimes useful, and even beautiful. (a) Dry ice sublimates directly to carbon dioxide gas. The visible vapor is made of water droplets. (credit: Windell Oskay) (b) Frost forms patterns on a very cold window, an example of a solid formed directly from a vapor. (credit: Liz West)

All phase transitions involve heat. In the case of direct solid-vapor transitions, the energy required is given by the equation $Q = mL_s$, where L_s is the **heat of sublimation**, which is the energy required to change 1.00 kg of a substance from the solid phase to the vapor phase. L_s is analogous to L_f and L_v, and its value depends on the substance. Sublimation requires energy input, so that dry ice is an effective coolant, whereas the reverse process (i.e., frosting) releases energy. The amount of energy required for sublimation is of the same order of magnitude as that for other phase transitions.

The material presented in this section and the preceding section allows us to calculate any number of effects related to temperature and phase change. In each case, it is necessary to identify which temperature and phase changes are taking place and then to apply the appropriate equation. Keep in mind that heat transfer and work can cause both temperature and phase changes.

Problem-Solving Strategies for the Effects of Heat Transfer

1. *Examine the situation to determine that there is a change in the temperature or phase. Is there heat transfer into or out of the system?* When the presence or absence of a phase change is not obvious, you may wish to first solve the problem as if there were no phase changes, and examine the temperature change obtained. If it is sufficient to take you past a boiling or melting point, you should then go back and do the problem in steps—temperature change, phase change, subsequent temperature change, and so on.

2. *Identify and list all objects that change temperature and phase.*

3. *Identify exactly what needs to be determined in the problem (identify the unknowns).* A written list is useful.

4. *Make a list of what is given or what can be inferred from the problem as stated (identify the knowns).*

5. *Solve the appropriate equation for the quantity to be determined (the unknown).* If there is a temperature change, the transferred heat depends on the specific heat (see **Table 14.1**) whereas, for a phase change, the transferred heat depends on the latent heat. See **Table 14.2**.

6. *Substitute the knowns along with their units into the appropriate equation and obtain numerical solutions complete with units.* You will need to do this in steps if there is more than one stage to the process (such as a temperature change followed by a phase change).

7. *Check the answer to see if it is reasonable: Does it make sense?* As an example, be certain that the temperature change does not also cause a phase change that you have not taken into account.

Check Your Understanding

Why does snow remain on mountain slopes even when daytime temperatures are higher than the freezing temperature?

Solution
Snow is formed from ice crystals and thus is the solid phase of water. Because enormous heat is necessary for phase changes, it takes a certain amount of time for this heat to be accumulated from the air, even if the air is above $0°C$. The warmer the air is, the faster this heat exchange occurs and the faster the snow melts.

14.4 Heat Transfer Methods

Learning Objectives

By the end of this section, you will be able to:

• Discuss the different methods of heat transfer.

Equally as interesting as the effects of heat transfer on a system are the methods by which this occurs. Whenever there is a temperature difference, heat transfer occurs. Heat transfer may occur rapidly, such as through a cooking pan, or slowly, such as through the walls of a picnic ice chest. We can control rates of heat transfer by choosing materials (such as thick wool clothing for the winter), controlling air movement (such as the use of weather stripping around doors), or by choice of color (such as a white roof to reflect summer sunlight). So many processes involve heat transfer, so that it is hard to imagine a situation where no heat transfer occurs. Yet every process involving heat transfer takes place by only three methods:

1. **Conduction** is heat transfer through stationary matter by physical contact. (The matter is stationary on a macroscopic scale—we know there is thermal motion of the atoms and molecules at any temperature above absolute zero.) Heat transferred between the electric burner of a stove and the bottom of a pan is transferred by conduction.

2. **Convection** is the heat transfer by the macroscopic movement of a fluid. This type of transfer takes place in a forced-air furnace and in weather systems, for example.

3. Heat transfer by **radiation** occurs when microwaves, infrared radiation, visible light, or another form of electromagnetic radiation is emitted or absorbed. An obvious example is the warming of the Earth by the Sun. A less obvious example is thermal radiation from the human body.

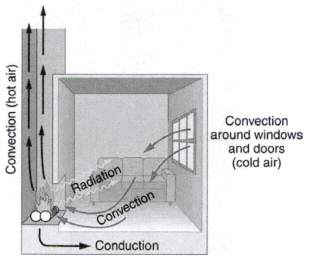

Figure 14.12 In a fireplace, heat transfer occurs by all three methods: conduction, convection, and radiation. Radiation is responsible for most of the heat transferred into the room. Heat transfer also occurs through conduction into the room, but at a much slower rate. Heat transfer by convection also occurs through cold air entering the room around windows and hot air leaving the room by rising up the chimney.

We examine these methods in some detail in the three following modules. Each method has unique and interesting characteristics, but all three do have one thing in common: they transfer heat solely because of a temperature difference **Figure 14.12**.

Check Your Understanding

Name an example from daily life (different from the text) for each mechanism of heat transfer.

Solution

Conduction: Heat transfers into your hands as you hold a hot cup of coffee.

Convection: Heat transfers as the barista "steams" cold milk to make hot *cocoa*.

Radiation: Reheating a cold cup of coffee in a microwave oven.

14.5 Conduction

Learning Objectives

By the end of this section, you will be able to:

- Calculate thermal conductivity.
- Observe conduction of heat in collisions.
- Study thermal conductivities of common substances.

The information presented in this section supports the following AP® learning objectives and science practices:

- **1.E.3.1** The student is able to design an experiment and analyze data from it to examine thermal conductivity. **(S.P. 4.1, 4.2, 5.1)**
- **5.B.6.1** The student is able to describe the models that represent processes by which energy can be transferred between a system and its environment because of differences in temperature: conduction, convection, and radiation. **(S.P. 1.2)**

Figure 14.13 Insulation is used to limit the conduction of heat from the inside to the outside (in winters) and from the outside to the inside (in summers). (credit: Giles Douglas)

Your feet feel cold as you walk barefoot across the living room carpet in your cold house and then step onto the kitchen tile floor. This result is intriguing, since the carpet and tile floor are both at the same temperature. The different sensation you feel is explained by the different rates of heat transfer: the heat loss during the same time interval is greater for skin in contact with the tiles than with the carpet, so the temperature drop is greater on the tiles.

Some materials conduct thermal energy faster than others. In general, good conductors of electricity (metals like copper, aluminum, gold, and silver) are also good heat conductors, whereas insulators of electricity (wood, plastic, and rubber) are poor heat conductors. Figure 14.14 shows molecules in two bodies at different temperatures. The (average) kinetic energy of a molecule in the hot body is higher than in the colder body. If two molecules collide, an energy transfer from the molecule with greater kinetic energy to the molecule with less kinetic energy occurs. The cumulative effect from all collisions results in a net flux of heat from the hot body to the colder body. The heat flux thus depends on the temperature difference $\Delta T = T_{\text{hot}} - T_{\text{cold}}$.

Therefore, you will get a more severe burn from boiling water than from hot tap water. Conversely, if the temperatures are the same, the net heat transfer rate falls to zero, and equilibrium is achieved. Owing to the fact that the number of collisions increases with increasing area, heat conduction depends on the cross-sectional area. If you touch a cold wall with your palm, your hand cools faster than if you just touch it with your fingertip.

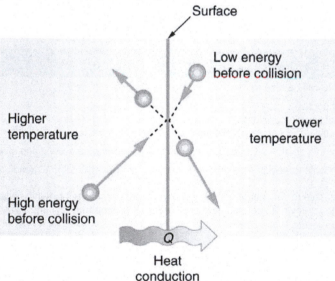

Figure 14.14 The molecules in two bodies at different temperatures have different average kinetic energies. Collisions occurring at the contact surface tend to transfer energy from high-temperature regions to low-temperature regions. In this illustration, a molecule in the lower temperature region (right side) has low energy before collision, but its energy increases after colliding with the contact surface. In contrast, a molecule in the higher temperature region (left side) has high energy before collision, but its energy decreases after colliding with the contact surface.

A third factor in the mechanism of conduction is the thickness of the material through which heat transfers. The figure below shows a slab of material with different temperatures on either side. Suppose that T_2 is greater than T_1, so that heat is

transferred from left to right. Heat transfer from the left side to the right side is accomplished by a series of molecular collisions. The thicker the material, the more time it takes to transfer the same amount of heat. This model explains why thick clothing is warmer than thin clothing in winters, and why Arctic mammals protect themselves with thick blubber.

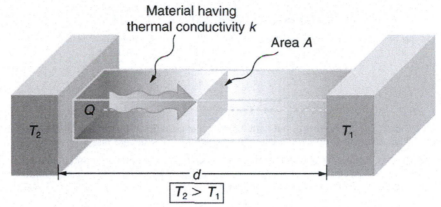

Figure 14.15 Heat conduction occurs through any material, represented here by a rectangular bar, whether window glass or walrus blubber. The temperature of the material is T_2 on the left and T_1 on the right, where T_2 is greater than T_1. The rate of heat transfer by conduction is directly proportional to the surface area A, the temperature difference $T_2 - T_1$, and the substance's conductivity k. The rate of heat transfer is inversely proportional to the thickness d.

Lastly, the heat transfer rate depends on the material properties described by the coefficient of thermal conductivity. All four factors are included in a simple equation that was deduced from and is confirmed by experiments. The **rate of conductive heat transfer** through a slab of material, such as the one in **Figure 14.15**, is given by

$$\frac{Q}{t} = \frac{kA(T_2 - T_1)}{d},$$

(14.27)

where Q/t is the rate of heat transfer in watts or kilocalories per second, k is the **thermal conductivity** of the material, A and d are its surface area and thickness, as shown in **Figure 14.15**, and $(T_2 - T_1)$ is the temperature difference across the slab. **Table 14.3** gives representative values of thermal conductivity.

Example 14.5 Calculating Heat Transfer Through Conduction: Conduction Rate Through an Ice Box

A Styrofoam ice box has a total area of $0.950 \ \text{m}^2$ and walls with an average thickness of 2.50 cm. The box contains ice, water, and canned beverages at $0°C$. The inside of the box is kept cold by melting ice. How much ice melts in one day if the ice box is kept in the trunk of a car at $35.0°C$?

Strategy

This question involves both heat for a phase change (melting of ice) and the transfer of heat by conduction. To find the amount of ice melted, we must find the net heat transferred. This value can be obtained by calculating the rate of heat transfer by conduction and multiplying by time.

Solution

1. Identify the knowns.

$$A = 0.950 \ \text{m}^2; d = 2.50 \ \text{cm} = 0.0250 \ \text{m}; T_1 = 0°C; T_2 = 35.0°C, t = 1 \ \text{day} = 24 \ \text{hours} = 86{,}400 \ \text{s}.$$ (14.28)

2. Identify the unknowns. We need to solve for the mass of the ice, m. We will also need to solve for the net heat transferred to melt the ice, Q.

3. Determine which equations to use. The rate of heat transfer by conduction is given by

$$\frac{Q}{t} = \frac{kA(T_2 - T_1)}{d}.$$ (14.29)

4. The heat is used to melt the ice: $Q = mL_f$.

5. Insert the known values:

$$\frac{Q}{t} = \frac{(0.010 \ \text{J/s} \cdot \text{m·°C})(0.950 \ \text{m}^2)(35.0°C - 0°C)}{0.0250 \ \text{m}} = 13.3 \ \text{J/s}.$$ (14.30)

6. Multiply the rate of heat transfer by the time ($1 \ \text{day} = 86{,}400 \ \text{s}$):

$$Q = (Q/t)t = (13.3 \text{ J/s})(86,400 \text{ s}) = 1.15 \times 10^6 \text{ J}. \qquad (14.31)$$

7.   Set this equal to the heat transferred to melt the ice: $Q = mL_f$. Solve for the mass m:

$$m = \frac{Q}{L_f} = \frac{1.15 \times 10^6 \text{ J}}{334 \times 10^3 \text{ J/kg}} = 3.44 \text{kg}. \qquad (14.32)$$

Discussion

The result of 3.44 kg, or about 7.6 lbs, seems about right, based on experience. You might expect to use about a 4 kg (7–10 lb) bag of ice per day. A little extra ice is required if you add any warm food or beverages.

Inspecting the conductivities in Table 14.3 shows that Styrofoam is a very poor conductor and thus a good insulator. Other good insulators include fiberglass, wool, and goose-down feathers. Like Styrofoam, these all incorporate many small pockets of air, taking advantage of air's poor thermal conductivity.

Table 14.3 Thermal Conductivities of Common Substances[7]

Substance	Thermal conductivity k (J/s·m·°C)
Silver	420
Copper	390
Gold	318
Aluminum	220
Steel iron	80
Steel (stainless)	14
Ice	2.2
Glass (average)	0.84
Concrete brick	0.84
Water	0.6
Fatty tissue (without blood)	0.2
Asbestos	0.16
Plasterboard	0.16
Wood	0.08–0.16
Snow (dry)	0.10
Cork	0.042
Glass wool	0.042
Wool	0.04
Down feathers	0.025
Air	0.023
Styrofoam	0.010

A combination of material and thickness is often manipulated to develop good insulators—the smaller the conductivity k and the larger the thickness d, the better. The ratio of d/k will thus be large for a good insulator. The ratio d/k is called the **R factor**. The rate of conductive heat transfer is inversely proportional to R. The larger the value of R, the better the insulation. R factors are most commonly quoted for household insulation, refrigerators, and the like—unfortunately, it is still in non-metric units of ft^2·°F·h/Btu, although the unit usually goes unstated (1 British thermal unit [Btu] is the amount of energy needed to change the temperature of 1.0 lb of water by 1.0 °F). A couple of representative values are an R factor of 11 for 3.5-in-thick fiberglass batts (pieces) of insulation and an R factor of 19 for 6.5-in-thick fiberglass batts. Walls are usually insulated with 3.5-in batts, while ceilings are usually insulated with 6.5-in batts. In cold climates, thicker batts may be used in ceilings and walls.

---

7. At temperatures near 0°C.

Figure 14.16 The fiberglass batt is used for insulation of walls and ceilings to prevent heat transfer between the inside of the building and the outside environment.

Note that in **Table 14.3**, the best thermal conductors—silver, copper, gold, and aluminum—are also the best electrical conductors, again related to the density of free electrons in them. Cooking utensils are typically made from good conductors.

Example 14.6 Calculating the Temperature Difference Maintained by a Heat Transfer: Conduction Through an Aluminum Pan

Water is boiling in an aluminum pan placed on an electrical element on a stovetop. The sauce pan has a bottom that is 0.800 cm thick and 14.0 cm in diameter. The boiling water is evaporating at the rate of 1.00 g/s. What is the temperature difference across (through) the bottom of the pan?

Strategy

Conduction through the aluminum is the primary method of heat transfer here, and so we use the equation for the rate of heat transfer and solve for the temperature difference.

$$T_2 - T_1 = \frac{Q}{t}\left(\frac{d}{kA}\right). \tag{14.33}$$

Solution

1. Identify the knowns and convert them to the SI units.

   The thickness of the pan, $d = 0.800 \text{ cm} = 8.0 \times 10^{-3}$ m, the area of the pan,

   $A = \pi(0.14/2)^2 \text{ m}^2 = 1.54 \times 10^{-2} \text{ m}^2$, and the thermal conductivity, $k = 220 \text{ J/s} \cdot \text{m} \cdot °\text{C}$.

2. Calculate the necessary heat of vaporization of 1 g of water:

$$Q = mL_v = \left(1.00 \times 10^{-3} \text{ kg}\right)\left(2256 \times 10^3 \text{ J/kg}\right) = 2256 \text{ J}. \tag{14.34}$$

3. Calculate the rate of heat transfer given that 1 g of water melts in one second:

$$Q/t = 2256 \text{ J/s or } 2.26 \text{ kW}. \tag{14.35}$$

4. Insert the knowns into the equation and solve for the temperature difference:

$$T_2 - T_1 = \frac{Q}{t}\left(\frac{d}{kA}\right) = (2256 \text{ J/s})\frac{8.00 \times 10^{-3}\text{m}}{(220 \text{ J/s} \cdot \text{m} \cdot °\text{C})\left(1.54 \times 10^{-2} \text{ m}^2\right)} = 5.33 °\text{C}. \tag{14.36}$$

Discussion

The value for the heat transfer $Q/t = 2.26\text{kW}$ or 2256 J/s is typical for an electric stove. This value gives a remarkably small temperature difference between the stove and the pan. Consider that the stove burner is red hot while the inside of the pan is nearly $100°\text{C}$ because of its contact with boiling water. This contact effectively cools the bottom of the pan in spite of its proximity to the very hot stove burner. Aluminum is such a good conductor that it only takes this small temperature difference to produce a heat transfer of 2.26 kW into the pan.

Conduction is caused by the random motion of atoms and molecules. As such, it is an ineffective mechanism for heat transport over macroscopic distances and short time distances. Take, for example, the temperature on the Earth, which would be unbearably cold during the night and extremely hot during the day if heat transport in the atmosphere was to be only through conduction. In another example, car engines would overheat unless there was a more efficient way to remove excess heat from the pistons.

Check Your Understanding

How does the rate of heat transfer by conduction change when all spatial dimensions are doubled?

Solution

Because area is the product of two spatial dimensions, it increases by a factor of four when each dimension is doubled

$\left(A_{\text{fina}} = (2d)^2 = 4d^2 = 4A_{\text{initial}}\right)$. The distance, however, simply doubles. Because the temperature difference and the coefficient of thermal conductivity are independent of the spatial dimensions, the rate of heat transfer by conduction increases by a factor of four divided by two, or two:

$$\left(\frac{Q}{t}\right)_{\text{fina}} = \frac{kA_{\text{fina}}(T_2 - T_1)}{d_{\text{fina}}} = \frac{k(4A_{\text{initial}})(T_2 - T_1)}{2d_{\text{initial}}} = 2\frac{kA_{\text{initial}}(T_2 - T_1)}{d_{\text{initial}}} = 2\left(\frac{Q}{t}\right)_{\text{initial}}. \tag{14.37}$$

Applying the Science Practices: Estimating Thermal Conductivity

The following equipment and materials are available to you for a thermal conductivity experiment:

1 high-density polyethylene cylindrical container
1 steel cylindrical container
1 glass cylindrical container
3 cork stoppers
3 glass thermometers
1 small incubator
1 cork base (2 cm thick)
crushed ice
1 digital timer
1 metric balance
1 meter stick or ruler
1 Vernier caliper
1 micrometer

Notes: The three cylindrical containers have equal volumes and are tested in sequence. All cork stoppers fit snugly into the open tops of the containers and have small holes through which a thermometer can be placed securely. There is enough ice to fill each of the containers. Each container with thermometer fits inside the incubator on the cork base. The incubator has been uniformly pre-heated to a temperature of 40°C. The thermometers can be observed through the incubator window

Exercise 14.1

Describe an experimental procedure to estimate the thermal conductivity (k) for each of the container materials. Point out what properties need to be measured, and how the available equipment can be used to make all of the necessary measurements. Identify sources of error in the measurements. Explain the purpose of the cork stoppers and base, the reason for using the incubator, and when the timer should be started and stopped. Draw a labeled diagram of your setup to help in your description. Include enough detail so that another student could carry out your procedure. For assistance, review the information and analysis in 'Example 14.5: Calculating Heat Transfer through Conduction.'

Solution
The dimensions of each container are measured, so that the side surface area (A) and the thickness of the sides (d) are determined. Weigh an empty container, fill it with ice, weigh it again, insert the cork stopper, and insert the thermometer through the stopper so that the bulb is near the bottom of the container. Place the container in the incubator on the cork base. The incubator provides a uniform high temperature, which evenly surrounds the container and will melt the ice within 20 minutes. Because the cork is an effective insulator, most of the heat transfer will occur through the sides of the containers. By using the incubator temperature of 40° (T_2) and the temperature of the ice $(T_1 = 0°\ C)$, the

temperature difference is uniform until all of the ice melts, and the temperature of the water in the container rises. During the time (t) the container is placed in the incubator and the ice completely melts, the amount of heat transferred into the container is almost all of the heat needed to melt the ice, which equals the mass of the ice (m) multiplied by the latent heat of ice (L_f). By using all the measured quantities in the rearranged equation (14.26) for thermal conductivity,

$k = \dfrac{d(mL_f)}{tA(T_2 - T_1)}$, the thermal conductivity (k) can be estimated for each of the container materials. Sources of error

include the measurements of the length and radius of the container, which are affected by the precision of the calipers and meter stick, and the measurement of the container thickness, which is made with the micrometer. The mass of the ice, the measured temperatures, and the time interval are also subject to precision limits of the balance, thermometers, and timer, respectively.

14.6 Convection

Learning Objectives
By the end of this section, you will be able to: • Discuss the method of heat transfer by convection. The information presented in this section supports the following AP® learning objectives and science practices: • **5.B.6.1** The student is able to describe the models that represent processes by which energy can be transferred between a system and its environment because of differences in temperature: conduction, convection, and radiation. **(S.P. 1.2)**

Convection is driven by large-scale flow of matter. In the case of Earth, the atmospheric circulation is caused by the flow of hot air from the tropics to the poles, and the flow of cold air from the poles toward the tropics. (Note that Earth's rotation causes the observed easterly flow of air in the northern hemisphere). Car engines are kept cool by the flow of water in the cooling system, with the water pump maintaining a flow of cool water to the pistons. The circulatory system is used the body: when the body overheats, the blood vessels in the skin expand (dilate), which increases the blood flow to the skin where it can be cooled by sweating. These vessels become smaller when it is cold outside and larger when it is hot (so more fluid flows, and more energy is transferred).

The body also loses a significant fraction of its heat through the breathing process.

While convection is usually more complicated than conduction, we can describe convection and do some straightforward, realistic calculations of its effects. Natural convection is driven by buoyant forces: hot air rises because density decreases as temperature increases. The house in **Figure 14.17** is kept warm in this manner, as is the pot of water on the stove in **Figure 14.18**. Ocean currents and large-scale atmospheric circulation transfer energy from one part of the globe to another. Both are examples of natural convection.

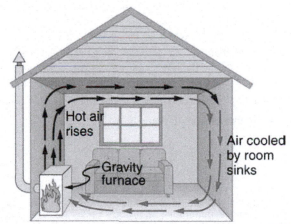

Figure 14.17 Air heated by the so-called gravity furnace expands and rises, forming a convective loop that transfers energy to other parts of the room. As the air is cooled at the ceiling and outside walls, it contracts, eventually becoming denser than room air and sinking to the floor. A properly designed heating system using natural convection, like this one, can be quite efficient in uniformly heating a home.

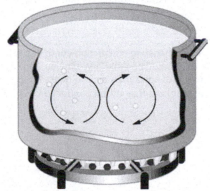

Figure 14.18 Convection plays an important role in heat transfer inside this pot of water. Once conducted to the inside, heat transfer to other parts of the pot is mostly by convection. The hotter water expands, decreases in density, and rises to transfer heat to other regions of the water, while colder water sinks to the bottom. This process keeps repeating.

Take-Home Experiment: Convection Rolls in a Heated Pan

Take two small pots of water and use an eye dropper to place a drop of food coloring near the bottom of each. Leave one on a bench top and heat the other over a stovetop. Watch how the color spreads and how long it takes the color to reach the top. Watch how convective loops form.

Example 14.7 Calculating Heat Transfer by Convection: Convection of Air Through the Walls of a House

Most houses are not airtight: air goes in and out around doors and windows, through cracks and crevices, following wiring to switches and outlets, and so on. The air in a typical house is completely replaced in less than an hour. Suppose that a moderately-sized house has inside dimensions $12.0 \text{m} \times 18.0 \text{m} \times 3.00 \text{m}$ high, and that all air is replaced in 30.0 min. Calculate the heat transfer per unit time in watts needed to warm the incoming cold air by $10.0°C$, thus replacing the heat transferred by convection alone.

Strategy

Heat is used to raise the temperature of air so that $Q = mc\Delta T$. The rate of heat transfer is then Q/t, where t is the time for air turnover. We are given that ΔT is $10.0°C$, but we must still find values for the mass of air and its specific heat before we can calculate Q. The specific heat of air is a weighted average of the specific heats of nitrogen and oxygen, which gives $c = c_\text{p} \cong 1000 \text{ J/kg·°C}$ from **Table 14.4** (note that the specific heat at constant pressure must be used for this process).

Solution

1. Determine the mass of air from its density and the given volume of the house. The density is given from the density ρ and the volume

$$m = \rho V = \left(1.29 \text{ kg/m}^3\right)(12.0 \text{ m} \times 18.0 \text{ m} \times 3.00 \text{ m}) = 836 \text{ kg}. \tag{14.38}$$

2. Calculate the heat transferred from the change in air temperature: $Q = mc\Delta T$ so that

$$Q = (836 \text{ kg})(1000 \text{ J/kg·°C})(10.0°C) = 8.36 \times 10^6 \text{ J}. \tag{14.39}$$

3. Calculate the heat transfer from the heat Q and the turnover time t. Since air is turned over in $t = 0.500 \text{ h} = 1800 \text{ s}$, the heat transferred per unit time is

$$\frac{Q}{t} = \frac{8.36 \times 10^6 \text{ J}}{1800 \text{ s}} = 4.64 \text{ kW}. \tag{14.40}$$

Discussion

This rate of heat transfer is equal to the power consumed by about forty-six 100-W light bulbs. Newly constructed homes are designed for a turnover time of 2 hours or more, rather than 30 minutes for the house of this example. Weather stripping, caulking, and improved window seals are commonly employed. More extreme measures are sometimes taken in very cold (or hot) climates to achieve a tight standard of more than 6 hours for one air turnover. Still longer turnover times are unhealthy, because a minimum amount of fresh air is necessary to supply oxygen for breathing and to dilute household pollutants. The term used for the process by which outside air leaks into the house from cracks around windows, doors, and the foundation is called "air infiltration."

A cold wind is much more chilling than still cold air, because convection combines with conduction in the body to increase the rate at which energy is transferred away from the body. The table below gives approximate wind-chill factors, which are the temperatures of still air that produce the same rate of cooling as air of a given temperature and speed. Wind-chill factors are a dramatic reminder of convection's ability to transfer heat faster than conduction. For example, a 15.0 m/s wind at $0°C$ has the chilling equivalent of still air at about $-18°C$.

Table 14.4 Wind-Chill Factors

Moving air temperature	Wind speed (m/s)				
(°C)	2	5	10	15	20
5	3	−1	−8	−10	−12
2	0	−7	−12	−16	−18
0	−2	−9	−15	−18	−20
−5	−7	−15	−22	−26	−29
−10	−12	−21	−29	−34	−36
−20	−23	−34	−44	−50	−52
−40	−44	−59	−73	−82	−84

Although air can transfer heat rapidly by convection, it is a poor conductor and thus a good insulator. The amount of available space for airflow determines whether air acts as an insulator or conductor. The space between the inside and outside walls of a house, for example, is about 9 cm (3.5 in) —large enough for convection to work effectively. The addition of wall insulation prevents airflow, so heat loss (or gain) is decreased. Similarly, the gap between the two panes of a double-paned window is about 1 cm, which prevents convection and takes advantage of air's low conductivity to prevent greater loss. Fur, fiber, and fiberglass also take advantage of the low conductivity of air by trapping it in spaces too small to support convection, as shown in the figure. Fur and feathers are lightweight and thus ideal for the protection of animals.

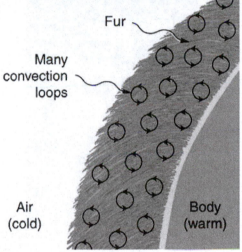

Figure 14.19 Fur is filled with air, breaking it up into many small pockets. Convection is very slow here, because the loops are so small. The low conductivity of air makes fur a very good lightweight insulator.

Some interesting phenomena happen *when convection is accompanied by a phase change*. It allows us to cool off by sweating, even if the temperature of the surrounding air exceeds body temperature. Heat from the skin is required for sweat to evaporate from the skin, but without air flow, the air becomes saturated and evaporation stops. Air flow caused by convection replaces the saturated air by dry air and evaporation continues.

Example 14.8 Calculate the Flow of Mass during Convection: Sweat-Heat Transfer away from the Body

The average person produces heat at the rate of about 120 W when at rest. At what rate must water evaporate from the body to get rid of all this energy? (This evaporation might occur when a person is sitting in the shade and surrounding temperatures are the same as skin temperature, eliminating heat transfer by other methods.)

Strategy

Energy is needed for a phase change ($Q = mL_v$). Thus, the energy loss per unit time is

$$\frac{Q}{t} = \frac{mL_v}{t} = 120 \ \text{W} = 120 \ \text{J/s}.$$

(14.41)

We divide both sides of the equation by L_v to find that the mass evaporated per unit time is

$$\frac{m}{t} = \frac{120 \text{ J/s}}{L_v}. \tag{14.42}$$

Solution

(1) Insert the value of the latent heat from **Table 14.2**, $L_v = 2430 \text{ kJ/kg} = 2430 \text{ J/g}$. This yields

$$\frac{m}{t} = \frac{120 \text{ J/s}}{2430 \text{ J/g}} = 0.0494 \text{ g/s} = 2.96 \text{ g/min}. \tag{14.43}$$

Discussion

Evaporating about 3 g/min seems reasonable. This would be about 180 g (about 7 oz) per hour. If the air is very dry, the sweat may evaporate without even being noticed. A significant amount of evaporation also takes place in the lungs and breathing passages.

Another important example of the combination of phase change and convection occurs when water evaporates from the oceans. Heat is removed from the ocean when water evaporates. If the water vapor condenses in liquid droplets as clouds form, heat is released in the atmosphere. Thus, there is an overall transfer of heat from the ocean to the atmosphere. This process is the driving power behind thunderheads, those great cumulus clouds that rise as much as 20.0 km into the stratosphere. Water vapor carried in by convection condenses, releasing tremendous amounts of energy. This energy causes the air to expand and rise, where it is colder. More condensation occurs in these colder regions, which in turn drives the cloud even higher. Such a mechanism is called positive feedback, since the process reinforces and accelerates itself. These systems sometimes produce violent storms, with lightning and hail, and constitute the mechanism driving hurricanes.

Figure 14.20 Cumulus clouds are caused by water vapor that rises because of convection. The rise of clouds is driven by a positive feedback mechanism. (credit: Mike Love)

Figure 14.21 Convection accompanied by a phase change releases the energy needed to drive this thunderhead into the stratosphere. (credit: Gerardo García Moretti)

Figure 14.22 The phase change that occurs when this iceberg melts involves tremendous heat transfer. (credit: Dominic Alves)

The movement of icebergs is another example of convection accompanied by a phase change. Suppose an iceberg drifts from Greenland into warmer Atlantic waters. Heat is removed from the warm ocean water when the ice melts and heat is released to the land mass when the iceberg forms on Greenland.

Check Your Understanding

Explain why using a fan in the summer feels refreshing!

Solution
Using a fan increases the flow of air: warm air near your body is replaced by cooler air from elsewhere. Convection increases the rate of heat transfer so that moving air "feels" cooler than still air.

14.7 Radiation

Learning Objectives

By the end of this section, you will be able to:

- Discuss heat transfer by radiation.
- Explain the radiant power of different materials.

The information presented in this section supports the following AP® learning objectives and science practices:

- **5.B.6.1** The student is able to describe the models that represent processes by which energy can be transferred between a system and its environment because of differences in temperature: conduction, convection, and radiation. **(S.P. 1.2)**

You can feel the heat transfer from a fire and from the Sun. Similarly, you can sometimes tell that the oven is hot without touching its door or looking inside—it may just warm you as you walk by. The space between the Earth and the Sun is largely empty, without any possibility of heat transfer by convection or conduction. In these examples, heat is transferred by radiation. That is, the hot body emits electromagnetic waves that are absorbed by our skin: no medium is required for electromagnetic waves to propagate. Different names are used for electromagnetic waves of different wavelengths: radio waves, microwaves, infrared **radiation**, visible light, ultraviolet radiation, X-rays, and gamma rays.

Figure 14.23 Most of the heat transfer from this fire to the observers is through infrared radiation. The visible light, although dramatic, transfers relatively little thermal energy. Convection transfers energy away from the observers as hot air rises, while conduction is negligibly slow here. Skin is very sensitive to infrared radiation, so that you can sense the presence of a fire without looking at it directly. (credit: Daniel X. O'Neil)

The energy of electromagnetic radiation depends on the wavelength (color) and varies over a wide range: a smaller wavelength (or higher frequency) corresponds to a higher energy. Because more heat is radiated at higher temperatures, a temperature change is accompanied by a color change. Take, for example, an electrical element on a stove, which glows from red to orange, while the higher-temperature steel in a blast furnace glows from yellow to white. The radiation you feel is mostly infrared, which corresponds to a lower temperature than that of the electrical element and the steel. The radiated energy depends on its intensity, which is represented in the figure below by the height of the distribution.

Electromagnetic Waves explains more about the electromagnetic spectrum and **Introduction to Quantum Physics** discusses how the decrease in wavelength corresponds to an increase in energy.

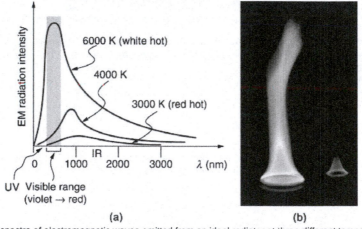

Figure 14.24 (a) A graph of the spectra of electromagnetic waves emitted from an ideal radiator at three different temperatures. The intensity or rate of radiation emission increases dramatically with temperature, and the spectrum shifts toward the visible and ultraviolet parts of the spectrum. The shaded portion denotes the visible part of the spectrum. It is apparent that the shift toward the ultraviolet with temperature makes the visible appearance shift from red to white to blue as temperature increases. (b) Note the variations in color corresponding to variations in flame temperature. (credit: Tuohirulla)

All objects absorb and emit electromagnetic radiation. The rate of heat transfer by radiation is largely determined by the color of the object. Black is the most effective, and white is the least effective. People living in hot climates generally avoid wearing black clothing, for instance (see **Take-Home Experiment: Temperature in the Sun**). Similarly, black asphalt in a parking lot will be hotter than adjacent gray sidewalk on a summer day, because black absorbs better than gray. The reverse is also true—black radiates better than gray. Thus, on a clear summer night, the asphalt will be colder than the gray sidewalk, because black radiates the energy more rapidly than gray. An *ideal radiator* is the same color as an *ideal absorber*, and captures all the radiation that falls on it. In contrast, white is a poor absorber and is also a poor radiator. A white object reflects all radiation, like a mirror. (A perfect, polished white surface is mirror-like in appearance, and a crushed mirror looks white.)

Figure 14.25 This illustration shows that the darker pavement is hotter than the lighter pavement (much more of the ice on the right has melted), although both have been in the sunlight for the same time. The thermal conductivities of the pavements are the same.

Gray objects have a uniform ability to absorb all parts of the electromagnetic spectrum. Colored objects behave in similar but more complex ways, which gives them a particular color in the visible range and may make them special in other ranges of the nonvisible spectrum. Take, for example, the strong absorption of infrared radiation by the skin, which allows us to be very sensitive to it.

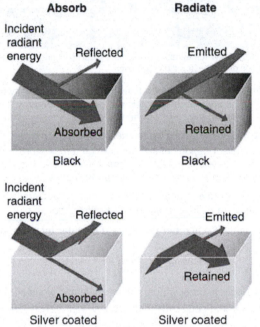

Figure 14.26 A black object is a good absorber and a good radiator, while a white (or silver) object is a poor absorber and a poor radiator. It is as if radiation from the inside is reflected back into the silver object, whereas radiation from the inside of the black object is "absorbed" when it hits the surface and finds itself on the outside and is strongly emitted.

The rate of heat transfer by emitted radiation is determined by the **Stefan-Boltzmann law of radiation**:

$$\frac{Q}{t} = \sigma e A T^4,$$

(14.44)

where $\sigma = 5.67 \times 10^{-8} \text{ J/s} \cdot \text{m}^2 \cdot \text{K}^4$ is the Stefan-Boltzmann constant, A is the surface area of the object, and T is its absolute temperature in kelvin. The symbol e stands for the **emissivity** of the object, which is a measure of how well it radiates. An ideal jet-black (or black body) radiator has $e = 1$, whereas a perfect reflector has $e = 0$. Real objects fall between these two values. Take, for example, tungsten light bulb filaments which have an e of about 0.5, and carbon black (a material used in printer toner), which has the (greatest known) emissivity of about 0.99.

The radiation rate is directly proportional to the *fourth power* of the absolute temperature—a remarkably strong temperature dependence. Furthermore, the radiated heat is proportional to the surface area of the object. If you knock apart the coals of a fire, there is a noticeable increase in radiation due to an increase in radiating surface area.

Figure 14.27 A thermograph of part of a building shows temperature variations, indicating where heat transfer to the outside is most severe. Windows are a major region of heat transfer to the outside of homes. (credit: U.S. Army)

Skin is a remarkably good absorber and emitter of infrared radiation, having an emissivity of 0.97 in the infrared spectrum. Thus, we are all nearly (jet) black in the infrared, in spite of the obvious variations in skin color. This high infrared emissivity is why we can so easily feel radiation on our skin. It is also the basis for the use of night scopes used by law enforcement and the military to detect human beings. Even small temperature variations can be detected because of the T^4 dependence. Images, called *thermographs*, can be used medically to detect regions of abnormally high temperature in the body, perhaps indicative of disease. Similar techniques can be used to detect heat leaks in homes **Figure 14.27**, optimize performance of blast furnaces, improve comfort levels in work environments, and even remotely map the Earth's temperature profile.

All objects emit and absorb radiation. The *net* rate of heat transfer by radiation (absorption minus emission) is related to both the temperature of the object and the temperature of its surroundings. Assuming that an object with a temperature T_1 is surrounded by an environment with uniform temperature T_2, the **net rate of heat transfer by radiation** is

$$\frac{Q_{\text{net}}}{t} = \sigma e A\left(T_2^4 - T_1^4\right),$$

(14.45)

where e is the emissivity of the object alone. In other words, it does not matter whether the surroundings are white, gray, or black; the balance of radiation into and out of the object depends on how well it emits and absorbs radiation. When $T_2 > T_1$, the quantity Q_{net}/t is positive; that is, the net heat transfer is from hot to cold.

Take-Home Experiment: Temperature in the Sun

Place a thermometer out in the sunshine and shield it from direct sunlight using an aluminum foil. What is the reading? Now remove the shield, and note what the thermometer reads. Take a handkerchief soaked in nail polish remover, wrap it around the thermometer and place it in the sunshine. What does the thermometer read?

Example 14.9 Calculate the Net Heat Transfer of a Person: Heat Transfer by Radiation

What is the rate of heat transfer by radiation, with an unclothed person standing in a dark room whose ambient temperature is $22.0°C$. The person has a normal skin temperature of $33.0°C$ and a surface area of 1.50 m^2. The emissivity of skin is 0.97 in the infrared, where the radiation takes place.

Strategy

We can solve this by using the equation for the rate of radiative heat transfer.

Solution

Insert the temperatures values $T_2 = 295 \text{ K}$ and $T_1 = 306 \text{ K}$, so that

$$\frac{Q}{t} = \sigma e A\left(T_2^4 - T_1^4\right)$$

(14.46)

$$= \left(5.67 \times 10^{-8} \text{ J/s} \cdot \text{m}^2 \cdot \text{K}^4\right)(0.97)\left(1.50 \text{ m}^2\right)\left[(295 \text{ K})^4 - (306 \text{ K})^4\right]$$

(14.47)

$$= -99 \text{ J/s} = -99 \text{ W}.$$

(14.48)

Discussion

This value is a significant rate of heat transfer to the environment (note the minus sign), considering that a person at rest may produce energy at the rate of 125 W and that conduction and convection will also be transferring energy to the environment. Indeed, we would probably expect this person to feel cold. Clothing significantly reduces heat transfer to the environment by many methods, because clothing slows down both conduction and convection, and has a lower emissivity (especially if it is white) than skin.

The Earth receives almost all its energy from radiation of the Sun and reflects some of it back into outer space. Because the Sun is hotter than the Earth, the net energy flux is from the Sun to the Earth. However, the rate of energy transfer is less than the equation for the radiative heat transfer would predict because the Sun does not fill the sky. The average emissivity (e) of the Earth is about 0.65, but the calculation of this value is complicated by the fact that the highly reflective cloud coverage varies greatly from day to day. There is a negative feedback (one in which a change produces an effect that opposes that change) between clouds and heat transfer; greater temperatures evaporate more water to form more clouds, which reflect more radiation back into space, reducing the temperature. The often mentioned **greenhouse effect** is directly related to the variation of the Earth's emissivity with radiation type (see the figure given below). The greenhouse effect is a natural phenomenon responsible for providing temperatures suitable for life on Earth. The Earth's relatively constant temperature is a result of the energy balance between the incoming solar radiation and the energy radiated from the Earth. Most of the infrared radiation emitted from the Earth is absorbed by carbon dioxide (CO_2) and water (H_2O) in the atmosphere and then re-radiated back to the Earth or into outer space. Re-radiation back to the Earth maintains its surface temperature about $40°C$ higher than it would be if there was no atmosphere, similar to the way glass increases temperatures in a greenhouse.

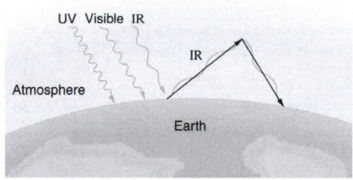

Figure 14.28 The greenhouse effect is a name given to the trapping of energy in the Earth's atmosphere by a process similar to that used in greenhouses. The atmosphere, like window glass, is transparent to incoming visible radiation and most of the Sun's infrared. These wavelengths are absorbed by the Earth and re-emitted as infrared. Since Earth's temperature is much lower than that of the Sun, the infrared radiated by the Earth has a much longer wavelength. The atmosphere, like glass, traps these longer infrared rays, keeping the Earth warmer than it would otherwise be. The amount of trapping depends on concentrations of trace gases like carbon dioxide, and a change in the concentration of these gases is believed to affect the Earth's surface temperature.

The greenhouse effect is also central to the discussion of global warming due to emission of carbon dioxide and methane (and other so-called greenhouse gases) into the Earth's atmosphere from industrial production and farming. Changes in global climate could lead to more intense storms, precipitation changes (affecting agriculture), reduction in rain forest biodiversity, and rising sea levels.

Heating and cooling are often significant contributors to energy use in individual homes. Current research efforts into developing environmentally friendly homes quite often focus on reducing conventional heating and cooling through better building materials, strategically positioning windows to optimize radiation gain from the Sun, and opening spaces to allow convection. It is possible to build a zero-energy house that allows for comfortable living in most parts of the United States with hot and humid summers and cold winters.

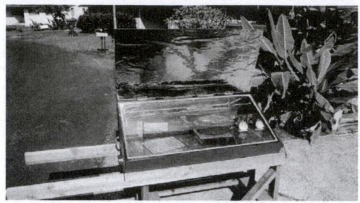

Figure 14.29 This simple but effective solar cooker uses the greenhouse effect and reflective material to trap and retain solar energy. Made of inexpensive, durable materials, it saves money and labor, and is of particular economic value in energy-poor developing countries. (credit: E.B. Kauai)

Conversely, dark space is very cold, about $3\text{K}(-454°\text{F})$, so that the Earth radiates energy into the dark sky. Owing to the fact that clouds have lower emissivity than either oceans or land masses, they reflect some of the radiation back to the surface, greatly reducing heat transfer into dark space, just as they greatly reduce heat transfer into the atmosphere during the day. The rate of heat transfer from soil and grasses can be so rapid that frost may occur on clear summer evenings, even in warm latitudes.

Check Your Understanding

What is the change in the rate of the radiated heat by a body at the temperature $T_1 = 20°\text{C}$ compared to when the body is at the temperature $T_2 = 40°\text{C}$?

Solution
The radiated heat is proportional to the fourth power of the *absolute temperature*. Because $T_1 = 293$ K and $T_2 = 313$ K, the rate of heat transfer increases by about 30 percent of the original rate.

Career Connection: Energy Conservation Consultation

The cost of energy is generally believed to remain very high for the foreseeable future. Thus, passive control of heat loss in both commercial and domestic housing will become increasingly important. Energy consultants measure and analyze the flow of energy into and out of houses and ensure that a healthy exchange of air is maintained inside the house. The job prospects for an energy consultant are strong.

Problem-Solving Strategies for the Methods of Heat Transfer

1. *Examine the situation to determine what type of heat transfer is involved.*

2. *Identify the type(s) of heat transfer—conduction, convection, or radiation.*

3. *Identify exactly what needs to be determined in the problem (identify the unknowns). A written list is very useful.*

4. *Make a list of what is given or can be inferred from the problem as stated (identify the knowns).*

5. *Solve the appropriate equation for the quantity to be determined (the unknown).*

6. *For conduction, equation $\dfrac{Q}{t} = \dfrac{kA(T_2 - T_1)}{d}$ is appropriate.* **Table 14.3** *lists thermal conductivities. For convection, determine the amount of matter moved and use equation $Q = mc\Delta T$, to calculate the heat transfer involved in the temperature change of the fluid. If a phase change accompanies convection, equation $Q = mL_f$ or $Q = mL_v$ is appropriate to find the heat transfer involved in the phase change.* **Table 14.2** *lists information relevant to phase change. For radiation, equation $\dfrac{Q_{net}}{t} = \sigma e A\left(T_2^4 - T_1^4\right)$ gives the net heat transfer rate.*

7. *Insert the knowns along with their units into the appropriate equation and obtain numerical solutions complete with units.*

8. *Check the answer to see if it is reasonable. Does it make sense?*

Glossary

conduction: heat transfer through stationary matter by physical contact

This OpenStax book is available for free at http://cnx.org/content/col11844/1.14   Download for free at https://openstax.org/details/books/college-physics-ap-courses

convection: heat transfer by the macroscopic movement of fluid

emissivity: measure of how well an object radiates

greenhouse effect: warming of the Earth that is due to gases such as carbon dioxide and methane that absorb infrared radiation from the Earth's surface and reradiate it in all directions, thus sending a fraction of it back toward the surface of the Earth

heat: the spontaneous transfer of energy due to a temperature difference

heat of sublimation: the energy required to change a substance from the solid phase to the vapor phase

kilocalorie: $1 \text{ kilocalorie} = 1000 \text{ calories}$

latent heat coefficient: a physical constant equal to the amount of heat transferred for every 1 kg of a substance during the change in phase of the substance

mechanical equivalent of heat: the work needed to produce the same effects as heat transfer

net rate of heat transfer by radiation: is $\dfrac{Q_{net}}{t} = \sigma e A \left(T_2^4 - T_1^4 \right)$

R factor: the ratio of thickness to the conductivity of a material

radiation: heat transfer which occurs when microwaves, infrared radiation, visible light, or other electromagnetic radiation is emitted or absorbed

radiation: energy transferred by electromagnetic waves directly as a result of a temperature difference

rate of conductive heat transfer: rate of heat transfer from one material to another

specific heat: the amount of heat necessary to change the temperature of 1.00 kg of a substance by 1.00 ºC

Stefan-Boltzmann law of radiation: $\dfrac{Q}{t} = \sigma e A T^4$, where σ is the Stefan-Boltzmann constant, A is the surface area of the object, T is the absolute temperature, and e is the emissivity

sublimation: the transition from the solid phase to the vapor phase

thermal conductivity: the property of a material's ability to conduct heat

Section Summary

14.1 Heat
- Heat and work are the two distinct methods of energy transfer.
- Heat is energy transferred solely due to a temperature difference.
- Any energy unit can be used for heat transfer, and the most common are kilocalorie (kcal) and joule (J).
- Kilocalorie is defined to be the energy needed to change the temperature of 1.00 kg of water between $14.5°C$ and $15.5°C$.
- The mechanical equivalent of this heat transfer is $1.00 \text{ kcal} = 4186 \text{ J}$.

14.2 Temperature Change and Heat Capacity
- The transfer of heat Q that leads to a change ΔT in the temperature of a body with mass m is $Q = mc\Delta T$, where c is the specific heat of the material. This relationship can also be considered as the definition of specific heat.

14.3 Phase Change and Latent Heat
- Most substances can exist either in solid, liquid, and gas forms, which are referred to as "phases."
- Phase changes occur at fixed temperatures for a given substance at a given pressure, and these temperatures are called boiling and freezing (or melting) points.
- During phase changes, heat absorbed or released is given by:
$$Q = mL,$$
where L is the latent heat coefficient.

14.4 Heat Transfer Methods

- Heat is transferred by three different methods: conduction, convection, and radiation.

14.5 Conduction

- Heat conduction is the transfer of heat between two objects in direct contact with each other.
- The rate of heat transfer Q/t (energy per unit time) is proportional to the temperature difference $T_2 - T_1$ and the contact area A and inversely proportional to the distance d between the objects:

$$\frac{Q}{t} = \frac{kA(T_2 - T_1)}{d}.$$

14.6 Convection

- Convection is heat transfer by the macroscopic movement of mass. Convection can be natural or forced and generally transfers thermal energy faster than conduction. Table 14.4 gives wind-chill factors, indicating that moving air has the same chilling effect of much colder stationary air. *Convection that occurs along with a phase change* can transfer energy from cold regions to warm ones.

14.7 Radiation

- Radiation is the rate of heat transfer through the emission or absorption of electromagnetic waves.
- The rate of heat transfer depends on the surface area and the fourth power of the absolute temperature:

$$\frac{Q}{t} = \sigma e A T^4,$$

where $\sigma = 5.67 \times 10^{-8} \, \text{J/s} \cdot \text{m}^2 \cdot \text{K}^4$ is the Stefan-Boltzmann constant and e is the emissivity of the body. For a black body, $e = 1$ whereas a shiny white or perfect reflector has $e = 0$, with real objects having values of e between 1 and 0. The net rate of heat transfer by radiation is

$$\frac{Q_{net}}{t} = \sigma e A \left(T_2^4 - T_1^4 \right)$$

where T_1 is the temperature of an object surrounded by an environment with uniform temperature T_2 and e is the emissivity of the *object*.

Conceptual Questions

14.1 Heat

1. How is heat transfer related to temperature?

2. Describe a situation in which heat transfer occurs. What are the resulting forms of energy?

3. When heat transfers into a system, is the energy stored as heat? Explain briefly.

14.2 Temperature Change and Heat Capacity

4. What three factors affect the heat transfer that is necessary to change an object's temperature?

5. The brakes in a car increase in temperature by ΔT when bringing the car to rest from a speed v. How much greater would ΔT be if the car initially had twice the speed? You may assume the car to stop sufficiently fast so that no heat transfers out of the brakes.

14.3 Phase Change and Latent Heat

6. Heat transfer can cause temperature and phase changes. What else can cause these changes?

7. How does the latent heat of fusion of water help slow the decrease of air temperatures, perhaps preventing temperatures from falling significantly below $0°C$, in the vicinity of large bodies of water?

8. What is the temperature of ice right after it is formed by freezing water?

9. If you place $0°C$ ice into $0°C$ water in an insulated container, what will happen? Will some ice melt, will more water freeze, or will neither take place?

10. What effect does condensation on a glass of ice water have on the rate at which the ice melts? Will the condensation speed up the melting process or slow it down?

11. In very humid climates where there are numerous bodies of water, such as in Florida, it is unusual for temperatures to rise above about $35°C(95°F)$. In deserts, however, temperatures can rise far above this. Explain how the evaporation of water helps limit high temperatures in humid climates.

12. In winters, it is often warmer in San Francisco than in nearby Sacramento, 150 km inland. In summers, it is nearly always hotter in Sacramento. Explain how the bodies of water surrounding San Francisco moderate its extreme temperatures.

13. Putting a lid on a boiling pot greatly reduces the heat transfer necessary to keep it boiling. Explain why.

14. Freeze-dried foods have been dehydrated in a vacuum. During the process, the food freezes and must be heated to facilitate dehydration. Explain both how the vacuum speeds up dehydration and why the food freezes as a result.

15. When still air cools by radiating at night, it is unusual for temperatures to fall below the dew point. Explain why.

16. In a physics classroom demonstration, an instructor inflates a balloon by mouth and then cools it in liquid nitrogen. When cold, the shrunken balloon has a small amount of light blue liquid in it, as well as some snow-like crystals. As it warms up, the liquid boils, and part of the crystals sublimate, with some crystals lingering for awhile and then producing a liquid. Identify the blue liquid and the two solids in the cold balloon. Justify your identifications using data from **Table 14.2**.

14.4 Heat Transfer Methods

17. What are the main methods of heat transfer from the hot core of Earth to its surface? From Earth's surface to outer space?

When our bodies get too warm, they respond by sweating and increasing blood circulation to the surface to transfer thermal energy away from the core. What effect will this have on a person in a $40.0°C$ hot tub?

Figure 14.30 shows a cut-away drawing of a thermos bottle (also known as a Dewar flask), which is a device designed specifically to slow down all forms of heat transfer. Explain the functions of the various parts, such as the vacuum, the silvering of the walls, the thin-walled long glass neck, the rubber support, the air layer, and the stopper.

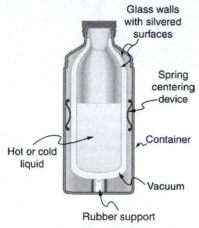

Figure 14.30 The construction of a thermos bottle is designed to inhibit all methods of heat transfer.

14.5 Conduction

18. Some electric stoves have a flat ceramic surface with heating elements hidden beneath. A pot placed over a heating element will be heated, while it is safe to touch the surface only a few centimeters away. Why is ceramic, with a conductivity less than that of a metal but greater than that of a good insulator, an ideal choice for the stove top?

19. Loose-fitting white clothing covering most of the body is ideal for desert dwellers, both in the hot Sun and during cold evenings. Explain how such clothing is advantageous during both day and night.

Figure 14.31 A jellabiya is worn by many men in Egypt. (credit: Zerida)

14.6 Convection

20. One way to make a fireplace more energy efficient is to have an external air supply for the combustion of its fuel. Another is to have room air circulate around the outside of the fire box and back into the room. Detail the methods of heat transfer involved in each.

21. On cold, clear nights horses will sleep under the cover of large trees. How does this help them keep warm?

14.7 Radiation

22. When watching a daytime circus in a large, dark-colored tent, you sense significant heat transfer from the tent. Explain why this occurs.

23. Satellites designed to observe the radiation from cold (3 K) dark space have sensors that are shaded from the Sun, Earth, and Moon and that are cooled to very low temperatures. Why must the sensors be at low temperature?

24. Why are cloudy nights generally warmer than clear ones?

25. Why are thermometers that are used in weather stations shielded from the sunshine? What does a thermometer measure if it is shielded from the sunshine and also if it is not?

26. On average, would Earth be warmer or cooler without the atmosphere? Explain your answer.

Problems & Exercises

14.2 Temperature Change and Heat Capacity

1. On a hot day, the temperature of an 80,000-L swimming pool increases by $1.50°C$. What is the net heat transfer during this heating? Ignore any complications, such as loss of water by evaporation.

2. Show that $1 \text{ cal/g} \cdot °C = 1 \text{ kcal/kg} \cdot °C$.

3. To sterilize a 50.0-g glass baby bottle, we must raise its temperature from $22.0°C$ to $95.0°C$. How much heat transfer is required?

4. The same heat transfer into identical masses of different substances produces different temperature changes. Calculate the final temperature when 1.00 kcal of heat transfers into 1.00 kg of the following, originally at $20.0°C$: (a) water; (b) concrete; (c) steel; and (d) mercury.

5. Rubbing your hands together warms them by converting work into thermal energy. If a woman rubs her hands back and forth for a total of 20 rubs, at a distance of 7.50 cm per rub, and with an average frictional force of 40.0 N, what is the temperature increase? The mass of tissues warmed is only 0.100 kg, mostly in the palms and fingers.

6. A 0.250-kg block of a pure material is heated from $20.0°C$ to $65.0°C$ by the addition of 4.35 kJ of energy. Calculate its specific heat and identify the substance of which it is most likely composed.

7. Suppose identical amounts of heat transfer into different masses of copper and water, causing identical changes in temperature. What is the ratio of the mass of copper to water?

8. (a) The number of kilocalories in food is determined by calorimetry techniques in which the food is burned and the amount of heat transfer is measured. How many kilocalories per gram are there in a 5.00-g peanut if the energy from burning it is transferred to 0.500 kg of water held in a 0.100-kg aluminum cup, causing a $54.9°C$ temperature increase? (b) Compare your answer to labeling information found on a package of peanuts and comment on whether the values are consistent.

9. Following vigorous exercise, the body temperature of an 80.0-kg person is $40.0°C$. At what rate in watts must the person transfer thermal energy to reduce the the body temperature to $37.0°C$ in 30.0 min, assuming the body continues to produce energy at the rate of 150 W? $(1 \text{ watt} = 1 \text{ joule/second or } 1 \text{ W} = 1 \text{ J/s})$.

10. Even when shut down after a period of normal use, a large commercial nuclear reactor transfers thermal energy at the rate of 150 MW by the radioactive decay of fission products. This heat transfer causes a rapid increase in temperature if the cooling system fails $(1 \text{ watt} = 1 \text{ joule/second or } 1 \text{ W} = 1 \text{ J/s and } 1 \text{ MW} = 1 \text{ megawatt})$. (a) Calculate the rate of temperature increase in degrees Celsius per second $(°C/s)$ if the mass of the reactor core is $1.60 \times 10^5 \text{ kg}$ and it has an average specific heat of $0.3349 \text{ kJ/kg}° \cdot C$. (b) How long would it take to obtain a temperature increase of $2000°C$, which could cause some metals holding the radioactive materials to melt? (The initial rate of temperature increase would be greater than that calculated here because the heat transfer is concentrated in a smaller mass. Later, however, the temperature increase would slow down because the 5×10^5-kg steel containment vessel would also begin to heat up.)

Figure 14.32 Radioactive spent-fuel pool at a nuclear power plant. Spent fuel stays hot for a long time. (credit: U.S. Department of Energy)

14.3 Phase Change and Latent Heat

11. How much heat transfer (in kilocalories) is required to thaw a 0.450-kg package of frozen vegetables originally at $0°C$ if their heat of fusion is the same as that of water?

12. A bag containing $0°C$ ice is much more effective in absorbing energy than one containing the same amount of $0°C$ water.

  a. How much heat transfer is necessary to raise the temperature of 0.800 kg of water from $0°C$ to $30.0°C$?

  b. How much heat transfer is required to first melt 0.800 kg of $0°C$ ice and then raise its temperature?

  c. Explain how your answer supports the contention that the ice is more effective.

13. (a) How much heat transfer is required to raise the temperature of a 0.750-kg aluminum pot containing 2.50 kg of water from $30.0°C$ to the boiling point and then boil away 0.750 kg of water? (b) How long does this take if the rate of heat transfer is 500 W

$1 \text{ watt} = 1 \text{ joule/second } (1 \text{ W} = 1 \text{ J/s})$?

14. The formation of condensation on a glass of ice water causes the ice to melt faster than it would otherwise. If 8.00 g of condensation forms on a glass containing both water and 200 g of ice, how many grams of the ice will melt as a result? Assume no other heat transfer occurs.

15. On a trip, you notice that a 3.50-kg bag of ice lasts an average of one day in your cooler. What is the average power in watts entering the ice if it starts at $0°C$ and completely melts to $0°C$ water in exactly one day

$1 \text{ watt} = 1 \text{ joule/second } (1 \text{ W} = 1 \text{ J/s})$?

16. On a certain dry sunny day, a swimming pool's temperature would rise by $1.50°C$ if not for evaporation. What fraction of the water must evaporate to carry away precisely enough energy to keep the temperature constant?

17. (a) How much heat transfer is necessary to raise the temperature of a 0.200-kg piece of ice from $-20.0°C$ to $130°C$, including the energy needed for phase changes?

(b) How much time is required for each stage, assuming a constant 20.0 kJ/s rate of heat transfer?

(c) Make a graph of temperature versus time for this process.

18. In 1986, a gargantuan iceberg broke away from the Ross Ice Shelf in Antarctica. It was approximately a rectangle 160 km long, 40.0 km wide, and 250 m thick.

(a) What is the mass of this iceberg, given that the density of ice is 917 kg/m^3?

(b) How much heat transfer (in joules) is needed to melt it?

(c) How many years would it take sunlight alone to melt ice this thick, if the ice absorbs an average of 100 W/m^2, 12.00 h per day?

19. How many grams of coffee must evaporate from 350 g of coffee in a 100-g glass cup to cool the coffee from $95.0°C$ to $45.0°C$? You may assume the coffee has the same thermal properties as water and that the average heat of vaporization is 2340 kJ/kg (560 cal/g). (You may neglect the change in mass of the coffee as it cools, which will give you an answer that is slightly larger than correct.)

20. (a) It is difficult to extinguish a fire on a crude oil tanker, because each liter of crude oil releases $2.80 \times 10^7 \text{ J}$ of energy when burned. To illustrate this difficulty, calculate the number of liters of water that must be expended to absorb the energy released by burning 1.00 L of crude oil, if the water has its temperature raised from $20.0°C$ to $100°C$, it boils, and the resulting steam is raised to $300°C$. (b) Discuss additional complications caused by the fact that crude oil has a smaller density than water.

21. The energy released from condensation in thunderstorms can be very large. Calculate the energy released into the atmosphere for a small storm of radius 1 km, assuming that 1.0 cm of rain is precipitated uniformly over this area.

22. To help prevent frost damage, 4.00 kg of $0°C$ water is sprayed onto a fruit tree.

(a) How much heat transfer occurs as the water freezes?

(b) How much would the temperature of the 200-kg tree decrease if this amount of heat transferred from the tree? Take the specific heat to be 3.35 kJ/kg·°C, and assume that no phase change occurs.

23. A 0.250-kg aluminum bowl holding 0.800 kg of soup at $25.0°C$ is placed in a freezer. What is the final temperature if 377 kJ of energy is transferred from the bowl and soup, assuming the soup's thermal properties are the same as that of water? Explicitly show how you follow the steps in **Problem-Solving Strategies for the Effects of Heat Transfer.**

24. A 0.0500-kg ice cube at $-30.0°C$ is placed in 0.400 kg of $35.0°C$ water in a very well-insulated container. What is the final temperature?

25. If you pour 0.0100 kg of $20.0°C$ water onto a 1.20-kg block of ice (which is initially at $-15.0°C$), what is the final temperature? You may assume that the water cools so rapidly that effects of the surroundings are negligible.

26. Indigenous people sometimes cook in watertight baskets by placing hot rocks into water to bring it to a boil. What mass of $500°C$ rock must be placed in 4.00 kg of $15.0°C$ water to bring its temperature to $100°C$, if 0.0250 kg of water escapes as vapor from the initial sizzle? You may neglect the effects of the surroundings and take the average specific heat of the rocks to be that of granite.

27. What would be the final temperature of the pan and water in **Example 14.3** if 0.260 kg of water was placed in the pan and 0.0100 kg of the water evaporated immediately, leaving the remainder to come to a common temperature with the pan?

28. In some countries, liquid nitrogen is used on dairy trucks instead of mechanical refrigerators. A 3.00-hour delivery trip requires 200 L of liquid nitrogen, which has a density of 808 kg/m^3.

(a) Calculate the heat transfer necessary to evaporate this amount of liquid nitrogen and raise its temperature to $3.00°C$. (Use c_p and assume it is constant over the temperature range.) This value is the amount of cooling the liquid nitrogen supplies.

(b) What is this heat transfer rate in kilowatt-hours?

(c) Compare the amount of cooling obtained from melting an identical mass of $0°C$ ice with that from evaporating the liquid nitrogen.

29. Some gun fanciers make their own bullets, which involves melting and casting the lead slugs. How much heat transfer is needed to raise the temperature and melt 0.500 kg of lead, starting from $25.0°C$?

14.5 Conduction

30. (a) Calculate the rate of heat conduction through house walls that are 13.0 cm thick and that have an average thermal conductivity twice that of glass wool. Assume there are no windows or doors. The surface area of the walls is 120 m^2 and their inside surface is at $18.0°C$, while their outside surface is at $5.00°C$. (b) How many 1-kW room heaters would be needed to balance the heat transfer due to conduction?

31. The rate of heat conduction out of a window on a winter day is rapid enough to chill the air next to it. To see just how rapidly the windows transfer heat by conduction, calculate the rate of conduction in watts through a 3.00-m^2 window that is 0.635 cm thick (1/4 in) if the temperatures of the inner and outer surfaces are $5.00°C$ and $-10.0°C$, respectively. This rapid rate will not be maintained—the inner surface will cool, and even result in frost formation.

32. Calculate the rate of heat conduction out of the human body, assuming that the core internal temperature is $37.0°C$, the skin temperature is $34.0°C$, the thickness of the tissues between averages 1.00 cm, and the surface area is 1.40 m^2.

33. Suppose you stand with one foot on ceramic flooring and one foot on a wool carpet, making contact over an area of 80.0 cm^2 with each foot. Both the ceramic and the carpet are 2.00 cm thick and are $10.0°C$ on their bottom sides. At what rate must heat transfer occur from each foot to keep the top of the ceramic and carpet at $33.0°C$?

34. A man consumes 3000 kcal of food in one day, converting most of it to maintain body temperature. If he loses half this energy by evaporating water (through breathing and sweating), how many kilograms of water evaporate?

35. (a) A firewalker runs across a bed of hot coals without sustaining burns. Calculate the heat transferred by conduction into the sole of one foot of a firewalker given that the bottom of the foot is a 3.00-mm-thick callus with a conductivity at the low end of the range for wood and its density is 300 kg/m^3. The area of contact is 25.0 cm^2, the temperature of the coals is $700°C$, and the time in contact is 1.00 s.

(b) What temperature increase is produced in the 25.0 cm^3 of tissue affected?

(c) What effect do you think this will have on the tissue, keeping in mind that a callus is made of dead cells?

36. (a) What is the rate of heat conduction through the 3.00-cm-thick fur of a large animal having a 1.40-m^2 surface area? Assume that the animal's skin temperature is $32.0°C$, that the air temperature is $-5.00°C$, and that fur has the same thermal conductivity as air. (b) What food intake will the animal need in one day to replace this heat transfer?

37. A walrus transfers energy by conduction through its blubber at the rate of 150 W when immersed in $-1.00°C$ water. The walrus's internal core temperature is $37.0°C$, and it has a surface area of 2.00 m^2. What is the average thickness of its blubber, which has the conductivity of fatty tissues without blood?

Figure 14.33 Walrus on ice. (credit: Captain Budd Christman, NOAA Corps)

38. Compare the rate of heat conduction through a 13.0-cm-thick wall that has an area of 10.0 m^2 and a thermal conductivity twice that of glass wool with the rate of heat conduction through a window that is 0.750 cm thick and that has an area of 2.00 m^2, assuming the same temperature difference across each.

39. Suppose a person is covered head to foot by wool clothing with average thickness of 2.00 cm and is transferring energy by conduction through the clothing at the rate of 50.0 W. What is the temperature difference across the clothing, given the surface area is 1.40 m^2?

40. Some stove tops are smooth ceramic for easy cleaning. If the ceramic is 0.600 cm thick and heat conduction occurs through the same area and at the same rate as computed in Example 14.6, what is the temperature difference across it? Ceramic has the same thermal conductivity as glass and brick.

41. One easy way to reduce heating (and cooling) costs is to add extra insulation in the attic of a house. Suppose the house already had 15 cm of fiberglass insulation in the attic and in all the exterior surfaces. If you added an extra 8.0 cm of fiberglass to the attic, then by what percentage would the heating cost of the house drop? Take the single story house to be of dimensions 10 m by 15 m by 3.0 m. Ignore air infiltration and heat loss through windows and doors.

42. (a) Calculate the rate of heat conduction through a double-paned window that has a 1.50-m^2 area and is made of two panes of 0.800-cm-thick glass separated by a 1.00-cm air gap. The inside surface temperature is $15.0°C$, while that on the outside is $-10.0°C$. (Hint: There are identical temperature drops across the two glass panes. First find these and then the temperature drop across the air gap. This problem ignores the increased heat transfer in the air gap due to convection.)

(b) Calculate the rate of heat conduction through a 1.60-cm-thick window of the same area and with the same temperatures. Compare your answer with that for part (a).

43. Many decisions are made on the basis of the payback period: the time it will take through savings to equal the capital cost of an investment. Acceptable payback times depend upon the business or philosophy one has. (For some industries, a payback period is as small as two years.) Suppose you wish to install the extra insulation in Exercise 14.41. If energy cost $1.00 per million joules and the insulation was $4.00 per square meter, then calculate the simple payback time. Take the average ΔT for the 120 day heating season to be $15.0°C$.

44. For the human body, what is the rate of heat transfer by conduction through the body's tissue with the following conditions: the tissue thickness is 3.00 cm, the change in temperature is $2.00°C$, and the skin area is $1.50 \ m^2$. How does this compare with the average heat transfer rate to the body resulting from an energy intake of about 2400 kcal per day? (No exercise is included.)

14.6 Convection

45. At what wind speed does $-10°C$ air cause the same chill factor as still air at $-29°C$?

46. At what temperature does still air cause the same chill factor as $-5°C$ air moving at 15 m/s?

47. The "steam" above a freshly made cup of instant coffee is really water vapor droplets condensing after evaporating from the hot coffee. What is the final temperature of 250 g of hot coffee initially at $90.0°C$ if 2.00 g evaporates from it? The coffee is in a Styrofoam cup, so other methods of heat transfer can be neglected.

48. (a) How many kilograms of water must evaporate from a 60.0-kg woman to lower her body temperature by $0.750°C$?

(b) Is this a reasonable amount of water to evaporate in the form of perspiration, assuming the relative humidity of the surrounding air is low?

49. On a hot dry day, evaporation from a lake has just enough heat transfer to balance the $1.00 \ kW/m^2$ of incoming heat from the Sun. What mass of water evaporates in 1.00 h from each square meter? Explicitly show how you follow the steps in the **Problem-Solving Strategies for the Methods of Heat Transfer**.

50. One winter day, the climate control system of a large university classroom building malfunctions. As a result, $500 \ m^3$ of excess cold air is brought in each minute. At what rate in kilowatts must heat transfer occur to warm this air by $10.0°C$ (that is, to bring the air to room temperature)?

51. The Kilauea volcano in Hawaii is the world's most active, disgorging about $5\times10^5 \ m^3$ of $1200°C$ lava per day. What is the rate of heat transfer out of Earth by convection if this lava has a density of $2700 \ kg/m^3$ and eventually cools to $30°C$? Assume that the specific heat of lava is the same as that of granite.

Figure 14.34 Lava flow on Kilauea volcano in Hawaii. (credit: J. P. Eaton, U.S. Geological Survey)

52. During heavy exercise, the body pumps 2.00 L of blood per minute to the surface, where it is cooled by $2.00°C$. What is the rate of heat transfer from this forced convection alone, assuming blood has the same specific heat as water and its density is $1050 \ kg/m^3$?

53. A person inhales and exhales 2.00 L of $37.0°C$ air, evaporating $4.00\times10^{-2} \ g$ of water from the lungs and breathing passages with each breath.

(a) How much heat transfer occurs due to evaporation in each breath?

(b) What is the rate of heat transfer in watts if the person is breathing at a moderate rate of 18.0 breaths per minute?

(c) If the inhaled air had a temperature of $20.0°C$, what is the rate of heat transfer for warming the air?

(d) Discuss the total rate of heat transfer as it relates to typical metabolic rates. Will this breathing be a major form of heat transfer for this person?

54. A glass coffee pot has a circular bottom with a 9.00-cm diameter in contact with a heating element that keeps the coffee warm with a continuous heat transfer rate of 50.0 W

(a) What is the temperature of the bottom of the pot, if it is 3.00 mm thick and the inside temperature is $60.0°C$?

(b) If the temperature of the coffee remains constant and all of the heat transfer is removed by evaporation, how many grams per minute evaporate? Take the heat of vaporization to be 2340 kJ/kg.

14.7 Radiation

55. At what net rate does heat radiate from a 275-m^2 black roof on a night when the roof's temperature is $30.0°C$ and the surrounding temperature is $15.0°C$? The emissivity of the roof is 0.900.

56. (a) Cherry-red embers in a fireplace are at $850°C$ and have an exposed area of $0.200 \ \text{m}^2$ and an emissivity of 0.980. The surrounding room has a temperature of $18.0°C$. If 50% of the radiant energy enters the room, what is the net rate of radiant heat transfer in kilowatts? (b) Does your answer support the contention that most of the heat transfer into a room by a fireplace comes from infrared radiation?

57. Radiation makes it impossible to stand close to a hot lava flow. Calculate the rate of heat transfer by radiation from $1.00 \ \text{m}^2$ of $1200°C$ fresh lava into $30.0°C$ surroundings, assuming lava's emissivity is 1.00.

58. (a) Calculate the rate of heat transfer by radiation from a car radiator at $110°C$ into a $50.0°C$ environment, if the radiator has an emissivity of 0.750 and a 1.20-m^2 surface area. (b) Is this a significant fraction of the heat transfer by an automobile engine? To answer this, assume a horsepower of $200 \ \text{hp} \ (1.5 \ \text{kW})$ and the efficiency of automobile engines as 25%.

59. Find the net rate of heat transfer by radiation from a skier standing in the shade, given the following. She is completely clothed in white (head to foot, including a ski mask), the clothes have an emissivity of 0.200 and a surface temperature of $10.0°C$, the surroundings are at $-15.0°C$, and her surface area is $1.60 \ \text{m}^2$.

60. Suppose you walk into a sauna that has an ambient temperature of $50.0°C$. (a) Calculate the rate of heat transfer to you by radiation given your skin temperature is $37.0°C$, the emissivity of skin is 0.98, and the surface area of your body is $1.50 \ \text{m}^2$. (b) If all other forms of heat transfer are balanced (the net heat transfer is zero), at what rate will your body temperature increase if your mass is 75.0 kg?

61. Thermography is a technique for measuring radiant heat and detecting variations in surface temperatures that may be medically, environmentally, or militarily meaningful.(a) What is the percent increase in the rate of heat transfer by radiation from a given area at a temperature of $34.0°C$ compared with that at $33.0°C$, such as on a person's skin? (b) What is the percent increase in the rate of heat transfer by radiation from a given area at a temperature of $34.0°C$ compared with that at $20.0°C$, such as for warm and cool automobile hoods?

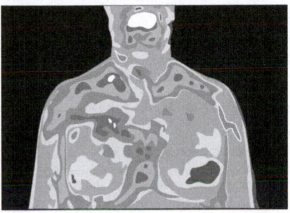

Figure 14.35 Artist's rendition of a thermograph of a patient's upper body, showing the distribution of heat represented by different colors.

62. The Sun radiates like a perfect black body with an emissivity of exactly 1. (a) Calculate the surface temperature of the Sun, given that it is a sphere with a 7.00×10^8-m radius that radiates $3.80 \times 10^{26} \ \text{W}$ into 3-K space. (b) How much power does the Sun radiate per square meter of its surface? (c) How much power in watts per square meter is that value at the distance of Earth, $1.50 \times 10^{11} \ \text{m}$ away? (This number is called the solar constant.)

63. A large body of lava from a volcano has stopped flowing and is slowly cooling. The interior of the lava is at $1200°C$, its surface is at $450°C$, and the surroundings are at $27.0°C$. (a) Calculate the rate at which energy is transferred by radiation from $1.00 \ \text{m}^2$ of surface lava into the surroundings, assuming the emissivity is 1.00. (b) Suppose heat conduction to the surface occurs at the same rate. What is the thickness of the lava between the $450°C$ surface and the $1200°C$ interior, assuming that the lava's conductivity is the same as that of brick?

64. Calculate the temperature the entire sky would have to be in order to transfer energy by radiation at $1000 \ \text{W/m}^2$ —about the rate at which the Sun radiates when it is directly overhead on a clear day. This value is the effective temperature of the sky, a kind of average that takes account of the fact that the Sun occupies only a small part of the sky but is much hotter than the rest. Assume that the body receiving the energy has a temperature of $27.0°C$.

65. (a) A shirtless rider under a circus tent feels the heat radiating from the sunlit portion of the tent. Calculate the temperature of the tent canvas based on the following information: The shirtless rider's skin temperature is $34.0°C$ and has an emissivity of 0.970. The exposed area of skin is 0.400 m^2. He receives radiation at the rate of 20.0 W—half what you would calculate if the entire region behind him was hot. The rest of the surroundings are at $34.0°C$. (b) Discuss how this situation would change if the sunlit side of the tent was nearly pure white and if the rider was covered by a white tunic.

66. Integrated Concepts

One $30.0°C$ day the relative humidity is 75.0%, and that evening the temperature drops to $20.0°C$, well below the dew point. (a) How many grams of water condense from each cubic meter of air? (b) How much heat transfer occurs by this condensation? (c) What temperature increase could this cause in dry air?

67. Integrated Concepts

Large meteors sometimes strike the Earth, converting most of their kinetic energy into thermal energy. (a) What is the kinetic energy of a 10^9 kg meteor moving at 25.0 km/s? (b) If this meteor lands in a deep ocean and 80% of its kinetic energy goes into heating water, how many kilograms of water could it raise by $5.0°C$? (c) Discuss how the energy of the meteor is more likely to be deposited in the ocean and the likely effects of that energy.

68. Integrated Concepts

Frozen waste from airplane toilets has sometimes been accidentally ejected at high altitude. Ordinarily it breaks up and disperses over a large area, but sometimes it holds together and strikes the ground. Calculate the mass of $0°C$ ice that can be melted by the conversion of kinetic and gravitational potential energy when a 20.0 kg piece of frozen waste is released at 12.0 km altitude while moving at 250 m/s and strikes the ground at 100 m/s (since less than 20.0 kg melts, a significant mess results).

69. Integrated Concepts

(a) A large electrical power facility produces 1600 MW of "waste heat," which is dissipated to the environment in cooling towers by warming air flowing through the towers by $5.00°C$. What is the necessary flow rate of air in m^3/s? (b) Is your result consistent with the large cooling towers used by many large electrical power plants?

70. Integrated Concepts

(a) Suppose you start a workout on a Stairmaster, producing power at the same rate as climbing 116 stairs per minute. Assuming your mass is 76.0 kg and your efficiency is 20.0%, how long will it take for your body temperature to rise $1.00°C$ if all other forms of heat transfer in and out of your body are balanced? (b) Is this consistent with your experience in getting warm while exercising?

71. Integrated Concepts

A 76.0-kg person suffering from hypothermia comes indoors and shivers vigorously. How long does it take the heat transfer to increase the person's body temperature by $2.00°C$ if all other forms of heat transfer are balanced?

72. Integrated Concepts

In certain large geographic regions, the underlying rock is hot. Wells can be drilled and water circulated through the rock for heat transfer for the generation of electricity. (a) Calculate the heat transfer that can be extracted by cooling 1.00 km^3 of granite by $100°C$. (b) How long will it take for heat transfer at the rate of 300 MW, assuming no heat transfers back into the 1.00 km^3 of rock by its surroundings?

73. Integrated Concepts

Heat transfers from your lungs and breathing passages by evaporating water. (a) Calculate the maximum number of grams of water that can be evaporated when you inhale 1.50 L of $37°C$ air with an original relative humidity of 40.0%. (Assume that body temperature is also $37°C$.) (b) How many joules of energy are required to evaporate this amount? (c) What is the rate of heat transfer in watts from this method, if you breathe at a normal resting rate of 10.0 breaths per minute.

74. Integrated Concepts

(a) What is the temperature increase of water falling 55.0 m over Niagara Falls? (b) What fraction must evaporate to keep the temperature constant?

75. Integrated Concepts

Hot air rises because it has expanded. It then displaces a greater volume of cold air, which increases the buoyant force on it. (a) Calculate the ratio of the buoyant force to the weight of $50.0°C$ air surrounded by $20.0°C$ air. (b) What energy is needed to cause 1.00 m^3 of air to go from $20.0°C$ to $50.0°C$? (c) What gravitational potential energy is gained by this volume of air if it rises 1.00 m? Will this cause a significant cooling of the air?

76. Unreasonable Results

(a) What is the temperature increase of an 80.0 kg person who consumes 2500 kcal of food in one day with 95.0% of the energy transferred as heat to the body? (b) What is unreasonable about this result? (c) Which premise or assumption is responsible?

77. Unreasonable Results

A slightly deranged Arctic inventor surrounded by ice thinks it would be much less mechanically complex to cool a car engine by melting ice on it than by having a water-cooled system with a radiator, water pump, antifreeze, and so on. (a) If 80.0% of the energy in 1.00 gal of gasoline is converted into "waste heat" in a car engine, how many kilograms of $0°C$ ice could it melt? (b) Is this a reasonable amount of ice to carry around to cool the engine for 1.00 gal of gasoline consumption? (c) What premises or assumptions are unreasonable?

78. Unreasonable Results

(a) Calculate the rate of heat transfer by conduction through a window with an area of $1.00 \ m^2$ that is 0.750 cm thick, if its inner surface is at $22.0°C$ and its outer surface is at $35.0°C$. (b) What is unreasonable about this result? (c) Which premise or assumption is responsible?

79. Unreasonable Results

A meteorite 1.20 cm in diameter is so hot immediately after penetrating the atmosphere that it radiates 20.0 kW of power. (a) What is its temperature, if the surroundings are at $20.0°C$ and it has an emissivity of 0.800? (b) What is unreasonable about this result? (c) Which premise or assumption is responsible?

80. Construct Your Own Problem

Consider a new model of commercial airplane having its brakes tested as a part of the initial flight permission procedure. The airplane is brought to takeoff speed and then stopped with the brakes alone. Construct a problem in which you calculate the temperature increase of the brakes during this process. You may assume most of the kinetic energy of the airplane is converted to thermal energy in the brakes and surrounding materials, and that little escapes. Note that the brakes are expected to become so hot in this procedure that they ignite and, in order to pass the test, the airplane must be able to withstand the fire for some time without a general conflagration.

81. Construct Your Own Problem

Consider a person outdoors on a cold night. Construct a problem in which you calculate the rate of heat transfer from the person by all three heat transfer methods. Make the initial circumstances such that at rest the person will have a net heat transfer and then decide how much physical activity of a chosen type is necessary to balance the rate of heat transfer. Among the things to consider are the size of the person, type of clothing, initial metabolic rate, sky conditions, amount of water evaporated, and volume of air breathed. Of course, there are many other factors to consider and your instructor may wish to guide you in the assumptions made as well as the detail of analysis and method of presenting your results.

Test Prep for AP® Courses

14.1 Heat

1. An ice cube is placed in a cup of hot water. Which of the following statements correctly describes energy transfer at the molecular level?
   a. Kinetic energy is transferred only from the hot water molecules to the water molecules in the ice.
   b. Kinetic energy is transferred only from the water molecules in the ice to the hot water molecules.
   c. Kinetic energy is transferred mostly from the hot water molecules to the water molecules in the ice.
   d. Kinetic energy is transferred mostly from the water molecules in the ice to the hot water molecules.

2. The molecular description of heat transfer from higher to lower temperatures applies to 'spontaneous' processes, that is, processes in which no energy is added to or removed or from the systems by work or heat. Refrigeration is an example of work being done to remove energy from air within a given space, and thus lower the temperature of the air. Assume a typical kitchen refrigerator, where the air inside the unit forms the system with a temperature of 25°C, and the walls are kept at a constant temperature of 10°C. In terms of molecules and average kinetic energy, describe how the air is made colder.

14.5 Conduction

3. For the experiment that you devised Example 14.1, which variables can be changed, and how should they be changed, so as to shorten the time in which a measurement is made?
   a. Use a smaller quantity of ice (smaller m).
   b. Use containers with greater thickness (larger d).
   c. Use containers with smaller surface areas (smaller A).
   d. Use a lower ambient temperature outside the container (smaller T_2).

4. With reference to Example 14.1, why does the change in the temperature of the ice indicate that it has entirely melted?

5. Which of the following correctly describes the rate of conductive heat transfer for a substance?

   a.   The rate decreases with increased surface area and
        decreased thickness.
   b.   The rate increases with increased temperature
        difference and surface area.
   c.   The rate increases with decreased temperature
        difference and increased thickness.
   d.   The rate decreases with decreased temperature
        difference and increased surface area.

6. You wish to design a saucepan that has the same rate of
thermal conduction as a pan made of silver. You need to use
a less costly material, and limit the size of the pan so that the
surface area in contact with a range heating element is no
more than 50% greater than that of the hypothetical silver
pan. Explain what other factor(s) can be adjusted, and by how
much, so that your design will be successful. Use **Table 14.3**
to obtain thermal conductivity values for different substances.

14.6 Convection

7. Under which two conditions would convection in a fluid be
greatest?
   a.   The gravitational acceleration is large, and the fluid
        density varies greatly for a given temperature change.
   b.   The gravitational acceleration is small, and the fluid
        density varies greatly for a given temperature change.
   c.   The gravitational acceleration is large, and the fluid
        density varies slightly for a given temperature change.
   d.   The gravitational acceleration is small, and the fluid
        density varies slightly for a given temperature change.

8. Sea breezes occur along coastlines, and consist of cool air
moving toward the shore from the ocean. However, this only
occurs during the day, and is a stronger effect when the air
temperature on the land is greatest and the air temperature
above the water is coldest. At night, the breezes are
reversed, moving from the land toward the ocean. Taking into
consideration the specific heat capacities of water and sand
(which is about the same as that of concrete), explain how
sea breezes form during the day and change direction at
night.

14.7 Radiation

9. Two ideal, black-body radiators have temperatures of 275
K and 1100 K. The rate of heat transfer from the latter
radiator is how many times greater than the rate of the former
radiator?
   a.   4
   b.   16
   c.   64
   d.   256

10. On a warm, sunny day, a car is parked along a street
where there is no shade. The car's windows are closed. The
air inside the car becomes noticeably warmer than the air
outside. What factors contribute to the higher temperature?

15 THERMODYNAMICS

Figure 15.1 A steam engine uses heat transfer to do work. Tourists regularly ride this narrow-gauge steam engine train near the San Juan Skyway in Durango, Colorado, part of the National Scenic Byways Program. (credit: Dennis Adams)

Chapter Outline

Connection for AP® Courses

Heat is energy in transit, and it can be used to do work. It can also be converted to any other form of energy. When a car engine burns fuel, for example, heat transfers into a gas. Work is done by the heated gas as it exerts a force through a distance (Essential Knowledge 5.B.5), converting its energy into a variety of other forms—into the car's kinetic or gravitational potential energy; into electrical energy to run the spark plugs, radio, and lights; and into stored energy in the car's battery. But most of the heat produced from burning fuel in the engine does not do work. Rather, the heat is released into the environment, implying that the engine is quite inefficient.

It is often said that modern gasoline engines cannot be made to be significantly more efficient. We hear the same about heat transfer to electrical energy in large power stations, whether they are coal, oil, natural gas, or nuclear powered. Why is that the case? Is the inefficiency caused by design problems that could be solved with better engineering and superior materials? Is it part of some money-making conspiracy by those who sell energy? Actually, the truth is more interesting, and reveals much about the nature of heat transfer. Basic physical laws govern how heat transfer for doing work takes place and place insurmountable limits onto its efficiency. This chapter will explore these laws as well as many applications and concepts associated with them. These topics are part of *thermodynamics*—the study of heat transfer and its relationship to doing work.

This chapter discusses thermodynamics in practical contexts including heat engines, heat pumps, and refrigerators, which support Big Idea 4, and that interactions between systems can result in changes in those systems. As systems either do work or have work done on them, the total energy of a system can change (Enduring Understanding 4.C). These ideas are based on the previous understanding of heat as the process of energy transfer from a higher temperature system to a lower temperature system (Essential Knowledge 4.C.3). You will learn about the first law of thermodynamics, which supports Big Idea 5, that changes that occur as a result of interactions are constrained by conservation laws. The first law of thermodynamics is a special case of energy conservation that explains the relationship between changes in the internal energy of a system (Essential Knowledge 5.B.4) and energy transfer in the form of heat or work (Essential Knowledge 5.B.7). Note that the energy of a system is conserved (Enduring Understanding 5.B). You will also learn about the second law of thermodynamics and entropy. These are applications of Big Idea 7, that the mathematics of probability can be used to describe the behavior of complex systems. For example, an isolated system will reach thermal equilibrium (Enduring Understanding 7.B), a state with higher disorder. This process has a probabilistic nature (Essential Knowledge 7.B.1) and is described by the second law of thermodynamics. The second law of thermodynamics describes the change of entropy for reversible and irreversible processes (Essential Knowledge 7.B.2). Entropy is considered qualitatively at this level.

Big Idea 4 Interactions between systems can result in changes in those systems.

Enduring Understanding 4.C Interactions with other objects or systems can change the total energy of a system.

Essential Knowledge 4.C.3 Energy is transferred spontaneously from a higher temperature system to a lower temperature system. The process through which energy is transferred between systems at different temperatures is called heat.

Big Idea 5 Changes that occur as a result of interactions are constrained by conservation laws.

Enduring Understanding 5.B The energy of a system is conserved.

Essential Knowledge 5.B.4 The internal energy of a system includes the kinetic energy of the objects that make up the system and the potential energy of the configuration of the objects that make up the system.

Essential Knowledge 5.B.5 Energy can be transferred by an external force exerted on an object or system that moves the object or system through a distance; this energy transfer is called work. Energy transfer in mechanical or electrical systems may occur at different rates. Power is defined as the rate of energy transfer into, out of, or within a system. [A piston filled with gas getting compressed or expanded is treated in Physics 2 as a part of thermodynamics.]

Essential Knowledge 5.B.7 The first law of thermodynamics is a specific case of the law of conservation of energy involving the internal energy of a system and the possible transfer of energy through work and/or heat. Examples should include P–V diagrams — isochoric process, isothermal process, isobaric process, adiabatic process. No calculations of heat or internal energy from temperature change; and in this course, examples of these relationships are qualitative and/or semi–quantitative.

Big Idea 7 The mathematics of probability can be used to describe the behavior of complex systems and to interpret the behavior of quantum mechanical systems.

Enduring Understanding 7.B The tendency of isolated systems to move toward states with higher disorder is described by probability.

Essential Knowledge 7.B.1 The approach to thermal equilibrium is a probability process.

Essential Knowledge 7.B.2 The second law of thermodynamics describes the change in entropy for reversible and irreversible processes. Only a qualitative treatment is considered in this course.

15.1 The First Law of Thermodynamics

Learning Objectives

By the end of this section, you will be able to:

- Define the first law of thermodynamics.
- Describe how conservation of energy relates to the first law of thermodynamics.
- Identify instances of the first law of thermodynamics working in everyday situations, including biological metabolism.
- Calculate changes in the internal energy of a system, after accounting for heat transfer and work done.

The information presented in this section supports the following AP® learning objectives and science practices:

- **4.C.3.1** The student is able to make predictions about the direction of energy transfer due to temperature differences based on interactions at the microscopic level. **(S.P. 6.1)**
- **5.B.4.1** The student is able to describe and make predictions about the internal energy of systems. **(S.P. 6.4, 7.2)**
- **5.B.7.1** The student is able to predict qualitative changes in the internal energy of a thermodynamic system involving transfer of energy due to heat or work done and justify those predictions in terms of conservation of energy principles. **(S.P. 6.4, 7.2)**

Figure 15.2 This boiling tea kettle represents energy in motion. The water in the kettle is turning to water vapor because heat is being transferred from the stove to the kettle. As the entire system gets hotter, work is done—from the evaporation of the water to the whistling of the kettle. (credit: Gina Hamilton)

If we are interested in how heat transfer is converted into doing work, then the conservation of energy principle is important. The first law of thermodynamics applies the conservation of energy principle to systems where heat transfer and doing work are the methods of transferring energy into and out of the system. The **first law of thermodynamics** states that the change in internal energy of a system equals the net heat transfer *into* the system minus the net work done *by* the system. In equation form, the first law of thermodynamics is

$$\Delta U = Q - W. \tag{15.1}$$

Here ΔU is the *change in internal energy* U of the system. Q is the *net heat transferred into the system*—that is, Q is the sum of all heat transfer into and out of the system. W is the *net work done by the system*—that is, W is the sum of all work done on or by the system. We use the following sign conventions: if Q is positive, then there is a net heat transfer into the system; if W is positive, then there is net work done by the system. So positive Q adds energy to the system and positive W takes energy from the system. Thus $\Delta U = Q - W$. Note also that if more heat transfer into the system occurs than work done, the difference is stored as internal energy. Heat engines are a good example of this—heat transfer into them takes place so that they can do work. (See **Figure 15.3**.) We will now examine Q, W, and ΔU further.

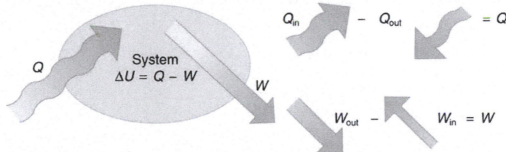

Figure 15.3 The first law of thermodynamics is the conservation-of-energy principle stated for a system where heat and work are the methods of transferring energy for a system in thermal equilibrium. Q represents the net heat transfer—it is the sum of all heat transfers into and out of the system. Q is positive for net heat transfer *into* the system. W is the total work done on and by the system. W is positive when more work is done *by* the system than on it. The change in the internal energy of the system, ΔU, is related to heat and work by the first law of thermodynamics, $\Delta U = Q - W$.

Making Connections: Law of Thermodynamics and Law of Conservation of Energy

The first law of thermodynamics is actually the law of conservation of energy stated in a form most useful in thermodynamics. The first law gives the relationship between heat transfer, work done, and the change in internal energy of a system.

Heat Q and Work W

Heat transfer (Q) and doing work (W) are the two everyday means of bringing energy into or taking energy out of a system.

The processes are quite different. Heat transfer, a less organized process, is driven by temperature differences. Work, a quite organized process, involves a macroscopic force exerted through a distance. Nevertheless, heat and work can produce identical results. For example, both can cause a temperature increase. Heat transfer into a system, such as when the Sun warms the air in a bicycle tire, can increase its temperature, and so can work done on the system, as when the bicyclist pumps air into the tire. Once the temperature increase has occurred, it is impossible to tell whether it was caused by heat transfer or by doing work. This uncertainty is an important point. Heat transfer and work are both energy in transit—neither is stored as such in a system. However, both can change the internal energy U of a system. Internal energy is a form of energy completely different from either heat or work.

Internal Energy U

We can think about the internal energy of a system in two different but consistent ways. The first is the atomic and molecular view, which examines the system on the atomic and molecular scale. The **internal energy** U of a system is the sum of the kinetic and potential energies of its atoms and molecules. Recall that kinetic plus potential energy is called mechanical energy. Thus internal energy is the sum of atomic and molecular mechanical energy. Because it is impossible to keep track of all individual atoms and molecules, we must deal with averages and distributions. A second way to view the internal energy of a system is in terms of its macroscopic characteristics, which are very similar to atomic and molecular average values.

Macroscopically, we define the change in internal energy ΔU to be that given by the first law of thermodynamics:

$$\Delta U = Q - W. \tag{15.2}$$

Many detailed experiments have verified that $\Delta U = Q - W$, where ΔU is the change in total kinetic and potential energy of all atoms and molecules in a system. It has also been determined experimentally that the internal energy U of a system depends only on the state of the system and *not how it reached that state*. More specifically, U is found to be a function of a few macroscopic quantities (pressure, volume, and temperature, for example), independent of past history such as whether there has been heat transfer or work done. This independence means that if we know the state of a system, we can calculate changes in its internal energy U from a few macroscopic variables.

Real World Connections: Pistons

In a previous chapter, temperature was related to the average kinetic energy of the individual molecules in the material. This relates to the concept of total internal energy in a system. Recall that the total internal energy of a system is defined as the sum of all the kinetic energies of all the elements of the system, plus the sum of all the potential energies of interactions between all of the pairs of elements in the system. For example, consider an internal combustion engine utilizing a piston in a cylinder. First, the piston compresses the gas in the cylinder, forcing the molecules closer to each other and changing their potential energy by an outside force doing work on the system. This is the compression stroke, Figure 15.4(b). Then fuel is burned in the cylinder, raising the temperature and hence kinetic energy of all of the molecules. Therefore, the internal energy of the system has had chemical potential energy converted into kinetic energy. This occurs between the compression and power strokes in the figure. Then the piston is pushed back out, using some of the internal energy of the system to do work on an outside system, as shown in the power stroke in the figure.

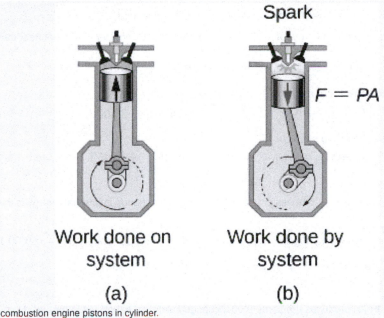

Figure 15.4 Internal combustion engine pistons in cylinder.

Making Connections: Macroscopic and Microscopic

In thermodynamics, we often use the macroscopic picture when making calculations of how a system behaves, while the atomic and molecular picture gives underlying explanations in terms of averages and distributions. We shall see this again in later sections of this chapter. For example, in the topic of entropy, calculations will be made using the atomic and molecular view.

To get a better idea of how to think about the internal energy of a system, let us examine a system going from State 1 to State 2. The system has internal energy U_1 in State 1, and it has internal energy U_2 in State 2, no matter how it got to either state. So the change in internal energy $\Delta U = U_2 - U_1$ is independent of what caused the change. In other words, ΔU *is independent of path*. By path, we mean the method of getting from the starting point to the ending point. Why is this independence important? Note that $\Delta U = Q - W$. Both Q and W *depend on path*, but ΔU does not. This path independence means that internal energy U is easier to consider than either heat transfer or work done.

Example 15.1 Calculating Change in Internal Energy: The Same Change in U is Produced by Two Different Processes

(a) Suppose there is heat transfer of 40.00 J to a system, while the system does 10.00 J of work. Later, there is heat transfer of 25.00 J out of the system while 4.00 J of work is done on the system. What is the net change in internal energy of the system?

(b) What is the change in internal energy of a system when a total of 150.00 J of heat transfer occurs out of (from) the system and 159.00 J of work is done on the system? (See **Figure 15.5**).

Strategy

In part (a), we must first find the net heat transfer and net work done from the given information. Then the first law of thermodynamics $(\Delta U = Q - W)$ can be used to find the change in internal energy. In part (b), the net heat transfer and work done are given, so the equation can be used directly.

Solution for (a)

The net heat transfer is the heat transfer into the system minus the heat transfer out of the system, or

$$Q = 40.00 \text{ J } - 25.00 \text{ J} = 15.00 \text{ J}. \qquad (15.3)$$

Similarly, the total work is the work done by the system minus the work done on the system, or

$$W = 10.00 \text{ J } - 4.00 \text{ J } = 6.00 \text{ J}. \qquad (15.4)$$

Thus the change in internal energy is given by the first law of thermodynamics:

$$\Delta U = Q - W = 15.00 \text{ J} - 6.00 \text{ J} = 9.00 \text{ J}. \tag{15.5}$$

We can also find the change in internal energy for each of the two steps. First, consider 40.00 J of heat transfer in and 10.00 J of work out, or

$$\Delta U_1 = Q_1 - W_1 = 40.00 \text{ J} - 10.00 \text{ J} = 30.00 \text{ J}. \tag{15.6}$$

Now consider 25.00 J of heat transfer out and 4.00 J of work in, or

$$\Delta U_2 = Q_2 - W_2 = -25.00 \text{ J} - (-4.00 \text{ J}) = -21.00 \text{ J}. \tag{15.7}$$

The total change is the sum of these two steps, or

$$\Delta U = \Delta U_1 + \Delta U_2 = 30.00 \text{ J} + (-21.00 \text{ J}) = 9.00 \text{ J}. \tag{15.8}$$

Discussion on (a)

No matter whether you look at the overall process or break it into steps, the change in internal energy is the same.

Solution for (b)

Here the net heat transfer and total work are given directly to be $Q = -150.00 \text{ J}$ and $W = -159.00 \text{ J}$, so that

$$\Delta U = Q - W = -150.00 \text{ J} - (-159.00 \text{ J}) = 9.00 \text{ J}. \tag{15.9}$$

Discussion on (b)

A very different process in part (b) produces the same 9.00-J change in internal energy as in part (a). Note that the change in the system in both parts is related to ΔU and not to the individual Qs or Ws involved. The system ends up in the *same* state in both (a) and (b). Parts (a) and (b) present two different paths for the system to follow between the same starting and ending points, and the change in internal energy for each is the same—it is independent of path.

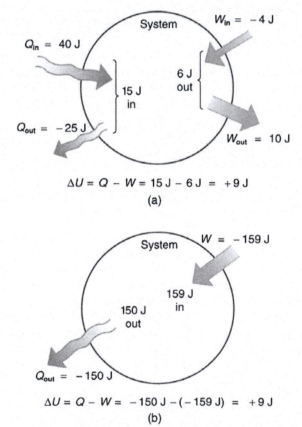

Figure 15.5 Two different processes produce the same change in a system. (a) A total of 15.00 J of heat transfer occurs into the system, while work takes out a total of 6.00 J. The change in internal energy is $\Delta U = Q - W = 9.00 \text{ J}$. (b) Heat transfer removes 150.00 J from the system while work puts 159.00 J into it, producing an increase of 9.00 J in internal energy. If the system starts out in the same state in (a) and (b), it will end up in the same final state in either case—its final state is related to internal energy, not how that energy was acquired.

Plan and design an experiment to measure the total energy out of a potato cannon. How will you measure how much work was done? How will you calculate the energy put in? How can you then estimate how much heat was output? What variables do you need to hold constant over multiple trials for the best results?

The trajectory of the projectile should be measurable and provide a means of calculating the work done. The energy input should be calculable from the type of fuel used. Using the same amount of fuel in each trial would be helpful. The heat output can be estimated by comparing the energy input with the energy required to do the work.

Human Metabolism and the First Law of Thermodynamics

Human metabolism is the conversion of food into heat transfer, work, and stored fat. Metabolism is an interesting example of the first law of thermodynamics in action. We now take another look at these topics via the first law of thermodynamics. Considering the body as the system of interest, we can use the first law to examine heat transfer, doing work, and internal energy in activities ranging from sleep to heavy exercise. What are some of the major characteristics of heat transfer, doing work, and energy in the body? For one, body temperature is normally kept constant by heat transfer to the surroundings. This means Q is negative. Another fact is that the body usually does work on the outside world. This means W is positive. In such situations, then, the body loses internal energy, since $\Delta U = Q - W$ is negative.

Now consider the effects of eating. Eating increases the internal energy of the body by adding chemical potential energy (this is an unromantic view of a good steak). The body *metabolizes* all the food we consume. Basically, metabolism is an oxidation process in which the chemical potential energy of food is released. This implies that food input is in the form of work. Food energy is reported in a special unit, known as the Calorie. This energy is measured by burning food in a calorimeter, which is how the units are determined.

In chemistry and biochemistry, one calorie (spelled with a *lowercase* c) is defined as the energy (or heat transfer) required to raise the temperature of one gram of pure water by one degree Celsius. Nutritionists and weight-watchers tend to use the *dietary* calorie, which is frequently called a Calorie (spelled with a *capital* C). One food Calorie is the energy needed to raise the temperature of one *kilogram* of water by one degree Celsius. This means that one dietary Calorie is equal to one kilocalorie for the chemist, and one must be careful to avoid confusion between the two.

Again, consider the internal energy the body has lost. There are three places this internal energy can go—to heat transfer, to doing work, and to stored fat (a tiny fraction also goes to cell repair and growth). Heat transfer and doing work take internal energy out of the body, and food puts it back. If you eat just the right amount of food, then your average internal energy remains constant. Whatever you lose to heat transfer and doing work is replaced by food, so that, in the long run, $\Delta U = 0$. If you overeat repeatedly, then ΔU is always positive, and your body stores this extra internal energy as fat. The reverse is true if you eat too little. If ΔU is negative for a few days, then the body metabolizes its own fat to maintain body temperature and do work that takes energy from the body. This process is how dieting produces weight loss.

Life is not always this simple, as any dieter knows. The body stores fat or metabolizes it only if energy intake changes for a period of several days. Once you have been on a major diet, the next one is less successful because your body alters the way it responds to low energy intake. Your basal metabolic rate (BMR) is the rate at which food is converted into heat transfer and work done while the body is at complete rest. The body adjusts its basal metabolic rate to partially compensate for over-eating or under-eating. The body will decrease the metabolic rate rather than eliminate its own fat to replace lost food intake. You will chill more easily and feel less energetic as a result of the lower metabolic rate, and you will not lose weight as fast as before. Exercise helps to lose weight, because it produces both heat transfer from your body and work, and raises your metabolic rate even when you are at rest. Weight loss is also aided by the quite low efficiency of the body in converting internal energy to work, so that the loss of internal energy resulting from doing work is much greater than the work done. It should be noted, however, that living systems are not in thermalequilibrium.

The body provides us with an excellent indication that many thermodynamic processes are *irreversible*. An irreversible process can go in one direction but not the reverse, under a given set of conditions. For example, although body fat can be converted to do work and produce heat transfer, work done on the body and heat transfer into it cannot be converted to body fat. Otherwise, we could skip lunch by sunning ourselves or by walking down stairs. Another example of an irreversible thermodynamic process is photosynthesis. This process is the intake of one form of energy—light—by plants and its conversion to chemical potential energy. Both applications of the first law of thermodynamics are illustrated in **Figure 15.6**. One great advantage of conservation laws such as the first law of thermodynamics is that they accurately describe the beginning and ending points of complex processes, such as metabolism and photosynthesis, without regard to the complications in between. **Table 15.1** presents a summary of terms relevant to the first law of thermodynamics.

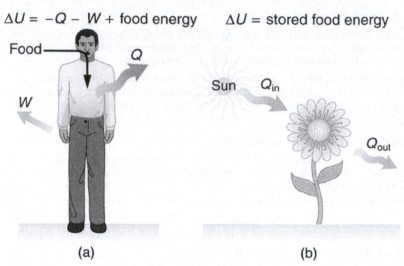

$$\Delta U = -Q - W + \text{food energy} \qquad \Delta U = \text{stored food energy}$$

Figure 15.6 (a) The first law of thermodynamics applied to metabolism. Heat transferred out of the body (Q) and work done by the body (W)
remove internal energy, while food intake replaces it. (Food intake may be considered as work done on the body.) (b) Plants convert part of the radiant
heat transfer in sunlight to stored chemical energy, a process called photosynthesis.

Table 15.1 Summary of Terms for the First Law of Thermodynamics, $\Delta U = Q - W$

Term	Definition
U	Internal energy—the sum of the kinetic and potential energies of a system's atoms and molecules. Can be divided into many subcategories, such as thermal and chemical energy. Depends only on the state of a system (such as its P, V, and T), not on how the energy entered the system. Change in internal energy is path independent.
Q	Heat—energy transferred because of a temperature difference. Characterized by random molecular motion. Highly dependent on path. Q entering a system is positive.
W	Work—energy transferred by a force moving through a distance. An organized, orderly process. Path dependent. W done by a system (either against an external force or to increase the volume of the system) is positive.

15.2 The First Law of Thermodynamics and Some Simple Processes

Learning Objectives

By the end of this section, you will be able to:

- Describe the processes of a simple heat engine.
- Explain the differences among the simple thermodynamic processes—isobaric, isochoric, isothermal, and adiabatic.
- Calculate total work done in a cyclical thermodynamic process.

The information presented in this section supports the following AP® learning objectives and science practices:

- **5.B.5.6** The student is able to design an experiment and analyze graphical data in which interpretations of the area under a pressure-volume curve are needed to determine the work done on or by the object or system. **(S.P. 4.2, 5.1)**
- **5.B.7.2** The student is able to create a plot of pressure versus volume for a thermodynamic process from given data. **(S.P. 1.1)**
- **5.B.7.3** The student is able to use a plot of pressure versus volume for a thermodynamic process to make calculations of internal energy changes, heat, or work, based upon conservation of energy principles (i.e., the first law of thermodynamics). **(S.P. 1.1, 1.4, 2.2)**

Figure 15.7 Beginning with the Industrial Revolution, humans have harnessed power through the use of the first law of thermodynamics, before we even understood it completely. This photo, of a steam engine at the Turbinia Works, dates from 1911, a mere 61 years after the first explicit statement of the first law of thermodynamics by Rudolph Clausius. (credit: public domain; author unknown)

One of the most important things we can do with heat transfer is to use it to do work for us. Such a device is called a **heat engine**. Car engines and steam turbines that generate electricity are examples of heat engines. **Figure 15.8** shows schematically how the first law of thermodynamics applies to the typical heat engine.

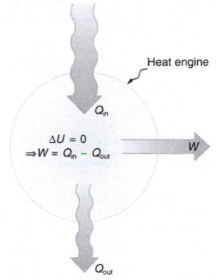

Figure 15.8 Schematic representation of a heat engine, governed, of course, by the first law of thermodynamics. It is impossible to devise a system where $Q_{\text{out}} = 0$, that is, in which no heat transfer occurs to the environment.

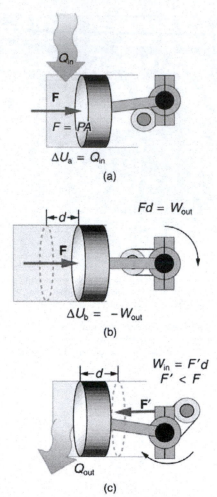

Figure 15.9 (a) Heat transfer to the gas in a cylinder increases the internal energy of the gas, creating higher pressure and temperature. (b) The force exerted on the movable cylinder does work as the gas expands. Gas pressure and temperature decrease when it expands, indicating that the gas's internal energy has been decreased by doing work. (c) Heat transfer to the environment further reduces pressure in the gas so that the piston can be more easily returned to its starting position.

The illustrations above show one of the ways in which heat transfer does work. Fuel combustion produces heat transfer to a gas in a cylinder, increasing the pressure of the gas and thereby the force it exerts on a movable piston. The gas does work on the outside world, as this force moves the piston through some distance. Heat transfer to the gas cylinder results in work being done. To repeat this process, the piston needs to be returned to its starting point. Heat transfer now occurs from the gas to the surroundings so that its pressure decreases, and a force is exerted by the surroundings to push the piston back through some distance. Variations of this process are employed daily in hundreds of millions of heat engines. We will examine heat engines in detail in the next section. In this section, we consider some of the simpler underlying processes on which heat engines are based.

PV Diagrams and their Relationship to Work Done on or by a Gas

A process by which a gas does work on a piston at constant pressure is called an **isobaric process**. Since the pressure is constant, the force exerted is constant and the work done is given as

$$P\Delta V. \tag{15.10}$$

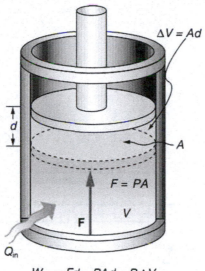

$$W_{out} = Fd = PAd = P\Delta V$$

Figure 15.10 An isobaric expansion of a gas requires heat transfer to keep the pressure constant. Since pressure is constant, the work done is $P\Delta V$.

$$W = Fd \tag{15.11}$$

See the symbols as shown in **Figure 15.10**. Now $F = PA$, and so

$$W = PAd. \tag{15.12}$$

Because the volume of a cylinder is its cross-sectional area A times its length d , we see that $Ad = \Delta V$, the change in volume; thus,

$$W = P\Delta V \text{ (isobaric process).} \tag{15.13}$$

Note that if ΔV is positive, then W is positive, meaning that work is done *by* the gas on the outside world.

(Note that the pressure involved in this work that we've called P is the pressure of the gas *inside* the tank. If we call the pressure outside the tank P_{ext} , an expanding gas would be working *against* the external pressure; the work done would therefore be $W = -P_{ext}\Delta V$ (isobaric process). Many texts use this definition of work, and not the definition based on internal pressure, as the basis of the First Law of Thermodynamics. This definition reverses the sign conventions for work, and results in a statement of the first law that becomes $\Delta U = Q + W$.)

It is not surprising that $W = P\Delta V$, since we have already noted in our treatment of fluids that pressure is a type of potential energy per unit volume and that pressure in fact has units of energy divided by volume. We also noted in our discussion of the ideal gas law that PV has units of energy. In this case, some of the energy associated with pressure becomes work.

Figure 15.11 shows a graph of pressure versus volume (that is, a PV diagram for an isobaric process. You can see in the figure that the work done is the area under the graph. This property of PV diagrams is very useful and broadly applicable: *the work done on or by a system in going from one state to another equals the area under the curve on a PV diagram.*

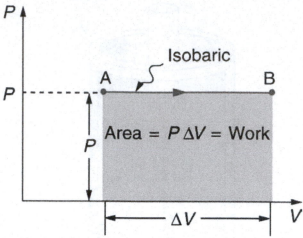

Figure 15.11 A graph of pressure versus volume for a constant-pressure, or isobaric, process, such as the one shown in **Figure 15.10**. The area under the curve equals the work done by the gas, since $W = P\Delta V$.

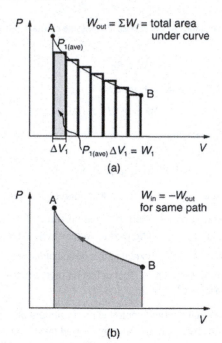

Figure 15.12 (a) A PV diagram in which pressure varies as well as volume. The work done for each interval is its average pressure times the change in volume, or the area under the curve over that interval. Thus the total area under the curve equals the total work done. (b) Work must be done on the system to follow the reverse path. This is interpreted as a negative area under the curve.

We can see where this leads by considering **Figure 15.12**(a), which shows a more general process in which both pressure and volume change. The area under the curve is closely approximated by dividing it into strips, each having an average constant pressure $P_{i(\text{ave})}$. The work done is $W_i = P_{i(\text{ave})}\Delta V_i$ for each strip, and the total work done is the sum of the W_i. Thus the total work done is the total area under the curve. If the path is reversed, as in **Figure 15.12**(b), then work is done on the system. The area under the curve in that case is negative, because ΔV is negative.

PV diagrams clearly illustrate that *the work done depends on the path taken and not just the endpoints*. This path dependence is seen in **Figure 15.13**(a), where more work is done in going from A to C by the path via point B than by the path via point D. The vertical paths, where volume is constant, are called **isochoric** processes. Since volume is constant, $\Delta V = 0$, and no work is done in an isochoric process. Now, if the system follows the cyclical path ABCDA, as in **Figure 15.13**(b), then the total work done is the area inside the loop. The negative area below path CD subtracts, leaving only the area inside the rectangle. In fact, the work done in any cyclical process (one that returns to its starting point) is the area inside the loop it forms on a PV diagram, as **Figure 15.13**(c) illustrates for a general cyclical process. Note that the loop must be traversed in the clockwise direction for work to be positive—that is, for there to be a net work output.

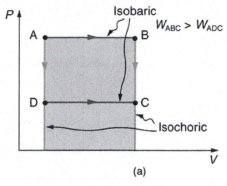

(a)

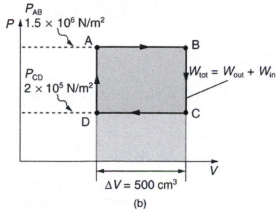

(b)

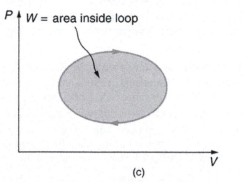

(c)

Figure 15.13 (a) The work done in going from A to C depends on path. The work is greater for the path ABC than for the path ADC, because the former is at higher pressure. In both cases, the work done is the area under the path. This area is greater for path ABC. (b) The total work done in the cyclical process ABCDA is the area inside the loop, since the negative area below CD subtracts out, leaving just the area inside the rectangle. (The values given for the pressures and the change in volume are intended for use in the example below.) (c) The area inside any closed loop is the work done in the cyclical process. If the loop is traversed in a clockwise direction, W is positive—it is work done on the outside environment. If the loop is traveled in a counter-clockwise direction, W is negative—it is work that is done to the system.

Example 15.2 Total Work Done in a Cyclical Process Equals the Area Inside the Closed Loop on a *PV* Diagram

Calculate the total work done in the cyclical process ABCDA shown in **Figure 15.13**(b) by the following two methods to verify that work equals the area inside the closed loop on the PV diagram. (Take the data in the figure to be precise to three significant figures.) (a) Calculate the work done along each segment of the path and add these values to get the total work. (b) Calculate the area inside the rectangle ABCDA.

Strategy

To find the work along any path on a PV diagram, you use the fact that work is pressure times change in volume, or $W = P\Delta V$. So in part (a), this value is calculated for each leg of the path around the closed loop.

Solution for (a)

The work along path AB is

$$W_{AB} = P_{AB}\Delta V_{AB} \tag{15.14}$$
$$= (1.50\times10^6 \text{ N/m}^2)(5.00\times10^{-4} \text{ m}^3) = 750 \text{ J}.$$

Since the path BC is isochoric, $\Delta V_{BC} = 0$, and so $W_{BC} = 0$. The work along path CD is negative, since ΔV_{CD} is negative (the volume decreases). The work is

$$W_{CD} = P_{CD}\Delta V_{CD} \tag{15.15}$$
$$= (2.00\times10^5 \text{ N/m}^2)(-5.00\times10^{-4} \text{ m}^3) = -100 \text{ J}.$$

Again, since the path DA is isochoric, $\Delta V_{DA} = 0$, and so $W_{DA} = 0$. Now the total work is

$$W = W_{AB} + W_{BC} + W_{CD} + W_{DA} \tag{15.16}$$
$$= 750 \text{ J} + 0 + (-100\text{J}) + 0 = 650 \text{ J}.$$

Solution for (b)

The area inside the rectangle is its height times its width, or

$$\text{area} = (P_{AB} - P_{CD})\Delta V \tag{15.17}$$
$$= \left[(1.50\times10^6 \text{ N/m}^2) - (2.00\times10^5 \text{ N/m}^2) \right](5.00\times10^{-4} \text{ m}^3)$$
$$= 650 \text{ J}.$$

Thus,

$$\text{area} = 650 \text{ J} = W. \tag{15.18}$$

Discussion

The result, as anticipated, is that the area inside the closed loop equals the work done. The area is often easier to calculate than is the work done along each path. It is also convenient to visualize the area inside different curves on PV diagrams in order to see which processes might produce the most work. Recall that work can be done to the system, or by the system, depending on the sign of W. A positive W is work that is done by the system on the outside environment; a negative W represents work done by the environment on the system.

Figure 15.14(a) shows two other important processes on a PV diagram. For comparison, both are shown starting from the same point A. The upper curve ending at point B is an **isothermal** process—that is, one in which temperature is kept constant. If the gas behaves like an ideal gas, as is often the case, and if no phase change occurs, then $PV = nRT$. Since T is constant, PV is a constant for an isothermal process. We ordinarily expect the temperature of a gas to decrease as it expands, and so we correctly suspect that heat transfer must occur from the surroundings to the gas to keep the temperature constant during an isothermal expansion. To show this more rigorously for the special case of a monatomic ideal gas, we note that the average kinetic energy of an atom in such a gas is given by

$$\frac{1}{2}m\bar{v}^2 = \frac{3}{2}kT. \tag{15.19}$$

The kinetic energy of the atoms in a monatomic ideal gas is its only form of internal energy, and so its total internal energy U is

$$U = N\frac{1}{2}m\bar{v}^2 = \frac{3}{2}NkT, \text{ (monatomic ideal gas)}, \tag{15.20}$$

where N is the number of atoms in the gas. This relationship means that the internal energy of an ideal monatomic gas is constant during an isothermal process—that is, $\Delta U = 0$. If the internal energy does not change, then the net heat transfer into the gas must equal the net work done by the gas. That is, because $\Delta U = Q - W = 0$ here, $Q = W$. We must have just enough heat transfer to replace the work done. An isothermal process is inherently slow, because heat transfer occurs continuously to keep the gas temperature constant at all times and must be allowed to spread through the gas so that there are no hot or cold regions.

Also shown in **Figure 15.14**(a) is a curve AC for an **adiabatic** process, defined to be one in which there is no heat transfer—that is, $Q = 0$. Processes that are nearly adiabatic can be achieved either by using very effective insulation or by performing the process so fast that there is little time for heat transfer. Temperature must decrease during an adiabatic expansion process, since work is done at the expense of internal energy:

$$U = \frac{3}{2}NkT. \tag{15.21}$$

(You might have noted that a gas released into atmospheric pressure from a pressurized cylinder is substantially colder than the gas in the cylinder.) In fact, because $Q = 0$, $\Delta U = -W$ for an adiabatic process. Lower temperature results in lower pressure along the way, so that curve AC is lower than curve AB, and less work is done. If the path ABCA could be followed by cooling the gas from B to C at constant volume (isochorically), **Figure 15.14**(b), there would be a net work output.

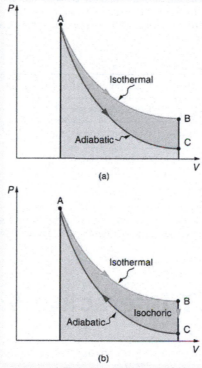

(a)

(b)

Figure 15.14 (a) The upper curve is an isothermal process ($\Delta T = 0$), whereas the lower curve is an adiabatic process ($Q = 0$). Both start from the same point A, but the isothermal process does more work than the adiabatic because heat transfer into the gas takes place to keep its temperature constant. This keeps the pressure higher all along the isothermal path than along the adiabatic path, producing more work. The adiabatic path thus ends up with a lower pressure and temperature at point C, even though the final volume is the same as for the isothermal process. (b) The cycle ABCA produces a net work output.

Applying the Science Practices: Work in a Potato Cannon

Plan an experiment using a potato cannon, meter stick, and pressure gauge to measure the work done by a potato cannon. Your experiment should produce P–V diagrams to analyze and determine the work done on a gas or by a gas. What do you need to measure? How will you measure it? Can you modify the potato cannon to make your measurements easier? When you perform multiple trials, what variables do you need to keep fixed between each trial? Which variables will you change?

One class decides to use a heavy piston, capable of being latched in place, to replace the potato. They latch the piston in place so that the contained volume is 0.50 L, load the cannon with fuel, and close the cannon with their pressure gauge, which reads 101 kPa. Then they light the fuel, and the pressure jumps to 405 kPa. Next, they release the latch, and the piston moves out until the internal volume is 2.0 L. The pressure is measured at this point to be 101 kPa again. Finally, they release the pressure gauge, and move the piston back down to 0.50 L, still at atmospheric pressure. Draw a diagram of this process, and calculate the net work performed by this system. Can you think of any ways to improve the measurements?

You should find that you have a right triangle on a P–V diagram, the area of which is the net work done by the system. Using a pressure gauge that can take continuous measurements during the expansion phase might be useful, as it is unlikely that this would actually be a linear process.

Reversible Processes

Both isothermal and adiabatic processes such as shown in **Figure 15.14** are reversible in principle. A **reversible process** is one in which both the system and its environment can return to exactly the states they were in by following the reverse path. The reverse isothermal and adiabatic paths are BA and CA, respectively. Real macroscopic processes are never exactly reversible. In the previous examples, our system is a gas (like that in **Figure 15.10**), and its environment is the piston, cylinder, and the rest of the universe. If there are any energy-dissipating mechanisms, such as friction or turbulence, then heat transfer to the environment occurs for either direction of the piston. So, for example, if the path BA is followed and there is friction, then the gas will be returned to its original state but the environment will not—it will have been heated in both directions. Reversibility requires the direction of heat transfer to reverse for the reverse path. Since dissipative mechanisms cannot be completely eliminated, real processes cannot be reversible.

There must be reasons that real macroscopic processes cannot be reversible. We can imagine them going in reverse. For example, heat transfer occurs spontaneously from hot to cold and never spontaneously the reverse. Yet it would not violate the first law of thermodynamics for this to happen. In fact, all spontaneous processes, such as bubbles bursting, never go in reverse. There is a second thermodynamic law that forbids them from going in reverse. When we study this law, we will learn something about nature and also find that such a law limits the efficiency of heat engines. We will find that heat engines with the greatest possible theoretical efficiency would have to use reversible processes, and even they cannot convert all heat transfer into doing work. **Table 15.2** summarizes the simpler thermodynamic processes and their definitions.

Table 15.2 Summary of Simple
Thermodynamic Processes

Isobaric	Constant pressure $W = P\Delta V$
Isochoric	Constant volume $W = 0$
Isothermal	Constant temperature $Q = W$
Adiabatic	No heat transfer $Q = 0$

PhET Explorations: States of Matter

Watch different types of molecules form a solid, liquid, or gas. Add or remove heat and watch the phase change. Change the temperature or volume of a container and see a pressure-temperature diagram respond in real time. Relate the interaction potential to the forces between molecules.

Figure 15.15 States of Matter (http://cnx.org/content/m55259/1.3/states-of-matter_en.jar)

15.3 Introduction to the Second Law of Thermodynamics: Heat Engines and Their Efficiency

Learning Objectives

By the end of this section, you will be able to:

- State the expressions of the second law of thermodynamics.
- Calculate the efficiency and carbon dioxide emission of a coal-fired electricity plant, using second law characteristics.
- Describe and define the Otto cycle.

Figure 15.16 These ice floes melt during the Arctic summer. Some of them refreeze in the winter, but the second law of thermodynamics predicts that it would be extremely unlikely for the water molecules contained in these particular floes to reform the distinctive alligator-like shape they formed when the picture was taken in the summer of 2009. (credit: Patrick Kelley, U.S. Coast Guard, U.S. Geological Survey)

The second law of thermodynamics deals with the direction taken by spontaneous processes. Many processes occur spontaneously in one direction only—that is, they are irreversible, under a given set of conditions. Although irreversibility is seen in day-to-day life—a broken glass does not resume its original state, for instance—complete irreversibility is a statistical statement that cannot be seen during the lifetime of the universe. More precisely, an **irreversible process** is one that depends on path. If the process can go in only one direction, then the reverse path differs fundamentally and the process cannot be reversible. For example, as noted in the previous section, heat involves the transfer of energy from higher to lower temperature. A cold object in contact with a hot one never gets colder, transferring heat to the hot object and making it hotter. Furthermore, mechanical energy, such as kinetic energy, can be completely converted to thermal energy by friction, but the reverse is impossible. A hot stationary object never spontaneously cools off and starts moving. Yet another example is the expansion of a puff of gas introduced into one corner of a vacuum chamber. The gas expands to fill the chamber, but it never regroups in the corner. The random motion of the gas molecules could take them all back to the corner, but this is never observed to happen. (See **Figure 15.17**.)

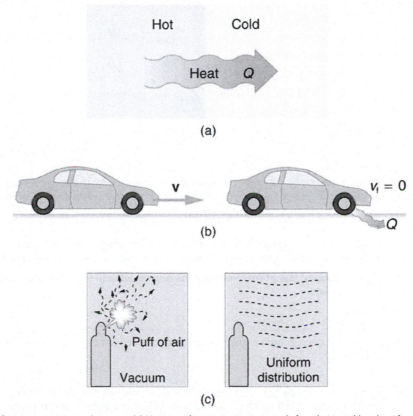

Figure 15.17 Examples of one-way processes in nature. (a) Heat transfer occurs spontaneously from hot to cold and not from cold to hot. (b) The brakes of this car convert its kinetic energy to heat transfer to the environment. The reverse process is impossible. (c) The burst of gas let into this vacuum chamber quickly expands to uniformly fill every part of the chamber. The random motions of the gas molecules will never return them to the corner.

The fact that certain processes never occur suggests that there is a law forbidding them to occur. The first law of thermodynamics would allow them to occur—none of those processes violate conservation of energy. The law that forbids these processes is called the second law of thermodynamics. We shall see that the second law can be stated in many ways that may seem different, but which in fact are equivalent. Like all natural laws, the second law of thermodynamics gives insights into nature, and its several statements imply that it is broadly applicable, fundamentally affecting many apparently disparate processes.

The already familiar direction of heat transfer from hot to cold is the basis of our first version of the **second law of thermodynamics**.

The Second Law of Thermodynamics (first expression)

Heat transfer occurs spontaneously from higher- to lower-temperature bodies but never spontaneously in the reverse direction.

Another way of stating this: It is impossible for any process to have as its sole result heat transfer from a cooler to a hotter object.

Heat Engines

Now let us consider a device that uses heat transfer to do work. As noted in the previous section, such a device is called a heat engine, and one is shown schematically in **Figure 15.18**(b). Gasoline and diesel engines, jet engines, and steam turbines are all heat engines that do work by using part of the heat transfer from some source. Heat transfer from the hot object (or hot reservoir) is denoted as Q_h, while heat transfer into the cold object (or cold reservoir) is Q_c, and the work done by the engine is W. The temperatures of the hot and cold reservoirs are T_h and T_c, respectively.

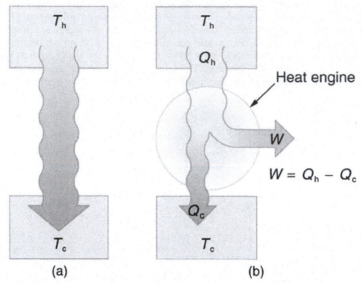

Figure 15.18 (a) Heat transfer occurs spontaneously from a hot object to a cold one, consistent with the second law of thermodynamics. (b) A heat engine, represented here by a circle, uses part of the heat transfer to do work. The hot and cold objects are called the hot and cold reservoirs. Q_h is the heat transfer out of the hot reservoir, W is the work output, and Q_c is the heat transfer into the cold reservoir.

Because the hot reservoir is heated externally, which is energy intensive, it is important that the work is done as efficiently as possible. In fact, we would like W to equal Q_h, and for there to be no heat transfer to the environment ($Q_c = 0$).

Unfortunately, this is impossible. The **second law of thermodynamics** also states, with regard to using heat transfer to do work (the second expression of the second law):

The Second Law of Thermodynamics (second expression)

It is impossible in any system for heat transfer from a reservoir to completely convert to work in a cyclical process in which the system returns to its initial state.

A **cyclical process** brings a system, such as the gas in a cylinder, back to its original state at the end of every cycle. Most heat engines, such as reciprocating piston engines and rotating turbines, use cyclical processes. The second law, just stated in its second form, clearly states that such engines cannot have perfect conversion of heat transfer into work done. Before going into the underlying reasons for the limits on converting heat transfer into work, we need to explore the relationships among W, Q_h, and Q_c, and to define the efficiency of a cyclical heat engine. As noted, a cyclical process brings the system back to its original condition at the end of every cycle. Such a system's internal energy U is the same at the beginning and end of every cycle—that is, $\Delta U = 0$. The first law of thermodynamics states that

$$\Delta U = Q - W, \tag{15.22}$$

where Q is the *net* heat transfer during the cycle ($Q = Q_h - Q_c$) and W is the net work done by the system. Since $\Delta U = 0$ for a complete cycle, we have

$$0 = Q - W, \tag{15.23}$$

so that

$$W = Q. \tag{15.24}$$

Thus the net work done by the system equals the net heat transfer into the system, or

$$W = Q_h - Q_c \text{ (cyclical process)}, \tag{15.25}$$

just as shown schematically in **Figure 15.18**(b). The problem is that in all processes, there is some heat transfer Q_c to the environment—and usually a very significant amount at that.

In the conversion of energy to work, we are always faced with the problem of getting less out than we put in. We define *conversion efficiency* Eff to be the ratio of useful work output to the energy input (or, in other words, the ratio of what we get to what we spend). In that spirit, we define the efficiency of a heat engine to be its net work output W divided by heat transfer to the engine Q_h; that is,

$$Eff = \frac{W}{Q_h}.$$ (15.26)

Since $W = Q_h - Q_c$ in a cyclical process, we can also express this as

$$Eff = \frac{Q_h - Q_c}{Q_h} = 1 - \frac{Q_c}{Q_h} \text{ (cyclical process)},$$ (15.27)

making it clear that an efficiency of 1, or 100%, is possible only if there is no heat transfer to the environment ($Q_c = 0$). Note that all Q s are positive. The direction of heat transfer is indicated by a plus or minus sign. For example, Q_c is out of the system and so is preceded by a minus sign.

Example 15.3 Daily Work Done by a Coal-Fired Power Station, Its Efficiency and Carbon Dioxide Emissions

A coal-fired power station is a huge heat engine. It uses heat transfer from burning coal to do work to turn turbines, which are used to generate electricity. In a single day, a large coal power station has 2.50×10^{14} J of heat transfer from coal and 1.48×10^{14} J of heat transfer into the environment. (a) What is the work done by the power station? (b) What is the efficiency of the power station? (c) In the combustion process, the following chemical reaction occurs: $C + O_2 \rightarrow CO_2$. This implies that every 12 kg of coal puts 12 kg + 16 kg + 16 kg = 44 kg of carbon dioxide into the atmosphere. Assuming that 1 kg of coal can provide 2.5×10^6 J of heat transfer upon combustion, how much CO_2 is emitted per day by this power plant?

Strategy for (a)

We can use $W = Q_h - Q_c$ to find the work output W, assuming a cyclical process is used in the power station. In this process, water is boiled under pressure to form high-temperature steam, which is used to run steam turbine-generators, and then condensed back to water to start the cycle again.

Solution for (a)

Work output is given by:

$$W = Q_h - Q_c.$$ (15.28)

Substituting the given values:

$$\begin{aligned} W &= 2.50 \times 10^{14} \text{ J} - 1.48 \times 10^{14} \text{ J} \\ &= 1.02 \times 10^{14} \text{ J}. \end{aligned}$$ (15.29)

Strategy for (b)

The efficiency can be calculated with $Eff = \frac{W}{Q_h}$ since Q_h is given and work W was found in the first part of this example.

Solution for (b)

Efficiency is given by: $Eff = \frac{W}{Q_h}$. The work W was just found to be 1.02×10^{14} J, and Q_h is given, so the efficiency is

$$\begin{aligned} Eff &= \frac{1.02 \times 10^{14} \text{ J}}{2.50 \times 10^{14} \text{ J}} \\ &= 0.408, \text{ or } 40.8\% \end{aligned}$$ (15.30)

Strategy for (c)

The daily consumption of coal is calculated using the information that each day there is 2.50×10^{14} J of heat transfer from coal. In the combustion process, we have $C + O_2 \rightarrow CO_2$. So every 12 kg of coal puts 12 kg + 16 kg + 16 kg = 44 kg of CO_2 into the atmosphere.

Solution for (c)

The daily coal consumption is

$$\frac{2.50 \times 10^{14} \text{ J}}{2.50 \times 10^6 \text{ J/kg}} = 1.0 \times 10^8 \text{ kg.} \tag{15.31}$$

Assuming that the coal is pure and that all the coal goes toward producing carbon dioxide, the carbon dioxide produced per day is

$$1.0 \times 10^8 \text{ kg coal} \times \frac{44 \text{ kg CO}_2}{12 \text{ kg coal}} = 3.7 \times 10^8 \text{ kg CO}_2. \tag{15.32}$$

This is 370,000 metric tons of CO_2 produced every day.

Discussion

If all the work output is converted to electricity in a period of one day, the average power output is 1180 MW (this is left to you as an end-of-chapter problem). This value is about the size of a large-scale conventional power plant. The efficiency found is acceptably close to the value of 42% given for coal power stations. It means that fully 59.2% of the energy is heat transfer to the environment, which usually results in warming lakes, rivers, or the ocean near the power station, and is implicated in a warming planet generally. While the laws of thermodynamics limit the efficiency of such plants—including plants fired by nuclear fuel, oil, and natural gas—the heat transfer to the environment could be, and sometimes is, used for heating homes or for industrial processes. The generally low cost of energy has not made it economical to make better use of the waste heat transfer from most heat engines. Coal-fired power plants produce the greatest amount of CO_2 per unit energy output (compared to natural gas or oil), making coal the least efficient fossil fuel.

With the information given in **Example 15.3**, we can find characteristics such as the efficiency of a heat engine without any knowledge of how the heat engine operates, but looking further into the mechanism of the engine will give us greater insight. **Figure 15.19** illustrates the operation of the common four-stroke gasoline engine. The four steps shown complete this heat engine's cycle, bringing the gasoline-air mixture back to its original condition.

The **Otto cycle** shown in **Figure 15.20**(a) is used in four-stroke internal combustion engines, although in fact the true Otto cycle paths do not correspond exactly to the strokes of the engine.

The adiabatic process AB corresponds to the nearly adiabatic compression stroke of the gasoline engine. In both cases, work is done on the system (the gas mixture in the cylinder), increasing its temperature and pressure. Along path BC of the Otto cycle, heat transfer Q_h into the gas occurs at constant volume, causing a further increase in pressure and temperature. This process corresponds to burning fuel in an internal combustion engine, and takes place so rapidly that the volume is nearly constant. Path CD in the Otto cycle is an adiabatic expansion that does work on the outside world, just as the power stroke of an internal combustion engine does in its nearly adiabatic expansion. The work done by the system along path CD is greater than the work done on the system along path AB, because the pressure is greater, and so there is a net work output. Along path DA in the Otto cycle, heat transfer Q_c from the gas at constant volume reduces its temperature and pressure, returning it to its original state. In an internal combustion engine, this process corresponds to the exhaust of hot gases and the intake of an air-gasoline mixture at a considerably lower temperature. In both cases, heat transfer into the environment occurs along this final path.

The net work done by a cyclical process is the area inside the closed path on a PV diagram, such as that inside path ABCDA in **Figure 15.20**. Note that in every imaginable cyclical process, it is absolutely necessary for heat transfer from the system to occur in order to get a net work output. In the Otto cycle, heat transfer occurs along path DA. If no heat transfer occurs, then the return path is the same, and the net work output is zero. The lower the temperature on the path AB, the less work has to be done to compress the gas. The area inside the closed path is then greater, and so the engine does more work and is thus more efficient. Similarly, the higher the temperature along path CD, the more work output there is. (See **Figure 15.21**.) So efficiency is related to the temperatures of the hot and cold reservoirs. In the next section, we shall see what the absolute limit to the efficiency of a heat engine is, and how it is related to temperature.

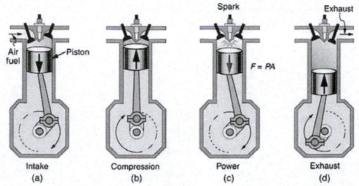

Figure 15.19 In the four-stroke internal combustion gasoline engine, heat transfer into work takes place in the cyclical process shown here. The piston is connected to a rotating crankshaft, which both takes work out of and does work on the gas in the cylinder. (a) Air is mixed with fuel during the intake stroke. (b) During the compression stroke, the air-fuel mixture is rapidly compressed in a nearly adiabatic process, as the piston rises with the valves closed. Work is done on the gas. (c) The power stroke has two distinct parts. First, the air-fuel mixture is ignited, converting chemical potential energy into thermal energy almost instantaneously, which leads to a great increase in pressure. Then the piston descends, and the gas does work by exerting a force through a distance in a nearly adiabatic process. (d) The exhaust stroke expels the hot gas to prepare the engine for another cycle, starting again with the intake stroke.

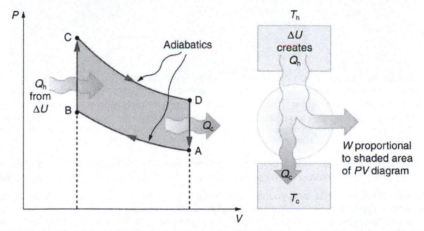

Figure 15.20 PV diagram for a simplified Otto cycle, analogous to that employed in an internal combustion engine. Point A corresponds to the start of the compression stroke of an internal combustion engine. Paths AB and CD are adiabatic and correspond to the compression and power strokes of an internal combustion engine, respectively. Paths BC and DA are isochoric and accomplish similar results to the ignition and exhaust-intake portions, respectively, of the internal combustion engine's cycle. Work is done on the gas along path AB, but more work is done by the gas along path CD, so that there is a net work output.

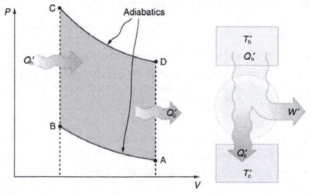

Figure 15.21 This Otto cycle produces a greater work output than the one in Figure 15.20, because the starting temperature of path CD is higher and the starting temperature of path AB is lower. The area inside the loop is greater, corresponding to greater net work output.

15.4 Carnot's Perfect Heat Engine: The Second Law of Thermodynamics Restated

Learning Objectives
By the end of this section, you will be able to:
• Identify a Carnot cycle.

- Calculate maximum theoretical efficiency of a nuclear reactor.
- Explain how dissipative processes affect the ideal Carnot engine.

Figure 15.22 This novelty toy, known as the drinking bird, is an example of Carnot's engine. It contains methylene chloride (mixed with a dye) in the abdomen, which boils at a very low temperature—about $100°F$. To operate, one gets the bird's head wet. As the water evaporates, fluid moves up into the head, causing the bird to become top-heavy and dip forward back into the water. This cools down the methylene chloride in the head, and it moves back into the abdomen, causing the bird to become bottom heavy and tip up. Except for a very small input of energy—the original head-wetting—the bird becomes a perpetual motion machine of sorts. (credit: Arabesk.nl, Wikimedia Commons)

We know from the second law of thermodynamics that a heat engine cannot be 100% efficient, since there must always be some heat transfer Q_c to the environment, which is often called waste heat. How efficient, then, can a heat engine be? This question was answered at a theoretical level in 1824 by a young French engineer, Sadi Carnot (1796–1832), in his study of the then-emerging heat engine technology crucial to the Industrial Revolution. He devised a theoretical cycle, now called the **Carnot cycle**, which is the most efficient cyclical process possible. The second law of thermodynamics can be restated in terms of the Carnot cycle, and so what Carnot actually discovered was this fundamental law. Any heat engine employing the Carnot cycle is called a **Carnot engine**.

What is crucial to the Carnot cycle—and, in fact, defines it—is that only reversible processes are used. Irreversible processes involve dissipative factors, such as friction and turbulence. This increases heat transfer Q_c to the environment and reduces the efficiency of the engine. Obviously, then, reversible processes are superior.

Carnot Engine

Stated in terms of reversible processes, the **second law of thermodynamics** has a third form:

A Carnot engine operating between two given temperatures has the greatest possible efficiency of any heat engine operating between these two temperatures. Furthermore, all engines employing only reversible processes have this same maximum efficiency when operating between the same given temperatures.

Figure 15.23 shows the PV diagram for a Carnot cycle. The cycle comprises two isothermal and two adiabatic processes. Recall that both isothermal and adiabatic processes are, in principle, reversible.

Carnot also determined the efficiency of a perfect heat engine—that is, a Carnot engine. It is always true that the efficiency of a cyclical heat engine is given by:

$$Eff = \frac{Q_h - Q_c}{Q_h} = 1 - \frac{Q_c}{Q_h}.$$

(15.33)

What Carnot found was that for a perfect heat engine, the ratio Q_c / Q_h equals the ratio of the absolute temperatures of the heat reservoirs. That is, $Q_c / Q_h = T_c / T_h$ for a Carnot engine, so that the maximum or **Carnot efficiency** Eff_C is given by

$$Eff_C = 1 - \frac{T_c}{T_h},$$

(15.34)

where T_h and T_c are in kelvins (or any other absolute temperature scale). No real heat engine can do as well as the Carnot efficiency—an actual efficiency of about 0.7 of this maximum is usually the best that can be accomplished. But the ideal Carnot engine, like the drinking bird above, while a fascinating novelty, has zero power. This makes it unrealistic for any applications.

Carnot's interesting result implies that 100% efficiency would be possible only if $T_c = 0 \, \text{K}$ —that is, only if the cold reservoir were at absolute zero, a practical and theoretical impossibility. But the physical implication is this—the only way to have all heat transfer go into doing work is to remove *all* thermal energy, and this requires a cold reservoir at absolute zero.

It is also apparent that the greatest efficiencies are obtained when the ratio T_c / T_h is as small as possible. Just as discussed for the Otto cycle in the previous section, this means that efficiency is greatest for the highest possible temperature of the hot reservoir and lowest possible temperature of the cold reservoir. (This setup increases the area inside the closed loop on the PV diagram; also, it seems reasonable that the greater the temperature difference, the easier it is to divert the heat transfer to work.) The actual reservoir temperatures of a heat engine are usually related to the type of heat source and the temperature of the environment into which heat transfer occurs. Consider the following example.

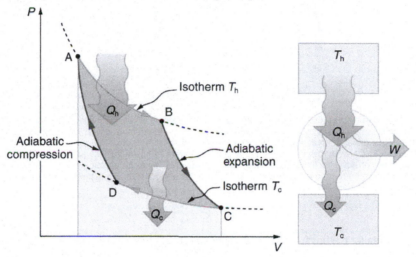

Figure 15.23 PV diagram for a Carnot cycle, employing only reversible isothermal and adiabatic processes. Heat transfer Q_h occurs into the working substance during the isothermal path AB, which takes place at constant temperature T_h. Heat transfer Q_c occurs out of the working substance during the isothermal path CD, which takes place at constant temperature T_c. The net work output W equals the area inside the path ABCDA. Also shown is a schematic of a Carnot engine operating between hot and cold reservoirs at temperatures T_h and T_c. Any heat engine using reversible processes and operating between these two temperatures will have the same maximum efficiency as the Carnot engine.

Example 15.4 Maximum Theoretical Efficiency for a Nuclear Reactor

A nuclear power reactor has pressurized water at $300°C$. (Higher temperatures are theoretically possible but practically not, due to limitations with materials used in the reactor.) Heat transfer from this water is a complex process (see **Figure 15.24**). Steam, produced in the steam generator, is used to drive the turbine-generators. Eventually the steam is condensed to water at $27°C$ and then heated again to start the cycle over. Calculate the maximum theoretical efficiency for a heat engine operating between these two temperatures.

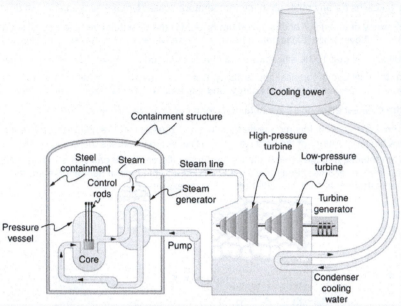

Figure 15.24 Schematic diagram of a pressurized water nuclear reactor and the steam turbines that convert work into electrical energy. Heat exchange is used to generate steam, in part to avoid contamination of the generators with radioactivity. Two turbines are used because this is less expensive than operating a single generator that produces the same amount of electrical energy. The steam is condensed to liquid before being returned to the heat exchanger, to keep exit steam pressure low and aid the flow of steam through the turbines (equivalent to using a lower-temperature cold reservoir). The considerable energy associated with condensation must be dissipated into the local environment; in this example, a cooling tower is used so there is no direct heat transfer to an aquatic environment. (Note that the water going to the cooling tower does not come into contact with the steam flowing over the turbines.)

Strategy

Since temperatures are given for the hot and cold reservoirs of this heat engine, $Eff_C = 1 - \dfrac{T_c}{T_h}$ can be used to calculate the Carnot (maximum theoretical) efficiency. Those temperatures must first be converted to kelvins.

Solution

The hot and cold reservoir temperatures are given as $300°C$ and $27.0°C$, respectively. In kelvins, then, $T_h = 573\ \text{K}$ and $T_c = 300\ \text{K}$, so that the maximum efficiency is

$$Eff_C = 1 - \frac{T_c}{T_h}. \tag{15.35}$$

Thus,

$$
\begin{aligned}
Eff_C &= 1 - \frac{300\ \text{K}}{573\ \text{K}} \\
&= 0.476,\ \text{or}\ 47.6\%.
\end{aligned}
\tag{15.36}
$$

Discussion

A typical nuclear power station's actual efficiency is about 35%, a little better than 0.7 times the maximum possible value, a tribute to superior engineering. Electrical power stations fired by coal, oil, and natural gas have greater actual efficiencies (about 42%), because their boilers can reach higher temperatures and pressures. The cold reservoir temperature in any of these power stations is limited by the local environment. **Figure 15.25** shows (a) the exterior of a nuclear power station and (b) the exterior of a coal-fired power station. Both have cooling towers into which water from the condenser enters the tower near the top and is sprayed downward, cooled by evaporation.

(a)

(b)

Figure 15.25 (a) A nuclear power station (credit: BlatantWorld.com) and (b) a coal-fired power station. Both have cooling towers in which water evaporates into the environment, representing Q_c. The nuclear reactor, which supplies Q_h, is housed inside the dome-shaped containment buildings. (credit: Robert & Mihaela Vicol, publicphoto.org)

Since all real processes are irreversible, the actual efficiency of a heat engine can never be as great as that of a Carnot engine, as illustrated in **Figure 15.26**(a). Even with the best heat engine possible, there are always dissipative processes in peripheral equipment, such as electrical transformers or car transmissions. These further reduce the overall efficiency by converting some of the engine's work output back into heat transfer, as shown in **Figure 15.26**(b).

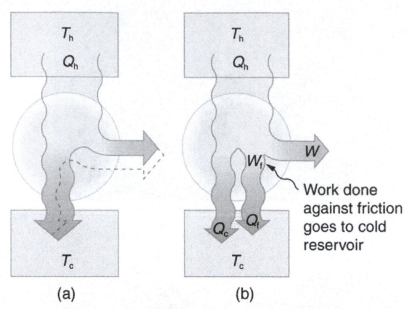

Figure 15.26 Real heat engines are less efficient than Carnot engines. (a) Real engines use irreversible processes, reducing the heat transfer to work. Solid lines represent the actual process; the dashed lines are what a Carnot engine would do between the same two reservoirs. (b) Friction and other dissipative processes in the output mechanisms of a heat engine convert some of its work output into heat transfer to the environment.

15.5 Applications of Thermodynamics: Heat Pumps and Refrigerators

Learning Objectives

By the end of this section, you will be able to:

- Describe the use of heat engines in heat pumps and refrigerators.
- Demonstrate how a heat pump works to warm an interior space.
- Explain the differences between heat pumps and refrigerators.
- Calculate a heat pump's coefficient of performance.

Figure 15.27 Almost every home contains a refrigerator. Most people don't realize they are also sharing their homes with a heat pump. (credit: ld1337x, Wikimedia Commons)

Heat pumps, air conditioners, and refrigerators utilize heat transfer from cold to hot. They are heat engines run backward. We say backward, rather than reverse, because except for Carnot engines, all heat engines, though they can be run backward, cannot truly be reversed. Heat transfer occurs from a cold reservoir Q_c and into a hot one. This requires work input W, which is also converted to heat transfer. Thus the heat transfer to the hot reservoir is $Q_h = Q_c + W$. (Note that Q_h, Q_c, and W are positive, with their directions indicated on schematics rather than by sign.) A heat pump's mission is for heat transfer Q_h to occur into a warm environment, such as a home in the winter. The mission of air conditioners and refrigerators is for heat transfer Q_c to occur from a cool environment, such as chilling a room or keeping food at lower temperatures than the environment. (Actually, a heat pump can be used both to heat and cool a space. It is essentially an air conditioner and a heating unit all in one.

In this section we will concentrate on its heating mode.)

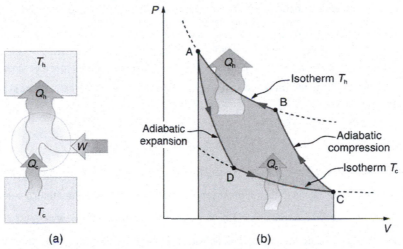

(a)   (b)

Figure 15.28 Heat pumps, air conditioners, and refrigerators are heat engines operated backward. The one shown here is based on a Carnot (reversible) engine. (a) Schematic diagram showing heat transfer from a cold reservoir to a warm reservoir with a heat pump. The directions of W, Q_h, and Q_c are opposite what they would be in a heat engine. (b) PV diagram for a Carnot cycle similar to that in **Figure 15.29** but reversed, following path ADCBA. The area inside the loop is negative, meaning there is a net work input. There is heat transfer Q_c into the system from a cold reservoir along path DC, and heat transfer Q_h out of the system into a hot reservoir along path BA.

Heat Pumps

The great advantage of using a heat pump to keep your home warm, rather than just burning fuel, is that a heat pump supplies $Q_h = Q_c + W$. Heat transfer is from the outside air, even at a temperature below freezing, to the indoor space. You only pay for W, and you get an additional heat transfer of Q_c from the outside at no cost; in many cases, at least twice as much energy is transferred to the heated space as is used to run the heat pump. When you burn fuel to keep warm, you pay for all of it. The disadvantage is that the work input (required by the second law of thermodynamics) is sometimes more expensive than simply burning fuel, especially if the work is done by electrical energy.

The basic components of a heat pump in its heating mode are shown in **Figure 15.29**. A working fluid such as a non-CFC refrigerant is used. In the outdoor coils (the evaporator), heat transfer Q_c occurs to the working fluid from the cold outdoor air, turning it into a gas.

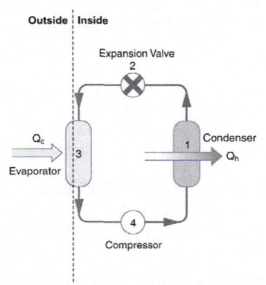

Figure 15.29 A simple heat pump has four basic components: (1) condenser, (2) expansion valve, (3) evaporator, and (4) compressor. In the heating mode, heat transfer Q_c occurs to the working fluid in the evaporator (3) from the colder outdoor air, turning it into a gas. The electrically driven compressor (4) increases the temperature and pressure of the gas and forces it into the condenser coils (1) inside the heated space. Because the temperature of the gas is higher than the temperature in the room, heat transfer from the gas to the room occurs as the gas condenses to a liquid. The working fluid is then cooled as it flows back through an expansion valve (2) to the outdoor evaporator coils.

The electrically driven compressor (work input W) raises the temperature and pressure of the gas and forces it into the condenser coils that are inside the heated space. Because the temperature of the gas is higher than the temperature inside the room, heat transfer to the room occurs and the gas condenses to a liquid. The liquid then flows back through a pressure-reducing valve to the outdoor evaporator coils, being cooled through expansion. (In a cooling cycle, the evaporator and condenser coils exchange roles and the flow direction of the fluid is reversed.)

The quality of a heat pump is judged by how much heat transfer Q_h occurs into the warm space compared with how much work input W is required. In the spirit of taking the ratio of what you get to what you spend, we define a **heat pump's coefficient of performance** (COP_{hp}) to be

$$COP_{hp} = \frac{Q_h}{W}.$$

(15.37)

Since the efficiency of a heat engine is $Eff = W / Q_h$, we see that $COP_{hp} = 1 / Eff$, an important and interesting fact. First, since the efficiency of any heat engine is less than 1, it means that COP_{hp} is always greater than 1—that is, a heat pump always has more heat transfer Q_h than work put into it. Second, it means that heat pumps work best when temperature differences are small. The efficiency of a perfect, or Carnot, engine is $Eff_C = 1 - (T_c / T_h)$; thus, the smaller the temperature difference, the smaller the efficiency and the greater the COP_{hp} (because $COP_{hp} = 1 / Eff$). In other words, heat pumps do not work as well in very cold climates as they do in more moderate climates.

Friction and other irreversible processes reduce heat engine efficiency, but they do *not* benefit the operation of a heat pump—instead, they reduce the work input by converting part of it to heat transfer back into the cold reservoir before it gets into the heat pump.

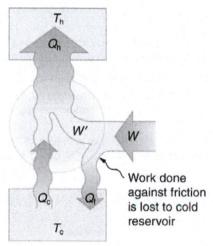

Figure 15.30 When a real heat engine is run backward, some of the intended work input (W) goes into heat transfer before it gets into the heat engine, thereby reducing its coefficient of performance COP_{hp}. In this figure, W' represents the portion of W that goes into the heat pump, while the remainder of W is lost in the form of frictional heat (Q_f) to the cold reservoir. If all of W had gone into the heat pump, then Q_h would have been greater. The best heat pump uses adiabatic and isothermal processes, since, in theory, there would be no dissipative processes to reduce the heat transfer to the hot reservoir.

Example 15.5 The Best COP_{hp} of a Heat Pump for Home Use

A heat pump used to warm a home must employ a cycle that produces a working fluid at temperatures greater than typical indoor temperature so that heat transfer to the inside can take place. Similarly, it must produce a working fluid at temperatures that are colder than the outdoor temperature so that heat transfer occurs from outside. Its hot and cold reservoir temperatures therefore cannot be too close, placing a limit on its COP_{hp}. (See **Figure 15.31**.) What is the best coefficient of performance possible for such a heat pump, if it has a hot reservoir temperature of $45.0°C$ and a cold reservoir temperature of $-15.0°C$?

Strategy

A Carnot engine reversed will give the best possible performance as a heat pump. As noted above, $COP_{hp} = 1 / Eff$, so that we need to first calculate the Carnot efficiency to solve this problem.

Solution

Carnot efficiency in terms of absolute temperature is given by:

$$Eff_C = 1 - \frac{T_c}{T_h}.$$

(15.38)

The temperatures in kelvins are $T_h = 318 \text{ K}$ and $T_c = 258 \text{ K}$, so that

$$Eff_C = 1 - \frac{258 \text{ K}}{318 \text{ K}} = 0.1887.$$

(15.39)

Thus, from the discussion above,

$$COP_{hp} = \frac{1}{Eff} = \frac{1}{0.1887} = 5.30,$$

(15.40)

or

$$COP_{hp} = \frac{Q_h}{W} = 5.30,$$

(15.41)

so that

$$Q_h = 5.30 \text{ W}.$$

(15.42)

Discussion

This result means that the heat transfer by the heat pump is 5.30 times as much as the work put into it. It would cost 5.30 times as much for the same heat transfer by an electric room heater as it does for that produced by this heat pump. This is not a violation of conservation of energy. Cold ambient air provides 4.3 J per 1 J of work from the electrical outlet.

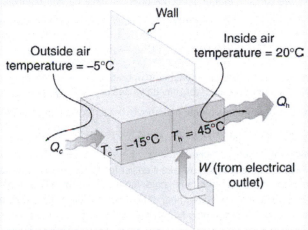

Figure 15.31 Heat transfer from the outside to the inside, along with work done to run the pump, takes place in the heat pump of the example above. Note that the cold temperature produced by the heat pump is lower than the outside temperature, so that heat transfer into the working fluid occurs. The pump's compressor produces a temperature greater than the indoor temperature in order for heat transfer into the house to occur.

Real heat pumps do not perform quite as well as the ideal one in the previous example; their values of COP_{hp} range from about 2 to 4. This range means that the heat transfer Q_h from the heat pumps is 2 to 4 times as great as the work W put into them. Their economical feasibility is still limited, however, since W is usually supplied by electrical energy that costs more per joule than heat transfer by burning fuels like natural gas. Furthermore, the initial cost of a heat pump is greater than that of many furnaces, so that a heat pump must last longer for its cost to be recovered. Heat pumps are most likely to be economically superior where winter temperatures are mild, electricity is relatively cheap, and other fuels are relatively expensive. Also, since they can cool as well as heat a space, they have advantages where cooling in summer months is also desired. Thus some of the best locations for heat pumps are in warm summer climates with cool winters. **Figure 15.32** shows a heat pump, called a "*reverse cycle*" or "*split-system cooler*" in some countries.

Figure 15.32 In hot weather, heat transfer occurs from air inside the room to air outside, cooling the room. In cool weather, heat transfer occurs from air outside to air inside, warming the room. This switching is achieved by reversing the direction of flow of the working fluid.

Air Conditioners and Refrigerators

Air conditioners and refrigerators are designed to cool something down in a warm environment. As with heat pumps, work input is required for heat transfer from cold to hot, and this is expensive. The quality of air conditioners and refrigerators is judged by how much heat transfer Q_c occurs from a cold environment compared with how much work input W is required. What is considered the benefit in a heat pump is considered waste heat in a refrigerator. We thus define the **coefficient of performance** (COP_{ref}) of an air conditioner or refrigerator to be

$$COP_{ref} = \frac{Q_c}{W}.$$

(15.43)

Noting again that $Q_h = Q_c + W$, we can see that an air conditioner will have a lower coefficient of performance than a heat pump, because $COP_{hp} = Q_h / W$ and Q_h is greater than Q_c. In this module's Problems and Exercises, you will show that

$$COP_{ref} = COP_{hp} - 1$$

(15.44)

for a heat engine used as either an air conditioner or a heat pump operating between the same two temperatures. Real air conditioners and refrigerators typically do remarkably well, having values of COP_{ref} ranging from 2 to 6. These numbers are better than the COP_{hp} values for the heat pumps mentioned above, because the temperature differences are smaller, but they are less than those for Carnot engines operating between the same two temperatures.

A type of COP rating system called the "energy efficiency rating" (EER) has been developed. This rating is an example where non-SI units are still used and relevant to consumers. To make it easier for the consumer, Australia, Canada, New Zealand, and the U.S. use an Energy Star Rating out of 5 stars—the more stars, the more energy efficient the appliance. EERs are expressed in mixed units of British thermal units (Btu) per hour of heating or cooling divided by the power input in watts. Room air conditioners are readily available with EERs ranging from 6 to 12. Although not the same as the COPs just described, these EERs are good for comparison purposes—the greater the EER, the cheaper an air conditioner is to operate (but the higher its purchase price is likely to be).

The EER of an air conditioner or refrigerator can be expressed as

$$EER = \frac{Q_c / t_1}{W / t_2},$$

(15.45)

where Q_c is the amount of heat transfer from a cold environment in British thermal units, t_1 is time in hours, W is the work input in joules, and t_2 is time in seconds.

Problem-Solving Strategies for Thermodynamics

1. *Examine the situation to determine whether heat, work, or internal energy are involved.* Look for any system where the primary methods of transferring energy are heat and work. Heat engines, heat pumps, refrigerators, and air conditioners are examples of such systems.

2. *Identify the system of interest and draw a labeled diagram of the system showing energy flow.*

3. *Identify exactly what needs to be determined in the problem (identify the unknowns).* A written list is useful. Maximum efficiency means a Carnot engine is involved. Efficiency is not the same as the coefficient of performance.

4. *Make a list of what is given or can be inferred from the problem as stated (identify the knowns).* Be sure to distinguish heat transfer into a system from heat transfer out of the system, as well as work input from work output. In many situations, it is useful to determine the type of process, such as isothermal or adiabatic.

5. *Solve the appropriate equation for the quantity to be determined (the unknown).*

6. *Substitute the known quantities along with their units into the appropriate equation and obtain numerical solutions complete with units.*

7. *Check the answer to see if it is reasonable: Does it make sense?* For example, efficiency is always less than 1, whereas coefficients of performance are greater than 1.

15.6 Entropy and the Second Law of Thermodynamics: Disorder and the Unavailability of Energy

Learning Objectives

By the end of this section, you will be able to:

* Define entropy.
* Calculate the increase of entropy in a system with reversible and irreversible processes.
* Explain the expected fate of the universe in entropic terms.
* Calculate the increasing disorder of a system.

The information presented in this section supports the following AP® learning objectives and science practices:

* **7.B.2.1:** The student is able to connect qualitatively the second law of thermodynamics in terms of the state function called entropy and how it (entropy) behaves in reversible and irreversible processes. **(S.P. 7.1)**

Figure 15.33 The ice in this drink is slowly melting. Eventually the liquid will reach thermal equilibrium, as predicted by the second law of thermodynamics. (credit: Jon Sullivan, PDPhoto.org)

There is yet another way of expressing the second law of thermodynamics. This version relates to a concept called **entropy**. By examining it, we shall see that the directions associated with the second law—heat transfer from hot to cold, for example—are related to the tendency in nature for systems to become disordered and for less energy to be available for use as work. The entropy of a system can in fact be shown to be a measure of its disorder and of the unavailability of energy to do work.

Making Connections: Entropy, Energy, and Work

Recall that the simple definition of energy is the ability to do work. Entropy is a measure of how much energy is not available to do work. Although all forms of energy are interconvertible, and all can be used to do work, it is not always possible, even in principle, to convert the entire available energy into work. That unavailable energy is of interest in thermodynamics, because the field of thermodynamics arose from efforts to convert heat to work.

We can see how entropy is defined by recalling our discussion of the Carnot engine. We noted that for a Carnot cycle, and hence for any reversible processes, $Q_c / Q_h = T_c / T_h$. Rearranging terms yields

$$\frac{Q_c}{T_c} = \frac{Q_h}{T_h}$$

(15.46)

for any reversible process. Q_c and Q_h are absolute values of the heat transfer at temperatures T_c and T_h, respectively. This ratio of Q/T is defined to be the **change in entropy** ΔS for a reversible process,

$$\Delta S = \left(\frac{Q}{T}\right)_{\text{rev}},$$ (15.47)

where Q is the heat transfer, which is positive for heat transfer into and negative for heat transfer out of, and T is the absolute temperature at which the reversible process takes place. The SI unit for entropy is joules per kelvin (J/K). If temperature changes during the process, then it is usually a good approximation (for small changes in temperature) to take T to be the average temperature, avoiding the need to use integral calculus to find ΔS.

The definition of ΔS is strictly valid only for reversible processes, such as used in a Carnot engine. However, we can find ΔS precisely even for real, irreversible processes. The reason is that the entropy S of a system, like internal energy U, depends only on the state of the system and not how it reached that condition. Entropy is a property of state. Thus the change in entropy ΔS of a system between state 1 and state 2 is the same no matter how the change occurs. We just need to find or imagine a reversible process that takes us from state 1 to state 2 and calculate ΔS for that process. That will be the change in entropy for any process going from state 1 to state 2. (See **Figure 15.34**.)

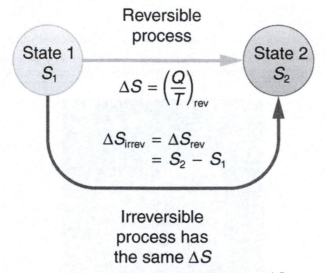

Figure 15.34 When a system goes from state 1 to state 2, its entropy changes by the same amount ΔS, whether a hypothetical reversible path is followed or a real irreversible path is taken.

Now let us take a look at the change in entropy of a Carnot engine and its heat reservoirs for one full cycle. The hot reservoir has a loss of entropy $\Delta S_{\text{h}} = -Q_{\text{h}}/T_{\text{h}}$, because heat transfer occurs out of it (remember that when heat transfers out, then Q has a negative sign). The cold reservoir has a gain of entropy $\Delta S_{\text{c}} = Q_{\text{c}}/T_{\text{c}}$, because heat transfer occurs into it. (We assume the reservoirs are sufficiently large that their temperatures are constant.) So the total change in entropy is

$$\Delta S_{\text{tot}} = \Delta S_{\text{h}} + \Delta S_{\text{c}}.$$ (15.48)

Thus, since we know that $Q_{\text{h}}/T_{\text{h}} = Q_{\text{c}}/T_{\text{c}}$ for a Carnot engine,

$$\Delta S_{\text{tot}} = \frac{Q_{\text{h}}}{T_{\text{h}}} + \frac{Q_{\text{c}}}{T_{\text{c}}} = 0.$$ (15.49)

This result, which has general validity, means that *the total change in entropy for a system in any reversible process is zero.*

The entropy of various parts of the system may change, but the total change is zero. Furthermore, the system does not affect the entropy of its surroundings, since heat transfer between them does not occur. Thus the reversible process changes neither the total entropy of the system nor the entropy of its surroundings. Sometimes this is stated as follows: *Reversible processes do not affect the total entropy of the universe.* Real processes are not reversible, though, and they do change total entropy. We can, however, use hypothetical reversible processes to determine the value of entropy in real, irreversible processes. The following example illustrates this point.

Example 15.6 Entropy Increases in an Irreversible (Real) Process

Spontaneous heat transfer from hot to cold is an irreversible process. Calculate the total change in entropy if 4000 J of heat transfer occurs from a hot reservoir at $T_{\text{h}} = 600 \text{ K}(327° \text{ C})$ to a cold reservoir at $T_{\text{c}} = 250 \text{ K}(-23° \text{ C})$, assuming there is no temperature change in either reservoir. (See **Figure 15.35**.)

Strategy

How can we calculate the change in entropy for an irreversible process when $\Delta S_{tot} = \Delta S_h + \Delta S_c$ is valid only for

reversible processes? Remember that the total change in entropy of the hot and cold reservoirs will be the same whether a reversible or irreversible process is involved in heat transfer from hot to cold. So we can calculate the change in entropy of the hot reservoir for a hypothetical reversible process in which 4000 J of heat transfer occurs from it; then we do the same for a hypothetical reversible process in which 4000 J of heat transfer occurs to the cold reservoir. This produces the same changes in the hot and cold reservoirs that would occur if the heat transfer were allowed to occur irreversibly between them, and so it also produces the same changes in entropy.

Solution

We now calculate the two changes in entropy using $\Delta S_{tot} = \Delta S_h + \Delta S_c$. First, for the heat transfer from the hot reservoir,

$$\Delta S_h = \frac{-Q_h}{T_h} = \frac{-4000 \text{ J}}{600 \text{ K}} = -6.67 \text{ J/K}. \tag{15.50}$$

And for the cold reservoir,

$$\Delta S_c = \frac{-Q_c}{T_c} = \frac{4000 \text{ J}}{250 \text{ K}} = 16.0 \text{ J/K}. \tag{15.51}$$

Thus the total is

$$
\begin{aligned}
\Delta S_{tot} &= \quad \Delta S_h + \Delta S_c \\
&= \quad (-6.67 + 16.0) \text{ J/K} \\
&= \quad 9.33 \text{ J/K}.
\end{aligned}
\tag{15.52}
$$

Discussion

There is an *increase* in entropy for the system of two heat reservoirs undergoing this irreversible heat transfer. We will see that this means there is a loss of ability to do work with this transferred energy. Entropy has increased, and energy has become unavailable to do work.

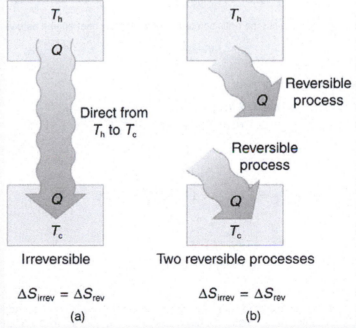

Figure 15.35 (a) Heat transfer from a hot object to a cold one is an irreversible process that produces an overall increase in entropy. (b) The same final state and, thus, the same change in entropy is achieved for the objects if reversible heat transfer processes occur between the two objects whose temperatures are the same as the temperatures of the corresponding objects in the irreversible process.

It is reasonable that entropy increases for heat transfer from hot to cold. Since the change in entropy is Q/T, there is a larger change at lower temperatures. The decrease in entropy of the hot object is therefore less than the increase in entropy of the cold object, producing an overall increase, just as in the previous example. This result is very general:

There is an increase in entropy for any system undergoing an irreversible process.

With respect to entropy, there are only two possibilities: entropy is constant for a reversible process, and it increases for an irreversible process. There is a fourth version of **the second law of thermodynamics stated in terms of entropy**:

The total entropy of a system either increases or remains constant in any process; it never decreases.

For example, heat transfer cannot occur spontaneously from cold to hot, because entropy would decrease.

Entropy is very different from energy. Entropy is *not* conserved but increases in all real processes. Reversible processes (such as in Carnot engines) are the processes in which the most heat transfer to work takes place and are also the ones that keep entropy constant. Thus we are led to make a connection between entropy and the availability of energy to do work.

Entropy and the Unavailability of Energy to Do Work

What does a change in entropy mean, and why should we be interested in it? One reason is that entropy is directly related to the fact that not all heat transfer can be converted into work. The next example gives some indication of how an increase in entropy results in less heat transfer into work.

Example 15.7 Less Work is Produced by a Given Heat Transfer When Entropy Change is Greater

(a) Calculate the work output of a Carnot engine operating between temperatures of 600 K and 100 K for 4000 J of heat transfer to the engine. (b) Now suppose that the 4000 J of heat transfer occurs first from the 600 K reservoir to a 250 K reservoir (without doing any work, and this produces the increase in entropy calculated above) before transferring into a Carnot engine operating between 250 K and 100 K. What work output is produced? (See **Figure 15.36**.)

Strategy

In both parts, we must first calculate the Carnot efficiency and then the work output.

Solution (a)

The Carnot efficiency is given by

$$Eff_C = 1 - \frac{T_c}{T_h}. \tag{15.53}$$

Substituting the given temperatures yields

$$Eff_C = 1 - \frac{100 \text{ K}}{600 \text{ K}} = 0.833. \tag{15.54}$$

Now the work output can be calculated using the definition of efficiency for any heat engine as given by

$$Eff = \frac{W}{Q_h}. \tag{15.55}$$

Solving for W and substituting known terms gives

$$\begin{aligned} W &= Eff_C Q_h \\ &= (0.833)(4000 \text{ J}) = 3333 \text{ J}. \end{aligned} \tag{15.56}$$

Solution (b)

Similarly,

$$Eff'_C = 1 - \frac{T_c}{T'_c} = 1 - \frac{100 \text{ K}}{250 \text{ K}} = 0.600, \tag{15.57}$$

so that

$$\begin{aligned} W &= Eff'_C Q_h \\ &= (0.600)(4000 \text{ J}) = 2400 \text{ J}. \end{aligned} \tag{15.58}$$

Discussion

There is 933 J less work from the same heat transfer in the second process. This result is important. The same heat transfer into two perfect engines produces different work outputs, because the entropy change differs in the two cases. In the second case, entropy is greater and less work is produced. Entropy is associated with the *un*availability of energy to do work.

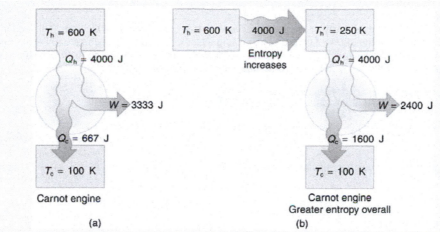

Figure 15.36 (a) A Carnot engine working at between 600 K and 100 K has 4000 J of heat transfer and performs 3333 J of work. (b) The 4000 J of heat transfer occurs first irreversibly to a 250 K reservoir and then goes into a Carnot engine. The increase in entropy caused by the heat transfer to a colder reservoir results in a smaller work output of 2400 J. There is a permanent loss of 933 J of energy for the purpose of doing work.

When entropy increases, a certain amount of energy becomes *permanently* unavailable to do work. The energy is not lost, but its character is changed, so that some of it can never be converted to doing work—that is, to an organized force acting through a distance. For instance, in the previous example, 933 J less work was done after an increase in entropy of 9.33 J/K occurred in the 4000 J heat transfer from the 600 K reservoir to the 250 K reservoir. It can be shown that the amount of energy that becomes unavailable for work is

$$W_{unavail} = \Delta S \cdot T_0,$$ (15.59)

where T_0 is the lowest temperature utilized. In the previous example,

$$W_{unavail} = (9.33 \text{ J/K})(100 \text{ K}) = 933 \text{ J}$$ (15.60)

as found.

Heat Death of the Universe: An Overdose of Entropy

In the early, energetic universe, all matter and energy were easily interchangeable and identical in nature. Gravity played a vital role in the young universe. Although it may have *seemed* disorderly, and therefore, superficially entropic, in fact, there was enormous potential energy available to do work—all the future energy in the universe.

As the universe matured, temperature differences arose, which created more opportunity for work. Stars are hotter than planets, for example, which are warmer than icy asteroids, which are warmer still than the vacuum of the space between them.

Most of these are cooling down from their usually violent births, at which time they were provided with energy of their own—nuclear energy in the case of stars, volcanic energy on Earth and other planets, and so on. Without additional energy input, however, their days are numbered.

As entropy increases, less and less energy in the universe is available to do work. On Earth, we still have great stores of energy such as fossil and nuclear fuels; large-scale temperature differences, which can provide wind energy; geothermal energies due to differences in temperature in Earth's layers; and tidal energies owing to our abundance of liquid water. As these are used, a certain fraction of the energy they contain can never be converted into doing work. Eventually, all fuels will be exhausted, all temperatures will equalize, and it will be impossible for heat engines to function, or for work to be done.

Entropy increases in a closed system, such as the universe. But in parts of the universe, for instance, in the Solar system, it is not a locally closed system. Energy flows from the Sun to the planets, replenishing Earth's stores of energy. The Sun will continue to supply us with energy for about another five billion years. We will enjoy direct solar energy, as well as side effects of solar energy, such as wind power and biomass energy from photosynthetic plants. The energy from the Sun will keep our water at the liquid state, and the Moon's gravitational pull will continue to provide tidal energy. But Earth's geothermal energy will slowly run down and won't be replenished.

But in terms of the universe, and the very long-term, very large-scale picture, the entropy of the universe is increasing, and so the availability of energy to do work is constantly decreasing. Eventually, when all stars have died, all forms of potential energy have been utilized, and all temperatures have equalized (depending on the mass of the universe, either at a very high temperature following a universal contraction, or a very low one, just before all activity ceases) there will be no possibility of doing work.

Either way, the universe is destined for thermodynamic equilibrium—maximum entropy. This is often called the *heat death of the universe*, and will mean the end of all activity. However, whether the universe contracts and heats up, or continues to expand and cools down, the end is not near. Calculations of black holes suggest that entropy can easily continue for at least 10^{100} years.

Order to Disorder

Entropy is related not only to the unavailability of energy to do work—it is also a measure of disorder. This notion was initially postulated by Ludwig Boltzmann in the 1800s. For example, melting a block of ice means taking a highly structured and orderly system of water molecules and converting it into a disorderly liquid in which molecules have no fixed positions. (See **Figure 15.37**.) There is a large increase in entropy in the process, as seen in the following example.

Example 15.8 Entropy Associated with Disorder

Find the increase in entropy of 1.00 kg of ice originally at $0°$ C that is melted to form water at $0°$ C.

Strategy

As before, the change in entropy can be calculated from the definition of ΔS once we find the energy Q needed to melt the ice.

Solution

The change in entropy is defined as:

$$\Delta S = \frac{Q}{T}. \tag{15.61}$$

Here Q is the heat transfer necessary to melt 1.00 kg of ice and is given by

$$Q = mL_f, \tag{15.62}$$

where m is the mass and L_f is the latent heat of fusion. $L_f = 334$ kJ/kg for water, so that

$$Q = (1.00 \text{ kg})(334 \text{ kJ/kg}) = 3.34\times10^5 \text{ J}. \tag{15.63}$$

Now the change in entropy is positive, since heat transfer occurs into the ice to cause the phase change; thus,

$$\Delta S = \frac{Q}{T} = \frac{3.34\times10^5 \text{ J}}{T}. \tag{15.64}$$

T is the melting temperature of ice. That is, $T = 0°\text{C} = 273$ K. So the change in entropy is

$$\begin{aligned} \Delta S &= \frac{3.34\times10^5 \text{ J}}{273 \text{ K}} \\ &= 1.22\times10^3 \text{ J/K}. \end{aligned} \tag{15.65}$$

Discussion

This is a significant increase in entropy accompanying an increase in disorder.

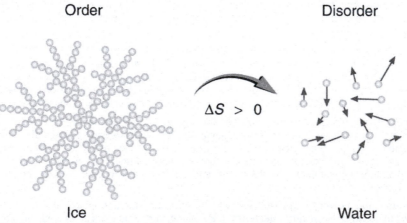

Figure 15.37 When ice melts, it becomes more disordered and less structured. The systematic arrangement of molecules in a crystal structure is replaced by a more random and less orderly movement of molecules without fixed locations or orientations. Its entropy increases because heat transfer occurs into it. Entropy is a measure of disorder.

In another easily imagined example, suppose we mix equal masses of water originally at two different temperatures, say $20.0°$ C and $40.0°$ C. The result is water at an intermediate temperature of $30.0°$ C. Three outcomes have resulted: entropy has increased, some energy has become unavailable to do work, and the system has become less orderly. Let us think about each of these results.

First, entropy has increased for the same reason that it did in the example above. Mixing the two bodies of water has the same effect as heat transfer from the hot one and the same heat transfer into the cold one. The mixing decreases the entropy of the hot water but increases the entropy of the cold water by a greater amount, producing an overall increase in entropy.

Second, once the two masses of water are mixed, there is only one temperature—you cannot run a heat engine with them. The energy that could have been used to run a heat engine is now unavailable to do work.

Third, the mixture is less orderly, or to use another term, less structured. Rather than having two masses at different temperatures and with different distributions of molecular speeds, we now have a single mass with a uniform temperature.

These three results—entropy, unavailability of energy, and disorder—are not only related but are in fact essentially equivalent.

Life, Evolution, and the Second Law of Thermodynamics

Some people misunderstand the second law of thermodynamics, stated in terms of entropy, to say that the process of the evolution of life violates this law. Over time, complex organisms evolved from much simpler ancestors, representing a large decrease in entropy of the Earth's biosphere. It is a fact that living organisms have evolved to be highly structured, and much lower in entropy than the substances from which they grow. But it is *always* possible for the entropy of one part of the universe to decrease, provided the total change in entropy of the universe increases. In equation form, we can write this as

$$\Delta S_{\text{tot}} = \Delta S_{\text{syst}} + \Delta S_{\text{envir}} > 0. \tag{15.66}$$

Thus ΔS_{syst} can be negative as long as ΔS_{envir} is positive and greater in magnitude.

How is it possible for a system to decrease its entropy? Energy transfer is necessary. If I pick up marbles that are scattered about the room and put them into a cup, my work has decreased the entropy of that system. If I gather iron ore from the ground and convert it into steel and build a bridge, my work has decreased the entropy of that system. Energy coming from the Sun can decrease the entropy of local systems on Earth—that is, ΔS_{syst} is negative. But the overall entropy of the rest of the universe

increases by a greater amount—that is, ΔS_{envir} is positive and greater in magnitude. Thus, $\Delta S_{\text{tot}} = \Delta S_{\text{syst}} + \Delta S_{\text{envir}} > 0$,

and the second law of thermodynamics is *not* violated.

Every time a plant stores some solar energy in the form of chemical potential energy, or an updraft of warm air lifts a soaring bird, the Earth can be viewed as a heat engine operating between a hot reservoir supplied by the Sun and a cold reservoir supplied by dark outer space—a heat engine of high complexity, causing local decreases in entropy as it uses part of the heat transfer from the Sun into deep space. There is a large total increase in entropy resulting from this massive heat transfer. A small part of this heat transfer is stored in structured systems on Earth, producing much smaller local decreases in entropy. (See **Figure 15.38**.)

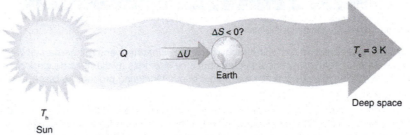

Figure 15.38 Earth's entropy may decrease in the process of intercepting a small part of the heat transfer from the Sun into deep space. Entropy for the entire process increases greatly while Earth becomes more structured with living systems and stored energy in various forms.

PhET Explorations: Reversible Reactions

Watch a reaction proceed over time. How does total energy affect a reaction rate? Vary temperature, barrier height, and potential energies. Record concentrations and time in order to extract rate coefficients. Do temperature dependent studies to extract Arrhenius parameters. This simulation is best used with teacher guidance because it presents an analogy of chemical reactions.

Figure 15.39 Reversible Reactions (http://cnx.org/content/m55267/1.2/reversible-reactions_en.jar)

15.7 Statistical Interpretation of Entropy and the Second Law of Thermodynamics: The Underlying Explanation

Figure 15.40 When you toss a coin a large number of times, heads and tails tend to come up in roughly equal numbers. Why doesn't heads come up 100, 90, or even 80% of the time? (credit: Jon Sullivan, PDPhoto.org)

The various ways of formulating the second law of thermodynamics tell what happens rather than why it happens. Why should heat transfer occur only from hot to cold? Why should energy become ever less available to do work? Why should the universe become increasingly disorderly? The answer is that it is a matter of overwhelming probability. Disorder is simply vastly more likely than order.

When you watch an emerging rain storm begin to wet the ground, you will notice that the drops fall in a disorganized manner both in time and in space. Some fall close together, some far apart, but they never fall in straight, orderly rows. It is not impossible for rain to fall in an orderly pattern, just highly unlikely, because there are many more disorderly ways than orderly ones. To illustrate this fact, we will examine some random processes, starting with coin tosses.

Coin Tosses

What are the possible outcomes of tossing 5 coins? Each coin can land either heads or tails. On the large scale, we are concerned only with the total heads and tails and not with the order in which heads and tails appear. The following possibilities exist:

$$5 \text{ heads, } 0 \text{ tails} \tag{15.67}$$
$$4 \text{ heads, } 1 \text{ tail}$$
$$3 \text{ heads, } 2 \text{ tails}$$
$$2 \text{ heads, } 3 \text{ tails}$$
$$1 \text{ head, } 4 \text{ tails}$$
$$0 \text{ head, } 5 \text{ tails}$$

These are what we call macrostates. A **macrostate** is an overall property of a system. It does not specify the details of the system, such as the order in which heads and tails occur or which coins are heads or tails.

Using this nomenclature, a system of 5 coins has the 6 possible macrostates just listed. Some macrostates are more likely to occur than others. For instance, there is only one way to get 5 heads, but there are several ways to get 3 heads and 2 tails, making the latter macrostate more probable. **Table 15.3** lists of all the ways in which 5 coins can be tossed, taking into account the order in which heads and tails occur. Each sequence is called a **microstate**—a detailed description of every element of a system.

Table 15.3 5-Coin Toss

	Individual microstates	Number of microstates
5 heads, 0 tails	HHHHH	1
4 heads, 1 tail	HHHHT, HHHTH, HHTHH, HTHHH, THHHH	5
3 heads, 2 tails	HTHTH, THTHH, HTHHT, THHTH, THHHT HTHTH, THTHH, HTHHT, THHHT, THHHT	10
2 heads, 3 tails	TTTHH, TTHHT, THHTT, HHTTT, TTHTH, THTHT, HTHTT, THTTH, HTTHT, HTTTH	10
1 head, 4 tails	TTTTH, TTTHT, TTHTT, THTTT, HTTTT	5
0 heads, 5 tails	TTTTT	1
		Total: 32

The macrostate of 3 heads and 2 tails can be achieved in 10 ways and is thus 10 times more probable than the one having 5 heads. Not surprisingly, it is equally probable to have the reverse, 2 heads and 3 tails. Similarly, it is equally probable to get 5 tails as it is to get 5 heads. Note that all of these conclusions are based on the crucial assumption that each microstate is equally probable. With coin tosses, this requires that the coins not be asymmetric in a way that favors one side over the other, as with loaded dice. With any system, the assumption that all microstates are equally probable must be valid, or the analysis will be erroneous.

The two most orderly possibilities are 5 heads or 5 tails. (They are more structured than the others.) They are also the least likely, only 2 out of 32 possibilities. The most disorderly possibilities are 3 heads and 2 tails and its reverse. (They are the least structured.) The most disorderly possibilities are also the most likely, with 20 out of 32 possibilities for the 3 heads and 2 tails and its reverse. If we start with an orderly array like 5 heads and toss the coins, it is very likely that we will get a less orderly array as a result, since 30 out of the 32 possibilities are less orderly. So even if you start with an orderly state, there is a strong tendency to go from order to disorder, from low entropy to high entropy. The reverse can happen, but it is unlikely.

Table 15.4 100-Coin Toss

Macrostate		Number of microstates
Heads	Tails	(W)
100	0	1
99	1	1.0×10^2
95	5	7.5×10^7
90	10	1.7×10^{13}
75	25	2.4×10^{23}
60	40	1.4×10^{28}
55	45	6.1×10^{28}
51	49	9.9×10^{28}
50	50	1.0×10^{29}
49	51	9.9×10^{28}
45	55	6.1×10^{28}
40	60	1.4×10^{28}
25	75	2.4×10^{23}
10	90	1.7×10^{13}
5	95	7.5×10^7
1	99	1.0×10^2
0	100	1
		Total: 1.27×10^{30}

This result becomes dramatic for larger systems. Consider what happens if you have 100 coins instead of just 5. The most orderly arrangements (most structured) are 100 heads or 100 tails. The least orderly (least structured) is that of 50 heads and 50 tails. There is only 1 way (1 microstate) to get the most orderly arrangement of 100 heads. There are 100 ways (100 microstates) to get the next most orderly arrangement of 99 heads and 1 tail (also 100 to get its reverse). And there are 1.0×10^{29} ways to get 50 heads and 50 tails, the least orderly arrangement. **Table 15.4** is an abbreviated list of the various macrostates and the number of microstates for each macrostate. The total number of microstates—the total number of different ways 100 coins can be tossed—is an impressively large 1.27×10^{30}. Now, if we start with an orderly macrostate like 100 heads and toss the coins, there is a virtual certainty that we will get a less orderly macrostate. If we keep tossing the coins, it is possible, but exceedingly unlikely, that we will ever get back to the most orderly macrostate. If you tossed the coins once each second, you could expect to get either 100 heads or 100 tails once in 2×10^{22} years! This period is 1 trillion (10^{12}) times longer than the age of the universe, and so the chances are essentially zero. In contrast, there is an 8% chance of getting 50 heads, a 73% chance of getting from 45 to 55 heads, and a 96% chance of getting from 40 to 60 heads. Disorder is highly likely.

Disorder in a Gas

The fantastic growth in the odds favoring disorder that we see in going from 5 to 100 coins continues as the number of entities in the system increases. Let us now imagine applying this approach to perhaps a small sample of gas. Because counting microstates and macrostates involves statistics, this is called **statistical analysis**. The macrostates of a gas correspond to its macroscopic properties, such as volume, temperature, and pressure; and its microstates correspond to the detailed description of the positions and velocities of its atoms. Even a small amount of gas has a huge number of atoms: 1.0 cm^3 of an ideal gas at 1.0 atm and $0°$ C has 2.7×10^{19} atoms. So each macrostate has an immense number of microstates. In plain language, this means that there are an immense number of ways in which the atoms in a gas can be arranged, while still having the same pressure, temperature, and so on.

The most likely conditions (or macrostates) for a gas are those we see all the time—a random distribution of atoms in space with a Maxwell-Boltzmann distribution of speeds in random directions, as predicted by kinetic theory. This is the most disorderly and least structured condition we can imagine. In contrast, one type of very orderly and structured macrostate has all of the atoms in one corner of a container with identical velocities. There are very few ways to accomplish this (very few microstates corresponding to it), and so it is exceedingly unlikely ever to occur. (See **Figure 15.41**(b).) Indeed, it is so unlikely that we have a law saying that it is impossible, which has never been observed to be violated—the second law of thermodynamics.

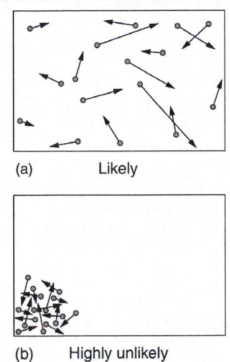

(a)   Likely

(b)   Highly unlikely

Figure 15.41 (a) The ordinary state of gas in a container is a disorderly, random distribution of atoms or molecules with a Maxwell-Boltzmann distribution of speeds. It is so unlikely that these atoms or molecules would ever end up in one corner of the container that it might as well be impossible. (b) With energy transfer, the gas can be forced into one corner and its entropy greatly reduced. But left alone, it will spontaneously increase its entropy and return to the normal conditions, because they are immensely more likely.

The disordered condition is one of high entropy, and the ordered one has low entropy. With a transfer of energy from another system, we could force all of the atoms into one corner and have a local decrease in entropy, but at the cost of an overall increase in entropy of the universe. If the atoms start out in one corner, they will quickly disperse and become uniformly distributed and will never return to the orderly original state (**Figure 15.41**(b)). Entropy will increase. With such a large sample of atoms, it is possible—but unimaginably unlikely—for entropy to decrease. Disorder is vastly more likely than order.

The arguments that disorder and high entropy are the most probable states are quite convincing. The great Austrian physicist Ludwig Boltzmann (1844–1906)—who, along with Maxwell, made so many contributions to kinetic theory—proved that the entropy of a system in a given state (a macrostate) can be written as

$$S = k \ln W,$$ (15.68)

where $k = 1.38 \times 10^{-23}$ J/K is Boltzmann's constant, and $\ln W$ is the natural logarithm of the number of microstates W corresponding to the given macrostate. W is proportional to the probability that the macrostate will occur. Thus entropy is directly related to the probability of a state—the more likely the state, the greater its entropy. Boltzmann proved that this expression for S is equivalent to the definition $\Delta S = Q / T$, which we have used extensively.

Thus the second law of thermodynamics is explained on a very basic level: entropy either remains the same or increases in every process. This phenomenon is due to the extraordinarily small probability of a decrease, based on the extraordinarily larger number of microstates in systems with greater entropy. Entropy *can* decrease, but for any macroscopic system, this outcome is so unlikely that it will never be observed.

Example 15.9 Entropy Increases in a Coin Toss

Suppose you toss 100 coins starting with 60 heads and 40 tails, and you get the most likely result, 50 heads and 50 tails. What is the change in entropy?

Strategy

Noting that the number of microstates is labeled W in **Table 15.4** for the 100-coin toss, we can use

$\Delta S = S_{\text{f}} - S_{\text{i}} = k \ln W_{\text{f}} - k \ln W_{\text{i}}$ to calculate the change in entropy.

Solution

The change in entropy is

$$\Delta S = S_f - S_i = k\ln W_f - k\ln W_i, \tag{15.69}$$

where the subscript i stands for the initial 60 heads and 40 tails state, and the subscript f for the final 50 heads and 50 tails state. Substituting the values for W from **Table 15.4** gives

$$\begin{aligned} \Delta S &= (1.38\times10^{-23}\ \text{J/K})[\ln(1.0\times10^{29}) - \ln(1.4\times10^{28})] \\ &= 2.7\times10^{-23}\ \text{J/K} \end{aligned} \tag{15.70}$$

Discussion

This increase in entropy means we have moved to a less orderly situation. It is not impossible for further tosses to produce the initial state of 60 heads and 40 tails, but it is less likely. There is about a 1 in 90 chance for that decrease in entropy (-2.7×10^{-23} J/K) to occur. If we calculate the decrease in entropy to move to the most orderly state, we get $\Delta S = -92\times10^{-23}$ J/K. There is about a 1 in 10^{30} chance of this change occurring. So while very small decreases in entropy are unlikely, slightly greater decreases are impossibly unlikely. These probabilities imply, again, that for a macroscopic system, a decrease in entropy is impossible. For example, for heat transfer to occur spontaneously from 1.00 kg of $0°C$ ice to its $0°C$ environment, there would be a decrease in entropy of 1.22×10^3 J/K. Given that a ΔS of 10^{-21} J/K corresponds to about a 1 in 10^{30} chance, a decrease of this size (10^3 J/K) is an *utter* impossibility. Even for a milligram of melted ice to spontaneously refreeze is impossible.

Problem-Solving Strategies for Entropy

1. *Examine the situation to determine if entropy is involved.*
2. *Identify the system of interest and draw a labeled diagram of the system showing energy flow.*
3. *Identify exactly what needs to be determined in the problem (identify the unknowns). A written list is useful.*
4. *Make a list of what is given or can be inferred from the problem as stated (identify the knowns). You must carefully identify the heat transfer, if any, and the temperature at which the process takes place. It is also important to identify the initial and final states.*
5. *Solve the appropriate equation for the quantity to be determined (the unknown). Note that the change in entropy can be determined between any states by calculating it for a reversible process.*
6. *Substitute the known value along with their units into the appropriate equation, and obtain numerical solutions complete with units.*
7. *To see if it is reasonable: Does it make sense? For example, total entropy should increase for any real process or be constant for a reversible process. Disordered states should be more probable and have greater entropy than ordered states.*

Exercise 15.1

(a) If you toss 10 coins, what percent of the time will you get the three most likely macrostates (6 heads and 4 tails, 5 heads and 5 tails, 4 heads and 6 tails)? (b) You can realistically toss 10 coins and count the number of heads and tails about twice a minute. At that rate, how long will it take on average to get either 10 heads and 0 tails or 0 heads and 10 tails?

Exercise 15.2

(a) Construct a table showing the macrostates and all of the individual microstates for tossing 6 coins. (Use **Table 15.5** as a guide.) (b) How many macrostates are there? (c) What is the total number of microstates? (d) What percent chance is there of tossing 5 heads and 1 tail? (e) How much more likely are you to toss 3 heads and 3 tails than 5 heads and 1 tail? (Take the ratio of the number of microstates to find out.)

Solution

(b) 7

(c) 64

(d) 9.38%

(e) 3.33 times more likely (20 to 6)

Exercise 15.3

In an air conditioner, 12.65 MJ of heat transfer occurs from a cold environment in 1.00 h. (a) What mass of ice melting would involve the same heat transfer? (b) How many hours of operation would be equivalent to melting 900 kg of ice? (c) If ice costs 20 cents per kg, do you think the air conditioner could be operated more cheaply than by simply using ice? Describe in detail how you evaluate the relative costs.

Glossary

adiabatic process: a process in which no heat transfer takes place

Carnot cycle: a cyclical process that uses only reversible processes, the adiabatic and isothermal processes

Carnot efficiency: the maximum theoretical efficiency for a heat engine

Carnot engine: a heat engine that uses a Carnot cycle

change in entropy: the ratio of heat transfer to temperature Q / T

coefficient of performance: for a heat pump, it is the ratio of heat transfer at the output (the hot reservoir) to the work supplied; for a refrigerator or air conditioner, it is the ratio of heat transfer from the cold reservoir to the work supplied

cyclical process: a process in which the path returns to its original state at the end of every cycle

entropy: a measurement of a system's disorder and its inability to do work in a system

first law of thermodynamics: states that the change in internal energy of a system equals the net heat transfer *into* the system minus the net work done *by* the system

heat engine: a machine that uses heat transfer to do work

heat pump: a machine that generates heat transfer from cold to hot

human metabolism: conversion of food into heat transfer, work, and stored fat

internal energy: the sum of the kinetic and potential energies of a system's atoms and molecules

irreversible process: any process that depends on path direction

isobaric process: constant-pressure process in which a gas does work

isochoric process: a constant-volume process

isothermal process: a constant-temperature process

macrostate: an overall property of a system

microstate: each sequence within a larger macrostate

Otto cycle: a thermodynamic cycle, consisting of a pair of adiabatic processes and a pair of isochoric processes, that converts heat into work, e.g., the four-stroke engine cycle of intake, compression, ignition, and exhaust

reversible process: a process in which both the heat engine system and the external environment theoretically can be returned to their original states

second law of thermodynamics: heat transfer flows from a hotter to a cooler object, never the reverse, and some heat energy in any process is lost to available work in a cyclical process

second law of thermodynamics stated in terms of entropy: the total entropy of a system either increases or remains constant; it never decreases

statistical analysis: using statistics to examine data, such as counting microstates and macrostates

Section Summary

15.1 The First Law of Thermodynamics

- The first law of thermodynamics is given as $\Delta U = Q - W$, where ΔU is the change in internal energy of a system, Q is the net heat transfer (the sum of all heat transfer into and out of the system), and W is the net work done (the sum of all work done on or by the system).

- Both Q and W are energy in transit; only ΔU represents an independent quantity capable of being stored.
- The internal energy U of a system depends only on the state of the system and not how it reached that state.
- Metabolism of living organisms, and photosynthesis of plants, are specialized types of heat transfer, doing work, and internal energy of systems.

15.2 The First Law of Thermodynamics and Some Simple Processes

- One of the important implications of the first law of thermodynamics is that machines can be harnessed to do work that humans previously did by hand or by external energy supplies such as running water or the heat of the Sun. A machine that uses heat transfer to do work is known as a heat engine.
- There are several simple processes, used by heat engines, that flow from the first law of thermodynamics. Among them are the isobaric, isochoric, isothermal and adiabatic processes.
- These processes differ from one another based on how they affect pressure, volume, temperature, and heat transfer.
- If the work done is performed on the outside environment, work (W) will be a positive value. If the work done is done to the heat engine system, work (W) will be a negative value.
- Some thermodynamic processes, including isothermal and adiabatic processes, are reversible in theory; that is, both the thermodynamic system and the environment can be returned to their initial states. However, because of loss of energy owing to the second law of thermodynamics, complete reversibility does not work in practice.

15.3 Introduction to the Second Law of Thermodynamics: Heat Engines and Their Efficiency

- The two expressions of the second law of thermodynamics are: (i) Heat transfer occurs spontaneously from higher- to lower-temperature bodies but never spontaneously in the reverse direction; and (ii) It is impossible in any system for heat transfer from a reservoir to completely convert to work in a cyclical process in which the system returns to its initial state.
- Irreversible processes depend on path and do not return to their original state. Cyclical processes are processes that return to their original state at the end of every cycle.
- In a cyclical process, such as a heat engine, the net work done by the system equals the net heat transfer into the system, or $W = Q_h - Q_c$, where Q_h is the heat transfer from the hot object (hot reservoir), and Q_c is the heat transfer into the cold object (cold reservoir).
- Efficiency can be expressed as $Eff = \dfrac{W}{Q_h}$, the ratio of work output divided by the amount of energy input.
- The four-stroke gasoline engine is often explained in terms of the Otto cycle, which is a repeating sequence of processes that convert heat into work.

15.4 Carnot's Perfect Heat Engine: The Second Law of Thermodynamics Restated

- The Carnot cycle is a theoretical cycle that is the most efficient cyclical process possible. Any engine using the Carnot cycle, which uses only reversible processes (adiabatic and isothermal), is known as a Carnot engine.
- Any engine that uses the Carnot cycle enjoys the maximum theoretical efficiency.
- While Carnot engines are ideal engines, in reality, no engine achieves Carnot's theoretical maximum efficiency, since dissipative processes, such as friction, play a role. Carnot cycles without heat loss may be possible at absolute zero, but this has never been seen in nature.

15.5 Applications of Thermodynamics: Heat Pumps and Refrigerators

- An artifact of the second law of thermodynamics is the ability to heat an interior space using a heat pump. Heat pumps compress cold ambient air and, in so doing, heat it to room temperature without violation of conservation principles.
- To calculate the heat pump's coefficient of performance, use the equation $COP_{hp} = \dfrac{Q_h}{W}$.
- A refrigerator is a heat pump; it takes warm ambient air and expands it to chill it.

15.6 Entropy and the Second Law of Thermodynamics: Disorder and the Unavailability of Energy

- Entropy is the loss of energy available to do work.
- Another form of the second law of thermodynamics states that the total entropy of a system either increases or remains constant; it never decreases.
- Entropy is zero in a reversible process; it increases in an irreversible process.
- The ultimate fate of the universe is likely to be thermodynamic equilibrium, where the universal temperature is constant and no energy is available to do work.
- Entropy is also associated with the tendency toward disorder in a closed system.

15.7 Statistical Interpretation of Entropy and the Second Law of Thermodynamics: The Underlying Explanation

- Disorder is far more likely than order, which can be seen statistically.

- The entropy of a system in a given state (a macrostate) can be written as

$$S = k \ln W,$$

where $k = 1.38 \times 10^{-23}$ J/K is Boltzmann's constant, and $\ln W$ is the natural logarithm of the number of microstates W corresponding to the given macrostate.

Conceptual Questions

15.1 The First Law of Thermodynamics

1. Describe the photo of the tea kettle at the beginning of this section in terms of heat transfer, work done, and internal energy. How is heat being transferred? What is the work done and what is doing it? How does the kettle maintain its internal energy?

2. The first law of thermodynamics and the conservation of energy, as discussed in **Conservation of Energy**, are clearly related. How do they differ in the types of energy considered?

3. Heat transfer Q and work done W are always energy in transit, whereas internal energy U is energy stored in a system. Give an example of each type of energy, and state specifically how it is either in transit or resides in a system.

4. How do heat transfer and internal energy differ? In particular, which can be stored as such in a system and which cannot?

5. If you run down some stairs and stop, what happens to your kinetic energy and your initial gravitational potential energy?

6. Give an explanation of how food energy (calories) can be viewed as molecular potential energy (consistent with the atomic and molecular definition of internal energy).

7. Identify the type of energy transferred to your body in each of the following as either internal energy, heat transfer, or doing work: (a) basking in sunlight; (b) eating food; (c) riding an elevator to a higher floor.

15.2 The First Law of Thermodynamics and Some Simple Processes

8. A great deal of effort, time, and money has been spent in the quest for the so-called perpetual-motion machine, which is defined as a hypothetical machine that operates or produces useful work indefinitely and/or a hypothetical machine that produces more work or energy than it consumes. Explain, in terms of heat engines and the first law of thermodynamics, why or why not such a machine is likely to be constructed.

9. One method of converting heat transfer into doing work is for heat transfer into a gas to take place, which expands, doing work on a piston, as shown in the figure below. (a) Is the heat transfer converted directly to work in an isobaric process, or does it go through another form first? Explain your answer. (b) What about in an isothermal process? (c) What about in an adiabatic process (where heat transfer occurred prior to the adiabatic process)?

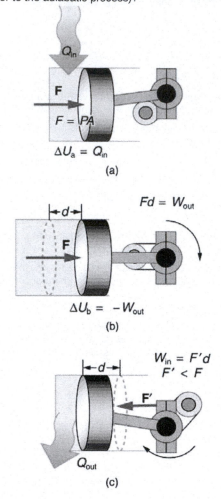

Figure 15.42

10. Would the previous question make any sense for an isochoric process? Explain your answer.

11. We ordinarily say that $\Delta U = 0$ for an isothermal process. Does this assume no phase change takes place? Explain your answer.

12. The temperature of a rapidly expanding gas decreases. Explain why in terms of the first law of thermodynamics. (Hint: Consider whether the gas does work and whether heat transfer occurs rapidly into the gas through conduction.)

13. Which cyclical process represented by the two closed loops, ABCFA and ABDEA, on the PV diagram in the figure below produces the greatest *net* work? Is that process also the one with the smallest work input required to return it to point A? Explain your responses.

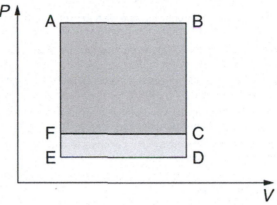

Figure 15.43 The two cyclical processes shown on this PV diagram start with and return the system to the conditions at point A, but they follow different paths and produce different amounts of work.

14. A real process may be nearly adiabatic if it occurs over a very short time. How does the short time span help the process to be adiabatic?

15. It is unlikely that a process can be isothermal unless it is a very slow process. Explain why. Is the same true for isobaric and isochoric processes? Explain your answer.

15.3 Introduction to the Second Law of Thermodynamics: Heat Engines and Their Efficiency

16. Imagine you are driving a car up Pike's Peak in Colorado. To raise a car weighing 1000 kilograms a distance of 100 meters would require about a million joules. You could raise a car 12.5 kilometers with the energy in a gallon of gas. Driving up Pike's Peak (a mere 3000-meter climb) should consume a little less than a quart of gas. But other considerations have to be taken into account. Explain, in terms of efficiency, what factors may keep you from realizing your ideal energy use on this trip.

17. Is a temperature difference necessary to operate a heat engine? State why or why not.

18. Definitions of efficiency vary depending on how energy is being converted. Compare the definitions of efficiency for the human body and heat engines. How does the definition of efficiency in each relate to the type of energy being converted into doing work?

19. Why—other than the fact that the second law of thermodynamics says reversible engines are the most efficient—should heat engines employing reversible processes be more efficient than those employing irreversible processes? Consider that dissipative mechanisms are one cause of irreversibility.

15.4 Carnot's Perfect Heat Engine: The Second Law of Thermodynamics Restated

20. Think about the drinking bird at the beginning of this section (Figure 15.22). Although the bird enjoys the theoretical maximum efficiency possible, if left to its own devices over time, the bird will cease "drinking." What are some of the dissipative processes that might cause the bird's motion to cease?

21. Can improved engineering and materials be employed in heat engines to reduce heat transfer into the environment? Can they eliminate heat transfer into the environment entirely?

22. Does the second law of thermodynamics alter the conservation of energy principle?

15.5 Applications of Thermodynamics: Heat Pumps and Refrigerators

23. Explain why heat pumps do not work as well in very cold climates as they do in milder ones. Is the same true of refrigerators?

24. In some Northern European nations, homes are being built without heating systems of any type. They are very well insulated and are kept warm by the body heat of the residents. However, when the residents are not at home, it is still warm in these houses. What is a possible explanation?

25. Why do refrigerators, air conditioners, and heat pumps operate most cost-effectively for cycles with a small difference between T_h and T_c? (Note that the temperatures of the cycle employed are crucial to its COP.)

26. Grocery store managers contend that there is *less* total energy consumption in the summer if the store is kept at a *low* temperature. Make arguments to support or refute this claim, taking into account that there are numerous refrigerators and freezers in the store.

27. Can you cool a kitchen by leaving the refrigerator door open?

15.6 Entropy and the Second Law of Thermodynamics: Disorder and the Unavailability of Energy

28. A woman shuts her summer cottage up in September and returns in June. No one has entered the cottage in the meantime. Explain what she is likely to find, in terms of the second law of thermodynamics.

29. Consider a system with a certain energy content, from which we wish to extract as much work as possible. Should the system's entropy be high or low? Is this orderly or disorderly? Structured or uniform? Explain briefly.

30. Does a gas become more orderly when it liquefies? Does its entropy change? If so, does the entropy increase or decrease? Explain your answer.

31. Explain how water's entropy can decrease when it freezes without violating the second law of thermodynamics. Specifically, explain what happens to the entropy of its surroundings.

32. Is a uniform-temperature gas more or less orderly than one with several different temperatures? Which is more structured? In which can heat transfer result in work done without heat transfer from another system?

33. Give an example of a spontaneous process in which a system becomes less ordered and energy becomes less available to do work. What happens to the system's entropy in this process?

34. What is the change in entropy in an adiabatic process? Does this imply that adiabatic processes are reversible? Can a process be precisely adiabatic for a macroscopic system?

35. Does the entropy of a star increase or decrease as it radiates? Does the entropy of the space into which it radiates (which has a temperature of about 3 K) increase or decrease? What does this do to the entropy of the universe?

36. Explain why a building made of bricks has smaller entropy than the same bricks in a disorganized pile. Do this by considering the number of ways that each could be formed (the number of microstates in each macrostate).

15.7 Statistical Interpretation of Entropy and the Second Law of Thermodynamics: The Underlying Explanation

37. Explain why a building made of bricks has smaller entropy than the same bricks in a disorganized pile. Do this by considering the number of ways that each could be formed (the number of microstates in each macrostate).

Problems & Exercises

15.1 The First Law of Thermodynamics

1. What is the change in internal energy of a car if you put 12.0 gal of gasoline into its tank? The energy content of gasoline is 1.3×10^8 J/gal. All other factors, such as the car's temperature, are constant.

2. How much heat transfer occurs from a system, if its internal energy decreased by 150 J while it was doing 30.0 J of work?

3. A system does 1.80×10^8 J of work while 7.50×10^8 J of heat transfer occurs to the environment. What is the change in internal energy of the system assuming no other changes (such as in temperature or by the addition of fuel)?

4. What is the change in internal energy of a system which does 4.50×10^5 J of work while 3.00×10^6 J of heat transfer occurs into the system, and 8.00×10^6 J of heat transfer occurs to the environment?

5. Suppose a woman does 500 J of work and 9500 J of heat transfer occurs into the environment in the process. (a) What is the decrease in her internal energy, assuming no change in temperature or consumption of food? (That is, there is no other energy transfer.) (b) What is her efficiency?

6. (a) How much food energy will a man metabolize in the process of doing 35.0 kJ of work with an efficiency of 5.00%? (b) How much heat transfer occurs to the environment to keep his temperature constant? Explicitly show how you follow the steps in the Problem-Solving Strategy for thermodynamics found in **Problem-Solving Strategies for Thermodynamics**.

7. (a) What is the average metabolic rate in watts of a man who metabolizes 10,500 kJ of food energy in one day? (b) What is the maximum amount of work in joules he can do without breaking down fat, assuming a maximum efficiency of 20.0%? (c) Compare his work output with the daily output of a 187-W (0.250-horsepower) motor.

8. (a) How long will the energy in a 1470-kJ (350-kcal) cup of yogurt last in a woman doing work at the rate of 150 W with an efficiency of 20.0% (such as in leisurely climbing stairs)? (b) Does the time found in part (a) imply that it is easy to consume more food energy than you can reasonably expect to work off with exercise?

9. (a) A woman climbing the Washington Monument metabolizes 6.00×10^2 kJ of food energy. If her efficiency is 18.0%, how much heat transfer occurs to the environment to keep her temperature constant? (b) Discuss the amount of heat transfer found in (a). Is it consistent with the fact that you quickly warm up when exercising?

15.2 The First Law of Thermodynamics and Some Simple Processes

10. A car tire contains 0.0380 m^3 of air at a pressure of 2.20×10^5 N/m^2 (about 32 psi). How much more internal energy does this gas have than the same volume has at zero gauge pressure (which is equivalent to normal atmospheric pressure)?

11. A helium-filled toy balloon has a gauge pressure of 0.200 atm and a volume of 10.0 L. How much greater is the internal energy of the helium in the balloon than it would be at zero gauge pressure?

12. Steam to drive an old-fashioned steam locomotive is supplied at a constant gauge pressure of 1.75×10^6 N/m^2 (about 250 psi) to a piston with a 0.200-m radius. (a) By calculating $P\Delta V$, find the work done by the steam when the piston moves 0.800 m. Note that this is the net work output, since gauge pressure is used. (b) Now find the amount of work by calculating the force exerted times the distance traveled. Is the answer the same as in part (a)?

13. A hand-driven tire pump has a piston with a 2.50-cm diameter and a maximum stroke of 30.0 cm. (a) How much work do you do in one stroke if the average gauge pressure is 2.40×10^5 N/m^2 (about 35 psi)? (b) What average force do you exert on the piston, neglecting friction and gravitational force?

14. Calculate the net work output of a heat engine following path ABCDA in the figure below.

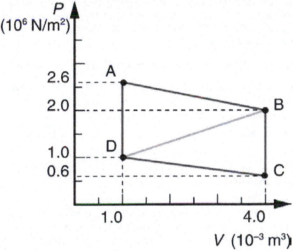

Figure 15.44

15. What is the net work output of a heat engine that follows path ABDA in the figure above, with a straight line from B to D? Why is the work output less than for path ABCDA? Explicitly show how you follow the steps in the **Problem-Solving Strategies for Thermodynamics**.

16. Unreasonable Results

What is wrong with the claim that a cyclical heat engine does 4.00 kJ of work on an input of 24.0 kJ of heat transfer while 16.0 kJ of heat transfers to the environment?

17. (a) A cyclical heat engine, operating between temperatures of $450°$ C and $150°$ C produces 4.00 MJ of work on a heat transfer of 5.00 MJ into the engine. How much heat transfer occurs to the environment? (b) What is unreasonable about the engine? (c) Which premise is unreasonable?

18. Construct Your Own Problem

Consider a car's gasoline engine. Construct a problem in which you calculate the maximum efficiency this engine can have. Among the things to consider are the effective hot and cold reservoir temperatures. Compare your calculated efficiency with the actual efficiency of car engines.

19. Construct Your Own Problem

Consider a car trip into the mountains. Construct a problem in which you calculate the overall efficiency of the car for the trip as a ratio of kinetic and potential energy gained to fuel consumed. Compare this efficiency to the thermodynamic efficiency quoted for gasoline engines and discuss why the thermodynamic efficiency is so much greater. Among the factors to be considered are the gain in altitude and speed, the mass of the car, the distance traveled, and typical fuel economy.

15.3 Introduction to the Second Law of Thermodynamics: Heat Engines and Their Efficiency

20. A certain heat engine does 10.0 kJ of work and 8.50 kJ of heat transfer occurs to the environment in a cyclical process. (a) What was the heat transfer into this engine? (b) What was the engine's efficiency?

21. With 2.56×10^6 J of heat transfer into this engine, a given cyclical heat engine can do only 1.50×10^5 J of work. (a) What is the engine's efficiency? (b) How much heat transfer to the environment takes place?

22. (a) What is the work output of a cyclical heat engine having a 22.0% efficiency and 6.00×10^9 J of heat transfer into the engine? (b) How much heat transfer occurs to the environment?

23. (a) What is the efficiency of a cyclical heat engine in which 75.0 kJ of heat transfer occurs to the environment for every 95.0 kJ of heat transfer into the engine? (b) How much work does it produce for 100 kJ of heat transfer into the engine?

24. The engine of a large ship does 2.00×10^8 J of work with an efficiency of 5.00%. (a) How much heat transfer occurs to the environment? (b) How many barrels of fuel are consumed, if each barrel produces 6.00×10^9 J of heat transfer when burned?

25. (a) How much heat transfer occurs to the environment by an electrical power station that uses 1.25×10^{14} J of heat transfer into the engine with an efficiency of 42.0%? (b) What is the ratio of heat transfer to the environment to work output? (c) How much work is done?

26. Assume that the turbines at a coal-powered power plant were upgraded, resulting in an improvement in efficiency of 3.32%. Assume that prior to the upgrade the power station had an efficiency of 36% and that the heat transfer into the engine in one day is still the same at 2.50×10^{14} J. (a) How much more electrical energy is produced due to the upgrade? (b) How much less heat transfer occurs to the environment due to the upgrade?

27. This problem compares the energy output and heat transfer to the environment by two different types of nuclear power stations—one with the normal efficiency of 34.0%, and another with an improved efficiency of 40.0%. Suppose both have the same heat transfer into the engine in one day, 2.50×10^{14} J. (a) How much more electrical energy is produced by the more efficient power station? (b) How much less heat transfer occurs to the environment by the more efficient power station? (One type of more efficient nuclear power station, the gas-cooled reactor, has not been reliable enough to be economically feasible in spite of its greater efficiency.)

15.4 Carnot's Perfect Heat Engine: The Second Law of Thermodynamics Restated

28. A certain gasoline engine has an efficiency of 30.0%. What would the hot reservoir temperature be for a Carnot engine having that efficiency, if it operates with a cold reservoir temperature of $200°C$?

29. A gas-cooled nuclear reactor operates between hot and cold reservoir temperatures of $700°C$ and $27.0°C$. (a) What is the maximum efficiency of a heat engine operating between these temperatures? (b) Find the ratio of this efficiency to the Carnot efficiency of a standard nuclear reactor (found in **Example 15.4**).

30. (a) What is the hot reservoir temperature of a Carnot engine that has an efficiency of 42.0% and a cold reservoir temperature of $27.0°C$? (b) What must the hot reservoir temperature be for a real heat engine that achieves 0.700 of the maximum efficiency, but still has an efficiency of 42.0% (and a cold reservoir at $27.0°C$)? (c) Does your answer imply practical limits to the efficiency of car gasoline engines?

31. Steam locomotives have an efficiency of 17.0% and operate with a hot steam temperature of $425°C$. (a) What would the cold reservoir temperature be if this were a Carnot engine? (b) What would the maximum efficiency of this steam engine be if its cold reservoir temperature were $150°C$?

32. Practical steam engines utilize $450°C$ steam, which is later exhausted at $270°C$. (a) What is the maximum efficiency that such a heat engine can have? (b) Since $270°C$ steam is still quite hot, a second steam engine is sometimes operated using the exhaust of the first. What is the maximum efficiency of the second engine if its exhaust has a temperature of $150°C$? (c) What is the overall efficiency of the two engines? (d) Show that this is the same efficiency as a single Carnot engine operating between $450°C$ and $150°C$. Explicitly show how you follow the steps in the **Problem-Solving Strategies for Thermodynamics**.

33. A coal-fired electrical power station has an efficiency of 38%. The temperature of the steam leaving the boiler is $550°C$. What percentage of the maximum efficiency does this station obtain? (Assume the temperature of the environment is $20°C$.)

34. Would you be willing to financially back an inventor who is marketing a device that she claims has 25 kJ of heat transfer at 600 K, has heat transfer to the environment at 300 K, and does 12 kJ of work? Explain your answer.

35. Unreasonable Results

(a) Suppose you want to design a steam engine that has heat transfer to the environment at $270°C$ and has a Carnot efficiency of 0.800. What temperature of hot steam must you use? (b) What is unreasonable about the temperature? (c) Which premise is unreasonable?

36. Unreasonable Results

Calculate the cold reservoir temperature of a steam engine that uses hot steam at $450°C$ and has a Carnot efficiency of 0.700. (b) What is unreasonable about the temperature? (c) Which premise is unreasonable?

15.5 Applications of Thermodynamics: Heat Pumps and Refrigerators

37. What is the coefficient of performance of an ideal heat pump that has heat transfer from a cold temperature of $-25.0°C$ to a hot temperature of $40.0°C$?

38. Suppose you have an ideal refrigerator that cools an environment at $-20.0°C$ and has heat transfer to another environment at $50.0°C$. What is its coefficient of performance?

39. What is the best coefficient of performance possible for a hypothetical refrigerator that could make liquid nitrogen at $-200°C$ and has heat transfer to the environment at $35.0°C$?

40. In a very mild winter climate, a heat pump has heat transfer from an environment at $5.00°C$ to one at $35.0°C$. What is the best possible coefficient of performance for these temperatures? Explicitly show how you follow the steps in the **Problem-Solving Strategies for Thermodynamics**.

41. (a) What is the best coefficient of performance for a heat pump that has a hot reservoir temperature of $50.0°C$ and a cold reservoir temperature of $-20.0°C$? (b) How much heat transfer occurs into the warm environment if $3.60×10^7$ J of work ($10.0 kW \cdot h$) is put into it? (c) If the cost of this work input is 10.0 cents/kW $\cdot$ h, how does its cost compare with the direct heat transfer achieved by burning natural gas at a cost of 85.0 cents per therm. (A therm is a common unit of energy for natural gas and equals $1.055×10^8$ J.)

42. (a) What is the best coefficient of performance for a refrigerator that cools an environment at $-30.0°C$ and has heat transfer to another environment at $45.0°C$? (b) How much work in joules must be done for a heat transfer of 4186 kJ from the cold environment? (c) What is the cost of doing this if the work costs 10.0 cents per $3.60×10^6$ J (a kilowatt-hour)? (d) How many kJ of heat transfer occurs into the warm environment? (e) Discuss what type of refrigerator might operate between these temperatures.

43. Suppose you want to operate an ideal refrigerator with a cold temperature of $-10.0°C$, and you would like it to have a coefficient of performance of 7.00. What is the hot reservoir temperature for such a refrigerator?

44. An ideal heat pump is being considered for use in heating an environment with a temperature of $22.0°C$. What is the cold reservoir temperature if the pump is to have a coefficient of performance of 12.0?

45. A 4-ton air conditioner removes $5.06×10^7$ J (48,000 British thermal units) from a cold environment in 1.00 h. (a) What energy input in joules is necessary to do this if the air conditioner has an energy efficiency rating (EER) of 12.0? (b) What is the cost of doing this if the work costs 10.0 cents per $3.60×10^6$ J (one kilowatt-hour)? (c) Discuss whether this cost seems realistic. Note that the energy efficiency rating (EER) of an air conditioner or refrigerator is defined to be the number of British thermal units of heat transfer from a cold environment per hour divided by the watts of power input.

46. Show that the coefficients of performance of refrigerators and heat pumps are related by $COP_{ref} = COP_{hp} - 1$.

Start with the definitions of the COP s and the conservation of energy relationship between Q_h, Q_c, and W.

15.6 Entropy and the Second Law of Thermodynamics: Disorder and the Unavailability of Energy

47. (a) On a winter day, a certain house loses $5.00×10^8$ J of heat to the outside (about 500,000 Btu). What is the total change in entropy due to this heat transfer alone, assuming an average indoor temperature of $21.0° C$ and an average outdoor temperature of $5.00° C$? (b) This large change in entropy implies a large amount of energy has become unavailable to do work. Where do we find more energy when such energy is lost to us?

48. On a hot summer day, $4.00×10^6$ J of heat transfer into a parked car takes place, increasing its temperature from $35.0° C$ to $45.0° C$. What is the increase in entropy of the car due to this heat transfer alone?

49. A hot rock ejected from a volcano's lava fountain cools from $1100° C$ to $40.0° C$, and its entropy decreases by 950 J/K. How much heat transfer occurs from the rock?

50. When $1.60×10^5$ J of heat transfer occurs into a meat pie initially at $20.0° C$, its entropy increases by 480 J/K. What is its final temperature?

51. The Sun radiates energy at the rate of $3.80×10^{26}$ W from its $5500° C$ surface into dark empty space (a negligible fraction radiates onto Earth and the other planets). The effective temperature of deep space is $-270° C$. (a) What is the increase in entropy in one day due to this heat transfer? (b) How much work is made unavailable?

52. (a) In reaching equilibrium, how much heat transfer occurs from 1.00 kg of water at $40.0° C$ when it is placed in contact with 1.00 kg of $20.0° C$ water in reaching equilibrium? **(b)** What is the change in entropy due to this heat transfer? **(c)** How much work is made unavailable, taking the lowest temperature to be $20.0° C$? Explicitly show how you follow the steps in the **Problem-Solving Strategies for Entropy**.

53. What is the decrease in entropy of 25.0 g of water that condenses on a bathroom mirror at a temperature of $35.0° C$, assuming no change in temperature and given the latent heat of vaporization to be 2450 kJ/kg?

54. Find the increase in entropy of 1.00 kg of liquid nitrogen that starts at its boiling temperature, boils, and warms to $20.0° C$ at constant pressure.

55. A large electrical power station generates 1000 MW of electricity with an efficiency of 35.0%. **(a)** Calculate the heat transfer to the power station, Q_h, in one day. **(b)** How much heat transfer Q_c occurs to the environment in one day? **(c)** If the heat transfer in the cooling towers is from $35.0° C$ water into the local air mass, which increases in temperature from $18.0° C$ to $20.0° C$, what is the total increase in entropy due to this heat transfer? **(d)** How much energy becomes unavailable to do work because of this increase in entropy, assuming an $18.0° C$ lowest temperature? (Part of Q_c could be utilized to operate heat engines or for simply heating the surroundings, but it rarely is.)

56. (a) How much heat transfer occurs from 20.0 kg of $90.0° C$ water placed in contact with 20.0 kg of $10.0° C$ water, producing a final temperature of $50.0° C$? **(b)** How much work could a Carnot engine do with this heat transfer, assuming it operates between two reservoirs at constant temperatures of $90.0° C$ and $10.0° C$? **(c)** What increase in entropy is produced by mixing 20.0 kg of $90.0° C$ water with 20.0 kg of $10.0° C$ water? **(d)** Calculate the amount of work made unavailable by this mixing using a low temperature of $10.0° C$, and compare it with the work done by the Carnot engine. Explicitly show how you follow the steps in the **Problem-Solving Strategies for Entropy**. **(e)** Discuss how everyday processes make increasingly more energy unavailable to do work, as implied by this problem.

15.7 Statistical Interpretation of Entropy and the Second Law of Thermodynamics: The Underlying Explanation

57. Using **Table 15.4**, verify the contention that if you toss 100 coins each second, you can expect to get 100 heads or 100 tails once in 2×10^{22} years; calculate the time to two-digit accuracy.

58. What percent of the time will you get something in the range from 60 heads and 40 tails through 40 heads and 60 tails when tossing 100 coins? The total number of microstates in that range is 1.22×10^{30}. (Consult **Table 15.4**.)

59. (a) If tossing 100 coins, how many ways (microstates) are there to get the three most likely macrostates of 49 heads and 51 tails, 50 heads and 50 tails, and 51 heads and 49 tails? **(b)** What percent of the total possibilities is this? (Consult **Table 15.4**.)

60. (a) What is the change in entropy if you start with 100 coins in the 45 heads and 55 tails macrostate, toss them, and get 51 heads and 49 tails? **(b)** What if you get 75 heads and 25 tails? **(c)** How much more likely is 51 heads and 49 tails than 75 heads and 25 tails? **(d)** Does either outcome violate the second law of thermodynamics?

61. (a) What is the change in entropy if you start with 10 coins in the 5 heads and 5 tails macrostate, toss them, and get 2 heads and 8 tails? **(b)** How much more likely is 5 heads and 5 tails than 2 heads and 8 tails? (Take the ratio of the number of microstates to find out.) **(c)** If you were betting on 2 heads and 8 tails would you accept odds of 252 to 45? Explain why or why not.

Table 15.5 10-Coin Toss

Macrostate		Number of Microstates (*W*)
Heads	Tails	
10	0	1
9	1	10
8	2	45
7	3	120
6	4	210
5	5	252
4	6	210
3	7	120
2	8	45
1	9	10
0	10	1
		Total: 1024

15.1 The First Law of Thermodynamics

1. A cylinder is divided in half by a movable disk in the middle. Each half is filled with an equal number of gas molecules, but one half is at a higher temperature than the other. Which choice best describes what happens next?

    a.   Nothing.
    b.   The high temperature side expands, compressing the low temperature side.
    c.   Heat moves from hot to cold, so the low temperature side will gradually increase in temperature and expand
    d.   (b) happens quickly, but after that (c) happens more slowly.

2. Imagine a solid material at the molecular level as consisting of a bunch of billiard balls connected to each other by springs (this is actually a surprisingly useful approximation). If we have two blocks of the same material, but in one the billiard balls are shaking back and forth on their springs a great deal, and in the other they are barely moving, which block is at the higher temperature? Using what you know about conservation of momentum in collisions, describe which block will transfer energy to the other, and justify your answer.

3. A system has 300 J of work done on it, and has a heat transfer of -320 J. Compared to prior to these processes, the internal energy is:

    a.   20 J less
    b.   20 J more
    c.   620 J more
    d.   620 J less

4. Find a snack or drink item in the classroom, or at your next meal. Find the total Calories (kilocalories) in the item, and calculate how long it would take exercising at 150 W (moderately, climbing stairs) at 20% efficiency to burn off this energy.

5. A potato cannon has the fuel combusted, generating a lot of heat and pressure, which launch a potato. The combustion process _____ the internal energy, while launching the potato _____ the internal energy of the potato cannon.

    a.   increases, increases
    b.   increases, decreases
    c.   decreases, increases
    d.   decreases, decreases

6. Describe what happens to the system inside of a refrigerator or freezer in terms of heat transfer, work, and conservation of energy. Confine yourself to time periods in which the door is closed.

15.2 The First Law of Thermodynamics and Some Simple Processes

7. In **Figure 15.44**, how much work is done by the system in process AB?

    a. 4.5×10^3 J
    b. 6.0×10^3 J
    c. 6.9×10^3 J
    d. 7.8×10^3 J

8. Consider process CD in **Figure 15.44**. Does this represent work done by or on the system, and how much?

9. A thermodynamic process begins at 1.2×10^6 N/m^2 and 5 L. The state then changes to 1.2×10^6 N/m^2 and 2 L. Next it

becomes 2.2×10^6 N/m^2 and 2 L. The next change is 2.2×10^6 N/m^2 and 5 L. Finally, the system ends at 1.0×10^6 N/m^2 and 5 L.

On **Figure 15.43**, this process is best described by

    a.   EFCDB
    b.   DEFCD
    c.   CFABC
    d.   CFABD

10. The first step of a thermodynamic cycle is an isobaric process with increasing volume. The second is an isochoric process, with decreasing pressure. The last step may be either an isothermal or adiabatic process, ending at the starting point of the isobaric process. Sketch a graph of these two possibilities, and comment on which will have greater net work per cycle.

11. In **Figure 15.44**, which of the following cycles has the greatest net work output?

    a.   ABDA
    b.   BCDB
    c.   (a) and (b) are equal
    d.   ADCBA

12. Look at **Figure 15.43**, and assign values to the three pressures and two volumes given in the graph. Then calculate the net work for the cycle ABCFEDCFA using those values. How does this work compare to the heat output or input of the system? Which value(s) would you change to maximize the net work per cycle?

15.6 Entropy and the Second Law of Thermodynamics: Disorder and the Unavailability of Energy

13. Equal masses of steam (100 degrees C) and ice (0 degrees C) are placed in contact with each other in an otherwise insulated container. They both end up as liquid water at a common temperature. The steam ___ entropy and ___ order, while the ice ___ entropy and ___ order.

    a.   gained, gained, lost, lost
    b.   gained, lost, lost, gained
    c.   lost, gained, gained, lost
    d.   lost, lost, gained, gained

14. A high temperature reservoir losing heat and hence entropy is a reversible process. A low temperature reservoir gaining a certain amount of heat and hence entropy is a reversible process. But a high temperature reservoir losing heat to a low temperature reservoir is irreversible. Why?

15.7 Statistical Interpretation of Entropy and the Second Law of Thermodynamics: The Underlying Explanation

15. A piston is resting halfway into a cylinder containing gas in thermal equilibrium. The layer of molecules next to the closed end of the cylinder is suddenly flash-heated to a very high temperature. Which best describes what happens next?

a. The high temperature molecules push out the piston until their energy is reduced enough that the system is in equilibrium.
b. The molecules with the highest temperature bounce off their neighbors, losing energy to them, and so on until the system is at a new equilibrium with the piston moved out.
c. The molecules with the highest temperature bounce off their neighbors, losing energy to them, and so on until the system is at a new equilibrium with the piston where it started.
d. The high temperature molecules push out the piston until their energy is reduced enough that the system is in equilibrium, and then the piston gets sucked back in.

16. Design a macroscopic simulation using reasonably common materials to represent one very high energy particle gradually transferring energy to a bunch of lower energy particles, and determine if you end up with some sort of equilibrium.

16   OSCILLATORY MOTION AND WAVES

Figure 16.1 There are at least four types of waves in this picture—only the water waves are evident. There are also sound waves, light waves, and waves on the guitar strings. (credit: John Norton)

Connection for AP® Courses

In this chapter, students are introduced to oscillation, the regular variation in the position of a system about a central point accompanied by transfer of energy and momentum, and to **waves**. A child's swing, a pendulum, a spring, and a vibrating string are all examples of oscillations. This chapter will address simple harmonic motion and periods of vibration, aspects of oscillation that produce waves, a common phenomenon in everyday life. Waves carry energy from one place to another." This chapter will show how harmonic oscillations produce waves that transport energy across space and through time. The information and examples presented support Big Ideas 1, 2, and 3 of the AP® Physics Curriculum Framework.

The chapter opens by discussing the forces that govern oscillations and waves. It goes on to discuss important concepts such as **simple harmonic motion**, uniform harmonic motion, and damped harmonic motion. You will also learn about energy in simple harmonic motion and how it changes from kinetic to potential, and how the total sum, which would be the mechanical energy of the oscillator, remains constant or conserved at all times. The chapter also discusses characteristics of waves, such as their frequency, period of oscillation, and the forms in which they can exist, i.e., transverse or longitudinal. The chapter ends by discussing what happens when two or more waves overlap and how the **amplitude** of the resultant wave changes, leading to the phenomena of **superposition** and **interference**.

The concepts in this chapter support:

Big Idea 3 The interactions of an object with other objects can be described by forces.

Enduring Understanding 3.B Classically, the acceleration of an object interacting with other objects can be predicted by using

$$\vec{a} = \sum \frac{\vec{F}}{m} .$$

Essential Knowledge 3.B.3 Restoring forces can result in oscillatory motion. When a linear restoring force is exerted on an object displaced from an equilibrium position, the object will undergo a special type of motion called simple harmonic motion. Examples should include gravitational force exerted by the Earth on a simple pendulum and a mass-spring oscillator.

Big Idea 4 Interactions between systems can result in changes in those systems.

Enduring Understanding 4.C Interactions with other objects or systems can change the total energy of a system.

Essential Knowledge 4.C.1 The energy of a system includes its kinetic energy, potential energy, and microscopic internal energy. Examples should include gravitational potential energy, elastic potential energy, and kinetic energy.

Essential Knowledge 4.C.2 Mechanical energy (the sum of kinetic and potential energy) is transferred into or out of a system when an external force is exerted on a system such that a component of the force is parallel to its displacement. The process through which the energy is transferred is called work.

Big Idea 5 Changes that occur as a result of interactions are constrained by conservation laws.

Enduring Understanding 5.B The energy of a system is conserved.

Essential Knowledge 5.B.2 A system with internal structure can have internal energy, and changes in a system's internal structure can result in changes in internal energy. [Physics 1: includes mass-spring oscillators and simple pendulums. Physics 2: includes charged object in electric fields and examining changes in internal energy with changes in configuration.]

Big Idea 6 Waves can transfer energy and momentum from one location to another without the permanent transfer of mass and serve as a mathematical model for the description of other phenomena.

Enduring Understanding 6.A A wave is a traveling disturbance that transfers energy and momentum.

Essential Knowledge 6.A.1 Waves can propagate via different oscillation modes such as transverse and longitudinal.

Essential Knowledge 6.A.2 For propagation, mechanical waves require a medium, while electromagnetic waves do not require a physical medium. Examples should include light traveling through a vacuum and sound not traveling through a vacuum.

Essential Knowledge 6.A.3 The amplitude is the maximum displacement of a wave from its equilibrium value.

Essential Knowledge 6.A.4 Classically, the energy carried by a wave depends on and increases with amplitude. Examples should include sound waves.

Enduring Understanding 6.B A periodic wave is one that repeats as a function of both time and position and can be described by its amplitude, frequency, wavelength, speed, and energy.

Essential Knowledge 6.B.1 The period is the repeat time of the wave. The frequency is the number of repetitions over a period of time.

Essential Knowledge 6.B.2 The wavelength is the repeat distance of the wave.

Essential Knowledge 6.B.3 A simple wave can be described by an equation involving one sine or cosine function involving the wavelength, amplitude, and frequency of the wave.

Essential Knowledge 6.B.4 The wavelength is the ratio of speed over frequency.

Enduring Understanding 6.C Only waves exhibit interference and diffraction.

Essential Knowledge 6.C.1 When two waves cross, they travel through each other; they do not bounce off each other. Where the waves overlap, the resulting displacement can be determined by adding the displacements of the two waves. This is called superposition.

Enduring Understanding 6.D Interference and superposition lead to standing waves and beats.

Essential Knowledge 6.D.1 Two or more wave pulses can interact in such a way as to produce amplitude variations in the resultant wave. When two pulses cross, they travel through each other; they do not bounce off each other. Where the pulses overlap, the resulting displacement can be determined by adding the displacements of the two pulses. This is called superposition.

Essential Knowledge 6.D.2 Two or more traveling waves can interact in such a way as to produce amplitude variations in the resultant wave.

Essential Knowledge 6.D.3 Standing waves are the result of the addition of incident and reflected waves that are confined to a region and have nodes and antinodes. Examples should include waves on a fixed length of string, and sound waves in both closed and open tubes.

Essential Knowledge 6.D.4 The possible wavelengths of a standing wave are determined by the size of the region to which it is confined.

Essential Knowledge 6.D.5 Beats arise from the addition of waves of slightly different frequency.

16.1 Hooke's Law: Stress and Strain Revisited

Learning Objectives

By the end of this section, you will be able to:

- Explain Newton's third law of motion with respect to stress and deformation.
- Describe the restoring force and displacement.
- Use Hooke's law of deformation, and calculate stored energy in a spring.

The information presented in this section supports the following AP® learning objectives and science practices:

- **3.B.3.3** The student can analyze data to identify qualitative or quantitative relationships between given values and variables (i.e., force, displacement, acceleration, velocity, period of motion, frequency, spring constant, string length, mass) associated with objects in oscillatory motion to use that data to determine the value of an unknown. **(S.P. 2.2, 5.1)**
- **3.B.3.4** The student is able to construct a qualitative and/or a quantitative explanation of oscillatory behavior given evidence of a restoring force. **(S.P. 2.2, 6.2)**

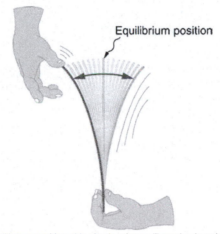

Figure 16.2 When displaced from its vertical equilibrium position, this plastic ruler oscillates back and forth because of the restoring force opposing displacement. When the ruler is on the left, there is a force to the right, and vice versa.

Newton's first law implies that an object oscillating back and forth is experiencing forces. Without force, the object would move in a straight line at a constant speed rather than oscillate. Consider, for example, plucking a plastic ruler to the left as shown in **Figure 16.2**. The deformation of the ruler creates a force in the opposite direction, known as a **restoring force**. Once released, the restoring force causes the ruler to move back toward its stable equilibrium position, where the net force on it is zero. However, by the time the ruler gets there, it gains momentum and continues to move to the right, producing the opposite deformation. It is then forced to the left, back through equilibrium, and the process is repeated until dissipative forces dampen the motion. These forces remove mechanical energy from the system, gradually reducing the motion until the ruler comes to rest.

The simplest oscillations occur when the restoring force is directly proportional to displacement. When stress and strain were covered in **Newton's Third Law of Motion**, the name was given to this relationship between force and displacement was Hooke's law:

$$F = -kx. \tag{16.1}$$

Here, F is the restoring force, x is the displacement from equilibrium or **deformation**, and k is a constant related to the difficulty in deforming the system. The minus sign indicates the restoring force is in the direction opposite to the displacement.

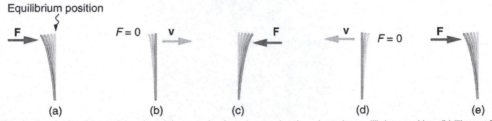

Figure 16.3 (a) The plastic ruler has been released, and the restoring force is returning the ruler to its equilibrium position. (b) The net force is zero at the equilibrium position, but the ruler has momentum and continues to move to the right. (c) The restoring force is in the opposite direction. It stops the ruler and moves it back toward equilibrium again. (d) Now the ruler has momentum to the left. (e) In the absence of damping (caused by frictional forces), the ruler reaches its original position. From there, the motion will repeat itself.

The **force constant** k is related to the rigidity (or stiffness) of a system—the larger the force constant, the greater the restoring force, and the stiffer the system. The units of k are newtons per meter (N/m). For example, k is directly related to Young's modulus when we stretch a string. **Figure 16.4** shows a graph of the absolute value of the restoring force versus the displacement for a system that can be described by Hooke's law—a simple spring in this case. The slope of the graph equals the force constant k in newtons per meter. A common physics laboratory exercise is to measure restoring forces created by springs, determine if they follow Hooke's law, and calculate their force constants if they do.

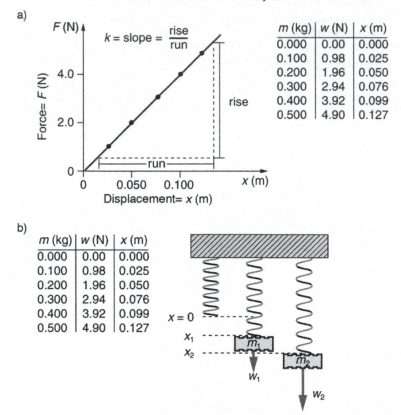

b)

m (kg)	w (N)	x (m)
0.000	0.00	0.000
0.100	0.98	0.025
0.200	1.96	0.050
0.300	2.94	0.076
0.400	3.92	0.099
0.500	4.90	0.127

Figure 16.4 (a) A graph of absolute value of the restoring force versus displacement is displayed. The fact that the graph is a straight line means that the system obeys Hooke's law. The slope of the graph is the force constant k. (b) The data in the graph were generated by measuring the displacement of a spring from equilibrium while supporting various weights. The restoring force equals the weight supported, if the mass is stationary.

Example 16.1 How Stiff Are Car Springs?

Figure 16.5 The mass of a car increases due to the introduction of a passenger. This affects the displacement of the car on its suspension system. (credit: exfordy on Flickr)

What is the force constant for the suspension system of a car that settles 1.20 cm when an 80.0-kg person gets in?

Strategy

Consider the car to be in its equilibrium position $x = 0$ before the person gets in. The car then settles down 1.20 cm, which means it is displaced to a position $x = -1.20 \times 10^{-2}$ m . At that point, the springs supply a restoring force F equal to the person's weight $w = mg = (80.0 \text{ kg})(9.80 \text{ m/s}^2) = 784 \text{ N}$. We take this force to be F in Hooke's law. Knowing F and x , we can then solve the force constant k .

Solution

1. Solve Hooke's law, $F = -kx$, for k :

$$k = -\frac{F}{x}. \tag{16.2}$$

Substitute known values and solve k :

$$
\begin{aligned}
k &= -\frac{784 \text{ N}}{-1.20 \times 10^{-2} \text{ m}} \\
&= 6.53 \times 10^4 \text{ N/m}.
\end{aligned} \tag{16.3}
$$

Discussion

Note that F and x have opposite signs because they are in opposite directions—the restoring force is up, and the displacement is down. Also, note that the car would oscillate up and down when the person got in if it were not for damping (due to frictional forces) provided by shock absorbers. Bouncing cars are a sure sign of bad shock absorbers.

Energy in Hooke's Law of Deformation

In order to produce a deformation, work must be done. That is, a force must be exerted through a distance, whether you pluck a guitar string or compress a car spring. If the only result is deformation, and no work goes into thermal, sound, or kinetic energy, then all the work is initially stored in the deformed object as some form of potential energy. The potential energy stored in a spring is $\text{PE}_{el} = \frac{1}{2}kx^2$. Here, we generalize the idea to elastic potential energy for a deformation of any system that can be described by Hooke's law. Hence,

$$\text{PE}_{el} = \frac{1}{2}kx^2, \tag{16.4}$$

where PE_{el} is the **elastic potential energy** stored in any deformed system that obeys Hooke's law and has a displacement x from equilibrium and a force constant k .

It is possible to find the work done in deforming a system in order to find the energy stored. This work is performed by an applied force F_{app} . The applied force is exactly opposite to the restoring force (action-reaction), and so $F_{app} = kx$. **Figure 16.6** shows a graph of the applied force versus deformation x for a system that can be described by Hooke's law. Work done on the system is force multiplied by distance, which equals the area under the curve or $(1/2)kx^2$ (Method A in the figure). Another way to determine the work is to note that the force increases linearly from 0 to kx , so that the average force is $(1/2)kx$, the distance moved is x , and thus $W = F_{app}d = [(1/2)kx](x) = (1/2)kx^2$ (Method B in the figure).

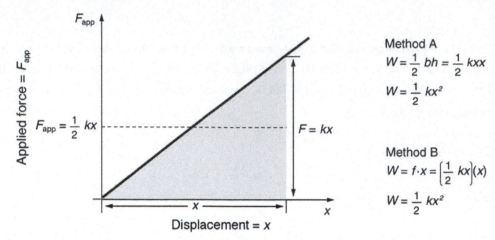

Method A
$$W = \frac{1}{2} bh = \frac{1}{2} kxx$$
$$W = \frac{1}{2} kx^2$$

Method B
$$W = f \cdot x = \left(\frac{1}{2} kx\right)(x)$$
$$W = \frac{1}{2} kx^2$$

Figure 16.6 A graph of applied force versus distance for the deformation of a system that can be described by Hooke's law is displayed. The work done on the system equals the area under the graph or the area of the triangle, which is half its base multiplied by its height, or $W = (1/2)kx^2$.

Example 16.2 Calculating Stored Energy: A Tranquilizer Gun Spring

We can use a toy gun's spring mechanism to ask and answer two simple questions: (a) How much energy is stored in the spring of a tranquilizer gun that has a force constant of 50.0 N/m and is compressed 0.150 m? (b) If you neglect friction and the mass of the spring, at what speed will a 2.00-g projectile be ejected from the gun?

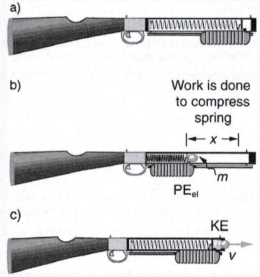

Figure 16.7 (a) In this image of the gun, the spring is uncompressed before being cocked. (b) The spring has been compressed a distance x, and the projectile is in place. (c) When released, the spring converts elastic potential energy $\mathrm{PE}_{\mathrm{el}}$ into kinetic energy.

Strategy for a

(a): The energy stored in the spring can be found directly from elastic potential energy equation, because k and x are given.

Solution for a

Entering the given values for k and x yields

$$
\begin{aligned}
\mathrm{PE}_{\mathrm{el}} &= \tfrac{1}{2}kx^2 = \tfrac{1}{2}(50.0 \text{ N/m})(0.150 \text{ m})^2 = 0.563 \text{ N} \cdot \text{m} \\
&= 0.563 \text{ J}
\end{aligned}
\tag{16.5}
$$

Strategy for b

Because there is no friction, the potential energy is converted entirely into kinetic energy. The expression for kinetic energy can be solved for the projectile's speed.

Solution for b

1. Identify known quantities:

$$\text{KE}_f = \text{PE}_{el} \text{ or } 1/2mv^2 = (1/2)kx^2 = \text{PE}_{el} = 0.563 \text{ J} \qquad (16.6)$$

2. Solve for v:

$$v = \left[\frac{2\text{PE}_{el}}{m}\right]^{1/2} = \left[\frac{2(0.563 \text{ J})}{0.002 \text{ kg}}\right]^{1/2} = 23.7(\text{J/kg})^{1/2} \qquad (16.7)$$

3. Convert units: 23.7 m/s

Discussion

(a) and (b): This projectile speed is impressive for a tranquilizer gun (more than 80 km/h). The numbers in this problem seem reasonable. The force needed to compress the spring is small enough for an adult to manage, and the energy imparted to the dart is small enough to limit the damage it might do. Yet, the speed of the dart is great enough for it to travel an acceptable distance.

Check your Understanding

Envision holding the end of a ruler with one hand and deforming it with the other. When you let go, you can see the oscillations of the ruler. In what way could you modify this simple experiment to increase the rigidity of the system?

Solution

You could hold the ruler at its midpoint so that the part of the ruler that oscillates is half as long as in the original experiment.

Check your Understanding

If you apply a deforming force on an object and let it come to equilibrium, what happened to the work you did on the system?

Solution

It was stored in the object as potential energy.

16.2 Period and Frequency in Oscillations

Learning Objectives

By the end of this section, you will be able to:

- Relate recurring mechanical vibrations to the frequency and period of harmonic motion, such as the motion of a guitar string.
- Compute the frequency and period of an oscillation.

The information presented in this section supports the following AP® learning objectives and science practices:

- **3.B.3.3** The student can analyze data to identify qualitative or quantitative relationships between given values and variables (i.e., force, displacement, acceleration, velocity, period of motion, frequency, spring constant, string length, mass) associated with objects in oscillatory motion to use that data to determine the value of an unknown. **(S.P. 2.2, 5.1)**

Figure 16.8 The strings on this guitar vibrate at regular time intervals. (credit: JAR)

When you pluck a guitar string, the resulting sound has a steady tone and lasts a long time. Each successive vibration of the string takes the same time as the previous one. We define **periodic motion** to be a motion that repeats itself at regular time intervals, such as exhibited by the guitar string or by an object on a spring moving up and down. The time to complete one oscillation remains constant and is called the **period** T. Its units are usually seconds, but may be any convenient unit of time.

The word period refers to the time for some event whether repetitive or not; but we shall be primarily interested in periodic motion, which is by definition repetitive. A concept closely related to period is the frequency of an event. For example, if you get a paycheck twice a month, the frequency of payment is two per month and the period between checks is half a month. **Frequency** f is defined to be the number of events per unit time. For periodic motion, frequency is the number of oscillations per unit time.

The relationship between frequency and period is

$$f = \frac{1}{T}.$$ (16.8)

The SI unit for frequency is the *cycle per second*, which is defined to be a *hertz* (Hz):

$$1 \text{ Hz} = 1\frac{\text{cycle}}{\text{sec}} \text{ or } 1 \text{ Hz} = \frac{1}{\text{s}}$$ (16.9)

A cycle is one complete oscillation. Note that a vibration can be a single or multiple event, whereas oscillations are usually repetitive for a significant number of cycles.

Example 16.3 Determine the Frequency of Two Oscillations: Medical Ultrasound and the Period of Middle C

We can use the formulas presented in this module to determine both the frequency based on known oscillations and the oscillation based on a known frequency. Let's try one example of each. (a) A medical imaging device produces ultrasound by oscillating with a period of 0.400 µs. What is the frequency of this oscillation? (b) The frequency of middle C on a typical musical instrument is 264 Hz. What is the time for one complete oscillation?

Strategy

Both questions (a) and (b) can be answered using the relationship between period and frequency. In question (a), the period T is given and we are asked to find frequency f. In question (b), the frequency f is given and we are asked to find the period T.

Solution a

1. Substitute 0.400 µs for T in $f = \frac{1}{T}$:

$$f = \frac{1}{T} = \frac{1}{0.400 \times 10^{-6} \text{ s}}.$$ (16.10)

Solve to find

$$f = 2.50 \times 10^{6} \text{ Hz}.$$ (16.11)

Discussion a

The frequency of sound found in (a) is much higher than the highest frequency that humans can hear and, therefore, is called ultrasound. Appropriate oscillations at this frequency generate ultrasound used for noninvasive medical diagnoses, such as observations of a fetus in the womb.

Solution b

1. Identify the known values:
   The time for one complete oscillation is the period T:

$$f = \frac{1}{T}.$$ (16.12)

2. Solve for T:

$$T = \frac{1}{f}.$$ (16.13)

3. Substitute the given value for the frequency into the resulting expression:

$$T = \frac{1}{f} = \frac{1}{264 \text{ Hz}} = \frac{1}{264 \text{ cycles/s}} = 3.79 \times 10^{-3} \text{ s} = 3.79 \text{ ms}.$$ (16.14)

Discussion

The period found in (b) is the time per cycle, but this value is often quoted as simply the time in convenient units (ms or milliseconds in this case).

Identify an event in your life (such as receiving a paycheck) that occurs regularly. Identify both the period and frequency of this event.

Solution

I visit my parents for dinner every other Sunday. The frequency of my visits is 26 per calendar year. The period is two weeks.

16.3 Simple Harmonic Motion: A Special Periodic Motion

Learning Objectives

By the end of this section, you will be able to:

- Describe a simple harmonic oscillator.
- Relate physical characteristics of a vibrating system to aspects of simple harmonic motion and any resulting waves.

The information presented in this section supports the following AP® learning objectives and science practices:

- **3.B.3.1** The student is able to predict which properties determine the motion of a simple harmonic oscillator and what the dependence of the motion is on those properties. **(S.P. 6.4, 7.2)**
- **3.B.3.4** The student is able to construct a qualitative and/or a quantitative explanation of oscillatory behavior given evidence of a restoring force. **(S.P. 2.2, 6.2)**
- **6.A.3.1** The student is able to use graphical representation of a periodic mechanical wave to determine the amplitude of the wave. **(S.P. 1.4)**
- **6.B.1.1** The student is able to use a graphical representation of a periodic mechanical wave (position versus time) to determine the period and frequency of the wave and describe how a change in the frequency would modify features of the representation. **(S.P. 1.4, 2.2)**

The oscillations of a system in which the net force can be described by Hooke's law are of special importance, because they are very common. They are also the simplest oscillatory systems. **Simple Harmonic Motion** (SHM) is the name given to oscillatory motion for a system where the net force can be described by Hooke's law, and such a system is called a **simple harmonic oscillator**. If the net force can be described by Hooke's law and there is no *damping* (by friction or other non-conservative forces), then a simple harmonic oscillator will oscillate with equal displacement on either side of the equilibrium position, as shown for an object on a spring in **Figure 16.9**. The maximum displacement from equilibrium is called the **amplitude** X. The units for amplitude and displacement are the same, but depend on the type of oscillation. For the object on the spring, the units of amplitude and displacement are meters; whereas for sound oscillations, they have units of pressure (and other types of oscillations have yet other units). Because amplitude is the maximum displacement, it is related to the energy in the oscillation.

Take-Home Experiment: SHM and the Marble

Find a bowl or basin that is shaped like a hemisphere on the inside. Place a marble inside the bowl and tilt the bowl periodically so the marble rolls from the bottom of the bowl to equally high points on the sides of the bowl. Get a feel for the force required to maintain this periodic motion. What is the restoring force and what role does the force you apply play in the simple harmonic motion (SHM) of the marble?

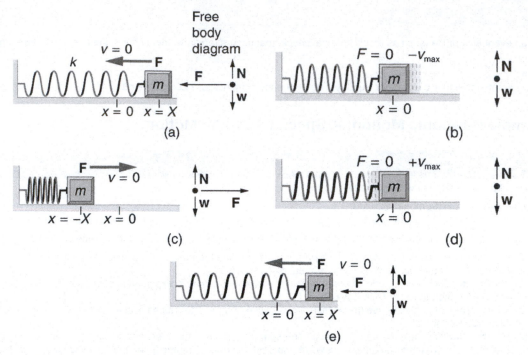

Figure 16.9 An object attached to a spring sliding on a frictionless surface is an uncomplicated simple harmonic oscillator. When displaced from equilibrium, the object performs simple harmonic motion that has an amplitude X and a period T. The object's maximum speed occurs as it passes through equilibrium. The stiffer the spring is, the smaller the period T. The greater the mass of the object is, the greater the period T.

What is so significant about simple harmonic motion? One special thing is that the period T and frequency f of a simple harmonic oscillator are independent of amplitude. The string of a guitar, for example, will oscillate with the same frequency whether plucked gently or hard. Because the period is constant, a simple harmonic oscillator can be used as a clock.

Two important factors do affect the period of a simple harmonic oscillator. The period is related to how stiff the system is. A very stiff object has a large force constant k, which causes the system to have a smaller period. For example, you can adjust a diving board's stiffness—the stiffer it is, the faster it vibrates, and the shorter its period. Period also depends on the mass of the oscillating system. The more massive the system is, the longer the period. For example, a heavy person on a diving board bounces up and down more slowly than a light one.

In fact, the mass m and the force constant k are the *only* factors that affect the period and frequency of simple harmonic motion.

Period of Simple Harmonic Oscillator

The *period of a simple harmonic oscillator* is given by

$$T = 2\pi\sqrt{\frac{m}{k}}\qquad\qquad(16.15)$$

and, because $f = 1/T$, the *frequency of a simple harmonic oscillator* is

$$f = \frac{1}{2\pi}\sqrt{\frac{k}{m}}.\qquad\qquad(16.16)$$

Note that neither T nor f has any dependence on amplitude.

Example 16.4 Mechanical Waves

What do sound waves, water waves, and seismic waves have in common? They are all governed by Newton's laws and they can exist only when traveling in a medium, such as air, water, or rocks. Waves that require a medium to travel are collectively known as "mechanical waves."

Take-Home Experiment: Mass and Ruler Oscillations

Find two identical wooden or plastic rulers. Tape one end of each ruler firmly to the edge of a table so that the length of each ruler that protrudes from the table is the same. On the free end of one ruler tape a heavy object such as a few large coins. Pluck the ends of the rulers at the same time and observe which one undergoes more cycles in a time period, and measure the period of oscillation of each of the rulers.

Example 16.5 Calculate the Frequency and Period of Oscillations: Bad Shock Absorbers in a Car

If the shock absorbers in a car go bad, then the car will oscillate at the least provocation, such as when going over bumps in the road and after stopping (See **Figure 16.10**). Calculate the frequency and period of these oscillations for such a car if the car's mass (including its load) is 900 kg and the force constant (k) of the suspension system is 6.53×10^4 N/m.

Strategy

The frequency of the car's oscillations will be that of a simple harmonic oscillator as given in the equation $f = \frac{1}{2\pi}\sqrt{\frac{k}{m}}$. The mass and the force constant are both given.

Solution

1. Enter the known values of k and m:

$$f = \frac{1}{2\pi}\sqrt{\frac{k}{m}} = \frac{1}{2\pi}\sqrt{\frac{6.53 \times 10^4 \text{ N/m}}{900 \text{ kg}}}.$$

(16.17)

2. Calculate the frequency:

$$\frac{1}{2\pi}\sqrt{72.6 / \text{s}^{-2}} = 1.3656 / \text{s}^{-1} \approx 1.36 / \text{s}^{-1} = 1.36 \text{ Hz}.$$

(16.18)

3. You could use $T = 2\pi\sqrt{\frac{m}{k}}$ to calculate the period, but it is simpler to use the relationship $T = 1/f$ and substitute the value just found for f:

$$T = \frac{1}{f} = \frac{1}{1.356 \text{ Hz}} = 0.738 \text{ s}.$$

(16.19)

Discussion

The values of T and f both seem about right for a bouncing car. You can observe these oscillations if you push down hard on the end of a car and let go.

The Link between Simple Harmonic Motion and Waves

If a time-exposure photograph of the bouncing car were taken as it drove by, the headlight would make a wavelike streak, as shown in **Figure 16.10**. Similarly, **Figure 16.11** shows an object bouncing on a spring as it leaves a wavelike "trace of its position on a moving strip of paper. Both waves are sine functions. All simple harmonic motion is intimately related to sine and cosine waves.

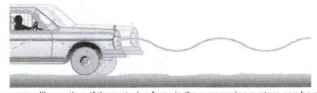

Figure 16.10 The bouncing car makes a wavelike motion. If the restoring force in the suspension system can be described only by Hooke's law, then the wave is a sine function. (The wave is the trace produced by the headlight as the car moves to the right.)

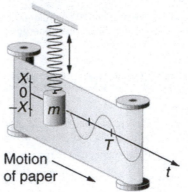

Figure 16.11 The vertical position of an object bouncing on a spring is recorded on a strip of moving paper, leaving a sine wave.

The displacement as a function of time t in any simple harmonic motion—that is, one in which the net restoring force can be described by Hooke's law, is given by

$$x(t) = X \cos\frac{2\pi t}{T}, \tag{16.20}$$

where X is amplitude. At $t = 0$, the initial position is $x_0 = X$, and the displacement oscillates back and forth with a period T. (When $t = T$, we get $x = X$ again because $\cos 2\pi = 1$.). Furthermore, from this expression for x, the velocity v as a function of time is given by:

$$v(t) = -v_{\text{max}} \sin\left(\frac{2\pi t}{T}\right), \tag{16.21}$$

where $v_{\text{max}} = 2\pi X / T = X\sqrt{k/m}$. The object has zero velocity at maximum displacement—for example, $v = 0$ when $t = 0$, and at that time $x = X$. The minus sign in the first equation for $v(t)$ gives the correct direction for the velocity. Just after the start of the motion, for instance, the velocity is negative because the system is moving back toward the equilibrium point. Finally, we can get an expression for acceleration using Newton's second law. [Then we have $x(t)$, $v(t)$, t, and $a(t)$, the quantities needed for kinematics and a description of simple harmonic motion.] According to Newton's second law, the acceleration is $a = F/m = kx/m$. So, $a(t)$ is also a cosine function:

$$a(t) = -\frac{kX}{m}\cos\frac{2\pi t}{T}. \tag{16.22}$$

Hence, $a(t)$ is directly proportional to and in the opposite direction to $x(t)$.

Figure 16.12 shows the simple harmonic motion of an object on a spring and presents graphs of $x(t), v(t)$, and $a(t)$ versus time.

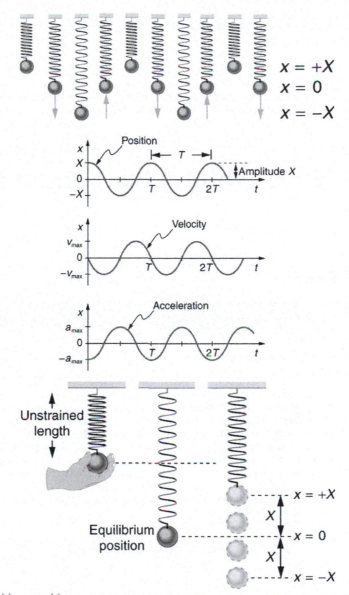

Figure 16.12 Graphs of $x(t)$, $v(t)$, and $a(t)$ versus t for the motion of an object on a spring. The net force on the object can be described by

Hooke's law, and so the object undergoes simple harmonic motion. Note that the initial position has the vertical displacement at its maximum value X; v is initially zero and then negative as the object moves down; and the initial acceleration is negative, back toward the equilibrium position and becomes zero at that point.

The most important point here is that these equations are mathematically straightforward and are valid for all simple harmonic motion. They are very useful in visualizing waves associated with simple harmonic motion, including visualizing how waves add with one another.

Check Your Understanding

Suppose you pluck a banjo string. You hear a single note that starts out loud and slowly quiets over time. Describe what happens to the sound waves in terms of period, frequency and amplitude as the sound decreases in volume.

Solution
Frequency and period remain essentially unchanged. Only amplitude decreases as volume decreases.

Check Your Understanding

A babysitter is pushing a child on a swing. At the point where the swing reaches x, where would the corresponding point on a wave of this motion be located?

Solution

x is the maximum deformation, which corresponds to the amplitude of the wave. The point on the wave would either be at the very top or the very bottom of the curve.

PhET Explorations: Masses and Springs

A realistic mass and spring laboratory. Hang masses from springs and adjust the spring stiffness and damping. You can even slow time. Transport the lab to different planets. A chart shows the kinetic, potential, and thermal energy for each spring.

Figure 16.13 Masses and Springs (http://cnx.org/content/m55273/1.3/mass-spring-lab_en.jar)

16.4 The Simple Pendulum

Learning Objectives
By the end of this section, you will be able to: • Determine the period of oscillation of a hanging pendulum. The information presented in this section supports the following AP® learning objectives and science practices: • **3.B.3.1** The student is able to predict which properties determine the motion of a simple harmonic oscillator and what the dependence of the motion is on those properties. **(S.P. 6.4, 7.2)** • **3.B.3.2** The student is able to design a plan and collect data in order to ascertain the characteristics of the motion of a system undergoing oscillatory motion caused by a restoring force. **(S.P. 4.2)** • **3.B.3.3** The student can analyze data to identify qualitative or quantitative relationships between given values and variables (i.e., force, displacement, acceleration, velocity, period of motion, frequency, spring constant, string length, mass) associated with objects in oscillatory motion to use that data to determine the value of an unknown. **(S.P. 2.2, 5.1)** • **3.B.3.4** The student is able to construct a qualitative and/or a quantitative explanation of oscillatory behavior given evidence of a restoring force. **(S.P. 2.2, 6.2)**

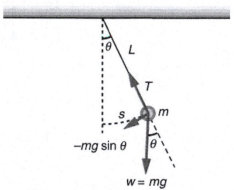

Figure 16.14 A simple pendulum has a small-diameter bob and a string that has a very small mass but is strong enough not to stretch appreciably. The linear displacement from equilibrium is s, the length of the arc. Also shown are the forces on the bob, which result in a net force of $-mg \sin\theta$ toward the equilibrium position—that is, a restoring force.

Pendulums are in common usage. Some have crucial uses, such as in clocks; some are for fun, such as a child's swing; and some are just there, such as the sinker on a fishing line. For small displacements, a pendulum is a simple harmonic oscillator. A **simple pendulum** is defined to have an object that has a small mass, also known as the pendulum bob, which is suspended from a light wire or string, such as shown in **Figure 16.14**. Exploring the simple pendulum a bit further, we can discover the conditions under which it performs simple harmonic motion, and we can derive an interesting expression for its period.

We begin by defining the displacement to be the arc length s. We see from **Figure 16.14** that the net force on the bob is tangent to the arc and equals $-mg \sin\theta$. (The weight mg has components $mg \cos\theta$ along the string and $mg \sin\theta$ tangent to the arc.) Tension in the string exactly cancels the component $mg \cos\theta$ parallel to the string. This leaves a *net* restoring force back toward the equilibrium position at $\theta = 0$.

Now, if we can show that the restoring force is directly proportional to the displacement, then we have a simple harmonic oscillator. In trying to determine if we have a simple harmonic oscillator, we should note that for small angles (less than about $15°$), $\sin\theta \approx \theta$ ($\sin\theta$ and θ differ by about 1% or less at smaller angles). Thus, for angles less than about $15°$, the restoring force F is

$$F \approx -mg\theta. \tag{16.23}$$

The displacement s is directly proportional to θ. When θ is expressed in radians, the arc length in a circle is related to its radius (L in this instance) by:

$$s = L\theta, \tag{16.24}$$

so that

$$\theta = \frac{s}{L}. \tag{16.25}$$

For small angles, then, the expression for the restoring force is:

$$F \approx -\frac{mg}{L}s \tag{16.26}$$

This expression is of the form:

$$F = -kx, \tag{16.27}$$

where the force constant is given by $k = mg/L$ and the displacement is given by $x = s$. For angles less than about $15°$, the restoring force is directly proportional to the displacement, and the simple pendulum is a simple harmonic oscillator.

Using this equation, we can find the period of a pendulum for amplitudes less than about $15°$. For the simple pendulum:

$$T = 2\pi\sqrt{\frac{m}{k}} = 2\pi\sqrt{\frac{m}{mg/L}}. \tag{16.28}$$

Thus,

$$T = 2\pi\sqrt{\frac{L}{g}} \tag{16.29}$$

for the period of a simple pendulum. This result is interesting because of its simplicity. The only things that affect the period of a simple pendulum are its length and the acceleration due to gravity. The period is completely independent of other factors, such as mass. As with simple harmonic oscillators, the period T for a pendulum is nearly independent of amplitude, especially if θ is less than about $15°$. Even simple pendulum clocks can be finely adjusted and accurate.

Note the dependence of T on g. If the length of a pendulum is precisely known, it can actually be used to measure the acceleration due to gravity. Consider the following example.

Example 16.6 Measuring Acceleration due to Gravity: The Period of a Pendulum

What is the acceleration due to gravity in a region where a simple pendulum having a length 75.000 cm has a period of 1.7357 s?

Strategy

We are asked to find g given the period T and the length L of a pendulum. We can solve $T = 2\pi\sqrt{\frac{L}{g}}$ for g, assuming only that the angle of deflection is less than $15°$.

Solution

1. Square $T = 2\pi\sqrt{\frac{L}{g}}$ and solve for g:

$$g = 4\pi^2\frac{L}{T^2}. \tag{16.30}$$

2. Substitute known values into the new equation:

$$g = 4\pi^2\frac{0.75000\ \text{m}}{(1.7357\ \text{s})^2}. \tag{16.31}$$

3. Calculate to find g:

$$g = 9.8281 \text{ m/s}^2. \tag{16.32}$$

Discussion

This method for determining g can be very accurate. This is why length and period are given to five digits in this example. For the precision of the approximation $\sin\theta \approx \theta$ to be better than the precision of the pendulum length and period, the maximum displacement angle should be kept below about $0.5°$.

Making Career Connections

Knowing g can be important in geological exploration; for example, a map of g over large geographical regions aids the study of plate tectonics and helps in the search for oil fields and large mineral deposits.

Take Home Experiment: Determining g

Use a simple pendulum to determine the acceleration due to gravity g in your own locale. Cut a piece of a string or dental floss so that it is about 1 m long. Attach a small object of high density to the end of the string (for example, a metal nut or a car key). Starting at an angle of less than $10°$, allow the pendulum to swing and measure the pendulum's period for 10 oscillations using a stopwatch. Calculate g. How accurate is this measurement? How might it be improved?

Check Your Understanding

An engineer builds two simple pendula. Both are suspended from small wires secured to the ceiling of a room. Each pendulum hovers 2 cm above the floor. Pendulum 1 has a bob with a mass of 10 kg. Pendulum 2 has a bob with a mass of 100 kg. Describe how the motion of the pendula will differ if the bobs are both displaced by $12°$.

Solution
The movement of the pendula will not differ at all because the mass of the bob has no effect on the motion of a simple pendulum. The pendula are only affected by the period (which is related to the pendulum's length) and by the acceleration due to gravity.

PhET Explorations: Pendulum Lab

Play with one or two pendulums and discover how the period of a simple pendulum depends on the length of the string, the mass of the pendulum bob, and the amplitude of the swing. It's easy to measure the period using the photogate timer. You can vary friction and the strength of gravity. Use the pendulum to find the value of g on planet X. Notice the anharmonic behavior at large amplitude.

Figure 16.15 Pendulum Lab (http://cnx.org/content/m55274/1.2/pendulum-lab_en.jar)

16.5 Energy and the Simple Harmonic Oscillator

Learning Objectives
By the end of this section, you will be able to: • Describe the changes in energy that occur while a system undergoes simple harmonic motion.

To study the energy of a simple harmonic oscillator, we first consider all the forms of energy it can have We know from **Hooke's Law: Stress and Strain Revisited** that the energy stored in the deformation of a simple harmonic oscillator is a form of potential energy given by:

$$PE_{el} = \frac{1}{2}kx^2. \tag{16.33}$$

Because a simple harmonic oscillator has no dissipative forces, the other important form of energy is kinetic energy KE. Conservation of energy for these two forms is:

$$KE + PE_{el} = \text{constant} \tag{16.34}$$

or

$$\tfrac{1}{2}mv^2 + \tfrac{1}{2}kx^2 = \text{constant.} \tag{16.35}$$

This statement of conservation of energy is valid for *all* simple harmonic oscillators, including ones where the gravitational force plays a role

Namely, for a simple pendulum we replace the velocity with $v = L\omega$, the spring constant with $k = mg/L$, and the displacement term with $x = L\theta$. Thus

$$\tfrac{1}{2}mL^2\omega^2 + \tfrac{1}{2}mgL\theta^2 = \text{constant.} \tag{16.36}$$

In the case of undamped simple harmonic motion, the energy oscillates back and forth between kinetic and potential, going completely from one to the other as the system oscillates. So for the simple example of an object on a frictionless surface attached to a spring, as shown again in **Figure 16.16**, the motion starts with all of the energy stored in the spring. As the object starts to move, the elastic potential energy is converted to kinetic energy, becoming entirely kinetic energy at the equilibrium position. It is then converted back into elastic potential energy by the spring, the velocity becomes zero when the kinetic energy is completely converted, and so on. This concept provides extra insight here and in later applications of simple harmonic motion, such as alternating current circuits.

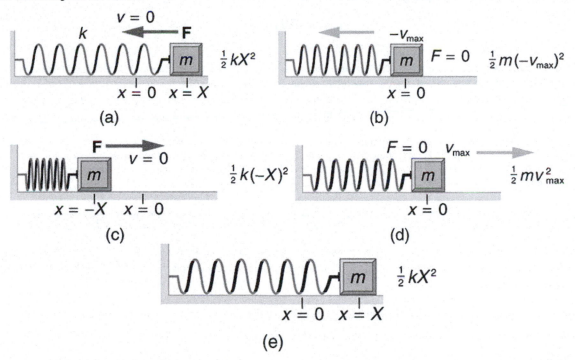

Figure 16.16 The transformation of energy in simple harmonic motion is illustrated for an object attached to a spring on a frictionless surface.

The conservation of energy principle can be used to derive an expression for velocity v. If we start our simple harmonic motion with zero velocity and maximum displacement ($x = X$), then the total energy is

$$\tfrac{1}{2}kX^2. \tag{16.37}$$

This total energy is constant and is shifted back and forth between kinetic energy and potential energy, at most times being shared by each. The conservation of energy for this system in equation form is thus:

$$\tfrac{1}{2}mv^2 + \tfrac{1}{2}kx^2 = \tfrac{1}{2}kX^2. \tag{16.38}$$

Solving this equation for v yields:

$$v = \pm\sqrt{\tfrac{k}{m}\left(X^2 - x^2\right)}. \tag{16.39}$$

Manipulating this expression algebraically gives:

$$v = \pm\sqrt{\frac{k}{m}}X\sqrt{1 - \frac{x^2}{X^2}}$$

(16.40)

and so

$$v = \pm v_{max}\sqrt{1 - \frac{x^2}{X^2}},$$

(16.41)

where

$$v_{max} = \sqrt{\frac{k}{m}}X.$$

(16.42)

From this expression, we see that the velocity is a maximum (v_{max}) at $x = 0$, as stated earlier in $v(t) = -v_{max}\sin\frac{2\pi t}{T}$.

Notice that the maximum velocity depends on three factors. Maximum velocity is directly proportional to amplitude. As you might guess, the greater the maximum displacement the greater the maximum velocity. Maximum velocity is also greater for stiffer systems, because they exert greater force for the same displacement. This observation is seen in the expression for v_{max}; it is proportional to the square root of the force constant k. Finally, the maximum velocity is smaller for objects that have larger masses, because the maximum velocity is inversely proportional to the square root of m. For a given force, objects that have large masses accelerate more slowly.

A similar calculation for the simple pendulum produces a similar result, namely:

$$\omega_{max} = \sqrt{\frac{g}{L}}\theta_{max}.$$

(16.43)

Making Connections: Mass Attached to a Spring

Consider a mass m attached to a spring, with spring constant k, fixed to a wall. When the mass is displaced from its equilibrium position and released, the mass undergoes simple harmonic motion. The spring exerts a force $F = -kv$ on the mass. The potential energy of the system is stored in the spring. It will be zero when the spring is in the equilibrium position. All the internal energy exists in the form of kinetic energy, given by $KE = \frac{1}{2}mv^2$. As the system oscillates, which means that the spring compresses and expands, there is a change in the structure of the system and a corresponding change in its internal energy. Its kinetic energy is converted to potential energy and vice versa. This occurs at an equal rate, which means that a loss of kinetic energy yields a gain in potential energy, thus preserving the work-energy theorem and the law of conservation of energy.

Example 16.7 Determine the Maximum Speed of an Oscillating System: A Bumpy Road

Suppose that a car is 900 kg and has a suspension system that has a force constant $k = 6.53\times10^4$ N/m. The car hits a bump and bounces with an amplitude of 0.100 m. What is its maximum vertical velocity if you assume no damping occurs?

Strategy

We can use the expression for v_{max} given in $v_{max} = \sqrt{\frac{k}{m}}X$ to determine the maximum vertical velocity. The variables m and k are given in the problem statement, and the maximum displacement X is 0.100 m.

Solution

1. Identify known.

2. Substitute known values into $v_{max} = \sqrt{\frac{k}{m}}X$:

$$v_{max} = \sqrt{\frac{6.53\times10^4 \text{ N/m}}{900 \text{ kg}}}(0.100 \text{ m}).$$

(16.44)

3. Calculate to find $v_{max}= 0.852$ m/s.

Discussion

This answer seems reasonable for a bouncing car. There are other ways to use conservation of energy to find v_{max}. We could use it directly, as was done in the example featured in **Hooke's Law: Stress and Strain Revisited**.

The small vertical displacement y of an oscillating simple pendulum, starting from its equilibrium position, is given as

$$y(t) = a \sin \omega t, \tag{16.45}$$

where a is the amplitude, ω is the angular velocity and t is the time taken. Substituting $\omega = \frac{2\pi}{T}$, we have

$$yt = a \sin\!\left(\frac{2\pi t}{T}\right). \tag{16.46}$$

Thus, the displacement of pendulum is a function of time as shown above.

Also the velocity of the pendulum is given by

$$v(t) = \frac{2a\pi}{T} \cos\left(\frac{2\pi t}{T}\right), \tag{16.47}$$

so the motion of the pendulum is a function of time.

Check Your Understanding

Why does it hurt more if your hand is snapped with a ruler than with a loose spring, even if the displacement of each system is equal?

Solution
The ruler is a stiffer system, which carries greater force for the same amount of displacement. The ruler snaps your hand with greater force, which hurts more.

Check Your Understanding

You are observing a simple harmonic oscillator. Identify one way you could decrease the maximum velocity of the system.

Solution
You could increase the mass of the object that is oscillating.

16.6 Uniform Circular Motion and Simple Harmonic Motion

Learning Objectives

By the end of this section, you will be able to:

* Compare simple harmonic motion with uniform circular motion.

Figure 16.17 The horses on this merry-go-round exhibit uniform circular motion. (credit: Wonderlane, Flickr)

There is an easy way to produce simple harmonic motion by using uniform circular motion. **Figure 16.18** shows one way of using this method. A ball is attached to a uniformly rotating vertical turntable, and its shadow is projected on the floor as shown. The shadow undergoes simple harmonic motion. Hooke's law usually describes uniform circular motions (ω constant) rather than systems that have large visible displacements. So observing the projection of uniform circular motion, as in **Figure 16.18**, is often easier than observing a precise large-scale simple harmonic oscillator. If studied in sufficient depth, simple harmonic motion produced in this manner can give considerable insight into many aspects of oscillations and waves and is very useful mathematically. In our brief treatment, we shall indicate some of the major features of this relationship and how they might be useful.

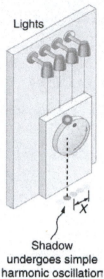

Figure 16.18 The shadow of a ball rotating at constant angular velocity ω on a turntable goes back and forth in precise simple harmonic motion.

Figure 16.19 shows the basic relationship between uniform circular motion and simple harmonic motion. The point P travels around the circle at constant angular velocity ω. The point P is analogous to an object on the merry-go-round. The projection of the position of P onto a fixed axis undergoes simple harmonic motion and is analogous to the shadow of the object. At the time shown in the figure, the projection has position x and moves to the left with velocity v. The velocity of the point P around the circle equals $\bar{v}_{max}$. The projection of $\bar{v}_{max}$ on the x-axis is the velocity v of the simple harmonic motion along the x-axis.

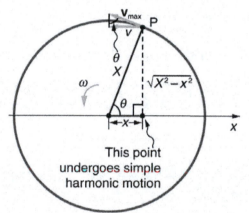

Figure 16.19 A point P moving on a circular path with a constant angular velocity ω is undergoing uniform circular motion. Its projection on the x-axis undergoes simple harmonic motion. Also shown is the velocity of this point around the circle, $\bar{v}_{max}$, and its projection, which is v. Note that these velocities form a similar triangle to the displacement triangle.

To see that the projection undergoes simple harmonic motion, note that its position x is given by

$$x = X \cos \theta, \tag{16.48}$$

where $\theta = \omega t$, ω is the constant angular velocity, and X is the radius of the circular path. Thus,

$$x = X \cos \omega t. \tag{16.49}$$

The angular velocity ω is in radians per unit time; in this case 2π radians is the time for one revolution T. That is, $\omega = 2\pi / T$. Substituting this expression for ω, we see that the position x is given by:

$$x(t) = \cos\left(\frac{2\pi t}{T}\right). \tag{16.50}$$

This expression is the same one we had for the position of a simple harmonic oscillator in **Simple Harmonic Motion: A Special Periodic Motion**. If we make a graph of position versus time as in **Figure 16.20**, we see again the wavelike character (typical of simple harmonic motion) of the projection of uniform circular motion onto the x-axis.

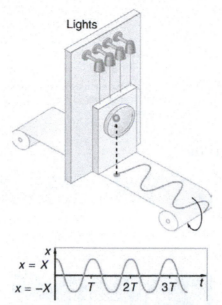

Figure 16.20 The position of the projection of uniform circular motion performs simple harmonic motion, as this wavelike graph of x versus t indicates.

Now let us use **Figure 16.19** to do some further analysis of uniform circular motion as it relates to simple harmonic motion. The triangle formed by the velocities in the figure and the triangle formed by the displacements (X, x, and $\sqrt{X^2 - x^2}$) are similar right triangles. Taking ratios of similar sides, we see that

$$\frac{v}{v_{\max}} = \frac{\sqrt{X^2 - x^2}}{X} = \sqrt{1 - \frac{x^2}{X^2}}. \tag{16.51}$$

We can solve this equation for the speed v or

$$v = v_{\max}\sqrt{1 - \frac{x^2}{X^2}}. \tag{16.52}$$

This expression for the speed of a simple harmonic oscillator is exactly the same as the equation obtained from conservation of energy considerations in **Energy and the Simple Harmonic Oscillator**. You can begin to see that it is possible to get all of the characteristics of simple harmonic motion from an analysis of the projection of uniform circular motion.

Finally, let us consider the period T of the motion of the projection. This period is the time it takes the point P to complete one revolution. That time is the circumference of the circle $2\pi X$ divided by the velocity around the circle, $v_{\max}$. Thus, the period T is

$$T = \frac{2\pi X}{v_{\max}}. \tag{16.53}$$

We know from conservation of energy considerations that

$$v_{\max} = \sqrt{\frac{k}{m}}X. \tag{16.54}$$

Solving this equation for $X/v_{\max}$ gives

$$\frac{X}{v_{\max}} = \sqrt{\frac{m}{k}}. \tag{16.55}$$

Substituting this expression into the equation for T yields

$$T = 2\pi\sqrt{\frac{m}{k}}. \tag{16.56}$$

Thus, the period of the motion is the same as for a simple harmonic oscillator. We have determined the period for any simple harmonic oscillator using the relationship between uniform circular motion and simple harmonic motion.

Some modules occasionally refer to the connection between uniform circular motion and simple harmonic motion. Moreover, if you carry your study of physics and its applications to greater depths, you will find this relationship useful. It can, for example, help to analyze how waves add when they are superimposed.

Identify an object that undergoes uniform circular motion. Describe how you could trace the simple harmonic motion of this object as a wave.

Solution

A record player undergoes uniform circular motion. You could attach dowel rod to one point on the outside edge of the turntable and attach a pen to the other end of the dowel. As the record player turns, the pen will move. You can drag a long piece of paper under the pen, capturing its motion as a wave.

16.7 Damped Harmonic Motion

Learning Objectives
By the end of this section, you will be able to: • Compare and discuss underdamped and overdamped oscillating systems. • Explain critically damped systems.

Figure 16.21 In order to counteract dampening forces, this dad needs to keep pushing the swing. (credit: Erik A. Johnson, Flickr)

A guitar string stops oscillating a few seconds after being plucked. To keep a child happy on a swing, you must keep pushing. Although we can often make friction and other non-conservative forces negligibly small, completely undamped motion is rare. In fact, we may even want to damp oscillations, such as with car shock absorbers.

For a system that has a small amount of damping, the period and frequency are nearly the same as for simple harmonic motion, but the amplitude gradually decreases as shown in **Figure 16.22**. This occurs because the non-conservative damping force removes energy from the system, usually in the form of thermal energy. In general, energy removal by non-conservative forces is described as

$$W_{nc} = \Delta(KE + PE), \tag{16.57}$$

where W_{nc} is work done by a non-conservative force (here the damping force). For a damped harmonic oscillator, W_{nc} is negative because it removes mechanical energy (KE + PE) from the system.

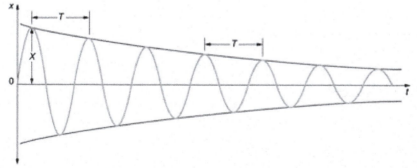

Figure 16.22 In this graph of displacement versus time for a harmonic oscillator with a small amount of damping, the amplitude slowly decreases, but the period and frequency are nearly the same as if the system were completely undamped.

If you gradually *increase* the amount of damping in a system, the period and frequency begin to be affected, because damping opposes and hence slows the back and forth motion. (The net force is smaller in both directions.) If there is very large damping, the system does not even oscillate—it slowly moves toward equilibrium. **Figure 16.23** shows the displacement of a harmonic oscillator for different amounts of damping. When we want to damp out oscillations, such as in the suspension of a car, we may want the system to return to equilibrium as quickly as possible **Critical damping** is defined as the condition in which the damping of an oscillator results in it returning as quickly as possible to its equilibrium position The critically damped system may overshoot the equilibrium position, but if it does, it will do so only once. Critical damping is represented by Curve A in **Figure 16.23**. With less-than critical damping, the system will return to equilibrium faster but will overshoot and cross over one or more times. Such a system is **underdamped**; its displacement is represented by the curve in **Figure 16.22**. Curve B in **Figure 16.23** represents an **overdamped** system. As with critical damping, it too may overshoot the equilibrium position, but will reach equilibrium over a longer period of time.

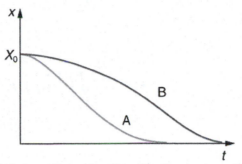

Figure 16.23 Displacement versus time for a critically damped harmonic oscillator (A) and an overdamped harmonic oscillator (B). The critically damped oscillator returns to equilibrium at $X = 0$ in the smallest time possible without overshooting.

Critical damping is often desired, because such a system returns to equilibrium rapidly and remains at equilibrium as well. In addition, a constant force applied to a critically damped system moves the system to a new equilibrium position in the shortest time possible without overshooting or oscillating about the new position. For example, when you stand on bathroom scales that have a needle gauge, the needle moves to its equilibrium position without oscillating. It would be quite inconvenient if the needle oscillated about the new equilibrium position for a long time before settling. Damping forces can vary greatly in character. Friction, for example, is sometimes independent of velocity (as assumed in most places in this text). But many damping forces depend on velocity—sometimes in complex ways, sometimes simply being proportional to velocity.

Making Connections: Damped Oscillator

Consider a mass attached to a spring. This system oscillates when in air because air exerts almost no force on the spring. Now put this system in a liquid, say, water. You will see that the system hardly oscillates when in water. When the system is submerged in water an external force acts on the oscillator. This force is exerted by the liquid against the motion of the spring-mass oscillator and is responsible for the inhibition of oscillations. A force that inhibits oscillations is called a "damping force," and the system that experiences it is called a "damped oscillator." A damped oscillator sees a change in its energy. With time the total energy of the oscillator, which would be its mechanical energy, decreases. Since energy has to be conserved, the energy gets converted into thermal energy, which is stored in the water and the spring.

Example 16.8 Damping an Oscillatory Motion: Friction on an Object Connected to a Spring

Damping oscillatory motion is important in many systems, and the ability to control the damping is even more so. This is generally attained using non-conservative forces such as the friction between surfaces, and viscosity for objects moving through fluids. The following example considers friction. Suppose a 0.200-kg object is connected to a spring as shown in **Figure 16.24**, but there is simple friction between the object and the surface, and the coefficient of friction μ_k is equal to 0.0800. (a) What is the frictional force between the surfaces? (b) What total distance does the object travel if it is released 0.100 m from equilibrium, starting at $v = 0$? The force constant of the spring is $k = 50.0 \text{ N/m}$.

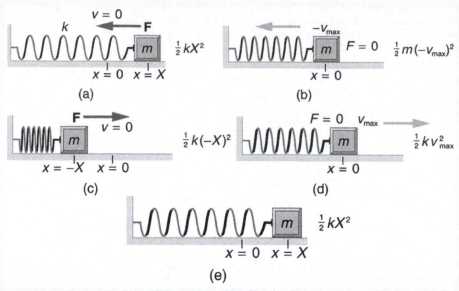

Figure 16.24 The transformation of energy in simple harmonic motion is illustrated for an object attached to a spring on a frictionless surface.

Strategy

This problem requires you to integrate your knowledge of various concepts regarding waves, oscillations, and damping. To solve an integrated concept problem, you must first identify the physical principles involved. Part (a) is about the frictional force. This is a topic involving the application of Newton's Laws. Part (b) requires an understanding of work and conservation of energy, as well as some understanding of horizontal oscillatory systems.

Now that we have identified the principles we must apply in order to solve the problems, we need to identify the knowns and unknowns for each part of the question, as well as the quantity that is constant in Part (a) and Part (b) of the question.

Solution a

1. Choose the proper equation: Friction is $f = \mu_k mg$.

2. Identify the known values.

3. Enter the known values into the equation:

$$f = (0.0800)(0.200 \text{ kg})(9.80 \text{ m/s}^2).$$
(16.58)

4. Calculate and convert units: $f = 0.157 \text{ N}$.

Discussion a

The force here is small because the system and the coefficients are small.

Solution b

Identify the known:

*   The system involves elastic potential energy as the spring compresses and expands, friction that is related to the work done, and the kinetic energy as the body speeds up and slows down.
*   Energy is not conserved as the mass oscillates because friction is a non-conservative force.
*   The motion is horizontal, so gravitational potential energy does not need to be considered.
*   Because the motion starts from rest, the energy in the system is initially $\text{PE}_{\text{el,i}} = (1/2)kX^2$. This energy is removed by work done by friction $W_{\text{nc}} = -fd$, where d is the total distance traveled and $f = \mu_k mg$ is the force of friction.

    When the system stops moving, the friction force will balance the force exerted by the spring, so $\text{PE}_{\text{el,f}} = (1/2)kx^2$ where x is the final position and is given by

$$
\begin{aligned}
F_{\text{el}} &= f \\
kx &= \mu_k mg \\
x &= \frac{\mu_k mg}{k}
\end{aligned}
$$
(16.59)

1. By equating the work done to the energy removed, solve for the distance d .

2. The work done by the non-conservative forces equals the initial, stored elastic potential energy. Identify the correct

equation to use:

$$W_{nc} = \Delta(KE + PE) = PE_{el,f} - PE_{el,i} = \frac{1}{2}k\left(\left(\frac{\mu_k mg}{k}\right)^2 - X^2\right).$$ (16.60)

3. Recall that $W_{nc} = -fd$.

4. Enter the friction as $f = \mu_k mg$ into $W_{nc} = -fd$, thus

$$W_{nc} = -\mu_k mgd.$$ (16.61)

5. Combine these two equations to find

$$\frac{1}{2}k\left(\left(\frac{\mu_k mg}{k}\right)^2 - X^2\right) = -\mu_k mgd.$$ (16.62)

6. Solve the equation for d :

$$d = \frac{k}{2\mu_k mg}\left(X^2 - \left(\frac{\mu_k mg}{k}\right)^2\right).$$ (16.63)

7. Enter the known values into the resulting equation:

$$d = \frac{50.0\ \text{N/m}}{2(0.0800)(0.200\ \text{kg})(9.80\ \text{m/s}^2)}\left((0.100\ \text{m})^2 - \left(\frac{(0.0800)(0.200\ \text{kg})(9.80\ \text{m/s}^2)}{50.0\ \text{N/m}}\right)^2\right).$$ (16.64)

8. Calculate d and convert units:

$$d = 1.59\ \text{m}.$$ (16.65)

Discussion b

This is the total distance traveled back and forth across $x = 0$, which is the undamped equilibrium position. The number of oscillations about the equilibrium position will be more than $d/X = (1.59\ \text{m})/(0.100\ \text{m}) = 15.9$ because the amplitude of the oscillations is decreasing with time. At the end of the motion, this system will not return to $x = 0$ for this type of damping force, because static friction will exceed the restoring force. This system is underdamped. In contrast, an overdamped system with a simple constant damping force would not cross the equilibrium position $x = 0$ a single time. For example, if this system had a damping force 20 times greater, it would only move 0.0484 m toward the equilibrium position from its original 0.100-m position.

This worked example illustrates how to apply problem-solving strategies to situations that integrate the different concepts you have learned. The first step is to identify the physical principles involved in the problem. The second step is to solve for the unknowns using familiar problem-solving strategies. These are found throughout the text, and many worked examples show how to use them for single topics. In this integrated concepts example, you can see how to apply them across several topics. You will find these techniques useful in applications of physics outside a physics course, such as in your profession, in other science disciplines, and in everyday life.

Check Your Understanding

Why are completely undamped harmonic oscillators so rare?

Solution
Friction often comes into play whenever an object is moving. Friction causes damping in a harmonic oscillator.

Check Your Understanding

Describe the difference between overdamping, underdamping, and critical damping.

Solution
An overdamped system moves slowly toward equilibrium. An underdamped system moves quickly to equilibrium, but will oscillate about the equilibrium point as it does so. A critically damped system moves as quickly as possible toward equilibrium without oscillating about the equilibrium.

16.8 Forced Oscillations and Resonance

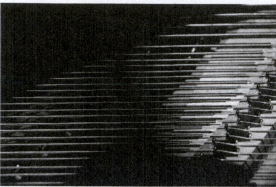

Figure 16.25 You can cause the strings in a piano to vibrate simply by producing sound waves from your voice. (credit: Matt Billings, Flickr)

Sit in front of a piano sometime and sing a loud brief note at it with the dampers off its strings. It will sing the same note back at you—the strings, having the same frequencies as your voice, are resonating in response to the forces from the sound waves that you sent to them. Your voice and a piano's strings is a good example of the fact that objects—in this case, piano strings—can be forced to oscillate but oscillate best at their natural frequency. In this section, we shall briefly explore applying a *periodic driving force* acting on a simple harmonic oscillator. The driving force puts energy into the system at a certain frequency, not necessarily the same as the natural frequency of the system. The **natural frequency** is the frequency at which a system would oscillate if there were no driving and no damping force.

Most of us have played with toys involving an object supported on an elastic band, something like the paddle ball suspended from a finger in Figure 16.26. Imagine the finger in the figure is your finger. At first you hold your finger steady, and the ball bounces up and down with a small amount of damping. If you move your finger up and down slowly, the ball will follow along without bouncing much on its own. As you increase the frequency at which you move your finger up and down, the ball will respond by oscillating with increasing amplitude. When you drive the ball at its natural frequency, the ball's oscillations increase in amplitude with each oscillation for as long as you drive it. The phenomenon of driving a system with a frequency equal to its natural frequency is called **resonance**. A system being driven at its natural frequency is said to **resonate**. As the driving frequency gets progressively higher than the resonant or natural frequency, the amplitude of the oscillations becomes smaller, until the oscillations nearly disappear and your finger simply moves up and down with little effect on the ball.

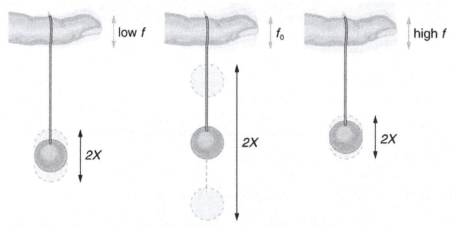

Figure 16.26 The paddle ball on its rubber band moves in response to the finger supporting it. If the finger moves with the natural frequency f_0 of the ball on the rubber band, then a resonance is achieved, and the amplitude of the ball's oscillations increases dramatically. At higher and lower driving frequencies, energy is transferred to the ball less efficiently, and it responds with lower-amplitude oscillations.

Figure 16.27 shows a graph of the amplitude of a damped harmonic oscillator as a function of the frequency of the periodic force driving it. There are three curves on the graph, each representing a different amount of damping. All three curves peak at the point where the frequency of the driving force equals the natural frequency of the harmonic oscillator. The highest peak, or greatest response, is for the least amount of damping, because less energy is removed by the damping force.

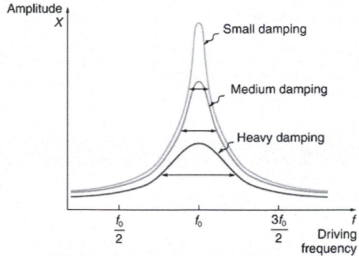

Figure 16.27 Amplitude of a harmonic oscillator as a function of the frequency of the driving force. The curves represent the same oscillator with the same natural frequency but with different amounts of damping. Resonance occurs when the driving frequency equals the natural frequency, and the greatest response is for the least amount of damping. The narrowest response is also for the least damping.

It is interesting that the widths of the resonance curves shown in **Figure 16.27** depend on damping: the less the damping, the narrower the resonance. The message is that if you want a driven oscillator to resonate at a very specific frequency, you need as little damping as possible. Little damping is the case for piano strings and many other musical instruments. Conversely, if you want small-amplitude oscillations, such as in a car's suspension system, then you want heavy damping. Heavy damping reduces the amplitude, but the tradeoff is that the system responds at more frequencies.

These features of driven harmonic oscillators apply to a huge variety of systems. When you tune a radio, for example, you are adjusting its resonant frequency so that it only oscillates to the desired station's broadcast (driving) frequency. The more selective the radio is in discriminating between stations, the smaller its damping. Magnetic resonance imaging (MRI) is a widely used medical diagnostic tool in which atomic nuclei (mostly hydrogen nuclei) are made to resonate by incoming radio waves (on the order of 100 MHz). A child on a swing is driven by a parent at the swing's natural frequency to achieve maximum amplitude. In all of these cases, the efficiency of energy transfer from the driving force into the oscillator is best at resonance. Speed bumps and gravel roads prove that even a car's suspension system is not immune to resonance. In spite of finely engineered shock absorbers, which ordinarily convert mechanical energy to thermal energy almost as fast as it comes in, speed bumps still cause a large-amplitude oscillation. On gravel roads that are corrugated, you may have noticed that if you travel at the "wrong" speed, the bumps are very noticeable whereas at other speeds you may hardly feel the bumps at all. **Figure 16.28** shows a photograph of a famous example (the Tacoma Narrows Bridge) of the destructive effects of a driven harmonic oscillation. The Millennium Bridge in London was closed for a short period of time for the same reason while inspections were carried out.

In our bodies, the chest cavity is a clear example of a system at resonance. The diaphragm and chest wall drive the oscillations of the chest cavity which result in the lungs inflating and deflating. The system is critically damped and the muscular diaphragm oscillates at the resonant value for the system, making it highly efficient.

Figure 16.28 In 1940, the Tacoma Narrows Bridge in Washington state collapsed. Heavy cross winds drove the bridge into oscillations at its resonant frequency. Damping decreased when support cables broke loose and started to slip over the towers, allowing increasingly greater amplitudes until the structure failed (credit: PRI's *Studio 360*, via Flickr)

Check Your Understanding

A famous magic trick involves a performer singing a note toward a crystal glass until the glass shatters. Explain why the trick works in terms of resonance and natural frequency.

Solution

The performer must be singing a note that corresponds to the natural frequency of the glass. As the sound wave is directed at the glass, the glass responds by resonating at the same frequency as the sound wave. With enough energy introduced into the system, the glass begins to vibrate and eventually shatters.

16.9 Waves

Learning Objectives
By the end of this section, you will be able to: • Describe various characteristics associated with a wave. • Differentiate between transverse and longitudinal waves.

Figure 16.29 Waves in the ocean behave similarly to all other types of waves. (credit: Steve Jurveston, Flickr)

What do we mean when we say something is a wave? The most intuitive and easiest wave to imagine is the familiar water wave. More precisely, a **wave** is a disturbance that propagates, or moves from the place it was created. For water waves, the disturbance is in the surface of the water, perhaps created by a rock thrown into a pond or by a swimmer splashing the surface repeatedly. For sound waves, the disturbance is a change in air pressure, perhaps created by the oscillating cone inside a speaker. For earthquakes, there are several types of disturbances, including disturbance of Earth's surface and pressure disturbances under the surface. Even radio waves are most easily understood using an analogy with water waves. Visualizing water waves is useful because there is more to it than just a mental image. Water waves exhibit characteristics common to all waves, such as amplitude, period, frequency and energy. All wave characteristics can be described by a small set of underlying principles.

A wave is a disturbance that propagates, or moves from the place it was created. The simplest waves repeat themselves for several cycles and are associated with simple harmonic motion. Let us start by considering the simplified water wave in **Figure 16.30**. The wave is an up and down disturbance of the water surface. It causes a sea gull to move up and down in simple harmonic motion as the wave crests and troughs (peaks and valleys) pass under the bird. The time for one complete up and down motion is the wave's period T. The wave's frequency is $f = 1/T$, as usual. The wave itself moves to the right in the figure. This movement of the wave is actually the disturbance moving to the right, not the water itself (or the bird would move to the right). We define **wave velocity** v_{w} to be the speed at which the disturbance moves. Wave velocity is sometimes also called the *propagation velocity or propagation speed,* because the disturbance propagates from one location to another.

Misconception Alert

Many people think that water waves push water from one direction to another. In fact, the particles of water tend to stay in one location, save for moving up and down due to the energy in the wave. The energy moves forward through the water, but the water stays in one place. If you feel yourself pushed in an ocean, what you feel is the energy of the wave, not a rush of water.

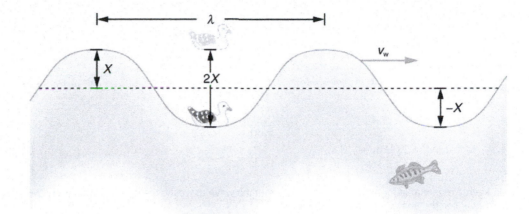

Figure 16.30 An idealized ocean wave passes under a sea gull that bobs up and down in simple harmonic motion. The wave has a wavelength λ, which is the distance between adjacent identical parts of the wave. The up and down disturbance of the surface propagates parallel to the surface at a speed v_w.

The water wave in the figure also has a length associated with it, called its **wavelength** λ, the distance between adjacent identical parts of a wave. (λ is the distance parallel to the direction of propagation.) The speed of propagation v_w is the distance the wave travels in a given time, which is one wavelength in the time of one period. In equation form, that is

$$v_w = \frac{\lambda}{T} \tag{16.66}$$

or

$$v_w = f\lambda. \tag{16.67}$$

This fundamental relationship holds for all types of waves. For water waves, v_w is the speed of a surface wave; for sound, v_w is the speed of sound; and for visible light, v_w is the speed of light, for example.

Applying the Science Practices: Different Types of Waves

Consider a spring fixed to a wall with a mass connected to its end. This fixed point on the wall exerts a force on the complete spring-and-mass system, and this implies that the momentum of the complete system is not conserved. Now, consider energy. Since the system is fixed to a point on the wall, it does not do any work; hence, the total work done is conserved, which means that the energy is conserved. Consequently, we have an oscillator in which energy is conserved but momentum is not. Now, consider a system of two masses connected to each other by a spring. This type of system also forms an oscillator. Since there is no fixed point, momentum is conserved as the forces acting on the two masses are equal and opposite. Energy for such a system will be conserved, because there are no external forces acting on the spring-two-masses system. It is clear from above that, for momentum to be conserved, momentum needs to be carried by waves. This is a typical example of a mechanical oscillator producing mechanical waves that need a medium in which to propagate. Sound waves are also examples of mechanical waves. There are some waves that can travel in the absence of a medium of propagation. Such waves are called "electromagnetic waves." Light waves are examples of electromagnetic waves. Electromagnetic waves are created by the vibration of electric charge. This vibration creates a wave with both electric and magnetic field components.

Take-Home Experiment: Waves in a Bowl

Fill a large bowl or basin with water and wait for the water to settle so there are no ripples. Gently drop a cork into the middle of the bowl. Estimate the wavelength and period of oscillation of the water wave that propagates away from the cork. Remove the cork from the bowl and wait for the water to settle again. Gently drop the cork at a height that is different from the first drop. Does the wavelength depend upon how high above the water the cork is dropped?

Example 16.9 Calculate the Velocity of Wave Propagation: Gull in the Ocean

Calculate the wave velocity of the ocean wave in **Figure 16.30** if the distance between wave crests is 10.0 m and the time for a sea gull to bob up and down is 5.00 s.

Strategy

We are asked to find v_w. The given information tells us that $\lambda = 10.0 \text{ m}$ and $T = 5.00 \text{ s}$. Therefore, we can use

$v_W = \frac{\lambda}{T}$ to find the wave velocity.

Solution

1. Enter the known values into $v_W = \frac{\lambda}{T}$:

$$v_W = \frac{10.0 \text{ m}}{5.00 \text{ s}}.$$
(16.68)

2. Solve for v_W to find $v_W = 2.00$ m/s.

Discussion

This slow speed seems reasonable for an ocean wave. Note that the wave moves to the right in the figure at this speed, not the varying speed at which the sea gull moves up and down.

Transverse and Longitudinal Waves

A simple wave consists of a periodic disturbance that propagates from one place to another. The wave in **Figure 16.31** propagates in the horizontal direction while the surface is disturbed in the vertical direction. Such a wave is called a **transverse wave** or shear wave; in such a wave, the disturbance is perpendicular to the direction of propagation. In contrast, in a **longitudinal wave** or compressional wave, the disturbance is parallel to the direction of propagation. **Figure 16.32** shows an example of a longitudinal wave. The size of the disturbance is its amplitude X and is completely independent of the speed of propagation v_W.

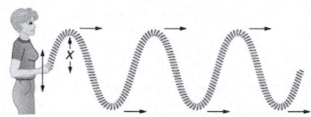

Figure 16.31 In this example of a transverse wave, the wave propagates horizontally, and the disturbance in the cord is in the vertical direction.

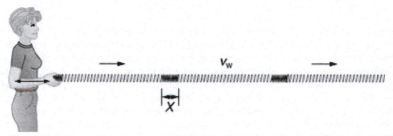

Figure 16.32 In this example of a longitudinal wave, the wave propagates horizontally, and the disturbance in the cord is also in the horizontal direction.

Waves may be transverse, longitudinal, or *a combination of the two*. (Water waves are actually a combination of transverse and longitudinal. The simplified water wave illustrated in **Figure 16.30** shows no longitudinal motion of the bird.) The waves on the strings of musical instruments are transverse—so are electromagnetic waves, such as visible light.

Sound waves in air and water are longitudinal. Their disturbances are periodic variations in pressure that are transmitted in fluids. Fluids do not have appreciable shear strength, and thus the sound waves in them must be longitudinal or compressional. Sound in solids can be both longitudinal and transverse.

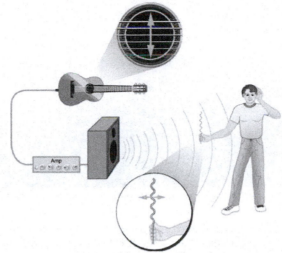

Figure 16.33 The wave on a guitar string is transverse. The sound wave rattles a sheet of paper in a direction that shows the sound wave is longitudinal.

Earthquake waves under Earth's surface also have both longitudinal and transverse components (called compressional or P-waves and shear or S-waves, respectively). These components have important individual characteristics—they propagate at different speeds, for example. Earthquakes also have surface waves that are similar to surface waves on water.

Applying the Science Practices: Electricity in Your Home

The source of electricity is of a sinusoidal nature. If we appropriately probe using an oscilloscope (an instrument used to display and analyze electronic signals), we can precisely determine the frequency and wavelength of the waveform. Inquire about the maximum voltage current that you get in your house and plot a sinusoidal waveform representing the frequency, wavelength, and period for it.

Check Your Understanding

Why is it important to differentiate between longitudinal and transverse waves?

Solution
In the different types of waves, energy can propagate in a different direction relative to the motion of the wave. This is important to understand how different types of waves affect the materials around them.

PhET Explorations: Wave on a String

Watch a string vibrate in slow motion. Wiggle the end of the string and make waves, or adjust the frequency and amplitude of an oscillator. Adjust the damping and tension. The end can be fixed, loose, or open.

Figure 16.34 Wave on a String (http://cnx.org/content/m55281/1.3/wave-on-a-string_en.jar)

16.10 Superposition and Interference

Learning Objectives

By the end of this section, you will be able to:

- Determine the resultant waveform when two waves act in superposition relative to each other.
- Explain standing waves.
- Describe the mathematical representation of overtones and beat frequency.

Figure 16.35 These waves result from the superposition of several waves from different sources, producing a complex pattern. (credit: waterborough, Wikimedia Commons)

Most waves do not look very simple. They look more like the waves in **Figure 16.35** than like the simple water wave considered in **Waves**. (Simple waves may be created by a simple harmonic oscillation, and thus have a sinusoidal shape). Complex waves are more interesting, even beautiful, but they look formidable. Most waves appear complex because they result from several simple waves adding together. Luckily, the rules for adding waves are quite simple.

When two or more waves arrive at the same point, they superimpose themselves on one another. More specifically, the disturbances of waves are superimposed when they come together—a phenomenon called **superposition**. Each disturbance corresponds to a force, and forces add. If the disturbances are along the same line, then the resulting wave is a simple addition of the disturbances of the individual waves—that is, their amplitudes add. **Figure 16.36** and **Figure 16.37** illustrate superposition in two special cases, both of which produce simple results.

Figure 16.36 shows two identical waves that arrive at the same point exactly in phase. The crests of the two waves are precisely aligned, as are the troughs. This superposition produces pure **constructive interference**. Because the disturbances add, pure constructive interference produces a wave that has twice the amplitude of the individual waves, but has the same wavelength.

Figure 16.37 shows two identical waves that arrive exactly out of phase—that is, precisely aligned crest to trough—producing pure **destructive interference**. Because the disturbances are in the opposite direction for this superposition, the resulting amplitude is zero for pure destructive interference—the waves completely cancel.

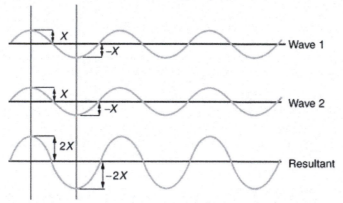

Figure 16.36 Pure constructive interference of two identical waves produces one with twice the amplitude, but the same wavelength.

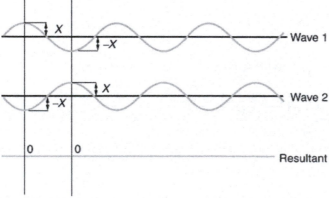

Figure 16.37 Pure destructive interference of two identical waves produces zero amplitude, or complete cancellation.

While pure constructive and pure destructive interference do occur, they require precisely aligned identical waves. The superposition of most waves produces a combination of constructive and destructive interference and can vary from place to place and time to time. Sound from a stereo, for example, can be loud in one spot and quiet in another. Varying loudness means the sound waves add partially constructively and partially destructively at different locations. A stereo has at least two speakers creating sound waves, and waves can reflect from walls. All these waves superimpose. An example of sounds that vary over time from constructive to destructive is found in the combined whine of airplane jets heard by a stationary passenger. The combined sound can fluctuate up and down in volume as the sound from the two engines varies in time from constructive to destructive. These examples are of waves that are similar.

An example of the superposition of two dissimilar waves is shown in Figure 16.38. Here again, the disturbances add and subtract, producing a more complicated looking wave.

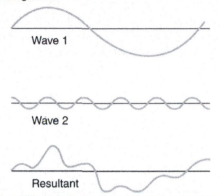

Figure 16.38 Superposition of non-identical waves exhibits both constructive and destructive interference.

Standing Waves

Sometimes waves do not seem to move; rather, they just vibrate in place. Unmoving waves can be seen on the surface of a glass of milk in a refrigerator, for example. Vibrations from the refrigerator motor create waves on the milk that oscillate up and down but do not seem to move across the surface. These waves are formed by the superposition of two or more moving waves, such as illustrated in Figure 16.39 for two identical waves moving in opposite directions. The waves move through each other with their disturbances adding as they go by. If the two waves have the same amplitude and wavelength, then they alternate between constructive and destructive interference. The resultant looks like a wave standing in place and, thus, is called a **standing wave**. Waves on the glass of milk are one example of standing waves. There are other standing waves, such as on guitar strings and in organ pipes. With the glass of milk, the two waves that produce standing waves may come from reflections from the side of the glass.

A closer look at earthquakes provides evidence for conditions appropriate for resonance, standing waves, and constructive and destructive interference. A building may be vibrated for several seconds with a driving frequency matching that of the natural frequency of vibration of the building—producing a resonance resulting in one building collapsing while neighboring buildings do not. Often buildings of a certain height are devastated while other taller buildings remain intact. The building height matches the condition for setting up a standing wave for that particular height. As the earthquake waves travel along the surface of Earth and reflect off denser rocks, constructive interference occurs at certain points. Often areas closer to the epicenter are not damaged while areas farther away are damaged.

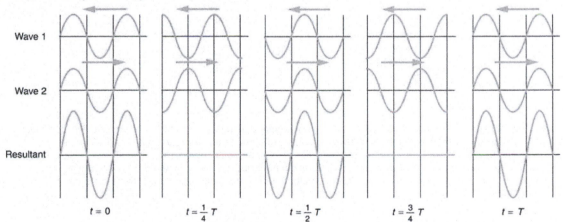

Figure 16.39 Standing wave created by the superposition of two identical waves moving in opposite directions. The oscillations are at fixed locations in space and result from alternately constructive and destructive interference.

Standing waves are also found on the strings of musical instruments and are due to reflections of waves from the ends of the string. Figure 16.40 and Figure 16.41 show three standing waves that can be created on a string that is fixed at both ends. **Nodes** are the points where the string does not move; more generally, nodes are where the wave disturbance is zero in a standing wave. The fixed ends of strings must be nodes, too, because the string cannot move there. The word **antinode** is used to denote the location of maximum amplitude in standing waves. Standing waves on strings have a frequency that is related to the propagation speed v_w of the disturbance on the string. The wavelength λ is determined by the distance between the points where the string is fixed in place.

The lowest frequency, called the **fundamental frequency**, is thus for the longest wavelength, which is seen to be $\lambda_1 = 2L$. Therefore, the fundamental frequency is $f_1 = v_w / \lambda_1 = v_w / 2L$. In this case, the **overtones** or harmonics are multiples of the fundamental frequency. As seen in **Figure 16.41**, the first harmonic can easily be calculated since $\lambda_2 = L$. Thus, $f_2 = v_w / \lambda_2 = v_w / 2L = 2f_1$. Similarly, $f_3 = 3f_1$, and so on. All of these frequencies can be changed by adjusting the tension in the string. The greater the tension, the greater v_w is and the higher the frequencies. This observation is familiar to anyone who has ever observed a string instrument being tuned. We will see in later chapters that standing waves are crucial to many resonance phenomena, such as in sounding boxes on string instruments.

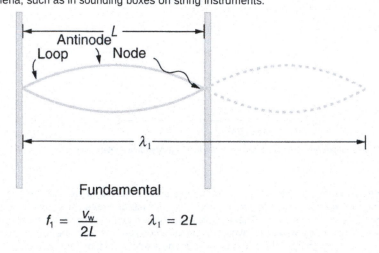

Figure 16.40 The figure shows a string oscillating at its fundamental frequency.

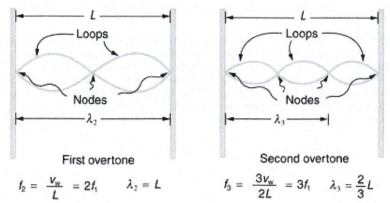

Figure 16.41 First and second harmonic frequencies are shown.

Beats

Striking two adjacent keys on a piano produces a warbling combination usually considered to be unpleasant. The superposition of two waves of similar but not identical frequencies is the culprit. Another example is often noticeable in jet aircraft, particularly the two-engine variety, while taxiing. The combined sound of the engines goes up and down in loudness. This varying loudness happens because the sound waves have similar but not identical frequencies. The discordant warbling of the piano and the fluctuating loudness of the jet engine noise are both due to alternately constructive and destructive interference as the two waves go in and out of phase. **Figure 16.42** illustrates this graphically.

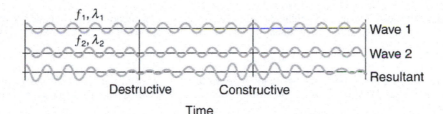

Figure 16.42 Beats are produced by the superposition of two waves of slightly different frequencies but identical amplitudes. The waves alternate in time between constructive interference and destructive interference, giving the resulting wave a time-varying amplitude.

The wave resulting from the superposition of two similar-frequency waves has a frequency that is the average of the two. This wave fluctuates in amplitude, or *beats*, with a frequency called the **beat frequency**. We can determine the beat frequency by adding two waves together mathematically. Note that a wave can be represented at one point in space as

$$x = X \cos\left(\frac{2\pi t}{T}\right) = X \cos(2\pi \, ft),$$ (16.69)

where $f = 1/T$ is the frequency of the wave. Adding two waves that have different frequencies but identical amplitudes produces a resultant

$$x = x_1 + x_2.$$ (16.70)

More specifically,

$$x = X \cos(2\pi \, f_1 \, t) + X \cos(2\pi \, f_2 \, t).$$ (16.71)

Using a trigonometric identity, it can be shown that

$$x = 2X \cos(\pi \, f_\text{B} \, t)\cos(2\pi \, f_\text{ave} \, t),$$ (16.72)

where

$$f_\text{B} = |\, f_1 - f_2 \,|$$ (16.73)

is the beat frequency, and f_ave is the average of f_1 and f_2. These results mean that the resultant wave has twice the amplitude and the average frequency of the two superimposed waves, but it also fluctuates in overall amplitude at the beat frequency f_B. The first cosine term in the expression effectively causes the amplitude to go up and down. The second cosine term is the wave with frequency f_ave. This result is valid for all types of waves. However, if it is a sound wave, providing the two frequencies are similar, then what we hear is an average frequency that gets louder and softer (or warbles) at the beat frequency.

Real World Connections: Tuning Forks

The **MIT physics demo (http://openstaxcollege.org/l/31tuningforks/)** entitled "Tuning Forks: Resonance and Beat Frequency" provides a qualitative picture of how wave interference produces beats.

Description: Two identical forks and sounding boxes are placed next to each other. Striking one tuning fork will cause the other to resonate at the same frequency. When a weight is attached to one tuning fork, they are no longer identical. Thus, one will not cause the other to resonate. When two different forks are struck at the same time, the interference of their pitches produces beats.

Real World Connections: Jump Rop

This is a fun activity with which to learn about interference and superposition. Take a jump rope and hold it at the two ends with one of your friends. While each of you is holding the rope, snap your hands to produce a wave from each side. Record your observations and see if they match with the following:

a. One wave starts from the right end and travels to the left end of the rope.

b. Another wave starts at the left end and travels to the right end of the rope.

c. The waves travel at the same speed.

d. The shape of the waves depends on the way the person snaps his or her hands.

e. There is a region of overlap.

f. The shapes of the waves are identical to their original shapes after they overlap.

Now, snap the rope up and down and ask your friend to snap his or her end of the rope sideways. The resultant that one sees here is the vector sum of two individual displacements.

This activity illustrates superposition and interference. When two or more waves interact with each other at a point, the disturbance at that point is given by the sum of the disturbances each wave will produce in the absence of the other. This is the principle of superposition. Interference is a result of superposition of two or more waves to form a resultant wave of greater or lower amplitude.

While beats may sometimes be annoying in audible sounds, we will find that beats have many applications. Observing beats is a very useful way to compare similar frequencies. There are applications of beats as apparently disparate as in ultrasonic imaging and radar speed traps.

Check Your Understanding

Imagine you are holding one end of a jump rope, and your friend holds the other. If your friend holds her end still, you can move your end up and down, creating a transverse wave. If your friend then begins to move her end up and down, generating a wave in the opposite direction, what resultant wave forms would you expect to see in the jump rope?

Solution
The rope would alternate between having waves with amplitudes two times the original amplitude and reaching equilibrium with no amplitude at all. The wavelengths will result in both constructive and destructive interference

Check Your Understanding

Define nodes and antinodes.

Solution
Nodes are areas of wave interference where there is no motion. Antinodes are areas of wave interference where the motion is at its maximum point.

Check Your Understanding

You hook up a stereo system. When you test the system, you notice that in one corner of the room, the sounds seem dull. In another area, the sounds seem excessively loud. Describe how the sound moving about the room could result in these effects.

Solution
With multiple speakers putting out sounds into the room, and these sounds bouncing off walls, there is bound to be some wave interference. In the dull areas, the interference is probably mostly destructive. In the louder areas, the interference is probably mostly constructive.

PhET Explorations: Wave Interference

Make waves with a dripping faucet, audio speaker, or laser! Add a second source or a pair of slits to create an interference pattern.

Figure 16.43 Wave Interference (http://cnx.org/content/m55282/1.2/wave-interference_en.jar)

16.11 Energy in Waves: Intensity

Learning Objectives

By the end of this section, you will be able to:

- Calculate the intensity and the power of rays and waves.

Figure 16.44 The destructive effect of an earthquake is palpable evidence of the energy carried in these waves. The Richter scale rating of earthquakes is related to both their amplitude and the energy they carry. (credit: Petty Officer 2nd Class Candice Villarreal, U.S. Navy)

All waves carry energy. The energy of some waves can be directly observed. Earthquakes can shake whole cities to the ground, performing the work of thousands of wrecking balls.

Loud sounds pulverize nerve cells in the inner ear, causing permanent hearing loss. Ultrasound is used for deep-heat treatment of muscle strains. A laser beam can burn away a malignancy. Water waves chew up beaches.

The amount of energy in a wave is related to its amplitude. Large-amplitude earthquakes produce large ground displacements. Loud sounds have higher pressure amplitudes and come from larger-amplitude source vibrations than soft sounds. Large ocean breakers churn up the shore more than small ones. More quantitatively, a wave is a displacement that is resisted by a restoring force. The larger the displacement x, the larger the force $F = kx$ needed to create it. Because work W is related to force multiplied by distance (Fx) and energy is put into the wave by the work done to create it, the energy in a wave is related to amplitude. In fact, a wave's energy is directly proportional to its amplitude squared because

$$W \propto Fx = kx^2.$$
(16.74)

The energy effects of a wave depend on time as well as amplitude. For example, the longer deep-heat ultrasound is applied, the more energy it transfers. Waves can also be concentrated or spread out. Sunlight, for example, can be focused to burn wood. Earthquakes spread out, so they do less damage the farther they get from the source. In both cases, changing the area the waves cover has important effects. All these pertinent factors are included in the definition of **intensity** I as power per unit area:

$$I = \frac{P}{A}$$
(16.75)

where P is the power carried by the wave through area A. The definition of intensity is valid for any energy in transit, including that carried by waves. The SI unit for intensity is watts per square meter (W/m^2). For example, infrared and visible energy from the Sun impinge on Earth at an intensity of 1300 W/m^2 just above the atmosphere. There are other intensity-related units in use, too. The most common is the decibel. For example, a 90 decibel sound level corresponds to an intensity of 10^{-3} W/m^2. (This quantity is not much power per unit area considering that 90 decibels is a relatively high sound level. Decibels will be discussed in some detail in a later chapter.

Example 16.10 Calculating intensity and power: How much energy is in a ray of sunlight?

The average intensity of sunlight on Earth's surface is about 700 W/m^2.

(a) Calculate the amount of energy that falls on a solar collector having an area of 0.500 m^2 in 4.00 h.

(b) What intensity would such sunlight have if concentrated by a magnifying glass onto an area 200 times smaller than its own?

Strategy a

Because power is energy per unit time or $P = \frac{E}{t}$, the definition of intensity can be written as $I = \frac{P}{A} = \frac{E/t}{A}$, and this equation can be solved for E with the given information.

Solution a

1. Begin with the equation that states the definition of intensity:

$$I = \frac{P}{A}.$$
(16.76)

2. Replace P with its equivalent E/t:

$$I = \frac{E/t}{A}.$$

(16.77)

3. Solve for E:

$$E = IAt.$$

(16.78)

4. Substitute known values into the equation:

$$E = \left(700 \text{ W/m}^2\right)\left(0.500 \text{ m}^2\right)[(4.00 \text{ h})(3600 \text{ s/h})].$$

(16.79)

5. Calculate to find E and convert units:

$$5.04{\times}10^6 \text{ J},$$

(16.80)

Discussion a

The energy falling on the solar collector in 4 h in part is enough to be useful—for example, for heating a significant amount of water.

Strategy b

Taking a ratio of new intensity to old intensity and using primes for the new quantities, we will find that it depends on the ratio of the areas. All other quantities will cancel.

Solution b

1. Take the ratio of intensities, which yields:

$$\frac{I'}{I} = \frac{P'/A'}{P/A} = \frac{A}{A'} \left(\text{The powers cancel because } P' = P\right).$$

(16.81)

2. Identify the knowns:

$$A = 200A',$$

(16.82)

$$\frac{I'}{I} = 200.$$

(16.83)

3. Substitute known quantities:

$$I' = 200I = 200\left(700 \text{ W/m}^2\right).$$

(16.84)

4. Calculate to find I':

$$I' = 1.40{\times}10^5 \text{ W/m}^2.$$

(16.85)

Discussion b

Decreasing the area increases the intensity considerably. The intensity of the concentrated sunlight could even start a fire.

Example 16.11 Determine the combined intensity of two waves: Perfect constructive interference

If two identical waves, each having an intensity of 1.00 W/m^2, interfere perfectly constructively, what is the intensity of the resulting wave?

Strategy

We know from **Superposition and Interference** that when two identical waves, which have equal amplitudes X, interfere perfectly constructively, the resulting wave has an amplitude of $2X$. Because a wave's intensity is proportional to amplitude squared, the intensity of the resulting wave is four times as great as in the individual waves.

Solution

1. Recall that intensity is proportional to amplitude squared.

2. Calculate the new amplitude:

$$I' \propto (X')^2 = (2X)^2 = 4X^2.$$

(16.86)

3. Recall that the intensity of the old amplitude was:

$$I \propto X^2.$$

(16.87)

4. Take the ratio of new intensity to the old intensity. This gives:

$$\frac{I\prime}{I} = 4. \tag{16.88}$$

5. Calculate to find $I\prime$:

$$I\prime = 4I = 4.00 \text{ W/m}^2. \tag{16.89}$$

Discussion

The intensity goes up by a factor of 4 when the amplitude doubles. This answer is a little disquieting. The two individual waves each have intensities of 1.00 W/m^2, yet their sum has an intensity of 4.00 W/m^2, which may appear to violate conservation of energy. This violation, of course, cannot happen. What does happen is intriguing. The area over which the intensity is 4.00 W/m^2 is much less than the area covered by the two waves before they interfered. There are other areas where the intensity is zero. The addition of waves is not as simple as our first look in **Superposition and Interference** suggested. We actually get a pattern of both constructive interference and destructive interference whenever two waves are added. For example, if we have two stereo speakers putting out 1.00 W/m^2 each, there will be places in the room where the intensity is 4.00 W/m^2, other places where the intensity is zero, and others in between. **Figure 16.45** shows what this interference might look like. We will pursue interference patterns elsewhere in this text.

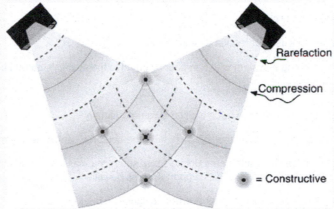

Figure 16.45 These stereo speakers produce both constructive interference and destructive interference in the room, a property common to the superposition of all types of waves. The shading is proportional to intensity.

Check Your Understanding

Which measurement of a wave is most important when determining the wave's intensity?

Solution
Amplitude, because a wave's energy is directly proportional to its amplitude squared.

Glossary

amplitude: the maximum displacement from the equilibrium position of an object oscillating around the equilibrium position

antinode: the location of maximum amplitude in standing waves

beat frequency: the frequency of the amplitude fluctuations of a wave

constructive interference: when two waves arrive at the same point exactly in phase; that is, the crests of the two waves are precisely aligned, as are the troughs

critical damping: the condition in which the damping of an oscillator causes it to return as quickly as possible to its equilibrium position without oscillating back and forth about this position

deformation: displacement from equilibrium

destructive interference: when two identical waves arrive at the same point exactly out of phase; that is, precisely aligned crest to trough

elastic potential energy: potential energy stored as a result of deformation of an elastic object, such as the stretching of a spring

force constant: a constant related to the rigidity of a system: the larger the force constant, the more rigid the system; the force constant is represented by *k*

frequency: number of events per unit of time

fundamental frequency: the lowest frequency of a periodic waveform

intensity: power per unit area

longitudinal wave: a wave in which the disturbance is parallel to the direction of propagation

natural frequency: the frequency at which a system would oscillate if there were no driving and no damping forces

nodes: the points where the string does not move; more generally, nodes are where the wave disturbance is zero in a standing wave

oscillate: moving back and forth regularly between two points

over damping: the condition in which damping of an oscillator causes it to return to equilibrium without oscillating; oscillator moves more slowly toward equilibrium than in the critically damped system

overtones: multiples of the fundamental frequency of a sound

period: time it takes to complete one oscillation

periodic motion: motion that repeats itself at regular time intervals

resonance: the phenomenon of driving a system with a frequency equal to the system's natural frequency

resonate: a system being driven at its natural frequency

restoring force: force acting in opposition to the force caused by a deformation

simple harmonic motion: the oscillatory motion in a system where the net force can be described by Hooke's law

simple harmonic oscillator: a device that implements Hooke's law, such as a mass that is attached to a spring, with the other end of the spring being connected to a rigid support such as a wall

simple pendulum: an object with a small mass suspended from a light wire or string

superposition: the phenomenon that occurs when two or more waves arrive at the same point

transverse wave: a wave in which the disturbance is perpendicular to the direction of propagation

under damping: the condition in which damping of an oscillator causes it to return to equilibrium with the amplitude gradually decreasing to zero; system returns to equilibrium faster but overshoots and crosses the equilibrium position one or more times

wave: a disturbance that moves from its source and carries energy

wave velocity: the speed at which the disturbance moves. Also called the propagation velocity or propagation speed

wavelength: the distance between adjacent identical parts of a wave

Section Summary

16.1 Hooke's Law: Stress and Strain Revisited
- An oscillation is a back and forth motion of an object between two points of deformation.
- An oscillation may create a wave, which is a disturbance that propagates from where it was created.
- The simplest type of oscillations and waves are related to systems that can be described by Hooke's law:
$$F = -kx,$$
  where F is the restoring force, x is the displacement from equilibrium or deformation, and k is the force constant of the system.
- Elastic potential energy PE_{el} stored in the deformation of a system that can be described by Hooke's law is given by
$$PE_{el} = (1/2)kx^2.$$

16.2 Period and Frequency in Oscillations
- Periodic motion is a repetitious oscillation.
- The time for one oscillation is the period T.

- The number of oscillations per unit time is the frequency f.
- These quantities are related by

$$f = \frac{1}{T}.$$

16.3 Simple Harmonic Motion: A Special Periodic Motion
- Simple harmonic motion is oscillatory motion for a system that can be described only by Hooke's law. Such a system is also called a simple harmonic oscillator.
- Maximum displacement is the amplitude X. The period T and frequency f of a simple harmonic oscillator are given by

$T = 2\pi\sqrt{\frac{m}{k}}$ and $f = \frac{1}{2\pi}\sqrt{\frac{k}{m}}$, where m is the mass of the system.

- Displacement in simple harmonic motion as a function of time is given by $x(t) = X \cos\frac{2\pi t}{T}$.

- The velocity is given by $v(t) = -v_{max}\sin\frac{2\pi t}{T}$, where $v_{max} = \sqrt{k/m}X$.

- The acceleration is found to be $a(t) = -\frac{kX}{m}\cos\frac{2\pi t}{T}$.

16.4 The Simple Pendulum
- A mass m suspended by a wire of length L is a simple pendulum and undergoes simple harmonic motion for amplitudes less than about $15°$.
  The period of a simple pendulum is

$$T = 2\pi\sqrt{\frac{L}{g}},$$

  where L is the length of the string and g is the acceleration due to gravity.

16.5 Energy and the Simple Harmonic Oscillator
- Energy in the simple harmonic oscillator is shared between elastic potential energy and kinetic energy, with the total being constant:

$$\frac{1}{2}mv^2 + \frac{1}{2}kx^2 = \text{constant}.$$

- Maximum velocity depends on three factors: it is directly proportional to amplitude, it is greater for stiffer systems, and it is smaller for objects that have larger masses:

$$v_{max} = \sqrt{\frac{k}{m}}X.$$

16.6 Uniform Circular Motion and Simple Harmonic Motion

A projection of uniform circular motion undergoes simple harmonic oscillation.

16.7 Damped Harmonic Motion
- Damped harmonic oscillators have non-conservative forces that dissipate their energy.
- Critical damping returns the system to equilibrium as fast as possible without overshooting.
- An underdamped system will oscillate through the equilibrium position.
- An overdamped system moves more slowly toward equilibrium than one that is critically damped.

16.8 Forced Oscillations and Resonance
- A system's natural frequency is the frequency at which the system will oscillate if not affected by driving or damping forces.
- A periodic force driving a harmonic oscillator at its natural frequency produces resonance. The system is said to resonate.
- The less damping a system has, the higher the amplitude of the forced oscillations near resonance. The more damping a system has, the broader response it has to varying driving frequencies.

16.9 Waves
- A wave is a disturbance that moves from the point of creation with a wave velocity v_w.

- A wave has a wavelength λ, which is the distance between adjacent identical parts of the wave.

- Wave velocity and wavelength are related to the wave's frequency and period by $v_w = \frac{\lambda}{T}$ or $v_w = f\lambda$.

- A transverse wave has a disturbance perpendicular to its direction of propagation, whereas a longitudinal wave has a

disturbance parallel to its direction of propagation.

16.10 Superposition and Interference
- Superposition is the combination of two waves at the same location.
- Constructive interference occurs when two identical waves are superimposed in phase.
- Destructive interference occurs when two identical waves are superimposed exactly out of phase.
- A standing wave is one in which two waves superimpose to produce a wave that varies in amplitude but does not propagate.
- Nodes are points of no motion in standing waves.
- An antinode is the location of maximum amplitude of a standing wave.
- Waves on a string are resonant standing waves with a fundamental frequency and can occur at higher multiples of the fundamental, called overtones or harmonics.
- Beats occur when waves of similar frequencies f_1 and f_2 are superimposed. The resulting amplitude oscillates with a beat frequency given by

$$f_B = |f_1 - f_2|.$$

16.11 Energy in Waves: Intensity
Intensity is defined to be the power per unit area:

$I = \frac{P}{A}$ and has units of W/m^2.

Conceptual Questions

16.1 Hooke's Law: Stress and Strain Revisited
1. Describe a system in which elastic potential energy is stored.

16.3 Simple Harmonic Motion: A Special Periodic Motion
2. What conditions must be met to produce simple harmonic motion?

3. (a) If frequency is not constant for some oscillation, can the oscillation be simple harmonic motion?

(b) Can you think of any examples of harmonic motion where the frequency may depend on the amplitude?

4. Give an example of a simple harmonic oscillator, specifically noting how its frequency is independent of amplitude.

5. Explain why you expect an object made of a stiff material to vibrate at a higher frequency than a similar object made of a spongy material.

6. As you pass a freight truck with a trailer on a highway, you notice that its trailer is bouncing up and down slowly. Is it more likely that the trailer is heavily loaded or nearly empty? Explain your answer.

7. Some people modify cars to be much closer to the ground than when manufactured. Should they install stiffer springs? Explain your answer.

16.4 The Simple Pendulum
8. Pendulum clocks are made to run at the correct rate by adjusting the pendulum's length. Suppose you move from one city to another where the acceleration due to gravity is slightly greater, taking your pendulum clock with you, will you have to lengthen or shorten the pendulum to keep the correct time, other factors remaining constant? Explain your answer.

16.5 Energy and the Simple Harmonic Oscillator
9. Explain in terms of energy how dissipative forces such as friction reduce the amplitude of a harmonic oscillator. Also explain how a driving mechanism can compensate. (A pendulum clock is such a system.)

16.7 Damped Harmonic Motion
10. Give an example of a damped harmonic oscillator. (They are more common than undamped or simple harmonic oscillators.)

11. How would a car bounce after a bump under each of these conditions?
- overdamping
- underdamping
- critical damping

12. Most harmonic oscillators are damped and, if undriven, eventually come to a stop. How is this observation related to the second law of thermodynamics?

16.8 Forced Oscillations and Resonance
13. Why are soldiers in general ordered to "route step" (walk out of step) across a bridge?

16.9 Waves

14. Give one example of a transverse wave and another of a longitudinal wave, being careful to note the relative directions of the disturbance and wave propagation in each.

15. What is the difference between propagation speed and the frequency of a wave? Does one or both affect wavelength? If so, how?

16.10 Superposition and Interference

16. Speakers in stereo systems have two color-coded terminals to indicate how to hook up the wires. If the wires are reversed, the speaker moves in a direction opposite that of a properly connected speaker. Explain why it is important to have both speakers connected the same way.

16.11 Energy in Waves: Intensity

17. Two identical waves undergo pure constructive interference. Is the resultant intensity twice that of the individual waves? Explain your answer.

18. Circular water waves decrease in amplitude as they move away from where a rock is dropped. Explain why.

Problems & Exercises

16.1 Hooke's Law: Stress and Strain Revisited

1. Fish are hung on a spring scale to determine their mass (most fishermen feel no obligation to truthfully report the mass).

(a) What is the force constant of the spring in such a scale if it the spring stretches 8.00 cm for a 10.0 kg load?

(b) What is the mass of a fish that stretches the spring 5.50 cm?

(c) How far apart are the half-kilogram marks on the scale?

2. It is weigh-in time for the local under-85-kg rugby team. The bathroom scale used to assess eligibility can be described by Hooke's law and is depressed 0.75 cm by its maximum load of 120 kg. (a) What is the spring's effective spring constant? (b) A player stands on the scales and depresses it by 0.48 cm. Is he eligible to play on this under-85 kg team?

3. One type of BB gun uses a spring-driven plunger to blow the BB from its barrel. (a) Calculate the force constant of its plunger's spring if you must compress it 0.150 m to drive the 0.0500-kg plunger to a top speed of 20.0 m/s. (b) What force must be exerted to compress the spring?

4. (a) The springs of a pickup truck act like a single spring with a force constant of $1.30 \times 10^5 \ \text{N/m}$. By how much will the truck be depressed by its maximum load of 1000 kg?

(b) If the pickup truck has four identical springs, what is the force constant of each?

5. When an 80.0-kg man stands on a pogo stick, the spring is compressed 0.120 m.

(a) What is the force constant of the spring? (b) Will the spring be compressed more when he hops down the road?

6. A spring has a length of 0.200 m when a 0.300-kg mass hangs from it, and a length of 0.750 m when a 1.95-kg mass hangs from it. (a) What is the force constant of the spring? (b) What is the unloaded length of the spring?

16.2 Period and Frequency in Oscillations

7. What is the period of $60.0 \ \text{Hz}$ electrical power?

8. If your heart rate is 150 beats per minute during strenuous exercise, what is the time per beat in units of seconds?

9. Find the frequency of a tuning fork that takes 2.50×10^{-3} s to complete one oscillation.

10. A stroboscope is set to flash every 8.00×10^{-5} s. What is the frequency of the flashes?

11. A tire has a tread pattern with a crevice every 2.00 cm. Each crevice makes a single vibration as the tire moves. What is the frequency of these vibrations if the car moves at 30.0 m/s?

12. Engineering Application

Each piston of an engine makes a sharp sound every other revolution of the engine. (a) How fast is a race car going if its eight-cylinder engine emits a sound of frequency 750 Hz, given that the engine makes 2000 revolutions per kilometer? (b) At how many revolutions per minute is the engine rotating?

16.3 Simple Harmonic Motion: A Special Periodic Motion

13. A type of cuckoo clock keeps time by having a mass bouncing on a spring, usually something cute like a cherub in a chair. What force constant is needed to produce a period of 0.500 s for a 0.0150-kg mass?

14. If the spring constant of a simple harmonic oscillator is doubled, by what factor will the mass of the system need to change in order for the frequency of the motion to remain the same?

15. A 0.500-kg mass suspended from a spring oscillates with a period of 1.50 s. How much mass must be added to the object to change the period to 2.00 s?

16. By how much leeway (both percentage and mass) would you have in the selection of the mass of the object in the previous problem if you did not wish the new period to be greater than 2.01 s or less than 1.99 s?

17. Suppose you attach the object with mass m to a vertical spring originally at rest, and let it bounce up and down. You release the object from rest at the spring's original rest length. (a) Show that the spring exerts an upward force of $2.00 \ mg$ on the object at its lowest point. (b) If the spring has a force constant of $10.0 \ \text{N/m}$ and a 0.25-kg-mass object is set in motion as described, find the amplitude of the oscillations. (c) Find the maximum velocity.

18. A diver on a diving board is undergoing simple harmonic motion. Her mass is 55.0 kg and the period of her motion is 0.800 s. The next diver is a male whose period of simple harmonic oscillation is 1.05 s. What is his mass if the mass of the board is negligible?

19. Suppose a diving board with no one on it bounces up and down in a simple harmonic motion with a frequency of 4.00 Hz. The board has an effective mass of 10.0 kg. What is the frequency of the simple harmonic motion of a 75.0-kg diver on the board?

20.

Figure 16.46 This child's toy relies on springs to keep infants entertained. (credit: By Humboldthead, Flickr)

The device pictured in **Figure 16.46** entertains infants while keeping them from wandering. The child bounces in a harness suspended from a door frame by a spring constant.

(a) If the spring stretches 0.250 m while supporting an 8.0-kg child, what is its spring constant?

(b) What is the time for one complete bounce of this child? (c) What is the child's maximum velocity if the amplitude of her bounce is 0.200 m?

21. A 90.0-kg skydiver hanging from a parachute bounces up and down with a period of 1.50 s. What is the new period of oscillation when a second skydiver, whose mass is 60.0 kg, hangs from the legs of the first, as seen in **Figure 16.47**.

Figure 16.47 The oscillations of one skydiver are about to be affected by a second skydiver. (credit: U.S. Army, www.army.mil)

16.4 The Simple Pendulum

As usual, the acceleration due to gravity in these problems is taken to be $g = 9.80 \text{ m/s}^2$, unless otherwise specified.

22. What is the length of a pendulum that has a period of 0.500 s?

23. Some people think a pendulum with a period of 1.00 s can be driven with "mental energy" or psycho kinetically, because its period is the same as an average heartbeat. True or not, what is the length of such a pendulum?

24. What is the period of a 1.00-m-long pendulum?

25. How long does it take a child on a swing to complete one swing if her center of gravity is 4.00 m below the pivot?

26. The pendulum on a cuckoo clock is 5.00 cm long. What is its frequency?

27. Two parakeets sit on a swing with their combined center of mass 10.0 cm below the pivot. At what frequency do they swing?

28. (a) A pendulum that has a period of 3.00000 s and that is located where the acceleration due to gravity is 9.79 m/s^2 is moved to a location where it the acceleration due to gravity is 9.82 m/s^2. What is its new period? (b) Explain why so many digits are needed in the value for the period, based on the relation between the period and the acceleration due to gravity.

29. A pendulum with a period of 2.00000 s in one location $\left(g = 9.80 \text{ m/s}^2\right)$ is moved to a new location where the period is now 1.99796 s. What is the acceleration due to gravity at its new location?

30. (a) What is the effect on the period of a pendulum if you double its length?

(b) What is the effect on the period of a pendulum if you decrease its length by 5.00%?

31. Find the ratio of the new/old periods of a pendulum if the pendulum were transported from Earth to the Moon, where the acceleration due to gravity is 1.63 m/s^2.

32. At what rate will a pendulum clock run on the Moon, where the acceleration due to gravity is 1.63 m/s^2, if it keeps time accurately on Earth? That is, find the time (in hours) it takes the clock's hour hand to make one revolution on the Moon.

33. Suppose the length of a clock's pendulum is changed by 1.000%, exactly at noon one day. What time will it read 24.00 hours later, assuming it the pendulum has kept perfect time before the change? Note that there are two answers, and perform the calculation to four-digit precision.

34. If a pendulum-driven clock gains 5.00 s/day, what fractional change in pendulum length must be made for it to keep perfect time?

16.5 Energy and the Simple Harmonic Oscillator

35. The length of nylon rope from which a mountain climber is suspended has a force constant of $1.40 \times 10^4 \text{ N/m}$.

(a) What is the frequency at which he bounces, given his mass plus and the mass of his equipment are 90.0 kg?

(b) How much would this rope stretch to break the climber's fall if he free-falls 2.00 m before the rope runs out of slack? Hint: Use conservation of energy.

(c) Repeat both parts of this problem in the situation where twice this length of nylon rope is used.

36. Engineering Application

Near the top of the Citigroup Center building in New York City, there is an object with mass of $4.00 \times 10^5 \text{ kg}$ on springs that have adjustable force constants. Its function is to dampen wind-driven oscillations of the building by oscillating at the same frequency as the building is being driven—the driving force is transferred to the object, which oscillates instead of the entire building. (a) What effective force constant should the springs have to make the object oscillate with a period of 2.00 s? (b) What energy is stored in the springs for a 2.00-m displacement from equilibrium?

16.6 Uniform Circular Motion and Simple Harmonic Motion

37. (a)What is the maximum velocity of an 85.0-kg person bouncing on a bathroom scale having a force constant of $1.50 \times 10^6 \text{ N/m}$, if the amplitude of the bounce is 0.200 cm? (b)What is the maximum energy stored in the spring?

38. A novelty clock has a 0.0100-kg mass object bouncing on a spring that has a force constant of 1.25 N/m. What is the maximum velocity of the object if the object bounces 3.00 cm above and below its equilibrium position? (b) How many joules of kinetic energy does the object have at its maximum velocity?

39. At what positions is the speed of a simple harmonic oscillator half its maximum? That is, what values of x / X give $v = \pm v_{\text{max}} / 2$, where X is the amplitude of the motion?

40. A ladybug sits 12.0 cm from the center of a Beatles music album spinning at 33.33 rpm. What is the maximum velocity of its shadow on the wall behind the turntable, if illuminated parallel to the record by the parallel rays of the setting Sun?

16.7 Damped Harmonic Motion

41. The amplitude of a lightly damped oscillator decreases by 3.0% during each cycle. What percentage of the mechanical energy of the oscillator is lost in each cycle?

16.8 Forced Oscillations and Resonance

42. How much energy must the shock absorbers of a 1200-kg car dissipate in order to damp a bounce that initially has a velocity of 0.800 m/s at the equilibrium position? Assume the car returns to its original vertical position.

43. If a car has a suspension system with a force constant of $5.00 \times 10^4 \text{ N/m}$, how much energy must the car's shocks remove to dampen an oscillation starting with a maximum displacement of 0.0750 m?

44. (a) How much will a spring that has a force constant of 40.0 N/m be stretched by an object with a mass of 0.500 kg when hung motionless from the spring? (b) Calculate the decrease in gravitational potential energy of the 0.500-kg object when it descends this distance. (c) Part of this gravitational energy goes into the spring. Calculate the energy stored in the spring by this stretch, and compare it with the gravitational potential energy. Explain where the rest of the energy might go.

45. Suppose you have a 0.750-kg object on a horizontal surface connected to a spring that has a force constant of 150 N/m. There is simple friction between the object and surface with a static coefficient of friction $\mu_s = 0.100$. (a) How far can the spring be stretched without moving the mass? (b) If the object is set into oscillation with an amplitude twice the distance found in part (a), and the kinetic coefficient of friction is $\mu_k = 0.0850$, what total distance does it travel before stopping? Assume it starts at the maximum amplitude.

46. Engineering Application: A suspension bridge oscillates with an effective force constant of $1.00 \times 10^8 \text{ N/m}$. (a) How much energy is needed to make it oscillate with an amplitude of 0.100 m? (b) If soldiers march across the bridge with a cadence equal to the bridge's natural frequency and impart $1.00 \times 10^4 \text{ J}$ of energy each second, how long does it take for the bridge's oscillations to go from 0.100 m to 0.500 m amplitude?

16.9 Waves

47. Storms in the South Pacific can create waves that travel all the way to the California coast, which are 12,000 km away. How long does it take them if they travel at 15.0 m/s?

48. Waves on a swimming pool propagate at 0.750 m/s. You splash the water at one end of the pool and observe the wave go to the opposite end, reflect, and return in 30.0 s. How far away is the other end of the pool?

49. Wind gusts create ripples on the ocean that have a wavelength of 5.00 cm and propagate at 2.00 m/s. What is their frequency?

50. How many times a minute does a boat bob up and down on ocean waves that have a wavelength of 40.0 m and a propagation speed of 5.00 m/s?

51. Scouts at a camp shake the rope bridge they have just crossed and observe the wave crests to be 8.00 m apart. If they shake it the bridge twice per second, what is the propagation speed of the waves?

52. What is the wavelength of the waves you create in a swimming pool if you splash your hand at a rate of 2.00 Hz and the waves propagate at 0.800 m/s?

53. What is the wavelength of an earthquake that shakes you with a frequency of 10.0 Hz and gets to another city 84.0 km away in 12.0 s?

54. Radio waves transmitted through space at $3.00 \times 10^8 \text{ m/s}$ by the Voyager spacecraft have a wavelength of 0.120 m. What is their frequency?

55. Your ear is capable of differentiating sounds that arrive at the ear just 1.00 ms apart. What is the minimum distance between two speakers that produce sounds that arrive at noticeably different times on a day when the speed of sound is 340 m/s?

56. (a) Seismographs measure the arrival times of earthquakes with a precision of 0.100 s. To get the distance to the epicenter of the quake, they compare the arrival times of S- and P-waves, which travel at different speeds. **Figure 16.48**) If S- and P-waves travel at 4.00 and 7.20 km/s, respectively, in the region considered, how precisely can the distance to the source of the earthquake be determined? (b) Seismic waves from underground detonations of nuclear bombs can be used to locate the test site and detect violations of test bans. Discuss whether your answer to (a) implies a serious limit to such detection. (Note also that the uncertainty is greater if there is an uncertainty in the propagation speeds of the S- and P-waves.)

Figure 16.48 A seismograph as described in above problem.(credit: Oleg Alexandrov)

16.10 Superposition and Interference

57. A car has two horns, one emitting a frequency of 199 Hz and the other emitting a frequency of 203 Hz. What beat frequency do they produce?

58. The middle-C hammer of a piano hits two strings, producing beats of 1.50 Hz. One of the strings is tuned to 260.00 Hz. What frequencies could the other string have?

59. Two tuning forks having frequencies of 460 and 464 Hz are struck simultaneously. What average frequency will you hear, and what will the beat frequency be?

60. Twin jet engines on an airplane are producing an average sound frequency of 4100 Hz with a beat frequency of 0.500 Hz. What are their individual frequencies?

61. A wave traveling on a Slinky® that is stretched to 4 m takes 2.4 s to travel the length of the Slinky and back again. (a) What is the speed of the wave? (b) Using the same Slinky stretched to the same length, a standing wave is created which consists of three antinodes and four nodes. At what frequency must the Slinky be oscillating?

62. Three adjacent keys on a piano (F, F-sharp, and G) are struck simultaneously, producing frequencies of 349, 370, and 392 Hz. What beat frequencies are produced by this discordant combination?

16.11 Energy in Waves: Intensity

63. Medical Application

Ultrasound of intensity 1.50×10^2 W/m^2 is produced by the rectangular head of a medical imaging device measuring 3.00 by 5.00 cm. What is its power output?

64. The low-frequency speaker of a stereo set has a surface area of 0.05 m^2 and produces 1W of acoustical power. What is the intensity at the speaker? If the speaker projects sound uniformly in all directions, at what distance from the speaker is the intensity 0.1 W/m^2?

65. To increase intensity of a wave by a factor of 50, by what factor should the amplitude be increased?

66. Engineering Application

A device called an insolation meter is used to measure the intensity of sunlight has an area of 100 cm^2 and registers 6.50 W. What is the intensity in W/m^2?

67. Astronomy Application

Energy from the Sun arrives at the top of the Earth's atmosphere with an intensity of 1.30 kW/m^2. How long does it take for 1.8×10^9 J to arrive on an area of 1.00 m^2?

68. Suppose you have a device that extracts energy from ocean breakers in direct proportion to their intensity. If the device produces 10.0 kW of power on a day when the breakers are 1.20 m high, how much will it produce when they are 0.600 m high?

69. Engineering Application

(a) A photovoltaic array of (solar cells) is 10.0% efficient in gathering solar energy and converting it to electricity. If the average intensity of sunlight on one day is 700 W/m^2, what area should your array have to gather energy at the rate of 100 W? (b) What is the maximum cost of the array if it must pay for itself in two years of operation averaging 10.0 hours per day? Assume that it earns money at the rate of 9.00 ¢ per kilowatt-hour.

70. A microphone receiving a pure sound tone feeds an oscilloscope, producing a wave on its screen. If the sound intensity is originally 2.00×10^{-5} W/m^2, but is turned up until the amplitude increases by 30.0%, what is the new intensity?

71. Medical Application

(a) What is the intensity in W/m^2 of a laser beam used to burn away cancerous tissue that, when 90.0% absorbed, puts 500 J of energy into a circular spot 2.00 mm in diameter in 4.00 s? (b) Discuss how this intensity compares to the average intensity of sunlight (about 700 W/m^2) and the implications that would have if the laser beam entered your eye. Note how your answer depends on the time duration of the exposure.

Test Prep for AP® Courses

16.1 Hooke's Law: Stress and Strain Revisited

1. Which of the following represents the distance (how much ground the particle covers) moved by a particle in a simple harmonic motion in one time period? (Here, *A* represents the amplitude of the oscillation.)
   a. 0 cm
   b. *A* cm
   c. 2*A* cm
   d. 4*A* cm

2. A spring has a spring constant of 80 N·m^{-1}. What is the force required to (a) compress the spring by 5 cm and (b) expand the spring by 15 cm?

3. In the formula $F = -kx$, what does the minus sign indicate?
   a. It indicates that the restoring force is in the direction of the displacement.
   b. It indicates that the restoring force is in the direction opposite the displacement.
   c. It indicates that mechanical energy in the system decreases when a system undergoes oscillation.
   d. None of the above

4. The splashing of a liquid resembles an oscillation. The restoring force in this scenario will be due to which of the following?
   a. Potential energy
   b. Kinetic energy
   c. Gravity
   d. Mechanical energy

16.2 Period and Frequency in Oscillations

5. A mass attached to a spring oscillates and completes 50 full cycles in 30 s. What is the time period and frequency of this system?

16.3 Simple Harmonic Motion: A Special Periodic Motion

6. Use these figures to answer the following questions.

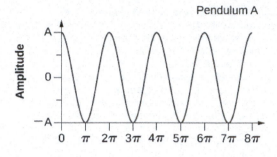

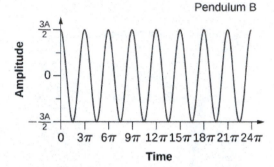

Figure 16.49
   a. Which of the two pendulums oscillates with larger amplitude?
   b. Which of the two pendulums oscillates at a higher frequency?

7. A particle of mass 100 g undergoes a simple harmonic motion. The restoring force is provided by a spring with a spring constant of 40 N·m^{-1}. What is the period of oscillation?

   a. 10π
   b. 0.5π
   c. 0.1π
   d. 1π

8. The graph shows the simple harmonic motion of a mass *m* attached to a spring with spring constant *k*.

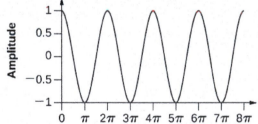

Figure 16.50

What is the displacement at time 8π?

   a. 1 m
   b. 0 m
   c. Not defined
   d. −1 m

9. A pendulum of mass 200 g undergoes simple harmonic motion when acted upon by a force of 15 N. The pendulum crosses the point of equilibrium at a speed of 5 m·s^{-1}. What is the energy of the pendulum at the center of the oscillation?

16.4 The Simple Pendulum

10. A ball is attached to a string of length 4 m to make a pendulum. The pendulum is placed at a location that is away from the Earth's surface by twice the radius of the Earth. What is the acceleration due to gravity at that height and what is the period of the oscillations?

11. Which of the following gives the correct relation between the acceleration due to gravity and period of a pendulum?

a. $g = \dfrac{2\pi L}{T^2}$

b. $g = \dfrac{4\pi^2 L}{T^2}$

c. $g = \dfrac{2\pi L}{T}$

d. $g = \dfrac{2\pi^2 L}{T}$

12. Tom has two pendulums with him. Pendulum 1 has a ball of mass 0.1 kg attached to it and has a length of 5 m. Pendulum 2 has a ball of mass 0.5 kg attached to a string of length 1 m. How does mass of the ball affect the frequency of the pendulum? Which pendulum will have a higher frequency and why?

16.5 Energy and the Simple Harmonic Oscillator

13. A mass of 1 kg undergoes simple harmonic motion with amplitude of 1 m. If the period of the oscillation is 1 s, calculate the internal energy of the system.

16.6 Uniform Circular Motion and Simple Harmonic Motion

14. In the equation $x = A \sin wt$, what values can the position x take?

a. −1 to +1
b. −A to +A
c. 0
d. −t to t

16.7 Damped Harmonic Motion

15. The non-conservative damping force removes energy from a system in which form?
a. Mechanical energy
b. Electrical energy
c. Thermal energy
d. None of the above

16. The time rate of change of mechanical energy for a damped oscillator is always:
a. 0
b. Negative
c. Positive
d. Undefined

17. A 0.5-kg object is connected to a spring that undergoes oscillatory motion. There is friction between the object and the surface it is kept on given by coefficient of friction $\mu_k = 0.06$. If the object is released 0.2 m from equilibrium, what is the distance that the object travels? Given that the force constant of the spring is 50 N m^{-1} and the frictional force between the objects is 0.294 N.

16.8 Forced Oscillations and Resonance

18. How is constant amplitude sustained in forced oscillations?

16.9 Waves

19. What is the difference between the waves coming from a tuning fork and electromagnetic waves?

20. Represent longitudinal and transverse waves in a graphical form.

21. Why is the sound produced by a tambourine different from that produced by drums?

22. A transverse wave is traveling left to right. Which of the following is correct about the motion of particles in the wave?
a. The particles move up and down when the wave travels in a vacuum.
b. The particles move left and right when the wave travels in a medium.
c. The particles move up and down when the wave travels in a medium.
d. The particles move right and left when the wave travels in a vacuum.

23.

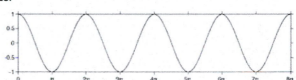

Figure 16.51 The graph shows propagation of a mechanical wave. What is the wavelength of this wave?

16.10 Superposition and Interference

24. A guitar string has a number of frequencies at which it vibrates naturally. Which of the following is true in this context?
a. The resonant frequencies of the string are integer multiples of fundamental frequencies.
b. The resonant frequencies of the string are not integer multiples of fundamental frequencies.
c. They have harmonic overtones.
d. None of the above

25. Explain the principle of superposition with figures that show the changes in the wave amplitude.

26. In this figure which points represent the points of constructive interference?

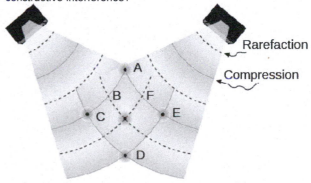

Figure 16.52
a. A, B, F
b. A, B, C, D, E, F
c. A, C, D, E
d. A, B, D

27. A string is fixed on both sides. It is snapped from both

ends at the same time by applying an equal force. What happens to the shape of the waves generated in the string? Also, will you observe an overlap of waves?

28. In the preceding question, what would happen to the amplitude of the waves generated in this way? Also, consider another scenario where the string is snapped up from one end and down from the other end. What will happen in this situation?

29. Two sine waves travel in the same direction in a medium. The amplitude of each wave is A, and the phase difference between the two is 180°. What is the resultant amplitude?
   a. $2A$
   b. $3A$
   c. 0
   d. $9A$

30. Standing wave patterns consist of nodes and antinodes formed by repeated interference between two waves of the same frequency traveling in opposite directions. What are nodes and antinodes and how are they produced?

17   PHYSICS OF HEARING

Figure 17.1 This tree fell some time ago. When it fell, atoms in the air were disturbed. Physicists would call this disturbance sound whether someone was around to hear it or not. (credit: B.A. Bowen Photography)

Chapter Outline

17.1. Sound

17.2. Speed of Sound, Frequency, and Wavelength

17.3. Sound Intensity and Sound Level

17.4. Doppler Effect and Sonic Booms

17.5. Sound Interference and Resonance: Standing Waves in Air Columns

17.6. Hearing

17.7. Ultrasound

Connection for AP® Courses

In this chapter, the concept of waves is specifically applied to the phenomena of sound. As such, Big Idea 6 continues to be supported, as sound waves carry energy and momentum from one location to another without the permanent transfer of mass. This energy is carried through vibrations caused by disturbances in air pressure (Enduring Understanding 6.A). As air pressure increases, amplitudes of vibration and energy transfer do as well. This idea (Enduring Understanding 6.A.4) explains why a very loud sound can break glass.

The chapter continues the fundamental analysis of waves addressed in Chapter 16. Sound waves are periodic, and can therefore be expressed as a function of position and time. Furthermore, sound waves are described by amplitude, frequency, wavelength, and speed (Enduring Understanding 6.B). The relationship between speed and frequency is analyzed further in Section 17.4, as the frequency of sound depends upon the relative motion between the source and observer. This concept, known as the Doppler effect, supports Essential Knowledge 6.B.5.

Like all other waves, sound waves can overlap. When they do so, their interaction will produce an amplitude variation within the resultant wave. This amplitude can be determined by adding the displacement of the two pulses, through a process called superposition. This process, covered in Section 17.5, reinforces the content in Enduring Understanding 6.D.1.

In situations where the interfering waves are confined, such as on a fixed length of string or in a tube, standing waves can result. These waves are the result of interference between the incident and reflecting wave. Standing waves are described using nodes and antinodes, and their wavelengths are determined by the size of the region to which they are confined. This chapter's description of both standing waves and the concept of beats strongly support Enduring Understanding 6.D, as well as Essential Knowledge 6.D.1, 6.D.3, and 6.D.4.

The concepts in this chapter support:

Big Idea 6 Waves can transfer energy and momentum from one location to another without the permanent transfer of mass and serve as a mathematical model for the description of other phenomena.

Enduring Understanding 6.B A periodic wave is one that repeats as a function of both time and position and can be described by its amplitude, frequency, wavelength, speed, and energy.

Essential Knowledge 6.B.5 The observed frequency of a wave depends on the relative motion of the source and the observer. This is a qualitative measurement only.

Enduring Understanding 6.D Interference and superposition lead to standing waves and beats.

Essential Knowledge 6.D.1 Two or more wave pulses can interact in such a way as to produce amplitude variations in the resultant wave. When two pulses cross, they travel through each other; they do not bounce off each other. Where the pulses overlap, the resulting displacement can be determined by adding the displacements of the two pulses. This is called superposition.

Essential Knowledge 6.D.3 Standing waves are the result of the addition of incident and reflected waves that are confined to a region and have nodes and antinodes. Examples should include waves on a fixed length of string, and sound waves in both closed and open tubes.

Essential Knowledge 6.D.4 The possible wavelengths of a standing wave are determined by the size of the region in which it is confined.

17.1 Sound

Learning Objectives
By the end of this section, you will be able to: • Define sound and hearing. • Describe sound as a longitudinal wave.

Figure 17.2 This glass has been shattered by a high-intensity sound wave of the same frequency as the resonant frequency of the glass. While the sound is not visible, the effects of the sound prove its existence. (credit: ||read||, Flickr)

Sound can be used as a familiar illustration of waves. Because hearing is one of our most important senses, it is interesting to see how the physical properties of sound correspond to our perceptions of it. **Hearing** is the perception of sound, just as vision is the perception of visible light. But sound has important applications beyond hearing. Ultrasound, for example, is not heard but can be employed to form medical images and is also used in treatment.

The physical phenomenon of **sound** is defined to be a disturbance of matter that is transmitted from its source outward. Sound is a wave. On the atomic scale, it is a disturbance of atoms that is far more ordered than their thermal motions. In many instances, sound is a periodic wave, and the atoms undergo simple harmonic motion. In this text, we shall explore such periodic sound waves.

A vibrating string produces a sound wave as illustrated in **Figure 17.3**, **Figure 17.4**, and **Figure 17.5**. As the string oscillates back and forth, it transfers energy to the air, mostly as thermal energy created by turbulence. But a small part of the string's energy goes into compressing and expanding the surrounding air, creating slightly higher and lower local pressures. These compressions (high pressure regions) and rarefactions (low pressure regions) move out as longitudinal pressure waves having the same frequency as the string—they are the disturbance that is a sound wave. (Sound waves in air and most fluids are longitudinal, because fluids have almost no shear strength. In solids, sound waves can be both transverse and longitudinal.) **Figure 17.5** shows a graph of gauge pressure versus distance from the vibrating string.

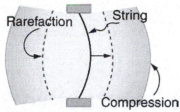

Figure 17.3 A vibrating string moving to the right compresses the air in front of it and expands the air behind it.

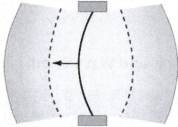

Figure 17.4 As the string moves to the left, it creates another compression and rarefaction as the ones on the right move away from the string.

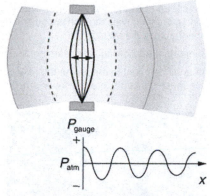

Figure 17.5 After many vibrations, there are a series of compressions and rarefactions moving out from the string as a sound wave. The graph shows gauge pressure versus distance from the source. Pressures vary only slightly from atmospheric for ordinary sounds.

The amplitude of a sound wave decreases with distance from its source, because the energy of the wave is spread over a larger and larger area. But it is also absorbed by objects, such as the eardrum in **Figure 17.6**, and converted to thermal energy by the viscosity of air. In addition, during each compression a little heat transfers to the air and during each rarefaction even less heat transfers from the air, so that the heat transfer reduces the organized disturbance into random thermal motions. (These processes can be viewed as a manifestation of the second law of thermodynamics presented in **Introduction to the Second Law of Thermodynamics: Heat Engines and Their Efficiency**.) Whether the heat transfer from compression to rarefaction is significant depends on how far apart they are—that is, it depends on wavelength. Wavelength, frequency, amplitude, and speed of propagation are important for sound, as they are for all waves.

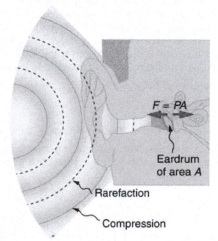

Figure 17.6 Sound wave compressions and rarefactions travel up the ear canal and force the eardrum to vibrate. There is a net force on the eardrum, since the sound wave pressures differ from the atmospheric pressure found behind the eardrum. A complicated mechanism converts the vibrations to nerve impulses, which are perceived by the person.

Make waves with a dripping faucet, audio speaker, or laser! Add a second source or a pair of slits to create an interference pattern.

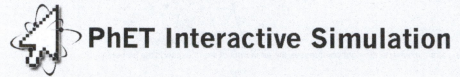

Figure 17.7 Wave Interference (http://cnx.org/content/m55288/1.2/wave-interference_en.jar)

17.2 Speed of Sound, Frequency, and Wavelength

Learning Objectives
By the end of this section, you will be able to: • Define pitch. • Describe the relationship between the speed of sound, its frequency, and its wavelength. • Describe the effects on the speed of sound as it travels through various media. • Describe the effects of temperature on the speed of sound. The information presented in this section supports the following AP® learning objectives and science practices: • **6.B.4.1** The student is able to design an experiment to determine the relationship between periodic wave speed, wavelength, and frequency, and relate these concepts to everyday examples. **(S.P. 4.2, 5.1, 7.2)**

Figure 17.8 When a firework explodes, the light energy is perceived before the sound energy. Sound travels more slowly than light does. (credit: Dominic Alves, Flickr)

Sound, like all waves, travels at a certain speed and has the properties of frequency and wavelength. You can observe direct evidence of the speed of sound while watching a fireworks display. The flash of an explosion is seen well before its sound is heard, implying both that sound travels at a finite speed and that it is much slower than light. You can also directly sense the frequency of a sound. Perception of frequency is called **pitch**. The wavelength of sound is not directly sensed, but indirect evidence is found in the correlation of the size of musical instruments with their pitch. Small instruments, such as a piccolo, typically make high-pitch sounds, while large instruments, such as a tuba, typically make low-pitch sounds. High pitch means small wavelength, and the size of a musical instrument is directly related to the wavelengths of sound it produces. So a small instrument creates short-wavelength sounds. Similar arguments hold that a large instrument creates long-wavelength sounds.

The relationship of the speed of sound, its frequency, and wavelength is the same as for all waves:

$$v_{\mathrm{w}} = f\lambda, \tag{17.1}$$

where v_{w} is the speed of sound, f is its frequency, and λ is its wavelength. The wavelength of a sound is the distance between adjacent identical parts of a wave—for example, between adjacent compressions as illustrated in **Figure 17.9**. The frequency is the same as that of the source and is the number of waves that pass a point per unit time.

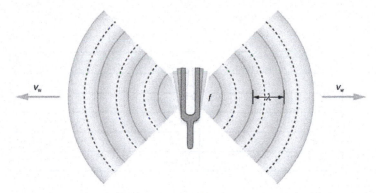

Figure 17.9 A sound wave emanates from a source vibrating at a frequency f , propagates at v_{w} , and has a wavelength λ .

Table 17.4 makes it apparent that the speed of sound varies greatly in different media. The speed of sound in a medium is determined by a combination of the medium's rigidity (or compressibility in gases) and its density. The more rigid (or less compressible) the medium, the faster the speed of sound. This observation is analogous to the fact that the frequency of a simple harmonic motion is directly proportional to the stiffness of the oscillating object. The greater the density of a medium, the slower the speed of sound. This observation is analogous to the fact that the frequency of a simple harmonic motion is inversely proportional to the mass of the oscillating object. The speed of sound in air is low, because air is compressible. Because liquids and solids are relatively rigid and very difficult to compress, the speed of sound in such media is generally greater than in gases.

Applying the Science Practices: Bottle Music

When liquid is poured into a small-necked container like a soda bottle, it can make for a fun musical experience! Find a small-necked bottle and pour water into it. When you blow across the surface of the bottle, a musical pitch should be created. This pitch, which corresponds to the resonant frequency of the air remaining in the bottle, can be determined using **Equation 17.1**. Your task is to design an experiment and collect data to confirm this relationship between the frequency created by blowing into the bottle and the depth of air remaining.

1. Use the explanation above to design an experiment that will yield data on depth of air column and frequency of pitch. Use the data table below to record your data.

Table 17.1

Depth of air column (λ)	Frequency of pitch generated (f)

2. Construct a graph using the information collected above. The graph should include all five data points and should display frequency on the dependent axis.

3. What type of relationship is displayed on your graph? (direct, inverse, quadratic, etc.)

4. Does your graph align with equation 17.1, given earlier in this section? Explain.

Note: For an explanation of why a frequency is created when you blow across a small-necked container, explore **Section 17.5** later in this chapter.

Answer

1. As the depth of the air column increases, the frequency values must decrease. A sample set of data is displayed below.

Table 17.2

Depth of air column (λ)	Frequency of pitch generated (f)
24 cm	689.6 Hz
22 cm	752.3 Hz
20 cm	827.5 Hz
18 cm	919.4 Hz
16 cm	1034.4 Hz

2. The graph drawn should have frequency on the vertical axis, contain five data points, and trend downward and to the right. A graph using the sample data from above is displayed below.

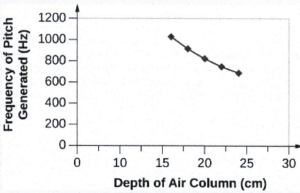

Figure 17.10 A graph of the depth of air column versus the frequency of pitch generated.

3. Inverse relationship.

Table 17.3

Depth of air column (λ)	Frequency of pitch generated (f)	Product of wavelength and frequency
24 cm	689.6 Hz	165.5
22 cm	752.3 Hz	165.5
20 cm	827.5 Hz	165.5
18 cm	919.4 Hz	165.5
16 cm	1034.4 Hz	165.5

4. The graph does align with the equation $v = f\lambda$. As the wavelength decreases, the frequency of the pitch generated increases. This relationship is validated by both the sample data table and the sample graph. Additionally, as Table 17.1 demonstrates, the product of λ and f is constant across all five data points.

    In addition to these explanations, the student may use the formula as given in the problem statement to show that the product $f \times$ *air column height* is consistently 165.5.

Table 17.4 Speed of Sound in
Various Media

Medium	v_W(m/s)
Gases at 0^oC	
Air	331
Carbon dioxide	259
Oxygen	316
Helium	965
Hydrogen	1290
Liquids at 20^oC	
Ethanol	1160
Mercury	1450
Water, fresh	1480
Sea water	1540
Human tissue	1540
Solids (longitudinal or bulk)	
Vulcanized rubber	54
Polyethylene	920
Marble	3810
Glass, Pyrex	5640
Lead	1960
Aluminum	5120
Steel	5960

Earthquakes, essentially sound waves in Earth's crust, are an interesting example of how the speed of sound depends on the rigidity of the medium. Earthquakes have both longitudinal and transverse components, and these travel at different speeds. The bulk modulus of granite is greater than its shear modulus. For that reason, the speed of longitudinal or pressure waves (P-waves) in earthquakes in granite is significantly higher than the speed of transverse or shear waves (S-waves). Both components of earthquakes travel slower in less rigid material, such as sediments. P-waves have speeds of 4 to 7 km/s, and S-waves correspondingly range in speed from 2 to 5 km/s, both being faster in more rigid material. The P-wave gets progressively farther ahead of the S-wave as they travel through Earth's crust. The time between the P- and S-waves is routinely used to determine the distance to their source, the epicenter of the earthquake.

The speed of sound is affected by temperature in a given medium. For air at sea level, the speed of sound is given by

$$v_{\text{w}} = (331 \text{ m/s})\sqrt{\frac{T}{273 \text{ K}}}, \qquad (17.2)$$

where the temperature (denoted as T) is in units of kelvin. The speed of sound in gases is related to the average speed of particles in the gas, v_{rms}, and that

$$v_{\text{rms}} = \sqrt{\frac{3kT}{m}}, \qquad (17.3)$$

where k is the Boltzmann constant (1.38×10^{-23} J/K) and m is the mass of each (identical) particle in the gas. So, it is reasonable that the speed of sound in air and other gases should depend on the square root of temperature. While not negligible, this is not a strong dependence. At 0^oC, the speed of sound is 331 m/s, whereas at 20.0^oC it is 343 m/s, less than a 4% increase. **Figure 17.11** shows a use of the speed of sound by a bat to sense distances. Echoes are also used in medical imaging.

Figure 17.11 A bat uses sound echoes to find its way about and to catch prey. The time for the echo to return is directly proportional to the distance.

One of the more important properties of sound is that its speed is nearly independent of frequency. This independence is certainly true in open air for sounds in the audible range of 20 to 20,000 Hz. If this independence were not true, you would certainly notice it for music played by a marching band in a football stadium, for example. Suppose that high-frequency sounds traveled faster—then the farther you were from the band, the more the sound from the low-pitch instruments would lag that from the high-pitch ones. But the music from all instruments arrives in cadence independent of distance, and so all frequencies must travel at nearly the same speed. Recall that

$$v_{\text{w}} = f\lambda. \tag{17.4}$$

In a given medium under fixed conditions, v_{w} is constant, so that there is a relationship between f and λ; the higher the frequency, the smaller the wavelength. See **Figure 17.12** and consider the following example.

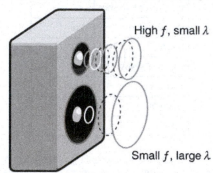

High f, small λ

Small f, large λ

Figure 17.12 Because they travel at the same speed in a given medium, low-frequency sounds must have a greater wavelength than high-frequency sounds. Here, the lower-frequency sounds are emitted by the large speaker, called a woofer, while the higher-frequency sounds are emitted by the small speaker, called a tweeter.

Example 17.1 Calculating Wavelengths: What Are the Wavelengths of Audible Sounds?

Calculate the wavelengths of sounds at the extremes of the audible range, 20 and 20,000 Hz, in $30.0°\text{C}$ air. (Assume that the frequency values are accurate to two significant figures.)

Strategy

To find wavelength from frequency, we can use $v_{\text{w}} = f\lambda$.

Solution

1. Identify knowns. The value for v_{w}, is given by

$$v_{\text{w}} = (331 \text{ m/s})\sqrt{\frac{T}{273 \text{ K}}}. \tag{17.5}$$

2. Convert the temperature into kelvin and then enter the temperature into the equation

$$v_{\text{w}} = (331 \text{ m/s})\sqrt{\frac{303 \text{ K}}{273 \text{ K}}} = 348.7 \text{ m/s}. \tag{17.6}$$

3. Solve the relationship between speed and wavelength for λ:

$$\lambda = \frac{v_{\text{w}}}{f}. \tag{17.7}$$

4. Enter the speed and the minimum frequency to give the maximum wavelength:

$$\lambda_{\text{max}} = \frac{348.7 \text{ m/s}}{20 \text{ Hz}} = 17 \text{ m}. \tag{17.8}$$

5. Enter the speed and the maximum frequency to give the minimum wavelength:

$$\lambda_{min} = \frac{348.7 \text{ m/s}}{20,000 \text{ Hz}} = 0.017 \text{ m} = 1.7 \text{ cm}.$$ (17.9)

Discussion

Because the product of f multiplied by λ equals a constant, the smaller f is, the larger λ must be, and vice versa.

The speed of sound can change when sound travels from one medium to another. However, the frequency usually remains the same because it is like a driven oscillation and has the frequency of the original source. If v_w changes and f remains the same, then the wavelength λ must change. That is, because $v_w = f\lambda$, the higher the speed of a sound, the greater its wavelength for a given frequency.

Making Connections: Take-Home Investigation—Voice as a Sound Wave

Suspend a sheet of paper so that the top edge of the paper is fixed and the bottom edge is free to move. You could tape the top edge of the paper to the edge of a table. Gently blow near the edge of the bottom of the sheet and note how the sheet moves. Speak softly and then louder such that the sounds hit the edge of the bottom of the paper, and note how the sheet moves. Explain the effects.

Check Your Understanding

Imagine you observe two fireworks explode. You hear the explosion of one as soon as you see it. However, you see the other firework for several milliseconds before you hear the explosion. Explain why this is so.

Solution
Sound and light both travel at definite speeds. The speed of sound is slower than the speed of light. The first firework is probably very close by, so the speed difference is not noticeable. The second firework is farther away, so the light arrives at your eyes noticeably sooner than the sound wave arrives at your ears.

Check Your Understanding

You observe two musical instruments that you cannot identify. One plays high-pitch sounds and the other plays low-pitch sounds. How could you determine which is which without hearing either of them play?

Solution
Compare their sizes. High-pitch instruments are generally smaller than low-pitch instruments because they generate a smaller wavelength.

17.3 Sound Intensity and Sound Level

Learning Objectives

By the end of this section, you will be able to:

*   Define intensity, sound intensity, and sound pressure level.
*   Calculate sound intensity levels in decibels (dB).

The information presented in this section supports the following AP® learning objectives and science practices:

*   **6.A.4.1** The student is able to explain and/or predict qualitatively how the energy carried by a sound wave relates to the amplitude of the wave, and/or apply this concept to a real-world example. **(S.P. 6.4)**

Figure 17.13 Noise on crowded roadways like this one in Delhi makes it hard to hear others unless they shout. (credit: Lingaraj G J, Flickr)

In a quiet forest, you can sometimes hear a single leaf fall to the ground. After settling into bed, you may hear your blood pulsing through your ears. But when a passing motorist has his stereo turned up, you cannot even hear what the person next to you in your car is saying. We are all very familiar with the loudness of sounds and aware that they are related to how energetically the source is vibrating. In cartoons depicting a screaming person (or an animal making a loud noise), the cartoonist often shows an open mouth with a vibrating uvula, the hanging tissue at the back of the mouth, to suggest a loud sound coming from the throat **Figure 17.14**. High noise exposure is hazardous to hearing, and it is common for musicians to have hearing losses that are sufficiently severe that they interfere with the musicians' abilities to perform. The relevant physical quantity is sound intensity, a concept that is valid for all sounds whether or not they are in the audible range.

Intensity is defined to be the power per unit area carried by a wave. Power is the rate at which energy is transferred by the wave. In equation form, **intensity** I is

$$I = \frac{P}{A}, \tag{17.10}$$

where P is the power through an area A. The SI unit for I is $\mathrm{W/m^2}$. The intensity of a sound wave is related to its amplitude squared by the following relationship:

$$I = \frac{(\Delta p)^2}{2\rho v_{\mathrm{w}}}. \tag{17.11}$$

Here Δp is the pressure variation or pressure amplitude (half the difference between the maximum and minimum pressure in the sound wave) in units of pascals (Pa) or $\mathrm{N/m^2}$. (We are using a lower case p for pressure to distinguish it from power, denoted by P above.) The energy (as kinetic energy $\frac{mv^2}{2}$) of an oscillating element of air due to a traveling sound wave is proportional to its amplitude squared. In this equation, ρ is the density of the material in which the sound wave travels, in units of $\mathrm{kg/m^3}$, and v_{w} is the speed of sound in the medium, in units of m/s. The pressure variation is proportional to the amplitude of the oscillation, and so I varies as $(\Delta p)^2$ (**Figure 17.14**). This relationship is consistent with the fact that the sound wave is produced by some vibration; the greater its pressure amplitude, the more the air is compressed in the sound it creates.

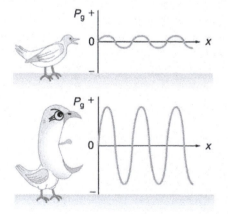

Figure 17.14 Graphs of the gauge pressures in two sound waves of different intensities. The more intense sound is produced by a source that has larger-amplitude oscillations and has greater pressure maxima and minima. Because pressures are higher in the greater-intensity sound, it can exert larger forces on the objects it encounters.

Sound intensity levels are quoted in decibels (dB) much more often than sound intensities in watts per meter squared. Decibels are the unit of choice in the scientific literature as well as in the popular media. The reasons for this choice of units are related to how we perceive sounds. How our ears perceive sound can be more accurately described by the logarithm of the intensity rather than directly to the intensity. The **sound intensity level** β in decibels of a sound having an intensity I in watts per meter squared is defined to be

$$\beta \, (\text{dB}) = 10 \log_{10}\!\left(\frac{I}{I_0}\right),$$

(17.12)

where $I_0 = 10^{-12} \, \text{W/m}^2$ is a reference intensity. In particular, I_0 is the lowest or threshold intensity of sound a person with normal hearing can perceive at a frequency of 1000 Hz. Sound intensity level is not the same as intensity. Because β is defined in terms of a ratio, it is a unitless quantity telling you the *level* of the sound relative to a fixed standard ($10^{-12} \, \text{W/m}^2$, in this case). The units of decibels (dB) are used to indicate this ratio is multiplied by 10 in its definition. The bel, upon which the decibel is based, is named for Alexander Graham Bell, the inventor of the telephone.

Table 17.5 Sound Intensity Levels and Intensities

Sound intensity level β (dB)	Intensity I(W/m^2)	Example/effect
0	1×10^{-12}	Threshold of hearing at 1000 Hz
10	1×10^{-11}	Rustle of leaves
20	1×10^{-10}	Whisper at 1 m distance
30	1×10^{-9}	Quiet home
40	1×10^{-8}	Average home
50	1×10^{-7}	Average office, soft music
60	1×10^{-6}	Normal conversation
70	1×10^{-5}	Noisy office, busy traffic
80	1×10^{-4}	Loud radio, classroom lecture
90	1×10^{-3}	Inside a heavy truck; damage from prolonged exposure[1]
100	1×10^{-2}	Noisy factory, siren at 30 m; damage from 8 h per day exposure
110	1×10^{-1}	Damage from 30 min per day exposure
120	1	Loud rock concert, pneumatic chipper at 2 m; threshold of pain
140	1×10^{2}	Jet airplane at 30 m; severe pain, damage in seconds
160	1×10^{4}	Bursting of eardrums

The decibel level of a sound having the threshold intensity of $10^{-12} \, \text{W/m}^2$ is $\beta = 0 \, \text{dB}$, because $\log_{10} 1 = 0$. That is, the threshold of hearing is 0 decibels. **Table 17.5** gives levels in decibels and intensities in watts per meter squared for some familiar sounds.

One of the more striking things about the intensities in **Table 17.5** is that the intensity in watts per meter squared is quite small for most sounds. The ear is sensitive to as little as a trillionth of a watt per meter squared—even more impressive when you realize that the area of the eardrum is only about $1 \, \text{cm}^2$, so that only 10^{-16} W falls on it at the threshold of hearing! Air molecules in a sound wave of this intensity vibrate over a distance of less than one molecular diameter, and the gauge pressures involved are less than 10^{-9} atm.

Another impressive feature of the sounds in **Table 17.5** is their numerical range. Sound intensity varies by a factor of 10^{12} from

---

1.   Several government agencies and health-related professional associations recommend that 85 dB not be exceeded for 8-hour daily exposures in the absence of hearing protection.

threshold to a sound that causes damage in seconds. You are unaware of this tremendous range in sound intensity because how your ears respond can be described approximately as the logarithm of intensity. Thus, sound intensity levels in decibels fit your experience better than intensities in watts per meter squared. The decibel scale is also easier to relate to because most people are more accustomed to dealing with numbers such as 0, 53, or 120 than numbers such as 1.00×10^{-11} .

One more observation readily verified by examining **Table 17.5** or using $I = \dfrac{(\Delta p)^2}{2\rho v_{\text{w}}}$ is that each factor of 10 in intensity corresponds to 10 dB. For example, a 90 dB sound compared with a 60 dB sound is 30 dB greater, or three factors of 10 (that is, 10^3 times) as intense. Another example is that if one sound is 10^7 as intense as another, it is 70 dB higher. See **Table 17.6**.

Table 17.6 Ratios of Intensities and Corresponding Differences in Sound Intensity Levels

I_2/I_1	$\beta_2 - \beta_1$
2.0	3.0 dB
5.0	7.0 dB
10.0	10.0 dB

Example 17.2 Calculating Sound Intensity Levels: Sound Waves

Calculate the sound intensity level in decibels for a sound wave traveling in air at $0°C$ and having a pressure amplitude of 0.656 Pa.

Strategy

We are given Δp , so we can calculate I using the equation $I = (\Delta p)^2 / (2\rho v_{\text{w}})^2$. Using I , we can calculate β straight from its definition in $\beta \text{ (dB)} = 10 \log_{10}(I/I_0)$.

Solution

(1) Identify knowns:

Sound travels at 331 m/s in air at $0°C$.

Air has a density of 1.29 kg/m^3 at atmospheric pressure and $0°C$.

(2) Enter these values and the pressure amplitude into $I = (\Delta p)^2 / (2\rho v_{\text{w}})$:

$$I = \frac{(\Delta p)^2}{2\rho v_{\text{w}}} = \frac{(0.656 \text{ Pa})^2}{2(1.29 \text{ kg/m}^3)(331 \text{ m/s})} = 5.04 \times 10^{-4} \text{ W/m}^2. \qquad (17.13)$$

(3) Enter the value for I and the known value for I_0 into $\beta \text{ (dB)} = 10 \log_{10}(I/I_0)$. Calculate to find the sound intensity level in decibels:

$$10 \log_{10}(5.04 \times 10^8) = 10 \,(8.70) \text{ dB} = 87 \text{ dB}. \qquad (17.14)$$

Discussion

This 87 dB sound has an intensity five times as great as an 80 dB sound. So a factor of five in intensity corresponds to a difference of 7 dB in sound intensity level. This value is true for any intensities differing by a factor of five.

Example 17.3 Change Intensity Levels of a Sound: What Happens to the Decibel Level?

Show that if one sound is twice as intense as another, it has a sound level about 3 dB higher.

Strategy

You are given that the ratio of two intensities is 2 to 1, and are then asked to find the difference in their sound levels in decibels. You can solve this problem using of the properties of logarithms.

Solution

(1) Identify knowns:

The ratio of the two intensities is 2 to 1, or:

$$\frac{I_2}{I_1} = 2.00. \tag{17.15}$$

We wish to show that the difference in sound levels is about 3 dB. That is, we want to show:

$$\beta_2 - \beta_1 = 3 \text{ dB}. \tag{17.16}$$

Note that:

$$\log_{10} b - \log_{10} a = \log_{10}\left(\frac{b}{a}\right). \tag{17.17}$$

(2) Use the definition of β to get:

$$\beta_2 - \beta_1 = 10 \log_{10}\left(\frac{I_2}{I_1}\right) = 10 \log_{10} 2.00 = 10\,(0.301) \text{ dB}. \tag{17.18}$$

Thus,

$$\beta_2 - \beta_1 = 3.01 \text{ dB}. \tag{17.19}$$

Discussion

This means that the two sound intensity levels differ by 3.01 dB, or about 3 dB, as advertised. Note that because only the ratio I_2/I_1 is given (and not the actual intensities), this result is true for any intensities that differ by a factor of two. For example, a 56.0 dB sound is twice as intense as a 53.0 dB sound, a 97.0 dB sound is half as intense as a 100 dB sound, and so on.

It should be noted at this point that there is another decibel scale in use, called the **sound pressure level**, based on the ratio of the pressure amplitude to a reference pressure. This scale is used particularly in applications where sound travels in water. It is beyond the scope of most introductory texts to treat this scale because it is not commonly used for sounds in air, but it is important to note that very different decibel levels may be encountered when sound pressure levels are quoted. For example, ocean noise pollution produced by ships may be as great as 200 dB expressed in the sound pressure level, where the more familiar sound intensity level we use here would be something under 140 dB for the same sound.

Take-Home Investigation: Feeling Sound

Find a CD player and a CD that has rock music. Place the player on a light table, insert the CD into the player, and start playing the CD. Place your hand gently on the table next to the speakers. Increase the volume and note the level when the table just begins to vibrate as the rock music plays. Increase the reading on the volume control until it doubles. What has happened to the vibrations?

Check Your Understanding

Describe how amplitude is related to the loudness of a sound.

Solution

Amplitude is directly proportional to the experience of loudness. As amplitude increases, loudness increases.

Check Your Understanding

Identify common sounds at the levels of 10 dB, 50 dB, and 100 dB.

Solution

10 dB: Running fingers through your hair.

50 dB: Inside a quiet home with no television or radio.

100 dB: Take-off of a jet plane.

17.4 Doppler Effect and Sonic Booms

By the end of this section, you will be able to:

- Define Doppler effect, Doppler shift, and sonic boom.
- Calculate the frequency of a sound heard by someone observing Doppler shift.
- Describe the sounds produced by objects moving faster than the speed of sound.

The information presented in this section supports the following AP® learning objectives and science practices:

- **6.B.5.1** The student is able to create or use a wave front diagram to demonstrate or interpret qualitatively the observed frequency of a wave, dependent upon relative motions of source and observer. **(S.P. 1.4)**

The characteristic sound of a motorcycle buzzing by is an example of the **Doppler effect**. The high-pitch scream shifts dramatically to a lower-pitch roar as the motorcycle passes by a stationary observer. The closer the motorcycle brushes by, the more abrupt the shift. The faster the motorcycle moves, the greater the shift. We also hear this characteristic shift in frequency for passing race cars, airplanes, and trains. It is so familiar that it is used to imply motion and children often mimic it in play.

The Doppler effect is an alteration in the observed frequency of a sound due to motion of either the source or the observer. Although less familiar, this effect is easily noticed for a stationary source and moving observer. For example, if you ride a train past a stationary warning bell, you will hear the bell's frequency shift from high to low as you pass by. The actual change in frequency due to relative motion of source and observer is called a **Doppler shift**. The Doppler effect and Doppler shift are named for the Austrian physicist and mathematician Christian Johann Doppler (1803–1853), who did experiments with both moving sources and moving observers. Doppler, for example, had musicians play on a moving open train car and also play standing next to the train tracks as a train passed by. Their music was observed both on and off the train, and changes in frequency were measured.

What causes the Doppler shift? **Figure 17.15**, **Figure 17.16**, and **Figure 17.17** compare sound waves emitted by stationary and moving sources in a stationary air mass. Each disturbance spreads out spherically from the point where the sound was emitted. If the source is stationary, then all of the spheres representing the air compressions in the sound wave centered on the same point, and the stationary observers on either side see the same wavelength and frequency as emitted by the source, as in **Figure 17.15**. If the source is moving, as in **Figure 17.16**, then the situation is different. Each compression of the air moves out in a sphere from the point where it was emitted, but the point of emission moves. This moving emission point causes the air compressions to be closer together on one side and farther apart on the other. Thus, the wavelength is shorter in the direction the source is moving (on the right in **Figure 17.16**), and longer in the opposite direction (on the left in **Figure 17.16**). Finally, if the observers move, as in **Figure 17.17**, the frequency at which they receive the compressions changes. The observer moving toward the source receives them at a higher frequency, and the person moving away from the source receives them at a lower frequency.

X   Y

Figure 17.15 Sounds emitted by a source spread out in spherical waves. Because the source, observers, and air are stationary, the wavelength and frequency are the same in all directions and to all observers.

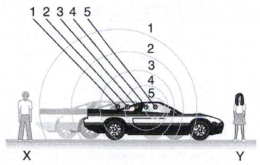

Figure 17.16 Sounds emitted by a source moving to the right spread out from the points at which they were emitted. The wavelength is reduced and, consequently, the frequency is increased in the direction of motion, so that the observer on the right hears a higher-pitch sound. The opposite is true for the observer on the left, where the wavelength is increased and the frequency is reduced.

Figure 17.17 The same effect is produced when the observers move relative to the source. Motion toward the source increases frequency as the observer on the right passes through more wave crests than she would if stationary. Motion away from the source decreases frequency as the observer on the left passes through fewer wave crests than he would if stationary.

We know that wavelength and frequency are related by $v_w = f\lambda$, where v_w is the fixed speed of sound. The sound moves in a medium and has the same speed v_w in that medium whether the source is moving or not. Thus f multiplied by λ is a constant. Because the observer on the right in **Figure 17.16** receives a shorter wavelength, the frequency she receives must be higher. Similarly, the observer on the left receives a longer wavelength, and hence he hears a lower frequency. The same thing happens in **Figure 17.17**. A higher frequency is received by the observer moving toward the source, and a lower frequency is received by an observer moving away from the source. In general, then, relative motion of source and observer toward one another increases the received frequency. Relative motion apart decreases frequency. The greater the relative speed is, the greater the effect.

The Doppler Effect

The Doppler effect occurs not only for sound but for any wave when there is relative motion between the observer and the source. There are Doppler shifts in the frequency of sound, light, and water waves, for example. Doppler shifts can be used to determine velocity, such as when ultrasound is reflected from blood in a medical diagnostic. The recession of galaxies is determined by the shift in the frequencies of light received from them and has implied much about the origins of the universe. Modern physics has been profoundly affected by observations of Doppler shifts.

For a stationary observer and a moving source, the frequency f_{obs} received by the observer can be shown to be

$$f_{obs} = f_s\left(\frac{v_w}{v_w \pm v_s}\right),$$ (17.20)

where f_s is the frequency of the source, v_s is the speed of the source along a line joining the source and observer, and v_w is the speed of sound. The minus sign is used for motion toward the observer and the plus sign for motion away from the observer, producing the appropriate shifts up and down in frequency. Note that the greater the speed of the source, the greater the effect. Similarly, for a stationary source and moving observer, the frequency received by the observer f_{obs} is given by

$$f_{obs} = f_s\left(\frac{v_w \pm v_{obs}}{v_w}\right),$$ (17.21)

where v_{obs} is the speed of the observer along a line joining the source and observer. Here the plus sign is for motion toward the source, and the minus is for motion away from the source.

Example 17.4 Calculate Doppler Shift: A Train Horn

Suppose a train that has a 150-Hz horn is moving at 35.0 m/s in still air on a day when the speed of sound is 340 m/s.

(a) What frequencies are observed by a stationary person at the side of the tracks as the train approaches and after it passes?

(b) What frequency is observed by the train's engineer traveling on the train?

Strategy

To find the observed frequency in (a), $f_{obs} = f_s \left(\dfrac{v_w}{v_w \pm v_s} \right)$, must be used because the source is moving. The minus sign is used for the approaching train, and the plus sign for the receding train. In (b), there are two Doppler shifts—one for a moving source and the other for a moving observer.

Solution for (a)

(1) Enter known values into $f_{obs} = f_s \left(\dfrac{v_w}{v_w - v_s} \right)$.

$$f_{obs} = f_s \left(\frac{v_w}{v_w - v_s} \right) = (150 \text{ Hz}) \left(\frac{340 \text{ m/s}}{340 \text{ m/s} - 35.0 \text{ m/s}} \right) \qquad (17.22)$$

(2) Calculate the frequency observed by a stationary person as the train approaches.

$$f_{obs} = (150 \text{ Hz})(1.11) = 167 \text{ Hz} \qquad (17.23)$$

(3) Use the same equation with the plus sign to find the frequency heard by a stationary person as the train recedes.

$$f_{obs} = f_s \left(\frac{v_w}{v_w + v_s} \right) = (150 \text{ Hz}) \left(\frac{340 \text{ m/s}}{340 \text{ m/s} + 35.0 \text{ m/s}} \right) \qquad (17.24)$$

(4) Calculate the second frequency.

$$f_{obs} = (150 \text{ Hz})(0.907) = 136 \text{ Hz} \qquad (17.25)$$

Discussion on (a)

The numbers calculated are valid when the train is far enough away that the motion is nearly along the line joining train and observer. In both cases, the shift is significant and easily noticed. Note that the shift is 17.0 Hz for motion toward and 14.0 Hz for motion away. The shifts are not symmetric.

Solution for (b)

(1) Identify knowns:

- It seems reasonable that the engineer would receive the same frequency as emitted by the horn, because the relative velocity between them is zero.
- Relative to the medium (air), the speeds are $v_s = v_{obs} = 35.0$ m/s.
- The first Doppler shift is for the moving observer; the second is for the moving source.

(2) Use the following equation:

$$f_{obs} = \left[f_s \left(\frac{v_w \pm v_{obs}}{v_w} \right) \right] \left(\frac{v_w}{v_w \pm v_s} \right). \qquad (17.26)$$

The quantity in the square brackets is the Doppler-shifted frequency due to a moving observer. The factor on the right is the effect of the moving source.

(3) Because the train engineer is moving in the direction toward the horn, we must use the plus sign for v_{obs}; however, because the horn is also moving in the direction away from the engineer, we also use the plus sign for v_s. But the train is carrying both the engineer and the horn at the same velocity, so $v_s = v_{obs}$. As a result, everything but f_s cancels, yielding

$$f_{obs} = f_s. \qquad (17.27)$$

Discussion for (b)

We may expect that there is no change in frequency when source and observer move together because it fits your experience. For example, there is no Doppler shift in the frequency of conversations between driver and passenger on a motorcycle. People talking when a wind moves the air between them also observe no Doppler shift in their conversation. The crucial point is that source and observer are not moving relative to each other.

Sonic Booms to Bow Wakes

What happens to the sound produced by a moving source, such as a jet airplane, that approaches or even exceeds the speed of sound? The answer to this question applies not only to sound but to all other waves as well.

Suppose a jet airplane is coming nearly straight at you, emitting a sound of frequency f_s. The greater the plane's speed v_s, the greater the Doppler shift and the greater the value observed for f_{obs}. Now, as v_s approaches the speed of sound, f_{obs} approaches infinity, because the denominator in $f_{obs} = f_s\left(\dfrac{v_w}{v_w \pm v_s}\right)$ approaches zero. At the speed of sound, this result means that in front of the source, each successive wave is superimposed on the previous one because the source moves forward at the speed of sound. The observer gets them all at the same instant, and so the frequency is infinite. (Before airplanes exceeded the speed of sound, some people argued it would be impossible because such constructive superposition would produce pressures great enough to destroy the airplane.) If the source exceeds the speed of sound, no sound is received by the observer until the source has passed, so that the sounds from the approaching source are mixed with those from it when receding. This mixing appears messy, but something interesting happens—a sonic boom is created. (See **Figure 17.18**.)

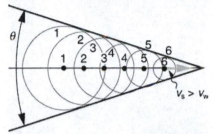

Figure 17.18 Sound waves from a source that moves faster than the speed of sound spread spherically from the point where they are emitted, but the source moves ahead of each. Constructive interference along the lines shown (actually a cone in three dimensions) creates a shock wave called a sonic boom. The faster the speed of the source, the smaller the angle θ.

There is constructive interference along the lines shown (a cone in three dimensions) from similar sound waves arriving there simultaneously. This superposition forms a disturbance called a **sonic boom**, a constructive interference of sound created by an object moving faster than sound. Inside the cone, the interference is mostly destructive, and so the sound intensity there is much less than on the shock wave. An aircraft creates two sonic booms, one from its nose and one from its tail. (See **Figure 17.19**.) During television coverage of space shuttle landings, two distinct booms could often be heard. These were separated by exactly the time it would take the shuttle to pass by a point. Observers on the ground often do not see the aircraft creating the sonic boom, because it has passed by before the shock wave reaches them, as seen in **Figure 17.19**. If the aircraft flies close by at low altitude, pressures in the sonic boom can be destructive and break windows as well as rattle nerves. Because of how destructive sonic booms can be, supersonic flights are banned over populated areas of the United States.

Figure 17.19 Two sonic booms, created by the nose and tail of an aircraft, are observed on the ground after the plane has passed by.

Sonic booms are one example of a broader phenomenon called bow wakes. A **bow wake**, such as the one in **Figure 17.20**, is created when the wave source moves faster than the wave propagation speed. Water waves spread out in circles from the point where created, and the bow wake is the familiar V-shaped wake trailing the source. A more exotic bow wake is created when a subatomic particle travels through a medium faster than the speed of light travels in that medium. (In a vacuum, the maximum speed of light will be $c = 3.00 \times 10^8$ m/s ; in the medium of water, the speed of light is closer to $0.75c$. If the particle creates light in its passage, that light spreads on a cone with an angle indicative of the speed of the particle, as illustrated in **Figure 17.21**. Such a bow wake is called Cerenkov radiation and is commonly observed in particle physics.

Figure 17.20 Bow wake created by a duck. Constructive interference produces the rather structured wake, while there is relatively little wave action inside the wake, where interference is mostly destructive. (credit: Horia Varlan, Flickr)

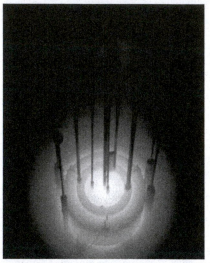

Figure 17.21 The blue glow in this research reactor pool is Cerenkov radiation caused by subatomic particles traveling faster than the speed of light in water. (credit: U.S. Nuclear Regulatory Commission)

Doppler shifts and sonic booms are interesting sound phenomena that occur in all types of waves. They can be of considerable use. For example, the Doppler shift in ultrasound can be used to measure blood velocity, while police use the Doppler shift in radar (a microwave) to measure car velocities. In meteorology, the Doppler shift is used to track the motion of storm clouds; such "Doppler Radar" can give velocity and direction and rain or snow potential of imposing weather fronts. In astronomy, we can examine the light emitted from distant galaxies and determine their speed relative to ours. As galaxies move away from us, their light is shifted to a lower frequency, and so to a longer wavelength—the so-called red shift. Such information from galaxies far, far away has allowed us to estimate the age of the universe (from the Big Bang) as about 14 billion years.

Check Your Understanding

Why did scientist Christian Doppler observe musicians both on a moving train and also from a stationary point not on the train?

Solution
Doppler needed to compare the perception of sound when the observer is stationary and the sound source moves, as well as when the sound source and the observer are both in motion.

Check Your Understanding

Describe a situation in your life when you might rely on the Doppler shift to help you either while driving a car or walking near traffic.

Solution
If I am driving and I hear Doppler shift in an ambulance siren, I would be able to tell when it was getting closer and also if it has passed by. This would help me to know whether I needed to pull over and let the ambulance through.

17.5 Sound Interference and Resonance: Standing Waves in Air Columns

Learning Objectives

By the end of this section, you will be able to:

- Define antinode, node, fundamental, overtones, and harmonics.
- Identify instances of sound interference in everyday situations.
- Describe how sound interference occurring inside open and closed tubes changes the characteristics of the sound, and how this applies to sounds produced by musical instruments.
- Calculate the length of a tube using sound wave measurements.

The information presented in this section supports the following AP® learning objectives and science practices:

- **6.D.1.1** The student is able to use representations of individual pulses and construct representations to model the interaction of two wave pulses to analyze the superposition of two pulses. **(S.P. 1.1, 1.4)**
- **6.D.1.2** The student is able to design a suitable experiment and analyze data illustrating the superposition of mechanical waves (only for wave pulses or standing waves). **(S.P. 4.2, 5.1)**
- **6.D.1.3** The student is able to design a plan for collecting data to quantify the amplitude variations when two or more traveling waves or wave pulses interact in a given medium. **(S.P. 4.2)**
- **6.D.3.1** The student is able to refine a scientific question related to standing waves and design a detailed plan for the experiment that can be conducted to examine the phenomenon qualitatively or quantitatively. **(S.P. 2.1, 2.2, 4.2)**
- **6.D.3.2** The student is able to predict properties of standing waves that result from the addition of incident and reflected waves that are confined to a region and have nodes and antinodes. **(S.P. 6.4)**
- **6.D.3.3** The student is able to plan data collection strategies, predict the outcome based on the relationship under test, perform data analysis, evaluate evidence compared to the prediction, explain any discrepancy and, if necessary, revise the relationship among variables responsible for establishing standing waves on a string or in a column of air. **(S.P. 3.2, 4.1, 5.1, 5.2, 5.3)**
- **6.D.3.4** The student is able to describe representations and models of situations in which standing waves result from the addition of incident and reflected waves confined to a region. **(S.P. 1.2)**
- **6.D.4.2** The student is able to calculate wavelengths and frequencies (if given wave speed) of standing waves based on boundary conditions and length of region within which the wave is confined, and calculate numerical values of wavelengths and frequencies. Examples should include musical instruments. **(S.P. 2.2)**

Figure 17.22 Some types of headphones use the phenomena of constructive and destructive interference to cancel out outside noises. (credit: JVC America, Flickr)

Interference is the hallmark of waves, all of which exhibit constructive and destructive interference exactly analogous to that seen for water waves. In fact, one way to prove something "is a wave" is to observe interference effects. So, sound being a wave, we expect it to exhibit interference; we have already mentioned a few such effects, such as the beats from two similar notes played simultaneously.

Figure 17.23 shows a clever use of sound interference to cancel noise. Larger-scale applications of active noise reduction by destructive interference are contemplated for entire passenger compartments in commercial aircraft. To obtain destructive interference, a fast electronic analysis is performed, and a second sound is introduced with its maxima and minima exactly reversed from the incoming noise. Sound waves in fluids are pressure waves and consistent with Pascal's principle; pressures from two different sources add and subtract like simple numbers; that is, positive and negative gauge pressures add to a much smaller pressure, producing a lower-intensity sound. Although completely destructive interference is possible only under the simplest conditions, it is possible to reduce noise levels by 30 dB or more using this technique.

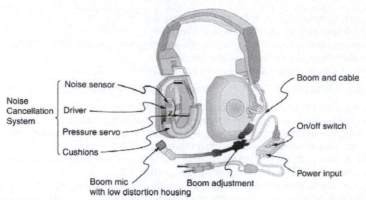

Figure 17.23 Headphones designed to cancel noise with destructive interference create a sound wave exactly opposite to the incoming sound. These headphones can be more effective than the simple passive attenuation used in most ear protection. Such headphones were used on the record-setting, around the world nonstop flight of the Voyager aircraft to protect the pilots' hearing from engine noise.

Where else can we observe sound interference? All sound resonances, such as in musical instruments, are due to constructive and destructive interference. Only the resonant frequencies interfere constructively to form standing waves, while others interfere destructively and are absent. From the toot made by blowing over a bottle, to the characteristic flavor of a violin's sounding box, to the recognizability of a great singer's voice, resonance and standing waves play a vital role.

Interference

Interference is such a fundamental aspect of waves that observing interference is proof that something is a wave. The wave nature of light was established by experiments showing interference. Similarly, when electrons scattered from crystals exhibited interference, their wave nature was confirmed to be exactly as predicted by symmetry with certain wave characteristics of light.

Applying the Science Practices: Standing Wave

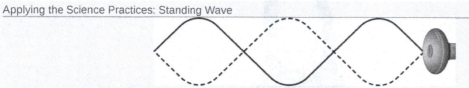

Figure 17.24 The standing wave pattern of a rubber tube attached to a doorknob.

Tie one end of a strip of long rubber tubing to a stable object (doorknob, fence post, etc.) and shake the other end up and down until a standing wave pattern is achieved. Devise a method to determine the frequency and wavelength generated by your arm shaking. Do your results align with the equation? Do you find that the velocity of the wave generated is consistent for each trial? If not, explain why this is the case.

Answer

This task will likely require two people. The frequency of the wave pattern can be found by timing how long it takes the student shaking the rubber tubing to move his or her hand up and down one full time. (It may be beneficial to time how long it takes the student to do this ten times, and then divide by ten to reduce error.) The wavelength of the standing wave can be measured with a meter stick by measuring the distance between two nodes and multiplying by two. This information should be gathered for standing wave patterns of multiple different wavelengths. As students collect their data, they can use the equation to determine if the wave velocity is consistent. There will likely be some error in the experiment yielding velocities of slightly different value. This error is probably due to an inaccuracy in the wavelength and/or frequency measurements.

Suppose we hold a tuning fork near the end of a tube that is closed at the other end, as shown in **Figure 17.25**, **Figure 17.26**, **Figure 17.27**, and **Figure 17.28**. If the tuning fork has just the right frequency, the air column in the tube resonates loudly, but at most frequencies it vibrates very little. This observation just means that the air column has only certain natural frequencies. The figures show how a resonance at the lowest of these natural frequencies is formed. A disturbance travels down the tube at the speed of sound and bounces off the closed end. If the tube is just the right length, the reflected sound arrives back at the tuning fork exactly half a cycle later, and it interferes constructively with the continuing sound produced by the tuning fork. The incoming and reflected sounds form a standing wave in the tube as shown.

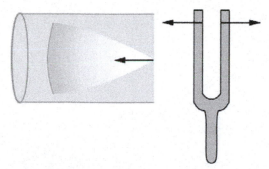

Figure 17.25 Resonance of air in a tube closed at one end, caused by a tuning fork. A disturbance moves down the tube.

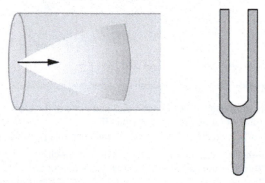

Figure 17.26 Resonance of air in a tube closed at one end, caused by a tuning fork. The disturbance reflects from the closed end of the tube.

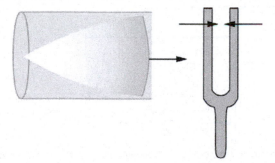

Figure 17.27 Resonance of air in a tube closed at one end, caused by a tuning fork. If the length of the tube L is just right, the disturbance gets back to the tuning fork half a cycle later and interferes constructively with the continuing sound from the tuning fork. This interference forms a standing wave, and the air column resonates.

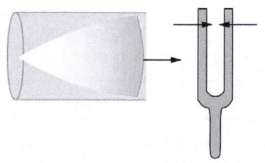

Figure 17.28 Resonance of air in a tube closed at one end, caused by a tuning fork. A graph of air displacement along the length of the tube shows none at the closed end, where the motion is constrained, and a maximum at the open end. This standing wave has one-fourth of its wavelength in the tube, so that $\lambda = 4L$.

The standing wave formed in the tube has its maximum air displacement (an **antinode**) at the open end, where motion is unconstrained, and no displacement (a **node**) at the closed end, where air movement is halted. The distance from a node to an antinode is one-fourth of a wavelength, and this equals the length of the tube; thus, $\lambda = 4L$. This same resonance can be produced by a vibration introduced at or near the closed end of the tube, as shown in **Figure 17.29**. It is best to consider this a natural vibration of the air column independently of how it is induced.

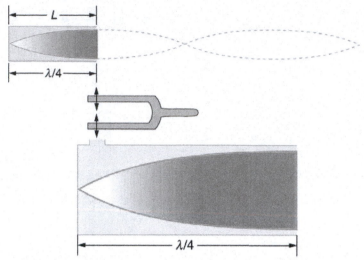

Figure 17.29 The same standing wave is created in the tube by a vibration introduced near its closed end.

Given that maximum air displacements are possible at the open end and none at the closed end, there are other, shorter wavelengths that can resonate in the tube, such as the one shown in **Figure 17.30**. Here the standing wave has three-fourths of its wavelength in the tube, or $L = (3/4)\lambda'$, so that $\lambda' = 4L/3$. Continuing this process reveals a whole series of shorter-

wavelength and higher-frequency sounds that resonate in the tube. We use specific terms for the resonances in any system. The lowest resonant frequency is called the **fundamental**, while all higher resonant frequencies are called **overtones**. All resonant frequencies are integral multiples of the fundamental, and they are collectively called **harmonics**. The fundamental is the first harmonic, the first overtone is the second harmonic, and so on. **Figure 17.31** shows the fundamental and the first three overtones (the first four harmonics) in a tube closed at one end.

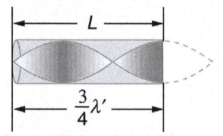

Figure 17.30 Another resonance for a tube closed at one end. This has maximum air displacements at the open end, and none at the closed end. The wavelength is shorter, with three-fourths λ' equaling the length of the tube, so that $\lambda' = 4L/3$. This higher-frequency vibration is the first overtone.

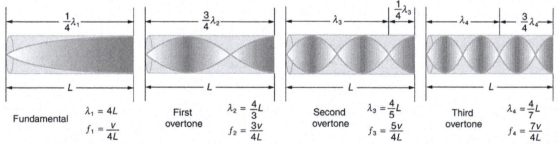

Figure 17.31 The fundamental and three lowest overtones for a tube closed at one end. All have maximum air displacements at the open end and none at the closed end.

The fundamental and overtones can be present simultaneously in a variety of combinations. For example, middle C on a trumpet has a sound distinctively different from middle C on a clarinet, both instruments being modified versions of a tube closed at one end. The fundamental frequency is the same (and usually the most intense), but the overtones and their mix of intensities are different and subject to shading by the musician. This mix is what gives various musical instruments (and human voices) their distinctive characteristics, whether they have air columns, strings, sounding boxes, or drumheads. In fact, much of our speech is determined by shaping the cavity formed by the throat and mouth and positioning the tongue to adjust the fundamental and combination of overtones. Simple resonant cavities can be made to resonate with the sound of the vowels, for example. (See **Figure 17.32**.) In boys, at puberty, the larynx grows and the shape of the resonant cavity changes giving rise to the difference in predominant frequencies in speech between men and women.

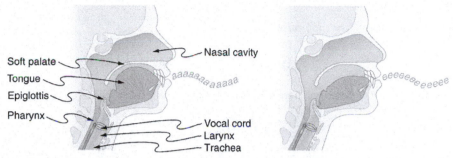

Figure 17.32 The throat and mouth form an air column closed at one end that resonates in response to vibrations in the voice box. The spectrum of overtones and their intensities vary with mouth shaping and tongue position to form different sounds. The voice box can be replaced with a mechanical vibrator, and understandable speech is still possible. Variations in basic shapes make different voices recognizable.

Now let us look for a pattern in the resonant frequencies for a simple tube that is closed at one end. The fundamental has $\lambda = 4L$, and frequency is related to wavelength and the speed of sound as given by:

$$v_\text{w} = f\lambda. \tag{17.28}$$

Solving for f in this equation gives

$$f = \frac{v_\text{w}}{\lambda} = \frac{v_\text{w}}{4L}, \tag{17.29}$$

where v_w is the speed of sound in air. Similarly, the first overtone has $\lambda' = 4L/3$ (see **Figure 17.31**), so that

$$f' = 3\frac{v_\text{w}}{4L} = 3f. \tag{17.30}$$

Because $f' = 3f$, we call the first overtone the third harmonic. Continuing this process, we see a pattern that can be generalized in a single expression. The resonant frequencies of a tube closed at one end are

$$f_n = n\frac{v_\text{w}}{4L}, n = 1,3,5, \tag{17.31}$$

where f_1 is the fundamental, f_3 is the first overtone, and so on. It is interesting that the resonant frequencies depend on the speed of sound and, hence, on temperature. This dependence poses a noticeable problem for organs in old unheated cathedrals, and it is also the reason why musicians commonly bring their wind instruments to room temperature before playing them.

Example 17.5 Find the Length of a Tube with a 128 Hz Fundamental

(a) What length should a tube closed at one end have on a day when the air temperature, is $22.0°\text{C}$, if its fundamental frequency is to be 128 Hz (C below middle C)?

(b) What Is the frequency of its fourth overtone?

Strategy

The length L can be found from the relationship in $f_n = n\frac{v_\text{w}}{4L}$, but we will first need to find the speed of sound v_w.

Solution for (a)

(1) Identify knowns:

- the fundamental frequency is 128 Hz
- the air temperature is $22.0°\text{C}$

(2) Use $f_n = n\frac{v_\text{w}}{4L}$ to find the fundamental frequency ($n = 1$).

$$f_1 = \frac{v_\text{w}}{4L} \tag{17.32}$$

(3) Solve this equation for length.

$$L = \frac{v_\text{w}}{4f_1} \tag{17.33}$$

(4) Find the speed of sound using $v_w = (331 \text{ m/s})\sqrt{\dfrac{T}{273 \text{ K}}}$.

$$v_w = (331 \text{ m/s})\sqrt{\frac{295 \text{ K}}{273 \text{ K}}} = 344 \text{ m/s} \tag{17.34}$$

(5) Enter the values of the speed of sound and frequency into the expression for L .

$$L = \frac{v_w}{4f_1} = \frac{344 \text{ m/s}}{4(128 \text{ Hz})} = 0.672 \text{ m} \tag{17.35}$$

Discussion on (a)

Many wind instruments are modified tubes that have finger holes, valves, and other devices for changing the length of the resonating air column and hence, the frequency of the note played. Horns producing very low frequencies, such as tubas, require tubes so long that they are coiled into loops.

Solution for (b)

(1) Identify knowns:

- the first overtone has $n = 3$
- the second overtone has $n = 5$
- the third overtone has $n = 7$
- the fourth overtone has $n = 9$

(2) Enter the value for the fourth overtone into $f_n = n\dfrac{v_w}{4L}$.

$$f_9 = 9\frac{v_w}{4L} = 9f_1 = 1.15 \text{ kHz} \tag{17.36}$$

Discussion on (b)

Whether this overtone occurs in a simple tube or a musical instrument depends on how it is stimulated to vibrate and the details of its shape. The trombone, for example, does not produce its fundamental frequency and only makes overtones.

Another type of tube is one that is *open* at both ends. Examples are some organ pipes, flutes, and oboes. The resonances of tubes open at both ends can be analyzed in a very similar fashion to those for tubes closed at one end. The air columns in tubes open at both ends have maximum air displacements at both ends, as illustrated in **Figure 17.33**. Standing waves form as shown.

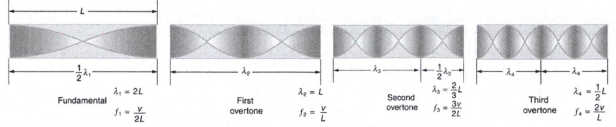

Figure 17.33 The resonant frequencies of a tube open at both ends are shown, including the fundamental and the first three overtones. In all cases the maximum air displacements occur at both ends of the tube, giving it different natural frequencies than a tube closed at one end.

Based on the fact that a tube open at both ends has maximum air displacements at both ends, and using **Figure 17.33** as a guide, we can see that the resonant frequencies of a tube open at both ends are:

$$f_n = n\frac{v_w}{2L}, \quad n = 1, 2, 3..., \tag{17.37}$$

where f_1 is the fundamental, f_2 is the first overtone, f_3 is the second overtone, and so on. Note that a tube open at both

ends has a fundamental frequency twice what it would have if closed at one end. It also has a different spectrum of overtones than a tube closed at one end. So if you had two tubes with the same fundamental frequency but one was open at both ends and the other was closed at one end, they would sound different when played because they have different overtones. Middle C, for example, would sound richer played on an open tube, because it has even multiples of the fundamental as well as odd. A closed tube has only odd multiples.

Applying the Science Practices: Closed- and Open-Ended Tubes

Strike an open-ended length of plastic pipe while holding it in the air. Now place one end of the pipe on a hard surface, sealing one opening, and strike it again. How does the sound change? Further investigate the sound created by the pipe by striking pipes of different lengths and composition.

Answer

When the pipe is placed on the ground, the standing wave within the pipe changes from being open on both ends to being closed on one end. As a result, the fundamental frequency will change from $f = \frac{v}{2L}$ to $f = \frac{v}{4L}$. This decrease in frequency results in a decrease in observed pitch.

Real-World Applications: Resonance in Everyday Systems

Resonance occurs in many different systems, including strings, air columns, and atoms. Resonance is the driven or forced oscillation of a system at its natural frequency. At resonance, energy is transferred rapidly to the oscillating system, and the amplitude of its oscillations grows until the system can no longer be described by Hooke's law. An example of this is the distorted sound intentionally produced in certain types of rock music.

Wind instruments use resonance in air columns to amplify tones made by lips or vibrating reeds. Other instruments also use air resonance in clever ways to amplify sound. **Figure 17.34** shows a violin and a guitar, both of which have sounding boxes but with different shapes, resulting in different overtone structures. The vibrating string creates a sound that resonates in the sounding box, greatly amplifying the sound and creating overtones that give the instrument its characteristic flavor. The more complex the shape of the sounding box, the greater its ability to resonate over a wide range of frequencies. The marimba, like the one shown in **Figure 17.35** uses pots or gourds below the wooden slats to amplify their tones. The resonance of the pot can be adjusted by adding water.

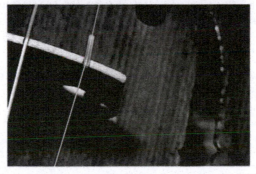

Figure 17.34 String instruments such as violins and guitars use resonance in their sounding boxes to amplify and enrich the sound created by their vibrating strings. The bridge and supports couple the string vibrations to the sounding boxes and air within. (credits: guitar, Feliciano Guimares, Fotopedia; violin, Steve Snodgrass, Flickr)

Figure 17.35 Resonance has been used in musical instruments since prehistoric times. This marimba uses gourds as resonance chambers to amplify its sound. (credit: APC Events, Flickr)

We have emphasized sound applications in our discussions of resonance and standing waves, but these ideas apply to any system that has wave characteristics. Vibrating strings, for example, are actually resonating and have fundamentals and overtones similar to those for air columns. More subtle are the resonances in atoms due to the wave character of their electrons. Their orbitals can be viewed as standing waves, which have a fundamental (ground state) and overtones (excited states). It is fascinating that wave characteristics apply to such a wide range of physical systems.

Check Your Understanding

Describe how noise-canceling headphones differ from standard headphones used to block outside sounds.

Solution
Regular headphones only block sound waves with a physical barrier. Noise-canceling headphones use destructive interference to reduce the loudness of outside sounds.

Check Your Understanding

How is it possible to use a standing wave's node and antinode to determine the length of a closed-end tube?

Solution
When the tube resonates at its natural frequency, the wave's node is located at the closed end of the tube, and the antinode is located at the open end. The length of the tube is equal to one-fourth of the wavelength of this wave. Thus, if we know the wavelength of the wave, we can determine the length of the tube.

PhET Explorations: Sound

This simulation lets you see sound waves. Adjust the frequency or volume and you can see and hear how the wave changes. Move the listener around and hear what she hears.

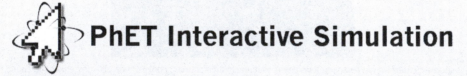

Figure 17.36 Sound (http://cnx.org/content/m55293/1.3/sound_en.jar)

Applying the Science Practices: Variables Affecting Superposition

In the PhET Interactive Simulation above, select the tab titled 'Two Source Interference.' Within this tab, manipulate the variables present (frequency, amplitude, and speaker separation) to investigate the relationship the variables have with the superposition pattern constructed on the screen. Record all observations.

Answer

As frequency is increased, the wavelength within the standing wave pattern will decrease. This results in an increase in nodes and antinodes, as represented in the applet by black and white surfaces. As amplitude is decreased, the contrast between black and white surfaces decreases, in demonstration of the decrease in sound level. There is no impact on sound wavelength, or number of nodes and antinodes shown. Increasing the speaker separation will affect the wavelength constructed. However, by separating the speakers, the number of nodes and antinodes within the applet will increase, as the waves are able to interfere over a greater distance.

17.6 Hearing

Learning Objectives
By the end of this section, you will be able to: • Define hearing, pitch, loudness, timbre, note, tone, phon, ultrasound, and infrasound. • Compare loudness to frequency and intensity of a sound. • Identify structures of the inner ear and explain how they relate to sound perception.

Figure 17.37 Hearing allows this vocalist, his band, and his fans to enjoy music. (credit: West Point Public Affairs, Flickr)

The human ear has a tremendous range and sensitivity. It can give us a wealth of simple information—such as pitch, loudness, and direction. And from its input we can detect musical quality and nuances of voiced emotion. How is our hearing related to the physical qualities of sound, and how does the hearing mechanism work?

Hearing is the perception of sound. (Perception is commonly defined to be awareness through the senses, a typically circular definition of higher-level processes in living organisms.) Normal human hearing encompasses frequencies from 20 to 20,000 Hz, an impressive range. Sounds below 20 Hz are called **infrasound**, whereas those above 20,000 Hz are **ultrasound**. Neither is perceived by the ear, although infrasound can sometimes be felt as vibrations. When we do hear low-frequency vibrations, such as the sounds of a diving board, we hear the individual vibrations only because there are higher-frequency sounds in each. Other animals have hearing ranges different from that of humans. Dogs can hear sounds as high as 30,000 Hz, whereas bats and dolphins can hear up to 100,000-Hz sounds. You may have noticed that dogs respond to the sound of a dog whistle which produces sound out of the range of human hearing. Elephants are known to respond to frequencies below 20 Hz.

The perception of frequency is called **pitch**. Most of us have excellent relative pitch, which means that we can tell whether one sound has a different frequency from another. Typically, we can discriminate between two sounds if their frequencies differ by 0.3% or more. For example, 500.0 and 501.5 Hz are noticeably different. Pitch perception is directly related to frequency and is not greatly affected by other physical quantities such as intensity. Musical **notes** are particular sounds that can be produced by most instruments and in Western music have particular names. Combinations of notes constitute music. Some people can identify musical notes, such as A-sharp, C, or E-flat, just by listening to them. This uncommon ability is called perfect pitch.

The ear is remarkably sensitive to low-intensity sounds. The lowest audible intensity or threshold is about 10^{-12} W/m^2 or 0 dB. Sounds as much as 10^{12} more intense can be briefly tolerated. Very few measuring devices are capable of observations over a range of a trillion. The perception of intensity is called **loudness**. At a given frequency, it is possible to discern differences of about 1 dB, and a change of 3 dB is easily noticed. But loudness is not related to intensity alone. Frequency has a major effect on how loud a sound seems. The ear has its maximum sensitivity to frequencies in the range of 2000 to 5000 Hz, so that sounds in this range are perceived as being louder than, say, those at 500 or 10,000 Hz, even when they all have the same intensity. Sounds near the high- and low-frequency extremes of the hearing range seem even less loud, because the ear is even less sensitive at those frequencies. **Table 17.7** gives the dependence of certain human hearing perceptions on physical quantities.

Table 17.7 Sound Perceptions

Perception	Physical quantity
Pitch	Frequency
Loudness	Intensity and Frequency
Timbre	Number and relative intensity of multiple frequencies. Subtle craftsmanship leads to non-linear effects and more detail.
Note	Basic unit of music with specific names, combined to generate tunes
Tone	Number and relative intensity of multiple frequencies.

When a violin plays middle C, there is no mistaking it for a piano playing the same note. The reason is that each instrument produces a distinctive set of frequencies and intensities. We call our perception of these combinations of frequencies and intensities **tone** quality, or more commonly the **timbre** of the sound. It is more difficult to correlate timbre perception to physical quantities than it is for loudness or pitch perception. Timbre is more subjective. Terms such as dull, brilliant, warm, cold, pure, and rich are employed to describe the timbre of a sound. So the consideration of timbre takes us into the realm of perceptual psychology, where higher-level processes in the brain are dominant. This is true for other perceptions of sound, such as music and noise. We shall not delve further into them; rather, we will concentrate on the question of loudness perception.

A unit called a **phon** is used to express loudness numerically. Phons differ from decibels because the phon is a unit of loudness perception, whereas the decibel is a unit of physical intensity. **Figure 17.38** shows the relationship of loudness to intensity (or intensity level) and frequency for persons with normal hearing. The curved lines are equal-loudness curves. Each curve is labeled with its loudness in phons. Any sound along a given curve will be perceived as equally loud by the average person. The curves were determined by having large numbers of people compare the loudness of sounds at different frequencies and sound intensity levels. At a frequency of 1000 Hz, phons are taken to be numerically equal to decibels. The following example helps illustrate how to use the graph:

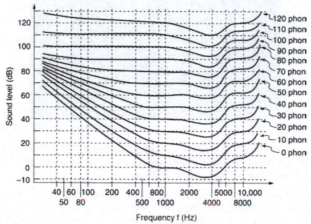

Figure 17.38 The relationship of loudness in phons to intensity level (in decibels) and intensity (in watts per meter squared) for persons with normal hearing. The curved lines are equal-loudness curves—all sounds on a given curve are perceived as equally loud. Phons and decibels are defined to be the same at 1000 Hz.

Example 17.6 Measuring Loudness: Loudness Versus Intensity Level and Frequency

(a) What is the loudness in phons of a 100-Hz sound that has an intensity level of 80 dB? (b) What is the intensity level in decibels of a 4000-Hz sound having a loudness of 70 phons? (c) At what intensity level will an 8000-Hz sound have the same loudness as a 200-Hz sound at 60 dB?

Strategy for (a)

The graph in **Figure 17.38** should be referenced in order to solve this example. To find the loudness of a given sound, you must know its frequency and intensity level and locate that point on the square grid, then interpolate between loudness curves to get the loudness in phons.

Solution for (a)

(1) Identify knowns:

- The square grid of the graph relating phons and decibels is a plot of intensity level versus frequency—both physical quantities.
- 100 Hz at 80 dB lies halfway between the curves marked 70 and 80 phons.

(2) Find the loudness: 75 phons.

Strategy for (b)

The graph in **Figure 17.38** should be referenced in order to solve this example. To find the intensity level of a sound, you must have its frequency and loudness. Once that point is located, the intensity level can be determined from the vertical axis.

Solution for (b)

(1) Identify knowns:

- Values are given to be 4000 Hz at 70 phons.

(2) Follow the 70-phon curve until it reaches 4000 Hz. At that point, it is below the 70 dB line at about 67 dB.

(3) Find the intensity level:

67 dB

Strategy for (c)

The graph in **Figure 17.38** should be referenced in order to solve this example.

Solution for (c)

(1) Locate the point for a 200 Hz and 60 dB sound.

(2) Find the loudness: This point lies just slightly above the 50-phon curve, and so its loudness is 51 phons.

(3) Look for the 51-phon level is at 8000 Hz: 63 dB.

Discussion

These answers, like all information extracted from **Figure 17.38**, have uncertainties of several phons or several decibels, partly due to difficulties in interpolation, but mostly related to uncertainties in the equal-loudness curves.

Further examination of the graph in **Figure 17.38** reveals some interesting facts about human hearing. First, sounds below the 0-phon curve are not perceived by most people. So, for example, a 60 Hz sound at 40 dB is inaudible. The 0-phon curve represents the threshold of normal hearing. We can hear some sounds at intensity levels below 0 dB. For example, a 3-dB, 5000-Hz sound is audible, because it lies above the 0-phon curve. The loudness curves all have dips in them between about 2000 and 5000 Hz. These dips mean the ear is most sensitive to frequencies in that range. For example, a 15-dB sound at 4000 Hz has a loudness of 20 phons, the same as a 20-dB sound at 1000 Hz. The curves rise at both extremes of the frequency range, indicating that a greater-intensity level sound is needed at those frequencies to be perceived to be as loud as at middle frequencies. For example, a sound at 10,000 Hz must have an intensity level of 30 dB to seem as loud as a 20 dB sound at 1000 Hz. Sounds above 120 phons are painful as well as damaging.

We do not often utilize our full range of hearing. This is particularly true for frequencies above 8000 Hz, which are rare in the environment and are unnecessary for understanding conversation or appreciating music. In fact, people who have lost the ability to hear such high frequencies are usually unaware of their loss until tested. The shaded region in **Figure 17.39** is the frequency and intensity region where most conversational sounds fall. The curved lines indicate what effect hearing losses of 40 and 60 phons will have. A 40-phon hearing loss at all frequencies still allows a person to understand conversation, although it will seem very quiet. A person with a 60-phon loss at all frequencies will hear only the lowest frequencies and will not be able to understand speech unless it is much louder than normal. Even so, speech may seem indistinct, because higher frequencies are not as well perceived. The conversational speech region also has a gender component, in that female voices are usually characterized by higher frequencies. So the person with a 60-phon hearing impediment might have difficulty understanding the normal conversation of a woman.

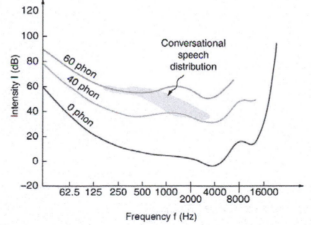

Figure 17.39 The shaded region represents frequencies and intensity levels found in normal conversational speech. The 0-phon line represents the normal hearing threshold, while those at 40 and 60 represent thresholds for people with 40- and 60-phon hearing losses, respectively.

Hearing tests are performed over a range of frequencies, usually from 250 to 8000 Hz, and can be displayed graphically in an audiogram like that in **Figure 17.40**. The hearing threshold is measured in dB *relative to the normal threshold*, so that normal hearing registers as 0 dB at all frequencies. Hearing loss caused by noise typically shows a dip near the 4000 Hz frequency, irrespective of the frequency that caused the loss and often affects both ears. The most common form of hearing loss comes with age and is called *presbycusis*—literally elder ear. Such loss is increasingly severe at higher frequencies, and interferes with music appreciation and speech recognition.

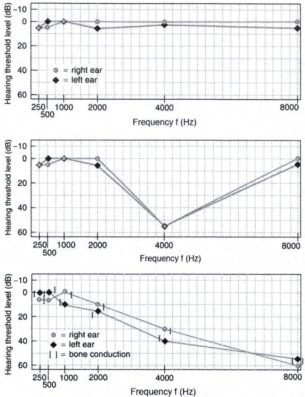

Figure 17.40 Audiograms showing the threshold in intensity level versus frequency for three different individuals. Intensity level is measured relative to the normal threshold. The top left graph is that of a person with normal hearing. The graph to its right has a dip at 4000 Hz and is that of a child who suffered hearing loss due to a cap gun. The third graph is typical of presbycusis, the progressive loss of higher frequency hearing with age. Tests performed by bone conduction (brackets) can distinguish nerve damage from middle ear damage.

The Hearing Mechanism

The hearing mechanism involves some interesting physics. The sound wave that impinges upon our ear is a pressure wave. The ear is a transducer that converts sound waves into electrical nerve impulses in a manner much more sophisticated than, but analogous to, a microphone. **Figure 17.41** shows the gross anatomy of the ear with its division into three parts: the outer ear or ear canal; the middle ear, which runs from the eardrum to the cochlea; and the inner ear, which is the cochlea itself. The body part normally referred to as the ear is technically called the pinna.

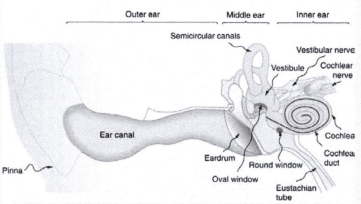

Figure 17.41 The illustration shows the gross anatomy of the human ear.

The outer ear, or ear canal, carries sound to the recessed protected eardrum. The air column in the ear canal resonates and is partially responsible for the sensitivity of the ear to sounds in the 2000 to 5000 Hz range. The middle ear converts sound into mechanical vibrations and applies these vibrations to the cochlea. The lever system of the middle ear takes the force exerted on the eardrum by sound pressure variations, amplifies it and transmits it to the inner ear via the oval window, creating pressure waves in the cochlea approximately 40 times greater than those impinging on the eardrum. (See **Figure 17.42**.) Two muscles in the middle ear (not shown) protect the inner ear from very intense sounds. They react to intense sound in a few milliseconds and reduce the force transmitted to the cochlea. This protective reaction can also be triggered by your own voice, so that humming while shooting a gun, for example, can reduce noise damage.

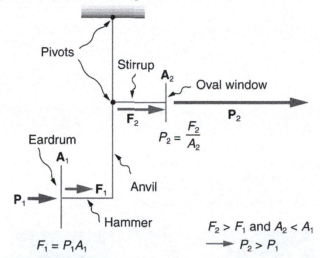

Figure 17.42 This schematic shows the middle ear's system for converting sound pressure into force, increasing that force through a lever system, and applying the increased force to a small area of the cochlea, thereby creating a pressure about 40 times that in the original sound wave. A protective muscle reaction to intense sounds greatly reduces the mechanical advantage of the lever system.

Figure 17.43 shows the middle and inner ear in greater detail. Pressure waves moving through the cochlea cause the tectorial membrane to vibrate, rubbing cilia (called hair cells), which stimulate nerves that send electrical signals to the brain. The membrane resonates at different positions for different frequencies, with high frequencies stimulating nerves at the near end and low frequencies at the far end. The complete operation of the cochlea is still not understood, but several mechanisms for sending information to the brain are known to be involved. For sounds below about 1000 Hz, the nerves send signals at the same frequency as the sound. For frequencies greater than about 1000 Hz, the nerves signal frequency by position. There is a structure to the cilia, and there are connections between nerve cells that perform signal processing before information is sent to the brain. Intensity information is partly indicated by the number of nerve signals and by volleys of signals. The brain processes the cochlear nerve signals to provide additional information such as source direction (based on time and intensity comparisons of sounds from both ears). Higher-level processing produces many nuances, such as music appreciation.

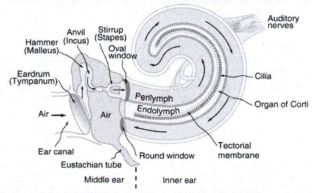

Figure 17.43 The inner ear, or cochlea, is a coiled tube about 3 mm in diameter and 3 cm in length if uncoiled. When the oval window is forced inward, as shown, a pressure wave travels through the perilymph in the direction of the arrows, stimulating nerves at the base of cilia in the organ of Corti.

Hearing losses can occur because of problems in the middle or inner ear. Conductive losses in the middle ear can be partially overcome by sending sound vibrations to the cochlea through the skull. Hearing aids for this purpose usually press against the bone behind the ear, rather than simply amplifying the sound sent into the ear canal as many hearing aids do. Damage to the nerves in the cochlea is not repairable, but amplification can partially compensate. There is a risk that amplification will produce further damage. Another common failure in the cochlea is damage or loss of the cilia but with nerves remaining functional. Cochlear implants that stimulate the nerves directly are now available and widely accepted. Over 100,000 implants are in use, in about equal numbers of adults and children.

The cochlear implant was pioneered in Melbourne, Australia, by Graeme Clark in the 1970s for his deaf father. The implant consists of three external components and two internal components. The external components are a microphone for picking up sound and converting it into an electrical signal, a speech processor to select certain frequencies and a transmitter to transfer the signal to the internal components through electromagnetic induction. The internal components consist of a receiver/transmitter secured in the bone beneath the skin, which converts the signals into electric impulses and sends them through an internal cable to the cochlea and an array of about 24 electrodes wound through the cochlea. These electrodes in turn send the impulses directly into the brain. The electrodes basically emulate the cilia.

Are ultrasound and infrasound imperceptible to all hearing organisms? Explain your answer.

Solution
No, the range of perceptible sound is based in the range of human hearing. Many other organisms perceive either infrasound or ultrasound.

17.7 Ultrasound

Learning Objectives

By the end of this section, you will be able to:

- Define acoustic impedance and intensity reflection coefficient.
- Describe medical and other uses of ultrasound technology.
- Calculate acoustic impedance using density values and the speed of ultrasound.
- Calculate the velocity of a moving object using Doppler-shifted ultrasound.

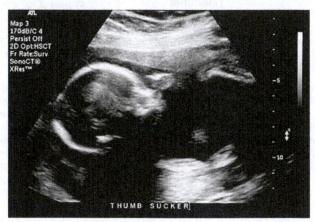

Figure 17.44 Ultrasound is used in medicine to painlessly and noninvasively monitor patient health and diagnose a wide range of disorders. (credit: abbybatchelder, Flickr)

Any sound with a frequency above 20,000 Hz (or 20 kHz)—that is, above the highest audible frequency—is defined to be ultrasound. In practice, it is possible to create ultrasound frequencies up to more than a gigahertz. (Higher frequencies are difficult to create; furthermore, they propagate poorly because they are very strongly absorbed.) Ultrasound has a tremendous number of applications, which range from burglar alarms to use in cleaning delicate objects to the guidance systems of bats. We begin our discussion of ultrasound with some of its applications in medicine, in which it is used extensively both for diagnosis and for therapy.

Characteristics of Ultrasound

The characteristics of ultrasound, such as frequency and intensity, are wave properties common to all types of waves. Ultrasound also has a wavelength that limits the fineness of detail it can detect. This characteristic is true of all waves. We can never observe details significantly smaller than the wavelength of our probe; for example, we will never see individual atoms with visible light, because the atoms are so small compared with the wavelength of light.

Ultrasound in Medical Therapy

Ultrasound, like any wave, carries energy that can be absorbed by the medium carrying it, producing effects that vary with intensity. When focused to intensities of 10^3 to 10^5 W/m^2, ultrasound can be used to shatter gallstones or pulverize cancerous tissue in surgical procedures. (See **Figure 17.45**.) Intensities this great can damage individual cells, variously causing their protoplasm to stream inside them, altering their permeability, or rupturing their walls through *cavitation*. Cavitation is the creation of vapor cavities in a fluid—the longitudinal vibrations in ultrasound alternatively compress and expand the medium, and at sufficient amplitudes the expansion separates molecules. Most cavitation damage is done when the cavities collapse, producing even greater shock pressures.

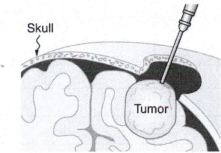

Figure 17.45 The tip of this small probe oscillates at 23 kHz with such a large amplitude that it pulverizes tissue on contact. The debris is then aspirated. The speed of the tip may exceed the speed of sound in tissue, thus creating shock waves and cavitation, rather than a smooth simple harmonic oscillator–type wave.

Most of the energy carried by high-intensity ultrasound in tissue is converted to thermal energy. In fact, intensities of 10^3 to 10^4 W/m^2 are commonly used for deep-heat treatments called ultrasound diathermy. Frequencies of 0.8 to 1 MHz are typical. In both athletics and physical therapy, ultrasound diathermy is most often applied to injured or overworked muscles to relieve pain and improve flexibility. Skill is needed by the therapist to avoid "bone burns" and other tissue damage caused by overheating and cavitation, sometimes made worse by reflection and focusing of the ultrasound by joint and bone tissue.

In some instances, you may encounter a different decibel scale, called the sound *pressure* level, when ultrasound travels in water or in human and other biological tissues. We shall not use the scale here, but it is notable that numbers for sound pressure levels range 60 to 70 dB higher than you would quote for β, the sound intensity level used in this text. Should you encounter a sound pressure level of 220 decibels, then, it is not an astronomically high intensity, but equivalent to about 155 dB—high enough to destroy tissue, but not as unreasonably high as it might seem at first.

Ultrasound in Medical Diagnostics

When used for imaging, ultrasonic waves are emitted from a transducer, a crystal exhibiting the piezoelectric effect (the expansion and contraction of a substance when a voltage is applied across it, causing a vibration of the crystal). These high-frequency vibrations are transmitted into any tissue in contact with the transducer. Similarly, if a pressure is applied to the crystal (in the form of a wave reflected off tissue layers), a voltage is produced which can be recorded. The crystal therefore acts as both a transmitter and a receiver of sound. Ultrasound is also partially absorbed by tissue on its path, both on its journey away from the transducer and on its return journey. From the time between when the original signal is sent and when the reflections from various boundaries between media are received, (as well as a measure of the intensity loss of the signal), the nature and position of each boundary between tissues and organs may be deduced.

Reflections at boundaries between two different media occur because of differences in a characteristic known as the **acoustic impedance** Z of each substance. Impedance is defined as

$$Z = \rho v, \tag{17.38}$$

where ρ is the density of the medium (in kg/m^3) and v is the speed of sound through the medium (in m/s). The units for Z are therefore $\text{kg/(m}^2 \cdot \text{s)}$.

Table 17.8 shows the density and speed of sound through various media (including various soft tissues) and the associated acoustic impedances. Note that the acoustic impedances for soft tissue do not vary much but that there is a big difference between the acoustic impedance of soft tissue and air and also between soft tissue and bone.

Table 17.8 The Ultrasound Properties of Various Media, Including Soft Tissue Found in the Body

Medium	Density (kg/m³)	Speed of Ultrasound (m/s)	Acoustic Impedance $\left(\mathrm{kg}/\left(\mathrm{m}^2 \cdot \mathrm{s}\right)\right)$
Air	1.3	330	429
Water	1000	1500	1.5×10^6
Blood	1060	1570	1.66×10^6
Fat	925	1450	1.34×10^6
Muscle (average)	1075	1590	1.70×10^6
Bone (varies)	1400–1900	4080	5.7×10^6 to 7.8×10^6
Barium titanate (transducer material)	5600	5500	30.8×10^6

At the boundary between media of different acoustic impedances, some of the wave energy is reflected and some is transmitted. The greater the *difference* in acoustic impedance between the two media, the greater the reflection and the smaller the transmission.

The **intensity reflection coefficient** a is defined as the ratio of the intensity of the reflected wave relative to the incident (transmitted) wave. This statement can be written mathematically as

$$a = \frac{(Z_2 - Z_1)^2}{(Z_1 + Z_2)^2},$$

(17.39)

where Z_1 and Z_2 are the acoustic impedances of the two media making up the boundary. A reflection coefficient of zero (corresponding to total transmission and no reflection) occurs when the acoustic impedances of the two media are the same. An impedance "match" (no reflection) provides an efficient coupling of sound energy from one medium to another. The image formed in an ultrasound is made by tracking reflections (as shown in **Figure 17.46**) and mapping the intensity of the reflected sound waves in a two-dimensional plane.

Example 17.7 Calculate Acoustic Impedance and Intensity Reflection Coefficient: Ultrasound and Fat Tissue

(a) Using the values for density and the speed of ultrasound given in **Table 17.8**, show that the acoustic impedance of fat tissue is indeed 1.34×10^6 kg/(m²·s).

(b) Calculate the intensity reflection coefficient of ultrasound when going from fat to muscle tissue.

Strategy for (a)

The acoustic impedance can be calculated using $Z = \rho v$ and the values for ρ and v found in **Table 17.8**.

Solution for (a)

(1) Substitute known values from **Table 17.8** into $Z = \rho v$.

$$Z = \rho v = \left(925 \text{ kg/m}^3\right)(1450 \text{ m/s})$$

(17.40)

(2) Calculate to find the acoustic impedance of fat tissue.

$$1.34 \times 10^6 \text{ kg/(m}^2 \cdot \text{s)}$$

(17.41)

This value is the same as the value given for the acoustic impedance of fat tissue.

Strategy for (b)

The intensity reflection coefficient for any boundary between two media is given by $a = \frac{(Z_2 - Z_1)^2}{(Z_1 + Z_2)^2}$, and the acoustic impedance of muscle is given in **Table 17.8**.

Solution for (b)

Substitute known values into $a = \dfrac{(Z_2 - Z_1)^2}{(Z_1 + Z_2)^2}$ to find the intensity reflection coefficient:

$$a = \frac{(Z_2 - Z_1)^2}{(Z_1 + Z_2)^2} = \frac{\left(1.34 \times 10^6 \text{ kg/(m}^2 \cdot \text{s)} - 1.70 \times 10^6 \text{ kg/(m}^2 \cdot \text{s)}\right)^2}{\left(1.70 \times 10^6 \text{ kg/(m}^2 \cdot \text{s)} + 1.34 \times 10^6 \text{ kg/(m}^2 \cdot \text{s)}\right)^2} = 0.014 \tag{17.42}$$

Discussion

This result means that only 1.4% of the incident intensity is reflected, with the remaining being transmitted.

The applications of ultrasound in medical diagnostics have produced untold benefits with no known risks. Diagnostic intensities are too low (about 10^{-2} W/m^2) to cause thermal damage. More significantly, ultrasound has been in use for several decades and detailed follow-up studies do not show evidence of ill effects, quite unlike the case for x-rays.

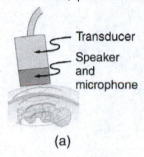

(a)

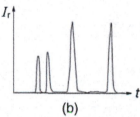

(b)

Figure 17.46 (a) An ultrasound speaker doubles as a microphone. Brief bleeps are broadcast, and echoes are recorded from various depths. (b) Graph of echo intensity versus time. The time for echoes to return is directly proportional to the distance of the reflector, yielding this information noninvasively.

The most common ultrasound applications produce an image like that shown in **Figure 17.47**. The speaker-microphone broadcasts a directional beam, sweeping the beam across the area of interest. This is accomplished by having multiple ultrasound sources in the probe's head, which are phased to interfere constructively in a given, adjustable direction. Echoes are measured as a function of position as well as depth. A computer constructs an image that reveals the shape and density of internal structures.

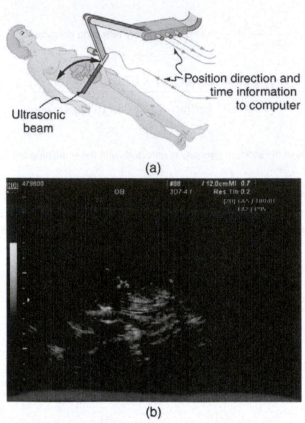

(a)

(b)

Figure 17.47 (a) An ultrasonic image is produced by sweeping the ultrasonic beam across the area of interest, in this case the woman's abdomen. Data are recorded and analyzed in a computer, providing a two-dimensional image. (b) Ultrasound image of 12-week-old fetus. (credit: Margaret W. Carruthers, Flickr)

How much detail can ultrasound reveal? The image in **Figure 17.47** is typical of low-cost systems, but that in **Figure 17.48** shows the remarkable detail possible with more advanced systems, including 3D imaging. Ultrasound today is commonly used in prenatal care. Such imaging can be used to see if the fetus is developing at a normal rate, and help in the determination of serious problems early in the pregnancy. Ultrasound is also in wide use to image the chambers of the heart and the flow of blood within the beating heart, using the Doppler effect (echocardiology).

Whenever a wave is used as a probe, it is very difficult to detect details smaller than its wavelength λ. Indeed, current technology cannot do quite this well. Abdominal scans may use a 7-MHz frequency, and the speed of sound in tissue is about 1540 m/s—so the wavelength limit to detail would be $\lambda = \frac{v_w}{f} = \frac{1540 \text{ m/s}}{7 \times 10^6 \text{ Hz}} = 0.22 \text{ mm}$. In practice, 1-mm detail is attainable, which is sufficient for many purposes. Higher-frequency ultrasound would allow greater detail, but it does not penetrate as well as lower frequencies do. The accepted rule of thumb is that you can effectively scan to a depth of about 500λ into tissue. For 7 MHz, this penetration limit is $500 \times 0.22 \text{ mm}$, which is 0.11 m. Higher frequencies may be employed in smaller organs, such as the eye, but are not practical for looking deep into the body.

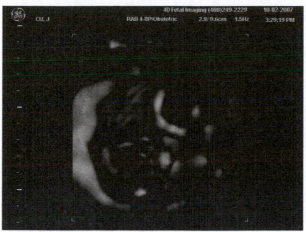

Figure 17.48 A 3D ultrasound image of a fetus. As well as for the detection of any abnormalities, such scans have also been shown to be useful for strengthening the emotional bonding between parents and their unborn child. (credit: Jennie Cu, Wikimedia Commons)

In addition to shape information, ultrasonic scans can produce density information superior to that found in X-rays, because the intensity of a reflected sound is related to changes in density. Sound is most strongly reflected at places where density changes are greatest.

Another major use of ultrasound in medical diagnostics is to detect motion and determine velocity through the Doppler shift of an echo, known as **Doppler-shifted ultrasound**. This technique is used to monitor fetal heartbeat, measure blood velocity, and detect occlusions in blood vessels, for example. (See **Figure 17.49**.) The magnitude of the Doppler shift in an echo is directly proportional to the velocity of whatever reflects the sound. Because an echo is involved, there is actually a double shift. The first occurs because the reflector (say a fetal heart) is a moving observer and receives a Doppler-shifted frequency. The reflector then acts as a moving source, producing a second Doppler shift.

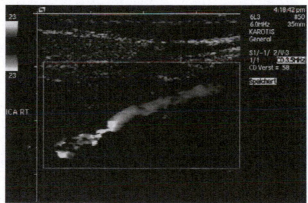

Figure 17.49 This Doppler-shifted ultrasonic image of a partially occluded artery uses color to indicate velocity. The highest velocities are in red, while the lowest are blue. The blood must move faster through the constriction to carry the same flow. (credit: Arning C, Grzyska U, Wikimedia Commons)

A clever technique is used to measure the Doppler shift in an echo. The frequency of the echoed sound is superimposed on the broadcast frequency, producing beats. The beat frequency is $F_B = |f_1 - f_2|$, and so it is directly proportional to the Doppler shift ($f_1 - f_2$) and hence, the reflector's velocity. The advantage in this technique is that the Doppler shift is small (because the reflector's velocity is small), so that great accuracy would be needed to measure the shift directly. But measuring the beat frequency is easy, and it is not affected if the broadcast frequency varies somewhat. Furthermore, the beat frequency is in the audible range and can be amplified for audio feedback to the medical observer.

Uses for Doppler-Shifted Radar

Doppler-shifted radar echoes are used to measure wind velocities in storms as well as aircraft and automobile speeds. The principle is the same as for Doppler-shifted ultrasound. There is evidence that bats and dolphins may also sense the velocity of an object (such as prey) reflecting their ultrasound signals by observing its Doppler shift.

Example 17.8 Calculate Velocity of Blood: Doppler-Shifted Ultrasound

Ultrasound that has a frequency of 2.50 MHz is sent toward blood in an artery that is moving toward the source at 20.0 cm/s, as illustrated in **Figure 17.50**. Use the speed of sound in human tissue as 1540 m/s. (Assume that the frequency of 2.50 MHz is accurate to seven significant figures.)

a. What frequency does the blood receive?

b. What frequency returns to the source?

c. What beat frequency is produced if the source and returning frequencies are mixed?

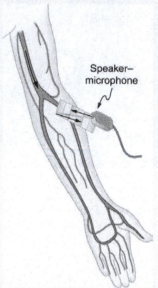

Figure 17.50 Ultrasound is partly reflected by blood cells and plasma back toward the speaker-microphone. Because the cells are moving, two Doppler shifts are produced—one for blood as a moving observer, and the other for the reflected sound coming from a moving source. The magnitude of the shift is directly proportional to blood velocity.

Strategy

The first two questions can be answered using $f_{obs} = f_s\left(\dfrac{v_w}{v_w \pm v_s}\right)$ and $f_{obs} = f_s\left(\dfrac{v_w \pm v_{obs}}{v_w}\right)$ for the Doppler shift.

The last question asks for beat frequency, which is the difference between the original and returning frequencies.

Solution for (a)

(1) Identify knowns:

*   The blood is a moving observer, and so the frequency it receives is given by

$$f_{obs} = f_s\left(\frac{v_w \pm v_{obs}}{v_w}\right). \tag{17.43}$$

* v_b is the blood velocity (v_{obs} here) and the plus sign is chosen because the motion is toward the source.

(2) Enter the given values into the equation.

$$f_{obs} = (2{,}500{,}000 \text{ Hz})\left(\frac{1540 \text{ m/s} + 0.2 \text{ m/s}}{1540 \text{ m/s}}\right) \tag{17.44}$$

(3) Calculate to find the frequency: 2,500,325 Hz.

Solution for (b)

(1) Identify knowns:

*   The blood acts as a moving source.
*   The microphone acts as a stationary observer.
*   The frequency leaving the blood is 2,500,325 Hz, but it is shifted upward as given by

$$f_{obs} = f_s\left(\frac{v_w}{v_w - v_b}\right). \tag{17.45}$$

f_{obs} is the frequency received by the speaker-microphone.

*   The source velocity is v_b.

*   The minus sign is used because the motion is toward the observer.

The minus sign is used because the motion is toward the observer.

(2) Enter the given values into the equation:

$$f_{obs} = (2{,}500{,}325 \text{ Hz})\left(\frac{1540 \text{ m/s}}{1540 \text{ m/s} - 0.200 \text{ m/s}}\right) \tag{17.46}$$

(3) Calculate to find the frequency returning to the source: 2,500,649 Hz.

Solution for (c)

(1) Identify knowns:

- The beat frequency is simply the absolute value of the difference between f_s and f_{obs}, as stated in:

$$f_B = |f_{obs} - f_s|. \tag{17.47}$$

(2) Substitute known values:

$$|2,500,649 \text{ Hz} - 2,500,000 \text{ Hz}| \tag{17.48}$$

(3) Calculate to find the beat frequency: 649 Hz.

Discussion

The Doppler shifts are quite small compared with the original frequency of 2.50 MHz. It is far easier to measure the beat frequency than it is to measure the echo frequency with an accuracy great enough to see shifts of a few hundred hertz out of a couple of megahertz. Furthermore, variations in the source frequency do not greatly affect the beat frequency, because both f_s and f_{obs} would increase or decrease. Those changes subtract out in $f_B = |f_{obs} - f_s|$.

Industrial and Other Applications of Ultrasound

Industrial, retail, and research applications of ultrasound are common. A few are discussed here. Ultrasonic cleaners have many uses. Jewelry, machined parts, and other objects that have odd shapes and crevices are immersed in a cleaning fluid that is agitated with ultrasound typically about 40 kHz in frequency. The intensity is great enough to cause cavitation, which is responsible for most of the cleansing action. Because cavitation-produced shock pressures are large and well transmitted in a fluid, they reach into small crevices where even a low-surface-tension cleaning fluid might not penetrate.

Sonar is a familiar application of ultrasound. Sonar typically employs ultrasonic frequencies in the range from 30.0 to 100 kHz. Bats, dolphins, submarines, and even some birds use ultrasonic sonar. Echoes are analyzed to give distance and size information both for guidance and finding prey. In most sonar applications, the sound reflects quite well because the objects of interest have significantly different density than the medium in which they travel. When the Doppler shift is observed, velocity information can also be obtained. Submarine sonar can be used to obtain such information, and there is evidence that some bats also sense velocity from their echoes.

Similarly, there are a range of relatively inexpensive devices that measure distance by timing ultrasonic echoes. Many cameras, for example, use such information to focus automatically. Some doors open when their ultrasonic ranging devices detect a nearby object, and certain home security lights turn on when their ultrasonic rangers observe motion. Ultrasonic "measuring tapes" also exist to measure such things as room dimensions. Sinks in public restrooms are sometimes automated with ultrasound devices to turn faucets on and off when people wash their hands. These devices reduce the spread of germs and can conserve water.

Ultrasound is used for nondestructive testing in industry and by the military. Because ultrasound reflects well from any large change in density, it can reveal cracks and voids in solids, such as aircraft wings, that are too small to be seen with x-rays. For similar reasons, ultrasound is also good for measuring the thickness of coatings, particularly where there are several layers involved.

Basic research in solid state physics employs ultrasound. Its attenuation is related to a number of physical characteristics, making it a useful probe. Among these characteristics are structural changes such as those found in liquid crystals, the transition of a material to a superconducting phase, as well as density and other properties.

These examples of the uses of ultrasound are meant to whet the appetites of the curious, as well as to illustrate the underlying physics of ultrasound. There are many more applications, as you can easily discover for yourself.

Check Your Understanding

Why is it possible to use ultrasound both to observe a fetus in the womb and also to destroy cancerous tumors in the body?

Solution

Ultrasound can be used medically at different intensities. Lower intensities do not cause damage and are used for medical imaging. Higher intensities can pulverize and destroy targeted substances in the body, such as tumors.

Glossary

acoustic impedance: property of medium that makes the propagation of sound waves more difficult

antinode: point of maximum displacement

bow wake: V-shaped disturbance created when the wave source moves faster than the wave propagation speed

Doppler effect: an alteration in the observed frequency of a sound due to motion of either the source or the observer

Doppler shift: the actual change in frequency due to relative motion of source and observer

Doppler-shifted ultrasound: a medical technique to detect motion and determine velocity through the Doppler shift of an echo

fundamental: the lowest-frequency resonance

harmonics: the term used to refer collectively to the fundamental and its overtones

hearing: the perception of sound

infrasound: sounds below 20 Hz

intensity: the power per unit area carried by a wave

intensity reflection coefficient: a measure of the ratio of the intensity of the wave reflected off a boundary between two media relative to the intensity of the incident wave

loudness: the perception of sound intensity

node: point of zero displacement

note: basic unit of music with specific names, combined to generate tunes

overtones: all resonant frequencies higher than the fundamental

phon: the numerical unit of loudness

pitch: the perception of the frequency of a sound

sonic boom: a constructive interference of sound created by an object moving faster than sound

sound: a disturbance of matter that is transmitted from its source outward

sound intensity level: a unitless quantity telling you the level of the sound relative to a fixed standard

sound pressure level: the ratio of the pressure amplitude to a reference pressure

timbre: number and relative intensity of multiple sound frequencies

tone: number and relative intensity of multiple sound frequencies

ultrasound: sounds above 20,000 Hz

Section Summary

17.1 Sound
- Sound is a disturbance of matter that is transmitted from its source outward.
- Sound is one type of wave.
- Hearing is the perception of sound.

17.2 Speed of Sound, Frequency, and Wavelength

The relationship of the speed of sound v_w, its frequency f, and its wavelength λ is given by

$$v_w = f\lambda,$$

which is the same relationship given for all waves.

In air, the speed of sound is related to air temperature T by

$$v_w = (331 \text{ m/s})\sqrt{\frac{T}{273 \text{ K}}}.$$

v_w is the same for all frequencies and wavelengths.

17.3 Sound Intensity and Sound Level
- Intensity is the same for a sound wave as was defined for all waves; it is

$$I = \frac{P}{A},$$

where P is the power crossing area A. The SI unit for I is watts per meter squared. The intensity of a sound wave is also related to the pressure amplitude Δp

$$I = \frac{(\Delta p)^2}{2\rho v_{\mathrm{w}}},$$

where ρ is the density of the medium in which the sound wave travels and v_{w} is the speed of sound in the medium.

- Sound intensity level in units of decibels (dB) is

$$\beta \ (\mathrm{dB}) = 10 \log_{10}\left(\frac{I}{I_0}\right),$$

where $I_0 = 10^{-12} \ \mathrm{W/m}^2$ is the threshold intensity of hearing.

17.4 Doppler Effect and Sonic Booms

- The Doppler effect is an alteration in the observed frequency of a sound due to motion of either the source or the observer.
- The actual change in frequency is called the Doppler shift.
- A sonic boom is constructive interference of sound created by an object moving faster than sound.
- A sonic boom is a type of bow wake created when any wave source moves faster than the wave propagation speed.
- For a stationary observer and a moving source, the observed frequency f_{obs} is:

$$f_{\mathrm{obs}} = f_{\mathrm{s}}\left(\frac{v_{\mathrm{w}}}{v_{\mathrm{w}} \pm v_{\mathrm{s}}}\right),$$

where f_{s} is the frequency of the source, v_{s} is the speed of the source, and v_{w} is the speed of sound. The minus sign is used for motion toward the observer and the plus sign for motion away.

- For a stationary source and moving observer, the observed frequency is:

$$f_{\mathrm{obs}} = f_{\mathrm{s}}\left(\frac{v_{\mathrm{w}} \pm v_{\mathrm{obs}}}{v_{\mathrm{w}}}\right),$$

where v_{obs} is the speed of the observer.

17.5 Sound Interference and Resonance: Standing Waves in Air Columns

- Sound interference and resonance have the same properties as defined for all waves.
- In air columns, the lowest-frequency resonance is called the fundamental, whereas all higher resonant frequencies are called overtones. Collectively, they are called harmonics.
- The resonant frequencies of a tube closed at one end are:

$$f_n = n\frac{v_{\mathrm{w}}}{4L}, n = 1, 3, 5...,$$

f_1 is the fundamental and L is the length of the tube.

- The resonant frequencies of a tube open at both ends are:

$$f_n = n\frac{v_{\mathrm{w}}}{2L}, n = 1, 2, 3...$$

17.6 Hearing

- The range of audible frequencies is 20 to 20,000 Hz.
- Those sounds above 20,000 Hz are ultrasound, whereas those below 20 Hz are infrasound.
- The perception of frequency is pitch.
- The perception of intensity is loudness.
- Loudness has units of phons.

17.7 Ultrasound

- The acoustic impedance is defined as:

$$Z = \rho v,$$

ρ is the density of a medium through which the sound travels and v is the speed of sound through that medium.

- The intensity reflection coefficient a, a measure of the ratio of the intensity of the wave reflected off a boundary between two media relative to the intensity of the incident wave, is given by

$$a = \frac{(Z_2 - Z_1)^2}{(Z_1 + Z_2)^2}.$$

- The intensity reflection coefficient is a unitless quantity.

Conceptual Questions

17.2 Speed of Sound, Frequency, and Wavelength

1. How do sound vibrations of atoms differ from thermal motion?

2. When sound passes from one medium to another where its propagation speed is different, does its frequency or wavelength change? Explain your answer briefly.

17.3 Sound Intensity and Sound Level

3. Six members of a synchronized swim team wear earplugs to protect themselves against water pressure at depths, but they can still hear the music and perform the combinations in the water perfectly. One day, they were asked to leave the pool so the dive team could practice a few dives, and they tried to practice on a mat, but seemed to have a lot more difficulty. Why might this be?

4. A community is concerned about a plan to bring train service to their downtown from the town's outskirts. The current sound intensity level, even though the rail yard is blocks away, is 70 dB downtown. The mayor assures the public that there will be a difference of only 30 dB in sound in the downtown area. Should the townspeople be concerned? Why?

17.4 Doppler Effect and Sonic Booms

5. Is the Doppler shift real or just a sensory illusion?

6. Due to efficiency considerations related to its bow wake, the supersonic transport aircraft must maintain a cruising speed that is a constant ratio to the speed of sound (a constant Mach number). If the aircraft flies from warm air into colder air, should it increase or decrease its speed? Explain your answer.

7. When you hear a sonic boom, you often cannot see the plane that made it. Why is that?

17.5 Sound Interference and Resonance: Standing Waves in Air Columns

8. How does an unamplified guitar produce sounds so much more intense than those of a plucked string held taut by a simple stick?

9. You are given two wind instruments of identical length. One is open at both ends, whereas the other is closed at one end. Which is able to produce the lowest frequency?

10. What is the difference between an overtone and a harmonic? Are all harmonics overtones? Are all overtones harmonics?

17.6 Hearing

11. Why can a hearing test show that your threshold of hearing is 0 dB at 250 Hz, when **Figure 17.39** implies that no one can hear such a frequency at less than 20 dB?

17.7 Ultrasound

12. If audible sound follows a rule of thumb similar to that for ultrasound, in terms of its absorption, would you expect the high or low frequencies from your neighbor's stereo to penetrate into your house? How does this expectation compare with your experience?

13. Elephants and whales are known to use infrasound to communicate over very large distances. What are the advantages of infrasound for long distance communication?

14. It is more difficult to obtain a high-resolution ultrasound image in the abdominal region of someone who is overweight than for someone who has a slight build. Explain why this statement is accurate.

15. Suppose you read that 210-dB ultrasound is being used to pulverize cancerous tumors. You calculate the intensity in watts per centimeter squared and find it is unreasonably high (10^5 W/cm^2). What is a possible explanation?

Problems & Exercises

17.2 Speed of Sound, Frequency, and Wavelength

1. When poked by a spear, an operatic soprano lets out a 1200-Hz shriek. What is its wavelength if the speed of sound is 345 m/s?

2. What frequency sound has a 0.10-m wavelength when the speed of sound is 340 m/s?

3. Calculate the speed of sound on a day when a 1500 Hz frequency has a wavelength of 0.221 m.

4. (a) What is the speed of sound in a medium where a 100-kHz frequency produces a 5.96-cm wavelength? (b) Which substance in **Table 17.4** is this likely to be?

5. Show that the speed of sound in $20.0°C$ air is 343 m/s, as claimed in the text.

6. Air temperature in the Sahara Desert can reach $56.0°C$ (about $134°F$). What is the speed of sound in air at that temperature?

7. Dolphins make sounds in air and water. What is the ratio of the wavelength of a sound in air to its wavelength in seawater? Assume air temperature is $20.0°C$.

8. A sonar echo returns to a submarine 1.20 s after being emitted. What is the distance to the object creating the echo? (Assume that the submarine is in the ocean, not in fresh water.)

9. (a) If a submarine's sonar can measure echo times with a precision of 0.0100 s, what is the smallest difference in distances it can detect? (Assume that the submarine is in the ocean, not in fresh water.)

(b) Discuss the limits this time resolution imposes on the ability of the sonar system to detect the size and shape of the object creating the echo.

10. A physicist at a fireworks display times the lag between seeing an explosion and hearing its sound, and finds it to be 0.400 s. (a) How far away is the explosion if air temperature is $24.0°C$ and if you neglect the time taken for light to reach the physicist? (b) Calculate the distance to the explosion taking the speed of light into account. Note that this distance is negligibly greater.

11. Suppose a bat uses sound echoes to locate its insect prey, 3.00 m away. (See **Figure 17.11**.) (a) Calculate the echo times for temperatures of $5.00°C$ and $35.0°C$. (b) What percent uncertainty does this cause for the bat in locating the insect? (c) Discuss the significance of this uncertainty and whether it could cause difficulties for the bat. (In practice, the bat continues to use sound as it closes in, eliminating most of any difficulties imposed by this and other effects, such as motion of the prey.)

17.3 Sound Intensity and Sound Level

12. What is the intensity in watts per meter squared of 85.0-dB sound?

13. The warning tag on a lawn mower states that it produces noise at a level of 91.0 dB. What is this in watts per meter squared?

14. A sound wave traveling in $20°C$ air has a pressure amplitude of 0.5 Pa. What is the intensity of the wave?

15. What intensity level does the sound in the preceding problem correspond to?

16. What sound intensity level in dB is produced by earphones that create an intensity of $4.00×10^{-2}$ W/m^2?

17. Show that an intensity of 10^{-12} W/m^2 is the same as 10^{-16} W/cm^2.

18. (a) What is the decibel level of a sound that is twice as intense as a 90.0-dB sound? (b) What is the decibel level of a sound that is one-fifth as intense as a 90.0-dB sound?

19. (a) What is the intensity of a sound that has a level 7.00 dB lower than a $4.00×10^{-9}$ W/m^2 sound? (b) What is the intensity of a sound that is 3.00 dB higher than a $4.00×10^{-9}$ W/m^2 sound?

20. (a) How much more intense is a sound that has a level 17.0 dB higher than another? (b) If one sound has a level 23.0 dB less than another, what is the ratio of their intensities?

21. People with good hearing can perceive sounds as low in level as -8.00 dB at a frequency of 3000 Hz. What is the intensity of this sound in watts per meter squared?

22. If a large housefly 3.0 m away from you makes a noise of 40.0 dB, what is the noise level of 1000 flies at that distance, assuming interference has a negligible effect?

23. Ten cars in a circle at a boom box competition produce a 120-dB sound intensity level at the center of the circle. What is the average sound intensity level produced there by each stereo, assuming interference effects can be neglected?

24. The amplitude of a sound wave is measured in terms of its maximum gauge pressure. By what factor does the amplitude of a sound wave increase if the sound intensity level goes up by 40.0 dB?

25. If a sound intensity level of 0 dB at 1000 Hz corresponds to a maximum gauge pressure (sound amplitude) of 10^{-9} atm, what is the maximum gauge pressure in a 60-dB sound? What is the maximum gauge pressure in a 120-dB sound?

26. An 8-hour exposure to a sound intensity level of 90.0 dB may cause hearing damage. What energy in joules falls on a 0.800-cm-diameter eardrum so exposed?

27. (a) Ear trumpets were never very common, but they did aid people with hearing losses by gathering sound over a large area and concentrating it on the smaller area of the eardrum. What decibel increase does an ear trumpet produce if its sound gathering area is 900 cm^2 and the area of the eardrum is 0.500 cm^2, but the trumpet only has an efficiency of 5.00% in transmitting the sound to the eardrum? (b) Comment on the usefulness of the decibel increase found in part (a).

28. Sound is more effectively transmitted into a stethoscope by direct contact than through the air, and it is further intensified by being concentrated on the smaller area of the eardrum. It is reasonable to assume that sound is transmitted into a stethoscope 100 times as effectively compared with transmission though the air. What, then, is the gain in decibels produced by a stethoscope that has a sound gathering area of 15.0 cm^2, and concentrates the sound onto two eardrums with a total area of 0.900 cm^2 with an efficiency of 40.0%?

29. Loudspeakers can produce intense sounds with surprisingly small energy input in spite of their low efficiencies. Calculate the power input needed to produce a 90.0-dB sound intensity level for a 12.0-cm-diameter speaker that has an efficiency of 1.00%. (This value is the sound intensity level right at the speaker.)

17.4 Doppler Effect and Sonic Booms

30. (a) What frequency is received by a person watching an oncoming ambulance moving at 110 km/h and emitting a steady 800-Hz sound from its siren? The speed of sound on this day is 345 m/s. (b) What frequency does she receive after the ambulance has passed?

31. (a) At an air show a jet flies directly toward the stands at a speed of 1200 km/h, emitting a frequency of 3500 Hz, on a day when the speed of sound is 342 m/s. What frequency is received by the observers? (b) What frequency do they receive as the plane flies directly away from them?

32. What frequency is received by a mouse just before being dispatched by a hawk flying at it at 25.0 m/s and emitting a screech of frequency 3500 Hz? Take the speed of sound to be 331 m/s.

33. A spectator at a parade receives an 888-Hz tone from an oncoming trumpeter who is playing an 880-Hz note. At what speed is the musician approaching if the speed of sound is 338 m/s?

34. A commuter train blows its 200-Hz horn as it approaches a crossing. The speed of sound is 335 m/s. (a) An observer waiting at the crossing receives a frequency of 208 Hz. What is the speed of the train? (b) What frequency does the observer receive as the train moves away?

35. Can you perceive the shift in frequency produced when you pull a tuning fork toward you at 10.0 m/s on a day when the speed of sound is 344 m/s? To answer this question, calculate the factor by which the frequency shifts and see if it is greater than 0.300%.

36. Two eagles fly directly toward one another, the first at 15.0 m/s and the second at 20.0 m/s. Both screech, the first one emitting a frequency of 3200 Hz and the second one emitting a frequency of 3800 Hz. What frequencies do they receive if the speed of sound is 330 m/s?

37. What is the minimum speed at which a source must travel toward you for you to be able to hear that its frequency is Doppler shifted? That is, what speed produces a shift of 0.300% on a day when the speed of sound is 331 m/s?

17.5 Sound Interference and Resonance: Standing Waves in Air Columns

38. A "showy" custom-built car has two brass horns that are supposed to produce the same frequency but actually emit 263.8 and 264.5 Hz. What beat frequency is produced?

39. What beat frequencies will be present: (a) If the musical notes A and C are played together (frequencies of 220 and 264 Hz)? (b) If D and F are played together (frequencies of 297 and 352 Hz)? (c) If all four are played together?

40. What beat frequencies result if a piano hammer hits three strings that emit frequencies of 127.8, 128.1, and 128.3 Hz?

41. A piano tuner hears a beat every 2.00 s when listening to a 264.0-Hz tuning fork and a single piano string. What are the two possible frequencies of the string?

42. (a) What is the fundamental frequency of a 0.672-m-long tube, open at both ends, on a day when the speed of sound is 344 m/s? (b) What is the frequency of its second harmonic?

43. If a wind instrument, such as a tuba, has a fundamental frequency of 32.0 Hz, what are its first three overtones? It is closed at one end. (The overtones of a real tuba are more complex than this example, because it is a tapered tube.)

44. What are the first three overtones of a bassoon that has a fundamental frequency of 90.0 Hz? It is open at both ends. (The overtones of a real bassoon are more complex than this example, because its double reed makes it act more like a tube closed at one end.)

45. How long must a flute be in order to have a fundamental frequency of 262 Hz (this frequency corresponds to middle C on the evenly tempered chromatic scale) on a day when air temperature is $20.0°C$? It is open at both ends.

46. What length should an oboe have to produce a fundamental frequency of 110 Hz on a day when the speed of sound is 343 m/s? It is open at both ends.

47. What is the length of a tube that has a fundamental frequency of 176 Hz and a first overtone of 352 Hz if the speed of sound is 343 m/s?

48. (a) Find the length of an organ pipe closed at one end that produces a fundamental frequency of 256 Hz when air temperature is $18.0°C$. (b) What is its fundamental frequency at $25.0°C$?

49. By what fraction will the frequencies produced by a wind instrument change when air temperature goes from $10.0°C$ to $30.0°C$? That is, find the ratio of the frequencies at those temperatures.

50. The ear canal resonates like a tube closed at one end. (See **Figure 17.41**.) If ear canals range in length from 1.80 to 2.60 cm in an average population, what is the range of fundamental resonant frequencies? Take air temperature to be $37.0°C$, which is the same as body temperature. How does this result correlate with the intensity versus frequency graph (**Figure 17.39** of the human ear?

51. Calculate the first overtone in an ear canal, which resonates like a 2.40-cm-long tube closed at one end, by taking air temperature to be $37.0°C$. Is the ear particularly sensitive to such a frequency? (The resonances of the ear canal are complicated by its nonuniform shape, which we shall ignore.)

52. A crude approximation of voice production is to consider the breathing passages and mouth to be a resonating tube closed at one end. (See **Figure 17.32**.) (a) What is the fundamental frequency if the tube is 0.240-m long, by taking air temperature to be $37.0°C$? (b) What would this frequency become if the person replaced the air with helium? Assume the same temperature dependence for helium as for air.

53. (a) Students in a physics lab are asked to find the length of an air column in a tube closed at one end that has a fundamental frequency of 256 Hz. They hold the tube vertically and fill it with water to the top, then lower the water while a 256-Hz tuning fork is rung and listen for the first resonance. What is the air temperature if the resonance occurs for a length of 0.336 m? (b) At what length will they observe the second resonance (first overtone)?

54. What frequencies will a 1.80-m-long tube produce in the audible range at $20.0°C$ if: (a) The tube is closed at one end? (b) It is open at both ends?

17.6 Hearing

55. The factor of 10^{-12} in the range of intensities to which the ear can respond, from threshold to that causing damage after brief exposure, is truly remarkable. If you could measure distances over the same range with a single instrument and the smallest distance you could measure was 1 mm, what would the largest be?

56. The frequencies to which the ear responds vary by a factor of 10^3. Suppose the speedometer on your car measured speeds differing by the same factor of 10^3, and the greatest speed it reads is 90.0 mi/h. What would be the slowest nonzero speed it could read?

57. What are the closest frequencies to 500 Hz that an average person can clearly distinguish as being different in frequency from 500 Hz? The sounds are not present simultaneously.

58. Can the average person tell that a 2002-Hz sound has a different frequency than a 1999-Hz sound without playing them simultaneously?

59. If your radio is producing an average sound intensity level of 85 dB, what is the next lowest sound intensity level that is clearly less intense?

60. Can you tell that your roommate turned up the sound on the TV if its average sound intensity level goes from 70 to 73 dB?

61. Based on the graph in **Figure 17.38**, what is the threshold of hearing in decibels for frequencies of 60, 400, 1000, 4000, and 15,000 Hz? Note that many AC electrical appliances produce 60 Hz, music is commonly 400 Hz, a reference frequency is 1000 Hz, your maximum sensitivity is near 4000 Hz, and many older TVs produce a 15,750 Hz whine.

62. What sound intensity levels must sounds of frequencies 60, 3000, and 8000 Hz have in order to have the same loudness as a 40-dB sound of frequency 1000 Hz (that is, to have a loudness of 40 phons)?

63. What is the approximate sound intensity level in decibels of a 600-Hz tone if it has a loudness of 20 phons? If it has a loudness of 70 phons?

64. (a) What are the loudnesses in phons of sounds having frequencies of 200, 1000, 5000, and 10,000 Hz, if they are all at the same 60.0-dB sound intensity level? (b) If they are all at 110 dB? (c) If they are all at 20.0 dB?

65. Suppose a person has a 50-dB hearing loss at all frequencies. By how many factors of 10 will low-intensity sounds need to be amplified to seem normal to this person? Note that smaller amplification is appropriate for more intense sounds to avoid further hearing damage.

66. If a woman needs an amplification of $5.0×10^{12}$ times the threshold intensity to enable her to hear at all frequencies, what is her overall hearing loss in dB? Note that smaller amplification is appropriate for more intense sounds to avoid further damage to her hearing from levels above 90 dB.

67. (a) What is the intensity in watts per meter squared of a just barely audible 200-Hz sound? (b) What is the intensity in watts per meter squared of a barely audible 4000-Hz sound?

68. (a) Find the intensity in watts per meter squared of a 60.0-Hz sound having a loudness of 60 phons. (b) Find the intensity in watts per meter squared of a 10,000-Hz sound having a loudness of 60 phons.

69. A person has a hearing threshold 10 dB above normal at 100 Hz and 50 dB above normal at 4000 Hz. How much more intense must a 100-Hz tone be than a 4000-Hz tone if they are both barely audible to this person?

70. A child has a hearing loss of 60 dB near 5000 Hz, due to noise exposure, and normal hearing elsewhere. How much more intense is a 5000-Hz tone than a 400-Hz tone if they are both barely audible to the child?

71. What is the ratio of intensities of two sounds of identical frequency if the first is just barely discernible as louder to a person than the second?

17.7 Ultrasound

Unless otherwise indicated, for problems in this section, assume that the speed of sound through human tissues is 1540 m/s.

72. What is the sound intensity level in decibels of ultrasound of intensity $10^5 \ W/m^2$, used to pulverize tissue during surgery?

73. Is 155-dB ultrasound in the range of intensities used for deep heating? Calculate the intensity of this ultrasound and compare this intensity with values quoted in the text.

74. Find the sound intensity level in decibels of $2.00×10^{-2} \ W/m^2$ ultrasound used in medical diagnostics.

75. The time delay between transmission and the arrival of the reflected wave of a signal using ultrasound traveling through a piece of fat tissue was 0.13 ms. At what depth did this reflection occur?

76. In the clinical use of ultrasound, transducers are always coupled to the skin by a thin layer of gel or oil, replacing the air that would otherwise exist between the transducer and the skin. (a) Using the values of acoustic impedance given in **Table 17.8** calculate the intensity reflection coefficient between transducer material and air. (b) Calculate the intensity reflection coefficient between transducer material and gel (assuming for this problem that its acoustic impedance is identical to that of water). (c) Based on the results of your calculations, explain why the gel is used.

77. (a) Calculate the minimum frequency of ultrasound that will allow you to see details as small as 0.250 mm in human tissue. (b) What is the effective depth to which this sound is effective as a diagnostic probe?

78. (a) Find the size of the smallest detail observable in human tissue with 20.0-MHz ultrasound. (b) Is its effective penetration depth great enough to examine the entire eye (about 3.00 cm is needed)? (c) What is the wavelength of such ultrasound in $0°C$ air?

79. (a) Echo times are measured by diagnostic ultrasound scanners to determine distances to reflecting surfaces in a patient. What is the difference in echo times for tissues that are 3.50 and 3.60 cm beneath the surface? (This difference is the minimum resolving time for the scanner to see details as small as 0.100 cm, or 1.00 mm. Discrimination of smaller time differences is needed to see smaller details.) (b) Discuss whether the period T of this ultrasound must be smaller than the minimum time resolution. If so, what is the minimum frequency of the ultrasound and is that out of the normal range for diagnostic ultrasound?

80. (a) How far apart are two layers of tissue that produce echoes having round-trip times (used to measure distances) that differ by $0.750~\mu s$? (b) What minimum frequency must the ultrasound have to see detail this small?

81. (a) A bat uses ultrasound to find its way among trees. If this bat can detect echoes 1.00 ms apart, what minimum distance between objects can it detect? (b) Could this distance explain the difficulty that bats have finding an open door when they accidentally get into a house?

82. A dolphin is able to tell in the dark that the ultrasound echoes received from two sharks come from two different objects only if the sharks are separated by 3.50 m, one being that much farther away than the other. (a) If the ultrasound has a frequency of 100 kHz, show this ability is not limited by its wavelength. (b) If this ability is due to the dolphin's ability to detect the arrival times of echoes, what is the minimum time difference the dolphin can perceive?

83. A diagnostic ultrasound echo is reflected from moving blood and returns with a frequency 500 Hz higher than its original 2.00 MHz. What is the velocity of the blood? (Assume that the frequency of 2.00 MHz is accurate to seven significant figures and 500 Hz is accurate to three significant figures.)

84. Ultrasound reflected from an oncoming bloodstream that is moving at 30.0 cm/s is mixed with the original frequency of 2.50 MHz to produce beats. What is the beat frequency? (Assume that the frequency of 2.50 MHz is accurate to seven significant figures.)

Test Prep for AP® Courses

17.2 Speed of Sound, Frequency, and Wavelength

1. A teacher wants to demonstrate that the speed of sound is not a constant value. Considering her regular classroom voice as the control, which of the following will increase the speed of sound leaving her mouth?
  I. Submerge her mouth underwater and speak at the same volume.
  II. Increase the temperature of the room and speak at the same volume.
  III. Increase the pitch of her voice and speak at the same volume.
  a. I only
  b. I and II only
  c. I, II and III
  d. II and III
  e. III only

2. All members of an orchestra begin tuning their instruments at the same time. While some woodwind instruments play high frequency notes, other stringed instruments play notes of lower frequency. Yet an audience member will hear all notes simultaneously, in apparent contrast to the equation.

Explain how a student could demonstrate the flaw in the above logic, using a slinky, stopwatch, and meter stick. Make sure to explain what relationship is truly demonstrated in the above equation, in addition to what would be necessary to get the speed of the slinky to actually change. You may include diagrams and equations as part of your explanation.

17.3 Sound Intensity and Sound Level

3. In order to waken a sleeping child, the volume on an alarm clock is tripled. Under this new scenario, how much more energy will be striking the child's ear drums each second?
  a. twice as much
  b. three times as much
  c. approximately 4.8 times as much
  d. six times as much
  e. nine times as much

4. A musician strikes the strings of a guitar such that they vibrate with twice the amplitude.

a. Explain why this requires an energy input greater than twice the original value.
b. Explain why the sound leaving the string will not result in a decibel level that is twice as great.

17.4 Doppler Effect and Sonic Booms

5. A baggage handler stands on the edge of a runway as a landing plane approaches. Compared to the pitch of the plane as heard by the plane's pilot, which of the following correctly describes the sensation experienced by the handler?
a. The frequency of the plane will be lower pitched according to the baggage handler and will become even lower pitched as the plane slows to a stop.
b. The frequency of the plane will be lower pitched according to the baggage handler but will increase in pitch as the plane slows to a stop.
c. The frequency of the plane will be higher pitched according to the baggage handler but will decrease in pitch as the plane slows to a stop.
d. The frequency of the plane will be higher pitched according to the baggage handler and will further increase in pitch as the plane slows to a stop.

6. The following graph represents the perceived frequency of a car as it passes a student.

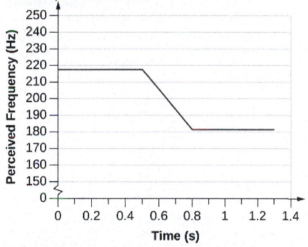

Figure 17.51 Plot of time versus perceived frequency to illustrate the Doppler effect.

a. If the true frequency of the car's horn is 200 Hz, how fast was the car traveling?
b. On the graph above, draw a line demonstrating the perceived frequency for a car traveling twice as fast. Label all intercepts, maximums, and minimums on the graph.

17.5 Sound Interference and Resonance: Standing Waves in Air Columns

7. A common misconception is that two wave pulses traveling in opposite directions will reflect off each other. Outline a procedure that you would use to convince someone that the two wave pulses do not reflect off each other, but instead travel through each other. You may use sketches to represent your understanding. Be sure to provide evidence to not only refute the original claim, but to support yours as well.

8. Two wave pulses are traveling toward each other on a string, as shown below. Which of the following representations correctly shows the string as the two pulses overlap?

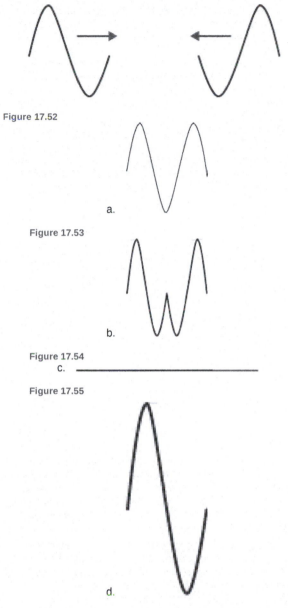

Figure 17.52

Figure 17.53

a.

Figure 17.54

b.

c.

Figure 17.55

d.

Figure 17.56

9. A student sends a transverse wave pulse of amplitude A along a rope attached at one end. As the pulse returns to the student, a second pulse of amplitude $3A$ is sent along the opposite side of the rope. What is the resulting amplitude when the two pulses interact?
a. $4A$
b. A
c. $2A$, on the side of the original wave pulse
d. $2A$, on the side of the second wave pulse

10. A student would like to demonstrate destructive interference using two sound sources. Explain how the student could set up this demonstration and what restrictions they would need to place upon their sources. Be sure to consider both the layout of space and the sounds created in your explanation.

11. A student is shaking a flexible string attached to a wooden board in a rhythmic manner. Which of the following choices will decrease the wavelength within the rope?

I. The student could shake her hand back and forth with greater frequency.
II. The student could shake her hand back in forth with a greater amplitude.
III. The student could increase the tension within the rope by stepping backwards from the board.
a. I only
b. I and II
c. I and III
d. II and III
e. I, II, and III

12. A ripple tank has two locations (L1 and L2) that vibrate in tandem as shown below. Both L1 and L2 vibrate in a plane perpendicular to the page, creating a two-dimensional interference pattern.

Figure 17.57

Describe an experimental procedure to determine the speed of the waves created within the water, including all additional equipment that you would need. You may use the diagram below to help your description, or you may create one of your own. Include enough detail so that another student could carry out your experiment.

13. A string is vibrating between two posts as shown above. Students are to determine the speed of the wave within this string. They have already measured the amount of time necessary for the wave to oscillate up and down. The students must also take what other measurements to determine the speed of the wave?
a. The distance between the two posts.
b. The amplitude of the wave
c. The tension in the string
d. The amplitude of the wave and the tension in the string
e. The distance between the two posts, the amplitude of the wave, and the tension in the string

14. The accepted speed of sound in room temperature air is 346 m/s. Knowing that their school is colder than usual, a group of students is asked to determine the speed of sound in their room. They are permitted to use any materials necessary; however, their lab procedure must utilize standing wave patterns. The students collect the information **Table 17.9**.

Table 17.9

Trial Number	Wavelength (m)	Frequency (Hz)	
1	3.45	95	
2	2.32	135	
3	1.70	190	
4	1.45	240	
5	1.08	305	

a. Describe an experimental procedure the group of students could have used to obtain this data. Include diagrams of the experimental setup and any equipment used in the process.
b. Select a set of data points from the table and plot those points on a graph to determine the speed of sound within the classroom. Fill in the blank column in the table for any quantities you graph other than the given data. Label the axes and indicate the scale for each. Draw a best-fit line or curve through your data points.
c. Using information from the graph, determine the speed of sound within the student's classroom, and explain what characteristic of the graph provides this evidence.
d. Determine the temperature of the classroom.

15. A tube is open at one end. If the fundamental frequency f is created by a wavelength λ, then which of the following describes the frequency and wavelength associated with the tube's fourth overtone?

	f	λ
(a)	$4f$	$\lambda/4$
(b)	$4f$	λ
(c)	$9f$	$\lambda/9$
(d)	$9f$	λ

16. A group of students were tasked with collecting information about standing waves. **Table 17.10** a series of their data, showing the length of an air column and a resonant frequency present when the column is struck.

Table 17.10

Length (m)	Resonant Frequency (Hz)
1	85.75
2	43
3	29
4	21.5

a. From their data, determine whether the air column was open or closed on each end.
b. Predict the resonant frequency of the column at a length of 2.5 meters.

17. When a student blows across a glass half-full of water, a resonant frequency is created within the air column remaining in the glass. Which of the following can the student do to increase this resonant frequency?
I. Add more water to the glass.
II. Replace the water with a more dense fluid.
III. Increase the temperature of the room.

a. I only
b. I and III
c. II and III
d. all of the above

18. A student decides to test the speed of sound through wood using a wooden ruler. The student rests the ruler on a desk with half of its length protruding off the desk edge. The student then holds one end in place and strikes the protruding end with his other hand, creating a musical sound, and counts the number of vibrations of the ruler. Explain why the student would not be able to measure the speed of sound through wood using this method.

19. A musician stands outside in a field and plucks a string on an acoustic guitar. Standing waves will most likely occur in which of the following media? Select two answers.
a. The guitar string
b. The air inside the guitar
c. The air surrounding the guitar
d. The ground beneath the musician

20.

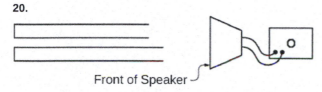

Front of Speaker

Figure 17.58 This figure shows two tubes that are identical except for their slightly different lengths. Both tubes have one open end and one closed end. A speaker connected to a variable frequency generator is placed in front of the tubes, as shown. The speaker is set to produce a note of very low frequency when turned on. The frequency is then slowly increased to produce resonances in the tubes. Students observe that at first only one of the tubes resonates at a time. Later, as the frequency gets very high, there are times when both tubes resonate.

In a clear, coherent, paragraph-length answer, explain why there are some high frequencies, but no low frequencies, at which both tubes resonate. You may include diagrams and/or equations as part of your explanation.

Oscillator

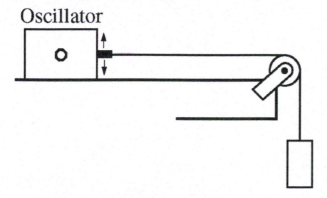

Figure 17.59

21. A student connects one end of a string with negligible mass to an oscillator. The other end of the string is passed over a pulley and attached to a suspended weight, as shown above. The student finds that a standing wave with one antinode is formed on the string when the frequency of the oscillator is f_0. The student then moves the oscillator to shorten the horizontal segment of string to half its original length. At what frequency will a standing wave with one

antinode now be formed on the string?
a. $f_0/2$
b. f_0
c. $2f_0$
d. There is no frequency at which a standing wave will be formed.

22. A guitar string of length L is bound at both ends. **Table 17.11** shows the string's harmonic frequencies when struck.

Table 17.11

Harmonic Number	Frequency
1	225/L
2	450/L
3	675/L
4	900/L

a. Based on the information above, what is the speed of the wave within the string?
b. The guitarist then slides her finger along the neck of the guitar, changing the string length as a result. Calculate the fundamental frequency of the string and wave speed present if the string length is reduced to 2/3 L.

APPENDIX A | ATOMIC MASSES

Table A1 Atomic Masses

Atomic Number, Z	Name	Atomic Mass Number, A	Symbol	Atomic Mass (u)	Percent Abundance or Decay Mode	Half-life, $t_{1/2}$
0	neutron	1	n	1.008 665	β^-	10.37 min
1	Hydrogen	1	^1H	1.007 825	99.985%	
	Deuterium	2	^2H or D	2.014 102	0.015%	
	Tritium	3	^3H or T	3.016 050	β^-	12.33 y
2	Helium	3	^3He	3.016 030	$1.38 \times 10^{-4}\%$	
		4	^4He	4.002 603	$\approx 100\%$	
3	Lithium	6	^6Li	6.015 121	7.5%	
		7	^7Li	7.016 003	92.5%	
4	Beryllium	7	^7Be	7.016 928	EC	53.29 d
		9	^9Be	9.012 182	100%	
5	Boron	10	^{10}B	10.012 937	19.9%	
		11	^{11}B	11.009 305	80.1%	
6	Carbon	11	^{11}C	11.011 432	EC, β^+	
		12	^{12}C	12.000 000	98.90%	
		13	^{13}C	13.003 355	1.10%	
		14	^{14}C	14.003 241	β^-	5730 y
7	Nitrogen	13	^{13}N	13.005 738	β^+	9.96 min
		14	^{14}N	14.003 074	99.63%	
		15	^{15}N	15.000 108	0.37%	
8	Oxygen	15	^{15}O	15.003 065	EC, β^+	122 s
		16	^{16}O	15.994 915	99.76%	
		18	^{18}O	17.999 160	0.200%	
9	Fluorine	18	^{18}F	18.000 937	EC, β^+	1.83 h
		19	^{19}F	18.998 403	100%	
10	Neon	20	^{20}Ne	19.992 435	90.51%	
		22	^{22}Ne	21.991 383	9.22%	
11	Sodium	22	^{22}Na	21.994 434	β^+	2.602 y
		23	^{23}Na	22.989 767	100%	

Atomic Number, Z	Name	Atomic Mass Number, A	Symbol	Atomic Mass (u)	Percent Abundance or Decay Mode	Half-life, $t_{1/2}$
		24	^{24}Na	23.990 961	β^-	14.96 h
12	Magnesium	24	^{24}Mg	23.985 042	78.99%	
13	Aluminum	27	^{27}Al	26.981 539	100%	
14	Silicon	28	^{28}Si	27.976 927	92.23%	2.62h
		31	^{31}Si	30.975 362	β^-	
15	Phosphorus	31	^{31}P	30.973 762	100%	
		32	^{32}P	31.973 907	β^-	14.28 d
16	Sulfur	32	^{32}S	31.972 070	95.02%	
		35	^{35}S	34.969 031	β^-	87.4 d
17	Chlorine	35	^{35}Cl	34.968 852	75.77%	
		37	^{37}Cl	36.965 903	24.23%	
18	Argon	40	^{40}Ar	39.962 384	99.60%	
19	Potassium	39	^{39}K	38.963 707	93.26%	
		40	^{40}K	39.963 999	0.0117%, EC, β^-	1.28×10^9 y
20	Calcium	40	^{40}Ca	39.962 591	96.94%	
21	Scandium	45	^{45}Sc	44.955 910	100%	
22	Titanium	48	^{48}Ti	47.947 947	73.8%	
23	Vanadium	51	^{51}V	50.943 962	99.75%	
24	Chromium	52	^{52}Cr	51.940 509	83.79%	
25	Manganese	55	^{55}Mn	54.938 047	100%	
26	Iron	56	^{56}Fe	55.934 939	91.72%	
27	Cobalt	59	^{59}Co	58.933 198	100%	
		60	^{60}Co	59.933 819	β^-	5.271 y
28	Nickel	58	^{58}Ni	57.935 346	68.27%	
		60	^{60}Ni	59.930 788	26.10%	
29	Copper	63	^{63}Cu	62.939 598	69.17%	
		65	^{65}Cu	64.927 793	30.83%	
30	Zinc	64	^{64}Zn	63.929 145	48.6%	
		66	^{66}Zn	65.926 034	27.9%	
31	Gallium	69	^{69}Ga	68.925 580	60.1%	
32	Germanium	72	^{72}Ge	71.922 079	27.4%	

Atomic Number, Z	Name	Atomic Mass Number, A	Symbol	Atomic Mass (u)	Percent Abundance or Decay Mode	Half-life, $t_{1/2}$
		74	^{74}Ge	73.921 177	36.5%	
33	Arsenic	75	^{75}As	74.921 594	100%	
34	Selenium	80	^{80}Se	79.916 520	49.7%	
35	Bromine	79	^{79}Br	78.918 336	50.69%	
36	Krypton	84	^{84}Kr	83.911 507	57.0%	
37	Rubidium	85	^{85}Rb	84.911 794	72.17%	
38	Strontium	86	^{86}Sr	85.909 267	9.86%	
		88	^{88}Sr	87.905 619	82.58%	
		90	^{90}Sr	89.907 738	β^-	28.8 y
39	Yttrium	89	^{89}Y	88.905 849	100%	
		90	^{90}Y	89.907 152	β^-	64.1 h
40	Zirconium	90	^{90}Zr	89.904 703	51.45%	
41	Niobium	93	^{93}Nb	92.906 377	100%	
42	Molybdenum	98	^{98}Mo	97.905 406	24.13%	
43	Technetium	98	^{98}Tc	97.907 215	β^-	4.2×10^6y
44	Ruthenium	102	^{102}Ru	101.904 348	31.6%	
45	Rhodium	103	^{103}Rh	102.905 500	100%	
46	Palladium	106	^{106}Pd	105.903 478	27.33%	
47	Silver	107	^{107}Ag	106.905 092	51.84%	
		109	^{109}Ag	108.904 757	48.16%	
48	Cadmium	114	^{114}Cd	113.903 357	28.73%	
49	Indium	115	^{115}In	114.903 880	95.7%, β^-	4.4×10^{14}y
50	Tin	120	^{120}Sn	119.902 200	32.59%	
51	Antimony	121	^{121}Sb	120.903 821	57.3%	
52	Tellurium	130	^{130}Te	129.906 229	33.8%, β^-	2.5×10^{21}y
53	Iodine	127	^{127}I	126.904 473	100%	
		131	^{131}I	130.906 114	β^-	8.040 d
54	Xenon	132	^{132}Xe	131.904 144	26.9%	
		136	^{136}Xe	135.907 214	8.9%	
55	Cesium	133	^{133}Cs	132.905 429	100%	

Atomic Number, Z	Name	Atomic Mass Number, A	Symbol	Atomic Mass (u)	Percent Abundance or Decay Mode	Half-life, $t_{1/2}$
		134	^{134}Cs	133.906 696	EC, β^-	2.06 y
56	Barium	137	^{137}Ba	136.905 812	11.23%	
		138	^{138}Ba	137.905 232	71.70%	
57	Lanthanum	139	^{139}La	138.906 346	99.91%	
58	Cerium	140	^{140}Ce	139.905 433	88.48%	
59	Praseodymium	141	^{141}Pr	140.907 647	100%	
60	Neodymium	142	^{142}Nd	141.907 719	27.13%	
61	Promethium	145	^{145}Pm	144.912 743	EC, α	17.7 y
62	Samarium	152	^{152}Sm	151.919 729	26.7%	
63	Europium	153	^{153}Eu	152.921 225	52.2%	
64	Gadolinium	158	^{158}Gd	157.924 099	24.84%	
65	Terbium	159	^{159}Tb	158.925 342	100%	
66	Dysprosium	164	^{164}Dy	163.929 171	28.2%	
67	Holmium	165	^{165}Ho	164.930 319	100%	
68	Erbium	166	^{166}Er	165.930 290	33.6%	
69	Thulium	169	^{169}Tm	168.934 212	100%	
70	Ytterbium	174	^{174}Yb	173.938 859	31.8%	
71	Lutecium	175	^{175}Lu	174.940 770	97.41%	
72	Hafnium	180	^{180}Hf	179.946 545	35.10%	
73	Tantalum	181	^{181}Ta	180.947 992	99.98%	
74	Tungsten	184	^{184}W	183.950 928	30.67%	
75	Rhenium	187	^{187}Re	186.955 744	62.6%, β^-	4.6×10^{10} y
76	Osmium	191	^{191}Os	190.960 920	β^-	15.4 d
		192	^{192}Os	191.961 467	41.0%	
77	Iridium	191	^{191}Ir	190.960 584	37.3%	
		193	^{193}Ir	192.962 917	62.7%	
78	Platinum	195	^{195}Pt	194.964 766	33.8%	
79	Gold	197	^{197}Au	196.966 543	100%	
		198	^{198}Au	197.968 217	β^-	2.696 d
80	Mercury	199	^{199}Hg	198.968 253	16.87%	

Atomic Number, Z	Name	Atomic Mass Number, A	Symbol	Atomic Mass (u)	Percent Abundance or Decay Mode	Half-life, $t_{1/2}$
		202	^{202}Hg	201.970 617	29.86%	
81	Thallium	205	^{205}Tl	204.974 401	70.48%	
82	Lead	206	^{206}Pb	205.974 440	24.1%	
		207	^{207}Pb	206.975 872	22.1%	
		208	^{208}Pb	207.976 627	52.4%	
		210	^{210}Pb	209.984 163	α, β^-	22.3 y
		211	^{211}Pb	210.988 735	β^-	36.1 min
		212	^{212}Pb	211.991 871	β^-	10.64 h
83	Bismuth	209	^{209}Bi	208.980 374	100%	
		211	^{211}Bi	210.987 255	α, β^-	2.14 min
84	Polonium	210	^{210}Po	209.982 848	α	138.38 d
85	Astatine	218	^{218}At	218.008 684	α, β^-	1.6 s
86	Radon	222	^{222}Rn	222.017 570	α	3.82 d
87	Francium	223	^{223}Fr	223.019 733	α, β^-	21.8 min
88	Radium	226	^{226}Ra	226.025 402	α	1.60×10^3 y
89	Actinium	227	^{227}Ac	227.027 750	α, β^-	21.8 y
90	Thorium	228	^{228}Th	228.028 715	α	1.91 y
		232	^{232}Th	232.038 054	100%, α	1.41×10^{10} y
91	Protactinium	231	^{231}Pa	231.035 880	α	3.28×10^4 y
92	Uranium	233	^{233}U	233.039 628	α	1.59×10^3 y
		235	^{235}U	235.043 924	0.720%, α	7.04×10^8 y
		236	^{236}U	236.045 562	α	2.34×10^7 y
		238	^{238}U	238.050 784	99.2745%, α	4.47×10^9 y
		239	^{239}U	239.054 289	β^-	23.5 min
93	Neptunium	239	^{239}Np	239.052 933	β^-	2.355 d
94	Plutonium	239	^{239}Pu	239.052 157	α	2.41×10^4 y
95	Americium	243	^{243}Am	243.061 375	α, fission	7.37×10^3 y
96	Curium	245	^{245}Cm	245.065 483	α	8.50×10^3 y
97	Berkelium	247	^{247}Bk	247.070 300	α	1.38×10^3 y

Atomic Number, Z	Name	Atomic Mass Number, A	Symbol	Atomic Mass (u)	Percent Abundance or Decay Mode	Half-life, $t_{1/2}$
98	Californium	249	^{249}Cf	249.074 844	α	351 y
99	Einsteinium	254	^{254}Es	254.088 019	α, β^-	276 d
100	Fermium	253	^{253}Fm	253.085 173	EC, α	3.00 d
101	Mendelevium	255	^{255}Md	255.091 081	EC, α	27 min
102	Nobelium	255	^{255}No	255.093 260	EC, α	3.1 min
103	Lawrencium	257	^{257}Lr	257.099 480	EC, α	0.646 s
104	Rutherfordium	261	^{261}Rf	261.108 690	α	1.08 min
105	Dubnium	262	^{262}Db	262.113 760	α, fission	34 s
106	Seaborgium	263	^{263}Sg	263.11 86	α, fission	0.8 s
107	Bohrium	262	^{262}Bh	262.123 1	α	0.102 s
108	Hassium	264	^{264}Hs	264.128 5	α	0.08 ms
109	Meitnerium	266	^{266}Mt	266.137 8	α	3.4 ms

APPENDIX B | SELECTED RADIOACTIVE ISOTOPES

Decay modes are α, β^-, β^+, electron capture (EC) and isomeric transition (IT). EC results in the same daughter nucleus as would β^+ decay. IT is a transition from a metastable excited state. Energies for $\beta^{\pm}$ decays are the maxima; average energies are roughly one-half the maxima.

Table B1 Selected Radioactive Isotopes

Isotope	$t_{1/2}$	DecayMode(s)	Energy(MeV)	Percent	γ-Ray Energy(MeV)		Percent
^{3}H	12.33 y	β^-	0.0186	100%			
^{14}C	5730 y	β^-	0.156	100%			
^{13}N	9.96 min	β^+	1.20	100%			
^{22}Na	2.602 y	β^+	0.55	90%	γ	1.27	100%
^{32}P	14.28 d	β^-	1.71	100%			
^{35}S	87.4 d	β^-	0.167	100%			
^{36}Cl	3.00×10^5y	β^-	0.710	100%			
^{40}K	1.28×10^9y	β^-	1.31	89%			
^{43}K	22.3 h	β^-	0.827	87%	γs	0.373	87%
						0.618	87%
^{45}Ca	165 d	β^-	0.257	100%			
^{51}Cr	27.70 d	EC			γ	0.320	10%
^{52}Mn	5.59d	β^+	3.69	28%	γs	1.33	28%
						1.43	28%
^{52}Fe	8.27 h	β^+	1.80	43%		0.169	43%
						0.378	43%
^{59}Fe	44.6 d	β^- s	0.273	45%	γs	1.10	57%
			0.466	55%		1.29	43%
^{60}Co	5.271 y	β^-	0.318	100%	γs	1.17	100%
						1.33	100%
^{65}Zn	244.1 d	EC			γ	1.12	51%
^{67}Ga	78.3 h	EC			γs	0.0933	70%
						0.185	35%
						0.300	19%
						others	
^{75}Se	118.5 d	EC			γs	0.121	20%
						0.136	65%

Isotope	$t_{1/2}$	DecayMode(s)	Energy(MeV)	Percent		γ-Ray Energy(MeV)	Percent
						0.265	68%
						0.280	20%
						others	
^{86}Rb	18.8 d	β^- s	0.69	9%	γ	1.08	9%
			1.77	91%			
^{85}Sr	64.8 d	EC			γ	0.514	100%
^{90}Sr	28.8 y	β^-	0.546	100%			
^{90}Y	64.1 h	β^-	2.28	100%			
^{99m}Tc	6.02 h	IT			γ	0.142	100%
^{113m}In	99.5 min	IT			γ	0.392	100%
^{123}I	13.0 h	EC			γ	0.159	$\approx$100%
^{131}I	8.040 d	β^- s	0.248	7%	γ s	0.364	85%
			0.607	93%		others	
			others				
^{129}Cs	32.3 h	EC			γ s	0.0400	35%
						0.372	32%
						0.411	25%
						others	
^{137}Cs	30.17 y	β^- s	0.511	95%	γ	0.662	95%
			1.17	5%			
^{140}Ba	12.79 d	β^-	1.035	$\approx$100%	γ s	0.030	25%
						0.044	65%
						0.537	24%
						others	
^{198}Au	2.696 d	β^-	1.161	$\approx$100%	γ	0.412	$\approx$100%
^{197}Hg	64.1 h	EC			γ	0.0733	100%
^{210}Po	138.38 d	α	5.41	100%			
^{226}Ra	1.60×10^3 y	α s	4.68	5%	γ	0.186	100%
			4.87	95%			
^{235}U	7.038×10^8 y	α	4.68	$\approx$100%	γ s	numerous	<0.400%
^{238}U	4.468×10^9 y	α s	4.22	23%	γ	0.050	23%
			4.27	77%			
^{237}Np	2.14×10^6 y	α s	numerous		γ s	numerous	<0.250%
			4.96 (max.)				
^{239}Pu	2.41×10^4 y	α s	5.19	11%	γ s	7.5×10^{-5}	73%
			5.23	15%		0.013	15%

Isotope	$t_{1/2}$	DecayMode(s)	Energy(MeV)	Percent		γ-Ray Energy(MeV)	Percent
			5.24	73%		0.052	10%
						others	
^{243}Am	7.37×10^3y	αs	Max. 5.44		γs	0.075	
			5.37	88%		others	
			5.32	11%			
			others				

APPENDIX C | USEFUL INFORMATION

This appendix is broken into several tables.

- **Table C1**, Important Constants
- **Table C2**, Submicroscopic Masses
- **Table C3**, Solar System Data
- **Table C4**, Metric Prefixes for Powers of Ten and Their Symbols
- **Table C5**, The Greek Alphabet
- **Table C6**, SI units
- **Table C7**, Selected British Units
- **Table C8**, Other Units
- **Table C9**, Useful Formulae

Table C1 Important Constants [1]

Symbol	Meaning	Best Value	Approximate Value
c	Speed of light in vacuum	$2.99792458 \times 10^8 \, \mathrm{m/s}$	$3.00 \times 10^8 \, \mathrm{m/s}$
G	Gravitational constant	$6.67408(31) \times 10^{-11} \, \mathrm{N \cdot m^2/kg^2}$	$6.67 \times 10^{-11} \, \mathrm{N \cdot m^2/kg^2}$
N_A	Avogadro's number	$6.02214129(27) \times 10^{23}$	6.02×10^{23}
k	Boltzmann's constant	$1.3806488(13) \times 10^{-23} \, \mathrm{J/K}$	$1.38 \times 10^{-23} \, \mathrm{J/K}$
R	Gas constant	$8.3144621(75) \, \mathrm{J/mol \cdot K}$	$8.31 \, \mathrm{J/mol \cdot K} = 1.99 \, \mathrm{cal/mol \cdot K} = 0.0821 \, \mathrm{atm \cdot L/mol \cdot K}$
σ	Stefan-Boltzmann constant	$5.670373(21) \times 10^{-8} \, \mathrm{W/m^2 \cdot K}$	$5.67 \times 10^{-8} \, \mathrm{W/m^2 \cdot K}$
k	Coulomb force constant	$8.987551788... \times 10^9 \, \mathrm{N \cdot m^2/C^2}$	$8.99 \times 10^9 \, \mathrm{N \cdot m^2/C^2}$
q_e	Charge on electron	$-1.602176565(35) \times 10^{-19} \, \mathrm{C}$	$-1.60 \times 10^{-19} \, \mathrm{C}$
ε_0	Permittivity of free space	$8.854187817... \times 10^{-12} \, \mathrm{C^2/N \cdot m^2}$	$8.85 \times 10^{-12} \, \mathrm{C^2/N \cdot m^2}$
μ_0	Permeability of free space	$4\pi \times 10^{-7} \, \mathrm{T \cdot m/A}$	$1.26 \times 10^{-6} \, \mathrm{T \cdot m/A}$
h	Planck's constant	$6.62606957(29) \times 10^{-34} \, \mathrm{J \cdot s}$	$6.63 \times 10^{-34} \, \mathrm{J \cdot s}$

Table C2 Submicroscopic Masses [2]

Symbol	Meaning	Best Value	Approximate Value
m_e	Electron mass	$9.10938291(40) \times 10^{-31} \mathrm{kg}$	$9.11 \times 10^{-31} \mathrm{kg}$
m_p	Proton mass	$1.672621777(74) \times 10^{-27} \mathrm{kg}$	$1.6726 \times 10^{-27} \mathrm{kg}$
m_n	Neutron mass	$1.674927351(74) \times 10^{-27} \mathrm{kg}$	$1.6749 \times 10^{-27} \mathrm{kg}$

1. Stated values are according to the National Institute of Standards and Technology Reference on Constants, Units, and Uncertainty, www.physics.nist.gov/cuu (http://www.physics.nist.gov/cuu) (accessed May 18, 2012). Values in parentheses are the uncertainties in the last digits. Numbers without uncertainties are exact as defined.
2. Stated values are according to the National Institute of Standards and Technology Reference on Constants, Units, and Uncertainty, www.physics.nist.gov/cuu (http://www.physics.nist.gov/cuu) (accessed May 18, 2012). Values in parentheses are the uncertainties in the last digits. Numbers without uncertainties are exact as defined.

Symbol	Meaning	Best Value	Approximate Value
u	Atomic mass unit	$1.660538921(73) \times 10^{-27} kg$	$1.6605 \times 10^{-27} kg$

Table C3 Solar System Data

Sun	mass	$1.99 \times 10^{30} kg$
	average radius	$6.96 \times 10^{8} m$
	Earth-sun distance (average)	$1.496 \times 10^{11} m$
Earth	mass	$5.9736 \times 10^{24} kg$
	average radius	$6.376 \times 10^{6} m$
	orbital period	$3.16 \times 10^{7} s$
Moon	mass	$7.35 \times 10^{22} kg$
	average radius	$1.74 \times 10^{6} m$
	orbital period (average)	$2.36 \times 10^{6} s$
	Earth-moon distance (average)	$3.84 \times 10^{8} m$

Table C4 Metric Prefixes for Powers of Ten and Their Symbols

Prefix	Symbol	Value	Prefix	Symbol	Value
tera	T	10^{12}	deci	d	10^{-1}
giga	G	10^{9}	centi	c	10^{-2}
mega	M	10^{6}	milli	m	10^{-3}
kilo	k	10^{3}	micro	μ	10^{-6}
hecto	h	10^{2}	nano	n	10^{-9}
deka	da	10^{1}	pico	p	10^{-12}
—	—	$10^{0}(=1)$	femto	f	10^{-15}

Table C5 The Greek Alphabet

Alpha	A	α	Eta	H	η	Nu	N	ν	Tau	T	τ
Beta	B	β	Theta	Θ	θ	Xi	Ξ	ξ	Upsilon	Υ	υ
Gamma	Γ	γ	Iota	I	ι	Omicron	O	o	Phi	Φ	ϕ
Delta	Δ	δ	Kappa	K	κ	Pi	Π	π	Chi	X	χ
Epsilon	E	ε	Lambda	Λ	λ	Rho	P	ρ	Psi	Ψ	ψ
Zeta	Z	ζ	Mu	M	μ	Sigma	Σ	σ	Omega	Ω	ω

Table C6 SI Units

	Entity	Abbreviation	Name
Fundamental units	Length	m	meter
	Mass	kg	kilogram

	Entity	Abbreviation	Name
	Time	s	second
	Current	A	ampere
Supplementary unit	Angle	rad	radian
Derived units	Force	$N = kg \cdot m/s^2$	newton
	Energy	$J = kg \cdot m^2/s^2$	joule
	Power	$W = J/s$	watt
	Pressure	$Pa = N/m^2$	pascal
	Frequency	$Hz = 1/s$	hertz
	Electronic potential	$V = J/C$	volt
	Capacitance	$F = C/V$	farad
	Charge	$C = s \cdot A$	coulomb
	Resistance	$\Omega = V/A$	ohm
	Magnetic field	$T = N/(A \cdot m)$	tesla
	Nuclear decay rate	$Bq = 1/s$	becquerel

Table C7 Selected British Units

Length	1 inch (in.) = 2.54 cm (exactly)
	1 foot (ft) = 0.3048 m
	1 mile (mi) = 1.609 km
Force	1 pound (lb) = 4.448 N
Energy	1 British thermal unit (Btu) = 1.055×10^3 J
Power	1 horsepower (hp) = 746 W
Pressure	$1 \, lb/in^2 = 6.895 \times 10^3$ Pa

Table C8 Other Units

Length	1 light year (ly) = 9.46×10^{15} m
	1 astronomical unit (au) = 1.50×10^{11} m
	1 nautical mile = 1.852 km
	1 angstrom (Å) = 10^{-10} m
Area	1 acre (ac) = $4.05 \times 10^3 \, m^2$
	1 square foot (ft^2) = $9.29 \times 10^{-2} \, m^2$
	1 barn (b) = $10^{-28} \, m^2$
Volume	1 liter (L) = $10^{-3} \, m^3$
	1 U.S. gallon (gal) = $3.785 \times 10^{-3} \, m^3$

Mass	1 solar mass $= 1.99 \times 10^{30}$ kg
	1 metric ton $= 10^3$ kg
	1 atomic mass unit $(u) = 1.6605 \times 10^{-27}$ kg
Time	1 year $(y) = 3.16 \times 10^7$ s
	1 day $(d) = 86,400$ s
Speed	1 mile per hour (mph) $= 1.609$ km / h
	1 nautical mile per hour (naut) $= 1.852$ km / h
Angle	1 degree (°) $= 1.745 \times 10^{-2}$ rad
	1 minute of arc (') $= 1 / 60$ degree
	1 second of arc (") $= 1 / 60$ minute of arc
	1 grad $= 1.571 \times 10^{-2}$ rad
Energy	1 kiloton TNT (kT) $= 4.2 \times 10^{12}$ J
	1 kilowatt hour (kW $\cdot$ h) $= 3.60 \times 10^6$ J
	1 food calorie (kcal) $= 4186$ J
	1 calorie (cal) $= 4.186$ J
	1 electron volt (eV) $= 1.60 \times 10^{-19}$ J
Pressure	1 atmosphere (atm) $= 1.013 \times 10^5$ Pa
	1 millimeter of mercury (mm Hg) $= 133.3$ Pa
	1 torricelli (torr) $= 1$ mm Hg $= 133.3$ Pa
Nuclear decay rate	1 curie (Ci) $= 3.70 \times 10^{10}$ Bq

Table C9 Useful Formulae

Circumference of a circle with radius r or diameter d	$C = 2\pi r = \pi d$
Area of a circle with radius r or diameter d	$A = \pi r^2 = \pi d^2 / 4$
Area of a sphere with radius r	$A = 4\pi r^2$
Volume of a sphere with radius r	$V = (4 / 3)\left(\pi r^3\right)$

APPENDIX D | GLOSSARY OF KEY SYMBOLS AND NOTATION

In this glossary, key symbols and notation are briefly defined.

Table D1

Symbol	Definition
$\overline{\text{any symbol}}$	average (indicated by a bar over a symbol—e.g., $\bar{v}$ is average velocity)
°C	Celsius degree
°F	Fahrenheit degree
//	parallel
⊥	perpendicular
∝	proportional to
±	plus or minus
0	zero as a subscript denotes an initial value
α	alpha rays
α	angular acceleration
α	temperature coefficient(s) of resistivity
β	beta rays
β	sound level
β	volume coefficient of expansion
β^-	electron emitted in nuclear beta decay
β^+	positron decay
γ	gamma rays
γ	surface tension
$\gamma = 1/\sqrt{1 - v^2/c^2}$	a constant used in relativity
Δ	change in whatever quantity follows
δ	uncertainty in whatever quantity follows
ΔE	change in energy between the initial and final orbits of an electron in an atom
ΔE	uncertainty in energy
Δm	difference in mass between initial and final products
ΔN	number of decays that occur
Δp	change in momentum
Δp	uncertainty in momentum
ΔPE_g	change in gravitational potential energy
$\Delta \theta$	rotation angle

Symbol	Definition
Δs	distance traveled along a circular path
Δt	uncertainty in time
Δt_0	proper time as measured by an observer at rest relative to the process
ΔV	potential difference
Δx	uncertainty in position
ε_0	permittivity of free space
η	viscosity
θ	angle between the force vector and the displacement vector
θ	angle between two lines
θ	contact angle
θ	direction of the resultant
θ_b	Brewster's angle
θ_c	critical angle
κ	dielectric constant
λ	decay constant of a nuclide
λ	wavelength
λ_n	wavelength in a medium
μ_0	permeability of free space
μ_k	coefficient of kinetic friction
μ_s	coefficient of static friction
v_e	electron neutrino
π^+	positive pion
π^-	negative pion
π^0	neutral pion
ρ	density
ρ_c	critical density, the density needed to just halt universal expansion
ρ_{fl}	fluid density
$\bar{\rho}_{obj}$	average density of an object
ρ / ρ_w	specific gravity
τ	characteristic time constant for a resistance and inductance (RL) or resistance and capacitance (RC) circuit
τ	characteristic time for a resistor and capacitor (RC) circuit
τ	torque
Υ	upsilon meson
Φ	magnetic flux

Symbol	Definition
ϕ	phase angle
Ω	ohm (unit)
ω	angular velocity
A	ampere (current unit)
A	area
A	cross-sectional area
A	total number of nucleons
a	acceleration
a_B	Bohr radius
a_c	centripetal acceleration
a_t	tangential acceleration
AC	alternating current
AM	amplitude modulation
atm	atmosphere
B	baryon number
B	blue quark color
$\bar{B}$	antiblue (yellow) antiquark color
b	quark flavor bottom or beauty
B	bulk modulus
B	magnetic field strength
B_{int}	electron's intrinsic magnetic field
B_{orb}	orbital magnetic field
BE	binding energy of a nucleus—it is the energy required to completely disassemble it into separate protons and neutrons
BE/A	binding energy per nucleon
Bq	becquerel—one decay per second
C	capacitance (amount of charge stored per volt)
C	coulomb (a fundamental SI unit of charge)
C_p	total capacitance in parallel
C_s	total capacitance in series
CG	center of gravity
CM	center of mass
c	quark flavor charm
c	specific heat
c	speed of light
Cal	kilocalorie

Symbol	Definition
cal	calorie
COP_{hp}	heat pump's coefficient of performance
COP_{ref}	coefficient of performance for refrigerators and air conditioners
$\cos\theta$	cosine
$\cot\theta$	cotangent
$\csc\theta$	cosecant
D	diffusion constant
d	displacement
d	quark flavor down
dB	decibel
d_i	distance of an image from the center of a lens
d_o	distance of an object from the center of a lens
DC	direct current
E	electric field strength
ε	emf (voltage) or Hall electromotive force
emf	electromotive force
E	energy of a single photon
E	nuclear reaction energy
E	relativistic total energy
E	total energy
E_0	ground state energy for hydrogen
E_0	rest energy
EC	electron capture
E_{cap}	energy stored in a capacitor
Eff	efficiency—the useful work output divided by the energy input
Eff_C	Carnot efficiency
E_{in}	energy consumed (food digested in humans)
E_{ind}	energy stored in an inductor
E_{out}	energy output
e	emissivity of an object
e^+	antielectron or positron
eV	electron volt
F	farad (unit of capacitance, a coulomb per volt)
F	focal point of a lens
$\mathbf{F}$	force

Symbol	Definition
F	magnitude of a force
F	restoring force
F_B	buoyant force
F_c	centripetal force
F_i	force input
$\mathbf{F}_\text{net}$	net force
F_o	force output
FM	frequency modulation
f	focal length
f	frequency
f_0	resonant frequency of a resistance, inductance, and capacitance (RLC) series circuit
f_0	threshold frequency for a particular material (photoelectric effect)
f_1	fundamental
f_2	first overtone
f_3	second overtone
f_B	beat frequency
f_k	magnitude of kinetic friction
f_s	magnitude of static friction
G	gravitational constant
G	green quark color
$\bar{G}$	antigreen (magenta) antiquark color
g	acceleration due to gravity
g	gluons (carrier particles for strong nuclear force)
h	change in vertical position
h	height above some reference point
h	maximum height of a projectile
h	Planck's constant
hf	photon energy
h_i	height of the image
h_o	height of the object
I	electric current
I	intensity
I	intensity of a transmitted wave
I	moment of inertia (also called rotational inertia)

Symbol	Definition
I_0	intensity of a polarized wave before passing through a filter
I_{ave}	average intensity for a continuous sinusoidal electromagnetic wave
I_{rms}	average current
J	joule
J / Ψ	Joules/psi meson
K	kelvin
k	Boltzmann constant
k	force constant of a spring
K_α	x rays created when an electron falls into an $n = 1$ shell vacancy from the $n = 3$ shell
K_β	x rays created when an electron falls into an $n = 2$ shell vacancy from the $n = 3$ shell
kcal	kilocalorie
KE	translational kinetic energy
KE + PE	mechanical energy
KE_e	kinetic energy of an ejected electron
KE_{rel}	relativistic kinetic energy
KE_{rot}	rotational kinetic energy
$\overline{KE}$	thermal energy
kg	kilogram (a fundamental SI unit of mass)
L	angular momentum
L	liter
L	magnitude of angular momentum
L	self-inductance
ℓ	angular momentum quantum number
L_α	x rays created when an electron falls into an $n = 2$ shell from the $n = 3$ shell
L_e	electron total family number
L_μ	muon family total number
L_τ	tau family total number
L_f	heat of fusion
L_f and L_v	latent heat coefficients
L_{orb}	orbital angular momentum
L_s	heat of sublimation
L_v	heat of vaporization
L_z	z - component of the angular momentum
M	angular magnification

Symbol	Definition
M	mutual inductance
m	indicates metastable state
m	magnification
m	mass
m	mass of an object as measured by a person at rest relative to the object
m	meter (a fundamental SI unit of length)
m	order of interference
m	overall magnification (product of the individual magnifications)
$m\left(^{A}X\right)$	atomic mass of a nuclide
MA	mechanical advantage
m_e	magnification of the eyepiece
m_e	mass of the electron
m_ℓ	angular momentum projection quantum number
m_n	mass of a neutron
m_o	magnification of the objective lens
mol	mole
m_p	mass of a proton
m_s	spin projection quantum number
N	magnitude of the normal force
N	newton
$\mathbf{N}$	normal force
N	number of neutrons
n	index of refraction
n	number of free charges per unit volume
N_A	Avogadro's number
N_r	Reynolds number
N · m	newton-meter (work-energy unit)
N · m	newtons times meters (SI unit of torque)
OE	other energy
P	power
P	power of a lens
P	pressure
$\mathbf{p}$	momentum
p	momentum magnitude
p	relativistic momentum
$\mathbf{p}_{tot}$	total momentum

Symbol	Definition
$\mathbf{p}'_{tot}$	total momentum some time later
P_{abs}	absolute pressure
P_{atm}	atmospheric pressure
P_{atm}	standard atmospheric pressure
PE	potential energy
PE_{el}	elastic potential energy
PE_{elec}	electric potential energy
PE_s	potential energy of a spring
P_g	gauge pressure
P_{in}	power consumption or input
P_{out}	useful power output going into useful work or a desired, form of energy
Q	latent heat
Q	net heat transferred into a system
Q	flow rate—volume per unit time flowing past a point
$+Q$	positive charge
$-Q$	negative charge
q	electron charge
q_p	charge of a proton
q	test charge
QF	quality factor
R	activity, the rate of decay
R	radius of curvature of a spherical mirror
R	red quark color
$\bar{R}$	antired (cyan) quark color
R	resistance
R	resultant or total displacement
R	Rydberg constant
R	universal gas constant
r	distance from pivot point to the point where a force is applied
r	internal resistance
$r_\perp$	perpendicular lever arm
r	radius of a nucleus
r	radius of curvature
r	resistivity

Symbol	Definition
r or rad	radiation dose unit
rem	roentgen equivalent man
rad	radian
RBE	relative biological effectiveness
RC	resistor and capacitor circuit
rms	root mean square
r_n	radius of the nth H-atom orbit
R_p	total resistance of a parallel connection
R_s	total resistance of a series connection
R_s	Schwarzschild radius
S	entropy
S	intrinsic spin (intrinsic angular momentum)
S	magnitude of the intrinsic (internal) spin angular momentum
S	shear modulus
S	strangeness quantum number
s	quark flavor strange
s	second (fundamental SI unit of time)
s	spin quantum number
s	total displacement
$\sec \theta$	secant
$\sin \theta$	sine
s_z	z-component of spin angular momentum
T	period—time to complete one oscillation
T	temperature
T_c	critical temperature—temperature below which a material becomes a superconductor
T	tension
T	tesla (magnetic field strength B)
t	quark flavor top or truth
t	time
$t_{1/2}$	half-life—the time in which half of the original nuclei decay
$\tan \theta$	tangent
U	internal energy
u	quark flavor up
u	unified atomic mass unit
u	velocity of an object relative to an observer
u'	velocity relative to another observer

Symbol	Definition
V	electric potential
V	terminal voltage
V	volt (unit)
V	volume
$\mathbf{v}$	relative velocity between two observers
v	speed of light in a material
$\mathbf{v}$	velocity
$\bar{\mathbf{v}}$	average fluid velocity
$V_B - V_A$	change in potential
$\mathbf{v}_d$	drift velocity
V_p	transformer input voltage
V_{rms}	rms voltage
V_s	transformer output voltage
$\mathbf{v}_{tot}$	total velocity
v_w	propagation speed of sound or other wave
$\mathbf{v}_w$	wave velocity
W	work
W	net work done by a system
W	watt
w	weight
w_{fl}	weight of the fluid displaced by an object
W_c	total work done by all conservative forces
W_{nc}	total work done by all nonconservative forces
W_{out}	useful work output
X	amplitude
X	symbol for an element
$^{Z}X_{N}$	notation for a particular nuclide
x	deformation or displacement from equilibrium
x	displacement of a spring from its undeformed position
x	horizontal axis
X_C	capacitive reactance
X_L	inductive reactance
x_{rms}	root mean square diffusion distance
y	vertical axis
Y	elastic modulus or Young's modulus

Symbol	Definition
Z	atomic number (number of protons in a nucleus)
Z	impedance

ANSWER KEY

Chapter 1

Problems & Exercises

1

  a. 27.8 m/s

  b. 62.1 mph

3

$$\frac{1.0 \text{ m}}{\text{s}} = \frac{1.0 \text{ m}}{\text{s}} \times \frac{3600 \text{ s}}{1 \text{ hr}} \times \frac{1 \text{ km}}{1000 \text{ m}}$$

$$= 3.6 \text{ km/h}.$$

5

length: 377 ft; $4.53 \times 10^3 \text{ in}$. width: 280 ft; $3.3 \times 10^3 \text{ in}$.

7

8.847 km

9

(a) $1.3 \times 10^{-9} \text{ m}$

(b) 40 km/My

11

2 kg

13

  a. 85.5 to 94.5 km/h

  b. 53.1 to 58.7 mi/h

15

(a) 7.6×10^7 beats

(b) 7.57×10^7 beats

(c) 7.57×10^7 beats

17

  a. 3

  b. 3

  c. 3

19

a) 2.2%

(b) 59 to 61 km/h

21

80 ± 3 beats/min

23

2.8 h

25

$11 \pm 1 \text{ cm}^3$

27

$12.06 \pm 0.04 \text{ m}^2$

29

Sample answer: 2×10^9 heartbeats

31

Sample answer: 2×10^{31} if an average human lifetime is taken to be about 70 years.

33

Sample answer: 50 atoms

35

Sample answers:

(a) 10^{12} cells/hummingbird

(b) 10^{16} cells/human

Chapter 2

Problems & Exercises

1

(a) 7 m

(b) 7 m

(c) $+7$ m

3

(a) 13 m

(b) 9 m

(c) $+9$ m

5

(a) 3.0×10^4 m/s

(b) 0 m/s

7

2×10^7 years

9

34.689 m/s = 124.88 km/h

11

(a) 40.0 km/h

(b) 34.3 km/h, 25° S of E.

(c) average speed = 3.20 km/h, $\bar{v} = 0$.

13

384,000 km

15

(a) 6.61×10^{15} rev/s

(b) 0 m/s

16

4.29 m/s^2

18

(a) 1.43 s

(b) -2.50 m/s^2

20

(a) 10.8 m/s

(b)

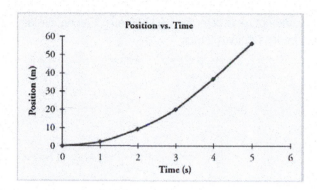

Figure 2.48.

21
38.9 m/s (about 87 miles per hour)
23

(a) 16.5 s

(b) 13.5 s

(c) -2.68 m/s^2

25

(a) 20.0 m

(b) -1.00 m/s

(c) This result does not really make sense. If the runner starts at 9.00 m/s and decelerates at 2.00 m/s^2, then she will have stopped after 4.50 s. If she continues to decelerate, she will be running backwards.

27
0.799 m
29

(a) 28.0 m/s

(b) 50.9 s

(c) 7.68 km to accelerate and 713 m to decelerate

31

(a) 51.4 m

(b) 17.1 s

33

(a) -80.4 m/s^2

(b) 9.33×10^{-2} s

35

(a) 7.7 m/s

(b) -15×10^2 m/s^2. This is about 3 times the deceleration of the pilots, who were falling from thousands of meters high!

37

(a) 32.6 m/s^2

(b) 162 m/s

(c) $v > v_{max}$, because the assumption of constant acceleration is not valid for a dragster. A dragster changes gears, and would have a greater acceleration in first gear than second gear than third gear, etc. The acceleration would be greatest at the beginning, so it would not be accelerating at 32.6 m/s^2 during the last few meters, but substantially less, and the final velocity would be less than 162 m/s.

39
104 s

40

(a) $v = 12.2 \text{ m/s}$; $a = 4.07 \text{ m/s}^2$

(b) $v = 11.2 \text{ m/s}$

41

(a) $y_1 = 6.28 \text{ m}$; $v_1 = 10.1 \text{ m/s}$

(b) $y_2 = 10.1 \text{ m}$; $v_2 = 5.20 \text{ m/s}$

(c) $y_3 = 11.5 \text{ m}$; $v_3 = 0.300 \text{ m/s}$

(d) $y_4 = 10.4 \text{ m}$; $v_4 = -4.60 \text{ m/s}$

43
$v_0 = 4.95 \text{ m/s}$

45

(a) $a = -9.80 \text{ m/s}^2$; $v_0 = 13.0 \text{ m/s}$; $y_0 = 0 \text{ m}$

(b) $v = 0 \text{m/s}$. Unknown is distance y to top of trajectory, where velocity is zero. Use equation $v^2 = v_0^2 + 2a(y - y_0)$ because it contains all known values except for y , so we can solve for y . Solving for y gives

$$
\begin{aligned}
v^2 - v_0^2 &= 2a(y - y_0) &\text{(2.100)}\\
\frac{v^2 - v_0^2}{2a} &= y - y_0 \\
y &= y_0 + \frac{v^2 - v_0^2}{2a} = 0 \text{ m} + \frac{(0 \text{ m/s})^2 - (13.0 \text{ m/s})^2}{2\left(-9.80 \text{ m/s}^2\right)} = 8.62 \text{ m}
\end{aligned}
$$

Dolphins measure about 2 meters long and can jump several times their length out of the water, so this is a reasonable result.

(c) 2.65 s

47

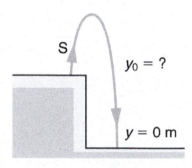

Figure 2.57.

(a) 8.26 m

(b) 0.717 s

49
1.91 s

51

(a) 94.0 m

(b) 3.13 s

53

(a) -70.0 m/s (downward)

(b) 6.10 s

55

(a) 19.6 m

(b) 18.5 m

57

(a) 305 m

(b) 262 m, -29.2 m/s

(c) 8.91 s

59

(a) 115 m/s

(b) 5.0 m/s^2

61

$$v = \frac{(11.7 - 6.95)\times 10^3 \text{ m}}{(40.0 - 20.0) \text{ s}} = 238 \text{ m/s}$$

63

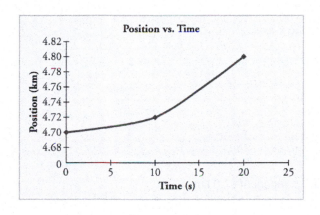

Figure 2.63.

65

(a) 6 m/s

(b) 12 m/s

(c) 3 m/s^2

(d) 10 s

Test Prep for AP® Courses

1
(b)

3

a. Use tape to mark off two distances on the track — one for cart *A* before the collision and one for the combined carts after the collision. Push cart *A* to give it an initial speed. Use a stopwatch to measure the time it takes for the cart(s) to cross the marked distances. The speeds are the distances divided by the times.

b. If the measurement errors are of the same magnitude, they will have a greater effect after the collision. The speed of the combined carts will be less than the initial speed of cart *A*. As a result, these errors will be a greater percentage of the actual velocity value after the collision occurs. (Note: Other arguments could properly be made for 'more error before the collision' and error that 'equally affects both sets of measurement.')

5

The position vs. time graph should be represented with a positively sloped line whose slope steadily decreases to zero. The *y*-intercept of the graph may be any value. The line on the velocity vs. time graph should have a positive *y*-intercept and a negative slope. Because the final velocity of the book is zero, the line should finish on the *x*-axis.

The position vs. time graph should be represented with a negatively sloped line whose slope steadily decreases to zero. The *y*-intercept of the graph may be any value. The line on the velocity vs. time graph should have a negative *y*-intercept and a positive slope. Because the final velocity of the book is zero, the line should finish on the *x*-axis.]

7

(c)

Chapter 3

Problems & Exercises

1

(a) $480 \, \text{m}$

(b) $379 \, \text{m}$, $18.4°$ east of north

3

north component 3.21 km, east component 3.83 km

5

$19.5 \, \text{m}$, $4.65°$ south of west

7

(a) $26.6 \, \text{m}$, $65.1°$ north of east

(b) $26.6 \, \text{m}$, $65.1°$ south of west

9

$52.9 \, \text{m}$, $90.1°$ with respect to the *x*-axis.

11

x-component 4.41 m/s

y-component 5.07 m/s

13

(a) 1.56 km

(b) 120 m east

15

North-component 87.0 km, east-component 87.0 km

17

30.8 m, 35.8 west of north

19

(a) $30.8 \, \text{m}$, $54.2°$ south of west

(b) $30.8 \, \text{m}$, $54.2°$ north of east

21

18.4 km south, then 26.2 km west(b) 31.5 km at $45.0°$ south of west, then 5.56 km at $45.0°$ west of north

23

$7.34 \, \text{km}$, $63.5°$ south of east

25

$x = 1.30 \, \text{m} \times 10^2$
$y = 30.9 \, \text{m}.$

27

(a) 3.50 s

(b) 28.6 m/s (c) 34.3 m/s

(d) 44.7 m/s, $50.2°$ below horizontal

29

(a) $18.4°$

(b) The arrow will go over the branch.

31

$$R = \frac{v_0}{\sin 2\theta_0\, g}$$

For $\theta = 45°$, $R = \frac{v_0}{g}$

$R = 91.8$ m for $v_0 = 30$ m/s ; $R = 163$ m for $v_0 = 40$ m/s ; $R = 255$ m for $v_0 = 50$ m/s .

33

(a) 560 m/s

(b) 8.00×10^3 m

(c) 80.0 m. This error is not significant because it is only 1% of the answer in part (b).

35

1.50 m, assuming launch angle of $45°$

37

$\theta = 6.1°$

yes, the ball lands at 5.3 m from the net

39

(a) −0.486 m

(b) The larger the muzzle velocity, the smaller the deviation in the vertical direction, because the time of flight would be smaller. Air resistance would have the effect of decreasing the time of flight, therefore increasing the vertical deviation.

41

4.23 m. No, the owl is not lucky; he misses the nest.

43

No, the maximum range (neglecting air resistance) is about 92 m.

45

15.0 m/s

47

(a) 24.2 m/s

(b) The ball travels a total of 57.4 m with the brief gust of wind.

49

$$y - y_0 = 0 = v_{0y}t - \frac{1}{2}gt^2 = (v_0\ \sin\ \theta)t - \frac{1}{2}gt^2 ,$$

so that $t = \dfrac{2(v_0\ \sin\ \theta)}{g}$

$x - x_0 = v_{0x}t = (v_0\ \cos\ \theta)t = R,$ and substituting for t gives:

$$R = v_0\ \cos\ \theta\left(\frac{2v_0\ \sin\ \theta}{g}\right) = \frac{2v_0^2\ \sin\ \theta\ \cos\ \theta}{g}$$

since $2\ \sin\ \theta\ \cos\ \theta = \sin\ 2\theta,$ the range is:

$$R = \frac{v_0^2\ \sin\ 2\theta}{g} .$$

52

(a) 35.8 km , $45°$ south of east

(b) 5.53 m/s , $45°$ south of east

(c) 56.1 km , $45°$ south of east

54

(a) 0.70 m/s faster

(b) Second runner wins

(c) 4.17 m

56
17.0 m/s , 22.1°

58

(a) 230 m/s , 8.0° south of west

(b) The wind should make the plane travel slower and more to the south, which is what was calculated.

60

(a) 63.5 m/s

(b) 29.6 m/s

62
6.68 m/s , 53.3° south of west

64

(a) $H_{\text{average}} = 14.9\dfrac{\text{km/s}}{\text{Mly}}$

(b) 20.2 billion years

66
1.72 m/s , 42.3° north of east

Test Prep for AP® Courses

1
(d)
3
We would need to know the horizontal and vertical positions of each ball at several times. From that data, we could deduce the velocities over several time intervals and also the accelerations (both horizontal and vertical) for each ball over several time intervals.
5

The graph of the ball's vertical velocity over time should begin at 4.90 m/s during the time interval 0 - 0.1 sec (there should be a data point at t = 0.05 sec, v = 4.90 m/s). It should then have a slope of -9.8 m/s², crossing through v = 0 at t = 0.55 sec and ending at v = -0.98 m/s at t = 0.65 sec.

The graph of the ball's horizontal velocity would be a constant positive value, a flat horizontal line at some positive velocity from t = 0 until t = 0.7 sec.

Chapter 4

Problems & Exercises

1
265 N
3
13.3 m/s²
7

(a) 12 m/s² .

(b) The acceleration is not one-fourth of what it was with all rockets burning because the frictional force is still as large as it was with all rockets burning.

9

(a) The system is the child in the wagon plus the wagon.

(b

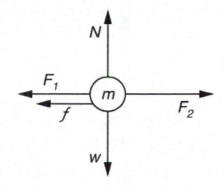

Figure 4.10.

(c) $a = 0.130 \text{ m/s}^2$ in the direction of the second child's push.

(d) $a = 0.00 \text{ m/s}^2$

11

(a) $3.68 \times 10^3 \text{ N}$. This force is 5.00 times greater than his weight.

(b) 3750 N; 11.3° above horizontal

13

$1.5 \times 10^3 \text{ N}, \ 150 \text{ kg}, \ 150 \text{ kg}$

15

Force on shell: $2.64 \times 10^7 \text{ N}$

Force exerted on ship = $-2.64 \times 10^7 \text{ N}$, by Newton's third law

17

a. 0.11 m/s^2

b. $1.2 \times 10^4 \text{ N}$

19

(a) $7.84 \times 10^{-4} \text{ N}$

(b) $1.89 \times 10^{-3} \text{ N}$. This is 2.41 times the tension in the vertical strand.

21

Newton's second law applied in vertical direction gives

$$F_y = F - 2T \sin \theta = 0 \tag{}$$
$$F = 2T \sin \theta \tag{}$$
$$T = \frac{F}{2 \sin \theta}. \tag{}$$

23

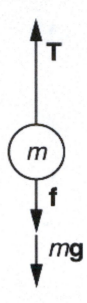

Figure 4.26.

Using the free-body diagram:

$$F_{\text{net}} = T - f - mg = ma \, ,$$

so that

$$a = \frac{T - f - mg}{m} = \frac{1.250 \times 10^7 \text{ N} - 4.50 \times 10^6 \text{ N} - (5.00 \times 10^5 \text{ kg})(9.80 \text{ m/s}^2)}{5.00 \times 10^5 \text{ kg}} = 6.20 \text{ m/s}^2 \, .$$

25
1. Use Newton's laws of motion.

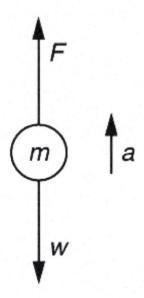

Figure 4.26.

2. Given : $a = 4.00g = (4.00)(9.80 \text{ m/s}^2) = 39.2 \text{ m/s}^2$; $m = 70.0 \text{ kg}$,
   Find: F.

3. $\sum F = +F - w = ma,$ so that $F = ma + w = ma + mg = m(a + g)$.

   $F = (70.0 \text{ kg})[(39.2 \text{ m/s}^2) + (9.80 \text{ m/s}^2)] \overset{=}{} 3.43 \times 10^3 \text{N}$. The force exerted by the high-jumper is actually down on the ground, but F is up from the ground and makes him jump.

4. This result is reasonable, since it is quite possible for a person to exert a force of the magnitude of 10^3 N .

27

(a) 4.41×10^5 N

(b) 1.50×10^5 N

29

(a) 910 N

(b) 1.11×10^3 N

31

$a = 0.139$ m/s , $\theta = 12.4°$ north of east

33

1. Use Newton's laws since we are looking for forces.

2. Draw a free-body diagram:

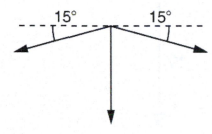

Figure 4.29.

3. The tension is given as $T = 25.0$ N. Find F_{app}. Using Newton's laws gives: $\Sigma F_y = 0,$ so that applied force is due to the y-components of the two tensions: $F_{app} = 2\,T\,\sin\theta = 2(25.0 \text{ N})\sin(15°) = 12.9$ N

   The x-components of the tension cancel. $\sum F_x = 0$.

4. This seems reasonable, since the applied tensions should be greater than the force applied to the tooth.

40

10.2 m/s^2 , $4.67°$ from vertical

42

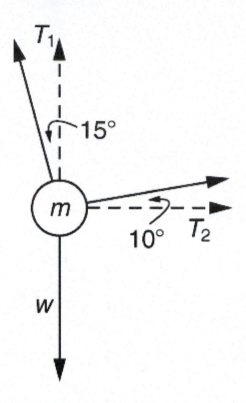

Figure 4.35.

$T_1 = 736 \text{ N}$

$T_2 = 194 \text{ N}$

44

(a) 7.43 m/s

(b) 2.97 m

46

(a) 4.20 m/s

(b) 29.4 m/s^2

(c) 4.31×10^3 N

48

(a) 47.1 m/s

(b) 2.47×10^3 m/s^2

(c) 6.18×10^3 N . The average force is 252 times the shell's weight.

52

(a) 1×10^{-13}

(b) 1×10^{-11}

54

10^2

Test Prep for AP® Courses

1

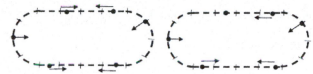

Figure 4.4. Car X is shown on the left, and Car Y is shown on the right.

i.

Car X takes longer to accelerate and does not spend any time traveling at top speed. Car Y accelerates over a shorter time and spends time going at top speed. So Car Y must cover the straightaways in a shorter time. Curves take the same time, so Car Y must overall take a shorter time.

ii.

The only difference in the calculations for the time of one segment of linear acceleration is the difference in distances. That shows that Car X takes longer to accelerate. The equation $\frac{d}{4v_c} = t_c$ corresponds to Car Y traveling for a time at top speed.

Substituting $a = \frac{v_c}{t_1}$ into the displacement equation in part (b) ii gives $D = \frac{3}{2}v_c t_1$. This shows that a car takes less time to reach its maximum speed when it accelerates over a shorter distance. Therefore, Car Y reaches its maximum speed more quickly, and spends more time at its maximum speed than Car X does, as argued in part (b) i.

3

A body cannot exert a force on itself. The hawk may accelerate as a result of several forces. The hawk may accelerate toward Earth as a result of the force due to gravity. The hawk may accelerate as a result of the additional force exerted on it by wind. The hawk may accelerate as a result of orienting its body to create less air resistance, thus increasing the net force forward.

5

(a) A soccer player, gravity, air, and friction commonly exert forces on a soccer ball being kicked.

(b) Gravity and the surrounding water commonly exert forces on a dolphin jumping. (The dolphin moves its muscles to exert a force on the water. The water exerts an equal force on the dolphin, resulting in the dolphin's motion.)

(c) Gravity and air exert forces on a parachutist drifting to Earth.

7

(c)

9

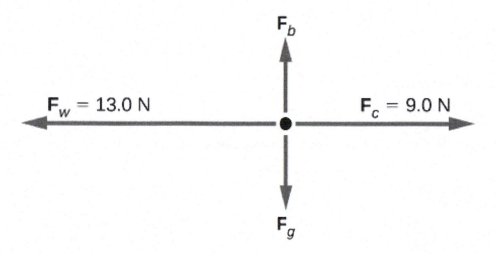

Figure 4.14.

The diagram consists of a black dot in the center and two small red arrows pointing up (Fb) and down (Fg) and two long red arrows pointing right (Fc = 9.0 N) and left (Fw=13.0 N).

In the diagram, F_g represents the force due to gravity on the balloon, and F_b represents the buoyant force. These two forces are equal in magnitude and opposite in direction. F_c represents the force of the current. F_w represents the force of the wind. The net force on the balloon will be $F_w - F_c = 4.0$ N and the balloon will accelerate in the direction the wind is blowing.

11

Since $m = F\,/\,a$, the parachutist has a mass of $539\,\text{N}/9.8\,\text{km/s}^2 = 55\,\text{kg}$.

For the first 2 s, the parachutist accelerates at 9.8 m/s^2.

$$v = at$$
$$= 9.8\frac{\text{m}}{\text{s}^2} \bullet 2\text{s}$$
$$= 17.6\frac{\text{m}}{\text{s}}$$

Her speed after 2 s is 19.6 m/s.

From 2 s to 10 s, the net force on the parachutist is 539 N – 615 N, or 76 N upward.

$$a = \frac{F}{m}$$
$$= \frac{-76\,\text{N}}{55\,\text{kg}}$$
$$= -1.4\frac{\text{m}}{\text{s}^2}$$

Since $v = v_0 + at$, $v = 17.6\,\text{m/s}^2 + (-1.4\,\text{m/s}^2)(8\text{s}) = 6.5\,\text{m/s}^2$.

At 10 s, the parachutist is falling to Earth at 8.4 m/s.

13

The system includes the gardener and the wheelbarrow with its contents. The following forces are important to include: the weight of the wheelbarrow, the weight of the gardener, the normal force for the wheelbarrow and the gardener, the force of the gardener pushing against the ground and the equal force of the ground pushing back against the gardener, and any friction in the wheelbarrow's wheels.

15

The system undergoing acceleration is the two figure skaters together.

Net force = $120\,\text{N} - 5.0\,\text{N} = 115\,\text{N}$.

Total mass = $40\,\text{kg} + 50\,\text{kg} = 90\,\text{kg}$.

Using Newton's second law, we have that

$$a = \frac{F}{m}$$
$$= \frac{115\,\text{N}}{90\,\text{kg}}$$
$$= 1.28\frac{\text{m}}{\text{s}^2}$$

The pair accelerates forward at 1.28 m/s^2.

17

The force of tension must equal the force of gravity plus the force necessary to accelerate the mass. $F = mg$ can be used to calculate the first, and $F = ma$ can be used to calculate the second.

For gravity:

$$F = mg$$
$$= (120.0\,\text{kg})(9.8\,\text{m/s}^2)$$
$$= 1205.4\,\text{N}$$

For acceleration:

$$F = ma$$
$$= (120.0\,\text{kg})(1.3\,\text{m/s}^2)$$
$$= 159.9\,\text{N}$$

The total force of tension in the cable is 1176 N + 156 N = 1332 N.

19

(b)

21

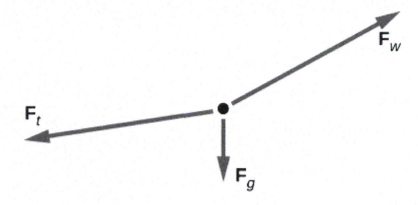

Figure 4.24.

The diagram has a black dot and three solid red arrows pointing away from the dot. Arrow Ft is long and pointing to the left and slightly down. Arrow Fw is also long and is a bit below a diagonal line halfway between pointing up and pointing to the right. A short arrow Fg is pointing down.

F$_g$ is the force on the kite due to gravity.

F$_w$ is the force exerted on the kite by the wind.

F$_t$ is the force of tension in the string holding the kite. It must balance the vector sum of the other two forces for the kite to float stationary in the air.

23
(b)
25
(d)
27

A free-body diagram would show a northward force of 64 N and a westward force of 38 N. The net force is equal to the sum of the two applied forces. It can be found using the Pythagorean theorem:

$$F_{net} = \sqrt{F_x{}^2 + F_y{}^w}$$
$$= \sqrt{(38\,N)^2 + (64\,N)^2}$$
$$= 74.4\,N$$

Since $a = \frac{F}{m}$,

$$a = \frac{74.4\,N}{825\,kg}$$
$$= 0.09\,m/s^2$$

The boulder will accelerate at 0.09 m/s^2.

29
(b)
31
(b)
33
(d)

Chapter 5

Problems & Exercises

1
5.00 N

4

(a) 588 N

(b) 1.96 m/s^2

6

(a) 3.29 m/s^2

(b) 3.52 m/s^2

(c) 980 N; 945 N

10

1.83 m/s^2

14

(a) 4.20 m/s^2

(b) 2.74 m/s^2

(c) −0.195 m/s^2

16

(a) 1.03×10^6 N

(b) 3.48×10^5 N

18

(a) 51.0 N

(b) 0.720 m/s^2

20

115 m/s; 414 km/hr

22

25 m/s; 9.9 m/s

24

2.9

26

$$[\eta] = \frac{[F_s]}{[r][v]} = \frac{\text{kg} \cdot \text{m/s}^2}{\text{m} \cdot \text{m/s}} = \frac{\text{kg}}{\text{m} \cdot \text{s}}$$

28

0.76 kg/m $\cdot$ s

29

$$1.90 \times 10^{-3} \text{ cm} \tag{5.57}$$

31

(a) 1 mm

(b) This does seem reasonable, since the lead does seem to shrink a little when you push on it.

33

(a) 9 cm

(b) This seems reasonable for nylon climbing rope, since it is not supposed to stretch that much.

35

8.59 mm

37

$$1.49 \times 10^{-7} \text{ m} \tag{5.58}$$

39

(a) 3.99×10^{-7} m

(b) 9.67×10^{-8} m

41

4×10^6 N/m^2. This is about 36 atm, greater than a typical jar can withstand.

43

1.4 cm

Test Prep for AP® Courses

1

(b)

3

(c)

Chapter 6

Problems & Exercises

1

723 km

3

5×10^7 rotations

5

117 rad/s

7

76.2 rad/s

728 rpm

8

(a) 33.3 rad/s

(b) 500 N

(c) 40.8 m

10

12.9 rev/min

12

4×10^{21} m

14

a) $3.47\times10^4 \, \mathrm{m/s^2}$, $3.55\times10^3 \, g$

b) $51.1 \, \mathrm{m/s}$

16

a) 31.4 rad/s

b) $118 \, \mathrm{m/s}$

c) $384 \, \mathrm{m/s}$

d)The centripetal acceleration felt by Olympic skaters is 12 times larger than the acceleration due to gravity. That's quite a lot of acceleration in itself. The centripetal acceleration felt by Button's nose was 39.2 times larger than the acceleration due to gravity. It is no wonder that he ruptured small blood vessels in his spins.

18

a) 0.524 km/s

b) 29.7 km/s

20

(a) 1.35×10^3 rpm

(b) $8.47\times10^3 \, \mathrm{m/s^2}$

(c) $8.47\times10^{-12} \, \mathrm{N}$

(d) 865

21

(a) $16.6 \, \mathrm{m/s}$

(b) $19.6 \, \mathrm{m/s^2}$

(c)

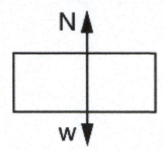

Figure 6.10.

(d) 1.76×10^3 N or $3.00\ w$, that is, the normal force (upward) is three times her weight.

(e) This answer seems reasonable, since she feels like she's being forced into the chair MUCH stronger than just by gravity.

22

a) $40.5\ \mathrm{m/s^2}$

b) 905 N

c) The force in part (b) is very large. The acceleration in part (a) is too much, about 4 g.

d) The speed of the swing is too large. At the given velocity at the bottom of the swing, there is enough kinetic energy to send the child all the way over the top, ignoring friction.

23

a) 483 N

b) 17.4 N

c) 2.24 times her weight, 0.0807 times her weight

25

 4.14°

27

a) 24.6 m

b) $36.6\ \mathrm{m/s^2}$

c) $a_c = 3.73\ g.$ This does not seem too large, but it is clear that bobsledders feel a lot of force on them going through sharply banked turns.

29

a) 2.56 rad/s

b) 5.71°

30

a) 16.2 m/s

b) 0.234

32

a) 1.84

b) A coefficient of friction this much greater than 1 is unreasonable .

c) The assumed speed is too great for the tight curve.

33

a) $5.979 \times 10^{24}\ \mathrm{kg}$

b) This is identical to the best value to three significant figures.

35

a) $1.62\ \mathrm{m/s^2}$

b) $3.75 \ m/s^2$

37

a) $3.42\times10^{-5} \ m/s^2$

b) $3.34\times10^{-5} \ m/s^2$

The values are nearly identical. One would expect the gravitational force to be the same as the centripetal force at the core of the system.

39

a) $7.01\times10^{-7} \ N$

b) $1.35\times10^{-6} \ N$, 0.521

41

a) $1.66\times10^{-10} \ m/s^2$

b) $2.17\times10^5 \ m/s$

42

a) $2.94\times10^{17} \ kg$

b) 4.92×10^{-8}

of the Earth's mass.

c) The mass of the mountain and its fraction of the Earth's mass are too great.

d) The gravitational force assumed to be exerted by the mountain is too great.

44

$1.98\times10^{30} \ kg$

46

$\dfrac{M_J}{M_E} = 316$

48

a) $7.4\times10^3 \ m/s$

b) $1.05\times10^3 \ m/s$

c) $2.86\times10^{-7} \ s$

d) $1.84\times10^7 \ N$

e) $2.76\times10^4 \ J$

49

a) $5.08\times10^3 \ km$

b) This radius is unreasonable because it is less than the radius of earth.

c) The premise of a one-hour orbit is inconsistent with the known radius of the earth.

Test Prep for AP® Courses

1
(a)
3
(b)
5
(b)

Chapter 7

Problems & Exercises

1

$3.00 \text{ J} = 7.17 \times 10^{-4} \text{ kcal}$

3

(a) $5.92 \times 10^5 \text{ J}$

(b) $-5.88 \times 10^5 \text{ J}$

(c) The net force is zero.

5

$3.14 \times 10^3 \text{ J}$

7

(a) -700 J

(b) 0

(c) 700 J

(d) 38.6 N

(e) 0

9

$1/250$

11

$1.1 \times 10^{10} \text{ J}$

13

$2.8 \times 10^3 \text{ N}$

15

102 N

16

(a) $1.96 \times 10^{16} \text{ J}$

(b) The ratio of gravitational potential energy in the lake to the energy stored in the bomb is 0.52. That is, the energy stored in the lake is approximately half that in a 9-megaton fusion bomb.

18

(a) 1.8 J

(b) 8.6 J

20

$$v_f = \sqrt{2gh + v_0{}^2} = \sqrt{2(9.80 \text{ m/s}^2)(-0.180 \text{ m}) + (2.00 \text{ m/s})^2} = 0.687 \text{ m/s}$$

22

$7.81 \times 10^5 \text{ N/m}$

24

9.46 m/s

26

4×10^4 molecules

27

Equating ΔPE_g and ΔKE, we obtain $v = \sqrt{2gh + v_0{}^2} = \sqrt{2(9.80 \text{ m/s}^2)(20.0 \text{ m}) + (15.0 \text{ m/s})^2} = 24.8 \text{ m/s}$

29

(a) 25×10^6 years

(b) This is much, much longer than human time scales.

30

2×10^{-10}

32

(a) 40

(b) 8 million

34
$149

36

(a) 208 W

(b) 141 s

38

(a) 3.20 s

(b) 4.04 s

40

(a) 9.46×10^7 J

(b) 2.54 y

42

Identify knowns: $m = 950$ kg , slope angle $\theta = 2.00°$, $v = 3.00$ m/s , $f = 600$ N

Identify unknowns: power P of the car, force F that car applies to road

Solve for unknown:

$$P = \frac{W}{t} = \frac{Fd}{t} = F\left(\frac{d}{t}\right) = Fv,$$

where F is parallel to the incline and must oppose the resistive forces and the force of gravity:

$$F = f + w = 600 \text{ N} + mg \sin \theta$$

Insert this into the expression for power and solve:

$$\begin{aligned} P &= (f + mg \sin \theta)v \\ &= \left[600 \text{ N} + (950 \text{ kg})(9.80 \text{ m/s}^2)\sin 2°\right](30.0 \text{ m/s}) \\ &= 2.77 \times 10^4 \text{ W} \end{aligned}$$

About 28 kW (or about 37 hp) is reasonable for a car to climb a gentle incline.

44

(a) 9.5 min

(b) 69 flights of stairs

46
641 W, 0.860 hp

48
31 g

50
14.3%

52

(a) 3.21×10^4 N

(b) 2.35×10^3 N

(c) Ratio of net force to weight of person is 41.0 in part (a); 3.00 in part (b)

54

(a) 108 kJ

(b) 599 W

56

(a) 144 J

(b) 288 W

58

(a) 2.50×10^{12} J

(b) 2.52%

(c) 1.4×10^4 kg (14 metric tons)

60

(a) 294 N

(b) 118 J

(c) 49.0 W

62

(a) 0.500 m/s^2

(b) 62.5 N

(c) Assuming the acceleration of the swimmer decreases linearly with time over the 5.00 s interval, the frictional force must therefore be increasing linearly with time, since $f = F - ma$. If the acceleration decreases linearly with time, the velocity will contain a term dependent on time squared (t^2). Therefore, the water resistance will not depend linearly on the velocity.

64

(a) 16.1×10^3 N

(b) 3.22×10^5 J

(c) 5.66 m/s

(d) 4.00 kJ

66

(a) 4.65×10^3 kcal

(b) 38.8 kcal/min

(c) This power output is higher than the highest value on **Table 7.5**, which is about 35 kcal/min (corresponding to 2415 watts) for sprinting.

(d) It would be impossible to maintain this power output for 2 hours (imagine sprinting for 2 hours!).

69

(a) 4.32 m/s

(b) 3.47×10^3 N

(c) 8.93 kW

Test Prep for AP® Courses

1
(b)
3
(d)
5
(a)
7
The kinetic energy should change in the form of –cos, with an initial value of 0 or slightly above, and ending at the same level.
9
Any force acting perpendicular will have no effect on kinetic energy. Obvious examples are gravity and the normal

force, but others include wind directly from the side and rain or other precipitation falling straight down.

11

Note that the wind is pushing from behind and one side, so your KE will increase. The net force has components of 1400 N in the direction of travel and 212 N perpendicular to the direction of travel. So the net force is 1420 N at 8.5 degrees from the direction of travel.

13

Gravity has a component perpendicular to the cannon (and to displacement, so it is irrelevant) and has a component parallel to the cannon. The latter is equal to 9.8 N. Thus the net force in the direction of the displacement is 50 N − 9.8 N, and the kinetic energy is 60 J.

15

The potato cannon (and many other projectile launchers) above is an option, with a force launching the projectile, friction, potentially gravity depending on the direction it is pointed, etc. A drag (or other) car accelerating is another possibility.

17

The kinetic energy of the rear wagon increases. The front wagon does not, until the rear wagon collides with it. The total system may be treated by its center of mass, halfway between the wagons, and its energy increases by the same amount as the sum of the two individual wagons.

19

(d)

21

0.049 J; 0.041 m, 0.25 m

23

20 m high, 20 m/s.

25

(a)

27

(d)

29

(c)

31

(b)

33

(c)

35

(c)

37

(c), (d)

39

(a)

41

(b)

Chapter 8

Problems & Exercises

1

(a) $1.50 \times 10^4 \text{ kg} \cdot \text{m/s}$

(b) 625 to 1

(c) $6.66 \times 10^2 \text{ kg} \cdot \text{m/s}$

3

(a) $8.00 \times 10^4 \text{ m/s}$

(b) $1.20 \times 10^6 \text{ kg} \cdot \text{m/s}$

(c) Because the momentum of the airplane is 3 orders of magnitude smaller than of the ship, the ship will not recoil very much. The recoil would be -0.0100 m/s, which is probably not noticeable.

5

54 s

7

$9.00{\times}10^3$ N

9

a) $2.40{\times}10^3$ N toward the leg

b) The force on each hand would have the same magnitude as that found in part (a) (but in opposite directions by Newton's third law) because the change in momentum and the time interval are the same.

11

a) 800 kg $\cdot$ m/s away from the wall

b) 1.20 m/s away from the wall

13

(a) $1.50{\times}10^6$ N away from the dashboard

(b) $1.00{\times}10^5$ N away from the dashboard

15

$4.69{\times}10^5$ N in the boat's original direction of motion

17

$2.10{\times}10^3$ N away from the wall

19

$$\mathbf{p} = m\mathbf{v} \Rightarrow p^2 = m^2v^2 \Rightarrow \frac{p^2}{m} = mv^2$$

$$\Rightarrow \frac{p^2}{2m} = \frac{1}{2}mv^2 = KE$$

$$KE = \frac{p^2}{2m}$$

21
60.0 g
23
0.122 m/s
25

In a collision with an identical car, momentum is conserved. Afterwards $v_f = 0$ for both cars. The change in momentum will be the same as in the crash with the tree. However, the force on the body is not determined since the time is not known. A padded stop will reduce injurious force on body.

27
22.4 m/s in the same direction as the original motion
29
0.250 m/s
31

(a) 86.4 N perpendicularly away from the bumper

(b) 0.389 J

(c) 64.0%

33

(a) 8.06 m/s

(b) -56.0 J

(c)(i) 7.88 m/s; (ii) -223 J

35

(a) 0.163 m/s in the direction of motion of the more massive satellite

(b) 81.6 J

(c) $8.70{\times}10^{-2}$ m/s in the direction of motion of the less massive satellite, 81.5 J. Because there are no external forces, the velocity of the center of mass of the two-satellite system is unchanged by the collision. The two velocities calculated above are the velocity of the center of mass in each of the two different individual reference frames. The loss in KE is the same in both reference frames because the KE lost to internal forces (heat, friction, etc.) is the same regardless of the coordinate system chosen.

37

0.704 m/s

−2.25 m/s

38

(a) 4.58 m/s away from the bullet

(b) 31.5 J

(c) −0.491 m/s

(d) 3.38 J

40

(a) 1.02×10^{-6} m/s

(b) 5.63×10^{20} J (almost all KE lost)

(c) Recoil speed is 6.79×10^{-17} m/s , energy lost is 6.25×10^{9} J . The plume will not affect the momentum result because the plume is still part of the Moon system. The plume may affect the kinetic energy result because a significant part of the initial kinetic energy may be transferred to the kinetic energy of the plume particles.

42
24.8 m/s
44

(a) 4.00 kg

(b) 210 J

(c) The clown does work to throw the barbell, so the kinetic energy comes from the muscles of the clown. The muscles convert the chemical potential energy of ATP into kinetic energy.

45

(a) 3.00 m/s, $60°$ below x-axis

(b) Find speed of first puck after collision: $0 = mv'_1 \sin 30° - mv'_2 \sin 60° \Rightarrow v'_1 = v'_2\dfrac{\sin 60°}{\sin 30°} = 5.196$ m/s

Verify that ratio of initial to final KE equals one:
$$\left.\begin{array}{l}\mathrm{KE} = \frac{1}{2}mv_1{}^2 = 18m \text{ J}\\ \mathrm{KE} = \frac{1}{2}mv'_1{}^2 + \frac{1}{2}mv'_2{}^2 = 18m \text{ J}\end{array}\right\}\dfrac{\mathrm{KE}}{\mathrm{KE}'} = 1.00$$

47

(a) −2.26 m/s

(b) 7.63×10^{3} J

(c) The ground will exert a normal force to oppose recoil of the cannon in the vertical direction. The momentum in the vertical direction is transferred to the earth. The energy is transferred into the ground, making a dent where the cannon is. After long barrages, cannon have erratic aim because the ground is full of divots.

49

(a) 5.36×10^{5} m/s at $-29.5°$

(b) 7.52×10^{-13} J

51

We are given that $m_1 = m_2 \equiv m$. The given equations then become:

$$v_1 = v_1 \cos\theta_1 + v_2 \cos\theta_2$$

and

$0 = v'_1 \sin \theta_1 + v'_2 \sin \theta_2.$

Square each equation to get

$$v_1^2 = v'^2_1 \cos^2 \theta_1 + v'^2_2 \cos^2 \theta_2 + 2v'_1 v'_2 \cos \theta_1 \cos \theta_2$$
$$0 = v'^2_1 \sin^2 \theta_1 + v'^2_2 \sin^2 \theta_2 + 2v'_1 v'_2 \sin \theta_1 \sin \theta_2.$$

Add these two equations and simplify:

$$v_1^2 = v'^2_1 + v'^2_2 + 2v'_1 v'_2 (\cos \theta_1 \cos \theta_2 + \sin \theta_1 \sin \theta_2)$$
$$= v'^2_1 + v'^2_2 + 2v'_1 v'_2 \left[\tfrac{1}{2} \cos (\theta_1 - \theta_2) + \tfrac{1}{2} \cos (\theta_1 + \theta_2) + \tfrac{1}{2} \cos (\theta_1 - \theta_2) - \tfrac{1}{2} \cos (\theta_1 + \theta_2) \right]$$
$$= v'^2_1 + v'^2_2 + 2v'_1 v'_2 \cos (\theta_1 - \theta_2).$$

Multiply the entire equation by $\frac{1}{2}m$ to recover the kinetic energy:

$$\tfrac{1}{2}mv_1^2 = \tfrac{1}{2}mv'^2_1 + \tfrac{1}{2}mv'^2_2 + mv'_1 v'_2 \cos(\theta_1 - \theta_2)$$

53

39.2 m/s^2

55

$4.16 \times 10^3 \text{ m/s}$

57

The force needed to give a small mass Δm an acceleration $a_{\Delta m}$ is $F = \Delta m a_{\Delta m}$. To accelerate this mass in the small time interval Δt at a speed v_e requires $v_e = a_{\Delta m} \Delta t$, so $F = v_e \frac{\Delta m}{\Delta t}$. By Newton's third law, this force is equal in magnitude to the thrust force acting on the rocket, so $F_{\text{thrust}} = v_e \frac{\Delta m}{\Delta t}$, where all quantities are positive. Applying Newton's second law to the rocket gives $F_{\text{thrust}} - mg = ma \Rightarrow a = \frac{v_e}{m} \frac{\Delta m}{\Delta t} - g$, where m is the mass of the rocket and unburnt fuel.

60

$2.63 \times 10^3 \text{ kg}$

61

(a) 0.421 m/s away from the ejected fluid.

(b) 0.237 J.

Test Prep for AP® Courses

1
(b)
3
(b)
5
(a)
7
(c) (based on calculation of $F = \frac{m\Delta v}{\Delta t}$)

9
(c)
11
(d)
13
(b)
15
(d)
17
(b)

19
(c)
21
(b)
23
(c)
25
(b)
27
(a)
29
(c)
31
(b)
33
(a)
35
(b)
37
(a)
39
(a)
41
(d)
43
(c). Because of conservation of momentum, the final velocity of the combined mass must be 4.286 m/s. The initial kinetic energy is $(0.5)(2.0)(15)^2 = 225 \text{ J}$. The final kinetic energy is $(0.5)(7.0)(4.286)^2 = 64 \text{ J}$, so the difference is -161 J.
45
(a)
47
(d)
49
(c)
51
(b)

Chapter 9

Problems & Exercises

1

a) 46.8 N·m

b) It does not matter at what height you push. The torque depends on only the magnitude of the force applied and the perpendicular distance of the force's application from the hinges. (Children don't have a tougher time opening a door because they push lower than adults, they have a tougher time because they don't push far enough from the hinges.)

3
23.3 N
5

Given:

$$m_1 \;=\; 26.0 \text{ kg}, \, m_2 = 32.0 \text{ kg}, \, m_s = 12.0 \text{ kg},$$
$$r_1 \;=\; 1.60 \text{ m}, \, r_s = 0.160 \text{ m}, \text{ find (a } r_2, \text{ (b) } F_p$$

a) Since children are balancing:

$$\text{net } \tau_{cw} = - \text{ net } \tau_{ccw}$$
$$\Rightarrow w_1 r_1 + m_s g r_s = w_2 r_2$$

So, solving for r_2 gives:

$$r_2 = \frac{w_1 r_1 + m_s g r_s}{w_2} = \frac{m_1 g r_1 + m_s g r_s}{m_2 g} = \frac{m_1 r_1 + m_s r_s}{m_2}$$

$$= \frac{(26.0 \text{ kg})(1.60 \text{ m}) + (12.0 \text{ kg})(0.160 \text{ m})}{32.0 \text{ kg}}$$

$$= 1.36 \text{ m}$$

b) Since the children are not moving:

$$\text{net } F = 0 = F_p - w_1 - w_2 - w_s$$
$$\Rightarrow F_p = w_1 + w_2 + w_s$$

So that

$$F_p = (26.0 \text{ kg} + 32.0 \text{ kg} + 12.0 \text{ kg})(9.80 \text{ m}/\text{s}^2)$$
$$= 686 \text{ N}$$

6

$$F_{\text{wall}} = 1.43 \times 10^3 \text{ N}$$

8

a) 2.55×10^3 N, 16.3° to the left of vertical (i.e., toward the wall)

b) 0.292

10

$$F_B = 2.12 \times 10^4 \text{ N}$$

12

a) 0.167, or about one-sixth of the weight is supported by the opposite shore.

b) $F = 2.0 \times 10^4$ N , straight up.

14

a) 21.6 N

b) 21.6 N

16

350 N directly upwards

19

25

50 N

21

a) $\text{MA} = 18.5$

b) $F_i = 29.1$ N

c) 510 N downward

23

1.3×10^3 N

25

a) $T = 299$ N

b) 897 N upward

26

F_B = 470 N; r_1 = 4.00 cm; w_a = 2.50 kg; r_2 = 16.0 cm; w_b = 4.00 kg; r_3 = 38.0 cm

$$F_E = w_a\left(\frac{r_2}{r_1} - 1\right) + w_b\left(\frac{r_3}{r_1} - 1\right)$$

$$= (2.50\text{ kg})(9.80\text{ m}/\text{s}^2)\left(\frac{16.0\text{ cm}}{4.0\text{ cm}} - 1\right)$$

$$+ (4.00\text{ kg})(9.80\text{ m}/\text{s}^2)\left(\frac{38.0\text{ cm}}{4.00\text{ cm}} - 1\right)$$

$$= 407\text{ N}$$

28

$$1.1\times10^3\text{ N}$$

$\theta = 190°$ ccw from positive x axis

30

$F_V = 97$ N, $\theta = 59°$

32

(a) 25 N downward

(b) 75 N upward

33

(a) $F_A = 2.21\times10^3$ N upward

(b) $F_B = 2.94\times10^3$ N downward

35

(a) $F_{\text{teeth on bullet}} = 1.2\times10^2$ N upward

(b) $F_J = 84$ N downward

37

(a) 147 N downward

(b) 1680 N, 3.4 times her weight

(c) 118 J

(d) 49.0 W

39

a) $\bar{x}_2 = 2.33$ m

b) The seesaw is 3.0 m long, and hence, there is only 1.50 m of board on the other side of the pivot. The second child is off the board.

c) The position of the first child must be shortened, i.e. brought closer to the pivot.

Test Prep for AP® Courses

1
(a)
3
Both objects are in equilibrium. However, they will respond differently if a force is applied to their sides. If the cone placed on its base is displaced to the side, its center of gravity will remain over its base and it will return to its original position. When the traffic cone placed on its tip is displaced to the side, its center of gravity will drift from its base, causing a torque that will accelerate it to the ground.
5
(d)
7

   a. F_L = 7350 N, F_R = 2450 N

   b.   As the car moves to the right side of the bridge, F_L will decrease and F_R will increase. (At exactly halfway across the bridge, F_L and F_R will both be 4900 N.)

9
The student should mention that the guiding principle behind simple machines is the second condition of equilibrium.

Though the torque leaving a machine must be equivalent to torque entering a machine, the same requirement does not exist for forces. As a result, by decreasing the lever arm to the existing force, the size of the existing force will be increased. The mechanical advantage will be equivalent to the ratio of the forces exiting and entering the machine.

11

a. The force placed on your bicep muscle will be greater than the force placed on the dumbbell. The bicep muscle is closer to your elbow than the downward force placed on your hand from the dumbbell. Because the elbow is the pivot point of the system, this results in a decreased lever arm for the bicep. As a result, the force on the bicep must be greater than that placed on the dumbbell. (How much greater? The ratio between the bicep and dumbbell forces is equal to the inverted ratio of their distances from the elbow. If the dumbbell is ten times further from the elbow than the bicep, the force on the bicep will be 200 pounds!)

b. The force placed on your bicep muscle will decrease. As the forearm lifts the dumbbell, it will get closer to the elbow. As a result, the torque placed on the arm from the weight will decrease and the countering torque created by the bicep muscle will do so as well.

Chapter 10

Problems & Exercises

1

$\omega = 0.737$ rev/s

3

(a) -0.26 rad/s^2

(b) 27 rev

5

(a) 80 rad/s^2

(b) 1.0 rev

7

(a) 45.7 s

(b) 116 rev

9

a) 600 rad/s^2

b) 450 rad/s

c) 21.0 m/s

10

(a) 0.338 s

(b) 0.0403 rev

(c) 0.313 s

12

0.50 kg $\cdot$ m^2

14

(a) 50.4 N $\cdot$ m

(b) 17.1 rad/s^2

(c) 17.0 rad/s^2

16

3.96×10^{18} s

or 1.26×10^{11} y

18

$$I_{end} = I_{center} + m\left(\frac{l}{2}\right)^2$$

Thus, $I_{center} = I_{end} - \frac{1}{4}ml^2 = \frac{1}{3}ml^2 - \frac{1}{4}ml^2 = \frac{1}{12}ml^2$

19

(a) 2.0 ms

(b) The time interval is too short.

(c) The moment of inertia is much too small, by one to two orders of magnitude. A torque of $500\ \text{N} \cdot \text{m}$ is reasonable.

20

(a) 17,500 rpm

(b) This angular velocity is very high for a disk of this size and mass. The radial acceleration at the edge of the disk is > 50,000 gs.

(c) Flywheel mass and radius should both be much greater, allowing for a lower spin rate (angular velocity).

21

(a) 185 J

(b) 0.0785 rev

(c) $W = 9.81\ \text{N}$

23

(a) 2.57×10^{29} J

(b) $KE_{rot} = 2.65 \times 10^{33}$ J

25

$KE_{rot} = 434$ J

27

(a) 128 rad/s

(b) 19.9 m

29

(a) 10.4 rad/s^2

(b) net $W = 6.11$ J

34

(a) 1.49 kJ

(b) 2.52×10^4 N

36

(a) 2.66×10^{40} kg $\cdot$ m^2/s

(b) 7.07×10^{33} kg $\cdot$ m^2/s

The angular momentum of the Earth in its orbit around the Sun is 3.77×10^6 times larger than the angular momentum of the Earth around its axis.

38

22.5 kg $\cdot$ m^2/s

40

25.3 rpm

43

(a) 0.156 rad/s

(b) 1.17×10^{-2} J

(c) $0.188 \; kg \cdot m/s$

45

(a) 3.13 rad/s

(b) Initial KE = 438 J, final KE = 438 J

47

(a) 1.70 rad/s

(b) Initial KE = 22.5 J, final KE = 2.04 J

(c) $1.50 \; kg \cdot m/s$

48

(a) $5.64 \times 10^{33} \; kg \cdot m^2 /s$

(b) $1.39 \times 10^{22} \; N \cdot m$

(c) 2.17×10^{15} N

Test Prep for AP® Courses

1
(b)
3
(d)
5
(d)
You are given a thin rod of length 1.0 m and mass 2.0 kg, a small lead weight of 0.50 kg, and a not-so-small lead weight of 1.0 kg. The rod has three holes, one in each end and one through the middle, which may either hold a pivot point or one of the small lead weights.

7
(a)
9
(c)
11
(a)
13
(a)
15
(b)
17
(c)
19
(b)
21
(b)
23
(c)
25
(d)
27
A door on hinges is a rotational system. When you push or pull on the door handle, the angular momentum of the system changes. If a weight is hung on the door handle, then pushing on the door with the same force will cause a different increase in angular momentum. If you push or pull near the hinges with the same force, the resulting angular momentum of the system will also be different.
29

Since the globe is stationary to start with,

$$\tau = \frac{\Delta L}{\Delta t}$$

$$\tau \cdot \Delta t = \Delta L$$

By substituting,

120 N•m • 1.2 s = 144 N•m•s.

The angular momentum of the globe after 1.2 s is 144 N•m•s.

Chapter 11

Problems & Exercises

1

1.610 cm^3

3

(a) 2.58 g

(b) The volume of your body increases by the volume of air you inhale. The average density of your body decreases when you take a deep breath, because the density of air is substantially smaller than the average density of the body before you took the deep breath.

4

2.70 g/cm^3

6

(a) 0.163 m

(b) Equivalent to 19.4 gallons, which is reasonable

8

$7.9 \times 10^2 \text{ kg/m}^3$

9

15.6 g/cm^3

10

(a) 10^{18} kg/m^3

(b) $2 \times 10^4 \text{ m}$

11

$3.59 \times 10^6 \text{ Pa}$; or 521 lb/in^2

13

$2.36 \times 10^3 \text{ N}$

14

0.760 m

16

$$(h\rho g)_{\text{units}} = (\text{m})(\text{kg/m}^3)(\text{m/s}^2) = (\text{kg} \cdot \text{m}^2)/(\text{m}^3 \cdot \text{s}^2)$$

$$= (\text{kg} \cdot \text{m/s}^2)(1/\text{m}^2)$$

$$= \text{N/m}^2$$

18

(a) 20.5 mm Hg

(b) The range of pressures in the eye is 12–24 mm Hg, so the result in part (a) is within that range

20

$1.09 \times 10^3 \text{ N/m}^2$

22

24.0 N

24

$2.55 \times 10^7 \text{ Pa}$; or 251 atm

26

5.76×10^3 N extra force

28

(a) $V = d_i A_i = d_o A_o \Rightarrow d_o = d_i \left(\dfrac{A_i}{A_o} \right)$.

Now, using equation:

$\dfrac{F_1}{A_1} = \dfrac{F_2}{A_2} \Rightarrow F_o = F_i \left(\dfrac{A_o}{A_i} \right)$.

Finally,

$W_o = F_o d_o = \left(\dfrac{F_i A_o}{A_i} \right) \left(\dfrac{d_i A_i}{A_o} \right) = F_i d_i = W_i$.

In other words, the work output equals the work input.

(b) If the system is not moving, friction would not play a role. With friction, we know there are losses, so that $W_{out} = W_{in} - W_f$; therefore, the work output is less than the work input. In other words, with friction, you need to push harder on the input piston than was calculated for the nonfriction case.

29

Balloon:

$P_g \quad = \quad 5.00 \text{ cm } H_2O,$

$P_{abs} \quad = \quad 1.035 \times 10^3 \text{ cm } H_2O.$

Jar:

$P_g \quad = \quad -50.0 \text{ mm Hg},$

$P_{abs} \quad = \quad 710 \text{ mm Hg}.$

31

4.08 m

33

$\quad \Delta P = 38.7 \text{ mm Hg},$

Leg blood pressure $= \dfrac{159}{119}$.

35

22.4 cm^2

36

91.7%

38

815 kg/m^3

40

(a) 41.4 g

(b) 41.4 cm^3

(c) 1.09 g/cm^3

42

(a) 39.5 g

(b) 50 cm^3

(c) 0.79 g/cm^3

It is ethyl alcohol.

44

8.21 N

46

(a) 960 kg/m^3

(b) 6.34%

She indeed floats more in seawater.

48

(a) 0.24

(b) 0.68

(c) Yes, the cork will float because $\rho_{obj} < \rho_{ethyl \ alcohol}(0.678 \text{ g/cm}^3 < 0.79 \text{ g/cm}^3)$

50

The difference is 0.006%.

52

$$F_{net} = F_2 - F_1 = P_2 A - P_1 A = (P_2 - P_1)A$$
$$= (h_2 \rho_{fl} g - h_1 \rho_{fl} g)A$$
$$= (h_2 - h_1)\rho_{fl} gA$$

where ρ_{fl} = density of fluid. Therefore,

$$F_{net} = (h_2 - h_1)A\rho_{fl} g = V_{fl}\rho_{fl} g = m_{fl} g = w_{fl}$$

where is w_{fl} the weight of the fluid displaced.

54

592 N/m^2

56

$2.23 \times 10^{-2} \text{ mm Hg}$

58

(a) $1.65 \times 10^{-3} \text{ m}$

(b) $3.71 \times 10^{-4} \text{ m}$

60

$6.32 \times 10^{-2} \text{ N/m}$

Based on the values in table, the fluid is probably glycerin.

62

$$P_w = 14.6 \text{ N/m}^2,$$
$$P_a = 4.46 \text{ N/m}^2,$$
$$P_{sw} = 7.40 \text{ N/m}^2.$$

Alcohol forms the most stable bubble, since the absolute pressure inside is closest to atmospheric pressure.

64

$5.1°$

This is near the value of $\theta = 0°$ for most organic liquids.

66

-2.78

The ratio is negative because water is raised whereas mercury is lowered.

68
479 N
70
1.96 N
71
$-63.0 \text{ cm H}_2\text{O}$
73

(a) $3.81 \times 10^3 \text{ N/m}^2$

(b) 28.7 mm Hg, which is sufficient to trigger micturition reflex

75

(a) 13.6 m water

(b) 76.5 cm water

77

(a) $3.98 \times 10^6 \text{ Pa}$

(b) $2.1 \times 10^{-3} \text{ cm}$

79

(a) 2.97 cm

(b) $3.39 \times 10^{-6} \text{ J}$

(c) Work is done by the surface tension force through an effective distance $h/2$ to raise the column of water.

81

(a) $2.01 \times 10^4 \text{ N}$

(b) $1.17 \times 10^{-3} \text{ m}$

(c) $2.56 \times 10^{10} \text{ N/m}^2$

83

(a) $1.38 \times 10^4 \text{ N}$

(b) $2.81 \times 10^7 \text{ N/m}^2$

(c) 283 N

85

(a) 867 N

(b) This is too much force to exert with a hand pump.

(c) The assumed radius of the pump is too large; it would be nearly two inches in diameter—too large for a pump or even a master cylinder. The pressure is reasonable for bicycle tires.

Test Prep for AP® Courses

1
(e)
3
(a) 100 kg/m^3 (b) 60% (c) yes; yes (76% will be submerged) (d) answers vary

Chapter 12

Problems & Exercises

1
2.78 cm^3/s
3
27 cm/s

5

(a) 0.75 m/s

(b) 0.13 m/s

7

(a) 40.0 cm^2

(b) 5.09×10^7

9

(a) 22 h

(b) 0.016 s

11

(a) 12.6 m/s

(b) $0.0800 \text{ m}^3/\text{s}$

(c) No, independent of density.

13

(a) 0.402 L/s

(b) 0.584 cm

15

(a) $127 \text{ cm}^3/\text{s}$

(b) 0.890 cm

17

$$P = \frac{\text{Force}}{\text{Area}},$$
$$(P)_{\text{units}} = \text{N/m}^2 = \text{N} \cdot \text{m/m}^3 = \text{J/m}^3$$
$$= \text{energy/volume}$$

19
184 mm Hg
21
$2.54 \times 10^5 \text{ N}$
23

(a) $1.58 \times 10^6 \text{ N/m}^2$

(b) 163 m

25

(a) $9.56 \times 10^8 \text{ W}$

(b) 1.4

27
1.26 W
29

(a) $3.02 \times 10^{-3} \text{ N}$

(b) 1.03×10^{-3}

31
$1.60 \text{ cm}^3/\text{min}$
33
$8.7 \times 10^{-11} \text{ m}^3/\text{s}$
35

0.316

37

(a) 1.52

(b) Turbulence will decrease the flow rate of the blood, which would require an even larger increase in the pressure difference, leading to higher blood pressure.

39

225 mPa · s

41

0.138 Pa · s,

or

Olive oil.

43

(a) $1.62 \times 10^4 \, \text{N/m}^2$

(b) $0.111 \, \text{cm}^3/\text{s}$

(c) 10.6 cm

45

1.59

47

$2.95 \times 10^6 \, \text{N/m}^2$ (gauge pressure)

51

$N_R = 1.99 \times 10^2 < 2000$

53

(a) nozzle: 1.27×10^5 , not laminar

(b) hose: 3.51×10^4 , not laminar.

55

2.54 << 2000, laminar.

57

1.02 m/s

$1.28 \times 10^{-2} \, \text{L/s}$

59

(a) ≥ 13.0 m

(b) $2.68 \times 10^{-6} \, \text{N/m}^2$

61

(a) 23.7 atm or $344 \, \text{lb/in}^2$

(b) The pressure is much too high.

(c) The assumed flow rate is very high for a garden hose.

(d) 5.27×10^6 >> 3000, turbulent, contrary to the assumption of laminar flow when using this equation.

62

1.41×10^{-3} m

64

1.3×10^2 s

66

0.391 s

Test Prep for AP® Courses

1
(c)
3
(a)
5
(a)
7
(a)
9
(d)

Chapter 13

Problems & Exercises

1
102°F
3
20.0°C and 25.6°C
5
9890°F
7

(a) 22.2°C

(b)
$$\begin{aligned}\Delta T(°F) &= T_2(°F) - T_1(°F) \\ &= \tfrac{9}{5}T_2(°C) + 32.0° - \left(\tfrac{9}{5}T_1(°C) + 32.0°\right) \\ &= \tfrac{9}{5}(T_2(°C) - T_1(°C)) = \tfrac{9}{5}\Delta T(°C)\end{aligned}$$

9
169.98 m
11
5.4×10^{-6} m
13
Because the area gets smaller, the price of the land DECREASES by ~$17,000.
15
$$\begin{aligned}V &= V_0 + \Delta V = V_0(1 + \beta\Delta T) \\ &= (60.00\text{ L})\left[1 + \left(950 \times 10^{-6}/°C\right)(35.0°C - 15.0°C)\right] \\ &= 61.1\text{ L}\end{aligned}$$

17

(a) 9.35 mL

(b) 7.56 mL

19
0.832 mm
21

We know how the length changes with temperature: $\Delta L = \alpha L_0 \Delta T$. Also we know that the volume of a cube is related to its length by $V = L^3$, so the final volume is then $V = V_0 + \Delta V = (L_0 + \Delta L)^3$. Substituting for ΔL gives

$$V = (L_0 + \alpha L_0 \Delta T)^3 = L_0^3(1 + \alpha\Delta T)^3.$$

Now, because $\alpha\Delta T$ is small, we can use the binomial expansion:

$$V \approx L_0^3(1 + 3\alpha\Delta T) = L_0^3 + 3\alpha L_0^3 \Delta T.$$

So writing the length terms in terms of volumes gives $V = V_0 + \Delta V \approx V_0 + 3\alpha V_0 \Delta T$, and so

$$\Delta V = \beta V_0 \Delta T \approx 3\alpha V_0 \Delta T, \text{ or } \beta \approx 3\alpha.$$

22
1.62 atm
24

(a) 0.136 atm

(b) 0.135 atm. The difference between this value and the value from part (a) is negligible.

26

(a) $nRT = (\text{mol})(\text{J/mol} \cdot \text{K})(\text{K}) = \text{J}$

(b) $nRT = (\text{mol})(\text{cal/mol} \cdot \text{K})(\text{K}) = \text{cal}$

(c)
$$\begin{aligned} nRT &= (\text{mol})(\text{L} \cdot \text{atm/mol} \cdot \text{K})(\text{K}) \\ &= \text{L} \cdot \text{atm} = (\text{m}^3)(\text{N/m}^2) \\ &= \text{N} \cdot \text{m} = \text{J} \end{aligned}$$

28
7.86×10^{-2} mol
30

(a) 6.02×10^5 km^3

(b) 6.02×10^8 km

32
$-73.9°$C
34

(a) 9.14×10^6 N/m^2

(b) 8.23×10^6 N/m^2

(c) 2.16 K

(d) No. The final temperature needed is much too low to be easily achieved for a large object.

36
41 km
38

(a) 3.7×10^{-17} Pa

(b) 6.0×10^{17} m^3

(c) 8.4×10^2 km

39
1.25×10^3 m/s
41

(a) 1.20×10^{-19} J

(b) 1.24×10^{-17} J

43
458 K
45
1.95×10^7 K
47
6.09×10^5 m/s
49
7.89×10^4 Pa
51

(a) 1.99×10^5 Pa

(b) 0.97 atm

53

3.12×10^4 Pa

55

78.3%

57

(a) 2.12×10^4 Pa

(b) 1.06 %

59

(a) 8.80×10^{-2} g

(b) 6.30×10^3 Pa ; the two values are nearly identical.

61

82.3%

63

4.77°C

65

38.3 m

67

$\dfrac{(F_B / w_{Cu})}{(F_B / w_{Cu})\prime} = 1.02$. The buoyant force supports nearly the exact same amount of force on the copper block in both

circumstances.

69

(a) 4.41×10^{10} mol/m^3

(b) It's unreasonably large.

(c) At high pressures such as these, the ideal gas law can no longer be applied. As a result, unreasonable answers come up when it is used.

71

(a) 7.03×10^8 m/s

(b) The velocity is too high—it's greater than the speed of light.

(c) The assumption that hydrogen inside a supernova behaves as an idea gas is responsible, because of the great temperature and density in the core of a star. Furthermore, when a velocity greater than the speed of light is obtained, classical physics must be replaced by relativity, a subject not yet covered.

Test Prep for AP® Courses

1
(a), (c)
3
(d)
5
(b)
7
(a) 7.29 × 10^{-21} J; (b) 352K or 79°C

Chapter 14

Problems & Exercises

1

5.02×10^8 J

3

3.07×10^3 J

5
0.171°C

7
10.8

9
617 W

11
35.9 kcal

13

(a) 591 kcal

(b) 4.94×10^3 s

15
13.5 W

17

(a) 148 kcal

(b) 0.418 s, 3.34 s, 4.19 s, 22.6 s, 0.456 s

19
33.0 g

20

(a) 9.67 L

(b) Crude oil is less dense than water, so it floats on top of the water, thereby exposing it to the oxygen in the air, which it uses to burn. Also, if the water is under the oil, it is less efficient in absorbing the heat generated by the oil.

22

a) 319 kcal

b) 2.00°C

24
20.6°C

26
4.38 kg

28

(a) 1.57×10^4 kcal

(b) 18.3 kW · h

(c) 1.29×10^4 kcal

30

(a) 1.01×10^3 W

(b) One

32
84.0 W

34
2.59 kg

36

(a) 39.7 W

(b) 820 kcal

38
35 to 1, window to wall

40
1.05×10^3 K

42

(a) 83 W

(b) 24 times that of a double pane window.

44
20.0 W, 17.2% of 2400 kcal per day
45
10 m/s
47
85.7°C
49
1.48 kg
51
$2 \times 10^4 \text{ MW}$
53

(a) 97.2 J

(b) 29.2 W

(c) 9.49 W

(d) The total rate of heat loss would be $29.2 \text{ W} + 9.49 \text{ W} = 38.7 \text{ W}$. While sleeping, our body consumes 83 W of power, while sitting it consumes 120 to 210 W. Therefore, the total rate of heat loss from breathing will not be a major form of heat loss for this person.

55
-21.7 kW
Note that the negative answer implies heat loss to the surroundings.
57
-266 kW
59
-36.0 W
61

(a) 1.31%

(b) 20.5%

63

(a) -15.0 kW

(b) 4.2 cm

65

(a) 48.5°C

(b) A pure white object reflects more of the radiant energy that hits it, so a white tent would prevent more of the sunlight from heating up the inside of the tent, and the white tunic would prevent that heat which entered the tent from heating the rider. Therefore, with a white tent, the temperature would be lower than 48.5°C, and the rate of radiant heat transferred to the rider would be less than 20.0 W.

67

(a) $3 \times 10^{17} \text{ J}$

(b) $1 \times 10^{13} \text{ kg}$

(c) When a large meteor hits the ocean, it causes great tidal waves, dissipating large amount of its energy in the form of kinetic energy of the water.

69

(a) $3.44 \times 10^5 \text{ m}^3/\text{s}$

(b) This is equivalent to 12 million cubic feet of air per second. That is tremendous. This is too large to be dissipated by heating the air by only 5°C. Many of these cooling towers use the circulation of cooler air over warmer water to increase the rate of evaporation. This would allow much smaller amounts of air necessary to remove such a large amount of heat because evaporation removes larger quantities of heat than was considered in part (a).

71
20.9 min
73

(a) 3.96×10^{-2} g

(b) 96.2 J

(c) 16.0 W

75

(a) 1.102

(b) 2.79×10^4 J

(c) 12.6 J. This will not cause a significant cooling of the air because it is much less than the energy found in part (b), which is the energy required to warm the air from $20.0°C$ to $50.0°C$.

76

(a) $36°C$

(b) Any temperature increase greater than about $3°C$ would be unreasonably large. In this case the final temperature of the person would rise to $73°C \, (163°F)$.

(c) The assumption of 95% heat retention is unreasonable.

78

(a) 1.46 kW

(b) Very high power loss through a window. An electric heater of this power can keep an entire room warm.

(c) The surface temperatures of the window do not differ by as great an amount as assumed. The inner surface will be warmer, and the outer surface will be cooler.

Test Prep for AP® Courses

1
(c)
3
(a)
5
(b)
7
(a)
9
(d)

Chapter 15

Problems & Exercises

1

1.6×10^9 J

3

-9.30×10^8 J

5

(a) -1.0×10^4 J , or -2.39 kcal

(b) 5.00%

7

(a) 122 W

(b) 2.10×10^6 J

(c) Work done by the motor is 1.61×10^7 J ;thus the motor produces 7.67 times the work done by the man

9

(a) 492 kJ

(b) This amount of heat is consistent with the fact that you warm quickly when exercising. Since the body is inefficient, the excess heat produced must be dissipated through sweating, breathing, etc.

10

6.77×10^3 J

12

(a) $W = P\Delta V = 1.76 \times 10^5$ J

(b) $W = Fd = 1.76 \times 10^5$ J . Yes, the answer is the same.

14

$W = 4.5 \times 10^3$ J

16

W is not equal to the difference between the heat input and the heat output.

20

(a) 18.5 kJ

(b) 54.1%

22

(a) 1.32×10^9 J

(b) 4.68×10^9 J

24

(a) 3.80×10^9 J

(b) 0.667 barrels

26

(a) 8.30×10^{12} J , which is 3.32% of 2.50×10^{14} J .

(b) -8.30×10^{12} J , where the negative sign indicates a reduction in heat transfer to the environment.

28

403°C

30

(a) 244°C

(b) 477°C

(c)Yes, since automobiles engines cannot get too hot without overheating, their efficiency is limited.

32

(a) $Eff_1 = 1 - \dfrac{T_{c,1}}{T_{h,1}} = 1 - \dfrac{543 \text{ K}}{723 \text{ K}} = 0.249$ or 24.9%

(b) $Eff_2 = 1 - \dfrac{423 \text{ K}}{543 \text{ K}} = 0.221$ or 22.1%

(c) $Eff_1 = 1 - \dfrac{T_{c,1}}{T_{h,1}} \Rightarrow T_{c,1} = T_{h,1}(1, \; -, \; eff_1)$ similarly, $T_{c,2} = T_{h,2}(1 - Eff_2)$

$$T_{c,2} = T_{h,1}(1 - Eff_1)(1 - Eff_2) \equiv T_{h,1}(1 - Eff_{\text{overall}})$$
using $T_{h,2} = T_{c,1}$ in above equation gives $?(1 - Eff_{\text{overall}}) = (1 - Eff_1)(1 - Eff_2)$
$$Eff_{\text{overall}} = 1 - (1 - 0.249)(1 - 0.221) = 41.5\%$$

(d) $Eff_{\text{overall}} = 1 - \dfrac{423 \text{ K}}{723 \text{ K}} = 0.415$ or 41.5%

34

The heat transfer to the cold reservoir is $Q_c = Q_h - W = 25 \text{ kJ} - 12 \text{ kJ} = 13 \text{ kJ}$, so the efficiency is

$Eff = 1 - \dfrac{Q_c}{Q_h} = 1 - \dfrac{13 \text{ kJ}}{25 \text{ kJ}} = 0.48$. The Carnot efficiency is $Eff_C = 1 - \dfrac{T_c}{T_h} = 1 - \dfrac{300 \text{ K}}{600 \text{ K}} = 0.50$. The actual

efficiency is 96% of the Carnot efficiency, which is much higher than the best-ever achieved of about 70%, so her scheme is likely to be fraudulent.

36

(a) –56.3°C

(b) The temperature is too cold for the output of a steam engine (the local environment). It is below the freezing point of water.

(c) The assumed efficiency is too high.

37
4.82
39
0.311
41

(a) 4.61

(b) $1.66 \times 10^8 \text{ J}$ or $3.97 \times 10^4 \text{ kcal}$

(c) To transfer $1.66 \times 10^8 \text{ J}$, heat pump costs $1.00, natural gas costs $1.34.

43
27.6°C
45

(a) $1.44 \times 10^7 \text{ J}$

(b) 40 cents

(c) This cost seems quite realistic; it says that running an air conditioner all day would cost $9.59 (if it ran continuously).

47

(a) $9.78 \times 10^4 \text{ J/K}$

(b) In order to gain more energy, we must generate it from things within the house, like a heat pump, human bodies, and other appliances. As you know, we use a lot of energy to keep our houses warm in the winter because of the loss of heat to the outside.

49
$8.01 \times 10^5 \text{ J}$
51

(a) $1.04 \times 10^{31} \text{ J/K}$

(b) $3.28 \times 10^{31} \text{ J}$

53
199 J/K
55

(a) $2.47 \times 10^{14} \text{ J}$

(b) $1.60 \times 10^{14} \text{ J}$

(c) $2.85 \times 10^{10} \text{ J/K}$

(d) $8.29 \times 10^{12} \text{ J}$

57
It should happen twice in every $1.27 \times 10^{30} \text{ s}$ or once in every $6.35 \times 10^{29} \text{ s}$

$$\left(6.35\times10^{29}\ \text{s}\right)\left(\frac{1\ \text{h}}{3600\ \text{s}}\right)\ \left(\frac{1\ \text{d}}{24\ \text{h}}\right)\left(\frac{1\ \text{y}}{365.25\ \text{d}}\right)$$

$$=\qquad\qquad 2.0\times10^{22}\ \text{y}$$

59

(a) 3.0×10^{29}

(b) 24%

61

(a) -2.38×10^{-23} J/K

(b) 5.6 times more likely

(c) If you were betting on two heads and 8 tails, the odds of breaking even are 252 to 45, so on average you would break even. So, no, you wouldn't bet on odds of 252 to 45.

Test Prep for AP® Courses

1
(d)
3
(a)
5
(b)
7
(c)
9
(d)
11
(a)
13
(c)
15
(b)

Chapter 16

Problems & Exercises

1

(a) 1.23×10^3 N/m

(b) 6.88 kg

(c) 4.00 mm

3

(a) 889 N/m

(b) 133 N

5

(a) 6.53×10^3 N/m

(b) Yes

7
16.7 ms
8
0.400 s / beats
9
400 Hz
10
12,500 Hz
11

1.50 kHz
12

(a) 93.8 m/s

(b) 11.3×10^3 rev/min

13
 2.37 N/m
15
0.389 kg
18
94.7 kg
21
1.94 s
22
6.21 cm
24
2.01 s
26
2.23 Hz
28

(a) 2.99541 s

(b) Since the period is related to the square root of the acceleration of gravity, when the acceleration changes by 1% the period changes by $(0.01)^2 = 0.01\%$ so it is necessary to have at least 4 digits after the decimal to see the changes.

30

(a) Period increases by a factor of 1.41 ($\sqrt{2}$)

(b) Period decreases to 97.5% of old period

32
Slow by a factor of 2.45
34
length must increase by 0.0116%.
35

(a) 1.99 Hz

(b) 50.2 cm

(c) 1.41 Hz, 0.710 m

36

(a) 3.95×10^6 N/m

(b) 7.90×10^6 J

37

a). 0.266 m/s

b). 3.00 J

39

$\pm \dfrac{\sqrt{3}}{2}$

42
384 J
44

(a). 0.123 m

(b). −0.600 J

(c). 0.300 J. The rest of the energy may go into heat caused by friction and other damping forces.

46

(a) 5.00×10^5 J

(b) 1.20×10^3 s

47
$t = 9.26$ d

49
$f = 40.0$ Hz

51
$v_w = 16.0$ m/s

53
$\lambda = 700$ m

55
$d = 34.0$ cm

57
$f = 4$ Hz

59

462 Hz,

4 Hz

61

(a) 3.33 m/s

(b) 1.25 Hz

63
0.225 W

65
7.07

67
16.0 d

68
2.50 kW

70
3.38×10^{-5} W/m^2

Test Prep for AP® Courses

1
(d)

3
(b)

5

The frequency is given by

$$f = \frac{1}{T} = \frac{50\,cycles}{30s} = 1.66\,Hz$$

Time period is:

$$T = \frac{1}{f} = \frac{1}{1.66} = 0.6\,s$$

7
(c)

9

The energy of the particle at the center of the oscillation is given by

$$E = \frac{1}{2}mv^2 = \frac{1}{2} \times 0.2\,kg \times (5\,m \cdot s^{-1})^2 = 2.5\,J$$

11

(b)
13
19.7 J
15
(c)
17

$$d = \frac{k}{2\mu_K mg}\left(X^2 - \frac{\mu_K mg}{k}\right)^2 \quad where\, k = 50\,\text{N} \cdot \text{m}^{-1} \quad \mu_k = 0.06 \quad m = 0.5\text{kg}$$

$$d = \frac{50\text{N} \cdot \text{m}^{-1}}{2\times0.06\times9.8\text{m} \cdot \text{s}^{-2}}\left((0.2)^2 - \left(\frac{\left(0.06\times0.5\text{kg}\times9.8\text{m} \cdot \text{s}^{-2}\right)^2}{(50\text{N} \cdot \text{m}^{-1})^2}\right)\right) = 1.698\,\text{m}$$

19
The waves coming from a tuning fork are mechanical waves that are longitudinal in nature, whereas electromagnetic waves are transverse in nature.
21
The sound energy coming out of an instrument depends on its size. The sound waves produced are relative to the size of the musical instrument. A smaller instrument such as a tambourine will produce a high-pitched sound (higher frequency, shorter wavelength), whereas a larger instrument such as a drum will produce a deeper sound (lower frequency, longer wavelength).
23
$2\pi\,\text{m}$
25
The student explains the principle of superposition and then shows two waves adding up to form a bigger wave when a crest adds with a crest and a trough with another trough. Also the student shows a wave getting cancelled out when a crest meets a trough and vice versa.
27
The student must note that the shape of the wave remains the same and there is first an overlap and then receding of the waves.
29
(c)

Chapter 17

Problems & Exercises

1
0.288 m
3
332 m/s
5

$$v_w = (331\,\text{m/s})\sqrt{\frac{T}{273\,\text{K}}} = (331\,\text{m/s})\sqrt{\frac{293\,\text{K}}{273\,\text{K}}}$$
$$= 343\,\text{m/s}$$

7
0.223
9

(a) 7.70 m

(b) This means that sonar is good for spotting and locating large objects, but it isn't able to resolve smaller objects, or detect the detailed shapes of objects. Objects like ships or large pieces of airplanes can be found by sonar, while smaller pieces must be found by other means.

11

(a) 18.0 ms, 17.1 ms

(b) 5.00%

(c) This uncertainty could definitely cause difficulties for the bat, if it didn't continue to use sound as it closed in on its prey. A 5% uncertainty could be the difference between catching the prey around the neck or around the chest, which means that it could miss grabbing its prey.

12
$3.16\times10^{-4}\,\text{W/m}^2$

14

3.04×10^{-4} W/m^2

16

106 dB

18

(a) 93 dB

(b) 83 dB

20

(a) 50.1

(b) 5.01×10^{-3} or $\frac{1}{200}$

22

70.0 dB

24

100

26

1.45×10^{-3} J

28

28.2 dB

30

(a) 878 Hz

(b) 735 Hz

32

3.79×10^3 Hz

34

(a) 12.9 m/s

(b) 193 Hz

36

First eagle hears 4.23×10^3 Hz

Second eagle hears 3.56×10^3 Hz

38

0.7 Hz

40

0.3 Hz, 0.2 Hz, 0.5 Hz

42

(a) 256 Hz

(b) 512 Hz

44

180 Hz, 270 Hz, 360 Hz

46

1.56 m

48

(a) 0.334 m

(b) 259 Hz

50

3.39 to 4.90 kHz

52

(a) 367 Hz

(b) 1.07 kHz

54

(a) $f_n = n(47.6 \text{ Hz})$, $n = 1, 3, 5,..., 419$

(b) $f_n = n(95.3 \text{ Hz})$, $n = 1, 2, 3,..., 210$

55

1×10^6 km

57

498.5 or 501.5 Hz

59

82 dB

61

approximately 48, 9, 0, –7, and 20 dB, respectively

63

(a) 23 dB

(b) 70 dB

65

Five factors of 10

67

(a) 2×10^{-10} W/m^2

(b) 2×10^{-13} W/m^2

69

2.5

71

1.26

72

170 dB

74

103 dB

76

(a) 1.00

(b) 0.823

(c) Gel is used to facilitate the transmission of the ultrasound between the transducer and the patient's body.

78

(a) 77.0 μm

(b) Effective penetration depth = 3.85 cm, which is enough to examine the eye.

(c) 16.6 μm

80

(a) 5.78×10^{-4} m

(b) 2.67×10^6 Hz

82

(a) $v_w = 1540 \text{ m/s} = f\lambda$ $\lambda = \dfrac{1540 \text{ m/s}}{100 \times 10^3 \text{ Hz}} = 0.0154 \text{ m} < 3.50 \text{ m}$. Because the wavelength is much shorter than

the distance in question, the wavelength is not the limiting factor.

(b) 4.55 ms

84

974 Hz

(Note: extra digits were retained in order to show the difference.)

Test Prep for AP® Courses

1

(b)

3

(e)

5

(c)

7

Answers vary. Students could include a sketch showing an increased amplitude when two waves occupy the same location. Students could also cite conceptual evidence such as sound waves passing through each other.

9

(d)

11

(c)

13

(a)

15

(c)

17

(b)

18

Striking the end of a protruding ruler would create transverse waves, not sound waves. Any measurable, audible sound would come from the repetitive striking of the ruler against the desk. The student is confused about the speed of sound through solids versus the speed of sound in air.

19

(a), (b)

21

(c)

Some items in the index may be found in Chapters 18-34, which are not in this book.

CPSIA information can be obtained
at www.ICGtesting.com
Printed in the USA
BVOW04s1552170917

495072BV00047B/363/P